Neurobiology: A Functional Approach

Neurobiology: A Functional Approach

Georg F. Striedter
University of California, Irvine

New York Oxford
OXFORD UNIVERSITY PRESS

Oxford University Press is a department of the University of Oxford.
It furthers the University's objective of excellence in research,
scholarship, and education by publishing worldwide.

Oxford New York
Auckland Cape Town Dar es Salaam Hong Kong Karachi
Kuala Lumpur Madrid Melbourne Mexico City Nairobi
New Delhi Shanghai Taipei Toronto

With offices in
Argentina Austria Brazil Chile Czech Republic France Greece
Guatemala Hungary Italy Japan Poland Portugal Singapore
South Korea Switzerland Thailand Turkey Ukraine Vietnam

For titles covered by Section 112 of the US Higher Education Opportunity Act, please visit www.oup.com/us/he for the latest information about pricing and alternate formats.

Published by Oxford University Press
198 Madison Avenue, New York, New York 10016
http://www.oup.com

ISBN: 978-0-19-539615-7

ABOUT THE COVER

The cover image of neurons in a mammalian hippocampus captures the complexity and elegance of neuronal circuits. As the book explains, hippocampal neurons participate in several functional systems: they regulate stress hormone levels, help animals navigate, and are essential for remembering what happened when and where. Greg Dunn created this image using an enamel and gold leaf technique. For more of the artist's work, please visit gregadunn.com.

Hippocampus II
Enamel on composition gold and aluminum
42" × 42"
Greg Dunn, 2011

Printing number: 9 8 7 6 5 4 3 2 1

Printed in the United States of America
on acid-free paper

Brief Contents

Contents

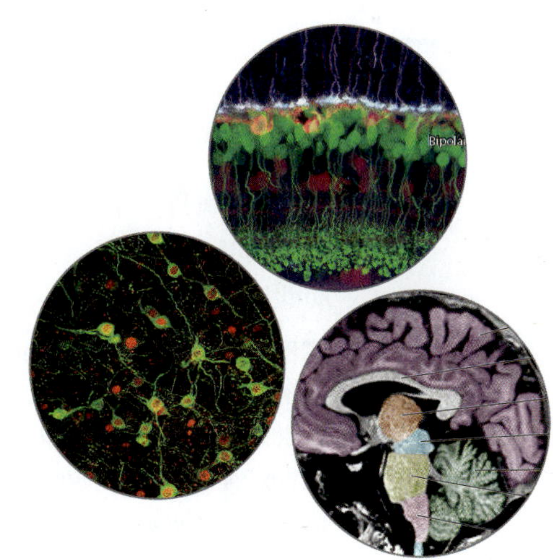

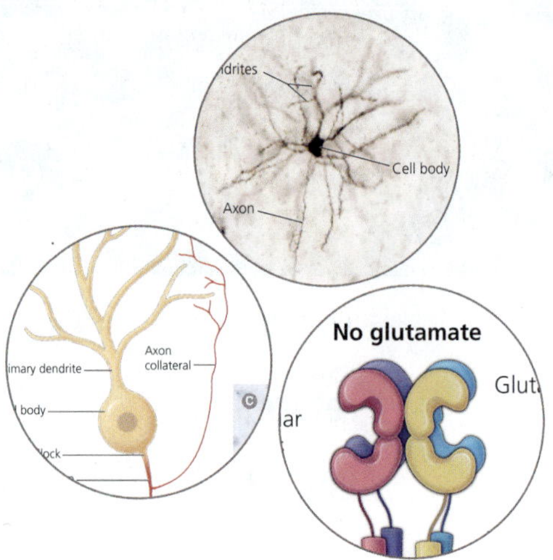

Cell body
Axon
Axon collateral
No glutamate

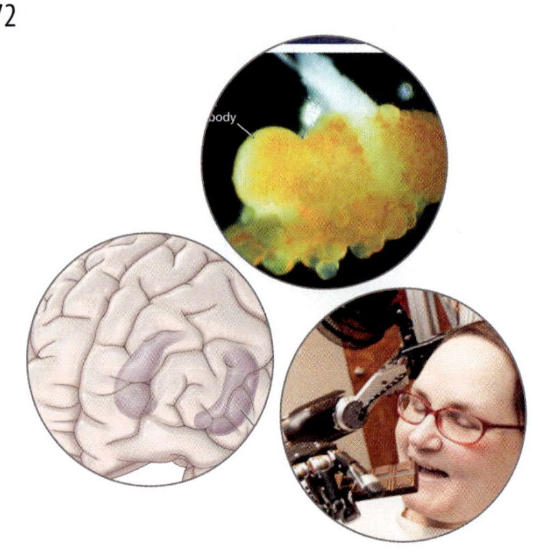

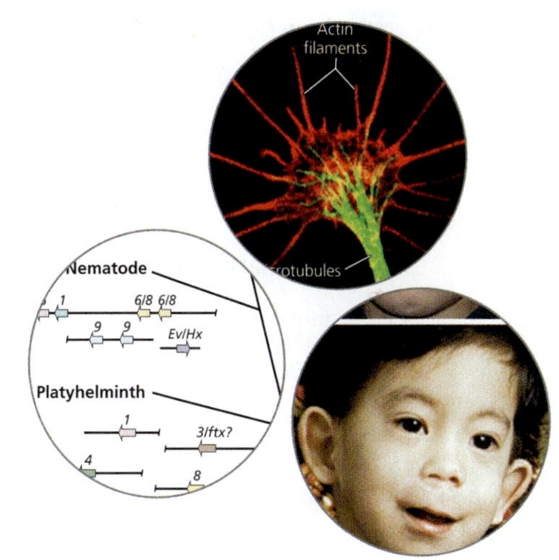
Actin
filaments
Nematode
Platyhelminth

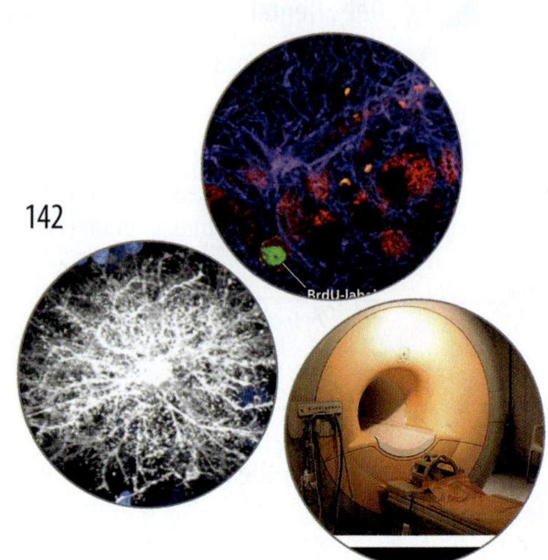

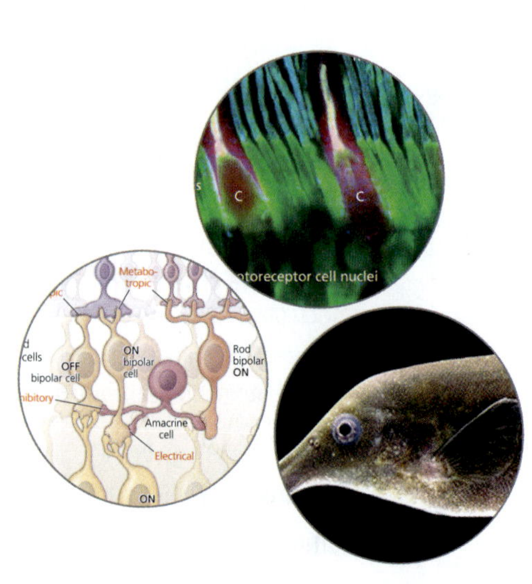

Metabo-
tropic
ON
bipolar
cell
OFF
bipolar cell
Rod
bipolar
ON
Amacrine
cell
Electrical
ON

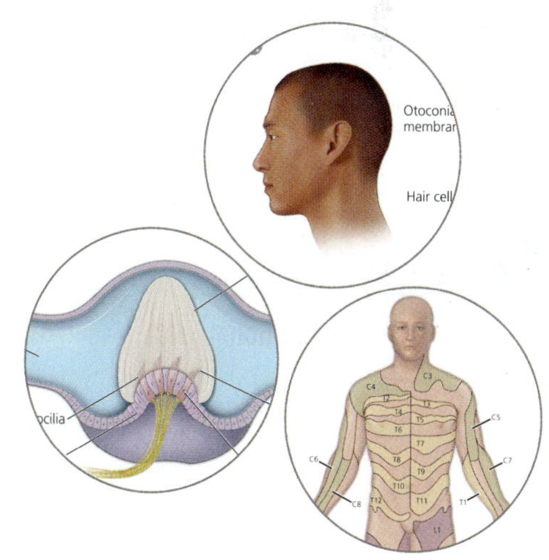
Hair cell

Chapter 8 Using Muscles and Glands 235

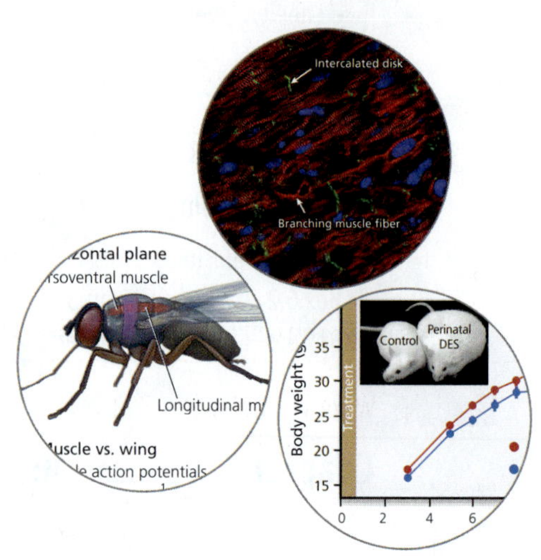

Chapter 9 Regulating Vital Bodily Functions 265

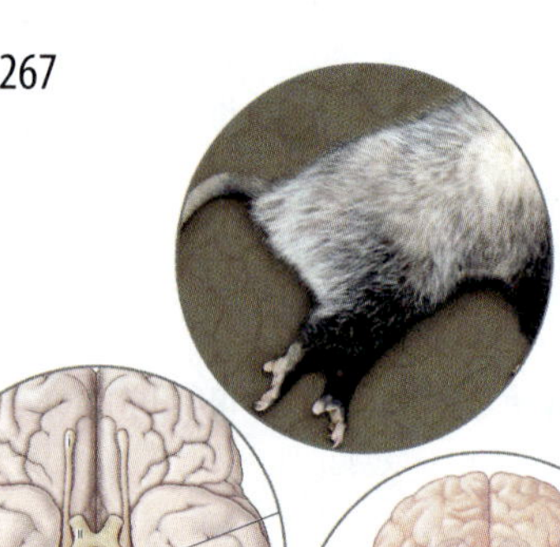

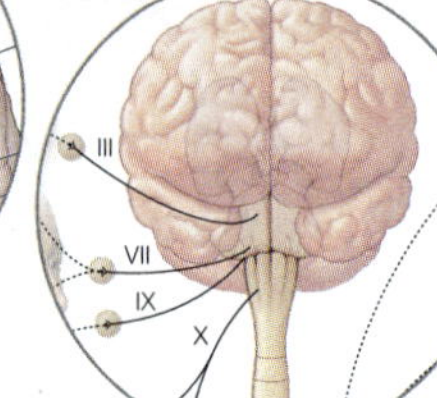

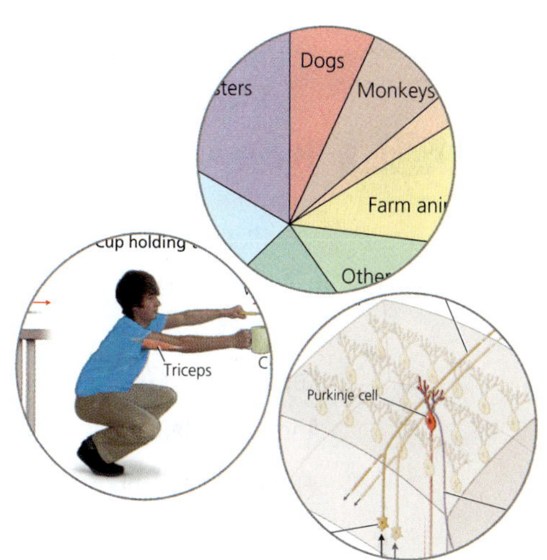
Dogs
Monkeys
Farm ani
Triceps
Purkinje cell

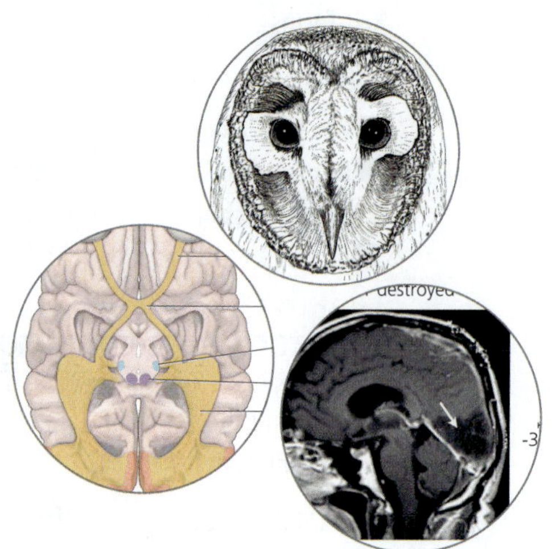

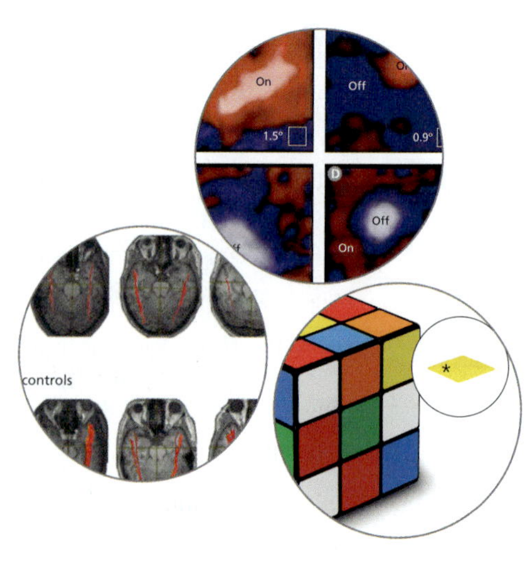
On
Off
1.5°
0.9°
Off
On
controls

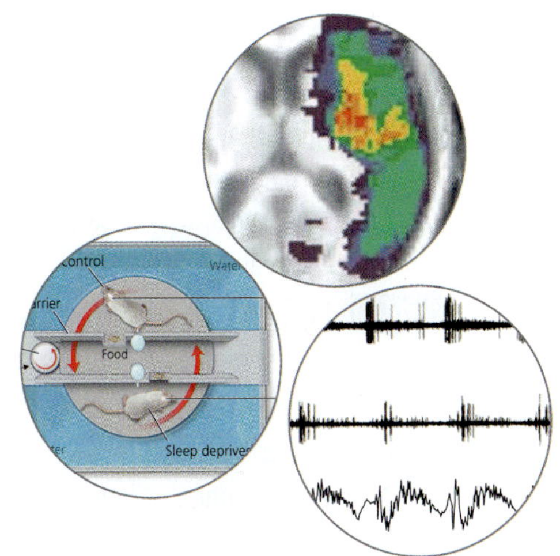
Control
Food
Sleep deprive

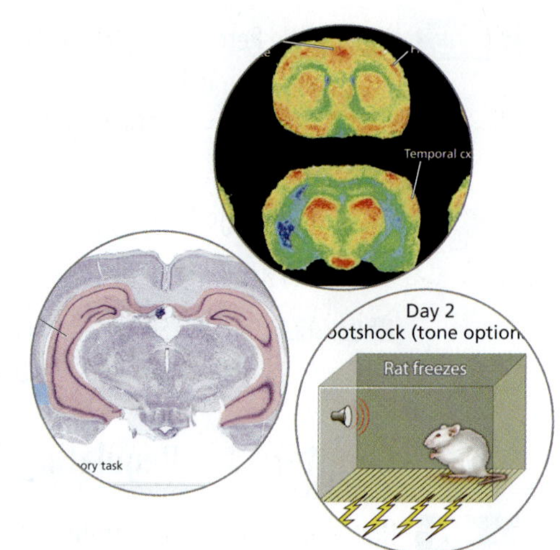
Temporal cx
Day 2
Rat freezes

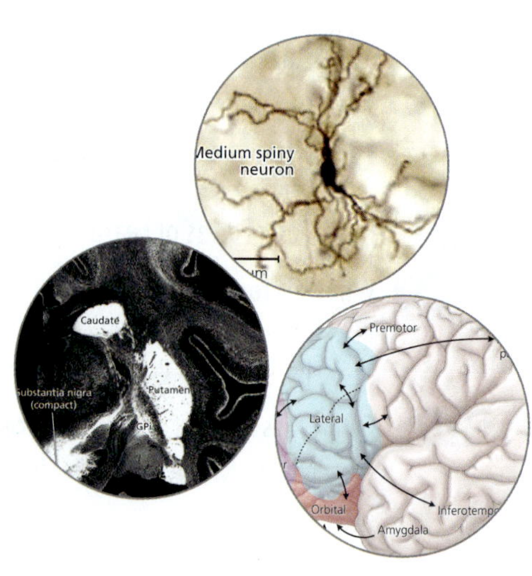
neuron
Caudate
Putamen
Premotor
Lateral
Orbital
Amygdala

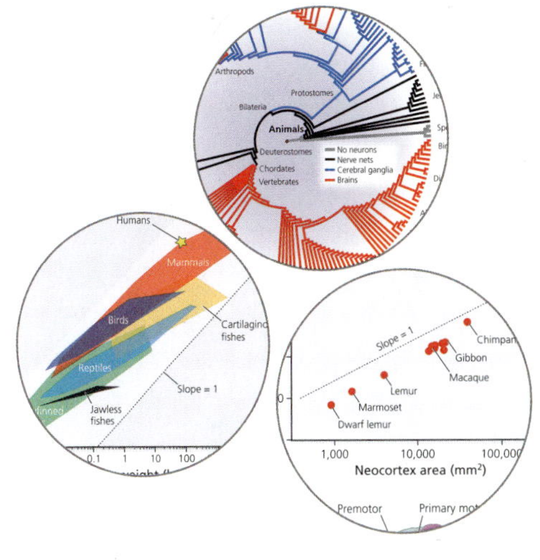

Arthropods
Protostomes
Bilateria
Animals
Deuterostomes
Chordates
Vertebrates
No neurons
Nerve nets
Cerebral ganglia
Brains
Humans
Mammals
Birds
Reptiles
Jawless fishes
Slope = 1
Gibbon
Macaque
Lemur
Marmoset
Dwarf lemur
1,000
10,000
Neocortex area (mm²)
Premotor

Preface

> To explicate the uses of the Brain seems as difficult a task as to paint the Soul, of which it is commonly said that it understands all things but itself.
>
> —Thomas Willis (1664)

Although inquisitive individuals have long sought to understand the neural basis of our thoughts, our emotions, our "soul," neurobiology is still a young and rapidly growing field. In 1971, the first meeting of the Society for Neuroscience had only 1,400 participants; by 2014, the list of attendees had swelled to 31,000. In parallel with this explosive growth, the number of US educational institutions offering undergraduate neurobiology majors increased from fewer than 10 in 1986 to more than 100 in 2008 (Ramos et al., 2011). Clearly, the quest to "explicate the uses of the Brain" is attracting an ever-growing number of bright minds.

The most fascinating aspect of neurobiological knowledge is that it spans numerous levels of analysis, ranging from DNA and other molecules to complex neural systems that interact with one another and the world. In contrast to other organs, the brain contains a vast number of distinct cell types, all performing different functions. Therefore, to understand the brain, it is not enough to know how a "generic neuron" works; we also have to figure out how those diverse neurons interact with one another and how they generate behavior. This task is enormous and, as you will see, still far from completion.

The Functional Approach

Neurobiology: A Functional Approach is unusual among introductory neurobiology textbooks in that it *asks not only how the nervous system works but also why it works as it does; it is focused on the problems that brains help organisms solve.*

For example, the book explains why neurons use all-or-nothing action potentials to communicate with other cells, and why the blood–brain barrier exists. Such "why" questions can often be answered by describing a specific feature's impact on survival or reproduction. The blood–brain barrier, for example, can be understood as a set of mechanisms that protect the brain, and thus the organism, from harmful compounds in the blood. Alternatively, "why" questions can be answered in terms of engineering principles. For example, the use of action potentials significantly improves the reliability of signal transmission along axons. Because many neurobiological "why" questions currently lack definitive answers, this book focuses on the strongest and most plausible hypotheses. Their primary purpose is to provoke thought and highlight what remains unknown.

Another unique aspect of the book is its emphasis on neural circuits and systems. It is relatively common for introductory neurobiology courses and texts to focus either on the cellular and molecular end of the neuroscience spectrum or on the cognitive end, neglecting the "middle ground." By emphasizing circuits and systems, the book allows readers to grasp the wide array of problems that nervous systems solve, such as sensing stimuli, controlling movement, learning what to fear or desire, and pursuing a goal. In keeping with this broad functional approach, the book covers molecular and cellular information not just at the beginning, but wherever it is relevant.

Organization

Neurobiology: A Functional Approach is organized around the major kinds of problems that nervous systems solve or help to solve. After an introductory chapter that provides an overview of nervous system organization, Chapters 2–5 focus on problems that are internal to the nervous system, such as the design of biological computing elements and the embryological development of complex brains. Chapters 6–8 deal with sensors that provide the brain with information about the outside world and with effectors, specifically muscles and glands. Chapters 9–12 explore problems that involve relatively simple interactions between the nervous system and its environment, such as regulating bodily functions and orienting the eyes toward external objects. Chapters 13–15 then focus on more complex interactions between organisms and their environment, including attention, arousal, memory, and decision making. The final chapter highlights how brains vary across species as well as within species.

If you are a student who is new to neurobiology, this book's organization should feel relatively natural because it progresses gradually from relatively simple functions, such as neuronal signaling, to ever more complex functions, including memory, decision making, and language. However, if you have previously studied neurobiology from other books, some aspects of this book will seem unusual to you.

Comparison to Other Textbooks

Neurobiology: A Functional Approach covers the cellular and molecular mechanisms of neuronal signaling in just one chapter, instead of the more usual 4–6 separate chapters. This decision was driven by my desire to keep this textbook relatively short. I also feel that it is important to familiarize students with the whole spectrum of modern neurobiological research, and to do so from the outset, not just in more advanced courses.

A second unusual feature of the present book is that it separates the question of how neurons sense stimuli from the question of how this sensory information is analyzed and used by the rest of the brain. This division makes it easier to compare and contrast the various senses. It also facilitates the discussion of multimodal sensory integration and low-level sensorimotor control, two topics that get short shrift in other books. Conventional treatments encourage you to think that the various sensory pathways are separate from one another and from motor systems until they reach the cerebral cortex, but this is not how real brains work.

The third big difference between this book and more traditional texts is that here the basal ganglia and dopamine are not discussed in the context of movement control, which is how most books deal with them. Instead, I cover them in Chapter 15, which explores the neural mechanisms of decision making, defined as the selection of behavioral goals, actions, and movements. In this larger context, the dopamine signals are viewed as helping organisms learn which behaviors are most appropriate for a given situation. This function is much broader than movement control.

Chapter 15 also covers the prefrontal cortex, which most textbooks discuss in a separate chapter on "executive functions." My decision to combine the basal ganglia with the prefrontal cortex was motivated by the dense anatomical connections between the prefrontal cortex and the striatum, the largest component of the basal ganglia. To understand the functions of these "frontostriatal loops," it helps to realize that both the striatum and the frontal lobe are hierarchically organized. This functional similarity is easier to grasp when both brain regions are covered together.

The book's last chapter discusses species, sex, and individual differences in brains. Other textbooks tend to ignore individual differences and cover sex differences only in the context of reproductive behaviors. However, both sorts of differences are increasingly recognized as being both widespread and significant. Species differences are likewise important. Most textbooks cover brain evolution in an opening chapter, but such discussions are more fruitful when you have already learned what brains are all about. Combining species, sex, and individual differences into a single chapter feels natural because, in all three cases, the differences exist on a background of widespread similarities.

As you read *Neurobiology: A Functional Approach*, you will find numerous links to neurological disorders and other clinical issues, but the book's primary focus is on the functioning of normal, intact brains. The rationale behind this emphasis is that knowing how a system normally functions makes it much easier to understand how the system can break. Of course, learning about neurological disorders is inherently fascinating and motivates many of us to learn about the brain. Therefore, instructors may want to supplement their lectures with stories and videos about disorders and treatments. Those in-class activities are especially useful if the students have read the book ahead of class and come prepared to contribute their own questions and ideas.

Special Features

Textbooks are most effective when they are fun to read and pull you into an unfolding "story." To achieve this goal, *Neurobiology: A Functional Approach* has been designed with several features that make it more enjoyable and easier to use.

User-friendly Style

Written by a single author, the text contains a minimum of jargon and is written with the beginning student in mind. A clear, consistent voice aims to engage you and make you curious about the brain and what it does. The book is more an epic narrative than an encyclopedic reference work. Yet it respects the brain's complexity and avoids oversimplification.

Experimental Emphasis

For any scientist, both young and old, it is important to know not only the key facts and concepts in a field, but also how those facts and concepts were obtained. Therefore, this book includes some historical information as well as descriptions of many key experiments. Data from these experiments are presented in many of the book's illustrations, making it easier to "see" how neuroscience works.

Special Topics Boxes

The book includes 46 "special topics boxes" that complement the central narrative. Some of the boxes focus on neurological disorders and therapies; others describe techniques or fascinating facts; and some highlight the power of evolution. The boxes can be skipped without breaking the text's main story line, but some are sure to pique your interest.

Brain Exercise Questions

Sprinkled throughout the book are "brain exercise questions" that encourage you to think about what you just read. These questions tend to be open-ended and lack simple answers, but trying to answer them will help you integrate the information and remember it. After all, none of us enjoy the feeling of having read large sections of a book only to realize that we remember none of it. Our time is much too valuable.

Chapter Summaries

Each chapter ends with a bulleted "chapter summary" that reprises the major points. These summaries can help you figure out which core concepts you remember—and which ones you should revisit. The summaries will also help you think about what you have learned and inscribe that knowledge in your long-term memory.

Annotated Bibliography

A list of additional readings at the end of each chapter provides interested readers with an entry into the scientific literature. Some of the listed articles and books review the field; others describe specific experiments. While some are broadly accessible, others are aimed at specialists. To gauge the scope and content of each listed publication, you can consult the annotated version of this bibliography, which is available on the book's website.

Key Terms and Glossary

Key terms are highlighted the first time they appear in the text and listed at the end of each chapter. Concise explanations of those terms are provided in the glossary at the end of the book.

Zoomable Human Brain Atlas

As you read about the brain, it helps to have a 3-dimensional image of what that brain looks like. Therefore, I have created a web-based atlas of Nissl- and fiber-stained tissue sections through real human brains. The user interface allows you to zoom in on individual neurons and then zoom out again. Only about 50 structures are identified by name so that you will not be overwhelmed with information unrelated to this book. This brain atlas is available through the course website.

Human Brain MRI Atlas

Contemporary neurobiologists rely heavily on the technique of magnetic resonance imaging (MRI). Therefore, I have created for this book three sets of horizontal, sagittal, and coronal MRI sections through a single human brain. The images are designed to be viewed as a presentation in PDF viewing software. The labeling of individual structures is kept to a minimum. This resource, too, is available through the book's website.

Visual Guide to the Book

Neurobiology: A Functional Approach is distinguished by the following approaches:

- Investigates not only how the nervous system works, but why it works as it does
- Focuses on the problems that brains help organisms solve
- Highlights evolutionary connections and real-world applications
- Emphasizes inquiry and the process of science

Chapter Introductions

- Each chapter opens with a summary of the **Core Questions** being discussed and the **Features** used to bring the topics home.

5.1 Are New Neurons Added to Adult Brains?

5.2 How Is the Brain Protected from Physical Trauma?

5.3 How Does the Brain Protect Itself against Toxins and Pathogens?

5.4 How Does the Nervous System Respond to an Attack?

5.5 How Do Neurons Get Their Energy?

5.6 What Links Body and Brain?

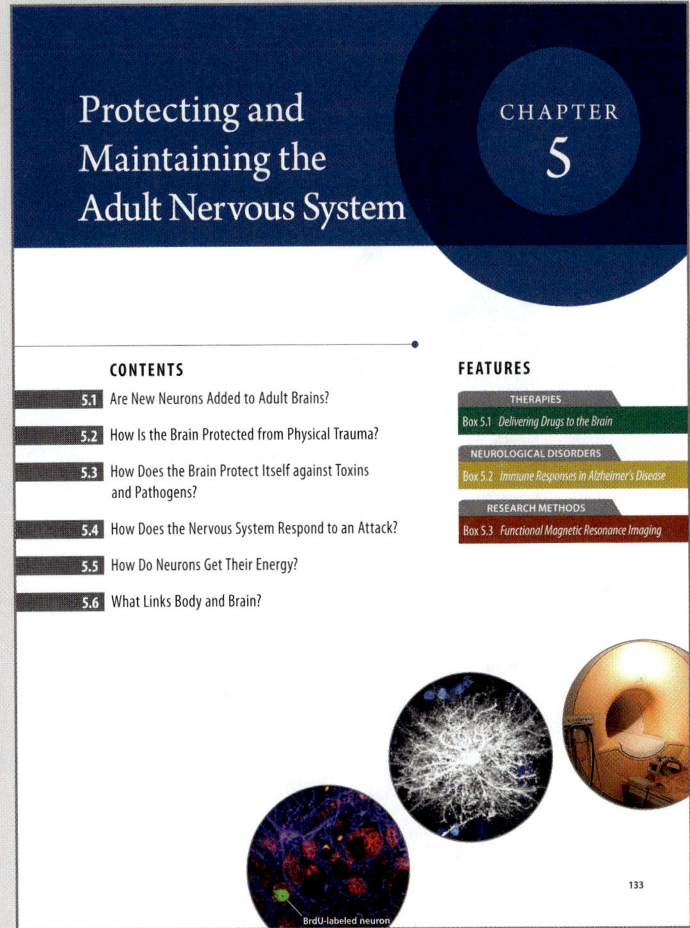

Protecting and Maintaining the Adult Nervous System

CHAPTER 5

CONTENTS

FEATURES

133

Inside the Chapter

- The **Core Questions** posed at the beginning of the chapter are reflected in the titles of each major section in the chapter; they pique student interest and mirror the spirit of scientific inquiry.
- **Brain Exercise Questions** integrated throughout the chapter prompt students to think critically and help them integrate and remember the chapter content.
- **Bulleted End-of-Chapter Summaries, Key Terms, and Suggested Further Readings** remind students of essential concepts and guide them to additional resources.

BRAIN EXERCISE

Imagine a piece of shrapnel entered your skull and killed a large number of neurons in your left auditory cortex, which is needed for speech production and comprehension. How might you and your brain respond to this injury?

5.3 How Does the Brain Protect Itself against Toxins and Pathogens?

Because the nervous system requires energy, derived mainly from glucose and oxygen, it must be supplied with blood. Unfortunately, blood also carries within it numerous substances that can damage neurons. Lead, mercury, pesticides, various other pollutants, pathogenic viruses, bacteria, and even the body's own immune cells would all wreak havoc in the CNS if they had free access to it. Fortunately, vertebrates evolved tight barriers between the blood and their neurons.

The Blood–brain Barrier

The discovery of a **blood–brain barrier** dates back to the 1880s, when Paul Ehrlich injected various aniline dyes into the veins of animals and noted that the dyes leaked out of the blood vessels into most tissues, but not into the brain and spinal cord. This was not what Ehrlich had expected. Struggling to interpret his results, Ehrlich concluded that the dyes must have leaked into the brain and spinal cord, just as they entered all other organs, but were incapable of staining neurons or glial cells. This hypothesis was wrong. In 1913, Edwin Goldman, a student of Ehrlich's, injected the same dyes Ehrlich had used directly into the CSF and noted that the brain and spinal cord were stained just fine. Therefore, Goldman concluded, neural tissue must be separated from the blood by some sort of barrier that keeps out aniline

Thematic Boxes

- A range of thematic boxes is used throughout the book to explore intriguing tangents and real-world applications.

RESEARCH METHODS

Box 5.3 *Functional Magnetic Resonance Imaging*

Functional magnetic resonance imaging (fMRI) has become the most widespread method of imaging activity in human brains. During a typical fMRI scan, patients are asked to place their head inside the open core of a large and noisy machine (Figure b5.3), to hold very still, and to perform some relatively simple tasks while the machine scans their brain. Scientists later process these data into images that show which regions of the brain were more (or less) active during the task than during a comparable rest period (or some alternative task). These images may seem easy to understand, but the fMRI technique is complex, and its data must be interpreted with care.

To produce an MRI brain scan, a person's head is placed inside a magnetic field that (at 1.5–7 Tesla) is more than 100,000 times as strong as the earth's magnetic field. In such a strong field, the protons of hydrogen atoms in the body tend to align their spin axes with the imposed magnetic field. Researchers then add an orthogonal, oscillating electromagnetic pulse that can, at the appropriate frequency, bump the spinning protons out of alignment. When the pulse is turned off, the protons move back into alignment and thereby generate a signal that the scanner can detect. Importantly, the frequency of the pulse that is most effective at perturbing the spinning protons (the resonant frequency) depends on the proton's chemical context (its atomic neighbors). This phenomenon has long formed the basis of nuclear magnetic resonance (NMR) spectroscopy, which chemists use to determine the molecular composition of homogeneous materials. In the 1970s, NMR spectroscopy was extended to let scientists visualize the chemical composition of nonhomogeneous materials, including brains. As the technique evolved, the spatial resolution of these images increased to the point where it is now smaller than a grain of rice. Scientists also learned to focus their images on diverse chemical signatures. Along the way, they changed the name of the technique from NMR to MRI. This helped alleviate patient concerns about being subjected to some sort of "nuclear" treatment when, in reality, no ionizing radiation is involved.

Early MRI scans were structural rather than functional. The early images focused on protons in gray matter versus white matter, ventricles, or bone; but Seiji Ogawa and his collaborators in the late 1980s turned their attention to the protons in oxygenated versus deoxygenated hemoglobin, developing what is now called *blood oxygen level-dependent fMRI (BOLD-fMRI)*. Ogawa reasoned that active neurons must pull some oxygen out of the local blood vessels and that this decrease in hemoglobin oxygenation should be detectable as a decrease in the MRI signal. Ogawa turned out to be correct, but surprisingly, the initial dip in the MRI signal is followed a few seconds later by a large increase signal strength. Because the late increase is larger and more reliable than the initial dip, it is the focus in most fMRI analyses. Why the delayed increase? Most likely it occurs because neuronal activity stimulates astrocytes, which then cause local blood vessels to dilate. This increases local blood flow and flushes the deoxygenated hemoglobin out of the local capillary beds.

An important caveat to interpreting fMRI results is that they usually depict differences in activity rather than absolute levels. They don't show you which regions were active during a task but only which regions were more active (or less active) during this task than during some other control task or resting period (Figure b5.3 B). This focus on differences greatly simplifies the analysis (because absolute signal levels depend on many factors that are difficult to control), but it easily misleads the uninitiated, who assume that the "hot spots" in an fMRI scan show all the brain areas that were active during a task. This is not true; they only indicate those areas that differed in activity between the task and its control. Keep this in mind when you come across fMRI results in upcoming chapters.

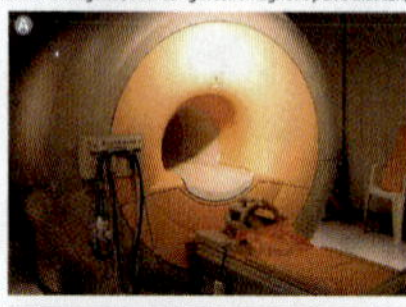

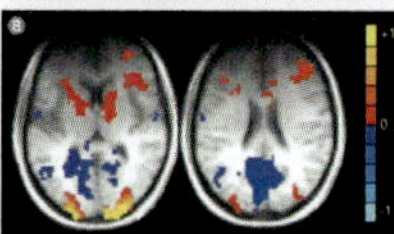

Figure b5.3 Functional MRI. Shown in (A) is a clinical MRI scanner. Panel (B) depicts BOLD-fMRI signals (colored) superimposed on structural MRI images (grayscale) for two horizontal sections through the "averaged brain" in the study. The color scale on the right shows the percent change in BOLD signal strength when people performed a simple task (detecting a letter hidden in visual noise) compared to staring at a screen that merely says "rest." As you can see, some brain regions became more active during the task; others became less active. [A courtesy of Kasuga Huang; B from Stark and Squire, 2001]

- **Research Methods:** Presents classic and modern experiments and techniques that have profoundly impacted current understanding.

EVOLUTION IN ACTION

Box 6.3 *Electroreception and Other Alien Senses*

Because we can perceive the world only through our own limited set of sensors, it is difficult to appreciate that other animals may well perceive the world quite differently. The average person is surprised to learn, for example, that many insects and birds can see intricate (and to us quite invisible) patterns of UV light reflected from flowers. Similarly, few people are aware of the enormous racket made by echolocating bats during their nightly flights because our auditory sensors are not tuned to ultrasonic frequencies. Still, we can at least imagine what it would be like to see UV light or to hear ultrasound. These sensory abilities are just extensions of senses that we already have. However, some animals possess senses that are truly alien to us. It is now clear, for example, that sea turtles can sense the earth's magnetic field and use it to guide them on their migrations. Homing pigeons can likewise orient by the terrestrial magnetic field, although they prefer to use the sun as their compass. Although the behavioral evidence for *magnetoreception* in these species is clear, the mechanisms underlying the transduction of magnetic fields remain somewhat mysterious. Crystals of magnetite likely play some role, but no one is sure how the physical or magnetic orientation of these crystals is sensed by neurons.

An equally strange sense is *electroreception*. We cannot detect weak electric fields, but many animals can, including lampreys, most cartilaginous fishes, and a few bony fishes (Figure b6.3). Many of these species can detect fields on the order of a few μV/cm, and some are sensitive down to ~10 nV/cm. In case you are not used to thinking in nanoVolts, this degree of sensitivity should, at least in theory, allow a shark to sense a difference of 1 Volt between two electrodes that are submerged in saltwater and separated by 10,000 km. Although electroreception is an evolutionarily ancient trait, it disappeared in the vertebrate lineages that left the water and invaded land. Aquatic amphibians can sense electric fields, but reptiles, birds, and most mammals cannot. This makes good sense, as electric fields require a conductive medium, such as water, to propagate. Remarkably, monotreme mammals have re-evolved electroreception. The duck-billed platypus, for one, has ~40,000 electrosensors on its bill and can, with some training, use them to find a 1.5V miniature battery hidden under a rock. The spiny anteater (or echidna), a close relative of the platypus, also has numerous electrosensors on its snout. It probably senses the weak electric fields produced by tiny prey tunneling through moist earth.

Indeed, the most widespread function of electroreceptors is the detection of electric fields generated by potential prey. As Ad Kalmijn showed in 1971, hungry sharks attack a flatfish buried under sand even if that flatfish is encased in agar, which blocks the spread of odors but is electrically transparent. Kalmijn further showed that hungry sharks have no interest in buried flatfish covered with electrically opaque plastic, but that they will attack when weak electrical current is passed between two buried stimulating electrodes (stronger currents tend to repel sharks!). These studies suggest that sharks, rays, and probably most electroreceptive animals use their electroreceptors to locate prey at night, in murky water, or hidden under ground.

In addition, electroreceptors may be used for navigation. Because some inanimate objects, especially metallic ones, generate weak electric fields under water, electroreceptive animals can find their way at night or in murky water by sensing the "electric landscape." Some electroreceptive fishes have gone a step further. They generate their own electric fields and then detect distortions in this field due to external objects. This sort of *electrolocation* is used by elephant nose fish (Figure b6.3) and by many of the knife fish that you can buy in pet stores. Some of these animals also communicate by means of electric signals.

How do electroreceptors work? The answer depends a bit on the specific type of electroreceptor you are talking about. The most widespread and sensitive type of electroreceptor sits at the bottom of slender tubes inside the skin. Because the tubes are often flask shaped, these receptors are called *ampullary receptors* (one type of ancient Roman flask was called an "ampulla"). Ampullary receptors are similar to hair cells in that they have a few microvillar "hairs" extending from their tops and form ribbon synapses with nerve terminals. The voltage across the receptor cell's basal membrane (the one facing inside the body) changes in response to external electric fields, which is thought to open or close (depending on the direction of the voltage change) voltage-gated calcium channels in the receptor cell's membrane. Changes in internal calcium levels then alter the rate of transmitter release, which modulates action potential firing in the postsynaptic nerve terminals.

Figure b6.3 A fish that can both generate and sense weak electric fields. The illustrated specimen of *Campylomormyrus rhynchophorus* belongs to the mormyrid (elephant nose) family of bony fishes. Modified muscle in its tail can generate weak electric fields, and electroreceptors on its head and snout can sense those fields as well as signals coming from moving prey. This specimen is approximately 14 cm long and was captured in an African river. [From Leal and Losos, 2010]

- **Evolution in Action:** Harnesses the explanatory power of the theme of evolution, allowing students to move beyond memorization and develop deep insight and intuition.

NEUROLOGICAL DISORDERS

Box 5.2 *Immune Responses in Alzheimer's Disease*

Alzheimer's disease is the most common form of dementia affecting the elderly. Roughly 3 out of every 1,000 people aged 65–69 have Alzheimer's disease, and this rate rises to ~56/1,000 in people over 90 years old. There are many aspects to Alzheimer's disease (see Chapter 16), but an especially intriguing one is its relationship to the immune system.

A key step in the development of Alzheimer's disease is the accumulation of amyloid β-protein (Aβ). This molecule is normally harmless, but Alzheimer's patients produce large amounts of an Aβ variant called Aβ42 that tends to form toxic aggregates. The largest of these aggregates are visible as *amyloid plaques* within the brain (Figure b5.2). In addition, Alzheimer's patients accumulate within their neurons incorrectly folded forms of a protein called tau. These tau aggregates are called *neurofibrillary tangles* (Figure b5.2). Although some Alzheimer's patients exhibit no plaques or tangles in their brain at autopsy, most neuroscientists believe that plaques or tangles, or both, are closely linked to the causes of Alzheimer's disease. Therefore, researchers have long been keen to prevent or reverse plaque and tangle formation, hoping that this might arrest or reverse the cognitive decline associated with Alzheimer's disease.

A major step toward this goal was taken in 1999 when researchers reported that an Aβ42 vaccine prevents plaque formation in a transgenic mouse model of Alzheimer's disease (Figure b5.2 B). The mice in the original study were injected with Aβ42 and then created their own antibodies. In later studies, the mice were injected with premanufactured Aβ42 antibodies. Either way, the antibodies consistently prevented amyloid aggregation and, in some cases, removed the aggregates that had already formed. In one study, Aβ42 antibody injections even cleared incipient tangles (Figure b5.2 D). These studies have raised hopes that similar vaccines might prevent or reverse Alzheimer's disease in humans as well. Indeed, several Alzheimer's patients injected with a synthetic form of Aβ42 as part of a clinical trial had few plaques in their brains at autopsy. Unfortunately, the treatment did not arrest the cognitive decline, as the patients still progressed to severe, final-stage Alzheimer's disease. Furthermore, the trial was aborted because some of the trial participants developed severe meningitis. This is frustrating and a bit of a challenge to the hypothesis that amyloid accumulations trigger Alzheimer's disease. However, the vaccines may have been given too late in the development of Alzheimer's disease, when the trigger for cognitive decline had already been pulled.

Another potential therapy for Alzheimer's disease is to manipulate the brain's innate immune response. The brains of Alzheimer's patients tend to be chronically inflamed, showing increased expression of cytokines and chemokines, as well as microglia around the edges of amyloid plaques. These findings suggest that the brain's own immune response may be responsible for some of the neu ron death and synapse loss in Alzheimer's disease. Alternatively, the brain's immune response may be a defensive measure, designed to prevent the further growth of amyloid plaques. To distinguish between these possibilities, one would want to know whether depressing the brain's immune response with drugs reduces the symptoms of Alzheimer's disease or worsens them. Unfortunately, the data on this point are conflicting. Overall, it seems likely that the brain's immune response to Alzheimer's disease is both good and bad. It may, for example, reduce the spread of plaques but kill some neurons as collateral damage. Future research will have to tease the various effects apart.

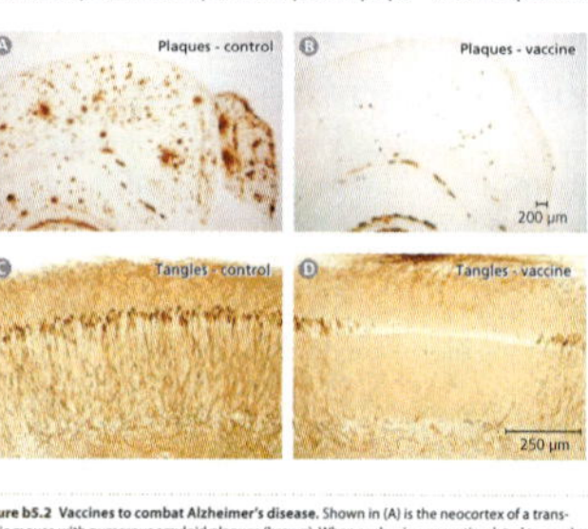

Figure b5.2 Vaccines to combat Alzheimer's disease. Shown in (A) is the neocortex of a transgenic mouse with numerous amyloid plaques (brown). When such mice were stimulated to produce antibodies against Aβ42, the number of plaques was reduced (B). In another study, Aβ42 antibodies were injected directly into the hippocampus of transgenic mice that develop neurofibrillary tangles as well as plaques. Image (C) shows a section through the hippocampus of an untreated control mouse, stained for tangles (brown). The photograph in (D) shows a comparable section from a mouse that received the Aβ42 antibodies. You can see that it contains fewer tangles at the injection site (in the center of the image). [A and B from Schenk et al., 1999; C and D from Oddo et al. 2004]

- **Neurological Disorders:** Motivates students through the presentation of fascinating examples of what happens when the nervous system doesn't function normally.

THERAPIES

Box 2.5 *Mood Molecules*

Among the most successful therapeutic molecules are drugs that improve or stabilize a person's mood. Most of these mood-boosting molecules target a distinct population of neurons that use *serotonin* (5-HT) as their neurotransmitter. The cell bodies of these neurons are located in a few hindbrain nuclei called the raphe nuclei, but the axons of these neurons terminate widely throughout the brain.

Many antidepressant drugs, notably Prozac, selectively inhibit the uptake of released serotonin back into presynaptic terminals. The effect of this reuptake inhibition is that released serotonin molecules remain in the synaptic cleft for a longer period of time and therefore exert a greater effect on their postsynaptic targets. The fact that antidepressants generally boost serotonin signaling suggests that clinical depression results from insufficient serotonin levels in the brain. It cannot be this simple, however, because reuptake inhibitors typically boost signaling within minutes or hours, whereas the positive effects on mood take weeks to manifest. To explain this discrepancy, it has been hypothesized that serotonin reuptake inhibitors cause gradual, long-term changes in the sensitivity and/or abundance of serotonin receptors. This hypothesis is controversial, but it is consistent with the general principle that chronic changes in transmitter abundance cause compensatory changes in the corresponding receptors.

One problem with antidepressant medications is that they tend to have side effects, including daytime sleepiness, nighttime restlessness, nausea, diarrhea, and sexual dysfunction. This is not particularly surprising because many of these drugs also inhibit the reuptake of norepinephrine, which regulates arousal, and because serotonin itself has multiple functions. One reason why there are so many different antidepressants on the market is that each drug represents a different compromise between benefits and side effects. Moreover, treating depression often involves a great deal of trial and error because some compounds work well in one person but not in another. A combination of several different drugs, at carefully adjusted dosages, usually works best.

Given how difficult it is to treat a diseased brain, is it reasonable to expect that we can alter the functions of a healthy brain and cause no harm? Probably not, but many people try. In particular, some people boost their mood by taking illegal drugs. One of these is *MDMA* (3,4-methylenedioxy-N-methamphetamine), also known as ecstasy (or, more recently, Molly). This drug is popular because it induces a strong sense of euphoria. Ecstasy does this mainly by altering serotonin transporter molecules so that instead of taking serotonin back up into the presynaptic terminal, they release it into the synaptic cleft. The result is a massive release of serotonin followed by serotonin depletion. The serotonin release feels good, but the depletion produces an ecstasy hangover that can last for several days and is characterized by a lack of motivation, focus, and appetite. Thus, the temporary boost in mood provided by ecstasy must be paid for later.

Research on rats and monkeys has shown that MDMA can irreversibly damage serotonergic axons in diverse brain regions (Figure b2.3). In these experiments, multiple doses of MDMA were injected into the animals, which is not how humans typically take MDMA. However, in squirrel monkeys even a single oral dose of MDMA reduces serotonin levels for at least 1 week in multiple brain areas. These data strongly suggest that MDMA has serious and long-lasting effects on brain function.

Indeed, studies with humans have shown that heavy ecstasy users have impaired verbal memory, display more impulsivity, and are more likely to be depressed than control subjects. These studies do not control for the fact that regular MDMA users also tend to take other drugs, such as amphetamines, that are already known to cause some brain damage. However, it certainly seems wise to find less risky, more creative ways to boost your mood. Finally, it is important to know that much of the so-called pure MDMA sold "on the street" (including Molly) is hardly pure and sometimes not MDMA at all. This lack of quality control is thought to account for at least some deaths associated with the taking of MDMA-related drugs.

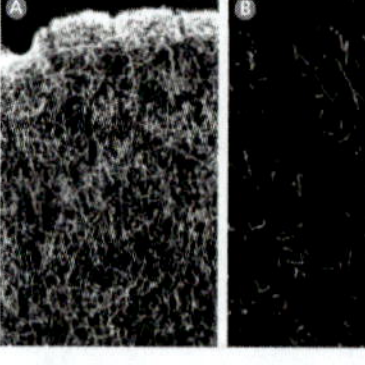

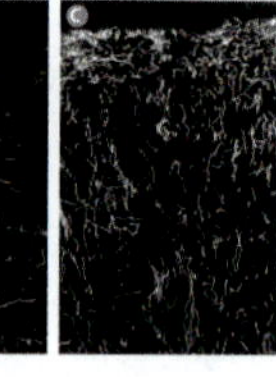

Figure b2.3 MDMA can destroy serotonergic axons. Serotonergic axons were stained with an antibody so that they appear as white lines on a dark background. Panel (A) shows a section through the frontal cortex of a control monkey. Panel (B) illustrates a similar section from a monkey 2 weeks after it received subcutaneous injections of MDMA (5 mg/kg twice daily for 4 days). A depletion of the serotonergic axons is evident. Panel (C) depicts an equivalent section from a monkey 7 years after receiving the MDMA injections. Despite some recovery, a deficit persists. [From Hatzidimitriou et al., 1999]

- **Therapies:** Highlights cutting-edge clinical applications derived from basic scientific research of the nervous system.

Visual Program

A rich ensemble of clear illustrations, vivid photographs, and informative graphs combine to highlight concepts and help students visualize complex neurological processes.

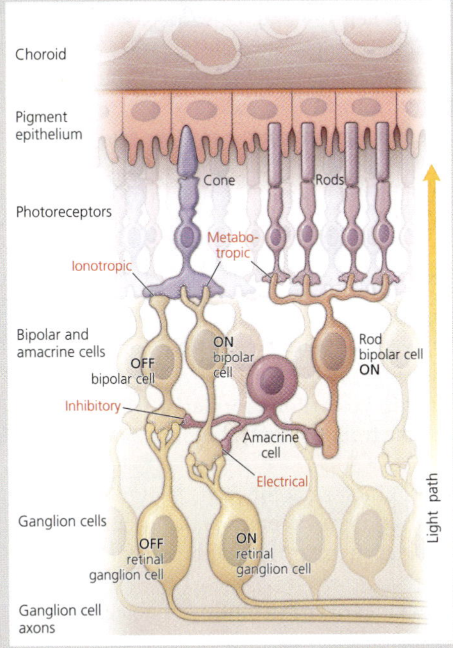

Clear, simple figures bring neurobiology to life.

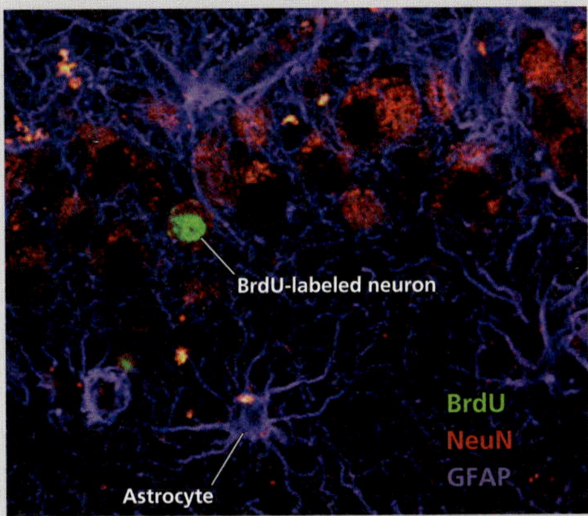

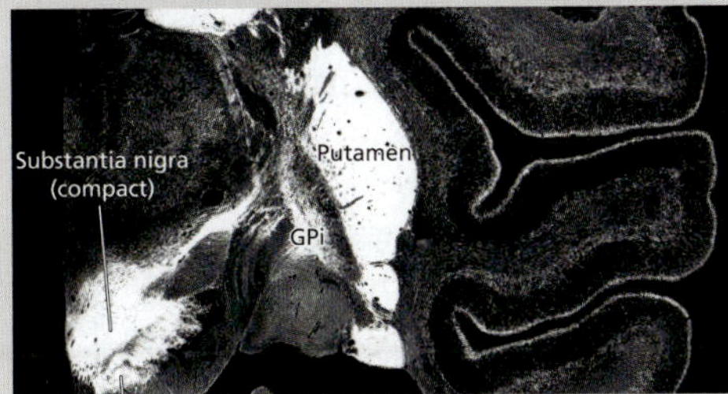

Vivid photographs bring students into the world of neuroscience as it is practiced today.

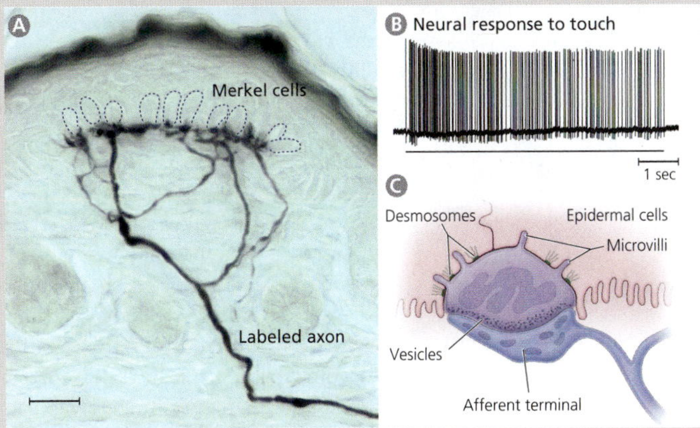

Photos and drawings work together, helping students connect concepts with real biological phenomena.

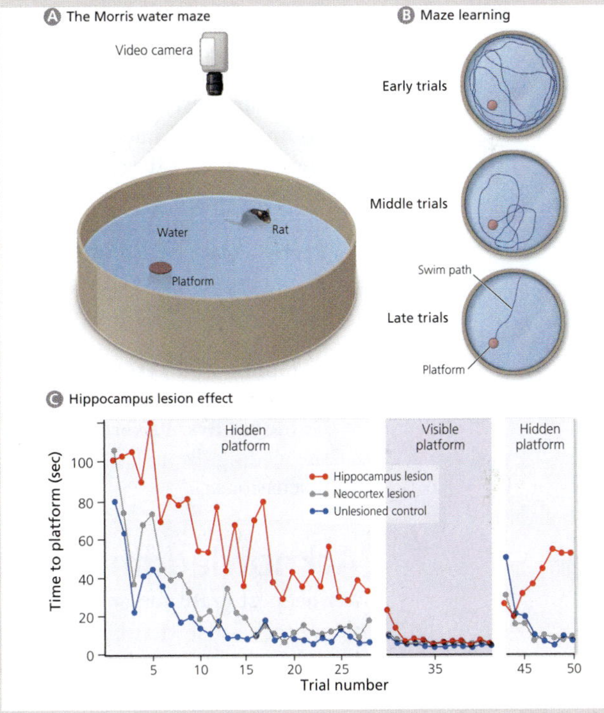

The illustration program consistently features real data and gets students thinking about the process of science.

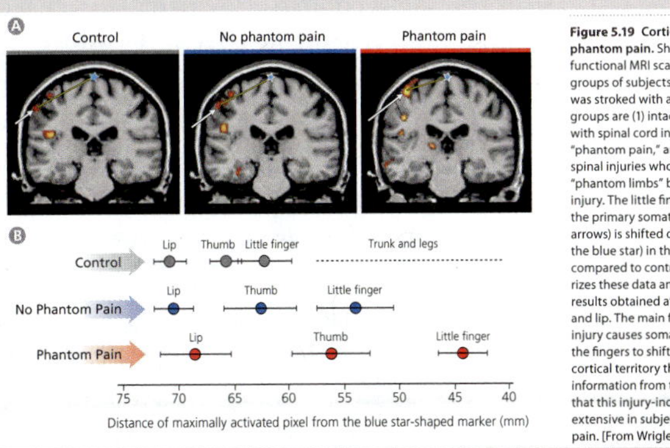

Figure 5.19 Cortical remapping and phantom pain. Shown in (A) are average functional MRI scans taken from three groups of subjects while their little finger was stroked with a brush. The three groups are (1) intact controls, (2) patients with spinal cord injury who feel no "phantom pain," and (3) people with spinal injuries who frequently feel pain in "phantom limbs" below the level of the injury. The little finger representation in the primary somatosensory cortex (white arrows) is shifted dorsomedially (toward the blue star) in the two injured groups compared to controls. Panel (B) summarizes these data and includes additional results obtained after stroking the thumb and lip. The main findings are that spinal injury causes somatosensory inputs from the fingers to shift dorsomedially, invading cortical territory that normally processes information from the trunk and legs, and that this injury-induced invasion is most extensive in subjects who feel phantom pain. [From Wrigley et al., 2009]

Multi-panel infographics with extended figure captions provide students the opportunity to explore topics in depth.

Support Package

Oxford University Press offers a comprehensive ancillary package for instructors and students using *Neurobiology: A Functional Approach.*

For Students

- **Zoomable Human Brain Atlas**: This resource is available for free on the companion website (www.oup.com/striedter). When reading about the brain, it helps to have a 3-dimensional image of what that brain looks like. Therefore, I have created a web-based atlas of Nissl- and fiber-stained sections through real human brains. This interface allows students to zoom in on individual neurons and then zoom out again.
- **Human Brain MRI Atlas**: Available for free on the companion website (www.oup.com/striedter). It includes horizontal, sagittal, and coronal MRI sections through a single human brain., viewable as presentations in most PDF readers.
- **Flashcards**: Electronic flashcards covering the key terms from the text are available for review.
- **Video Guide**: A curated index to hundreds of high-quality and freely-available animations and videos of neurobiological processes. This resource is available to both students and instructors.

For Instructors

The Ancillary Resource Center (ARC), located at www.oup-arc.com/striedter, contains the following teaching tools:

- **Digital Image Library**: Includes electronic files in PowerPoint format of every illustration, photo, graph, figure caption, and table from the text, both labeled and unlabeled versions.
- **Lecture Notes**: Editable lecture notes in PowerPoint format for each chapter help make preparing lectures faster and easier than ever. Each chapter's presentation includes a succinct outline of key concepts and incorporates the graphics from the chapter.
- **Computerized Test Bank**: The test item file includes over 750 multiple-choice, fill-in-the-blank, and short-answer questions.
- **Guide to "Brain Exercise" Questions**: Suggestions about how one might answer the reading "Brain Exercise" questions from the text serve as a guide to classroom discussions.

Acknowledgments

My deepest gratitude goes to my wife Anna, who has supported this project from the outset. I also thank my colleagues at UC Irvine. In particular, Jim McGaugh reassured me that this project was valuable and, more importantly, within my reach. Norm Weinberger's frequent reminders to challenge assumptions prodded me to seek another way of teaching neurobiology. I also thank the students in my own courses, who 7 years ago complained that they felt overwhelmed by facts and were starving for concepts; they prompted me to redesign my course. Subsequent student cohorts provided valuable feedback on endless drafts. This book is for them!

Given the vast and rapidly changing state of neurobiological knowledge, the idea of writing a single-author textbook on the subject seemed preposterous to me initially. Making it feasible required important advances in technology. The *Web of Science* software allowed me to search the literature very efficiently, and the *California Digital Library* made it possible to download most of the relevant papers immediately. Just as vital was the PDF file management program *Papers*, which helped me organize and read the more than 4,500 papers I gathered for this book.

I also thank the Higher Education division of Oxford University Press. Jason Noe guided this project through 6 years of hard work. He, along with editorial assistant Andrew Heaton, helped solicit and evaluate the more than 100 external reviews. Our discussions were vigorous at times, but his ideas have made the book stronger. Jason also selected the art studio Dragonfly Media Group, whose team, including Rob Duckwall, Caitlin Duckwall, and Craig Durant, did a superb job refining my rough drafts of the illustrations. My thanks also go to Editorial Director

Patrick Lynch and to John Haber, the developmental editor who helped me find my tone and style, which was no easy task. Barbara Mathieu oversaw the project's production; Michelle Laseau and Colleen Andrews created a beautiful design and cover; and Deanna Hegle meticulously corrected an alarming number of small mistakes in my last draft.

Institutional support was likewise critical. The University of California and the Department of Neurobiology & Behavior have been extremely supportive. A fellowship from the John Simon Guggenheim Memorial Foundation in 2009 came at the perfect time, boosting my confidence when it was frail. Meanwhile, the National Science Foundation funded my lab research and encouraged me to view this book as valuable outreach. To me, the book embodies the views and principles that the NSF promotes.

Manuscript Reviewers

Finally, I thank the many individuals who took the time to read and comment on my drafts. At UC Irvine, this list includes Michael Leon, John Marshall, James McGaugh, Raju Metherate, Ian Parker, and Norman Weinberger. Then there are the approximately 75 reviewers that OUP commissioned, whom I would like to thank for sharing their insights and suggestions; they contributed greatly to the published work. I am especially grateful to the following reviewers for their feedback and support:

Brenda Anderson
Stony Brook University

Marco Atzori
University of Texas at Dallas

Edwin Barea-Rodriguez
University of Texas at San Antonio

Andy Bass
Cornell University

Erik Bittman
University of Massachusetts Amherst

Martha Bosma
University of Washington

R. Thomas Boyd
Ohio State University

Jessica Brann
Loyola University Chicago

Catherine Carr
University of Maryland, College Park

Barbara Chapman
University of California, Davis

Corey Cleland
James Madison University

Lynwood Clemens
Michigan State University

Melissa Coleman
Claremont McKenna

Barry Condron
University of Virginia

Jeffrey Dean
Cleveland State University

Michael Denbow
Virginia Polytechnic Institute and State University

Harold B. Dowse
University of Maine

Gasper Farkas
The State University of New York, University at Buffalo

Wilma Friedman
Rutgers University

Evanna Gleason
Louisiana State University

Lincoln Gray
James Madison University

John Griffin
College of William and Mary

Adam Hall
Smith College

Melissa Harrington
Delaware State University

John Jellies
Western Michigan University

Glenn Kageyama
California State Polytechnic University, Pomona

Haig Keshishian
Yale University

Roger Knowles
Drew University

Esther Leise
The University of North Carolina at Greensboro

Barbara Lom
Davidson College

Zhongmin Lu
University of Miami

Jane Lubischer
North Carolina State University

Roger Mailler
University of Tulsa

Duane McPherson
State University of New York at Geneseo

Peter Meberg
University of North Dakota

Mary Morrison
Lycoming College

Riccardo Mozzachiodi
Texas A&M University

Alan Nighorn
University of Arizona

Judy Ochrietor
University of North Florida

Marliee Ogren-Balkema
Boston College

Tracie Paine
Oberlin College

Richard Payne
University of Maryland

Alexia Pollack
University of Massachusetts Boston

Astrid Prinz
Emory University

Princy Quadros Mennella
Delaware State University

Siobhan Robinson
Oberlin College

Michael Rowe
Ohio University

John M. Russell
Syracuse University

Donald Sakaguchi
Iowa State University

Damien Samways
Clarkson University

Joshua Sanes
Harvard University

Sara Sawyer
Glenville State College

Laura Schrader
Tulane University

Mark Segraves
Northwestern University

Sansar Sharma
New York Medical College

Annemarie Shibata
Creighton University

Wayne Silver
Wake Forest University

Daphne Soares
University of Maryland, College Park

Olaf Sporns
Indiana University

Malathi Srivatsan
Akransas State University

Thomas Terleph
Sacred Heart University

Karen Thompson
Agnes Scott College

Nathan Tublitz
University of Oregon

Kwoon Wong
University of Michigan

Michele Youakim
State University of New York–Amherst Campus

Heather Yu
Stonehill College

Jokubas Ziburkus
University of Houston

Thank you all! Although you often disagreed with one another and with me, I hope you will find that many of your suggestions have improved the book.

Georg Striedter—August 2015, Irvine, CA; georg.striedter@gmail.com

About the Author

Georg Striedter has been a professor of neurobiology and behavior at the University of California, Irvine, since 1995. For most of that time he has also been a fellow of UC Irvine's Center for the Neurobiology of Learning and Memory. Since 2010 he has served as editor-in-chief of the journal *Brain, Behavior and Evolution*. His research focused on fish brains during his graduate studies but then shifted to birds. In particular, he has studied the neural basis and behavioral functions of vocal imitation in parakeets. More recently, he has examined the connections between brain development and brain evolution, asking how the brains of different avian species diverge during development, and how one might experimentally manipulate neural development to simulate some species differences. Aside from research papers, Dr. Striedter has published a highly regarded book entitled *Principles of Brain Evolution*. Most of this book was written when Dr. Striedter was on sabbatical at the Institute for Advanced Studies in Berlin, Germany. In support of the present book, Dr. Striedter received a John Simon Guggenheim Memorial Fellowship in 2009. Outside of work, he enjoys interacting with animals, including until recently a lovely Bernese mountain dog. He also enjoys listening to classical music, an interest he likes to share with his wife and son, Anna and Ian.

CHAPTER 1

Nervous System Organization

CONTENTS

FEATURES

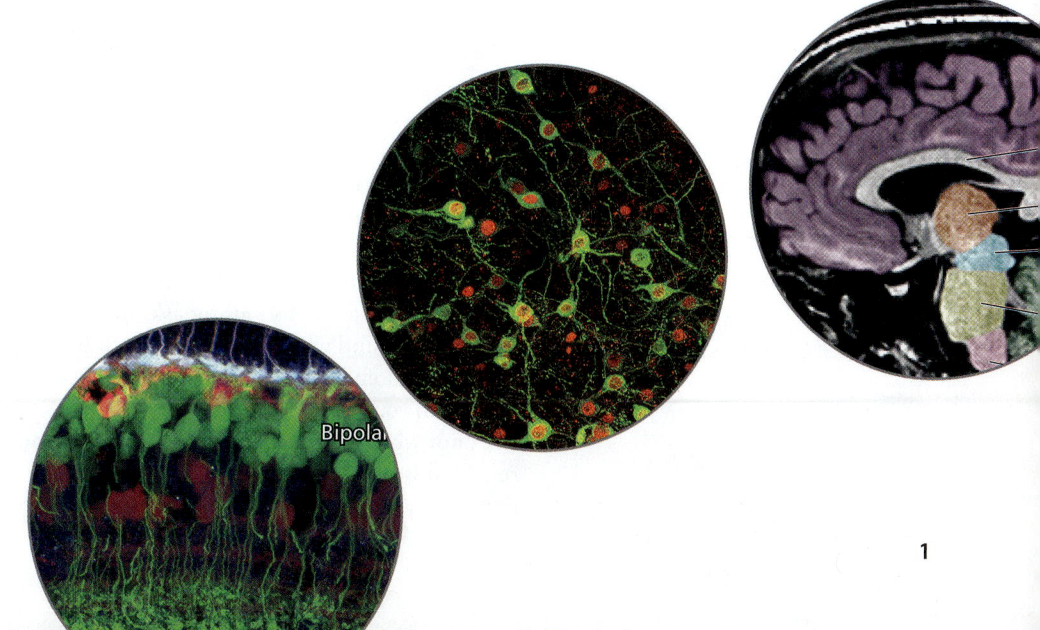

Aristotle, the ancient Greek philosopher-biologist, thought that our brain is used primarily to cool the blood and that the heart controls our thoughts. A very different view was put forth by Hippocrates, another famous Greek and "the father of medicine." After examining many patients with brain damage Hippocrates and his students concluded:

> Men ought to know that from nothing else but thence (*from the brain*) come joys, delights, laughter and sports, and sorrows, griefs, despondency, and lamentations. And by this, in an especial manner, we acquire wisdom and knowledge, and see and hear, and know what are foul and what are fair, what are bad and what are good, what are sweet, and what unsavory . . . And by the same organ we become mad and delirious, and fears and terrors assail us. . . . In these ways I am of the opinion that the brain exercises the greatest power in the man. This is the interpreter to us of those things which emanate from the air, when it (*the brain*) happens to be in a sound state.
>
> Hippocrates. From *The Genuine Works of Hippocrates*, translated by F. Adams (1886), Vol. 2, pp. 344–345.

Although these words sound amazingly modern, Hippocrates knew very little about brain structure or function. He wrote, for example, that behavioral problems arise when our brains are too humid or too dry, and he thought that inhaled air passes directly to the brain before it is distributed to the body. The latter views seem strange today, but Hippocrates lived more than 2,400 years ago. Much has been learned since then about how brains are built and how they work. This book will show you just how far neurobiologists have come.

1.1 How Do Neuroscientists Study the Brain?

Brains can be studied in a variety of ways. One time-honored method for studying complex systems, such as the brain, is to describe their structure, their anatomy. Part of understanding a computer, for example, is to know that it contains a central processor, some memory banks, a disk drive, and the fan. Knowing how these components are connected also helps. Similarly, if you want to explore the nervous system, you might begin your inquiry with a gross dissection. Next, you can cut a brain into thin slices, which can be stained to visualize brain cells and, possibly, their connections. You might also zoom in to examine the genes and proteins expressed within the brain. Using such neuroanatomical and molecular techniques, neurobiologists have gathered vast amounts of data on the nervous system's structural organization. Given this information, one can construct intricate diagrams of how the brain's main components are wired together. Unfortunately, such diagrams, no matter how detailed, do not reveal how nervous systems work. For that, you need some sort of physiological analysis. Thankfully, neurobiologists have been asking "how does the nervous system work" for nearly as long as they have studied neuroanatomy, and they have invented numerous methods to answer this question. We will get to these techniques shortly, but first we consider another sort of question about brains. Let us consider "why questions."

The Value of Why Questions

As you learn more about the anatomical and physiological intricacies of the brain, it is quite natural to wonder: Why? Why is the nervous system organized in this specific way? Why are its parts laid out as they are? Why do the parts act as they do? Young children are notorious for asking such "why questions," sometimes incessantly. Parents often find why questions difficult to answer, and so do most neurobiologists. But the children have a point. If you cannot explain why something works the way it does, then your understanding is incomplete. Knowing, for example, that a cell phone does not work if you remove the battery is good, but knowing why the battery is required would be better. Or consider an airplane's wing. Knowing that a wing's upper surface is more strongly curved than its lower surface is interesting, but a solid understanding of airplane wings requires understanding why that particular wing shape generates lift.

The trouble with answering "why questions" for biological systems is that these systems were not, in contrast to cell phones or airplane wings, designed by engineers. They evolved, and biological evolution isn't guided by any sort of designer. Evolution is based on random genetic variation, including mutations, gene duplications, and genomic rearrangements. These genetic alterations crop up accidentally, not because a struggling organism or some higher intelligence wills them into being. But if evolution involves no designer, how can we possibly explain why any biological features exist? This question has troubled biologists, philosophers, and students of all sorts for years. Yet it was answered long ago by Charles Darwin.

Charles Darwin's Core Idea

Darwin's simple but brilliant insight was that the origin of heritable variation (such as a genetic mutation) may be random, but whether an organism created by this variation survives and leaves its own offspring is decidedly non-random. As Thomas Malthus pointed out in the mid-1800s, organisms generally produce more offspring than their ecosystem can sustain. Therefore, only a fraction of the offspring lives long enough to create offspring of its own. Crucially, an offspring's chances for survival and reproduction in the Darwinian "struggle for existence" depend at least in part on the organism's traits. Some traits are adaptive, meaning that they boost an individual's chances of survival and reproduction; others are detrimental. Building on this key insight, Darwin realized that the adaptive traits, as long as they are heritable, will over time become more common in the population. Given enough time, these changes in trait frequency cause each species to be well suited to its niche. This is the essence of Darwin's concept of **natural selection**.

Because natural selection is not an agent but a process, we cannot say that natural selection explicitly designed organisms to live in their specific niche. However, it is acceptable to say that natural selection leads to improvements in organismal design. The end product may not be the optimal design, but it is generally good design. Design in this biological context is not an action but the outcome of a long process that is devoid of goals, intentions, or intelligence. Some biologists prefer to use the word "adaptation" for "good design" because it is less easily misunderstood, but the two terms are largely interchangeable. We can say, for example, that the gills of fish are adaptations for living in water, or that gills are well designed for gas exchange in an aquatic niche. Importantly, the notion of *evolutionary design* leads directly to the idea that biological structures often reflect clear **design principles**. These principles are often familiar to engineers. Vertebrate eyes, for example, elegantly demonstrate the laws of optics (Figure. 1.1). At other times, nature came up with striking new designs. Smart engineers are well aware of this and quick to learn from nature's accomplishments. Velcro, for example, was invented by a Swiss engineer who had been struck by how stubbornly burrs cling to dog fur.

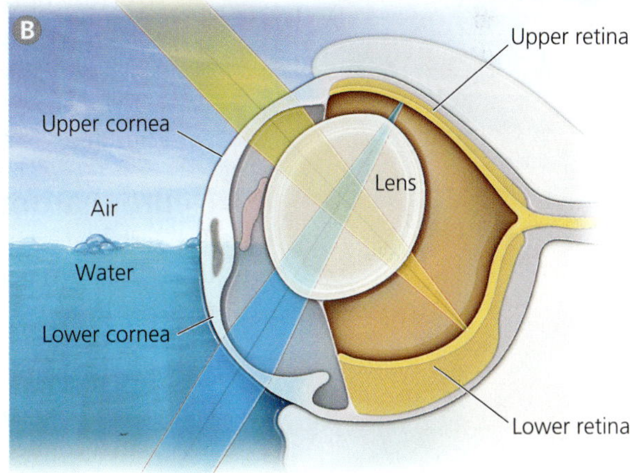

Figure 1.1 Eye morphology reflects the laws of optics. The "four-eyed" fish *Anableps* (A) spends most of its time at the water surface, simultaneously peeking out of the water and looking down into the water. It has only two eyes, but each cornea and retina are divided into top and bottom halves (B). The lenses of *Anableps* are ovoid, which is adaptive because (following Snell's Law) the cornea bends light much more when it interfaces with air than when it faces water. To compensate for this difference in corneal refraction, the lens is more curved where it passes light coming through the water and less curved where it passes light coming from the air. [A from Science Source; B after Sivak, 1976]

Principles of Nervous System Design

Biologists are often hesitant to talk about design features or principles in the nervous system. One reason for this reluctance is that we know much less about nervous systems than we know about eyes or most other organs. The brain really is biology's ultimate frontier. Another problem is that some well-known aspects of the nervous system seem, at first glance, poorly designed. For example, neurons transmit information with far more "noise" (less reliably) than we would tolerate in modern electronic devices. Some neurobiologists conclude from such observations that natural selection is a much feebler "designer" than Darwin had supposed. Yet evidence suggests that some aspects of neural noise perform useful functions that we do not yet fully understand. Furthermore, Darwin and his supporters never argued that natural selection optimizes all of an organism's traits. Since natural selection involves the survival and reproduction of entire organisms, it often "compromises" between various traits and needs. For example, the noisiness of neurons may represent a compromise between the need for reliable communication and the need to conserve energy (see Chapter 2). Such compromises make it difficult to identify design features and principles in the nervous system, but they do not make it impossible.

Therefore, in this book we do not avoid the why questions. Many of the offered answers will be hypotheses rather than absolute truths or demonstrated facts. Now you may ask, shouldn't textbooks be filled only with facts and avoid hypotheses? No. Hypotheses are good and essential; they help you think and learn. As Darwin once wrote, false facts impede scientific progress, but false hypotheses do little harm, for they inspire scientists to disprove them. Indeed, testing hypotheses is the main job of scientists. For students, too, hypotheses are valuable because they help make sense of otherwise disparate, unconnected facts. This is a major benefit of asking why questions: the answers help you integrate and synthesize the facts, making them easier to learn. Some of the answers will be wrong, but scientists accept that all their knowledge might eventually be proven wrong. Therefore, in this book we approach the nervous system from three complementary perspectives: anatomical, physiological, and evolutionary.

1.2 What Are the Basic Components of the Nervous System?

The structure of the nervous system can be described at macro- and microscopic scales. Here we begin with the macroscopic scale of gross anatomy, introducing the principles of neuroanatomical nomenclature and the nervous system's major divisions. Then we zoom in to look at the cells and molecules that are the nervous system's building blocks.

Neuroanatomical Nomenclature

The vertebrate nervous system is divided into central and peripheral nervous systems. The **central nervous system (CNS)** consists of brain and spinal cord. The brain needs little introduction: in humans it weighs about three pounds (1.3–1.4 kg) and sits atop the spinal cord. The spinal cord is a slender structure that weighs about 35 g and extends halfway down your vertebral canal. Any neural tissue outside of the brain and spinal cord belongs to the **peripheral nervous system**. This includes the nerves that connect the CNS to the various sense organs and muscles. The only exceptions to this clear division are the eye and optic nerve, which are considered part of the CNS because they develop from part of the brain. The peripheral nervous system also includes neurons that form distinct clusters, or ganglia, at various locations throughout the body. Most of these neurons belong to the **autonomic nervous system**, so named because it functions fairly independently (autonomously) of the CNS. The largest component of the autonomic system is the **enteric nervous system**. It lies within the lining of your gut and derives its name from the Greek word for "entrails." Although you have probably never heard of this enteric nervous system, it plays a crucial role in digestion and contains more neurons than the spinal cord (200–600 million).

As the term CNS implies, vertebrate nervous systems are highly centralized. Their neurons tend to be clustered in one location, rather than being scattered throughout the body, as they are in some invertebrates, such as anemones or comb jellies. Why are vertebrate nervous systems so highly centralized? Most likely because **centralization** makes neural communication faster and more efficient. If the neurons are close together, then the connections between them can be short, which makes them fast and metabolically cost-effective (more on this in Chapter 2).

Gray and White Matter

Even within the CNS, neurons are clustered, rather than randomly scattered. You may have heard of **gray matter** and **white matter**. Gray matter contains most parts of a neuron: its cell body, dendrites, and synapses (we will define these terms shortly). White matter, in contrast, contains almost exclusively the long processes that connect neurons with one another—the axons. Although some axons run through the gray matter, most of the longer axons are segregated into distinct bundles, called **fiber tracts**, which collectively make up the white matter. The term *white matter* derives from the observation that most long axons in the white matter are covered with a fatty substance (myelin) that appears white in fresh tissue (see Chapter 2). For now, what is important is that the existence of gray and white matter implies that most CNS neurons are clustered into specific regions (gray matter), separated by fiber tracts (white matter). This sort of clustering is analogous to people living in cities that are interconnected by roads. For neurons and people alike, clustering facilitates rapid interactions.

Looking more closely at the gray matter, slicing it thinly and staining it to visualize neurons (see Box 1.1: Neuroanatomical Techniques), one can see that gray matter itself is far from homogeneous. The cell bodies within the gray matter vary in size, shape, and staining properties. Importantly, neuronal cell bodies with similar features tend to cluster together, segregating themselves from cell bodies with different features. Neuroanatomists refer to such cell body clusters as **brain nuclei** (not to be confused with cell nuclei). Some brain nuclei contain only a single type of neuron; others contain several types. However, each brain nucleus contains a different mix of neuron types, and the neurons in each nucleus tend to have a unique set of connections and physiological properties. Therefore, brain nuclei are not just arbitrary collections of neurons, but fundamental units of brain structure and function. This is why neurobiologists consider them important enough to give them proper names. Even brain nuclei with fuzzy, vague boundaries tend to be given names; a good example is "nucleus ambiguus."

Zooming in further still, one can see that many brain nuclei contain multiple subdivisions. These subdivisions are sometimes named for the shape or size of the neuronal cell bodies within them. However, most subdivisions are named for their position within the larger nucleus. Although this is a sensible practice, it can be confusing because neuroanatomists have their own special terminology for spatial positions.

Neuroanatomical Orientations and Section Planes

Instead of using the terms upper, lower, front, or rear to indicate where cell groups are located, neuroanatomists use the terms **dorsal** (from the Latin word *dorsum*, meaning "back"), **ventral** (from *venter*, meaning "abdomen"), **rostral** (from *rostrum*, meaning "snout or beak"), and **caudal** (from *caudum*, meaning "tail"). These terms describe a structure's position relative to the main body axes in a typical four-legged animal such as a dog (see Figure 1.2).

Because humans rarely walk on all fours, neuroanatomists use four additional terms when talking about human brains. These are **superior** (toward the top of the head), **inferior** (toward the neck), **anterior** (toward the face), and **posterior** (toward the back of the head). Although this varied terminology seems awfully complex at first, one quickly gets used to it. A prominent neuroanatomist once told his mechanic about loud noises coming from the ventral side of his car.

Neuroanatomists also use the terms **medial** and **lateral**. Medial means toward the organism's midline, which is defined as the plane that bisects the organism into

RESEARCH METHODS

Box 1.1 *Neuroanatomical Techniques*

Fresh brain tissue is soft and decays quickly. This made it difficult for neuroanatomists to study the nervous system until they learned how to harden and preserve brain tissue in alcohol, various acids, or aldehydes. Neuroanatomy also got a boost in the late 1800s with the invention of *microtomes*, which allowed hardened tissue to be sliced into sections that are just a few microns thick. When such thin sections are stained with dyes, individual nerve cells become visible through good light microscopes, which were developed in the early 1800s. Collectively, these techniques and devices allowed neuroanatomists to make the transition from purely gross anatomy, which consists mainly of dissection, to neurohistology, which examines tissue in cellular detail.

Neuroanatomists have also developed techniques for looking at the molecular features of neurons and glial cells. One such method is *immunohistochemistry*. It uses antibodies against specific proteins to show where in a tissue the targeted proteins are located. By using multiple antibodies that fluoresce in different colors (Figure b1.1), neuroanatomists can determine which proteins coexist within specific areas or, if the spatial resolution is high enough, within specific cells. A closely related technique is *in situ hybridization* (ISH; see Chapter 4, Box 4.1). Collectively, immunohistochemistry and ISH have generated an impressive amount of data about the molecular neuroanatomy of the nervous system. They have shown, for example, which neurons use which neurotransmitters, where the receptors for those transmitters are located, and what kind of intracellular machinery diverse neurons contain.

Neuroanatomists in the 1870s developed the Golgi method, which labels individual neurons, including their axons. In the second half of the 20th century, they invented additional methods for tracing neural connections. Most of these involve injecting a molecule that is taken up by neurons through their dendrites or axon terminals and distributed throughout the neuron by intracellular transport. By visualizing these *axon tracer molecules* with special stains, neuroanatomists can visualize entire tracer-filled neurons whose cell bodies or axon terminals may be quite far from the tracer injection site. This is a powerful technique, especially when multiple tracers are injected simultaneously into different brain regions (see Chapter 8, Fig. 8.25C). Most of what we know about the brain's neural circuits derives from such neuroanatomical tracer experiments.

Some neuroanatomical techniques can be applied to human brains obtained at autopsy, but injecting neuroanatomical tracers into the brains of patients shortly before they die would obviously pose ethical problems. Fortunately, scientists have now developed several methods that can be used to image living human brains. Particularly exciting is the invention of *diffusion tensor imaging,* which allows the visualization and tracing of axon bundles in human brains.

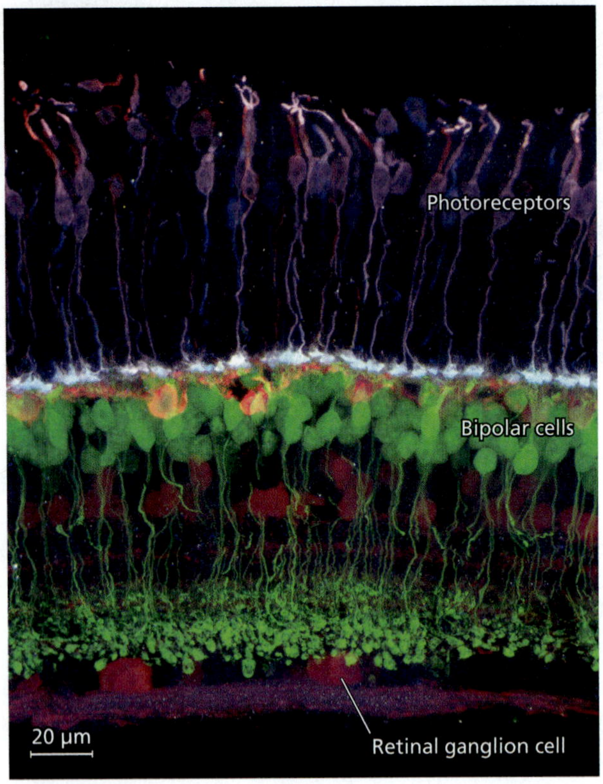

Figure b1.1 Immunohistochemistry reveals key neuroanatomical features. Shown here is a section through a mouse retina, processed with fluorescent antibodies. Cone photoreceptors are stained with an antibody against cone arrestin (purple). Amacrine, horizontal, and retinal ganglion cells are stained with anti-calbindin (orange/red). Bipolar cells fluoresce in green because the bipolar cells in this particular transgenic mouse strain express green fluorescent protein. [From Morgan et al., 2006]

left and right halves. This plane is also known as the **midsagittal** plane (Figure 1.3). It contrasts with two other important section planes, namely, the **horizontal** (axial) and **coronal** planes. The term "lateral" simply means away from the midsagittal plane. Because most neural structures are bilaterally symmetric, neurobiologists rarely label structures as left or right. However, they do use those labels whenever they suspect that structures on opposite sides of the body have different functions. In those cases, the terms left and right refer to the two sides of the organism, not of the observer.

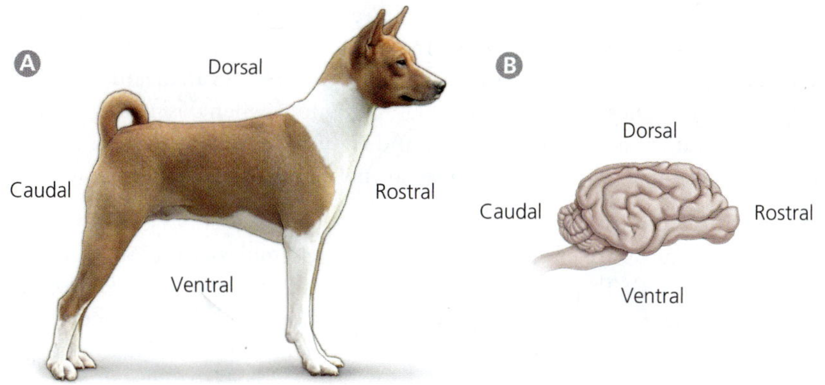

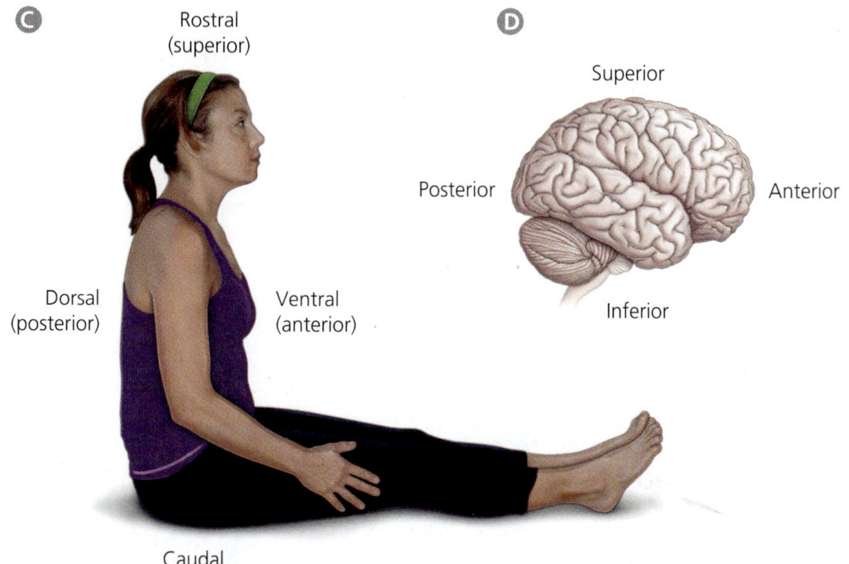

Figure 1.2 Major axes of body and brain. In four-legged vertebrates (A) the terms dorsal, ventral, rostral, and caudal are defined relative to the body's normal standing position. When applied to the brain (B), dorsal means toward the top of the head, ventral toward the bottom, rostral toward the front (tip of the snout), and caudal toward the tail. These terms get somewhat confusing when applied to humans. However, dorsal, ventral, rostral, and caudal are easily defined for humans if you imagine a human walking on four legs or sitting with the arms and legs out in front of them (C). In this sitting position, the terms posterior, anterior, superior, and inferior are synonymous with dorsal, ventral, rostral, and caudal, respectively. Panel (D) shows how these terms apply to human brains.

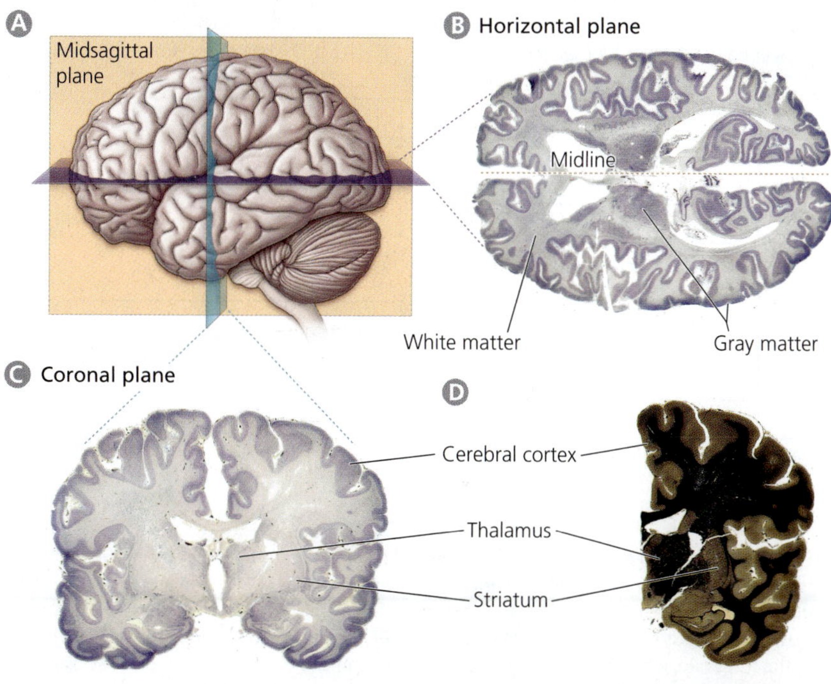

Figure 1.3 The human brain and its three major section planes. Panel (A) depicts a human brain transected by the three most common section planes, which are the midsagittal or midline plane (orange), the coronal or frontal plane (blue), and the horizontal plane (purple). Shown in (B) is a horizontal section stained so that cell bodies appear blue. Panel (C) presents a coronal section in which you can see the cerebral cortex, thalamus, and striatum. Shown in (D) is a coronal section treated with a "fiber stain" that turns white matter black. [Material from the Yakovlev-Haleem collection]

This usage avoids the ambiguities that would result when different observers look at a brain from different vantage points. In addition, neurobiologists use the terms **ipsilateral**, meaning "on or to the same side," and **contralateral**, meaning "on or to the opposite side." These terms are useful, for example, when describing neuronal connections that cross from left to right and right to left with mirror symmetry. In such cases one can say simply that the neuronal connections are contralateral.

Hierarchical Terminology

Individual brain nuclei are often lumped into larger assemblages, such as *forebrain* or *thalamus* (Figure 1.4). These nuclear assemblages are usually defined developmentally. As you will learn in Chapter 4, the brain begins as a single homogeneous entity that becomes progressively subdivided as development proceeds. Neurobiologists follow these developmental divisions and then group adult nuclei according to their developmental origins. Therefore, the hierarchy of names for adult structures is based on the hierarchical nature of brain development. For example, early in development the brain is divisible into three basic parts called hindbrain, midbrain, and forebrain. As these three regions grow, they differentiate into many adult brain nuclei. All the nuclei that develop from the embryonic hindbrain are called hindbrain nuclei. Those that develop from the embryonic midbrain are midbrain nuclei. What remains are

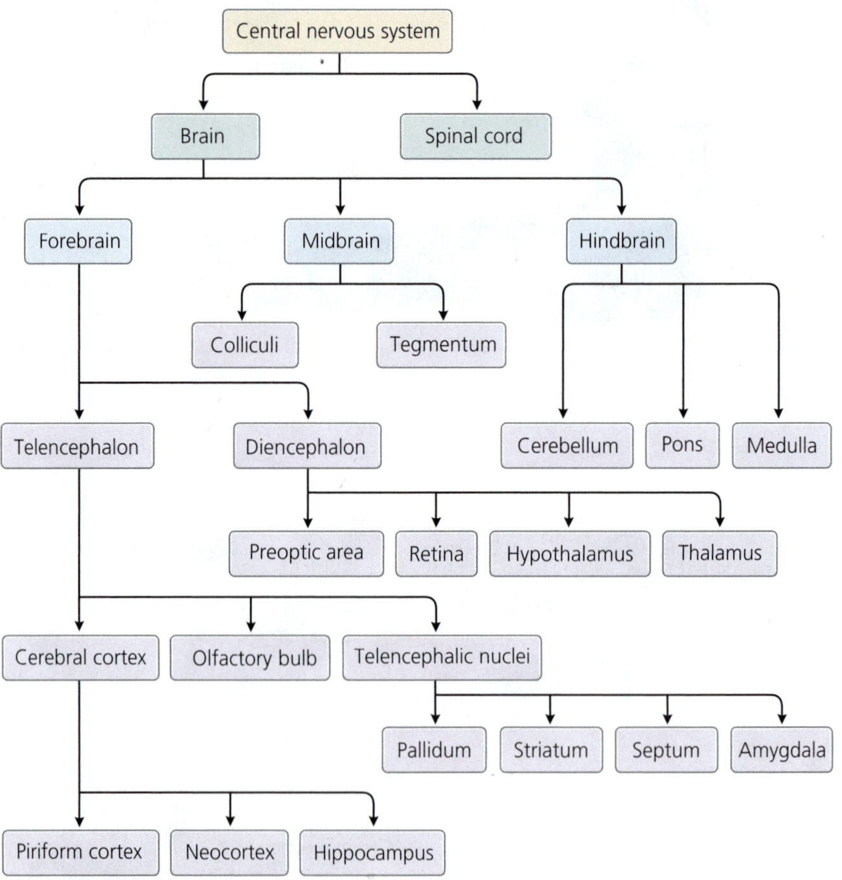

Figure 1.4 Hierarchical organization of the central nervous system (CNS). One of the difficult things about learning the names of the various components of the CNS is that almost every brain region is part of a larger region, which in turn is part of an even more inclusive division. For example, the hippocampus is part of the cerebral cortex, which is part of the telencephalon, which is part of the forebrain, which is part of the brain. This diagram shows the position of most major brain regions within this hierarchical scheme. A few major regions have been omitted, and each of the lower-level structures has still more subdivisions that are not shown.

the forebrain nuclei. Within the forebrain, neuroanatomists recognize thalamic nuclei, hypothalamic nuclei, and several additional aggregates. Again, these groupings are based on the observation that the embryonic forebrain divides into thalamic, hypothalamic, and several other regions before it gives rise to the individual brain nuclei in those divisions.

BRAIN EXERCISE

Why do neuroscientists use the terms "superior" and "inferior" to describe the location of structures in human (but not non-human) brains?

Major Divisions of the Nervous System

Having dealt with the fundamentals of brain nomenclature, let us quickly survey the large assemblages of nuclei you will encounter most often (see the online brain atlas for more detailed images). The **hindbrain** is, as you just learned, one of the three major brain divisions. It is sometimes referred to as the *rhombencephalon* because it is rhomboid in shape, especially in embryos. Within the hindbrain lie the **medulla**, the **cerebellum**, and the **pons**. Of these three structures, the cerebellum is by far the largest (Figure 1.5). Its surface is thrown into numerous tiny folds that make it look like a miniature version of the remaining brain. This explains the cerebellum's name, which means "little cerebrum" or "little brain." The medulla, in contrast, is relatively small, smooth, and oblong in shape. Its name in Latin means "marrow," which nicely describes its position deep underneath the other brain regions. The pons (Latin for "bridge") is named for the fact that it connects the cerebellum to various other brain regions.

The **midbrain** (or mesencephalon) lies rostral to the hindbrain in most animals and superior to it in humans. Its two main divisions are called **colliculus** (meaning "little hill") and **tegmentum**, respectively. The colliculus has superior and inferior subdivisions that are best known for processing visual and auditory information, respectively. The tegmentum contains diverse cell groups, including several that modulate activity in other brain regions. Although the midbrain is relatively small in humans (see Figure 1.5), it is quite large in many other species. For example, the superior colliculus (optic tectum) is large in most fishes, reptiles, and birds, and the inferior colliculus is large in dolphins and bats. Although the midbrain is functionally and embryologically distinct from the hindbrain, neurobiologists often lump these two brain regions together under the umbrella term **brainstem**. This name evokes a

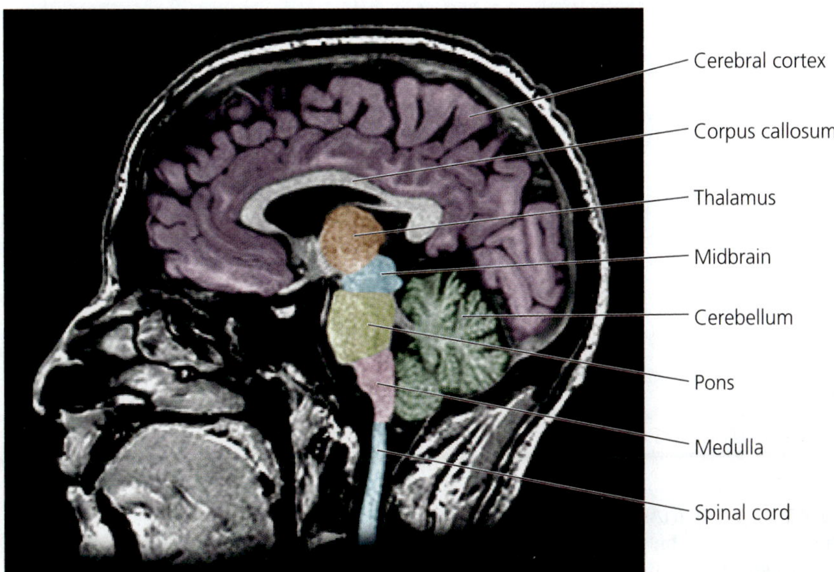

Figure 1.5 Midsagittal section through a human head, showing several of the brain's major divisions. This magnetic resonance imaging (MRI) scan passes down the middle of the head, along the midsagittal plane. The subject's nose and mouth are evident on the left. You can also see several of the brain's major divisions (color-coded). The corpus callosum is a large bundle of axons that interconnects the two cerebral hemispheres. [Courtesy of Craig Stark, PhD]

lovely image of the forebrain sitting like a flower on its stem, but it is ambiguous because some neurobiologists consider brainstem to be synonymous with medulla, while others think of it as the hindbrain, with or without the cerebellum. Some even include parts of the forebrain in their definition of brainstem. In this book, the term brainstem refers to medulla, pons, and tegmentum.

The **forebrain** (or prosencephalon) is by far the largest division of the adult human brain, and it contains numerous subdivisions (see Figure 1.4). The **diencephalon** abuts the midbrain and is divided into several parts, of which the **thalamus** and **hypothalamus** are the most prominent. The forebrain's other major division is the **telencephalon**, which is also divided into several major components. These include the **striatum**, named for its streaked (striated) appearance in tissue sections, and the **pallidum**, which in stained tissue sections often appears pale (pallid). Perched above and largely surrounding these structures is the **cerebral cortex**. The word *cerebral* means "of the brain," and *cortex* is Latin for "rind" or "bark." The most distinctive feature of the cerebral cortex is that its cells aren't segregated into brain nuclei but into discrete layers known as **laminae.** Similar laminae are also seen in the cerebellar cortex, but the number of layers is greater in the cerebral cortex.

Divisions of the Cerebral Cortex

How many laminae does the cerebral cortex contain? The answer depends largely on which part of the cortex you are examining. The most medial and the most lateral portions of the cerebral cortex contain only 2–3 layers. These relatively simple cortices are called **hippocampus** ("sea-horse" in Greek) and **piriform cortex**, respectively (Figure 1.6). In between the hippocampus and the piriform cortex (which is part of the olfactory cortex) lies a huge expanse of 5–6 layered cortex known as **neocortex**. The prefix *neo-* means "new" and indicates that this cortex is significantly more complex in mammals than in non-mammals. An important feature of the neocortex is that it is divisible into several **cortical areas** that differ from one another in their pattern of lamination (how thick and dense the various layers are), in their connections, and in their physiological functions. Small mammals have fewer than a dozen cortical areas. Their number in humans is debatable, but a reasonable estimate is that humans possess ~100 cortical areas.

One of the most obvious characteristics of the neocortex in large primates is that it is thrown into folds (Figure 1.7). The ridges of these folds are known as **gyri**

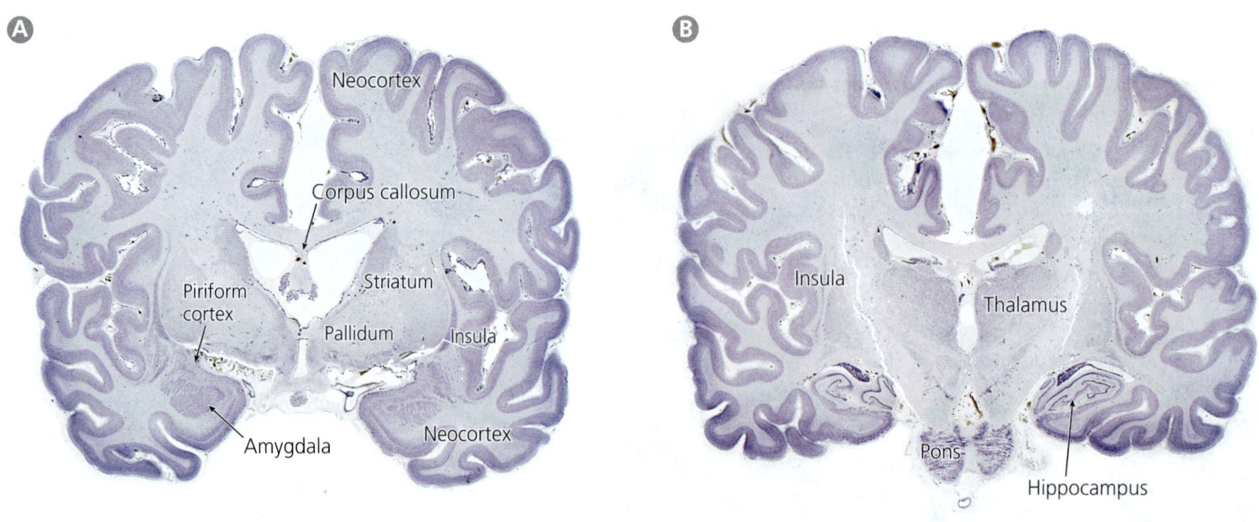

Figure 1.6 Coronal sections through a human brain. The section in (A) is anterior to that in (B). Both sections were stained so that neuronal and glial cell bodies are purplish blue. Only the major brain regions are labeled. For additional sections, higher magnification images, and additional labels, visit the Zoomable Human Brain Atlas (at http://zoomablebrain.bio.uci.edu/). [Material from the Yakovlev-Haleem Collection]

(singular: gyrus) and the valleys between them are **sulci** (singular: sulcus). Why all these folds? The simplest answer is that folding allows the large, sheet-like neocortex to fit inside a reasonably sized skull, much as a sheet of paper will fit inside a cup only if it is folded or crumpled. To better grasp this notion, consider that the surface area of a flattened human brain is about 1,900 cm^2 (~2 ft^2). Imagine trying to cram such a large sheet of tissue into your skull without folding it first! Thus, neocortical folding is an example of good biological design: it evolved because it elegantly solved the problem of how to cram a large cortical sheet into the skull. The same design principle explains the even more intricate folding of the cerebellar cortex.

The human neocortex is divided into four major lobes (Figure 1.8) that are named for the skull bones that cover them. The **frontal lobe** comprises the anterior one-third to one-half of the neocortex. Just posterior lies the **parietal lobe**, and just inferior to this is the **temporal lobe**. The **occipital lobe** occupies the posterior pole of the cerebral hemispheres. The frontal and parietal lobes are separated by a major sulcus called the **central sulcus**, but the boundaries between the other lobes are not so easily defined. In fact, only the temporal lobe is lobe-like in the sense that its anterior end can be pulled laterally away from the underlying neocortex, which is known as the **insular cortex** or insula (Figure 1.7). The medial aspect of the temporal lobe contains the piriform and hippocampal cortices, which were mentioned already. The anterior tip of the temporal lobe contains an almond-shaped structure called **amygdala** ("almond" in Latin and Greek), which comprises several brain nuclei and a few cortical areas.

Olfactory Bulb and Retina

This leaves us with two more forebrain structures to discuss, namely, the **olfactory bulb** and **retina**. The olfactory bulb protrudes rostrally from the cerebral hemispheres and is quite large in many vertebrates. In humans, it is relatively small (compared to the rest of the brain) and lies on the forebrain's inferior surface, overshadowed by the much larger frontal lobes. The retina isn't generally thought of as a component of the forebrain because it occupies the back of the eyeball and is connected to the brain through a major nerve, the optic nerve. However, the retina begins its development as part of the forebrain, positioned between the diencephalon and telencephalon. The embryonic retina gradually moves away from the other forebrain structures and toward the skin. When the migrating retina reaches the skin, a lens begins to form and the eyeball gradually takes shape. Thus, the retina is a forebrain derivative. This may explain why the retina, like many other forebrain components, is a complex structure with multiple laminae (see Box 1.1).

BRAIN EXERCISE

The formation of cortical folds saves space compared to having a smooth cortex. Why is this important to the organism?

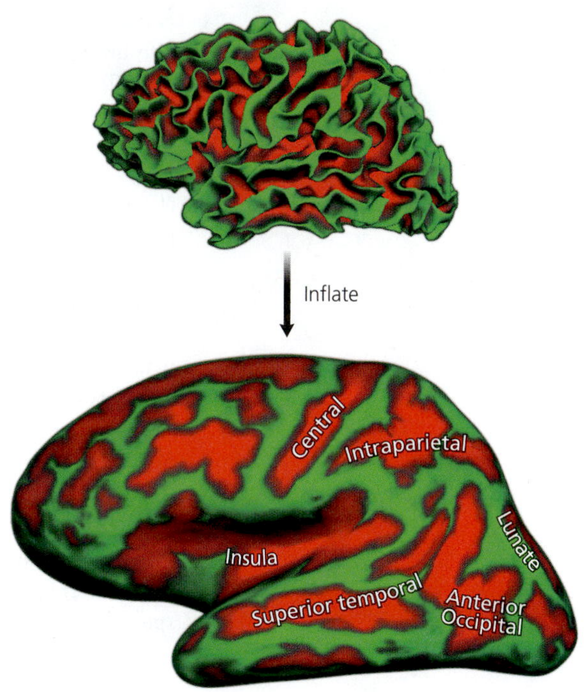

Figure 1.7 The benefits of cortical folding. Shown here is what would happen to a human brain if you inflated it and, thus, smoothed the neocortical folds (gyri are green, sulci are red). As you can see, the inflated brain would take up more volume and, therefore, be more difficult to fit into a skull. The white labels indicate major sulci and the insula, a cortical region that is normally hidden from view by the temporal lobe. [After Sereno and Tootel, 2005; image courtesy of Martin Sereno, PhD]

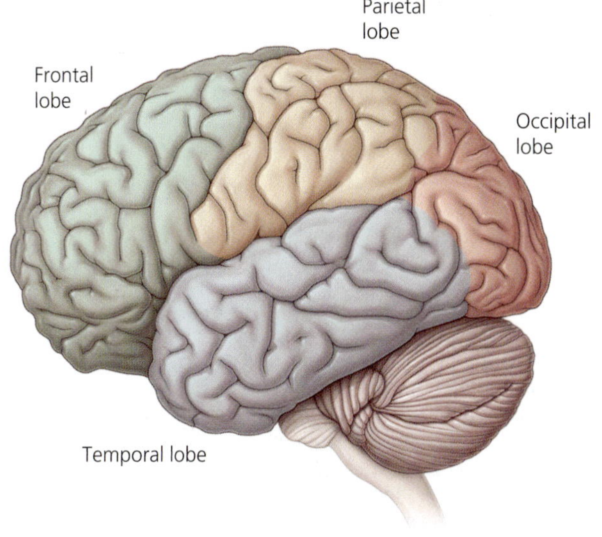

Figure 1.8 Lateral view of a human brain showing its major lobes. The primate neocortex is commonly divided into four major lobes named after the overlying frontal, parietal, occipital, and temporal bones. Hidden beneath the temporal lobe and part of the frontal lobes is the insula, which can be considered a fifth cerebral lobe (see Figure 1.7).

Neurons and Glial Cells

The human nervous system, considered in its entirety, is composed of more than 170 billion cells. Roughly half of them are the nervous system's most famous inhabitants, the neurons. Because neurons are discussed in detail in Chapter 2, we here limit ourselves to a brief survey.

Neurons and Synapses

Neurons are diverse but share a few common features. Virtually all neurons possess highly branched, tapering processes that extend away from the cell body: these are called **dendrites** (Figure 1.9). In addition, almost all neurons possess a long, thin process called the **axon**. It conducts electrical impulses known as **action potentials**, which propagate along axons toward the dendrites or cell bodies of other neurons. The small gaps between the axon of one neuron and the dendrites or cell body of the next neuron are called **synapses**. The human brain is estimated to contain about one quadrillion (10^{15}) synapses.

At most synapses, action potentials don't jump directly from one neuron to the next. Instead, when an action potential travels down an axon and approaches a synapse, the synapse releases molecules called **neurotransmitters**. These neurotransmitters diffuse across a tiny synaptic cleft to the other side of the synapse, where specialized receptor molecules await them. The binding of the neurotransmitters to their receptors elicits small electrical currents in the postsynaptic neuron. Depending on the neurotransmitters and receptors involved, these postsynaptic currents may be excitatory or inhibitory. The former increase the likelihood that the postsynaptic neuron will fire an action potential of its own; the latter decrease it. Excitatory synapses are more common than inhibitory synapses, at least in the cerebral cortex, but both are critical for normal brain function.

Glial Cells

Nervous systems also contain non-neuronal cells. Most of them are **glial cells** (glia). People used to think that the human brain contains 10 times as many glia as neurons, but recent studies have shown the glia-to-neuron ratio for the entire human brain to be closer to 1. To arrive at this estimate, researchers took human brains, put them

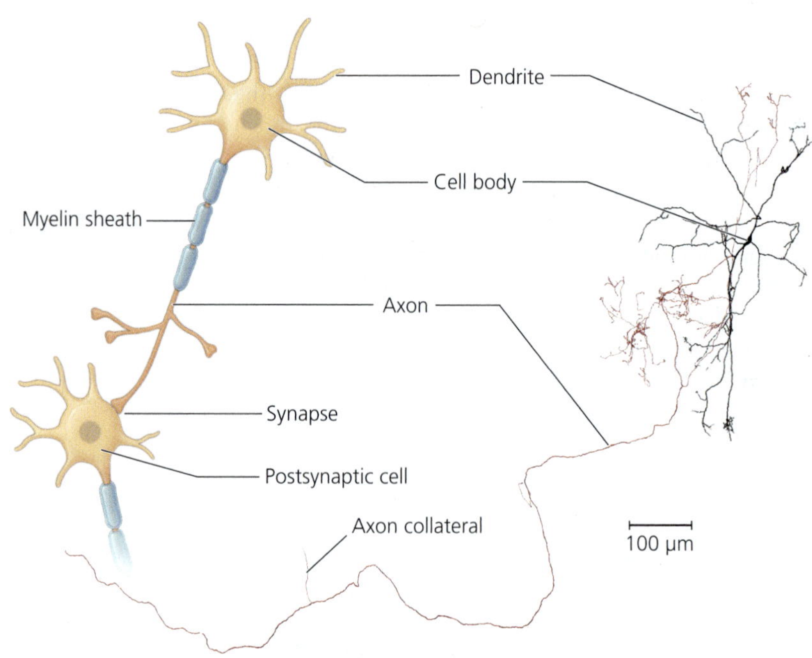

Figure 1.9 Neurons, schematic and real. Shown on the left is a schematic neuron that synapses onto a second neuron. Protruding from the neuronal cell body are several branching dendrites (only the thickest branches are shown) and an axon, which is thinner than the dendrites. Shown on the right is a drawing of a real neuron that was visualized by filling it with molecules of biocytin. This neuron's cell body and dendrites are shown in black. Its axon (red) emits a few side branches, known as collaterals, along its course. [Based on an image provided by Fulva Shah and James M. Tepper, 2013]

into a blender, and created a homogenized "brain soup" in which all cell membranes were ruptured, but the cell nuclei remained intact. The investigators then took a small sample from this homogenate and stained it with antibodies that specifically bind to neuronal cell nuclei. By counting the number of stained nuclei in the sample, the investigators could estimate how many neurons were in the entire homogenate; all the unlabeled nuclei were assumed to come from glial cells. Application of this *isotropic fractionator* method showed that human brains contain approximately 86 billion neurons and roughly the same number of glial cells (although the neuron:glia ratio varies considerably across the major brain regions).

Glial cells come in a variety of types, which collectively perform an impressive array of important functions. **Microglia**—the smallest type of glial cell—help control brain damage by engulfing cellular debris; they are sometimes called the macrophages of the brain. **Astrocytes** (astroglia) resemble microglia in also having a starburst shape (*astron* means "star" in ancient Greek), but they are larger and more diverse in function. Among other things, they play a role in protecting the brain from injury, controlling blood flow, and recycling neurotransmitters. **Oligodendrocytes** are the third major type of glial cell. They wrap themselves around axons, forming a "myelin sheath" (Figure 1.9) that enhances action potential propagation. **Schwann cells** perform an analogous function in the peripheral nervous system. In sum, glial cells perform a much more interesting set of functions than one would suppose from their name, which is derived from the Greek word for "glue."

Brain-specific Genes

Although neurons and glial cells are unique to the nervous system, they share molecular features with other cells of the body. Of the roughly 30,000 genes in the human genome, only 1–2% are expressed uniquely in the brain or at significantly higher levels in the brain than in other tissues. Most of these "brain-specific" genes are probably involved in electrical signaling or synaptic transmission. However, even physiological processes characteristic of neurons, such as synaptic transmission, involve numerous genes that are also used by other cells in different contexts and for different purposes. Similarly, neural development involves many genes that have additional functions outside of the nervous system. Therefore, the nervous system uses a molecular toolkit that is shared broadly with cells in other body parts.

BRAIN EXERCISE

Why do you think so few genes are uniquely expressed in brain tissue?

1.3 What Kinds of Circuits Do Neurons Form?

The most enigmatic aspect of the nervous system is the web of axons through which neurons communicate with one another. Neuroanatomists have used a variety of methods to study these axonal connections, and they have produced some fascinating diagrams of neuronal circuits. Such diagrams suggest that everything is connected to everything else. However, a closer look reveals that this is not the case, nor even possible. If every one of the brain's 86 billion neurons were connected to every other neuron, then the brain would contain almost 10^{22} connections, which would take up an enormous amount of space within the brain (Figure 1.10). As Mark Nelson and Jim Bower put it, if all the brain's neurons "were placed on the surface of a sphere and fully inter-connected by individual axons 0.1 μm in radius, the sphere would have to have a diameter of more than 20 km to accommodate the connections" (Nelson and Bower, 1990, p. 408).

Principles of Neural Circuit Organization

The brain's connections are far from random. Many neuronal pathways are highly stereotyped across individuals; and within individuals, most connections follow definite patterns. The most obvious of these patterns is that neighboring neurons in

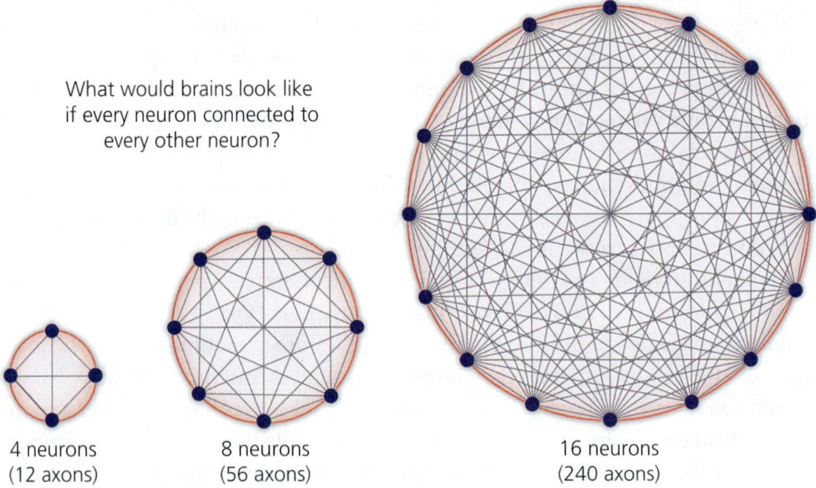

Figure 1.10 What would brains look like if every neuron were connected to every other neuron? With 4 neurons, you would have 12 axons; with 8 neurons, 56 axons; and with 16 neurons, 240 axons. In other words, the number of axons would increase exponentially with neuron number. You can also see that the axons would have to get longer as neuron number is increased. Nelson and Bower estimate that a human brain would have to be about 20 km in diameter if every neuron in it were connected to every other neuron. Thankfully, our neurons are less densely interconnected. [After Deacon, 1990]

one brain region often connect to neighboring neurons in another region. Such orderly, parallel connections are known as **topographic projections** (Figure 1.11). They are commonplace in all vertebrate nervous systems. Why are topographic projections so common? The most likely answer is that topographic projections ensure that neighboring neurons process similar information. This, in turn, allows the connections between neurons that process similar information to be as short as possible. In other words, topographic projections minimize overall connection lengths. Why is wire minimization a good design feature for brains? Because axons take up space, a scarce commodity in brains. Axons also require energy to build and operate; this metabolic energy is likewise limited.

Pathway Divergence

Many of the brain's connections are organized into multistep pathways. Often, these pathways are presented as fairly linear circuits that proceed from the sensory periphery to the neocortex, or from the neocortex to neurons that innervate muscles. This is a neat and tidy view, but reality is more complex. One complication is that axons projecting from one nucleus to another often send diverging branches also to other nuclei. Retinal neurons, for example, send axon branches to multiple targets (Figure 1.12). Motor pathways likewise diverge.

Why do neuronal pathways diverge? Probably because **divergence** allows information computed by one neuron to be transmitted to several other neurons, which can then use the information differently. In the auditory system, for example, one branch of the ascending pathway is specialized for comparing the intensity of sounds reaching the two ears, whereas a second branch is specialized for determining which ear receives a given sound earlier. These bits of information help us localize sounds. Importantly, comparing sound intensities in one branch of a pathway and sound arrival times in another is probably a better circuit design than performing both comparisons in a single linear pathway. In general, performing different kinds of computations in different branches of a diverging pathway is probably adaptive because specialization for one sort of computation generally makes it more difficult to perform other computations. A useful analogy is that training for one skill often interferes with training for another skill. This is why tennis pros rarely play Ping-Pong, and trombonists rarely play the flute.

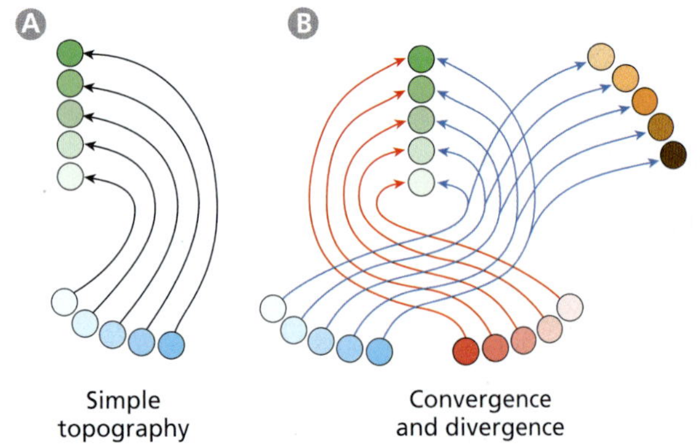

Figure 1.11 Neural circuits are often topographically organized. Panel (A) illustrates a neural projection that is topographically organized, meaning that adjacent neurons in one brain region tend to project to adjacent neurons in the target region. Shown in (B) is a more complex neural circuit that contains converging projections (axons from the red and the blue neurons converge onto the green neurons) and a diverging projection (individual axons from the blue neurons project to both the green and the brown neurons).

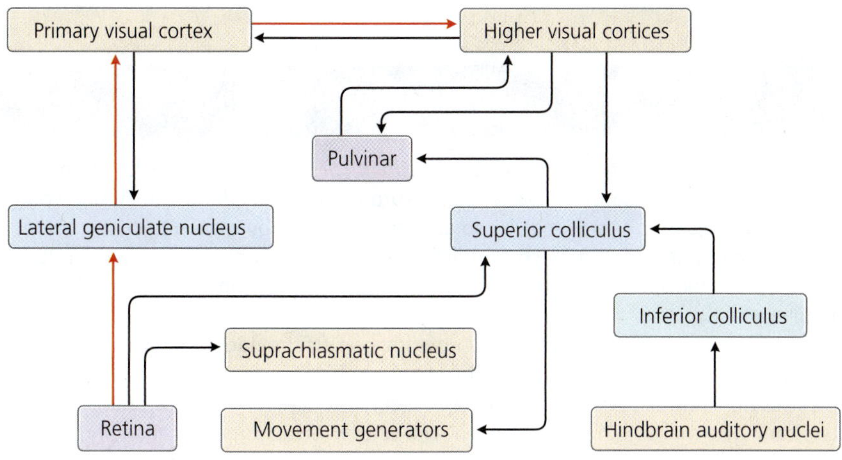

Figure 1.12 Complex visual pathways. The red arrows show how visual information is conveyed from the retina to the lateral geniculate nucleus, goes from there to the primary visual cortex, and ends in the higher order visual cortices. The pathway appears linear, but the primary visual cortex also projects back to the lateral geniculate nucleus, and some of the higher visual cortices project back to the primary visual cortex. In addition, the retina projects to the suprachiasmatic nucleus and to the midbrain's superior colliculus, which receives auditory inputs via the inferior colliculus and projects to motor control regions. Most neuronal pathways feature such diverging and converging projections.

Pathway Convergence

Convergence is the flip side of divergence. Pathways can only diverge so much before they must connect to some other pathways. A major benefit of pathway convergence is that it allows information from different neurons to be integrated. For example, some retinal neurons receive converging inputs from 15–30 photoreceptors. The retinal neurons average all these inputs and thereby filter out the random fluctuations in photoreceptor activity. This makes the retinal neurons less "noisy" and the overall system more sensitive to light. Convergence across sensory modalities can likewise be useful. For example, the convergence of visual and auditory inputs onto neurons in the midbrain's superior colliculus (Figure 1.12) improves an organism's ability to localize objects that can be seen as well as heard. Such cross-modal convergence makes it difficult to determine where one sensory modality ends and the other begins. Even the distinction between sensory and motor pathways is sometimes difficult to draw. Thus, the appealingly simple image of the brain as a collection of linear pathways must be replaced with the more accurate view of a tangled web in which pathway divergence and convergence are commonplace.

Feedback Loops

So far we have discussed neuronal pathways as if they were one-way paths. Indeed, most axons and synapses are unidirectional. However, many "forward" projections from one brain nucleus to another are reciprocated by "backward" projections. Often these backward projections are direct, from the target back to the origin; sometimes they involve relays in other brain regions. Regardless of how many regions are involved, the combination of forward and backward projections sets up potential feedback loops. If the forward connection is excitatory and the backward connection inhibitory, the result is a **negative feedback loop**. Such loops help to regulate excitatory activity within the forward pathway, keeping it within certain limits. If the forward and backward connections are both excitatory, the result is a **positive feedback loop**. As you can probably imagine, positive feedback can amplify activity to the point where it becomes dangerously high (we've all experienced what happens when microphones and loudspeakers form a positive feedback loop). However,

RESEARCH METHODS

Box 1.2 *Physiological Techniques*

The most ancient technique for studying nervous system function is to damage (or, as neuroscientists tend to say, to lesion) part of the brain and ask what the lesioned animals can and cannot do compared to controls. Brain lesions in humans typically result from strokes, which disrupt the brain's blood supply, or from physical injuries, such as a bullet wound. Because strokes and injuries typically encompass multiple brain regions (and parts of brain regions), their behavioral consequences can be difficult to interpret. This problem is minimized in animal studies, where experimenters can make more targeted brain lesions. They can, for example, kill specific types of cells within an area or temporarily inactivate specific neuron populations. Such experiments help neuroscientists identify which structures are necessary for a particular behavior. One caveat is that brain lesion effects are sometimes temporary, with the remaining brain "rewiring" itself to compensate for the damage (see Chapter 5). In such cases, we can infer that the lesioned region was necessary for the behavior shortly after the lesion, but not essential in the long term.

A second major method for studying nervous system function is to stimulate neurons electrically, chemically, or by some other means, and then observe what happens. If the organism alters its behavior, then the stimulated set of neurons is probably involved in producing that behavior. If the full behavior is triggered by the stimulation, we can conclude that the evoked activity is sufficient to generate the behavior. Of course, this does not mean that the stimulated neurons are *solely* responsible for the elicited behavior; they almost certainly work together with other brain regions. In addition, one must keep in mind that artificial stimulation is unlikely to mimic normal patterns of neuronal activity. Indeed, the induced patterns may be so abnormal that ongoing circuit activity comes to a crashing halt. These caveats aside, neuroscientists can combine lesion and stimulation experiments to show that a set of neurons is both necessary and sufficient for a particular behavior. This provides powerful evidence that the neurons are causally linked to the behavior.

The third major technique for learning how brains work is to "listen in" on neural activity while organisms are doing something of interest. To this end, neurobiologists may record the activity of neurons through small wire electrodes (microelectrodes) that are inserted into the brain. Such neurophysiological recording studies usually employ extracellular recording electrodes, which record small voltage changes just outside of the neuron. It is also possible to place recording electrodes inside of single neurons, but such intracellular recordings require very small electrode tips and are extremely difficult to perform in awake, behaving animals.

In humans, neuronal activity can be recorded with relatively large electrodes that are placed on the skull; the resulting data are called the *electroencephalogram* (EEG). Neurobiologists can also place a person's head into a *magnetic resonance imaging* (MRI) machine and monitor blood flow within the brain, which typically increases when neurons in the region are active. Yet another way to "listen in" on brain activity is to visualize the levels of *immediate early genes (IEGs)*, such as the *fos* gene, whose levels of expression increase whenever neurons are unusually active (Figure b1.2). Because the IEG technique requires sacrificing animals just after they performed the activity of interest, this method cannot be used in humans. Moreover, all techniques to monitor neural activity are limited by the fact that correlations between neuronal activity and an organism's behavior need not imply causality. Some other neurons may generate the behavior and simply take the recorded neurons along for the ride.

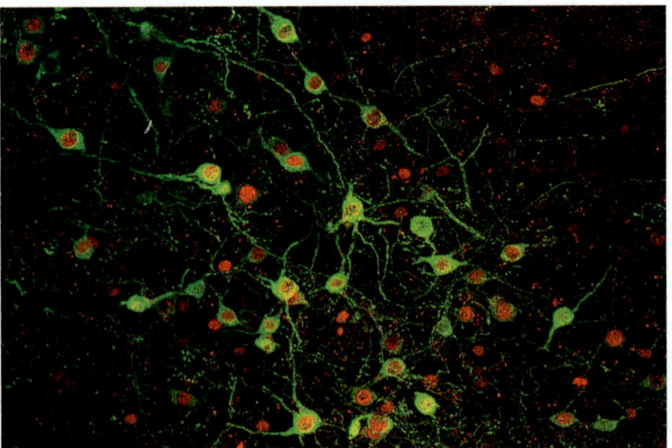

Figure b1.2 Immediate early gene expression reveals recent activity. In this section through the ventral forebrain the neuropeptide vasotocin appears green. The section is also stained with an antibody against the protein made by the immediate early gene *fos* (red). Because FOS levels typically increase a few minutes after a neuron has been electrically active, we can conclude that the neurons with the red nuclei in this image were active shortly before the animal was sacrificed. The animal in this case was a zebra finch that had just taken a water bath. [From Goodson et al., 2009]

positive feedback can also be useful. It can, for example, allow some neurons to maintain their activity in the face of inhibition by other neurons.

Analyses of Circuit Function

Electrical engineers can look at a simple electrical circuit and tell you what it does, but inferring a circuit's function from its anatomy is not so straightforward for neurobiologists. The physiological properties of the individual neurons must also be known.

Are the neurons inhibitory or excitatory? Do they fire rhythmically? How strong are all their connections? Answers to these questions require detailed physiological analyses (see Box 1.2: Physiological Techniques). Even then, the circuit's behavior is often hard to predict.

To tackle this problem, the circuits can be modeled *in silico* (via computer simulations). Such computer models are used to test ideas about how diverse circuit elements work together and can, in turn, suggest novel hypotheses that can be tested on living brains. In addition, neurobiologists can manipulate the circuit and examine how this manipulation affects other brain regions and the animal's behavior. In particular, they may test whether a circuit element is necessary and sufficient for a specific behavior. Such analyses have led, for example, to the discovery of large and powerful *command neurons* needed to trigger escape behavior in crayfish and other species. By combining all of these approaches, it is possible to decipher the nervous system's functional organization.

1.4 What Is the Brain's Functional Architecture?

If you ask someone today how the brain works, you are likely to hear that the brain works a bit like a computer, with inputs, outputs, long-term memory banks, short-term memory (RAM), and some sort of central processing unit that shuttles bits of data around, performing logical operations. You might even be told that brains contain both hardware and software: the former built by genes, the latter by a lifetime of experiences. These ideas contain kernels of truth. However, brains are only superficially similar to the sort of computer we use regularly. Thinking of the brain as a computer is a convenient analogy, but it is only an analogy. People in earlier times used different analogies to contemplate how brains might work.

Early Ideas on Brain Organization

In the17th century, the famous philosopher René Descartes proposed that brains are elaborate hydraulic machines. Specifically, he proposed that stimuli cause delicate fluids termed "animal spirits" to move through hollow nerves. The animal spirits then enter the brain's central cavity, or ventricle, and are reflected off the pineal gland, which Descartes depicted as a teardrop-shaped structure hanging from the top of the cerebral ventricle. According to Descartes, slight movements of the pineal gland change the direction of the reflected animal spirits and thus determine which motor nerves are activated in response to a given stimulus. Moreover, Descartes considered the pineal gland the "seat of the soul" because he thought it could also be moved, ever so slightly, by the immaterial human soul, which thereby modifies behavior. Descartes' model seems strange today because we now know that nerves are not hollow and that information processing occurs in brain tissue rather than the ventricles. Yet his model was sensible given the technology of the time. In Descartes' day, engineers had neither computers nor electricity, but they had learned to build elaborate, automated fountains that sprayed water in complex spatiotemporal patterns, controlled entirely by mechanical switches and valves.

Sherrington and Ramón y Cajal

With the discovery of bioelectricity in 1791 and the development of electrical engineering in the late 19th century, Descartes' hydraulic model of brain function was replaced with models that compared the brain to an electrical circuit. Particularly influential was the work of Charles Sherrington, who in 1906 proposed that the knee jerk reflex involves electrical signals flowing from the patellar tendon into the spinal cord, across one synapse, and out from there to the muscles. As you will see in Chapter 10, this proposal was prescient.

Around the same time, the world's most influential neuroanatomist, Santiago Ramón y Cajal, published a series of landmark papers using the Golgi method, which stains individual neurons in their entirety. Cajal inferred that electrical signals pass from dendrites to the cell body and, from there, down the axon to the synapses, where

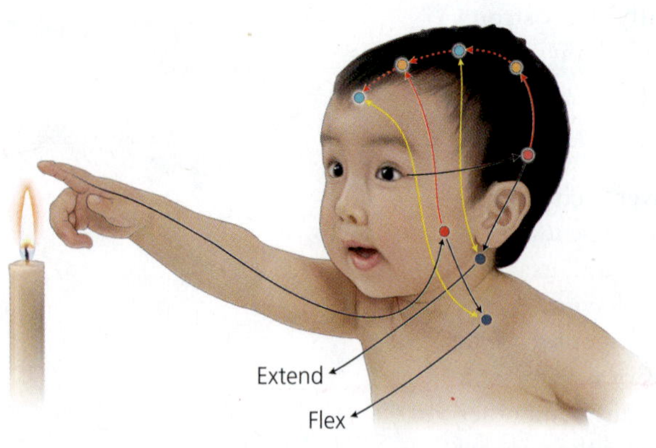

Figure 1.13 The Meynert–James model of brain organization. Activation of subcortical reflex arcs also activates cortical sensory and motor neurons. The near-simultaneous activation of these cortical neurons leads to a selective strengthening of the connections between them (dashed red lines). When the stimulus occurs a second time, a long reflex arc through the cortex is activated in addition to the subcortical reflex arc. William James illustrated this model using a child that learns to avoid a flame. The naïve child sees a flame and touches it reflexively; this causes pain, which leads to a reflexive withdrawal of the hand. After learning, seeing the flame activates a transcortical reflex arc that causes the hand to withdraw long before the child extends the arm. [After James, 1890]

the information is transmitted to the dendrites of other neurons. In addition, Cajal noted that the axons of many sensory neurons form two branches once they enter the spinal cord. One of the branches projects to the spinal motor neurons, as hypothesized by Sherrington. However, the other branch of the sensory axon ascends within the spinal cord toward the brain. Cajal interpreted this finding to mean that the sensory information that triggers simple reflexes is also sent up to the brain, where it can become incorporated into longer, more intricate reflex pathways (arcs).

Meynert and James

This **dual reflex arc model** was stated most explicitly by Theodor Meynert and William James in the late 1800s. They argued that information about sensory stimuli and motor responses is always sent from subcortical regions to the neocortex, even if the reflexive behavior is driven mainly through the subcortical, short reflex arc. Importantly, the cortical neurons are interconnected by a dense web of **association fibers**. According to Meynert and James, experience modifies these association fibers so that the connections between simultaneously activated cortical neurons are strengthened. As a result, subsequent activation of one set of cortical neurons triggers the activation of any other neurons that had previously been coactivated (associated) with them. The experience thus forges a novel transcortical reflex pathway that sits "on top" of the short subcortical reflex arc and modulates motor output through its descending connections.

To illustrate their model, Meynert and James considered how a young child learns not to touch a flame. They proposed that reaching for a flame and then withdrawing the hand in pain involves two short reflex arcs that course through the medulla and spinal cord (Figure 1.13). Activation of these short reflex arcs also activates cortical neurons. Some of the cortical neurons are activated by seeing the flame; others respond to extending the hand, feeling the pain, and withdrawing the hand. Because these groups of cortical neurons are activated closely together in time, the association fiber connections between them are selectively strengthened. Later, when the child sees another flame, the same groups of cortical neurons excite each other in rapid succession, starting with those that are activated by seeing the flame. The cortical chain reaction is so swift that the neurons commanding withdrawal of the hand become active long before the child has extended its hand or felt any pain. Activation of those neurons, in turn, causes the child to withdraw the hand before fully extending it. Hurrah! The child has learned not to touch the flame. According to Meynert and James, this learning involves the formation of a long cortical reflex arc that, once established, preempts the simple, subcortical reflexes that caused the naïve child so much distress.

The Meynert–James model of brain function was important because it wedded the idea that our minds are governed by laws of *mental association* to a concrete neurobiological substrate. It was a bold attempt to solve the mind–body problem (how minds and matter interact). It was also consistent with the available neurobiological data and with the observation that neocortex size appears roughly proportional to an animal's capacity for learned, intelligent behavior. All this explains why traces of the Meynert–James model persist today. For example, the term *association cortex* is still widely used to describe the cortical areas that lie between the main sensory and motor areas (Figure 1.14).

BRAIN EXERCISE

Why do you think Meynert and James used the example of a child seeing a flame to illustrate their dual reflex arc model?

RESEARCH METHODS

Box 1.3 *Opto- and Chemogenetic Techniques*

The classic brain lesion and stimulation techniques affect all neurons in a brain region indiscriminately, making it difficult to assign specific functions to the different types of neurons in the area. Deeply aware of this limitation, Francis Crick—the co-discoverer of DNA—challenged neuroscientists to develop "a method by which all neurons of just one type could be inactivated, leaving the others more or less unaltered" (Crick, 1979, p. 222). That challenge has now been met. Indeed, neuroscientists have developed several techniques for activating or inactivating specific cell types.

Optogenetic techniques use light-activated ion channels from algae or bacteria (prokaryotes) to alter the activity of eukaryotic cells. Most famous is an ion channel called channelrhodopsin-2 (ChR2), which depolarizes cells when it is opened by blue light. Designer viruses can be used to insert the DNA sequence of this microbial ChR2 into animal cells. Importantly, by coupling the ChR2 sequence to a cell-type specific promoter, scientists can restrict expression of the ChR2 to a specific subset of neurons (or glial cells). Once those cells express the foreign ChR2, they can be depolarized by illumination with blue light, which can be delivered deep into the brain through an implanted optical fiber (Figure b1.3). Carefully timed pulses of light will activate the neurons with millisecond precision. Because the procedure can be performed on awake animals, experimenters can observe how the neuronal activation alters behavior. Analogous experiments employing halorhodopsin, a bacterial chloride pump that hyperpolarizes cells, can be used to silence infected neurons. With this technique, modern neuroscientists have answered Crick's challenge.

The power of optogenetics is well illustrated by a recent study on the arcuate nucleus (Figure b1.3). This hypothalamic nucleus contains multiple types of neurons, some of which are thought to suppress feeding behavior while others increase it. Because these neuron types are intermingled within the arcuate nucleus, experiments that lesion or stimulate the entire nucleus often yield unclear results. To overcome this problem, experimenters used transgenic mice and a recombinant virus to express ChR2 selectively in arcuate neurons that express agouti-related peptide (AgRP). When those ChR2-expressing neurons were then activated with blue light, the mice began to feed voraciously until the light pulses were stopped (Figure b1.3B). In a complementary experiment, ChR2 was expressed in arcuate neurons that express a different peptide called proopiomelanocortin. Prolonged activation of these other neurons caused the mice to eat significantly less than control mice that did not express ChR2. Thus, optogenetics made it possible to analyze the circuits that control feeding with unprecedented precision.

Chemogenetics is conceptually similar to optogenetics but uses modified neurotransmitter receptors, rather than opsins, to modulate neural activity, and drugs, rather than light, to activate those receptors, which are often called Designer Receptors Exclusively Activated by Designer Drugs (DREADDs). For example, scientists have mutated a muscarinic acetylcholine receptor so that it can no longer be activated by acetylcholine but, instead, responds to *clozapine N oxide* (CNO), which is otherwise pharmacologically inert and easily administered to animals. Scientists have also created designer receptors that trigger specific intracellular signaling cascades when they are activated by CNO, other designer drugs, or, in some cases, light. Collectively, these approaches have allowed investigators to explore the functions of specific cell types (both neurons and glia) much more effectively than the traditional lesion and stimulation techniques had permitted.

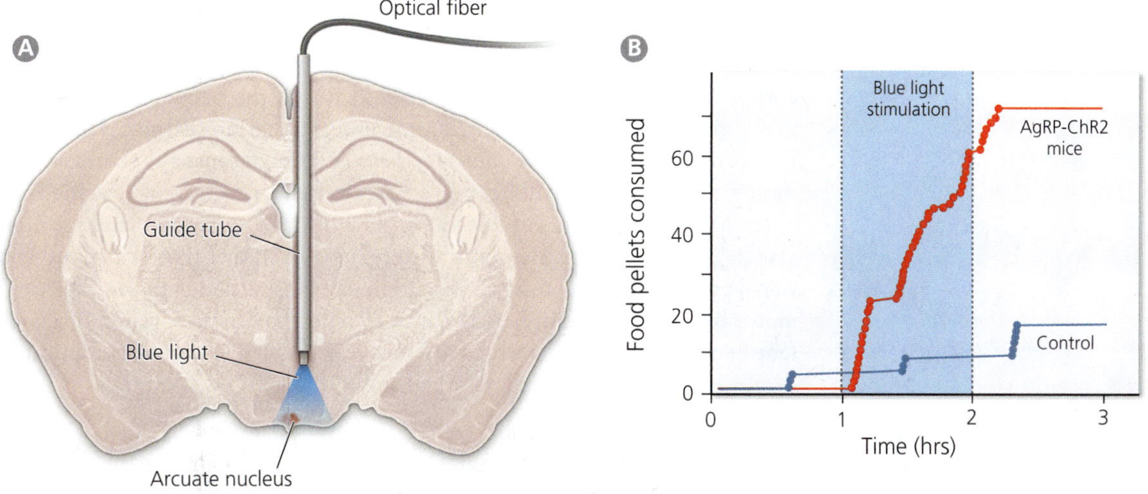

Figure b1.3 Optogenetic stimulation of feeding. Mice were engineered to express channelrhodopsin-2 (ChR2) in a subset of neurons of the hypothalamic arcuate nucleus that normally express agouti-related peptide (AgRP). The transgenic neurons were then activated by shining blue light onto them through an implanted optical fiber (A). When the blue light stimulation was turned on, the transgenic mice ate far more food pellets than control mice that did not express ChR2 in the AgRP neurons (B). [After Aponte et al., 2011]

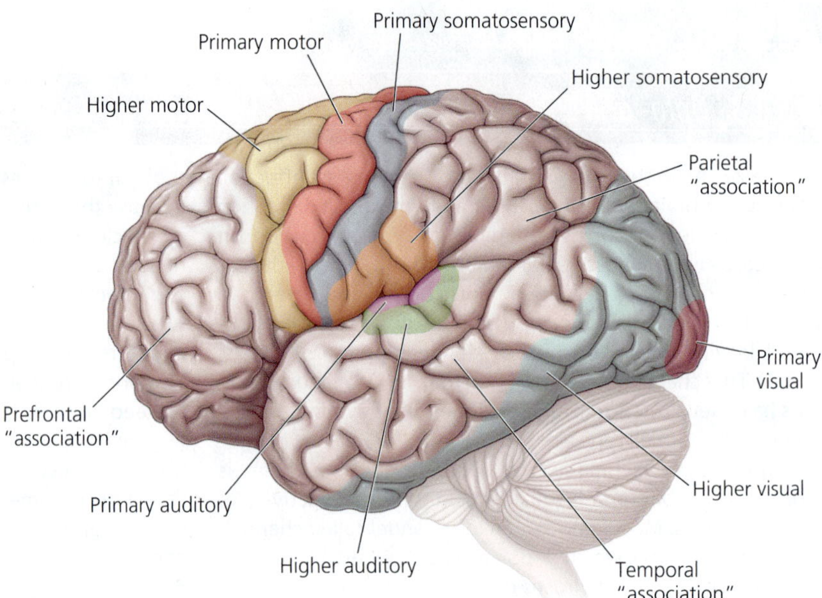

Figure 1.14 Major cortical areas. Human neocortex is divisible into primary sensory and motor areas, higher level sensory and motor cortices, and several "association" cortices. The word "association" is placed in quotation marks because the functions of the so-called association cortices are more complex and more varied than the word "association" implies. The primary auditory cortex is located on the superior surface of the temporal lobe and not visible from a strictly lateral view.

Modern Views of Brain Organization

Whereas some aspects of the Meynert–James model of brain organization seem remarkably modern, the model fails to account for much of the neuroanatomical and physiological data that have been gathered in the last 100 years. These newer findings present us with such a complex network of neuronal connections that it is difficult to glimpse a basic plan of how the brain is organized. Yet we should try. As Cajal noted in 1909, "to extend our understanding of neural function to the most complex human physiological and psychological activities, it is essential that we first generate a clear and accurate view of the structure of the . . . human brain itself, so that the basic plan—the overview—can be grasped in the blink of an eye." Although there is no consensus modern view on the brain's basic plan, and new techniques reveal ever more details (see Box 1.3: Opto- and Chemogenetic Techniques, p. 19), we can contemplate what features must be added to the Meynert–James model to bring it up to date.

The Shrinking Association Cortex

Meynert and James viewed the human neocortex as consisting of a few small sensory and motor areas separated by a large expanse of undifferentiated association cortex. More recent discoveries have shown that the neocortex actually contains many different sensory areas, including at least a dozen distinct visual cortices, as well as numerous premotor areas. As these additional sensory and motor areas were discovered, the regions thought of as association cortex shrunk. Furthermore, connectional and functional studies have shown that the various sensory, motor, and association cortices are hierarchically organized and that there are multiple bidirectional pathways along which information can be shuttled between the sensory and motor hierarchies (Figure 1.15). Thus, Meynert and James' vision of a single long loop coursing through the neocortex must be replaced by a more complicated image of multiple loops, all stacked on top of one another.

Widespread Synaptic Plasticity

Meynert and James believed that learning occurs primarily in association fibers that connect every cortical sensory neuron to every cortical motor neuron, at least in young children. With learning, the connections between simultaneously activated neurons are selectively strengthened, thereby forging new cortical reflex arcs. This idea is remarkably similar to modern theories of *synaptic plasticity* (see Chapter 3). However, we now know that synaptic plasticity is not limited to association cortex but exists even in primary sensory and motor cortices. In addition, regions outside of the neocortex play a role in learning and memory. These include the hippocampus, amygdala, striatum, and cerebellum. The hippocampus is particularly interesting because it is probably the only brain region that harbors the sort of dense all-to-all neuronal connections that Meynert and James considered crucial for associative learning.

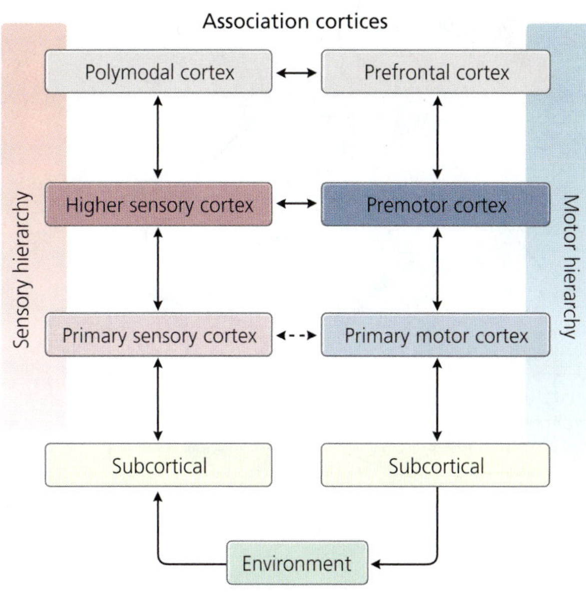

Figure 1.15 Parallel sensory and motor hierarchies. This diagram emphasizes that the cortical motor pathways are just as hierarchical as the sensory pathways. It also shows that sensory and motor areas are reciprocally interconnected, especially at the higher cortical levels. [After Fuster, 2003]

Central Pattern Generation

Another outdated aspect of the Meynert–James model is the notion that all behavior is triggered by sensory stimuli. James acknowledged that ideas and memories can drive behavior, but his model of the brain was based entirely on reflex arcs, which require sensory stimuli to elicit ideas and memories as well as simple motor responses. As subsequent research has shown, this emphasis on reflex arcs is excessive. Some behaviors, including locomotion and breathing, can be generated even without sensory inputs. These behaviors are driven by *central pattern generators* in the brainstem and spinal cord. Although the pattern generators receive sensory input, this merely modulates the intrinsically generated activity, adapting it to current conditions. As shown in Figure 1.16, the pattern generators can also be modulated by descending projections from the neocortex and from "behavior controllers" in the midbrain and hypothalamus. Thus, motor control involves not just motor neurons and motor cortex, as Meynert and James had thought, but a multitiered hierarchy of subcortical as well as cortical structures.

Behavioral State Modulation

A fourth important element that we must add to the Meynert–James model of brain organization is *behavioral state modulation*. As you know, organisms may be sleeping or awake, drowsy or alert, distracted or highly attentive to their surroundings. Modern neuroscience has shown that these variations in an animal's behavioral state are controlled by specialized neurons that tend to have widely dispersed projections and use distinctive neurotransmitters. Particularly important are midbrain neurons that release dopamine in response to unexpectedly rewarding stimuli, and neurons in the tegmentum that release norepinephrine when animals become aroused.

With these additions to the Meynert–James model, the possibilities for information flow within the brain become extremely diverse. Instead of dealing only with transcortical reflex arcs that modulate subcortical reflexes, we must now contemplate a great diversity of looped pathways, diffuse projection systems, and intrinsically generated activity, all modulating each other. These circuits are shown schematically in Figure 1.16. Although this diagram is hardly a basic plan that can be "grasped in the blink of an eye," it will seem less daunting once you have finished this book.

BRAIN EXERCISE

As modern technology evolves, how do you think this will affect our current understanding of the brain?

1.5 How Can Scientists "Reverse Engineer" the Brain?

Given the difficulty of grasping the nervous system's overall organization, it is natural for scientists to focus their attention on smaller, more manageable chunks of the overall system. Indeed, many neurobiologists focus their research on a specific brain nucleus or cortical area, or on a particular molecule. They ask: What does this

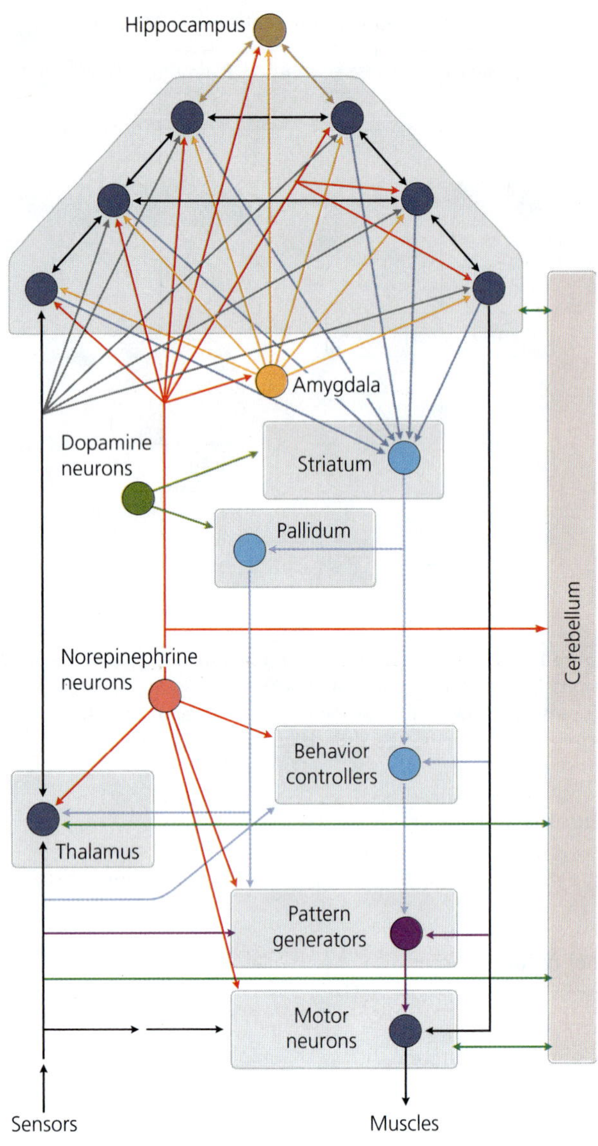

Figure 1.16 A model of the brain's overall organization. This schematic combines the Meynert–James model of brain organization (Figure 1.13) with the idea that the neocortex contains parallel sensory and motor hierarchies (Figure 1.15), central pattern generators, and modulatory systems. Even so, it is far from complete. [After Swanson, 2005]

structure do? What is its function in the larger scheme? To answer such questions, scientists typically inactivate or lesion (damage) the structure, stimulate it experimentally, or monitor its activity during some process or behavior. As scientists perform these tests, they must also ask themselves which specific behaviors or processes they're trying to explain in neural terms. This sounds like a simple question, but it can be surprisingly complex.

Consider, for example, how you might determine the effects of a specific brain lesion (e.g., of the hippocampus) on an animal's behavior and mental capacities. What is wrong with the brain-lesioned animals? To answer this question, you cannot simply observe the animals in their home cage or natural environment. You will have to test the animals somehow and then compare their performance to that of unlesioned controls. Which tests should you select? Does it matter which tests you choose? Yes it does because you will want to say more than simply that performance was impaired in some tests and not others. You will want to interpret the test results in terms of the behavioral or cognitive functions that are impaired. If the behavioral tests were well designed, they should let you draw these inferences. Therefore, your choice of tests should be governed by some hypotheses about which behavioral or cognitive functions you think might be impaired in the brain-lesioned animals. Those hypotheses would be based on a larger framework of ideas about the kinds of behavioral and cognitive processes that are at work within the animal and how those processes relate to one another. Whether or not you are fully aware of this underlying functional framework, it will influence which tests you choose and how you interpret the test results. Furthermore, a variety of different frameworks exist.

Functional Decomposition Strategies

To grasp the challenge posed by this problem of functional decomposition, as engineers call it, imagine that you were asked to build an intelligent robot. How would you proceed? Your first step would likely be to ask the client what they want the robot to do. Once you have this list of desired functions or capabilities, you need to ask yourself how these functions are interrelated (Figure 1.17). Next, you have to break these functions down into smaller processes and, most likely, add some functions that the client forgot to specify (e.g., "recharge when out of energy"). All these functions must then be assembled into a workable flow chart or control architecture. Importantly, different engineers might come up with very different flow charts (see Figure 1.17 A vs. B). Only after you have decided which architecture to implement would you start building the robot. From this engineering perspective, the study of functional neurobiology is an exercise in **reverse engineering**. You are presented with a brain and must deduce its various functions, their interrelationships (the flow chart), and all the implementation details.

How do neuroscientists come up with flow charts for the behaviors or processes that interest them? Two main approaches are used. The first is psychological; it attempts to deduce the functions of the mind through careful introspection and experiments. **Neuropsychologists** then attempt to link the psychological functions to neurobiology. The second approach is guided by ethology, which is the study of the behavior of animals in natural environments. At the core of the ethological approach is the task of identifying the discrete behaviors that organisms engage in and then determining how those behaviors are related to one another and the environment. Ethologists who ask how these behavioral processes are implemented in the brain are called **neuroethologists**. These scientists try to explain outwardly visible behavior, whereas neuropsychologists are interested in visible behavior mainly as a window into the mind.

These two approaches are explained more fully in the following sections. Because both approaches have benefits and limitations, we will not choose between them in this book. Instead, we'll seek to integrate the two. In our world today, the once-rigid boundaries between psychology and ethology are disappearing anyway, as is evident from the emergence of new research fields with names such as *evolutionary psychology* and *cognitive ethology*.

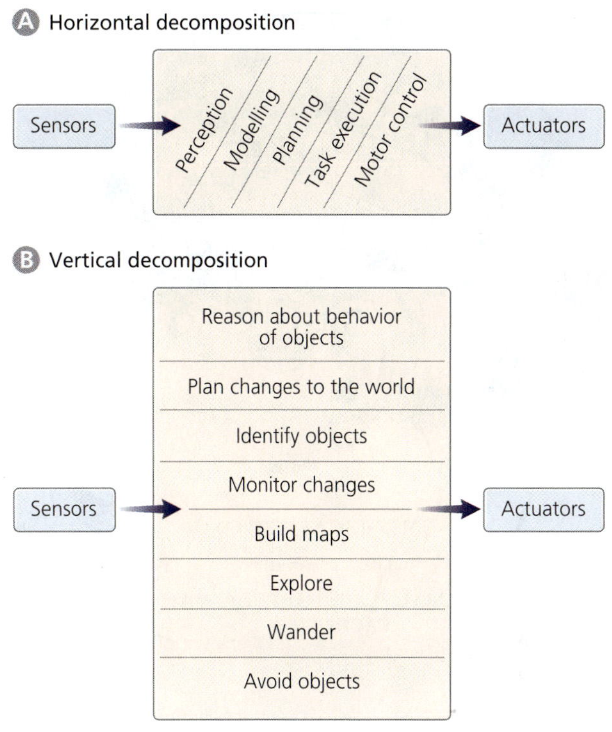

Figure 1.17 Functional decomposition. As you investigate how nervous systems work, you must ask yourself which processes you want to understand. You must decompose the organism's behavior into specific components (functions) and determine how they relate to one another. Engineers who build intelligent robots encounter an analogous problem: they must determine which functions they want to implement and how those functions are related to one another. Historically, robotics engineers have solved this problem in two very different ways. Some have opted for "horizontal decomposition" (A). In this approach perception comes before planning, which comes before action (produced by actuators). Other robotics engineers employ "vertical decomposition" (B), which allows different functions to occur in parallel. [After Brooks, 1999]

Neuropsychology

One of the earliest attempts to understand the mind–brain relationship was made by Franz Joseph Gall in the early 1800s. Gall was fascinated by the observation that talents and deficiencies are often highly specific and differentially distributed across a population. Some people are good at math, for example, but terrible at verbal communication. Based on such observations, Gall divided the human mind into 27 mental "faculties," including sense of place, sense of language, memory for people, cleverness, poetic talent, and arrogance. Gall believed that these mental faculties correspond to 27 "cerebral organs." Unfortunately, he had no way to visualize the postulated organs. Because of this limitation, Gall studied human skulls, which he assumed to be shaped by the brain. Specifically, Gall looked for persons who excelled or were deficient in some mental faculty and then asked whether their skulls had peculiar bumps or indentations. From these data, Gall and his fellow **phrenologists** constructed phrenological maps (Figure 1.18). These maps were wildly popular but controversial even in Gall's day. Today, they are mere curiosities because brain shape is only grossly related to skull shape and, in any case, has no clear link to brain function.

Psychology has come a long way since Gall invented phrenology. An important breakthrough came at the end of the nineteenth century, when German psychologist Hermann Ebbinghaus published a detailed experimental study of human memory, using himself as the only subject. This work convinced many psychologists that the mind could be studied experimentally. Before then, it had been studied almost exclusively by introspection or by the sort of comparative analysis that Gall had conducted. Experimental psychology then flourished in the twentieth century and, after behaviorism had run its course, psychologists began to break the mind down into its fundamental components. For example, memory was found to be divisible into several different types, including memory for motor skills, word meanings, and life events. Although debates continue about how best to carve up memory and other mental processes, psychologists now have a great deal of experimental evidence relevant to such questions. These data have led to good models and flow charts of how the human mind may work.

Neuropsychologists sometimes disagree about which models of the mind are best. For example, different psychologists may endorse different classifications of learning and memory, which then creates debates about which functions of the mind

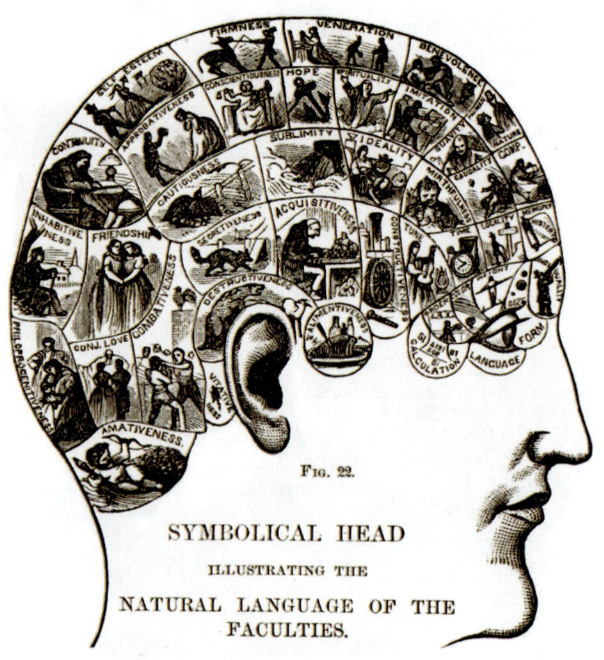

Figure 1.18 Phrenologists mapped mental faculties onto the surface of human skulls. Shown here is a phrenological map published by Samuel Wells (1870) in a little book called *How to Read Character*. In this book Wells summarizes, in old-fashioned language, the phrenologists' credo: "As is the brain, so is its bony casement, the cranium, on which may be read, in general forms and special elevations and depressions, and with unerring certainty, a correct outline of the intellectual and moral character of the man." [Wells, 1870, p. vi]

are implemented in the brain. Another problem is that the various boxes in the psychologists' flow charts may not be implemented by corresponding neuroanatomical boxes. Gall and his fellow phrenologists had hypothesized that each mental faculty is implemented by a specific brain region, but today's neuropsychologists realize that such simple function–structure correspondences cannot be taken for granted. For example, most models of the mind contain separate "boxes" for memory storage and memory retrieval, but neurobiological data suggest that the storage and retrieval of memories involves at least some overlapping brain regions. Such overlap immensely complicates the task of reverse engineering the brain.

BRAIN EXERCISE

Why do you think phrenology became wildly popular during the 19th century?

Neuroethology

As ethologists study the behavior of animals in their natural environment, they recognize that all animals face certain fundamental problems. To succeed evolutionarily, all organisms must survive long enough to leave viable offspring.

Staying Alive

One key to survival is eating and drinking. For most organisms, obtaining food and drink is a major challenge. It requires knowing when to eat or drink, which requires monitoring the body's internal state. It also involves knowing where to find food or drink. This, in turn, requires knowing where you are relative to the desired resources, as well as knowing when those resources are available. For example, waterholes dry out predictably, and fruit trees provide food only during specific seasons. Once food is located, it must be ingested and digested. Although digestion seems simple, it isn't. For example, food must be moved along the gut at a particular pace, and digestive enzymes must be secreted at the appropriate times, ideally before the food arrives. The nervous system plays a role in solving these and many other food-related challenges.

The second major key to survival is the avoidance of life-threatening situations. Organisms accomplish this, in part, by being instinctively afraid of heights, fire, low-frequency growls, and other ancient and foreseeable dangers. However, many dangers are not predictable enough to support the evolution of dedicated alarm circuits. Therefore, many organisms learn about dangerous objects or circumstances from experience. Because of the potential cost of making a mistake—not avoiding truly dangerous objects or situations—fear learning typically requires just one or a few trials. Many organisms even learn from the mistakes or fears of others. If dangerous objects or individuals cannot be avoided, organisms must determine whether they can afford to fight. This involves assessments of both the "enemy" and of the organism's own condition. Often survival depends on making the right choice in such situations. For highly social organisms, such as humans, assessing one's "fighting chance" is especially complex because such assessments must include estimates of how much help the individual may get from his or her "teammates." Thus, you can see that dealing with danger is a demanding task. It depends crucially on the nervous system.

Leaving Viable Offspring

In evolutionary terms, staying alive is not enough. Individuals must reproduce if their species is to persist. Therefore, organisms tend to be preoccupied with solving this particular problem. First, they must find potential mates. This is relatively simple if you are surrounded by many individuals of the right species and opposite sex. However, individuals are often very choosy about their mates and many do not get picked. As Darwin first realized, a vast number of physical and behavioral traits evolved specifically to solve the problem of mate choice. The sexual act itself is relatively simple in most species. However, in species where females are fertile only at certain times of the year, it makes sense for males to be attuned to the female's condition or "mood." Once offspring are produced, their survival has to be ensured. Some species do this by producing large numbers of offspring; others produce only a few but care for them diligently. For good parental care, it helps to form stable pair bonds and to recognize which babies are yours. Thus reproduction is complex, although commonly achieved. Again, the nervous system plays a major role in the entire process.

The preceding paragraphs described important real-life problems that the nervous system helps organisms solve. Each one of these problems is a function in the engineering sense of the term. Additional functions could be listed, and each problem can be broken down into more specific functions and processes. Ethologists have done precisely this for many species and many behaviors. Their flow charts are superficially similar to those of the neuropsychologists and pose similar challenges for neurobiologists. Specifically, it is often not clear whether individual boxes in an ethological flow chart correspond to discrete brain regions or neural mechanisms. For example, ethologists may hypothesize separate "modules" for learning to recognize dangerous predators and learning to recognize one's mate, but both forms of learning might well be implemented by overlapping brain systems. This uncertainty about how ethological functions map onto neural mechanisms is analogous to the uncertainty regarding the correspondence of psychological and neural processes that neuropsychologists confront.

BRAIN EXERCISE

Why do dogs so often bark at delivery people (e.g., mail carriers)? As you try to answer this question, are you thinking like an ethologist or more like a psychologist?

1.6 How Do Brains Evolve?

Our earth harbors roughly 60,000 vertebrate and more than 1 million invertebrate species. These differ in ancestry and in ecology. They also differ in the structure and function of their nervous systems. Nonetheless, many aspects of their nervous systems were conserved across evolutionary time and are, therefore, similar even in distantly related species. As a rule, the more closely two species are related, the more similarities in nervous system structure, function, and development they share. *Evolutionary conservation* usually predominates at the upper levels of biological organization, whereas *evolutionary change* occurs more frequently at the level of structural and physiological details.

Descent with Conservation and Modification

For example, an astonishing number of genes important for human brain function and development have **homologs** in other species—the genes can be traced back to a common ancestral species and were retained as evolution proceeded. Many mammalian genes have homologs in other vertebrates and in invertebrates. However, many of these genes duplicated in one lineage but not in another, and duplicated genes often diverge from one another in both DNA sequence and in function. Even homologous genes that have not duplicated may vary considerably across species, especially when the species being compared are distant relatives. Therefore, genetic differences abound at the level of individual genes, gene expression patterns, gene regulation, and protein interactions. A similar combination of high-level conservation with low-level change is apparent in the evolution of brain regions. High-level regions, such as the forebrain, tend to be broadly conserved. In contrast, small cell groups that are low in the structural hierarchy (see Figure 1.4) are often difficult to homologize across species, especially when those species are distant relatives. Even when homologs can be identified, they often vary in size or connectivity.

Which Species to Study?

Given this pattern of conservation and change, researchers must carefully select which species to study. Some phenomena may be easier to study in one species than another, but one should always ask: Can the findings be generalized across species? This is an important question. The history of neuroscience has shown rather convincingly that many findings can be generalized across species. Therefore, even if your interest is aimed primarily at human brains, you can learn valuable insights from work performed on non-humans. That is why, as you read on, you will learn about experiments performed on apes, monkeys, cats, rats, mice, as well as a few birds, some squid, and even snails.

Although many findings generalize across species, it is important to remember which findings came from which species and to keep in mind that species differences are likely to exist as well. For example, our brains differ considerably from other brains in size (Figure 1.19). We may not have the largest brains in absolute terms (that distinction goes to elephants and whales), but our brains are significantly larger than one would expect for primates of our body size. Do human brains differ from other brains also in structural details? Although research on chimpanzees,

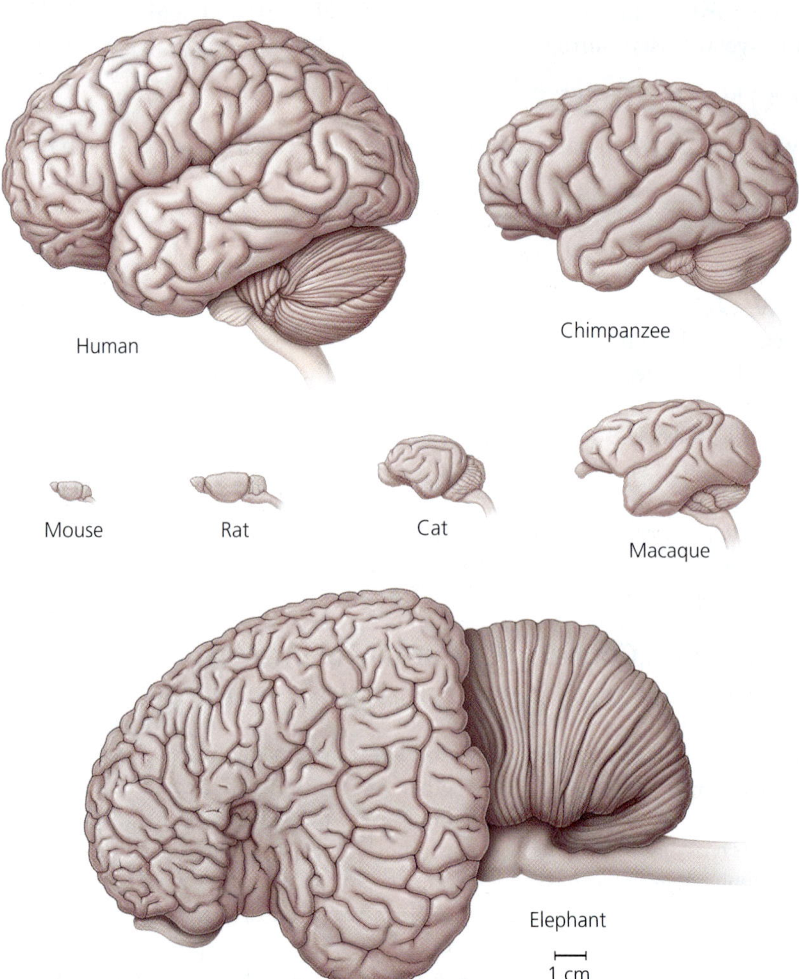

Figure 1.19 Seven mammalian brains, all shown to the same scale. The brain of a human is about three times larger than that of a chimpanzee. Elephant brains, in turn, are roughly three times larger than human brains. [After images from http://brainmuseum.org]

our closest relatives, is limited, their brains are very similar to our own, though just one-third the size. As we look to more distant relatives, including non-primates, more differences come into view. Rodents, for example, lack many of our visual cortical areas and some of the areas that one can find in primate prefrontal cortex. They also lack some primate specializations for color vision but excel at olfaction. Although keeping differences in mind, our focus in this book will be primarily on species similarities.

BRAIN EXERCISE

Given that elephants have larger brains than humans, why do we think humans are smarter? How might elephants be using their large brains?

SUMMARY

Section 1.1 - A major goal of neuroscience research is to discover general principles of brain organization. The search for these principles combines anatomical, physiological, and evolutionary perspectives.

Section 1.2 - The nervous system is hierarchically organized into molecules, cells, cell groups, major brain divisions, and central versus peripheral nervous systems.

- Learning neuroanatomy requires learning many new terms. Among the most important are orientation terms such as dorsal, ventral, superior, inferior, ipsilateral, contralateral, sagittal, and coronal.
- The central nervous system is divisible into brain and spinal cord. The brain comprises forebrain, midbrain, and hindbrain. These, in turn, consist of even smaller divisions.
- The nervous system contains neurons and glial cells. The neurons have distinctive dendrites, axons, and synapses. Despite these specializations, only 1–2% of human genes are "brain-specific."

Section 1.3 - Neural circuits tend to be replete with diverging, converging, and reciprocal connections. Although neural circuits may look as if everything is connected to everything else, this is rarely the case.

Section 1.4 - It is difficult to think about how the brain is organized without resorting to analogies with man-made systems, such as computers, but all analogies are flawed.

- Meynert and James in the late 1800s envisioned the brain as containing short (subcortical) and long (transcortical) reflex arcs. The latter were thought to exhibit plasticity and modulate activity in the subcortical circuits.
- Modern views of brain organization include not just long and short reflex arcs, but numerous looping pathways as well as central pattern generators and circuits that control an animal's behavioral state.

Section 1.5 - Studying the brain requires some sort of framework that describes the functions of the nervous system and how they are related to one another. Deriving such a framework amounts to reverse engineering the brain.

- For neurobiologists, the term "functional decomposition" means identifying which functions the system can perform and how those functions might be related to one another.
- Neuropsychology has a long history, beginning with phrenology in the early 1800s and blossoming today with cognitive neuroscience. It focuses on understanding the mechanisms of the mind.
- Neuroethologists aim to understand the neural mechanisms of naturally occurring behaviors. Their approach complements that of the neuropsychologists.

Section 1.6 - The more distantly related two species are, the more dissimilar their nervous systems are likely to be. Nonetheless, many aspects of nervous system organization are broadly conserved, especially at the higher levels of organization.

Box 1.1 - Anatomists study various aspects of brain structure, including histology, cellular and molecular details, as well as connectivity. Even functional methods, such as functional brain imaging, involve anatomical data.

Box 1.2 - Physiologists study brain function with electrodes and imaging methods to "listen in" on brain activity. They also lesion or inactivate brain areas and look for changes in behavior.

Box 1.3 - Optogenetic techniques involve infecting specific cell types with microbial opsins and then activating (or inactivating) the infected cells with light. Chemogenetic techniques are conceptually similar but use chemical compounds to activate mutated "designer receptors."

KEY TERMS

ADDITIONAL READINGS

1.1 - Approaches to Neuroscience Research

Striedter GF, Belgard TG, Chen C-C, Davis FP, Finlay BL, Güntürkün O, et al. 2014. NSF workshop report: discovering general principles of nervous system organization by comparing brain maps across species. *Brain Behav Evol* **83**:1–8.

1.2 - Basic Components of the Nervous System

Finger S. 2000. *Minds behind the brain.* Oxford, England: Oxford University Press.

Lee CR, Tepper JM. 2006. Morphological and physiological properties of parvalbumin- and calretinin-containing γ-aminobutyric acidergic neurons in the substantia nigra. *J Comp Neurol* **500**:958–972.

Nieuwenhuys R, Voogd J, van Huikzen C. 2006. *The human central nervous system: a synopsis and atlas.* Fourth Edition. Berlin: Springer-Verlag.

1.3 - Neuronal Circuits

Bassett DS, Bullmore E. 2006. Small-world brain networks. *Neuroscientist* **12**:512–523.

Clune J, Mouret, J-M, Lipson H. 2013. The evolutionary origins of modularity. *Proc Roy Soc B* **280**:20122863.

Nelson ME, Bower JM. 1990. Brain maps and parallel computers. *Trends Neurosci* **13**:403–408.

1.4 - Functional Organization of the Brain

Diamond IT. 1979. The subdivisions of neocortex: a proposal to revise the traditional view of sensory, motor, and association areas. *Prog Psychobiol Physiol Psychol* **8**:1–43.

Fuster J. 1997. Network memory. *Trends Neurosci* **20**:451–459.

Friston K. 2003. Learning and inference in the brain. *Neural Networks* **16**:1325–1352.

Grafton S, Dechamilton A. 2007. Evidence for a distributed hierarchy of action representation in the brain. *Hum Mov Sci* **26**:590–616.

James W. 1890. *The principles of psychology*. New York: H. Holt & Co.

Mesulam MM. 1998. From sensation to cognition. *Brain* **121**:1013–1052.

Young RM. 1970. *Mind, brain and adaptation in the nineteenth century: cerebral localization and its biological context from Gall to Ferrier*. Oxford: Clarendon Press.

1.5 - Functional Decomposition

Marr D. 1982. *Vision: a computational investigation into the human representation and processing of visual information*. San Francisco: W. H. Freeman.

Sherry DF. 2006. Neuroecology. *Annu Rev Psychol* **57**:167–197.

Tinbergen N. 1963. On the aims and methods of ethology. *Z Tierpsychol* **20**:410–440.

1.6 - Conservation and Change across Species

Sereno M, Tootell R. 2005. From monkeys to humans: what do we now know about brain homologies? *Curr Opin Neurobiol* **15**:135–144.

Shi P, Bakewell MA, Zhang J. 2006. Did brain-specific genes evolve faster in humans than in chimpanzees? *Trends Genet* **22**:608–613.

Striedter, GF. 2005. *Principles of brain evolution*. Sunderland, MA: Sinauer Associates.

Boxes

Aston-Jones G, Deisseroth K. 2013. Recent advances in optogenetics and pharmacogenetics. *Brain Res* **1511**:1–5.

Computing with Neurons

CHAPTER 2

CONTENTS

FEATURES

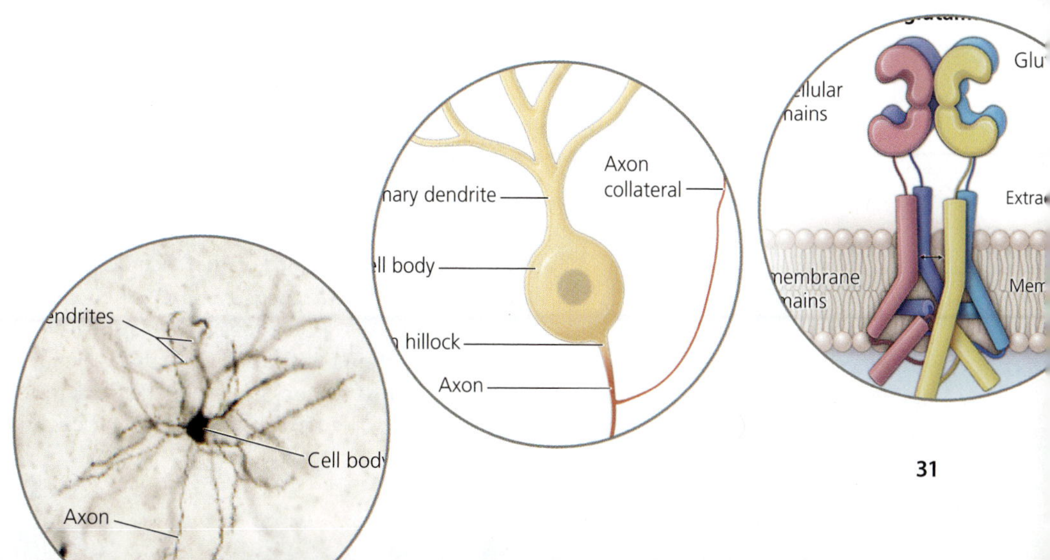

The human brain contains about 86 billion neurons. Although the nervous system also contains several other cell types, neurobiologists tend to be most interested in the neurons because neurons are thought to be the nervous system's principal computing elements. They gather information from other neurons or from sensory cells, process this information, and then send outputs to other neurons or to muscle cells and glands. Our goal in this chapter is to understand how neurons perform this computing function. Let us begin with a brief historical perspective on how neurons were discovered.

2.1 What Are Neurons?

In the mid-1600s, Descartes and others thought that the brain's critical functions were carried out by "animal spirits" coursing through hollow nerves and the cerebral ventricles. This notion now seems far-fetched, but scientists in Descartes' day could not examine neural tissue in detail because it is quite soft and quick to rot. Only in the mid-1800s did scientists discover ways to harden tissue with alcohol or other fixatives and cut it into thin slices that could be examined under a microscope. Even so, there was not much to see in such slices until the scientists discovered chemical dyes that could reveal some hints of the brain's inner texture. Essentially, these dyes showed that within the brain's gelatinous mass lay millions of tiny objects, which scientists initially referred to as *granules*. Seeing these granules was exciting but also raised many questions. Most importantly, it raised the question of whether these granules were cells.

History of the Neuron Doctrine

Theodore Schwann and Matthias Schleiden had postulated in 1839 that biological tissues were generally composed of distinct cells, but many scientists doubted that the nervous system was like other tissues in this respect. As these scientists looked carefully at the stained granules in the brain, they noticed that many granules had processes extending away from them. We now call these processes dendrites and axons, but only their stumps are visible with the dyes that were available in the 1850s and 1860s. This meant that the early investigators could not be sure where the processes of one granule end and those of another begin. Thus, they could not exclude the possibility that the unstained processes are all continuous with one another. This was a critical point because if the processes were continuous, then the individual granules would not be parts of distinct cells. Indeed, the brain would not consist of cells at all because cells by definition are distinct units with definitive boundaries. Instead, the brain would form a continuous web in which the stained granules are merely nodal points.

The Golgi Stain and Ramón y Cajal

A crucial first step toward resolving this uncertainty was taken in 1873 when Camillo Golgi invented a new stain. Golgi called his stain the "black reaction," but the world soon came to know it as the **Golgi stain**. The most exciting feature of the Golgi stain was that it stains the granule processes much more extensively. Moreover, Golgi's stain labels only a small fraction of all the granules and processes in a brain so that one can see the stained structures against a clear background. Did this new stain convince Golgi that each granule with its processes is a separate cell? No, it did not. Instead, Golgi came to believe that the processes of individual granules all fuse with one another to form a massive web (a reticulum). This notion was not as fanciful as it may seem today because the processes that Golgi wrote about are thin indeed and hard to see with the kinds of microscopes available in Golgi's time. In any case, Golgi concluded that the nervous system is not composed of discrete cells.

Golgi's view was challenged by Santiago Ramón y Cajal (Figure 2.1), who also used Golgi's stain but studied more diverse material. One of Cajal's key observations was that in most sensory pathways the thick processes we now call dendrites tend to be directed toward the sensory input, whereas the thin axonal processes are directed

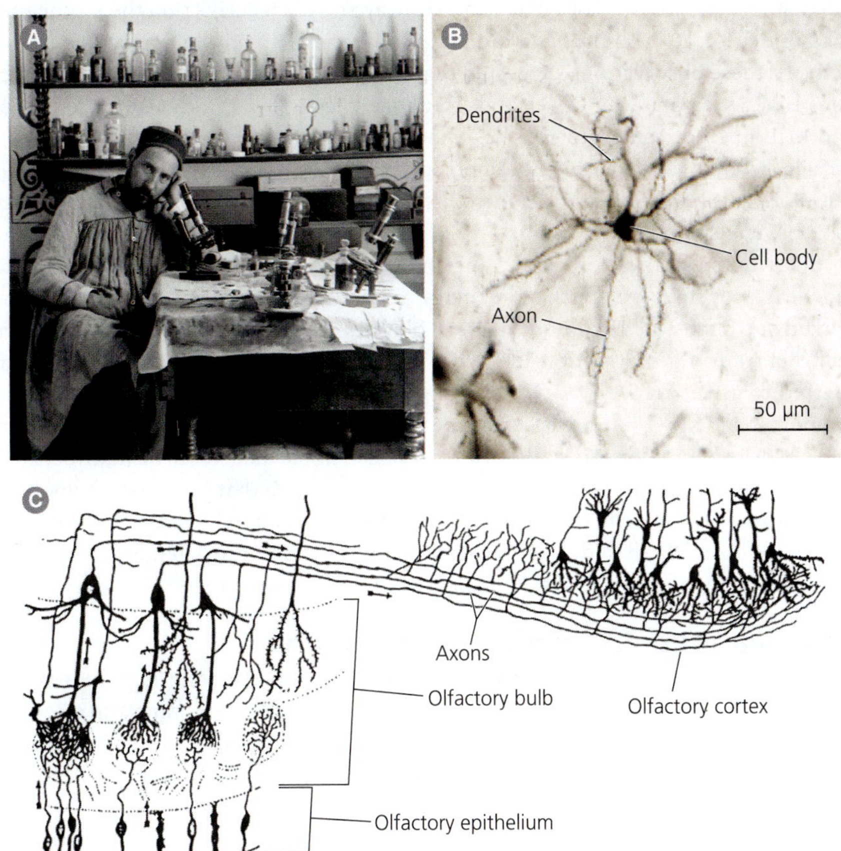

Figure 2.1 Ramón y Cajal used Golgi's stain. The photograph in (A) shows Cajal sitting in his lab, surrounded by microscopes, chemicals, and specimens. The photomicrograph in (B) depicts a Golgi-stained neuron. Shown in (C) is one of Cajal's detailed drawings of Golgi-stained neurons. The little arrows indicate the direction in which Cajal thought information flows; his inference turned out to be correct. [A courtesy of Cajal Legacy, Instituto Cajal, Madrid, Spain; B courtesy of Annie Vogel-Ciernia; C from Cajal, 1894]

toward the brain's center. This observation suggested to Cajal that neurons are functionally polarized, with inputs terminating on dendrites and outputs streaming through axons (Figure 2.1 C). It also implied that the axons of one neuron terminate on the dendrites of the next neuron(s) within a sensory pathway. Because Cajal never observed any thin axonal processes merging with thick dendritic processes, he concluded that axons and dendrites are separated by a gap. By implication, this meant that each granule with its processes is a distinct cell. This conclusion was directly opposed to Golgi's view.

Although Cajal's early work was slow to attract the attention of other scientists, Cajal gradually convinced most of his colleagues that neurons are functionally polarized and distinct cells, that Golgi's view was wrong. Nonetheless, Golgi and Cajal were both awarded the 1906 Nobel Prize in Physiology or Medicine. It may seem odd in retrospect to have the winner and the loser in a scientific dispute share the Nobel Prize, but Cajal could not have done his work without the Golgi stain. Furthermore, in 1906 no one had seen the tiny gaps between axons and dendrites that Cajal had hypothesized. They were not observed until the 1950s, when electron microscopes were invented. Because electrons vibrate at shorter wavelengths than the photons in visible light, the resolution of an electron microscope is roughly 1,000 times higher than that of a conventional light microscope. With this improved resolution, the gaps between neurons are visible.

Charles Sherrington and the Synapse

Around the time that Golgi and Cajal were studying neuron anatomy, the British scientist Charles Sherrington was studying the physiology of reflex pathways through the spinal cord. One of his key findings was that these reflexes took longer to complete than one would expect, given the available data on nerve conduction speeds. In addition, Sherrington found that the reflex pathways were unidirectional. He could stimulate the sensory end of a neuronal pathway and elicit activity on the motor end, but stimulation at the motor end did not produce activity on the sensory side. Collectively, these observations suggested that the spinal reflex pathways contain structures that act as one-way valves and introduce a time delay in the transmission of information from sensor to muscle. Sherrington summarized these findings and ideas in 1906, in a fascinating little book called *The Integrative Action of the Nervous System*. In this book Sherrington introduced the term **synapse** for the tiny one-way valves he envisioned.

Sherrington's physiological findings meshed well with Cajal's anatomical data and ideas. Not surprisingly, the two men were friends. Many details remained to be worked out, but by the 1920s most neurobiologists accepted the idea that neurons are cells that transmit information across tiny gaps called synapses. This key idea, often referred to as the **neuron doctrine**, has stood the test of time. Of course, as is so common in biology, there are exceptions to the norm. For example, some neurons are coupled by gap junctions that allow the cytoplasm in one neuron to be in direct contact with the cytoplasm of the other neuron. It is also true that information does not always flow from dendrite to axon, as Cajal had argued; some dendrites synapse onto other dendrites. We shall deal with such exceptions to the rule later. For now, let us consider the anatomy and physiology of stereotypical neurons.

Basic Features of a Stereotypical Neuron

As the neuron doctrine states, neurons are cells. This means that they share many features with other cells, such as a double-layered lipid cell membrane and a central nucleus where RNA is made from DNA. In addition, neurons exhibit numerous specializations that are not found in other cells. For example, all neurons contain

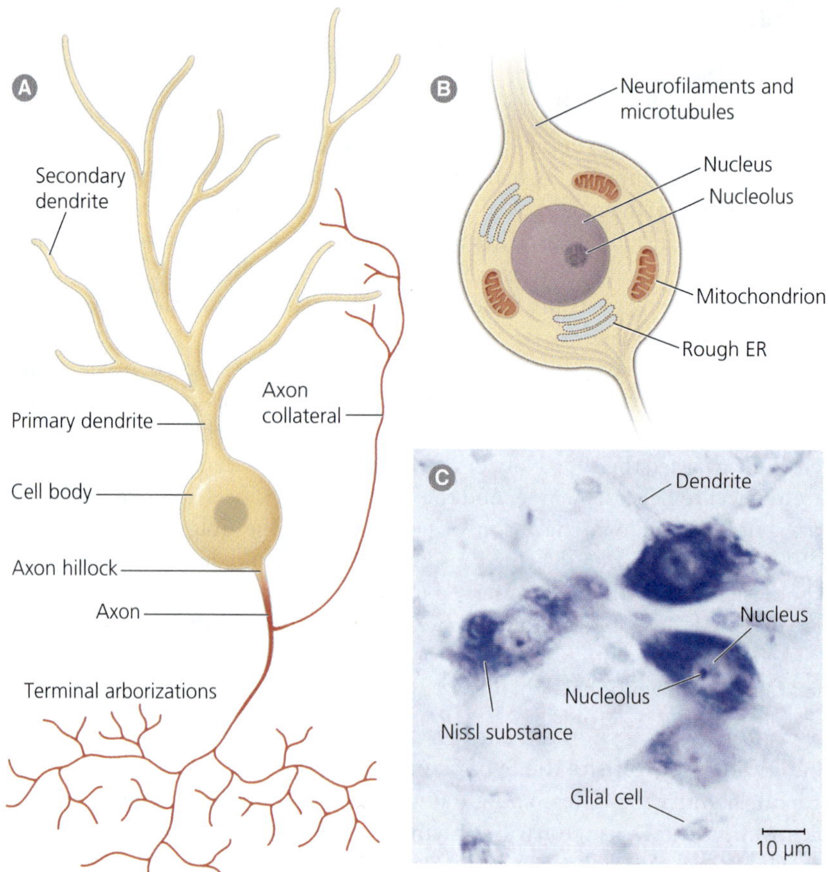

Figure 2.2 Neuron morphology. Panel (A) depicts an idealized neuron. The dendrites tend to become thinner with distance from the cell body. The axon gives off one major branch, called an axon collateral, and ends in a dense tangle of small branches. In contrast to dendrites, axons are uniformly thin and tend to branch at right angles. Panel (B) is a close-up of a typical neuronal cell body. Panel (C) shows several Nissl-stained neurons and glial cells.

a special kind of rough endoplasmic reticulum that is called **Nissl substance** after the German scientist who first discovered it. This fact is worth knowing, because neuroanatomists frequently use stains that label the Nissl substance (Figure 2.2 C). As Nissl substance is always located near a neuron's nucleus, Nissl stains tend to label only neuronal cell bodies (the "granules" of our earlier discussion) and those portions of neuronal dendrites that lie closest to the cell body (the stumps).

Axons and Dendrites

A distinctive anatomical feature of a typical neuron is that it has **dendrites**. These dendrites branch repeatedly and become progressively thinner as they extend away from the cell body (Figure 2.2). Most neurons also have an **axon**, which does not become thinner with distance from the cell body but is generally thinner than the main dendritic branches. In a typical neuron, the axon originates at a small mound, called the **axon hillock**, that protrudes from the cell body. From there, the axon wanders away, often splitting into a few major branches that are called **axon collaterals**. Near the end of these major branches, typical axons branch repeatedly to form a set of **terminal arborizations** (*arbor* is the Latin word for "tree").

Because axon terminals are typically far removed from the cell body, where most proteins are made, neurons contain an elaborate intracellular machinery for transporting proteins and organelles from the cell body to the terminals. Most proteins are transported down axons at speeds of ~1 mm/day, but some organelles, notably vesicles filled with neurotransmitter molecules (which we discuss shortly), are moved along axons 100 times faster than that (using molecular motors called dynein and kinesin). Most anatomical axon tracing techniques take advantage of this **axoplasmic transport**, as tracers are taken up at the cell body and transported down the axon to the terminals. Many tracers are also transported in the reverse directions, from terminal to cell body. This retrograde axoplasmic transport is also fast (at 100–200 mm/day) and is used to ship used bits of membrane and diverse molecules back to the cell body for recycling.

Resting and Action Potentials

Like most animal cells, neurons are electrically charged. If you insert a microelectrode (a finely tapered, electrolyte-filled capillary tube) into a neuron and hook it up to a voltmeter, you typically record a voltage change of roughly –70 mV as the microelectrode penetrates the cell membrane (Figure 2.3). This voltage difference between the inside and the outside of the neuron is called the neuron's resting membrane potential or, more simply, its **resting potential**. Muscle and glial cells also have strongly negative resting potentials, but they need not concern us now. What is important is that neurons occasionally abandon their negative resting potential and briefly become positively charged (Figure 2.3). These brief reversals of polarity are the **action potentials** that neurons use to transmit information along their axons.

Action potentials can be recorded both with intracellular electrodes, whose very sharp tip is inserted into the cell's interior (Figure 2.3), and with extracellular electrodes that are positioned close to a neuron but do not penetrate its cell membrane. Only intracellular electrodes allow experimenters to record the absolute value of a neuron's membrane potential. Moreover, action potentials recorded with intracellular electrodes are much larger than the same potentials recorded extracellularly (mainly because extracellular recordings only measure potential

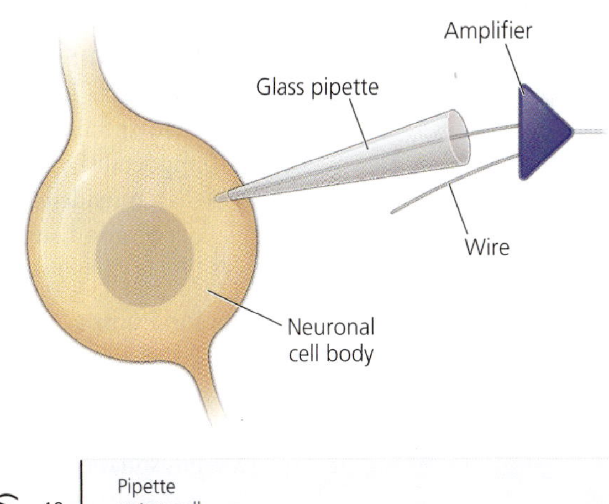

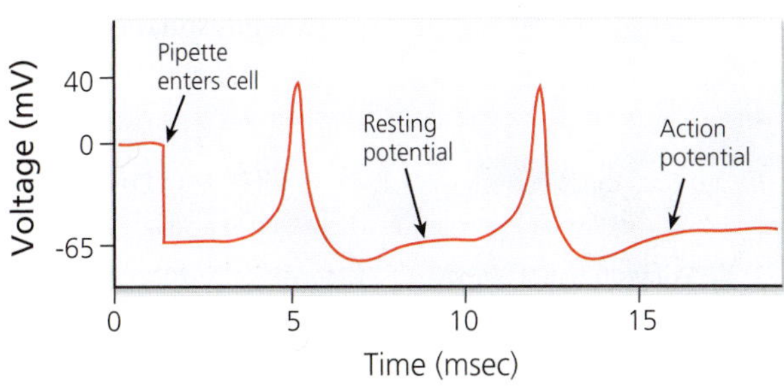

Figure 2.3 Intracellular recording. Shown at the top is the cell body of a neuron being probed with an intracellular recording electrode. The graph depicts an intracellular recording that reveals a neuron's membrane potential at rest and during two action potentials.

changes near the neuronal membrane, not across it). However, extracellular recordings are technically simpler to obtain, and they suffice for most of the research covered later in this book.

BRAIN EXERCISE

How did Ramón y Cajal infer the direction in which information flows through a stereotypical neuron (from dendrites to the tips of the axon)? Do you think it might be possible for information to flow also in the opposite direction? Why or why not?

2.2 What Mechanisms Generate Resting and Action Potentials?

Scientists had suspected since at least 1800 that animals can generate electric currents and that "animal electricity" can influence behavior. This kind of vague understanding inspired Mary Shelley in 1816 to write the novel *Frankenstein*, which features electricity as a life-giving force. However, no one knew how animals generate electricity or how they channel it productively.

This veil of mystery began to lift in the 1930s when a small band of scientists began to study the axons of giant neurons in the common squid. The axons of these neurons have such a large diameter that skilled experimenters could insert multiple electrodes into them and thus conduct experiments that would not have been possible on smaller neurons. **Squid giant axons** (note: *not* giant squid axons!) are also relatively easy to maintain outside of the animal, which means that one can easily manipulate the solutions in which the axons are bathed. Because of these advantages, most of what we know about the electrical properties of neurons was first discovered in the squid's giant axons. Fortunately, the key discoveries were later shown to generalize quite well to all neurons, including those of warm-blooded creatures like us.

Ionic Basis of the Resting Potential

One of the key early discoveries made on squid giant axons was that the axon's intracellular fluid differs from the extracellular fluid in ion composition (Table 2.1). Most importantly, the concentration of potassium ions (K^+) is much higher inside the axon than outside of it, whereas the opposite pattern holds for sodium ions (Na^+). This observation suggested that neurons are actively pumping sodium ions out across their cell membranes and pumping potassium in.

Ion Pumps

Later discoveries confirmed the existence of specialized transporter molecules that, indeed, pump ions into and out of neurons (Figure 2.4). Because moving ions against their concentration gradients is work, these transporters use up some adenosine triphosphate (ATP), the brain's main currency of metabolic energy. This is why the transporters are known as **Na^+/K^+-ATPases**. Intriguingly, the Na^+/K^+-ATPases

ION	EXTRACELLULAR CONCENTRATION (mM)	INTRACELLULAR CONCENTRATION (mM)	EXTRA-/INTRACELLULAR RATIO
Potassium (K^+)	20	400	1/20
Sodium (Na^+)	440	50	9/1
Chloride (Cl^-)	550	50	11/1
Calcium (Ca^{++})	10	0.0001	100,000/1
Organic Anions	~0	350	~0

Table 2.1 Ion concentrations inside and outside of a squid's giant axon. Sodium and calcium ions are more concentrated outside of neurons than inside of them, whereas the reverse is true for potassium and most organic anions. Intracellular free calcium concentration is extremely low.

move 2 potassium ions in for every 3 sodium ions that they move out. If you think about it, this net movement of positive ions out of a neuron should cause the neuron to become negatively charged, relative to its surroundings. Does this explain the negative resting potential of typical neurons? No, it doesn't, not by a long shot.

The principal reason why neurons are negatively charged at rest is that the resting neuronal membrane is slightly permeable to potassium ions but almost completely impermeable to other ions, including sodium. This means that some of the potassium ions pumped into the neuron by the Na^+/K^+-ATPases flow right back out, down their concentration gradient. As they do so, the neuron's extracellular space becomes positively charged relative to the neuron's interior, setting up a voltage difference across the neuronal membrane. More specifically, the net movement of potassium ions out of the neuron causes the neuron to develop their negative membrane potential.

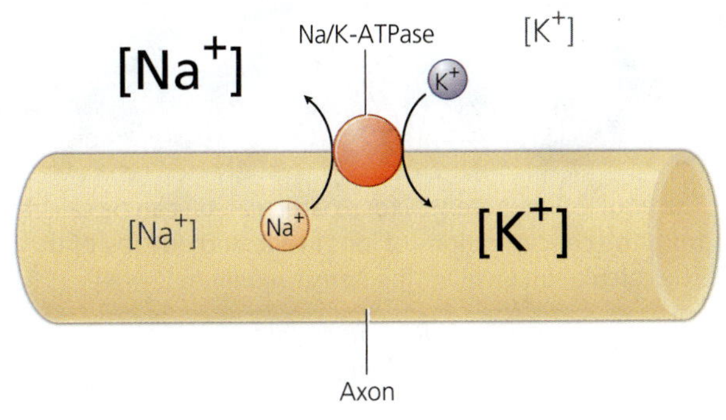

Figure 2.4 The sodium–potassium pump. The drawing shows how an Na/K-ATPase (red circle) pumps sodium ions out of the axons and potassium ions in. The process consumes metabolic energy provided by ATP.

Does the efflux of potassium ions undo the work performed by the Na^+/K^+-ATPases? Not really, at least in the short term, because the number of ions that move out of a neuron to generate its membrane potential is so small (barely a picomole per cm^2 of membrane or, more vividly, less than 0.001% of all the K^+ ions inside a large neuron) that it does not measurably change the intracellular potassium concentration. The change in extracellular potassium concentration is even smaller because the extracellular environment is relatively vast.

Electrochemical Equilibria

There is one more important twist to this story. As the neuron's inside becomes more negative, the positive potassium ions become ever more reluctant to exit the neuron because they are attracted to the net negative charge inside the cell and repelled by the net positive charge outside of the neuron. Thus, the movement of potassium ions across the neuronal membrane is governed by two opposing forces: (1) the **concentration gradient** for potassium tends to push potassium out of the neuron, and (2) the **electrical potential gradient** across the cell membrane tends to move potassium ions back into the neuron (Figure 2.5).

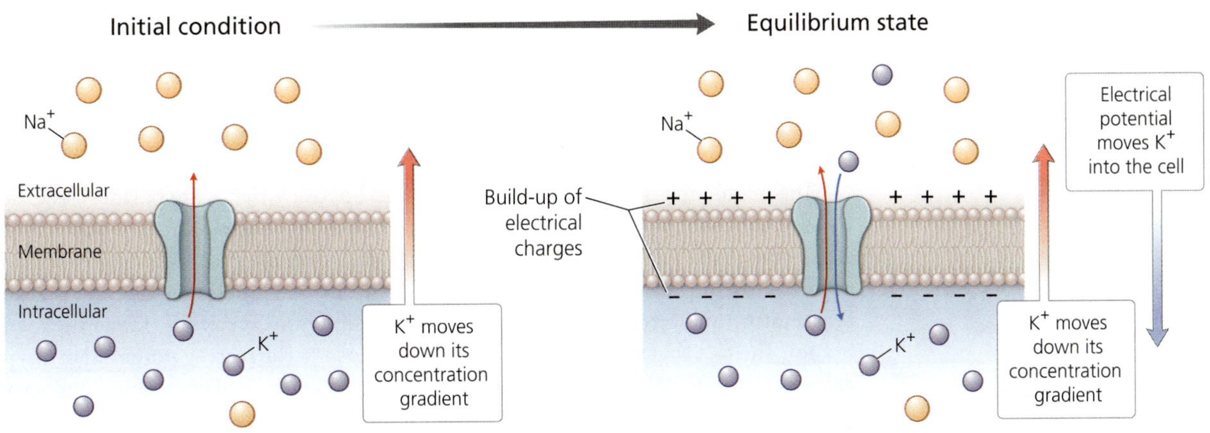

Figure 2.5 Electrochemical equilibrium. Shown on the left are the conditions created by Na^+/K^+-ATPases, which pump potassium ions (K^+) into the cell and sodium ions (Na^+) out. Because the neuronal cell membrane at rest is permeable only to potassium, some K^+ ions move down their concentration gradient out of the cell (red arrow). Over time, this movement causes positive and negative charges to build up on the extracellular and intracellular faces of the neuronal membrane, respectively (positive and negative charges attract one another across the membrane, which acts as an electrical insulator). As the charges build up, some K^+ ions move back into the cell, down the electrical potential gradient (blue arrow). The system soon equilibrates so that the outward flow of K^+ is precisely balanced by the inward flow. The voltage difference across the membrane when the system is in equilibrium is called the equilibrium potential.

NEUROBIOLOGY IN DEPTH

Box 2.1 *Neuronal Membrane Math*

The equilibrium potential for a particular ion can be calculated from the concentrations of ions inside and outside of the cell. The calculation involves the *Nernst equation*:

$$E = \frac{RT}{zF} \ln \frac{[ions\_outside]}{[ions\_inside]}$$

In this equation, *E* is the equilibrium potential; R, the ideal gas constant; *T*, the temperature in degrees Kelvin; *z*, the ion's valence; F, Faraday's constant; and ln, the natural logarithm. If the ion is monovalent (e.g., Na^+, K^+, or Cl^-) and the temperature is 37 °C, then the Nernst equation simplifies to the following:

$$E = \frac{61}{z} \log \frac{[ions\_outside]}{[ions\_inside]}$$

You can solve this equation separately for each of the ions that play a role in neuronal signaling; but if you want to consider all ions simultaneously, you need the *Goldman equation*:

$$E_m = \frac{RT}{F} \ln \left(\frac{\sum_i^N P_{M_i^+}[M_i^+]_{out} + \sum_j^M P_{A_j^-}[A_j^-]_{in}}{\sum_i^N P_{M_i^+}[M_i^+]_{in} + \sum_j^M P_{A_j^-}[A_j^-]_{out}} \right)$$

This equation looks intimidating, but the first part is the same as the first part of the Nernst equation. The part inside the parentheses simply says that you should multiply each ion's concentration (the positive ions here are called M^+ and the negative ions are called A^-) by the relative permeability of the membrane to that ion (P_M and P_A). It also says that for the positive ions, you should put the ion's concentration outside the cell in the numerator and the concentration inside the cell in the denominator; for negative ions, you should do the opposite. Finally, the summation symbols indicate that you need to calculate the permeability-by-concentration products separately for each ion and then sum the results. If this still sounds complicated, you'll be glad to know that for neurons you typically just have three ions to worry about, namely, K^+, Na^+, and Cl^-. In this case, Goldman's equation simplifies to:

$$E_m = \frac{RT}{F} \ln \left(\frac{P_{Na^+}[Na^+]_{out} + P_{K^+}[K^+]_{out} + P_{Cl^-}[Cl^-]_{in}}{P_{Na^+}[Na^+]_{in} + P_{K^+}[K^+]_{in} + P_{Cl^-}[Cl^-]_{out}} \right)$$

If you are interested in cold-blooded animals housed at a normal room temperature of 20 °C, you can rewrite the equation as the following:

$$E_m = 58 \log \left(\frac{P_{Na^+}[Na^+]_{out} + P_{K^+}[K^+]_{out} + P_{Cl^-}[Cl^-]_{in}}{P_{Na^+}[Na^+]_{in} + P_{K^+}[K^+]_{in} + P_{Cl^-}[Cl^-]_{out}} \right)$$

This equation is well loved by neurophysiologists because Alan Hodgkin and his collaborators used it to learn about changes in the ion permeability of neuronal membranes in squid axons. Specifically, they varied the concentrations of the various ions, recorded how that altered the membrane potential, and then fitted the data to the Goldman equation. This allowed them to determine that in a resting membrane, the ratio of P_K to P_{Na} to P_{Cl} is 1.0:0.04:0.45. In contrast, at the peak of the action potential, the corresponding ratios are 1.0:20:0.45. As these numbers reveal, the action potential involves an enormous increase in the membrane's permeability to sodium ions.

Finally, it is worth noting that the Goldman equation is just a special, more complicated form of the Nernst equation. For example, if you write the Goldman equation as if you had only sodium ions to consider and then simplify the equation by canceling out the sodium permeabilities, you obtain the Nernst equation for the sodium equilibrium potential:

$$E_m = \frac{RT}{F} \ln \left(\frac{P_{Na^+}[Na^+]_{out}}{P_{Na^+}[Na^+]_{in}} \right) = \frac{RT}{F} \ln \left(\frac{[Na^+]_{out}}{[Na^+]_{in}} \right)$$

The membrane potential at which these two opposing forces are balanced is called the **equilibrium potential** for potassium. Based on a formula called the **Nernst equation** (Box 2.1: Neuronal Membrane Math), one can calculate the potassium equilibrium potential to be approximately −80 mV (the specific value depends on temperature and on the concentrations of potassium ions on either side of the membrane). The fact that this value of −80 mV is so close to the resting potential of a typical neuron (−70 mV) strongly suggests that the push and pull of potassium ions across the neuronal membrane is indeed the major cause of the neuronal resting potential. Additional support for this hypothesis comes from the observation that systematic changes in the extracellular potassium concentration cause changes in the resting potential that correspond almost perfectly to what the Nernst equation predicts (Figure 2.6).

As you think about the ion movements underlying the neuronal resting potential, it is important to keep in mind that the neuronal cell membrane is only a few nanometers thick and, in general, a good electrical insulator (ions cannot cross pure lipid bilayers). Because the membrane is so thin, electrical charges on one side of the membrane can be electrostatically attracted to opposite charges on the other side (as you may recall from physics class, opposite charges attract one another). Therefore, when the inside of a neuron is more negative than the outside, a thin layer of negative charges develops on the inside of the neuronal membrane while a thin layer of positive charges accumulates on the membrane's extracellular side (Figure 2.5). This kind of arrangement—opposite charges lined up on either side of an electrical insulator—is what physicists call a capacitor. Most relevant to neurobiologists is that capacitors take time to charge (for the opposing charges to accumulate); and once capacitors are charged, they take time to discharge. As you will see later in this chapter, the fact that capacitors take time to change their charge affects how rapidly a neuron's membrane potential can change and, consequently, how rapidly neuronal signals can propagate along the neuronal membrane.

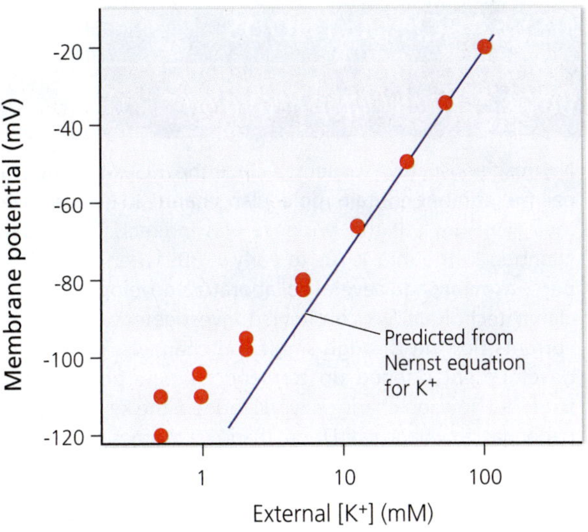

Figure 2.6 The role of potassium in setting the membrane potential. Hodgkin and Horowicz systematically varied the K^+ concentration $[K^+]$ outside of a frog muscle fiber and found that the fiber's membrane potential became less negative as extracellular $[K^+]$ increased. The data match what one would predict on the basis of the Nernst equation for K^+ (blue line) except at very low concentrations of K^+, when the membrane's permeability to Na^+ comes into play. [After Hodgkin and Horowicz, 1959]

Ion Channels

The ionic basis of the resting potential had become well understood by the late 1950s, but scientists remained unsure about the details of how ions flow through cell membranes. They suspected that neuronal cell membranes contain some sort of proteinaceous pores through which ions may flow, but the molecular identity and detailed function of those proteins remained unclear. This situation changed as new neurophysiological techniques (see Box 2.2: Patch Clamp Recording) allowed scientists to record electrical currents flowing through tiny patches of the cell membrane that often contained just one or two of the suspected pores. Advances in molecular biology then allowed the pore-forming proteins, which we now call **ion channels**, to be isolated, sequenced, and subjected to detailed structural and functional analyses. As a result of all this work, we now know an enormous amount of information about how ion channels open and close and what makes them selective to specific ions (see Box 2.3: Molecular Biology of Voltage-gated Ion Channels).

As research on ion channels progressed, scientists became more specific in their descriptions of ion flow and its regulation. For example, instead of saying simply that the neuronal membrane at rest is permeable to potassium ions, they can declare that it contains a specific type of potassium channel (potassium channels are quite diverse) that tends to be open near the resting membrane potential. This channel is generally called the **leak channel** because it allows potassium ions to "leak" out of neurons at rest. As you will discover in the next few sections, neuronal membranes also contain an assortment of additional ion channels, which collectively give neurons the ability to generate electrical signals.

Ionic Basis of the Action Potential

Although the electrical potential across the neuronal membrane tends to hover near –70 mV, it occasionally reverses polarity and briefly surges to about +40 mV. At those moments the neuron is no longer "at rest" but in the process of generating an action potential. This temporary reversal of the membrane potential had been described in the nineteenth century, but its causal basis remained unclear until research began on squid giant axons. This work involved scientists in several laboratories, but the most profound contributions were made by Alan Hodgkin and Andrew Huxley. Their work was so important that their core set of hypotheses came to be called the Hodgkin–Huxley model of neuronal excitability and earned both scientists the 1963 Nobel Prize in Physiology or Medicine.

RESEARCH METHODS

Box 2.2 *Patch Clamp Recording*

Neuroscientists had suspected since the 1950s that neuronal cell membranes contain molecular "channels" through which ions can flow, but the evidence was indirect. This situation changed in the mid-1970s to early 1980s when Erwin Neher, Bert Sakmann, and several collaborators developed the patch clamp technique, which allowed investigators to record the currents flowing through single ion channels. This technical development opened up tremendous new possibilities for studying how ion channels work. Indeed, the key patch clamp paper has now been cited more than 17,000 times, and Sakmann and Neher were awarded the 1991 Nobel Prize in Physiology or Medicine.

The core idea of patch clamp recording is to place the tip of an electrolyte-filled glass pipette directly over a tiny patch of cell membrane (Fig. b2.1). Ideally that patch is so small that it contains only a single ion channel. When this channel opens, ions flow through the channel into or out of the pipette tip. By connecting a very sensitive amplifier to the inside of the pipette, the resulting current can be recorded. This sounds simple enough, but the patch clamp technique requires an extremely tight seal between the rim of the pipette tip and the cell membrane; otherwise some ions can leak into the extracellular space instead of flowing through the pipette tip.

Fortunately, this problem can be solved by carefully cleaning both the pipette and the cell membrane (digesting away surrounding connective tissue) and then applying mild suction through the pipette. Under those conditions, the cell membrane tends to adhere very tightly to the rim of the pipette tip, forming a seal through which almost no current can leak (Figure b2.1 B). Because such seals typically have an electrical resistance of several gigaohms, they are nicknamed "gigaseals." Importantly, these seals reduce noise to such an extent that neurobiologists can see brief blips of current flow (measured in pico-amps, pA) that are variable in duration but very consistent in amplitude (Figure b2.1 C, E). Each blip represents the flow of ions through a single ion channel in its open state. When the channel closes, the current shuts down.

An unexpected but very important feature of the gigaseal is that the cell membrane adheres to the glass tip even when the pipette is withdrawn, detaching the membrane patch from the rest of the cell. When this happens, the originally intracellular surface of the membrane patch becomes exposed to the extracellular fluid creating an "inside-out" membrane patch. A slightly different technique generates "outside-out" patches in which the originally intracellular surface of the patch faces the inside of the pipette. In both cases, experimenters can control the fluid environment on both sides of the membrane patch. They can, for example, vary ion and transmitter concentrations both inside the pipette and in the extracellular fluid. By observing how these changes in fluid composition, as well as changes in membrane voltage, alter the frequency and duration of the recorded current blips, experimenters gained unprecedented knowledge of how ion channels work. Thus, the patch clamp technique revolutionized the study of electrically excitable membranes.

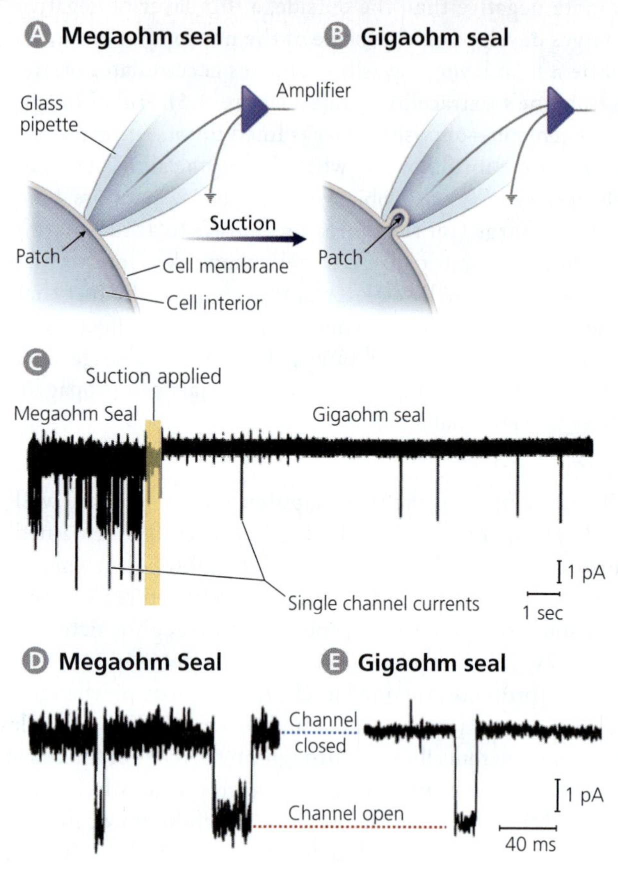

Figure b2.1 Patch clamp recording. When a glass pipette connected to an amplifier is placed directly onto a cell membrane, a small patch of membrane can be electrically isolated. The electrical seal may be on the order of megaohms (A) or, after mild suction, multiple gigaohms (B). After formation of a gigaohm seal (a gigaseal), the noise of the recording is dramatically reduced (C), allowing the experimenters to record with greater clarity the currents associated with the opening of single ion channels (D vs. E). The brief downward deflections (blips) in the recording traces represent depolarizing currents flowing through an open ion channel, which in this case is an activated acetylcholine receptor on a frog muscle fiber. [After Hamill et al., 1981]

NEUROBIOLOGY IN DEPTH

Box 2.3 *Molecular Biology of Voltage-gated Ion Channels*

Research over the last 30 years has revealed many molecular details of ion channel structure and function. A good example of this research is the work on voltage-gated sodium (Na_v) channels, which began with the isolation of Na_v channels from the electric organ of electric eels. The amino acid sequence of this purified protein led researchers to identify the DNA sequence for Na_v channels in diverse vertebrates, including mammals. Those channels turn out to be very large proteins, comprising ~2,000 amino acids, and they consist of 1 α and 2 β subunits (Figure b2.2 A). The primary sequence of the α subunit contains 4 very similar domains, each of which contains 6 hydrophobic α-helices that span the cell membrane. Between these helices, as well as between the 4 repeating modules, are protein loops that extend into the extracellular or intracellular space. To generate hypotheses about the functions of these structural domains, investigators used toxins that bind to specific sites in the protein and block specific aspects of channel function. They also mutated specific segments of the channel's DNA sequence, expressed the mutated channels in cells that do not normally express Na_v channels, and then examined how the mutations altered channel function.

Collectively, this research has revealed that the central pore of the Na_v channel is bordered by helices 5 and 6 of the 4 modules in the α subunit. The loops between helices 5 and 6 of all 4 domains penetrate the membrane and form part of the pore. Amino acids in these loops prevent ions other than Na^+ from passing through. Helix 4 is the channel's voltage sensor, moving within the membrane when the voltage exceeds the channel's activation threshold and thereby opening the channel's "activation gate." Strong membrane depolarization also causes an intracellular loop between the 3rd and 4th modules to move into the pore, closing it off. As long as this "inactivation gate" is closed, the channel remains impervious to sodium ions no matter how positive the membrane voltage is. Neurophysiologists had known for years that Na_v channels must have separate activation and inactivation gates, but now they can pinpoint where in the channel protein those gates are located.

Vertebrate Na_v channels have proven difficult to crystallize, which has thwarted determination of the channels' 3D structure through x-ray diffraction crystallography. However, bacterial Na_v channels are smaller and simpler than their vertebrate analogs. They consist of 4 identical subunits, each of which is very similar to one of the 4 modules in the α subunit of vertebrate Na_v channels. These features facilitate x-ray

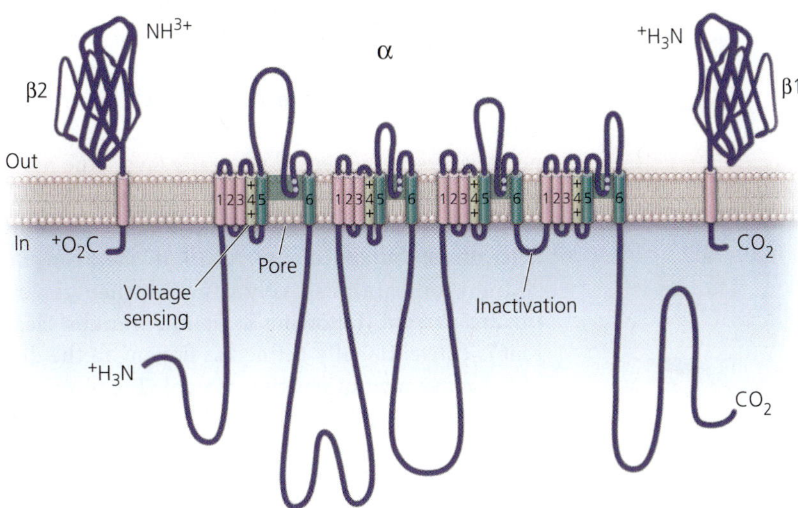

Figure b2.2 Voltage-gated sodium (Na_v) channels. Shown in (A) is the two-dimensional structure of the α and β subunits of a vertebrate Na_v channel, indicating key functional sites. The 3D crystal structure of the pore domain from a bacterial Na_v channel is shown in (B). The image on the left is a side view, sectioned through the central pore; the image on the right is taken from the intracellular perspective. [A after Caterall, 2000; B after McCusker et al., 2012]

Box 2.3 *Molecular Biology of Voltage-gated Ion Channels (continued)*

crystallography, allowing scientists to determine their 3D structure (Figure b2.2 B). Vertebrate Na_v channels are thought to look quite similar, at least in overall structure.

Molecular neurobiologists have also studied voltage-gated potassium (K_v) channels. This work began with the identification of the *shaker* mutation in fruit flies, which impairs K_v channel function. Subsequent research identified the DNA sequence of these channels, their structural details, and the functions of key domains. Intriguingly, K_v channels are similar to bacterial Na_v channels insofar as they are homotetramers (4 identical subunits that come together to form a single functional unit) and contain 6 membrane-spanning helices per subunit.

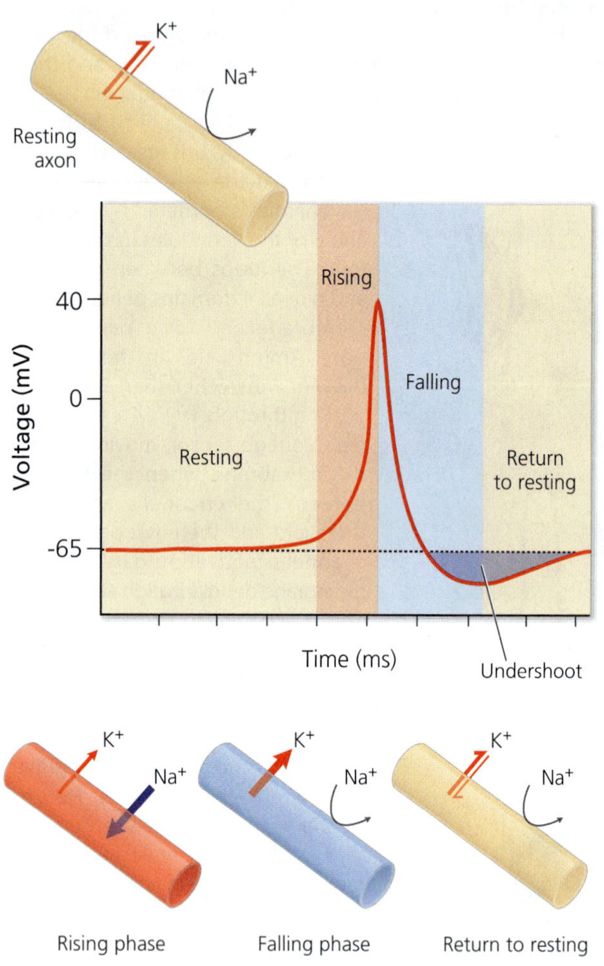

Figure 2.7 Phases of the action potential. The neuronal membrane at rest is relatively impermeable to Na^+, but some K^+ flows out; this keeps the membrane near its resting potential of −65 mV. During the rising phase of the action potential the membrane becomes highly permeable to Na^+, which rushes down its concentration gradient into the axon and thus depolarizes it to approximately +40 mV. Next comes the falling phase, during which the membrane becomes less permeable to Na^+ and much more permeable to K^+. This causes K^+ to rush out of the cell, repolarizing it to a level near the equilibrium potential for K^+ (approximately −80 mV). Soon thereafter, K^+ permeability returns to its resting level. When the membrane potential is more negative than its resting value, it is said to be hyperpolarized; this period is called the action potential undershoot.

The Hodgkin–Huxley Model

How did Hodgkin and Huxley explain neuronal action potentials? Their key discovery was that, at the beginning of an action potential, the axonal membrane suddenly becomes highly permeable to sodium ions (Figure 2.7). As later research showed, this sudden increase in sodium permeability results from the opening of specialized ion channels in the neuronal membrane that let sodium ions flow through their central pore (see Box 2.3). Because the concentration of sodium ions is much lower inside the axon than outside (Table 2.1) and because the neuron's negative resting potential generally favors the movement of positive ions into the neuron (down the electrical potential gradient), opening the membrane's sodium channels causes sodium ions to rush into the axon. Because sodium ions are positively charged, they generate an **inward current** (following Benjamin Franklin's lead, current is conventionally defined as flowing in the direction of net movement of positive charge). The inward current causes the inside of the axon to become more positive or, as a neurophysiologist would say, **depolarized**.

The influx of sodium does not continue indefinitely. As the axon becomes increasingly depolarized, the sodium ions become progressively more repelled by the excess of positive charges inside the cell (if you want to think about it the other way around, they become more attracted to excess of negative charges outside the cell). Thus, the movement of sodium ions across the neuronal membrane is governed by (1) the concentration gradient pushing them into the cell and (2) the voltage gradient pushing them out. The situation is similar to that for the potassium ions at rest (see Figure 2.5), but the corresponding forces are reversed in direction. Using the same equations Hodgkin and Huxley used to calculate the equilibrium potential for potassium, one can calculate the equilibrium potential for sodium (see Box 2.1). In squid giant axons, it is about +60 mV.

The idea that a sudden increase in membrane permeability might explain the onset of an action potential had been around for a few years before the work on squid giant axons began. However, the crucial role of sodium influx was not confirmed until Alan Hodgkin and his collaborator Bernard Katz showed that lowering the concentration of extracellular sodium ions decreases the size of the action potential precisely as one would predict from the Nernst equation for sodium ions. This was a great advance but still left open a crucial question: what causes the neuronal membrane to become more permeable to sodium during the rising phase of an action potential?

Voltage-clamp Recording

Hodgkin and Huxley were able to answer this question by using a clever new technique called **voltage-clamp recording**. This method employs two sets of electrodes. In Hodgkin and Huxley's experiments, those electrodes were arranged as shown in Figure 2.8. One set of electrodes is used to record the potential (the voltage difference) across the neuronal membrane. The second set is used to inject current (positive or negative charges) into the neuron so that its internal voltage can be set to any value the experimenter desires. Crucially, the first set of electrodes is connected to the second set in such a way that any time the first electrode detects a change in membrane potential, the second set of electrodes injects current into the cell to compensate precisely for the change in membrane potential. Because of this feedback mechanism, the voltage across the membrane can be held steady, or clamped, at any predetermined level even as ions flow into or out of the neuron (much as the cruise control in a car keeps the car's speed constant, even as the car goes up or down a hill). By keeping track of how much current is being injected into the cell, the experimenter can figure out how much current is being carried across the membrane by ions because these two currents are by design equal in magnitude (although opposite in sign).

The Action Potential Threshold

Using this voltage-clamp technique, Hodgkin and Huxley demonstrated that the membrane of the squid giant axon is relatively impermeable to sodium ions at rest. However, when the membrane potential is depolarized beyond some threshold voltage (typically 10–20 mV more positive than the resting potential), the inward current attributable to sodium influx temporarily increases (Figure 2.9). This observation allowed Hodgkin and Huxley to infer that strong depolarization makes the axonal membrane more permeable to sodium ions. Phrasing their conclusion in terms of ion channels, we can say that above-threshold depolarization causes sodium channels in the axonal membrane to increase their probability of being open rather than closed. Because these sodium channels open in response to a change in voltage across the

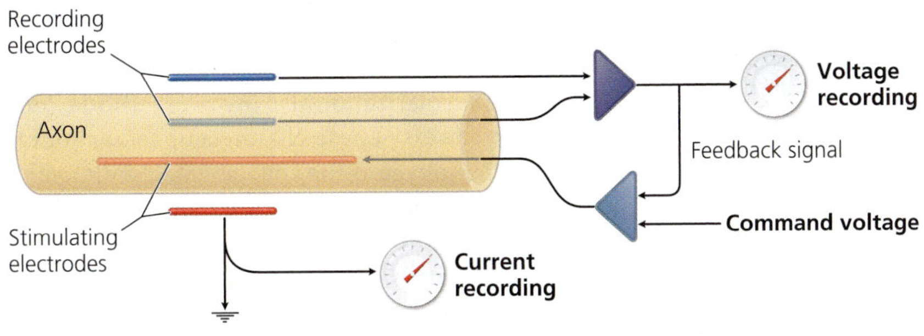

Figure 2.8 The voltage clamp technique. During voltage clamp recording, the voltage across the axonal membrane is set to an arbitrary value by a "command voltage" signal applied across a pair of stimulating electrodes (red). Ions flowing across the membrane tend to drive the membrane potential away from this desired value. These tiny deviations are recorded by two recording electrodes (blue). The output from these electrodes (feedback signal) is sent to the stimulating electrodes, which use it to cancel out the incipient deviations. Crucially, experimenters can infer how much ionic current is moving across the membrane because this current is exactly the inverse of the current needed to clamp the membrane at the desired voltage. By varying command voltage and ion concentrations, experimenters can deduce how much the different ions contribute to the overall current at various membrane voltages.

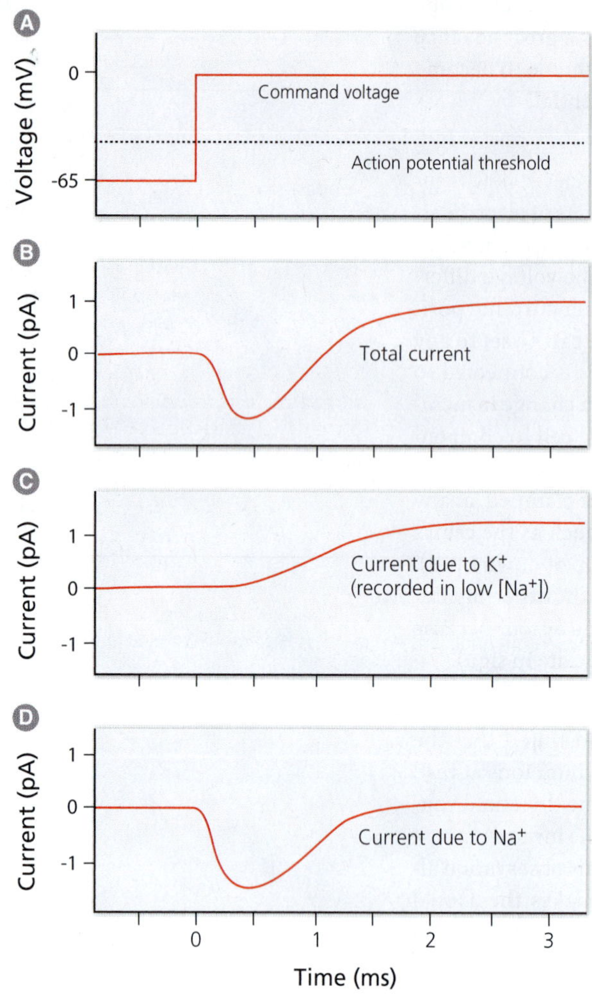

Figure 2.9 Changes in sodium and potassium permeabilities during the action potential. In a famous experiment, Hodgkin and Huxley depolarized a voltage-clamped squid axon to 0 mV, well above the level needed to trigger an action potential (A). They then measured current flow across the axonal membrane, both in artificial sea water (B) and in a solution lacking sodium ions (C), which eliminated sodium influx and thus isolated the current due to potassium. By comparing the currents observed in these two conditions (subtracting the curve in C from that in B), Hodgkin and Huxley could infer the current that is normally carried by sodium (D). Inward currents are depicted as downward deflections.

membrane (depolarization), neurobiologists refer to them as **voltage-gated sodium channels** (see Box 2.3). Also, the more depolarized the membrane is beyond that threshold, the greater the increase in sodium permeability (ever more sodium channels open). This is interesting because it implies a positive feedback loop: the more the membrane is depolarized, the more sodium flows in, which in turn causes further depolarization (Figure 2.10). The membrane potential at which the positive feedback between membrane depolarization and sodium influx kicks in is called the **action potential threshold**.

What causes a neuron's membrane potential to increase to the level where it reaches the action potential threshold? In studies using intracellular electrodes, the experimenter can inject the current needed to depolarize the axonal membrane to threshold. In an intact animal, of course, there is no stimulating electrode inside the neurons. Instead, neurons are normally depolarized by synaptic input. We discuss this in detail shortly. For now, we can note simply that synaptic inputs may gradually depolarize a neuron until its membrane voltage reaches the action potential threshold. You can see this if you look closely at the rising phase of an action potential. The rise in the membrane potential becomes significantly steeper after the membrane potential hits the action potential threshold (Figure 2.10, top right).

The Falling Phase of the Action Potential

The rising phase of an action potential typically lasts for less than 1 ms and is then followed by a decrease in the membrane potential during the falling phase (see Figure 2.7). This membrane **repolarization** occurs in part because the voltage-gated sodium channels automatically close about 1 ms after they open. This process, called **sodium channel inactivation** (see Box 2.3), explains why the current due to sodium influx in Hodgkin and Huxley's famous voltage-clamp experiment lasted only 2–3 ms, even though the membrane depolarization was maintained (Figure 2.9).

Membrane repolarization also involves **voltage-gated potassium channels**. These channels open at about the same membrane voltage as the voltage-gated sodium channels. Because the concentration of potassium ions is higher inside a neuron than outside of it, potassium ions start to rush out of the cell as the voltage-gated potassium channels open. This efflux of potassium generates an outward, repolarizing current (Figure 2.9C) that begins to reverse the membrane depolarization. However, the voltage-gated potassium channels open more slowly than the voltage-gated sodium channels that create the action potential's rising phase (Figures 2.9 C vs. D). This is important because if both channel types opened at the same rate, then potassium and sodium ions would be moving simultaneously in opposite directions across the membrane, causing their effects to cancel out. Some such overlap indeed occurs in squid giant axons (Figure 2.11) and is energetically inefficient. In mammals, however, the overlap between sodium influx and potassium efflux is minimal. Virtually all voltage-gated sodium channels are inactivated by the time the voltage-gated potassium channels have opened. Thus, evolution has found a way to make action potential generation more efficient in mammals than in squid.

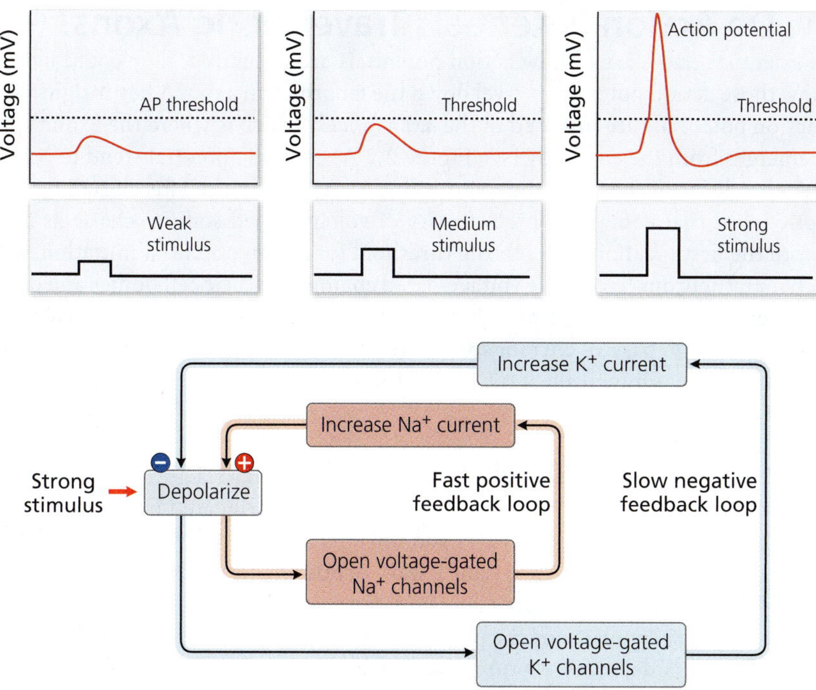

Figure 2.10 Feedback shapes the action potential. Shown across the top are the responses of a neuron to weak, medium, and strong depolarizing current pulses. The weak and medium stimuli cause sub-threshold changes in the membrane potential, whereas the strong stimulus triggers a full-blown action potential, whose dynamics are illustrated in the bottom diagram. Once the membrane is depolarized above threshold, the voltage-gated Na^+ channels open, which further increases membrane depolarization. Above-threshold depolarization also opens voltage-gated K^+ channels, which reduces the membrane potential. Because the K^+ channels open more slowly than the Na^+ channels, this repolarization comes after the depolarization, giving the action potential its distinct rising and falling phases.

Putting All the Pieces Together

Hodgkin and Huxley's most remarkable achievement was to model their data mathematically. Using a series of equations, they were able to reconstruct the shape of the action potential from the quantitative data they had obtained on sodium and potassium permeabilities in voltage-clamped axons (Figure 2.11). They even managed to reconstruct the **undershoot** of the membrane potential that occurs at the end of each action potential when the voltage-gated sodium channels have all become inactivated, but the voltage-gated potassium channels are still open. At that point, the number of open potassium channels is greater than it is at rest and, consequently, the membrane potential is closer to the equilibrium potential for potassium (–80 mV) than it would be in a neuron at rest. At these low membrane voltages, all voltage-gated potassium channels close and the neuronal membrane gradually returns to its resting potential. In other words, the action potential ends.

As you can now appreciate, action potentials are all-or-none phenomena. Once an action potential is triggered, the voltage-gated sodium and potassium channels follow a preset course, which determines both the amplitude and the duration of each action potential. For that reason, all the action potentials produced by a given neuron are normally invariant in size and shape; they are all-or-nothing events.

BRAIN EXERCISE

Why is the "overlap" region in Figure 2.11 "bad"? What causes this overlap, and how have mammals reduced it?

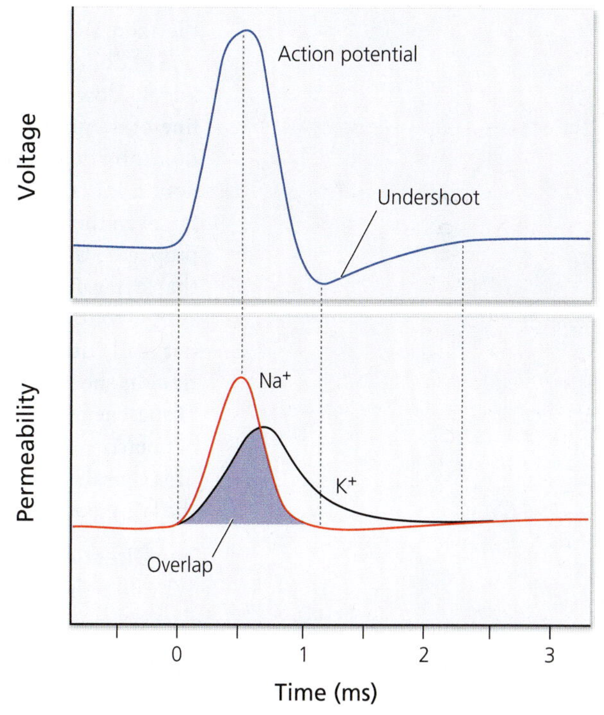

Figure 2.11 Hodgkin and Huxley's model of the action potential. Based on multiple experiments of the kind illustrated in Figure 2.9 (clamping squid axons at many different voltages) Hodgkin and Huxley inferred how the axon's permeability to Na^+ and K^+ changes during a normal action potential (when the membrane is not clamped). As the lower graph shows, the axon's membrane is permeable to both Na^+ and K^+ during a significant portion of the action potential (blue shading). This overlap is metabolically inefficient.

2.3 How Do Action Potentials Travel along Axons?

Once scientists had learned how action potentials are generated, they could figure out how those action potentials travel down the length of an axon. A key insight was that action potentials are initiated in the *axon hillock*, which is where the axon typically emerges from the cell body (see Figure 2.2 A). Action potentials tend to originate at that location because the axon hillock is typically the site closest to the synaptic input that contains a high density of voltage-gated sodium channels and therefore the first location to reach the threshold for action potential initiation. Although many neurons have some voltage-gated channels in their cell bodies and dendrites, membrane depolarizations that originate in the dendrites must typically propagate (through passive current spread) to the axon hillock before an action potential is triggered. Only in a few types of neurons can action potentials be initiated out in the dendrites.

A Traveling Wave of Membrane Depolarization

Now, consider what happens when an action potential is triggered in the axon hillock. First, sodium ions rush in. As these sodium ions accumulate in the axon hillock, they repel other positive ions, triggering a wave of positive ion movement away from the axon hillock. Some of this ionic current flows toward the cell body, but the rest flows into the axon, depolarizing the patch of axonal membrane right next to the axon hillock. When this depolarization reaches the action potential threshold, voltage-gated sodium channels open and sodium ions flow in. At that point the cycle repeats itself and the next patch of membrane becomes depolarized. In essence, you have a wave of membrane depolarization that spreads along the axon (Figure 2.12).

To better visualize the chain of events that propels an action potential down the axon, think of each patch of axonal membrane as an individual domino in a long line of dominoes! These dominoes are set up so that the falling of the first domino knocks down its neighbor, which then topples its neighbor, and so on until the entire line of dominoes lies flat. In this analogy the axon hillock corresponds to the first domino and the propagating action potential corresponds to the wave of falling dominoes. The core of this analogy is that the wave of falling dominos may travel far, even though no individual domino moves much, just as an action potential may propagate the entire length of an axon, even though individual ions move no more than a few microns.

Of course, the analogy is imperfect. Most importantly, the axonal membrane is not really divided into discrete patches. Instead, the axonal membrane forms a continuous sheet. Nonetheless, if you think about the individual ion channels within the axonal membrane, then you can say that sodium influx through one channel helps to depolarize the membrane around it and thus open some of its neighboring channels. This causal interplay sets up a chain reaction that propagates down the axon, much as the falling of one domino triggers a wave that travels down the line.

The Direction of Action Potential Propagation

Why does the action potential never reverse direction as it travels down the axon? The answer lies in the kinetics (time courses) of the voltage-gated sodium and potassium channels. As you may recall from our discussion of membrane repolarization during the action potential's falling phase, the voltage-gated sodium channels that open during an action potential's rising phase close automatically (inactivate) shortly after they open. In addition, voltage-gated potassium channels open slowly on depolarization and then stay open for a while. As long as the potassium channels remain open and the sodium channels are closed, it is impossible for the axonal membrane to generate another action potential. The membrane at this time is said to be in a **refractory period**. Therefore, as an action potential travels down an axon, it is inevitably followed by a wave of refractory periods that prevents the action potential from turning back on itself. If you like the domino analogy, you can think of it like this: once the dominoes have fallen, you need to set them up again before another wave can be triggered.

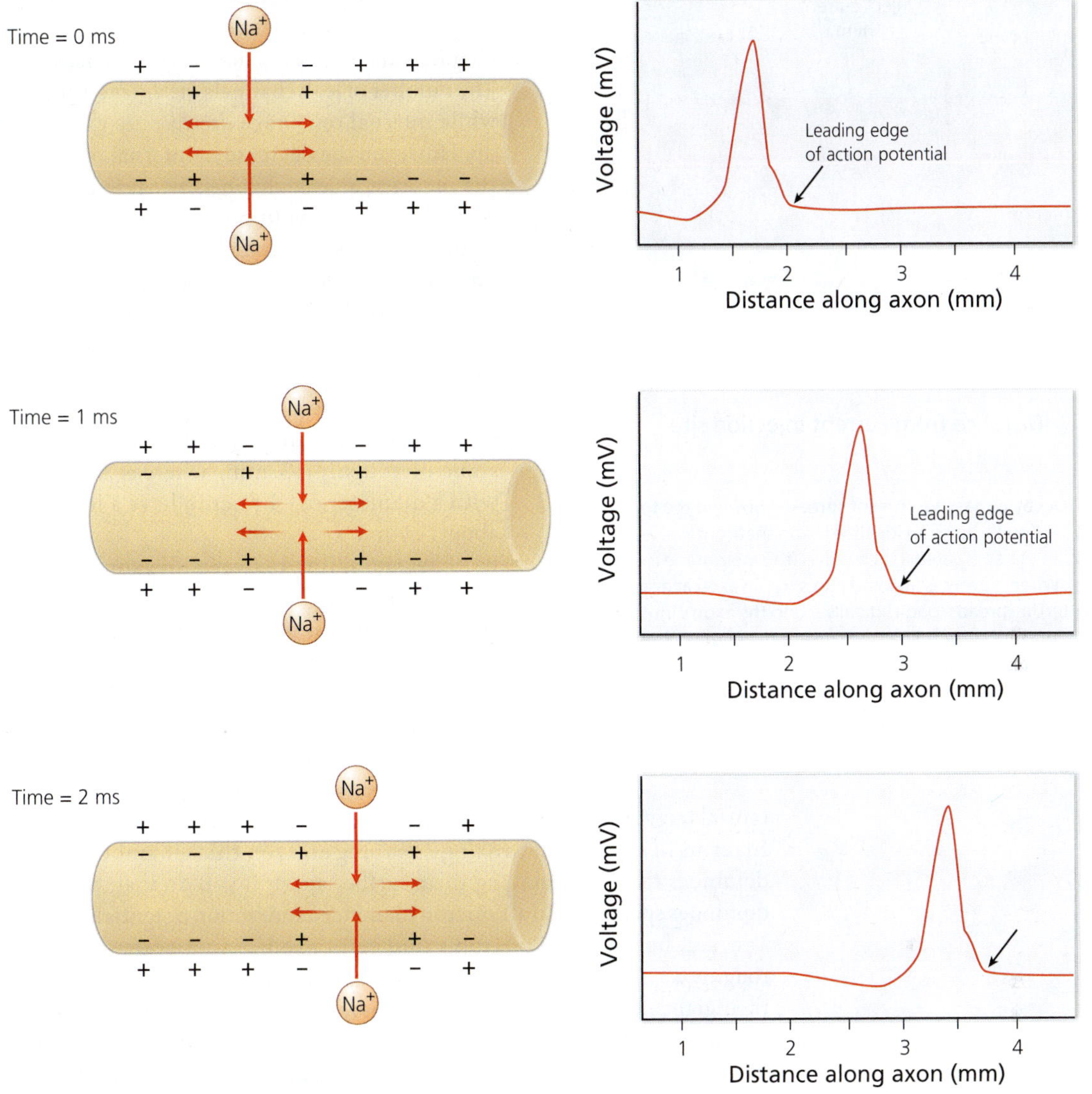

Figure 2.12 Action potential propagation. Action potentials travel along axons because the massive influx of Na^+ ions at one location of the axonal membrane tends to trigger Na^+ influx at adjacent locations, as long as those sites are not in the action potential's wake (refractory). The top, middle, and bottom panels in this figure illustrate successive time points. The plus and minus signs represent ionic charges that have accumulated on either side of the axonal membrane (a plus on the inside indicates depolarization). The red arrows indicate ion flow. The graphs on the right show how membrane voltage varies along the length of the illustrated axons. These traces may look odd at first, but it may help to think of them as time-reversed action potentials (compare to Figure 2.7).

The Speed of Action Potential Propagation

Action potentials travel down axons much more slowly than electrical signals race through typical copper wires. The main reason for the sluggishness of action potential propagation is that it takes time for voltage-gated sodium channels to open, for the sodium ions to flow into the axon, and for successive patches of axonal membrane to reach the threshold for action potential triggering.

One way for neurons to overcome this sluggishness is to increase axon diameter. To understand why this strategy works, you need to know that current flowing down an axon passively (without being amplified by voltage-gated ion channels) decays exponentially with distance, mainly because some of the current escapes through leak channels in the axonal membrane (Figure 2.13). The distance over which the voltage change generated by an injected current decays to ~37% of its maximum value is

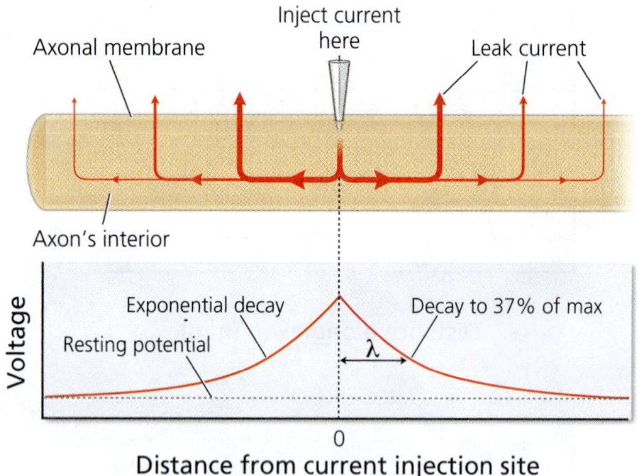

Figure 2.13 Decay of passive current spread. Shown at the top is a longitudinal section through an idealized axon that contains leak channels but no voltage-gated ion channels (the membrane is electrically passive). When current is injected into such an axon at a specific location (center), it spreads longitudinally along the axon's interior (black arrows indicate current flow). Because some of the current leaks out across the axonal membrane, the voltage change caused by the injected current decays exponentially with distance from the injection site (bottom graph). An axon's length constant (λ) is defined as the distance over which the generated voltage change decays to ~37% (1/e) of its maximum value.

called the axon's **length constant**. The value of an axon's length constant depends on both the number of leak channels and the axon's internal resistance to current flow. The number of leak channels *increases* with axon diameter, while internal resistance *decreases* as axon diameter goes up. However, the decrease in internal resistance outpaces the increase in leak channels because internal resistance varies with axon volume, whereas the number of leak channels varies with axonal surface area. Assuming (for now) that the density of leak channels is invariant, an axon's internal resistance is inversely proportional to its diameter. Put differently: the thicker an axon, the more easily current flows through it. This means that thicker axons should have a larger length constant. Indeed, squid giant axons with a diameter of ~1 mm have a length constant of roughly 13 mm, whereas mammalian axons with a diameter of ~1 micron have a length constant of about 0.2 mm.

The causal link between an axon's diameter and its length constant affects the speed of action potential propagation because, as noted earlier, the current generated by an action potential at one location must spread passively down the axon before it can depolarize another patch of axonal membrane sufficiently to trigger an action potential there. The less the current dissipates as it travels inside the axon, the farther down the axon that next action potential trigger site can be, and the faster the wave of action potentials propagates. In terms of our domino analogy, making an axon thicker amounts to making the dominoes taller and moving them farther apart; if you do that, the wave of falling dominoes speeds up. In a 1 micron diameter axon, action potentials travel at just over 1 m/sec, but in a 300 micron squid giant axon, action potentials zoom along at roughly 25 m/sec. This fast conduction speed is useful to the squid because they use their giant axons to trigger escape behaviors, jetting away from threats.

Increasing conduction speed by increasing axon diameter is a costly strategy, however, because larger diameter axons move more sodium and potassium ions across their membrane during an action potential. After a neuron has fired many action potentials, these ions must be moved back up their respective concentration gradients, a process that requires a considerable amount of metabolic energy. In addition, increasing axon diameter increases the amount of space those axons occupy. As noted in Chapter 1 (see Figure 1.10), space is at a premium in many nervous systems. Therefore, increasing axon diameter is not an ideal strategy for increasing axonal conduction speed.

The Effects of Myelination

A more cost-effective trick some neurons use to boost action potential propagation is to wrap their axons in a fatty substance called myelin, which is produced by specialized glial cells (Schwann cells and oligodendrocytes). This **myelin sheath** (Figure 2.14) changes both the axon's length constant and its capacitance. As you will learn in the next two sections, both of these effects increase the speed at which action potentials travel along axons.

Myelination Increases an Axon's Length Constant

The myelin sheath around an axon acts like a layer of electrical insulation that prevents the flow of ions through the previously mentioned leak channels, thereby increasing the resistance to current flow across the axonal membrane. In doing so, the myelin sheath increases the axon's length constant, allowing the positive current generated at the axon hillock to flow further down the axon before it fades away. If you think of the axon as a garden hose full of tiny holes, then the myelin is like an extra sheet of plastic around the hose that prevents water from leaking out. The less water leaks out, the more water (current) comes out the far end of the hose.

The analogy is imperfect, but the myelin does extend the distance over which currents flow within an axon. Thus, the myelin sheath allows an action potential in the axon hillock to trigger an action potential in a quite distant patch of axonal membrane (typically at least 100 μm away but varying with axon diameter). Of course, this second patch must be capable of triggering an action potential. That is, it must be free of myelin and packed with voltage-gated sodium channels. Otherwise the action potential wave would just die out. Indeed, careful examination shows that the myelin sheath is interrupted every few hundred microns (up to 2 mm in very thick, heavily myelinated axons). These gaps in the myelin sheath are known as **Nodes of Ranvier**, and they are packed with voltage-gated sodium and potassium channels.

Therefore, when an action potential is triggered at the axon hillock in a myelinated axon, positive current flows down the myelinated segment of the axon to the nearest Node of Ranvier. There, the positive current depolarizes the membrane until it exceeds the action potential threshold and lets sodium ions rush in (Figure 2.14 B). This depolarization then causes current to flow down the axon to the next Node of Ranvier, where the cycle repeats itself. Thus, in myelinated axons, the axonal membrane really is divisible into discrete patches between which current flows (in terms of the domino analogy, think of very tall dominoes spaced very far apart).

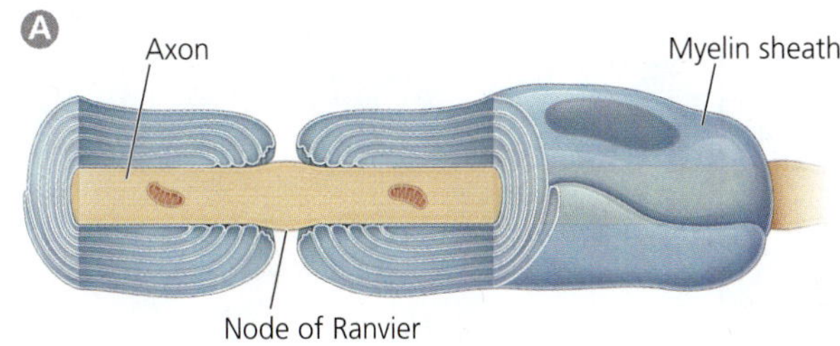

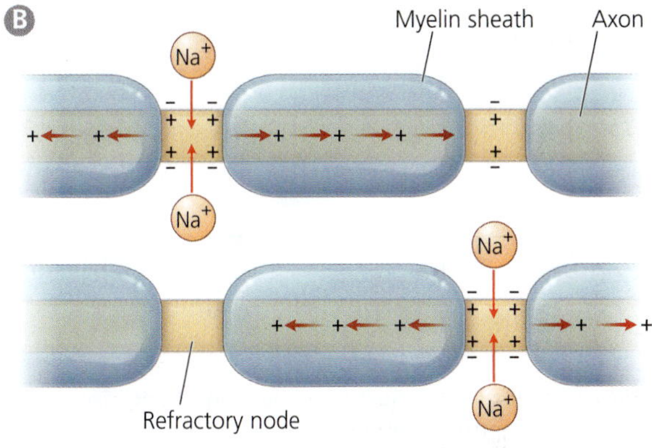

Figure 2.14 Axon myelination. Panel (A) shows how a myelin sheath wraps itself repeatedly around a myelinated axon. The axon remains unmyelinated at regularly spaced Nodes of Ranvier. Shown in (B) is a myelinated axon at two points in time. Initially (top), Na^+ rushes into the axon at a node of Ranvier, triggering a "pressure wave" of positive ions that travels down the axon. At the next node of Ranvier, the wave of current triggers another action potential (bottom). Na^+ influx due to this second action potential triggers another wave of positive current, which flows both up and down the axon. However, current flow back up the axon does not trigger another action potential at the first node because this node is now refractory.

Because the Nodes of Ranvier are separated by myelinated segments that do not support the generation of action potentials, action potentials in myelinated axons effectively leap from one node to the next. Indeed, action potentials typically leap across several nodes in a single bound, which ensures the propagation of those action potentials even if some nodes are malfunctioning. This **saltatory conduction** (saltatory means "jumping") of action potentials in myelinated axons increases conduction speed because, as you may recall, the most time-consuming aspect of action potential propagation is the depolarization of a membrane patch to the threshold for action potential triggering. In myelinated axons, significantly fewer patches must be depolarized (over a given distance) than in unmyelinated axons.

Myelination Decreases Membrane Capacitance

As noted earlier, charges of opposite sign may attract each other across the neuronal membrane, forming a capacitor. Because a glial sheath is an excellent electrical insulator, wrapping an axon in myelin increases the thickness of the electrical insulation between the intra- and extracellular fluids. This increase in physical separation between opposing charges on opposite sides of the neuronal membrane decreases the attractive force between those charges because the attractive force falls off with the square of distance. Another way of saying this is that myelination greatly *decreases* an axon's membrane capacitance.

Reducing membrane capacitance increases the speed of action potential propagation because, as mentioned earlier, capacitors take time to change their charge. As depolarizing current flows along the inside of an axon, it must dislodge some of the negative charges that had built up on the intracellular surface of the axonal membrane (it must discharge the membrane capacitance) before it can fully depolarize the next patch of membrane. The greater the membrane capacitance, the longer this

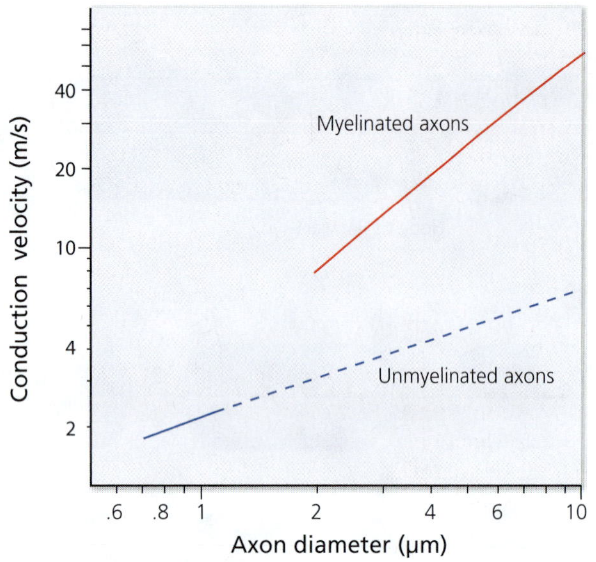

Figure 2.15 Factors that influence conduction velocity. The speed at which action potentials travel down an axon increases with axon diameter. This rule holds for both myelinated and unmyelinated axons, but the rate at which conduction velocity increases with axon diameter is significantly faster for myelinated axons than for unmyelinated ones. The illustrated data are from axons in cat peripheral nerves. The dashed line indicates predicted velocities of hypothetical thick unmyelinated axons. [After Hursh, 1939]

process takes. Therefore, increasing membrane capacitance reduces the rate at which a wave of depolarization spreads inside an axon. Conversely, the decrease in membrane capacitance created by a myelin sheath increases the rate at which depolarizing current spreads inside of an axon. This increase in the rate of current spread, in turn, speeds up the rate at which an action potential can be triggered at the next Node of Ranvier.

Myelination Boosts Conduction Velocity

As you have now seen, myelination increases action potential velocity through two distinct mechanisms—by increasing an axon's length constant and by decreasing its membrane capacitance. The two effects are synergistic; but for axons with thick myelin sheaths, the effect of decreasing membrane capacitance is more powerful than that of increasing the length constant (by reducing the leak current).

The degree to which action potential propagation is sped up by myelination depends on axon diameter (Figure 2.15). However, for mammalian axons between 2 and 10 μm in diameter, myelination boosts the speed of action potential propagation 5–10 fold. This amounts to a significant decrease in signal transmission time, especially through long axons (notably those from the spine to the toes). Myelination also saves metabolic energy because it drastically reduces the number of sodium and potassium ions that move across the axonal membrane (recall our leaky garden hose analogy). Given these advantages of myelination, it is not surprising that myelin evolved several times independently—in vertebrates and several invertebrate groups, including annelids (such as earthworms) and a few species of shrimp. Squid, though, lack myelin, which probably explains why some of their axons have such enormous diameters.

If myelination is so great for speeding up transmission and saving energy, why does the brain contain many axons that are not covered in myelin? One answer is that myelin sheaths are not entirely cost-free; they must be built during development and then maintained. A second answer is that myelin sheaths are bulky. Because myelinated axons take up more space than unmyelinated ones, they tend to be absent from areas where neurons are tightly packed (such as the gray matter of the cerebral cortex). If those tightly packed neurons all had myelinated axons, then they would have to be more widely spaced; and all the connections between them would have to be longer. To cut down on these wiring costs, the nervous system myelinates primarily its long axons, leaving the short ones bare.

BRAIN EXERCISE

Imagine a neuron that has one action potential initiation site in the axon hillock and another one in a distal dendrite. If action potentials were triggered simultaneously at both initiation zones, what would happen when the resulting action potential waves "collide"? What would happen if action potentials were triggered earlier at one initiation site than the other?

2.4 How Do Neurons Transmit and Integrate Information?

Having considered how action potentials propagate along axons, we can start to think about what happens when those action potentials reach the "end of the line," the axon terminal. As we have already discussed, an action potential that reaches the end of an axon does not simply reverse course and travel back up the axon. Nor can the electrical current associated with an action potential simply jump across the gap between neurons that we call the synapse (unless you are talking about "electrical synapses" in which ions may flow through gap junctions from one side of the synapse to the other). How, then, is information transmitted from a presynaptic axon terminal to a postsynaptic dendrite? The answer to this fundamental question was first worked out for synapses between motor neurons and muscle cells (see Chapter 8). Here, we focus on synaptic transmission at a stereotypical synapse within the brain or spinal cord.

Synaptic Transmission

Synapses are highly specialized both presynaptically and postsynaptically. One of the main presynaptic specializations is the presence of numerous **synaptic vesicles**, which are ~50 nanometers in diameter and filled with a few thousand neurotransmitter molecules each. For most excitatory neurons in the central nervous system, the neurotransmitter is L-glutamate, an ionic form of glutamic acid (which is one of the 20 standard amino acids). Neuroscientists generally refer to the L-glutamate in neurons simply as **glutamate**.

Transmitter Release and Postsynaptic Receptors

A second key specialization of presynaptic terminals is that their membranes are packed with **voltage-gated calcium channels**. When an action potential wave comes crashing into an axon terminal, these calcium channels open and let calcium ions rush down their concentration gradient into the terminal. The rise in calcium levels within the terminal sets in motion a complex molecular machinery that causes some synaptic vesicles to fuse with the cell membrane and thus (through a process called exocytosis) release their neurotransmitter content into the **synaptic cleft**, which is the tiny gap between the pre- and postsynaptic cells (Figure 2.16). Key evidence for the idea that calcium influx triggers neurotransmitter release was obtained

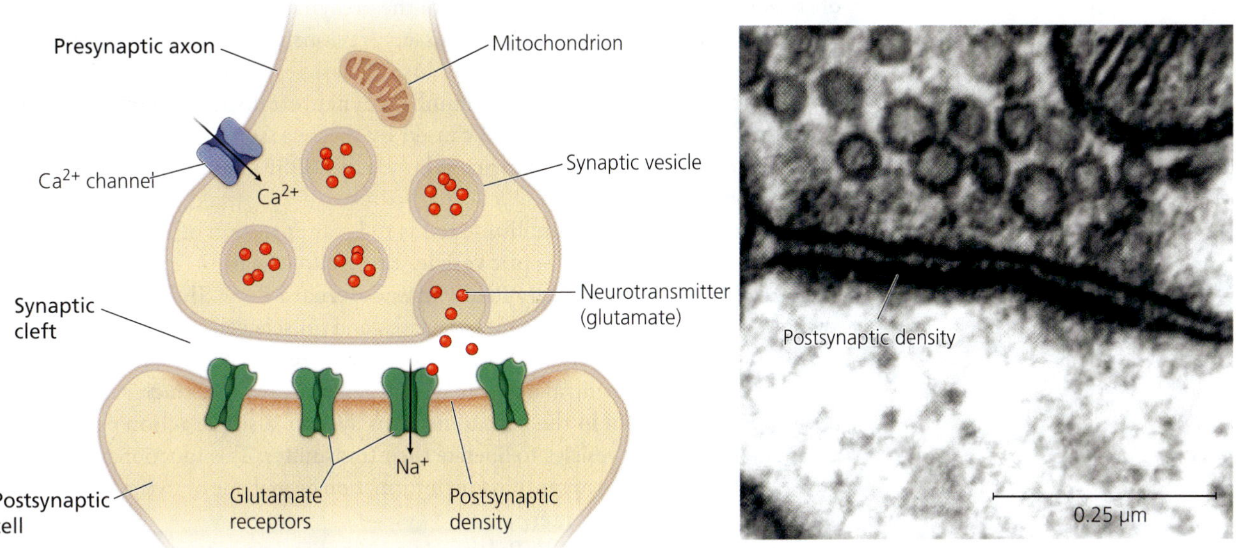

Figure 2.16 A synapse using glutamate. When an action potential comes down an axon, Ca^{2+} ions flow into the presynaptic terminal. This causes synaptic vesicles in the terminal to move toward the synaptic cleft and release glutamate. When glutamate binds to a postsynaptic glutamate receptor (of the AMPA[α-Amino-3-hydroxy-5-methyl-4-isoxazolepropionic acid] type), the receptor allows Na^+ ions to flow through its central pore into the postsynaptic cell. Shown on the right is an electron micrograph of a real synapse, showing presynaptic vesicles and the postsynaptic density, which is a dense meshwork of proteins that lines the postsynaptic side of the synapse. [Image courtesy of Kristen M. Harris, PhD; SynapseWeb, http://synapses.clm.utexas.edu/]

by Ricardo Miledi in 1973 when he injected calcium ions into terminals of squid giant axons and observed increased transmitter release.

After the neurotransmitter molecules have been released into the synaptic cleft, they diffuse across to the postsynaptic side of the synapse. Because a typical synaptic cleft is only 20–25 nm wide, the transsynaptic diffusion of neurotransmitter requires only about 0.05 ms, which is considerably less than the ~0.3 ms it takes for neurotransmitter to be released from the presynaptic terminal in response to an incoming action potential.

Once the neurotransmitter molecules reach the postsynaptic side, they tend to bind to specialized receptor molecules that are embedded in the postsynaptic membrane. These receptor molecules come in many different flavors (as you will see shortly), but one key receptor for glutamate is the **AMPA receptor**, which derives its name from the fact that it can be activated by an artificial glutamate agonist (a glutamate mimic) called AMPA (α-amino-3-hydroxy-5-methyl-4-isoxazolepropionic acid). This AMPA receptor consists of four molecular subunits that are arranged around a central pore through which ions can flow, but only when glutamate is bound to the receptor's subunits. This is why the AMPA receptor is referred to as a **ligand-gated ion channel** (with glutamate being the ligand).

Which ions pass through open AMPA receptors? Both potassium and sodium are capable of doing so. However, relatively few potassium ions flow through open AMPA receptors when a neuron is resting (not in the middle of firing an action potential) because a neuron's resting potential is near the equilibrium potential for potassium. In contrast, sodium ions do flow through open AMPA receptors when a neuron is at rest because the equilibrium potential for sodium is far more positive than a neuron's resting potential. Therefore, when glutamate binds to the AMPA receptors of a neuron at rest, the neuron will experience a net influx of positive ions (mainly the sodium) and, consequently, be depolarized. This depolarization is called an **excitatory postsynaptic potential (EPSP)**.

Excitatory Postsynaptic Potentials

How large is an EPSP? The answer to this question depends in part on the amount of neurotransmitter that is released from the presynaptic terminal in response to an incoming action potential. As we discussed previously, neurotransmitter molecules are released from synaptic vesicles. Because all vesicles within a terminal are similar in size, each vesicle releases similar numbers of neurotransmitter molecules into the synaptic cleft. Therefore, an EPSP caused by the release of neurotransmitter from two synaptic vesicles is roughly twice as large as an EPSP caused by neurotransmitter release from a single vesicle.

More generally, EPSP amplitude tends to be an integer multiple of the depolarization caused by a single synaptic vesicle. This **quantal** nature of synaptic transmission was discovered in 1952 by Paul Fatt and Bernard Katz. Their work focused on synaptic transmission between motor nerves and muscle cells where a typical EPSP is caused by the release of neurotransmitter from dozens, if not hundreds, of synaptic vesicles. EPSPs in the central nervous system generally involve much less transmitter release. However, even in the central nervous system, a single action potential may cause several synaptic vesicles to liberate their transmitter. This functional redundancy probably ensures that synapses transmit information even if one or two synaptic vesicles fail to release their transmitter.

The size of a recorded EPSP depends on the distance between the synapse and the tip of the microelectrode that is used to record this EPSP because the current underlying the EPSP fades away with distance from the synapse (Figure 2.17 A & B). As in the case of current spreading through an unmyelinated axon (see Figure 2.13), leakage of current across the dendritic membrane gradually reduces the amount of current that flows in parallel with the membrane. Therefore, EPSPs caused by synapses close to the cell body (proximal synapses) tend to generate more depolarization at the axon hillock than do EPSPs from synapses far out on the dendritic tree (distal synapses). This difference is important because the amount of depolarization at the axon

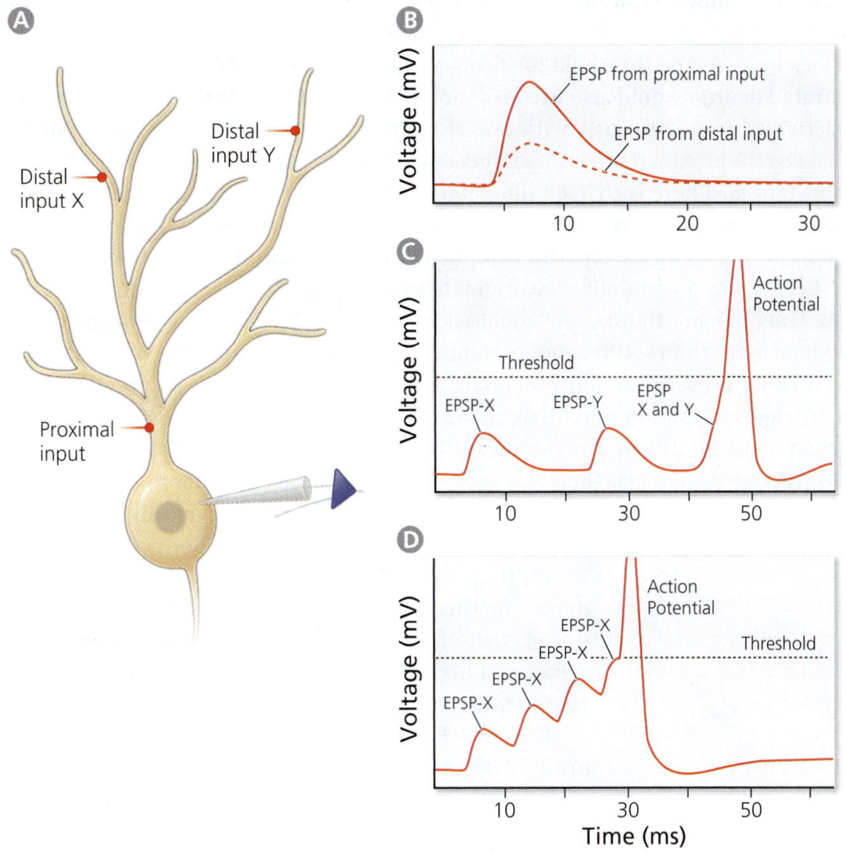

Figure 2.17 Integrating excitatory postsynaptic potentials (EPSPs). Dendritic EPSPs decrease in size as they propagate toward the cell body. When recording intracellularly at the cell body (A) and activating synapses either close to the cell body (proximally) or far out on the dendritic tree (distally), EPSPs resulting from proximal synapse activation are much larger than EPSPs due to distal input (B). Panel (C) shows that EPSPs elicited by the activation of distal synapses (EPSP-X or EPSP-Y) often fail to reach the threshold for action potential firing. However, if two distal inputs are activated simultaneously, then the two EPSPs tend to sum when they reach the cell body, producing a single large EPSP that can trigger an action potential. This phenomenon is known as spatial summation. Panel (D) depicts the summation of small, subthreshold EPSPs that occur in rapid succession, known as temporal summation.

hillock determines whether the neuron will fire an action potential (see Figure 2.10). In effect, proximal synapses are more likely to trigger an action potential in the postsynaptic neuron than distal synapses. This is an important principle to which we shall return.

For now, let us focus on the observation that even EPSPs generated at proximal synapses generally cause too little depolarization at the axon hillock to trigger an action potential all by themselves. This may seem odd at first. Would you not expect nervous systems to be designed such that each presynaptic action potential triggers a postsynaptic action potential, with the EPSP acting as go-between? No; you would expect this only if you think of neurons as forming a simple linear chain in which each neuron is connected to the next one through a single synapse. But this is not how typical neurons are wired together. Instead, a typical neuron is connected to many different neurons, with each connection involving at least a handful of synapses. Moreover, many neurons feed back onto themselves. Given this kind of weblike connectivity, it would be a disaster for the brain if every EPSP triggered an action potential. Action potentials would spread like wildfire throughout the brain, cause excessive neuronal firing and perhaps even neuronal cell death due to overexcitation (see Chapter 5). Given these considerations, it is good that EPSPs are relatively small.

Synaptic Integration

Because EPSPs tend to be small, many are needed to bring a postsynaptic neuron above the threshold for firing an action potential. This turns out to be crucial, as it allows neurons to respond differently to different spatial or temporal patterns of synaptic activity.

Spatial and Temporal Summation

Consider, for example, the neuron shown in Figure 2.17. Neither of its two distal synapses produce EPSPs that are large enough by themselves to trigger an action potential at the axon hillock. However, when both distal synapses are active simultaneously, or nearly so, then their EPSPs sum; and the total depolarization produced at the axon hillock exceeds the threshold for firing an action potential (Figure 2.17 C). The illustrated neuron would also fire an action potential whenever the proximal synapse is activated simultaneously with one of the distal synapses because the EPSP produced by the proximal synapse is larger than that produced by the distal synapse. The important idea here is that simultaneous EPSPs produced by synapses in different parts of the dendritic tree are additive in their effects on the postsynaptic membrane potential. The technical term for this phenomenon is **spatial summation**.

EPSPs may also sum if they originate from a single synapse, as long as that synapse fires repeatedly in rapid succession (Figure 2.17 D). Most neurons cannot produce more than ~400 action potentials per second because the same refractory period that prevents action potentials from reversing course also prevents axons from triggering action potentials that are separated in time by less than ~2.5 ms. However, most EPSPs are significantly longer than this minimum interval between action potentials. Their average duration depends on numerous factors (especially membrane capacitance) but typically ranges from 10–30 ms when recorded at the cell body. Therefore, any EPSPs that arrive at the axon hillock within 10–30 ms of one another will overlap in time and therefore be additive in their effect, even if they originated from the same synapse. This summing of EPSPs over time is called **temporal summation**.

Because EPSPs are typically mound shaped (Figure 2.17 B), the degree of temporal summation increases as the interval between successive EPSPs decreases. Therefore, the more rapidly a presynaptic axon fires, the more likely it is to push the postsynaptic cell above the action potential threshold.

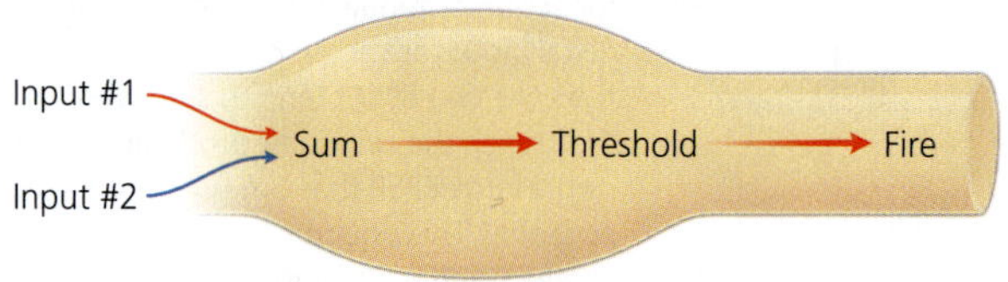

B Long integration time

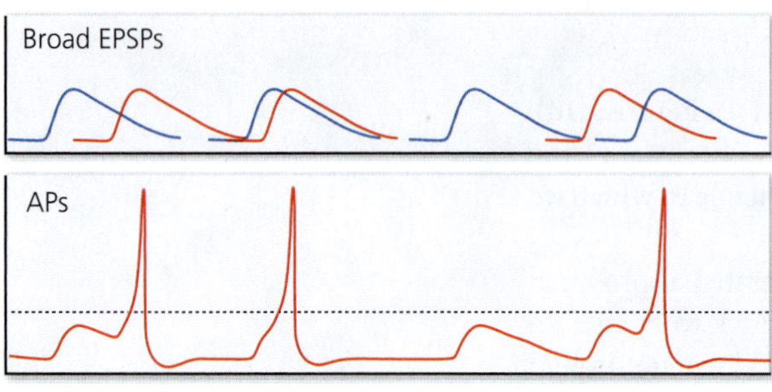

C Short integration time

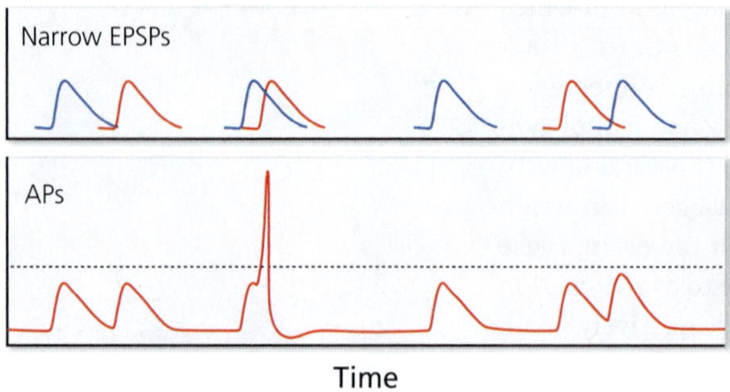

Figure 2.18 Integrate-and-fire devices. According to the integrate-and-fire model (A), neurons sum their inputs and then "fire" action potentials only if this sum exceeds a threshold. Because EPSPs typically last for many milliseconds (are broad), they overlap and sum even if they are not precisely coincident (B). Some neurons exhibit relatively brief (narrow) EPSPs (C), allowing them to function as precise coincidence detectors.

Neurons as Leaky Integrate-and-fire Devices

Temporal and spatial facilitation together make neurons sensitive to particular temporal and/or spatial patterns of synaptic input. Some input patterns depolarize the axon hillock enough to trigger an action potential; others do not. The key is that the membrane potential at the axon hillock at any given moment represents the sum of all the EPSPs that have arrived at the axon hillock within the last 10–30 ms.

Because of this EPSP summing, engineers like to think of neurons as leaky **integrate-and-fire devices** (Figure 2.18). The "integrate-and-fire" portion of this phrase refers to the idea that neurons sum their inputs and then fire an all-or-none action potential if the sum exceeds a particular threshold. The word "leaky" indicates that the EPSPs decay with time. This view of neurons as leaky integrate-and-fire devices is somewhat oversimplified because in real neurons some inputs count more than others (e.g., proximal inputs are weighted more heavily than distal ones). However, for some neurons, the leaky integrate-and-fire model fits reasonably well.

Transmitter Clearance

Before we leave the topic of synaptic transmission and integration, let us consider what happens to neurotransmitter molecules after they bind to the postsynaptic receptors. Most of them disengage quickly from their receptor and float back into the synaptic cleft. They are then free to bind to other receptors. However, if this process of dissociation and rebinding continued indefinitely, synaptic transmission would lose its temporal precision and neurons would quickly become hyperactive.

Some neurons deal with this potential problem by having enzymes in the synaptic cleft that break down and thus inactivate the neurotransmitter. Other neurons contain specialized proteins that actively transport neurotransmitter molecules from the synaptic cleft back into the presynaptic terminal. Some neurons combine inactivation and reuptake. Particularly interesting is that some glutamatergic (glutamate releasing) neurons use special glial cells, called astrocytes, to help with transmitter inactivation and recycling (see Chapter 5).

BRAIN EXERCISE

If an excitatory synaptic input to a neuron is activated hundreds of times in rapid succession, what (if anything) might happen to the amplitude of the EPSP you can record in the postsynaptic neuron? Why?

2.5 How Do Neurons Differ from One Another?

Thus far we have discussed stereotypical neurons, but real neurons exhibit astounding variety. The best way to deal with this diversity is to think of neurons as varying along several different dimensions, such as their size and shape, their transmitters and receptors, and the kinds of ion channels they express in their membranes. We discuss these dimensions of variation in the following sections.

Anatomical Variety

Neurons vary considerably in size (Figure 2.19). The cell bodies of the smallest mammalian neurons are no more than 4–5 μm in diameter, whereas those of the largest neurons are roughly 100 μm across. Mammalian axons exhibit a similar degree of variation, ranging in diameter from ~0.1 μm to about 15 μm. Some invertebrates have much thicker axons and larger cell bodies, but such enormous neurons are rare

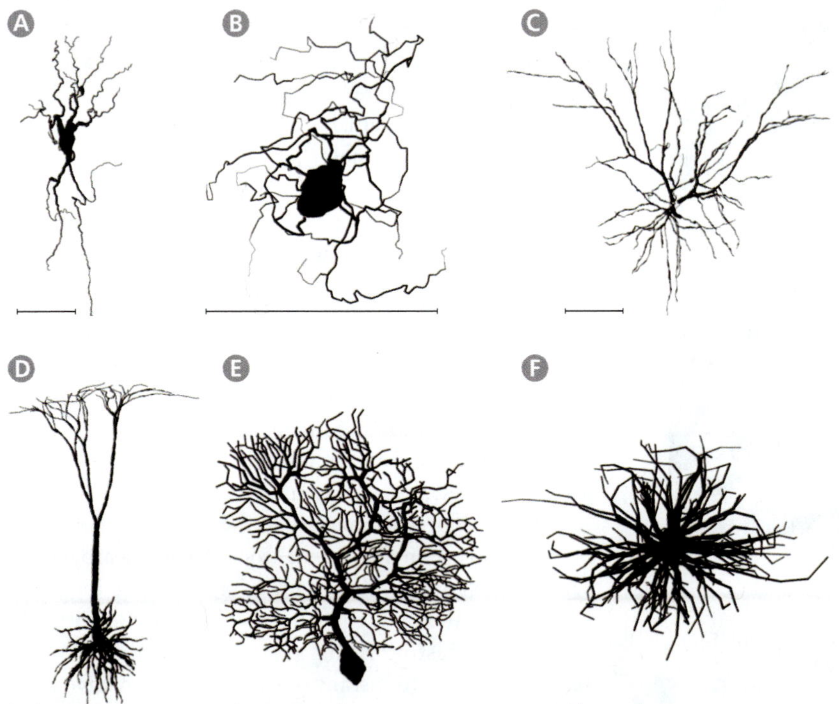

Figure 2.19 Neurons are amazingly diverse. Shown here are six neurons that differ dramatically in size and in the form of their dendritic trees (axons not shown). All scale bars equal 100 μm. The six neurons are (A) a vagal motor neuron, (B) an olivary nucleus neuron, (C) a neocortical neuron from the upper cortical layers, (D) a pyramidal neuron from layer 5 of the neocortex, (E) a cerebellar Purkinje cell, and (F) an alpha motor neuron from the spinal cord. [From Segev, 1998]

in vertebrates and not found in the human brain. Most likely, giant axons in mammals would require more metabolic energy than cell bodies could generate.

How small can neurons be? A lower limit on axon diameter is likely set by the fact that axons thinner than 0.1 μm in diameter would have so few ion channels per millimeter of length that ion channel noise (the random opening and closing of ion channels) would trigger many random action potentials, disrupting information processing. Neuronal cell bodies also have a lower size limit of 4–5 μm in diameter because all neurons need a nucleus, ribosomes, and endoplasmic reticulum to synthesize critical enzymes and other proteins. These organelles require space within the cell body. Indeed, the cell bodies of the smallest neurons contain little more than the nucleus and a bit of endoplasmic reticulum.

Dendritic Variation

Dendrites vary even more than neuronal cell bodies or axons (Figure 2.19). Some neurons have no dendrites at all, whereas others have huge dendritic trees that extend more than 1 mm away from the cell body. Some neurons have a single dendrite, extending in a single direction; others have multiple dendrites that extend in different directions. Some neurons have smooth dendrites; others cover their dendrites with numerous tiny **dendritic spines** (Figure 2.20), which tend to be the target of excitatory synapses. Models of spine function tend to stress that mature spines are often mushroom shaped and that the relatively thin neck of these spines acts as a diffusion barrier, biochemically isolating the spine head from the rest of the dendrite. This biochemical isolation allows synapses in one spine to be modified independently of synapses on other spines (see Chapter 3).

Variation in dendritic structure helps neuroanatomists classify neurons into distinct cell types, but what use is it to the neurons themselves? To answer this question, one may point out that neurons with large dendritic trees tend to receive more synaptic inputs than neurons with small dendrites. For example, the so-called **Purkinje cells** in the cerebellum have enormous dendritic trees (see Figure 2.19 E) and receive more synapses than most other neurons (>100,000 each). In contrast, neurons in some of the autonomic ganglia (you will learn about these structures in Chapter 9) possess simple dendrites and receive only a handful of synaptic inputs. Furthermore, the number of dendritic branches in these neurons correlates positively with the number of distinct inputs the cells receive. This correlation between the size of a neuron's dendritic tree and the number of synapses it receives is relatively easy to explain. Because synapses have a minimum size (~0.45 μm in diameter), you can increase the number of synapses onto a neuron only if you increase the size of the dendrites. So the more inputs a neuron receives, the larger its dendritic tree must be.

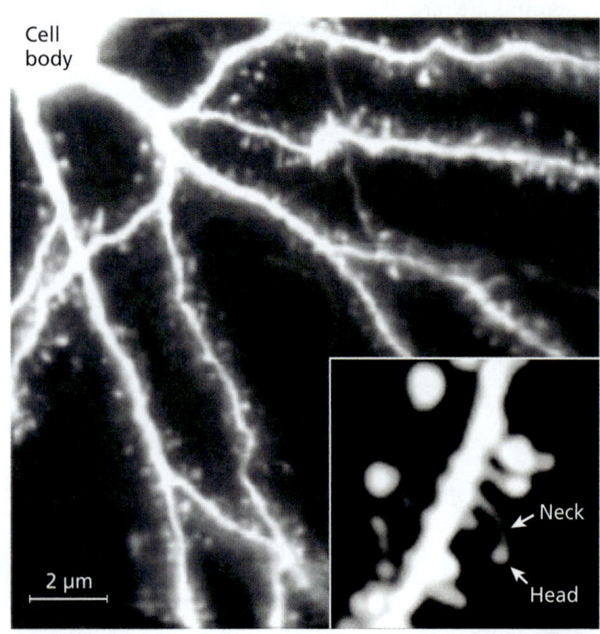

Figure 2.20 Dendritic spines. At medium magnification, the spines of a fluorescently labeled neuron look like "fuzz" on the dendrites. The insert shows a few spines at higher magnification, revealing that many spines have a thin "neck." [From Roberts et al., 2010; inset by T. M. Hoogland, public domain]

Variation in Axonal Output

Neurons also vary in the number of synaptic outputs. Some neurons synapse onto just a handful of other neurons; others contact thousands of cells. Some neurons contact other neurons through multiple, functionally redundant synapses; others contact their postsynaptic cells with just a single, sometimes enormous synapse.

In addition, neurons vary in the length of their axons. Some neurons contact only neurons that are nearby; others send their axons to distant brain regions. The former are called **interneurons**, the latter **projection neurons**. Particularly interesting is that many neurons do both. They send long axons to distant sites, but they also have collateral branches that synapse onto nearby neurons. Many neurons even send outputs back onto themselves through what are called **recurrent collaterals** (see Figure 2.2).

Neurotransmitter Variety

Another important dimension along which neurons vary is the identity of their neurotransmitter. More than half of all the neurons in a human brain use glutamate as their main neurotransmitter, but 15–40% of all neurons use a very different neurotransmitter called **gamma-aminobutyric acid**, abbreviated as **GABA**. Although GABA is synthesized from glutamate, it tends to have essentially the opposite effect on postsynaptic neurons. Instead of causing EPSPs, GABA causes **inhibitory postsynaptic potentials** (**IPSPs**). It does so by binding to receptors that allow negative chloride ions to diffuse down their concentration gradient into the postsynaptic cell. Because the equilibrium potential for chloride typically lies close to a neuron's resting potential, or even lower, the opening of chloride channels tends to counteract any depolarization caused by incoming EPSPs. Therefore, an IPSP pulls the neuronal membrane away from the threshold for triggering action potentials (Figure 2.21), thus reducing the neuron's probability of firing. This is why GABA's function is said to be inhibitory. It is interesting to note, however, that at very low membrane potentials, IPSPs can be depolarizing rather than hyperpolarizing (Figure 2.21). The point at which they switch polarity is called the IPSP **reversal potential**.

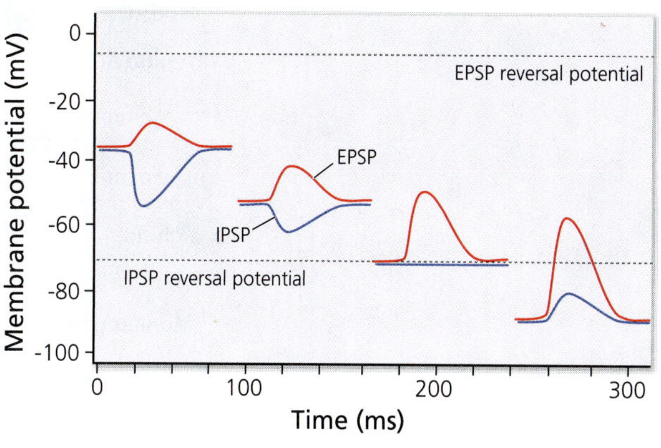

Figure 2.21 EPSPs and IPSPs vary with membrane voltage. EPSPs tend to become smaller at more positive membrane voltages and disappear at about −5 mV when the influx of Na^+ is balanced by an efflux of K^+ ions through open glutamate receptors. Similarly, IPSPs become smaller as the membrane potential approaches −70 mV, the equilibrium potential for chloride. At even more negative membrane potentials, IPSPs become depolarizing because chloride ions start moving up their concentration gradient and out of the cell. The membrane potential at which the IPSPs switch their polarity is called the IPSP reversal potential. Although IPSPs can be depolarizing at very negative membrane potentials, they cannot depolarize the cell membrane enough to trigger action potentials because the action potential threshold tends to lie above −60 mV. The reversal potential for EPSPs is near −5 mV.

Integrating IPSPs and EPSPs

As you might imagine, having synaptic inhibition as well as excitation expands the range of possibilities for synaptic integration. In our discussion a few paragraphs ago, we considered how multiple EPSPs arriving simultaneously or in quick succession may push the postsynaptic cell above threshold (see Figure 2.17). Adding IPSPs into this mix allows for the summing of both positive deflections of the membrane potential (EPSPs) and negative deflections (IPSPs).

If you think of neurons as integrate-and-fire devices, then postsynaptic cells should simply add up the number of EPSPs within a given time window, subtract the number of IPSPs, and then determine whether the resultant sum exceeds the firing threshold. This is a reasonable approximation of what real neurons do. However, IPSPs can be depolarizing at very low membrane potentials (Figure 2.21). Furthermore, inhibitory synapses are often located close to the axon hillock of the postsynaptic cell, whereas excitatory synapses are mainly found on the dendrites. Since inputs close to the axon hillock have more influence than distal inputs over the postsynaptic cell's probability of firing, IPSPs are frequently more powerful in their effect than EPSPs. In fact, a single IPSP delivered to a presynaptic terminal (yes, some axons terminate on the terminals of other axons!) may prevent an incoming action potential from causing transmitter release.

Acetylcholine and Monoamines

Aside from glutamate and GABA, neurons use a panoply of other neurotransmitters (Table 2.2). One of them is **acetylcholine**. This transmitter is employed by all mammalian motor neurons (neurons that innervate muscles), by many of the neurons that control our glands and internal organs, and by a few additional cell groups within the brain. Because it is such an important transmitter, acetylcholine is targeted by many of the toxins organisms have evolved to kill or paralyze one another (see Box 2.4: Nature's Neurotoxins).

NEUROACTIVE SUBSTANCE	PRINCIPAL FUNCTIONS OR CHARACTERISTICS
Amino Acid Transmitters	
L-glutamate	Main excitatory transmitter in CNS
γ-aminobutyric acid (GABA)	Main inhibitory transmitter in CNS
glycine	Inhibitory transmitter in some brainstem, spinal cord, and retinal neurons
Monoamines	
Dopamine	Involved in learning, depleted in Parkinson's disease
Norepinephrine	Involved in arousal, fight-or-flight response
Serotonin (5-HT)	Involved in mood control
Other Small Molecules	
Acetylcholine	Controls muscle contraction; modulator in the CNS
Nitric oxide	Promotes vasodilation; modulator in the CNS
Adenosine triphosphate (ATP)	Main transmitter for taste cells; co-transmitter at diverse synapses in the CNS
Selected Neuropeptides	
Vasopressin	Controls fluid balance, reproductive behaviors
Oxytocin	Induces labor; controls various reproductive behaviors
Dynorphin	Modulates pain; involved in drug addiction
Substance P	Modulates pain sensitivity
Neuropeptide Y	Regulates appetite

Table 2.2 The major neurotransmitters and some key neuromodulators

Three other important transmitters are **dopamine, norepinephrine,** and **serotonin**, which are collectively referred to as **monoamines** because they contain a single amino group connected to an aromatic carbon ring. Although monoamines are used by relatively few neurons, they are extremely important. The loss of dopamine neurons, for example, causes Parkinson's disease, which generally involves severe motor and cognitive deficits (see Chapter 15). Norepinephrine is used by neurons in the sympathetic nervous system, which mediates the fight-or-flight response, and by neurons that regulate arousal (see Chapter 13). Serotonin is released by neurons in the brainstem *raphe nuclei* and helps to regulate a whole host of processes, including sleep, appetite, and mood (see Box 2.5: Mood Molecules).

Nonclassical Transmitters

In addition to the classical transmitters, many neurons use a variety of other neuroactive substances. Most of these other substances are **neuropeptides**, which typically comprise 10 or more amino acids and thus are larger than the classical neurotransmitters. Importantly, neuropeptides are almost never used in isolation. They are used in combination with other neurotransmitters, including other neuropeptides. This makes it difficult to make general statements about their postsynaptic effects. However, one good rule is that neuropeptides modify the effect of other transmitters on postsynaptic cells.

EVOLUTION IN ACTION

Box 2.4 *Nature's Neurotoxins*

Many organisms have evolved toxins that incapacitate prey or ward off predators. Toxins that affect the nervous system are particularly common, mainly because they tend to act quickly. Even bacteria produce neurotoxins to discourage small animals from eating them (or larger animals from eating their hosts). The most powerful bacterial toxin is *botulinum toxin,* which derives its name from the Latin word for "blood sausage" (where the bacteria occasionally thrive). The potency of botulinum toxin depends on whether you eat it or inhale it, but 1 gram of botulinum toxin can kill about a million people. No other toxin is more powerful. How does it kill? Botulinum toxin is an enzyme that is taken up into the terminals of the axons that innervate muscle fibers. There the botulinum toxin digests key proteins needed for the release of synaptic vesicles filled with acetylcholine. Without synaptic vesicle release, you get no synaptic transmission, no muscle contractions and, in short order, respiratory arrest. Given its toxicity, it is ironic that botulinum toxin is often used to treat wrinkled skin. In this therapy, doctors inject a small amount of botulinum toxin (Botox) into the muscles under the skin. As the muscles are paralyzed, the wrinkles subside.

Another potent bacterial toxin is *tetrodotoxin* (TTX). Its name derives from the fact that TTX is often found in tetraodontiform fishes, including puffer fish. These fishes get infected with the TTX-making bacteria and then accumulate the toxin in some of their body parts, which is why sushi chefs must be trained to remove the toxic tissue. TTX also accumulates in the tissues of blue-ringed octopus and rough-skinned newt. These animals are relatively small yet contain enough toxin to kill dozens of adult humans within minutes. TTX accomplishes this gruesome task by blocking voltage-gated sodium channels. Without these channels, you get no action potential conduction, no centrally controlled muscle contractions, and, again, respiratory arrest. Curiously, the rough-skinned newts appear to be more toxic that they need to be. What predator of theirs requires a toxic dose sufficient to kill dozens of humans? The answer is a garter snake that likes to dine on rough-skinned newts. These snakes and the newts have been involved in an evolutionary arms race: As garter snakes became partially resistant to the TTX, the newts produced more TTX, which promoted the survival of garter snakes with even better resistance.

The most famous plant-derived neurotoxin is *curare* (d-tubocurarine). This molecule is harmless if you eat it because it is too large and charged to be absorbed through the gut. However, if curare is injected into your bloodstream, it blocks the acetylcholine receptors at the synapses between neurons and skeletal muscle, and paralysis ensues. This paralytic effect was discovered long ago by hunters in South America who dipped their arrowheads in curare. Animals wounded by such poisoned arrows quickly grew weak or died outright. Importantly, the hunters could eat their prey without being poisoned themselves because, as noted earlier, curare does not get absorbed through the gastrointestinal tract.

Poisonous snakes tend to be immune to their own venom, which is remarkable because snake venoms are typically a mix of several different neurotoxins. Some snake toxins are metalloproteases, which tend to break the connections between cells in the victim's body, leading to internal bleeding; they also interfere with blood clotting. Most snake venoms also contain toxins that irreversibly block the acetylcholine receptor at the neuromuscular junction. The most famous of these toxins is *alpha-bungarotoxin* (isolated from a snake called *Bungarus multicinctus*). The venom of Africa's black mamba snake includes toxins that interfere with acetylcholinesterase (which degrades acetylcholine), voltage-gated potassium channels, and muscarinic acetylcholine receptors. Curiously, many of the toxins in snake venom share a core molecular structure, suggesting that they all evolved from a common ancestral molecule. Another interesting fact is that snakes are not the only species immune to their venom. The mongoose, for example, is well known for dining on snakes, and its acetylcholine receptors are resistant to alpha-bungarotoxin.

Many scorpions and spiders are poisonous. Most scorpions are not life-threatening to humans, but the death stalker *Leiurus quinquestriatus,* in the deserts of Northern Africa and the Middle East, can kill a child. Its venom includes a paralyzing inhibitor of chloride channels. Most spiders likewise present no mortal danger to humans, but Australian funnel web spiders can be lethal. One of the neurotoxins in their deadly brew blocks sodium channel inactivation, which causes massive neuronal depolarization. In contrast, the principal toxin in the venom of American black widows is *latrotoxin,* which stimulates the uptake of calcium into synaptic terminals, causing massive transmitter release and, among other problems, rigid paralysis.

Cone snails are less widely known than spiders or scorpions, but some of them are almost as lethal. They kill fishes or mollusks with a toxic cocktail that includes acetylcholine receptor blockers and *omega-conotoxin,* which blocks a specific type of voltage-gated calcium channel and is >100 times more potent than morphine as a painkiller. This powerful analgesic eases the pain of death for the cone snail's victims, making them struggle less. It may also help scientists discover new treatments for human pain.

Neurons may also release some gases, notably nitric oxide and carbon monoxide. The latter will kill you if you inhale too much of it, but the brain uses carbon monoxide in small doses to modulate neural activity. As you will learn in Chapter 7, even ATP, the body's main carrier of metabolic energy, is used by some cells as a neurotransmitter. It seems that nature has been very inventive when it comes to using chemicals to exchange information between cells.

THERAPIES

Box 2.5 *Mood Molecules*

Among the most successful therapeutic molecules are drugs that improve or stabilize a person's mood. Most of these mood-boosting molecules target a distinct population of neurons that use *serotonin* (5-HT) as their neurotransmitter. The cell bodies of these neurons are located in a few hindbrain nuclei called the raphe nuclei, but the axons of these neurons terminate widely throughout the brain.

Many antidepressant drugs, notably Prozac, selectively inhibit the uptake of released serotonin back into presynaptic terminals. The effect of this reuptake inhibition is that released serotonin molecules remain in the synaptic cleft for a longer period of time and therefore exert a greater effect on their postsynaptic targets. The fact that antidepressants generally boost serotonin signaling suggests that clinical depression results from insufficient serotonin levels in the brain. It cannot be this simple, however, because reuptake inhibitors typically boost signaling within minutes or hours, whereas the positive effects on mood take weeks to manifest. To explain this discrepancy, it has been hypothesized that serotonin reuptake inhibitors cause gradual, long-term changes in the sensitivity and/or abundance of serotonin receptors. This hypothesis is controversial, but it is consistent with the general principle that chronic changes in transmitter abundance cause compensatory changes in the corresponding receptors.

One problem with antidepressant medications is that they tend to have side effects, including daytime sleepiness, nighttime restlessness, nausea, diarrhea, and sexual dysfunction. This is not particularly surprising because many of these drugs also inhibit the reuptake of norepinephrine, which regulates arousal, and because serotonin itself has multiple functions. One reason why there are so many different antidepressants on the market is that each drug represents a different compromise between benefits and side effects. Moreover, treating depression often involves a great deal of trial and error because some compounds work well in one person but not in another. A combination of several different drugs, at carefully adjusted dosages, usually works best.

Given how difficult it is to treat a diseased brain, is it reasonable to expect that we can alter the functions of a healthy brain and cause no harm? Probably not, but many people try. In particular, some people boost their mood by taking illegal drugs. One of these is *MDMA* (3,4-methylenedioxy-N-methamphetamine), also known as ecstasy (or, more recently, Molly). This drug is popular because it induces a strong sense of euphoria. Ecstasy does this mainly by altering serotonin transporter molecules so that instead of taking serotonin back up into the presynaptic terminal, they release it into the synaptic cleft. The result is a massive release of serotonin followed by serotonin depletion. The serotonin release feels good, but the depletion produces an ecstasy hangover that can last for several days and is characterized by a lack of motivation, focus, and appetite. Thus, the temporary boost in mood provided by ecstasy must be paid for later.

Research on rats and monkeys has shown that MDMA can irreversibly damage serotonergic axons in diverse brain regions (Figure b2.3). In these experiments, multiple doses of MDMA were injected into the animals, which is not how humans typically take MDMA. However, in squirrel monkeys even a single oral dose of MDMA reduces serotonin levels for at least 1 week in multiple brain areas. These data strongly suggest that MDMA has serious and long-lasting effects on brain function.

Indeed, studies with humans have shown that heavy ecstasy users have impaired verbal memory, display more impulsivity, and are more likely to be depressed than control subjects. These studies do not control for the fact that regular MDMA users also tend to take other drugs, such as amphetamines, that are already known to cause some brain damage. However, it certainly seems wise to find less risky, more creative ways to boost your mood. Finally, it is important to know that much of the so-called pure MDMA sold "on the street" (including Molly) is hardly pure and sometimes not MDMA at all. This lack of quality control is thought to account for at least some deaths associated with the taking of MDMA-related drugs.

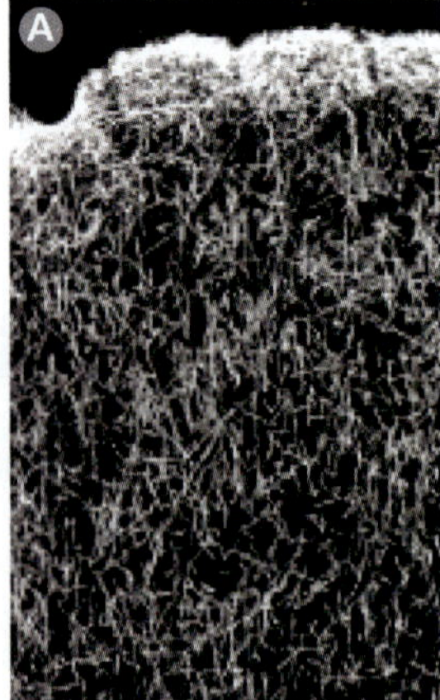

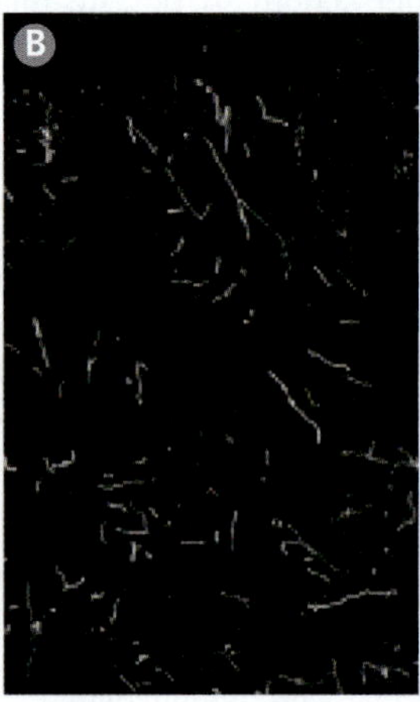

Figure b2.3 MDMA can destroy serotonergic axons. Serotonergic axons were stained with an antibody so that they appear as white lines on a dark background. Panel (A) shows a section through the frontal cortex of a control monkey. Panel (B) illustrates a similar section from a monkey 2 weeks after it received subcutaneous injections of MDMA (5 mg/kg twice daily for 4 days). A depletion of the serotonergic axons is evident. Panel (C) depicts an equivalent section from a monkey 7 years after receiving the MDMA injections. Despite some recovery, a deficit persists. [From Hatzidimitriou et al., 1999]

Although individual neurons may release more than one neurotransmitter, they typically release the same mix of substances at all their synapses. This rule of thumb is known as **Dale's principle**. This principle is sometimes confused with the idea that neurons use only one transmitter, which is not the case (e.g., some neurons use both dopamine and glutamate). However, Dale's principle does seem to be generally true, at least in vertebrates. Only some invertebrate neurons are known to use different combinations of neuroactive substances at different synapses.

Neurotransmitters versus Neuromodulators

Neuroscientists sometimes distinguish neurotransmitters from neuromodulators. Most commonly, they use the term **neuromodulation** to indicate that a neuron's response to transmitter released from a specific set of synapses can be altered (modulated) by the action of other substances—the neuromodulators—that were released from other sites and act on different receptors. For example, norepinephrine modulates how postsynaptic neurons respond to glutamatergic sensory input (see Chapter 13). Similarly, acetylcholine modifies how cortical neurons respond to inputs that are not cholinergic. One complication with this definition of neuromodulation is that some substances may be either a neurotransmitter or a neuromodulator, depending on the context of how they are released and what receptors they bind. Acetylcholine, for example, tends to act as a neuromodulator in the cortex but as a neurotransmitter at synapses onto skeletal muscle. Because the distinction between neurotransmitters and neuromodulators is so tricky to define, this book employs the term "neurotransmitter" quite loosely to denote all substances that neurons use to communicate with one another.

Receptor Variety

Given that neurons can use so many different neurotransmitters, it is not surprising that neurons express many different neurotransmitter receptors. However, it is somewhat startling to realize that neurotransmitter receptors are even more diverse than neurotransmitters.

Ionotropic versus Metabotropic Receptors

To manage this diversity we can sort all neurotransmitter receptors into two major classes. One class comprises the **ionotropic receptors**, which resemble the ion channels we discussed earlier in that they contain a central pore through which ions can flow (Figure 2.22). The AMPA receptor for glutamate is in this class. The effects of such ionotropic receptors may be excitatory or inhibitory, depending on which ions they let through, but the effects are usually quite fast and brief.

The other broad receptor class comprises the **metabotropic receptors**. They do not have a central pore through which ions can pass, but they can open or close nearby ion channels indirectly by modulating intracellular signaling cascades that ultimately affect the ion channels. Most metabotropic receptors belong to the large family of G protein-coupled receptors and affect neuronal activity by activating enzymes that control the production of second messengers, including cyclic AMP (adenosine monophosphate) and GMP (guanosine monophosphate). We will discuss these intracellular signaling systems more thoroughly in later chapters.

The postsynaptic effects of metabotropic receptors are significantly slower than those of ionotropic receptors but longer lasting. In many neurons, glutamate activates both the ionotropic AMPA receptor and several different

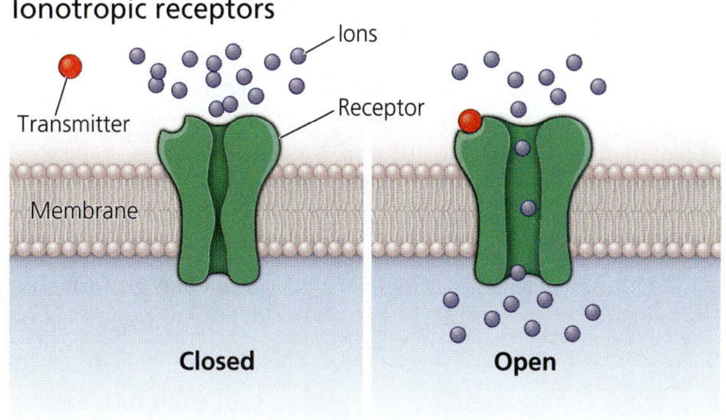

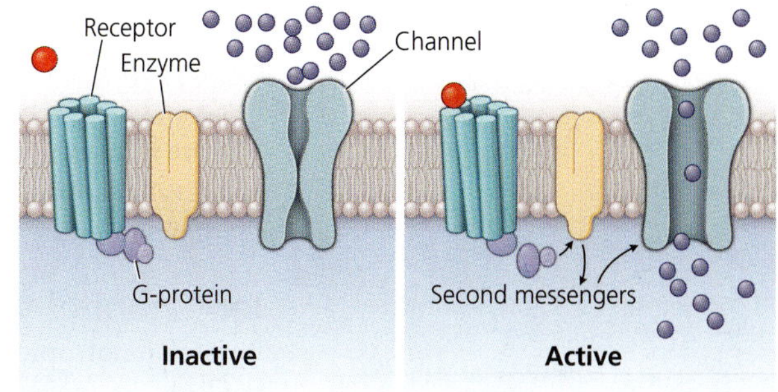

Figure 2.22 Ionotropic versus metabotropic receptors. Ionotropic receptors (top) let ions flow through their central pore when a transmitter is bound to them. In contrast, metabotropic receptors (bottom) do not contain an ion-passing pore. When metabotropic receptors are activated by a neurotransmitter, they activate enzymes that generate second messenger molecules which in turn can open or close nearby ion channels.

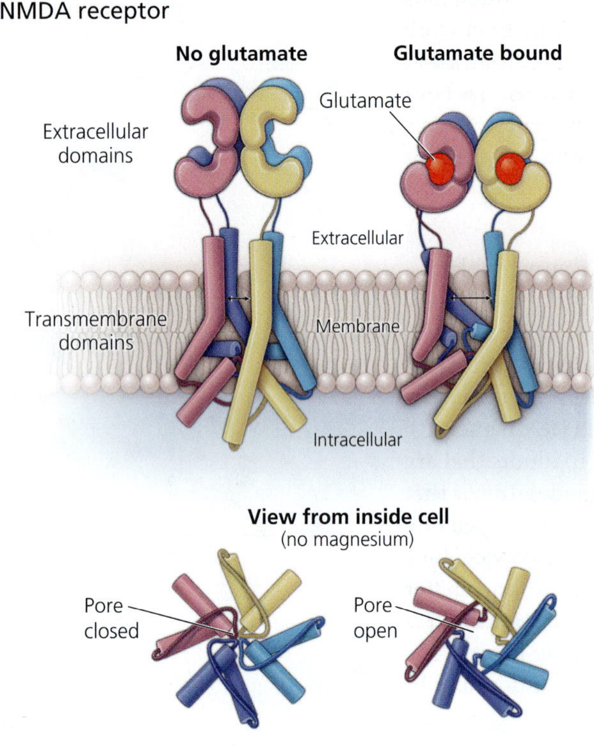

Figure 2.23 Structural model of the NMDA receptor. The NMDA-type glutamate receptor is assembled from four subunits (different colors). Each subunit has an extracellular domain, which contains the glutamate binding site; a transmembrane domain; and a small intracellular domain. When glutamate binds (right), the NMDA receptor changes shape to widen its central pore (bottom right) through which ions may flow. [After Wollmuth and Sobolevsky, 2004]

metabotropic receptors. Similarly, many neurons respond to GABA with both ionotropic and metabotropic receptors. Indeed, most neurons express a complex mix of ionotropic and metabotropic receptors.

The NMDA Receptor

One of the most intriguing neurotransmitter receptors is the **NMDA-type glutamate receptor** (Figure 2.23). This receptor gets its name from the observation that it can be activated not only by glutamate but also by the artificial compound N-methyl-D-aspartate (NMDA). In contrast, the AMPA receptor is not affected by NMDA. Both AMPA and NMDA receptors are ionotropic, but the NMDA receptor has an important additional feature. Its central pore lets current flow through it only when glutamate is bound *and* the postsynaptic membrane is already depolarized to levels well above the cell's resting potential. Thus, the NMDA receptor is both ligand gated and voltage gated.

Detailed studies have shown that the voltage dependence of the NMDA receptor is due to magnesium ions blocking the central pore when the membrane potential is negative. At positive membrane potentials, the magnesium ions are dislodged and NMDA receptors become permeable to positive ions, including sodium and calcium. The calcium inflow through open NMDA receptors is important because it activates some second messenger systems that play a role in synaptic plasticity (see Chapter 3).

Extrasynaptic Receptors

As you might expect, the receptors for a particular neurotransmitter are usually located at the synapses that release this transmitter. For example, NMDA and AMPA receptors tend to be located across from terminals that release glutamate. Sometimes, however, receptors and transmitters are mismatched. For example, receptors for dopamine, norepinephrine, and several neuropeptides are occasionally located far from synapses releasing these transmitters. The existence of such **extrasynaptic receptors** long posed a serious puzzle. However, recent work suggests that some neurons release their transmitter directly into the extracellular fluid. If these transmitter molecules are not degraded or taken up by cells, they may diffuse quite far and ultimately reach neurons that can respond to them through the appropriate extrasynaptic receptors. This kind of long-distance communication between neurons is called **volume transmission**. In contrast to synaptic transmission, volume transmission cannot communicate detailed spatial or temporal information. However, it allows relatively few neurons with relatively few transmitter release sites to influence many neurons.

Ion Channel Variety

Besides ionotropic receptors, neurons express many other ion channels, including the leak channels responsible for the resting potential and the various voltage-gated channels you read about in Section 2.2. These other ion channels are also incredibly diverse. For example, the mammalian nervous system contains at least five different calcium channels, 10 types of sodium channel, and upward of 30 different potassium channels. The potassium channels are each composed of multiple molecular subunits, which are themselves diverse and assembled in various combinations. Functionally these ion channels vary mainly in their degree of voltage sensitivity, in the time courses of channel opening and closing, and in how permeable they are to various ions. They are expressed in different combinations, to varying degrees, in different neurons. Some neurons even vary their ion channel mix during development.

What is the point of this variety? Neurobiologists are far from having a complete answer to this question, particularly as neurons expressing different combinations of ion channels may display nearly identical physiological properties. Nonetheless, variations in ion channel composition are likely to cause differences in how neurons respond to inputs. For example, a neuron's ion channel mix is likely to affect the threshold at which action potentials are triggered, the likelihood that a neuron responds to a given input with a burst of several action potentials (rather than just one), and a neuron's maximal rate of firing. The more you think about such differences in how neurons respond to their inputs, the more you realize that neurons are not simple integrate-and-fire devices, whose response depends only on the device's inputs and not on its intrinsic properties.

Dendritic Action Potentials

As noted earlier, many neurons have voltage-gated channels not only in their axons but also in their dendrites. Scientists discovered this fact soon after they learned how to record simultaneously from dendrites and the cell body. Such recordings showed that in some neurons, action potentials triggered in the axon hillock spread not only down the axon but also up into the dendritic tree. This **back-propagation** of action potentials is accomplished by voltage-gated channels and plays a major role in some forms of synaptic plasticity.

Some dendrites are even capable of triggering action potentials. Such dendritic spikes can propagate within the dendritic tree and may travel all the way down into the axon. Because dendritic spikes have a high triggering threshold, they tend to be rare. However, even sub-threshold activation of voltage-gated ion channels alters dendritic processing because it amplifies EPSPs as they travel within dendrites. Thus, the density of voltage-gated ion channels in a neuron's dendrites affects how neurons compute.

All this variation in ion channel composition, together with the previously described variations in neuronal anatomy, neurotransmitters, and neurotransmitter receptors, renders neurons incredibly diverse. They fit no stereotype. This is important because different kinds of neurons tend to perform different kinds of computations. Indeed, neuronal diversity is part of what makes the nervous system such a powerful computing device.

BRAIN EXERCISE

Neurobiologists may classify (and name) neuronal cell types according to the transmitters they use, the size and shape of their cell bodies, the targets of their axons, or many other features. In your view, which criteria for classifying (and naming) neurons would be most useful for research? Why?

2.6 Neuronal Information Processing

How do the details of neuronal structure, function, and plasticity help to explain how neural circuits regulate behavior? This question cannot be answered without delving into the particular circuits and behaviors that we want to explain. That task must wait. For now, let us ask a simpler, more focused question: how do the details of neuronal structure and function influence the way neurons transform their inputs into outputs? More succinctly, how do individual neurons process information? How do they compute?

How Neurons Encode Information

As we discussed, neurons integrate EPSPs and IPSPs over both space and time and then fire an action potential if the resultant depolarization at the axon hillock exceeds the firing threshold. An important additional feature that we have not yet discussed is that the number of action potentials a neuron fires in response to stimulation is generally proportional to the degree of depolarization at the axon hillock. A depolarization that is barely above threshold tends to evoke just a single action potential (or spike), whereas a strong depolarization tends to trigger several spikes in rapid succession. This means that the level of excitation coming into a neuron is reflected in the neuron's rate of firing. Even though action potentials are all-or-none events, the rate at which they are produced reflects the degree to which a neuron is depolarized.

Spike Rate versus Spike Timing

Using more formal terminology, we can say that the degree of depolarization is *encoded* in the neuron's spike rate. Furthermore, this **spike rate code** can be passed from one neuron to another because a high firing rate in one neuron tends to cause a strong depolarization in its postsynaptic neurons, which then fire their own series of action potentials. Of course, this is true only if the projections are excitatory, strong, and not outnumbered by other connections carrying different information. Despite these caveats, most neurobiologists agree that neurons generally use a spike rate code to process and transmit information.

Although most neurons use a spike rate code to encode information, **spike timing** can be important. As you will learn in Chapter 13, the synchronization of action potentials and synaptic activity across a population of neurons generates important rhythms in the electroencephalogram (EEG). Some studies have shown that the timing of neuronal spikes relative to such ensemble rhythms can encode information. In addition, there are reports of neuronal spike trains (sequences of action potentials fired by single neurons) that contain relatively long, repeating temporal patterns of activity. Such repeating patterns might encode information in much the same way as the repeating patterns of pulses and intervals encode information in the Morse code. Unfortunately, there is no evidence that neurons can decode such lengthy spike patterns. Therefore, it appears that neurons rely far more on spike rates than on spike timing to encode information.

Neuronal Response Preferences

Neurobiologists use the concept of a spike rate code to determine how individual neurons signal the presence or absence of particular sensory stimuli. In a typical experiment, an organism is presented with various stimuli while the experimenter records the firing rate of individual neurons. The experimenter then determines how the neuron's firing rate changes in response to each stimulus presentation (Figure 2.24). Such measurements generally show that individual neurons fire at some relatively low **background rate** when no stimuli are presented. Shortly after a stimulus appears, the neurons tend to increase their firing rate above background. Crucially, most neurons increase their firing rate much more for some stimuli than others.

When a neuron preferentially responds to stimuli at a specific location, the "preferred" location is called the neuron's **spatial receptive field**. If, within that receptive field, a neuron responds more strongly to some stimuli than to others, neurobiologists say that the neuron is **tuned** to the stimuli to which it responds with the highest firing rate. As you can see in Figure 2.24 C, neurons may also decrease their firing rate to below the background level for some stimuli. A neuronal **response profile** simply summarizes how a neuron responds to a variety of stimuli, revealing the neuron's stimulus preferences and, if you will, "dislikes."

The Metabolic Cost of Neuronal Information Processing

Given these characteristics, it is interesting that neurons rarely fire at their maximal rate, even when a preferred stimulus is presented. Moreover, most neurons respond only when stimuli appear or disappear; they rapidly habituate when stimuli persist. Why do neurons skimp on spikes? One reason is that organisms need to know about changes in their environment more than they need to know about what stays constant. Another reason for skimping on spikes is that spikes are energetically expensive to produce. Although action potentials are caused by ions flowing down their

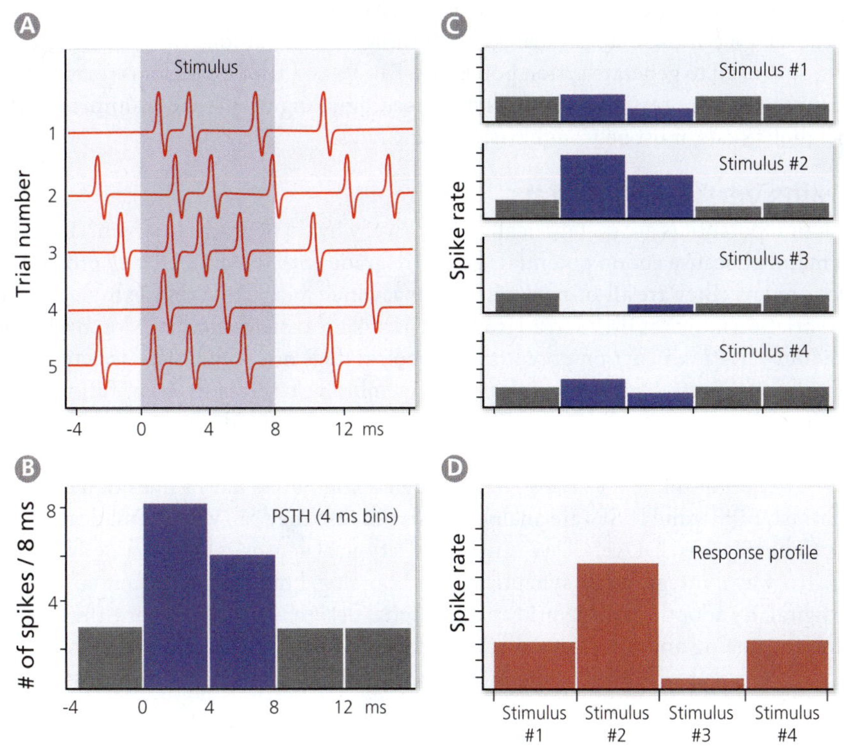

Figure 2.24 A neuronal response profile. Panel (A) shows extracellular recordings from a single neuron as it responds to an 8-ms long stimulus on 5 consecutive trials. Shown in (B) is a peri-stimulus time histogram (PSTH) of the spikes recorded during those trials. You can see that the spike rate increases during the stimulus. Panel (C) depicts four separate PSTHs that summarize the responses of a single neuron to 4 different stimuli. The neurons responds most strongly to stimulus #2 and is inhibited by stimulus #3. These variations in response strength are summarized in (D) as a "response profile," which is simply a histogram of the neuron's response to each of the presented stimuli.

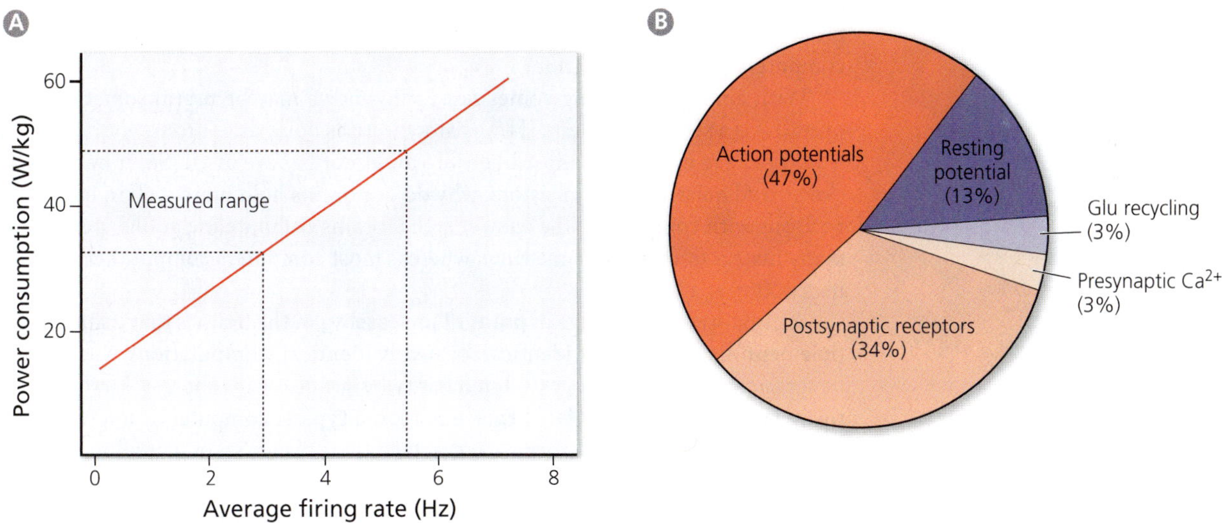

Figure 2.25 A neuron's energy budget. Rat cortical neurons consume about 14 watts/kg when they are at rest (A). Their power consumption increases with firing rate (red line). Because the neocortex in waking rats consumes 32–48 watts/kg, we can calculate that the average firing rate of neurons in the rat neocortex must be about 3–5.5 Hz. This is much lower than a neuron's maximum firing rate, suggesting that neurons economize their use of action potentials. Panel (B) shows the estimated energy budget of a rat cortical neuron firing at 4 Hz. These estimates involve assumptions that may not hold for all neurons. [After Laughlin and Sejnowski, 2003, and Attwell and Laughlin, 2001]

concentration gradients, they do produce a net influx of sodium ions, which must eventually be pumped back out by Na^+/K^+-ATPases. Detailed calculations indicate that reversing the ion fluxes generated by action potentials requires a significant amount of metabolic energy (Figure 2.25).

It has been estimated that the neocortex uses about as much ATP per gram of tissue as a leg muscle does during the running of a marathon. Up to 80% of that energy is used to generate action potentials, EPSPs and IPSPs, and for returning the membrane to the resting potential after those signaling events. To minimize these metabolic costs, most neurons have low firing rates.

Brains versus Computers

If action potentials are metabolically expensive, why do neurons use them? They do so mainly because action potentials do not degrade with distance as they propagate along axons. They are all-or-none phenomena, which means that they exhibit almost no variation in size as they spread from one part of an axon to another. Moreover, any random variation in action potential size (noise) does not accumulate as action potentials propagate because each patch of membrane triggers its own, full-fledged action potential when it has been depolarized above threshold.

Because of these features, we can say action potentials are **digital signals**. They are either on or off, just as the signals in digital computers have values of 0 or 1. In contrast, EPSPs and IPSPs are analog signals because they can vary dramatically and continuously in size. Overall, we can say that a typical neuron is both analog and digital. The way it integrates its synaptic inputs is analog, but its action potential output is digital. By adopting this hybrid analog/digital design neurons combine the best of both the analog and the digital worlds. The analog processing of synaptic inputs is fast and efficient, whereas the digital nature of action potentials makes it possible to transmit neural signals reliably, over long distances.

These considerations naturally lead us to compare neurons and transistors (the tiny digital switches inside all electronic devices), brains and computers. Arguably the biggest difference between neurons and transistors is that the latter are at least a million times faster. Neurons make up for some of their sluggishness by using analog processes to perform basic computations that require many transistors to execute in computers (e.g., multiplication).

With analog processing comes noise. This noise may be useful sometimes, but overall it is a serious problem. If transistors were as noisy as neurons, your computer would behave erratically; most likely, it would not behave at all. So, if neurons are slower and noisier than transistors, why do our brains hold up so well in most comparisons with computers? The answer is that brains compute in parallel, performing many instructions at the same time, whereas most computers compute serially (one instruction at a time).

An important advantage of **parallel processing** in the brain is that it allows multiple neurons to carry out identical, or nearly identical, computations. Such redundancy increases metabolic costs, but it has two significant advantages. First, it makes the nervous system more **fault tolerant** than a typical computer. If a few neurons die, there are likely to be others that can take their place. Second, it allows the nervous system to average the outputs of the various redundant computations. This averaging improves the precision of the overall computation because any noise that plagues the individual computing elements should cancel out. Thus, the brain's parallel computing architecture offers significant advantages over the serial design of a typical computer.

Perhaps the most intriguing difference between brains and computers is that the individual computing elements are more diverse in brains than in a computer. Transistors come in just a few flavors, but neurons are fantastically diverse. In Cajal's prescient words, "the functional superiority of the human brain is intimately bound up with the prodigious abundance and unusual wealth of forms of the so-called neurons with short axon[s] [interneurons]" (Cajal, 1917, pp. 345–346). Importantly, a neuron's anatomical and physiological characteristics often match the kinds of computations that the neuron is tasked with. In the auditory system, for example, neurons that are involved in localizing sounds by comparing when those sounds arrive at the two ears have anatomical and physiological specializations that help them detect minute differences in the arrival time of multiple inputs. This kind of functional specialization of the individual computing elements (the neurons) improves the brain's overall computing efficiency. Particularly useful is that many neurons acquire their anatomical and physiological specializations over time, with learning and experience. This neuronal plasticity is our focus in the next chapter (Chapter 3).

BRAIN EXERCISE

Why are brains so good at what they do, despite being relatively slow and "noisy" compared to computers?

SUMMARY

Section 2.1 - European scientists at the beginning of the 20th century discovered that neurons are discrete cells. Stereotypical neurons have long axons that synapse on dendrites, are electrically excitable, and use glutamate or GABA as their neurotransmitter.

Section 2.2 - Neurons have an inside-negative resting potential that is temporarily reversed during action potentials.

- Neurons use Na/K-ATPase to pump sodium out and potassium into the cell. A neuron's resting potential is determined mainly by the balance between potassium ions flowing down their concentration gradient and the prevailing voltage gradient.
- During an action potential, sodium rushes into an axon to depolarize its neuronal membrane. This is followed by a repolarizing efflux of potassium ions. These ion fluxes are governed by voltage-sensitive sodium channels that open rapidly and by more sluggish voltage-sensitive potassium channels.

Section 2.3 - Action potentials propagate by internal current flow that decays exponentially with distance but can trigger successive membrane patches to fire all-or-nothing action potentials. Other things being equal, axon diameter correlates with propagation speed. Myelin sheaths increase conduction speed and metabolic efficiency.

Section 2.4 - Synaptic glutamate release triggers EPSPs, which are integrated with one another and with IPSPs. A postsynaptic neuron fires an action potential only if the membrane depolarization at its axon hillock exceeds the action potential threshold.

Section 2.5 - Neurons vary along multiple dimensions.

- Neurons vary in the size and shape of their dendritic tree as well as in the length of their axons. Many neurons have dendrites covered with synapse-bearing spines.
- Glutamate and GABA are the main excitatory and inhibitory neurotransmitters, respectively. Most neurons use only one main neurotransmitter, but some also release modulatory neuropeptides.
- Most neurotransmitters activate multiple receptor types. Particularly important is the distinction between ionotropic and metabotropic receptors, as well as that between AMPA- and NMDA-type glutamate receptors.
- Neurons vary in the mix of ion channels they express; this can make them respond differently to identical input.

Section 2.6 - Neurons encode information primarily in their action potential firing rate. They combine analog with digital signaling and compute in a massively parallel manner.

Box 2.1 - The Nernst equation is a simplified form of the Goldman equation, which uses external and internal concentrations of various ions, as well as their permeabilities, to predict a cell's membrane potential.

Box 2.2 - The voltage clamp technique allows experimenters to record the currents flowing through individual ion channels.

Box 2.3 - Voltage-gated sodium channels contain distinct activation and inactivation gates, a voltage sensor, and a central pore through which sodium ions can flow.

Box 2.4 - Some plants and animals have evolved powerful neurotoxins to protect themselves from being eaten. Humans have learned to use some of these toxins productively.

Box 2.5 - Many antidepressant medications boost serotonin signaling. Ecstasy is a drug of abuse that acts on serotonin transporters and, at high doses, destroys serotonergic axons.

KEY TERMS

Golgi stain 32
synapse 34
neuron doctrine 34
Nissl substance 35
dendrite 35
axon 35
axon hillock 35
axon collateral 35
terminal arborization 35
axoplasmic transport 35
resting potential 35
action potential 35
squid giant axon 36
Na^+/K^+-ATPase 36
concentration gradient 37
electrical potential gradient 37
equilibrium potential 38
Nernst equation 38
ion channel 39
leak channel 39
inward current 42
depolarize 42
voltage-clamp recording 43
voltage-gated sodium channel 44
action potential threshold 44
repolarization 44
sodium channel inactivation 44
voltage-gated potassium channel 44
undershoot 45
refractory period 46
length constant 48
myelin sheath 48
Node of Ranvier 49
saltatory conduction 49
synaptic vesicle 51
glutamate 51
voltage-gated calcium channel 51
synaptic cleft 51
AMPA receptor 52
ligand-gated ion channel 52
excitatory postsynaptic potential (EPSP) 52
quantal 52
spatial summation 54
temporal summation 54
integrate-and-fire device 54
dendritic spine 56
Purkinje cell 56
interneuron 56
projection neuron 56
recurrent collateral 57
gamma-aminobutyric acid (GABA) 57
inhibitory postsynaptic potential (IPSP) 57
reversal potential 57
acetylcholine 57
dopamine 58
norepinephrine 58
serotonin 58
monoamine 58
neuropeptide 58
Dale's principle 61
neuromodulation 61
ionotropic receptor 61
metabotropic receptor 61
NMDA-type glutamate receptor 62
extrasynaptic receptor 62
volume transmission 62
back-propagation 63
spike rate code 64
spike timing 64
background rate 64
spatial receptive field 64
tuned 64
response profile 64
digital signal 66
parallel processing 66
fault tolerant 66

ADDITIONAL READINGS

2.1 - Neurons

Guillery RW. 2007. Relating the neuron doctrine to the cell theory: should contemporary knowledge change our view of the neuron doctrine? *Brain Res Rev* **55**:411–425.

Shepherd GM. 1991. *Foundations of the neuron doctrine.* New York: Oxford University Press.

2.2 - Resting and Action Potentials

Alle H, Roth A, Geiger JRP. 2009. Energy-efficient action potentials in hippocampal mossy fibers. *Science* **325**:1405–1408.

Hodgkin AL, Huxley AF. 1952. A quantitative description of membrane current and its application to conduction and excitation in nerve. *J Physiol* **117**:500–544.

Verkhratsky A, Krishtal O, Petersen O. 2006. From Galvani to patch clamp: the development of electrophysiology. *Pflügers Arch: Eur J Physiol* **453**:233–247.

2.4 - Synaptic Transmission and Integration

Heuser JS. 1979. Synaptic vesicle exocytosis captured by quick freezing and correlated with quantal transmitter release. *J Cell Biol* **81**:275–300.

Koch C, Segev I. 2000. The role of single neurons in information processing. *Nat Neurosci* **3(Suppl)**:1171–1177.

Lisman J, Raghavachari S, Tsien R. 2007. The sequence of events that underlie quantal transmission at central glutamatergic synapses. *Nat Rev Neurosci* **8**:597–609.

Magee JC. 2000. Dendritic integration of excitatory synaptic input. *Nat Rev Neurosci* **1**:181–190.

2.5 - Neuronal Diversity

LeBeau FE, El Manira A, Griller S. 2005. Tuning the network: modulation of neuronal microcircuits in the spinal cord and hippocampus. *Trends Neurosci* **28**:552–561.

Marder E, Goaillard JM. 2006. Variability, compensation and homeostasis in neuron and network function. *Nat Rev Neurosci* **7**:563–574.

Merighi A. 2002. Costorage and coexistence of neuropeptides in the mammalian CNS. *Prog Neurobiol* **66**:161–190.

Migliore M, Shepherd GM. 2002. Emerging rules for the distributions of active dendritic conductances. *Nat Rev Neurosci* **3**:362–370.

Miller C. 2000. An overview of the potassium channel family. *Genome Biol* **1**:reviews0004. http://genomebiology.com/2000/1/4/reviews/0004.1

Ramirez JM, Tryba AK, Peña F. 2004. Pacemaker neurons and neuronal networks: an integrative view. *Curr Opin Neurobiol* **14**:665–674.

Siegler Retchless B, Gao W, Johnson JW. 2012. A single GluN2 subunit residue controls NMDA receptor channel properties via intersubunit interaction. *Nature Neurosci* **15**:406–413.

Sterling P, Laughlin S. 2015. *Principles of neural design.* Cambridge, MA: MIT Press.

Traynelis SF, Wollmuth LP, McBain CJ, Menniti FS, Vance KM, Ogden KK, Hansen KB, Yuan H, Myers SJ, Dingledine R. 2010. Glutamate receptor ion channels: structure, regulation, and function. *Pharmacol Rev* **62**:405–496.

2.6 - Neuronal Information Processing

Faisal AA, Selen LP, Wolpert DM. 2008. Noise in the nervous system. *Nat Rev Neurosci* **9**:292–303.

Harris KD, Henze D, Hirase H, Leinekugel X, Dragoi G, Czurkó A, Buzsáki G. 2002. Spike train dynamics predicts theta-related phase precession in hippocampal pyramidal cells. *Nature* **417**:738–741.

Huxter J, Senior T, Allen K, Csicsvari J. 2008. Theta phase-specific codes for two-dimensional position, trajectory and heading in the hippocampus. *Nat Neurosci* **11**:587–594.

Laughlin SB, Sejnowski TJ. 2003. Communication in neuronal networks. *Science* **301**:1870–1874.

von Neumann J. 2000. *The computer and the brain.* Second Edition. Silliman Memorial Lectures. New Haven, CT: Yale University Press.

White JA, Rubinstein JT, Kay AR. 2000. Channel noise in neurons. *Trends Neurosci* **23**:131–137.

Boxes

Catterall WA. 2014. Structure and function of voltage-gated sodium channels at atomic resolution. *Exp Physiol* **99**(1):35–51.

Hille B. 2001. *Ion channels of excitable membranes.* Third Edition. Sunderland, MA: Sinauer Assoc.

Liebeskind BJ, Hillis DM, and Zakon HH. 2013. Independent acquisition of sodium selectivity in bacterial and animal sodium channels. *Curr Biol* **23**:R948–949.

Lyvers M. 2006. Recreational ecstasy use and the neurotoxic potential of MDMA: current status of the controversy and methodological issues. *Drug Alcohol Rev* **25**:269–276.

Mueller M, Yuan J, McCann UD, Hatzidimitriou G, Ricaurte GA. 2012. Single oral doses of (±) 3,4-methylenedioxymethamphetamine ("Ecstasy") produce lasting serotonergic deficits in non-human primates: relationship to plasma drug and metabolite concentrations. *Int J Neuropsychopharmacol* **16**:1–11.

Mebs D. 2001. Toxicity in animals: trends in evolution? *Toxicon* **39**:87–96.

Thase ME, Denko T. 2008. Pharmacotherapy of mood disorders. *Annu Rev Clin Psychol* **4**:53–91.

Vollenweider FX, Jones RT, Baggott MJ. 2001. Caveat emptor: editors beware. *Neuropsychopharmacology* **24**:461–463.

CHAPTER 3

Neuronal Plasticity

CONTENTS

FEATURES

Computers can store vast amounts of information by flipping tiny electronic switches (transistors) on or off, but they cannot change their own, internal wiring. Brains, in contrast, are plastic. They tend not to add more neurons once the organism has reached adulthood, but the connections between existing neurons can and typically do change as a result of experience. Good evidence for neuronal plasticity has accumulated over the last 50 years or so, but the idea itself is old. James and Meynert (see Chapter 1) had relied on the notion of neuronal plasticity to explain how a child learns not to touch a flame. They had proposed that mental associations (between the sight of a flame, pain in the finger, and withdrawal of the arm) result from the strengthening of connections between groups of cortical neurons that represent those percepts and movements. A more detailed vision of neuronal plasticity was proposed by Ramón y Cajal, who wrote

> One can admit as highly probable that mental exercise promotes in the most involved areas a greater development of the protoplasmic extensions [i.e., dendrites] and [axon] collaterals. As a result, associations that have already been created between certain cell groups strengthen themselves, notably by multiplying the terminal twigs of the protoplasmic extensions [i.e., dendrites] and [axon] collaterals. In addition, entirely new intercellular connections may establish themselves thanks to the formation of new collaterals and the expansion of [dendrites].
>
> Ramón y Cajal, 1894, pp. 466–467, author's translation.

Thus, the notion that brains can reorganize themselves as a result of "mental exercise" is hardly new. Yet direct evidence for neuronal plasticity did not accumulate until the second half of the twentieth century, when new methods made it possible to generate detailed maps of the brain's functional organization, trace neuronal connections at microscopic scales, and quantify the strength of synapses.

3.1 How Are Synapses Strengthened in the Marine Snail *Aplysia*?

How the brain rewires itself as a result of "mental exercise" is difficult to study in humans, for both practical and ethical reasons. Therefore, neurobiologists interested in neuronal plasticity tend to focus their research on monkeys, rodents, and, quite frequently, invertebrates. Especially important for our understanding of synaptic plasticity has been research on a marine snail called **Aplysia californica** (California sea hare). Research on this species began in the late 1960s and was led by Eric Kandel, who won the 2000 Nobel Prize in Physiology or Medicine.

Sensitization in *Aplysia*

Eric Kandel decided to study synaptic plasticity in *Aplysia* because this animal has a relatively small number of very large neurons (see Box 3.1: The Impact of Invertebrates on Neurobiology), which are much easier to study than neurons in the mammalian hippocampus that Kandel had studied previously. *Aplysia* also exhibits various behaviors that can be modified as a result of experience. Most important for our purposes is the **gill withdrawal reflex**. To understand this reflex, you need to know that sea hares use a fleshy tube (siphon) to draw water over their gills. When an *Aplysia* is relaxed, the siphon and gills are visible from above. However, when the siphon is touched, both structures are withdrawn and covered by protective flaps. In nature, this withdrawal reflex protects the delicate gills from rough seas or predators. In the laboratory, the reflex can be triggered by a puff of water aimed at the siphon. Importantly, the gill withdrawal reflex can be triggered even when much of the body is dissected away, leaving only siphon, gill, and tail as well as the neurons connecting those body parts. Touching the siphon in such a **semi-intact preparation** causes the gill to contract for a few seconds before it relaxes again. As shown in Figure 3.1, the neural circuit underlying this behavior consists of sensory neurons that innervate the siphon and synapse directly on large motor neurons, which innervate the gill muscles.

Neural Mechanisms of Sensitization

Sensitization of the gill withdrawal reflex occurs when an *Aplysia* receives a noxious (potentially harmful) stimulus on its external body surface, most commonly the tail. After such a stimulus, the gill withdrawal reflex is potentiated (strengthened) in the sense that a light touch on the siphon now causes the gill to be withdrawn for much longer than before (Figure 3.1). The animal becomes "sensitized" to future threats after experiencing the noxious stimulus. As Kandel and his collaborators discovered, one neural correlate of this behavioral sensitization is a slight increase in the duration (broadening) of the action potentials that sensory neurons generate in response to a touch of the siphon. Broadening the action potentials causes more calcium ions to flow into the presynaptic terminal, which increases the amount of neurotransmitter (glutamate) that is released onto the motor neuron. Increased transmitter release increases the amplitude of the excitatory postsynaptic potentials (EPSPs) in the motor neuron by more than 100% (Figure 3.1 C). Because the change in EPSP amplitude results mainly from an increase in transmitter release, the change is said to be presynaptic. Later in this chapter you will see that changes in synaptic strength often involve changes inside the postsynaptic cell, but sensitization of the gill withdrawal reflex in *Aplysia* involves mainly presynaptic alterations, notably action potential broadening and increased transmitter release.

Central to the mechanisms underlying sensitization is a set of neurons that are activated by noxious stimulation of the skin and then release serotonin at several locations in the *Aplysia* nervous system, including the presynaptic terminals of the sensory neurons in the gill withdrawal reflex circuit (Figure 3.1 A). Experiments have shown that blocking serotonin receptors on those sensory neurons prevents the ability of noxious stimuli, such as tail shocks, to induce sensitization. Conversely, pumping a bit of serotonin out of a micropipette onto the sensory neurons can sensitize the gill withdrawal reflex, even if no tail shocks are applied. Together, these studies indicate that serotonin release is *necessary and sufficient* for sensitization of *Aplysia's* gill withdrawal reflex. In general, showing that a neural mechanism is necessary and sufficient for a particular behavior (or change in behavior) goes a long way to establishing a causal link between the two.

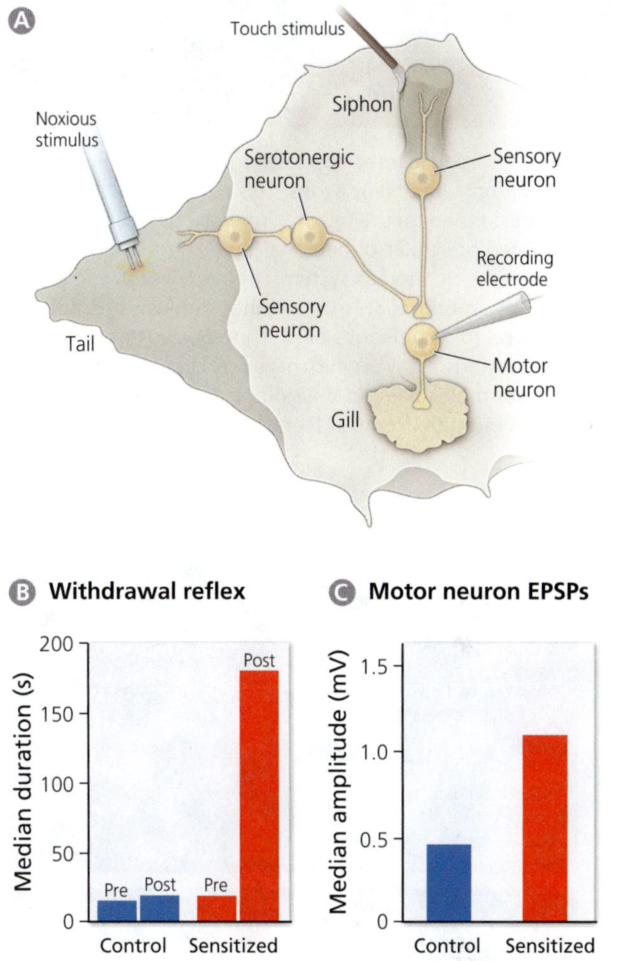

Figure 3.1 Sensitization of *Aplysia's* withdrawal reflex. Depicted in (A) is a semi-intact *Aplysia* preparation. The bar graph in (B) reveals that the mean duration of the gill withdrawal reflex increases after applying a painful stimulus to the animal's tail. No such enhancement is seen in control animals. Panel (C) shows that sensitization increases the amplitude of motor neuron EPSPs elicited by touch of the siphon. [After Kandel, 2001]

The Role of Serotonin in Sensitization

How does serotonin change the sensory neurons? First, it binds to G protein-coupled (see Figure 2.22) serotonin receptors on the sensory neuron. Activation of these metabotropic receptors activates the enzyme **adenylate cyclase**, which synthesizes **cyclic adenosine monophosphate (cAMP)** from ATP. Increased cAMP levels within the sensory neuron then activate **protein kinase A (PKA)**, an enzyme that phosphorylates voltage-gated potassium channels in the neuronal membrane (Figure 3.2). When these potassium channels are phosphorylated, their probability of being open decreases. This reduces potassium influx and therefore decreases the rate at which the cell membrane is repolarized during the falling phase of the action potential (see Chapter 2, Figure 2.7). As a result, the duration of the action potential increases and, as we just discussed, this broadening

EVOLUTION IN ACTION

Box 3.1 *The Impact of Invertebrates on Neurobiology*

In Chapter 2 you learned about Hodgkin and Huxley's influential work on squid giant axons. This research could not have been performed on the much thinner axons of mammals, yet it revealed principles of neuronal signaling that apply across all species with nervous systems. In this chapter, you learned about landmark studies on another mollusk, the sea hare *Aplysia californica* (Figure b3.1). This species has only about 20,000 neurons, many of which have very large cell bodies, which facilitates molecular and physiological analyses. *Aplysia* also exhibits multiple forms of learning, including sensitization, habituation, and classical (Pavlovian) conditioning. Although *Aplysia*'s behavior is less complex than that of vertebrates, the molecular mechanisms of learning and memory in this humble species have turned out to be broadly conserved.

Another invertebrate species that plays a major role in neurobiology is the fruit fly *Drosophila melanogaster*. Its main attraction for research is that mutant flies are fairly easy to generate and analyze. Taking advantage of these features, researchers generated multiple strains of learning-impaired fly mutants and then identified which genes had been altered. They found that many of the *Drosophila* genes involved in learning and memory are homologs of the genes that had been identified as being important for learning and memory in *Aplysia*. This discovery of molecular conservation across two invertebrate species that are relatively distant relatives suggested that the same mechanisms might be at work also in vertebrates. To a large extent, they are.

Lobsters and other crustaceans have also advanced neuroscience research significantly. In these species, small groups of neurons generate rhythmic outputs that control various aspects of digestion. By recording and manipulating the activity of these neurons, researchers discovered some general principles of how neurons generate rhythmic activity and, more generally, of neural network dynamics. They discovered, for example, that a neuron's physiological properties may be altered by serotonin and other neuromodulators, and that this modulation can profoundly alter network dynamics, even though the network's anatomy remains unchanged.

Invertebrates sometimes don't get enough credit for their contributions to neuroscience, presumably because human neurological disorders are difficult to study in invertebrates and because invertebrate nervous systems are organized quite differently from vertebrate nervous systems. Still, many problems are much easier to study in invertebrates, and some of the differences between vertebrates and invertebrate nervous systems may be smaller than they at first appear. Moreover, some invertebrates are cognitively quite complex. For example, honeybees can grasp simple concepts, such as "same versus different"; wasps can recognize each other by their facial markings; and octopuses can learn to solve puzzles by watching each other. Thus, many invertebrates deserve more recognition than they get.

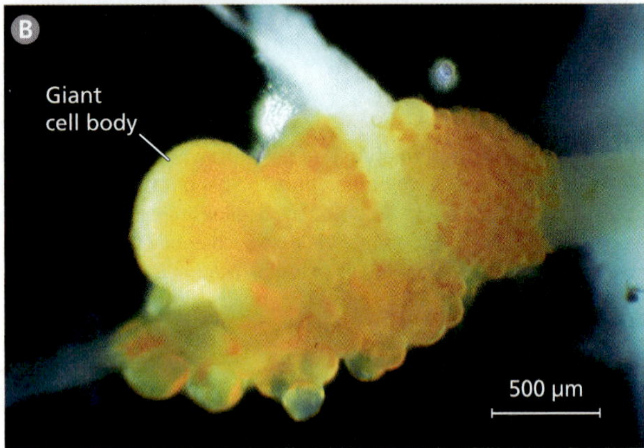

Figure b3.1 The large neurons of *Aplysia*. The photograph in (A) depicts the sea hare *Aplysia californica*. (B) shows its abdominal ganglion, dissected free of surrounding tissues, although several nerves and major axon tracts remain attached. Some of the cell bodies are gigantic. [Images courtesy of John H. Byrne]

of the action potential increases calcium influx into the axon terminal and boosts transmitter release. Voila! This is how serotonin release in response to a noxious stimulus strengthens the gill withdrawal reflex. Some additional mechanisms are known to be involved as well, but they need not concern us here. Instead, let us consider variations in how long sensitization lasts.

BRAIN EXERCISE

Can you think of an experimental manipulation that would prevent short-term sensitization in *Aplysia*?

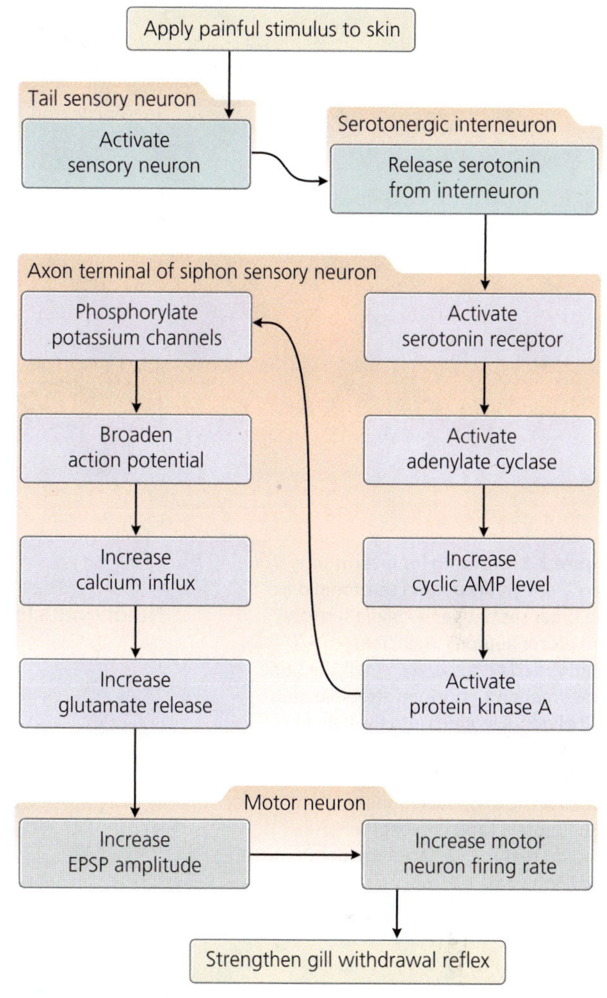

Figure 3.2 Mechanisms of short-term sensitization. Shown here is the principal chain of events leading to short-term sensitization of *Aplysia's* gill withdrawal reflex.

Making Sensitization Last for Days

The sensitization that we have discussed so far lasts a few minutes or hours. However, if an *Aplysia* is given a series of noxious stimuli, rather than just one, then the induced sensitization is much more prolonged. For example, giving the animal 5 tail shocks over a period of 4 days can enhance the gill withdrawal for several weeks. In a semi-intact preparation, a series of 5 serotonin pulses, spaced half an hour apart, can trigger sensitization lasting more than 24 hours. Incidentally, if the same number of shocks or serotonin pulses is given over a much shorter interval, then the sensitization is less persistent. This phenomenon has a clear parallel in your own life. Cramming for an exam the night before is not nearly as effective as spacing those learning sessions out over several days, at least if the goal is to remember the information long term. This is a general psychological principle: spaced learning is better than massed learning at creating long-lasting memories.

The mechanisms underlying long-term sensitization differ from those that cause short-term sensitization because long-term sensitization requires protein synthesis, whereas short-term sensitization does not. Applying a series of noxious stimuli to a semi-intact *Aplysia* that is bathed in a drug that inhibits protein synthesis fails to induce long-term sensitization, even though short-term sensitization is intact. Of course, short-term sensitization does involve proteins (enzymes and ion channels), but the proteins used in short-term sensitization were synthesized before the sensitization occurs. In contrast, long-term sensitization requires the translation of new proteins from mRNA.

The Role of CREB in Long-term Sensitization

Intrigued by this discovery, researchers set out to determine which proteins must be synthesized to generate long-term sensitization and what signals trigger their synthesis. They found that repeated noxious stimulation leads to repeated release of serotonin. As a result, the levels of activated PKA in the sensory neuron rise to such a high level that some of the activated PKA makes its way into the sensory neuron's nucleus, where it phosphorylates **cAMP response element binding protein (CREB)**. This protein is a transcription factor, which means that it helps regulate the expression (transcription from DNA) of other genes. Specifically, phosphorylated CREB binds to cAMP response element (CRE) sequences that are located upstream of many genes. When bound to such a CRE sequence, the phosphorylated CREB interacts with nearby RNA polymerases, which then begin their task of transcribing the downstream gene (Figure 3.3). Soon thereafter, the transcribed mRNAs are translated into proteins and shipped from the nucleus to other parts of the neuron.

To show that CREB is a critical link in the induction of long-term sensitization, experimenters injected synthetic oligonucleotides (short single-stranded RNA molecules) into an *Aplysia* sensory neuron and then applied several pulses of serotonin to its synapse onto a motor neuron (Figure 3.3). When the injected oligonucleotides contained the CRE sequence, and therefore bound any CREB molecules not yet attached to DNA, then only short-term sensitization was induced. In contrast, when

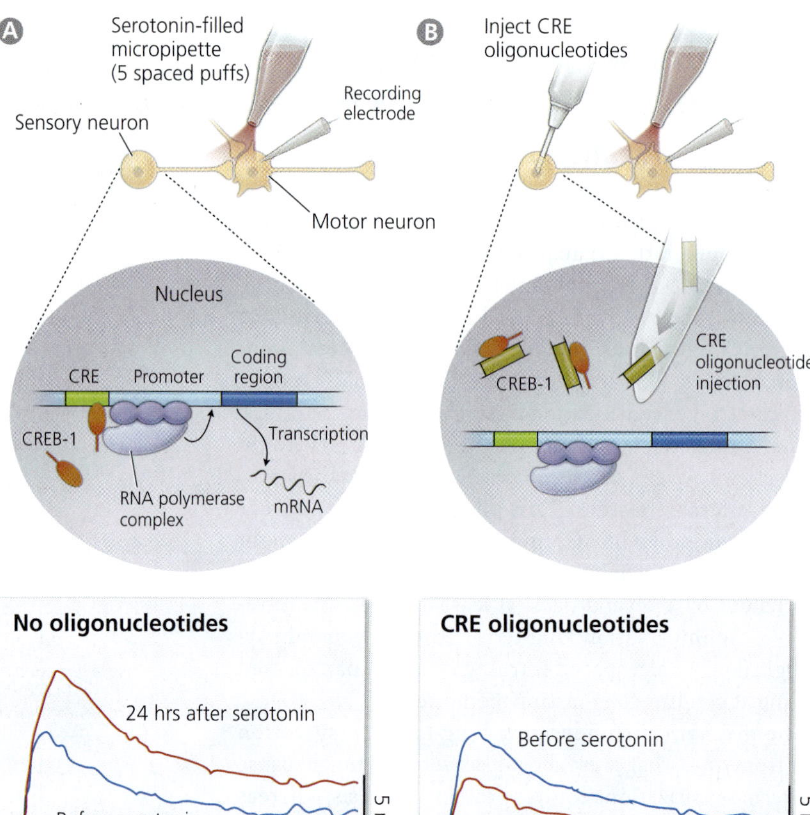

Figure 3.3 Blocking long-term sensitization. Repeatedly puffing serotonin onto the synapses between *Aplysia* sensory and motor neurons (A) increases EPSP amplitude 24 hours later. When the same experiment is performed after injecting CRE oligonucleotides into the cell body of the sensory neuron (B), long-term sensitization is blocked, presumably because the oligonucleotides bind to the CREB proteins and prevent them from binding to the DNA. [After Kandel, 2001, and Dash et al., 1990]

the injected oligonucleotides contained random sequences that do not bind CREB, then both short-term and long-term sensitization were observed. Therefore, CREB is necessary for the induction of long-term sensitization. In a complementary experiment, researchers showed that an injection of phosphorylated CREB into the sensory neuron is sufficient to trigger long-term sensitization. Overall, these studies show that phosphorylated CREB is both necessary and sufficient for long-term sensitization in *Aplysia*.

Many different genes have CRE sequences upstream of their promoters and are, therefore, upregulated by phosphorylated CREB. Among the most interesting is a gene that codes for **ubiquitin hydrolase**, which degrades the regulatory subunit of PKA and thereby renders PKA persistently active. This enzymatic modification prolongs synaptic potentiation because it ensures that PKA remains activated long after the serotonin release. Other genes induced by phosphorylated CREB are thought to be involved in increasing the physical size of sensory axon terminals and the sprouting of new synapses. Indeed, neuroanatomical studies have shown that the number of synapses that each sensory neuron makes on a motor neuron more than doubles with long-term sensitization.

Heterosynaptic versus Homosynaptic Potentiation

Before we leave this discussion of sensitization in *Aplysia*, it is important to note that the strengthened synapses in this form of learning are not the synapses that were initially activated. The noxious stimuli that trigger the sensitization activate the tail sensory neurons and, through them, the serotonin neurons. They do not activate the sensory neurons that carry information from the siphon to the motor neurons (see Figure 3.1 A). Therefore, the synaptic potentiation responsible for sensitization in *Aplysia* is **heterosynaptic** (involving different synapses), rather than **homosynaptic** (involving the same synapse). This distinction is worth making because Meynert, James, and Cajal had envisioned synaptic potentiation to be homosynaptic. They thought that only active synapses would be strengthened. As you will see, homosynaptic potentiation is, indeed, quite common in mammalian brains. Nonetheless, it is important to learn about sensitization in *Aplysia* because this work has revealed molecular mechanisms, such as the role of CREB in regulating transcription of learning-related genes, that are at work also in other species, including fruit flies and mammals (see Box 3.1).

BRAIN EXERCISE

In your view, why might new protein synthesis be needed for long-term sensitization but not for short-term sensitization?

3.2 How Are Synapses Strengthened in Mammals?

Research on the mammalian spinal cord in the 1950s revealed that high-frequency repetitive stimulation of sensory spinal nerves can strengthen the responses of spinal motor neurons to sensory inputs. One such experiment is shown schematically in Figure 3.4. First, the experimenters record the EPSP that is generated by the motor neuron in response to stimulating its sensory input once. Next, the experimenters stimulate the input pathway at a rate of a hundred or so pulses per second. This kind of high-frequency repetitive stimulation is called **tetanic stimulation** (derived from the word *tetanus*, which refers to prolonged muscle spasms). During the initial phase of the tetanic stimulation, the motor neuron's membrane potential becomes increasingly depolarized as a result of temporal summation (see Chapter 2). The membrane potential then becomes slightly less depolarized (depressed) as the presynaptic neurons run out of vesicles with neurotransmitter. At the end of the tetanic stimulation, the motor neuron's membrane potential gradually returns to its resting value. None of this is very surprising. Half a minute later, however, the motor neuron's response to a single stimulus is much larger than it had been before the tetanic stimulation (*post* vs. *pre* in Figure 3.4). This phenomenon is called **post-tetanic potentiation**, and it is thought to result mainly from a buildup of calcium within the presynaptic terminal. Although the discovery of post-tetanic potentiation was very exciting, the phenomenon only lasts a few minutes and is, therefore, insufficient to account for long-term memory.

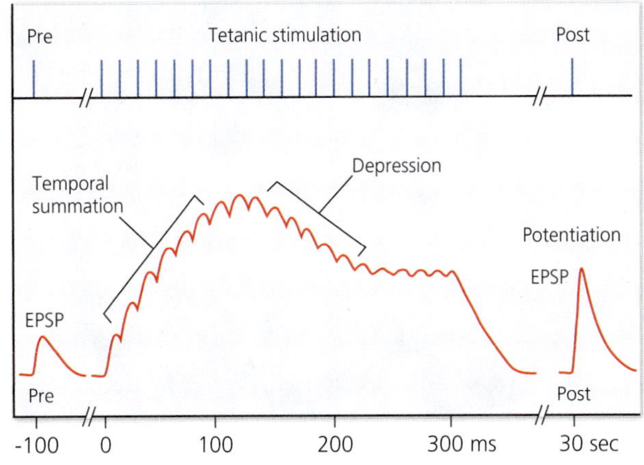

Figure 3.4 Effects of tetanic stimulation. In this idealized experiment, a series of electrical stimuli (blue vertical lines) is applied to an axon that synapses onto a neuron, whose responses are recorded intracellularly (red traces). During repetitive high-frequency (tetanic) stimulation the membrane potential increases because of temporal summation and then decreases as the presynaptic cell runs low on releasable transmitter. However, 30 seconds after the end of the tetanic stimulus, EPSP amplitude is larger than it had been before the tetanic stimulation, revealing post-tetanic potentiation.

Hippocampal Long-term Potentiation

In the 1960s, research on neuronal plasticity in mammals shifted its focus from the spinal cord to the hippocampus (see Box 3.2) because by then, researchers knew that lesions of the hippocampus cause profound memory loss. Two of the first neuroscientists looking for evidence of memory-related synaptic plasticity in the hippocampus were Timothy Bliss and Terje Lømo. In an influential experiment, they stimulated a set of axons in the **perforant path**, which projects from the neocortex to a part of the hippocampus termed **dentate gyrus** (Figure 3.5). Bliss and Lømo also recorded the postsynaptic responses of the dentate neurons to the perforant path stimulation. Their recordings reflected the sum of many different EPSPs but still indicated the strength of the synaptic connection between the axons of the perforant path and the dentate neurons. Bliss and Lømo discovered that tetanic stimulation of the perforant path in anesthetized rabbits increases the amplitude of the EPSPs in the dentate gyrus by more than 100% (Figure 3.6). Because this synaptic potentiation

NEUROBIOLOGY IN DEPTH

Box 3.2 *Hippocampal Structure and Functions*

The hippocampus is one of the most intensively studied brain areas. In rodents it is a relatively large, banana-shaped structure in the caudal half of the telencephalon, beneath the neocortex. In humans the hippocampus occupies a much smaller fraction of the telencephalon and looks somewhat like a seahorse (*hippocampus* means "sea-horse" in Ancient Greek). Compared to a rodent's hippocampus, the human hippocampus is shifted caudoventrally so that most of it lies within the medial aspect of the temporal lobe. Because of this shift in position, the ventral hippocampus in rodents is topologically equivalent to the anterior hippocampus in humans, and a rodent's dorsal hippocampus corresponds to the human posterior hippocampus (Figure b3.2).

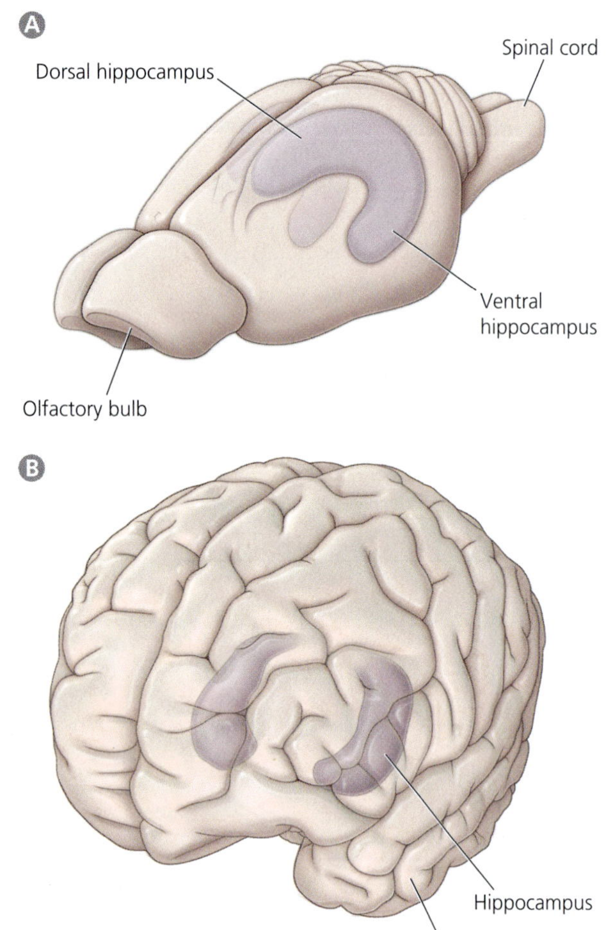

Figure b3.2 The hippocampus in a rat and a human. In a rat's brain, here shown from an anterolateral perspective (A), the hippocampus (purple) lies in the caudal telencephalon. In a human brain, here shown from the same anterolateral perspective (B), the hippocampus lies mainly in the temporal lobe.

Lesions of the hippocampus in rats or mice impair their ability to remember spatial locations relative to one another. Furthermore, many neurons in the rodent hippocampus are "place cells," which increase their firing rate when the rat is in a particular location, regardless of how the animal is oriented. Therefore, we can conclude that a major function of the hippocampus in rodents is spatial memory.

Humans with hippocampus lesions also have trouble forming new spatial memories, but they have a much more obvious general problem: they become unable to form new *episodic memories*, defined as memories of what happened when and where in a person's own life. For example, when a person with bilateral damage to the hippocampus meets someone new, they soon forget that they had ever met this person, even if they meet the same new person every day for several weeks. Given these data, it seems as if the hippocampus has a larger set of functions in humans than in rodents. In the latter, its function is spatial memory; in the former, it has the far more general function of remembering what happened when and where.

The hypothesis that hippocampal functions differ substantially across species is reasonable, but some scientists suggest that the differences may be more apparent than real. Because we cannot ask a rodent to tell us in words what it experienced yesterday, researchers must devise clever experiments to test whether rodents might nonetheless remember what they did where and when. Such experiments have shown that rodents can form some types of episodic memory. Hippocampal lesions impair those memories.

Another factor to consider is that scientists working on rodents have studied mainly the dorsal hippocampus, whereas studies on the human hippocampus focus mainly on the anterior hippocampus. As noted earlier, the anterior portion of the human hippocampus is topologically equivalent to the rodent ventral hippocampus. Therefore, it is possible that the suspected species differences between rodent and human hippocampi are not true species differences but rather a consequence of functional differences between different parts of the hippocampus (dorsal vs. ventral and anterior vs. posterior). This hypothesis is consistent with anatomical data showing that the posterior hippocampus in primates receives more refined spatial information than the anterior hippocampus and with physiological data indicating that in rodents the dorsal hippocampus encodes spatial information more precisely than the ventral hippocampus. Unfortunately, the functions of the ventral hippocampus in rodents and the posterior hippocampus in humans remain quite poorly understood. Therefore, the question of how the human hippocampus differs from that of rodents is still a subject of considerable debate.

lasts for many hours, it came to be known as **long-term potentiation (LTP)**. Some studies have reported that hippocampal LTP can last for several weeks or even months.

To understand the mechanisms underlying LTP, researchers employed the **brain slice** technique, which involves dissecting the brain out of a deeply anesthetized animal, slicing it into relatively thick slabs (~0.4 mm thick), and then bathing each slice in a highly oxygenated solution containing nutrients. Under these conditions, neurons can be kept alive for many hours. Stimulating and recording electrodes can be placed more precisely in brain slices than in intact brains, intracellular recordings are simpler to obtain, and drugs that block specific receptors or ion channels can be added to the solution in which a slice is bathed.

Using the brain slice technique, experimenters discovered that LTP can be elicited not only in the synapses of the perforant path, but also in several other hippocampal pathways. A second major feature of hippocampal LTP is that it tends to be input specific, meaning that only the stimulated, active synapses are strengthened. Therefore, in contrast to sensitization in *Aplysia*, hippocampal LTP is homosynaptic. Another difference between *Aplysia* and mammals is that LTP in the mammalian hippocampus involves mainly postsynaptic changes, whereas sensitization in *Aplysia* depends primarily on presynaptic modifications.

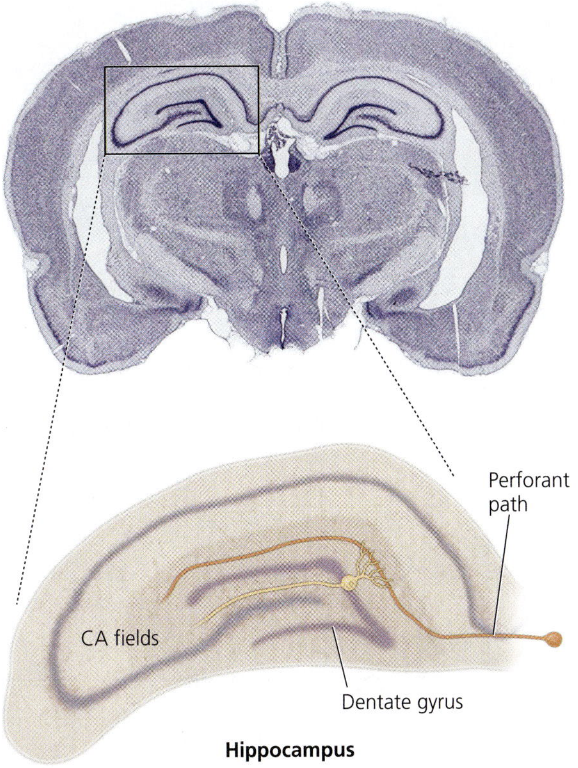

Figure 3.5 Section through a rat hippocampus. Shown at the top is a coronal section through a rat brain. The bottom diagram shows some major hippocampal divisions and connections. The yellow neuron projects from the dentate gyrus to one of the cornu ammonis (CA) fields. The orange neuron projects to the dentate gyrus through the perforant path. [Brain section image from brainmaps.org]

Hebbian Long-term Potentiation

So far we have discussed LTP as if strong, repetitive firing of a neuron strengthens *all* the synapses that this neuron makes onto other neurons. This is not what Meynert and James had in mind when they proposed their model of how a child learns not to touch a flame. They had proposed that strengthening occurs only in the connections between neurons whose activity is associated (linked) in time. This idea was formalized by Donald Hebb in 1949. As part of an ambitious effort to understand the neural basis of thought and memory, Hebb proposed the following:

> When an axon of cell A is near enough to excite a cell B and repeatedly or persistently takes part in firing it, some growth process or metabolic change takes place in one or both cells such that A's efficiency, as one of the cells firing B, is increased.
>
> (Hebb, 1949, p. 62).

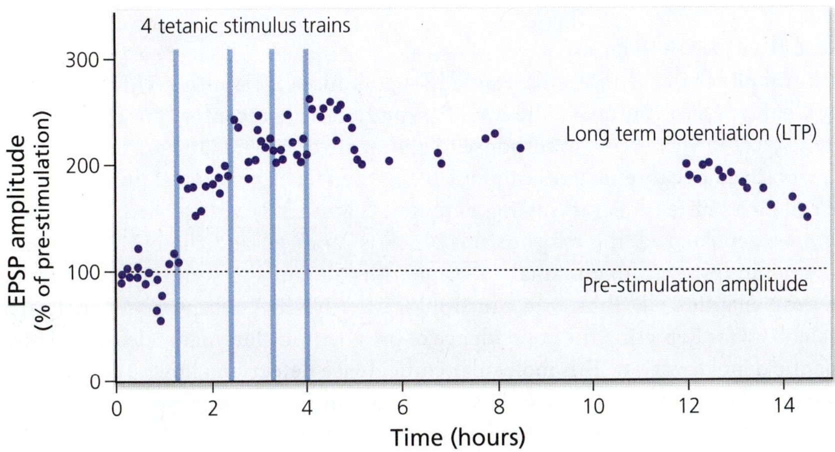

Figure 3.6 Long-term potentiation (LTP). Bliss and Lømo (1973) recorded synaptic responses in the dentate gyrus of a rabbit's hippocampus while applying four high-frequency (tetanic) trains of electrical stimuli to the perforant path (see Fig. 3.5). Already after the first tetanic stimulus, EPSP amplitude increased almost 100%. Importantly, some synaptic potentiation persisted for more than 10 hours after the stimulation. [After Bliss and Lømo, 1973]

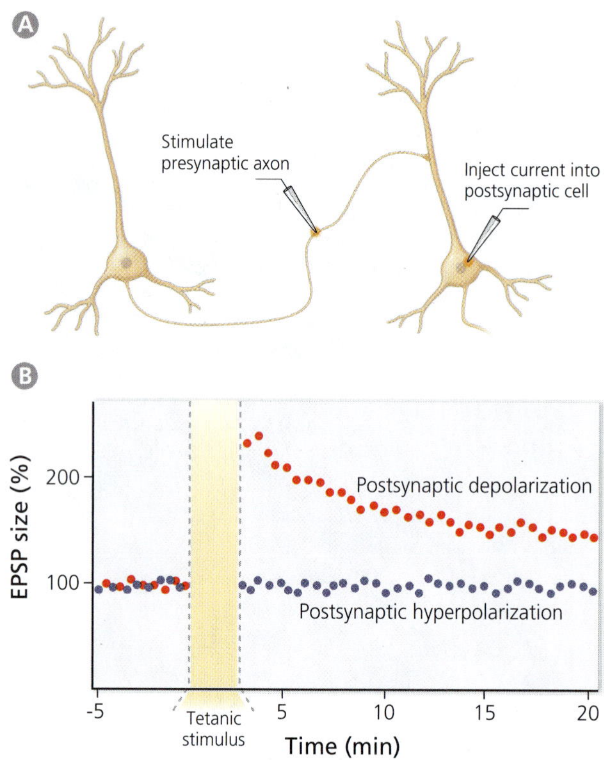

Figure 3.7 Testing whether LTP obeys Hebb's rule. In this idealized experiment, one electrode is used to stimulate presynaptic axons repeatedly (tetanically). A second electrode is used to inject either depolarizing or hyperpolarizing current into the postsynaptic cell during the presynaptic stimulation. Every 30 seconds the postsynaptic neuron's response to a single presynaptic stimulus is recorded. If the EPSPs increase in amplitude only when the stimulation is coupled with depolarizing current injections, and not when it is coupled with hyperpolarization, then the synaptic potentiation obeys Hebb's rule.

The key idea in this statement, commonly referred to as **Hebb's rule**, is that synapses should be strengthened only if they were active when the postsynaptic neuron was depolarized enough to fire an action potential. If a synapse is active but the postsynaptic cell does not fire, then this synapse should not be strengthened. Conversely, even weak synapses should be potentiated if they were active just before the postsynaptic neuron fires an action potential. A simplified version of Hebb's rule states: *Neurons that fire together, wire together.* Hebb had no direct evidence for such plasticity, but we now know that LTP at many synapses does follow Hebb's rule.

To test whether LTP at a specific set of synapses follows Hebb's rule, neuroscientists can stimulate presynaptic axons while depolarizing or hyperpolarizing the postsynaptic cell by means of intracellular current injections (Figure 3.7). Depolarizing the postsynaptic cell during the presynaptic stimulation mimics the situation in which the presynaptic cell "takes part in firing" the postsynaptic cell, whereas hyperpolarizing current injections mimic the situation in which the presynaptic stimulation *does not* participate in firing the postsynaptic cell. Hebb's rule predicts that LTP should be observed only when the presynaptic stimulation is accompanied by postsynaptic depolarization. Using such experiments, researchers have shown that LTP at many synapses does, indeed, obey Hebb's rule. This form of LTP is sometimes termed *Hebbian LTP* or *associative LTP*; but we can simply call it LTP, keeping in mind that not all instances of LTP are necessarily Hebbian.

BRAIN EXERCISE

In the statement "neurons that fire together, wire together," what exactly does "wire together" mean?

Mechanisms of LTP Induction

As mentioned earlier, short-term sensitization in *Aplysia* involves changes in existing proteins, whereas long-term sensitization requires new protein synthesis. A similar distinction applies to mechanisms underlying LTP in mammals, except that neuroscientists don't talk about short-term and long-term LTP (these terms would be confusing). Instead, neurobiologists distinguish between the *induction* (triggering) of LTP and its *stabilization*.

Role of the NMDA Receptor

A central player in the molecular cascade that induces LTP is the NMDA receptor (see Chapter 2). In contrast to the AMPA-type glutamate receptor, NMDA-type glutamate receptors can open in response to glutamate only when the postsynaptic cell is depolarized because magnesium ions block the receptor's central pore as long as the cell membrane is near its resting potential (Figure 3.8). As the postsynaptic cell becomes depolarized, the **magnesium block** is removed and the NMDA channel can open in response to glutamate, letting both calcium and sodium ions flow into the postsynaptic cell. Thus, you can think of the NMDA receptor as a molecular mechanism for detecting the coincidence of presynaptic glutamate release and postsynaptic depolarization. This molecular coincidence detection allows a cell to determine whether an active synapse has taken part in firing the postsynaptic cell. It would, therefore, be a good mechanism for triggering LTP.

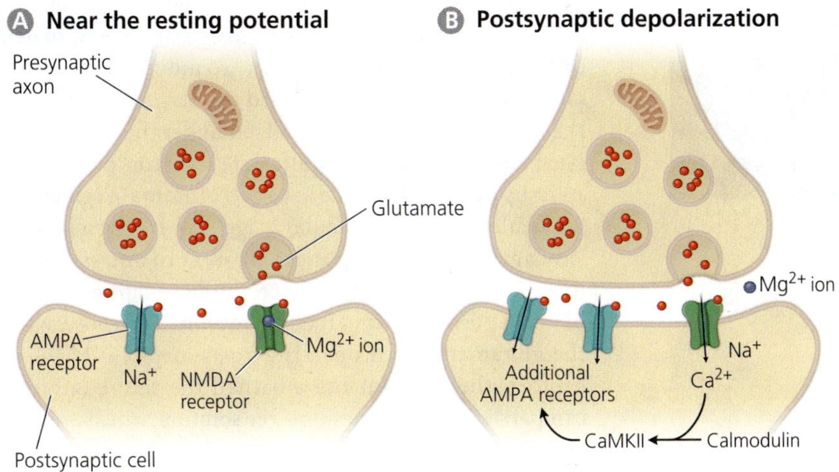

Figure 3.8 The NMDA receptor as a molecular trigger for LTP. If the postsynaptic membrane is near its resting potential (A), then NMDA receptors are blocked by magnesium (Mg^{2+}) ions. These ions are dislodged when the postsynaptic cell is strongly depolarized (B). Once the magnesium block is gone, Na^{+} and Ca^{2+} ions can flow through the NMDA receptor when glutamate is bound. An increase in postsynaptic calcium then triggers an intracellular signaling cascade that ultimately leads to the insertion of additional AMPA receptors into the postsynaptic membrane, which strengthens the synapse.

A potential problem with using NMDA receptors to trigger LTP is that a postsynaptic cell expressing only NMDA receptors would be unable to open in response to glutamate. As noted earlier, the NMDA receptors cannot open until the cell becomes depolarized; but as long as the NMDA receptors remain closed, the depolarization cannot get going. Such "silent synapses" have been observed in circuits that are still developing, but they would not work well in adult nervous systems. Most neurons solve this silent synapse problem by populating their postsynaptic membranes with AMPA as well as NMDA receptors. Because AMPA receptors have no magnesium block, they can open in response to glutamate even near the cell's resting potential. Thus, they allow the postsynaptic depolarization to get off the ground. Once the postsynaptic cell has become sufficiently depolarized, the magnesium block is removed from the NMDA receptors, allowing them to open as well.

How does the opening of postsynaptic NMDA receptors cause changes in synapse strength? Extensive studies, mainly on hippocampal slices, have shown that calcium influx through the open NMDA receptors is a critical factor. When postsynaptic calcium levels rise, a protein called **calmodulin** binds calcium, and the resulting calcium/calmodulin complex activates a protein kinase termed **calcium/calmodulin kinase II (CaMKII)**. Activated CaMKII promotes the insertion of additional AMPA receptors into the postsynaptic membrane (Figure 3.8 B). It also phosphorylates the AMPA receptors, which increases the rate at which ions can flow through them (the probability of the channels being open is reportedly unchanged). The upshot of these modifications is that the postsynaptic cell becomes more responsive to presynaptic glutamate release. For a given amount of released glutamate, the cell's EPSP increases in amplitude. Importantly, the synaptic potentiation is accomplished by changes in the postsynaptic cell, not by increased presynaptic transmitter release (which, as you may recall, is the basis of synaptic potentiation in the gill withdrawal reflex of *Aplysia*). There is some evidence for presynaptic changes also in LTP, but these effects are less well understood than the postsynaptic modifications.

The Role of Synapse Growth

The intracellular processes that induce LTP are the kind of "metabolic change" that Hebb proposed as one possible mechanism for strengthening synapses. The other mechanism Hebb envisioned is synapse growth. Indeed, an interesting study by Masanori Matsuzaki and his collaborators has shown that synapses may grow quite rapidly during LTP induction (Figure 3.9). To understand this somewhat complicated experiment, you need to know that the dendrites of many neurons are covered with tiny spines (see Figure 2.20). Such "spiny neurons" receive most of their excitatory input onto the tips of their spines, allowing you to think of each spine as the postsynaptic side of a synapse. Taking advantage of this arrangement, Matsuzaki

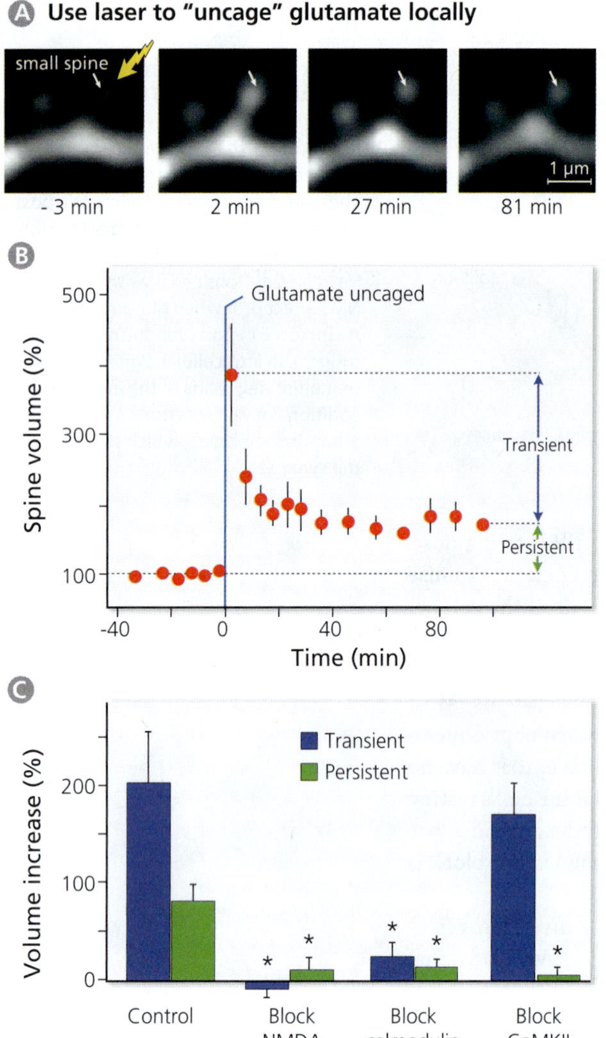

Figure 3.9 LTP and spine growth. A laser was used to release caged glutamate at one spine of a fluorescently labeled hippocampal neuron (A). The graph in (B) shows that spine volume increased dramatically right after the glutamate release. Although some of the volume increase was transient, some of it persisted for at least 100 minutes. The tissue was bathed in a magnesium-free solution to boost the ability of glutamate to open NMDA channels. Panel (C) shows that spine growth does not occur when NMDA receptors or calmodulin are blocked pharmacologically. Blocking CaMKII prevents persistent enlargement without blocking transient growth. [After Matsuzaki et al., 2004]

could determine whether synapses grow larger when they are strengthened by measuring changes in the size of dendritic spines after LTP induction. To induce LTP at a specific spine, the researchers labeled a postsynaptic neuron so that the spines could be visualized under the microscope. Next, they stimulated the glutamate receptors on a specific spine by bathing the tissue in **caged glutamate** (glutamate molecules inside a molecular "cage") and using a laser to "uncage" some of the glutamate next to one spine. The uncaged glutamate diffuses across the synaptic cleft and binds to postsynaptic glutamate receptors. To ensure that the glutamate could open the postsynaptic NMDA receptors, the experimenters bathed the tissue (a hippocampal slice) in a magnesium-free solution, which prevents magnesium block.

The results of Matsuzaki's experiment are shown in Figure 3.9. Within 1–5 minutes after the glutamate release, the volume of the stimulated spine doubles, on average. Spine volume then decreases, but the stimulated spines remain enlarged for more than an hour. Importantly, bathing the tissue in drugs that block NMDA receptors or calmodulin prevents spine growth. Therefore, spine growth seems to be triggered by the same molecular processes that are needed for LTP, suggesting that spine growth contributes causally to LTP induction.

BRAIN EXERCISE

Why is it important that Matsuzaki et al. in their spine growth experiment bathed the hippocampal slice in a magnesium-free solution?

Mechanisms of LTP Stabilization

As you learned in the previous section, LTP induction involves the activation of CaMKII, which then boosts AMPA receptor number and ion flow in the activated synapses. However, CaMKII stays active for only about a minute, and the AMPA receptors that were inserted into the postsynaptic membrane are soon removed again. Even spine growth is largely transient, especially when CaMKII is blocked (Figure 3.9 C). Yet studies with chronically implanted electrodes have shown that LTP can sometimes persist for months. We must wonder, therefore, how changes in synaptic strength are stabilized for the long term.

The Role of Protein Synthesis

The stabilization of LTP requires protein synthesis because protein synthesis inhibitors prevent LTP from developing, allowing only for transient synaptic strengthening. This is not surprising, given the need for protein synthesis in *Aplysia's* long-term sensitization, but ask yourself, how can proteins that are synthesized in the nucleus of the postsynaptic cell be used to strengthen only those synapses that contributed to firing the postsynaptic cell, as Hebb's rule requires? Protein synthesis usually occurs in a cell's nucleus, and the newly synthesized proteins are then shipped from the nucleus into the rest of the cell, including the dendrites. How do those proteins know which synapses they ought to fortify and which they ought to leave untouched? One answer to this question is that synapses that were active during a postsynaptic

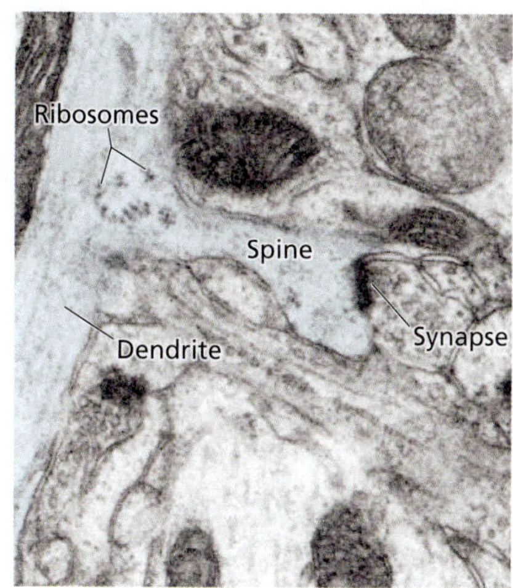

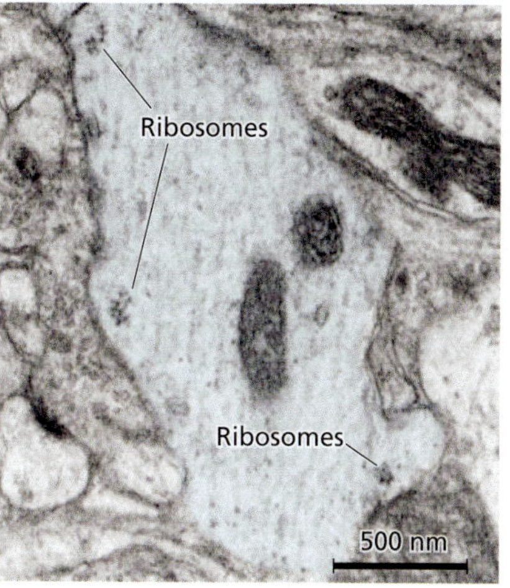

Figure 3.10 The machinery for protein synthesis in dendrites. These transmission electron micrographs show several small clusters of ribosomes in the dendrites (shaded light blue) of rat hippocampal neurons. Because ribosomes are used to generate proteins from mRNA, these data strongly suggest that proteins can be synthesized not only in the cell body but also in dendrites, close to the synapses. [From Steward and Levy, 1982]

depolarization can be marked with some sort of molecular "synaptic tag." Newly synthesized proteins are thought to recognize those tags and go to work selectively at those tagged synapses. Although there is some evidence for this **synaptic tagging hypothesis**, its details remain unclear.

Dendritic Protein Synthesis

Another mechanism for deploying newly synthesized proteins selectively at synapses that were active during a postsynaptic depolarization is to synthesize the needed proteins out in the dendrites, right next to the synapses that are to be strengthened. Early evidence for **dendritic protein synthesis** came from the anatomical observation that a few ribosomes, which translate mRNA into protein, are located inside the dendrites of hippocampal neurons (Figure 3.10). These ribosomes are typically found at the base of dendritic spines or sometimes inside a spine. To demonstrate that these dendritic ribosomes synthesize proteins that are needed to stabilize LTP, researchers used a knife to separate hippocampal dendrites from their cell bodies (Figure 3.11). In such isolated dendrites, LTP can be induced by tetanic stimulation of the input pathway. However, if the dendrites are bathed in a protein synthesis inhibitor, then only transient synaptic potentiation is seen (Figure 3.11 B). These findings imply that LTP induction causes ribosomes near the activated synapses to synthesize new proteins, which are then used to stabilize the induced changes in synaptic strength.

Although dendritic synthesis is known to play a role in stabilizing LTP, which proteins are involved remains unclear. One likely candidate is CaMKII, whose role in adding AMPA receptors to the postsynaptic membrane we already discussed. Another candidate is **activity-related cytoskeletal protein (ARC)**. Strong synaptic activity rapidly increases *Arc* gene expression in the cell nucleus. Much of the newly transcribed *Arc* mRNA is then shipped from the nucleus into the dendrites, where it is translated. Because blocking *Arc* translation impairs LTP and some forms of long-term memory, we can conclude that ARC is probably involved in stabilizing LTP.

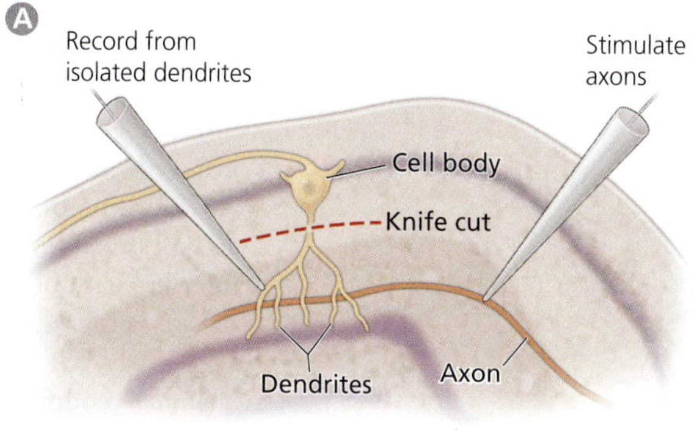

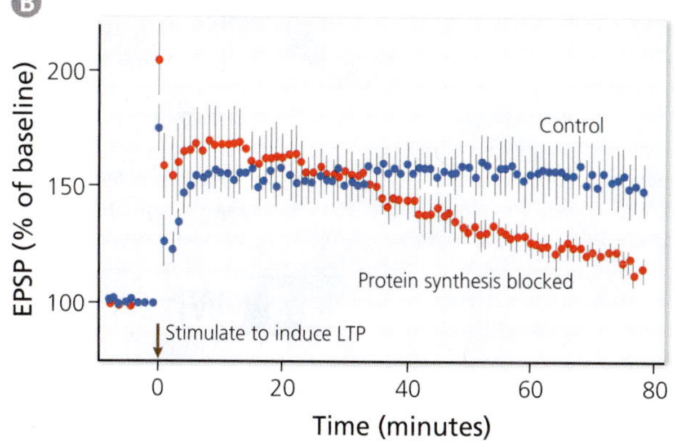

Figure 3.11 Dendritic protein synthesis is needed for LTP. The dendrites of hippocampal neurons were separated from their cell bodies by a knife cut (A). Axons terminating on those dendrites were stimulated repeatedly, leading to long-term potentiation (LTP) in the stimulated pathway, as demonstrated by a persistent increase in EPSP amplitude (B). Blocking protein synthesis (with emetine) permits transient potentiation (red) but prevents LTP stabilization. [After Cracco et al., 2005]

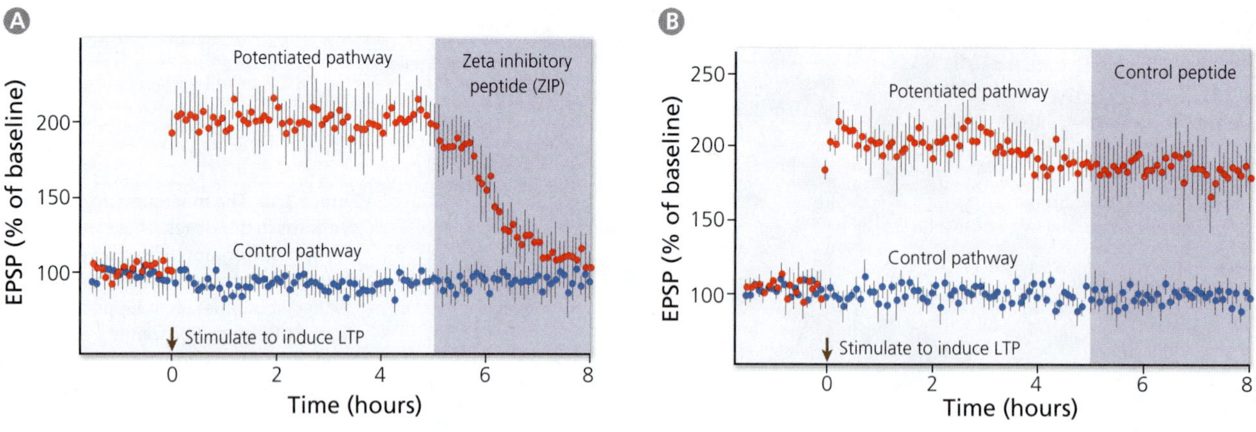

Figure 3.12 Reversing late-stage LTP with zeta inhibitory peptide (ZIP). Five hours after LTP induction, hippocampal neurons were bathed in ZIP (A). This treatment rapidly eliminated the LTP, whereas application of a scrambled (control) version of the same peptide had no effect on LTP (B). EPSPs elicited by test stimuli applied to a second (control) pathway that had not been stimulated tetanically was not potentiated and not affected by either peptide. [After Serrano et al., 2005]

Most likely, however, LTP stabilization involves a wide variety of proteins besides CaMKII and ARC.

Other Mechanisms Involved in LTP

Recent studies have shown that LTP can be destabilized by **zeta inhibitory peptide (ZIP).** When this peptide is applied to potentiated synapses as much as 5 hours after LTP induction, synaptic strengths rapidly return to their prestimulation levels of strength (Figure 3.12). In essence, ZIP destroys the synaptic "memory." No one is sure how ZIP accomplishes this feat. Some data indicate that ZIP interferes with **protein kinase M-zeta (PKM-zeta),** an enzyme that enhances synaptic transmission through AMPA receptors. However, mice engineered to lack PKM-zeta still exhibit LTP and are not impaired in several forms of learning and memory. Therefore, ZIP must also interfere with other mechanisms that are crucial for LTP stabilization. The identity of those other mechanisms is presently unknown.

Finally, the stabilization of LTP involves **epigenetic changes**, defined as the chemical modification of DNA and its associated proteins, which lead to long-lasting changes in gene expression. For example, drugs that prevent the de-acetylation of *histones*, around which DNA strands are wound, can convert transient synaptic enhancement into LTP and make some memories more persistent.

BRAIN EXERCISE

Are the "synaptic tagging" and "dendritic protein synthesis" hypotheses mutually exclusive? Why or why not?

3.3 When Are Synapses Weakened?

Although most of the research on synaptic plasticity has focused on the strengthening of synapses, brains must also have mechanisms for making synapses weaker. Otherwise, if synapses could only get stronger, neural activity would soon spread wildly through the brain, causing indiscriminate and excessive action potential firing. Indeed, neurobiologists have found numerous examples of synapses being weakened (depressed) for many hours after certain patterns of presynaptic stimulation. This phenomenon is called **long-term depression (LTD).**

Cerebellar Long-term Depression

A particularly interesting form of LTD does the opposite of what Hebb had predicted: it causes cells that "fire together" to become less strongly interconnected. Such

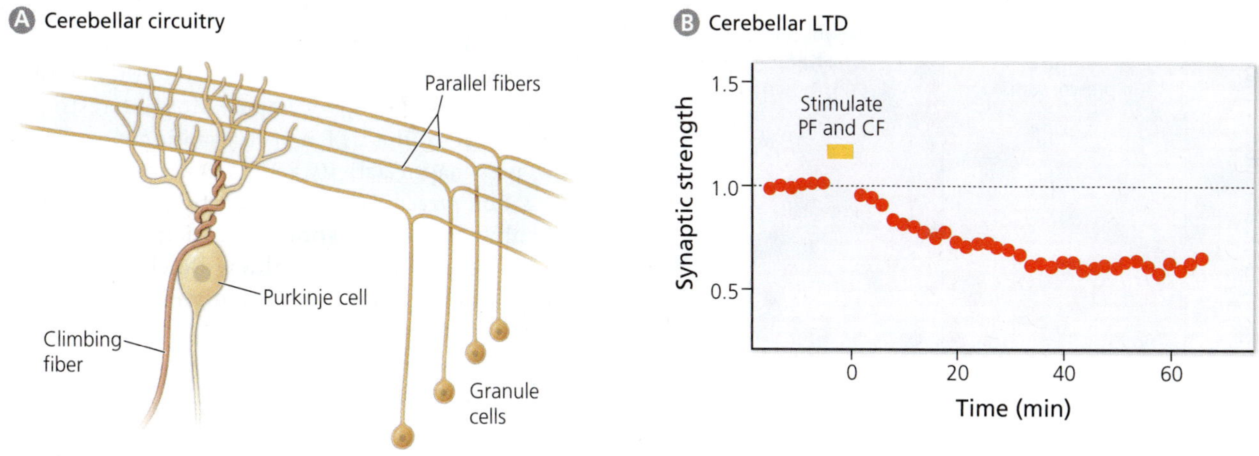

Figure 3.13 Long-term depression (LTD) in the cerebellum. Panel (A) depicts the climbing and parallel fiber inputs to a cerebellar Purkinje cell. Graph (B) shows that repetitive coincident stimulation of the parallel and climbing fiber (PF and CF, respectively) inputs leads to a persistent decrease in the strength of the stimulated parallel fiber-Purkinje cell synapses. This synaptic weakening is called long-term depression. [After Finch et al., 2012]

anti-Hebbian plasticity is found in the cerebellum, whose functions we will discuss extensively in Chapter 10. Briefly, the cerebellum fine-tunes movements (and, more controversially, some cognitive processes) by learning from errors and then adjusting its output accordingly.

As shown in Figure 3.13 A, the cerebellum contains some very large neurons, called *Purkinje cells*, and a much larger number of very small neurons, called cerebellar granule cells. Each Purkinje cell receives input from thousands of granule cells, whose axons are known as **parallel fibers**. In addition, each Purkinje cell receives input from a **climbing fiber**, which wraps itself around the Purkinje cell dendrites. When a climbing fiber fires an action potential and releases glutamate, the postsynaptic Purkinje cell becomes strongly depolarized and fires an action potential of its own. In contrast, the synapse between parallel fibers and Purkinje cells is relatively weak. Glutamate release at these synapses tends to evoke small postsynaptic depolarizations that trigger action potentials only when many parallel fibers are activated simultaneously. Most important for our present discussion is that activating the parallel and climbing fiber inputs simultaneously, and doing so repeatedly (~1/sec for several minutes), causes the parallel fiber input to decrease in strength (Fig. 3.13 B). This anti-Hebbian plasticity develops gradually and can last for many hours.

The mechanisms underlying cerebellar LTD resemble those of LTP but differ in some important respects. As in LTP, calcium influx into the postsynaptic cell plays a crucial role in cerebellar LTD. Specifically, the postsynaptic depolarization triggered by climbing fiber activation causes calcium influx through voltage-gated calcium channels and NMDA receptors at the climbing fiber synapses. Parallel fiber synapses do not contain NMDA receptors, but they do contain metabotropic glutamate receptors (see Chapter 2, Fig. 2.22). When these metabotropic receptors are activated by repetitive stimulation, they trigger an intracellular cascade (involving inositol trisphosphate, abbreviated IP3) that ultimately leads to the release of calcium ions from intracellular organelles. Although both climbing fiber and parallel fiber activation lead to increased calcium levels within the postsynaptic cell, coincident activation of both the parallel and the climbing fibers increases calcium levels much more than either form of stimulation can accomplish by itself. The sharp rise in postsynaptic calcium then triggers persistent activation of an enzyme called **protein kinase C**, which is involved in the removal of AMPA receptors from the postsynaptic side of the synapse. As you can imagine, the removal of AMPA receptors weakens the synapse.

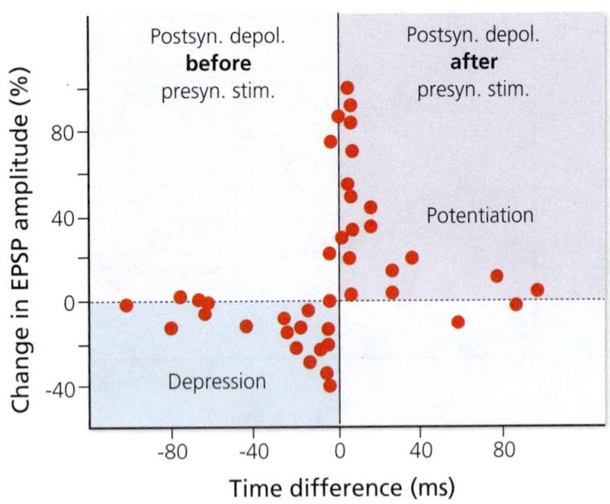

Figure 3.14 Spike timing-dependent plasticity. Using an experimental design similar to that shown in Figure 3.7, scientists discovered that whether the synapses between two cultured hippocampal neurons are strengthened (potentiation) or weakened (depression) depends on the relative timing of presynaptic stimulation and postsynaptic depolarization. [After Bi and Poo, 1998]

Spike Timing-dependent Plasticity

Although LTD has been studied most thoroughly in the cerebellum, LTD also occurs in several other brain regions, including the hippocampus. Most interesting are synapses that exhibit LTP if postsynaptic depolarization *follows* presynaptic activity and LTD if postsynaptic depolarization *precedes* presynaptic activity (Figure 3.14). Such **spike timing-dependent plasticity** has been observed in several different pathways, including some in the hippocampus, but it is not ubiquitous. In general, the type of plasticity that synapses can exhibit varies across cell types. This variability results from many factors, including differences in the transmitter receptors, ion channels, and intracellular enzymes. As we discussed in the previous chapter, neurons vary in many different respects. One of these is how their connections change in response to activity.

BRAIN EXERCISE

What does it mean for synaptic plasticity to be "anti-Hebbian"?

3.4 Can Inactive Neurons Strengthen Their Inputs?

So far we have discussed changes in synaptic strength that are linked to specific patterns of neural activity, but less specific forms of synaptic plasticity also occur. For example, neurons that have been inactive for some time tend to increase the strength of their excitatory synaptic inputs. In contrast, neurons that have been firing at a high rate tend to become less responsive. Strong evidence for such **synaptic scaling** comes from studies with cortical neurons that were maintained in cell culture for several days. Neurons cultured for 48 hours in a drug that blocks voltage-gated sodium channels (tetrodotoxin) exhibit larger **excitatory postsynaptic currents (EPSCs)**, which are analogous to EPSPs except that they measure current rather than membrane voltage, than neurons in control cultures without tetrodotoxin (Figure 3.15). Conversely, neurons cultured in a solution containing bicuculine, which blocks GABA receptors and therefore increases neuronal firing rates, exhibit reduced EPSCs (Figure 3.15 C). The mechanisms underlying these activity-dependent increases and decreases in synaptic strength are poorly understood but probably involve the addition or removal, respectively, of postsynaptic AMPA receptors.

Why do neurons scale up their sensitivity to synaptic input after they have been inactive and, conversely, scale it down after firing frequently? Because it would be wasteful to have neurons in your brain that are capable of generating action potentials but never receive enough excitatory input to reach the action potential threshold. Synaptic scaling ensures that such neurons do not remain silent for long. Conversely, it is dangerous for neurons to be overly sensitive to excitatory synaptic input and, therefore, firing too frequently. Such hypersensitivity can trigger epileptic seizures through runaway excitation. It can also kill neurons through excitotoxicity (see Chapter 5). Synaptic scaling mitigates these risks.

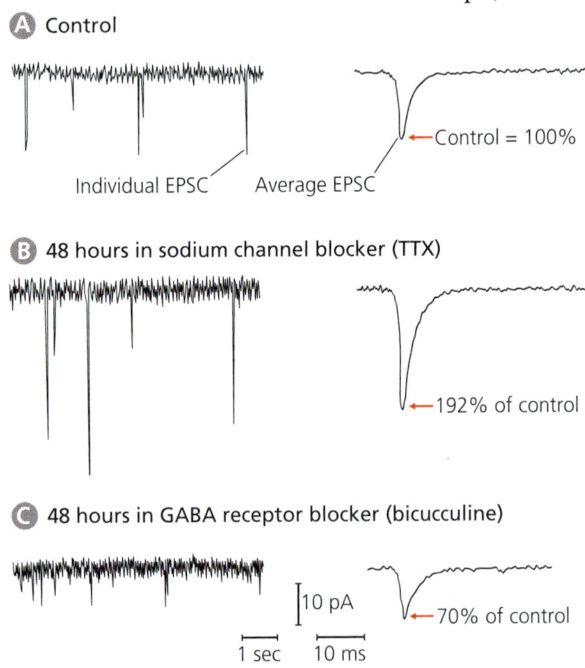

Figure 3.15 Synaptic scaling. Excitatory postsynaptic currents (EPSCs) were recorded from voltage-clamped cortical neurons maintained in cell culture (downward deflections indicate depolarizing current). After 48 hours of exposure to tetrodotoxin (TTX), which eliminates all action potentials, average EPSC amplitude was increased dramatically. In contrast, after blocking GABA receptors for 48 hours, EPSC amplitudes decreased. Thus, the neurons scale their responses to glutamate release up or down, depending on how much they have been firing. [From Turrigiano et al., 1998]

BRAIN EXERCISE

In what sense is synaptic scaling "anti-Hebbian"?

3.5 Can Experiences Rewire the Brain?

Neuronal plasticity may involve changes in the strength of existing synapses, but it may also involve the formation of entirely new dendritic spines, axon terminals, and major axonal branches. Such neuronal "rewiring" frequently occurs in the aftermath of brain damage, but it also occurs in the context of learning and memory.

Turnover of Dendritic Spines

Using fluorescent markers to label spiny neurons, researchers have found that dendritic spines are surprisingly dynamic. New spines are formed quite frequently and older spines are lost. This high rate of **spine turnover** implies that the synapses between neurons are more dynamic than people once believed. Most importantly, changes in the rate of spine turnover may be linked to learning. For example, spiny neurons that control singing in songbirds reduce their rate of spine turnover (Figure 3.16) when young birds learn to sing. Thus, the capacity for learning seems to be associated with a high degree of spine plasticity, whereas the act of learning involves the stabilization of newly formed spines.

Sprouting of Axonal Connections

If each spine represents the postsynaptic side of a synapse, then the existence of spine turnover implies that the presynaptic elements, the axon terminals, are also more dynamic than once thought. It is technically difficult to document small changes in axon terminals, but some studies have shown that axons can indeed sprout new branches, even in intact, adult brains.

A good example of axon sprouting comes from mice undergoing eye-blink conditioning. In this learning paradigm, animals are repeatedly presented with a tone, at the end of which a noxious puff of air is delivered to one of the eyes. The animals soon learn to associate the air puff with the tone and learn to blink when the tone is presented alone. This eye-blink conditioning involves the sprouting of new axon branches that terminate in the deep cerebellar nuclei, which lie beneath the cerebellar cortex (where the Purkinje cells are located).

To reach this conclusion, experimenters injected an axon tracer into a hindbrain region called the *pons*, which contains numerous auditory neurons. In untrained or pseudoconditioned animals (mice presented with unpaired air puffs and tones), the

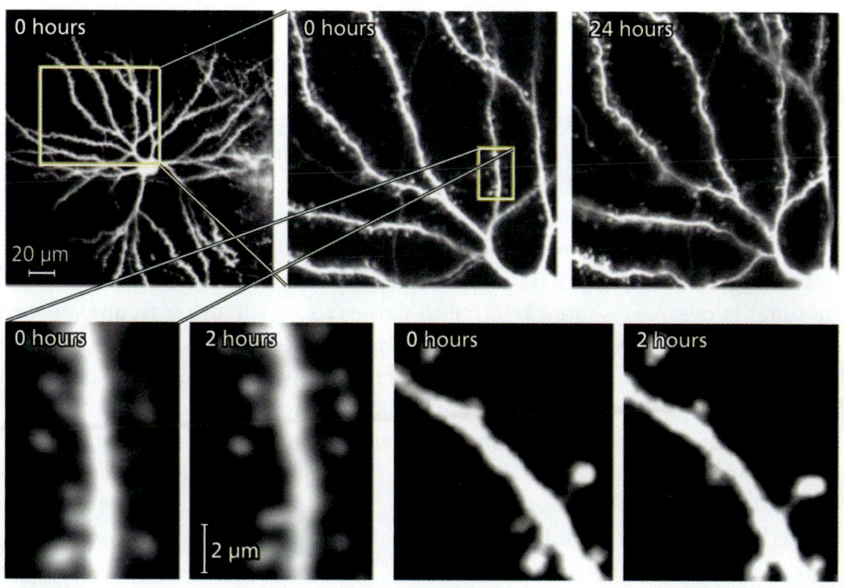

Figure 3.16 Dendritic spine turnover. Shown in the top row is a spiny neuron in the brain of an adult zebra finch, shown at different magnifications and different times. The left 2 images in the bottom row are close-ups of one dendrite, imaged 2 hours apart. You can see that one small spine was lost during this interval. The other images in the bottom row are from another neuron, imaged similarly. In this case, one spine disappeared while another one sprouted anew. As young birds learn to sing, the rate of spine turnover decreases (at least for neurons that had a high turnover rate to begin with). [From Roberts et al., 2010]

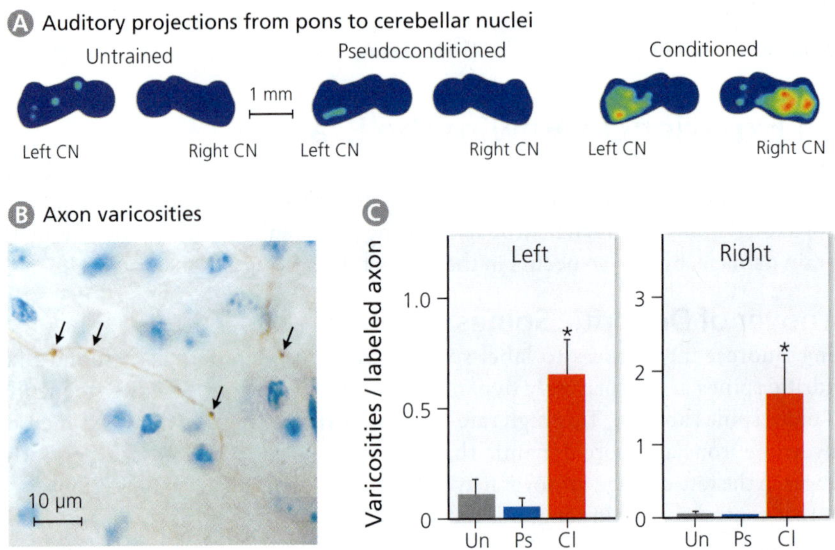

Figure 3.17 Axon sprouting during eye-blink conditioning. Axon tracers were used to visualize axons that project from the auditory region of the pons to the deep cerebellar nuclei (CN). Panel (A) shows that these projections are more extensive in eye-blink conditioned mice than in pseudoconditioned or untrained mice. Panel (B) depicts some labeled axons with varicosities, which are most likely synapses. The graphs in (C) show that the number of varicosities in the cerebellar nuclei is significantly increased in the conditioned animals, relative to the controls. To control for the size of the tracer injection, the number of varicosities was divided by the number of labeled axons in the main fiber tract connecting the pons to the cerebellum. [From Boele et al., 2013]

pons projects only weakly to the deep cerebellar nuclei. In contrast, after eye-blink conditioning, the projections from the pons to the deep cerebellar nuclei are much more extensive (Figure 3.17 A). They also exhibit more *varicosities*, which are most likely sites of neurotransmitter release (Figure 3.17 B). Overall, these data strongly suggest that eye-blink conditioning causes new axon branches to sprout and form new synapses. It is not yet clear, however, whether these learning-related changes are necessary or sufficient for eye-blink conditioning. They may be a correlate or consequence, but not a cause.

BRAIN EXERCISE

Go back to the quote from Ramón y Cajal at the beginning of this chapter. How well has his hypothesis been supported by modern data? What kind of plasticity did Cajal overlook in his statement?

Sensory Cortex Plasticity

In most cortical areas, a given stimulus tends to activate a spatially coherent cluster of neurons rather than a scattered set of isolated cells. This organizational feature allows neuroscientists to construct "maps" of how sensory stimuli are represented in the cortex by recording the stimulus preferences of neurons at several systematically varied locations across the cortex. One such map is shown in Figure 3.18. It shows that tones of a specific frequency tend to activate neurons in a specific part of the primary auditory cortex. Moreover, the sound frequency to which cortical neurons respond most strongly varies systematically across the auditory cortex, creating what is called a **tonotopic map**. This map turns out to be plastic. Strong evidence for this plasticity is that the tonotopic maps in naïve, untrained rats looks very different from those in thirsty rats that were trained to associate a specific tone with the availability of water (Figure 3.18). In such trained rats, the frequency of the conditioned stimulus (the tone) is overrepresented relative to naïve rats, while other frequencies are underrepresented. These learning-related changes in the tonotopic map are driven by changes in how individual neurons respond to auditory stimuli. In the case illustrated in Figure 3.18, neurons that responded most robustly to frequencies other than 6 kHz before training gradually shift their response preferences toward 6 kHz as the animals learn the association between the 6 kHz tone and water availability. The greater the number of cortical neurons that shift their response preference toward the conditioned stimulus, the more this stimulus becomes overrepresented

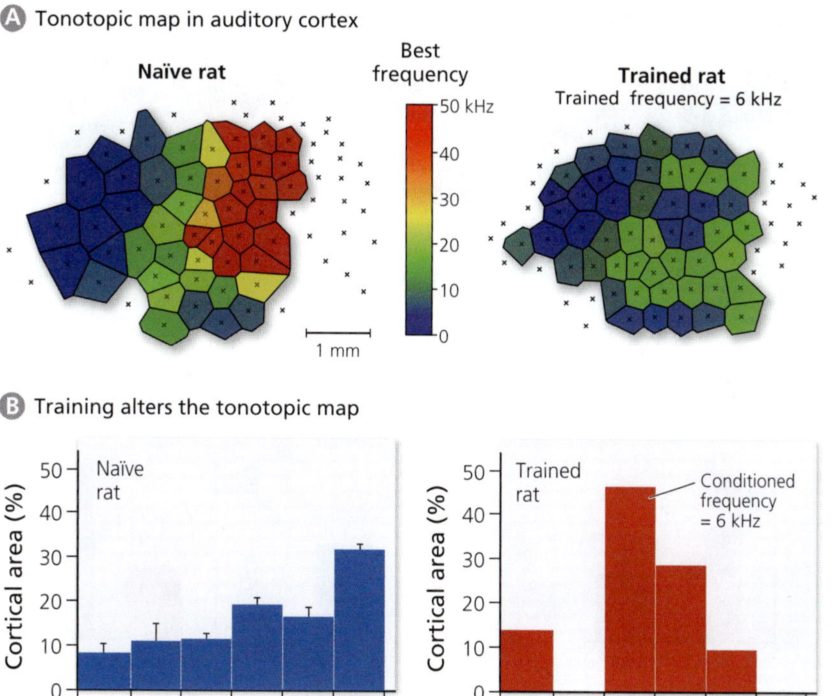

Figure 3.18 Learning-related plasticity in the auditory cortex. Thirsty rats were trained to press a bar for water whenever they heard a 6 kHz tone. The experimenters then compared the stimulus preferences of neurons in the primary auditory cortex of naïve (untrained) and trained rats (A). They found that the cortical territory containing neurons tuned to frequencies near 6 kHz (green zones) had expanded in the trained rats, relative to naïve rats, whereas the territory dedicated to higher frequencies (yellow to red shading) had shrunk. A quantitative summary is shown in (B). [From Weinberger, 2007]

in the map. Analogous forms of learning-related plasticity have been observed in the primary somatosensory and visual cortices.

The mechanisms underlying the overrepresentation of conditioned stimuli in the sensory cortex remain unclear. Learning-related changes in the auditory map require acetylcholine release in the neocortex, and repeatedly pairing acetylcholine release with stimulus presentation is sufficient to generate plasticity. However, it is not known whether the changes in functional organization involve axonal sprouting and spine turnover or just changes in the strength of existing synapses. Some learning-related changes in the sensory cortex occur within an hour of training and are therefore unlikely to require the sprouting of long axon branches. However, other changes develop over the course of several days, suggesting that they may involve axon sprouting. In any case, the data clearly show that neurons in the primary sensory cortices are not simply providing the raw material for learning and memory; they are part of the circuitry that is involved in learning and memory.

Motor Cortex Plasticity

The primary motor cortex (see Figure 1.14) exhibits learning-related plasticity very similar to that observed in the sensory cortex. This plasticity can be demonstrated by comparing cortical **motor maps** (showing which neurons are involved in which movements) between animals that learned a complex motor skill and animals that performed much simpler movements. In one experiment, adult rats were trained to reach through a small opening in their cage and grasp a slowly moving food pellet. At the end of the training period, the motor cortex of the rats was mapped and compared to the motor cortex of control rats that simply pressed a lever to get food. Remarkably, the rats that had learned the skilled movement dedicated significantly more of their motor cortex to movements of the forelimb digits and wrist, which were essential to grasping the food, than the rats that had simply pressed the lever repeatedly (Figure 3.19). The rats in the skilled learning group didn't just expand the areas required for the skilled movements; they also shrank the areas devoted to movements

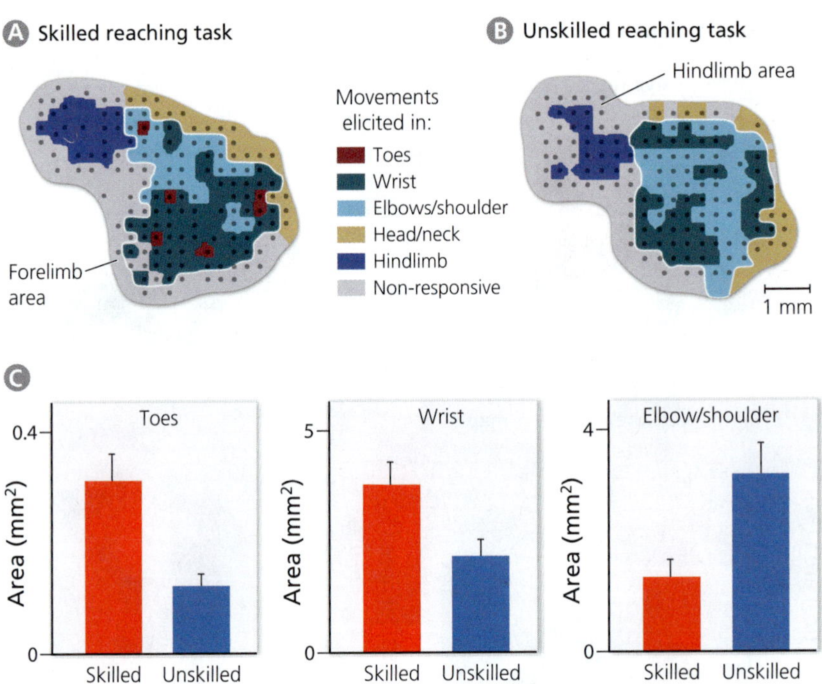

Figure 3.19 Training can alter the motor cortex map. Adult rats were trained to reach for moving food pellets (a task requiring skill) or simply press a bar for food (an unskilled task). Shortly after the last training session, the experimenters used electrical microstimulation to map the motor cortex in both groups of animals. They found that the representation of the forelimb's wrist and toes was larger in the skilled group (A) than the unskilled group (B). The bar graphs in (C) show that the wrist and toe representations expanded at the expense of the elbow and shoulder representation. [After Kleim et al., 2002]

of the elbow and shoulder, which are much less important for grasping a food pellet. The general conclusion is that learning a skilled motor task leads to an expansion of the cortical territory that is in involved in controlling the learned movements (at the expense of other areas). By contrast, repetitive performance of an easy, unskilled task causes little or no plasticity.

Similar findings have been reported for humans. For example, the hand region of the cortical motor map is expanded in highly skilled pianists and violinists, relative to nonmusicians. Because violinists make faster, more differentiated movements with the fingers of their left hand than their right, you would expect violinists to show a disproportionate enlargement of the hand area in the right motor cortex, which controls movements of the left hand. Indeed, they do! Given that fine motor control requires detailed sensory feedback, you might also expect the sensory representation of the hand to be enlarged in skilled violinists. Again, this prediction has been confirmed. These observations raise an interesting question: did the sensory and motor maps change *as a result* of extensive practice, or did the musicians become good at their craft *because* their cortex was unusual to begin with? A tentative answer to this question comes from the observation that the degree of cortical sensory and motor enlargement tends to correlate with the number of years of musical training. Thus, the data suggest that, the more you practice, the more the cortical areas related to the practiced skill will grow.

An interesting demonstration of motor cortex plasticity involved monkeys whose motor cortex had been chronically implanted with electrodes at two different locations, sites A and B. At the beginning of the experiment, electrical stimulation at these two sites evoked two very different movements. The experimenters then stimulated site B every time the neurons at site A fired an action potential. After two days of such stimulation, while the monkey was moving freely around its cage, the experimenters again stimulated sites A and B separately. They found that the evoked movements were now quite similar at the two sites, with the movements elicited by stimulation at site B having changed more than those evoked at site A. This observation suggests that the two days of stimulation caused the neurons at site A to project more strongly to the neurons at site B, just as Hebb's rule would predict (because the neurons at sites A and B were active simultaneously). More work is needed to determine whether this plasticity involves the sprouting of new connections or just

THERAPIES

Box 3.3 *Brain–Machine Interfaces*

Brain–machine interfaces (BMIs) make it possible for paralyzed subjects to control machines, such as a wheelchair or a robot arm, without contracting a muscle. The earliest studies on BMIs were conducted on non-humans. After implanting electrodes in the motor cortex of monkeys or rats, researchers recorded the activity of multiple neurons as the animals performed diverse movements. Using a computer to identify correlations between movement parameters and patterns of neural activity, the researchers built a "decoder" that can infer the subject's intended movements from the neural activity and do so in real time. The experimenters then restricted the animal's own movements, while continuing to record the neural activity, and used the decoder to infer the subject's intended movements. Another computer then translated the inferred intentions into commands that drive the robot arm.

Using this approach to help paralyzed humans is difficult because their inability to move means that you cannot use the subject's own movements to figure out how neural activity in the motor cortex correlates with movement. To solve this problem, researchers typically ask the subject to imagine moving the robot arm. Because neural activity during imagined movements correlates with activity during actual movements, researchers can use those correlations to construct the decoder. This is great, but there's another problem, namely, that the implanted electrodes tend to move within the brain over the course of several days or weeks, which means that the population of recorded neurons changes slowly over time. This "population drift" problem can be solved by regularly recalibrating the decoder using the imagined movement method. Alternatively, or in addition, experimenters can provide the decoder with feedback about how well it decoded the subject's intended movements. If the movement of the robot arm did not match the intention, the decoder's algorithm is altered; when performance is good, no changes are made. In this manner, the decoder learns to optimize its performance.

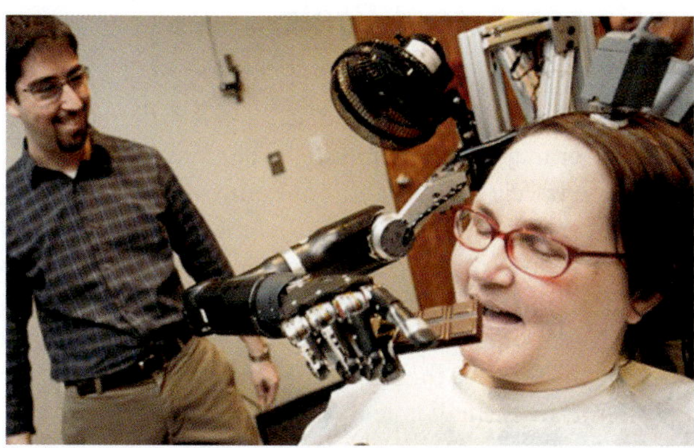

Figure b3.3 Paralyzed but able to control a robot arm. A paraplegic woman learned to control a robotic arm using a brain–machine interface. Two sets of electrodes implanted in the motor cortex are connected to a computer, which translates the subject's intended movements into signals that control the robot arm. Both the woman and the computer use error signals to improve their performance. It took several months of regular practice, but now the subject can feed herself a chocolate bar. Video at http://youtube/76lIQtE8oDY. [Image courtesy of University of Pittsburgh Medical Center]

What does this have to do with neuronal plasticity? The answer is that the computer is not the only one that learns. The paralyzed subject also learns to use the BMI. In fact, providing a subject with visual feedback on the success or failure of their attempts to use the robot arm greatly improves their rate of progress in controlling that arm. Over several days, such feedback leads to changes in the "movement tuning" of the recorded neurons. The mechanisms underlying this form of brain plasticity remain unclear but probably include the kind of remapping you learned about in this chapter.

Although changes in the brain might force the decoder to adjust its algorithm, and changes in the decoder might force the brain to tweak its own activity, brain and decoder tend to learn in concert with one another, coadapting their internal mechanisms to optimize performance. The effectiveness of this approach is best illustrated by a paraplegic subject who learned to feed herself with a robotic arm after just a few months of diligent practice (Figure b3.3). Remarking on her own progress, she said, "I used to have to think, up, clockwise, down, forward, back.... Now I just look at the target, and Hector [the arm] goes there."

changes in synaptic strength. Either way, the finding implies that the internal wiring of the motor cortex can be modified by electrical stimulation. Researchers are now exploring how this knowledge might be used to develop better neural prostheses, which allow patients with paralyzed limbs to control robot arms or other machines (see Box 3.3).

BRAIN EXERCISE

How would you expect the neocortex of professional baseball players to differ from that of people who have never played baseball? How might you determine whether those differences are the result of experience (learning)?

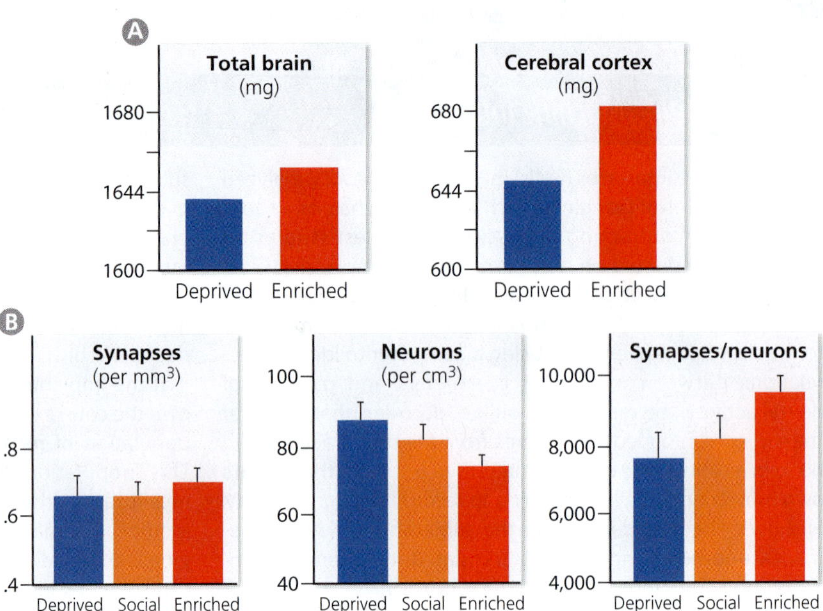

Figure 3.20 Environmental enrichment effects. Total brain and cerebral cortex mass are increased in rats that grew up in an enriched environment with access to various "toys," interactions with other rats, and daily maze training (A). Rats in the "deprived" comparison group are housed individually in standard rat cages (no toys). Panel (B) demonstrates that environmental enrichment decreases cortical neuron density without changing synapse density. Therefore, the number of synapses per neuron is increased. Rats housed with other rats but without toys or daily training (the "social" condition) exhibit only weak enrichment effects. [After Rosenzweig et al., 1962, and Turner and Greenough, 1985]

3.6 How Does Experience Affect Brain and Cortex Size?

One form of neuronal plasticity that we have not yet discussed is experience-dependent growth of the entire brain or major brain regions. To demonstrate this kind of large-scale plasticity, Mark Rosenzweig and his colleagues in the 1960s studied the brains of laboratory rats that were housed in two very different environments. Some rats were housed by themselves in small cages without toys or opportunities for exercise. A second group of rats was given daily maze training and housed in roomy cages with social companions, toys, ladders, and running wheels. Two months later the rats were sacrificed. As Rosenzweig discovered, the rats living in the impoverished environment ended up having a significantly smaller neocortex than the rats that had been living in the enriched conditions. Subsequent studies using electron microscopy revealed that the visual cortex in the enriched rats contains larger neurons, a larger number of synapses per neuron, and more glial cells (Figure 3.20).

Environmental enrichment or deprivation can also affect the size of human brains. Most dramatically, the brains of orphans who were institutionalized under atrociously deprived conditions in Romania (under the Ceauşescu regime) were 16% smaller, on average, than control brains. The orphans also suffered from a wide variety of cognitive defects. Whether these neural and behavioral effects of developmental deprivation can be reversed, at least to some extent, remains unclear. Similarly, it remains unclear why the effects of deprivation are more severe in some children than others, even when the deprivation was equally severe. Some of this residual variation is likely genetic.

BRAIN EXERCISE

Do you think laboratory rats and mice should be housed in enriched or deprived conditions (or somewhere in between)? What might be some pros and cons?

3.7 Does Neural Plasticity Cause Learning and Memory?

You have now learned that even adult brains exhibit diverse forms of neuronal plasticity. They may change the strength of some existing synapses, form novel connections or eliminate established ones, change the number of neurons representing a particular function, and grow or shrink in absolute size. This neuronal plasticity often accompanies the learning of new information or new skills, confirming what Ramón y Cajal and others had suspected back in the late 1800s. Based on a mountain of evidence, neuroscientists now know that experience-dependent neuronal plasticity may be found in many different brain regions, including the hippocampus, sensory and motor cortices, and the cerebellum. Although plasticity in these brain areas tends to accompany learning and memory, providing convincing evidence that the neural changes *cause* the learning or the memories is difficult.

To establish a causal link between learning and neural plasticity, researchers can ask whether preventing the plasticity impairs the learning. Such **loss-of-function experiments** have shown, for example, that CREB is *necessary* for long-term sensitization in *Aplysia* and that blocking LTP in the hippocampus can interfere with memory. It is important to note, however, that negative results in such experiments do not disprove a causal link because other mechanisms may compensate. For example, the auditory cortex is probably involved in tone conditioning, even though animals with lesions of the auditory cortex can still associate a tone with noxious stimuli, most likely using plasticity in some subcortical circuits (see Chapter 14).

Gain-of-function experiments that determine whether boosting plasticity is *sufficient* to enhance learning and memory can also provide evidence for causal links; but negative results are, once again, inconclusive. For example, you learned that injecting phosphorylated CREB into an *Aplysia* sensory neuron suffices to induce long-term sensitization at its synapse onto the gill motor neuron. Combined with the aforementioned loss-of-function results, these data provide compelling evidence that CREB is causally linked to long-term sensitization. However, negative results in such gain-of-function experiments would be difficult to interpret. The manipulation might have targeted an insufficient number of neurons, or additional mechanisms may have to be altered before a gain-of-function becomes evident. Even positive results must be interpreted with care. For example, boosting NMDA receptor function may facilitate learning and memory, but it may do so through a variety of mechanisms, some of which may have nothing to do with LTP. Boosting NMDA receptor function might, for instance, make an animal more sensitive to noxious stimuli, which would make the animal more likely to learn associations with noxious stimuli; but we would not be able to conclude that NMDA receptors are causally linked to memory. The difference in sensitivity would be a potential confound.

One other important way to test for causal links between neuronal plasticity and memory is to ask whether the degree of plasticity correlates with the amount of learning animals exhibit. If the correlation is strong, then a causal link is probable. For instance, the hypothesis that spine turnover is causally linked to song learning in birds is bolstered by the observation that changes in the rate of spine turnover correlate with the fidelity of the song imitations. Similarly, the hypothesis that changes in the tonotopic map of the auditory cortex are causally linked to learning and memory is supported by data indicating that the amount of map plasticity correlates with the strength of a rat's memory for the learned association. In the motor cortex, too, learning a difficult motor skill elicits more plasticity than learning a simple task and then performing it repeatedly. Given such correlations between learning and plasticity, a causal link between the two is highly probable. Finally, it is worth noting that both learning and neural plasticity are greater when the information being learned is important to the animal. As you know, it is difficult to learn new information if it seems irrelevant. The larger lesson here is this: if you want to learn something so well that it alters your neural circuitry, then try to figure out why the information is relevant and important to you.

BRAIN EXERCISE

In your view, what would be the best possible experiment to test whether extensive violin training changes the cortical representation for movements of the left hand? What would be some problems with this ideal experiment?

SUMMARY

Section 3.1 - The mechanisms underlying synaptic plasticity have been studied extensively in the marine snail *Aplysia californica.*

- Sensitization of the gill withdrawal reflex involves the release of serotonin onto presynaptic terminals in a simple sensorimotor circuit. Activation of the serotonin receptors triggers an intracellular signaling cascade that ultimately broadens the presynaptic action potentials, which then boosts transmitter release and increases EPSP amplitude.
- Long-term sensitization requires new protein synthesis. A crucial factor is the phosphorylation of CREB, which regulates the transcription of multiple genes involved in making changes in synaptic strength more persistent.

Section 3.2 - Synapse strengthening (potentiation) in mammals is commonly induced by high-frequency repetitive (tetanic) stimulation of an input pathway.

- Long-term potentiation (LTP) is characterized by an increase in EPSP amplitude that persists for several hours (or more) after the end of the tetanic stimulus. It is most often studied in hippocampal slices.
- LTP is associative when it obeys Hebb's rule, which states that synapses should only be strengthened when presynaptic activity is accompanied by postsynaptic depolarization, which may trigger action potentials.
- Glutamate can open NMDA receptors only when the postsynaptic membrane is depolarized enough to remove their magnesium block. NMDA receptor activation triggers an intracellular cascade that ultimately causes more AMPA receptors to be inserted into the postsynaptic membrane. It can also stimulate synapse growth.
- The stabilization of LTP requires protein synthesis. Dendritic protein synthesis and synaptic tagging are two possible mechanisms for ensuring that the newly synthesized proteins strengthen only those synapses that contributed to the postsynaptic depolarization (as required by Hebb's rule).

Section 3.3 - The best studied form of synaptic weakening is cerebellar long-term depression (LTD), which involves metabotropic glutamate receptors and the removal of AMPA receptors from the postsynaptic membrane.

Section 3.4 - Synaptic scaling refers to neurons increasing the average strength of all their synaptic inputs after they have been relatively inactive and decreasing the strength of those inputs when neurons have been very active.

Section 3.5 - Adult brains may rewire themselves as a result of experience.

- Dendritic spines may sprout or disappear in adult brains, and changes in the rate of spine turnover may correlate with learning. Axonal branches may also sprout as animals learn.
- As animals learn important sensory information, the cortical territory representing that information tends to expand at the expense of territory representing less important information.
- As animals learn a new motor skill, portions of the motor cortex related to that skill tend to expand at the expense of other areas. The more difficult the skill, the greater the expansion.

Section 3.6 - Animals living in an enriched environment tend to have a larger neocortex than deprived animals. The enlargement is due mainly to the growth of individual neurons and the addition of glial cells.

Section 3.7 - Establishing a causal link between neural phenomena and behavior is best done by combining gain-of-function and loss-of-function experiments with correlative data.

Box 3.1 - It is easy to underestimate invertebrates, both in their cognitive capacities and in their impact on neuroscience.

Box 3.2 - The hippocampus probably plays similar roles in spatial and episodic memory in both humans and rodents, but this idea remains debatable.

Box 3.3 - Paralyzed humans can control a robot arm through a brain–machine interface. Optimal performance is reached when both the machine and the human learn through feedback about motor performance.

KEY TERMS

***Aplysia californica* 72**
gill withdrawal reflex 72
semi-intact preparation 72
sensitization 73
adenylate cyclase 73
cyclic adenosine monophosphate (cAMP) 73
protein kinase A (PKA) 73
cAMP response element binding protein (CREB) 75
ubiquitin hydrolase 76
heterosynaptic 76
homosynaptic 76
tetanic stimulation 77
post-tetanic potentiation 77

perforant path 77
dentate gyrus 77
long-term potentiation 79
brain slice 79
Hebb's rule 80
magnesium block 80
calmodulin 81
calcium/calmodulin kinase II (CaMKII) 81
caged glutamate 82
synaptic tagging hypothesis 83
dendritic protein synthesis 83
activity-related cytoskeletal protein (ARC) 83
zeta inhibitory peptide (ZIP) 84
protein kinase M-zeta (PKM-zeta) 84
epigenetic change 84
long-term depression (LTD) 84
parallel fiber 85
climbing fiber 85
protein kinase C 85
spike timing-dependent plasticity 86
synaptic scaling 86
excitatory postsynaptic current (EPSC) 86
spine turnover 87
tonotopic map 88
motor map 89
loss-of-function experiment 93
gain-of-function experiment 93

ADDITIONAL READINGS

3.1 - Synaptic Plasticity in Aplysia

Kandel ER. 2001. The molecular biology of memory storage: a dialogue between genes and synapses. *Science* **294**:1030–1038.

3.2 - Synaptic Plasticity in Mammals

Cooke SF, Bliss TVP. 2006. Plasticity in the human central nervous system. *Brain* **129**:1659–1673.

Cooper S. 2005. Donald O. Hebb's synapse and learning rule: a history and commentary. *Neurosci Biobehav Rev* **28**:851–874.

Day JJ, Sweatt JD. 2011. Cognitive neuroepigenetics: a role for epigenetic mechanisms in learning and memory. *Neurobiol Learn Mem* **96**:2–12.

Derkach V, Barria A, Soderling TR. 1999. Ca2+/calmodulin-kinase II enhances channel conductance of alpha-amino-3-hydroxy-5-methyl-4-isoxazolepropionate type glutamate receptors. *Proc Natl Acad Sci U S A* **96**:3269–3274.

Gustafsson B, Wigström H, Abraham WC, Huang YY. 1987. Long-term potentiation in the hippocampus using depolarizing current pulses as the conditioning stimulus to single volley synaptic potentials. *J Neurosci* 7:774–780.

Korb E, Finkbeiner S. 2011. Arc in synaptic plasticity: from gene to behavior. *Trends Neurosci* **34**:591–598.

Lisman J, Yasuda R, Raghavachari S. 2012. Mechanisms of CaMKII action in long-term potentiation. *Nat Rev Neurosci* **13**:169–182.

Martin KC, Zukin RS. 2006. RNA trafficking and local protein synthesis in dendrites: an overview. *J Neurosci* **26**:7131–7134.

Martin KC, Kosik KS. 2002. Synaptic tagging—who's it? *Nat Rev Neurosci* **3**:813–820.

Murakoshi H, Yasuda R. 2012. Postsynaptic signaling during plasticity of dendritic spines. *Trends Neurosci* **35**:135–143.

Sacktor TC. 2011. How does PKMζ maintain long-term memory? *Nat Rev Neurosci* **12**:9–15.

Volk LJ, Bachman JL, Johnson R, Yu Y, Huganir RL. 2013. PKM-ζ is not required for hippocampal synaptic plasticity, learning and memory. *Nature* **493**:420–423.

3.3 - Synaptic Weakening

Gao Z, van Beugen BJ, De Zeeuw CI. 2012. Distributed synergistic plasticity and cerebellar learning. *Nat Rev Neurosci* **13**:619–635.

3.4 - Synaptic Scaling

Abbott LF, Nelson SB. 2000. Synaptic plasticity: taming the beast. *Nat Neurosci* **3**(Suppl):1178–1183.

3.5 - Synaptic Rewiring

Barnes SJ, Finnerty GT. 2010. Sensory experience and cortical rewiring. *Neuroscientist* **16**:186–198.

Bieszczad KM, Weinberger NM. 2010. Representational gain in cortical area underlies increase of memory strength. *Proc Natl Acad Sci U S A* **107**:3793–3798.

Galván VV, Weinberger NM. 2002. Long-term consolidation and retention of learning-induced tuning plasticity in the auditory cortex of the guinea pig. *Neurobiol Learn Mem* **77**:78–108.

Hihara S, Notoya T, Tanaka M, Ichinose S, Ojima H, Obayashi S, et al. 2006. Extension of corticocortical afferents into the anterior bank of the intraparietal sulcus by tool-use training in adult monkeys. *Neuropsychologia* **44**:2636–2646.

Münte TF, Altenmüller E, Jäncke L. 2002. The musician's brain as a model of neuroplasticity. *Nat Rev Neurosci* **3**:473–478.

Rosenkranz K, Williamon A, Rothwell JC. 2007. Motorcortical excitability and synaptic plasticity is enhanced in professional musicians. *J Neurosci* **27**:5200–5206.

3.6 - Changes in Brain and Cortex Size

Anderson BJ. 2011. Plasticity of gray matter volume: the cellular and synaptic plasticity that underlies volumetric change. *Dev Psychobiol* **53**:456–465.

Sheridan M, Drury S, Mclaughlin K, Almas A. 2010. Early institutionalization: neurobiological consequences and genetic modifiers. *Neuropsychol Rev* **20**:414–429.

3.7 - From Correlation to Causality

Destexhe A, Marder E. 2004. Plasticity in single neuron and circuit computations. *Nature* **431**:789–795.

Tang Y, Shimizu E, Tsien JZ. 2001. Do "smart" mice feel more pain, or are they just better learners? *Nat Neurosci* **4**:453–453.

Boxes

Burne T, Scott E, van Swinderen B, Hilliard M, Reinhard J, Claudianos C, et al. 2011. Big ideas for small brains: what can psychiatry learn from worms, flies, bees and fish? *Mol Psychiatry* **16**:7–16.

Nadel L, Hoscheidt S, Ryan LR. 2013. Spatial cognition and the hippocampus: the anterior-posterior axis. *J Cogn Neurosci* **25**:22–28.

Orsborn AL, Carmena JM. 2013. Creating new functional circuits for action via brain-machine interfaces. *Front Comput Neurosci* **7**:157.

Velliste M, Perel S, Spalding MC, Whitford AS, Schwartz AB. 2008. Cortical control of a prosthetic arm for self-feeding. *Nature* **453**:1098–1101.

CHAPTER

4

Developing a Nervous System

CONTENTS

FEATURES

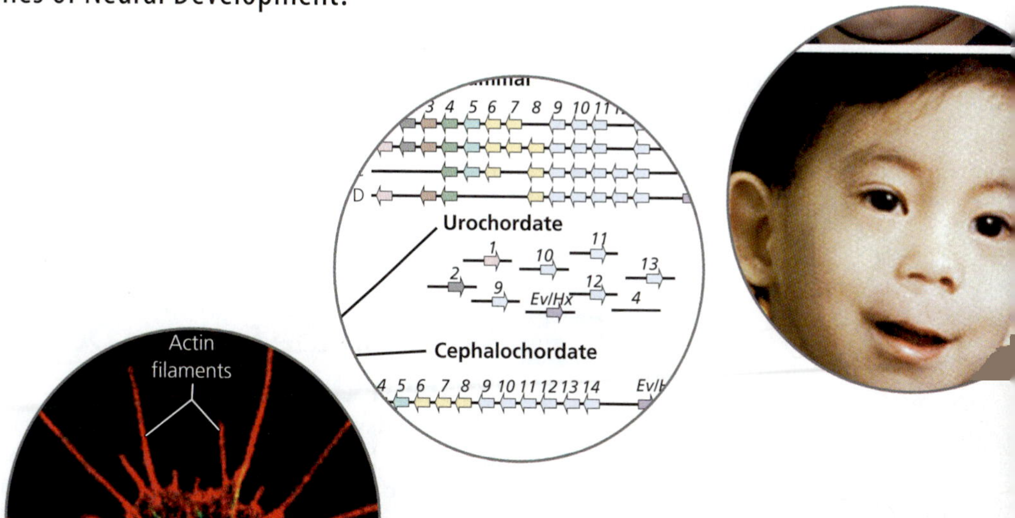

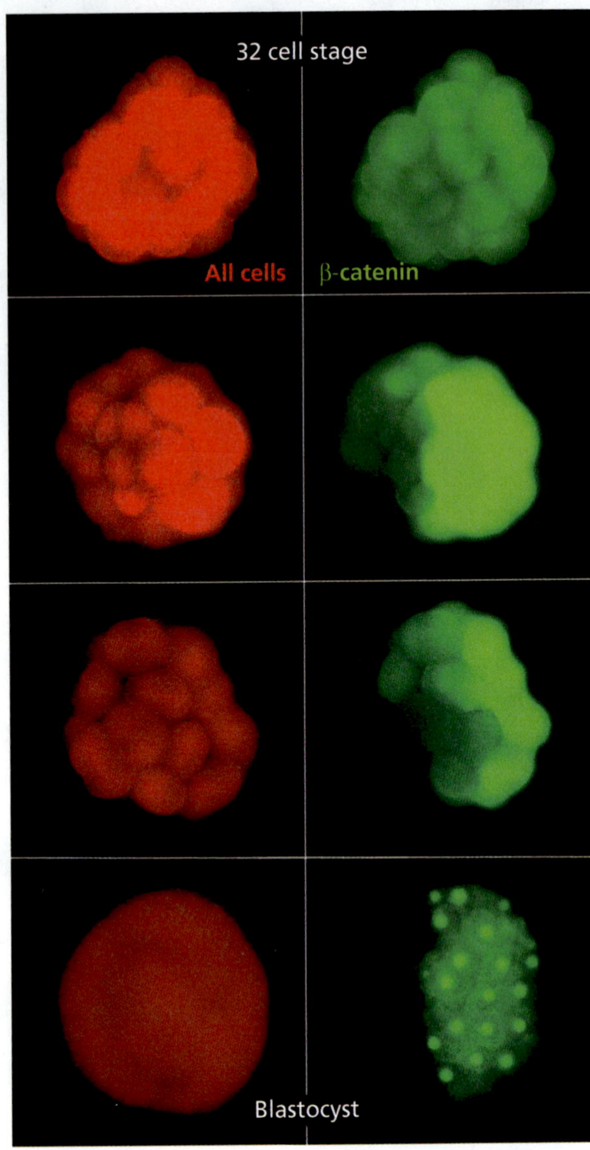

Figure 4.1 Breaking symmetry in embryos. In very young sea anemone embryos, every cell expresses beta-catenin (β-catenin; top). As development proceeds, β-catenin expression becomes restricted to one side of the embryo (the future endoderm). This change in gene expression is an example of symmetry breaking because the embryo is less symmetrical after the change. [From Wikramanayake et al., 2003]

Although it might appear that every neuron in a complex nervous system is connected to every other neuron, a closer look reveals that real neurons tend to have rather specific connections with select groups of other cells. Just as neurons differ in size, shape, transmitters, and ion channel types, so they differ in axonal connections. Furthermore, neurons with similar connections tend to cluster together, forming discrete brain areas, laminae, or nuclei. Thus, real nervous systems are heterogeneous (spatially variable) rather than homogeneous. But how does structural and functional heterogeneity arise during development? Given that development starts with a single fertilized egg cell, how do some cells of the growing embryo get specified to become the nervous system? How do the various cells in the nervous system become different from one another? How does each neuron "know" to which other neurons it should be connected?

On one level, the answer is relatively simple. As a fertilized egg divides again and again, its daughter cells come to express different sets of genes, which causes the cells to vary in size, shape, and other properties (Figure 4.1). This only begs the question, however. What causes different daughter cells to express different genes, given that they all contain the same DNA? The answer is that different cells contain different **transcription factors**, which contain specific DNA-binding domains that, when bound, promote (or repress) the expression of genes close to the binding site. In other words, they recognize specific DNA sequences, bind to them, and then regulate gene transcription at those selected locations (you may recall that CREB, which we discussed in Chapter 3, is one such transcription factor). Because transcription factors regulate many genes, some of which in turn encode other transcription factors or proteins that mediate cell–cell interactions, differences in transcription factor expression usually translate into long-lasting differences in how the cells behave, both intracellularly and in their interactions with other cells. Thus, differences in transcription factor expression tend to produce differences in **cell fate**.

Of course, this still doesn't answer the more fundamental question: what causes different cells to express different transcription factors? For some invertebrates, the answer lies with mom, who makes her eggs so that they already contain an asymmetric arrangement of transcription factors. In mammals, both the mother and the father's sperm cooperate to distribute transcription factors heterogeneously within the fertilized egg. This need not concern us here. What is important is that spatial differences in transcription factor expression are crucial to the early formation of the nervous system.

4.1 Where in the Embryo Does the Nervous System Originate?

Five to six days after fertilization, a human egg has grown into a hollow clump of cells that would fit comfortably on the head of a pin. At or before this **blastocyst stage**, embryonic cells that accidentally get separated from the others can form a complete embryo, an identical twin. After the blastocyst stage, twinning is rare because the embryonic cells become more specialized. In particular, the cells of the blastocyst rearrange themselves during gastrulation to form three distinct **germ layers**: namely, ectoderm, endoderm, and, sandwiched between them, mesoderm (Figure 4.2). The **ectoderm** is of special interest to neurobiologists because its cells differentiate into two seemingly very different tissues: epidermis (skin) and nervous system. It may seem strange that skin and nervous system are developmentally so closely related, but the earliest animals probably had their entire nervous system located within the skin. Centralized brains and spinal cords evolved later (see Chapter 16).

Induction of the Nervous System

Why do some ectodermal cells develop into skin while others form the nervous system? A key experiment addressing this question was performed by Hilde Mangold and her dissertation advisor Hans Spemann in the 1920s (Figure 4.3). They took a piece of mesoderm called the **dorsal blastopore lip** from the embryo of a white (lightly pigmented) amphibian just after the blastocyst stage and transplanted it into a darkly pigmented amphibian embryo of the same age. This modified embryo had two blastopore lips, one on each side of the embryo, and it developed into Siamese twins, joined at the belly but with two complete nervous systems (Figure 4.3). Crucially, Mangold noticed that the cells in both nervous systems were darkly pigmented,

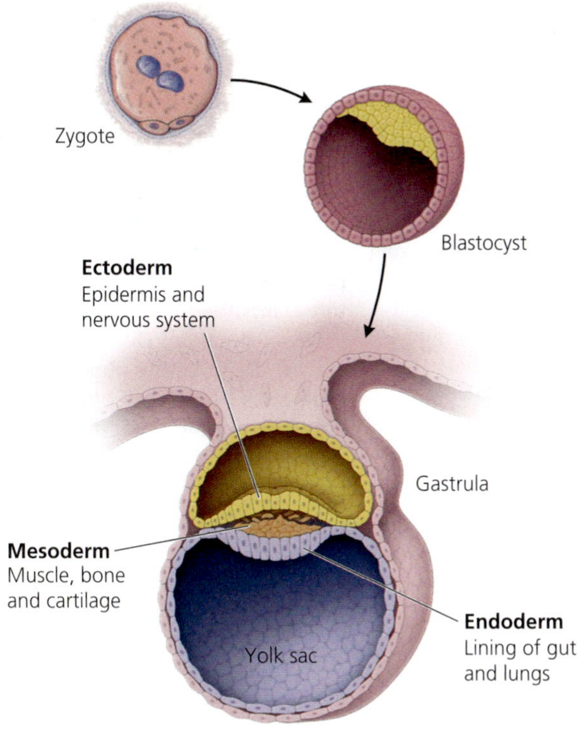

Figure 4.2 From zygote to gastrula. The zygote is a fertilized egg. The blastocyst (shown in cross section) is a hollow ball of cells. The gastrula (also sectioned) contains the three germ layers—endoderm, mesoderm, and ectoderm—suspended as a flat sheet between two fluid-filled spaces (blue and yellow). The ectoderm develops mainly into the epidermis (skin) and nervous system. The endoderm is fated to become the lining of the gastrointestinal tract and lungs. The mesoderm is sandwiched between the other two germ layers and spreads laterally to surround the yolk sac. It gives rise to muscle, connective tissue, and red blood cells.

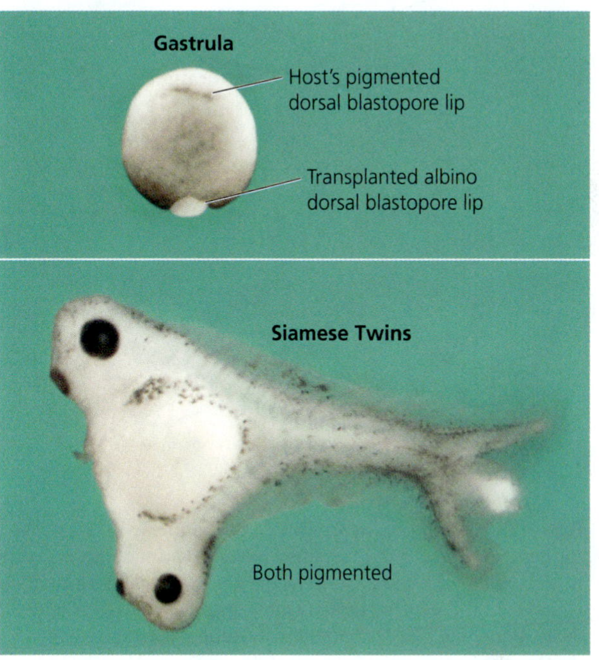

Figure 4.3 The Mangold–Spemann "organizer" experiment. Mangold and Spemann cut the dorsal blastopore lip out of an albino amphibian embryo and transplanted it into the ventral pole of a pigmented gastrula. When such a gastrula grows up, it produces Siamese twins (bottom). Because both twins are pigmented, the transplanted blastopore lip must have "induced" its host to form the second embryo. This experiment (originally performed on salamanders) led to the idea that the dorsal blastopore lip is an "organizer" that controls the fate of surrounding tissue. [From De Robertis and Kuroda, 2004]

RESEARCH METHODS

Box 4.1 *In Situ Hybridization*

The method of *in situ hybridization (ISH)* reveals when and where specific genes are expressed. It is a widely used technique that has played an exceptionally important role in developmental neurobiology. Although the method is complex, its basic principles are straightforward. The process begins with the construction of a "probe" for the gene of interest. One way to make such a probe is to intentionally damage the isolated gene's DNA in a few places with the enzyme Deoxyribonuclease I (DNase-I). This "nicked" DNA is then incubated with DNA Polymerase I (Pol-I), which replaces any removed nucleotides with new ones from the surrounding solution, based on the template provided by the other DNA strand (Figure b4.1). Crucially, some of the free nucleotides in the solution are conjugated to a molecule that researchers can later visualize. Often the labeled nucleotide is deoxyuridine triphosphate (dUTP) bound to biotin or digoxygenin. When the DNA incorporating these labeled nucleotides is separated into single strands (denatured), the labeled single strands become the desired "probe" that can bind to other samples of denatured DNA or to RNA.

To determine when and where in an embryo genes are transcribed, ISH probes must be brought into contact with the tissue's RNA. This is often accomplished by fixing the embryo (often by immersing it in paraformaldehyde) and slicing it into thin sections that are then mounted onto glass slides. Alternatively, a whole embryo can be treated with detergents and/or proteases to poke them full of holes through which ISH probes can pass. The latter method is called *whole mount in situ hybridization* because the tissue is placed (mounted) in the ISH reaction apparatus without having been sliced. Sectioned or not, the tissue containing the probes is processed through several solutions to wash away any probe that is not tightly bound to the intended target. Then the probe is visualized with antibodies that are bound to fluorescent molecules. These antibodies can be seen through microscopes with special light sources and optical filters; they end up glowing in the dark (typically in shades of red, yellow, or blue). Alternatively, ISH probes can be subjected to a series of chemical reactions that generate a colored product visible through a standard light microscope. Either way, you end up with a colorful image, where the presence of the color indicates the presence of the transcribed RNA.

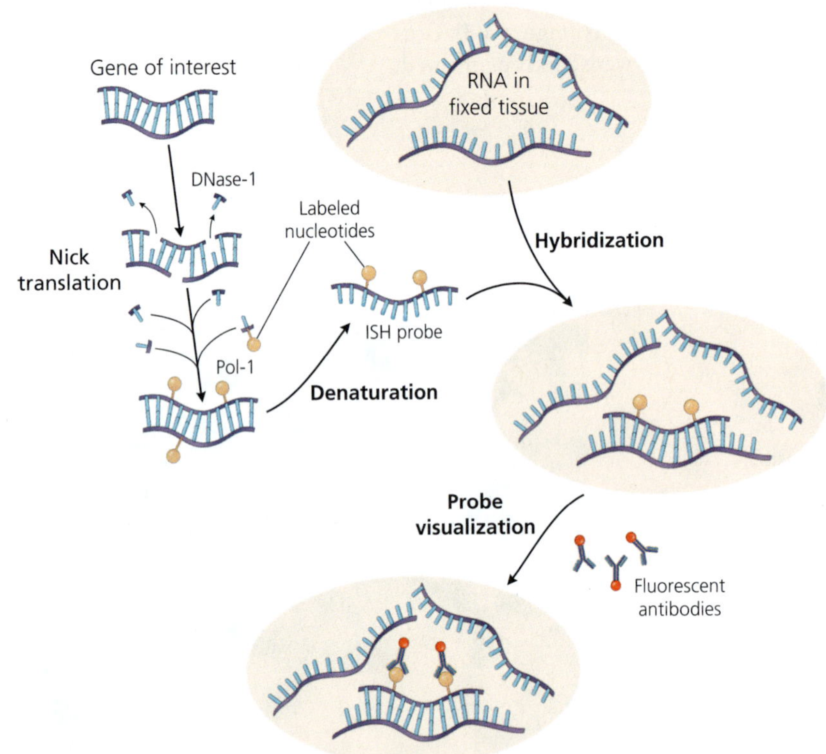

Figure b4.1 The in situ hybridization technique. To make an RNA probe, one takes DNA from the gene of interest and substitutes unlabeled nucleotides with nucleotides labeled with biotin or digoxygenin (yellow circles). The double stranded DNA is then denatured to generate a single-stranded probe, which is brought into contact with RNA in the tissue of interest (lightly shaded ovals). Under the proper conditions, the probe binds selectively to (hybridizes with) any RNA that has the complementary nucleotide sequence. After the hybridization, the labeled probe is exposed to antibodies that fluoresce at a specific wavelength (red circles linked to blue Y-shaped structures).

implying that they must have developed from cells that came from the "host" embryo. The transplanted, lightly pigmented cells developed into structures that were adjacent to the second nervous system. Based on these observations, Mangold and Spemann hypothesized that cells of the dorsal blastopore lip emit some sort of signal that induces neighboring cells to develop into a nervous system. This conclusion raised many questions about the cellular and molecular mechanisms underlying nervous system induction.

The search for the neural inducer (organizer) molecule hypothesized by Mangold and Spemann advanced significantly in the 1990s when experimenters used **in situ hybridization** (Box 4.1: In Situ Hybridization) and other molecular techniques to identify several genes that are expressed selectively in the dorsal blastopore lip. One of these genes is **chordin**. When injected into embryos, molecules of the chordin protein cause ectodermal cells to differentiate into neural tissue. Chordin does this by inhibiting **bone morphogenetic protein (BMP)**, which is secreted by cells in the ventral portion of the embryo (Figure 4.4). When BMP molecules diffuse away from their ventral source and bind to the ectoderm, they cause ectodermal cells to become skin. However, as BMP molecules approach the dorsal blastopore lip, they are inactivated by chordin and other BMP antagonists (other blockers of BMP function). What do ectodermal cells become if they do not receive the BMP signal? They become the nervous system.

The discovery that inhibiting BMP causes the ectoderm to develop into neural tissue was surprising because most people had expected that becoming the nervous system would be something special, not the ectoderm's "default" mode.

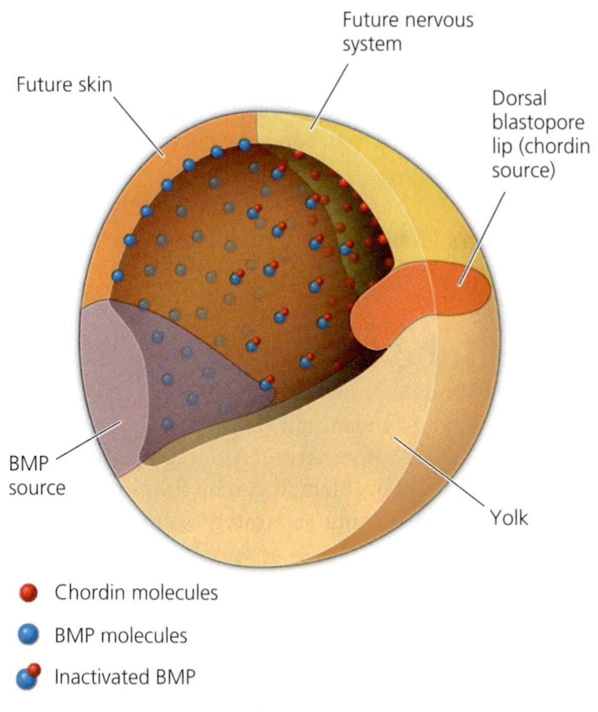

Figure 4.4 Neural induction. According to a widely accepted model of neural induction, the dorsal blastopore lip (red) secretes diffusible molecules of chordin (orange). When the chordin molecules encounter bone morphogenetic protein (BMP) molecules, which are secreted from a source at the other end of the embryo, they prevent the diffusible BMPs (blue) from binding to the ectoderm. Ectodermal cells that interact with BMP develop into skin. In contrast, ectodermal cells that are prevented by chordin from interacting with BMP develop into the nervous system (yellow). Although this model is simplified, it captures crucial aspects of neural induction, including the observation that forming the nervous system seems to be the ectoderm's default state. The model shown here is for a typical amphibian embryo.

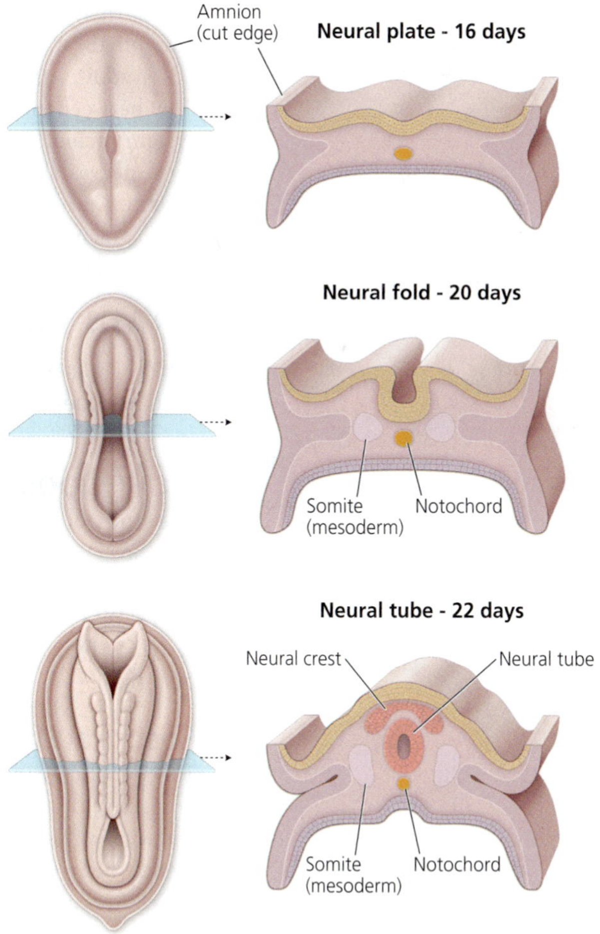

Figure 4.5 Formation of the neural tube. The neural plate (top) bends to form a neural groove (middle), which then closes up to form the neural tube (bottom). The images on the left depict human embryos at 16, 20, and 22 days of age. The images on the right illustrate sections through these embryos (at sites indicated by the light blue section planes). These images illustrate mainly the ectoderm, but notochord and somites are shown as well. The amnion and other extraembryonic membranes have been cut away. [After images from Marvin Sodicoff]

However, the evidence is pretty clear. Most convincingly, ectodermal cells grown individually in tissue culture, so that they receive no signals from other cells, adopt a neural fate. This finding shows that at the very root of nervous system development lies not some positive inductive signal, as Mangold and Spemann had thought, but an inhibitory signal that prevents the alternative outcome of becoming skin.

Forming the Neural Tube

After BMP and its inhibitors have divided the ectoderm into neural and skin-forming portions, the emerging nervous system is a flat sheet of cells referred to as the **neural plate** (Figure 4.5). Soon thereafter the left and right edges of this neural plate lift up, transforming the plate into a **neural groove**. As the future skin cells to the left and right of this groove proliferate, they push the groove's edges toward the midline until they meet. At this point, special cell adhesion molecules on the surface of the future skin cells cause the skin cells on both sides of the neural groove to stick to one another but not to other cells. Neural groove cells express different adhesion molecules, which make them stick to one another but not to the skin cells. The overall effect of this selective adhesion is that the neural groove becomes a **neural tube** that is separate from, and covered by, the skin. The neural tube soon closes at the front and at the rear. It then goes on to form the entire central nervous system, including both brain and spinal cord.

In addition, so-called **neural crest** cells migrate away from their original location right between the skin and the neural plate. They form much of the peripheral nervous system, including the neurons of the cranial and spinal nerves, the glia associated with those nerves, the ganglia of the sympathetic nervous system, and the enteric nervous system. The neural crest also gives rise to a number of non-neural structures, including skin pigment cells (melanocytes) and much of the skull. This, in brief, is the ectoderm's fate. As you can see, a large part of it is destined to become the nervous system.

BRAIN EXERCISE

Why is it fair to say that the ectoderm's "default" fate is to differentiate into neural tissue?

4.2 How Does the Neural Tube Get Subdivided?

The study of tissue patterning is the study of how tissue becomes heterogeneous. For the nervous system, it is the study of how the initially homogeneous neural tube becomes divided into a complex heterogeneous structure. Although patterning the neural tube is a complex three-dimensional problem, it can be simplified, at least initially, by considering rostrocaudal patterning separately from dorsoventral patterning.

Rostrocaudal Patterning

The spinal cord develops from the caudal portion of the neural tube, whereas the brain develops from its rostral end. The spinal cord is further subdivided into 31 segments; and the brain is subdivided into hindbrain, midbrain, and forebrain. Developmental neurobiologists have long wondered how these rostrocaudal divisions of the central nervous system come into existence. A full answer remains elusive, but most scientists agree that rostrocaudal neural tube patterning involves molecular signals that increase in concentration as you go from rostral to caudal along the neural tube. These molecules **caudalize** neural tube cells in a concentration-dependent manner. That is, they cause the affected cells to become caudal, rather than rostral, neural tissue. One likely candidate for such a caudalizing signal is **retinoic acid**. Interfering with retinoic acid signaling prevents caudal brain regions from forming normally. Conversely, artificial increases in retinoic acid concentration impair the differentiation of rostral brain regions (Figure 4.6). An important implication of

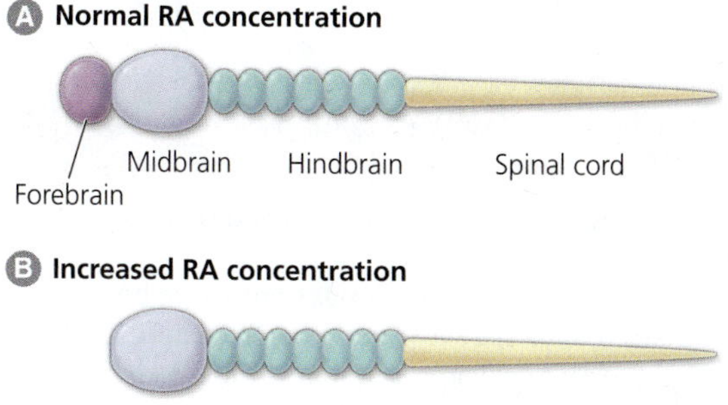

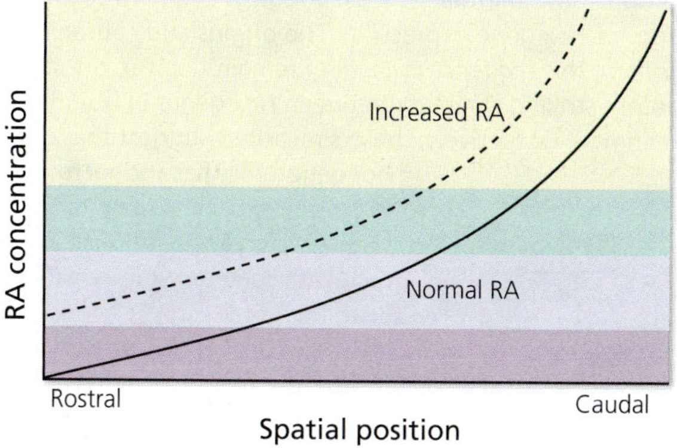

Figure 4.6 Retinoic acid affects brain patterning. Shown in (A) is a schematic dorsal view of an embryo that developed with the normal amount of retinoic acid (RA). Panel (B) shows an embryo that developed no forebrain because RA levels were artificially increased. Shown in (C) is a model of RA function, according to which RA concentration increases as you proceed caudally. Where RA levels are high (yellow), the tissue becomes spinal cord. Where they are very low (purple), the tissue becomes forebrain. At intermediate levels (blue, green) the neural tube adopts hindbrain and midbrain fates, respectively. This model predicts that an artificial increase in RA concentration (dashed curve) should yield the sort of embryo depicted in B. [After Maden, 2002]

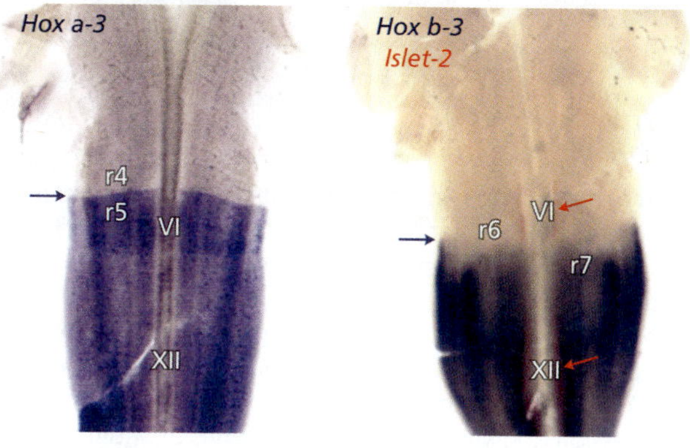

Figure 4.7 ***Hox*** **gene expression domains.** Shown here are dorsal views of the hindbrain from two chick embryos, stained with whole-mount in situ hybridization to reveal the expression patterns of *Hox* a-3 (left, purple), *Hox* b-3 (right, purple stain), and *Islet*-2 (right, red stain and arrows). The purple arrows indicate the rostral expression boundaries of the two *Hox* genes. The labels r4–r7 refer to numbered hindbrain segments (rhombomeres); and the Roman numerals VI and XII indicate motor neurons associated with the 6th and 12th cranial nerves, respectively. [From Guidato et al., 2003]

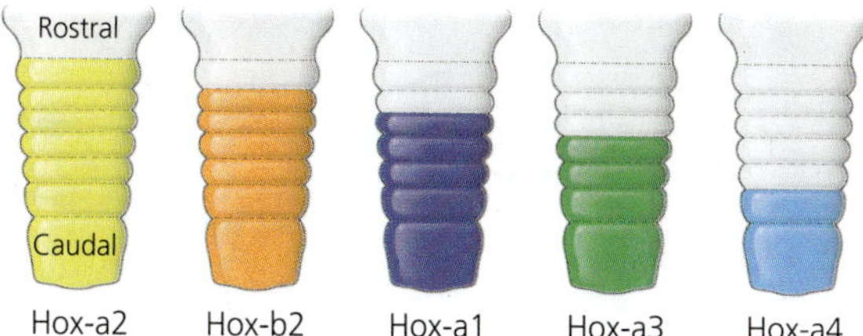

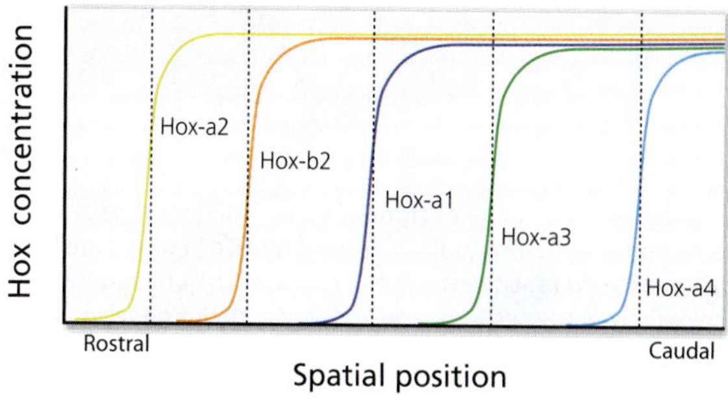

Figure 4.8 Nested expression of ***Hox*** **genes in the hindbrain.** Shown in (A) are dorsal views of vertebrate hindbrains in which individual segments are separated by dashed lines. The colors indicate that each *Hox* gene has a different rostral expression boundary. Panel (B) summarizes this finding in a single graph to show how the number of different *Hox* genes expressed within a hindbrain segment increases as you get to more caudal segments.

these findings is that the "default mode" of ectodermal cells is not just to become neural tissue, but to become rostral neural tissue (brain).

The *Hox* Gene Family

Rostrocaudal neural tube patterning also involves ***Hox* genes**. *Hox* genes are a family of transcription factors that is highly conserved across species (Box 4.2: *Hox* Genes in Evolution). Individual members of this family are expressed in various combinations at different rostrocaudal levels of the nervous system (Figure 4.7). Caudal hindbrain segments express many different *Hox* genes simultaneously, but more rostral segments express progressively fewer *Hox* genes (Figure 4.8). This nested expression pattern suggests that different *Hox* genes are activated at different concentrations of a caudalizing signal, such as retinoic acid. Specifically, *Hox* genes with the most extensive expression domain are thought to be activated by very low levels of the caudalizing

EVOLUTION IN ACTION

Box 4.2 Hox *Genes in Evolution*

Hox genes were first identified in fruit flies. Ed Lewis and others discovered that mutations in fruit fly *Hox* genes cause duplications of some body segments and the deletion of others. For example, some *Hox* gene mutations cause flies to develop legs where their antennae should be. Subsequent research revealed that fruit flies have multiple *Hox* genes that all share a conserved DNA-binding sequence (called the homeobox). Moreover, the expression domains of different *Hox* genes have different rostral boundaries during fruit fly development, which means that caudal body parts coexpress a larger number of *Hox* genes than rostral body parts. Additional data reveal that the specific combination of *Hox* genes that a body part expresses during development determines the identity of that body region later in life. It determines, for example, whether a body part will grow antennae or legs.

By now, homologs of the fruit fly *Hox* genes have been discovered in many animal species (Figure b4.2). As described in the main text, the mammalian *Hox* genes exhibit a rostrocaudally nested pattern of expression that is similar to that seen in fruit flies. Even more remarkable is that, in both fruit flies and rodents, the sequence in which the *Hox* genes are arranged on the chromosome mirrors (at least roughly) the rostrocaudal sequence of *Hox* gene expression. The origins and causal implications of this *sequence colinearity* remain unclear, but it is yet another striking similarity between *Hox* genes in fruit flies and mammals. Collectively, these similarities suggest that the *Hox* gene family has ancient roots, likely going back at least to the last common ancestor of all animals with bilateral symmetry (the Bilateria).

The discovery of conserved *Hox* gene expression patterns in both the body and the central nervous system has revitalized an old debate about whether the central nervous systems of insects and mammals are homologous. Are they descended from a central nervous system that evolved very early in animal phylogeny and was retained continuously since then, or did the insect and vertebrate central nervous systems evolve independently of one another? The conserved *Hox* gene expression pattern is consistent with the former hypothesis. However, it is also possible that an ancient set of *Hox* genes became involved in nervous system patterning long after the genes themselves evolved, and that it did so independently in multiple lineages. The debate is ongoing and unlikely to be resolved soon.

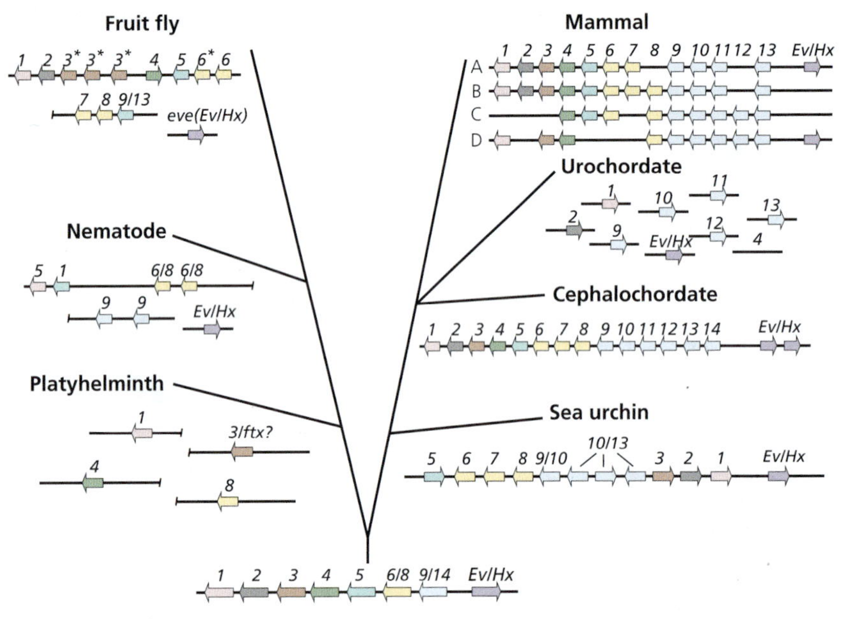

Figure b4.2 ***Hox* gene evolution.** Individual *Hox* genes are depicted as thick, colored arrows. The direction of each arrow indicates the direction in which the gene is transcribed. Most *Hox* genes are numbered in sequence (*Hox*-1, *Hox*-2, etc.). Homologous genes are given matching names and colors across species. The ancestral cluster of *Hox* genes is thought to have duplicated twice in the lineage leading to mammals, producing four *Hox* clusters (labeled A to D in the Mammal branch). Some genes in each cluster were lost. Curiously the presumed ancestral *Hox* cluster was disbanded in several taxonomic lineages, including flatworms (platyhelminths), round worms (nematodes), and tunicates (urochordates). The longest and most orderly *Hox* cluster is found in the cephalochordate *Amphioxus*. The fruit fly's *Hox* cluster is broken into two parts, and some of its genes have reversed their orientation relative to the ancestral condition. [From Lemons and McGinnis, 2006]

signal, whereas *Hox* genes expressed only in the caudal hindbrain require high levels of caudalizing signal. This hypothesis is supported by the finding that artificial increases in retinoic acid levels cause rostral hindbrain segments to express *Hox* gene combinations that are normally found only in more caudal segments. These "caudalized" segments also express non-*Hox* genes that are typically expressed only in caudal hindbrain segments of older animals, suggesting that the altered *Hox* gene expression pattern permanently alters cell fates.

The details of how *Hox* genes control cell fate remain a subject of debate. According to the **combinatorial *Hox* code model**, a cell's developmental fate is determined by the combination of *Hox* genes expressed within the cell. Alternatively, the **posterior prevalence model** states that some *Hox* genes are more important than others. Specifically, the *Hox* genes expressed in the more posterior hindbrain segments are thought to dominate the *Hox* genes with more anterior expression domains. According to this second model, a cell's fate depends on the most dominant *Hox* gene expressed within the cell. Experiments in which specific *Hox* genes were "knocked out" in transgenic mice tend to support the posterior prevalence model, but the matter is not settled yet. In any case, the data show that *Hox* genes are essential for rostrocaudal patterning of the vertebrate hindbrain and, to some extent, the spinal cord. As you will see shortly, rostrocaudal patterning in the midbrain and forebrain involves a different set of transcription factors.

BRAIN EXERCISE

How is the spatial expression of *Hox* genes linked to levels of retinoic acid in the embryo (as illustrated in Figure 4.6)?

Dorsoventral Patterning

The central nervous system is also patterned along the dorsoventral axis. In the spinal cord, for example, neurons that send their axons to muscles lie ventrally; whereas neurons that receive input from sensory nerves are located in the dorsal horn of the spinal cord. In between these motor and sensory neurons lie interneurons that connect to other neurons in a manner that varies with their dorsoventral position. As Thomas Jessel and his collaborators discovered, dorsoventral patterning in the spinal cord involves **sonic hedgehog (SHH)**. This protein (named after a once-popular video game character) is secreted by cells at the neural tube's ventral midline, which is called the **floor plate**. Because SHH diffuses freely away from its ventral source, it forms a ventral-to-dorsal concentration gradient within the developing spinal cord (Figure 4.9). This finding suggests that SHH is a **ventralizing** signal for the embryonic spinal cord. Indeed, injection of antibodies that block SHH also blocks the formation of motor neurons that lie ventrally in normal spinal cords. Conversely, the addition of SHH to embryonic spinal cords growing in tissue culture induces the cells to become motor neurons.

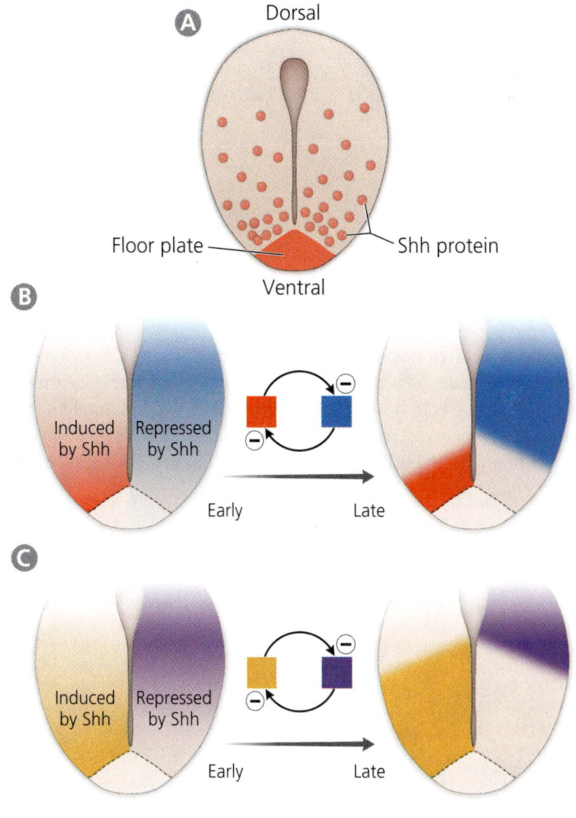

Figure 4.9 Dorsoventral patterning in the spinal cord. The diagram in (A) depicts an embryonic spinal cord in cross section. Floor plate cells secrete sonic hedgehog (SHH), which diffuses away (red circles), setting up an SHH concentration gradient. Panel (B) shows the expression of one gene that is induced by high levels of SHH (red), and one that is repressed by SHH (blue; for clarity, each protein is shown on just one side of the spinal cord although they are expressed on both). Initially these genes are expressed with spatial overlap (left), but their expression domains gradually become nonoverlapping (right side) because these two genes repress each other. Panel (C) is analogous to panel B but illustrates the expression of two other genes that develop a sharp boundary at a more dorsal location. [After Briscoe et al. 2000]

Boundary Formation

The ventralizing action of SHH involves two classes of transcription factors. Members of the first class are *induced* (turned on) by SHH. Importantly, different transcription factors within this class are induced at different concentrations of SHH, which causes them to have different dorsal expression boundaries. This is analogous to how different *Hox* genes are induced at different concentrations of retinoic acid. In contrast, the second class of transcription factors are *repressed* (turned off) by SHH rather than induced. Moreover, different transcription factors in this group are repressed at different concentrations of SHH, which causes them to have different ventral expression boundaries. In general, the expression domains of the SHH-induced genes and the SHH-repressed genes are complementary so that as one SHH-induced gene fades out, a SHH-repressed gene's expression fades in (Figure 4.9). This means that the interactions

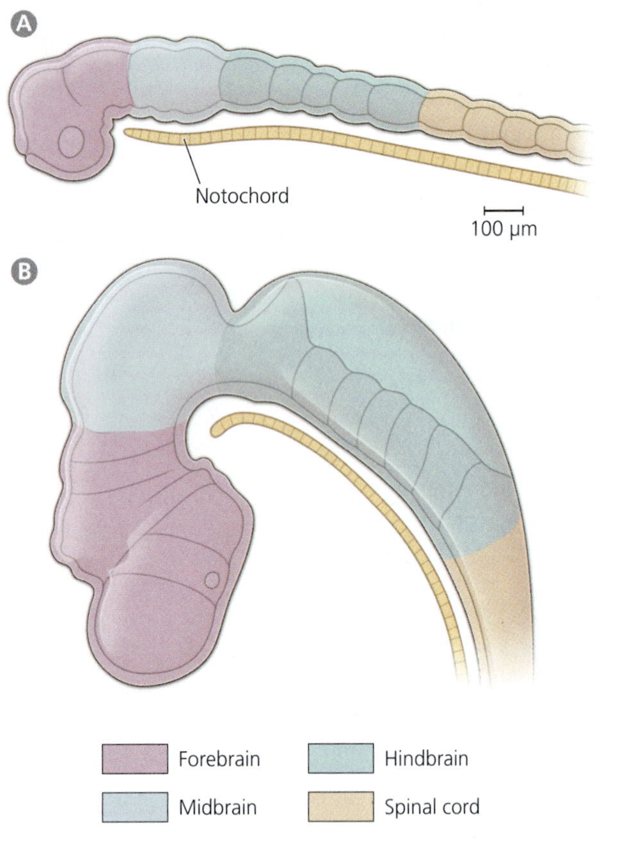

Figure 4.10 Development of the midbrain flexure. Shown here is the central nervous system at two stages of embryonic development. The tissue has been cut down the middle and you are looking at it from a medial perspective. You can see how the brain bends (flexes) around the anterior end of the notochord, with the midbrain at the apex of the flexure.

between SHH and its downstream transcription factors cause the spinal cord to become divided into genetically distinct dorsoventral domains.

Because the SHH-induced genes generally inhibit the expression of their SHH-repressed complements, and vice versa, individual cells tend to express one transcription factor or the other but not both. This mutual repression helps young cells make clear decisions about their fate. Think of it this way: if cells were to express transcription factors that prompt them to become motor neurons simultaneously with transcription factors that push them toward an interneuron fate, then the young cells would be "confused" about what to become. The mutual repression between the different sets of transcription factors clears up the confusion and forces the young neurons to choose just one of the alternative fates. It also leads to a gradual sharpening of the boundaries between the spinal cord's dorsoventral compartments.

Midbrain and Forebrain Patterning

Because the neural tube's rostral end bends dramatically during development (Figure 4.10), it is not immediately obvious what is rostral, caudal, dorsal, or ventral in the brain. However, if you mentally straighten the neural tube, then you can see that the midbrain lies rostral to the hindbrain and that the developing telencephalon lies at the rostral tip of the developing brain (together with the preoptic area and hypothalamus). You can also see that SHH is expressed along the ventral edge of the entire brain, just as it is expressed in the floor of the hindbrain and spinal cord (Figure 4.11). These data

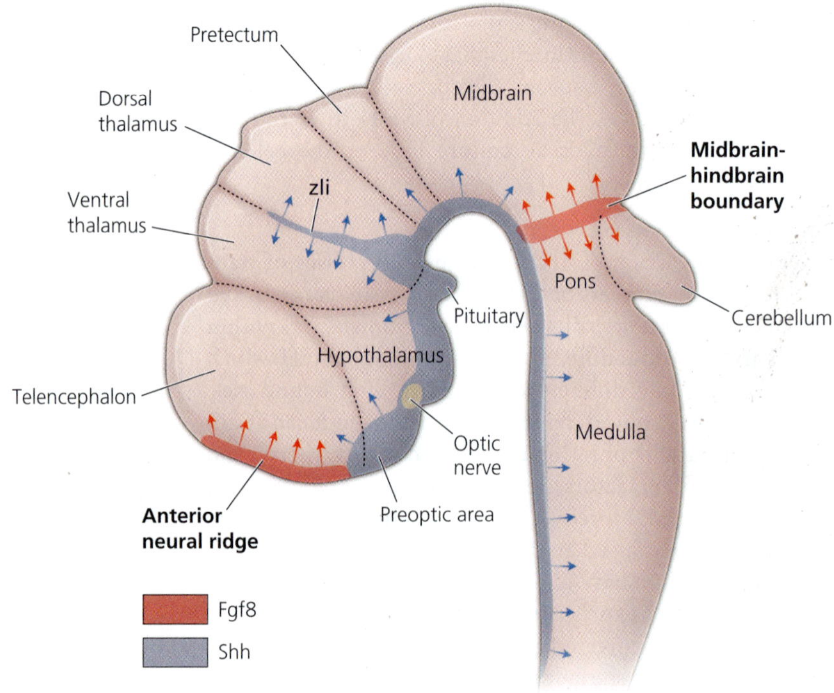

Figure 4.11 Brain patterning. In this schematic diagram of an embryonic brain, sonic hedgehog (SHH) expression is shown in blue. The spur of SHH expression extending between the dorsal and ventral thalamus is called the *zona limitans intrathalamica* (zli). Shown in red are two additional signaling centers, namely, the anterior neural ridge and the midbrain–hindbrain boundary. Both secrete (arrows) fibroblast growth factor 8 (Fgf8), which helps to pattern the developing brain.

suggest that SHH acts as a ventralizing signal throughout the neural tube, including both the brain and spinal cord. Indeed, blocking SHH prevents the formation of ventral tissues in all regions of the central nervous system.

The process of mentally straightening the neural tube also reveals that BMP proteins (yes, the same molecules that induce skin at earlier stages of development) are expressed dorsally throughout most of the neural tube, including the forebrain. Given this dorsal expression pattern, it is reasonable to hypothesize that BMP is a dorsalizing signal that counteracts the ventralizing activity of SHH. Experiments that boost or block BMP signaling are consistent with this hypothesis. Overall, these data indicate that dorsoventral patterning is fundamentally similar in the brain and spinal cord.

Rostrocaudal Patterning of the Brain

Rostrocaudal patterning is more complex in the midbrain and forebrain than in the hindbrain and spinal cord. This greater complexity arises because the brain contains multiple rostralizing and caudalizing factors secreted by multiple signaling centers. One of these signaling centers is the **midbrain–hindbrain boundary** (Figure 4.11). This ring-shaped region secretes several diffusible transcription factors, including **fibroblast growth factor 8 (FGF8)**. These signals are crucial for proper midbrain and cerebellum development.

A second important signaling center develops between the dorsal and ventral thalamus. This intrathalamic signaling center secretes SHH (Figure 4.11), just as the ventral forebrain does. However, SHH in this context does not act as a ventralizing signal but instead helps to pattern the thalamus. This is a recurring theme in developmental biology: molecules often play different roles in different locations and at different stages of development.

The third signaling center crucial for brain patterning is the **anterior neural ridge** (Figure 4.11). It secretes FGF8, just as the midbrain–hindbrain boundary does. However, FGF8 from the anterior neural ridge diffuses into the rostral forebrain, where it is responsible for rostrocaudal patterning. This was demonstrated in experiments that boosted FGF8 expression in the anterior forebrain, which caused the somatosensory cortex to develop in an abnormally caudal location (Figure 4.12). That is what you would expect if high levels of FGF8 promote the development of rostral cortical areas. In contrast, reducing FGF8 expression in the rostral forebrain causes somatosensory cortex to develop in an abnormally rostral location. Again, this is what you would expect if FGF8 rostralizes the cerebral cortex. In a third experiment, cells at the caudal edge of the neocortex were engineered to express FGF8, which they do not normally do. This manipulation caused the somatosensory cortex

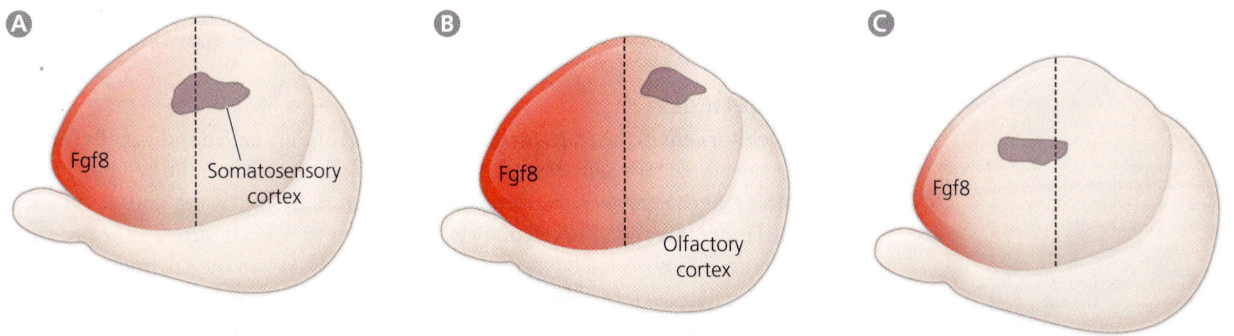

Figure 4.12 Rostrocaudal patterning of neocortex by Fgf8. Shown here are flattened mouse cortices (rostral to the left). Red shading indicates Fgf8 concentration, the primary somatosensory cortex is shown in purple, and the dashed line provides a rostrocaudal reference. Shown in (A) is the normal condition. Panel (B) shows that overexpression of Fgf8 in the rostral neocortex shifts the somatosensory cortex caudally. Conversely, reducing Fgf8 expression (C) shifts the somatosensory cortex rostrally. [After Assimacopoulos et al., 2012]

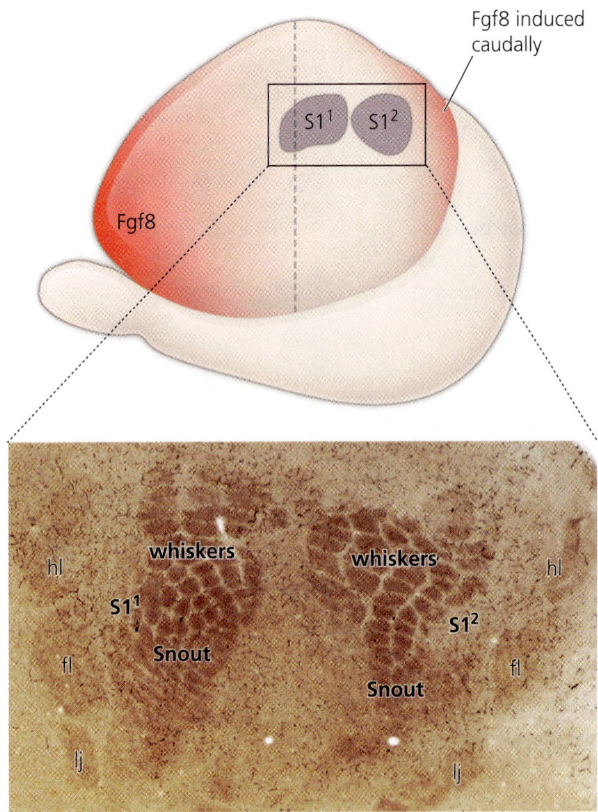

Figure 4.13 Duplicating a cortical map. In some mouse embryos, Fgf8 was artificially expressed caudally as well as rostrally (compare to Figure 4.12 A). These embryos developed a second primary somatosensory cortex (S1²) just caudal to the normal primary somatosensory cortex (S1¹). Shown here is a section through the two somatosensory areas, stained to reveal dark patches that represent distinct body parts (hl = hindlimb, fl = forelimb, lj = lower jaw). Note that the two somatosensory cortices are mirror images of one another. [Image from Assimacopoulos et al., 2012]

to appear in duplicate, once in its normal rostral location and then again more caudally (Figure 4.13).

In summary, brain and spinal cord patterning depends on transcription factors that are secreted from various signaling centers and then diffuse away so that each transcription factor forms a distinct concentration gradient. Secondary transcription factors are turned on (or off) at various specific concentrations of those primary, gradient-forming factors. The secondary transcription factors then interact with one another to sharpen the boundaries between their expression domains. They also modulate the expression of yet more genes. Some of these downstream genes themselves code for diffusible transcription factors that set up their own concentration gradients. Thus, as the neural tube develops and grows, an ever-increasing number of crisscrossing gradients cause an increasing number of other genes to be expressed in spatially restricted domains. Thus, the neural tube gradually becomes divided into ever more distinct territories.

BRAIN EXERCISE

How does rostrocaudal patterning in the midbrain and forebrain differ from rostrocaudal patterning in the hindbrain and spinal cord?

4.3 Where Do Neurons Come From?

At the early stages of embryonic development, the walls of the neural tube consist almost exclusively of rapidly dividing, undifferentiated cells that are called **progenitors**. As these cells divide, they make more progenitors just like themselves. However, the progenitors divide more often in some regions than in others, partly because of spatial variations in the transcription factors we just discussed. Combined with fluid pressure from the ventricle inside the tube, the differential proliferation of progenitors causes the neural tube to bulge in several places. Despite this bulging, the walls of the neural tube retain a uniform thickness (Figure 4.14). That is, the walls do not become thinner as a balloon does when you blow into it; nor do they thicken as the progenitors proliferate. Instead, the walls of the young neural tube expand only in surface area. They expand tangentially rather than radially.

Neurogenesis

The progenitor cells in the walls of the young neural tube are called **radial cells** (or radial glia) because they have a process that extends radially away from the cell body toward the brain's external surface. Each progenitor's cell body moves up and down within this radial process. When the cell body is near the external brain surface, the progenitor duplicates its chromosomal DNA (the cell is in S-phase). When the cell body sinks down to the ventricular surface, where brain tissue borders the ventricle, the cell divides by mitosis (Figure 4.14). After each cell division, the cell bodies of the two daughter cells move radially away from the ventricular surface and then repeat the cycle. At some point, however, the daughter cells do not reenter the cell cycle. Instead, one or both of them stop dividing and begin to differentiate into neurons (Figure 4.15 A).

This process of ceasing to divide and then becoming a neuron is called **neurogenesis**. The time when a daughter cell leaves the proliferative cell cycle is called its

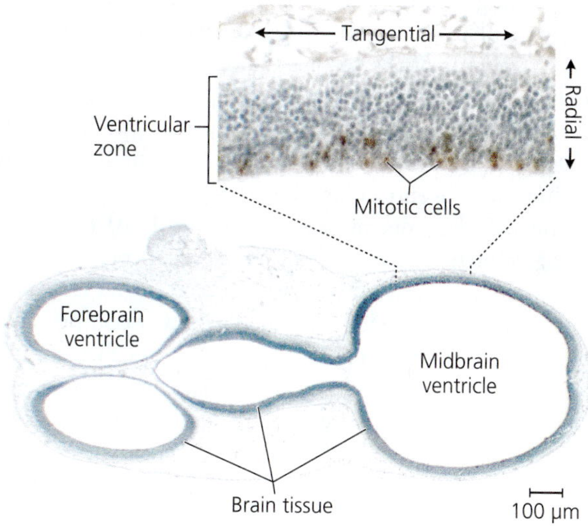

Figure 4.14 The walls of early embryonic brains are thin. Shown here is a horizontal section through the head of a young chicken embryo (rostral is to the left) stained so that brain cells are blue. As you can see, the brain at this stage consists of a relatively thin layer of cells, called the ventricular zone. This ventricular zone surrounds large fluid-filled ventricles. The illustrated section was also stained with an antibody that stains cells in mitosis brown. The insert at the top right shows that the mitotic nuclei are usually close to the ventricular surface. The black arrows indicate the radial and tangential dimensions in the ventricular zone.

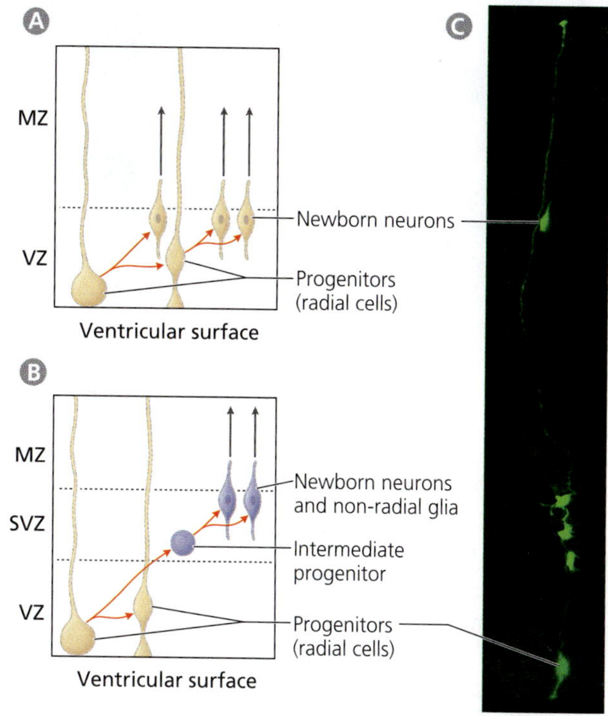

Figure 4.15 Neurogenesis and migration in the neocortex. Panel (A) shows two progenitors with their cell bodies in the ventricular zone (VZ) and a long, thin process extending radially through the mantle zone (MZ). When a progenitor divides (red arrows), two daughter cells result. One daughter migrates out of the VZ and becomes a newly born neuron. The other daughter remains a progenitor, which then divides into two newborn neurons. As shown in (B), some cells remain proliferative after they leave the VZ and accumulate in the subventricular zone (SVZ). When these "intermediate progenitors" divide, they generate neurons as well as astrocytes and oligodendrocytes. Panel (C) shows young neurons migrating radially away from the ventricular surface (bottom) along the slender processes of a radial cell. The newborn cells in this image are progeny of a single progenitor that was infected with a recombinant retrovirus. [A and B after Dehay and Kennedy, 2007; C from Noctor et al., 2001]

birth date. As you will learn in Chapter 5, a few neurons and many glial cells are born after the organism itself is born, and some neurogenesis persists well into adulthood. For now, let us focus on embryonic development, when the vast majority of neurons is produced. At the beginning of embryonic neurogenesis, only one of the daughter cells exits the cell cycle; the other daughter cell remains a progenitor. Toward the end of embryonic neurogenesis, each progenitor tends to produce two newborn cells. As you would predict, this leads to a rapid decrease in the number of progenitors.

Many of the transcription factors that partition the neural tube into its various subdivisions also affect neurogenesis timing. However, most transcription factors affect neurogenesis indirectly by acting on a few key molecules. One of these crucial molecules is **beta-catenin** (β-catenin; yes, the same molecule mentioned in Figure 4.1). Within proliferating brain cells, β-catenin tends to delay neurogenesis onset. This was demonstrated in experiments with transgenic mice in which β-catenin levels were artificially elevated. The progenitors in these manipulated brains divided more often than normal and therefore caused brains to grow much larger than normal. Another molecule that plays a major role in controlling neurogenesis is **delta-1**. At high levels of delta-1 expression, cells leave the cell cycle and begin neurogenesis. In a sense, this molecule has an effect opposite to that of β-catenin.

Radial Neuronal Migration

As shown in Figure 4.16, newborn cells tend to migrate radially away from their place of birth in the so-called **ventricular zone** and form a separate layer called the **mantle zone** that extends from the ventricular zone to the external brain surface. At first the mantle zone is thin, but it soon thickens enormously.

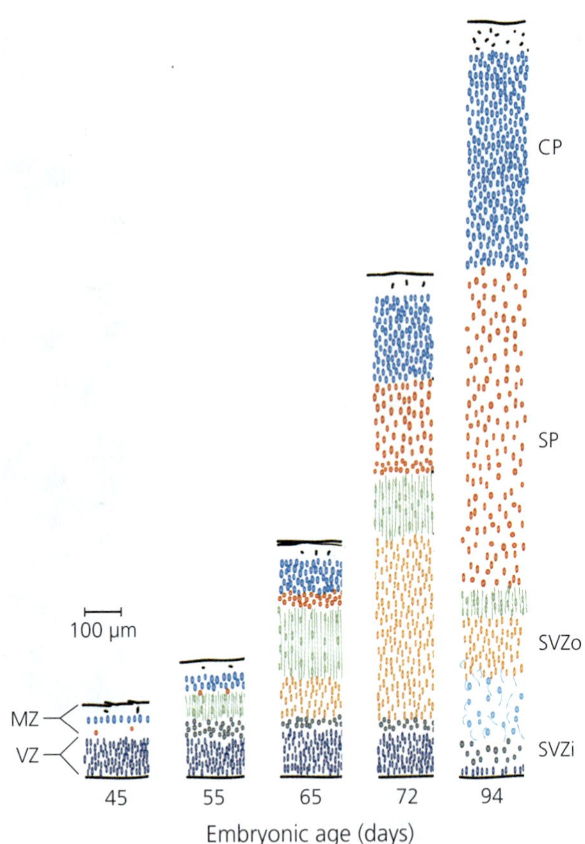

Figure 4.16 Brain tissue thickens as development proceeds. Shown here are drawings of radially oriented slivers through the neocortex of a monkey between 45 and 94 days of embryogenesis. Initially the neocortex consists almost entirely of proliferating cells that form the ventricular zone (VZ). As development proceeds, cells migrate out of the VZ to form more superficial layers, collectively referred to as the mantle zone (MZ). As the MZ thickens, it subdivides into multiple layers, including the internal and outer subventricular zones (SVZi and SVZo, respectively), the subplate (SP), and the cortical plate (CP). [From Dehay and Kennedy, 2007]

Although the mantle zone contains mainly postproliferative cells, there is one big exception to this rule. In several regions of the telencephalon, some cells leave the ventricular zone while they are still progenitors. These intermediate progenitors accumulate just outside of the ventricular in what is called the **subventricular zone**. Within the subventricular zone, the intermediate progenitors divide at least one more time (Figure 4.15 B). Eventually, they give birth to young neurons and glial cells. An interesting aspect of the subventricular zone is that it is much thicker in the neocortex of primates than of non-primates. Its enlargement probably explains, at least in part, why the neocortex is significantly larger in primates than in most other mammals, even after you account for differences in overall brain size.

After newborn cells leave the ventricular and subventricular zones, they tend to keep migrating radially. During this migration, the young cells maintain contact with the long processes of the radial cells, which extend all the way to the brain surface even as the mantle zone thickens (Figure 4.15 C). In fact, newborn cells tend to use the radial processes as a sort of guide rail or climbing rope.

Neurogenesis Timing and Cell Fate

In the neocortex, migrating young neurons burrow their way through the older cells above them. They do not leave their "radial monorail" until they approach the external brain surface, where they encounter a molecular "please disembark" signal called **reelin**. Because each generation of neocortical cells leaves the radial monorail after it has passed through most of the older cells, the adult neocortex is structured so that the first-born cells make up the deep cortical layers, whereas the younger cells occupy progressively more superficial layers (Figure 4.17 A). This orderly arrangement is unusual, however. In most regions of the nervous system, young cells do not consistently migrate

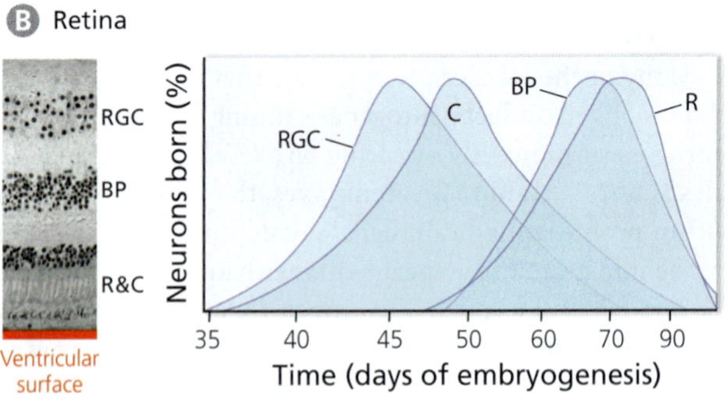

Figure 4.17 Neurogenesis timing in neocortex and retina. Illustrated in (A) are a stained section through the neocortex of an adult monkey (left; Roman numerals indicate distinct layers) and a graph showing that neurons in the upper cortical layers are born at progressively later times in development. This orderly relationship is not seen in the retina (B). Instead, retinal ganglion cells (RGC) are born before bipolar cells (BP) and end up further from the ventricular surface. Cone photoreceptors (C) tend to be born before rod photoreceptors (R), but both cell types are found in the same layer of the adult retina. Although the correlation between spatial location and birth date is less orderly in the retina than in the neocortex, time of birth in both regions plays at least some role in determining the neuron's fate. [A courtesy of Pasko Rakic, MD, PhD; B after Rapaport et al., 1996, and image from Anger et al., 2004]

past older cells. Instead, they tend to get off the radial monorail just before they reach the older cells. In the retina, another region in which neurogenesis has been studied extensively, some young neurons migrate past older cells, but others do not (Figure 4.17 B). Thus, different regions exhibit different patterns of radial neuronal migration.

The time at which a cell is born affects not only its position in the adult brain but also the type of cell it will become (its fate). Progenitors generally produce neurons before they produce glial cells, and they tend to make oligodendrocytes (the glia that insulate axons with myelin; see Chapter 2) before they make astrocytes (see Chapter 5). Moreover, progenitors give birth to different types of neurons at different times (Figure 4.17). Even in tissue culture, young progenitors typically produce different types of cells than old progenitors. This suggests that the progenitors themselves are changing over time. Consistent with this hypothesis, the mix of transcription factors expressed by various progenitors varies with age.

Studying Cell Type Specification with Transplantation Experiments

Important insights into cell type specification were obtained by Sue McConnell and her collaborators. They transplanted young neocortical progenitors, which normally generate cells of the deep cortical layers (see Fig 4.17 A), into the neocortex of older embryos and found that the offspring of the transplanted cells ended up in the upper cortical layers. The transplanted young progenitors acted like old progenitors. In contrast, when McConnell transplanted old progenitors into younger embryos, the transplanted progenitors acted just as old progenitors normally do: they generated neurons in the upper cortical layers. These experiments revealed that young progenitors are competent to generate a wide variety of different cell types; with age, their competence narrows. This probably means that some of the molecular changes occurring inside of the progenitors are irreversible. A major unresolved question is what causes these fateful changes in gene expression. Some evidence suggests that young neocortical neurons send signals back to the progenitors, instructing them to start making a different type of cell. This would explain why young progenitors transplanted into the older neocortex act like the old progenitors.

Recapping our discussion thus far, a cell's fate depends primarily on when and where it was born. In a way, young cells are like young people. Their behavior depends, at least in part, on when and where they were born. The analogy is also apt because, like humans, many neurons stay close to where they were born. They usually migrate along radial processes, but most of these migrations are relatively short. Of course, there are exceptions to this rule. For example, many young neurons in rodents migrate from the ventral telencephalon into the olfactory bulb and into the neocortex, where they become GABAergic interneurons. The human neocortex seems to harbor fewer of these long-distance "immigrants" than the rodent neocortex does, but human brains do exhibit some long **tangential migrations**. Specifically, in humans some young cells migrate tangentially out of the telencephalon into the dorsal thalamus. This seems to be a distinctly human trait. In any case, such long tangential migrations are rare. Most young brain cells migrate radially and do not wander far.

BRAIN EXERCISE

Given the data shown in Figure 4.17 B, which types of retinal cells must be migrating past which other types of retinal cells during development?

4.4 How Do Axons Find Their Targets?

So far we have discussed how the nervous systems develop a variety of distinct cell types and how those cells arrive at their proper locations. By and large, these cells do not yet have complete axons. How, then, do the connections between neurons form? This is a difficult problem because each neuron's axon must grow toward, and ultimately synapse with, a very specific subset of neurons. It must find a few needles in an

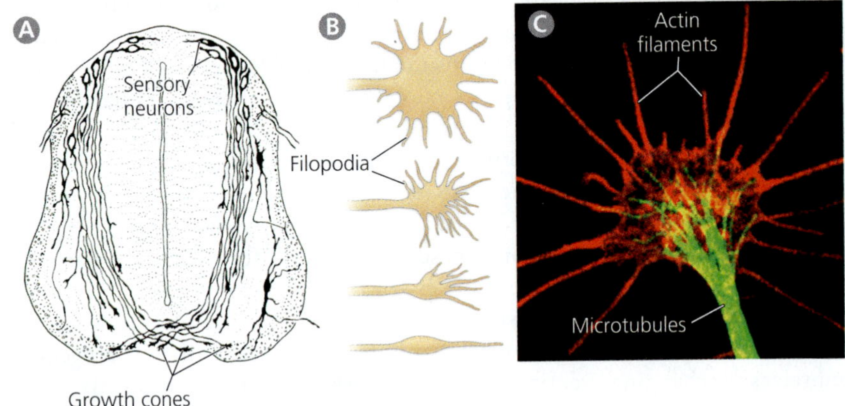

Figure 4.18 Growth cones. Panel (A) is a drawing by Ramón y Cajal showing axons of dorsal sensory neurons in the spinal cord sending their axons ventrally. Each growing axon is tipped with a growth cone, and some of the axons have crossed the ventral midline. Panel (B) shows that growth cones can have a variety of shapes, depending mainly on the kind of space they are growing through. Panel (C) shows actin filaments in red and microtubules in green. [A from Cajal, 1890; B after Bray, 1982; C courtesy of Paul Letourneau, PhD]

enormous proverbial haystack! To understand how neurons accomplish this task, you must first learn how axons grow.

Axonal Growth Cones

Axons typically grow out of a neuron's cell body shortly before the neuron ends its migration. Growing axons are filled with long and slender microtubules and tipped with **growth cones** (Figure 4.18). Ramón y Cajal originally described these growth cones as "a sort of club or battering ram, endowed with an exquisite chemical sensitivity, with rapid amoeboid movements, and with impulsive force by which it is able to proceed forward and overcome obstacles met in the way, forcing cellular interstices, until it arrives at its destination" (Ramon y Cajal, 1917, p. 599). This description is remarkably accurate, considering that Cajal based it entirely on what he saw in fixed and stained tissue sections. Modern scientists can watch time-lapse movies of growth cones wandering around a tissue culture dish or in a living brain slice, but their observations have generally confirmed Cajal's account.

Extending and Retracting Filopodia

Modern science also revealed that growth cones regularly extend and then retract slender protrusions called **filopodia** (Figure 4.18). These filopodia are filled with **actin filaments** that are linked to microtubules near the growth cone's center. The actin filaments are also anchored to the cell membrane near the filopodial tips. Individual actin subunits are regularly added to the end of the actin filament that extends into the filopodia. This tends to push the filopodia forward, lengthening them. However, even as the actin filaments extend, they are pulled back toward the growth cone's center by myosin molecules (you will learn about actin and myosin in Chapter 8). When the rate at which the actin filaments are pulled backward exceeds the rate at which actin is added at the tip, the filopodia retract. This sounds straightforward, but you might wonder: what happens to the basal ends of the actin filaments as they are pulled backward? Do they pile up inside the growth cone? No, they do not. Instead, actin subunits are regularly removed from the base of the actin filaments.

Filopodia can make a growth cone move, but only if the filopodia tips adhere to something in the external environment. Think of it like this: if a filopodium is retracted while its tip is stuck on some other cell or extracellular material, then the tip cannot move back toward the growth cone. Instead, the growth cone must move toward the filopodium tip. It is like pulling on a rope that is attached to the ceiling; when you pull on the rope, you end up pulling yourself off the floor. The force exerted by filopodial retraction can be visualized when axons are grown in tissue culture. Sometimes, when a filopodium from a cultured growth cone contacts the process of another neuron, it pulls on the other process (Figure 4.19). If you imagine this other process as an immovable extracellular object, then you can visualize the growth cone being pulled forward toward the filopodial tip. As the growth cone moves forward, the axon behind it elongates.

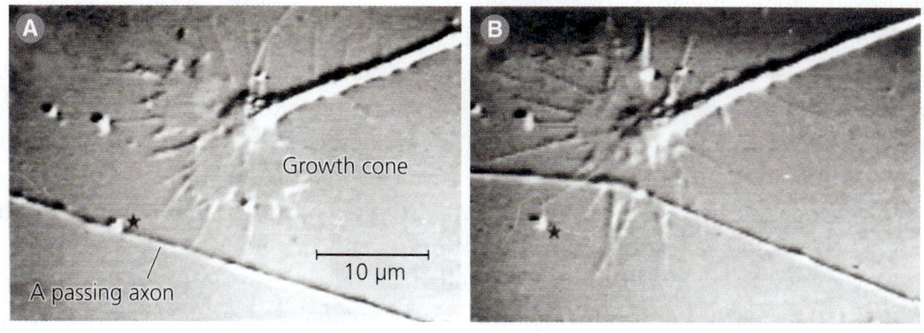

Figure 4.19 Growth cones can generate traction. Shown here are two frames from a time-lapse video of axons growing in culture. Some of the filopodia that have adhered to a passing axon in (A) have retracted by frame (B) and are now pulling the axon toward the growth cone's center. This demonstrates that retracting filopodia may exert considerable force. The asterisks in A and B indicate a speck of dirt that remains in a constant position between the two video frames. [From Heidemann et al., 1990]

Changing an Axon's Direction of Growth

Once you understand how axons elongate, you can grasp how they turn. Other things being equal, a growing axon will turn toward whatever locations offer the best "footholds" for its filopodia. This is only true, however, if the filopodia extend equally in all directions. If they extend mainly toward the left, for example, then the growth cone will probably turn left, even if the environment is slightly stickier on the right side. Thus, a growth cone's direction of movement is a product of (a) how well its filopodia stick to their surroundings and (b) how likely the growth cone is to extend its filopodia in a particular direction. This explains why growth cone guidance factors generally affect either the strength with which filopodia stick to their substrates or the likelihood of filopodia extending in a specific direction, or both.

Growth Cone Guidance

Ramón y Cajal proposed that growth cones "become oriented by chemical stimulation, and move toward the secreted products of certain cells" (Cajal, 1894, p. 146). This idea was appealing because many microorganisms exhibit such **chemotaxis** (movement toward a chemical). However, Cajal had no direct evidence for his hypothesis.

Evidence for Growth Cone Chemotaxis

The first convincing evidence for growth cone chemotaxis was obtained in the 1990s when researchers filled glass micropipettes with hypothesized guidance factors and then positioned the pipettes next to the growth cones of axons growing in tissue culture (Figure 4.20). As the molecules diffused out of the pipettes (or were pushed out slowly), a concentration gradient formed in the culture dish. Within about 30 minutes,

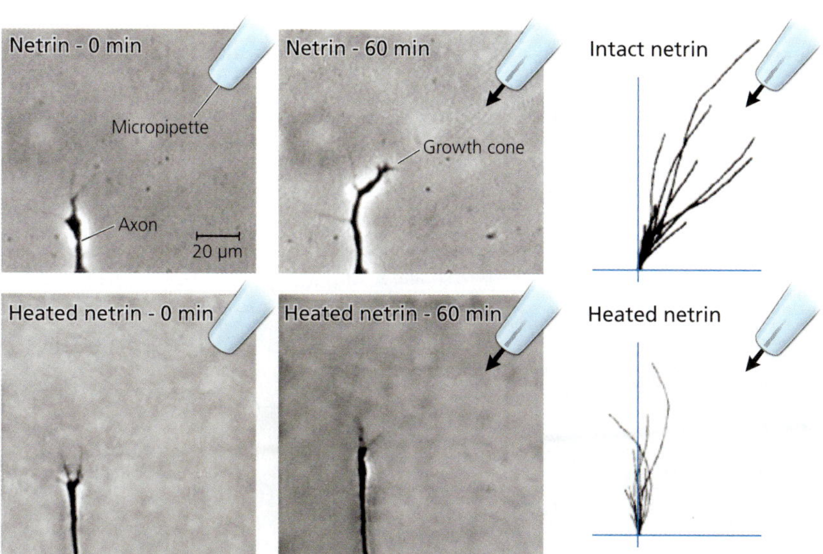

Figure 4.20 Axon chemotaxis. Shown in the top row are photographs of an axon growing in culture. After 60 minutes the axon has grown toward the tip of a glass micropipette filled with the molecule netrin. What you cannot see in these photographs is that the experimenters are slowly pushing the netrin out of the pipette, setting up a netrin concentration gradient within the culture dish. The graph on the right shows growth trajectories for 10 different axons. All of them grew toward the netrin-filled pipette. The bottom row illustrates the results from a control experiment in which the netrin was inactivated by heating it to 75°C. [From Ming et al. 1997]

the growth cones moved up this concentration gradient toward the pipette tip. This finding strongly supported Cajal's chemotaxis hypothesis.

Since those early studies, scientists have learned that growth cones can detect surprisingly small changes in molecule concentrations. Apparently, growth cones do this by moving slowly (<1 μm/minute), sampling their chemical environment repeatedly, and then integrating this information over time. The process is analogous to how you might locate a delicious meal by wandering around, sniffing repeatedly, and moving up the concentration gradient of odors given off by hot pizza. Scientists have also discovered that growth cones are repelled, rather than attracted, by some molecules. Proteins of the **semaphorin** family are particularly effective growth cone repellents. When growth cones contact such repellent molecules, the filopodia tend to collapse on the side with the highest concentration of the offending molecule. This differential collapse of filopodia forces the growth cone to turn away from the source of the repulsive molecules.

One problem for Cajal's chemotaxis hypothesis is that gradients of diffusible molecules tend to be steep, which means that growth cones can be guided by diffusion-based concentration gradients only over distances of less than a few hundred microns. Axons solve this problem in two ways. First, they use diffusion gradients mainly at early stages of development, when embryos are so small that a single concentration gradient spans a significant fraction of the embryo's nervous system. Second, growing axons use different guidance cues for different segments of their journey, much as driving directions are usually broken down into multiple segments (go North on the main road until you get to the stop sign, then turn right and go up the hill, etc.). This is a clever trick. However, journey segmentation creates a new problem: if an axon finds the first target in its journey by growing up a concentration gradient centered on the first target, how can it leave the first target and move on to the second one? It would seem that the growing axon would have to be reprogrammed when it reaches the first target. It must stop being attracted to the molecules secreted by the first target and start responding to new cues in novel ways.

Axon Reprogramming

Strong evidence for axon reprogramming has been obtained from research on commissural neurons in the spinal cord, which convey information from one side of the spinal cord to the other. The axons of these commissural neurons course ventrally toward the floor plate, cross the midline, and then ascend again on the other side of the spinal cord (Figure 4.21). In 1988, Marc Tessier-Lavigne and his collaborators discovered that the axons of commissural neurons grow ventrally because they are attracted to a diffusible substance secreted by floor plate cells. Six years later, this substance was identified as **netrin** (from the Sanskrit word for "he who guides"). Why do the commissural axons exit from the floor plate after crossing the midline? Because when commissural axons get to the floor plate, they start making a membrane receptor called **robo** (short for *roundabout*, which describes the abnormal growth of the commissural axons when this receptor is knocked out). The robo receptors make

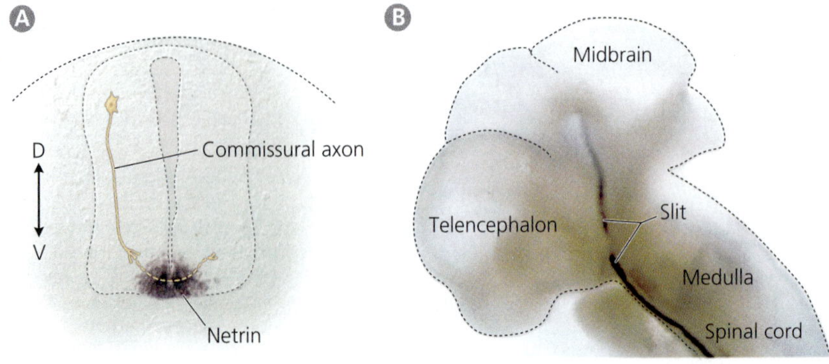

Figure 4.21 ***Netrin* and *slit* are expressed in the floor plate.** Shown in (A) is a section through an embryonic spinal cord of a chick, stained to reveal *netrin* expression in the floor plate. Shown in yellow is a commissural (midline crossing) neuron whose axon grows ventrally toward the *netrin* source, crosses to the other side, and then grows away from the floor plate. Shown in (B) is the expression pattern of *slit*, another molecule that is produced in the floor plate; the embryo shown here is from a mouse. [A from Kennedy et al., 1994; B from Holmes et al., 1998]

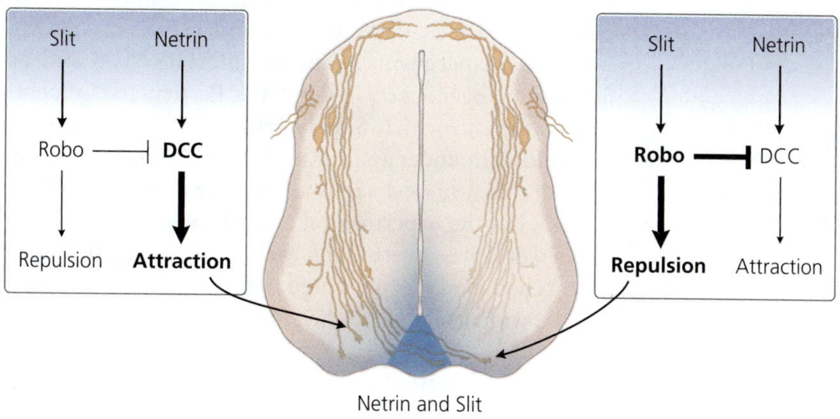

Figure 4.22 Axon reprogramming. The axons of commissural neurons initially grow toward the floor plate because they are attracted by netrin (left). This positive chemotaxis involves netrin activating its receptor Deleted in Colorectal Cancer (DCC) on the commissural axons. Once the axons have crossed the floor plate, they express high levels of robo, which is activated by slit. Robo represses the DCC receptor, making the growth cone less attracted to netrin. In addition, robo causes the growth cones to be repulsed by slit (right). As a result of all these molecular interactions, the axons are first attracted to the floor plate, and then repulsed by it. Arrows in the two flowcharts indicate positive interactions; T-shaped lines represent inhibition. Bold type indicates a high concentration of the specified molecule.

the growth cone less sensitive to netrin and, simultaneously, cause it to be repulsed by a molecule called **slit**, which is also secreted by floor plate cells. Because of these changes, growth cones that were initially attracted by netrin become indifferent to netrin and repulsed by slit, which they had hitherto ignored (Figure 4.22). Thus, the axons get reprogrammed from one step of their journey to the next.

Substrate-bound Axon Guidance Molecules

In the years since netrin was identified, neuroscientists have discovered many additional diffusible **axon guidance molecules**. They have also discovered guidance factors that do not diffuse freely but are, instead, attached to cell membranes or extracellular material. Many of these substrate-bound factors attract axons by offering the growth cones ideal levels of "stickiness" for growth; others promote filopodial extension. Some substrate-bound factors encourage growth cones to crawl along the surface of other axons that express the growth-promoting molecules on their surface. This is a good idea because it allows late-born neurons to find their targets by simply following the axons of neurons that were born earlier and have already blazed a trail. Because the trailblazing **pioneer axons** are growing out when the embryo is very small, they can navigate by the range-limited diffusion gradients. Later neurons cannot use the diffusion gradients, but they can follow the pioneers to the appropriate target. This helps explain why many axons travel in large bundles called nerves or axon tracts. Of course, follower axons must at some point leave the other axons and find their own, unique targets. This process of **defasciculation** (*fasciculus* means "bundle") also requires growth cone reprogramming, but has not yet been studied in detail.

The Retinotectal System

Extensive work on the mechanisms of axon guidance has been conducted on non-mammalian vertebrates, specifically on the projections from the retina to the midbrain's **optic tectum**, which is homologous to the mammalian superior colliculus and involved in orienting the eyes and head toward visual targets (see Chapter 11). The **retinotectal system** is well suited to research on axon guidance because both the retina and the optic tectum are relatively large and flat (sheet-like) in structure. Moreover, the retinotectal projections exhibit precise topography, with nasal retina projecting to the caudal tectum, temporal retina projecting to the rostral tectum,

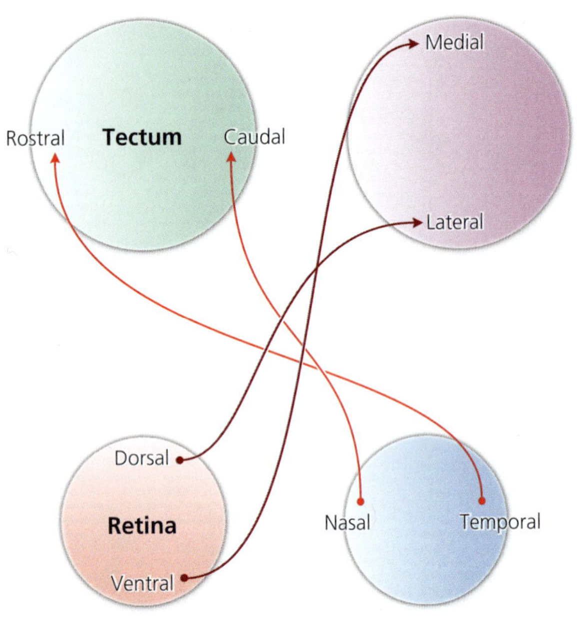

Figure 4.23 The retinotectal system. The projections from the retina to the optic tectum (the superior colliculus in mammals) are crossed and topographic. Nasal retina projects to caudal tectum, temporal retina to rostral tectum, dorsal retina to lateral tectum, and ventral retina to medial tectum.

ventral retina projecting to medial tectum, and dorsal retina projecting to the lateral tectum (Figure 4.23).

Classic experiments on the retinotectal system were performed by Roger Sperry in the 1940s. Particularly influential was a study in which Sperry crushed the optic nerve of an adult salamander and surgically rotated the affected eye by 180°. The crushed axons eventually grew back into the tectum, but the experimental animals now acted as if their visual world was inverted (on the side viewed by the rotated eye). When food was presented in front of the animal, it oriented toward its tail; and when food was presented above the salamander, the animal looked down. These behavioral observations led Sperry to conclude that the regenerating axons must have grown back to their original targets in the optic tectum but then provided incorrect, inverted information about the location of visual stimuli.

Based on this inference, Sperry developed his **chemoaffinity hypothesis**. Sperry proposed that each retinal axon expresses a distinct set of molecular markers, and the tectal neurons express matching (complementary) markers. As the retinal axons grow into the tectum (both in embryonic development and during regeneration after injury, as in Sperry's experiment), they seek out and selectively terminate on tectal neurons with the matching marker. In essence, each growing axon has a "chemical affinity" for a specific target neuron. Because it is improbable that every neuron expresses a unique chemical tag, Sperry proposed that the hypothesized markers are expressed in two or more intersecting gradients across the retina as well as the tectum (color gradients in Figure 4.23). Although Sperry's hypothesis was prescient, it took many years before neuroscientists discovered which molecules were expressed in the hypothesized gradients and how those molecules control axon guidance.

Crucial progress in the search for Sperry's hypothesized molecular tags came from in vitro experiments that used a so-called **stripe assay**. The first step in these experiments was to deposit membranes from cells in the rostral and caudal parts of the tectum onto a piece of filter paper so that membranes from rostral and caudal tectum formed alternating stripes (Figure 4.24). The experimenters then took a horizontal strip of embryonic retina, placed it right next to the alternating stripes in a tissue culture dish, and watched the retinal axons grow onto the tectal membrane stripes. Retinal neurons that normally sit close to the nose (in the nasal retina) grew indiscriminately onto both sorts of stripes. However, neurons that normally sit toward the side of the head (in the temporal retina) preferred growing onto stripes of membranes from the rostral tectum. When the tectal membranes were treated to remove or destroy most proteins, the temporal retinal axons lost their preference for rostral tectum. This showed that cells in the caudal tectum contain a membrane-bound protein that repels temporal, but not nasal, retinal axons.

The Role of Ephrins in Axon Guidance

The repulsive factor in the caudal tectum turned out to be a molecule of the **ephrin family** that is expressed in a smooth gradient across the tectum, with the highest ephrin levels in the most caudal tectum. It has also become apparent that axons from the nasal retina express low levels of the matching **ephrin receptor**, whereas axons from more temporal regions of the retina express progressively more ephrin receptor. This difference in ephrin receptor expression makes axons from the nasal retina less sensitive to ephrin's repulsive effects. Add to this the finding that ephrin expression is highest in the caudal tectum, and you come up with the following simple model: as retinal axons grow from the tectum's rostral edge toward the back, they simply grow until the ephrin-induced repulsion becomes too much for them to bear. Because the

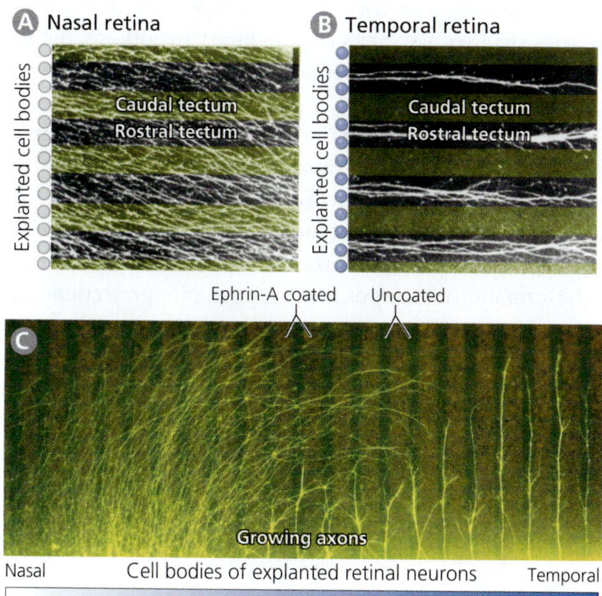

Figure 4.24 Ephrin guides retinal axons. Cultured neurons from nasal or temporal retina were prodded to extend their axons on top of cell membranes from either rostral or caudal tectum that had been deposited in alternating stripes on a piece of filter paper. The nasal retinal axons grew indiscriminately on both sets of membranes, but the temporal axons grew only on membranes from rostral tectum. A membrane-bound protein called ephrin helps to guide the retinal axons. Panel (C) shows what happens when nasal and temporal axons are prodded to grow on stripes that are coated with ephrin-A (faintly reddish lanes) or not (darker lanes). As you can see, the temporal axons avoid the ephrin molecules. [A and B from Walter et al., 1987; C from Monschau et al., 1997]

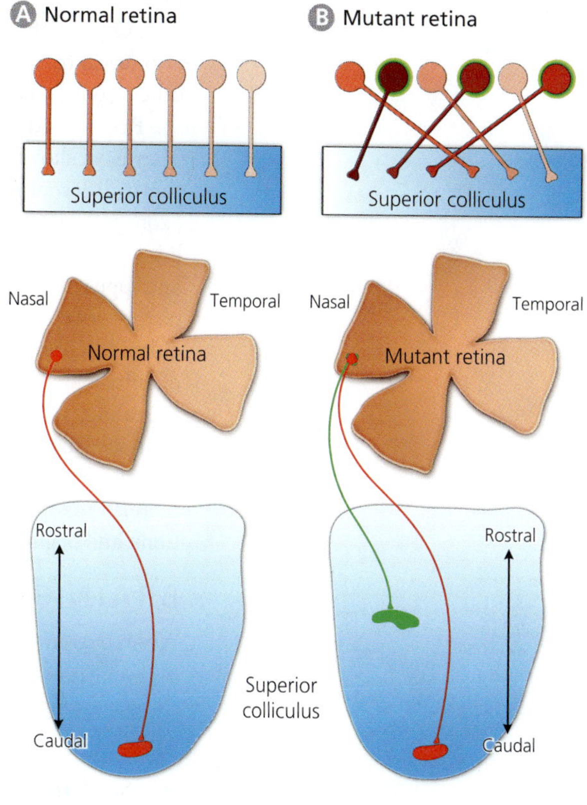

Figure 4.25 Ephrin gradients control map formation. In normal mice (A), nasal retina projects only to caudal tectum. This is not the case in transgenic mice (B) engineered so that a subset of retinal neurons (indicated by green rings) express more than their normal amount of ephrin receptor (concentration indicated by intensity of reddish-brown shading). The axons of these transgenic neurons target collicular cells that express very low levels of ephrin (light blue shading). Therefore, the retinal cells in the transgenic mice form two separate collicular maps. One map is formed by the normal retinal neurons, the other one by the transgenic cells. Consistent with this hypothesis, the nasal retina in the transgenic animals projects to two separate locations in the superior colliculus. The illustrated retina looks like a cross because it was cut before it was flattened. [After Brown et al., 2000]

nasal axons are less sensitive to ephrin than the temporal axons, they are more likely to end up growing all the way into the caudal tectum. Because both ephrin expression in the tectum and ephrin receptor expression in the retina follow smooth gradients, this model predicts the smooth rostrocaudal topography that the retinotectal projection normally exhibits.

To test this **ephrin gradient hypothesis**, researchers created transgenic mice that overexpress ephrin receptors in a randomly scattered subset of retinal neurons (Figure 4.25). These mice contain two superimposed gradients of retinal ephrin receptor expression: one normal gradient and one comprising the mutant cells. Importantly, when the scientists examined the retinal projections in these transgenic mice, they found two separate topographic sets of projections to the superior colliculus (the homolog of the optic tectum). Axons from the normal retinal neurons form a map in the caudal half of the superior colliculus, and axons from the mutant neurons form a second map in the rostral superior colliculus. This finding confirms that retinal axons expressing high levels of ephrin receptor abhor the caudal superior colliculus. The data also showed that a retinal axon's target is not determined by the absolute amount of ephrin receptor it expresses. If it were, then some of the mutant retinal neurons should have found no targets in the superior colliculus at all (because absolute ephrin receptor levels are higher in some of the mutant retinal cells than they are in any normal retinal neurons). Instead, a retinal axon's target in the superior colliculus is determined by the amount of ephrin receptor the axon expresses *relative to* the amount of ephrin receptor present in the other retinal axons. Essentially, the retinal axons are all jockeying for target space in the superior colliculus.

Eventually each retinal axon ends up with as good a target as it can get, given the competition. Overall, these data are consistent with Sperry's chemoaffinity hypothesis, but they also reveal that an axon's target is more flexible—more dependent on competition with other axons—than Sperry had originally envisioned.

Recent studies have filled in further details about the formation of topographic maps. They have shown, for example, that ephrins may be attractive as well as repulsive, and that a different member of the ephrin family is involved in establishing how retinal axons map onto the mediolateral (rather than rostrocaudal) axis of the superior colliculus. In fact, graded patterns of ephrin expression are seen all across the nervous system and appear to be crucial for the formation of many different topographic projections. This plethora of ephrin gradients is interesting because it suggests that substrate-bound molecules like ephrin are a powerful means of developing orderly axonal projections. The idea of diffusible signals such as netrin or semaphorin guiding axons at a distance is intuitively appealing, but such signals are not required. As long as the substrate-bound molecules are expressed in gradients, growing axons can use them to find their targets. The axons simply grow up or down such gradients until they find their "sweet spot" in the gradient. Once there, the axons slow their forward growth, branch repeatedly to form terminal arborizations, and begin to make some synapses.

BRAIN EXERCISE

When experimenters transplant embryonic neurons into adult nervous systems in the hope of repairing brain damage, the axons of the transplanted neurons usually fail to reach their normal targets. Why is that?

4.5 How Do Synapses Form?

Once an axon has found its general target, it needs to form synapses at specific locations within the target area. Because most presynaptic terminals are directly opposed to postsynaptic sites, the mechanisms of **synapse formation** must somehow assure that presynaptic terminals are located directly across from postsynaptic specializations. In addition, glutamatergic terminals must be matched up with postsynaptic sites that contain glutamate receptors; GABAergic terminals must be matched up with postsynaptic GABA receptors; and so forth for each kind of neurotransmitter. Finally, some axons terminate specifically on the cell bodies of their target neurons, whereas others prefer to end on their dendrites. How does such specificity arise?

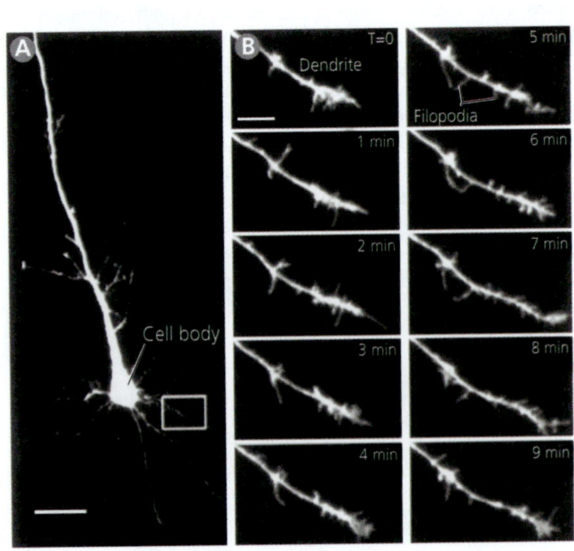

Figure 4.26 Time-lapse video reveals dendritic filopodia movements. The large image on the left (A) depicts a neuron from the neocortex of a 2-day-old mouse injected with a fluorescent dye so that its processes could be visualized in the living animal. The area in the small white rectangle is shown at higher magnification in (B), which shows 10 frames from a time-lapse video taken at one-minute intervals. Comparing successive frames, you can see dendritic filopodia extending from the growing dendrite and then retracting. Scale bars equal 25 μm in A and 5 μm in B. [From Portera-Cailliau et al., 2003]

Perhaps postsynaptic specializations exist before the presynaptic terminals arrive, and the presynaptic axons simply sniff them out by following some molecular trail. This idea is appealing but false. Instead, synapse formation begins with pre- and postsynaptic cells extending filopodial "feelers" that contact each other. Once contact has been established, the pre- and postsynaptic processes communicate. If the exchanged signals are positive, then pre- and postsynaptic specializations begin to form. Over time, these molecular interactions strengthen the bond between the pre- and postsynaptic cells, and a mature synapse is formed. Thus, postsynaptic specializations do not exist before the axon terminals arrive. Instead, they are laid down as synapses emerge. Now that you have the basic idea, let us get into details.

Filopodial Interactions

When young dendrites first grow out of the cell body, they are tipped with growth cones very similar to those of growing axons. Soon after the initial outgrowth, however, young dendrites distinguish themselves from young axons by growing thicker. Time-lapse movies further reveal that dendrites send out long and slender filopodia all along their length, not just at the tip (Figure 4.26). These **dendritic filopodia** either retract soon after forming, or they develop into new dendritic branches. Some such behind-the-tip filopodia are also seen in axons, particularly as they approach their target and form terminal arborizations, but they are more frequent in dendrites. Dendritic filopodia probably sample their chemical environment just as axonal filopodia do. Good evidence for this hypothesis comes from the observation that dendrites from different cells of the same cell type tend to exhibit little or no overlap (Figure 4.27). Most likely, this **dendritic tiling** arises because growing dendrites sense nearby dendrites of the same type and are repelled by them. This tendency to be repulsed by dendrites of the same type also ensures that the dendritic branches of a single neuron are spread far and wide rather than clumped. This spreading of the dendritic tree, which is very apparent in cerebellar Purkinje cells (Figure 4.28), maximizes a neuron's ability to gather inputs from many different axons.

Wild-type fly | *Furry* mutant

Figure 4.27 Dendritic tiling. Shown on the left are two sensory neurons in a normal, wild-type (wt) fruit fly. As you can see, the two dendrites don't overlap, a phenomenon referred to as dendritic tiling. Shown on the right are two neurons of the same type in a mutant fly with a defective furry (*Fry*) gene. In such mutants, dendritic tiling is disrupted. Even within a single dendritic tree, branches cross abnormally often (arrowheads). [After Emoto et al., 2004]

Axon-sensing Dendrites?

If dendrites can sense other dendrites, can they also sense axons? This question remains unanswered. However, some axons do release small amounts of neurotransmitter before proper synaptic contacts have formed, and some dendritic filopodia contain neurotransmitter receptors. Moreover, researchers have found that blocking transmitter release during development reduces the number of synapses eventually formed. These observations suggest, but hardly prove, that some sort of communication between axons and dendrites helps their respective filopodia find one another in the vast wilderness of brain space. Additional communication probably occurs after a dendritic filopodium has contacted an axonal filopodium. This would explain why filopodial contacts between GABAergic axons and GABA receptor-containing dendrites are more stable than contacts between axons and dendrites that are not so nicely matched. Apparently, axons and dendrites, having made contact, check each other out to see whether the axon's transmitter is a good match for the postsynaptic cell. If the two match, then synapse formation proceeds.

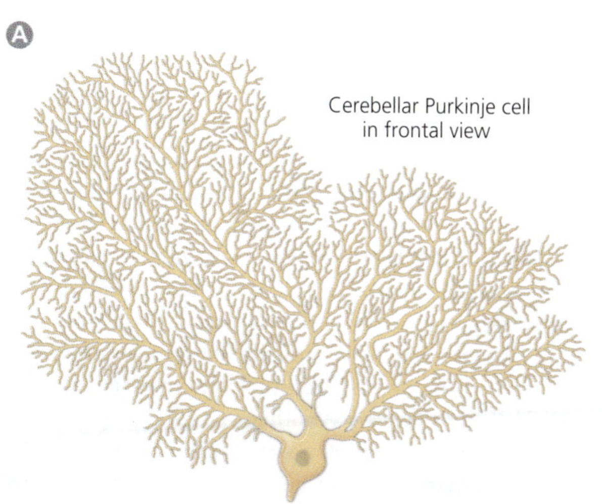

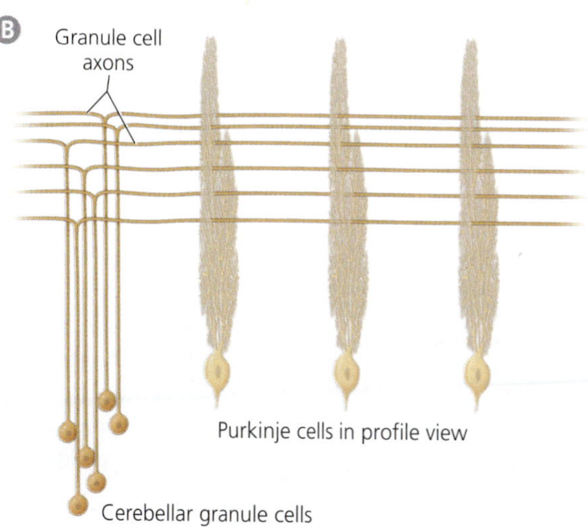

Figure 4.28 Purkinje cells' planar dendritic trees. Shown in (A) is a cerebellar Purkinje cell from an adult human. Its spectacular dendritic tree is largely confined to the plane of the paper (or screen) in front of you. Therefore, Purkinje cells look much skinnier in "profile," as shown in (B). A major source of input to the Purkinje cells are cerebellar granule cells, whose axons ascend to the level of the Purkinje cells, branch in a T-like fashion, and then course at right angles to the plane of the Purkinje cell dendrites. This orthogonal orientation, together with the large span of the Purkinje cell dendrites, allows each Purkinje cell to gather input from more than 100,000 granule cell axons. [A after Burns, 1911]

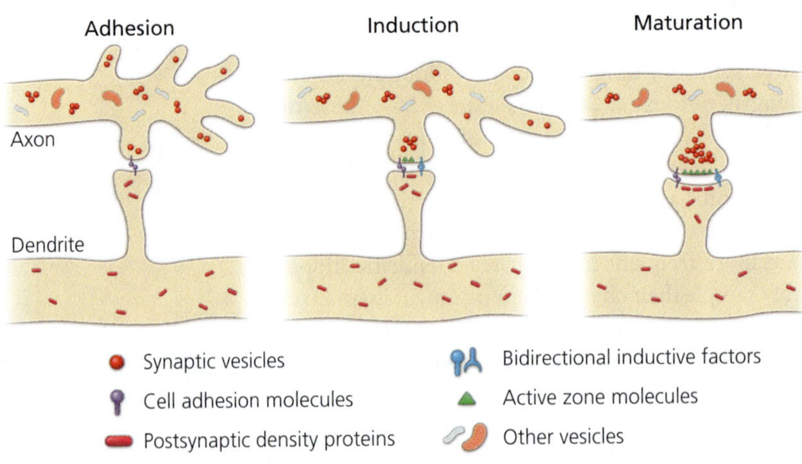

Figure 4.29 Bidirectional interactions during synapse formation. Shown at the left is what happens when an axonal filopodium first makes contact with a dendritic filopodium. The two stick together because of special cell adhesion molecules. Next (middle panel) comes a series of bidirectional molecular interactions that cause synaptic vesicles and various proteins needed for transmitter release to accumulate on the axonal side of the emerging synapse; postsynaptic proteins accumulate on the dendritic side. Many of the synaptic molecules are "prepackaged" into macromolecular assemblages that float around axons and dendrites and can be inserted rapidly into emerging synapses. Shown at the right is a maturing synapse. During this late stage of development, the strength of synaptic transmission increases considerably. [After Garner et al., 2006]

Synapse Formation

Once the pre- and postsynaptic elements are found to be compatible, synapse formation begins. This process involves a wide array of molecules (Figure 4.29). Adhesion molecules ensure that budding synapses do not get pulled apart. Other molecules cause pre- and postsynaptic specializations, such as synaptic vesicles and postsynaptic receptors, to gather at the new synapse. For example, the molecule **agrin** is released from axons when they contact a muscle. The agrin binds to a muscle-specific receptor that causes acetylcholine receptors in the muscle's membrane to cluster right beneath the axon's contact point. In the central nervous system, an analogous role is played by the molecule **neurexin**. It resides in the presynaptic membrane and binds to receptors in the postsynaptic cell. Activation of the neurexin receptor (neuroligin) then induces the postsynaptic side of the synapse to accumulate various scaffolding proteins and neurotransmitter receptors, which together form the postsynaptic density. It also induces presynaptic specializations. The details of these interactions between pre- and postsynaptic cells remain to be worked out, but the interactions are clearly bidirectional. When axons and dendrites interact, neither is passive; they stimulate one another.

Young synapses form within a few minutes or hours, but a typical synapse takes days to mature. During this **synapse maturation** phase, young synapses become larger and accumulate a greater variety of pre- and postsynaptic proteins. Particularly interesting is that AMPA-type glutamate receptors are recruited to glutamatergic synapses relatively late, after NMDA-type receptors have already been there for some time. Because NMDA-type receptors permit ion flow only after the postsynaptic cell has become depolarized (see Chapter 3), synapses that contain only NMDA receptors are called **silent synapses**. When AMPA receptors are inserted into these silent synapses, they become increasingly "vocal." Another important aspect of synapse maturation is that the excitatory and inhibitory postsynaptic potentials (EPSPs and IPSPs) generated by an active synapse tend to become faster as the synapses mature. These changes entail a variety of subtle changes in receptor composition. The biological function of these changes is unknown, but one good possibility is that young neurons with few synaptic inputs require broad synaptic potentials to push them past their firing threshold.

BRAIN EXERCISE

When dendritic and axonal filopodia contact each other, why don't the dendritic filopodia simply "instruct" the presynaptic filopodia on what kind of transmitters they should produce (e.g., GABA vs. glutamate)?

4.6 How Can a Neural Circuit Be Fine-Tuned?

Once synapses have formed and neurons are interconnected, you might think that brain development is essentially complete. You would be wrong. Nearly half of all the neurons created during development die off before the nervous system is fully mature, and the number of synapses declines by a similar fraction. A tempting analogy for this developmental reduction in neuron and synapse numbers is the chiseling of a fine sculpture from a block of marble. However, marble has no specific shape before the sculpting begins. In contrast, neural development is fairly specific from the start. Therefore, a better analogy for the elimination of neurons and synapses during development is the pruning of a plant. When you prune a tree, for example, you can only cut the branches that the tree produced. Moreover, as you cut off some branches, others continue to grow and new ones sprout. It is similar in the nervous system: the elimination of some connections in the brain is accompanied by the continued "sprouting" of connections elsewhere. Because the majority of axon terminals and synapses are eliminated after the major period of neuronal cell death has passed, we shall discuss the elimination of axonal branches and synapses after we have dealt with developmental neuron death.

Developmental Neuron Death

It may seem odd for neurons to die shortly after they were born, but **developmental neuron death** is a real and substantial phenomenon. In parts of the spinal cord, 3 times as many neurons are present in embryos than just a few weeks after birth. In the retina, about half of the neurons that project into the brain die off during development; and in the developing neocortex, 20–50% of all the neurons die during development. This great dying of neurons is not pathological; nor is developmental cell death limited to the nervous system. Programmed cell death, as it is often called, is part of normal development in many tissues and in most organisms. Importantly, the cells that die during normal development die peacefully. The neurons do not burst open and spill their various enzymes into extracellular space, where they could damage other cells. Instead, they self-destruct in an orderly manner, a process that is called **apoptosis** (from a Greek word used to describe when leaves "fall off" a tree). It involves the production of special cell death proteins that break down nucleic acids and other proteins. Ultimately, the apoptotic cells break into smaller pieces that are then "eaten" by other cells called microglia. You will learn more about microglia in Chapter 5.

Numerical Matching

Why do developing nervous systems produce more neurons than they need? Isn't this a waste of energy? Maybe it is, but consider this: would it not be good to have the number of neurons in one brain region be well matched to the number of neurons in its target? This kind of **numerical matching** between interconnected cell populations is difficult to achieve by controlling cell proliferation because precursor populations generally double with each round of cell division (2-4-8-16-32-64, etc.). One extra round of cell division, for example, would make a cell group twice as large as it normally is. Unless other components of the neuronal circuit are similarly enlarged, some of those extra neurons are likely to remain without proper inputs or outputs. This problem can be solved by overproducing neurons and then killing off any cells that fail to make proper connections. Before we see how this might work, let us review some evidence that numerical matching through developmental neuron death actually occurs.

A mutant strain of mice called **staggerer** has a specific deficit that kills all cerebellar Purkinje cells early in development. As you saw in Figure 4.28, Purkinje cells get much of their input from cerebellar granule cells. Indeed, the granule cells have no other target. So, what happens to the granule cells in staggerer mutants without any Purkinje cells? The granule cells all die. This is an extreme sort of numerical matching, where both populations go to zero. To explore this phenomenon further, scientists mixed together varying numbers of cells from embryonic staggerer and wild-type mice and thus created chimeric mice that had anywhere from 27,000 to 103,000 Purkinje cells (on one side of the brain). They then counted how many cerebellar granule cells these staggerer chimeras contain in adulthood. Remarkably, the number of granule cells was directly proportional to the number of Purkinje cells, forming an almost perfect linear correlation. This demonstrates that the sort of the numerical matching we discussed in the previous paragraph exists (at least in this pathway). Furthermore, because the staggerer mutation is known to have direct effects only on Purkinje cells, any adjustments in the number of granule cells must be due to indirect effects. The most likely mechanism for this indirect effect is that any granule cells that fail to make synaptic contact with a healthy Purkinje cell commit apoptotic suicide.

More evidence for the numerical matching hypothesis comes from experimental alterations of limb development (Figure 4.30). In a classic experiment, Viktor Hamburger removed one wing from a young chicken embryo and later counted the number of motor neurons in the portion of the spinal cord that normally innervates the wing. He found that the number of wing motor neurons was drastically reduced on the side missing the wing. The *dorsal root ganglia,* which house the cell bodies of neurons carrying sensory information from the wing into the CNS (see Chapter 7), were likewise smaller on the amputated side.

Hamburger initially believed that these effects were due to changes in cell proliferation (he hypothesized that tissues in the wing secrete a signal that promotes proliferation), but later work revealed that the decrease in neuron number on the amputated side is due to an increase in developmental neuron death. Another experiment showed that grafting an extra leg onto one side of a young chicken embryo increases the number of surviving leg motor neurons and dorsal root ganglia neurons on the side with the extra leg (Figure 4.30 B). Collectively, these experiments demonstrated that the number of neurons projecting to a set of limb muscles, as well as the number of sensory neurons carrying inputs from the limb, is matched during development to the amount of limb tissue available for innervation. They also showed that the numerical match is achieved primarily by modulating the extent of neuron death.

Trophic Factors

What kind of death-preventing signal might neurons obtain from other cells? A key answer was provided by Rita Levi-Montalcini and her collaborators. They had removed the leg of an embryonic chicken, replaced it with a sarcoma (a cancerous piece of connective tissue), and found that the transplanted sarcoma rescued many limb motor neurons and dorsal root ganglion cells from death. Importantly, the rescue happened even though none of the neural processes touched the sarcoma. This finding implied that the death-preventing signal must be a diffusible molecule produced by the sarcoma cells. Levi-Montalcini and Stanley Cohen isolated this diffusible molecule in 1954 by manipulating sarcoma extracts and testing whether they continued to promote neuron survival and axon outgrowth. The isolated

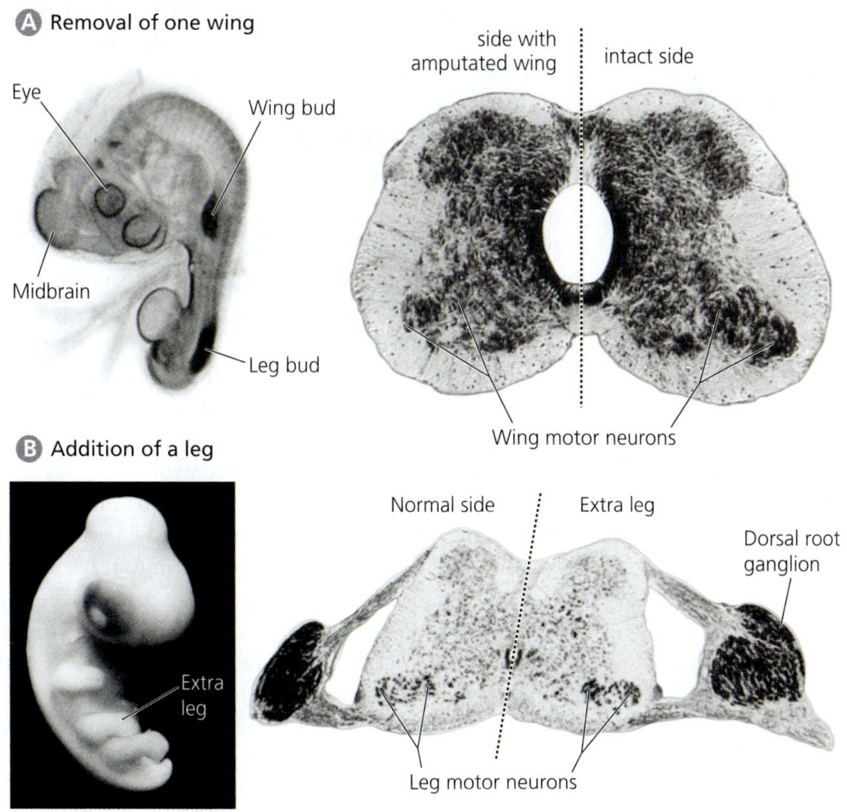

Figure 4.30 Manipulating neuron death. Shown in (A) is a lateral view of a chicken embryo (left). If one wing bud is cut away at this age, the chicken grows up with just one wing. A section through the spinal cord of such a one-winged chicken reveals that the side with the amputated wing contains fewer wing motor neurons. When experimenters graft an extra leg onto an embryo (B), the leg motor neurons are slightly more numerous; and the dorsal root ganglia (DRG) that ferry sensory information from the limb into the CNS (see Chapter 7) are much larger on the side with the extra leg. Both experiments reveal that the number of neurons innervating a limb can be altered by changing limb size and/or number. [A from Hamburger and Hamilton, 1992, and Hamburger, 1934; B from Hollyday and Hamburger, 1976]

molecule is called **nerve growth factor**, and its discoverers were awarded a Nobel prize in 1986.

Nerve growth factor (NGF) turned out to be only one of many neuron survival and growth promoting factors, collectively known as **trophic factors**. We now know, for example, that motor neurons are even more dependent for their survival on brain derived neurotrophic factor (BDNF) than NGF. The BDNF is made by muscle tissue and binds to special receptors on the axon terminal of motor neurons. It is then internalized through the receptor and transported back to the neuron's nucleus, where it prevents the apoptotic self-destruct sequence from running its course. Other neurons rely on different trophic factors. Some of these factors function as **retrograde signals**, meaning that they are (like BDNF) transported from a neuron's terminals back to the cell body (Figure 4.31). Others function as **anterograde signals**, which means that they originate from a neuron's inputs rather than its targets. The basic idea, however, is the same. Young neurons must obtain these trophic factors, whatever their source or type, to escape developmental death.

To understand the role of neurotrophic factors in numerical matching, you need to know one more thing: young neurons compete with one another for neurotrophic factors. This is certainly the simplest way to explain the results Hamburger obtained in his limb addition experiment. In normal embryos, the tissues of the limb secrete enough neurotrophic factor to support less than half of the motor and sensory neurons innervating the limb. However, in embryos with an extra limb, more neurotrophic factor is secreted and more can be transported back to the neuronal cell bodies to prevent apoptosis. Simply put, the more neurotrophic factor is available for uptake, the less the competition among neurons for that neurotrophic factor and the more neurons survive. It is a strange sort of economy: if demand for neurotrophic factor exceeds the supply, then the demand is trimmed. Thank goodness our own economies tend to adjust supply instead! In either case, however, the end result

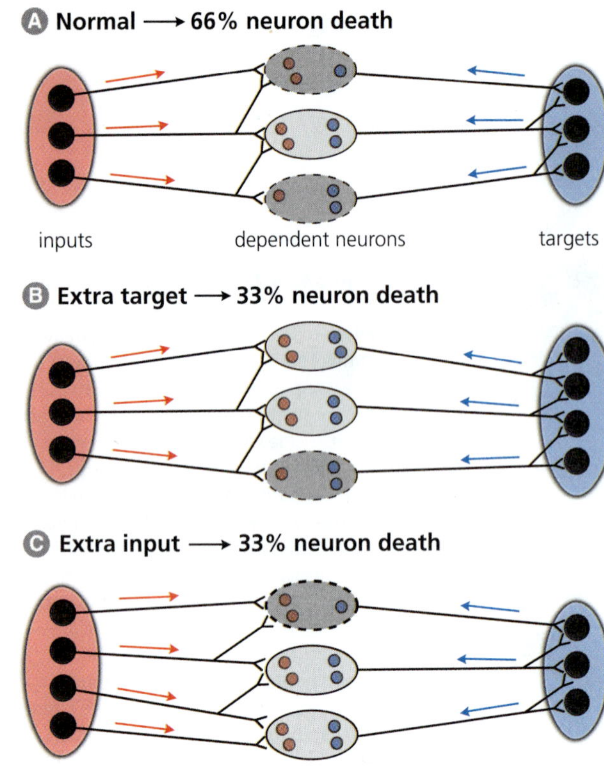

Figure 4.31 Numerical matching through trophic factors. Panel (A) illustrates three young neurons (gray ovals) that depend on neurotrophic factors secreted by their inputs (red ovals) and their targets (blue ovals). The trophic factor from the input side is transported anterogradely (red arrows) toward the neurons, which then internalize it (small red circles). The trophic factor from the target is transported retrogradely (blue arrows) and also internalized (small blue circles). In the illustrated example, each neuron requires two units of each trophic factor to survive. Because two of the neurons do not receive sufficient trophic factor, they die (dark gray with dashed outline). Panels (B) and (C) illustrate the effects of increasing target and input size, respectively. In both cases, more young neurons survive.

is that supply and demand are matched. This is "good design" in the construction of neural circuits as well as in economies.

Pruning and Sprouting Neuronal Connections

If neurons die, their axons disappear with them. However, major axon branches (collaterals) sometimes vanish during development while other parts of the neuron persist. This can be demonstrated by injecting a fluorescent axon tracer into a specific region of a developing brain and, some time later, injecting into the same region a second tracer that fluoresces in a different color. Both tracers should be transported back (retrogradely) to the cell bodies of any neurons that send axon branches into the area of the tracer injection. Any neurons that contain the first tracer but not the second one must have lost their projection to the injected area in the time between the two tracer injections. Such sequential double-labeling experiments have shown that individual neurons in the neocortex project to several different subcortical regions early in development but then eliminate a subset of their initial axon branches. The elimination of such **exuberant projections** has also been observed in other pathways, notably the pathways that interconnect the two cerebral hemispheres. However, even highly exuberant connections are at least somewhat specific rather than entirely random. To reinforce a point made earlier, the elimination of major axon collaterals during development is best compared to the pruning of a plant, not to the sculpting of marble.

Pruning Corticospinal Axons

Neurons prune not only their major axon collaterals but also smaller branches that spread within a target area. This was shown by injecting an anterograde tracer (a tracer that is transported from the cell body to axon terminals) into small groups of neurons in the primary motor cortex of young cats. After an appropriate survival time, the researchers sectioned the spinal cord of these animals and reconstructed the trajectory of labeled **corticospinal axons** (Figure 4.32). They found that in 25-day-old animals, individual corticospinal axons have widespread arborizations on both sides of the spinal cord. In contrast, when the tracer is injected into 35-day-old animals, individual axons have fewer long branches, exhibit a greater number of short branches, and are confined to just one side of the spinal cord. Because of these changes, the axons overlap significantly less at 35 days of age than 10 days earlier. By day 55, the corticospinal axons have even fewer long branches and more of the extremely short branches (Figure 4.32 B). Overall, these data show that corticospinal axons are pruned during development yet also sprout some new branches.

Pruning Retinogeniculate Axons

Axonal refinement has been studied extensively in the developing pathway from the retina to the thalamic **lateral geniculate nucleus (LGN)**, which sends visual information to the visual cortex (see Figure 1.12). The two retinas initially project to overlapping portions of the LGN on both sides of the brain, but the projections from the two eyes gradually segregate into nonoverlapping territories (Figure 4.33). By labeling individual retinal axons at various stages of development, researchers have shown that this segregation occurs because neurons eliminate axon branches that terminate in the "wrong" area and sprout additional branches in the "proper" location. Thus, axonal refinement in the retinogeniculate pathway also involves the loss of branches and the sprouting of new ones.

Research on the mechanisms underlying the segregation of retinal projections to the LGN revealed a crucial role for retinal activity. This is puzzling at first because the embryo at the time of axon segregation is still in the uterus and has its eyelids closed. How then can retinal activity be crucial for retinal axon segregation? The surprising answer is that the retina is intrinsically active long before it serves a visual function. Specifically, young retinas exhibit bursts of action potentials that spread like waves from one edge of the retina to the other. Blocking these **retinal activity waves** prevents the normal segregation of the retinogeniculate axons. Furthermore, boosting the retinal waves in one eye, relative to the other eye, causes the projection from the more active retina to expand at the expense of the less active projection. This implies that the elimination and sprouting of retinal axon branches in the LGN is an activity-dependent, competitive phenomenon.

Refinement through Competition

For a deeper understanding of how neuronal activity can drive competition between axon branches, let us revisit Hebb's postulate (see Chapter 3). It states that any synapses active when the postsynaptic cell fires are strengthened. Conversely, any synapses not contributing to the firing of a postsynaptic neuron are weakened. Stated more simply, neurons that fire together become more tightly wired together, whereas neurons that do not fire at the same time become less strongly interconnected. Applying these rules to the retinogeniculate pathway, we can see that a wave of neuronal activity spreading across one retina (but not into the other retina) would cause LGN neurons receiving synaptic input from that active retina to fire many action potentials. The synapses that were active just before or during those action potentials would be strengthened. In contrast, synaptic inputs from the other retina, which is unlikely to

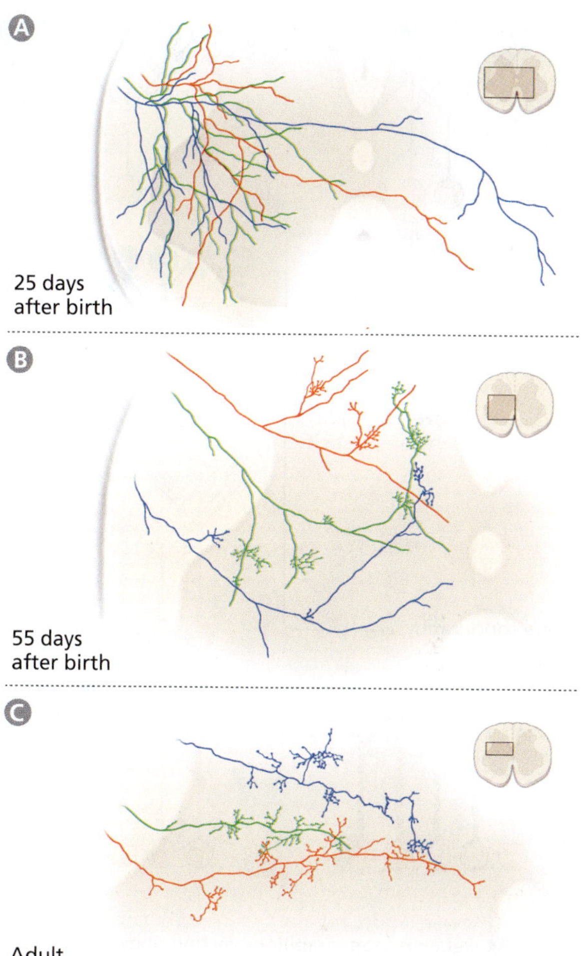

Figure 4.32 Refining corticospinal projections. Axonal tracers were injected into a small region of cat motor cortex at various stages of postnatal development. The authors followed labeled axons into the spinal cord and examined their terminal arborizations (each axon is a different color). At 25 days after birth (A), the axons have numerous long branches and overlap extensively. At 55 days (B), long branches have decreased in number, short branches are tightly clustered, and the axons overlap less than they did at 25 days. In adult cats (C), long branches have been reduced at the expense of short, tightly clustered arborizations. [After Li and Martin, 2002]

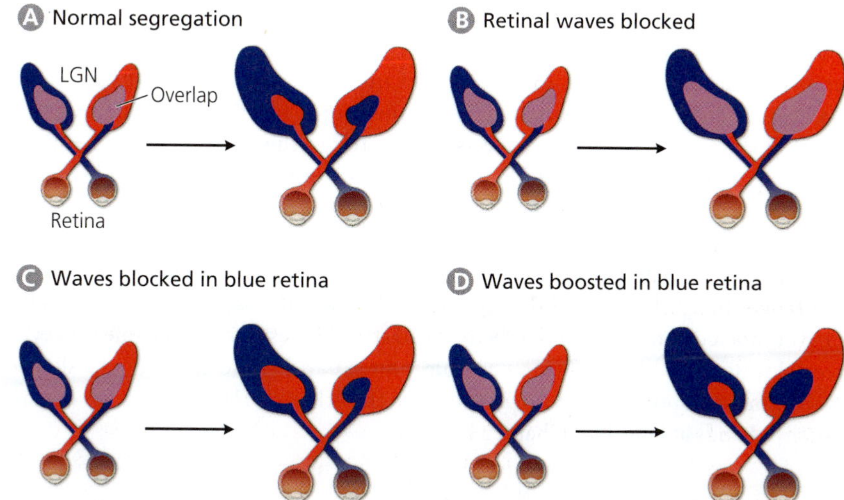

Figure 4.33 Eye-specific segregation of retinogeniculate projections. Panel (A) illustrates the normal development of retinal projections to the lateral geniculate nucleus (LGN) in a ferret. The two retinas (one red, the other blue) each project bilaterally to the LGN. The projections from the two eyes show a large amount of overlap (purple) within the LGN initially, but this overlap is eliminated as development goes on (arrow). When waves of retinal activity are blocked pharmacologically (B), the overlap persists into adulthood. Blocking retinal waves in just one retina (C) causes the more active retina to end up with a proportionately larger projection to the LGN. Boosting retinal waves (D) produces the opposite result. [After Huberman et al., 2008]

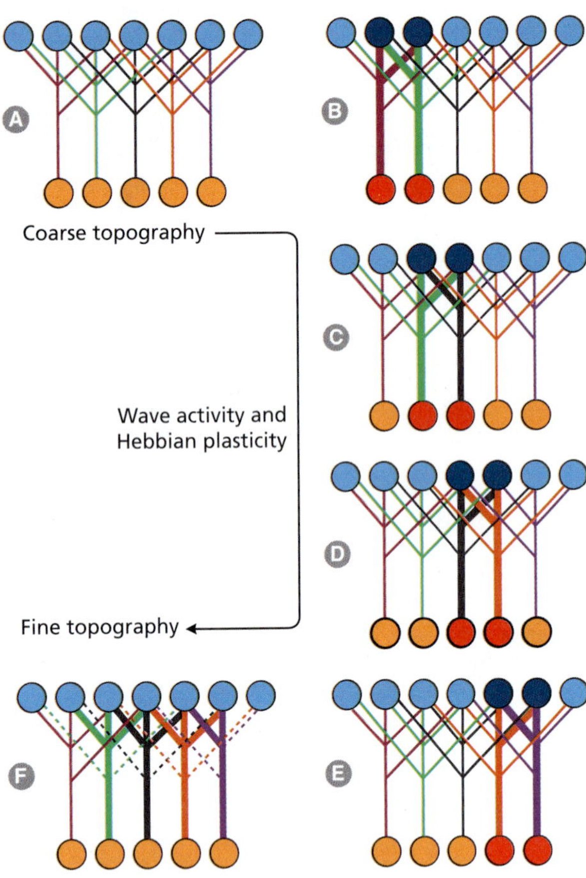

Figure 4.34 Refining a topographic projection. Shown in (A) is a coarse topographic projection in which each input neuron (orange) projects to 4 target neurons (blue). Panels (B) to (E) depict a wave of activity spreading from left to right across the input neurons (active neurons are red). A matching wave spreads across the target neurons, which fire only if they receive simultaneous input from at least two input neurons. According to Hebb's rule, synapses that contributed to firing a target cell are strengthened; all others are weakened. Panel (F) shows what you end up with after many such activity waves. Input lines that were never strengthened are lost (dashed lines), and the remaining inputs have become quite strong (bold lines). A comparison between panels (A) and (F) reveals that projection overlap has been reduced.

have been active at the same time, would be weakened. Sometime later, a wave of activity might spread across the other, previously inactive retina, leading to a strengthening of this retina's connections to any LGN neurons that fired in response to that retinal wave and to a weakening of the projections from the other retina. If these retinal waves and bouts of synapse strengthening and weakening happen repeatedly, as they would during development, then LGN neurons that initially received inputs from both retinas would gradually lose the inputs from one of the two retinas. Thus, the early developmental condition of LGN neurons receiving converging input from both retinas would gradually give way to the adult condition in which each LGN neuron receives input from just a single retina, as shown in Figure 4.33.

Refining Topographic Projections

Hebbian plasticity also contributes to the refinement of topographic, map-like projections. As we discussed in Section 4.4, molecular gradients play a major role in the initial formation of topographic projections. However, topographic maps based solely on molecular gradients tend to be relatively coarse, with the projections of individual neurons overlapping extensively. Waves of activity spreading across a brain region can refine this coarse topography because in such a wave neighboring neurons tend to fire simultaneously, whereas distant neurons fire asynchronously (Figure 4.34). When two adjacent neurons with overlapping projections are active simultaneously, they tend to trigger action potentials only in target neurons that receive synaptic input from both of them. According to Hebb's postulate, the synapses of the two input neurons onto the activated target neurons are strengthened; other synapses are weakened. Now imagine this process occurring repeatedly as wave after wave of activity sweeps across the input structure. Over time, many synapses would be weakened into oblivion, whereas others are strengthened progressively. In the end, each input neuron would project to a more limited set of target neurons, and the overlap in the projections of different input neurons would be reduced (Figure 4.34). This amounts to a refinement of the projection's topography.

Ocular Dominance Bands

As embryos mature, neural activity becomes increasingly controlled by external stimuli rather than intrinsic waves. Still, connectional plasticity persists. A good example of this persistent plasticity was discovered in the 1970s by David Hubel, Torsten Wiesel, and Simon Levay. They injected into the eyes of monkeys radioactive tracers that were transported from the retina to the LGN and from there to the visual cortex. After injecting such a tracer into one eye of an adult monkey, they found a zebra-like banding pattern of tracer in the primary visual cortex when it was sectioned tangentially (Figure 4.35). Physiological observations confirmed that cells within the labeled bands respond only to visual stimuli presented to the injected eye. Neurons located in between the **ocular dominance bands** (also called ocular dominance columns) respond to stimuli presented to the other eye. We will discuss ocular dominance bands at length in Chapter 11.

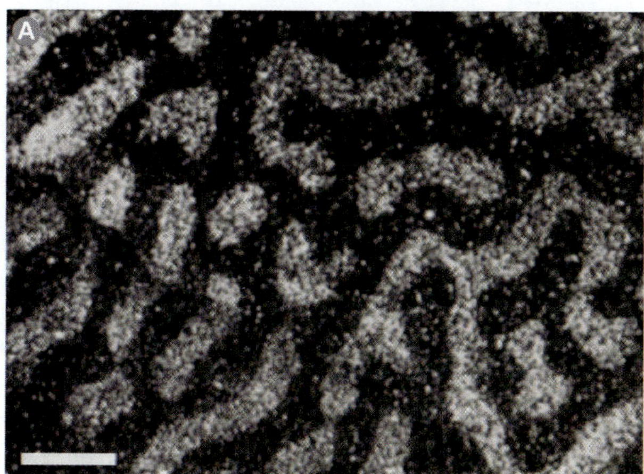

Figure 4.35 Cortical ocular dominance bands. Experimenters injected into one retina a radioactive tracer that is transported (across one synapse in the LGN) to the primary visual cortex (V1). Shown in (A) is a tangential section through V1 of an adult monkey that grew up normally. The tracer (white) is localized to discrete bands of relatively uniform thickness. Between these bands are unlabeled areas of similar width; they receive inputs from the other eye. Panel (B) shows analogous data from a monkey whose right eye was closed from 2 weeks after birth. Tracer was injected into the left eye 18 months later. The bands corresponding to the open left eye have expanded relative to the bands corresponding to the closed eye. Because monkeys at birth already have segregated ocular dominance bands, these findings indicate that visual activity in both eyes is required to maintain the normal pattern of ocular dominance bands. [From Levay et al., 1980]

Most interesting for present purposes is that in monkeys that grew up with one eye shut, the ocular dominance bands for the closed eye are much skinnier than the bands corresponding to the other eye (Figure 4.35 B). Because the bands for the two eyes are equally wide at birth, the asymmetry in width after monocular lid closure must have resulted from postnatal plasticity. Apparently, LGN neurons connected to the open eye expand their terminal arborizations in the visual cortex at the expense of the neurons that receive their input from the occluded eye.

Is the change in ocular dominance band width observed after suturing one eye shut an example of developmental plasticity, or is it learning? The answer depends on your definition of learning. The changes in the ocular dominance bands after eyelid closure are due to interactions between the organism and its external environment. They are also functional insofar as it is good design for nervous systems to expand connections that provide useful information about the world and to reduce those that do not. Therefore, the postnatal plasticity of ocular dominance bands is a form of learning, broadly defined. However, organisms exhibit many different forms of learning (which you will learn about in later chapters), and ocular dominance plasticity is at best an unsophisticated form of learning.

Beyond this definitional issue, the important lesson is that nervous systems are plastic throughout an organism's life. Only the form and extent of neuronal plasticity changes with age. Early in development, the factors controlling neuronal plasticity are mainly intrinsic to the organism (e.g., variations in neurotrophic factors and intrinsic neural activity). As development continues, neuronal plasticity becomes increasingly controlled by interactions between the organism and its environment. The late forms of plasticity also modify more subtle aspects of neuronal structure and function. In essence, developmental plasticity and learning are continuous with one another.

BRAIN EXERCISE

How would you define "learning," and how would you distinguish it from "developmental maturation"?

4.7 What Are the Major Themes of Neural Development?

The overarching theme of neural development is that everything—well, almost everything—changes. The earliest stages of neural development are dominated by cell proliferation and the emergence of progressively smaller tissue subdivisions. The proliferating cells then leave the cell cycle and differentiate. After migrating to where they need to be, young neurons send their axons out to specific targets. Once there, the axons branch and make some synapses. As a complement to these formative processes, many cells, axon branches, and synapses are pruned away. Underpinning these changes in neural structure are changes in molecular mechanisms. For example, gradients of diffusible transcription factors play a crucial role in early brain patterning and in the early phases of axon outgrowth, but they become less influential as embryos grow. The role of Hebbian plasticity likewise changes over time. It is initially driven by intrinsic waves of neuronal activity; but as the waves recede, Hebbian plasticity is increasingly guided by interactions between organisms and their environment. Thus, everything in neural development runs according to a grand and orderly schedule.

Sensitive Periods

An important corollary of neural development being so orderly is that the nervous system's sensitivity to environmental factors changes predictably with time. For example, the increased neuron death that is observed when embryonic limbs are cut (see Figure 4.30) occurs only up to a certain age. After this **sensitive period**, target-deprived neurons survive. Another good example is the nervous system's vulnerability to Accutane® (formally known as isotretinoin). This drug is commonly used to treat acne and is relatively safe for teenagers. However, Accutane is a form of retinoic acid that interferes with normal retinoic acid signaling in early embryos. Given the fundamental role of retinoic acid in nervous system patterning (see Section 4.2), it is not surprising that taking Accutane during pregnancy often leads to miscarriage or serious birth defects such as cleft palate, small brains, and intellectual disability. Of course, developing nervous systems are vulnerable to other chemicals as well. Alcohol, for example, may seriously damage embryonic brains, especially during very early stages of embryonic development (see Box 4.3). Even some of the medicines and dietary supplements that adults take in stride may impair brain development. Therefore, pregnant women must be very careful about the chemicals that they ingest.

Molecular Redeployment

Because the processes and mechanisms of neural development change over time, molecules that are used for one purpose at one point in development can be employed to different ends at other times. For example, some of the transcription factors crucial for early neural patterning later become involved in axon pathfinding, and some of these in turn are critical for later synaptic plasticity. This redeployment of molecules in different contexts, and to quite different ends, is surprising because we tend to think of molecules as having just a single function. However, even molecules that have a single molecular function (e.g., inhibiting BMP) may have quite different functions at the supramolecular level if the molecular function is performed in a different molecular and structural context. Even if a molecule interacts with the same small set of other molecules throughout development, the outcome of these interactions may differ across time just because the context of the interactions has changed. This solves an interesting riddle: When researchers first sequenced the human genome, many were surprised that it contains fewer than 30,000 genes. How can so few genes accomplish so much? The answer is that our genes can be deployed in countless combinations. It's the same combinatorial power that allows you to write a novel with just 26 letters, or to compose a symphony with a few dozen notes.

mature. These changes entail a variety of subtle changes in receptor composition. The biological function of these changes is unknown, but one good possibility is that young neurons with few synaptic inputs require broad synaptic potentials to push them past their firing threshold.

BRAIN EXERCISE

When dendritic and axonal filopodia contact each other, why don't the dendritic filopodia simply "instruct" the presynaptic filopodia on what kind of transmitters they should produce (e.g., GABA vs. glutamate)?

4.6 How Can a Neural Circuit Be Fine-Tuned?

Once synapses have formed and neurons are interconnected, you might think that brain development is essentially complete. You would be wrong. Nearly half of all the neurons created during development die off before the nervous system is fully mature, and the number of synapses declines by a similar fraction. A tempting analogy for this developmental reduction in neuron and synapse numbers is the chiseling of a fine sculpture from a block of marble. However, marble has no specific shape before the sculpting begins. In contrast, neural development is fairly specific from the start. Therefore, a better analogy for the elimination of neurons and synapses during development is the pruning of a plant. When you prune a tree, for example, you can only cut the branches that the tree produced. Moreover, as you cut off some branches, others continue to grow and new ones sprout. It is similar in the nervous system: the elimination of some connections in the brain is accompanied by the continued "sprouting" of connections elsewhere. Because the majority of axon terminals and synapses are eliminated after the major period of neuronal cell death has passed, we shall discuss the elimination of axonal branches and synapses after we have dealt with developmental neuron death.

Developmental Neuron Death

It may seem odd for neurons to die shortly after they were born, but **developmental neuron death** is a real and substantial phenomenon. In parts of the spinal cord, 3 times as many neurons are present in embryos than just a few weeks after birth. In the retina, about half of the neurons that project into the brain die off during development; and in the developing neocortex, 20–50% of all the neurons die during development. This great dying of neurons is not pathological; nor is developmental cell death limited to the nervous system. Programmed cell death, as it is often called, is part of normal development in many tissues and in most organisms. Importantly, the cells that die during normal development die peacefully. The neurons do not burst open and spill their various enzymes into extracellular space, where they could damage other cells. Instead, they self-destruct in an orderly manner, a process that is called **apoptosis** (from a Greek word used to describe when leaves "fall off" a tree). It involves the production of special cell death proteins that break down nucleic acids and other proteins. Ultimately, the apoptotic cells break into smaller pieces that are then "eaten" by other cells called microglia. You will learn more about microglia in Chapter 5.

Numerical Matching

Why do developing nervous systems produce more neurons than they need? Isn't this a waste of energy? Maybe it is, but consider this: would it not be good to have the number of neurons in one brain region be well matched to the number of neurons in its target? This kind of **numerical matching** between interconnected cell populations is difficult to achieve by controlling cell proliferation because precursor populations generally double with each round of cell division (2-4-8-16-32-64, etc.). One extra round of cell division, for example, would make a cell group twice as large as it normally is. Unless other components of the neuronal circuit are similarly enlarged, some of those extra neurons are likely to remain without proper inputs or outputs. This problem can be solved by overproducing neurons and then killing off any cells that fail to make proper connections. Before we see how this might work, let us review some evidence that numerical matching through developmental neuron death actually occurs.

A mutant strain of mice called **staggerer** has a specific deficit that kills all cerebellar Purkinje cells early in development. As you saw in Figure 4.28, Purkinje cells get much of their input from cerebellar granule cells. Indeed, the granule cells have no other target. So, what happens to the granule cells in staggerer mutants without any Purkinje cells? The granule cells all die. This is an extreme sort of numerical matching, where both populations go to zero. To explore this phenomenon further, scientists mixed together varying numbers of cells from embryonic staggerer and wild-type mice and thus created chimeric mice that had anywhere from 27,000 to 103,000 Purkinje cells (on one side of the brain). They then counted how many cerebellar granule cells these staggerer chimeras contain in adulthood. Remarkably, the number of granule cells was directly proportional to the number of Purkinje cells, forming an almost perfect linear correlation. This demonstrates that the sort of the numerical matching we discussed in the previous paragraph exists (at least in this pathway). Furthermore, because the staggerer mutation is known to have direct effects only on Purkinje cells, any adjustments in the number of granule cells must be due to indirect effects. The most likely mechanism for this indirect effect is that any granule cells that fail to make synaptic contact with a healthy Purkinje cell commit apoptotic suicide.

More evidence for the numerical matching hypothesis comes from experimental alterations of limb development (Figure 4.30). In a classic experiment, Viktor Hamburger removed one wing from a young chicken embryo and later counted the number of motor neurons in the portion of the spinal cord that normally innervates the wing. He found that the number of wing motor neurons was drastically reduced on the side missing the wing. The *dorsal root ganglia*, which house the cell bodies of neurons carrying sensory information from the wing into the CNS (see Chapter 7), were likewise smaller on the amputated side.

Hamburger initially believed that these effects were due to changes in cell proliferation (he hypothesized that tissues in the wing secrete a signal that promotes proliferation), but later work revealed that the decrease in neuron number on the amputated side is due to an increase in developmental neuron death. Another experiment showed that grafting an extra leg onto one side of a young chicken embryo increases the number of surviving leg motor neurons and dorsal root ganglia neurons on the side with the extra leg (Figure 4.30 B). Collectively, these experiments demonstrated that the number of neurons projecting to a set of limb muscles, as well as the number of sensory neurons carrying inputs from the limb, is matched during development to the amount of limb tissue available for innervation. They also showed that the numerical match is achieved primarily by modulating the extent of neuron death.

Trophic Factors

What kind of death-preventing signal might neurons obtain from other cells? A key answer was provided by Rita Levi-Montalcini and her collaborators. They had removed the leg of an embryonic chicken, replaced it with a sarcoma (a cancerous piece of connective tissue), and found that the transplanted sarcoma rescued many limb motor neurons and dorsal root ganglion cells from death. Importantly, the rescue happened even though none of the neural processes touched the sarcoma. This finding implied that the death-preventing signal must be a diffusible molecule produced by the sarcoma cells. Levi-Montalcini and Stanley Cohen isolated this diffusible molecule in 1954 by manipulating sarcoma extracts and testing whether they continued to promote neuron survival and axon outgrowth. The isolated

NEUROLOGICAL DISORDERS

Box 4.3 *Drugs and a Baby's Brain*

"SURGEON GENERAL'S WARNING: Smoking By Pregnant Women May Result in Fetal Injury, Premature Birth, and Low Birth Weight." You have probably seen this label on packs of cigarettes. Moreover, since 1988, every container of alcohol sold in the United States must warn that "According to the Surgeon General, women should not drink alcoholic beverages during pregnancy because of the risk of birth defects." Sadly, we are warned about so many things in life that it is easy to lose interest. Plus, we rarely understand the evidence on which the warnings are based. In the case of cigarettes and alcohol, the data are clear.

Cigarette smoke contains about 4,000 different chemicals, but *nicotine* is the most addictive and dangerous of them. In liquid form, nicotine is three times more toxic than arsenic, and a few drops on the skin can kill. The problems arise because nicotine binds to nicotinic acetylcholine receptors, which are crucial to central and peripheral nervous system function. They also play major roles in brain development. Most notably, exposure of newborn mice to nicotine interferes with synapse development and, later, sensory processing. Because newborn mice correspond to human fetuses in the third trimester, this suggests that late fetal human brains are vulnerable to nicotine. Consistent with this prediction, smoking during pregnancy correlates with various behavioral and cognitive problems in the offspring. Even right after birth, the babies of mothers who smoked in pregnancy experience nicotine withdrawal. Despite these concerns, roughly 16% of US women in 2006–2007 continued to smoke after they became pregnant.

Drinking alcohol during pregnancy is likewise dangerous. *Fetal alcohol syndrome (FAS)* occurs in 1–3 of every 1,000 live births and is associated with a variety of facial abnormalities (Figure b4.3) as well as low birth weight, reduced brain size, attention deficits, learning disabilities, and mood disorders. Because FAS also comes in milder varieties, scientists have coined the more inclusive term *Fetal Alcohol Spectrum Disorder (FASD)*. The incidence of FASD is estimated to be ~1% of all live births. This is not surprising, given that 10–12% of women report drinking during their pregnancy and that ~2% report having consumed 5 or more alcoholic drinks in a single "binge" while they were pregnant. The details of how alcohol damages brains remain largely unclear, but alcohol probably slows down cell proliferation at early stages of brain development and generally upsets the normal schedule of neurogenesis. It also interferes with the migration of some neuron types and with the forming of some neural connections. It also increases the incidence of developmental neuron death.

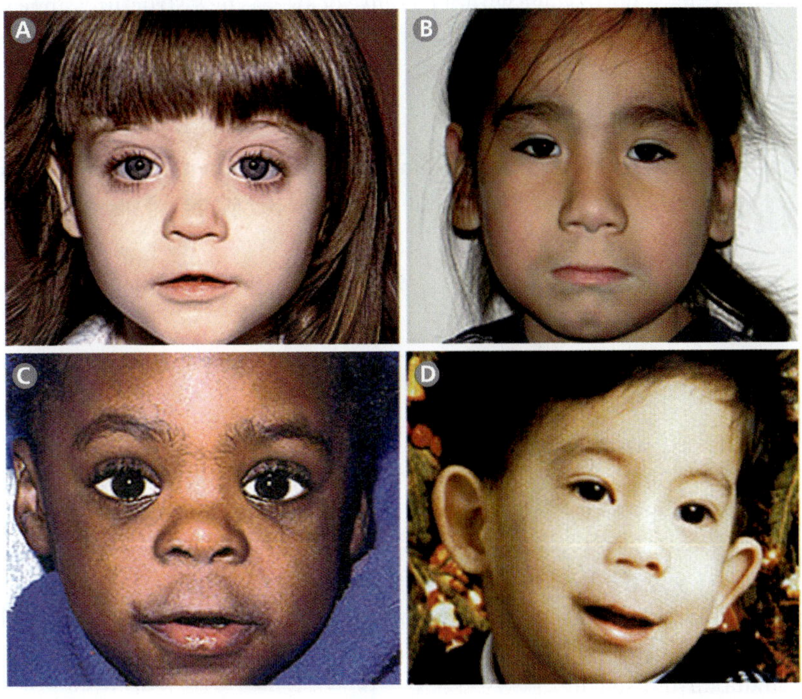

Figure b4.3 Fetal alcohol syndrome. Children born with fetal alcohol syndrome (FAS) display a unique combination of minor facial anomalies, namely small eyes, a thin upper lip, and a smooth philtrum (the vertical groove between the nose and upper lip). These features are diagnostic of FAS in (A) Caucasians, (B) Native Americans, (C) African Americans, and (D) Mexican Americans. Knowing these symptoms is useful because the mental problems associated with FAS tend not to emerge until later in life. [Copyright 2013, Susan J. Astley, PhD, University of Washington]

Given the severity and prevalence of FAS and FASD, it is odd that the phenomenon of *crack babies* is better known. The idea that babies of mothers who used crack cocaine would be doomed to a lifetime of severe mental problems originated in the late 1980s. It attracted significant media attention; but women who use cocaine during pregnancy also tend to drink and smoke, creating a potential confound. Furthermore, there had been no long-term studies to see how well crack babies did a few years after birth. These data are now in. They show that the children of cocaine-using mothers do have some lasting behavioral and cognitive problems, even after you control for alcohol consumption, smoking, and socioeconomic class. In addition, it is now well established that cocaine use can lead to premature birth and other pregnancy complications.

Some Regions Develop Faster Than Others

Although neural development generally progresses from cell proliferation to neurogenesis to axon growth, and from there to synapse formation, cell death, and connection refinement, this sequence plays out at different rates in different parts of the nervous systems. The caudal and ventral regions of the nervous system (notably the spinal cord and medulla) tend to mature before the more rostral and dorsal regions. The neocortex and the cerebellum are among the slowest to mature, and within the neocortex the prefrontal cortex does not fully develop until the teenage years. This asynchrony is adaptive because regions such as the spinal cord and medulla must, at least in most mammals, be functional by birth or shortly thereafter. In contrast, the neocortex and cerebellum are not essential for survival, at least while babies get their food from mom. Moreover, delaying neocortical and cerebellar maturation until after birth grants the environment a greater role in shaping that maturation. This allows more useful learning to take place. Thus, the fact that different brain regions develop and mature at different times is an effective compromise that allows organisms to survive outside the womb right after birth yet learn what's needed to be a successful adult.

Developmental Self-organization

The most amazing aspect of organismal development is that organisms construct themselves. Back in the 17th century, debates had raged about whether miniature adults lie hidden inside eggs or in the heads of sperm, waiting to unfold during development. Today we know that only DNA exists preformed in the gametes and that morphological structure emerges gradually out of a formless mass. We also have a vast amount of information on the molecular and cellular interactions that cause form to emerge in embryos. Still, it is difficult to watch an embryo develop from stage to stage without a sense of awe, as new structures appear continually, as if out of nowhere. Nothing in biology, except perhaps the working brain, is closer to magic.

BRAIN EXERCISE

Why do so many drugs contain warning labels telling women not to take them if they are pregnant or think they might be in the early stages of pregnancy?

SUMMARY

Section 4.1 - The embryo's ectoderm gives rise to both skin and the nervous system, the latter being the ectoderm's default fate. The neural plate folds and then closes to form the neural tube.

Section 4.2 - The mechanisms that pattern the neural tube along the rostrocaudal axis are distinct from those involved in dorsoventral patterning.

- Rostrocaudal patterning in the hindbrain and spinal cord involves retinoic acid and *Hox* genes. Without these caudalizing signals, the neural tube tends to form rostral brain regions.
- Floor plate cells secrete SHH, which induces some transcription factors and represses others in a concentration-dependent manner. Mutual repression between transcription factors creates distinct dorsoventral compartments with sharp boundaries.
- Midbrain and forebrain development involves SHH as well as several other transcription factors, including Fgf8 and BMP. Experimental manipulations of Ffg8 alter the rostrocaudal position of specific cortical areas.

Section 4.3 - On leaving the cell cycle, young cells migrate away from their place of birth, typically along radial processes. Where and when brain cells are born affects which transcription factors they express and thus their fate.

Section 4.4 - Axons typically grow long distances to find their targets.

- Developing axons are tipped with growth cones that forcefully extend and retract actin-filled filopodia.
- Growth cones can be guided by diffusible as well as substrate-bound molecules expressed in gradients. They may be attracted or repulsed by these guidance

factors, and their preferences may change as the axons grow.

- The topographic projections from the retina to the optic tectum (the retinotectal system) have been used to develop and test Sperry's chemoaffinity hypothesis as well as the more specific ephrin gradient hypotheses.

Section 4.5 - Dendrites extend filopodia that can sense other dendrites and axons. When dendrites and axons touch and are compatible, synapses are induced. As synapses mature, new types of receptors may be added.

Section 4.6 - Neural development involves a surprising amount of cell death and axon pruning.

- Many neurons depend on their survival for trophic factors obtained from postsynaptic cells or from their inputs. These trophic dependencies can match interconnected neuron populations for size.
- The selective elimination of axon branches during development tends to reduce projection overlap. It often involves activity-dependent competition that follows Hebb's rule.

Section 4.7 - Key themes of neural development are that (1) development is most easily disrupted during sensitive phases early in embryonic life, (2) the interactions and functional roles of developmentally important molecules vary with age, and (3) development proceeds at different rates in different brain regions.

Box 4.1 - In situ hybridization is used extensively in developmental neurobiology to study spatial and temporal changes in gene expression.

Box 4.2 - *Hox* gene evolution is a fascinating story of both conservation and change. In both insects and vertebrates, *Hox* genes are used in rostrocaudal body and brain patterning.

Box 4.3 - Nicotine and alcohol, as well as cocaine, can interfere with fetal brain development.

KEY TERMS

ADDITIONAL READINGS

4.1 - Developmental Origin of the Nervous System

De Robertis EM, Kuroda H. 2004. Dorsal-ventral patterning and neural induction in *Xenopus* embryos. *Annu Rev Cell Dev Biol* **20**:285–308.

4.2 - Patterning the Neural Tube

Briscoe J, Novitch B. 2008. Regulatory pathways linking progenitor patterning, cell fates and neurogenesis in the ventral neural tube. *Philos Trans R Soc Lond B* **363**:57–70.

Flames N, Pla R, Gelman D, Rubenstein J, Puelles L, Marin O. 2007. Delineation of multiple subpallial progenitor domains by the combinatorial expression of transcriptional codes. *J Neurosci* **27**:9682–9695.

O'Leary DD, Sahara S. 2008. Genetic regulation of arealization of the neocortex. *Curr Opin Neurobiol* **18**:90–100.

Puelles L. 2001. Brain segmentation and forebrain development in amniotes. *Brain Res Bull* **55**:695–710.

Scholpp S, Wolf O, Brand M, Lumsden A. 2006. Hedgehog signalling from the *zona limitans intrathalamica* orchestrates patterning of the zebrafish diencephalon. *Development* **133**:855–864.

Wurst W, Bally-Cuif L. 2001. Neural plate patterning: upstream and downstream of the isthmic organizer. *Nat Rev Neurosci* **2**:99–108.

4.3 - Neurogenesis and Migration

Desai AR, McConnell SK. 2000. Progressive restriction in fate potential by neural progenitors during cerebral cortical development. *Development* **127**:2863–2872.

Götz M, Huttner WB. 2005. The cell biology of neurogenesis. *Nat Rev Mol Cell Biol* **6**:777–788.

Kessaris N, Fogarty M, Iannarelli P, Grist M, Wegner M, Richardson W. 2006. Competing waves of oligodendrocytes in the forebrain and postnatal elimination of an embryonic lineage. *Nat Neurosci* **9**:173–179.

Zechner D, Müller T, Wende H, Walther I, Taketo MM, Crenshaw EB, Treier M, Birchmeier W, Birchmeier C. 2007. Bmp and Wnt/beta-catenin signals control expression of the transcription factor Olig3 and the specification of spinal cord neurons. *Dev Biol* **303**:181–190.

4.4 - Axon Growth and Targeting

Canty AJ, Murphy M. 2008. Molecular mechanisms of axon guidance in the developing corticospinal tract. *Prog Neurobiol* **85**:214–235.

Dickson BJ, Gilestro GF. 2006. Regulation of commissural axon pathfinding by slit and its Robo receptors. *Annu Rev Cell Dev Biol* **22**:651–675.

Guthrie S. 2007. Patterning and axon guidance of cranial motor neurons. *Nat Rev Neurosci* **8**:859–871.

Sperry RW. 1963. Chemoaffinity in the orderly growth of nerve fiber patterns and connections. *Proc Natl Acad Sci USA* **50**:703–10.

Wu K, Hengst U, Cox L, Macosko E, Jeromin A, Urquhart E, Jaffrey S. 2005. Local translation of RhoA regulates growth cone collapse. *Nature* **436**:1020–1024.

Young SH, Poo MM. 1983. Spontaneous release of transmitter from growth cones of embryonic neurones. *Nature* **305**:634–637.

4.5 - Synapse Formation

Jan YN, Jan LY. 2003. The control of dendrite development. *Neuron* **40**:229–242.

McAllister AK. 2007. Dynamic aspects of CNS synapse formation. *Annu Rev Neurosci* **30**:425–450.

Portera-Cailliau C, Pan DT, Yuste R. 2003. Activity-regulated dynamic behavior of early dendritic protrusions: evidence for different types of dendritic filopodia. *J Neurosci* **23**:7129–7142.

Wierenga C, Becker N, Bonhoeffer T. 2008. GABAergic synapses are formed without the involvement of dendritic protrusions. *Nat Neurosci* **11**:1044–1052.

4.6 - Refining the Neural Network

Cowan WM. 2001. Viktor Hamburger and Rita Levi-Montalcini: the path to the discovery of nerve growth factor. *Annu Rev Neurosci* **24**:551–600.

Huberman AD, Feller MB, Chapman B. 2008. Mechanisms underlying development of visual maps and receptive fields. *Annu Rev Neurosci* **31**:479–509.

Innocenti G, Price DJ. 2005. Exuberance in the development of cortical networks. *Nat Rev Neurosci* **6**:955–965.

Oppenheim RW. 1991. Cell death during development of the nervous system. *Annu Rev Neurosci* **14**:453–501.

Stellwagen D, Shatz CJ. 2002. An instructive role for retinal waves in the development of retinogeniculate connectivity. *Neuron* **33**:357–367.

4.7 - Major Themes of Neural Development

Liu A, Niswander L. 2005. Bone morphogenetic protein signaling and vertebrate nervous system development. *Nat Rev Neurosci* **6**:945–954.

Sanes DH, Reh TA, Harris WA. 2006. *Development of the nervous system.* Second edition. Amsterdam: Academic Press.

Boxes

Lewis MW, Misra S, Johnson HL, Rosen TS. 2004. Neurological and developmental outcomes of prenatally cocaine-exposed offspring from 12 to 36 months. *Am J Drug Alcohol Abuse* **30**:299–320.

Pauly JR, Slotkin TA. 2008. Maternal tobacco smoking, nicotine replacement and neurobehavioural development. *Acta Paediatr* **97**:1331–1337.

Wattendorf DJ, Muenke M. 2005. Fetal alcohol spectrum disorders. *Am Fam Physician* **72**:279–282, 285.

Denes AS, Jékely G, Steinmetz PR, Raible F, Snyman H, Prud'homme B, Ferrier DE, Balavoine G, Arendt D. 2007. Molecular architecture of annelid nerve cord supports common origin of nervous system centralization in bilateria. *Cell* **129**:277–288.

Protecting and Maintaining the Adult Nervous System

CHAPTER 5

CONTENTS

FEATURES

One strategy your body uses to maintain itself is to replace damaged or dying cells with new and healthy ones. Skin cells, for example, are replaced roughly every two months, and the cells lining your small intestine live only about a week on average. In contrast, most neurons are irreplaceable. They are born before the birth of the organism and, if they die, no new neurons are formed to take their place.

Although there are exceptions to this rule, the fact that human brains generate almost no new neurons after birth makes our nervous systems more vulnerable than other body parts. It means that neural damage, once incurred, cannot be readily repaired. The nervous system can compensate for damage by reorganizing what remains, but this capacity is limited. Therefore, our nervous systems must be protected well. Some of this protection is provided by the bones and membranes that surround the brain. Additional protection derives from the fluid in which the brain is bathed and from cellular barriers that keep out blood-borne toxins and pathogens. Should these protections fail and neurons start to die, the nervous system's immune response kicks in. In contrast to the body's general immune response, however, the nervous system's immune response is tightly controlled to minimize the chance that activated immune cells, in their battle against toxins or pathogens, kill neurons as "collateral damage."

We will discuss these mechanisms to protect and defend the brain in the following sections. Then we discuss how neurons are supplied with metabolic energy. These topics are related because a few seconds without energy causes neurons to stop working, and a few minutes of energy deprivation causes neurons to die. To minimize such disruptions in energy supply, the brain's vasculature is designed so that blood can be rerouted if a few blood vessels are blocked. In addition, neuronal activity can alter the diameter of nearby blood vessels and thus adjust the rate of blood flow through the area. The fact that neurons can control their local blood supply has been known for a while—and is widely used in functional brain imaging—but the mechanisms underlying these activity-dependent changes in blood flow are just now being revealed. As it turns out, neurons apparently control the blood vessels not directly but through a type of glial cell. Indeed, a major theme of this chapter is that neurons do not function in isolation from other types of cells.

5.1 Are New Neurons Added to Adult Brains?

We discussed neurogenesis in Chapter 4, but we did not review the methods used to determine when cells are "born." Several such methods have been developed. Virtually all of them involve injecting an animal with artificial analogs of thymidine, which is one of the nucleotides used to build DNA. One of these artificial thymidines is **tritiated thymidine**, which is weakly radioactive; another one is **bromodeoxyuridine (BrdU)**. Both compounds are taken up by dividing cells as if they were regular thymidine, and both are incorporated into newly synthesized DNA. Crucially, both compounds can be detected after the tissue is processed appropriately. The tritiated thymidine is visualized by coating thin slices of tissue with a photographic emulsion and then storing this "sandwich" in the dark for several days or weeks. The radioactivity coming from the tritiated thymidine reacts with silver grains in the emulsion, and a few additional processing steps will make the exposed silver grains turn black. Experimenters looking at such an **autoradiograph** through a microscope will see that some cells are covered with small black dots; these labeled cells contain the tritiated thymidine. BrdU, in contrast, is detected immunohistochemically. That is, tissue is sliced and exposed to antibodies against BrdU. Additional reactions are then used to visualize where those antibodies have bound. Any cells labeled with anti-BrdU are presumed to contain significant amounts of BrdU within their DNA.

Neuronal Birth-dating Experiments

If investigators sacrifice an animal within a half hour of when it was injected with BrdU or tritiated thymidine, then all the cells that duplicated their DNA in

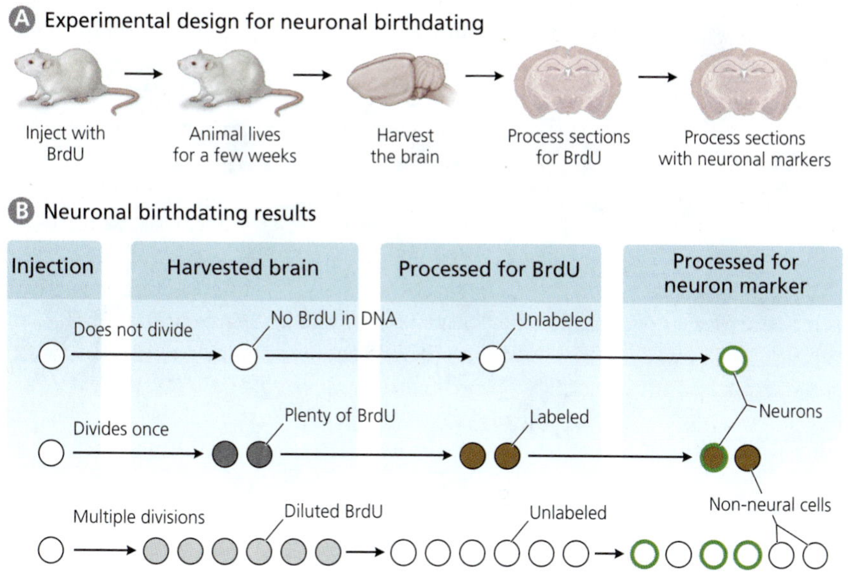

Figure 5.1 Birth-dating experiments. An animal is injected with the thymidine analog BrdU, which is available in the blood for about half an hour before being metabolized, and allowed to live for several weeks (A). Brain sections are then processed with antibodies against BrdU. The results of such an experiment are shown in (B). Any cells dividing shortly after the injection incorporate the BrdU into their DNA. Cells that divide only once retain this BrdU indefinitely, but dividing cells dilute the incorporated BrdU by a factor of two with each division. When the tissue is later processed, only the cells with high levels of BrdU are labeled (there is a labeling threshold). The labeled cells can be interpreted as having been "born" shortly after the BrdU injection. Staining with antibodies against neuron-specific proteins can reveal which of the newborn cells are neurons rather than glial cells.

preparation for cell division during that half hour will be labeled. This reveals where in the body dividing cells are located, but it does not reveal when individual cells are "born" (when they stop dividing). However, cellular birthdates can be obtained if the postinjection survival time is increased to several days (Figure 5.1). There are two reasons why this works. First, any BrdU or tritiated thymidine not incorporated into DNA is degraded by enzymes within about half an hour. Second, the amount of BrdU or tritiated thymidine that is incorporated into a cell's DNA decreases by half every time the cell divides. However, after a cell's last cell division, the DNA nucleotides are not exchanged (DNA is extremely stable). Therefore, after long survival times, any cells that are labeled by BrdU antibodies or radioactivity must have incorporated the BrdU or tritiated thymidine shortly after it was injected and then divided only once. In other words, they must have been born just after the thymidine analogs were injected. By injecting BrdU or tritiated thymidine into animals at different ages, experimenters can deduce when various cells are born.

Adult Neurogenesis in Mammals

Such **birth-dating experiments** have consistently revealed that, at least in monkeys and rodents, the vast majority of neurons are born before the organism itself is born. However, the studies also brought to light a few exceptions to this rule. Most importantly, Joseph Altman and Gopal Das discovered in the 1960s that the brains of rats injected with tritiated thymidine shortly after birth, and sacrificed a few weeks later, exhibit at least a few heavily labeled cells (Figure 5.2). Some of these labeled cells are clearly endothelial cells that line the blood vessels, and others may be glial cells. However, some labeled cells certainly look like neurons under the microscope. They are larger than typical glia and contain a large, pale nucleus. Altman and Das found these putative neurons primarily in the olfactory bulbs and in a small part of the hippocampus called the dentate gyrus. Although (or perhaps because) this finding ran counter to the accepted belief that all mammalian neurons are born prior to the organism's birth, it was widely ignored. Even after 1984, when researchers demonstrated neurogenesis in adult birds, most neuroscientists continued to deny the possibility of adult neurogenesis in mammals. They argued either that bird brains are odd or, in any case, that mammals are different.

In the 1990s, however, new discoveries revealed that mammals are not so different after all. First came a report by Elizabeth Gould and several collaborators confirming that new neurons are added to the dentate gyrus of full-grown, adult rats. Importantly, Gould went a step beyond Altman and Das by showing that the

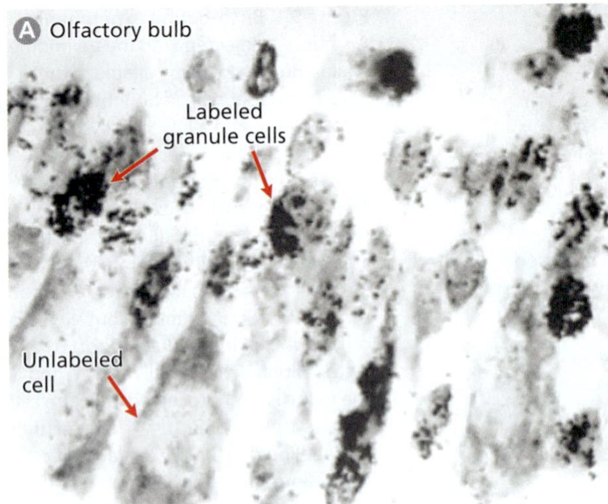

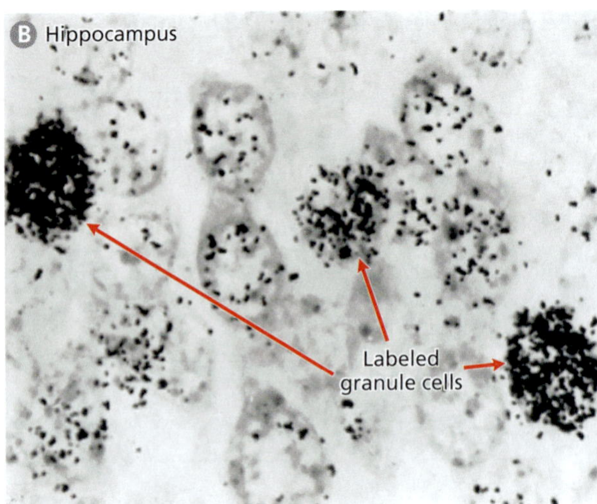

Figure 5.2 Postnatal neurogenesis in rat olfactory bulb and hippocampus. Newborn rats were injected with radioactive thymidine and sacrificed 2–3 weeks later. Any cells born shortly after the injection end up radioactive and can be detected by covering tissue sections with a photographic emulsion that, with proper processing, converts radioactive emissions into dark silver grains. Shown here are photographs of cells covered with such silver grains in (A) the olfactory bulb and (B) the dentate region of the hippocampus. Because these cells are small and contain little cytoplasm, they are called granule cells. [From Altman and Das, 1966]

newborn cells in the dentate of adult rats express a protein that is only expressed in neurons and do not express a protein that is expressed in many glial cells (glial fibrillary acidic protein, abbreviated GFAP). Soon thereafter, Gould and one of her students showed that stress increases neurogenesis as well as cell death in the dentate gyrus of adult rats. This is important because it answers why overall dentate volume does not increase dramatically throughout our adult lives: the new neurons are added to replace old neurons that died. In the same year, another team of researchers confirmed that new neurons are added to the olfactory bulbs in adult mice. All of these findings have now been corroborated by multiple laboratories, using BrdU as well as tritiated thymidine and many different markers for neurons and glial cells. Even in human brains, numerous neurons in the dentate gyrus are born in adulthood (Figure 5.3).

Despite this extensive body of work, uncertainties remain. For example, a recent study indicates that the olfactory bulb exhibits much less neuron turnover in humans than in rats. Are humans the outliers here, or are rodents unusual? The answer is unclear. Other studies question whether adult neurogenesis exists in the neocortex. Some researchers have described adult-born neurons in the neocortex of macaque monkeys, but these findings have not been replicated independently. A major obstacle to resolving this debate is that the detection of low levels of neurogenesis requires multiple injections of tritiated thymidine or BrdU, which raises the likelihood of artifactual labeling.

Figure 5.3 Adult neurogenesis in human hippocampus. A human cancer patient received BrdU injections for diagnostic purposes; on autopsy his brain was examined. Shown here is a section through the dentate gyrus of the patient's hippocampus stained with an antibody against BrdU (green), a second antibody against NeuN (which labels neurons red), and a third antibody that stains astrocytes purple. You can see one BrdU-labeled neuron. [From Eriksson et al., 1998]

Carbon Dating for Neurons

To circumvent this problem, an international team of collaborators invented "carbon dating for neurons." They took

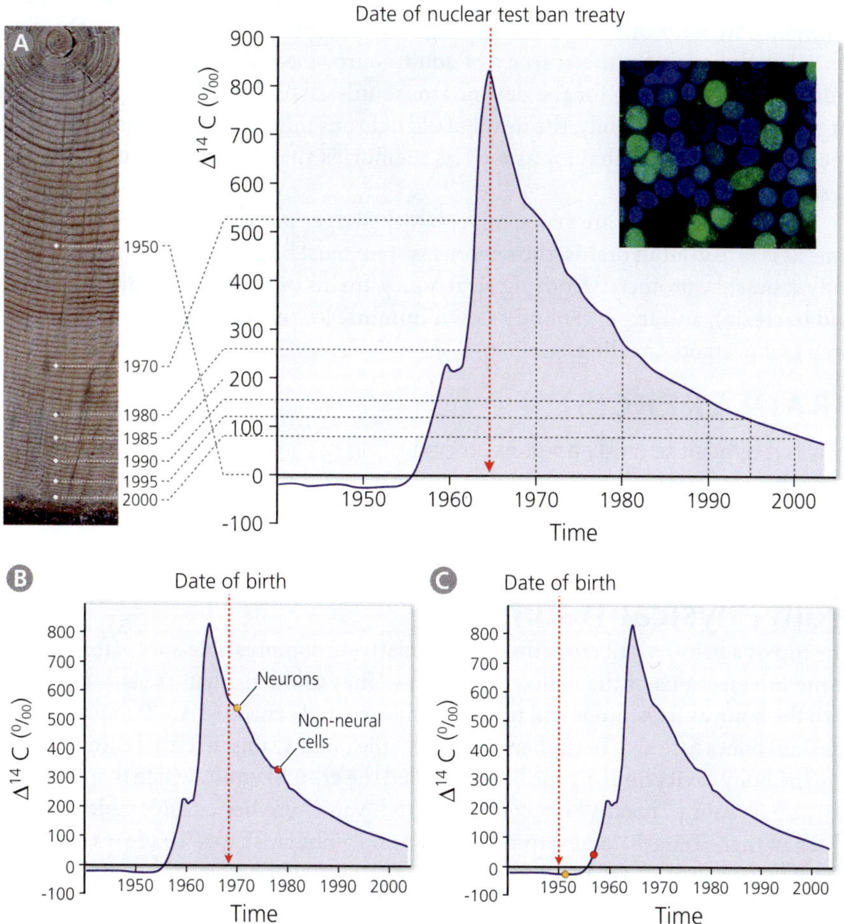

Figure 5.4 Carbon dating of human brain cells. (A) The concentration of radioactive carbon (^{14}C) in the atmosphere increased as a result of above-ground nuclear bomb tests and then decreased after 1963. This was verified by measuring the ^{14}C levels in tree rings (top left). Armed with this information, researchers homogenized pieces of neocortex from deceased humans and incubated the solution with a neuronal marker (green cell nuclei at top right). Neuronal nuclei were then separated from non-neuronal nuclei by a cell-sorting machine, and the amount of ^{14}C in each solution was measured by accelerator mass spectrometry. From this, the scientists calculated the level of ^{14}C per average neuron and average non-neuronal cell. Next, the scientists asked when in history (but after the person's birth) atmospheric ^{14}C levels corresponded to the level of ^{14}C in the cell nuclei. Because DNA is very stable, this look-up procedure reveals when the selected cells were born. Examination of patients born either after of before the test ban treaty (B and C) showed that the average neuron in the neocortex is born around the time the person is born, whereas the average non-neuronal cell (mainly glia) is born many years later. [From Spalding et al., 2005]

advantage of the fact that above-ground explosions of nuclear bombs during the Cold War dramatically increased the amount of **radioactive carbon** (^{14}C) in the atmosphere (Figure 5.4 A). After the Partial Test Ban Treaty was signed in 1963, ^{14}C levels declined again, not because the isotope decayed (its half-life is almost 6,000 years) but because the atmospheric ^{14}C was absorbed into the oceans and incorporated into living cells through photosynthesis. Given these observations, the experimenters reasoned that the amount of ^{14}C in the DNA of mature brain cells should reflect the level of atmospheric ^{14}C that existed when those cells were born. For example, if you were born in 1969 and the amount of ^{14}C in your neurons matches the atmospheric ^{14}C levels from 1969, then the vast majority of your neurons must have been born prior to your birth or shortly thereafter (Figure 5.4 B). Similarly, if you were born in 1955 and your neurons contain essentially no ^{14}C, then your neurons must have been born before atmospheric ^{14}C levels increased (before ~1956; Figure 5.4 C).

After studying numerous individuals of various ages, and using antibodies to discriminate between neurons and glial cells, the scientists concluded that virtually all neurons in the human neocortex are born around the time of birth. Non-neural cells, in contrast, continue to be added to neocortex in adulthood. Most of these late-born, non-neural cells are glia, but some may be endothelial cells in the lining of blood vessels.

Why is adult neurogenesis so scarce in most parts of the human brain? The likely answer is that most neurons have long axons that require, as we discussed in Chapter 4, complex molecular signals to reach their targets. If these axon guidance cues exist only in embryos, then neurons born in adulthood would have difficulty forming proper connections. Indeed, the adult-born neurons in the mammalian dentate gyrus have relatively short axons, and those in the olfactory bulb have no axons at all; they communicate with other neurons through dendrodendritic synapses. Another potential explanation for the scarcity of adult neurogenesis is that some old neurons must die to make room for the new neurons, unless you can afford to grow an ever larger brain. Unfortunately, the death of old neurons might destroy neuronal circuits needed for ongoing behavior as well as memories that those old neurons helped to store.

These explanations are speculative, but the larger point is clear: because neurogenesis is rare in adult brains, the nervous system must be guarded diligently. Specifically, it must be protected from physical injury, toxins or pathogens (harmful viruses and bacteria), and from the body's own immune system, which tends to kill cells even as it promotes healing.

BRAIN EXERCISE

Why is it difficult to study adult neurogenesis in humans, and how can those difficulties be overcome?

5.2 How Is the Brain Protected from Physical Trauma?

The top of a baby's skull contains several relatively compressible spots (fontanelles) at the intersections of the major skull bones. They allow the baby's skull to expand with the brain as its neurons add myelin, dendrites, and synapses. At ~9 months of age, the skull bones fuse and, from then on, protect the brain against external compression.

The bony cavity housing the brain is called the **cranial vault**, which is appropriate because "cranium" means helmet in Latin and "vault" implies considerable strength. Much of that strength comes from the cranium's globular shape. To see why globular structures are strong, try breaking an egg by squeezing it between your palms (wear an apron). Breaking an egg in this manner is harder than you might expect, mainly because the compressive forces are distributed along the egg shell's surface, which means that less of your applied force is directed toward the egg's center. Much of the

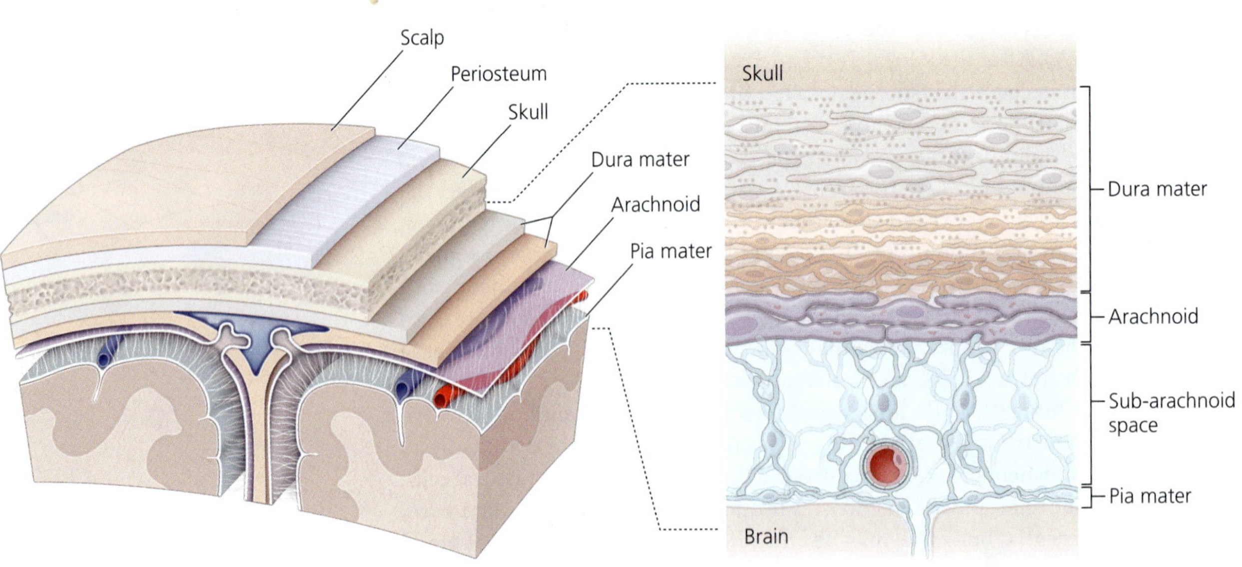

Figure 5.5 The brain's protective envelopes. Shown on the left are the layers that surround and protect the brain, starting with the scalp and ending with the thin pia mater. Blood vessels course mainly through the subarachnoid space (between the pia mater and the arachnoid membrane) and within the dura mater, which is the toughest of the three meningeal membranes. Shown on the right is a more detailed section through the meninges. The subarachnoid space is "spongy" and filled with cerebrospinal fluid. [After Haines, 1991]

cranium's strength derives from its egg-like shape. Think about it: a strong blow to the head is much more likely to break your jaw or nose than crack your skull: unless, of course, the force is focused on a tiny spot, as it would be with a bullet.

Meninges and Cerebrospinal Fluid

Between the cranial vault and the brain lie several protective membranes called the **meninges** (Figure 5.5). The outermost of these meninges is the **dura mater**, which means "tough mother" in Latin. This name is quite appropriate because the dura mater is tough indeed: when doctors want to cut it during surgery, they have to use a very sharp scalpel. The innermost of the meninges is the **pia mater.** It is much more delicate than the dura and tightly covers the central nervous system's external surface. Between the pia mater and the dura mater lies the **arachnoid membrane,** and just below it a spongy layer called the **subarachnoid space**. The "arachnoid" in both these terms refers to the fact that the subarachnoid space looks as if it were replete with tiny spider webs, made primarily of collagen.

Although the subarachnoid space and its two adjacent membranes are rather flimsy, they do protect the brain and spinal cord because they form a double-layered sack that holds a salty, watery substance called **cerebrospinal fluid (CSF)**. The density of CSF is nearly identical to that of nervous tissue. Therefore, the brain floats in the CSF, just as you tend to float in salt water. Without this buoyancy, the brain would slump against the bottom of your cranium, compress the medulla, and kill neurons.

Buoyancy also decreases the force with which the brain slams against the cranial wall when you bang your head against a hard surface. The fundamental physics behind this protection go back to Archimedes, who discovered that a fluid always pushes against a body immersed in it with a force equal to the weight of the displaced fluid (Archimedes' principle). If physics is not your thing, do this: put an egg in a zip-lock bag and drop it from waist-high. The egg will break. Now repeat the experiment, but fill the bag with salt water. Most likely, your egg will stay intact (though the bag might burst open). This is a graphic illustration of how the meninges, together with the CSF, protect your brain from physical trauma.

Ventricles and Choroid Plexus

The brain and spinal cord are bathed in CSF not only externally but also from within, through the cerebral ventricles (Figure 5.6 A). The largest portion of the ventricles lies in the paired cerebral hemispheres. Neurobiologists typically refer to these paired ventricular spaces as the left and right **lateral ventricles**. They are connected with one another through narrow tubes (the foramina of Monro), which lead into a single, slit-like ventricle that lies between the left and right sides of the diencephalon. This is the **third ventricle**. It transitions caudally into a thin tube called the **cerebral aqueduct**. The cerebral aqueduct passes through the midbrain and opens up into a more spacious **fourth ventricle**, which lies within the medulla. From the fourth ventricle extends another narrow tube that reaches down into the spinal cord; this is the **central canal**.

Most of the CSF is produced by the **choroid plexus**, which is found in the floor of the two lateral ventricles (Figure 5.6 B), as well as in the roof of the third and fourth ventricles.

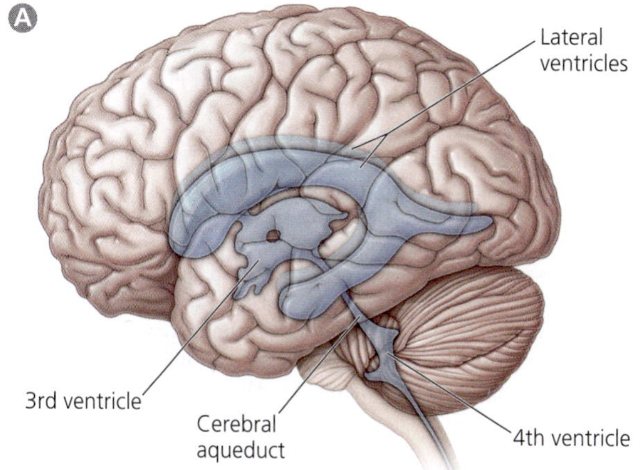

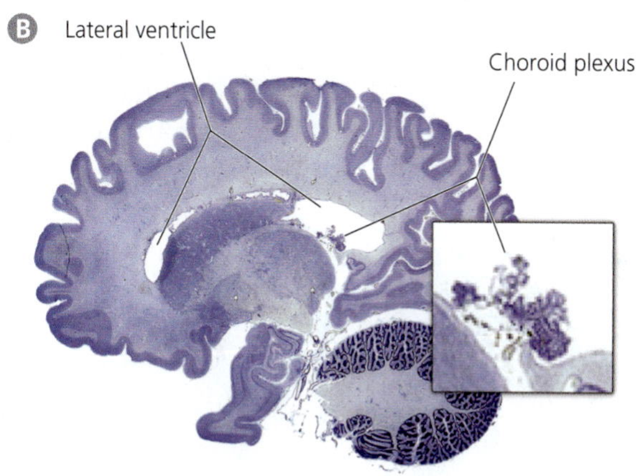

Figure 5.6 Cerebral ventricles and choroid plexus. (A) The ventricle that extends into the telencephalon is divisible into left and right halves, referred to as the left and right lateral ventricles. The thalamic portion of the ventricle is the third ventricle. The fourth ventricle lies in the hindbrain. Importantly, all the ventricles are continuous with one another. The narrow canal joining the third and fourth ventricles is called the cerebral aqueduct. The bottom image depicts a sagittal section through a human brain, highlighting the location of the choroid plexus, which is critical for producing the CSF. [B from brainmuseum.org]

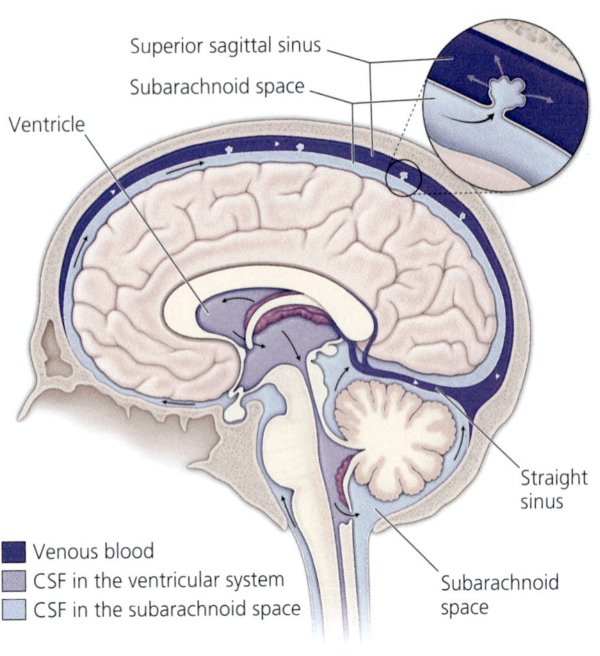

Figure 5.7 Circulation of cerebrospinal fluid (CSF). CSF is secreted primarily by cells of the choroid plexus (purple). Because the choroid plexus is largest in the lateral ventricles (LV), the CSF tends to flow from the forebrain toward the hindbrain. It exits the brain through small holes in the roof of the fourth ventricle, ending up in the subarachnoid space that surrounds the brain and spinal cord. The CSF eventually leaves the subarachnoid space through arachnoid granulations, which protrude into the veins and venous sinuses that drain blood from the brain (see insert). [From a supplement to Masdeu et al., 2009]

The choroid plexus is very thin, extensively folded, and associated with a dense capillary bed. Blood plasma seeps through the relatively porous walls of the choroid plexus capillaries but is then contained within the extracapillary space by epithelial cells that are linked by tight junctions. Only small lipophilic molecules can cross the choroid epithelium to enter the ventricles without some special assistance. However, **membrane transporters** ferry some large proteins across the choroid epithelium into the CSF. In addition, water molecules can enter the ventricles through specialized water channels called **aquaporins**. Therefore, you can think of CSF as salt water to which select proteins have been added. According to some estimates, adult humans produce ~500 ml of CSF per day.

Circulation of the CSF

Where does all this CSF go after it has been produced? It flows at first into the ventricles. From there the CSF flows into the subarachnoid space surrounding the brain and spinal cord (Figure 5.7). The journey to the subarachnoid space passes through the roof of the fourth ventricle, which contains several large holes through which the CSF can flow.

In addition, CSF seeps into the subarachnoid space through the brain tissue and pia mater. This seepage occurs because the CSF in the ventricles is under pressure, and the **ependymal cells** that line the ventricles are relatively permeable to CSF in adult brains. As CSF flows through the brain **parenchyma** (the bulk of all brain tissue), it gathers neuronal waste products and carries them toward the brain surface. There it crosses the relatively permeable pia mater and enters the subarachnoid space.

Where does the CSF go from there? It passes through the arachnoid membrane into large venous sinuses coursing inside the dura mater (Figures 5.7 and 5.8). Much of this CSF transport occurs at specialized protrusions of the arachnoid membrane into the veins and venous sinuses near the brain's dorsal midline. The largest of these protrusions are known as **arachnoid granulations**. They allow the CSF to be recycled, if you will, into the blood.

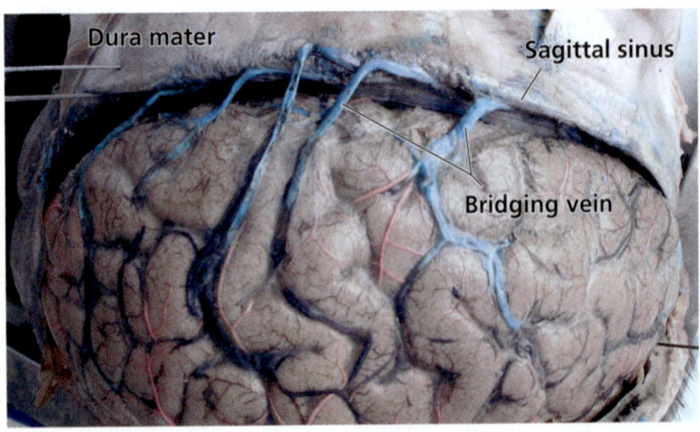

Figure 5.8 The brain's bridging veins. In this human brain, the veins and arteries colored blue and red, respectively. The dura mater was peeled back from the right hemisphere, revealing the bridging veins that drain blood from the brain into the superior sagittal sinus, which runs within the dura. [From Han et al., 2007]

Pressure Kills Neurons

Although the CSF helps to protect the brain and spinal cord, it can also cause harm. Because CSF is produced at a steady rate, any blockage in the flow of CSF from the choroid plexus to the subarachnoid space increases intraventricular pressure. This causes the ventricles to expand and creates a condition known as "water on the brain" or **hydrocephalus**. If this happens during embryonic development, the embryos develop with enormous ventricles and, typically, smooth cortices. Such embryos often abort spontaneously or die soon after birth unless the pressure in the ventricles is surgically reduced. Only rarely are the blockages resolved without intervention (Figure 5.9).

More commonly, reductions in intraventricular pressure are achieved by implanting a tube that runs from the ventricles under the skin via a one-way valve into a major body cavity (usually the peritoneal cavity). This sounds surreal, but works. Cerebral shunts are also used in adult

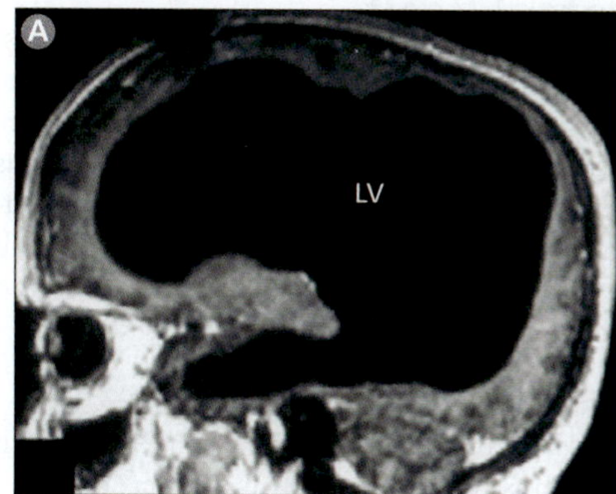

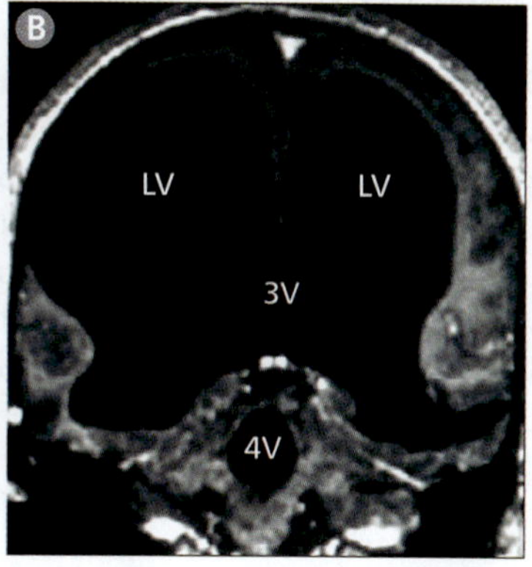

Figure 5.9 Hydrocephalus in an adult human. Shown here are sagittal (A) and coronal (B) MRI scans through the head of a 44-year-old man who developed hydrocephalus shortly after birth, was treated with an intraventricular shunt, and then led a relatively normal life. He was married, had two children, worked as a civil servant, and had an IQ of 75. His lateral, third, and fourth ventricles (LV, 3V, and 4V, respectively) were dramatically enlarged. [From Feuillet et al., 2007]

brains if CSF outflow is blocked (often by a tumor compressing the cerebral aqueduct). Without a shunt, such blockages cause rises in intraventricular pressure, which then lead to headaches, nausea, and destructive changes in brain shape. Intraventricular pressures greater than 25 torr (half a pound per square inch) are usually lethal.

Subdural Hematoma

Brain damage also results when strong blows to the head or rapid decelerations (as might occur in a car crash) violently push the brain against the cranial vault. Such head trauma may cause some neurons to be squashed, and it may rupture blood vessels. In particular, a severe blow may tear the **bridging veins** that course from the cerebral surface to the dural veins and venous sinuses (see Figure 5.8). Such tears cause blood to flow into the subarachnoid space and form what is called a **subdural hematoma**. Some of the smaller veins within the dura may also rupture, causing blood to accumulate inside the dura. A third terrifying possibility is that blood may spill into the space between the dura and the cranium from vessels that supply the meninges. This happens most often when a person is hit in the temple area, where a major meningeal artery runs through the skull. No matter where the leaking blood accumulates, it compresses the brain. The pressure in this case comes from the outside, rather than from inside as in hydrocephaly, but it is just as dangerous. Any type of pressure may kill neurons directly, or it may alter brain shape to the point that many axons are severed.

Concussions

Because the brain's protection against physical trauma is not perfect, helmet laws are generally a good idea. They have already saved thousands of lives. Furthermore, any time you get a concussion you ought to visit a neurologist, even if you think everything is fine. Intracranial bleeding is sometimes slow, which means that pressure on the brain increases gradually and symptoms are delayed. Fortunately, three-dimensional brain X-rays, commonly known as **computed tomography (CT) scans**, let doctors identify even small hematomas with relative ease. Even if those scans reveal no obvious damage, repeated concussions can cause long-term damage. This has long been apparent in professional boxers, who frequently develop cognitive impairments late in their careers or after retirement. However, similar problems are now recognized also in other contact sports, especially American football. The general syndrome caused by repeated head trauma is called **chronic traumatic encephalopathy (CTE)**. It involves substantial inflammation of the brain, diverse forms of neurodegeneration, and progressive dementia.

BRAIN EXERCISE

What would happen to the brain if the choroid plexus were damaged (not functioning) in adulthood?

5.3 How Does the Brain Protect Itself against Toxins and Pathogens?

Because the nervous system requires energy, derived mainly from glucose and oxygen, it must be supplied with blood. Unfortunately, blood also carries within it numerous substances that can damage neurons. Lead, mercury, pesticides, various other pollutants, pathogenic viruses, bacteria, and even the body's own immune cells would all wreak havoc in the CNS if they had free access to it. Fortunately, vertebrates evolved tight barriers between the blood and their neurons.

The Blood–brain Barrier

The discovery of a **blood–brain barrier** dates back to the 1880s, when Paul Ehrlich injected various aniline dyes into the veins of animals and noted that the dyes leaked out of the blood vessels into most tissues, but not into the brain and spinal cord. This was not what Ehrlich had expected. Struggling to interpret his results, Ehrlich concluded that the dyes must have leaked into the brain and spinal cord, just as they entered all other organs, but were incapable of staining neurons or glial cells. This hypothesis was wrong. In 1913, Edwin Goldman, a student of Ehrlich's, injected the same dyes Ehrlich had used directly into the CSF and noted that the brain and spinal cord were stained just fine. Therefore, Goldman concluded, neural tissue must be separated from the blood by some sort of barrier that keeps out aniline dyes and, most likely, some other molecules.

The mechanistic basis of the blood–brain barrier remained obscure until 1967 when scientists injected into the blood vessels of mice some tracer molecules (horseradish peroxidase or, later, biotinylated dextrans) that could be visualized with an electron microscope. They found that the tracer molecules remained, just like Paul Ehrlich's dyes, within the blood vessels. However, closer inspection revealed that tracer molecules did penetrate the spaces between the endothelial cells lining those vessels, up to the point where the cell membranes of neighboring endothelial cells became closely apposed (Figure 5.10).

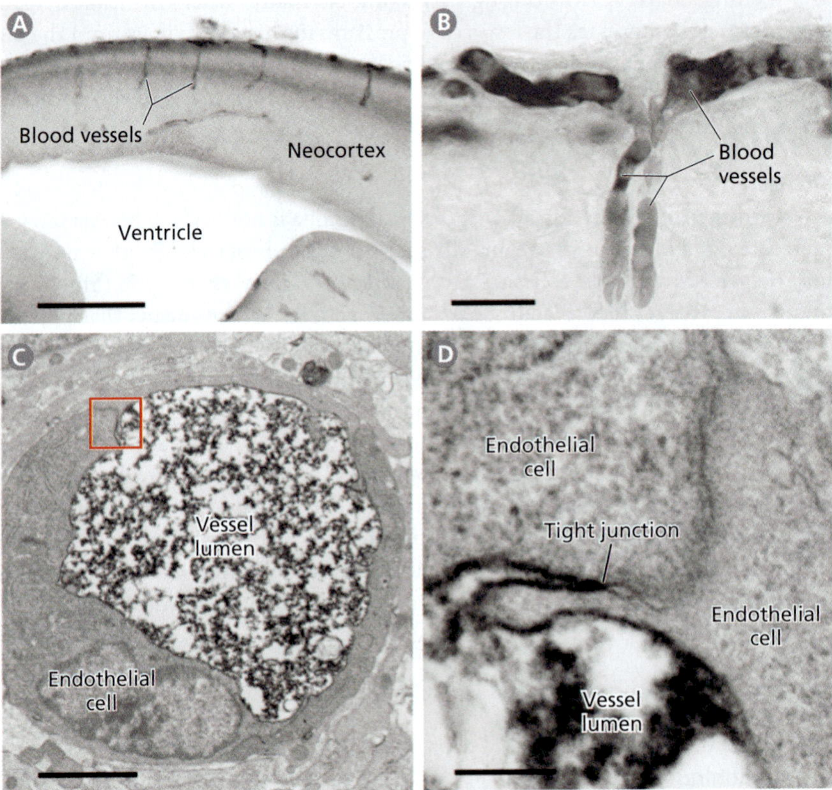

Figure 5.10 The blood–brain barrier. Small tracer molecules (biotinylated dextrans) were injected into the bloodstream of newborn opossums, which were then sacrificed. The image in (A) shows how the blood vessels containing the darkly stained tracer dip radially into the cerebral cortex from the brain surface. Image (B) is a close-up of two penetrating blood vessels. Panel (C) depicts a section through a penetrating blood vessel imaged with an electron microscope. It shows how a vessel's interior, or lumen, is surrounded by the vessel's endothelial cells. Image (D) is a higher magnification view of the area marked by the red square in (C). You can see that the tracer molecules spread from the vessel lumen into the spaces between the endothelial cells, but only to the point where the endothelial cell membranes are joined by tight junctions. Scale bars equal 200 μm in A, 25 μm in B, 2μm in C, and 200 nm in D. [From Ek et al., 2006]

THERAPIES

Box 5.1 *Delivering Drugs to the Brain*

The blood–brain barrier protects the brain from circulating toxins and pathogens, but it also prevents beneficial compounds from getting to the brain. Even compounds that reach the brain are often pumped back out or quickly degraded. This explains, for example, why brain cancers are difficult to treat: the chemotherapy drugs that attack most other cancers have trouble reaching the tumors. Similarly, drugs that might be used to treat Alzheimer's or other neurodegenerative diseases have a hard time crossing the blood–brain barrier. Nor are the RNA or DNA molecules that might be used for brain "gene therapy" easily delivered to the brain.

To overcome these problems, one can inject drugs or genes directly into the brain parenchyma or ventricles, but this entails some serious risks, especially of infection. Alternatively, one can temporarily weaken the blood–brain barrier. For example, one can inject a concentrated sugar solution into a cerebral artery. This causes water to diffuse out of the capillary endothelial cells, making them shrink and pulling the tight junctions between the cells apart. Another approach is to inject an artery with tiny air bubbles and then focus a beam of ultrasound onto a brain region fed by that artery. The ultrasound interacts with the microbubbles, and the resulting mechanical forces temporarily open the blood–brain barrier. While the blood–brain barrier is open, the therapeutic drug can enter the brain. So can toxins and pathogens, of course.

A more recent strategy for shuttling drugs into the brain is to use drug-filled nanoparticles (Figure b5.1). Some types of nanoparticles are solid balls suffused with drug; others are tiny hollow spheres surrounding a drug-filled core. Some are made from lipid membranes; others are made from polymers. To protect the nanoparticles from attack by the body's immune system, they are usually coated with "camouflaging" molecules such as polyethylene glycol. The nanoparticles may also be coated with molecules that are actively transported into the brain, such as transferrin or antibodies to the transferrin receptor (Figure b5.1). Because this receptor's normal function is to shuttle transferrin across the blood–brain barrier, it can transport the nanoparticles into the brain as well. In essence, the membrane transporters are tricked into ferrying a secret payload across the blood–brain barrier. Drug designers refer to such nanoparticles as "Trojan horses."

A less flashy means of getting drugs into the brain is to inject a precursor that crosses the blood–brain barrier. An example of this strategy is the use of L-dopa to treat the dopamine deficiency associated with Parkinson's disease (see Chapter 15). Dopamine does not cross the blood–brain barrier, but its precursor L-dopa is transported across the barrier by the same transporter molecules that bring large amino acids into the brain. Once in the brain, L-dopa is converted to dopamine. Therefore, oral doses of L-dopa boost dopamine levels inside the brain. Although L-dopa is not a cure for Parkinson's disease, it does alleviate its symptoms for a while.

Because many therapeutic drugs do not have lipophilic precursors, scientists may try to make the drugs themselves more lipophilic. This sometimes works, but it also works for drugs of abuse. Heroin, for example, is more lipophilic than its natural cousin morphine and therefore enters the brain 100 times more easily. Once in the brain, heroin is converted to morphine and has similar analgesic and euphoric effects.

Speaking of drugs, you ought to know that methamphetamine causes a significant breakdown of the blood–brain barrier. This observation has prompted suggestions that methamphetamine might be used to facilitate the delivery of therapeutic drugs into the brain. For fairly obvious reasons, this idea has not caught on. For one thing, methamphetamine is highly addictive. For another, any prolonged and uncontrolled disruption of the blood–brain barrier leaves the brain extremely vulnerable to toxins and pathogens. It is probably no accident that HIV-infected methamphetamine users are more likely than non-users to have the HIV virus damage the brain. The larger lesson here is simple: be kind to your blood–brain barrier; it exists for a reason!

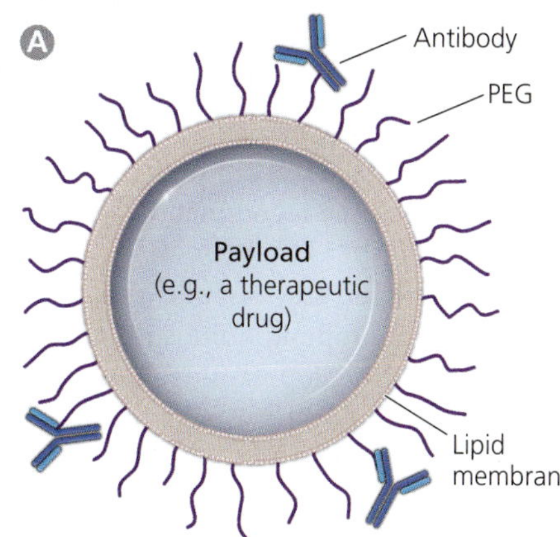

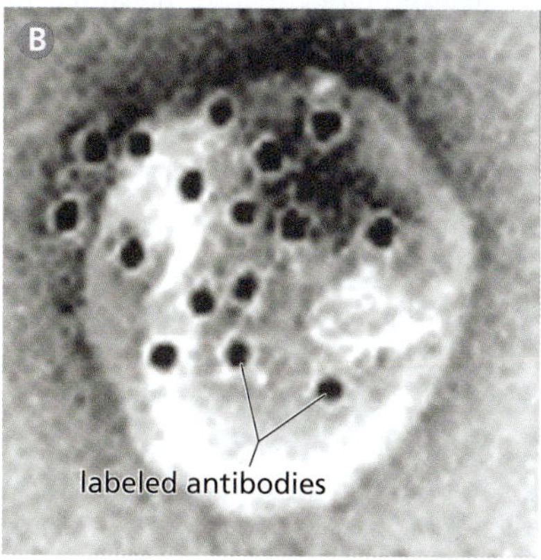

Figure b5.1 Trojan horse nanoparticles. Shown in (A) is a schematic section through a hollow nanoparticle that is covered with strands of polyethylene glycol (PEG) to prevent its detection and removal by the immune system. Bound to the tips of some PEG strands are antibodies against the transferrin receptor. Once such a nanoparticle binds to transferrin receptors at the blood-brain barrier, it is taken up by capillary endothelial cells and released into the brain. Shown in (B) is a scanning electron micrograph of an actual nanoparticle (a liposome) covered with antibodies that appear black because the experimenters reacted them with secondary antibodies that, after processing, became electron dense. The liposome is $<$ 200 nm in diameter. [B from Pardridge, 2003]

Such membrane appositions are called **tight junctions**, and they are formed by special molecules called claudins and occludins. In essence, the experiments revealed that endothelial cells lining the capillaries in the brain and spinal cord are stitched together by numerous tight junctions that block the passage of many molecules. In contrast, the endothelial cells surrounding capillaries in other organs are joined by fewer tight junctions and, therefore, more leaky. We already encountered some of these leaky capillaries in our discussion of the choroid plexus. The capillaries in the brain and spinal cord retain their contents much better.

Of course, brain capillaries must allow some substances to cross; otherwise their existence would be pointless. Indeed, oxygen and carbon dioxide diffuse readily across the blood–brain barrier, as do small lipophilic molecules, such as alcohol and most anesthetics. Large hydrophilic molecules, such as glucose and some amino acids, cannot diffuse across the blood–brain barrier, but the endothelial cells lining CNS capillaries are full of membrane transporters for glucose and other molecules. Some of these transporters move various ions across the endothelial membranes, which requires metabolic energy because the ions must be moved against their concentration gradients. For example, the transport of sodium ions into the brain involves the same Na^+/K^+-ATPases that help to generate a neuron's membrane potential (see Chapter 2). Other transporters move metabolic waste products and other compounds out of the CNS into the blood. Last, but not least, water molecules traverse the endothelial cell walls through the same sort of aquaporin channels found in the epithelial cells of the choroid plexus. In summary, the blood–brain barrier is not a simple barrier at all; it is a complex mix of semipermeable membranes, channels, and transporters.

Drugs and the Blood–brain Barrier

Understanding this complexity is crucial if you want to deliver drugs from blood to brain (see Box 5.1: Delivering Drugs to the Brain) or, conversely, if you want to keep a therapeutic drug out of the brain. A good example of the latter is provided by histamine antagonists, commonly known as **antihistamines**. They reduce allergic reactions to irritants in the eyes, nose, and sinuses; but when antihistamines enter the brain, they interfere with the ability of the neurotransmitter histamine to promote wakefulness (see Chapter 13). This explains why first-generation antihistamines, such as diphenhydramine (Benadryl), make you so sleepy; they cross the blood–brain barrier and block histamine signaling inside the brain. In contrast, the second generation of antihistamines, including loratadine (Claritin), are less lipophilic and, thus, less likely to enter the brain. Even if some of their molecules do get into the brain's capillaries, they are pumped back out by a transporter molecule called P-glycoprotein. Therefore, the newer antihistamines reduce allergic reactions in peripheral organs without making you drowsy. Of course, if you exceed the recommended dose, even the newer drugs have a sedating influence.

Another good example of a drug that is designed to stay out of the brain is the anti-diarrhea drug **loperamide** (Immodium). It is an opioid quite similar to morphine (see Chapter 7); but in contrast to the latter drug, loperamide does not cross the blood–brain barrier. Therefore, loperamide causes constipation, as most opioids do, but it does not share morphine's ability to reduce pain.

The Blood-CSF and Arachnoid Barriers

The blood–brain barrier is complemented by an equally tight **blood-CSF barrier**. This makes sense, because it would be pointless to bar toxins from crossing the brain's capillaries if they can enter the CSF, and from there the brain, through the choroid plexus. However, the tight junctions that form the blood-CSF barrier are located not between the leaky endothelial cells of the choroidal capillaries but between the choroid epithelial cells that face the ventricle. This was demonstrated by injecting a dye into the blood and noting that the dye leaks out of the choroidal blood vessels and penetrates the space between choroidal epithelial cells until it reaches the impenetrable tight junctions (Figure 5.11).

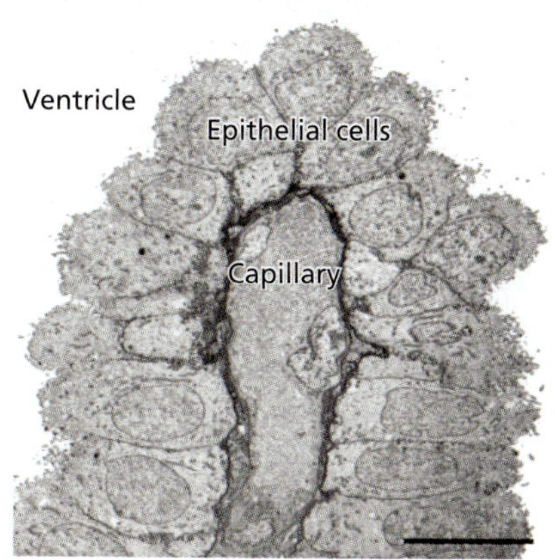

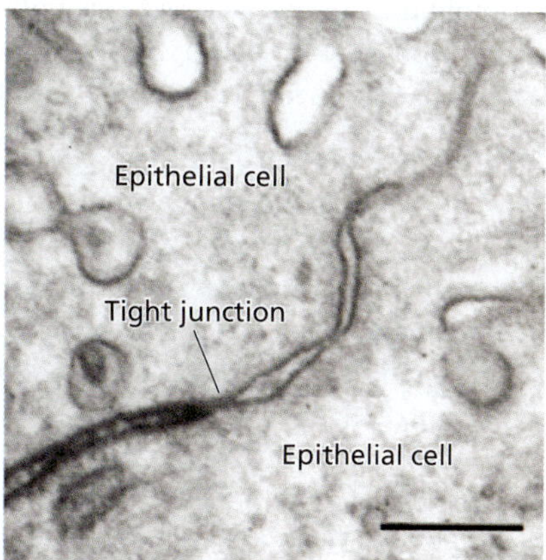

Figure 5.11 The blood-CSF barrier. Shown here are electron micrographs of sections through the choroid plexus of a newborn opossum that received an intravenous injection of darkly staining tracer. Shown on the left is a choroid plexus capillary. The tracer molecules were washed out of the capillary lumen but remain evident in the surrounding intercellular spaces (dark stain). This is shown at higher magnification on the right. The tracer has diffused between two epithelial cells up to where their cell membranes are coupled by a tight junction. Such tight junctions link all epithelial cells around choroidal capillaries and thereby keep the tracer molecules from leaking into the ventricle. Scale bars equal 10 μm and 200 nm. [From Ek et al., 2003]

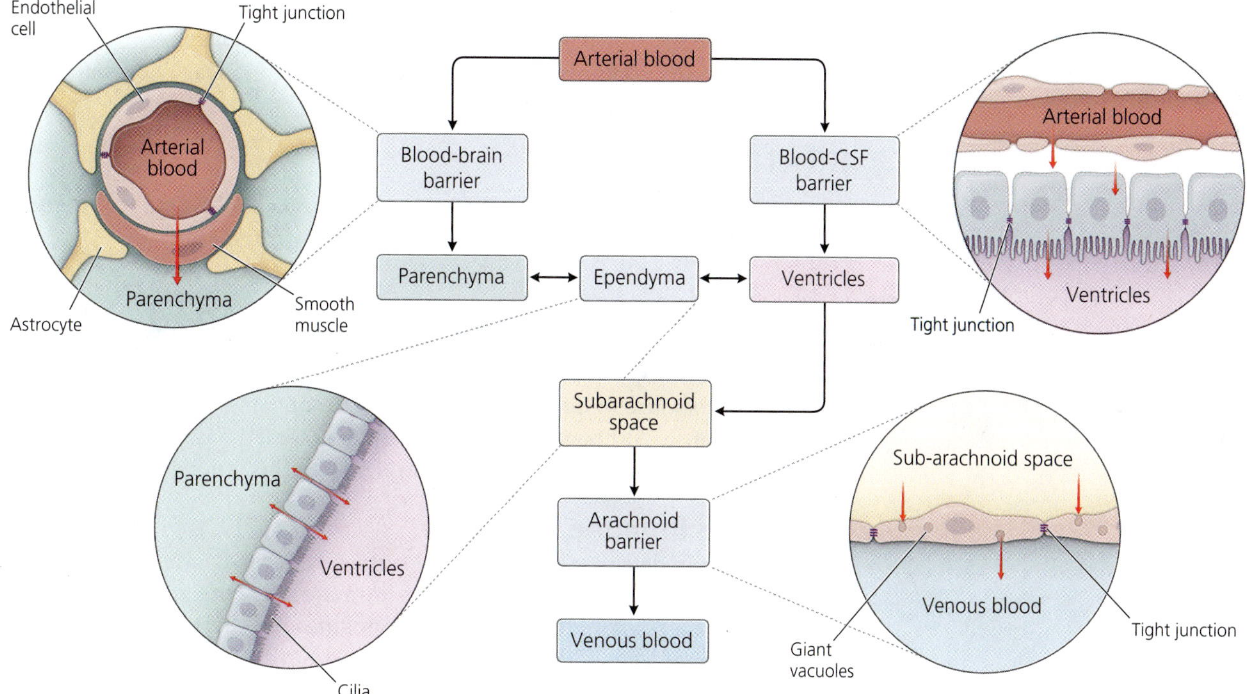

Figure 5.12 Barriers between blood, brain, and ventricles. The blood–brain barrier restricts the movement of molecules across brain capillary walls. Because the capillaries' endothelial cells are joined by tight junctions (small purple ovals), large lipophobic molecules can leave the capillaries and enter the brain's parenchyma only if they are transported across the endothelial cell walls. A similar barrier exists in the choroid plexus. However, the tight junctions in the blood-CSF barrier are located not between endothelial cells but between the epithelial cells of the choroid plexus. CSF from the ventricles readily enters the brain parenchyma, and vice versa, at least in adults. CSF from the ventricles ultimately collects in the subarachnoid space around the brain. It is transported from there into veins through "giant vacuoles" that can move through the arachnoid membrane. Not shown is the barrier between parenchyma and subarachnoid space, which is less well understood.

Further thought suggests that some sort of barrier ought also to exist where the CSF drains back into the blood, namely, at the arachnoid granulations, and more generally, the arachnoid membrane (Figure 5.12). Indeed, cells of the arachnoid membrane are sewn together by innumerable tight junctions, just like the cells at the blood–brain and blood-CSF barriers. In contrast to those other barriers, however, the **arachnoid barrier** contains few transporter molecules. Instead, CSF moves through its cells in **giant vacuoles**. These vacuoles engulf CSF when they form on the subarachnoid side of the arachnoid membrane, float through the arachnoid cells, and dump their contents into the veins and venous sinuses on the dural side of the arachnoid membrane. Because this bulk transport of CSF is unidirectional, always moving from CSF to blood, there is no danger of blood-borne toxins or pathogens invading the CSF from the venous side of the circulation.

Circumventricular Organs

Although the CNS is well insulated from the blood, there are a few places close to the ventricle where the blood–brain barrier is imperfect. At these **circumventricular organs** dyes injected into the blood can leak into neural tissue. This seems like a serious breach of security, but the circumventricular organs have tight junctions "behind them" as a backup barrier that protects the remaining brain. Why have the circumventricular organs at all? They exist so that some neurons can sense the blood's composition and initiate appropriate responses, such as drinking more water when the blood is too dehydrated or breathing faster when blood CO_2 levels are high. We will discuss these topics in Chapters 9 and 10.

In summary, the brain and spinal cord are well protected behind a series of barriers. Unfortunately, infection of the meninges (meningitis), physical trauma, some diseases (such as multiple sclerosis or cancer), and some drugs can weaken these barriers, creating further risks. However, a weakening of the brain's protective barriers can also be good for the brain because it allows some immune system cells to enter the brain. To fully comprehend this assertion, you need to learn how our nervous system responds to ongoing attacks.

BRAIN EXERCISE

Why do you think the brain is protected by such tight cellular barriers, whereas other vital organs, such as the liver, lack this kind of protection?

5.4 How Does the Nervous System Respond to an Attack?

One crucial response of our nervous system to an ongoing assault is to mount an immune response. Although the nervous system is sometimes said to lack such an immune response, in the next section, we explain that this is not quite true.

The Brain's Immune Response

Our body's main defender is the immune system, which is commonly divided into innate and adaptive subsystems. The **innate immune response** generates inflammation, which consists mainly in the release of cytokines and chemokines, and prompts some white blood cells to kill damaged cells and devour debris. These innate responses are triggered rapidly in response to injury. In contrast, the **adaptive immune system** is slower to respond. Its cells bind pathogens, migrate to the lymph nodes, and there present the pathogens to special white blood cells called **lymphocytes**. Some of these lymphocytes respond by releasing cytokines, which kick-start the aforementioned innate immune response. Other specialized lymphocytes, called B-cells, respond by making antibodies against the pathogens (antigens). Each B-cell responds only to a few specific antigens; but when it does respond, the B-cell divides and thus makes more cells of its kind. Therefore, the body soon swarms with B-cells making antibodies against the presented antigen. When these antibodies bind to their matching antigens, macrophages and special "killer cells" derived from bone marrow are stimulated to remove the threat.

This summary of the body's immune response reveals an obvious problem for the brain as it attempts to fight off attackers: if the nervous system lies behind impenetrable barriers, how can antigen-presenting cells get from the nervous system to the lymph nodes, and how can killer cells and macrophages get into the nervous system to remove threatening cells or debris? One possible answer is that the body's immune cells do not, in fact, cross the brain's various barriers. This hypothesis was supported by experiments in which skin or tumor cells were transplanted from one animal into the brain parenchyma of a second animal. Such grafts were not attacked by immune cells. That is, the grafts were rejected slowly or not at all. In contrast, when the same cells were transplanted under the skin of the second animal, they evoked a strong immune response and were rejected rapidly. Such findings suggested that the central nervous system is **immune privileged**. This notion fits well with the idea of the nervous system sitting behind tight barriers, and it is still widely believed. However, it is not quite correct, for the nervous system is capable of mounting a significant, albeit unorthodox, immune response.

One line of evidence against the nervous system being immune privileged is that tissue grafts into the ventricles, rather than the brain parenchyma, do trigger rapid rejection. This shows that antigens or antigen-presenting cells must be able to travel from the CSF into the lymphatic system. This hypothesis was confirmed when tracers injected into the subarachnoid space were subsequently found within the cervical lymph nodes. More detailed studies showed that some CSF seeps along the outside of the cerebral blood vessels and drains into lymphatic ducts within the nose. The finding of intraventricular graft rejection also implies that the immune system's effector cells—the B-cells, macrophages, and killer cells—must be capable of crossing the blood-CSF barrier. Precisely how this works remains unclear.

Astrocytes and Microglia

At this point you might ask, if white blood cells can enter the CSF, why do they not enter the brain parenchyma? After all, the brain parenchyma is separated from the ventricles by the relatively permeable ependyma (see Figure 5.12). The most likely answer to this question is that white blood cells do enter the parenchyma, only to be transformed or killed by glial cells, specifically by **astrocytes**. Astrocytes (named for their starburst shape) are quite diverse, but many of them have long processes with bulbous endings, called **end-feet**, that line blood vessels (Figure 5.13) as well as the

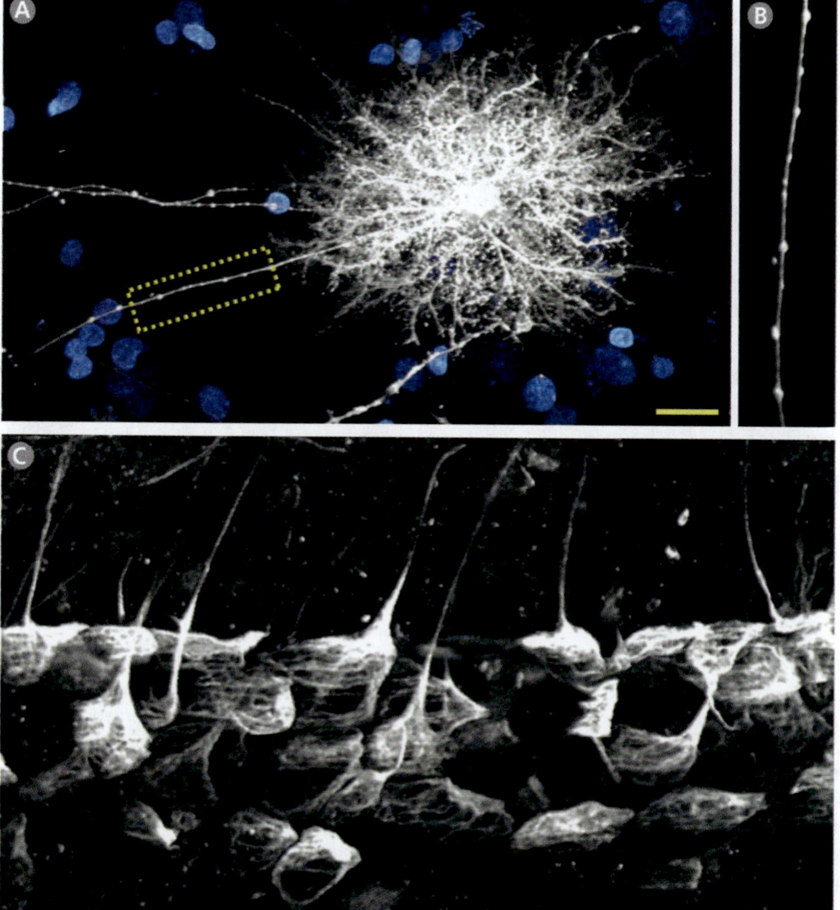

Figure 5.13 Human astrocytes. The photograph in (A) shows an astrocyte that was labeled with a fluorescent dye. It has numerous fine branching processes, one of which is long and exhibits small swellings along its length, shown at higher magnification in (B). Panel (C) depicts the "end-feet" of human astrocytes lining the surface of a blood vessel. The end-feet in this image were stained with an antibody against glial fibrillary acidic protein (GFAP), which is a marker for astrocytes. Scale bars are 50 μm in A and 20 μm in C. [From Oberheim et al., 2009]

pia mater. One function of these astrocytes is to secrete molecules that kill invading white blood cells or cause them to become much less aggressive. In essence, the astrocytes ensure that any immune effector cells that come into the brain parenchyma from the blood are quickly neutralized. This is why grafts into the brain parenchyma are rejected so slowly. One minor wrinkle in this elegant "story" is that grafts into the brain parenchyma can be rejected rapidly if the host animal previously received a graft of the same cells under the skin. In this case, the astrocyte "defenders at the gate" are simply overwhelmed by the invading immune cells, which multiplied as a result of the first graft.

Omitted from our discussion thus far has been one critical player in the neural response to injury: the microglial cell. **Microglia** comprise roughly 12% of all cells in a healthy human brain, but they are odd cells indeed. Early in development, they migrate into the central nervous system from the yolk sack and, possibly, other sources. In healthy adult brains, a few additional microglia develop from white blood cells that entered the parenchyma and were transformed by astrocytes (as discussed in the preceding paragraph). Regardless of origin, microglia look like tiny starbursts because they have complex, branching processes. The microglia use these processes to monitor the extracellular environment for signals of cellular damage. If such signals are detected, the microglia migrate toward the signal source (Figure 5.14). Often, the microglia divide along the way and thereby multiply. When this microglial defense force arrives at the site of damage, the microglia engulf and digest cellular debris, secrete cytokines, and kill some cells outright. Thus, the microglia perform most of the functions that are performed by macrophages and killer cells in other parts of the body.

Pros and Cons of Mounting an Immune Response

Now let us revisit an issue raised at the end of the previous section: why do the blood–brain and blood-CSF barriers become more leaky when the brain or its meninges are under attack? At first glance, opening the barriers seems counterproductive because it simply exposes the brain to further threats. However, opening the barriers also lets additional white blood cells enter the CSF and, from there, the brain parenchyma.

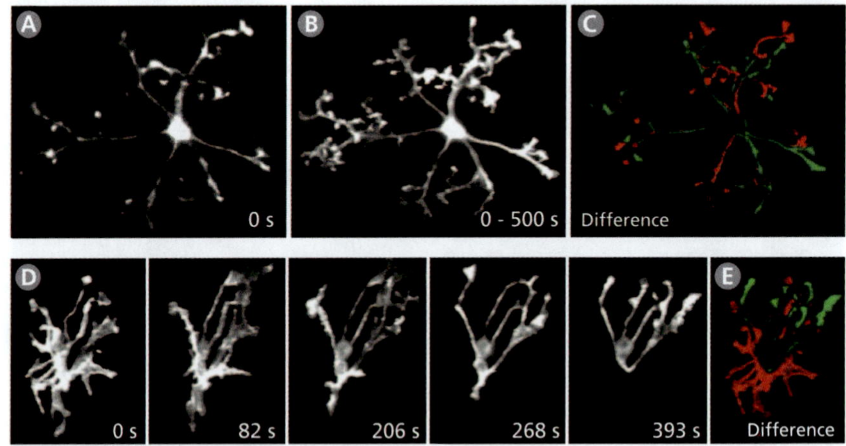

Figure 5.14 Microglia on the move. Shown here are images from time-lapse movies of retinas from transgenic mice whose microglia were made to fluoresce. Image (A) shows a microglial cell at time zero. Image (B) shows the same cell over the next 500 seconds (the sum of 50 images taken 10 sec apart). Image (C) depicts the difference in cell shape between A and B. Retracted processes are shown in red, additions in green. Shown along the bottom (D) are selected frames from a second time-lapse movie. At time zero, a laser was used to kill some cells near the top-right corner of the image. As you can see, the microglial processes soon orient and move toward the damage. Panel (E) shows the difference between the first and last images of this second movie. Again, retractions are in red, additions in green. Scale bars equal 20 μm. [From Lee et al., 2008]

Even if the white blood cells in the parenchyma are transformed into microglia, this still boosts the supply of cells available to fight the threat. Thus, opening the barriers strengthens the brain's immune response. This is a risky move because the heightened immune response may kill some innocent bystanders (healthy neurons), but the risk may be worth taking if it eliminates the threat or keeps it from spreading.

More generally, you can now appreciate that the unusual design of our brain's immune response has pluses and minuses. For example, the weak neural immune response allows some viruses (notably those of the *Herpes* family) to hide out in neurons, from where they emerge sporadically. This seems like bad design, but having the body's immune cells pursuing viruses into the nervous system might cause irreparable harm. Similarly, the low-level inflammation that accompanies many neurodegenerative diseases may not suffice to clear the threat entirely, but a stronger immune response might kill even more neurons.

Increasingly, neuroscientists have realized that the brain's immune response is finely tuned to maximize its benefits and minimize its costs. This explains why efforts to treat Alzheimer's disease by suppressing neural inflammation have not been very successful and why boosting the brain's immune response to beta-amyloid (see Box 5.2: Immune Responses in Alzheimer's Disease) has had some serious side effects. This does not mean that scientists should give up hope of treating neurodegenerative diseases by manipulating the brain's immune system. It simply means that efforts to do so require a careful appreciation of how complex and finely tuned that neural immune response already is.

Minimizing Neuron Death

Neural damage must be contained because dying neurons are a potential threat to neighboring neurons. If neurons die by **necrosis**, which is essentially just death by catastrophic rupture, then they spill into the extracellular space not only enzymes that might harm neighboring cells, but also all of their stored neurotransmitter. If this transmitter is glutamate, it is likely to excite most neighboring neurons. In fact, spilled glutamate may excite nearby neurons so much that they are liable to die. This glutamate-induced **excitotoxicity** is caused by excessive calcium influx, which overwhelms a cell's ability to regulate internal calcium and ultimately leads to death. Indeed, neuroscientists sometimes use glutamate or glutamate agonists (such as ibotenic acid) to kill neurons selectively and then examine the effects of those lesions on an animal's behavior. Now consider this: if the overexcited cells were themselves to suffer death by necrosis, then they would also spill their glutamate and would, in turn, kill their neighbors. A chain reaction of cell death would spread.

Fortunately, excitotoxicity tends not to kill neurons by necrosis. Unless the glutamate exposure is extreme, excitotoxicity kills neurons quietly. Specifically, overexcited neurons systematically dismantle their intracellular contents, including their neurotransmitters. The dying neurons then divide into smaller membrane-encased globules that are readily digested by microglia. Thus, a dying neuron's innards are recycled rather than spilled. You already learned about such *programmed cell death* occurring in embryos (see Chapter 4), but it also occurs in mature brains. It may result from overexposure to glutamate, from attacks by killer immune cells, or from disruptions in blood supply. Programmed cell death, including *apoptosis*, is also seen with normal aging in all neurodegenerative disorders, such as Alzheimer's and Parkinson's disease or amyotrophic lateral sclerosis (which entails the degeneration of motor neurons). Except in cases of direct physical trauma, neuronal necrosis is rare. It probably is minimized precisely because necrosis in the brain would tend to have such explosive ripple effects. In contrast, programmed cell death minimizes collateral damage to neighboring cells.

Astrocytes and Glial Scars

Astrocytes also help to minimize brain damage. We already discussed the role of astrocytes in dealing with invading immune cells, but there are multiple types of

NEUROLOGICAL DISORDERS

Box 5.2 *Immune Responses in Alzheimer's Disease*

Alzheimer's disease is the most common form of dementia affecting the elderly. Roughly 3 out of every 1,000 people aged 65–69 have Alzheimer's disease, and this rate rises to ~56/1,000 in people over 90 years old. There are many aspects to Alzheimer's disease (see Chapter 16), but an especially intriguing one is its relationship to the immune system.

A key step in the development of Alzheimer's disease is the accumulation of amyloid β-protein (Aβ). This molecule is normally harmless, but Alzheimer's patients produce large amounts of an Aβ variant called Aβ42 that tends to form toxic aggregates. The largest of these aggregates are visible as *amyloid plaques* within the brain (Figure b5.2). In addition, Alzheimer's patients accumulate within their neurons incorrectly folded forms of a protein called tau. These tau aggregates are called *neurofibrillary tangles* (Figure b5.2). Although some Alzheimer's patients exhibit no plaques or tangles in their brain at autopsy, most neuroscientists believe that plaques or tangles, or both, are closely linked to the causes of Alzheimer's disease. Therefore, researchers have long been keen to prevent or reverse plaque and tangle formation, hoping that this might arrest or reverse the cognitive decline associated with Alzheimer's disease.

A major step toward this goal was taken in 1999 when researchers reported that an Aβ42 vaccine prevents plaque formation in a transgenic mouse model of Alzheimer's disease (Figure b5.2 B). The mice in the original study were injected with Aβ42 and then created their own antibodies. In later studies, the mice were injected with premanufactured Aβ42 antibodies. Either way, the antibodies consistently prevented amyloid aggregation and, in some cases, removed the aggregates that had already formed. In one study, Aβ42 antibody injections even cleared incipient tangles (Figure b5.2 D). These studies have raised hopes that similar vaccines might prevent or reverse Alzheimer's disease in humans as well. Indeed, several Alzheimer's patients injected with a synthetic form of Aβ42 as part of a clinical trial had few plaques in their brains at autopsy. Unfortunately, the treatment did not arrest the cognitive decline, as the patients still progressed to severe, final-stage Alzheimer's disease. Furthermore, the trial was aborted because some of the trial participants developed severe meningitis. This is frustrating and a bit of a challenge to the hypothesis that amyloid accumulations trigger Alzheimer's disease. However, the vaccines may have been given too late in the development of Alzheimer's disease, when the trigger for cognitive decline had already been pulled.

Another potential therapy for Alzheimer's disease is to manipulate the brain's innate immune response. The brains of Alzheimer's patients tend to be chronically inflamed, showing increased expression of cytokines and chemokines, as well as microglia around the edges of amyloid plaques. These findings suggest that the brain's own immune response may be responsible for some of the neu ron death and synapse loss in Alzheimer's disease. Alternatively, the brain's immune response may be a defensive measure, designed to prevent the further growth of amyloid plaques. To distinguish between these possibilities, one would want to know whether depressing the brain's immune response with drugs reduces the symptoms of Alzheimer's disease or worsens them. Unfortunately, the data on this point are conflicting. Overall, it seems likely that the brain's immune response to Alzheimer's disease is both good and bad. It may, for example, reduce the spread of plaques but kill some neurons as collateral damage. Future research will have to tease the various effects apart.

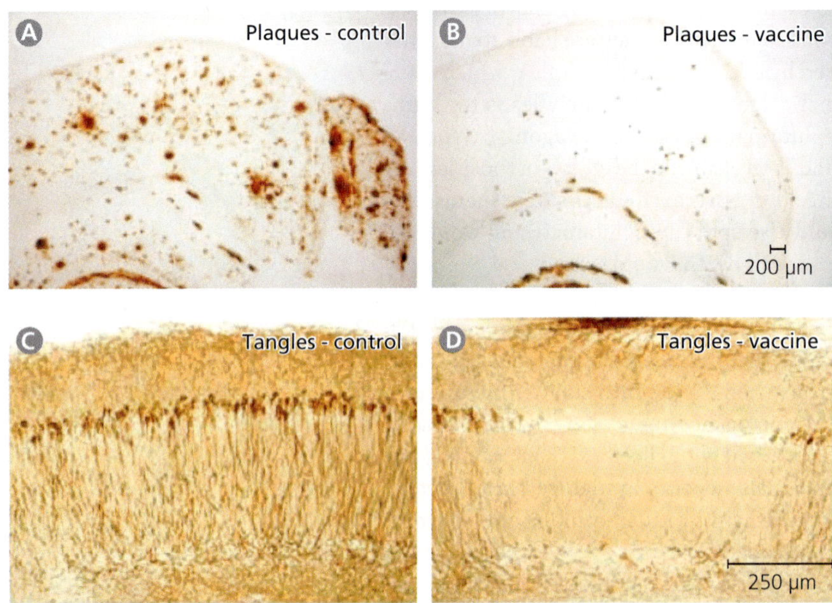

Figure b5.2 Vaccines to combat Alzheimer's disease. Shown in (A) is the neocortex of a transgenic mouse with numerous amyloid plaques (brown). When such mice were stimulated to produce antibodies against Aβ42, the number of plaques was reduced (B). In another study, Aβ42 antibodies were injected directly into the hippocampus of transgenic mice that develop neurofibrillary tangles as well as plaques. Image (C) shows a section through the hippocampus of an untreated control mouse, stained for tangles (brown). The photograph in (D) shows a comparable section from a mouse that received the Aβ42 antibodies. You can see that it contains fewer tangles at the injection site (in the center of the image). [A and B from Schenk et al., 1999; C and D from Oddo et al. 2004]

astrocytes. Some of them possess potassium channels that tend to be open at the astrocyte's resting potential and preferentially let potassium flow into the cell rather than out of it. Through these specialized potassium channels, astrocytes absorb much of the potassium that builds up in the extracellular space around highly active neurons. Without this astrocyte-mediated **potassium buffering**, active neurons would become ever more depolarized (recall that both the resting membrane potential and repolarization after an action potential require potassium concentration to be much lower outside of neurons than inside of them; see Chapter 2), making them likely to die from excitotoxicity.

The central nervous system also contains **reactive astrocytes**, which form a **glial scar** around sites of brain injury. When called to action, reactive astrocytes proliferate, move to the site of injury, and soak up glutamate as well as other substances released through necrosis. This protects the remaining neurons (Figure 5.15). Unfortunately, some of the molecules secreted by reactive astrocytes repel axonal growth cones, which prevents regenerating axons from growing across a glial scar. This is a major problem for patients whose spinal cord was cut. Researchers have tried to circumvent this problem by preventing scar formation, but this creates its own problems because, as we discussed, the glial scar is not all bad. Recognizing these complications, some scientists are trying to remove the glial scar just after it has performed damage control but before it impedes axonal regeneration. As you might expect, research in this area is moving rapidly.

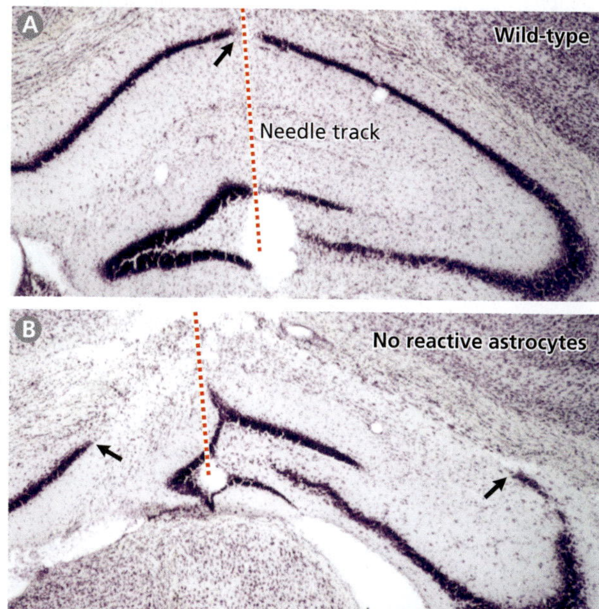

Figure 5.15 Protective role of astrocytes. Shown in (A) is a Nissl-stained section through the hippocampus of a normal mouse several days after the hippocampus was stabbed with a needle (along the red dashed line). Some cells next to the needle track have died, as evidenced by interruptions in the cell-dense layers. Shown in (B) is what happens when you perform the same experiment in transgenic mice that lack reactive astrocytes (dividing astrocytes are engineered to commit suicide). In such animals, the needle causes far more extensive neuron death (between the arrows). Scale bar equals 150 μm. [From Bush et al., 1999]

Axonal Regeneration

Although damaged axons in the central nervous system tend not to regenerate, axons in the peripheral nervous system often do regrow. If peripheral axons are cut, their distal, severed ends quickly degenerate. The proximal ends may withdraw, but they rarely degenerate. Instead, they develop growth cones and grow back. The regenerating axons are aided in their journey by *Schwann cells,* which ensheath peripheral axons in myelin (see Chapter 2). When a peripheral axon segment degenerates, the surrounding Schwann cells lose some of their morphological specializations (de-differentiate), re-enter the cell cycle (proliferate), and secrete a variety of molecules that promote axonal regeneration. Importantly, the Schwann cells and their secretions form a hollow tube that guides the regenerating axons back to their original target. Once the axons have successfully re-innervated their target, the Schwann cells re-differentiate and start to myelinate the regenerated parts of the axons. Although peripheral axons regenerate fairly slowly (a few mm per day), the regeneration can be amazingly complete, as long as the regenerating nerves can find the path laid down by the Schwann cells. This explains why surgeons who want to reattach a severed hand must carefully match the cut nerve ends before they sew them together.

Why do peripheral axons regenerate after an injury, whereas central axons do not? As mentioned earlier, the astrocytes in glial scars impede axonal regeneration, and glial scars tend not to form in the peripheral nervous system. At least as important, axons in the CNS are myelinated by oligodendrocytes rather than Schwann cells. In contrast to Schwann cells, oligodendrocytes express molecules that inhibit, rather than promote, axon outgrowth. One of these outgrowth-inhibiting molecules is called **Nogo**. When this molecule (or other, similar compounds) binds to Nogo receptors on a growing axon, the axon's growth cone collapses, halting the axon's ability to grow in length (see Chapter 4).

Researchers are working hard to understand what makes oligodendrocytes so different from Schwann cells, hoping that such an understanding will allow them to facilitate the regeneration of damaged axons in the central nervous system. A better understanding of Schwann cells and oligodendrocytes may also lead to treatments for **multiple sclerosis** and **Guillain-Barre syndrome**, which involve the loss of myelin in the CNS and PNS, respectively.

As noted earlier, damaged neurons in the CNS tend not to be replaced. In response to injury, some cells near the injury may give birth to a few neurons, but most of these appear to die within a month. Even the survivors are unlikely to make connections that match those of the deceased cells. There are no tubes to guide their axons, and the molecular signals that directed their embryonic counterparts disappeared long ago. This problem also plagues efforts to treat brain damage by grafting fetal neurons or neuronal stem cells from other animals into the damaged site. Some of the grafted cells may survive and mature, but they rarely replicate the missing connections. This is not to say that such grafts are futile. The grafted cells may, for example, secrete important growth factors that rescue neurons from further damage or death. Nonetheless, the injured CNS tends not to repair itself by regenerating its damaged or missing components. Instead, damaged brains tend to rewire themselves so that entirely different connections can substitute for the lost circuitry.

Functional Recovery through Brain Rewiring

As we discussed in Chapter 3, adult brains are capable of forming new connections when individuals learn new sensory associations or motor skills. Neuronal rewiring is even more dramatic after brain injury. This rewiring often helps to restore behavioral function, at least to some extent.

Compensating for a Lost Sensory System

You may have heard of blind people whose sense of hearing is so exquisite that they can navigate by making clicking noises and listening for echoes bouncing off solid surfaces. The stories are true, although most blind people do not echolocate. More commonly, the blind compensate for their lack of vision by developing more acute senses of hearing, touch, and smell.

An interesting clue about the neural mechanisms underlying such sensory compensation comes from the case of a woman who was born blind and then became a proficient reader of Braille, the writing system in which patterns of raised dots correspond to individual letters. At 63 years of age, the woman suffered a stroke (a disruption of her cerebral blood supply) that irrevocably damaged her occipital cortex, which contains most of the visual cortical areas (Figure 5.16). Remarkably, this

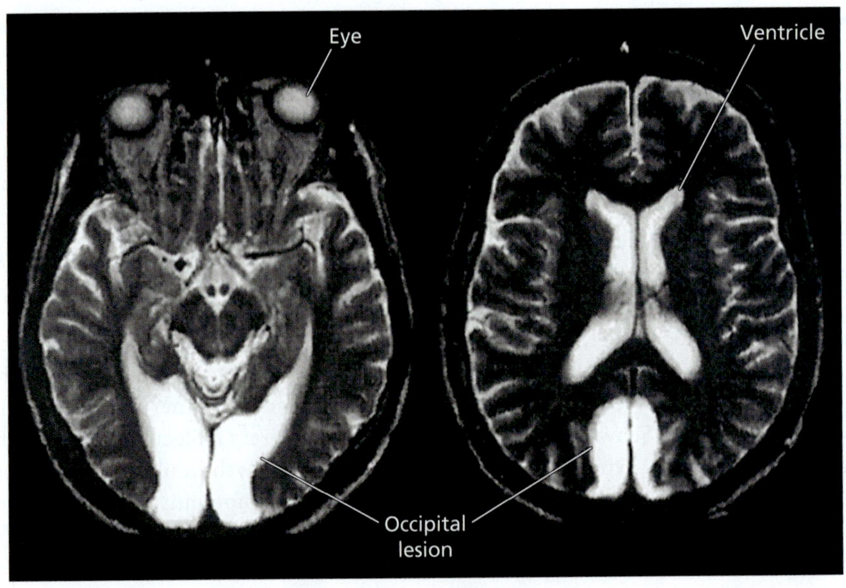

Figure 5.16 Braille alexia. Shown here are horizontal MRI scans through the brain of a woman who was blind from birth, learned to read Braille at age 6, and used it extensively thereafter. One morning, at age 63, she complained of lightheadedness and motor incoordination. Later that day the woman collapsed and was admitted to the hospital. The next day she was unable to read Braille. Although she could feel the raised dots of the Braille characters, she could no longer decipher them. A brain scan revealed that the woman had suffered a stroke, destroying her occipital cortex bilaterally, which suggests that the woman had been using her occipital (visual) cortex to read Braille. On these MRI scans, the damaged brain regions are white, as are the ventricles. [From Hamilton et al., 2000]

stroke destroyed her Braille reading ability. In technical terms, she developed **Braille alexia** (*alexia* is Latin for "without words"). Although the woman could still feel the dots, she could no longer interpret their meaning. This shows that the woman's occipital cortex had been involved in reading Braille. Because the occipital cortex normally plays little role in processing somatosensory (touch) information, we can infer that the woman's early loss of vision must have caused her occipital cortex to be rewired with some new, unusual inputs from the somatosensory system.

Braille alexia is rare, but functional brain imaging has shown that in many blind people, the occipital cortex is surprisingly active during the processing of complex somatosensory information (Figure 5.17). Such studies have further shown that the degree to which the occipital cortex responds to somatosensory stimuli depends on the age at which the person became blind. The earlier in life eyesight is lost, the greater the rewiring.

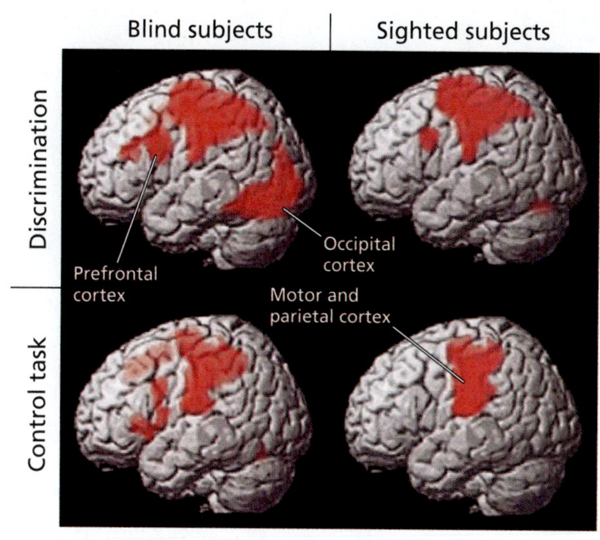

Figure 5.17 Blindness alters occipital cortex function. A functional MRI study compared brain activity in blind versus sighted subjects as they engaged in two different tasks. In the first task, subjects were asked whether two sequentially presented Braille characters were the same or different. The second (control) task involved similar Braille stimuli and hand movements, but no discrimination. Shown in red are cortical areas that were significantly activated in these tasks, compared to rest, in blind and sighted subjects. As you can see, occipital cortex was significantly activated in blind but not sighted subjects during the discrimination task (but not the control task). This shows that blindness causes occipital cortex, which is typically considered visual cortex, to become more involved in high-level somatosensory processing. [From Sadato et al., 2002]

This appears to be a general rule: the earlier a disruption occurs, the greater the ability of the nervous system to compensate for the damage. Scientists who study such compensatory rewiring in animals are well aware of this. They usually create the brain damage early in an animal's life and then look for rewired connections when the animal is an adult. More often than not, they find that the intact brain systems have "invaded" the brain regions that lost their normal inputs on account of the early damage. In mice that are born blind (due to a mutation), somatosensory pathways project to parts of the thalamus that normally receive only visual inputs. Such dramatic rewiring is not observed when mice are blinded as adults. This is consistent with the more general idea that the nervous system's degree of plasticity tends to decrease with age (see Chapter 4).

Remapping Within a Sensory System

Brain rewiring occurs even more readily within a sensory modality. For example, it has been shown that cutting the sensory nerves in the arm of 3- to 4-year-old monkeys led to significant rewiring in the monkey's somatosensory cortex by the time the monkey was examined 12 years later. Specifically, neurons that normally process sensory information from the hand were found to respond to stimulation of the face. Apparently, axons of neurons that respond to facial stimulation had "invaded" the neighboring cortical territory, which normally processes inputs from the hand (Figure 5.18) but was deprived of this input in the monkeys with the cut sensory pathways.

More recent work revealed that this sort of **remapping** of sensory inputs happens within a year or two and can extend over many millimeters. Unfortunately, the mechanisms by which inputs to one region of the cortex invade those of another remain unclear. One possibility is that cortical neurons always receive more inputs than physiological mapping studies suggest and that removal of inputs to one area simply reveals those hidden inputs. Although this sort of input unmasking is a real phenomenon, it cannot explain the relatively slow time course of the remapping that occurs after grave injuries. Furthermore, anatomical studies have shown that remapping involves the sprouting of at least some new axon branches. Some of the new axons belong to cortical neurons; others may lie at various points along the path from skin to the somatosensory cortex. In short, sensory remapping after injury probably involves multiple mechanisms.

Motor Rehabilitation

The fact that even adult brains can rewire themselves after damage gives hope to those who lost some functional capacities because of brain damage. However, substantial

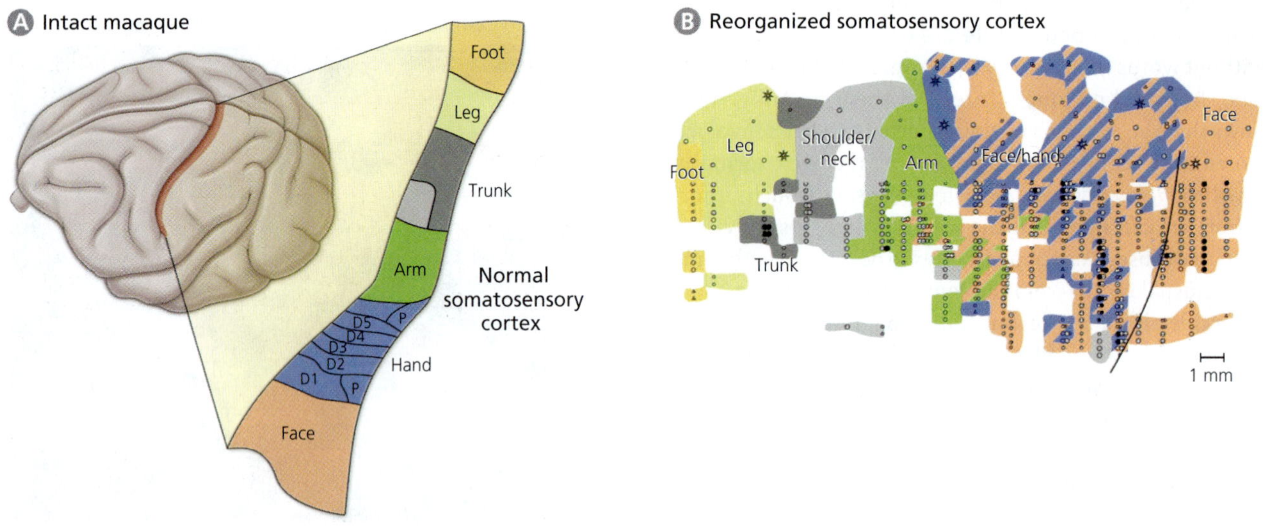

Figure 5.18 Remapping after sensory deprivation. Shown in (A) is the primary somatosensory cortex in an intact macaque. Tactile information from the body surface is represented topographically. Shown in (B) is the cortical body map of a monkey that had lost much (but not all) inputs from its hand 22 months before the somatosensory cortex was mapped. Much of the cortical territory that normally responds to touches on the hand now responds to touches of the face. The dots and other symbols indicate where neural responses were recorded. The curved line indicates where the face representation would end in an intact monkey. [From Jain et al., 2008]

recovery usually requires enormous effort. Edward Taub showed this back in the 1970s. He cut the sensory nerves in one arm of macaque monkeys and found that, not surprisingly, the monkeys rarely used the deafferented arm. However, when Taub prevented monkeys from using their other, intact arm, they did eventually learn to move the once so useless arm in useful ways. These experiments were quite controversial, partly because the experimental manipulations were unusually severe but also because the standards of animal care in Taub's research facility seem to have been relatively lax by today's standards.

Nonetheless, Taub's basic finding, that recovery from damage is improved if the organism is forced to use the damaged limb, has stood the test of time. It forms the core of **constraint-induced movement therapy** in which stroke victims or wounded veterans with damaged limbs are constrained to use them anyway. This is an extremely frustrating therapy, but it appears to work. A recent clinical trial found that, compared to "usual and customary care," constraint-induced movement therapy significantly increases a patient's ability to move the injured limb. Importantly, the improvement reportedly persists for at least two years.

Potential Downsides of Rewiring the Brain

Although neural plasticity following brain or nerve damage is usually beneficial, it need not be entirely benign. For example, amputees often feel terrible pain in parts of their lost limbs. Such **phantom limb pain** is associated with the sort of rewiring we have been discussing because in patients who lost an arm, some somatosensory inputs from the face seem to be rerouted into the cortical region that used to process information coming from the arm. This means that touches to the face will be perceived, at least sometimes, as stimulation of the missing arm. Sadly, this phantom stimulation is frequently perceived as pain.

A similar phenomenon occurs in patients with fractured spinal cords. Many such patients occasionally feel pain below the level of injury, which cannot be "real" because the sensory fibers carrying sensory information from those lower regions to the brain were cut. However, the somatosensory cortex in these patients is reorganized such that somatosensory inputs from above the spinal cut are rerouted into the adjacent deafferented (input-deprived) regions of cortex. The greater the extent of this rerouting, the greater the phantom pain (Figure 5.19).

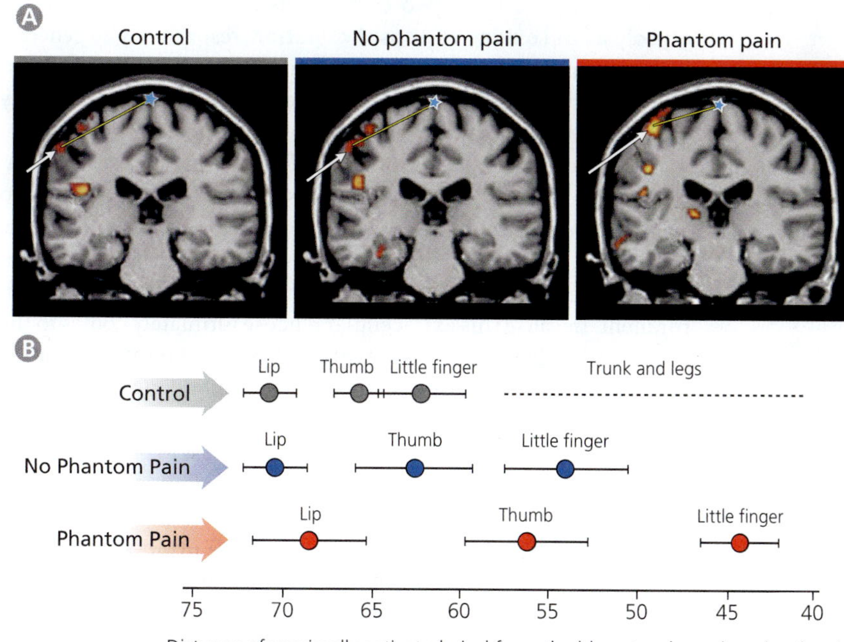

Figure 5.19 Cortical remapping and phantom pain. Shown in (A) are average functional MRI scans taken from three groups of subjects while their little finger was stroked with a brush. The three groups are (1) intact controls, (2) patients with spinal cord injury who feel no "phantom pain," and (3) people with spinal injuries who frequently feel pain in "phantom limbs" below the level of the injury. The little finger representation in the primary somatosensory cortex (white arrows) is shifted dorsomedially (toward the blue star) in the two injured groups compared to controls. Panel (B) summarizes these data and includes additional results obtained after stroking the thumb and lip. The main findings are that spinal injury causes somatosensory inputs from the fingers to shift dorsomedially, invading cortical territory that normally processes information from the trunk and legs, and that this injury-induced invasion is most extensive in subjects who feel phantom pain. [From Wrigley et al., 2009]

Overall, we can conclude that the central nervous system recovers fairly well, if not quite perfectly, from limited damage. Indeed, most brain damage is subtle enough that it goes unnoticed. Even Parkinson's disease (see Chapter 15) becomes symptomatic only after more than 80 percent of the dopaminergic neurons in the ventral midbrain have died. In your brain, too, a few thousand neurons die every day. Yet you rarely notice. Only after traumatic damage do we become aware of the brain's limited capacity for self-repair. Thankfully, many injuries that would historically have been lethal can now be managed medically, in part by boosting the brain's intrinsic capacity for self-healing and remapping.

BRAIN EXERCISE

Imagine a piece of shrapnel entered your skull and killed a large number of neurons in your left auditory cortex, which is needed for speech production and comprehension. How might you and your brain respond to this injury?

5.5 How Do Neurons Get Their Energy?

Thus far in this chapter, we have discussed various ways in which nervous systems are protected from harm and how they deal with injury. Now let us turn to a related problem, namely, how neurons are supplied with metabolic energy. Providing the brain with energy is important because neurons deprived of metabolic energy soon lose the ability to maintain their membrane resting potential. This causes them to fire excessively, release abnormal amounts of glutamate, and suffer from excitotoxicity. In short, neurons deprived of energy do not just stop working, they die. As we have seen, dead neurons are generally not replaced. Thus, providing the brain with energy is paramount.

Sources of Metabolic Energy

Although the brain occupies only about 2% of an adult human's body, it accounts for ~20% of the body's total energy consumption at rest. This is why neurobiologists like to say that the brain is an energetically expensive organ. More important, neurons are rather picky about the metabolic fuel they use. Many cells can produce energy from various carbohydrates, lipids, and even proteins, which they convert to glucose

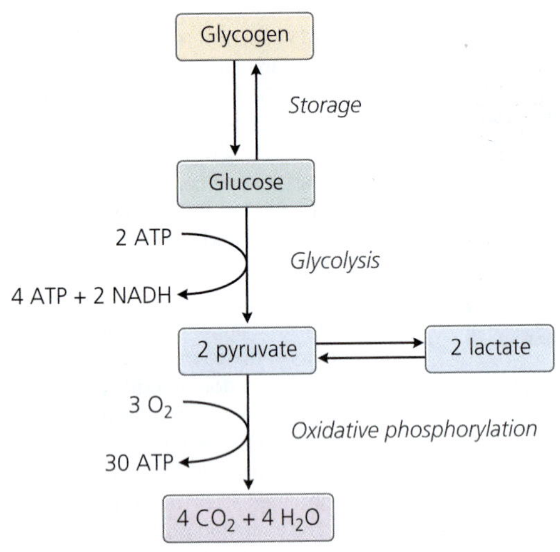

Figure 5.20 Glycolysis and oxidative phosphorylation. The glucose required for glycolysis comes mainly from the blood, but astrocytes can store some glucose in the form of glycogen, which is readily converted back to glucose when needed. The pyruvate created through glycolysis either enters the oxidative phosphorylation pathway or is converted to lactate. Lactate is either released into the blood or converted back to pyruvate. Compared to glycolysis, oxidative phosphorylation produces more ATP, which is the fundamental currency of metabolic energy. However, when oxygen levels are low, glycolysis becomes more important. [After Raichle and Mintun, 2006]

or pyruvate. The latter two compounds are then passed through **glycolysis** and **oxidative phosphorylation**, respectively, to generate molecules of ATP (Figure 5.20), which are the basic currency of metabolic energy. Neurons are like other cells in that they, too, employ glycolysis and oxidative phosphorylation to generate ATPs.

However, neurons feed these metabolic pathways almost exclusively with glucose, most likely because burning lipids and amino acids for energy would damage the connections between neurons irreparably. In contrast to most other cells, neurons cannot synthesize glucose and must instead obtain it from the extracellular environment. Because this extracellular glucose ultimately comes to the brain *via* the blood, neurons are totally dependent on a steady blood supply. This is why disruptions of cerebral blood flow cause you to faint within about 15 seconds and why disruptions longer than 5–10 minutes cause extensive neuron death.

Given the importance of cerebral blood flow, it is little wonder that the human brain contains about 100 billion capillaries, totaling an estimated 650 km in total length and 20 m^2 in surface area. Roughly 750 ml of blood course through these capillaries every minute. Although the average distance from a neuron to the nearest capillary is probably no more than 20 μm, there is a serious problem with getting glucose from these blood vessels to the neurons, namely, the blood–brain barrier. Oxygen can readily diffuse across the blood–brain barrier, but glucose cannot. This explains why the endothelial cells lining the blood vessels, as well as the neuronal and glial cell membranes, are loaded with **glucose transporter** molecules. They shuttle glucose molecules across the blood–brain barrier and into neurons and glial cells. The glia, primarily the astrocytes, convert some of this glucose to **glycogen**, a polysaccharide that contains as many as 30,000 tightly linked glucose molecules.

Astrocytes can store the glycogen and later convert it back into glucose, if needed (Figure 5.21). Neurons, however, do not make glycogen. They must use the glucose immediately and then restock their supply from the extracellular environment. Fortunately, when blood glucose is low, astrocytes convert glycogen to glucose and (through glycolysis) to lactate. Some of the lactate produced in this process may be shuttled to neurons, which can use it to drive oxidative phosphorylation (Figure 5.21).

Cerebral Blood Flow

Clogged arteries can threaten cerebral blood flow. Fortunately, the brain is fairly well protected against obstructions in its arterial supply. A key protective device is the

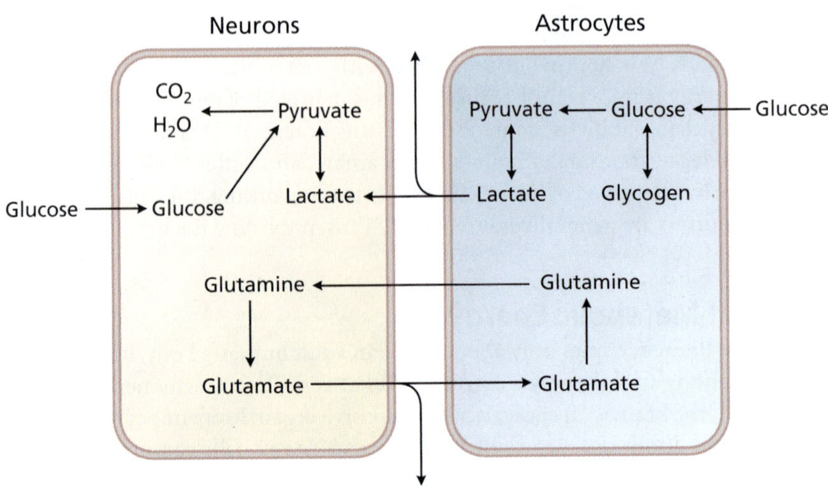

Figure 5.21 Metabolic interactions between neurons and astrocytes. According to the "lactate shuttle hypothesis," astrocytes convert glucose to lactate and then shuttle some of it across to neurons, which can use the lactate to drive oxidative phosphorylation. Astrocytes also recycle much of the glutamate released by neurons. The astrocytes convert this glutamate to glutamine and ship it over to neurons, which then convert it back to glutamate.

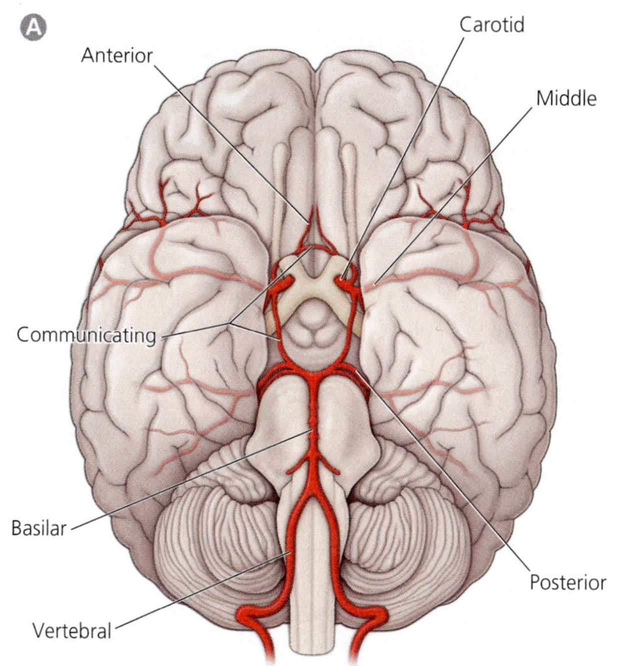

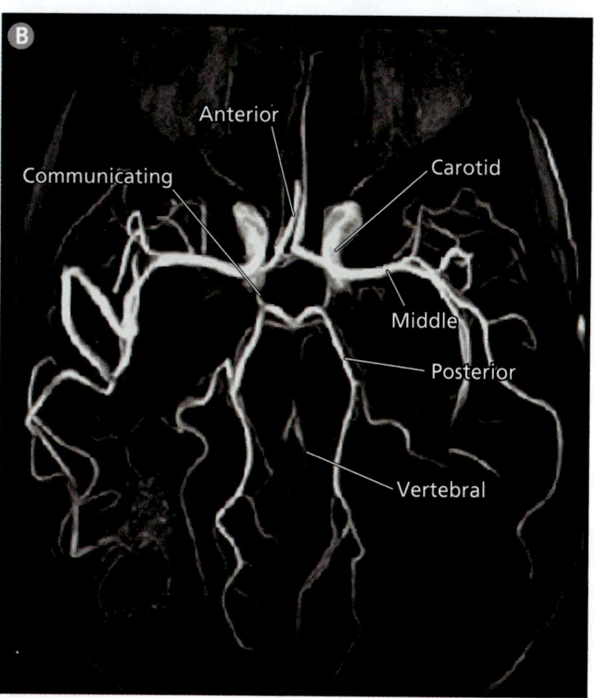

Figure 5.22 The circle of Willis. The drawing in (A) depicts a human brain from an inferior view, highlighting the major blood vessels. Shown in (B) is an MRI image of the cerebral arteries. The circle of Willis consists of segments of the two posterior cerebral arteries, which connect through "communicating arteries" to the two middle cerebral arteries, which are continuous with the anterior cerebral arteries, which are in turn connected across the midline by another communicating artery. Thomas Willis recognized that these "lateral" connections between the major arteries are functionally important. If one of the major arteries feeding into the circle of Willis (the carotid and basilar arteries) is blocked, then blood can flow "laterally" around the circle to compensate for the blockage. [B courtesy of Harald H. Quick, University of Erlangen, Erlangen, Germany]

circle of Willis, which is a ring of arteries at the base of the brain (Figure 5.22). This arterial ring forms because the major arteries that feed the brain are linked by small **communicating arteries**. Two of these join the left and right posterior cerebral arteries to the two middle cerebral arteries; the third connects the two anterior cerebral arteries across the midline. The upshot of all this "communication" between the main cerebral arteries is that blockage in any one of them causes no significant disruption in cerebral blood flow as long as the blockage occurs upstream of the circle of Willis (i.e., in the carotid or basilar arteries). As Sir Thomas Willis noted back in the 1660s, if one of the arteries feeding into the circle of arteries bearing his name is blocked, the other arteries take up the slack and, through the circle, supply the downstream arteries with blood. Although the circle of Willis is found in all textbooks dealing with the cerebral blood supply, it is quite variable across humans. Indeed, the circle of Willis is complete in only about 35% of humans. As you might expect, the probability of having a severe **ischemic stroke** (a major disruption in the cerebral blood supply) increases with the number of missing communicating arteries.

Additional layers of defense for the cerebral blood supply are provided by linkages between arteries and arterioles (small arterial branches) downstream of the circle of Willis. Some of the major cerebral arteries are linked by **collateral branches** through which blood may flow in either direction, depending on the blood pressures at the two ends (Figure 5.23). The small arterioles at the brain's surface are also linked by vessel fusions called **anastomoses** (from the Greek word *stoma*, meaning "mouth"), through which blood can flow between the connected vessels. Because of these anastomoses, the direction in which blood flows through any particular segment of the arteriole network is variable. Most importantly, if one of the arteriole segments is blocked, then blood flow through the other arterioles is rerouted automatically

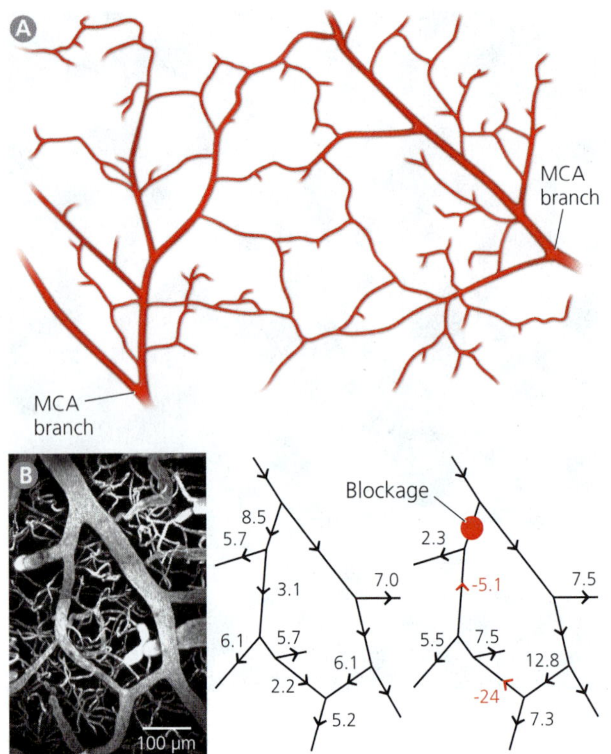

Figure 5.23 Compensatory blood flow through surface arterioles. The arterioles that run along the surface of a rat's parietal cortex (A) interconnect two branches of the middle cerebral artery (MCA). Short branches that seem to end abruptly are diving down into the cortex (into the page). Shown in (B) is a photograph of the blood vessels in another part of the cortex. Researchers used a two-photon microscope to measure the direction and speed of blood flowing through these vessels. Shown to the right are data obtained before and after one of the arterioles was experimentally blocked (red circle). The arrowheads indicate the direction of flow and the numbers indicate velocity in mm/sec. As you can see, blood flow through some of the arterioles reverses (red arrows) after the blockage, rerouting the blood. [From Schaffer et al., 2006]

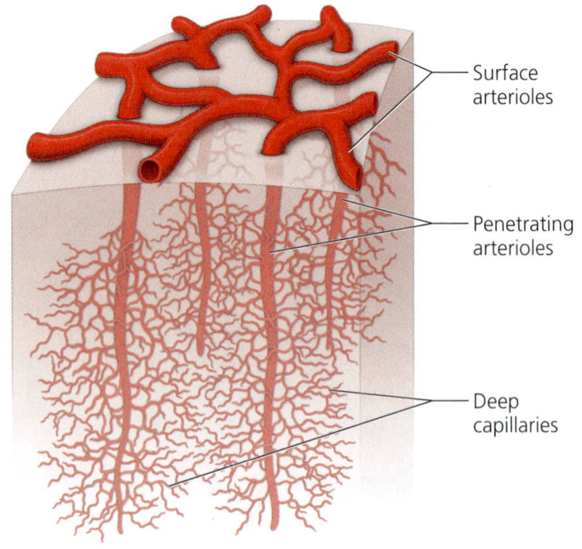

Figure 5.24 Bottlenecks in the cerebral blood supply. Penetrating arterioles dive from the brain surface down into the brain, where they give rise to a dense network of capillaries. Because penetrating arterioles are not connected with one another, a clot in one of them deprives the associated brain tissue of blood supply.

so that it flows around the blockage (Figure 5.23 B). Thus, the communicating arteries, collateral branches, and anastomoses of the cerebral vasculature provide a robust, multilayered defense against blocked blood vessels.

From the brain surface, blood is carried deep into the brain through **penetrating arterioles** (Figure 5.24). These arterioles run perpendicularly into the brain, where they branch to form a dense network of capillaries. In contrast to the surface arterioles, penetrating arterioles do not anastomose with one another. Therefore, penetrating arterioles are bottlenecks in the brain's blood supply. Measurements in rats have shown that blockage of a single penetrating arteriole significantly reduces blood flow over a cortical volume up to 700 μm in diameter.

This is quite odd, when you first think about it: why would the brain's principal arteries and surface arterioles be designed to minimize the effects of localized blockages but then allow the penetrating arterioles to be such bottlenecks? Why not connect the penetrating arterioles through some collateral branches or anastomoses? The answer seems to be that the penetrating arterioles are optimized for a very different function, namely the regional control of blood supply. Because they are such bottlenecks, changes in the diameter of the penetrating arterioles cause major changes in the amount of blood flowing into their downstream capillaries (flow rate increases with the 4th power of vessel diameter). Indeed, as you will learn in the next section, the amount of blood flowing through each region of the brain varies, at least somewhat, depending on the tasks the brain is performing. The harder the neurons work, the more blood flows to them.

Linking Blood Flow to Neuronal Activity

The link between blood flow and neural activity was first explored by Angelo Mosso in 1881. He reported that blood flow-induced pulsations of the brain increased when a subject performs a mathematical calculation, whereas pulsations in the arm remained unchanged. This finding implies that blood flow to the brain can be regulated independently of blood flow to the rest of the body.

Later research revealed that we can control not only how much blood flows to our brain as a whole but how much flows to each brain region. Specifically, it is now known that blood flow to a region of the brain tends to increase about 4–6 seconds after the synapses in that region increased their level of activity. Because synaptic activity tends to correlate with postsynaptic firing, we can say that regional blood flow correlates with local brain activity. This is why modern researchers are so keen to measure cerebral blood flow in awake human beings: it indicates which parts of the brain increase (or decrease) their level of activity during a particular behavior or cognitive task. Indeed, measures of local blood flow underlie many current techniques for visualizing human brain activity.

Functional Brain Imaging

The principal method currently employed to study brain function in awake humans is **functional magnetic resonance imaging** (**fMRI**; see Box 5.3: Functional Magnetic Resonance Imaging), which usually records **blood oxygen level dependent (BOLD) signals**. This is quite a mouthful, and *BOLD-fMRI* is a fairly complicated technique. In essence, the method compares the ratio of oxygenated (oxygen-carrying) hemoglobin to deoxygenated hemoglobin across various regions of the brain and across multiple time points. When arterial blood flow into a brain region increases, then the BOLD signal for that region increases because arterial blood is more oxygenated than venous blood. Importantly, the increased BOLD signal is generally interpreted as an increase in neuronal activity.

This description of fMRI might be somewhat confusing because you would expect more active neurons to suck more oxygen out of the blood, producing a *lower* BOLD signal. Indeed, this does happen, but the decrease in the BOLD signal due to neuronal oxygen consumption is highly localized and difficult to detect with fMRI. The "initial dip" in the BOLD signal is followed by a spatially more extensive and reliable increase in the BOLD signal that is produced by the increase in arterial blood flow (Figure 5.25). This extra blood is needed not because the neurons are starved for oxygen but because they need additional glucose. Under normal conditions, glucose is a more limited resource than oxygen. About 60% of the blood's oxygen never even enters the brain.

Local Regulation of Blood Flow

How does an increase in neuronal activity increase local blood flow? Part of the answer is that the penetrating arterioles are surrounded by **smooth muscle cells**.

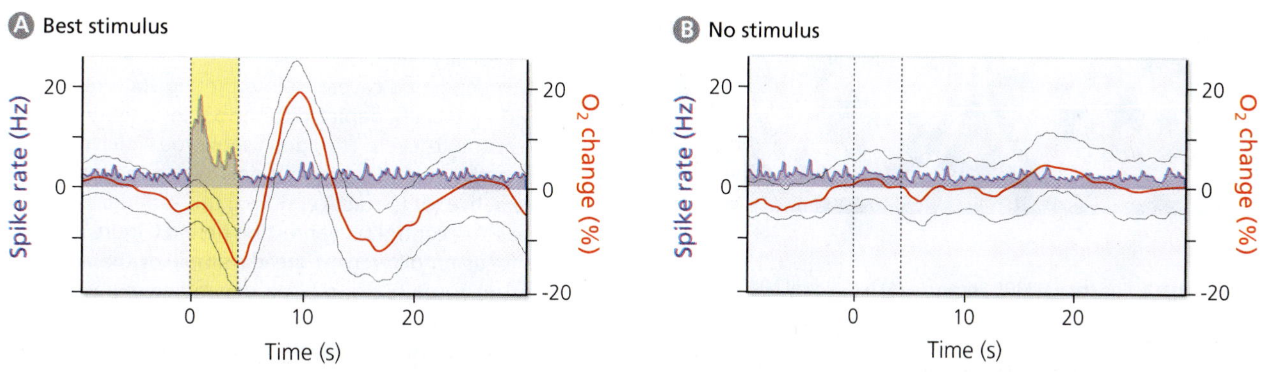

Figure 5.25 Neural activity correlates with local oxygen concentration. An electrode capable of simultaneously recording neural activity and tissue oxygen levels was placed into the visual cortex of anesthetized cats presented with various visual stimuli. Shown in (A) are data obtained with a stimulus (yellow shading) that strongly activates the cortical neurons. The histogram (blue shading) reflects the neuron's firing rate, which increases during the stimulus presentation. The red curve indicates the oxygen level, which initially dips but then increases. Panel (B) shows analogous recordings obtained when no stimulus is presented. [After Viswanathan and Freeman, 2007]

RESEARCH METHODS

Box 5.3 *Functional Magnetic Resonance Imaging*

Functional magnetic resonance imaging (fMRI) has become the most widespread method of imaging activity in human brains. During a typical fMRI scan, patients are asked to place their head inside the open core of a large and noisy machine (Figure b5.3), to hold very still, and to perform some relatively simple tasks while the machine scans their brain. Scientists later process these data into images that show which regions of the brain were more (or less) active during the task than during a comparable rest period (or some alternative task). These images may seem easy to understand, but the fMRI technique is complex, and its data must be interpreted with care.

To produce an MRI brain scan, a person's head is placed inside a magnetic field that (at 1.5–7 Tesla) is more than 100,000 times as strong as the earth's magnetic field. In such a strong field, the protons of hydrogen atoms in the body tend to align their spin axes with the imposed magnetic field. Researchers then add an orthogonal, oscillating electromagnetic pulse that can, at the appropriate frequency, bump the spinning protons out of alignment. When the pulse is turned off, the protons move back into alignment and thereby generate a signal that the scanner can detect. Importantly, the frequency of the pulse that is most effective at perturbing the spinning protons (the resonant frequency) depends on the proton's chemical context (its atomic neighbors). This phenomenon has long formed the basis of nuclear magnetic resonance (NMR) spectroscopy, which chemists use to determine the molecular composition of homogeneous materials. In the 1970s, NMR spectroscopy was extended to let scientists visualize the chemical composition of nonhomogeneous materials, including brains. As the technique evolved, the spatial resolution of these images increased to the point where it is now smaller than a grain of rice. Scientists also learned to focus their images on diverse chemical signatures. Along the way, they changed the name of the technique from NMR to MRI. This helped alleviate patient concerns about being subjected to some sort of "nuclear" treatment when, in reality, no ionizing radiation is involved.

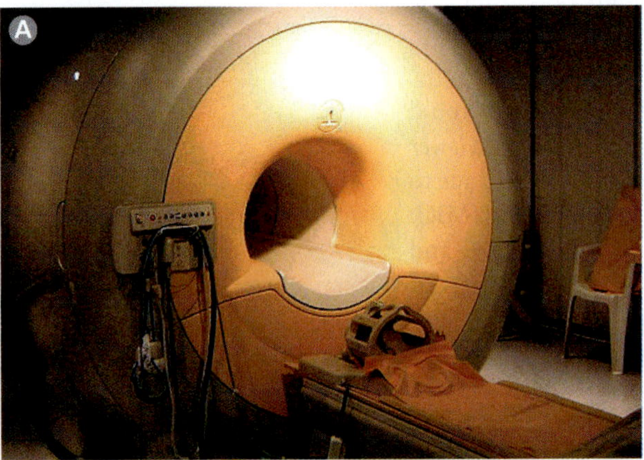

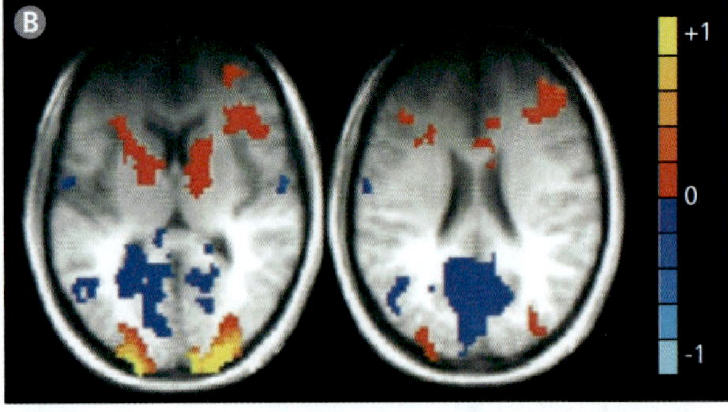

Figure b5.3 Functional MRI. Shown in (A) is a clinical MRI scanner. Panel (B) depicts BOLD-fMRI signals (colored) superimposed on structural MRI images (grayscale) for two horizontal sections through the "averaged brain" in the study. The color scale on the right shows the percent change in BOLD signal strength when people performed a simple task (detecting a letter hidden in visual noise) compared to staring at a screen that merely says "rest." As you can see, some brain regions became more active during the task; others became less active. [A courtesy of Kasuga Huang; B from Stark and Squire, 2001]

Early MRI scans were structural rather than functional. The early images focused on protons in gray matter versus white matter, ventricles, or bone; but Seiji Ogawa and his collaborators in the late 1980s turned their attention to the protons in oxygenated versus deoxygenated hemoglobin, developing what is now called *blood oxygen level-dependent fMRI (BOLD-fMRI).* Ogawa reasoned that active neurons must pull some oxygen out of the local blood vessels and that this decrease in hemoglobin oxygenation should be detectable as a decrease in the MRI signal. Ogawa turned out to be correct, but surprisingly, the initial dip in the MRI signal is followed a few seconds later by a large increase signal strength. Because the late increase is larger and more reliable than the initial dip, it is the focus in most fMRI analyses. Why the delayed increase? Most likely it occurs because neuronal activity stimulates astrocytes, which then cause local blood vessels to dilate. This increases local blood flow and flushes the deoxygenated hemoglobin out of the local capillary beds.

An important caveat to interpreting fMRI results is that they usually depict differences in activity rather than absolute levels. They don't show you which regions were active during a task but only which regions were more active (or less active) during this task than during some other control task or resting period (Figure b5.3 B). This focus on differences greatly simplifies the analysis (because absolute signal levels depend on many factors that are difficult to control), but it easily misleads the uninitiated, who assume that the "hot spots" in an fMRI scan show all the brain areas that were active during a task. This is not true; they only indicate those areas that differed in activity between the task and its control. Keep this in mind when you come across fMRI results in upcoming chapters.

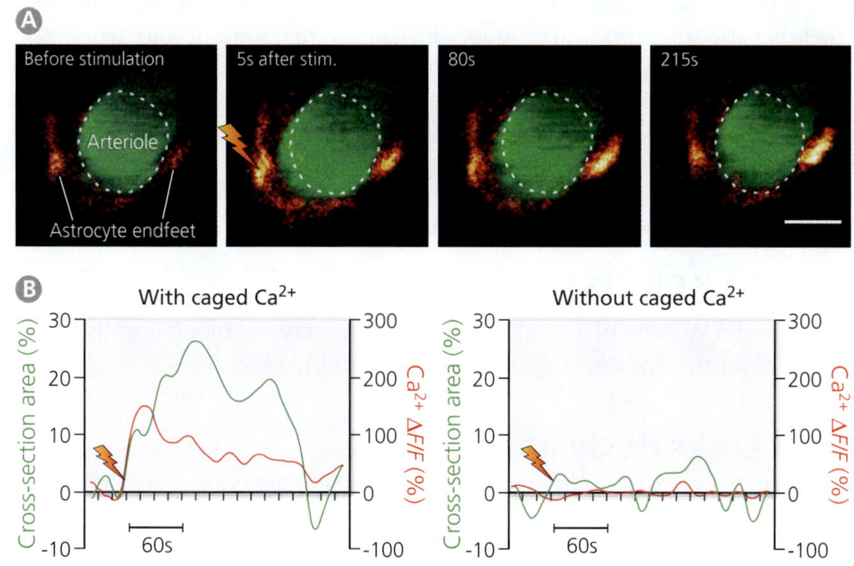

Figure 5.26 Astrocyte end-feet regulate arteriole diameter. Shown in (A) are frames from a time-lapse movie of an arteriole in cross section, bounded by two astrocyte end-feet. The blood is stained with green fluorescent dye, and the astrocytic end-feet were filled with a dye that glows orange when Ca^{2+} levels are high. The endfeet were filled with caged Ca^{2+} ions that can be freed from a molecular "cage" by application of a laser beam. The image at the left was taken before the laser stimulation. The other images were taken at the indicated intervals after the laser stimulation, which was aimed at the left end-foot. The dashed white line indicates the vessel's original diameter. As you can see, the vessel's diameter increases within 5 seconds after the laser stimulation. Quantitative data are shown in (B). The green traces track blood vessel diameter; the red traces show the concentration of uncaged Ca^{2+} in the astrocyte end-feet. [From Takano et al., 2006]

When these smooth muscle cells contract, the vessel's lumen shrinks. Conversely, when the muscles relax, the vessels dilate (for more on smooth muscle, see Chapter 8).

Of course, this raises a new question: what controls the muscle cells lining the blood vessels? The likely culprits here are astrocytes. As you saw in Figure 5.13, many astrocytes have long processes that end on blood vessels. In human brains, the astrocytic endfeet cover virtually the entire surface of each cerebral blood vessel in a cobblestone pattern (Figure 5.13 C). As we discussed, one function of these endfeet is to kill or transform invading immune cells. The endfeet also help induce and maintain the blood–brain barrier (especially in the capillaries). However, a crucial third role of astrocyte endfeet is the control of regional blood flow. Evidence for this function is provided by the observation that calcium levels inside the endfeet correlate with arteriole diameter. Furthermore, experimental increases in calcium levels within an astrocyte endfoot causes the adjacent blood vessel to dilate (Figure 5.26). The signals passing from astrocytes to the perivascular muscle cells are still being worked out.

What controls the calcium levels within the astrocytic endfeet? The answer is neurons. When glutamatergic synapses are active, some of the released glutamate diffuses out of the synaptic clefts. Much of that extrasynaptic glutamate binds to metabotropic glutamate receptors on astrocytes, which then causes transient increases in calcium levels within the stimulated astrocyte, including its endfeet. Thus, astrocytes detect how much glutamate is released from excitatory synapses and adjust blood flow accordingly. Importantly, astrocytes do not measure the activity of individual neurons but of thousands of synapses in their vicinity. In rodents, an individual astrocyte is estimated to contact 20,000–120,000 synapses. Human astrocytes are much larger and are estimated to contact 270,000 to 2 million synapses. This remarkable degree of convergence makes sense because blood flow cannot be targeted at single neurons anyway. Changing the diameter of a penetrating arteriole alters blood flow to a relatively small volume of tissue (~0.1 mm^3 in a rat neocortex), but this volume still harbors thousands of neurons.

All this learning about cerebral blood flow and metabolic energy may make you wonder how much energy it takes to think. Hardly more than not to think. The changes in total energy consumption typically visualized in fMRI studies are only on the order of 1 percent, and some brain areas consume less energy as others consume more. This shows that the brain isn't idly waiting for stimulation but always humming along. Occasionally, of course, the brain runs low on energy. This happens, for instance, when you have been running so hard that your muscles have consumed much of the available glucose. In such instances, you may suffer not only from muscle fatigue but also from **cerebral fatigue**, which means that some of your neurons slow to a crawl. This impairs your ability to think, but it also makes you less capable of keeping up the exercise. This negative feedback loop is annoying if you love sports, but it serves a valuable protective function. **Hyperventilation** is another protective measure that you may be familiar with. If you hyperventilate after running too hard, blood CO_2 levels decrease. This causes cerebral blood flow to slow down (for more on this, see Chapter 9), which then prompts the body to rest.

BRAIN EXERCISE

How does the location of a blockage in one of the cerebral blood vessels influence whether the blockage will cause neurons to die?

5.6 What Links Body and Brain?

It is tempting to regard the brain as a pure information processing device, independent of other body parts. This brain–body separation is clearly on display in the classic thought experiment of the **brain-in-a-vat** (Figure 5.27). Countless philosophers have wondered what would happen if you took a human brain out of its skull, put it in a jar containing a life-sustaining solution, and then connected all the sensory nerves to a supercomputer that presents them with artificial stimuli. The interesting question first asked by Descartes (who stipulated a malicious demon instead of the supercomputer) is this: would the disembodied brain be able to tell that its sensory inputs are artificial rather than real? Moreover, if the brain-in-a-vat cannot draw this distinction, then how can we ever be sure about reality? Because this book is not about philosophy, we need not deal with these questions. However, let us use some of what we learned to contemplate the perils a disembodied brain would face.

The first, most serious problem for our hypothetical brain-in-a-vat is that no solution can possibly keep the isolated brain alive as long as it only surrounds the brain. The oxygen and other nutrients in the solution cannot reach cells deep in the brain if they must get there by diffusion alone. This limitation is well known to neuroscientists who study brain slices. If the slices are thicker than about 0.5 mm, the cells deep in the slice's interior die off, no matter what the slice floats in. Therefore, to keep our potted brain alive, we have to pump a solution with nutrients through the cerebral blood vessels. We need an artificial heart.

Figure 5.27 What would a brain think if it were sitting in a vat and getting its inputs from a computer? This thought experiment is often used by philosophers to point out how difficult it is for a person to know what is real and what might be an illusion due to misleading sensory information. But ask yourself a different question: how long could a brain-in-a-vat survive?

A second concern is how to drain the fluid that we pumped into the brain. Perhaps we can discard the meninges and let the pumped-in fluid pour out of the veins as they emerge at the brain surface. The fluid could then accumulate in the space around the brain and drain from the bottom of our vessel to be discarded or recycled. Although this design would be relatively simple, it would severely compromise the blood–brain barrier. Any toxins or pathogens that manage to find their way into the fluid would freely enter the brain and start to kill neurons. A much better design would be to leave the meninges intact and have the veins drain into the space surrounding the meninges. This would leave the blood–brain barrier intact and thus protect the neurons.

What happens if you have an earthquake or someone bumps the vat roughly? What if the blood vessels feeding our hypothetical brain are ruptured or the brain suffers concussive compression? Would such a damaged brain reorganize itself as real brains do? Before you answer this question, consider how a disembodied brain could know that anything is wrong? An embodied brain with damage in the motor cortex can tell that grabbing a desirable object became impossible after the injury. It can then try to develop alternate means of reaching the object. That is, it can attempt to rewire itself. However, a disembodied brain with the same injury would be informed by the computer that everything is fine. Any damage that affects the brain's motor output would not be detected. Therefore, the injured brain-in-a-vat can never learn to compensate, to rewire itself. The only way out of this dilemma is to tell the computer about the injury and how it alters the body's reaction to the brain's commands. In short, the computer would have to simulate both the body and the brain. At that point, one could argue that our suspended brain is no longer disembodied. It would interact with a virtual body.

This prolonged thought experiment shows how intertwined our brains and bodies are. An isolated brain would be extremely vulnerable to injury and starvation; it could not avoid most threats, nor recover from injury. A brain-in-a-vat might be duped into believing that the computer's simulated stimuli are real, but the deception would be brief. Reality would bite Descartes' demon. Philosophy aside, the major point is simply that any serious discussion of brain structure and function must consider how brains interact with both their bodies and the outside world. This theme recurs throughout this book.

BRAIN EXERCISE

The brain helps to control the rest of the body, but how does the body control what happens in the brain?

SUMMARY

Section 5.1 - Most neurons in adult human brains are irreplaceable, but adult neurogenesis is evident in the olfactory bulb and the dentate gyrus. In addition, glial cells continue to be born in adulthood.

Section 5.2 - The brain is physically protected by the cranial vault, the meninges, and the cerebrospinal fluid (CSF). The CSF is produced mainly by the choroid plexus, circulates through the ventricles, and is eventually recycled into the blood. Hydrocephalus results when CSF drainage is blocked.

Section 5.3 - The brain is protected from toxins and pathogens by multiple barriers. The blood–brain barrier is formed by tight junctions between capillary endothelial cells, whereas the blood-CSF barrier is created by tight junctions in the epithelial layer of the choroid plexus.

Section 5.4 - When damaged or attacked, the brain responds in a variety of ways.

- Although the brain is often said to be immune privileged, lymphocytes may enter the brain under some conditions, although most are converted to microglia. Microglia protect the brain by moving to damaged areas, neutralizing threats, and clearing cellular debris.
- Programmed cell death recycles a dying cell's internal components, thus minimizing harm to neighboring cells. Astrocytes also reduce the spread of brain damage, mainly by forming glial scars.
- Removal of sensory input to cortical areas can result in significant remapping of the remaining inputs, especially when the affected individuals are young. Excessive cortical rewiring is linked to phantom pain.

Section 5.5 - Without glucose and oxygen, neurons quickly die.

- Neurons derive most of their energy from glucose, which they obtain from blood. They can also obtain lactate from nearby astrocytes. In contrast to astrocytes, neurons cannot store metabolic energy as glycogen.
- The brain's blood supply features connections between different arteries (and surface arterioles) that allow for lateral flow around localized arterial blockages.
- Astrocytes can match local blood flow to levels of neural activity by sensing synaptic glutamate release and adjusting the diameter of penetrating arterioles. These blood flow adjustments are the basis of functional brain imaging using the BOLD signal.

Section 5.6 - To appreciate how dependent the brain is on protective barriers and metabolic energy supplies, it helps to consider how fragile and helpless an isolated brain would be.

Box 5.1 - Many therapeutic drugs do not pass through the blood–brain barrier. To get them into the brain, one can temporarily weaken the barrier or encase the drugs in nanoparticles that are transported across the barrier.

Box 5.2 - Alzheimer's disease involves accumulations of amyloid β-protein (Aβ) and tau as well as brain inflammation. Therapies using antibodies against Aβ have been successful in animal models but less so in humans.

Box 5.3 - Functional magnetic resonance imaging (fMRI) usually measures changes in the ratio of oxygen-rich and oxygen-poor blood. This ratio increases as active brain regions are supplied with more blood.

KEY TERMS

tritiated thymidine 134
bromodeoxyuridine (BrdU) 134
autoradiograph 134
radioactive carbon (^{14}C) 137
cranial vault 138
meninges 139
dura mater 139
pia mater 139
arachnoid membrane 139
subarachnoid space 139
cerebrospinal fluid (CSF) 139
lateral ventricles 139
cerebral aqueduct 139
fourth ventricle 139
central canal 139
choroid plexus 139
membrane transporters 140
aquaporins 140
ependymal cells 140
parenchyma 140
arachnoid granulations 140
hydrocephalus 140
bridging veins 141
subdural hematoma 141
computed tomography (CT) scan 141
chronic traumatic encephalopathy (CTE) 141
blood–brain barrier 142
tight junction 144
antihistamines 144
loperamide 144
blood-CSF barrier 144
arachnoid barrier 145
giant vacuoles 145
circumventricular organs 146
innate immune response 146
adaptive immune system 146
lymphocytes 146
immune privileged 147
astrocytes 147
end-feet 147
microglia 148
necrosis 149
excitotoxicity 149
potassium buffering 151
reactive astrocyte 151
glial scar 151
Nogo 151
multiple sclerosis 152
Guillain-Barre syndrome 152
Braille alexia 153
remapping 153
constraint-induced movement therapy 154
phantom limb pain 154
glycolysis 156
oxidative phosphorylation 156
glucose transporter 156
glycogen 156
circle of Willis 157
communicating arteries 157
ischemic stroke 157
collateral branches 157
anastomoses 157
penetrating arterioles 158
functional magnetic resonance imaging (fMRI) 159
blood oxygen level dependent (BOLD) signals 159
smooth muscle cells 159
cerebral fatigue 162
hyperventilation 162
brain-in-a-vat 162

ADDITIONAL READINGS

5.1 - Cell Turnover in Adult Brains

Taupin P. 2007. BrdU immunohistochemistry for studying adult neurogenesis: paradigms, pitfalls, limitations, and validation. *Brain Res Rev* **53**:198–214.

5.2 - Protection from Physical Trauma

Brodbelt A, Stoodley M. 2007. CSF pathways: a review. *Br J Neurosurg* **21**:510–520.

Mack J, Squier W, Eastman JT. 2009. Anatomy and development of the meninges: implications for subdural collections and CSF circulation. *Pediatr Radiol* **39**:200–210.

5.3 - Protection Against Toxins and Pathogens

Bechmann I, Galea I, Perry VH. 2007. What is the blood-brain barrier (not)? *Trends Immunol* **28**:5–11.

Johansson P, Dziegielewska K, Liddelow S, Saunders N. 2008. The blood-CSF barrier explained: when development is not immaturity. *Bioessays* **30**:237–248.

Wolburg H, Noell S, Mack A, Wolburg-Buchholz K, Fallier-Becker P. 2009. Brain endothelial cells and the glio-vascular complex. *Cell Tissue Res* **335**:75–96.

5.4 - Responses to Attacks on the Nervous System

Ajami B, Bennett J, Krieger C, Tetzlaff W, Rossi F. 2007. Local self-renewal can sustain CNS microglia maintenance and function throughout adult life. *Nat Neurosci* **10**:1538–1543.

Bredesen D, Rao R, Mehlen P. 2006. Cell death in the nervous system. *Nature* **443**:796–802.

Chen R, Cohen LG, Hallett M. 2002. Nervous system reorganization following injury. *Neuroscience* **111**:761–773.

Galea I, Bechmann I, Perry VH. 2007. What is immune privilege (not)? *Trends Immunol* **28**:12–18.

Hanisch U, Kettenmann H. 2007. Microglia: active sensor and versatile effector cells in the normal and pathologic brain. *Nat Neurosci* **10**:1387–1394.

Noppeney U. 2007. The effects of visual deprivation on functional and structural organization of the human brain. *Neurosci Biobehav Rev* **31**:1169–1180.

Pons TP, Garraghty PE, Ommaya AK, Kaas JH, Taub E, Mishkin M. 1991. Massive cortical reorganization after sensory deafferentation in adult macaques. *Science* **252**:1857–1860.

Tambuyzer BR, Ponsaerts P, Nouwen EJ. 2009. Microglia: gatekeepers of central nervous system immunology. *J Leukoc Biol* **85**:352–370.

Wolf SL, Winstein CJ, Miller JP, Thompson PA, Taub E, Uswatte G, Morris D, Blanton S, Nichols-Larsen D, Clark PC. 2008. Retention of upper limb function in stroke survivors who have received constraint-induced movement therapy: the EXCITE randomised trial. *Lancet Neurol* **7**:33–40.

5.5 - Supplying the Brain with Energy

Iadecola C, Nedergaard M. 2007. Glial regulation of the cerebral microvasculature. *Nat Neurosci* **10**:1369–1376.

Nybo L, Secher NH. 2004. Cerebral perturbations provoked by prolonged exercise. *Prog Neurobiol* **72**:223–261.

Zonta M, Angulo MC, Gobbo S, Rosengarten B, Hossmann KA, Pozzan T, Carmignoto G. 2003. Neuron-to-astrocyte signaling is central to the dynamic control of brain microcirculation. *Nat Neurosci* **6**:43–50.

Boxes

Bennewitz MF, Saltzman WM. 2009. Nanotechnology for delivery of drugs to the brain for epilepsy. *Neurotherapeutics* **6**:323–336.

Deeken JF, Löscher W. 2007. The blood-brain barrier and cancer: transporters, treatment, and Trojan horses. *Clin Cancer Res* **13**:1663–1674.

Holmes C, Boche D, Wilkinson D, Yadegarfar G, Hopkins V, Bayer A, Jones RW, Bullock R, Love S, Neal JW, Zotova E, Nicoll JAR. 2008. Long-term effects of Aβ42 immunisation in Alzheimer's disease: follow-up of a randomised, placebo-controlled phase I trial. *Lancet* **372**:216–223.

Huettel SA, Song AW, McCarthy G. 2009. *Functional magnetic resonance imaging*. Sunderland, MA: Sinauer Associates.

Neuroinflammation Working Group. 2000. Inflammation and Alzheimer's disease. *Neurobiol Aging* **21**:383–421.

Niewoehner J, Bohrmann B, Collin L, Urich E, Sade H, Maier P, Rueger P, et al. 2014. Increased brain penetration and potency of a therapeutic antibody using a monovalent molecular shuttle. *Neuron* **81**:49–60.

Park D, Jeon JH, Shin S, Jang JY, Choi B, Nahm SS. 2008. Debilitating stresses do not increase blood–brain barrier permeability: Lack of the involvement of corticosteroids. *Environ Toxicol Pharmacol* **26**:30–37.

CHAPTER 6

Sensors I:

Remote Sensing

CONTENTS

FEATURES

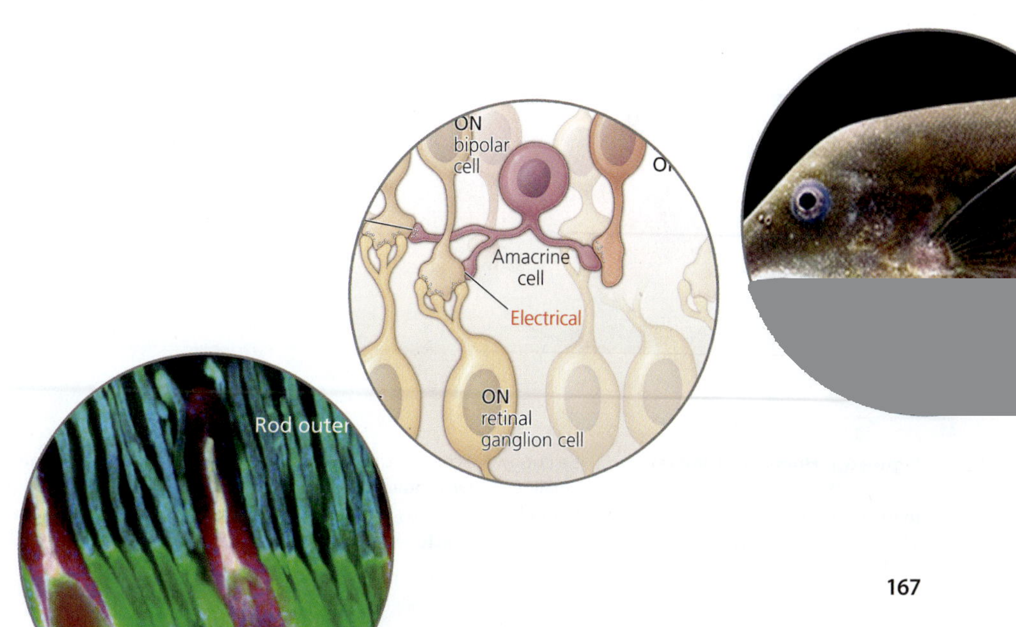

All animals use their nervous systems to gather useful information about the world. In this chapter, we focus on sensors that animals use to obtain information about distant stimuli; these are the sensors used in vision, hearing, and smell. In Chapter 7, we will discuss an array of other sensors that require direct contact with stimuli and provide animals with information about what is going on inside and on the surface of their own bodies. The division of sensors into remote sensors and contact sensors is somewhat arbitrary, as sensors in the skin can, for example, detect the low-frequency vibrations emanating from a bass drum or the heat coming off the sun. Still, the categories are functionally meaningful because the ability to detect distant threats or opportunities is very different from sensing stimuli that are already in contact with us. Just imagine how limited your own behavior would be if you were blind, deaf, and unable to smell. How would you find food, court a mate, or avoid threats?

All biological sensors must solve the problem of **sensory transduction**, which is to convert various forms of energy located outside of the nervous system (light rays, sound waves, mechanical forces, or chemicals) into a common set of neural signals—namely, membrane depolarization and hyperpolarization—that modulate the rate of neurotransmitter release. Another common theme is that all sensors respond to a limited range of stimuli and are adjustable in sensitivity. Although these features are clearly useful, they cause our perception of reality to be biased and variable. This sensor variability explains why individuals of different species often perceive the world quite differently and why physically identical stimuli may generate different perceptions at different times in different contexts. We will discuss these general principles of sensory transduction at the end of the chapter after we have explored the detection of light, odors, and sound.

6.1 How Do We Sense Darkness and Light?

The light sensors, or **photoreceptors**, of humans are all found in the retina (Figure 6.1 A). As light passes through the eye, it is bent by the cornea at the front of the eye and by the lens. Due to this bending, the incoming rays of light cross just behind the lens (Figure 6.1 B) and project an inverted image of the outside world onto the retina at the back of the eyeball. If you have normal vision, this projected image is focused on the retina. If you are near- or farsighted, the image is focused in front or just behind the retina, respectively. Fortunately, these vision problems can usually be corrected

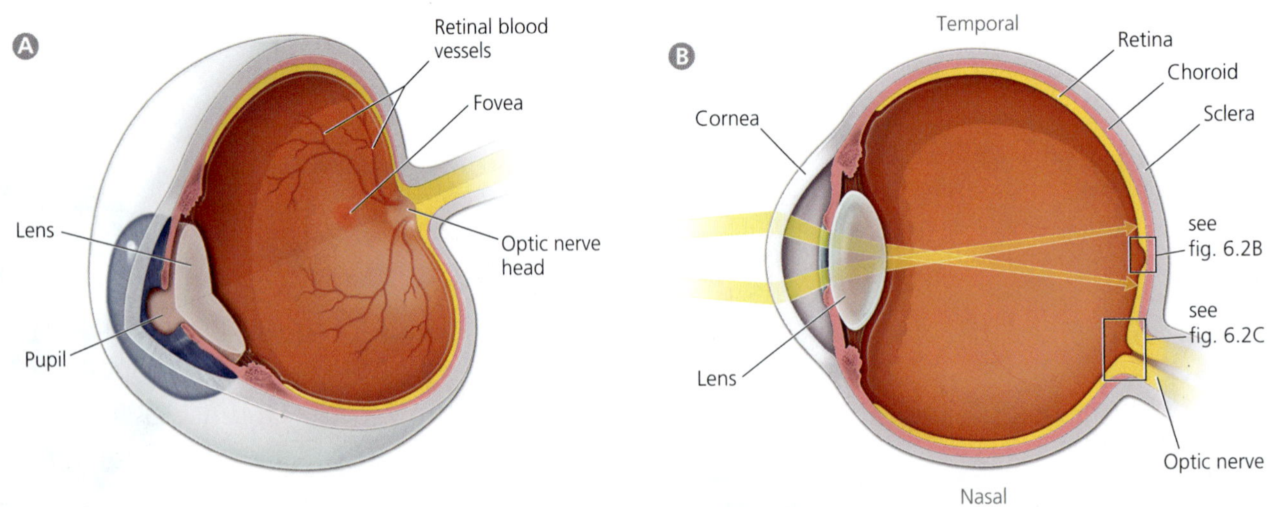

Figure 6.1 Human eye and retina. (A) is a cut-away schematic of a human eye. As shown in (B), light entering the eye is bent by the cornea and by the lens. Due to this refraction, light rays (yellow arrows) coming from a single, relatively distant location in the world are focused onto a small spot in the retina. Because light rays from different locations cross in (or just behind) the lens, the retinal image is inverted relative to the outside world. The black rectangles in (B) indicate sections through the fovea and optic nerve head, which are shown at higher magnification in Figure 6.2 B and C.

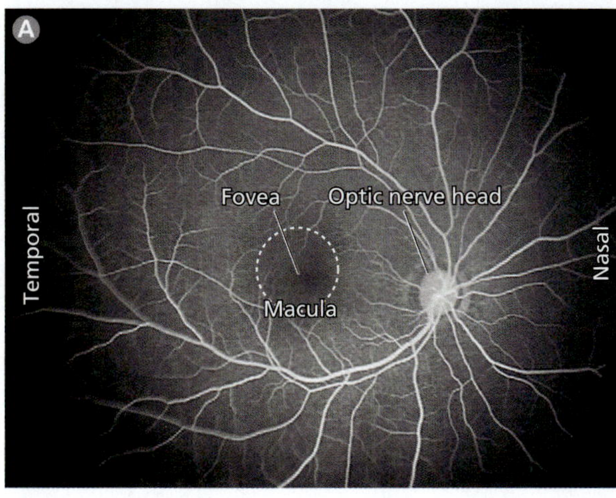

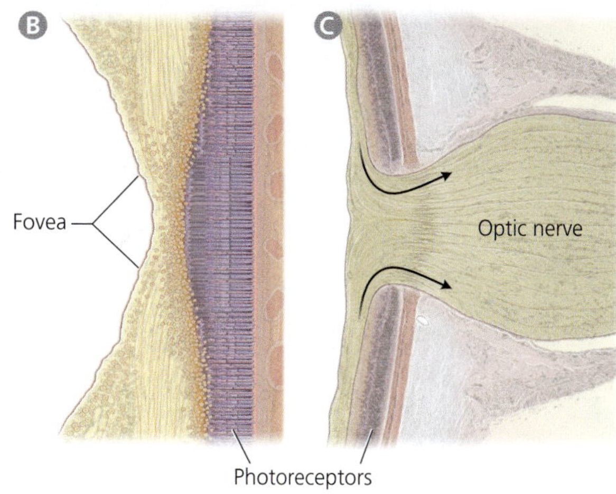

Figure 6.2 Retinal blood vessels, fovea, and optic nerve head. (A) is a photograph of a human retina taken through an ophthalmoscope shortly after an intravenous injection of a fluorescing substance. You can see blood vessels emanating from the center of the optic nerve head where axons leave the retina. Roughly 3 mm temporal to the optic nerve head lies a darker region called the macula (dashed circle). Within the macula lies the fovea. Shown in (B) and (C) are sections through the fovea and the optic nerve head, respectively. The black arrows in C indicate the path of axons leaving the retina to enter the optic nerve. [A from Witmer & Kiss, 2013; B after Polyak, 1957; C after Sugiyama et al., 1999]

with prescription eyeglasses, contact lenses, or a special form of laser surgery that alters the cornea's curvature. The aim in all these corrective procedures is to create a tightly focused projection of the outside world within the layer of photoreceptor cells. These photoreceptors are the visual system's sensory transducers; they convert light energy (photons) into neural signals.

Special Regions of the Retina

Before we discuss the photoreceptors, consider the retina as a whole. An ophthalmologist peering into a human eye sees a retina covered with blood vessels (Figure 6.2 A). All these vessels can be traced back to a circular white spot that is about 2 mm in diameter. This is the **blind spot**. It is where the retinal blood vessels enter and exit the eye. It is also the only portion of the retina devoid of photoreceptors. To understand why the blind spot exists, you must know that the axons carrying information out of the retina into the brain initially run along the inner side of the retina (the side facing the lens). For these axons to leave the retina and enter the optic nerve, they must cross through the depth of the retina. They perform this crossing at a single location, which is called the **optic nerve head** (Figure 6.2 A and C). Because the optic nerve head consists entirely of axons crossing through the retina, it cannot contain any photoreceptors and must, therefore, be "blind." Fortunately, we are typically unaware of our retinal blind spots because the blind spots in our two eyes are aimed at different regions of external space. In addition, we somehow "fill in" the parts of the retinal image that fall onto the blind spot even if we peer through just one eye (Figure 6.3).

Macula and Fovea

A few millimeters lateral (temporal) to the blind spot lies the **macula**, a roughly circular, more darkly pigmented region of the retina (Figure 6.2) that is devoid of overlying blood vessels. Deterioration of this region is called **macular degeneration**, a major cause of vision problems in old age, often brought on by diabetes. The main concern with macular degeneration is that the macula contains the **fovea**, which occupies the center of your visual field. Whenever you move your eyes to look at some object, you are placing the retinal image of that object onto the fovea. Because the fovea contains a very high density of photoreceptors, aiming the fovea at objects allows you to see them more sharply. Before we can discuss what kind of photoreceptors predominate within the fovea, you need to know that retinal photoreceptors

Figure 6.3 Mapping your blind spot. Close your left eye and focus the right eye on the black X, holding it 16 inches from your face. Now place the tip of a pencil over the arrow so that you can see its tip. Slowly move the pencil toward the top of the figure while maintaining your fixation on the X. At some point, the tip will seem to disappear because it has entered your retina's blind spot. Now move the pencil left, right, up, and down, and note where the tip disappears. Your blind spot will probably be oval in shape and surprisingly large. Finally, lay the pencil vertically across your blind spot map so that the pencil's tip protrudes beyond it. When you now fixate on the X with your right eye, the pencil does not display a gaping hole because the brain "fills in" the "missing" piece.

come in two basic flavors: rods and cones. Because rods are simpler than cones, we discuss them first.

Rod Photoreceptors

Rod photoreceptors are unusual neurons that fire no action potentials, have unusually short axons, and possess, instead of typical dendrites, a rod-shaped **outer segment** (Figure 6.4). Each outer segment contains a stack of membranous disks filled with **rhodopsin**, which consists of two linked molecules. The larger of these molecules is **opsin**, which has 7 transmembrane domains and belongs to the large family of G protein-coupled receptors. The smaller molecule, called **retinal**, lies deep within the opsin molecule. Retinal is bent in the dark but straightens out after activation by light (Figure 6.4 C). More specifically, absorption of a photon converts retinal from its *cis* isomer, which has a bent backbone, to the all-*trans* isomer, which has a straight backbone. This change in shape "activates" the rhodopsin molecule. Soon after rhodopsin activation, retinal loses its covalent attachment to the opsin molecule, thus allowing the retinal to drift away. Eventually, as we discuss later, the opsin molecule is recombined with a *cis*-retinal, regenerating rhodopsin.

The Phototransduction Cascade

The activation of rhodopsin initiates a complex chain of intracellular processes termed the **phototransduction cascade** (Figure 6.5). When an activated rhodopsin molecule bumps into the G protein **transducin**, it activates that transducin, which means that the GDP on one of the three transducin subunits is replaced with GTP, causing this subunit to dissociate from the others. The solitary subunit then activates a **phosphodiesterase** that degrades cyclic GMP (cGMP) to 5'-GMP. Each activated

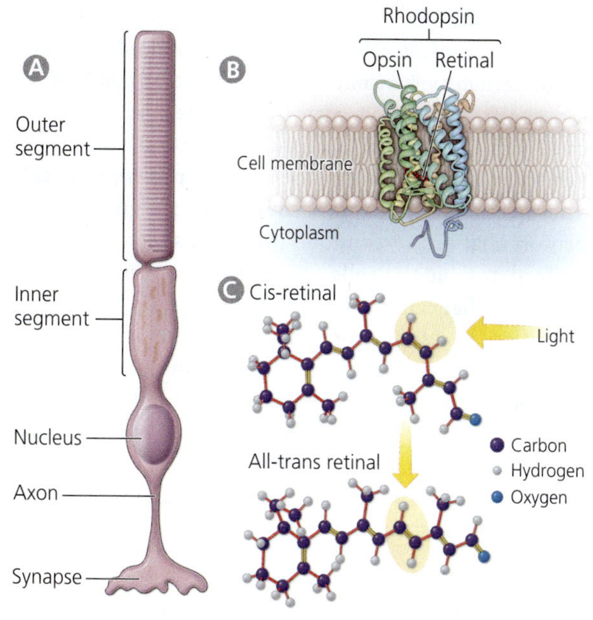

Figure 6.4 Rods and rhodopsin. (A) depicts a rod photoreceptor. Its outer segment contains numerous membranous disks that are loaded with rhodopsin molecules. Shown in (B) is a model of a rhodopsin molecule embedded in a disk's membrane. Its major component is opsin, a molecule with 7 helical transmembrane domains. Deep within the opsin molecule, and covalently bound to it, lies *cis*-retinal (red). As shown in (C), *cis*-retinal isomerizes to all-*trans*-retinal on interaction with photons. [Rhodopsin model created by Roland Deschain]

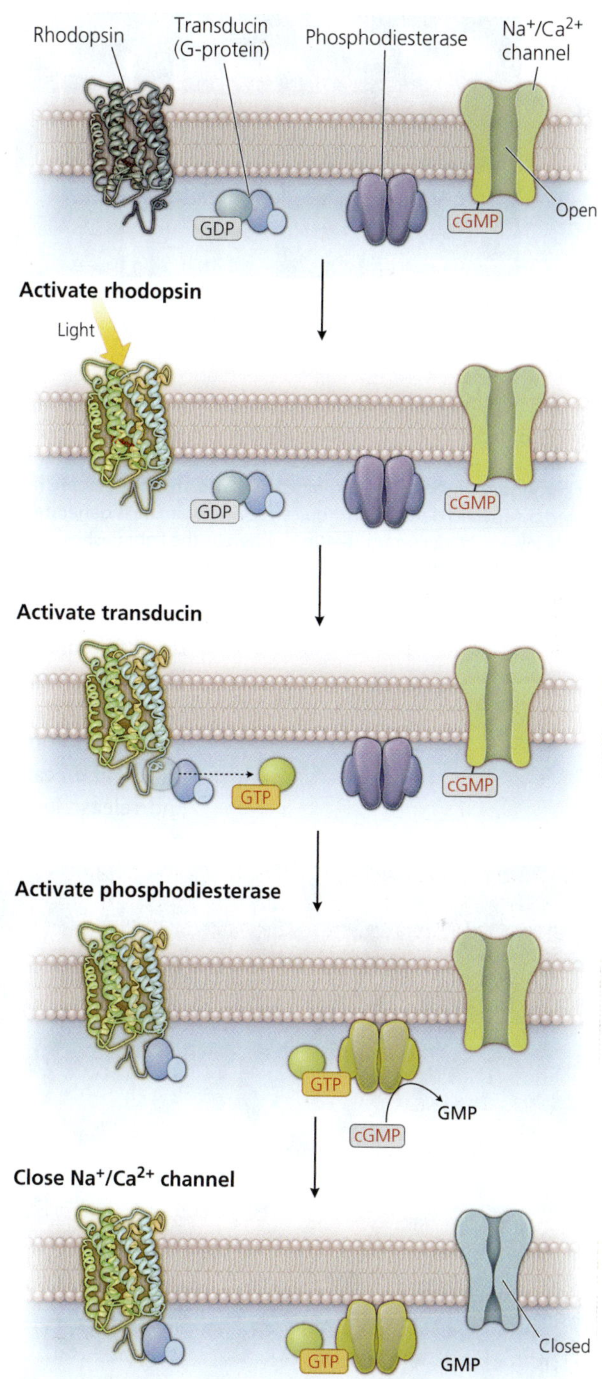

Figure 6.5 The phototransduction cascade. Light-activated rhodopsin causes the G protein transducin to split into multiple subunits, one of which then activates a phosphodiesterase; activated phosphodiesterase, in turn, lowers intracellular cyclic GMP (cGMP) levels, which causes cGMP-gated Na^+/Ca^{2+} channels to close.

rhodopsin can activate several transducin molecules (the number depends on how long the molecules interact, which no one knows for sure). Each activated transducin can activate numerous phosphodiesterases, which in turn can degrade hundreds of molecules of cGMP per second. Thus, the phototransduction cascade entails a series of signal amplification steps. Because of this amplification, even a few photons can dramatically decrease cGMP levels within a photoreceptor cell.

The light-induced decrease in cGMP concentration causes cGMP-gated cation (positive ion) channels to close, allowing fewer Na^+ and Ca^{2+} ions to enter the photoreceptor cell. This change in ion flux creates a membrane **hyperpolarization** (the opposite of depolarization). Importantly, this hyperpolarization occurs against a background of membrane depolarization that is generated by the open cGMP-gated cation channels when it is dark. Because of these channels, photoreceptors in the dark have resting potentials near −40 mV, which is significantly more depolarized than the resting potential of a typical neuron (see Chapter 2). In essence, light stimulation reduces the depolarizing "dark current" caused by open cGMP-gated cation channels (Figure 6.6). The light-induced membrane hyperpolarization rapidly spreads across the photoreceptor's cell membrane and (by closing some voltage-gated calcium channels) reduces Ca^{2+} influx into the rod's axon terminal. The decrease in intracellular calcium then causes fewer glutamate-filled vesicles to be released. Overall, we can say that illumination hyperpolarizes rod photoreceptors and decreases their rate of glutamate release.

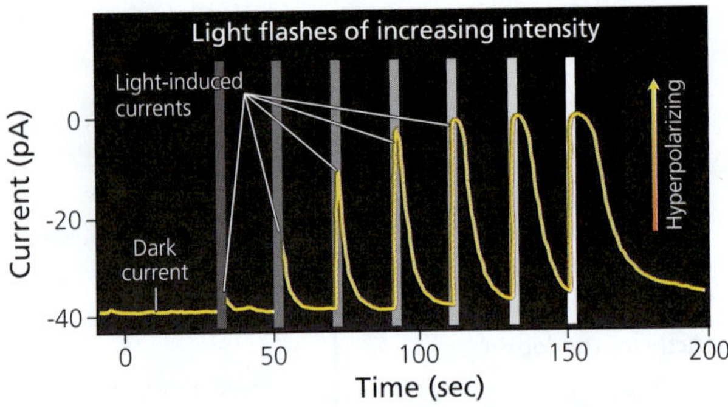

Figure 6.6 Dark-and light-induced currents. Shown in yellow are currents recorded from a rod in a salamander's retina. Light flashes of increasing intensity (vertical gray lines) were presented against a dark background. Outward currents, which hyperpolarize the cell, are represented by upward deflections. You can see that darkness generates a steady inward "dark current" (–40pA), whereas the light flashes generate transient outward (hyperpolarizing) currents whose amplitude and duration vary with light intensity. [After Matthews, 1990]

Rod Bipolar Cells

Because we intuitively expect sensory stimulation to *increase* excitatory transmitter release, it seems odd that light causes a *decrease* in glutamate release. The puzzle is resolved once you realize that rod photoreceptors synapse onto very unusual postsynaptic cells called **rod bipolar cells.** These cells do not express the ionotropic glutamate receptors that are so common in neurons (see Chapter 2). Instead, rod bipolar cells express metabotropic glutamate receptors that cause hyperpolarization, rather than depolarization, in response to glutamate. Therefore, when rods are hyperpolarized by light, their postsynaptic bipolar cells become depolarized (less hyperpolarized) and consequently increase their rate of neurotransmitter release. This makes the neural response to light consistent with our expectations: More light means more neurotransmitter release from rod bipolar cells.

Dark Adaptation

Still, you might wonder why photoreceptors evolved to release so much transmitter in the dark; it seems like a waste of metabolic energy. However, when rods are in constant darkness, they accumulate so much intracellular Ca^{2+} that the enzyme synthesizing cGMP becomes inhibited. As a result, cGMP levels decline and some Na^+ channels close, just as they would if photons were flittering about. Because of the decreased Na^+ influx, the rods become more hyperpolarized and release less neurotransmitter from their synapses. This conserves transmitter molecules and thus saves energy. It also makes the photoreceptors more sensitive to changes in light intensity when background illumination is low. Collectively, these processes are called **dark adaptation** because they "adapt" the retina to low-light conditions. An additional component of dark adaptation is that after prolonged exposure to darkness, most of the retinal in the photoreceptors has been returned to its *cis* form and recombined with opsin to form rhodopsin, which stands ready to be activated by photons. In contrast, after prolonged exposure to light, most of the retinal is "bleached," meaning that it has been converted to its all-*trans* form and separated from opsin. Overall, dark adaptation increases rod sensitivity to the point where rods are able to detect single photons—an amazing level of sensitivity when you consider that vision also works in bright sunlight, which is at least ten million times more intense than faint starlight.

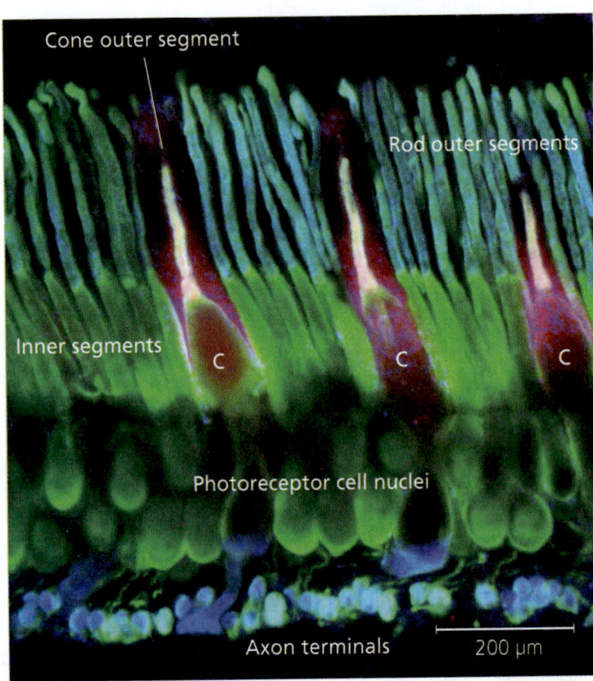

Figure 6.7 Rods and cones. In this cross section through the photoreceptor layer of a monkey's retina you can see three cones (c) and many rods. The cone outer segments are shorter and more conical than rod outer segments. The different colors were obtained by staining the tissue with multiple fluorescent antibodies. [Courtesy of Nicolás Cuenca, www.retinalmicroscopy.com]

Cone Photoreceptors

The second major type of vertebrate photoreceptor cell is called a cone because its outer segment is cone shaped (Figure 6.7). Although humans have far more rods than cones (~120 million vs. ~6 million), the cones are crucial because they predominate within the fovea (Figure 6.8). Because cones are much less sensitive to light than rods, vision in very dim light is mediated mainly by rods in the retina's periphery. This explains why dim stars are easier to see if you do not look directly at them with your cone-dominated fovea.

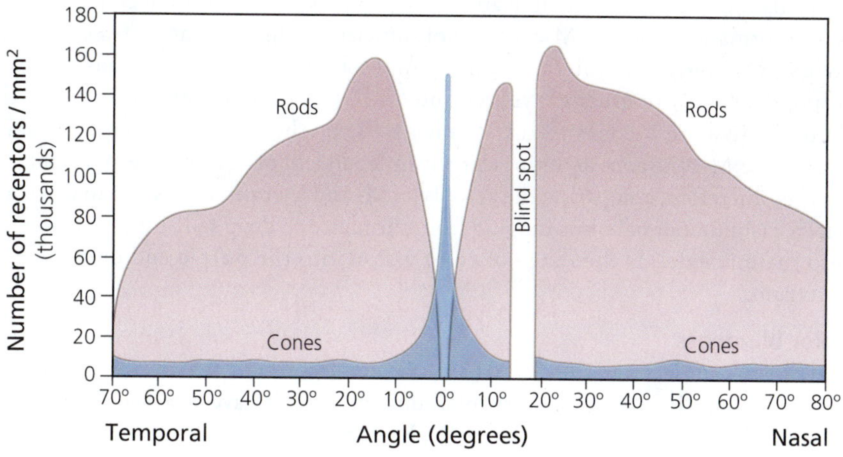

Figure 6.8 Rod and cone distributions. This graph shows the density of rods and cones across a typical human retina. The center of the fovea, which contains only cones, is defined as zero degrees. [After Osterberg, 1935]

Cone Tuning Curves

Have you ever noticed that it's harder to see colors in dim light? This phenomenon arises because cones differ from rods not only in their light sensitivity but also in their response to light of different wavelengths. Rods are most readily activated by light with a wavelength of about 500 nm, which most of us perceive as bluish green. The further the incoming light deviates from this optimal wavelength, the weaker the elicited response. In technical language, we say that rod rhodopsin is *tuned* to wavelengths around 500 nm, much as your radio or TV tuner is tuned to a particular wavelength or frequency.

The three different types of opsin molecules that are expressed in cones are tuned to different wavelengths. As you can see in Figure 6.9, **S-cones** respond maximally to 425 nm light, which we perceive as dark blue, **M-cones** respond maximally to green light at ~530 nm, and **L-cones** respond best to yellow light with a wavelength of ~560 nm. The letters S, M, and L indicate that these cones are tuned to short (blue), middle (green), and long (yellow and red) wavelengths, respectively. However, the response profiles of the three cone types overlap considerably. Therefore, it is more accurate to say that the three cone types preferentially respond (are tuned) to different wavelengths of light. These preference differences are due to relatively minor differences in the molecular structure of the opsin molecules expressed within the three cone types.

Why does your retina contain three differently tuned cone types? Because having those cones makes color vision possible. The neural circuits used to discriminate colors are complicated (see Chapter 12), but the basic principle is fairly simple. Imagine you are looking at two equally bright spots of light: one is a bluish green, the other

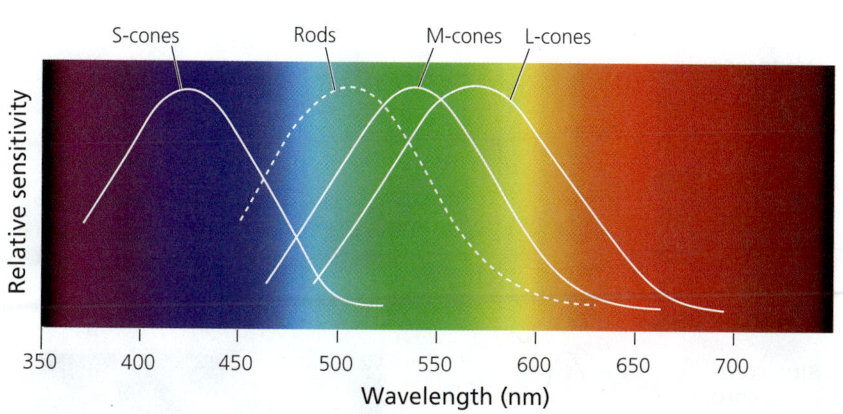

Figure 6.9 Photoreceptor tuning curves. S-cones are most sensitive to short wavelengths, M-cones respond best to medium wavelengths, and L-cones respond best to the longest wavelengths in the visible spectrum. Because the color sensitivity (tuning) curves are different for each type of cone, each color in the visible spectrum evokes a different pattern of activity across the three cone types. Orange light, for example, will activate L-cones more strongly than M-cones and the latter more strongly than S-cones. Rods are maximally sensitive to blue-green wavelengths, but they are not used in color vision.

a greenish yellow. Both spots activate your M-cones suboptimally but equally. Therefore, information from the M-cones is not sufficient to discriminate between the two colors. However, activity in the L-cones can resolve this ambiguity because L-cones respond strongly to greenish yellow and much less to bluish green. In short, the ratio of activity in L-cones versus M-cones tells you the color of the spot. Crucially, the L-cone:M-cone activity ratio is largely independent of how intense, or bright, the color is. Therefore, comparing the activity of M- and L-cones allows you to discriminate a vast number of colors ranging from turquoise to deep red. Throwing S-cones into the mix extends the range of color vision into the purple end of the visible spectrum.

Color-blindness

You can now understand the neural basis of **color-blindness**, which affects 8–12% of the population (Figure 6.10). Most colorblind individuals have defective (or missing) M-cones or L-cones. This makes them unable to resolve the color ambiguities we just discussed. Colorblind individuals find it difficult, for example, to discriminate red from green when the two colors are equally bright. They have even greater difficulties when the brightness of a colored spot is unknown or difficult to infer from other cues. Because of these challenges, colorblind individuals pay more attention to the position of the light in a traffic signal than to its color and get annoyed when city engineers mount traffic lights horizontally, thus removing key positional cues. Even more confusing are single-position caution lights that flash either yellow or red. Without the normal complement of M- and L-cones those colors are very difficult to discriminate.

Curiously, males are far more likely than females to be colorblind. This gender difference arises because the M- and L-opsin genes are both on the X chromosome. As you probably recall from basic biology, females have two X chromosomes and males have only one. Therefore, any deleterious mutations on an X chromosome will impact males disproportionately. Because the S-opsin gene lies on a different chromosome, mutations in this gene rarely lead to serious color vision problems. Also

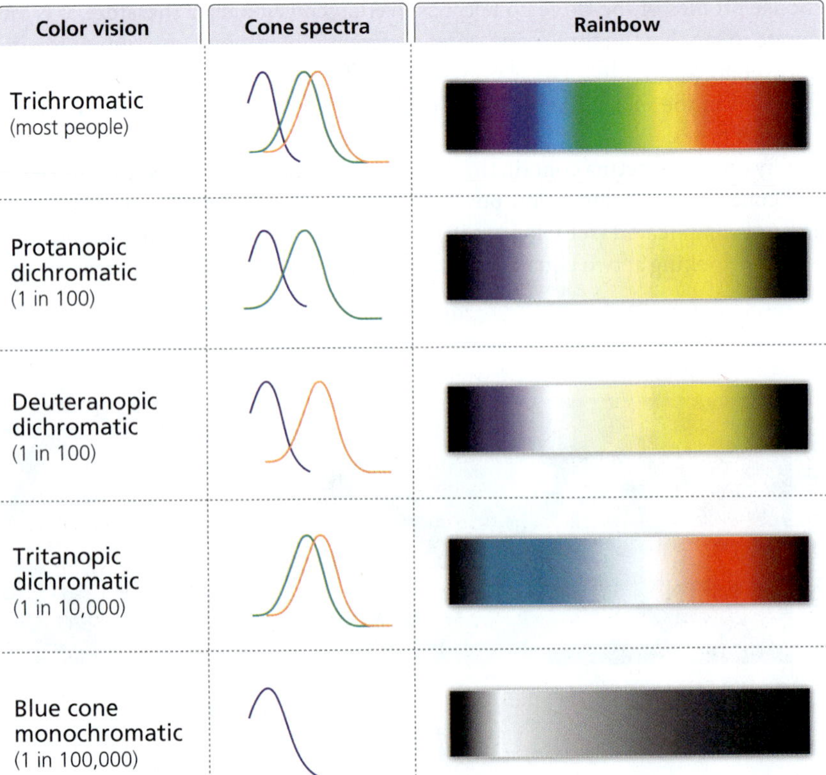

Figure 6.10 Forms of color blindness. Shown in the top row are the tuning curves of the three cone types in normal vision and a representation of how the visible spectrum is normally perceived. The other rows illustrate various forms of color blindness in which one or two cone types are nonfunctional. An additional 1–5% of people (depending on their ethnic group) have milder forms of color blindness in which one cone type has an abnormal color sensitivity curve. [From Deeb, 2006]

interesting is that most mammals do not possess L-cones and therefore are red-green colorblind. Genome comparisons reveal that the L-cone opsin gene was lost with the evolution of early mammals and that a new, slightly different L-opsin gene evolved in primates. The re-evolution of L-opsins in primates probably helped them identify ripe fruits, which tend to be yellow or red rather than green.

Pathways through the Retina

Cones, like rods, decrease their rate of glutamate release on exposure to light and synapse on bipolar cells with metabotropic glutamate receptors that elicit hyperpolarization. However, cones make additional synapses onto a second type of bipolar cell that expresses ionotropic rather than metabotropic glutamate receptors (Figure 6.11). This second type of cone bipolar cell is called an **OFF bipolar cell** because it releases less glutamate when the cone is exposed to light and more when the light turns off (it has an OFF response to light). Correspondingly, the first type of cone bipolar cell is called an **ON bipolar cell**. One may debate why cones synapse onto both ON and OFF bipolar cells, whereas rods synapse only onto ON-type bipolar cells, but the difference clearly reveals that the retinal circuits processing color are more complex than those that deal with low-light signals from the rods.

All cone bipolar cells release glutamate directly onto **retinal ganglion cells** (Figure 6.11), which have long axons that project through the optic nerve to multiple targets in the brain, including the superior colliculus (optic tectum) and lateral geniculate nucleus, both of which we have discussed in Chapter 4 and will discuss more thoroughly in Chapters 11 and 12. In general, each retinal ganglion cell receives input from either ON bipolar cells or OFF bipolar cells but not from both. Therefore, retinal ganglion cells can also be divided into ON and OFF types, respectively (Figure 6.11).

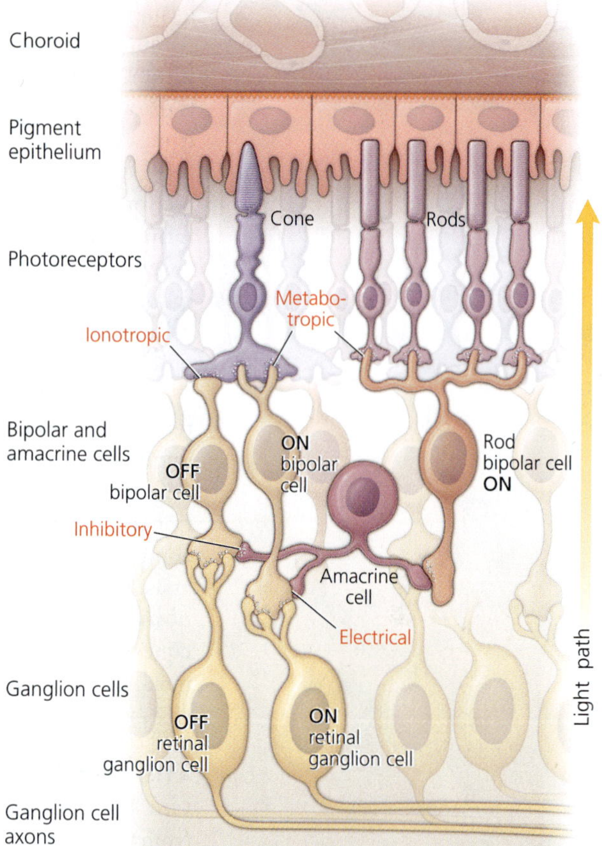

Figure 6.11 Circuits through the retina. Cone photoreceptors synapse on the dendrites of bipolar cells, which synapse onto retinal ganglion cells whose axons leave the retina. Rod photoreceptors transmit information to retinal ganglion cells through rod bipolar and specialized amacrine cells, which then synapse onto cone bipolar cells using electrical or inhibitory synapses. The photoreceptor outer segments are embedded in the pigment epithelium, which is covered by a dense capillary bed called the choroid.

Rod bipolar cells, in contrast, do not contact retinal ganglion cells directly. Instead, they synapse onto a type of retinal **amacrine cell** that synapses onto cone bipolar cells (Figure 6.11). The amacrine cells use the inhibitory transmitter glycine to communicate with OFF bipolar cells and electrical synapses (gap junctions that allow action potentials to propagate directly from one cell to the next) to contact ON bipolar cells. The upshot of these connections is that information from rod bipolar cells must pass through amacrine and cone bipolar cells to get out of the retina.

Convergence in the Retinal Pathways

A related difference between the rod and cone pathways is that the cone pathway exhibits a much higher degree of convergence. Only one or a few cones converge onto each cone bipolar cell, and each cone bipolar cell synapses on just a few retinal ganglion cells. In contrast, 30 or more rod photoreceptors converge onto each rod bipolar cell, each amacrine cell gathers input from numerous rod bipolar cells, and each cone bipolar cell gathers inputs from several amacrine cells. This high degree of connectional convergence allows the output of many rods to be summed as it is passed to a single retinal ganglion cell, making the rod system even more sensitive to light than it would be without this convergence (recall that rods are more sensitive to light than cones). However, the increased light sensitivity of the rod system comes at a cost, namely, a dramatically reduced spatial resolution.

As we discuss more thoroughly in Chapters 11 and 12, each retinal ganglion cell has a specific **spatial receptive field**, defined as the region of space where visual stimuli must be presented if they are to modify the neuron's activity. Using this definition, retinal ganglion cells that are activated mainly through the rod system (i.e., outside the fovea) have much larger spatial receptive fields (gather information from a much larger region of visual space) than those that are activated through the cone system, simply because rod-driven retinal ganglion cells combine visual information from a much larger set of photoreceptors. In essence, much of the spatial information captured by the rods is thrown away as signals are summed along the pathway through the rod bipolar and amacrine cells. Overall, you can think of the rod system as being specialized for vision in low light, sacrificing spatial detail and color; whereas the cone system is specialized for high-resolution color vision but goes offline when it gets dark.

Although you now understand the basics of how visual information traverses the retina, you should be aware that the retina is more complex than our discussion has implied. For example, researchers have identified roughly 30 distinct types of amacrine cells (the type we have discussed is most common), and it includes a few additional cell types (most notably horizontal cells; see Chapter 12, Figure 12.5). Another complication is that some retinal ganglion cells can sense photons even without the assistance of rods or cones. These light-sensitive retinal ganglion cells express **melanopsin**, a photosensitive pigment that is related to the opsins expressed in rods and cones but responds to light much more slowly. As you will learn in later chapters, the melanopsin-expressing retinal ganglion cells play a major role in controlling pupil size and the sleep-wake cycle.

The Puzzle of the Inverted Vertebrate Retina

One of the oddest aspects of retinal organization is that the rods and cones face away from the incoming light (yellow arrow in Figure 6.11). This seems backward because you would expect much of the incoming light to be scattered and absorbed by the overlying photoreceptor inner segments, bipolar cells, ganglion cells, axons, and blood vessels at the front of the retina (see Figure 6.2 A) long before it reaches the rods and cones. Furthermore, placing the photoreceptors at the back of the retina forces the retinal ganglion axons to cross through the retina on their way out of the eye, thereby creating the blind spot (see Figure 6.2 C).

Given this apparently backward construction, you may wonder whether the vertebrate retina is an example of "bad design." Some biologists have answered "yes" and argued that this poor design is evidence against the notion that biological structures were designed by an intelligent creator who supposedly designed all things optimally. However, the theory of evolution predicts neither optimal design nor poor design; it simply states that natural selection is capable of yielding structures and processes that give some organisms an advantage over competitors. Evolution by natural selection usually yields good designs in the sense that they serve organisms well, but those designs may be suboptimal in some respects. In any case, let us explore whether the vertebrate retina is really as poorly "designed" as it appears to be.

Retinoid Recycling

As we discussed before, light activation ultimately causes retinal molecules to float away from the opsins to which they had been bound. This raises an important question: how do photoreceptors recover their lost retinal and regain their sensitivity to light? The answer lies in the **pigment epithelium**, a layer of cells that covers the rod and cone outer segments (Figure 6.11). After *cis*-retinal is converted to all-*trans*-retinal and separated from opsin, it is transported out of the photoreceptors and into the pigment epithelium, where it is re-isomerized to *cis*-retinal. The *cis*-retinal is then transported back into the outer segments of the photoreceptors and recombined with opsin to form rhodopsin or one of the cone opsins. This **retinoid recycling** consumes

a significant amount of metabolic energy, which explains why the pigment epithelium is covered with a dense capillary bed called the **choroid** (Figure 6.11).

Once you realize that the photoreceptor outer segments must be covered by the pigment epithelium and the choroid to function efficiently, it becomes apparent why the outer segments evolved to face away from the incoming light: if the photoreceptors were facing the light, then the red blood cells in the choroid and the dark cells of the pigment epithelium would absorb most of the incoming photons before they reach the photoreceptors. In contrast, the other parts of the retina, including the axons of the retinal ganglion cells, are relatively transparent.

Light Paths through the Retina

Placing the pigment epithelium and choroid away from the incoming light may be the better of two options, but the retinal cells lying between the incoming light and the photoreceptors in the vertebrate retina do cause some photons to be scattered or absorbed. Fortunately, evolution has minimized these costs. For one thing, the cell bodies of bipolar and retinal ganglion cells in the fovea are pushed toward the fovea's perimeter (Figure 6.2 B), thereby clearing the path for photons from the central portion of the visual field.

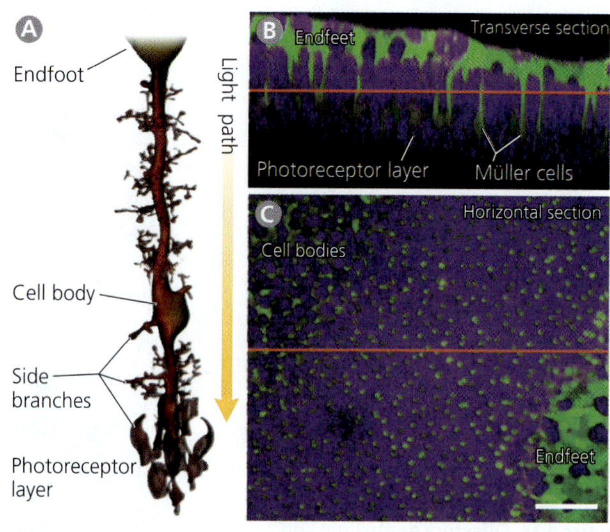

Figure 6.12 Light guides through the retina. The main trunk of a retinal Müller cell (A) has a relatively constant diameter of just less than 3 μm, but it widens into a prominent "endfoot" at the retina's inner surface. This endfoot has a lower refractive index than the rest of the cell, which facilitates the entry of light. Shown on the right is a retina in which Müller cells are labeled green. Panel (B) shows a transverse section, with the endfeet at the top and the photoreceptor layer at the bottom of the photograph. Shown in (C) is a roughly horizontal section through the same retina (the red horizontal line in B shows the section plane for C and vice versa). [From Franze et al., 2007]

In addition, the retina is studded with specialized glia, called **Müller cells** (Figure 6.12), which collect incoming light at the retina's inner surface and channel it, just as a fiber optic cable would, past all the retinal cells straight to the outer segments of the photoreceptors. Thus, Müller cells overcome a principal drawback of having an inverted retina. Müller cells also perform several additional functions, such as helping to degrade neurotransmitters and, at least in zebrafish, serving as potential stem cells for regenerating damaged portions of the retina.

Speaking of light paths, have you ever wondered why the eyes of some animals appear to glow when they are photographed at night? The answer is that many nocturnal animals have a highly reflective layer, called the **tapetum lucidum** (Latin for "bright tapestry"), either in their pigment epithelium or in the choroid. Photons that get past the photoreceptors without being captured by opsin molecules tend to bounce off this tapetum lucidum and, thus, get another chance at interacting with the retinal opsins. This mechanism increases the animal's ability to see in low light conditions. Of course, bright light from a camera flash also bounces off the tapetum lucidum, exits through the pupil, and generates the eerie glow we call "eye shine." Incidentally, the "red eye" you often see in flash photography of human eyes is not caused by a tapetum lucidum (we don't have one) but by light reflecting off the reddish blood vessels at the front of the retina.

BRAIN EXERCISE

In what sense might it be better to think of photoreceptors as darkness sensors rather than light sensors?

6.2 How Do We Sense Odors?

The sense of smell allows animals to detect and identify odor-generating objects, often over considerable distances. Prey can, for example, detect a predator by smell long before the predator is visible or audible. Odors may also be used to find potential food or to communicate with other members of one's species.

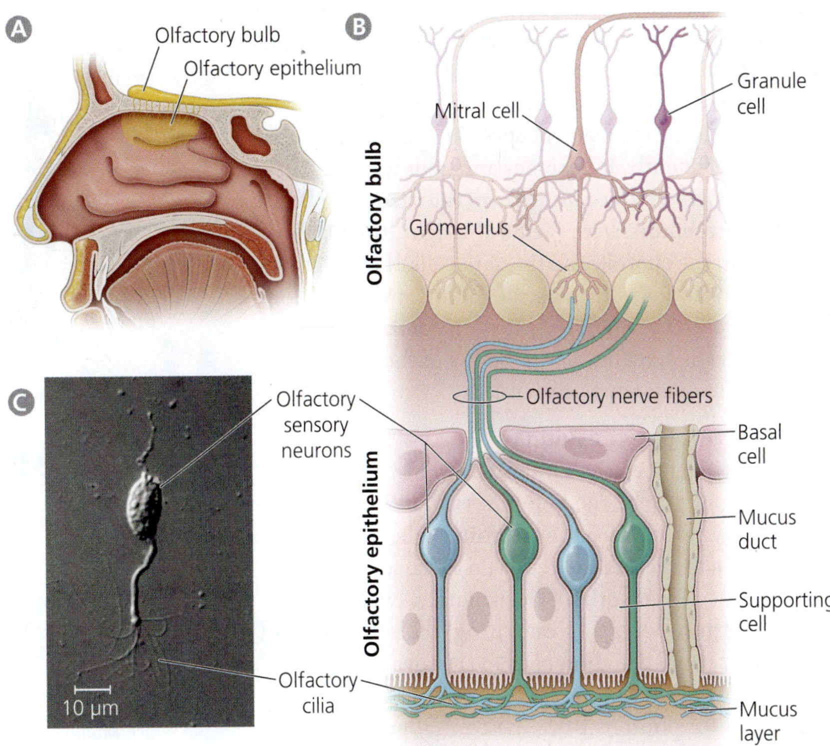

Figure 6.13 Olfactory epithelium and bulb (A) The olfactory epithelium contains olfactory sensory neurons that send their axons through a bony plate into the olfactory bulb. As shown in (B), olfactory sensory neurons extend their cilia into a layer of mucus, where they encounter odorants. Panel (C) shows an isolated olfactory sensory neuron. [C from Kleene & Gesteland,1981]

Olfactory Epithelium

Odor-sensing cells are located in the **olfactory epithelium**, which is located in the roof of the nasal cavity (Figure 6.13). The external surface of the olfactory epithelium is covered in mucus secreted from mucus glands whose ducts pass through the epithelium. Sticking into the mucus layer are the cilia of **olfactory sensory neurons**. The membranes of these cilia are packed with olfactory receptor molecules that can bind odorants (molecules that can be smelled) and trigger changes in the membrane potential of the olfactory sensory neurons. We will discuss the olfactory receptor molecules shortly; but for now, we remain at the cellular level.

Why are the olfactory cilia covered in mucus? Probably because the cilia would rapidly dry out if they were suspended in air. In addition, the mucus protects the cilia from pathogens. To this end, the olfactory mucus is enriched in antioxidants, as well as antimicrobial and anti-inflammatory proteins. The downside of all this mucus is that the airborne odor molecules must get through the mucus before they can contact the olfactory cilia. Fortunately, many odorants are ferried through the mucus by specialized **odorant binding proteins**. Others dissolve in the mucus and reach the cilia by simple diffusion. All this works fairly well until a cold or allergy attack makes you secrete so much nasal mucus that odorants have trouble reaching the cilia. Thus, the mucus that protects your olfactory system may, at times, prevent its normal functioning.

An intriguing aspect of olfactory sensory neurons is that they can be replaced in adulthood, which is why complete destruction of all olfactory sensory neurons impairs olfaction for only 2–3 months. Even under normal conditions, olfactory sensory neurons die and are replaced regularly, with an average life span of about 1 month. However, some olfactory sensory neurons live considerably longer. This cellular longevity was demonstrated by injecting into the olfactory bulb (of mice) a tracer that is transported along the axons back to the cell bodies of the olfactory sensory neurons. When the investigators examined the olfactory epithelium 3 months later, some labeled cell bodies could still be found in the olfactory epithelium. Because tracers are

taken up by neurons only shortly after an injection, the labeled cell bodies observed in the olfactory epithelium must have belonged to neurons that were present at the time of the tracer injection. That is, they must have lived at least 3 months.

Olfactory Receptor Molecules

The receptors responsible for odor sensing in vertebrates were identified in 1991 by Linda Buck and Richard Axel, who received the Nobel Prize in Medicine or Physiology for their efforts. Buck and Axel showed that the olfactory cilia of rats selectively express a large family of novel G protein-coupled receptors, each containing 7 transmembrane domains (Figure 6.14).

Although the **olfactory receptor molecules** are surprisingly similar to opsins, they are far more diverse. According to current estimates, humans possess ~350 different functional olfactory receptor types (vs. just 5 different opsins). Although 350 seems like a large number, some other mammals possess an even greater diversity of olfactory receptor proteins. Mice and rats are estimated to have ~1,200 functional olfactory receptor genes, and dogs appear to have at least 900 such genes. These species differences arose because the olfactory receptor genes proliferated and diversified independently in the various mammalian lineages. In addition, in the lineage leading to humans, an unusually high fraction of olfactory receptor genes (~60%) mutated into nonfunctional **pseudogenes** (Figure 6.15).

Given these differences, you might expect humans to have a poor sense of smell. They do, at least compared to dogs (which also have up to twenty times as many olfactory sensory neurons as humans), but humans can sense

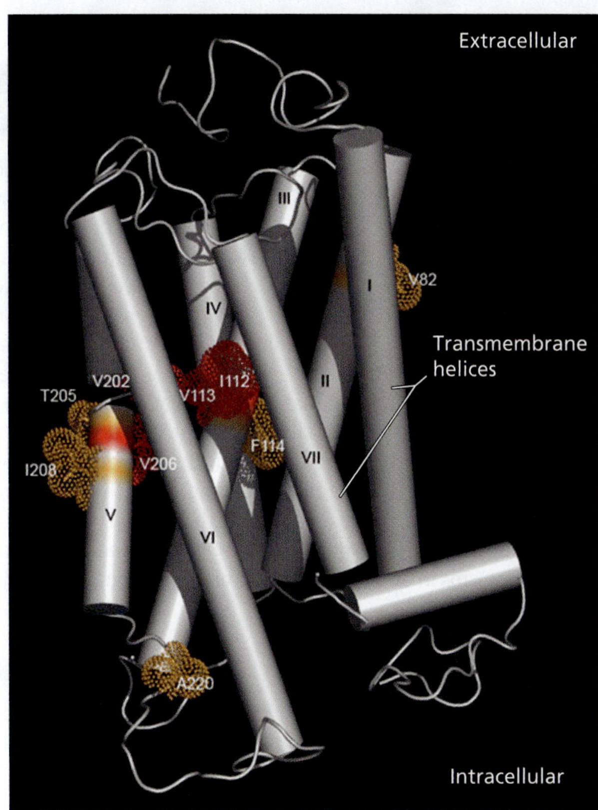

Figure 6.14 An olfactory receptor molecule. In this model of a mouse olfactory receptor, helical portions of the protein are depicted as gray cylinders. The 7 helices that span the cell membrane are labeled with Roman numerals. The dotted areas indicate regions that differ between different receptor types, and the white labels identify specific residues. The red regions deep within the molecule are known to be involved in the specific binding of odorant molecules. [From Abaffy et al., 2007]

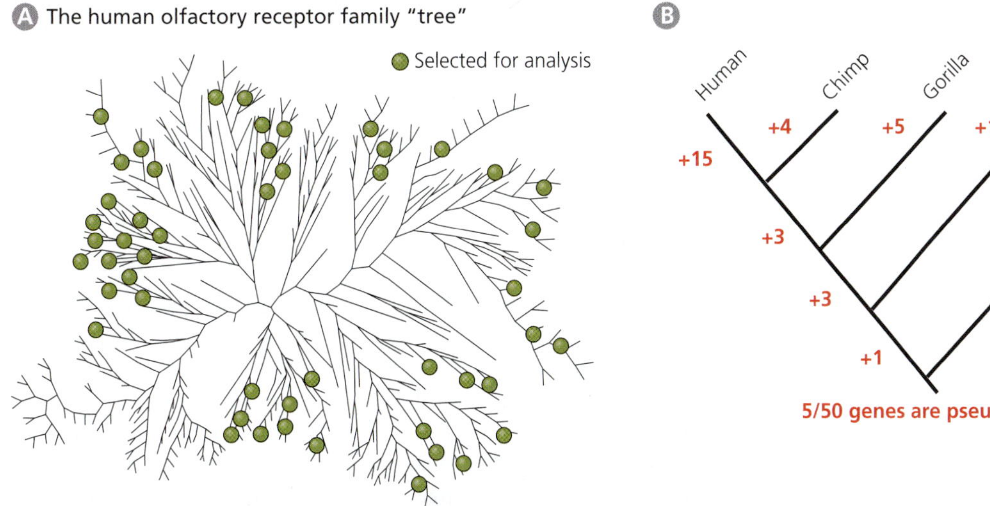

Figure 6.15 Olfactory receptor evolution. Of the roughly 1,000 genes in the human olfactory receptor family, more than 60% are pseudogenes, which means that they do not form functional proteins. A comparative study of 50 randomly selected olfactory receptor genes (A) revealed that many of these genes became nonfunctional within the primate lineage (B). Roughly 10% (5/50) were pseudogenes at the root of the primate family tree. An additional 22 genes became nonfunctional in the lineage leading to humans, 15 of them mutating after humans diverged from chimpanzees. [From Gilad et al., 2003]

NEUROBIOLOGY IN DEPTH

Box 6.1 *Anosmias: Does Your World Smell the Same as Mine?*

Roughly 5% of the general population are *anosmic*, meaning that they lack a sense of smell. This kind of general anosmia may result from various causes, including head trauma, nasal inflammation, and intranasal application of zinc gluconate (Zicam), as well as Alzheimer's and Parkinson's disease. However, approximately 1% of anosmics were born without a sense of smell. In subjects with Kallmann's syndrome, the olfactory bulb never developed. Other anosmics have mutations in a type of voltage-gated sodium channel that makes them incapable of smelling odorants and insensitive to pain (see Chapter 7).

Even people with a generally sound sense of smell may have one or more *specific anosmias*, meaning that they are much less sensitive to certain odorants than the general population. Several of these specific anosmias have been linked to polymorphisms in an olfactory receptor gene. A good example is the anosmia for *androstenone*. Male pigs use androstenone as a pheromone to impress female pigs, but most humans find its smell sickening (which makes them consider pork from male pigs distasteful). Androstenone is also found in human sweat, which most people find unpleasant at high concentrations. However, a fraction of the population cannot smell androstenone at all (Figure b6.1; incidence estimates vary widely across studies). Twin studies have shown that this androstenone anosmia is heritable, and genetic analysis have linked it to two amino acid substitutions in the olfactory receptor gene OR7D4. Finally, in vitro studies have confirmed that OR7D4 with the two substitutions does not respond to androstenone.

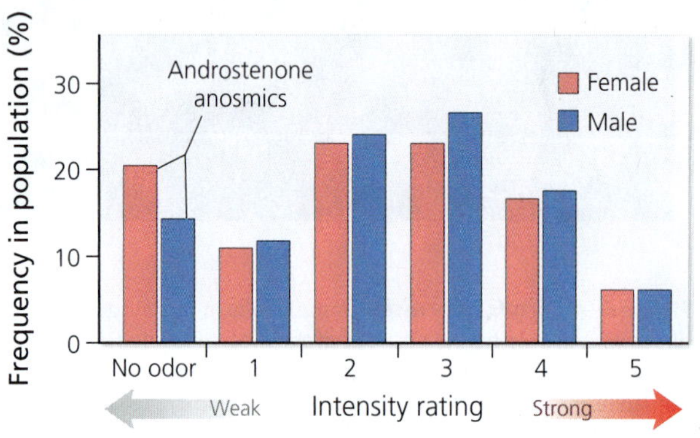

Figure b6.1 Variations in androstenone sensitivity. A sample of 555 women and 435 men were given scratch-and-sniff tests of several odors, including androstenone. They were asked whether they smelled anything and, if so, how strong the odor was. Roughly 18% of the individuals did not detect the androstenone at all, whereas the others displayed a normally distributed range of sensitivities. Sex differences were not statistically significant. [After Knaapila et al., 2012]

Another specific anosmia concerns *cis-3-hexen-1-ol* (C3HEX), which is found in diverse fruits and vegetables (such as raspberries and broccoli). This molecule smells like fresh-cut grass to most people, but roughly 12% of the general population have trouble smelling C3HEX. This deficit is genetically associated with two amino acid substitutions in the olfactory receptor gene OR2J3. Although cells expressing OR2J3 do not respond to C3HEX, humans possess two other olfactory receptors that can respond to C3HEX at high concentrations. Therefore, people with the modified OR2J3 allele are not completely anosmic for C3HEX; they can smell this compound when the odor is intense.

Have you ever noticed the sulfuric smell that taints your urine within a few hours of eating asparagus? Different people have different degrees of sensitivity to this "asparagus urine" odor, and at least 8% of people cannot detect it at all. What about you? Do you have this specific anosmia? It can be difficult to know. If your urine doesn't smell strange to you several hours after eating asparagus, then you might have the anosmia, or you might not, because production of the metabolized asparagus odor is itself a variable (and heritable) trait. Therefore, proving that you have the asparagus urine anosmia, you must conduct a more extensive experiment, involving multiple subjects.

some odors very well. For example, 80% of mothers can smell whether a shirt was worn by their own baby. Also interesting is that humans vary considerably in which odors they can smell, at least in part because of variation in their olfactory receptor genes (see Box 6.1: Anosmias: Does Your World Smell the Same as Mine?).

Olfactory receptor molecules bind odorants deep in their interior, similar to how opsins bind retinal. However, because of variations in amino acid sequence, each olfactory receptor type tends to bind a different set of odorants. Specifically, each olfactory receptor type binds to a specific molecular feature or **epitope**. Because most odorants have more than one of these epitopes, most odorants are bound by multiple olfactory receptor types.

The Olfactory Transduction Cascade

The binding of an olfactory receptor molecule to one of its "preferred" odorants activates a G protein called Golf (Figure 6.16). Activated Golf then activates an olfactory-specific *adenylate cyclase,* which raises cAMP levels within the olfactory cilium. Rising cAMP levels open cAMP-gated ion channels that allow Na^+ and Ca^{2+} to flow into the cell. The rise in intracellular Ca^{2+}, in turn, opens calcium-gated chloride channels. Through these channels Cl^- ions flow out of the cilia, which contain unusually high levels of intracellular Cl^-, and into the mucus, which harbors few chloride ions. The upshot of this complex signaling cascade is that odorant binding depolarizes an olfactory sensory neuron. If the depolarization exceeds threshold, action potentials are triggered in the olfactory sensory neuron's axon hillock and sent to the olfactory bulb.

The molecular mechanisms underlying sensory transduction in the olfactory system are similar to the phototransduction cascade (compare Figures 6.16 and 6.5). Both involve G proteins (G_{olf} and transducin), G protein-activated enzymes (adenylate cyclase and phosphodiesterase), and cyclic nucleotides (cAMP and cGMP) that open ion channels permeable to both Na^+ and Ca^{2+}. The major difference is that activating an olfactory receptor with an odorant *increases* the number of open Na^+/Ca^{2+} channels, thereby causing membrane depolarization; whereas activating opsins with light *decreases* the number of open Na^+/Ca^{2+} channels, thus triggering membrane hyperpolarization.

The difference in the physiological effect of opsin and olfactory receptor activation is surprising, given that opsins and olfactory receptors are structurally similar. To resolve the riddle, it helps to think of retinal as a ligand for opsin (a molecule that binds opsin), just as odorants are ligands for olfactory receptors. From that perspective, the physiological difference between the two sensory transduction cascades arises because light-induced activation leads to the gradual release of retinal from its opsin, whereas odor stimulation promotes the binding of an odorant to its respective receptor molecule. If you can get yourself to think of darkness as a visual stimulus, then the enigma disappears: darkness causes photoreceptor depolarization, just as smelling an odor causes depolarization in olfactory sensory neurons.

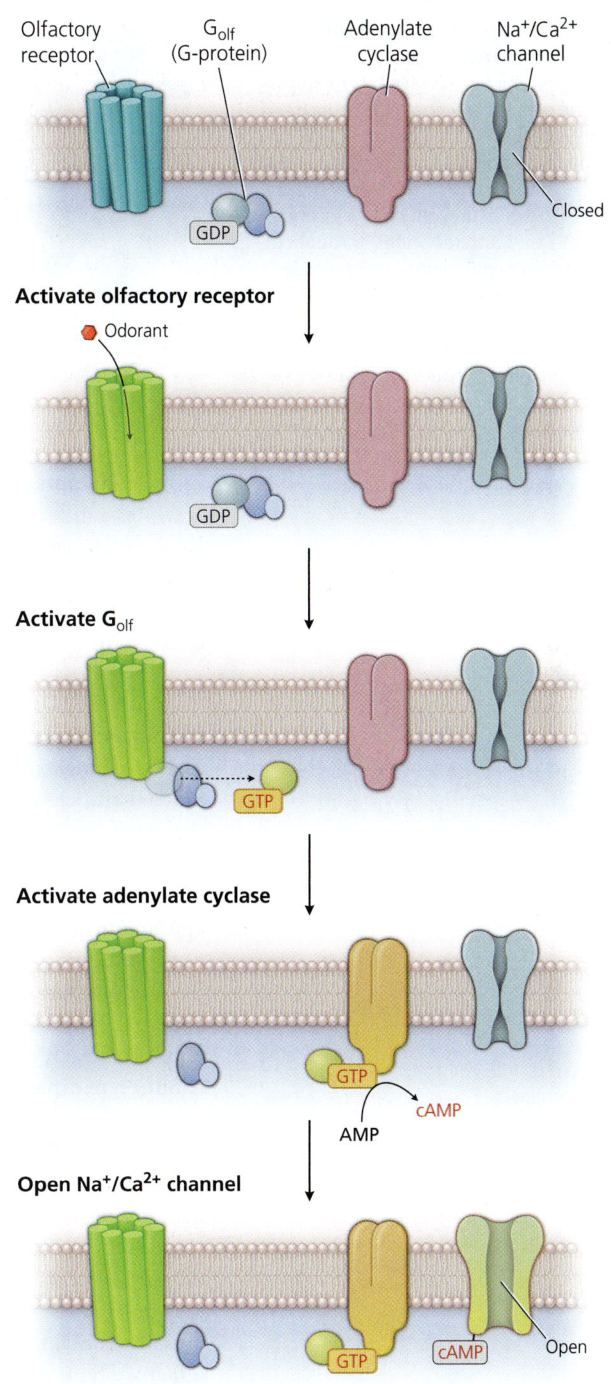

Figure 6.16 Olfactory transduction. The binding of an odorant to an olfactory receptor activates a G protein called Golf, which then activates adenylate cyclase. Activated adenylate cyclase raises intracellular levels of cAMP, which opens cAMP-gated Na^+/Ca^{2+} channels. Opening the Na^+/Ca^{2+} channels depolarizes the cell and elevates intracellular Ca^{2+} levels.

Odorant Discrimination

Imagine what would happen if each olfactory sensory neuron in the human olfactory epithelium expressed all 350 olfactory receptor types. Every odorant for which you have an olfactory receptor would depolarize every olfactory sensory neuron, making it impossible to discriminate between odors because all the olfactory sensory neurons would have an equal probability of responding to any given odor. Because we obviously can discriminate odors, something must be wrong with our imaginary scenario.

Indeed, each olfactory sensory neuron expresses only one type of olfactory receptor protein. Therefore, each odorant activates only a subset of olfactory sensory neurons, namely, the subset that expresses olfactory receptor molecules capable of binding one of the odorant's epitopes. This means that odorants can be discriminated from one another by examining which sets of olfactory sensory neurons are activated. Even complex mixtures of

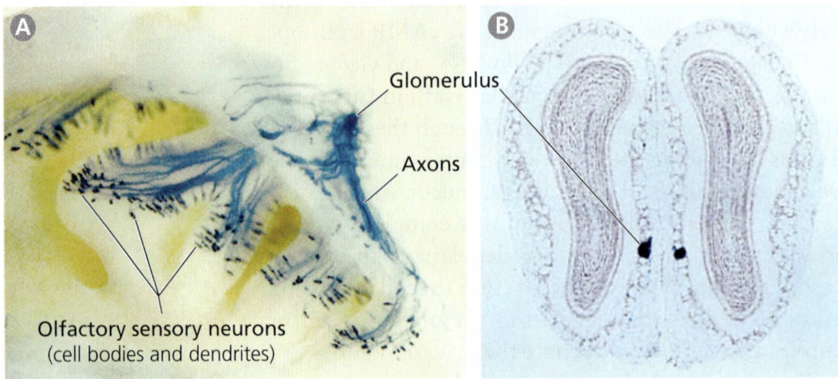

Figure 6.17 Projections to the olfactory bulb. Mombaerts et al. (1996) created transgenic mice in which all cells expressing the olfactory receptor molecule P2 also express an axonal marker that a simple histological procedure stains bright blue. Panel (A) shows that the cell bodies of the P2-expressing olfactory sensory neurons are scattered widely throughout the olfactory epithelium, whereas their axons converge onto a single glomerulus in the olfactory bulb. Panel (B) depicts a coronal section through the left and right olfactory bulbs of such a transgenic mouse, showing two of the four glomeruli (two in each bulb) in which the P2-expressing axons terminate. [From Mombaerts et al., 1996]

several odorants, as in most natural odors, are discriminable because each mixture activates a different set of olfactory sensory neurons. Thus, olfaction employs a *combinatorial scheme* to encode odor identity. We use analogous encoding schemes when we combine a limited set of letters into a profusion of words or combine thousands of pixels into pictures on a computer screen. In all of these examples, information is encoded in the combinatorial pattern of elements.

The Olfactory Bulb

In contrast to photoreceptors, the olfactory sensory neurons generate action potentials and have long axons, which exit the olfactory epithelium at the back (away from the mucus) and pass through tiny holes in a bony plate that separates the olfactory epithelium from the **olfactory bulb**. Once inside the olfactory bulb, the olfactory axons terminate in small globular structures called **glomeruli** (Figure 6.17). Each human olfactory bulb contains about 5,500 of these glomeruli. Given that each olfactory epithelium in humans contains ~7 million olfactory sensory neurons, we can infer that each glomerulus receives inputs from more than 1,000 olfactory sensory neurons.

The projections from the olfactory epithelium to the olfactory bulb are remarkably specific. Although the olfactory epithelium is divisible into several longitudinal bands that each contain a different assortment of olfactory sensory neurons, the distribution of different olfactory sensory neuron types within each band is essentially random. However, the axons of the olfactory sensory neurons "descramble" this random pattern as they project to the olfactory bulb. This orderly projection pattern was demonstrated by creating transgenic mice that produce a protein that fills axons and can be stained blue in every cell expressing a specific olfactory receptor gene (Figure 6.17). Examination of these mice revealed that all the neurons expressing the targeted olfactory receptor gene send their axons to just two glomeruli in each olfactory bulb. It is still unclear precisely how the axons of widely scattered neurons expressing the same olfactory receptor type find one another as they grow toward the olfactory bulb. However, because of this axonal sorting out, odorants that activate olfactory sensory neurons throughout much of the olfactory epithelium cause axonal transmitter release in just a few glomeruli within the olfactory bulb.

Chemotopic Organization of the Olfactory Bulb

To examine which glomeruli are activated by which odorants, rats can be injected with radioactive **2-deoxyglucose (2-DG)**, which neurons partially metabolize as if it were glucose. Shortly after receiving such an injection, the rats are exposed to a specific odorant and, half an hour later, the olfactory bulb is processed so that the pattern of radioactivity can be observed. In such experiments, highly radioactive areas must have been highly activated during the odor presentation. The results are clear: different odorants activate distinct, although often overlapping, subsets of the olfactory glomeruli (Figure 6.18).

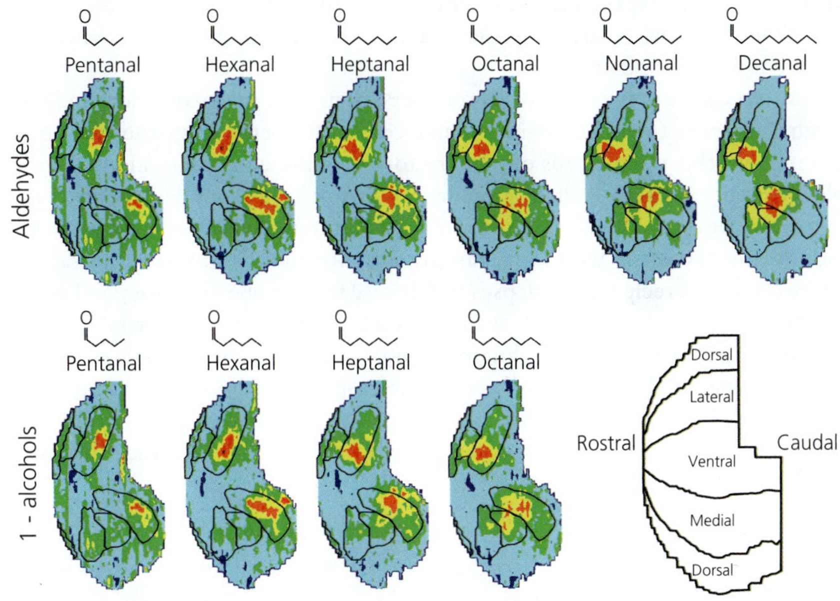

Figure 6.18 Mapping responses to odorants. Rats were injected with radioactive 2-DG and allowed to smell one of several odorants (pentanal, hexanal, etc.). Shortly thereafter, the rats were sacrificed and processed to determine where in the olfactory bulb the radioactive 2-DG had accumulated, indicating regions of high neuronal activity. The pseudocolored images reveal 2-DG uptake in a digitally flattened glomerular layer after exposure to aldehydes and alcohols with varying numbers of carbon atoms. The areas of highest 2-DG uptake move systematically across the bulb as carbon number increases, revealing a chemotopic map. [From Johnson et al., 2004]

Additional experiments have shown that structurally similar odorants activate similar subsets of glomeruli in spatially restricted regions of the olfactory bulb. Furthermore, the location of activated glomeruli varies systematically with the number of carbon molecules in the odorant's backbone (Figure 6.18). This orderly mapping between odorant structure and the spatial location of activated glomeruli is called **chemotopy** (the Greek word *topos* means "place"). Knowing the chemotopic map, scientists can look at the pattern of 2-DG activity within a rat's olfactory bulb and predict, at least to some extent, what the rat has smelled. Moreover, rats trained to discriminate between multiple odors tend to confuse odors that evoke very similar patterns of glomerular activity. This finding strongly suggests that the chemotopy in the olfactory bulb is involved in odor discrimination. We will return to this topic in Chapter 12 where we discuss the important contributions of the olfactory cortex to the identification of natural odors, which typically comprise a mix of many odorants and therefore elicit complex patterns of activity across the olfactory bulb.

The chemotopic organization of the olfactory bulb is similar to the retina's **retinotopic** organization, which is defined as the tendency of adjacent retinal neurons to be activated by visual stimuli from adjacent locations in space. In other words, the visual environment is systematically "mapped" onto the retina in a manner that is analogous to how olfactory stimuli are mapped onto the olfactory bulb. We explore **sensory maps** more thoroughly at the end of this chapter and later in the book. For now, it suffices to say that most brain areas involved in sensory processing contain sensory maps.

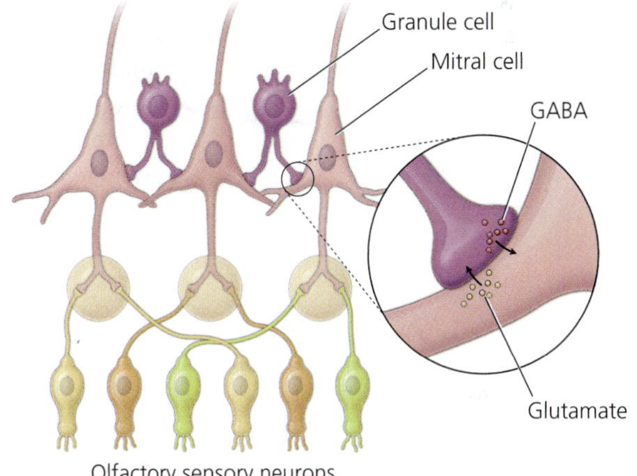

Figure 6.19 Circuits within the olfactory bulb. The diagram shows olfactory sensory neurons synapsing onto mitral cell dendrites in three glomeruli (yellow spheres). Mitral cells have axons that project out of the olfactory bulb, but they also release glutamate from their dendrites onto the dendrites of granule cells. The granule cells, in turn, release GABA back onto the dendrites of mitral cells. These reciprocal synapses (see inset) provide feedback inhibition to activated mitral cells. They also provide a mechanism through which an active mitral cell can inhibit less active neighbors, a phenomenon called lateral inhibition. [After Yokoi et al., 1995]

Circuits within the Olfactory Bulb

The largest neurons in the olfactory bulb are the **mitral cells**, which project to the olfactory cortex and extend long dendrites into the olfactory glomeruli (Figure 6.19). Because each mitral cell receives input in just one glomerulus, we would expect mitral cells to respond to odorants much as the olfactory sensory neurons do, but this is not the case. Instead, the mitral cells are more selective than the sensory neurons. To understand the basis of this increased selectivity, we need to consider an additional set of neurons that connects the mitral cells to one another.

Granule cells are the most numerous type of neuron in the olfactory bulb. Although they share their name with granule cells in the cerebellum and some other brain regions, the granule cells of the olfactory bulb are quite unusual. As Camillo Golgi first reported, the olfactory granule cells do not have axons. How then do they affect other neurons? The answer is that olfactory granule cells form **reciprocal dendrodendritic synapses** with long, tangentially directed dendrites of mitral cells (Figure 6.19). The reciprocal synapses in this case feature one excitatory and one inhibitory synapse, located right next to one another. At the excitatory synapse, the dendrite of a mitral cell releases glutamate onto the dendrite of a granule cell. At the inhibitory synapse, the granule cell dendrite releases GABA back onto the mitral cell dendrite. One likely function of these reciprocal synapses is negative feedback. That is, excitation of a mitral cell excites granule cells, which then inhibit the same mitral cell, shutting down its activity shortly after it began.

Lateral Inhibition

A second function of the reciprocal synapses between granule and mitral cells is **lateral inhibition**. It allows an activated mitral cell to excite granule cells, which then inhibit *neighboring* mitral cells. For lateral inhibition to work, the depolarization caused by glutamate release onto a granule cell must spread through the granule cell's dendritic tree to other synapses that contact other mitral cells. It can do so through passive current spread and dendritic action potentials (see Chapter 2).

To grasp the functional significance of lateral inhibition in the olfactory bulb, consider two mitral cells that can inhibit each other through their connections with the GABAergic granule cells. Of the two mitral cells, the more active one will inhibit the less active one, which means that the less active mitral cell will become even less active and less able to inhibit the more active mitral cell. You can think of it as a winner-take-all competition between the two mitral cells in which the more active cell wins and the less active one falls completely silent. How does lateral inhibition in the olfactory bulb affect the sense of smell? This question continues to be debated, but lateral inhibition probably narrows the range of odorants to which a mitral cell responds. In essence, lateral inhibition suppresses any responses to odorants that evoke stronger responses in neighboring mitral cells (Figure 6.20). As we discuss in Chapters 11 and 12, lateral inhibition also occurs in the visual and somatosensory systems, where it likewise suppresses weak responses in favor of stronger ones.

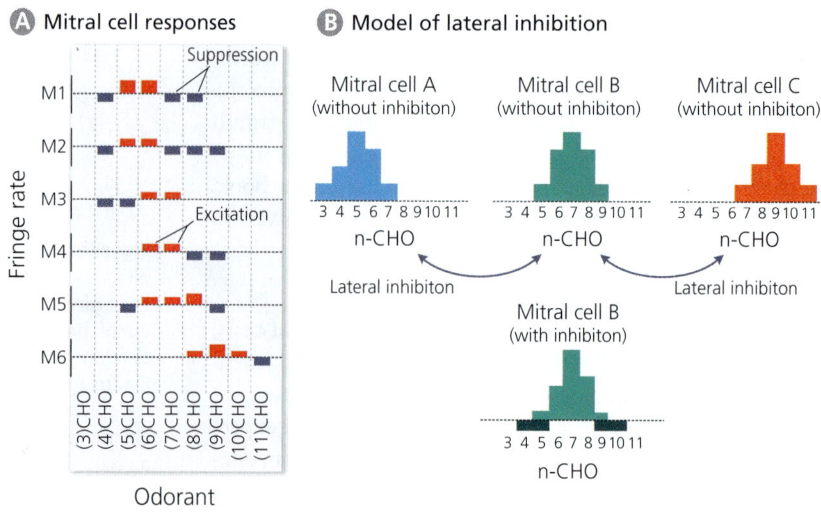

Figure 6.20 Lateral inhibition in the olfactory bulb. The histograms in (A) show the response profiles of 6 mitral cells (M1–6) for 9 odorants that vary in the length of their carbon backbone. Panel (B) presents a model of how lateral inhibition shapes the mitral cell response profiles. Shown along the top are the response profiles of three hypothetical mitral cells (A–C) in the absence of lateral inhibition. The bottom histogram depicts how cell B would respond to the odorants if lateral inhibition is present. For odorants with 4, 5, 9, and 10 carbon atoms, inhibition from the neighboring mitral cells would outweigh the excitatory inputs from the olfactory sensory neurons. [After Yokoi et al., 1995]

Besides mitral and granule cells, the olfactory bulb contains several other types of neurons, including some that allow activated mitral cells to inhibit mitral cells outside of their immediate neighborhood. These long-range inhibitory projections are thought to keep the average firing rate of all mitral cells relatively low, even when the air is full of odorants. Without this long-range inhibition, high odor concentrations would cause many mitral cells to fire at very high rates, which would be energetically costly. Keeping average firing rates low in odor-rich environments also makes it easier to detect subtle changes in odor concentrations or composition. To better understand this idea, consider how difficult it is to understand what someone is saying when many people are talking simultaneously. Having too many mitral cells generating too many action potentials presents an analogous challenge to the rest of the brain. Keeping the average firing rates low allows the activity of a few highly active mitral cells to stand out more clearly.

BRAIN EXERCISE

What are some similarities between olfaction and vision? What are the key differences?

6.3 How Do We Hear Sounds?

Sounds are longitudinal pressure waves produced by vibrating surfaces or turbulent streams. The main parameters of sound are intensity (power), which is typically measured in decibels (dB), and frequency (pitch), measured in Hertz (Hz) or Kilohertz (kHz). Children and young adults can hear sound frequencies from ~20 Hz to ~20 kHz. As humans age, they become progressively less capable of hearing high frequencies, which is why teenagers sometimes try to use high-frequency ring tones in classes with older instructors (as first reported in British newspapers). In terms of sound intensity, human hearing has an amazingly broad range. The sound of a flying mosquito rates just above 0 dB, which is defined as the threshold of human hearing. In contrast, the sound of a jet engine at 100 feet measures roughly 140 dB. Because the decibel scale is logarithmic (base 10), the value of 140 dB means that the jet's sound is 10^{14} times as intense as sounds that hover at the threshold of human hearing. We will cover the mechanisms underlying this broad range of sensitivity later. For now, we ask, what makes us capable of hearing sounds at all?

Outer and Middle Ears

We hear primarily with our ears, which are divisible into outer, middle, and inner ears (Figure 6.21). The outer ear consists of the ear flap, or **pinna**, and the **ear canal**. Together, they act as a funnel that directs sound waves toward the eardrum, or **tympanic membrane**. The sound waves cause the tympanic membrane to vibrate at the same frequency as the incoming sound. Most microphones are based on the same principle: they contain a thin membrane that vibrates in response to sound. As a rule, the larger and thinner the membrane, the more sensitive the microphone. Given this general principle, it is not surprising that the human tympanic membrane is fairly large (9 mm in diameter) and thin (~100 μm in thickness). Of course, there is a cost to making the tympanic membrane so thin and sensitive: sounds exceeding ~140 dB will cause it to rupture.

The Middle Ear Bones

Behind the tympanic membrane lies the **middle ear**, which consists of three small bones, suspended in air by tiny tendons and ligaments. The bones are called **malleus**, **incus**, and **stapes** (Latin for hammer, anvil, and stirrup, respectively). One end of the malleus is attached to the back of the tympanic membrane, right near its center, and the other end is coupled to the incus, which in turn attaches to one end of the stapes. Because of these linkages, movement of the tympanic membrane causes all three middle ear bones to move as a unit such that movement of the tympanic membrane in one direction causes the stapes to move in the opposite direction. Movement of the stapes, finally, pushes and pulls on the **oval window**, which is a thin membrane between the middle and inner ears. Thus, the middle ear bones convert vibrations of the tympanic membrane into matching (but opposite in direction) vibrations of the oval window.

Why do you need the middle ear bones? Wouldn't the vibrations of the tympanic membrane automatically induce the oval window to vibrate? Yes they would, but very inefficiently because the **inner ear** behind the oval window is filled with fluid rather than air, and water is much denser than air. Without the middle ear bones, 99.9% of the energy coming from the vibrating tympanic membrane would bounce right off the oval window. The middle ear bones focus the forces created by the relatively large tympanic membrane onto the much smaller membrane of the oval window (the same force-focusing principle explains why thumbtacks can easily be pushed into drywall or wood). In fact, the force with which the stapes pushes on the oval window membrane is at least 13 times as large as the force with which the tympanic membrane pushes on the malleus. In addition, the malleus acts as a force-amplifying lever, analogous to a crowbar, because its rotational axis (Figure 6.21) lies closer to the incus than to the tympanic membrane.

In reaction to very loud sounds, reflexive contraction of two small **middle ear muscles** (attached to the tympanic membrane and the stapes) can greatly reduce the degree to which vibrations are amplified within the middle ear. This mechanism helps protect the inner ear from sound-induced damage.

The Cochlea

The portion of our inner ear involved in hearing is called the **cochlea** (from the ancient Greek word for "snail with spiral shell"). It is a 6 cm long, fluid-filled tube that is folded back on itself, creating two parallel tubes that are continuous with one another

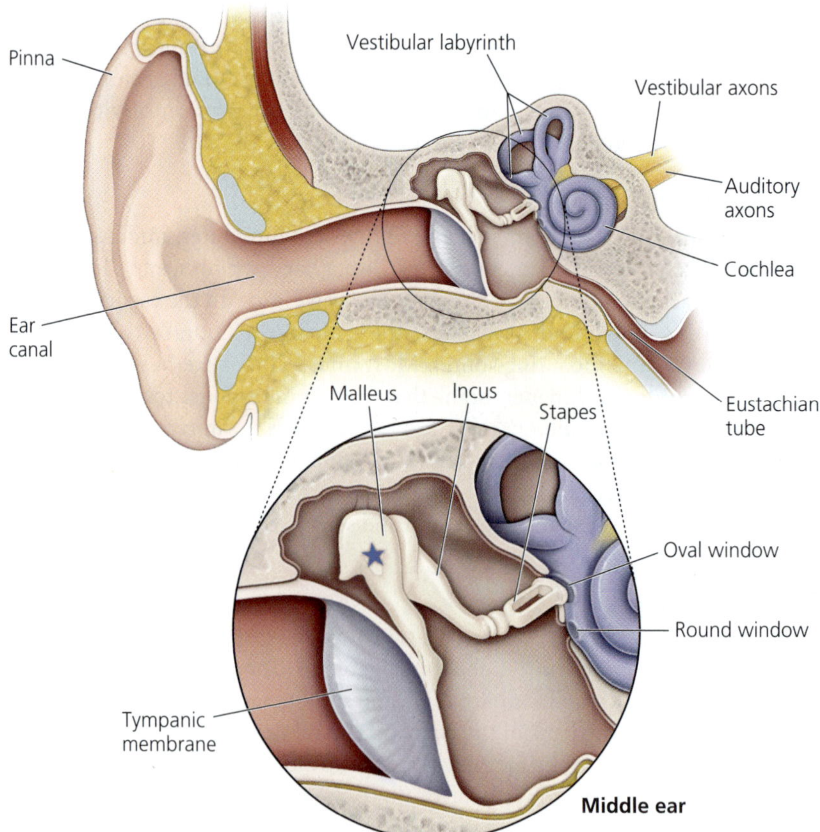

Figure 6.21 The human ear. The middle ear starts at the tympanic membrane and includes the middle ear bones (malleus, incus, and stapes). The malleus is attached to the tympanic membrane and tends to rotate around its center of mass (blue star) when the tympanic membrane moves. Flexible joints link the malleus to the incus, and the latter to the stapes; the stapes, in turn, is attached to the thin membrane of the oval window. Because of these linkages, any movement of the tympanic membrane causes an opposite movement of the oval window. The inner ear comprises the cochlea and vestibular labyrinth.

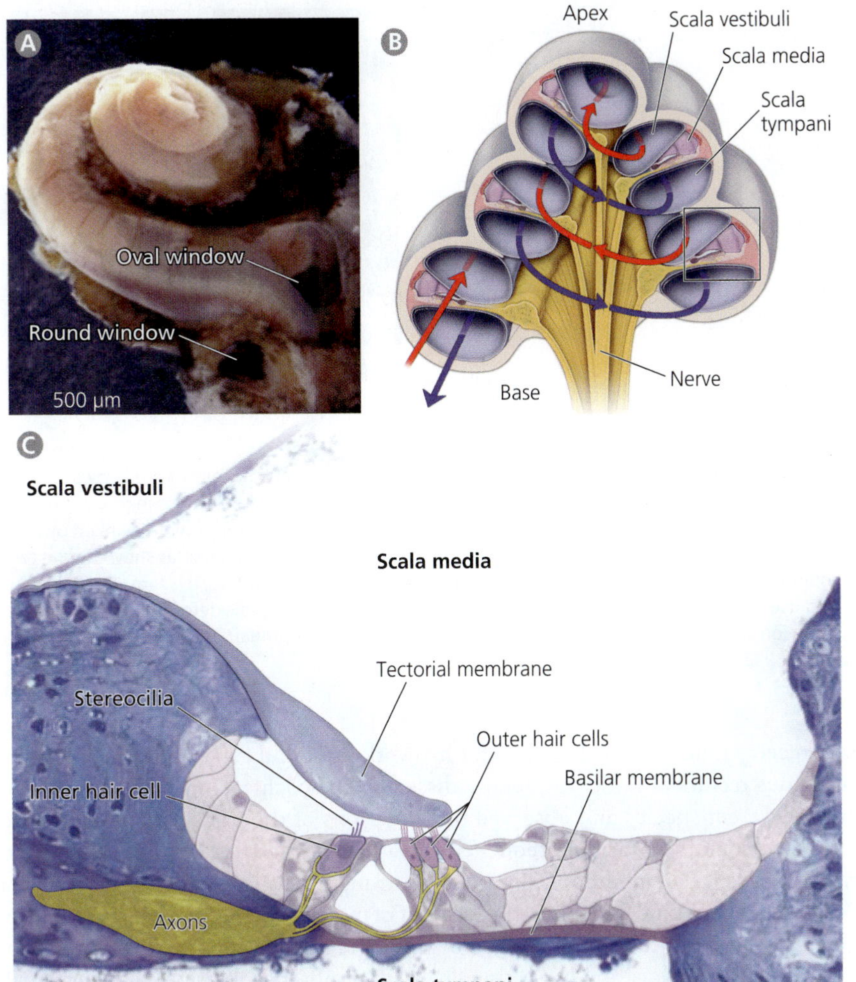

Figure 6.22 The cochlea. Shown in (A) is a cochlea from a 5-month-old human fetus. The base of the cochlea was opened so you can see the oval and round windows. Panel (B) illustrates how sound pressure waves travel up the scala vestibuli toward the apex and back down again through the scala tympani. Panel (C) depicts a section through the scala media (indicated by the rectangle in B). Inner and outer hair cells, as well as numerous supporting cells, are located on top of the thin basilar membrane. [A and B courtesy of Rémy Pujol, Association NeurOreille; C after Russel, 2007]

at one end and twisted into 2.5 spiraling coils (Figure 6.22). The half of the cochlear tube that extends from the oval window to the apex of the cochlea is called the **scala vestibuli**; the other half is called the **scala tympani**. It extends from the apex to the **round window**, which is a thin membrane quite similar to that of the oval window.

Why is the round window a thin membrane rather than bone? Consider that vibrations of the oval window membrane generate sound waves within the cochlear tube. Because sound waves travel much faster in salty water than air (1,560 vs. 343 m/s), the cochlear sound wave travels almost instantaneously up the scala vestibuli and then back down the scala tympani toward the round window. The resulting pressure changes at the round window cause it to vibrate just as the oval window does (although the two membranes move in opposite directions). Now imagine what would happen if the membrane of the round window were replaced with inflexible bone. Because the cochlear fluid is incompressible and the walls of the cochlea are stiff, any attempts to displace the cochlear fluid by pushing on the oval window would be futile. In short, the round window enables sound waves to propagate within the cochlea.

Hair Cells on the Basilar Membrane

Sandwiched between the scala vestibuli and the scala tympani (except at the apex) lies another tube, the **scala media** (Figure 6.22 B). It is separated from the scala tympani by the **basilar membrane**. Because the basilar membrane is flexible, it tends to vibrate at right angles to the longitudinal sound waves passing through the scala tympani. The physics of these vibrations were studied in detail by Georg von Békésy,

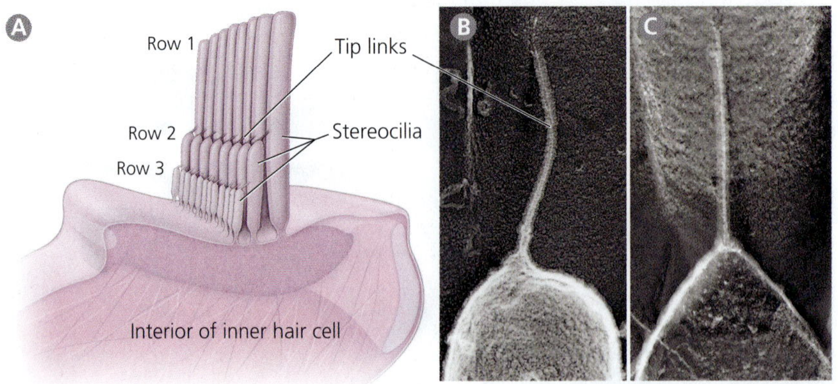

Figure 6.23 Inner hair cells. Protruding from the top of each inner hair cell are three rows of modified microvilli called stereocilia. The stereocilia in row 1 are taller than those in row 2, which are taller than those in row 3 (A). The tips of adjacent stereocilia in different rows are linked by thin filaments referred to as tip links. Deflections toward row 1 pull on the tip links, as shown in the two scanning electron micrographs on the right. In (B), the tip link is slack; in (C), it is taut. Tip link tension causes the tip of the lower stereocilium to become tent shaped. This deformation is thought to open ion channels, which changes the hair cell's membrane potential. [A after Hackney and Furness, 1995; B and C from Kachar et al., 2000]

who received the 1961 Nobel prize for his work on sound transduction. Displaying enormous technical skill, von Békésy dissected the cochleae out of cadavers from many different species and observed the vibrations of the basilar membrane in response to sounds at various frequencies. He discovered that low-frequency tones cause the basilar membrane to vibrate most strongly near the cochlea's apex. As von Békésy increased tone frequency, the location of maximal membrane vibration moved progressively closer to the cochlea's base. This systematic relationship between sound frequency and the location of maximal basilar membrane vibration arises primarily because the basilar membrane's width, tension, and stiffness decrease progressively along the cochlea.

The sound-induced vibrations in the cochlea are sensed by specialized **hair cells** that sit on top of the basilar membrane and extend small "hairs" into the scala media (Figure 6.22 C). These hairs, called **stereocilia**, are neither hairs nor cilia, but modified microvilli. Odd terminology aside, each cochlear hair cell is adorned with three rows of stereocilia that differ in height (Figure 6.23). As you can see in Figure 6.22 C, the cochlea contains inner and outer hair cells. The ~3,500 **inner hair cells** of the human cochlea form a single, spiraling row near the inner edge of the basilar membrane. The **outer hair cells**, in contrast, form 3 spiraling rows further away from the cochlea's central axis. Inner and outer hair cells differ in several respects. Most important is that inner hair cells are purely sensory, whereas outer hair cells perform both sensory and motor functions. As you will learn in Chapter 8, the outer hair cells are capable of causing basilar membrane vibrations and, thereby, amplifying some of the vibrations caused by sound.

Converting Membrane Vibrations into Ion Currents

Whenever sounds cause the basilar membrane to vibrate up and down, the stereocilia of the inner hair cells vibrate sideways. What causes these sideways vibrations? As you can see from Figure 6.22 C, a **tectorial membrane** lies just above the inner hair cell stereocilia. As sound waves travel up and down the cochlea, this tectorial membrane vibrates up and down, just as the basilar membrane does. However, the tectorial membrane's outer edge is not attached to the cochlear wall but instead rests on the basilar membrane where the outer hair cells are located. As a result of this arrangement, the basilar and tectorial membranes slide sideways against one another as they vibrate up and down. Because of this displacement, up–down vibrations of the two membranes cause the fluid between them to slosh back and forth sideways (slightly). These fluid movements, in turn, deflect the stereocilia sideways, usually by less than 1 μm. The vibrations of the stereocilia then cause oscillations in a hair cell's membrane potential.

To understand the origin of these membrane potential oscillations, you need to know that a hair cell's stereocilia are bathed in a special fluid called **endolymph** that contains an unusually high concentration of K^+ ions and is, therefore, positively charged relative to the inside of the hair cell. You also need to know that the tips of the shorter stereocilia (in rows 2 and 3) contain ion channels that, in their open state, allow K^+ and Ca^{2+} ions to flow into the stereocilia and depolarize them. Finally, you need to know that these ion channels open whenever the stereocilia bend toward the tallest stereocilia and close whenever they bend the opposite way.

This opening and closing of the stereocilial ion channels depends on slender filaments called **tip links** that extend from the tip of one stereocilium to the side of the next tallest stereocilium (Figure 6.23). When the stereocilia bend away from the tallest stereocilia, the tip links are slack; when they move in the other direction, the tip links are tight and pull on the stereocilial tips. This mechanical pulling opens the stereocilial ion channels. When the tension is relieved, the ion channels close. Although the molecular identity of the mechanosensitive ion channels in stereocilia remains debatable (transmembrane channel-like proteins appear to be involved), experimental evidence supports their existence. It has been shown, for example, that the chemical destruction of tip links blocks all sound-related oscillations in a hair cell's membrane potential. In addition, studies using calcium indicator dyes to monitor intracellular Ca^{2+} levels have revealed that deflections toward the tallest stereocilia trigger Ca^{2+} influx into the stereocilia of row 2 and 3 long before calcium levels rise in row 1 (Figure 6.24). These data are consistent with the hypothesis that tip links open calcium channels by actively pulling on the tips of row 2 and 3 stereocilia.

Hair Cell Neurotransmitter Release

The depolarization of hair cell stereocilia opens voltage-gated Ca^{2+} channels that then allow more Ca^{2+} ions to flow down the electrical potential and concentration gradients into the hair cell. As a result, Ca^{2+} levels quickly become elevated throughout the hair cell. What does this do? As you may recall from Chapter 2, increases in intracellular Ca^{2+} concentration trigger synaptic vesicle release in axon terminals. However, hair cells are not neurons. They develop from extraneural ectoderm, generate no action potentials, and lack axons. Nonetheless, hair cells possess elaborate presynaptic specializations, called **ribbon synapses**, that contain numerous synaptic

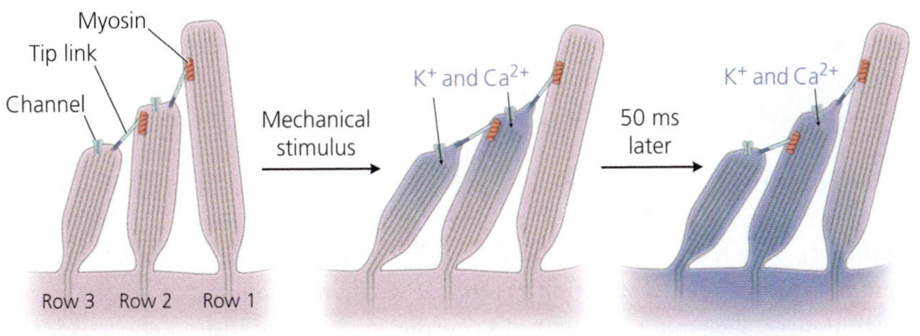

Figure 6.24 Bending the stereocilia causes Ca^{2+} and Na^+ influx. Bending the stereocilia toward row 1 opens Ca^{2+} and K^+ permeable ion channels in the tips of row 2 and 3 stereocilia. Over time, this influx causes membrane depolarization and elevates Ca^{2+} concentration throughout much of the hair cell, including the row 1 stereocilia. Myosin molecules (red rectangles) act as molecular motors that move along the stereocilium, adjusting tip link tension. [After Spinelli and Gillespie, 2009]

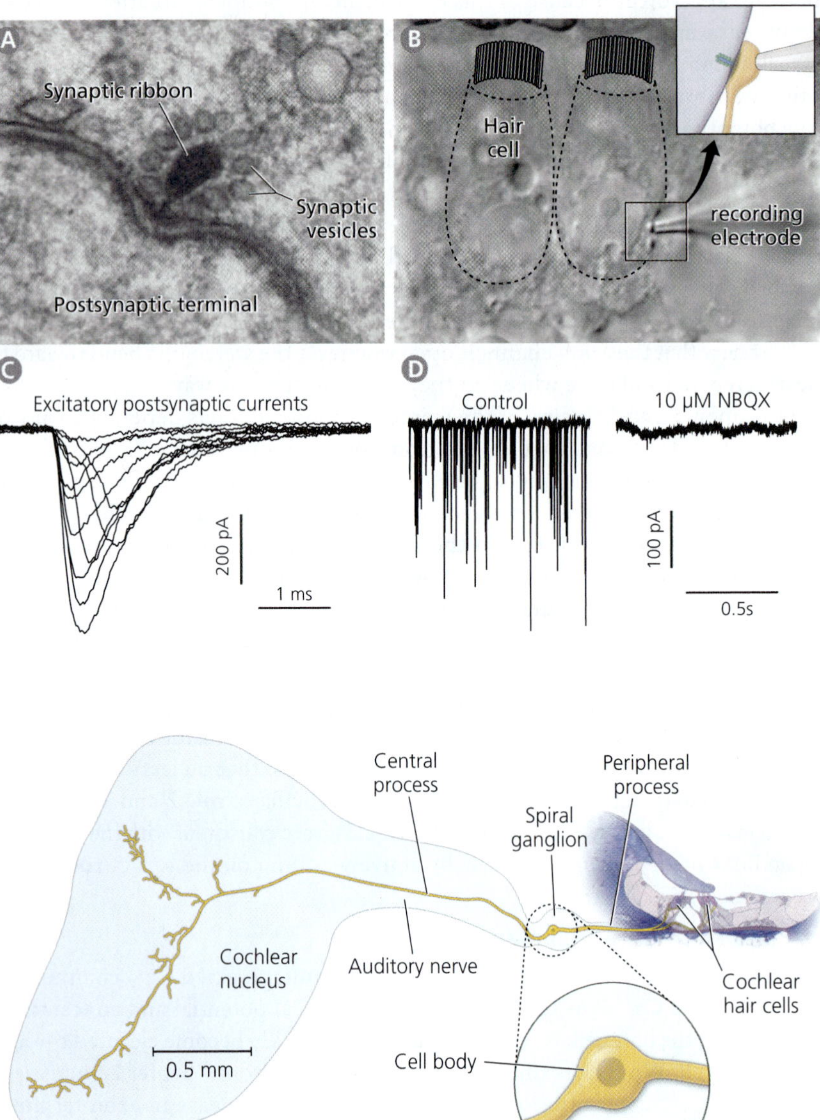

Figure 6.25 Synaptic transmission at inner hair cells. A typical hair cell synapse (A) contains numerous synaptic vesicles surrounding a "synaptic ribbon." Shown in (B) are two inner hair cells and an electrode attached to a postsynaptic terminal (yellow in the inset). The recordings in (C) are spontaneous excitatory postsynaptic currents (EPSCs) recorded from a postsynaptic terminal (voltage clamped at −94 mV; downward deflections represent an influx of positive ions). The enormous variation in EPSC amplitude implies that the synapse is releasing anywhere from 1 to 22 vesicles at a time (mean = 3–6). The EPSCs disappear when the hair cells are bathed in NBQX (6-nitro-7-sulfamoylbenzo(f)quinoxaline-2, 3-dione), which blocks AMPA-type glutamate receptors (D). [A from Meyer et al., 2009; B and C from Glowatzki and Fuchs, 2002; D courtesy of Elisabeth Glowatzki, Johns Hopkins School of Medicine]

Figure 6.26 The auditory nerve. The cell bodies of auditory nerve axons are located in the spiral ganglion and emit two axonal processes. The peripheral process forms postsynaptic terminals at cochlear hair cells (see Figure 6.25), whereas the central process forms the auditory nerve and, ultimately, presynaptic terminals in the cochlear nucleus of the medulla. As the close-up (bottom right) reveals, the peripheral and central processes emerge from opposite sides of the neuronal cell body, giving it a bipolar shape. [After Nayagam et al., 2011]

vesicles filled with glutamate and are positioned very close to the postsynaptic cell (Figure 6.25). When hair cells are depolarized and Ca^{2+} levels increase, those vesicles are released and elicit EPSPs in the postsynaptic terminals. Indeed, the ribbon synapses of hair cells can release substantial amounts of glutamate for prolonged periods of time.

The postsynaptic terminals onto which inner hair cells release their neurotransmitter are the tips of processes that originate from cell bodies in the cochlea's **spiral ganglion** (Figure 6.26). When the terminals become sufficiently depolarized, action potentials are triggered. The action potentials are then conducted centrally, traverse the cell body, travel along the neuron's central process, and ultimately cause transmitter release in a brainstem region called the **cochlear nucleus** (Figure 6.26). This arrangement does not fit the neuronal stereotype you learned about in Chapter 2 because these neurons appear to have two axons emerging from their cell body. The tip of the peripheral process functions as the neuron's dendrite, but the rest of the peripheral process looks and acts like an axon; for example, it is ensheathed in myelin. As it turns out, most peripheral sensory neurons also have such axon-like dendrites.

Hair Cell Repolarization

Returning to the topic of hair cells, let us consider how hair cells return to their resting membrane potential after being depolarized. As we discussed in Chapter 2, membrane depolarization in most neurons leads to a gradual opening of voltage-gated K^+ channels through which K^+ ions move out of the cell and repolarize the membrane. How can this work in hair cells if, as we just learned, K^+ tends to rush into (rather than out of) hair cells? The answer to this riddle is that the body of a hair cell is bathed in **perilymph**, which has a much lower K^+ concentration than the scala media's endolymph (the two fluid compartments are separated by cells with tight junctions between them). Therefore, the opening of K^+-permeable channels causes K^+ to flow from the endolymph into the stereocilia, down into the body of the hair cells, and then out into the perilymph. At every step of this pathway, K^+ ions are flowing down their concentration gradient, which means that the hair cells do not need to spend much metabolic energy pumping K^+ ions back out. Hair cells also differ from typical neurons in that their depolarization involves essentially no Na^+ influx. This, too, saves energy because there is no need to pump Na^+ ions back out across the hair cell membrane after a depolarization.

The fact that cochlear hair cells have little need for metabolic energy probably explains why they, in contrast to most sensory cells, are not adjacent to a dense capillary bed. However, a great deal of metabolic energy is needed to create the unusually high K^+ concentration within the scala media's endolymph. This work is not performed by the cochlear hair cells but by specialized epithelial cells in a structure called the **stria vascularis**. These cells are located on the peripheral wall of the scala media, contain a variety of ATP-driven ion pumps and are, as the name "stria vascularis" implies, supplied by a rich capillary bed. When the blood supply to the stria vascularis is disrupted, or when ion transport in these cells is blocked, then the endolymph becomes depleted of K^+ ions and cochlear hair cells rapidly cease functioning. Thus, the cochlear hair cells have essentially "outsourced" much of the metabolic work required for auditory transduction to the stria vascularis cells. This is a good design feature because placing a dense capillary bed on top of the basilar membrane (which is where it would have to be to feed the cochlear hair cells) would dampen the basilar membrane vibrations that are so critical for sensing sound.

Encoding Sound Parameters

As noted earlier, high-frequency tones tend to vibrate the basilar membrane near the base of the cochlea, whereas low-frequency tones cause maximal basilar membrane vibrations further up the cochlea, closer to its apex (Figure 6.27). Now that you have learned about inner hair cells, we can restate this principle, called cochlear **tonotopy**, in terms of hair cell activation: the higher the frequency of a tone, the closer to the cochlea's base will be the hair cells that are most activated by that tone.

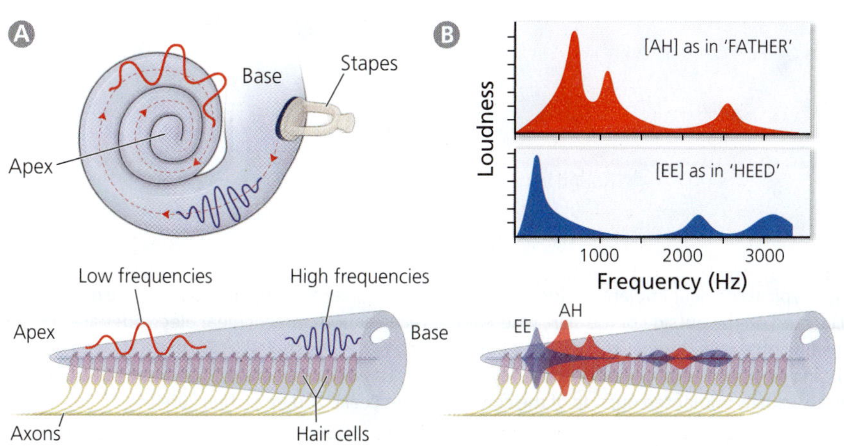

Figure 6.27 Cochlear tonopy. High sound frequencies cause maximal vibration of the basilar membrane near the cochlea's base, whereas low frequencies cause maximal vibrations near the cochlea's apex (A). Shown in (B) are the sound spectrum differences between the vowels AH and EE. Each vowel excites hair cells at several locations along the cochlea, much as a piano chord involves the pressing of multiple piano keys. Because the two vowels contain a different mix of frequencies, they excite a different (although overlapping) set of cochlear hair cells. [After hyperphysics .phy-astr.gsu.edu/hbase/HFrame.html]

THERAPIES

Box 6.2 *Cochlear Implants*

Alessandro Volta, the 18th-century inventor of the battery, stuck metal rods into his ears, connected them to a ~50-volt battery, and heard a boom inside his head. In the late 1950s and 1960s, scientists extended Volta's work by building cochlear implants to help the hearing impaired. Virtually all cochlear implants contain an external microphone and a sound processor that transmits its output via radio frequency waves to a receiver surgically implanted under the skin (Figure b6.2). The receiver then sends current pulses along one or more insulated wires into the cochlea's scala tympani. This current stimulates the sensory axons that normally innervate the inner hair cells. Thus, a cochlear implant bypasses the outer and middle ears as it translates sound waves into patterns of neural activity in the auditory nerve. Use of such cochlear implants can help the deaf understand speech, especially if they were born deaf (~1/2,000 live births) and received the implants as babies, or if they became deaf in adulthood, long after they learned to speak. According to some estimates, more than 150,000 people currently wear some type of cochlear implant.

Despite differences across manufacturers, most cochlear implants are fundamentally similar. For example, most employ some sort of volume control to boost soft sounds and attenuate loud ones. In addition, most cochlear implants decompose incoming sound into several frequency bands. The implant's processor then determines the amount of sound energy present in each of these frequency bands, generates current pulses that are proportional to this amount of energy, and sends the pulses out to one of the implanted electrodes. Importantly, the current pulses for each frequency band, or channel, are applied to different electrodes in keeping with the principle of cochlear tonotopy. Stimuli representing the highest frequencies are sent to the cochlea's most basal portion, and pulses representing progressively lower frequencies are sent to progressively more apical regions.

Although modern cochlear implants can restore some level of speech perception, they do have serious limitations. Surgical complications during the initial implantation are relatively rare, but replacing old or damaged implants is difficult. A related problem is that inserting the electrodes into the scala tympani usually damages any remaining cochlear hair cells. A third weakness is that unilateral cochlear implants do not provide the sort of cues needed for precise sound localization. Yet another worry is that children who grow up with cochlear implants may feel estranged from the sign-language-using deaf community but not be full members of the hearing community either. This quandary may be resolved by having the children with implants learn sign language as well.

A surprising but common complaint of cochlear implant users is that music sounds terrible. Simple tonal melodies are hardly recognizable, and harmonies are not perceived as such. The most likely explanation for these shortcomings is that the placement of the implanted stimulating electrodes fails to match the cochlea's natural tonotopy. To resolve this problem, some new cochlear implants allow audiologists to change which electrodes are assigned to which frequency channel. This can be done by trial and error or, if some residual hearing exists, by matching the pitch of a heard sound to the perception created by stimulating a particular electrode. In addition, it is possible to fine-tune the rate at which stimulus pulses are presented through the various electrodes and to disable any electrodes that are ineffective or, on stimulation, create unpleasant sensations. This kind of postimplantation fine-tuning often leads to significant improvements in the perception of music as well as speech.

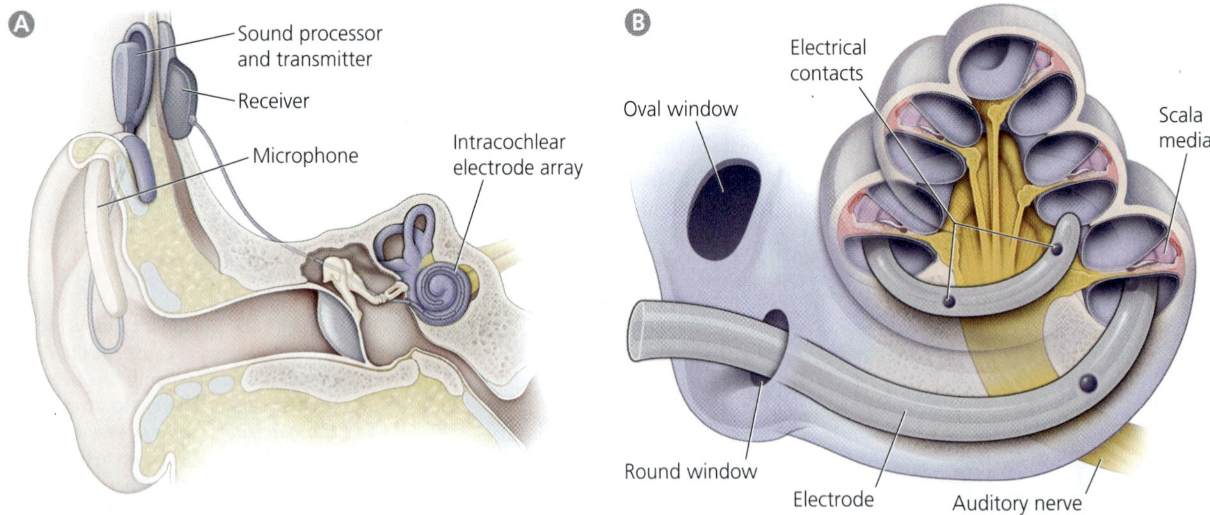

Figure b6.2 A cochlear implant. Shown in (A) is a typical cochlear implant. A close-up of the intracochlear array is shown in (B). The microphone, sound processor, and transmitter are all located on the outside of the skin, whereas the receiver and intracochlear electrodes are implanted under the skin. Signal and power transmission across the skin occurs via radiofrequency waves. As shown in (B), the electrode contacts face the underside of the basilar membrane to maximize stimulation efficiency. Current is passed between each of these electrode contacts and a ground electrode (not shown) that is usually implanted outside of the cochlea. [From Leake and Rebscher, 2004]

Cochlear tonotopy is also key to the design of electronic cochlear implants that are used to restore hearing to the deaf (see Box 6.2: Cochlear Implants).

So far so good, but natural sounds never comprise only a single frequency. Speech sounds, for example, contain a mix of many frequencies (Figure 6.27 B). How can we hear so many frequencies at the same time? The answer is simple. Sounds containing multiple frequencies activate multiple groups of hair cells at several locations along the cochlea. All our central nervous system has to do is to monitor which hair cells are activated. If you want an analogy, imagine a very clever deaf person who can infer what sounds a piano is producing by noting carefully which piano keys a pianist is depressing.

What about sound intensity? How is that aspect of sound encoded by hair cells? To answer this question, recall that tip links open ion channels when the stereocilia bend toward the tallest stereocilia. The louder the sound, the greater the degree of stereocilial bending, the greater the degree of hair cell depolarization, and the more neurotransmitter is released during each cycle of the sound wave. Furthermore, the more neurotransmitter is released by a hair cell, the greater the odds that action potentials are triggered in the postsynaptic axons. Other things being equal, this means that the average firing rate of axons in the auditory nerve increases with sound amplitude. The louder the sound, the higher the average firing rate.

A somewhat troublesome observation is that increases in sound intensity tend to shift the cochlear tonotopic "map" toward the cochlea's base. This change in activation pattern is a problem for the notion that our brains decode a sound's frequency simply by determining which hair cells are most active. After all, we generally perceive sound frequency as being independent of sound amplitude (imagine how musicians would feel if the pitch of their instrument depended on how loud they played).

How do our brains resolve this enigma? Mainly by analyzing the temporal pattern of action potentials in the auditory nerve. Because only one direction of stereocilial bending generates hair cell depolarization, auditory nerve fibers tend to fire only during one half of each cycle in the sound pressure wave. Therefore, the temporal pattern of action potentials in the auditory nerve correlates with sound frequency. This arrangement, in turn, allows auditory neurons in the brain to determine the frequency of an incoming sound by analyzing the temporal pattern of firing in auditory nerve fibers. The details of how this temporal analysis might be performed are controversial and complex (e.g., temporal analysis probably works only for low-frequency sounds). However, none of these complexities invalidate the notion that cochlear tonotopy plays a major role in the perception of sound frequencies, especially for complex sounds. The point is merely that neuroscientists still have some mysteries to solve.

BRAIN EXERCISE

How are your sound sensors protected from the external environment, and how can they, despite the protections, be harmed by loud sounds?

6.4 Are There Some Principles of Sensor Organization?

As you have seen, biological sensors vary in numerous respects. It is useful, therefore, to step back and highlight a few common principles of sensor organization. One such principle is that the information sensors provide is incomplete and variable. Another common theme is that, in most sensory systems, key stimulus parameters are represented in an orderly, map-like fashion.

Variability in Sensor Range

Biological sensors generally respond to a limited subset of all possible stimuli within their **sensory modality** (vision, hearing, somatosensation, etc.). Photoreceptors, for instance, respond most strongly to light within a relatively narrow range of wavelengths, and auditory hair cells respond most strongly to sounds of a particular frequency. Neuroscientists refer to this kind of stimulus selectivity as sensors being "tuned" to a specific range of stimuli, or they may speak anthropocentrically of sensors "preferring" a particular subset of stimuli. What causes such stimulus preferences? Sometimes the preference is due to the specificity of the receptor molecules expressed by a sensor; in other instances, the preference arises from the physical environment in which a sensor is embedded. Either way, sensors are limited in the kinds of stimuli they can detect. Sensors also encode only a limited range of stimulus intensities. Below some threshold level of intensity, a stimulus cannot be sensed at all; and once a sensor is activated maximally, further increases in stimulus intensity cannot be discerned.

From a mechanistic perspective, these limitations arise from how the sensors and their associated structures are built. However, many sensor limitations exist not because it was physically impossible for evolution to overcome them but because evolution favored sensors tuned to the subset of all possible stimuli that is important for survival or reproduction. Given that sensing is metabolically expensive, it is adaptive for organisms to focus on what is most important for members of their species. Moreover, flooding the brain with all sorts of irrelevant sensory signals would hardly be beneficial; as it is, we struggle to keep our sensory world from becoming "one great blooming, buzzing confusion" (as William James put it in 1890).

The strongest evidence in favor of the hypothesis that sensors evolved to focus on biologically important stimuli is that sensor limitations differ across species, often dramatically. For example, many insects and birds can detect ultraviolet radiation; some snakes find prey by means of infrared sensors; and bats, as well as dolphins, can hear ultrasound. Turtles and homing pigeons can sense magnetic fields; and many fishes, as well as the platypus, are able to detect weak electric fields (see Box 6.3: Electroreception and Other Alien Senses). Of course, each species has its own sensory limitations. For example, most fishes and salamanders cannot hear frequencies above ~2 kHz; and dolphins and their relatives, the whales, have lost their main olfactory sense. Thus, the point is not that other animals perceive a broader range of stimuli than we do but that each species senses a different range of stimuli. This conclusion is important to keep in mind when you are designing (or interpreting) behavioral experiments on non-humans. They often perceive the world quite differently than we do.

Variability in Sensor Sensitivity

Even within an individual, sensor sensitivity can vary dramatically as it is adjusted to current conditions. As we discussed, dark adaptation makes rods and cones more sensitive to light. In the auditory system, the sensitivity of individual hair cells can be tweaked by adjusting tip link tension by means of small molecular motors (Figure 6.24); and basilar membrane vibrations can be amplified by oscillatory changes in the length of the outer hair cells (more on this in Chapter 8). Adaptation to prolonged stimuli also occurs in the olfactory system, where Ca^{2+} accumulation in the olfactory sensory neurons reduces the signal amplification provided by the olfactory transduction cascade (through a variety of mechanisms). As a rule, such changes in sensitivity ensure that sensors do not respond to steady "background" stimuli. A major benefit of such **background adaptation** is that it leaves the sensors exquisitely sensitive to changes in stimulation. This is good design because changes in the external or internal environment are more likely than constancies to be important to an animal's survival or reproduction.

There is, however, a downside to the changeability of sensor sensitivity, namely the loss of information about absolute stimulus levels. If you look at the headlights of a car at night, for instance, they will seem brighter than they would in daylight, largely because your vision is more sensitive when it is generally dark. To determine the absolute intensity of those car lights, you have to use a light meter. Fortunately, in most real-world situations absolute stimulus intensities are less informative than relative

EVOLUTION IN ACTION

Box 6.3 *Electroreception and Other Alien Senses*

Because we can perceive the world only through our own limited set of sensors, it is difficult to appreciate that other animals may well perceive the world quite differently. The average person is surprised to learn, for example, that many insects and birds can see intricate (and to us quite invisible) patterns of UV light reflected from flowers. Similarly, few people are aware of the enormous racket made by echolocating bats during their nightly flights because our auditory sensors are not tuned to ultrasonic frequencies. Still, we can at least imagine what it would be like to see UV light or to hear ultrasound. These sensory abilities are just extensions of senses that we already have. However, some animals possess senses that are truly alien to us. It is now clear, for example, that sea turtles can sense the earth's magnetic field and use it to guide them on their migrations. Homing pigeons can likewise orient by the terrestrial magnetic field, although they prefer to use the sun as their compass. Although the behavioral evidence for *magnetoreception* in these species is clear, the mechanisms underlying the transduction of magnetic fields remain somewhat mysterious. Crystals of magnetite likely play some role, but no one is sure how the physical or magnetic orientation of these crystals is sensed by neurons.

An equally strange sense is *electroreception*. We cannot detect weak electric fields, but many animals can, including lampreys, most cartilaginous fishes, and a few bony fishes (Figure b6.3). Many of these species can detect fields on the order of a few μV/cm, and some are sensitive down to ~10 nV/cm. In case you are not used to thinking in nanoVolts, this degree of sensitivity should, at least in theory, allow a shark to sense a difference of 1 Volt between two electrodes that are submerged in saltwater and separated by 10,000 km. Although electroreception is an evolutionarily ancient trait, it disappeared in the vertebrate lineages that left the water and invaded land. Aquatic amphibians can sense electric fields, but reptiles, birds, and most mammals cannot. This makes good sense, as electric fields require a conductive medium, such as water, to propagate. Remarkably, monotreme mammals have re-evolved electroreception. The duck-billed platypus, for one, has ~40,000 electrosensors on its bill and can, with some training, use them to find a 1.5V miniature battery hidden under a rock. The spiny anteater (or echidna), a close relative of the platypus, also has numerous electrosensors on its snout. It probably senses the weak electric fields produced by tiny prey tunneling through moist earth.

Figure b6.3 A fish that can both generate and sense weak electric fields. The illustrated specimen of *Campylomormyrus rhynchophorus* belongs to the mormyrid (elephant nose) family of bony fishes. Modified muscle in its tail can generate weak electric fields, and electroreceptors on its head and snout can sense those fields as well as signals coming from moving prey. This specimen is approximately 14 cm long and was captured in an African river. [From Leal and Losos, 2010]

Indeed, the most widespread function of electroreceptors is the detection of electric fields generated by potential prey. As Ad Kalmijn showed in 1971, hungry sharks attack a flatfish buried under sand even if that flatfish is encased in agar, which blocks the spread of odors but is electrically transparent. Kalmijn further showed that hungry sharks have no interest in buried flatfish covered with electrically opaque plastic, but that they will attack when weak electrical current is passed between two buried stimulating electrodes (stronger currents tend to repel sharks!). These studies suggest that sharks, rays, and probably most electroreceptive animals use their electroreceptors to locate prey at night, in murky water, or hidden under ground.

In addition, electroreceptors may be used for navigation. Because some inanimate objects, especially metallic ones, generate weak electric fields under water, electroreceptive animals can find their way at night or in murky water by sensing the "electric landscape." Some electroreceptive fishes have gone a step further. They generate their own electric fields and then detect distortions in this field due to external objects. This sort of *electrolocation* is used by elephant nose fish (Figure b6.3) and by many of the knife fish that you can buy in pet stores. Some of these animals also communicate by means of electric signals.

How do electroreceptors work? The answer depends a bit on the specific type of electroreceptor you are talking about. The most widespread and sensitive type of electroreceptor sits at the bottom of slender tubes inside the skin. Because the tubes are often flask shaped, these receptors are called *ampullary receptors* (one type of ancient Roman flask was called an "ampulla"). Ampullary receptors are similar to hair cells in that they have a few microvillar "hairs" extending from their tops and form ribbon synapses with nerve terminals. The voltage across the receptor cell's basal membrane (the one facing inside the body) changes in response to external electric fields, which is thought to open or close (depending on the direction of the voltage change) voltage-gated calcium channels in the receptor cell's membrane. Changes in internal calcium levels then alter the rate of transmitter release, which modulates action potential firing in the postsynaptic nerve terminals.

intensities. For example, the absolute intensity of light reflected from an object often tells you more about the intensity of the light source than about the object. In contrast, the light reflected off an object relative to the same light bouncing off other surfaces facilitates object identification (see Chapter 12). It is adaptive, therefore, for organisms to prioritize relative stimulus intensity over absolute intensity.

Labeled Lines

Johannes Müller noted in 1838 that electrical stimulation of different sensory nerves generates different sensory perceptions. For instance, current applied to the optic nerve makes you see lights, whereas the same current applied to the auditory nerve makes you perceive a sound. Müller inferred that the sensation of sound must be due to the peculiar "energy" or "quality" of the auditory nerve, that the sensations of color and light are caused by special aspects of the optic nerve, and so on for each nerve. However, Müller was unclear on what those specific nerve energies or qualities might be.

In a way, the specificity of nerves derives from the sensors from which they receive their input. Activation of the optic nerve generates visual perceptions because photoreceptors transduce light; activation of the auditory nerve generates auditory perceptions because auditory hair cells transduce sounds, and so on for the other senses. However, Müller's point is more subtle. Imagine what would happen if you stimulated an auditory nerve fiber electrically, as you can do with cochlear implants. If the electrical stimulation is triggered by a microphone, then the person perceives the stimulation as a sound. This much is obvious. But what would happen if you replaced the microphone with a light sensor (a camera)? If you now activate the auditory nerve fiber by illuminating the light sensor, would the perception be auditory or visual? An answer to this question comes from blind people who were outfitted with a digital camera that converts visual images to patterns of somatosensory stimuli that are applied to the skin of the person's back or, in more recent experiments, the tongue. These patients gradually learn to interpret the optically driven somatosensory stimulation as information about objects that are located at a distance from the body, rather than on the skin. However, they do not perceive the stimuli as light (at least not until they have had extensive experience with the device). Therefore, we can conclude that the brain interprets activity in a set of axons as representing the kind of information that the axons normally carry.

Another way to phrase Müller's insight is to say that the brain interprets nerve activity according to a **labeled line** code. That is, action potentials in a particular set of axons are interpreted by other neurons according to the "label" carried by those axons. If the axons normally carry visual information, then they are labeled "visual" and their activity is interpreted as such. Importantly, the labels can be quite specific. For example, axons receiving input from cochlear hair cells that respond selectively to 1 kHz sounds would be labeled as "1 kHz." Axons from other portions of the cochlea would carry different frequency labels. In general, we can say that each sensory axon represents a specific "labeled line" that can be active to varying degrees but always represents a specific type of information. Sometimes the information is sufficient to identify a stimulus object. Male moths, for example, have sensors that respond selectively to odor molecules released by female moths (sex pheromones); when these sensors are active, males can be fairly certain that a female of their species is nearby. In most cases, however, the information carried by individual neurons is insufficient to identify external objects. Therefore, organisms must usually analyze the pattern of activity in many neurons at once, using a combinatorial code. We will discuss this kind of combinatorial coding—often called "population coding"—at length in later chapters.

Sensory Maps

An intriguing aspect of the labeled lines in our brains is that they tend to exhibit an orderly, map-like organization. In the retina, adjacent sensors convey information about stimuli presented at adjacent locations in space, which means that external space is "mapped" onto the retina. In the cochlea, the mapped stimulus parameter is frequency, rather than space. These retinotopic and tonotopic maps are found not only in the sensor arrays but also in most of the brain regions that process information from those arrays. Many parts of the mammalian visual system, for instance, retain a retinotopic organization (see Chapter 11). In addition, the brain constructs map-like representations that are not carried over from the sensor arrays. A good example of such centrally derived maps is the chemotopic mapping of odorants onto the olfactory bulb, which results from the descrambling of olfactory sensory axons. Another good example comes from the auditory system, which constructs a map of auditory space in the midbrain. You will learn more about such centrally generated maps later. For now, let us focus on the fundamental question of why sensory maps exist. Are they functionally significant or merely accidents of evolution and development?

This question is difficult to answer, but one likely reason for the existence of sensory maps is that neurons gathering inputs from such maps can efficiently sample the activity of many cells with similar stimulus preferences. If cells with similar preferences were scattered randomly about, then axonal connections required for such information gathering (polling, if you like) would need to be considerably longer (Figure 6.28). By integrating the activity of many similarly tuned cells, downstream neurons can minimize input noise (random activity not driven by the stimulus), making them more reliable and sensitive (recall our discussion of convergence in the rod system). A related benefit of sensory maps is that inhibitory connections between neurons that respond to similar stimuli can be much shorter than they would be if those neurons were not topographically arranged.

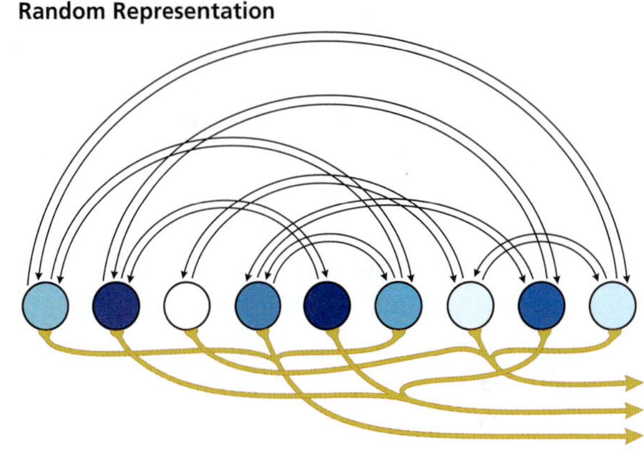

Figure 6.28 The benefits of topographic maps. Shown at the top are 9 neurons (circles) that are topographically organized so that adjacent neurons encode similar information (similar colors). The neurons at the bottom are arranged randomly. The figure also shows lateral connections between neurons that encode very similar information (black arrows) and postsynaptic axons that integrate information from multiple neurons carrying similar information (yellow arrows). The topographic arrangement minimizes axonal connection lengths.

BRAIN EXERCISE

Do cameras and microphones provide more "realistic," objectively precise information about the world than your own eyes and ears?

SUMMARY

Section 6.1 - Light sensing involves mainly rod- and cone-type photoreceptors in the retina. The highest density of cone photoreceptors is found in the retina's fovea.

- When light hits rhodopsin, retinal changes its shape and eventually floats away. Activated rhodopsin initiates an intracellular signaling cascade that ultimately hyperpolarizes the photoreceptor.
- Cone photoreceptors are less light sensitive than rods. S-, M-, and L-type cones are tuned to different but overlapping wavelengths of light. Color vision is based on comparing activity levels across the different cone types.
- Placing the photoreceptors at the back of the retina, away from the incoming light, seems like bad design until you realize that the photoreceptors must be covered with pigment epithelium and a dense capillary bed to recycle light-activated retinal. Specialized glial cells minimize the costs of an inverted retina.

Section 6.2 - The sense of smell is based on the binding of odorants to olfactory receptor molecules in the cilia of olfactory sensory neurons, which are located in the olfactory epithelium.

- Olfactory sensory neurons regularly die despite being protected by nasal mucus, but they can be replaced even in adulthood. The axons of olfactory sensory neurons project to the olfactory bulb.
- Olfactory receptor molecules are G protein-coupled receptors resembling opsins. They are extremely diverse, especially in non-humans, with different receptor types binding different odorant epitopes.
- Olfactory sensory neurons expressing the same type of olfactory receptor are scattered across the olfactory epithelium, but their axons converge onto just a few glomeruli in the olfactory bulb. Adjacent glomeruli tend to be activated by similar odorants.
- Reciprocal dendrodendritic synapses between mitral cells and granule cells mediate feedback inhibition and lateral inhibition in the olfactory bulb.

Section 6.3 - Hearing involves the outer, middle, and inner ears. The inner ear harbors the cochlea, where sound transduction takes place.

- The middle ear bones form a lever system that converts airborne vibrations into pressure waves that travel through the fluid-filled cochlea. The pressure waves induce vibrations of the basilar membrane.
- Pressure waves in the cochlea bend the stereocilia of hair cells on the basilar membrane, depolarizing them. In response, the hair cells release glutamate onto the axons of neurons in the auditory nerve.
- High-frequency sounds excite hair cells near the base of the cochlea; low frequencies excite hair cells near the cochlea's apex. This tonotopic organization facilitates sound frequency discrimination.

Section 6.4 - Despite the great diversity of biological sensors, some common principles are evident.

- All sensors respond to a specific subset of stimuli, and their sensitivity to these stimuli is typically adjustable. Therefore, our sensors provide incomplete and biased (but useful!) information about the external and internal environments.
- Adjacent sensory neurons tend to convey similar information. The brain's interpretation of activity in sensory neurons depends mainly on the type of sensor that normally activates the neurons.

Box 6.1 - Olfactory sensitivity varies not only across species but also across individual humans. Several specific anosmias have been linked to polymorphisms in specific olfactory receptor genes.

Box 6.2 - Cochlear implants can restore some hearing ability by converting signals from a microphone into electrical impulses delivered to auditory nerve fibers distributed along the cochlea.

Box 6.3 - Many insects and birds can see ultraviolet radiation, many bats hear ultrasonic sounds, and many fishes can detect weak electric fields. Some fishes can even generate their own electric fields, using them to electrolocate.

KEY TERMS

scala vestibuli 187
scala tympani 187
round window 187
scala media 187
basilar membrane 187
hair cell 188
stereocilia 188
inner hair cell 188
outer hair cell 188
tectorial membrane 188
endolymph 189
tip link 189
ribbon synapse 189
spiral ganglion 190
cochlear nucleus 190
perilymph 191
stria vascularis 191
tonotopy 191
sensory modality 193
background adaptation 194
labeled line 196

ADDITIONAL READINGS

6.1 - Light Sensors

Balasubramanian V, Sterling P. 2009. Receptive fields and functional architecture in the retina. *J Physiol (Lond)* **587**:2753–2767.

Fain GL, Matthews HR, Cornwall MC, Koutalos Y. 2001. Adaptation in vertebrate photoreceptors. *Physiol Rev* **81**:117–151.

Luo D-G, Xue T, Yau KW. 2008. How vision begins: an odyssey. *Proc Natl Acad Sci U S A* **105**:9855–9862.

Masland RH. 2001. The fundamental plan of the retina. *Nat Neurosci* **4**:877–886.

Tang PH, Kono M, Koutalos Y, Ablonczy Z, Crouch RK. 2013. New insights into retinoid metabolism and cycling within the retina. *Prog Retin Eye Res* **32**:48–63.

6.2 - Odor Sensors

Araneda RC, Kini AD, Firestein S. 2000. The molecular receptive range of an odorant receptor. *Nat Neurosci* **3**:1248–1255.

Kleene SJ. 2008. The electrochemical basis of odor transduction in vertebrate olfactory cilia. *Chem Senses* **33**:839–859.

Mackay-Sim A, Kittel PW. 1991. On the life span of olfactory receptor neurons. *Eur J Neurosci* **3**:209–215.

Mori K, Takahashi YK, Igarashi KM, Yamaguchi M. 2006. Maps of odorant molecular features in the mammalian olfactory bulb. *Physiol Rev* **86**:409–433.

Shepherd GM, Chen WR, Willhite D, Migliore M, Greer CA. 2007. The olfactory granule cell: from classical enigma to central role in olfactory processing. *Brain Res Rev* **55**:373–382.

Zhu P, Frank T, Friedrich RW. 2013. Equalization of odor representations by a network of electrically coupled inhibitory interneurons. *Nat Neurosci* **16**:1678–1686.

6.3 - Sound Sensors

Beurg M, Fettiplace R, Nam J, Ricci A. 2009. Localization of inner hair cell mechanotransducer channels using high-speed calcium imaging. *Nat Neurosci* **12**:553–558.

LeMasurier M, Gillespie P. 2005. Hair-cell mechanotransduction and cochlear amplification. *Neuron* **48**:403–415.

Mammano F, Bortolozzi M, Ortolano S, Anselmi F. 2007. Ca2+ signaling in the inner ear. *Physiology* **22**:131–144.

Mustafi D, Engel AH, Palczewski K. 2009. Structure of cone photoreceptors. *Prog Retin Eye Res* **28**:289–302.

Oxenham AJ, Bernstein JG, Penagos H. 2004. Correct tonotopic representation is necessary for complex pitch perception. *Proc Natl Acad Sci U S A* **101**:1421–1425.

6.4 - Principles of Sensor Organization

Block SM. 1992. Biophysical principles of sensory transduction. In "Sensory Transduction." *Soc General Physiologists Series* **47**:1–17.

Bockaert J, Pin JP. 1999. Molecular tinkering of G protein-coupled receptors: an evolutionary success. *EMBO J* **18**:1723–1729.

Clapham DE. 2003. TRP channels as cellular sensors. *Nature* **426**:517–524.

Danilov YP, Tyler ME, Skinner KL, Hogle RA, Bach-y-Rita P. 2007. Efficacy of electrotactile vestibular substitution in patients with peripheral and central vestibular loss. *J Vestib Res* **17**:119–130.

Ma Q. 2010. Labeled lines meet and talk: population coding of somatic sensations. *J Clin Invest* **120**:3773–3778.

Proulx MJ. 2010. Synthetic synaesthesia and sensory substitution. *Conscious Cogn* **19**:501–503.

Ward J, Meijer P. 2010. Visual experiences in the blind induced by an auditory sensory substitution device. *Conscious Cogn* **19**:492–500.

Webb B. 2006. Transformation, encoding and representation. *Curr Biol* **16**:R184–R185.

Boxes

Johnson S, Lohmann KJ. 2005. The physics and neurobiology of magnetoreception. *Nat Rev Neurosci* **6**:703–712.

Lunde K, Egelandsdal B, Skuterud E, Mainland JD, Lea T, Hersleth M, Matsunami H. 2012. Genetic variation of an odorant receptor OR7D4 and sensory perception of cooked meat containing androstenone. *PLoS ONE* **7**:e35259.

McRae JF, Mainland JD, Jaeger SR, Adipietro KA, Matsunami H, Newcomb RD. 2012. Genetic variation in the odorant receptor OR2J3 is associated with the ability to detect the "grassy" smelling odor, cis-3-hexen-1-ol. *Chem Senses* **37**:585–593.

Niparko JN. 2009. *Cochlear implants: principles and practices.* Philadelphia: Lippincott, Williams & Wilkins.

Pelchat ML, Bykowski C, Duke FF, Reed DR. 2011. Excretion and perception of a characteristic odor in urine after asparagus ingestion: a psychophysical and genetic study. *Chem Senses* **36**:9–17.

Pettigrew JD. 1999. Electroreception in monotremes. *J Exp Biol* **202**:1447–1454.

Sensors II: Sensing on Contact

CHAPTER 7

CONTENTS

FEATURES

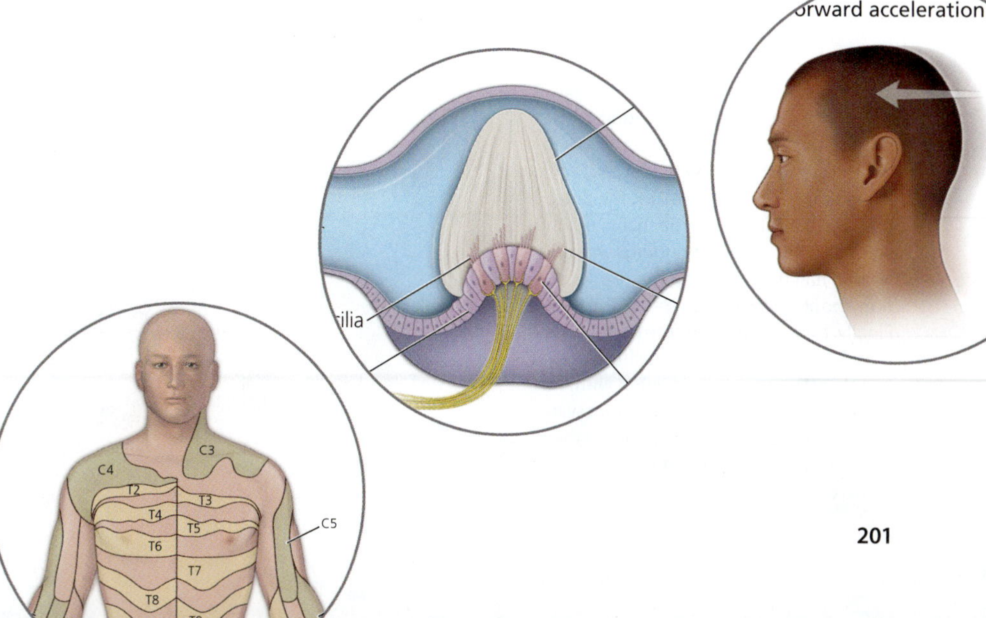

The three main senses we discussed in Chapter 6—vision, smell, and hearing—are ideal for sensing objects far away from the body. Now we turn to senses that require physical contact between the sensor and the object being sensed. The distinction is not hard and fast, and one can think of other ways to categorize biological sensors, but all sensor classification schemes are somewhat arbitrary. The important point is that animals possess a wide variety of sensors that are specialized for sensing stimuli at the body surface or inside of it.

Close-range sensors are located in the skin, in muscles, and in many internal organs. Most of them provide inputs to the **somatosensory system**, which processes information about the state of the body (*soma* being the Greek word for "body"). Another system that uses contact-dependent sensors is the **gustatory system**, which we use to taste our food (*gustatio* means "taste" in Latin). Toward the end of this chapter we also cover the sensory cells of the **vestibular system**, which informs our brain about the movements of the head through space. As you read about these various sensors, and how they send their information to the central nervous system, keep in mind the issues we discussed at the end of Chapter 6 relating to sensor range and variability, labeled lines, and sensory maps. We will revisit these principles of sensor design at the end of this chapter.

7.1 How Do We Sense Touch and Vibration?

The skin contains a variety of sensors (Figure 7.1). Many sensors in the skin are **mechanosensory**, meaning that they are specialized for sensing mechanical stimuli such as physical touch and vibration. Some of these mechanosensors are simply sensory axons that terminate as **free nerve endings** in the skin. Their sensitivity to mechanical stimuli results from special ion channels that depolarize the axon when its membrane is physically deformed (stretched or squeezed). The molecular identity of these mechanosensory ion channels remains controversial, but diverse candidate molecules have been identified. Unfortunately, many of these candidates respond also to nonmechanical stimuli, such as tissue acidity or temperature, making their mechanosensory function more difficult to prove.

Encapsulated Nerve Endings

In addition to free nerve endings, the skin contains mechanosensory axons that terminate on, or among, specialized skin cells. Some of these axons wrap around the base of hairs (Figure 7.1) and respond selectively to movements of those hairs. Other axons terminate in close association with specialized cells that are named after their

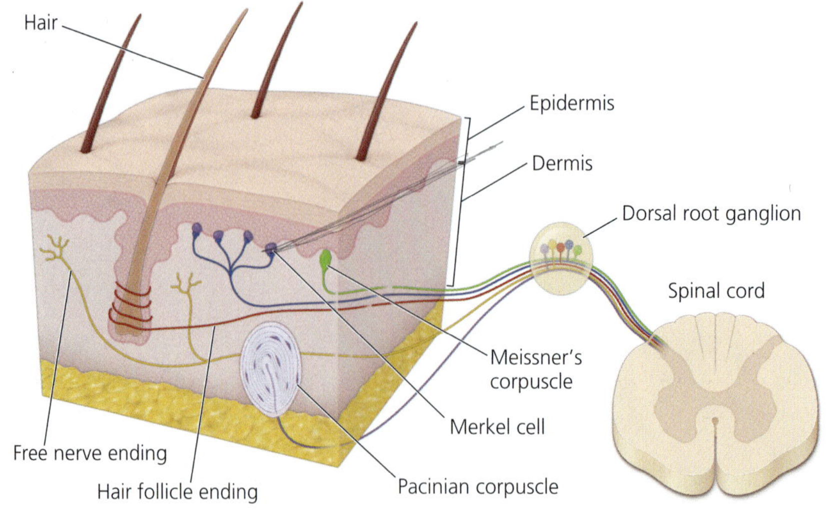

Figure 7.1 Sensors in the skin. The dermis and epidermis contain many nerve endings that are coupled to modified skin cells, notably Merkel cells, Pacinian corpuscles, and Meissner's corpuscles. Each of these sensors is specialized for the detection of different sorts of mechanical stimulation. Additional nerve endings wrap around hair follicles and provide information about slight movements of the hairs. Finally, the skin contains free nerve endings, which derive their name from the fact that they are not associated with modified skin cells. Most axons innervating the skin have their cell bodies in the dorsal root ganglion and project into the spinal cord.

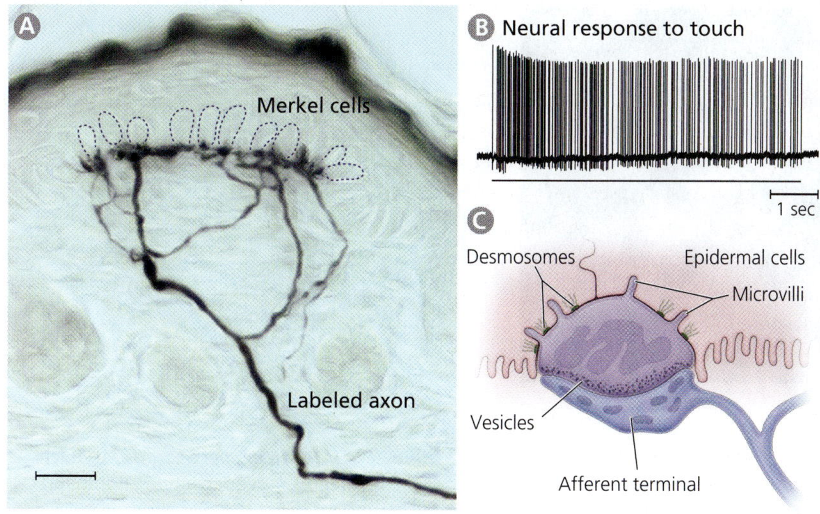

Figure 7.2 Merkel cell innervation. Shown in (A) is a section through the skin of a newborn mouse in which an axon innervating multiple Merkel cells (outlined with dashed lines) was labeled with an intracellular tracer. The responses of such an axon to prolonged touch is shown in (B). The diagram in (C) shows that the axons innervating a Merkel cell (purple) tend to have large axon terminals (blue). The drawing also shows that Merkel cells extend microvilli into the overlying epidermal cells and are coupled to those epidermal cells by desmosomes (which "glue" cells together). These features ensure that mechanical deformation of the epidermis stretches or compresses the membrane of the underlying Merkel cells. In response to such changes in membrane tension, Merkel cells modify their rate of neurotransmitter release. Scale bar in (A) equals 10 μm. [A and B from Woodbury & Koerber, 2007; C after Haeberle & Lumpkin, 2008]

discoverers: Merkel, Meissner, and Pacini (who also discovered the bacterium responsible for cholera). These three types of **encapsulated nerve endings**, together with the nerve endings around hair follicles, account for our ability to sense skin touch and vibration.

Merkel Cells

Merkel cells lie deep within the epidermis (Figure 7.1) and are most common in highly touch-sensitive areas such as the fingertips. The axons innervating Merkel cells are very sensitive to indentation of the overlying skin, exhibit excellent spatial resolution, reliably encode stimulus magnitude (depth of indentation), and respond even to sustained stimuli, which is to say that they adapt slowly to prolonged stimuli (Figure 7.2). As a result of these features, information coming into the nervous system through Merkel cells can be used to identify the shape and texture of objects in contact with the skin.

How do Merkel cells transmit sensory information to the brain? By analogy with auditory hair cells, we might expect Merkel cells to release neurotransmitter molecules onto the axon terminal in response to a mechanical deformation. Indeed, Merkel cells contain synaptic vesicles (Figure 7.2 C) and much of the molecular machinery required for glutamatergic transmission. However, when Merkel cells are killed experimentally, the associated axons can still respond to mechanical stimuli. This finding suggests that mechanosensory transduction takes place inside the axon terminals, using mechanosensory ion channels in the axon's membrane, and that Merkel cells play merely an accessory role in mechanotransduction. Still, other experiments have shown that even isolated Merkel cells are capable of responding to mechanical stimulation. Therefore, it seems prudent to conclude that mechanotransduction involves both Merkel cells and the axon terminals that contact them. The two work closely together.

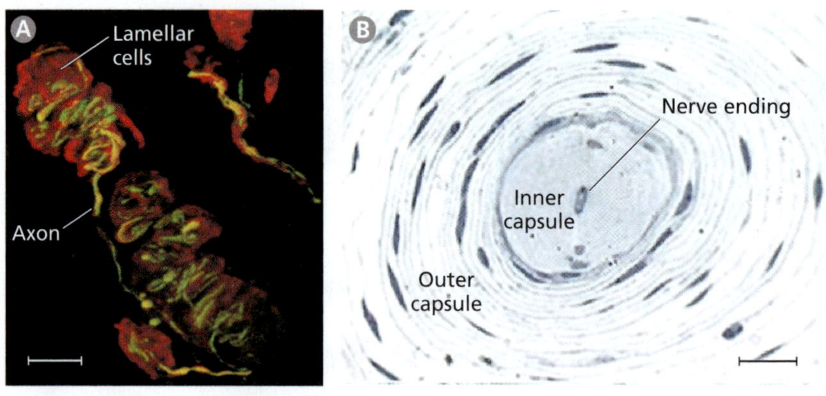

Figure 7.3 Meissner's and Pacinian corpuscles. Shown in (A) is a Meissner's corpuscle, which consists of a spiral-shaped sensory nerve ending (yellow/green) embedded among non-neural lamellar cells (red). Meissner's corpuscles lie relatively superficial in the skin and respond well to low-frequency vibration. Pacinian corpuscles (B) lie deeper in the skin, are more sensitive, and respond to higher frequency vibrations. They consist of a sensory nerve that extends into the center of an onion-like arrangement of lamellar cells forming distinct inner and outer capsules. Both scale bars equal 20 μm. [A from Guinard et al., 2000; B from Pawson et al., 2009]

Meissner's Corpuscles

Meissner's corpuscles consist of a spiraling nerve ending inside a stack of disk-shaped *lamellar cells* resembling glia (Figure 7.3 A). Because there is no evidence that the lamellar cells release neurotransmitters onto the axonal ending in their midst, it is likely that the transduction of the mechanical stimuli is performed by mechanically gated ion channels on the nerve ending itself. The principal function of the lamellar cells is probably to filter the incoming stimuli and protect the nerve ending from overstimulation.

Because Meissner's corpuscles lie just beneath the epidermis (Figure 7.1), their associated axons respond to mechanical stimuli in a relatively small patch of overlying skin. In this respect they resemble the axons contacting Merkel cells. However, the axons that end in Meissner's corpuscles exhibit lower spatial resolution, less sensitivity, and more rapid adaptation to steady stimuli. These features make the axons innervating Meissner's corpuscles well suited for detecting low-frequency vibration, as might occur when objects move across the skin. It has been suggested that Meissner's corpuscles are especially important for controlling grip strength because they signal when a grasped object begins to slip away and gripping force should be increased.

Pacinian Corpuscles

Pacinian corpuscles consist of numerous glia-like cells arranged like the layers of an onion around a central nerve ending (Figure 7.3 B). In contrast to Merkel cells and Meissner's corpuscles, Pacinian corpuscles lie deep within the skin and in some internal organs. Accordingly, they respond to stimuli applied over large areas. Although Pacinian corpuscles resemble Meissner's corpuscles in being rapidly adapting, they are ~30 times as sensitive and respond to vibrations of higher frequency. Electron micrographs reveal that some of the non-neural cells in Pacinian corpuscles contain "synapse-like" structures, suggesting that at least some aspects of mechanotransduction in these corpuscles occur outside of the axon. However, the fact that Pacinian corpuscles contain so many layers of non-neural cells means that most of the non-neural cells cannot be releasing any molecules directly onto the nerve ending. Therefore, their principal function is probably to protect the sensitive nerve ending from the kinds of intense low-frequency stimuli that would be produced by tapping your finger or gripping an object.

An intriguing potential function of Pacinian corpuscles is to sense the subtle forces applied to the working surfaces of handheld tools. As you may have noticed, tool use generally requires sensing when the tool's working surface (usually its tip) makes contact with the targeted object. This remote contact creates mechanical forces that are transmitted through the tool to the hand and ultimately sensed by the Pacinian corpuscles. Learning to interpret such information greatly facilitates tool use.

Central Projections of Mechanosensory Axons

Mechanosensory axons that innervate the skin project into the spinal cord through **spinal nerves** that are named after the segments of the vertebral column through which they pass (Figure 7.4 A). Humans possess 8 cervical, 12 thoracic, 5 lumbar, 5 sacral, and 1 coccygeal spinal nerves. Near the point where these nerves connect to the spinal cord, they separate into dorsal (or posterior) and ventral (or anterior) branches (Figure 7.4 B). The ventral branch is termed the **ventral root** and contains the axons of motor neurons that innervate skeletal muscles (see Chapter 8). The **dorsal root** contains the axons of sensory neurons that project into the spinal cord.

Dorsal Root Ganglion Cells

The cell bodies of axons in the dorsal roots form a distinct swelling along the dorsal root called the **dorsal root ganglion** (*ganglion* means "swelling" or "knot" in Greek). The cell bodies in the dorsal root ganglia are unusual in that they do not have dendrites emerging from them. In this respect they resemble the bipolar neurons in the spiral ganglion of the auditory nerve (see Chapter 6). However, the cell bodies in the dorsal root ganglia give rise to just one axon, which then splits in two, forming a T-shaped junction (Figure 7.5 A). One of the two axon branches passes through the spinal nerve to the periphery, where it ends in muscles or skin. The other (central) process heads into the spinal cord. Because of this unusual arrangement, the dorsal root ganglion cells are said to be **pseudounipolar** in shape. Functionally, the unipolar shape is probably adaptive because forcing action potentials to traverse the cell body (which tends to have few voltage-gated sodium channels) would risk having the conduction fail. This potential problem is largely eliminated by locating the cell body off the main line for action potential propagation.

The peripheral segments of dorsal root ganglion cell axons can be several meters long, and so can the central processes (e.g., in giraffes), but the cell bodies are rarely more than 50 μm in diameter (Figure 7.5 B). This size differential

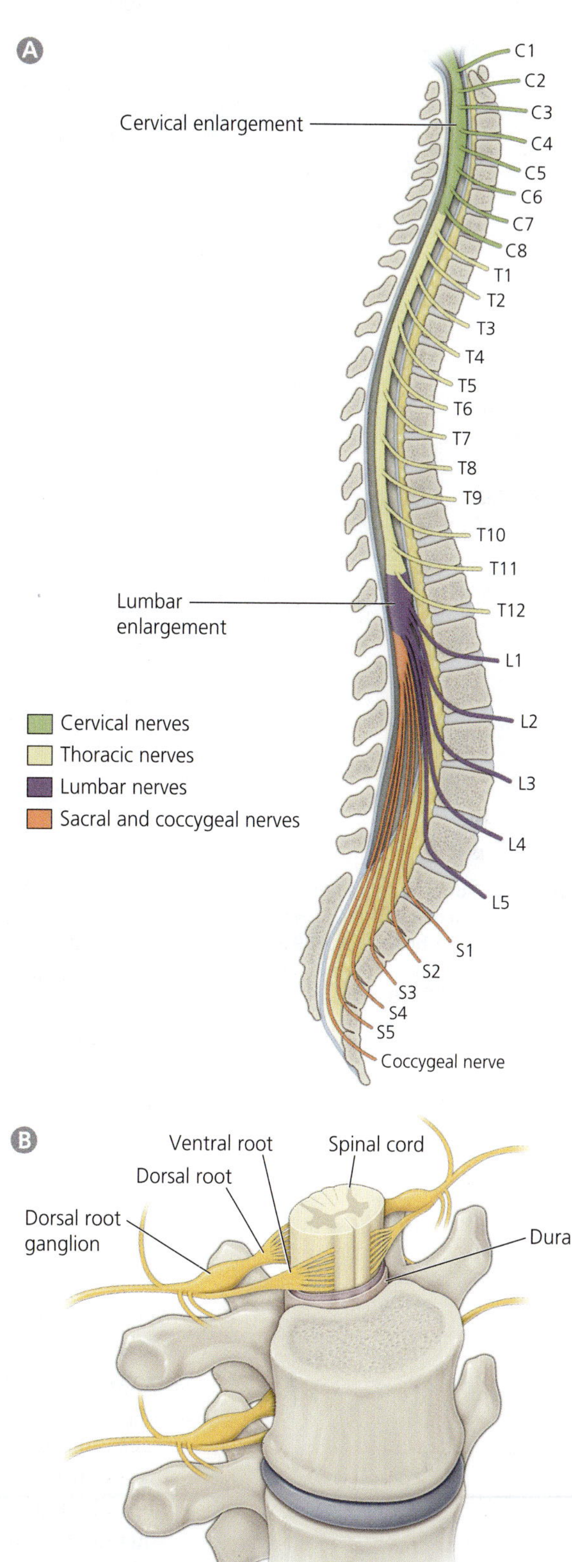

Figure 7.4 Spinal nerves and spinal roots. As shown in (A), humans have 31 pairs of spinal nerves that connect to the spinal cord, which is located inside of the vertebral canal. Although relatively uniform in structure, the spinal cord is thickened (enlarged) where the cervical and lumbar nerves emerge. Panel (B) illustrates the dorsal and ventral roots (called posterior and anterior roots, respectively, in humans). Sensory information comes into the spinal cord mainly through the dorsal roots. The cell bodies of dorsal root axons are located in dorsal root ganglia.

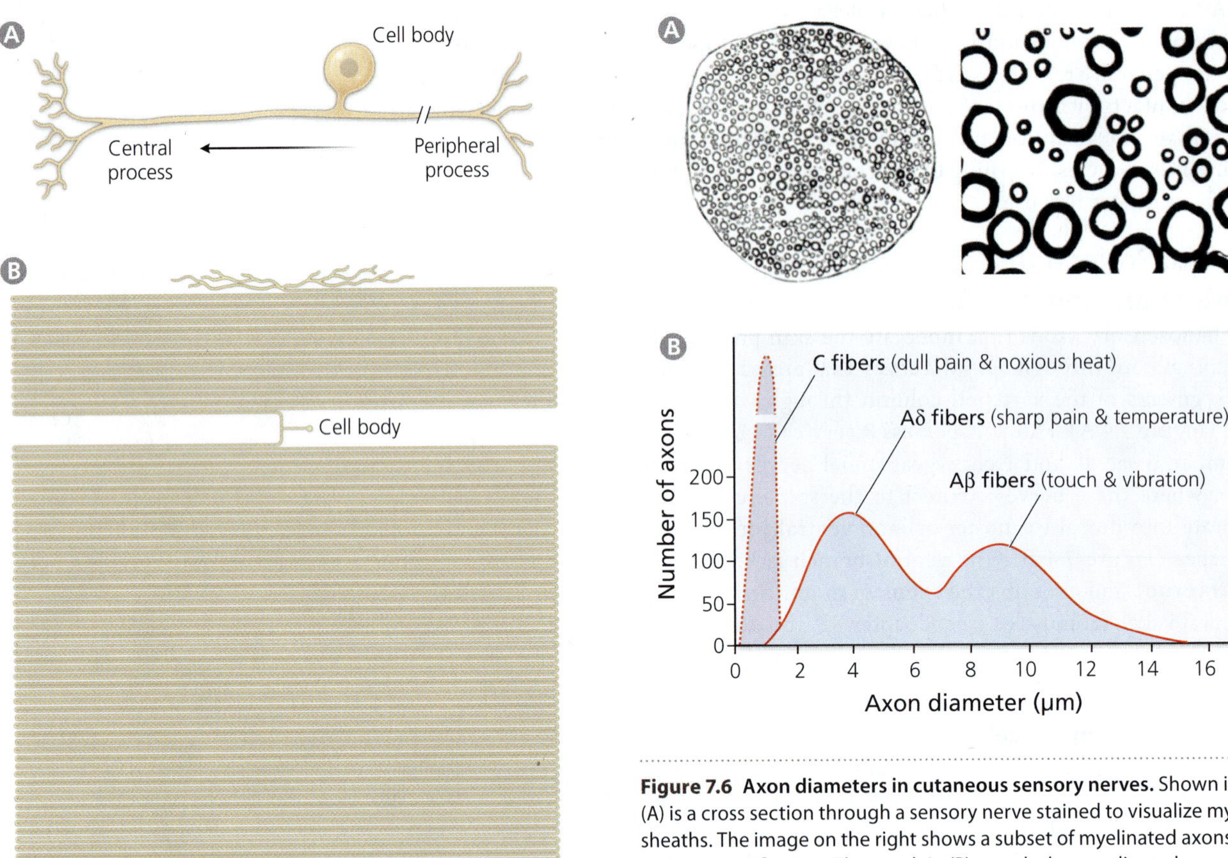

Figure 7.5 A dorsal root ganglion cell. Dorsal root ganglion cells are usually depicted as in (A), but in reality 99.8% of the cell's cytoplasm is in the axon. For the neuron illustrated in (B), the peripheral axon segment is 120 cm long, while the centrally projecting segment is 50 cm in length. [After Devor, 1999]

Figure 7.6 Axon diameters in cutaneous sensory nerves. Shown in (A) is a cross section through a sensory nerve stained to visualize myelin sheaths. The image on the right shows a subset of myelinated axons at higher magnification. The graph in (B) reveals that myelinated axon diameters in skin sensory nerves exhibit two broad peaks, corresponding to Aβ and Aδ fibers. Unmyelinated C fibers are more numerous than myelinated axons but very thin and not visible in (A). [A from Matsumoto & Mori, 1975; B after Boyd & Davey, 1968]

poses a serious metabolic challenge for the dorsal root ganglion cells. Because most of the molecules in axon terminals must be replaced regularly, and most proteins are synthesized in the cell body, large numbers of molecules must be shipped over long distances through the axons of dorsal root ganglion cells. This intracellular transport requires large amounts of metabolic energy, mostly in the form of ATP. Because ATP is synthesized primarily in the cell body (through oxidative phosphorylation), one would expect dorsal root ganglion cells to have large cell bodies. Indeed, the cell bodies of dorsal root ganglion cells are among the largest in the vertebrate nervous system. It is only relative to their extremely long axons that the cell bodies of these cells are small. One way in which dorsal root ganglion cells may minimize their metabolic challenges is to keep average firing rates relatively low, thus minimizing the need to replenish neurotransmitter stores.

Sensory Nerve Fiber Classification

Dorsal root sensory axons vary in their diameter and degree of myelination. The axons that innervate the hair follicles, Merkel cells, and Meissner's and Pacinian corpuscles tend to be encased in thick myelin sheaths and have a total diameter of 7–12 μm. They are often referred to as **Aβ fibers** to distinguish them from thinner myelinated axons (Aδ fibers) and unmyelinated axons (C fibers). The distribution of fiber diameters for these spinal nerve axons is shown in Figure 7.6. At this point you might suspect that spinal nerves also contain Aα and Aγ fibers. Indeed, Aα fibers are very thick myelinated axons that project to and from muscles, and Aγ fibers are myelinated axons of medium thickness that innervate specialized muscle fibers. Neither of these fiber types are evident in Figure 7.6 because those data were gathered from a sensory nerve that innervates only the skin (and no muscles). The Aα, Aβ, Aδ, and C fibers are sometimes referred to as class I, II, III, and IV fibers, respectively; but the latter terminology applies only to sensory fibers and is no longer widely used.

As you may recall from Chapter 2, the speed of action potential propagation varies with both myelination and axon diameter. This explains why Aα axons can conduct action potentials at speeds of up to 120 m/sec, while Aβ and Aδ fibers are limited to conduction velocities of 35–70 and 5–30 m/sec, respectively. The unmyelinated C fibers are even slower, clocking in at 0.5–2 m/sec. This variation in action potential conduction speed has important functional implications, which we discuss later. For now, suffice it to say that information about skin touch and vibration is conveyed to the central nervous system through a relatively rapid pathway consisting of the relatively thick and myelinated Aβ fibers.

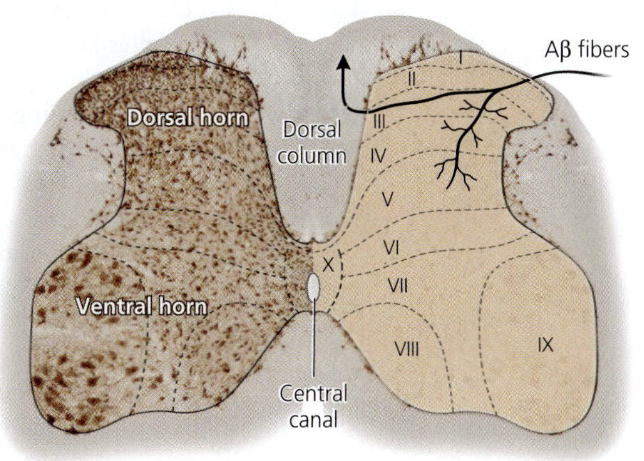

Figure 7.7 Termination of Aβ fibers in the spinal cord. Aβ fibers, which tend to carry information about touch and vibration of the skin, branch as they enter the dorsal horn of the spinal cord. One branch ascends to the medulla in the dorsal column fiber tract. The other branch forms terminal branches that synapse on neurons in the deeper layers of the dorsal horn, called Rexed laminae III, IV, and V. [After Todd, 2010]

Central Pathways for Touch and Vibration

The touch and vibration-sensitive Aβ fibers enter the gray matter of the spinal cord from its dorsal (posterior) aspect. Soon after entering the spinal cord, these axons split into two major branches. One branch enters the **dorsal column** fiber tract, which forms the dorsomedial portion of the spinal cord's white matter (Figure 7.7) and ascends to the caudal medulla, where the axons terminate in the ipsilateral **dorsal column nuclei**. We will discuss these nuclei and their projections to the thalamus in Chapter 11. The other branch of the incoming Aβ fibers terminates in the **dorsal horn** of the gray matter, close to where the fibers entered the spinal cord (Figure 7.7).

To specify the location of the axon terminals more precisely, neuroscientists divide the spinal gray matter into 10 distinct zones, called Rexed's laminae (after Bror Rexed). Laminae I–VI form the dorsal horn; laminae VIII and IX comprise the spinal cord's **ventral horn** and contain most of the spinal motor neurons; laminae VII and X are called the *intermediate zone*. Using this terminology, we can say that Aβ fibers terminate mainly in laminae III, IV, and V of the ipsilateral dorsal horn. The neurons in these laminae are quite diverse and project to a wide variety of targets both within the spinal cord and in the brain. Accordingly, they serve a wide variety of functions, including the control of locomotion through sensory feedback.

So far we have discussed how information from touch and vibration sensors in the skin of the body reaches the central nervous system. What about vibration and touch sensors in the skin of the head? They convey their information not through spinal nerves but through cranial nerve V, which is called the **trigeminal nerve** (*trigeminus* means "triplets" in Latin and refers to this nerve having three major branches). The touch- and vibration-sensitive Aβ fibers in the trigeminal nerve project to the **principal nucleus of the trigeminal nerve**, which is located in the caudal pons.

Somatotopy

Each touch sensor in the skin responds to somatosensory stimuli in a limited portion of the body surface. This region is called the sensor's *spatial receptive field*, closely analogous to the spatial receptive fields of retinal neurons that we discussed in Chapter 6. Meissner corpuscles and Merkel cells tend to have smaller spatial receptive fields than the Pacinian corpuscles, mainly because the latter are more sensitive and therefore respond to stimuli presented at more distant locations. As you might suspect, the neurons that receive information from the various somatosensory sensors also have spatial receptive fields, their size and location being determined primarily by the location of the sensors from which they gather information. In general, neurons that respond to touches on the fingertips and face have smaller spatial receptive fields than neurons that gather information from the back.

These differences in receptive field size (which we discuss more thoroughly in Chapter 11) largely explain why we can identify and localize touch stimuli much better on our fingertips and face than on our back. In analogy to the visual system, we can think of our fingertips as the "fovea" of the somatosensory system because we often aim our fingertips at portions of the outside world that warrant further scrutiny.

An important feature of the central pathways for touch and vibration sensors in the skin is that they are topographically organized (see Box 7.1: Shingles and Dermatomes), meaning that adjacent neurons in those pathways tend to have adjacent spatial receptive fields. This **somatotopic** organization is analogous to the retinotopic and tonotopic patterns of organization that we discussed in Chapter 6. In all three cases, the sensory surface is mapped topographically onto groups of neurons in the central nervous system. Inside the brain, the somatosensory maps of the body and head are located in two separate areas (dorsal column nuclei and principal nucleus of the trigeminal nerve, respectively); but these two partial body maps are combined into a single whole-body map at later stages of the somatosensory pathway (see Chapter 11).

BRAIN EXERCISE

In what way are the various types of sensors in the skin analogous to the various types of sensors in the retina?

7.2 How Do We Sense Pain?

Pain appears to be distinct from other sensations, but painful sensations can be triggered by a variety of stimuli. Pinching or poking the skin can generate a painful sensation, as can extreme temperatures, acid burn, or tissue inflammation. The sensors involved in detecting such noxious (potentially pain-inducing) stimuli are called **nociceptors** (from *nocere*, which means "to harm" in Latin). These nociceptors tend to be free nerve endings. Some are unimodal, meaning that they are involved in just a single kind of pain. Others are polymodal in the sense that they respond to a variety of noxious stimuli. Because there are so many different kinds of noxious stimuli, the molecules involved in transducing these stimuli are likewise diverse; often they remain unknown. Importantly, nociceptors tend to be activated only by fairly intense stimuli that are capable of creating tissue damage. Less intense stimuli may activate other sensors, but they tend to leave nociceptors silent.

Closely related to nociceptors are itch sensors, which trigger a desire to scratch the itchy area. The last few years have seen great progress in identifying various molecules that are involved in generating itch, but it remains unclear to what extent the nerve fibers expressing these itch-related molecules are separate from those responsible for pain. Considerable overlap is suggested by the observation that pain tends to inhibit itching, which, incidentally, explains why it feels good to scratch an itch.

Axons That Transmit Pain

Most of the axons responsible for sensing pain are thin unmyelinated C fibers, but roughly 30% are Aδ fibers (Figure 7.6 B). The fact that pain is sensed by two different classes of sensory fibers, with different conduction velocities (0.5–2 m/sec vs. 5–30 m/sec), explains why pain comes in both slow and fast varieties. Stubbing a toe, for example, generates a sharp initial pain, followed a few seconds later by a duller pain. The first, fast pain is relatively easy to localize and tends to trigger a withdrawal response. The second, slower pain is more difficult to localize and prompts a desire to protect the affected area from movement or contact.

Pathways for Pain at the Body Surface

Pain-sensitive C and Aδ fibers from the body surface enter the spinal cord through the dorsal roots, just as Aβ fibers do, but they behave quite differently once they enter the dorsal horn (compare Figures 7.7 and 7.8). For one thing, the nociceptive fibers do not send a long axon branch into the dorsal column tract. For another, they terminate

NEUROLOGICAL DISORDERS

Box 7.1 *Shingles and Dermatomes*

Chickenpox (varicella) is a highly infectious childhood disease characterized by fever and a blistering rash all over the skin. It is caused by the *varicella-zoster virus*, which belongs to the herpes family of DNA viruses. It is a neurotropic virus, meaning that it has a special affinity for neurons. Indeed, the virus "hides out" in neurons of the dorsal root and cranial nerve ganglia after the initial varicella infection, evading destruction by the immune system for years, often decades. The virus lies dormant in these structures until it is reactivated in some people (as many as 1% of people over 65 every year) by poorly understood factors, including a weakened immune system.

On reactivation, the varicella-zoster virus produces large numbers of virus particles that are transported along sensory axons to the skin, where they cause inflammation and a blistering rash. The rash may disappear within a week or so, but the patient tends to feel intense pain in the affected area, both before the rash develops and for weeks (sometimes months) afterward. This pain stems at least in part from inflammation-induced hyperalgesia, which makes even the lightest touch feel extremely painful. If you've ever had a blistering sunburn, imagine that kind of pain but more intense and lasting for weeks. Importantly, the rashes caused by this secondary infection remain limited to the portion of the skin that is innervated by the infected nerve. These skin regions, called *dermatomes*, tend to have sharp boundaries, especially at the midline, and sometimes appear belt shaped, which is why this disease is called *shingles* or *zoster* (these names come from the Latin and Greek words for "belt," respectively). Most vividly, the Norwegian name for shingles is "belt of roses from Hell."

After analyzing more than 470 cases of shingles, Henry Head and Alfred W. Campbell in 1900 published a comprehensive map of human dermatomes (Figure b7.1). Subsequent studies largely confirmed their findings (although alternative, less accurate dermatome maps still appear in numerous textbooks). Head and Campbell's dermatome map shows that each spinal nerve innervates a discrete region of the skin. However, most dermatomes overlap with one another (except at the midline), which is why damage to a single spinal nerve tends to have little effect on sensation.

Antiviral drugs can shorten the duration of shingles and reduce its severity as long as the drugs are given early enough. Unfortunately, the development of better treatments is hampered by the fact that the varicella-zoster virus infects almost exclusively human cells. Only recently have researchers begun to study the virus in mice with transplanted human cells or in monkeys that are infected with the monkey homolog of the varicella-zoster virus (simian varicella virus). Although treatments remain elusive, vaccinations against the varicella-zoster virus are quite effective. A comprehensive child vaccination campaign has reduced the incidence of chickenpox by more than 80% since 1995. A similar vaccine managed to reduce the incidence of shingles by 51% in a clinical trial. It was approved as an antishingles vaccine for older adults in 2005.

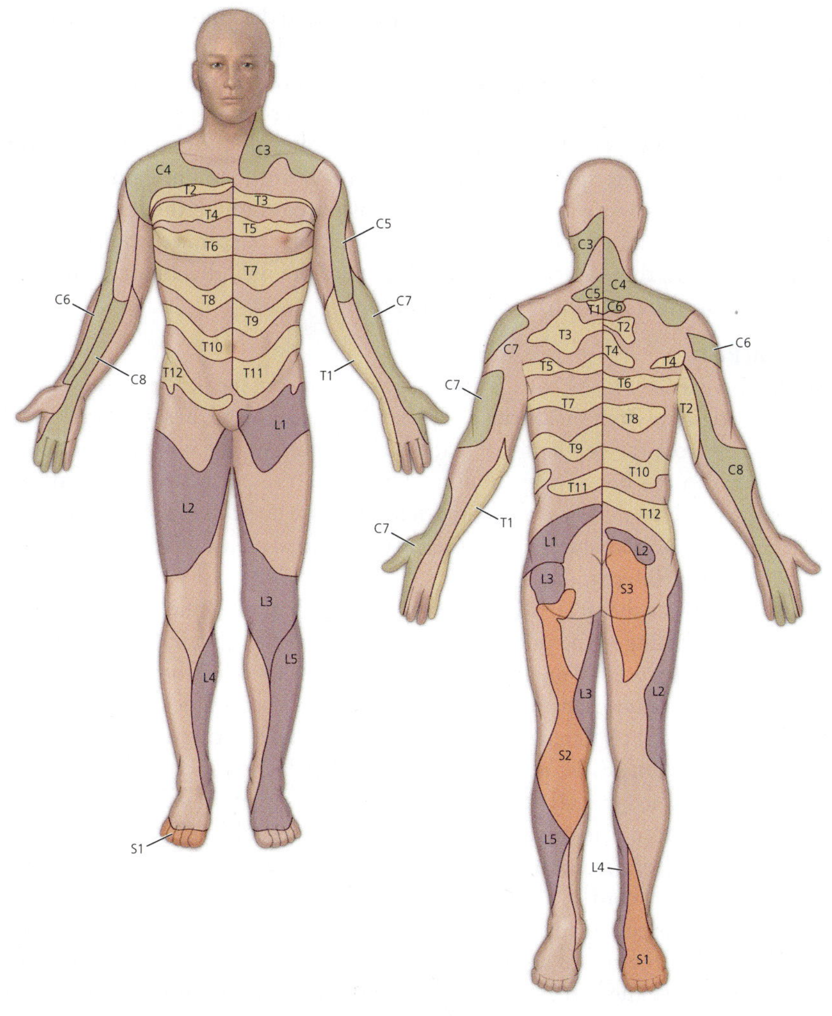

Figure b7.1 A shingles-based dermatome map. Having analyzed the pattern of rashes in numerous patients with shingles, Henry Head and A.W. Campbell inferred the illustrated dermatome map. Each dermatome is shown on just one side of the body and labeled with the spinal nerve that innervates it. [After Lee et al., 2008]

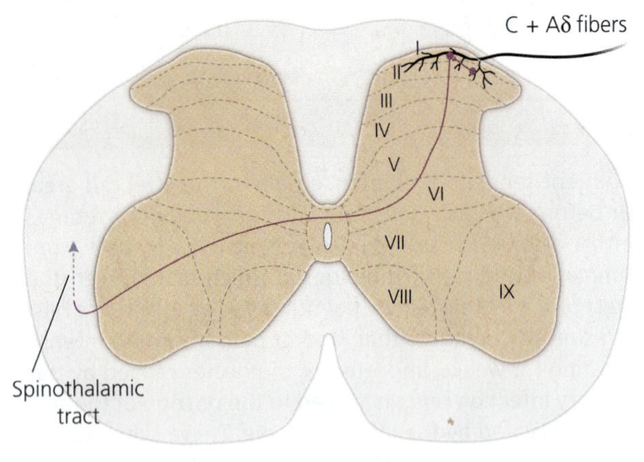

Figure 7.8 Pain pathway through the spinal cord. Nociceptive C and Aδ fibers terminate mainly in Rexed laminae I and II of the spinal cord. Neurons in lamina II project mainly to lamina I, but lamina I neurons have long axons that cross the midline and ascend to the brain in the contralateral spinothalamic tract. [After Todd, 2010]

more superficially within the dorsal horn. Specifically, nociceptive axons terminate primarily in Rexed's laminae I and II, which are collectively called the **substantia gelatinosa** (their gelatinous appearance results from a paucity of myelinated axons). The postsynaptic neurons in lamina II project mainly to lamina I, but the lamina I neurons have much longer projections. They send their axons across the midline to the contralateral spinal cord, where they ascend in the **spinothalamic tract** (Figure 7.8). As the name implies, axons in the spinothalamic tract terminate in part of the dorsal thalamus. However, axons in the spinothalamic tract also terminate in parts of the midbrain and medulla, where they mediate pain-triggered reflexes and modulate other behaviors.

The fact that pain-related signals ascend to the brain in the contralateral spinothalamic tract, whereas information about touch and vibration ascends in the ipsilateral dorsal column tract, has an interesting clinical consequence. If a person incurs damage to one half of the spinal cord, then that person becomes unable to sense pain below the level of the injury on the side of the body that is *contralateral* to the injury. However, the same patient will lose sensations of touch and vibration on the *ipsilateral* side. This clinical phenomenon, called **Brown-Séquard syndrome**, can be used to diagnose spinal cord injuries without the use of expensive CT or MRI scans.

The pain pathways we just discussed carry signals from the body surface through spinal nerves into the spinal cord and, from there, to the brain. Pain signals from the head follow a different path. They are transmitted through the trigeminal nerve directly to the brain, specifically to the caudal part of the **spinal nucleus of the trigeminal nerve**. This elongated nucleus is part of the medulla, but it is continuous with the substantia gelatinosa of the spinal cord.

Visceral Pain

Although we tend to think of pain as coming mainly from the skin, painful sensations can also originate from inside the body and the head. Headaches, for example, tend to originate from the cerebral vasculature and meninges; breaking a bone activates nociceptors in the bone's surrounding membrane (the periosteum); and digestive discomfort is generated by stretch- and acid-sensitive nociceptors in the lining of the gastrointestinal tract. Neuroscientists refer such internal pain as **visceral pain** (*viscerus* being Latin for "internal organ").

An important characteristic of visceral pain is that it is difficult to localize. Particularly interesting is the phenomenon of **referred pain**, which arises when pain originating in an internal organ is falsely identified as coming from (referred to) part of the skin. For example, patients having a heart attack sometimes report pain on the inside surface of their left arm, left hand, and left jaw. Another example of referred pain is an *ice cream headache*, which feels like a headache but arises when very cold stimuli (such as ice cream) contact the back of the throat. The neural mechanisms underlying referred pain remain obscure, but they probably include the convergence of nociceptive axons from the skin and from internal organs onto single neurons in the brain or spinal cord.

What good is pain if you cannot properly identify its source? This is a good question, but before the rise of modern medicine we could do little to alleviate internal pains in any case. The best approach for dealing with visceral pain in prehistoric days would have been to avoid repeating whatever actions preceded the pain. For such pain-mediated negative reinforcement learning, a diffuse or misplaced pain signal suffices. Now, of course, it would be nice if we could tell a doctor more precisely where it hurts.

Pain Modulation

As you know from personal experience, injuries that barely trigger a response when you are in the middle of a competitive game or other form of excitement can be quite painful once the excitement stops. You may also have discovered that touches of the skin that aren't usually painful can become quite painful when you have a serious sunburn. Furthermore, you know of several drugs that reduce pain. In sum, it is quite obvious that perceptions of pain can be modulated by a wide variety of mechanisms.

Pharmacological Alleviation of Pain

Morphine and various codeine derivatives, such as hydrocodone and oxycodone, are powerful *analgesics* (pain relievers) that are derived from opium in poppy seeds or synthesized from scratch. In addition to these **exogenous opioids**, animals make their own **endogenous opioids**, such as *enkephalin* or *endorphin*, which lessen pain during vigorous exercise or in response to acute stress. All these opioids bind to G protein-coupled **opioid receptors**. The intracellular mechanisms triggered by the activation of these receptors cause voltage-gated calcium channels to close and a specific subset of potassium channels to favor their open state. As a result, the neurons that express the activated opioid receptors become much less excitable. Which neurons are involved? Opioid receptors are located in many nociceptive axons that terminate in the substantia gelatinosa as well as in some brain regions that project back to the substantia gelatinosa. In both cases, activation of the opioid receptors tends to impede the transmission of nociceptive signals. If these signals fail to reach the higher levels of the brain, then the pain is not perceived.

A promising new direction in the search for better analgesics is based on the discovery that nociceptive axons express a unique complement of voltage-gated sodium channels. Of the 9 different types of voltage-gated Na^+ channels that vertebrates possess, nociceptive axons preferentially express $Na_v1.7$ (Figure 7.9 A). Genetic studies have shown that mutations in $Na_v1.7$ are associated with **congenital insensitivity to pain**, defined as the inability to perceive noxious stimuli as pain. Before you think feeling no pain is great, consider that people with such mutations frequently suffer from internal injuries without knowing that anything is wrong, which often makes the problem worse. Still, the discovery that people with mutations in $Na_v1.7$ are insensitive to pain has prompted an intensive search for drugs that selectively block this particular ion channel. Amazingly enough, a Chinese centipede has already evolved such a compound and uses it in its venom. Injecting this centipede-derived $Na_v1.7$ blocker into mice makes them less sensitive to pain (Figure 7.9).

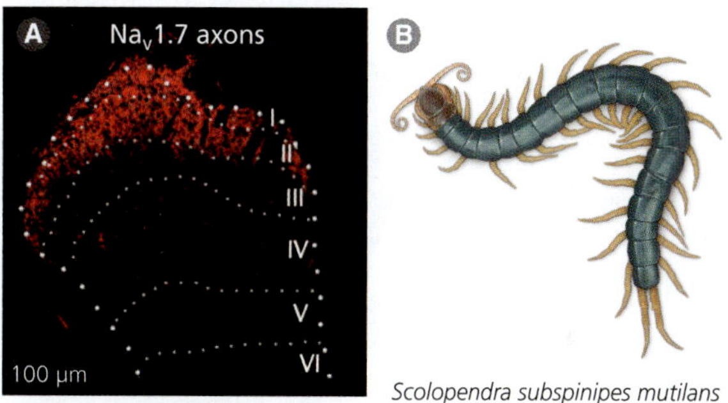

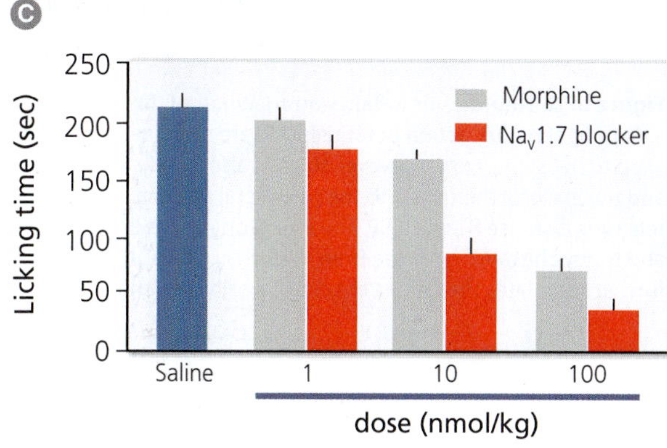

Figure 7.9 Reducing pain with a sodium channel blocker. Shown in (A) are axons that express $Na_v1.7$ and terminate in laminae I and II of the spinal cord, suggesting that they are nociceptive. Drawing (B) depicts a Chinese red-headed centipede whose toxin includes a compound that blocks the $Na_v1.7$ channel. Injecting this toxin into mice reduces their sensitivity to pain more than morphine does (C). To quantify pain, experimenters measured how long a mouse licked its paw after receiving an injection of formalin there. [A from Dib-Hajj et al., 2013; B and C after Yang et al., 2013]

Competition between Pain and Touch

Although pain is most commonly reduced with analgesic drugs, it can also be alleviated by activating touch-sensitive axons that innervate the same general area. As you may have noticed, rubbing a mosquito bite tends to reduce the pain. Most likely, the rubbing activates touch-sensitive axons that excite inhibitory neurons in the spinal cord that, in turn, reduce the ability of the nociceptive axons to transmit signals to higher brain regions. In essence, the neural activity caused by rubbing the skin "closes the gate" through which nociceptive activity can reach the brain and generate conscious pain perception. This idea is called the **gate control theory of pain**.

NEUROLOGICAL DISORDERS

Box 7.2 *Neurogenic Inflammation*

Tissue damage activates pain-sensitive axons and causes the release of various inflammatory compounds, some of which further heighten pain fiber activity (hyperalgesia). Tissue damage also causes redness, tissue swelling, and a rush of white blood cells into the affected area. Curiously, when the damaged area is treated with lidocaine, which blocks action potential propagation, the inflammatory response is markedly reduced. Therefore, the inflammation must be partly *neurogenic*, meaning "caused by neurons."

A key mechanism underlying neurogenic inflammation was discovered more than a century ago when William Bayliss cut a dog's spinal nerve dorsal root, electrically stimulated the peripheral side of the cut, and observed swelling of the body part innervated by that sensory nerve. Bayliss proposed that the electrical stimulation of the dorsal root causes action potentials to travel *antidromically* (opposite to their normal direction of travel) toward the skin where they trigger the release of substances that trigger capillary dilation, which causes the swelling. Subsequent research confirmed this core idea and led to the following, more elaborate hypothesis.

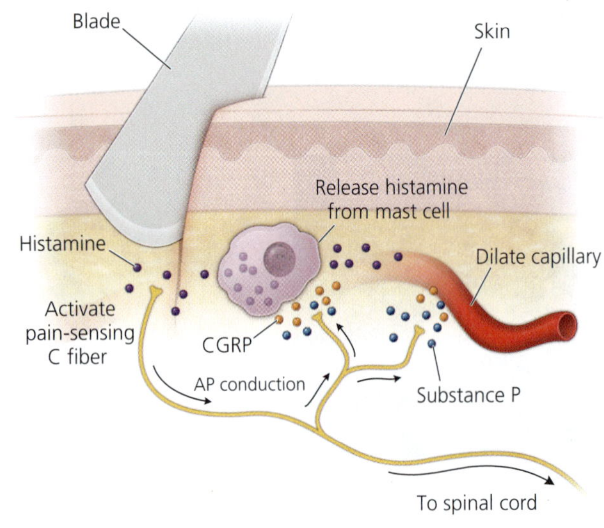

Figure b7.2 Neurogenic inflammation. When a C fiber is activated by a noxious stimulus, action potentials (AP) are propagated along the axon to the spinal cord. However, the APs also invade axon collaterals and travel along them back to the peripheral terminals, where they release substance P and CGRP. These neuropeptides dilate capillaries and trigger histamine release from mast cells. The histamine then further stimulates the C fiber, creating a positive feedback loop.

In neurogenic inflammation, noxious stimuli activate C and Aβ fibers and trigger action potentials that travel toward the spinal cord (Figure b7.2). Those action potentials also invade other peripheral axon branches (collaterals) and travel antidromically toward the peripheral axon terminals, where they cause the release of the neuropeptide transmitters *substance P* (the P apparently stands for "powder," rather than "pain") and *calcitonin gene-related peptide* (CGRP). Thus, the peripheral nerve endings of the pain-sensitive fibers are both *afferent* (receiving input) and *efferent* (generating output). Once released into the skin, substance P and CGRP cause local capillaries to dilate, thereby increasing blood flow. Substance P and CGRP also prompt mast cells in the skin to release histamine, which further increases activation of the pain-sensing fibers, setting up a positive feedback loop. As if that weren't enough, histamine also disrupts the cellular bonds between vascular endothelial cells, making the capillaries more "leaky," which contributes to tissue swelling and promotes the efflux of white blood cells. Given these mechanisms, you can see why blocking action potentials in the wound area reduces the inflammatory response.

Although neurogenic inflammation is most commonly studied in skin, it can also affect other tissues. For example, neurogenic inflammation is thought to play a role in migraine headaches, which involve the activation of pain-sensitive fibers in the meninges and the dilation of meningeal blood vessels. Neurogenic inflammation may also be at work in asthma and anaphylactic shock. In these allergic attacks, histamine is released from mast cells, which activates pain (or itch) sensitive axons, which may then (through CGRP and substance P) cause even more histamine release and further swelling. According to this hypothesis, reducing neural activity should lessen the severity of the attacks. Reducing neural activity, of course, is simpler said than done.

The discovery that touch sensor activation can "gate out" pain has sparked the development of **Transcutaneous Electrical Nerve Stimulation (TENS)** devices, which allow patients with chronic pain to temporarily reduce their pain. When activated by the patient, these devices deliver a weak electric current to a patch of skin close to the site of chronic pain. The electrical stimulation creates a temporary tingling sensation in the skin and reduces the pain for several minutes at a time. Why doesn't the electrical stimulation activate nociceptive axons, making the pain worse? Because the ability to activate axons with externally applied currents is roughly proportional to axon diameter. Because nociceptive axons tend to have very small diameters, they are not excited by the TENS device. In contrast, the large-diameter touch- and vibration-sensitive axons are activated by the electrical stimulation.

Hypersensitivity to Pain

The opposite of analgesia is **hyperalgesia**, which includes both a heightened reaction to normally painful stimuli and feeling pain from stimuli that normally do not activate nociceptive axons. A good example is the hypersensitivity that develops when you have a sunburn or some other form of tissue inflammation (see Box 7.2: Neurogenic Inflammation). Compounds released by the damaged tissue lower the action potential threshold of the nociceptive fibers. In addition, molecules released as part of the inflammatory response, notably *prostaglandins*, reduce inhibition in the spinal cord dorsal horn, making it easier for pain signals to reach the brain. Given that inflammation tends to cause hyperalgesia, it is not surprising that anti-inflammatory compounds, such as aspirin and nonsteroidal anti-inflammatory drugs (NSAIDs), can reduce pain. Most of these analgesics inhibit *cyclooxygenase (COX)* enzymes, which are required for prostaglandin synthesis.

BRAIN EXERCISE

If you were insensitive to pain, how would you have to adjust your behavior? If you were taking care of a child that is insensitive to pain, how could you help that child stay safe?

7.3 How Do We Sense Temperature?

The brain contains some temperature-sensitive neurons (see Chapter 9), but most sensors for heat and cold are located in our skin. They help us maintain a stable body temperature and warn us if objects are so hot or cold that they might damage us. Like nociceptors, temperature sensors consist of free nerve endings that transmit impulses to the central nervous system through relatively thin C and Aδ fibers. Depending on the ion channels they express, temperature-sensitive axons are activated by noxious cold (<15°C), noxious heat (>43°C), or intermediate temperatures (Figure 7.10).

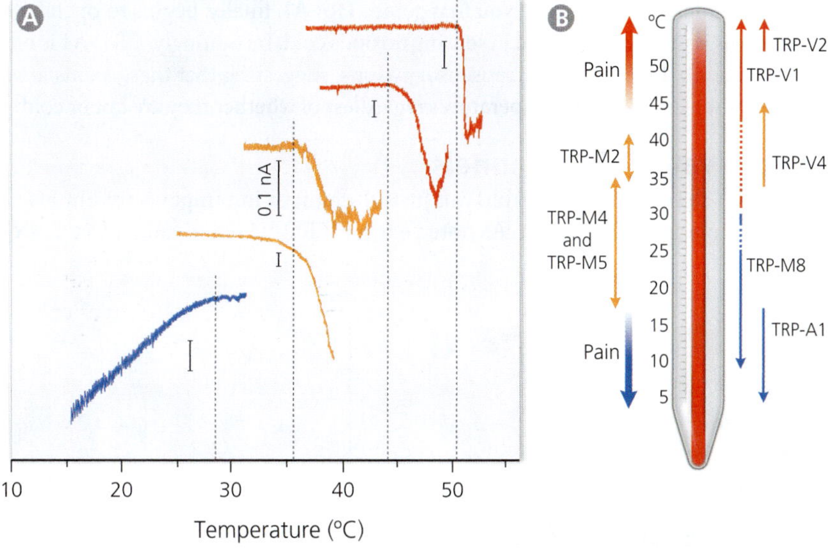

Figure 7.10 Temperature-sensitive TRP channels. Shown in (A) are recordings from embryonic kidney cells transfected with genes for temperature-sensitive transient receptor potential (TRP) channels. The colored traces are transmembrane currents recorded as the cells are gradually heated or cooled. The transfected cells respond to these temperature changes with inward (depolarizing) currents (all vertical scale bars represent 0.1 nA). As summarized in (B), cells expressing different TRP channels respond over different temperature ranges. The dashed lines in (B) indicate that the response thresholds for TRP-M8 and TRP-V1 are variable. Inflammation, for example, lowers the threshold for TRP-V1 activation. [After Tominaga and Caterina, 2004]

Temperature-sensitive TRP Channels

The various temperature-sensitive ion channels belong to a superfamily of about 30 different channels that are called **transient receptor potential (TRP) channels** (because mutations of its prototypical member cause fruit fly photoreceptors to have abnormally transient responses to light). The channels in this superfamily all have 6 transmembrane domains and are relatively nonselective cation channels that allow Ca^{2+}, Mg^{2+}, Na^{+}, and K^{+} to flow through them (permeabilities vary somewhat across the different TRP channels). Their functions are diverse, but several TRP channels are exquisitely sensitive to changes in temperature. This sensitivity is often studied by expressing individual TRP channel genes in cells that do not normally express these genes and then examining how the membrane potential of the transfected cells varies with temperature. Sample data of this kind are shown in Figure 7.10.

The first TRP channel discovered to be temperature sensitive is **TRP-V1** (the V1 refers to the fact that TRP-V1 can be activated by vanillin). The TRP-V1 channel is also known as the *capsaicin receptor* because it can be activated by capsaicin, the active ingredient of hot chili peppers. We will return to this odd property shortly. For now, the critical observation is that TRP-V1 channels begin to open at approximately 43°C, which is close to the point where heat becomes painful. Conduction through these channels increases further as temperatures rise to ~50°C, beyond which the tissue deteriorates. Mice lacking functional TRP-V1 channels exhibit diminished responses to noxious heat. Although this finding demonstrates that TRP-V1 is important for sensing noxious heat, mice also have some other channels that are sensitive to noxious heat. These other channels include TRP-V2 and TRP-M3, but their relative importance remains debated in the literature.

TRP-V3 and TRP-V4 channels sense temperatures just above normal body temperature, and **TRP-M8** opens below 30°C, reaching maximal activation around 8°C. Consistent with these data, mice lacking TRP-V3 or TRP-V4 have minor deficits in temperature sensing (although these appear to be strain dependent), and mice that lack TRP-M8 are less capable than wild-type mice of avoiding chilly environments (Figure 7.11). An interesting property of TRP-M8 channels is that they adapt to steady temperatures, which is why the ocean feels coldest when you first get in. **TRP-A1**, finally, begins to open below 17°C and thus is probably involved in sensing noxious cold. Intriguingly, TRP-A1 is often coexpressed with TRP-V1 in thermosensory axons, suggesting that these axons are responsible for sensing noxious temperatures regardless of whether they are hot or cold.

Food-activated TRP Channels

Several TRP channels are activated not just by changes in temperature but also by chemicals in plants we like to eat. As noted earlier, TRP-V1 is activated by **capsaicin**,

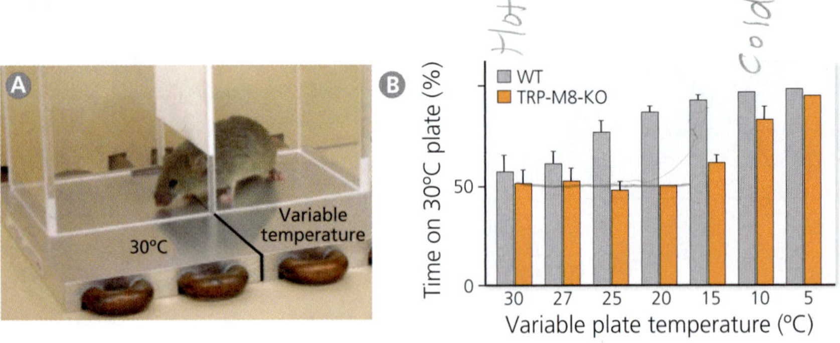

Figure 7.11 A behavioral function for TRP-M8. TRP-M8 knockout mice (TRP-M8-KO) or wild-type (WT) mice were placed in a chamber with two compartments (A). The floor of one compartment was a comfortable 30°C; the temperature of the other floor was adjustable from 5–30°C. Given a choice, WT mice preferred the 30°C compartment, as long as it was warmer than the other compartment by more than 3°C. In contrast, TRP-M8-KO mice preferred the warmer compartment only when the temperature in the variable compartment dropped below 15°C. [From Bautista et al., 2007]

the molecule that gives chili peppers their spice. Indeed, the "hotter" a pepper, the more it activates TRP-V1 expressing cells (Figure 7.12). The reason TRP-V1 responds to both hot temperature and hot chili peppers is that capsaicin lowers the temperature at which TRP-V1 channels open. Thus, on exposure to capsaicin, TRP-V1-expressing neurons signal "hot!" even at normal body temperatures. At high capsaicin concentrations, we perceive this signal as pain. After activation, the TRP-V1 receptor shuts down for several days or weeks, which explains why people who regularly eat chili peppers become desensitized to them and why capsaicin creams are used to relieve pain from arthritis and other painful conditions.

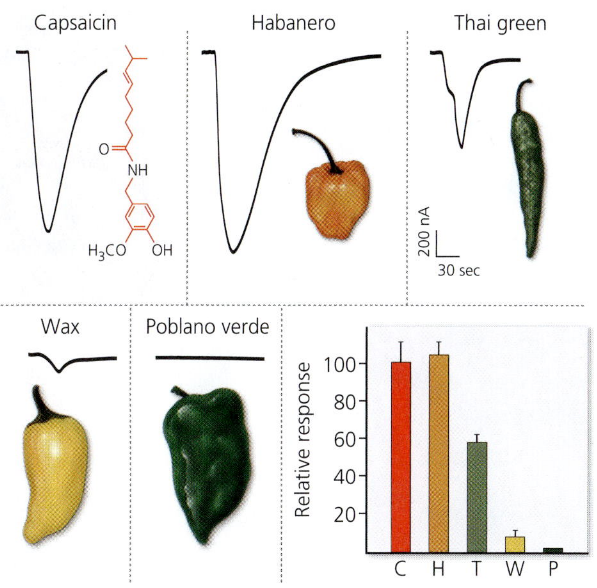

Figure 7.12 Capsaicin makes chili peppers hot. Intracellular responses (black traces) to synthetic capsaicin and extracts from various chili peppers were recorded from frog eggs that were transfected with the capsaicin receptor gene TRP-V1. The cells respond most strongly to synthetic capsaicin and to an extract of habanero peppers, which are among the spiciest hot peppers known. Their responses decline as extracts of less spicy peppers are applied, as illustrated in the histogram at bottom right in which all responses are expressed relative to the mean response to synthetic capsaicin. [From Caterina et al., 1997.]

Menthol, a major component of mint plants, affects temperature sensors in a manner opposite to that of capsaicin. It causes TRP-M8 channels to open at warmer temperatures than they normally do, reporting "cold" when it is still quite warm. This phenomenon explains why menthol is sometimes added to foods (or cigarettes) to make them taste "cool." Given that the interaction between TRP-M8 and menthol is opposite to that between TRP-V1 and capsaicin, you may wonder: do foods that combine mint leaves with hot chili peppers taste both cool and hot at the same time? Perhaps you can answer this question yourself, either in your own kitchen or a Thai restaurant.

Two additional TRP channels that can be activated by plant-derived compounds (phytochemicals) are TRP-A1 and TRP-V3. The former can be activated by the isothiocyanates in mustard oil, horseradish, and wasabi. The latter can be opened by camphor, thyme, oregano, and cloves.

BRAIN EXERCISE

Why do you think chili peppers, mustards, and several other plants evolved chemicals that activate specific mammalian TRP channels?

7.4 How Do We Taste Foods and Other Chemicals?

Judging the flavor of substances you eat or drink involves the sense of taste, or **gustation**, although it also draws on the sense of smell, which is why foods lose their flavor when they get cold (releasing fewer odorants) or when your nose is clogged. Flavor also depends on texture and spiciness, as we have just discussed. Still, taste sensors clearly make a significant contribution to flavor perception.

Taste Cells

The principal organs of taste are the **taste buds** on your tongue, the roof of the mouth, and the back of the throat. The human tongue contains approximately 10,000 taste buds, though this number varies enormously across individuals. Each taste bud consists of 50–150 **taste cells**. Like olfactory sensory neurons, taste cells die and are replaced regularly (every 10 days on average in rats). Within each bud, taste cells are clustered around a central pore that is continuous with the skin surface (Figure 7.13). Each taste cell extends into this pore several microvilli that are studded with **taste receptor** molecules. Activation of the taste receptors and cells generates sensations of the following basic tastes: sweet, umami (a Japanese word best translated as savory), bitter, salty, and sour. It was long thought that each basic taste is sensed in a different region of the tongue, but taste cells encoding each of the basic tastes are distributed throughout the tongue.

Although taste cells tend to be concentrated in the mouth, taste cells may also be located outside of the oral cavity. Catfish, for example, have taste buds on their

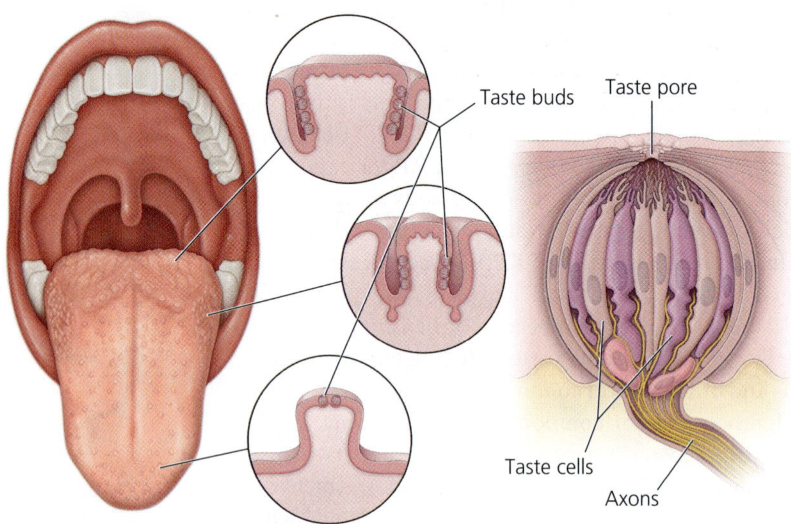

Figure 7.13 Taste buds and taste cells. Most taste buds are located in deep grooves at the back and sides of the tongue. The remaining taste buds sit on top of tiny mounds near the tip of the tongue. Shown at the right is a section through a single taste bud. It is shaped like a flask, with an opening at the top, and contains 50–150 taste cells. The taste cells contain the taste receptor molecules and are capable of releasing neurotransmitters onto postsynaptic terminals of axons in several cranial nerves that convey taste signals to the brain. [After Chandrashekar et al., 2006]

barbels and over much of their body surface (which they use at least in part to "taste" signaling molecules released by other catfish). In mammals, extra-oral taste cells are found in both the stomach and the intestines, where they help to sense the composition of ingested food and help to regulate eating behavior. For example, when intestinal taste cells sense that the ingested food contains a lot of fat, they help promote more food intake; when they detect bitter compounds, they activate circuits that cause the animal to stop eating. We will discuss this topic more in Chapter 9. Most recently, taste cells have been identified in the lining of the upper respiratory tract. These sensors are thought to be activated by bitter compounds, such as those secreted by invading bacteria, and then to trigger a variety of defensive actions including the secretion of antimicrobial peptides.

Taste Receptor Molecules

Scientists at the turn of the twenty-first century identified two major families of taste receptor genes. Receptors in the **T1 receptor family** bind molecules that taste either sweet or umami, whereas **T2 receptors** are activated by molecules that taste bitter. Both T1 and T2 receptors are G protein-coupled receptors with 7 transmembrane domains. Although these features make them similar to olfactory receptors, sequence analysis reveals that taste receptors are merely distant cousins of olfactory receptors and opsins.

The G protein coupled to T1 and T2 receptors is called **gustducin** and has two subunits. Gustducin-α is similar to transducin in that it activates a phosphodiesterase. Gustducin-β activates *phospholipase C*, which increases production of *inositol triphosphate*. The latter molecule causes calcium ions to be released from intracellular stores. In taste cells, the resulting rise in intracellular calcium concentration opens a TRP channel called **TRP-M5**, which allows positive ions to flow into the cell. Without a functional TRP-M5 gene, mice lose responsiveness to sweet, umami, and bitter stimuli.

T1 Receptors

The T1 receptor family has 3 members: T1R1, T1R2, and T1R3. These three proteins are never coexpressed with T2 receptors, but pairs of them are coexpressed with each other. When T1R2 and T1R3 are coexpressed in a single cell, the two proteins are assembled into a single compound receptor (a heterodimer) that preferentially binds sweet-tasting molecules, including natural sugars and various artificial sweeteners. Consistent with this idea, deletions of either T1R2 or T1R3 cause mice to become unresponsive to sweets (Figure 7.14) and lose their normal preference for sugar water. Moreover, transgenic mice that express human T1R2 prefer water laced with the artificial sweeteners aspartame and monellin, even though wild type mice show no such preference (artificial sweeteners are made for us, not mice).

Coexpression of T1R1 and T1R3 makes cells sensitive to the umami-tasting amino acids glutamate and aspartate (Figure 7.14). The predilection of the T1R1/T1R3 receptor for glutamate explains why **monosodium glutamate (MSG)** is such a potent flavor enhancer. On contact with saliva, MSG dissociates into sodium ions and glutamate, which then activates the umami taste cells. Importantly, the glutamate derived from MSG is the same glutamate you ingest with many savory foods (such as tomatoes and meat) and use as your brain's most abundant neurotransmitter. Nonetheless, some people react strongly to MSG, with symptoms that include headaches, flushing, facial numbness, tingling or burning around the mouth, heart palpitations, and nausea. In addition, MSG should always be consumed in moderation because high levels of dietary sodium can elevate a person's blood pressure.

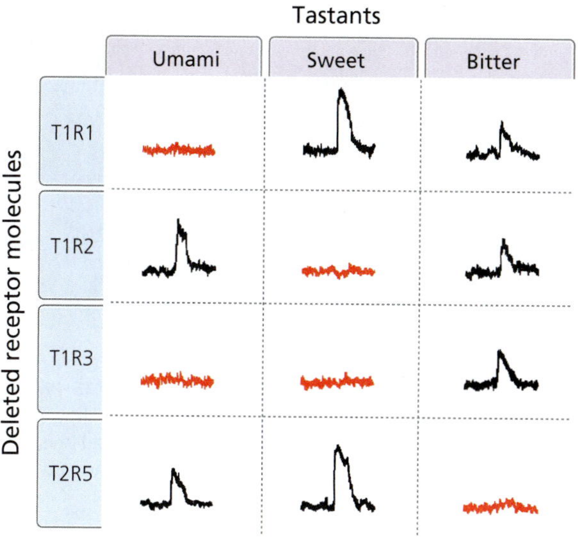

Figure 7.14 Deleting taste receptors. Shown here are responses of gustatory axons to 3 types of tastants in 4 transgenic mouse strains with specific deletions of individual taste receptors. Each deletion causes a specific deficit in the response to one or more tastants (red traces). Responses to umami and sweet tastes are each impaired by the deletion of two different taste receptor genes because these receptors are heterodimers. The umami receptor is a composite of T1R1 and T1R3, whereas the sweet receptor combines T1R2 with T1R3. Deletion of the T2R5 receptor makes mice insensitive to cycloheximide, a toxic antibiotic that normally tastes very bitter to rodents. [After Chandrashekar et al., 2006]

Is tasting umami in foods adaptive? The answer is probably yes. Because the umami taste is generally pleasant and triggered by two amino acids, it is reasonable to speculate that umami sensitivity evolved to promote the eating of protein. Unfortunately, our preference for sweets appears to be at least as strong as that for umami.

T2 Receptors

The T2 family of taste receptor molecules has 25 functional members that all bind bitter-tasting molecules. Transgenic mice that lack one of these T2 receptors lose their aversion to water flavored with whatever bitter tastants (molecules that can be tasted) the deleted T2 receptor normally binds (bottom traces in Figure 7.14). Furthermore, mice engineered to express the human T2 receptor for phenylthiocarbamide (PTC), which only humans find bitter, become averse to PTC-flavored water. Collectively, these studies show that the T2 receptors are necessary and sufficient for making compounds taste bitter.

Because most bitter compounds are at least potentially toxic (some produce cyanide when cleaved by enzymes), it is likely that the T2 receptors evolved to warn us of toxins. A related observation is that individual taste cells coexpress many, if not all, T2 receptor types, which is why we tend not to discriminate between different kinds of "bitter" sensations. Bitter is bitter and should be avoided (unless, of course, the bitterness is mild and comes from dark chocolate, coffee, or beer). Simply put, the great diversity of T2 receptors is typically not used for taste discrimination but to warn us before we ingest potentially harmful compounds.

Sour, Salt, and Fat Sensors

Sour foods tend to be spoiled or not yet ripe. To avoid eating such foods, many animals have evolved sensors of acidity, which is what makes foods taste sour. Taste cells sense acidity by means of a TRP channel called **PKD-2L1** (the PKD stands for polycystic kidney disease-like), which is expressed in a set of taste cells that do not express T1 or T2 receptors. Eliminating all the PKD-2L1-expressing taste cells makes mice unresponsive to sour tastes (Figure 7.15).

PKD-2L1-expressing cells also allow us to taste the carbon dioxide (CO2) in carbonated drinks. The way this works is that PKD-2L1 expressing taste cells also

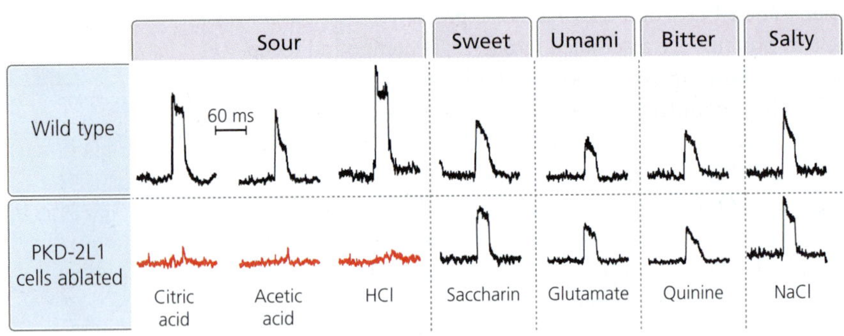

Figure 7.15 PKD-2L1 is needed to sense sour tastes. Mice engineered to lack taste cells expressing PKD-2L1 showed no neural responses to substances that normally taste sour, including citric acid, acetic acid (the acid in vinegar), and hydrochloric acid (HCl). Neural responses to sweet, umami, bitter, and salty substances were unaffected. [After Huang et al., 2006]

express, on the external side of their cell membrane, a carbonic anhydrase that converts dissolved CO_2 to H_2CO_3. The latter compound rapidly dissociates into HCO_3^- (bicarbonate) and H^+ (protons), creating an acidic solution that opens the PKD-2L1 channels. Thus, carbonation is detected by sensing the acidity created by the interactions between CO_2 and the carbonic anhydrase.

Our knowledge of the receptors for salty taste is meager. All mammals need some salt in their diet and have evolved salt-sensing taste cells to help them satisfy this need. The underlying mechanism includes an **epithelial sodium channel** that is expressed in many tissues, including the kidney, but also in a subset of taste cells. When extracellular sodium concentrations are high, sodium ions flow through this channel into the taste cells, depolarizing them. Consistent with this general hypothesis, transgenic mice without epithelial sodium channels are unable to taste salt. In humans, however, salt sensing is less dependent on the epithelial sodium channel, suggesting that other channels are involved as well.

Similarly uncertain is how mammals sense fat. It had long been thought that we sense dietary fat only by its smooth texture and, to a lesser degree, its smell. However, recent studies have shown that the molecule CD36, which binds fatty acids, is expressed in a subset of taste cells in both humans and mice and that mice lacking functional CD36 lose their normal preference for fatty foods. Intriguingly, CD36 is also expressed in the liver, in fat deposits, and in the lining of the small intestine, where it helps to regulate the secretion of dietary hormones in response to fat. Therefore, CD36 appears to be involved in a number of processes related to controlling fat intake and metabolism.

Central Taste Pathways

Most taste cells synapse directly onto the peripheral endings of gustatory axons. The principal transmitter at these synapses is **adenosine 5′-triphosphate (ATP)**. This may seem odd, given that ATP is better known as a key source of metabolic energy than as a neurotransmitter. However, gustatory axons express an ion channel called **P2X** that, after binding ATP, allows positive ions to flow into the axon terminal. Furthermore, transgenic mice that lack functional P2X receptors cannot taste a broad array of substances, including sour and salty foods. Although ATP is clearly critical for synaptic transmission at taste cell synapses, GABA and serotonin are also expressed in taste cells; their role is not yet clear.

Gustatory information is conveyed from the periphery into the brain through three cranial nerves. The **facial nerve** (cranial nerve VII) carries taste information from the anterior tip of the tongue, the **glossopharyngeal nerve** (IX) innervates the rest of the tongue, and the **vagus nerve** (X) innervates a few taste buds in the roof of

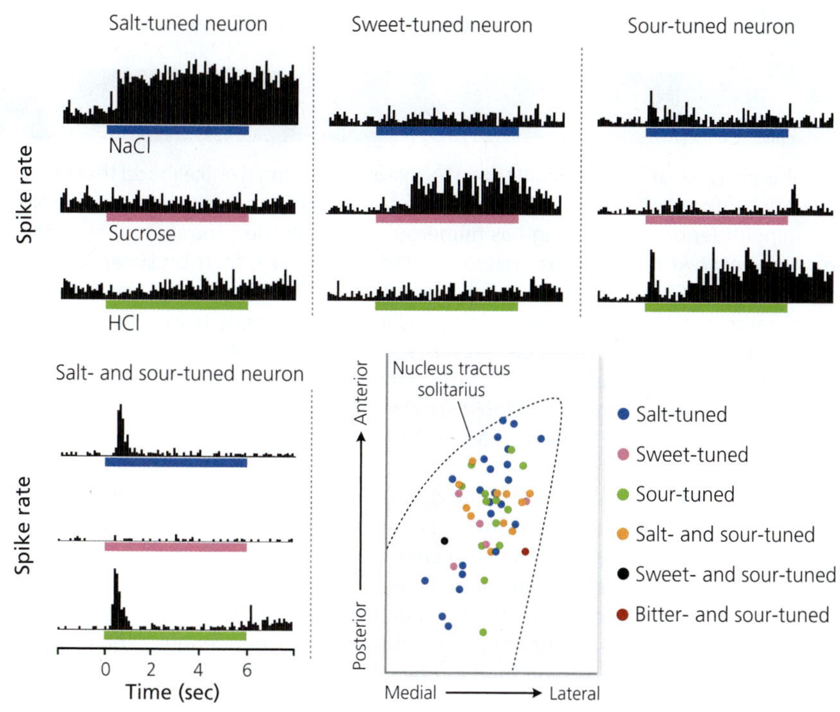

Figure 7.16 Taste coding in nucleus tractus solitarius (NTS). Experimenters recorded from individual neurons in a rat's NTS while the rat's tongue was stimulated with various tastants (the oral cavity was flushed with water between tastants). Shown here are the responses of four neurons to table salt (NaCl), sucrose, and hydrochloric acid (HCl). You can see that some NTS neurons respond strongly to just one tastant, while others respond to more than one. The diagram at the bottom right shows how neurons with different response profiles are distributed within the NTS. Neurons responsive to umami tastants are thought to overlap with those that are most responsive to sweet and salty stimuli, but they were not the focus of the illustrated study. [After Yokota et al., 2013]

the mouth and at the back of the throat. Taste-sensitive axons in these nerves project to a subdivision of **nucleus tractus solitarius (NTS)**, which lies in the medulla and receives several types of sensory input in addition to taste (see Chapter 9). Sweet, sour, salty, and bitter tastes are represented in somewhat different regions of NTS (Figure 7.16); but the topography of the primary gustatory projections is relatively imprecise. Indeed, some neurons in NTS respond to more than one kind of tastant, implying that they receive convergent input from at least two different classes of taste cells (Figure 7.16).

Variations in Tasting Ability

As we discussed in Chapter 6, sensory capacities vary widely across species; this is true also for taste. Birds, for example, are insensitive to capsaicin, which is why bird seed is sometimes laced with chili powder (the birds don't care, but squirrels won't touch it more than once). Another interesting example is that cats have a mutation in their T1R2 gene that makes them insensitive to sweet tastants. Now you know why your pet cat won't work for sweets! Birds also lost the T1R2 gene, making most birds insensitive to sweets. However, hummingbirds have modified their T1R1 and T1R3 genes so that their T1R1/T1R3 receptor responds much better to sugars than to umami. This change is probably adaptive because hummingbirds feed mainly on nectar. Yet another interesting example of evolutionary change in taste receptor repertoire is that giant pandas are insensitive to umami because they have a nonfunctional T1R1 gene. This is not a problem for pandas because they eat almost exclusively (99%) bamboo.

NEUROBIOLOGY IN DEPTH

Box 7.3 *Cannabinoid Effects on Food Intake*

The principal psychoactive component of cannabis (marijuana) is *tetrahydrocannabinol* (THC). Our bodies do not synthesize THC, but they do make the endogenous cannabinoids (endocannabinoids) *anandamide* and *2-arachidonoylglycerol* (2-AG). All these compounds bind to the cannabinoid receptors CB1 and CB2, which are widely expressed in the nervous and immune systems, respectively. In contrast to most other neurotransmitters, endocannabinoids are not stored in vesicles but made "on demand" when calcium levels rise or specific metabotropic receptors are activated. They are released mainly onto presynaptic axon terminals, changing the likelihood that those terminals will release some other transmitter. Cannabinoid signaling has numerous effects in the central and peripheral nervous systems, including diverse effects on synaptic plasticity. Its behavioral effects include increased food intake, decreased nausea, and a reduced sensitivity to pain. For these reasons, cannabis is sometimes prescribed to counteract the negative effects of chemotherapy. Most relevant to this chapter are cannabinoid effects on taste and food intake.

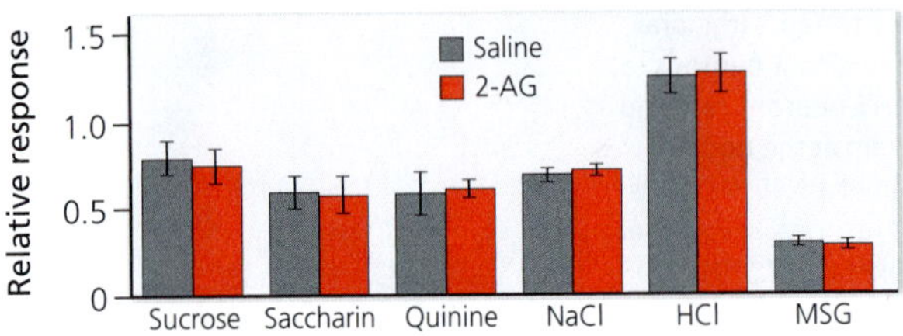

Figure b7.3 Cannabinoids selectively enhance sweet taste. Mice were injected with either the endocannabinoid 2-AG or saline (control) while investigators recorded electrical activity in one of the nerves that innervates taste buds. The 2-AG enhanced responses to sweets (sucrose and saccharine) without affecting responses to bitter, salty, sour, or umami tastants. The selective enhancement was not seen in CB1 knockout mice, confirming that the effect is due to cannabinoid signaling. [After Yoshida et al., 2010]

The cannabinoid receptor CB1 is coexpressed with T1R3 in sweet-sensing taste cells. When mice are injected with the endocannabinoid 2-AG, responses to sweet stimuli are selectively enhanced (Figure b7.3). Mice receiving such injections also show an increased preference for drinking sugar or saccharin-laced water. Neither effect is seen in transgenic mice without functional CB1 receptors (Figure b7.3). We can conclude that cannabis and the endogenous cannabinoids increase food intake at least in part by modulating taste cell responses.

Cannabinoids also make animals ingest more fat. The mechanism underlying this effect is not completely understood, but activating fat sensors in the oral cavity is known to increase endocannabinoid production in the small intestine; and blocking CB1 receptors in the small intestine reduces fat intake. These data suggest the existence of a positive feedback loop between the brain and the gut that makes animals increase their consumption of fatty foods.

Given these observations, it seems likely that cannabis plants evolved THC as way to increase the likelihood that animals eating those plants will go in search of sweet and fatty foods, thus ensuring that the plants' seeds would be dispersed widely. Similarly, endocannabinoids probably evolved to ensure that animals eat foods containing sugars and fat, which are nutritionally important but hard to obtain for animals in natural environments. Of course, contemporary humans in rich societies rarely have problems finding sweet and fatty foods, making the adaptive significance of "built-in" preferences for sweet and fatty foods harder to appreciate.

Tasting abilities vary also within a species. Best studied are genetic polymorphisms in humans that correlate with variations in the ability to taste bitter compounds. For example, variations in the taste receptor gene T2R38 account for most of the variation in the ability to perceive 6-*n*-propylthiouracil as tasting bitter. Such variations in bitter sensing ability may explain why some people really hate broccoli and Brussels sprouts, although the evidence underlying such claims is weak. Genetic variation in taste receptor genes may also explain why some people are connoisseurs of wine and food, whereas others are less discriminating. Again, however, the data supporting this hypothesis are relatively weak. Moreover, if food and wine critics really were atypical in the taste receptor genes that they express, then one might ask why "regular folk" should listen to their advice.

Some people love all kinds of sweets, while others dislike foods that are "too sweet." This variation in sweet-tooth proclivity is associated with variation in the T1R2 receptor gene. The T1R3 receptor is less variable than T1R2, but a significant fraction of the human population is insensitive to umami (just like the pandas). All this variation in tasting ability, combined with substantial variation in olfactory receptor genes (see Chapter 6, Box 6.1: Anosmias), helps to explain why different individuals prefer different kinds of foods and drink (see also Box 7.3: Cannabinoid Effects on Food Intake). As the Romans used to say, *de gustibus non est disputandum* (in matters of taste, there can be no disputes). Of course, differential experience and learning about foods (see Chapter 14) also contribute to variation in food preferences.

BRAIN EXERCISE

At the end of Chapter 6 we discussed the concept of labeled lines in sensory coding. To what extent do you think this concept applies to the gustatory system?

7.5 How Can We Sense Our Body's Physiological Condition?

The body contains a variety of sensors that monitor its physiological condition. For example, sensors in the lining of the gut tell the nervous system when our belly is too full or too empty, when to contract the intestines and push food toward the exit, and when to secrete which digestive enzymes. Sensors that detect tissue or blood dehydration help trigger thirst. The body also harbors sensors that monitor tissue pH and blood levels of oxygen, which help to regulate breathing. We will discuss the neural circuits underlying these regulatory reflexes in Chapter 9. For now, our focus is on the sensors.

Sensing Tissue Acidity

As mentioned earlier, sour-sensing taste cells use PKD-2L1 to measure the acidity of foods. Diverse other cells in the body express another kind of acid sensor, called **acid-sensing ion channel (ASIC),** that allows positive ions to flow into a cell when the extracellular pH is low (acidic). These ASICs were once thought to be involved in sour taste perception, but mice lacking these channels have normal sensitivity to sour foods and drinks. Instead, ASICs seem to play a major role in sensing the acidity of bodily tissues. This is useful because tissue acidity is related to carbon dioxide (CO_2) levels. Monitoring CO_2 levels, in turn, is important because high levels of CO_2 tend to indicate that oxygen is running low. How can an acid sensor monitor CO_2? As you learned a few paragraphs ago, CO_2 is converted by carbonic anhydrase to H_2CO_3, which quickly dissociates into bicarbonate and protons. By definition, the production of protons decreases the pH. Because the carbonic anhydrases required for this conversion are expressed throughout the body (not just in taste cells), ASICs can sense CO_2 levels in many locations. The signals generated by ASICs can then be used to increase breathing rate and clear the excess CO_2.

Another acid-sensing ion channel is TRP-V1. It is activated not only by capsaicin and heat but also by acidity. Although TRP-V1 is expressed at high levels in the tongue and oral cavity, it is also expressed in the rest of the body, where it acts as a general sensor for noxious stimuli, triggering pain. As you probably know, lactic acid buildup in muscles is painful, as is the spilling of stomach acids into the esophagus (acid reflux). Holding your breath for more than a minute or two also causes excruciating pain, presumably because the rising CO_2 level activates some TRP-V1 channels. Curiously, naked mole rats and hibernating mammals evolved a clever trick to circumvent this acid-induced pain, allowing them to live with high levels of tissue CO_2 (see Box 7.4: Feeling Less Pain in High CO_2).

EVOLUTION IN ACTION

Box 7.4 *Feeling Less Pain in High CO_2*

Naked mole rats are small, largely hairless rodents that live in underground tunnels. They live remarkably long (up to 21 years), partly because their cells cease dividing when crowded by other cells, thus preventing the growth of cancerous tumors. Naked mole rats are also unique among mammals in forming large colonies with a single breeding queen, several breeding males, and numerous nonbreeding workers. Living in large social groups deep underground does create a problem, though, namely, that naked mole rats must live in an oxygen-poor, CO_2-rich environment. To handle this challenge, naked mole rats evolved an unusually low metabolic rate and hemoglobin with an unusually high affinity for oxygen.

Naked mole rats also evolved a novel way to deal with the tissue acidosis that results when dissolved CO_2 comes into contact with carbonic anhydrase (see Section 7.4 of the main text). As revealed by molecular sequence comparisons, naked mole rats have modified the voltage-sensitive $Na_v1.7$ channels expressed in acid-sensing pain fibers. Specifically, they have converted a positively charged amino acid motif (Lys-Lys-Val) in this ion channel to a negatively charged motif (Glu-Lys-Glu), which increases the amount of depolarization that is required to trigger action potentials in the affected axons. Because of this evolutionary change, the acid-sensing axons of naked mole rats can sense tissue acidity, but they are less likely to send this information on to the central nervous system. Because activity in these acid-sensing fibers is generally interpreted as "pain," you can say that at a given tissue pH, naked mole rats feel less pain than typical rodents. Indeed, these animals seem generally oblivious to noxious stimuli.

An interesting twist on this story is that several groups of hibernating mammals have evolved a similar insensitivity to CO_2-induced acidosis (Figure b7.4). Because hibernating animals breathe at very low rates, their tissue accumulates high levels of CO_2, making it acidic. If this acidity activates nociceptors, the resulting pain might wake the animals from sleep and cut the hibernation short. To determine whether hibernators solve this problem in the same way as naked mole rats, researchers compared the amino acid sequences of the $Na_v1.7$ channel across 71 species of mammals, including 22 that hibernate. They discovered that the conversion of the positively charged amino acid motif into a negatively charged motif occurred at least 6 times independently, covering most of the hibernating species. Different taxonomic groups used slightly different amino acid substitutions to effect the changes in motif charge, but molecular modeling suggests that all the changes should make it more difficult to open $Na_v1.7$ channels in acidic environments.

These findings illustrate that the independent evolution of similar traits (convergent evolution) may occur quite frequently, even at the molecular level. They also illustrate that evolution often finds elegantly simple solutions to tricky biological problems.

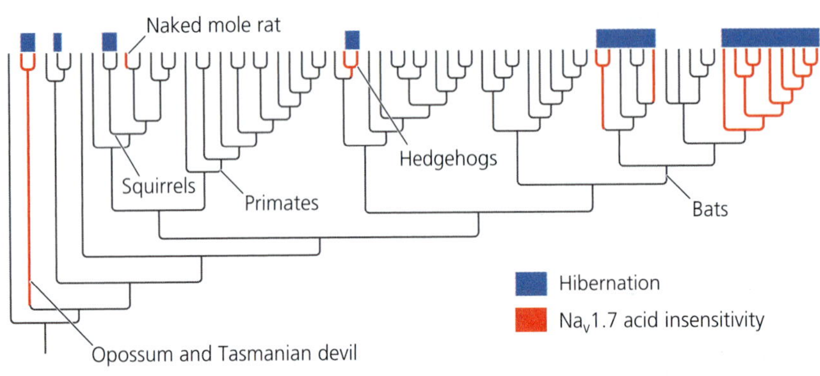

Figure b7.4 Acid insensitivity evolved repeatedly among mammals. This diagram shows the phylogenetic relationships of 71 mammalian species, 22 of which hibernate (blue bars). The groups indicated by the red lines all changed the amino acid sequence of their $Na_v1.7$ channel so that it becomes more difficult to open in acidic environments. As you can see, the molecular changes correlate well with the evolution of hibernation. Naked mole rats do not hibernate, but they live in a high-CO_2 environment that promotes tissue acidosis. [After Liu et al., 2013]

Sensing Blood Chemistry

Because changes in the body's general condition tend to be preceded by changes in blood chemistry, sensors of blood chemistry provide an "early warning" against poisons or other factors that might damage the body. Some of these blood chemistry sensors are located in parts of the brain where the blood-brain barrier is leaky. The sensors in these *circumventricular organs* (see Chapter 5) are involved in sensing blood osmolarity as well as various hormones, glucose, and, potentially, blood-borne toxins.

Sensors of blood chemistry are also located in the **carotid body**, a highly vascularized collection of cells located where the carotid artery divides into external and internal branches (Figure 7.17). A major function of the carotid body is to detect

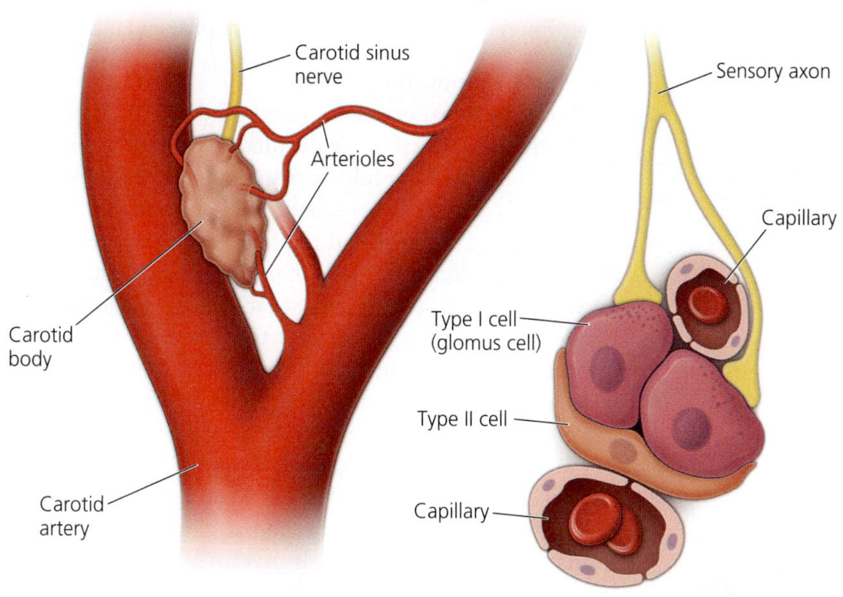

Figure 7.17 Sensing blood chemistry with glomus cells. The carotid body is located where the carotid artery divides into internal and external branches. It is penetrated by relatively leaky capillaries. Type I cells in the carotid body, known as glomus cells, can sense blood oxygen levels. When depolarized, they release neurotransmitters onto nerve terminals.

changes in blood oxygenation and then pass this information on to neural circuits that adjust breathing (see Chapter 9). Specifically, the carotid body contains **glomus cells** (type I cells) that are depolarized when blood oxygen levels are low. The details of this process are unclear but likely include O_2-sensitive potassium channels that close when oxygen levels drop. Also unclear is the identity of the neurotransmitter that glomus cells release. These cells contain both dopamine and acetylcholine, but recent evidence suggests that they, just like taste cells, use ATP as their main neurotransmitter.

Type II cells in the carotid body (Figure 7.17) are not directly involved in sensing blood oxygen. They are often considered "supporting cells," but recent evidence suggests that they are actually stem cells capable of dividing and giving rise to additional glomus cells. This capacity for cell division is triggered under conditions of chronically low oxygen levels (hypoxia) as may occur when living at high altitudes. Under those conditions, carotid body size increases substantially.

BRAIN EXERCISE

What might be the adaptive significance of being able to increase the number of glomus cells when living with chronic hypoxia?

7.6 How Do We Sense Body Position and Movement?

Moving through the world in a coordinated way requires knowing where your limbs and other body parts are located and how they are moving relative to one another, as well as relative to the external world. The ability to sense the position and movement of your body parts relative to one another is called **proprioception** (*proprius* being Latin for "one's own"). In contrast, sensing movements through external space is the domain of the vestibular system.

Proprioception

To reach for an object or place one foot in front of the other, you need to know the position of your limbs. Some of this information comes in through the eyes, but you

A Golgi tendon organ

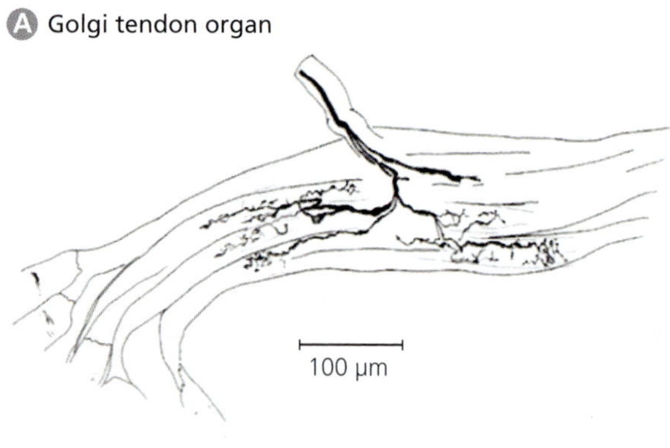

B Muscle spindles

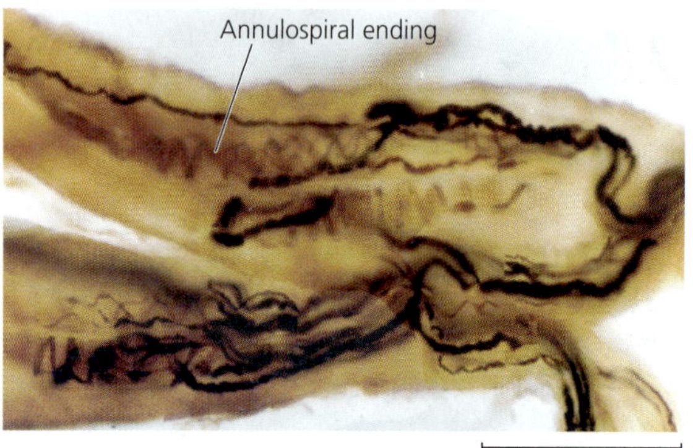

Figure 7.18 Sensors in tendons and muscles. Shown in (A) is a Golgi tendon organ, which consists of an axon that branches among the collagenous fibers of a tendon. Panel (B) depicts darkly stained annulospiral nerve endings inside two muscle spindles, which were rendered transparent. [From Banks et al., 2009]

can obviously move reasonably well in total darkness. For that, you rely on sensors in joints, tendons, and muscles.

Sensors in Tendons and Joints

Although mechanosensory nerve endings in limb joints could theoretically provide detailed information about limb position, inactivating **joint sensors** has relatively minor effects on an animal's movement. Furthermore, joint sensors often respond to multiple directions of joint rotation, implying that they don't encode limb position very precisely. Joint sensors also tend to fire only at rather extreme joint angles. Collectively, these data suggest that the major function of joint sensors in mammals is to protect against limb overextension or dislocation, not to provide detailed information about limb position.

Golgi tendon organs, named after Camillo Golgi (see Chapter 2), are found in the connective tissues that connect muscles to bone (i.e., tendons). They consist of myelinated axons that lose their myelin sheath as they enter the tendon and then branch among the tendon's collagen fibrils (Figure 7.18 A). As muscle fibers contract and pull on the fibrils, the tips of the axon branches are locally compressed, which opens mechanosensitive ion channels in the axon and triggers action potentials. Because of this arrangement, Golgi tendon organs convey information about muscle force (tension). Their functional significance remains somewhat obscure, but Golgi tendon organs are thought to be involved in regulating muscle force, which is what you must do, for example, when handling a delicate object.

Sensors in Muscle Spindles

Most skeletal muscles contain specialized structures called **muscle spindles**, which look like elongated seed pods sandwiched between the principal muscle fibers of skeletal muscle (see Chapter 8). Each muscle spindle is innervated by a few axons that lose their myelin sheath as they enter the spindle and terminate on slender (intrafusal) muscle fibers inside of the spindle. Some of the entering axons branch to form spiral-shaped endings, called **annulospiral endings**, that wrap around intrafusal fibers near their center (Figure 7.18 B). Others terminate off-center and do not form neat spirals; they form **flower-spray endings**.

Both types of muscle spindle endings are activated when the muscle is stretched. The annulospiral endings tend to respond most strongly during passive stretch (when someone pulls on your arm). The faster the stretch, the more they respond. Because of these response properties, annulospiral endings provide good information about the speed of muscle contraction. By contrast, the flower-spray endings are less sensitive to stretch velocity and encode mainly stretch magnitude. Both types of information are useful in proprioception: if you know the current length of all your muscles and the speed at which those lengths are changing, then you should be able to infer how your body is positioned and moving, assuming you also know some physics and musculoskeletal anatomy. Although the task may seem daunting, the nervous system performs it almost entirely subconsciously.

Central Targets of Muscle Sensors

Axons ending as flower sprays tend to be of the Aβ type and project to laminae III–V of the spinal cord, much as the touch and vibration-sensitive axons do. In contrast, annulospiral endings and Golgi tendon organs are associated with very thick (12–18 μm diameter) myelinated axons of the Aα type and project to deeper layers of the spinal

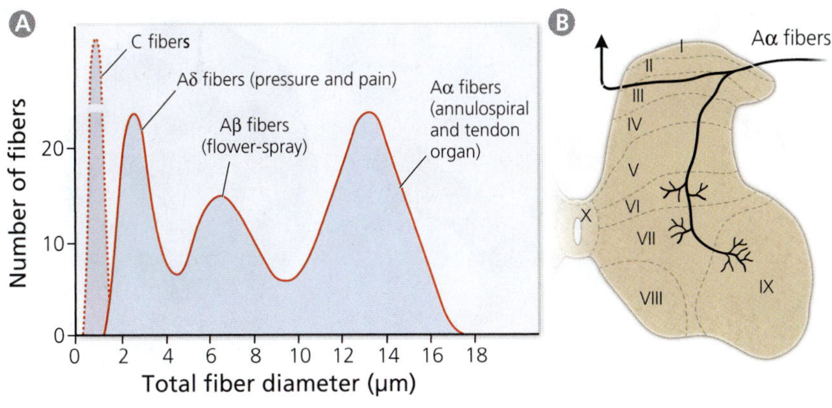

Figure 7.19 Sensory axons in a muscle nerve. The graph in (A) displays the spectrum of fiber diameters in a typical muscle nerve after all the motor axons were removed. The myelinated fibers form three peaks, corresponding to Aα, Aβ, and Aδ fibers. The dashed line shows that motor nerves also contain unmyelinated C fibers. The diagram in (B) shows that Aα axons tend to terminate in spinal laminae VI, VII, and IX. These axons also send a long branch through the dorsal columns to other parts of the spinal cord. [A after Boyd & Davey, 1968]

cord. Specifically, sensory Aα fibers terminate in Rexed's laminae VI, VII, and IX (Figure 7.19). The terminations in lamina IX are especially interesting because this region contains skeletal motor neurons. As you will learn in Chapter 10, these proprioceptive Aα fibers are links in vital reflex arcs that help maintain body posture.

Proprioceptive Aα fibers also send an axon branch into the dorsal column tracts, where they ascend with the touch- and vibration-sensitive axons we discussed earlier. However, the proprioceptive axons do not ascend all the way to the dorsal column nuclei. Instead, they terminate in two small structures in the spinal cord (called the accessory cuneate nucleus and Clarke's column), which then project to the ipsilateral cerebellum. These **spinocerebellar projections** are important for generating smooth and accurate limb movements.

BRAIN EXERCISE

If you have been lying still with your eyes closed for several minutes (e.g., shortly after waking up), can you feel where your hands or feet are located? Does moving those body parts help locate them in space? Why or why not?

Vestibular Sensors

Muscle spindle sensors can tell you about the location and movement of body parts relative to one another, but they provide no information about how the entire body is moving through space. On a roller coaster, for example, you may be sitting perfectly still yet moving through space along a tortuous trajectory. Even with your eyes closed, you can tell whether you are accelerating or decelerating, turning left or right, up or down, or tilting sideways. How are you tracking this whole body movement? The answer lies in the vestibular system, whose sensors are located adjacent to the cochlea (Figure 7.20) in a set of bony cavities called **vestibule** and **semicircular canals**. The fluid in these cavities is continuous with that in the cochlea and is, like cochlear endolymph, enriched in K^+ ions. The sensory cells of the vestibule and semicircular canals are very similar to cochlear hair cells, but they transduce different kinds of stimuli.

Sensing Head Tilt and Acceleration

Hair cells in the vestibule are clustered into two patches, or *maculae* (from the Latin word for "spot"). One macula sits in part of the vestibule that is called the **utricle**; the other lies within the nearby **sacculus**. Like cochlear hair cells, vestibular hair cells

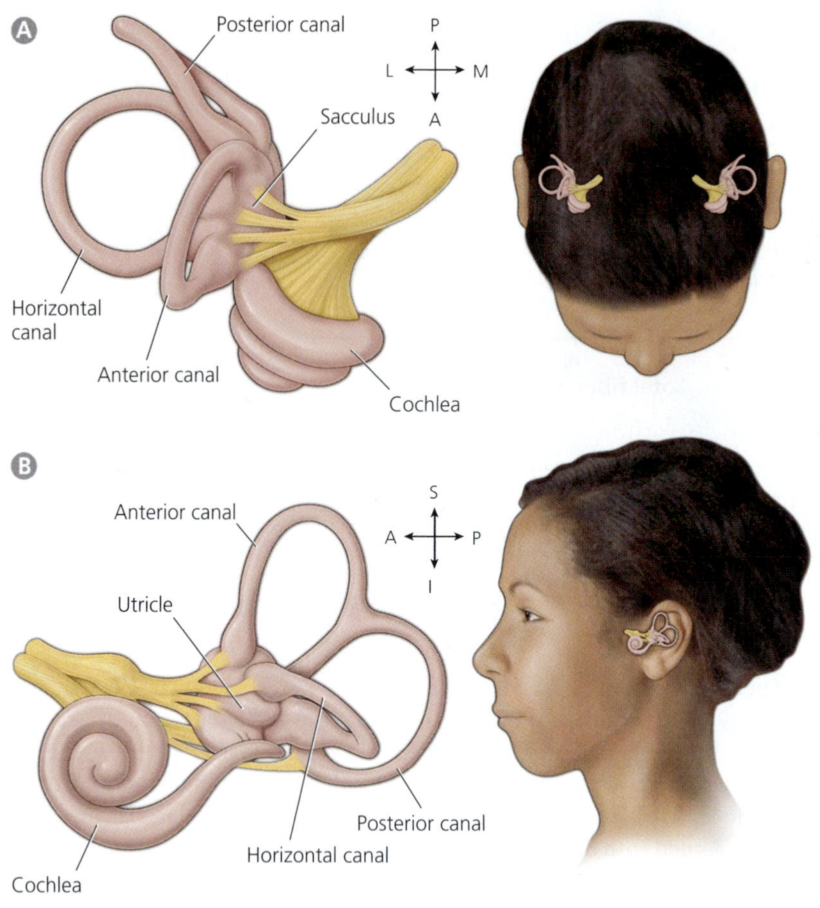

Figure 7.20 The vestibular apparatus. Panel (A) depicts the human semicircular canals, sacculus, and cochlea from a superior perspective. Panel (B) shows the vestibular apparatus from a lateral view. The horizontal, anterior, and posterior semicircular canals lie in three different, roughly orthogonal planes. The letters A, P, L, M, S, and I stand for anterior, posterior, lateral, medial, superior, and inferior, respectively. [After Ifediba et al., 2007]

have stereocilia of different lengths. In addition, vestibular hair cells have a long **kinocilium** adjacent to the tallest stereocilia (Figure 7.21). When the stereocilia are deflected toward this kinocilium, tip links open ion channels in the stereocilia. The open channels allow K^+ and Ca^{2+} entry, which causes membrane depolarization and neurotransmitter release. This is all quite similar to what happens in cochlear hair cells (except for the fact that cochlear hair cells lack kinocilia).

However, the vestibular hair cells sit on a substrate that does not vibrate in response to sound. Furthermore, the tips of the vestibular stereocilia are embedded in a dense, gelatinous membrane infused with thousands of calcium carbonate crystals called *otoconia* (ear dust). When this **otoconial membrane** moves sideways, the stereocilia bend. Importantly, only the hair cells in which the stereocilia bend toward the kinocilium are depolarized and release glutamate onto the axon terminals that innervate them (much as the cochlear hair cells that we discussed in Chapter 6). Because the orientation of the hair cells varies across the vestibular maculae, movements of the otoconial membrane in different directions cause different sets of hair cells to respond. Thus, the pattern of macular hair cell activation encodes the direction in which the otoconial membrane moves.

What causes the otoconial membranes to move? The answer is primarily inertia. Consider first the saccular macula. Its otoconial membrane lies parallel to the head's midsaggital plane (see Figure 1.3). If your head suddenly moves

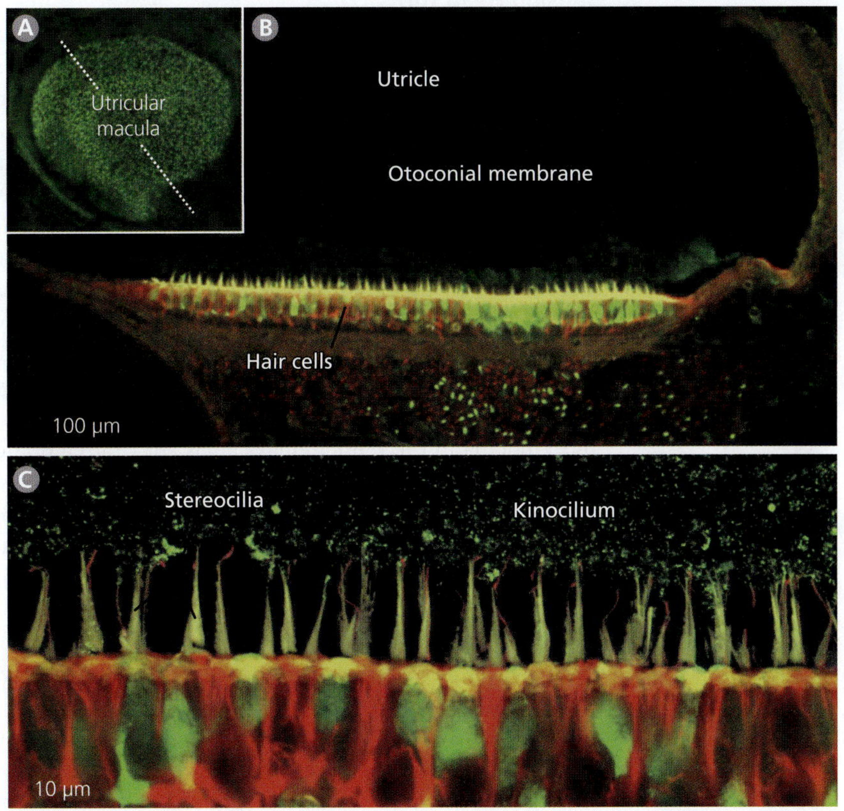

Figure 7.21 The utricle of the vestibular system. Panel (A) shows a top view of the utricular macula. Panel (B) depicts a transverse section through this macula (sectioned along the dashed line in A). The hair cell stereocilia are stained yellow. Their tips protrude into the otoconial membrane. A higher magnification view is shown in (C). In this image, you can see that each bundle of stereocilia is associated with a longer kinocilium (stained red). The cell bodies of some hair cells are stained green (with antibodies against calretinin); others remain unstained. [From Li et al., 2008]

forward, the otoconial membrane of the saccula lags behind because its inertial mass is greater than that of the surrounding fluid. This displacement of the membrane bends the saccular stereocilia and depolarizes a subset of saccular hair cells. Backward head movements bend the stereocilia in the opposite direction and therefore excite a different subset of saccular stereocilia. Other saccular hair cells respond to the sort of up or down movements you would experience in an elevator or a jump off a cliff. Importantly, saccular hair cells are activated mainly when the body is accelerating or decelerating and not when it is moving at a constant speed. This feature is a natural consequence of the physical fact that, at constant velocity, the otoconial membrane is no longer experiencing inertial lag.

The utricular macula is similar to the saccular macula except that it lies in the head's horizontal plane. Because of this difference in orientation, utricular hair cells respond to acceleration of the head in a different set of directions, namely, forward, backward, sideways, and anywhere in between. This much is relatively simple. However, utricular hair cells also encode head tilt (Figure 7.22). If the head is in its upright position, then the force of gravity pulling on the utricle's otoconial membrane runs parallel to the individual utricular stereocilia. In contrast, when the head is tilted, then the pull of gravity is no longer parallel to the stereocilia. This causes the stereocilia to be deflected in the direction of the tilt. Because head acceleration and head tilt activate utricular hair cells identically, head acceleration may be perceived as a head tilt. For example, when fighter jets take off, the pilots tend to feel as if their head is tilting backward, even when the head is stabilized by a headrest.

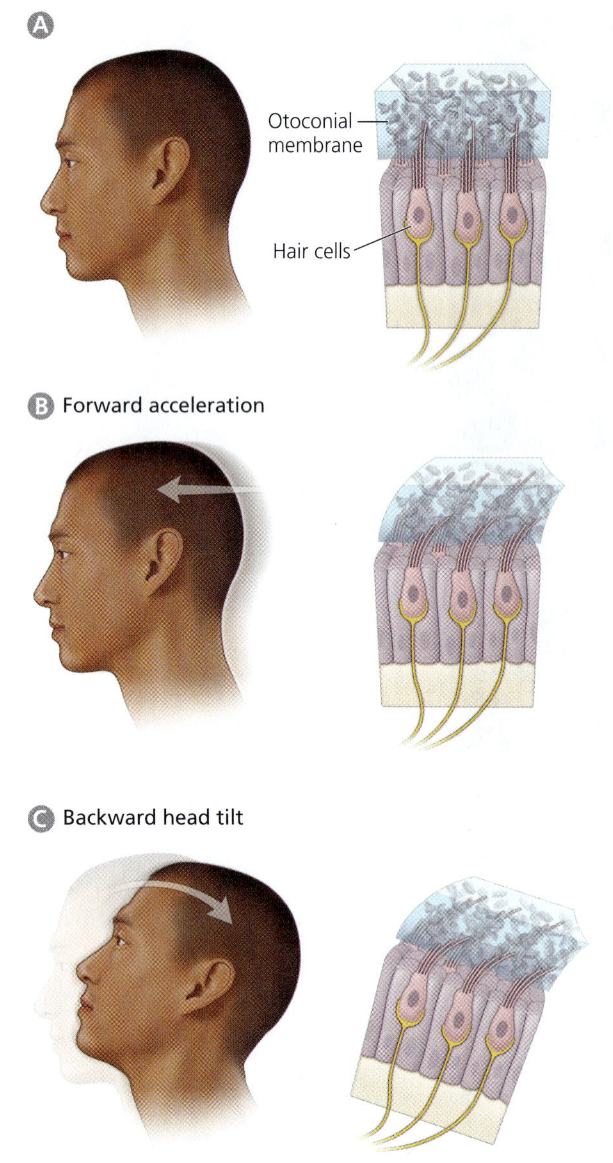

Figure 7.22 Activation of the utricle. When the head is in its normal position (A), the utricular hair cells are inactive. When the head accelerates forward (B), inertial forces displace the otoconial membrane backward, depolarizing the hair cells. The stereocilia bend in the same direction, however, when the head is tilted backward (C). Therefore, information from the utricle is insufficient to discriminate backward head tilt from forward acceleration.

Sensing Head Turns

Now let us turn to the largest part of the vestibular apparatus, the *semicircular canals*. As the name suggests, a semicircular canal consists largely of an endolymph-filled tube bent into a semicircle. The semicircle's open end is a more spacious cavity that, despite its less tubular shape, completes the circle (Figure 7.23). As you can see from Figure 7.20, each side of the head contains a set of three semicircular canals, each of which is oriented in a different plane. Moreover, the three planes are perpendicular to one another. Inside the tubular portion of each canal is a transverse ridge that is covered with hair cells. These hair cells are just like those of the utricular and saccular maculae except that they are not covered with an otoconial membrane. Instead, the stereocilia and kinocilia in the semicircular canals stick into a dome-shaped gelatinous mass called a **cupula**. Each cupula is so large that the canal's fluid cannot flow past it (Figure 7.23). However, because the cupula is flexible, fluid pressure can deflect it sideways (you may think of the cupula as a membrane that is stretched across each canal's interior). The deflection of the cupula then bends the stereocilia. Bending them toward the kinocilium causes the hair cells to depolarize and release glutamate.

What causes the fluid in a semicircular canal to exert pressure on the cupula? The answer is *angular acceleration* (changes in the velocity of head rotation). To understand how angular acceleration bends the cupula, conduct or imagine the following experiment: Fill a glass with water and sprinkle ground pepper on top. Now twist the glass around its vertical axis and observe the movement of the water with its floating particles. Because of rotational inertia, the water does not move with the glass, at least initially. The same phenomenon occurs inside the semicircular canals. When the head begins to rotate around its vertical axis, for instance, the fluid in the horizontal semicircular canals resists the rotation, presses on the cupula, and deflects the stereocilia. Because of how the hair cells are oriented (they all face the same direction within each canal, and the two sides of the body exhibit mirror symmetry), changing the speed of rightward head rotation activates hair cells only in the right horizontal canal. In contrast, leftward angular acceleration activates the hair cells in the left horizontal canal.

The anterior and posterior semicircular canals are similar to the horizontal canals but lie in different planes. If you think of the horizontal canals as encoding angular acceleration in the head's horizontal plane (around the head's vertical axis), then the anterior and posterior canals encode angular acceleration in the sagittal and coronal planes, respectively. By comparing activation patterns across all 6 canals, the central nervous system can determine quite precisely how the head is turning as long as the rotation includes a change in velocity (which it usually does). As you might expect, damage to one or more of your semicircular canals may lead to faulty inferences about your head's movement. Because of how the brain processes information coming from the left and right semicircular canals (see Chapter 10), such damage can make you feel as if your head were spinning endlessly. Fortunately, such nauseating vertigo is generally transient.

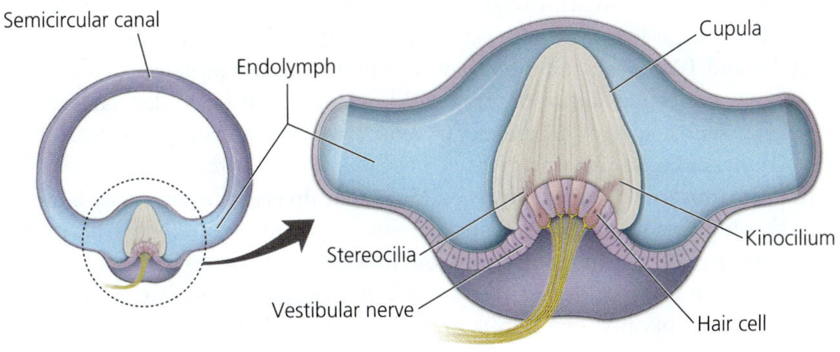

Figure 7.23 Hair cells in the semicircular canals. Each semicircular canal has a widened portion that contains a collection of hair cells whose stereocilia and kinocilia protrude into a large gelatinous mass called the cupula. When the cupula is bent by movement of the endolymph within the canal, stereocilia and kinocilia are deflected. Deflections toward the kinocilium cause the hair cells to increase their rate of transmitter release onto axons of the vestibular nerve.

Central Projections of the Vestibular Apparatus

The axons innervating hair cells of the vestibular apparatus all travel to the brain in the eighth cranial nerve (VIII). Because this nerve also contains auditory fibers from the cochlea, it is often called the **vestibulocochlear nerve**. The vestibular axons terminate mainly in the **vestibular complex**, which lies in the dorsal medulla and has several subdivisions. Within the vestibular complex, axons from the saccule, utricle, and semicircular canals terminate in separate, although overlapping, areas. Postsynaptic neurons in the vestibular complex project to a wide variety of targets including the spinal cord and brainstem areas that control eye movements. These projections tend to stabilize the head and eyes when the body moves in relation to the outside world.

BRAIN EXERCISE

Why do you feel an elevator's movement only for the first few seconds after the doors close and just before the doors open again at the destination?

7.7 What Are Some Common Themes of Contact Sensor Organization?

In this chapter, we have described a wide variety of sensors, most of which you probably had never thought about. Neuroscientists have amassed enormous amounts of information about these sensors, and remembering even the major points is challenging. Therefore, it is useful to ask, as we did at the end of Chapter 6, whether there are some common themes or general principles that can help us organize all this knowledge.

Variability in Sensor Range and Sensitivity

Most of the sensors we have discussed, both in this chapter and in Chapter 6, respond only to a specific subset of all possible stimuli. Even within a sensory modality, they generally respond to just a narrow range of stimuli. The various temperature sensors, for example, are activated over a limited range of temperatures. The various nerve endings in the skin respond to different types of stimuli, ranging from low-frequency vibration to light touches. The gustatory system, too, is characterized by a diverse set of sensors, each specialized to detect a different class of molecules.

The selectivity of an axon or sensory cell for a subset of stimuli is often based on the selectivity of the stimulus-sensing molecules that the cells or axons express. Taste cells, for example, derive their selectivity from the combination of taste receptor molecules that they express. Similarly, the range of temperatures over which an axon is depolarized depends on which temperature-sensitive ion channels that axon expresses. In both cases, the stimulus-sensing molecules are members of gene families (TRP, T1, and T2 genes); and the variation in their stimulus preferences is the result of divergent molecular evolution within those gene families. Molecular evolution also explains most species differences in sensory capacities, such as the inability of cats and chickens to sense sweets.

In mechanosensory axons, however, molecular divergence plays only a minor a role in shaping sensor selectivity. Instead, mechanosensory axons tend to derive their stimulus preferences from the environment in which they are embedded. For example, the nerve endings inside Meissner's and Pacinian corpuscles respond to different types of stimuli because the non-neural cells encapsulating those nerve endings affect which mechanical signals get through to the nerve terminals. Similarly, the annulospiral nerve endings in muscle spindles derive their response properties from the manner in which they are coiled around the intrafusal muscle fibers. Another example is provided by cochlear and vestibular hair cells, which are similar to one another but respond to different sorts of stimuli because of their physical environment. All these cells just sense the stretching of their cell membrane, which contains stretch-sensitive ion channels. The molecular identity of these mechanosensory channels remains hotly debated, but it is pretty clear that the range of mechanosensory stimuli to which the cells respond has more to do with the extracellular environment than with the molecules those cells express.

Another important principle of sensor organization is that sensors tend to adapt to steady stimulation. This is true also for the sensors we covered in this chapter, although some adapt more rapidly than others. Annulospiral endings, for example, adapt more quickly than flower-spray endings to steady muscle stretch, but both types of sensors respond more strongly to a changing stimulus than to a sustained one. Even pain sensors adapt, although usually more slowly than we'd like. These observations are consistent with the idea that nervous systems evolved to sense what is most important to the organism. Changes in the environment tend to be more important than constancies because the latter can be taken for granted, whereas the changes may be harbingers of opportunity or threat.

Sensor sensitivity can also be modulated by other factors. Tissue inflammation, for example, makes the inflamed region more sensitive to touch, often painfully so. Conversely, pain may be reduced by analgesic drugs, by rubbing the affected skin, or by electrical stimulation of touch-sensitive axons. These changes in pain sensitivity may involve changes in the sensors themselves or changes in the brain. Either way, this kind of modulation illustrates nicely that our sensory systems do not encode information truthfully. Sensation is influenced by many different factors, only one of which is objective reality (recall Box 7.3: Cannabinoid Effects on Food Intake).

Labeled Lines and Sensory Maps

As you may recall from Chapter 6, the "labeled line" hypothesis of neural coding states that activity in an axon or neural pathway always represents a specific type of information, regardless of how active the pathway is. The sensory systems we have discussed in this chapter also form some labeled lines. For example, the axons carrying information about noxious stimuli are distinct from the axons carrying information about touch and vibration, and they project to different targets in the brain and spinal cord. Indeed, most of the sensors we have discussed project to different targets in the brain, forming a multitude of separate pathways. However, not every stimulus has its own dedicated line. For example, painful heat, spice, and acidity all activate nociceptive axons expressing TRP-V1, making those stimuli difficult to discriminate from one another. Similarly, a wide variety of bitter tasting compounds all activate the same set of bitter-sensing taste cells. In this case, the "label" for the activated neural line is simply, "Warning! Something bitter was sensed."

Another twist on the labeled line concept is that several labeled lines may converge at very early stages of sensory processing. For example, sensory axons carrying information about salty and sour tastes sometimes converge onto single neurons in nucleus tractus solitarius, the first station in the brain's gustatory pathway. Such convergence does not mean that the "lines" for the five basic tastes become completely blurred within the brain, but it shows that integration across the labeled lines occurs even at surprisingly low levels of the brain.

Within each sensory modality, neural pathways tend to be topographically organized, with neighboring cells at one stage of the pathway projecting to neighboring cells at the next stage. This is true of the visual and auditory pathways we discussed in Chapter 6, and it applies quite well to the somatosensory pathway for touch and vibration of the skin. However, some of the other senses we discussed in this chapter exhibit at best an imprecise topography. Information about temperature, for example, is not obviously mapped inside the brain. That being said, most of the senses covered in this chapter have not been studied in as much detail as vision, hearing, or touch. It is possible, therefore, that central maps for some of these senses remain undiscovered.

BRAIN EXERCISE

Which types of sensors use mechanosensory ion channels to transduce sensory stimuli?

SUMMARY

Section 7.1 - The skin contains a variety of encapsulated nerve endings that are specialized for detecting touch and vibration of the skin.

- Axons innervating Merkel cells respond to skin indentation. The axons inside of Meissner's and Pacinian corpuscles respond to low- and high-frequency vibrations, respectively.
- Touch and vibration information from the body is carried by relatively thick myelinated axons of the Aβ type, which enter the spinal cord through spinal nerve dorsal roots and terminate in Rexed's laminae II–V.

Section 7.2 - Pain can be triggered by extreme mechanical stimulation, heat, or other stimuli that are capable of causing tissue damage.

- Nociceptive information is carried by C and Aδ fibers, which terminate in laminae I and II of the spinal cord; from there it ascends in the spinothalamic tract. Visceral pain is often localized incorrectly.
- Pain can be reduced by opioids and by activating touch and vibration sensors that compete with the pain signals for access to consciousness. Inflammation heightens pain sensitivity (hyperalgesia).

Section 7.3 - Temperature sensation involves multiple members of the transient receptor potential (TRP) family of ion channels. Several TRP channels also respond to plant-derived compounds.

- TRP channels are expressed by some C and Aδ fibers. TRPV1 and TRPV2 are activated at very hot temperatures; TRPA1 at very cold temperatures; and TRPV3, V4, and M8 at moderate temperatures.
- Capsaicin lowers the temperature at which TRPV1 channels open. Menthol raises the temperature at which TRPM8 channels open. Mustards and horseradish open TRPA1.

Section 7.4 - Taste buds contain taste cells that express various G protein-coupled taste receptor genes and release ATP onto sensory axons that project to nucleus tractus solitarius.

- Taste cells that express compound T1R1/T1R3 receptors sense umami, whereas those that express T1R2/T1R3 sense sweet. Taste cells expressing a variety of T2 receptors respond to various bitter compounds.
- Sensing that something is sour involves the acid-sensing TRP channel PKD-2L1. Sensing salt and fat involve epithelial sodium channels and CD36, respectively.

Section 7.5 - Various sensors provide the nervous system with information about the body's internal condition.

- Tissue acidity, often caused by high CO_2, can be sensed by ASICs and TRPV1.
- Blood oxygen levels are sensed by glomus cells in the carotid body, which use ATP as a neurotransmitter.

Section 7.6 - Proprioception is the sense of how our body parts are positioned and moving relative to one another.

The vestibular system senses how our head is moving through space.

- Golgi tendon organs sense the force of muscular contractions, whereas nerve endings inside muscle spindles sense muscle stretch. This information is sent to the intermediate zone and ventral horn of the spinal cord.
- Hair cells in the utricle and sacculus sense head tilt and linear acceleration; those in the semicircular canals sense head rotation. Vestibular hair cells project through cranial nerve VIII to the vestibular complex.

Section 7.7 - The sensors we discussed in this chapter respond to a limited range of stimuli and do so with variable degrees of sensitivity. Their central projections tend to be topographic, although the topography is sometimes rough.

Box 7.1 - After a primary chickenpox infection, the varicella-zoster virus hides out in dorsal root ganglia. When reactivated, the virus causes shingles (zoster), which consists of a painful rash that respects dermatome boundaries.

Box 7.2 - Neurogenic inflammation occurs when noxious stimuli trigger action potentials that invade peripheral axon collaterals, which then release substance P and CGRP. These peptides amplify the inflammatory response.

Box 7.4 - Both exogenous and endogenous cannabinoids increase your sensitivity to sweets. They also help promote the eating of fatty foods.

Box 7.4 - Naked mole rats and diverse hibernating mammals have adapted to living with high levels of CO_2 by modifying their $Na_v1.7$ channels in such a way that they are difficult to activate when tissue acidity is high.

KEY TERMS

ADDITIONAL READINGS

7.1 - Touch and Vibration

Johnson KO. 2001. The roles and functions of cutaneous mechanoreceptors. *Curr Opin Neurobiol* **11**:455–461.

Lumpkin EA, Caterina MJ. 2007. Mechanisms of sensory transduction in the skin. *Nature* **445**:858–865.

Mills LR, Diamond J. 1995. Merkel cells are not the mechanosensory transducers in the touch dome of the rat. *J Neurocytol* **24**:117–134.

Scheibert J, Leurent S, Prevost A, Debrégeas G. 2009. The role of fingerprints in the coding of tactile information probed with a biomimetic sensor. *Science* **323**:1503–1506.

Valle MED, Cobo T, Cobo JL, Vega JA. 2012. Mechanosensory neurons, cutaneous mechanoreceptors, and putative mechanoproteins. *Microsc Res Tech* **75**:1033–1043.

7.2 - Pain

Cox J, Reimann F, Nicholas A, Thornton G, Roberts E, Springell K, et al. 2006. An SCN9A channelopathy causes congenital inability to experience pain. *Nature* **444**: 894–898.

Dubin AE, Patapoutian A. 2010. Nociceptors: the sensors of the pain pathway. *J Clin Invest* **120**:3760–3772.

Liu T, Ji R-R. 2013. New insights into the mechanisms of itch: are pain and itch controlled by distinct mechanisms? *Pflugers Arch* **465**:1671–1685.

Sandkuhler J. 2009. Models and mechanisms of hyperalgesia and allodynia. *Physiol Rev* **89**:707–758.

7.3 - Temperature

Caterina MJ, Schumacher MA, Tominaga M, Rosen TA, Levine JD, Julius D. 1997. The capsaicin receptor: a heat-activated ion channel in the pain pathway. *Nature* **389**: 816–824.

Huang J, Zhang X, McNaughton PA. 2006. Modulation of temperature-sensitive TRP channels. *Sem Cell Dev Biol* **17**:638–645.

Xu H, Delling M, Jun JC, Clapham DE. 2006. Oregano, thyme and clove-derived flavors and skin sensitizers activate specific TRP channels. *Nat Neurosci* **9**:628–635.

7.4 - Taste

Adler E, Hoon MA, Mueller KL, Chandrashekar J, Ryba NJ, Zuker CS. 2000. A novel family of mammalian taste receptors. *Cell* **100**:693–702.

Baldwin MW, Toda Y, Nakagita T, O'Connell MJ, Klasing KC, Misaka T, Edwards SV, Liberles SD. 2014. Evolution of sweet taste perception in hummingbirds by transformation of the ancestral umami receptor. *Science* **345**:929–933.

Chandrashekar J, Hoon M, Ryba N, Zuker C. 2006. The receptors and cells for mammalian taste. *Nature* **444**:288–294.

Frank ME, Lundy RF, Contreras RJ. 2008. Cracking taste codes by tapping into sensory neuron impulse traffic. *Prog Neurobiol* **86**:245–263.

Bachmanov AA, Bosak NP, Floriano WB, Inoue M, Li X, Lin C, et al. 2011. Genetics of sweet taste preferences. *Flavour Fragr J* **26**:286–294.

Chandrashekar J, Yarmolinsky D, Buchholtz von L, Oka Y, Sly W, Ryba NJP, Zuker CS. 2009. The taste of carbonation. *Science* **326**:443–445.

Degrace-Passilly P, Besnard P. 2012. CD36 and taste of fat. *Curr Opin Clin Nutr Metab Care* **15**:107–111.

Hayes JE, Feeney EL, Allen AL. 2013. Do polymorphisms in chemosensory genes matter for human ingestive behavior? *Food Qual Prefer* **30**:202–216.

Kinnamon SC, Finger TE. 2013. A taste for ATP: neurotransmission in taste buds. *Front Cell Neurosci* **7**:264.

7.5 - Physiological Condition

Powley TL, Phillips RJ. 2002. Musings on the wanderer: what's new in our understanding of vago-vagal reflexes? I. morphology and topography of vagal afferents innervating the GI tract. *Am J Physiol Gastrointest Liver Physiol* **283**: G1217–G1225.

Kweon H-J, Suh B-C. 2013. Acid-sensing ion channels (ASICs): therapeutic targets for neurological diseases and their regulation. *BMB Rep* **46**:295–304.

Lopez-Barneo J, Ortega-Saenz P, Pardal R, Pascual A, Piruat JI. 2008. Carotid body oxygen sensing. *Eur Respir J* **32**: 1386–1398.

Mimee A, Smith PM, Ferguson AV. 2013. Circumventricular organs: targets for integration of circulating fluid and energy balance signals? *Physiol Behav* **121**:96–102.

7.6 - Body Position and Movement

Highstein SM, Rabbitt RD, Holstein GR, Boyle RD. 2005. Determinants of spatial and temporal coding by semicircular canal afferents. *J Neurophysiol* **93**:2359–2370.

Ovalle WK, Dow PR, Nahirney PC. 1999. Structure, distribution and innervation of muscle spindles in avian fast and slow skeletal muscle. *J Anat* **194**:381–394.

7.7 - General Principles

He C, Fitzpatrick DA, O'Halloran DM. 2013. A comparative study of the molecular evolution of signalling pathway members across olfactory, gustatory and photosensory modalities. *J Genet* **92**:327–334.

Boxes

Dib-Hajj SD, Yang Y, Black JA, Waxman SG. 2013. The Na(V)1.7 sodium channel: from molecule to man. *Nat Rev Neurosci* **14**:49–62.

Dipatrizio NV, Astarita G, Schwartz G, Li X, Piomelli D. 2011. Endocannabinoid signal in the gut controls dietary fat intake. *Proc Natl Acad Sci U S A* **108**:12904–12908.

Downs MB, Laporte C. 2011. Conflicting dermatome maps: educational and clinical implications. *J Orthop Sports Phys Ther* **41**:427–434.

Gilden D, Mahalingam R, Nagel MA, Pugazhenthi S, Cohrs RJ. 2011. Review: the neurobiology of varicella zoster virus infection. *Neuropathol Appl Neurobiol* **37**:441–463.

Harrison S, Geppetti P. 2001. Substance P. *Int J Biochem Cell Biol* **33**:555–576.

Smith ESJ, Omerbašić D, Lechner SG, Anirudhan G, Lapatsina L, Lewin GR. 2011. The molecular basis of acid insensitivity in the African naked mole-rat. *Science* **334**: 1557–1560.

Soria-Gómez E, Bellocchio L, Reguero L, Lepousez G, Martin C, Bendahmane M, et al. 2014. The endocannabinoid system controls food intake via olfactory processes. *Nat Neurosci* **17**:407–415.

Williamson DJ, Hargreaves RJ. 2001. Neurogenic inflammation in the context of migraine. *Microsc Res Tech* **53**:167–178.

Using Muscles and Glands

CHAPTER 8

CONTENTS

FEATURES

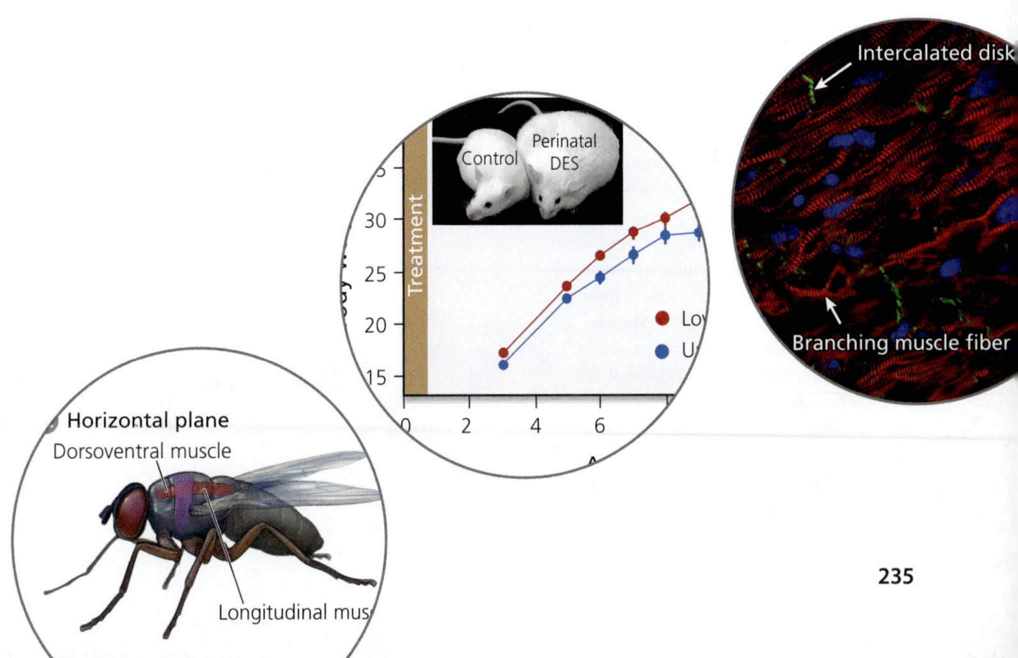

As we discussed in Chapter 5, a disembodied brain would quickly die. Even if one could provide such a brain-in-a-vat with protection, nutrients, and sensors to perceive the outside world, it would remain incapable of doing anything useful. To have a purpose, the nervous system needs some way of changing its environment. That is, it needs some **effectors**. The most obvious effectors in our bodies are the muscles. Although heart and intestinal muscles can contract independently of the CNS, all muscles are subject to some neural control. Thus, modulation of muscular activity is the major and most obvious way in which the nervous system can do things in the world. Besides working with muscles, the nervous system can change the body in more subtle ways by controlling glandular secretions. These include milk from mammary glands, sweat, saliva, tears, and gastric secretions as well as a variety of hormones circulating in the blood. Although hormone effects are less obvious than muscle contractions, they are quite powerful and often long lasting.

We begin this chapter with a review of the three major types of muscle: skeletal, cardiac, and smooth. All three have features in common, such as the basic contractile machinery, but they also differ in important respects. Because this is not a book on muscle physiology, many details are omitted. Instead, we focus on the control of muscle contraction by motor neurons. More complex aspects of neuromuscular control are discussed in Chapters 9 and 10. In the second half of this chapter, we focus on glands, especially the hormone-secreting endocrine glands. This, too, is a vast topic that we cannot discuss in much detail. Instead, we focus mainly on the pituitary gland, which is controlled by neurons in the hypothalamus and, in turn, controls the production and release of hormones that affect the brain, as well as other parts of the body.

8.1 How Do Neurons Control Skeletal Muscles?

The best known type of muscle is **skeletal muscle**. It moves our limbs and jaws, eyeballs, eyelids, facial skin, tongue, diaphragm, and vocal folds. In total, the human body contains about 320 bilateral pairs of skeletal muscles. Each of them contains many individual muscle fibers, bundled into fascicles (Figure 8.1). Individual skeletal muscle fibers are cylindrical, unbranched, and arranged in parallel. Each muscle fiber develops through the fusion of several precursor cells, which is why adult skeletal muscle fibers contain multiple cell nuclei and can be quite long (>30 cm in some thigh muscles). Because individual muscle fibers comprise multiple fused cells, biologists refer to the membrane surrounding each muscle fiber as **sarcolemma** (from the Greek words for "flesh" and "husk"), rather than plasmalemma (cell membrane).

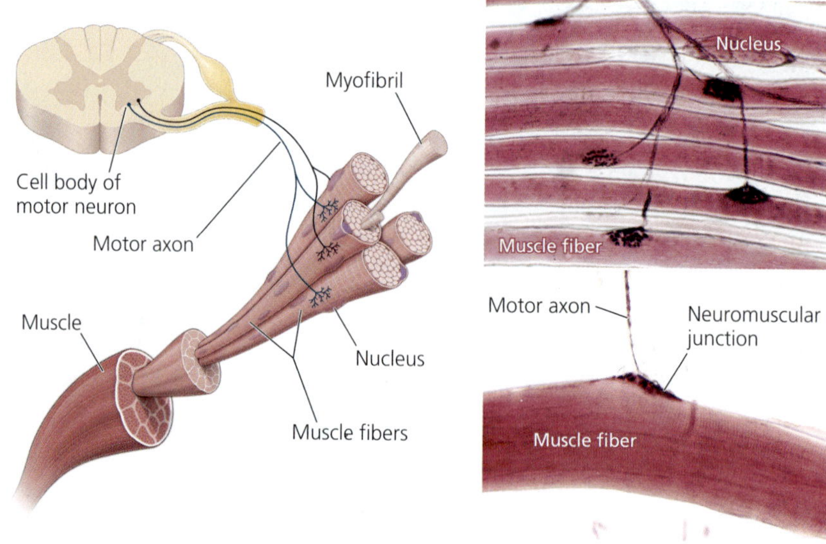

Figure 8.1 Skeletal muscle and its innervation. Shown on the left are two motor neurons in the spinal cord that innervate distinct sets of skeletal muscle fibers. Each muscle contains several bundles (fascicles) of muscle fibers, and each muscle fiber contains numerous myofibrils. Shown on the right are individual motor axons (brown). Each muscle fiber is innervated by only one axon, but each axon innervates multiple muscle fibers (top right). The large synapses between motor axons and muscle fibers are called neuromuscular junctions (bottom right). [Photographs courtesy of Thomas Caceci]

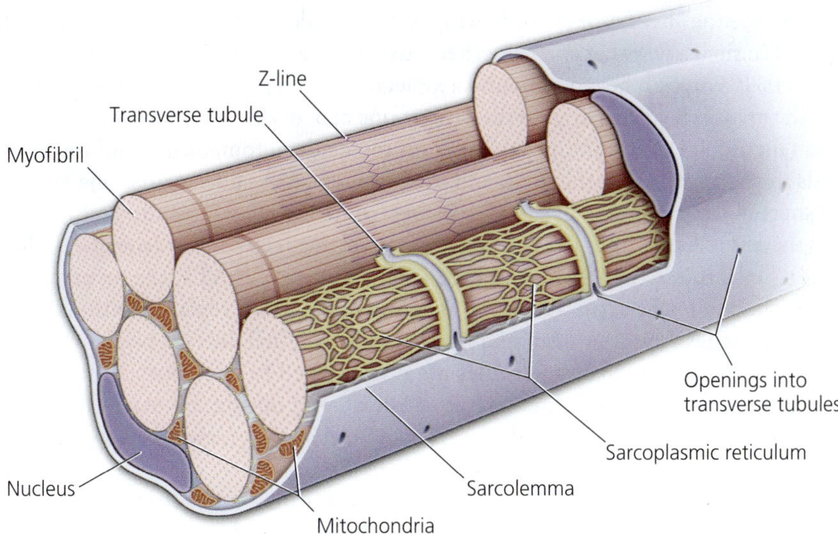

Figure 8.2 A muscle fiber's internal structure. In this cutaway view of a single muscle fiber, you can see two cell nuclei and eight myofibrils, arranged in parallel. Each myofibril consists of multiple sarcomeres, arranged in series. Surrounding the individual myofibrils is a dense mesh of sarcoplasmic reticulum (yellow). Transverse tubules (T-tubules) extend from small openings in the sarcolemma deep into the muscle fiber where they contact the sarcoplasmic reticulum.

Similarly, they refer to a muscle fiber's endoplasmic reticulum as **sarcoplasmic reticulum**.

The Contractile Machinery

An intriguing feature of skeletal muscle fibers is that they look striped under the microscope, even without special stains. Because of these stripes, or striations, skeletal muscle is classified as a type of **striated muscle** (cardiac muscle is also striated). To understand why skeletal muscle is striped, you have to know that each muscle fiber contains numerous slender **myofibrils**, all lying parallel to one another (Figures 8.1 and 8.2). Each myofibril is striped, and the stripes of neighboring myofibrils are all aligned (Figure 8.3A). Therefore, the entire muscle fiber looks striated. Why are the myofibrils striped? Because they contain a highly regular arrangement of filamentous molecules, notably actin and myosin, that repeat at regular intervals (every 2–3 μm). Each repeating unit is called a **sarcomere**. Each sarcomere is separated from its neighbors by a dense zig-zagging line called the **Z-line** (Figure 8.3B). Thin **actin** filaments emerge from these Z-lines at right angles, and interleaved between them are thick filaments of **myosin**. Seen through a microscope, the Z-lines and the regions where the actin and myosin filaments overlap look darker than the adjoining regions. This creates the striations.

Sliding Filament Model

When a skeletal muscle contracts, individual sarcomeres shorten by as much as 60%. Observation of muscle fibers during a contraction reveals that the myosin-containing bands remain constant in width, whereas the intervening (myosin-free) bands become narrower. This observation led to the **sliding filament model** of muscular contraction (Figure 8.3). According to this model, the alternating actin and myosin (thin and thick) filaments within a sarcomere slide past one another when a muscle fiber changes length.

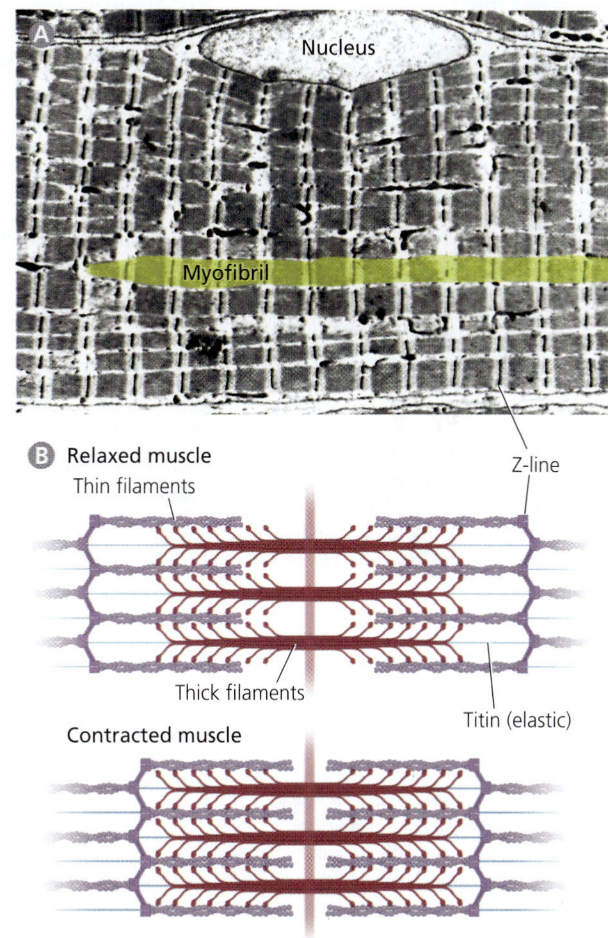

Figure 8.3 The sliding filament model. The electron micrograph in (A) is a longitudinal section through a skeletal muscle fiber showing one of the fiber's many cell nuclei and multiple myofibrils (one is shaded yellow). The diagram in (B) zooms in on one sarcomere. According to the sliding filament model, thin actin filaments slide past thick filaments of myosin. [A from Young and Heath, 2000]

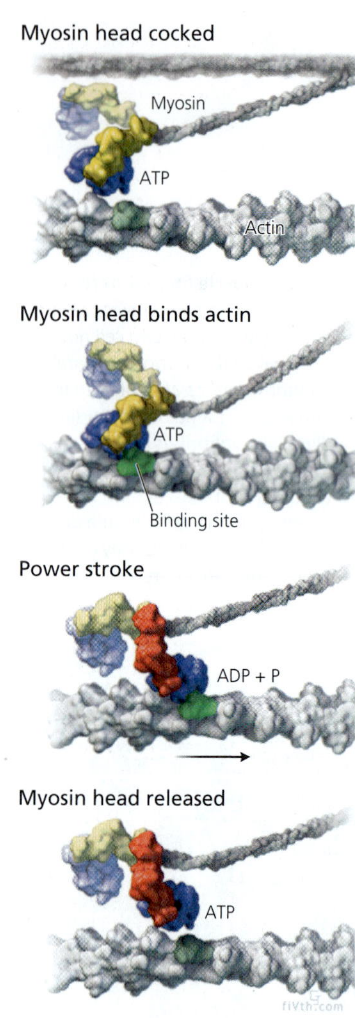

Figure 8.4 Actin-myosin interactions. When the head of a myosin molecule (colored) binds to an actin filament, the myosin changes its conformation so that the angle between the myosin head and the rest of the myosin decreases. The energy to generate this bending force comes from the release of a phosphate from ATP (adenosine triphosphate), creating ADP (adenosine diphosphate). To release the myosin head from the actin binding site, the ADP on the myosin head must be replaced by a new molecule of ATP. Each myosin molecule has two heads that act independently of one another; the second head is illustrated in paler colors. [From Vale & Milligan, 2000]

As actin and myosin filaments slide past one another, they interact. These interactions involve the "heads" of individual myosin molecules that stick out from the thick filament in such a way that the **myosin heads** come very close to the adjoining actin filaments. If a myosin head has bound a molecule of ATP, and if Ca^{2+} levels are high, then the myosin head can bind to the actin at one of many binding sites along the filament (Figure 8.4). On binding, the myosin molecule automatically and forcefully bends in such a way that the actin and myosin molecules slide against one another and shorten the sarcomere by a few nanometers. As the myosin bends, an inorganic phosphate is released from the bound ATP, converting it to ADP (adenosine diphosphate). Until this ADP is replaced with a new molecule of ATP, the myosin head remains bound to the actin. However, as soon as the ADP is replaced with ATP, the myosin detaches from the actin and returns to its original, less crooked shape. Because the actin and myosin molecules are not attached to one another during this second shape change, the Z-lines are not pushed apart during this stage (the sarcomere does not lengthen). Instead, the myosin head swings forward freely, ready to begin another contractile power stroke.

The repetition of these molecular interactions is called **cross-bridge cycling**. If the ends of the muscle are free to move, then the cross-bridge cycling shortens the muscle and generates a movement. If the ends of the muscle are fixed, then the actin–myosin interactions result in an **isometric contraction**, during which the muscle pulls on its attachment sites but does not change in length. Consider what happens when you make a fist and then clench it further: Bending the fingers to form a fist involves cross-bridge cycling in the muscle fibers. In contrast, the contractions that tighten a fully formed fist are primarily isometric.

Muscle Fiber Types

The contractile mechanisms we just discussed exist in all skeletal muscles. However, most animals have multiple myosin genes; and muscle fibers vary in which myosin genes they express. Different **myosin isoforms**, in turn, endow muscle fibers with different contractile properties. Because of these molecular differences, some muscle fibers contract more quickly than others, although the faster contractions tend to generate less force. The rapidly contracting "fast twitch" fibers predominate in small muscles controlling fast movements such as the muscles that move our eyes or vocal cords. Slow twitch fibers are more common in large muscles that tend to stay contracted for long periods of time, such as the antigravity muscles in our legs (see Chapter 10).

Muscle fibers also vary in **myoglobin** content. Myoglobin is similar to hemoglobin and ferries oxygen from the blood to the mitochondria. Because myoglobin is red, muscle that contains a high density of myoglobin (as well as many mitochondria and capillaries) is commonly referred to as dark meat. As long as oxygen is available, this kind of muscle derives most of its ATP from oxidative phosphorylation. In contrast, white meat (muscle with little myoglobin and few mitochondria) runs mainly on anaerobic glycolysis. Although glycolysis generates ATPs more rapidly than oxidative phosphorylation, it is much less efficient (2 vs. 30 ATPs) and therefore much more limited in how much energy it can provide. As a result, the muscle fibers in white meat can contract rapidly and forcefully but then fatigue within a couple of minutes.

Given this information, you may wonder why chicken breast meat is white. Wouldn't you want flight muscles to be fatigue resistant? For most birds the answer is yes. However, chickens rarely fly, and when they do, they do so for short distances. In ducks and geese, which fly for longer durations, breast meat is dark.

Excitation-contraction Coupling

Individual muscle fibers are innervated by the axons of motor neurons whose cell bodies are located either in the ventral (anterior) horn of the spinal cord or in the medulla. The synapse between a motor axon and a skeletal muscle fiber is called the **neuromuscular junction** (see Figure 8.1).

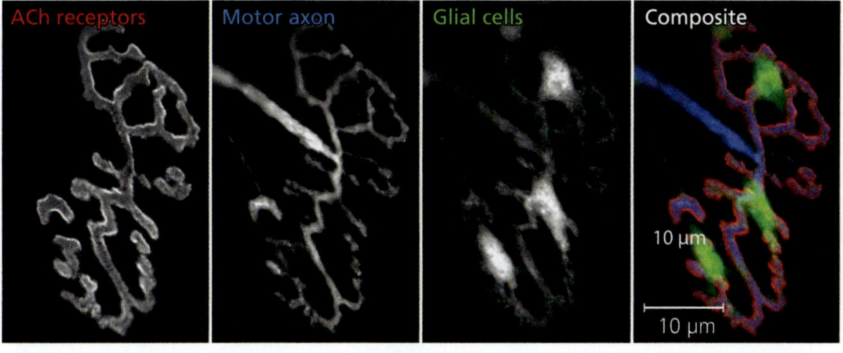

Figure 8.5 Acetylcholine receptors and glial cells at the neuromuscular junction. Shown here is a single neuromuscular junction from a transgenic mouse that expresses green fluorescent protein in all peripheral glial cells and cyan fluorescent protein in all motor axons (middle panels). The left-most image shows the distribution of postsynaptic acetylcholine (ACh) receptors, as revealed by binding with fluorescent alpha-bungarotoxin. The image on the right is a pseudocolored overlay of the other three images. [Images taken by Yi Zuo in Dr. Wes Thompson's lab at UT Austin]

The Neuromuscular Junction

With an average adult diameter of ~15 µm, a neuromuscular junction is much larger than most ordinary synapses. The axon terminal on the presynaptic side of a neuromuscular junction in rats typically contains 150,000–300,000 synaptic vesicles filled with the neurotransmitter acetylcholine. When an action potential enters the axon terminal, roughly 0.1% of these vesicles release their contents into the synaptic cleft. The released acetylcholine then diffuses to the postsynaptic side of the synapse, which is referred to as the *motor end plate*. The sarcolemma (muscle fiber membrane) at a motor endplate is studded with acetylcholine receptors (Figure 8.5). When acetylcholine binds to these receptors, they allow Na^+ and other positive ions to flow into the muscle fiber, thus depolarizing the sarcolemma. This depolarization, known as the **end plate potential**, opens nearby voltage-gated sodium channels, which then allow even more Na^+ ions to flow into the muscle fiber. At normal adult neuromuscular junctions, this ion influx almost always triggers a postsynaptic action potential. Most other synapses are not this reliable.

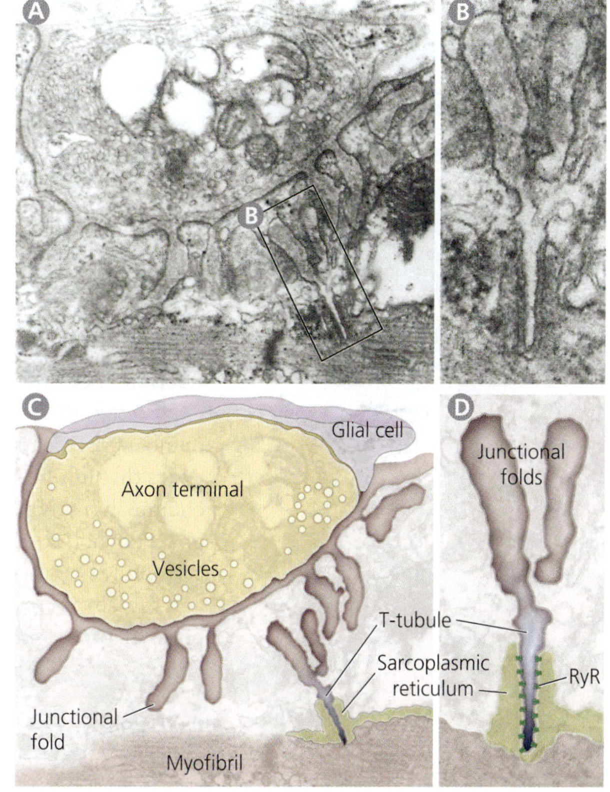

Figure 8.6 Section through a neuromuscular junction. The images in (A) and (B) are electron micrographs of a section through a neuromuscular junction in a mouse. Panels (C) and (D) indicate the various elements that can be seen in A and B, respectively. Junctional fold membranes are continuous with the T-tubules, which are linked to the sarcoplasmic reticulum through ryanodine receptors (RyR). [A and B from Dauber & Meister, 1986]

Spreading the Depolarization

Muscle action potentials begin at the motor end plate but spread from there across the muscle's sarcolemma. They also spread into the interior of a muscle fiber along deep folds in the sarcolemma just underneath each neuromuscular junction (Figure 8.6). These folds are continuous with a complex network of membranous tubes, called the **T-tubule system**, that penetrates deep into the muscle fiber. The T-tubules, in turn, contact the sarcoplasmic reticulum, which functions primarily as an internal calcium store. As an action potential travels down the T-tubules, it opens a special type of voltage-sensitive Ca^{2+} channel called the **ryanodine receptor**. These channels physically link the T-tubules to the sarcoplasmic reticulum (Figure 8.6 D) and ensure that action potentials conducted through the T-tubules cause Ca^{2+} to be released from the sarcoplasmic reticulum into the sarcoplasm.

Triggering a Muscle Contraction

Elevated calcium levels in the sarcoplasm open **calcium-activated calcium channels** in the membrane of the sarcoplasmic reticulum, causing even more Ca^{2+} release. Once sarcoplasmic Ca^{2+} levels have risen sufficiently, the binding sites on the actin filaments become accessible to the myosin heads. This unveiling of the binding sites involves a filamentous molecule called **tropomyosin**, which covers the binding sites when calcium levels are low, but (because of shape changes in some associated molecules) moves out of the way when calcium levels are high. As long as ATP levels are also high, muscle fiber contraction ensues.

Before a muscle can relax, its sarcoplasmic calcium levels must decrease to the point where the tropomyosin moves back to cover the binding sites on the actin filaments. This reduction in sarcoplasmic calcium levels is accomplished, at least in part, by the calcium-binding protein *parvalbumin*. In addition, sarcoplasmic calcium is pumped back into the sarcoplasmic reticulum by *calcium-ATPase*. The rate of this calcium transport varies between muscle fiber types, but it typically takes 25–200 ms to bring calcium levels back down after a muscle contraction. If action potentials recur at intervals shorter than this, the muscle fibers cannot relax fully between their contractions. At very high firing rates, this generates a state of tonic muscle contraction called **tetanus**. As you may remember, we discussed "tetanic stimulation" back in Chapter 3, in the context of synaptic plasticity. You may also have heard of "tetanus," a dangerous medical condition that involves uncontrolled, sustained muscle contractions. The condition is caused by **tetanus toxin**, which blocks GABA release from inhibitory synapses and, thereby, induces motor neuron hyperactivity and powerful muscle spasms. However, tetanic contraction of a muscle induced by rapid motor neuron firing is a normal aspect of any strong muscular exertion.

Muscle Contraction Gone Wrong

Now that you have learned about the mechanisms underlying muscular contraction, you can see why muscles cramp when they run out of energy. Without ATP, the myosin heads cannot detach from their actin binding sites. This impedes muscle lengthening as well as further contraction. In essence, the muscle fibers are stuck. Something similar happens in **rigor mortis** (which means "stiffness of death" in Latin). Two to three hours after death, the calcium stores in a muscle's interior disintegrate, releasing Ca^{2+} into the sarcoplasm and triggering a muscle contraction. Because ATP levels are negligible after death, these death-induced contractions cannot be reversed. Therefore, the rigor mortis lasts until the muscle cytoskeleton itself disintegrates, usually within 2–3 days after death.

Neuromuscular transmission is also impaired by **myasthenia gravis** (Latin for "serious muscular weakness"). In this autoimmune disease, the body generates antibodies against acetylcholine receptors. As the receptors are destroyed, muscular contractions become progressively weaker. Left untreated, patients with myasthenia gravis often die from respiratory arrest. Fortunately, such patients can be given immunosuppressants to reduce antibody production. They can also be treated with drugs that inhibit **acetylcholinesterase**, the enzyme that normally degrades acetylcholine in the synaptic cleft of neuromuscular junctions. By slowing the degradation of acetylcholine, acetylcholinesterase inhibitors create more opportunities for acetylcholine to bind their postsynaptic receptors. This amplifies the end plate potential and thereby increases the probability that a presynaptic action potential will elicit a postsynaptic contraction. Because of these treatments, the mortality rate for myasthenia gravis patients has dropped to less than 5%.

Controlling Muscle Force

Early in development, each muscle fiber is innervated by multiple motor neurons. Neuromuscular junctions are then eliminated through activity-dependent competition until, in adulthood, each skeletal muscle fiber is contacted by just one motor neuron. However, each motor neuron still innervates multiple muscle fibers.

Motor Unit Diversity

The set of muscle fibers innervated by a single motor neuron, together with the motor neuron itself, is called a **motor unit** (Figure 8.7). Some motor units are small, containing as few as 5 muscle fibers (in the delicate extraocular muscles); others include as many as 2,000 fibers. Even within a muscle, motor unit size varies considerably.

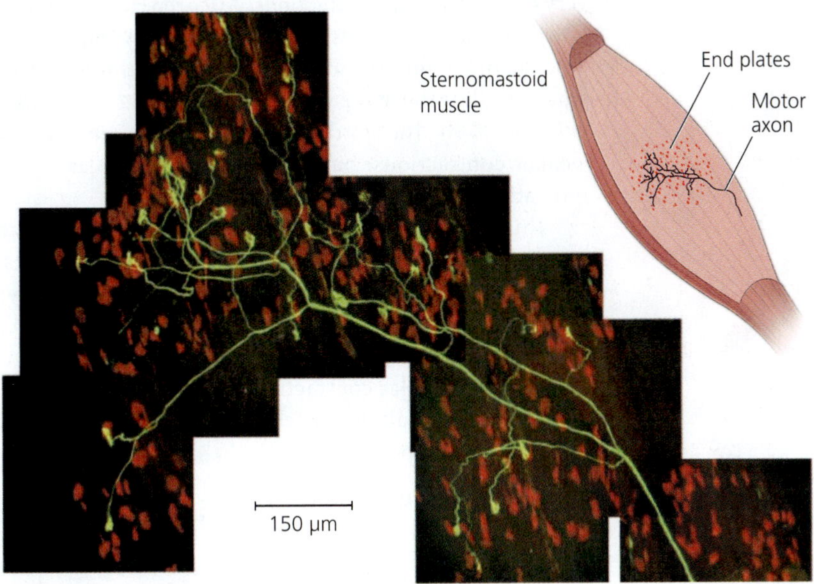

Figure 8.7 Anatomy of a motor unit. Shown here is a single motor axon (from an 8-day-old mouse) that is labeled with yellow fluorescent protein and innervates numerous muscle fibers in the sternomastoid, a superficial neck muscle. All motor end plates are stained with red fluorescent (rhodamine-conjugated) alpha-bungarotoxin, which binds acetylcholine receptors. You can see that the labeled axon innervates only a subset of the motor end plates, which tend to be located in the middle section of the muscle (inset). [Photograph from Keller-Peck et al., 2001]

Other things being equal, small motor units tend to elicit weak contractions; large motor units elicit stronger contractions.

The strength of the contraction elicited by the activity of a single motor unit also depends on the kinds of muscle fibers in that unit. As mentioned earlier, muscle fibers vary in the myosin genes they express and several other factors. Because of these differences, muscle fibers vary in contractile parameters. Some contract weakly on stimulation; others are strong. Some contract slowly, others fast. Some fatigue easily, but others can contract repeatedly without getting weaker. An important aspect of this variation is that individual motor neurons tend to innervate muscle fibers of the same type.

Evidence supporting this hypothesis comes mainly from experiments in which individual motor axons were stimulated electrically while muscle contractions were monitored with strain gauges. These experiments revealed three relatively distinct types of motor units (Figure 8.8). The fast, fatiguable type of motor unit generates

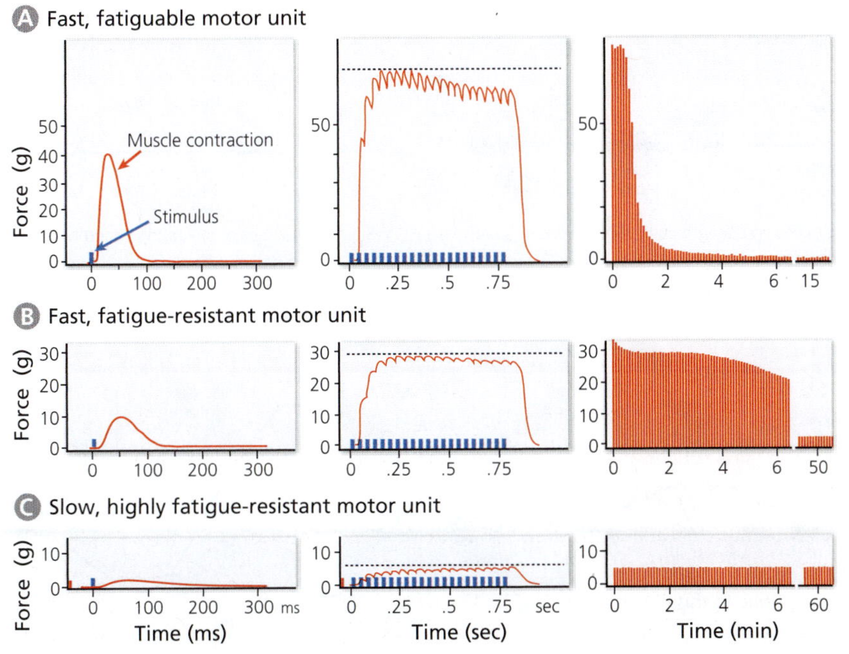

Figure 8.8 Motor unit variation. Electrical stimuli were applied to individual motor axons innervating a cat's calf muscle while muscle force was monitored by a strain gage. Shown in (A) are data for a fast, fatiguable type of motor unit. The muscle responds to a single stimulus (blue) with a brief but forceful twitch (red trace). A train of multiple stimuli generates a strong sustained contraction. Stimulating this type of motor unit repeatedly (brief trains of stimuli recurring every second for several minutes) causes the contractile force to decrease dramatically over time (right). Panel (B) shows analogous data for a motor unit that generates a weaker but less fatiguable contraction. Panel (C) presents data from a third type of motor unit that is highly fatigue resistant but generates very weak and slow muscle contractions. [After Burke et al., 1973]

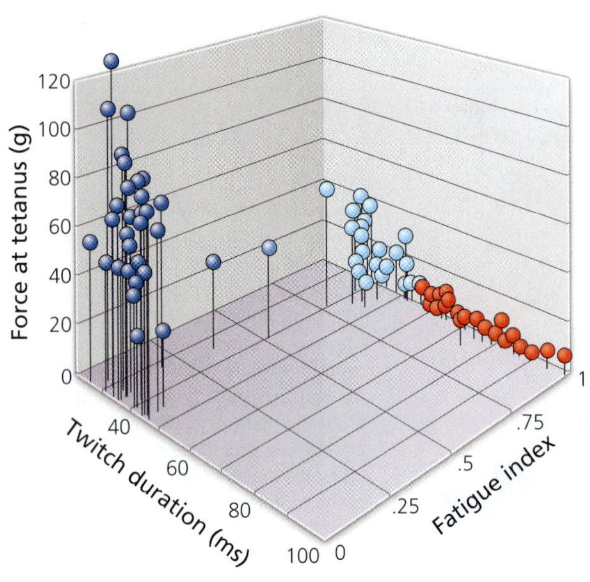

Figure 8.9 Motor unit classification. This 3-D graph summarizes physiological data for 81 calf muscle motor units. Dark blue circles represent fast, fatiguable motor units; light blue circles represent the fast, fatigue-resistant units; and red circles denote slow, highly fatigue-resistant units. The three types of motor units form relatively distinct clusters. [After Burke et al., 1973]

very fast and strong contractions. After repeated stimulation, however, the contractions weaken significantly (Figure 8.8 A). The second type of motor unit also generates fast contractions, but these are weaker and more fatigue resistant (Figure 8.8 B). The third type of motor unit generates even weaker contractions; but these are slow, long-lasting, and indefatigable (Figure 8.8 C).

This motor unit classification scheme is somewhat arbitrary in the sense that intermediate types of motor units do exist. Statistically speaking, however, the motor unit types are remarkably distinct (Figure 8.9). Therefore, we can conclude that the muscle fibers innervated by each motor neuron tend to have similar contractile properties. Another way of saying this is that all the muscle fibers in a motor unit tend to be of the same type.

Motor Unit Recruitment

How does the variation in motor unit size and type relate to variation in how strongly an entire muscle contracts? One possibility is that every contraction activates all of the motor units associated with that muscle but that these motor units fire more rapidly for strong contractions than for weak ones. Alternatively, weak and strong contractions might involve the activation of different sets of motor units. To distinguish between these two hypotheses, researchers have recorded from motor axons during various types of muscle contractions. In most of these experiments, recording electrodes are placed directly on (or into) the muscle in such a way that it is possible to record the activity of several axons simultaneously.

Figure 8.10 shows the result of one such experiment. It illustrates the activity of two different motor units during muscular contractions that vary in force. During a contraction that gradually increases in strength (Figure 8.10 A), one motor unit that fires relatively small action potentials (motor unit #1) begins to fire early in the contraction and then maintains its firing rate throughout the contraction. Once the contraction becomes strong, a second motor unit that fires larger action potentials (motor unit #2) becomes active as well. This orderly recruitment of additional motor

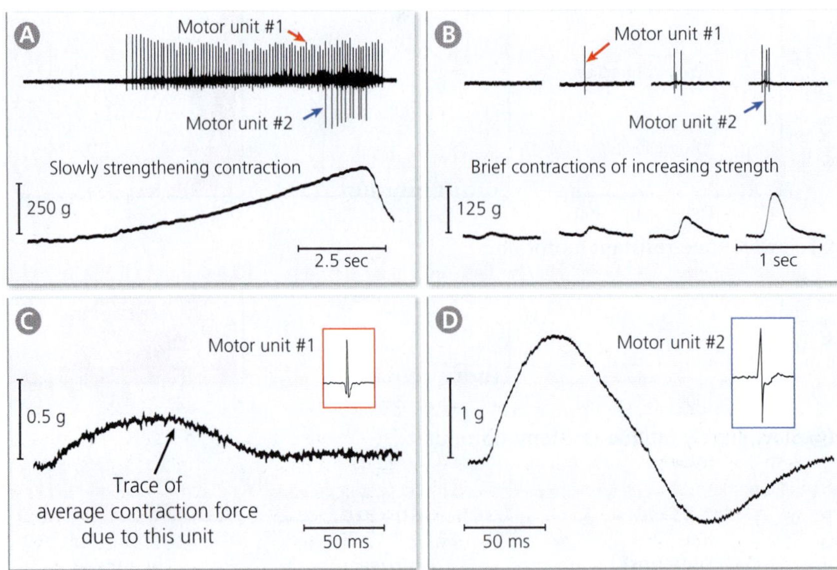

Figure 8.10 Motor unit recruitment. Wire electrodes were inserted into a shin muscle of humans as they pressed a pedal with their foot. The top trace in (A) shows the activity of two motor units. The bottom trace shows the force on the pedal. You can see that motor unit #2 does not begin to fire until the force exceeds ~250 g. Panel (B) shows how the same two motor units fire during a series of brief contractions of increasing force. Again, motor unit #2 fires at higher force levels than unit #1. Panels (C) and (D) show the contractile forces due to each motor unit's activity (averaging the forces recorded after individual action potentials). Motor unit #2 generates a much larger contractile force per action potential than unit #1. The inserts show the action potentials associated with each motor unit on a slower time scale. [After Desmedt and Godaux, 1977]

units as muscle contractions increase in strength is referred to as the **size principle of motor unit recruitment.** The principle has been confirmed in numerous studies, some of which recorded from up to 40 motor units simultaneously. The consistent finding is that more and more motor units become active as a contraction increases in strength. Once activated, the individual motor units maintain a rather steady firing rate throughout the contraction. As the contraction wanes, the motor units fall silent one by one, in the reverse of the sequence in which they were recruited (Figure 8.11).

Activation of the small, early recruited motor units produces only relatively weak muscle contractions, whereas activation of the large, late-recruited motor units elicits strong muscle twitches. The most likely explanation for this correlation is that the small, weak motor units innervate only a few muscle fibers, whereas the large, powerful motor units innervate a larger number of muscle fibers. This hypothesis has not been tested directly, but it is consistent with the observation that the more powerful motor units tend to fire larger action potentials (as recorded extracellularly) and have larger neuronal cell bodies (Figure 8.12). The larger neurons would be more capable of metabolically supporting the many axonal branches that are needed to innervate a large number of muscle fibers.

Exercise Effects

At this point, you may be asking yourself whether physical exercise can change the size of a motor unit. The answer is not known, but muscle biopsies have shown that limb muscles contain more slow, fatigue-resistant fibers in endurance athletes than in sprinters or weight lifters. These data suggest that exercise can change a muscle fiber's type. However, the variability in muscle fiber type composition within each group of athletes is large. Moreover, fiber type differences need not be caused by exercise. Perhaps endurance runners pursued their sport because they had more slow fibers to begin with and thus found it relatively easy to run for long periods of time.

To clarify this issue, researchers have turned to animal studies. In one classic experiment, the nerve from a muscle dominated by fast muscle fibers was experimentally rerouted to a muscle that usually contains mainly slow muscle fibers. Within a few weeks, the cross-innervated muscle had changed its muscle fiber composition to include

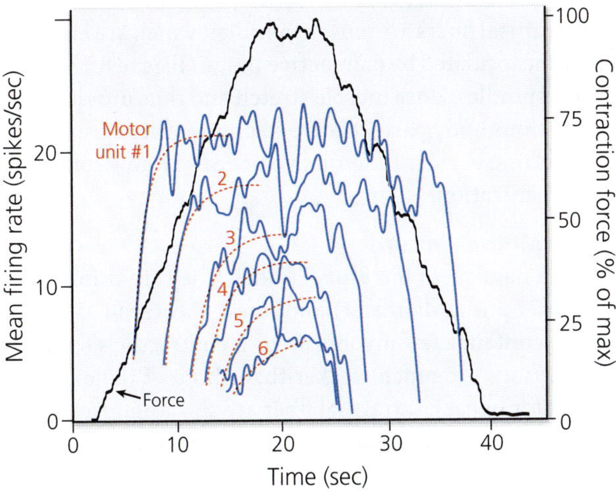

Figure 8.11 Motor unit recruitment in a human leg muscle. Shown here are the firing rates of 6 simultaneously recorded motor units (blue traces) while muscle force (black trace) was gradually increased and then decreased. The motor units were recruited successively (with unit #1 becoming active first) and fell silent in the opposite order. The later recruited units fired at lower maximum rates. The red lines are fitted curves. [After de Luca and Contessa, 2012]

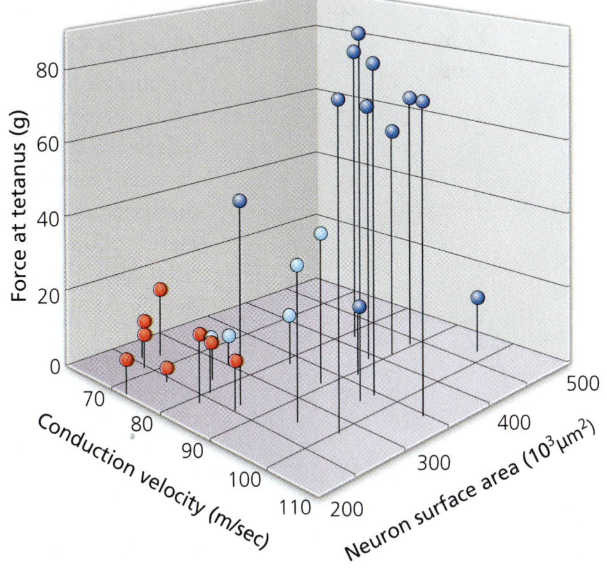

Figure 8.12 Larger neurons elicit stronger, faster contractions. A motor neuron's surface area (based on mean cell body and primary dendrite diameters) is positively correlated with axonal conduction velocity and with the contractile force elicited by activating the neuron. Motor neurons in slow motor units (red) tend to be smaller than neurons in fast, fatigue-resistant motor units (light blue), which tend to be smaller than the neurons in fast, fatiguable motor units (dark blue). [After Burke et al., 1982]

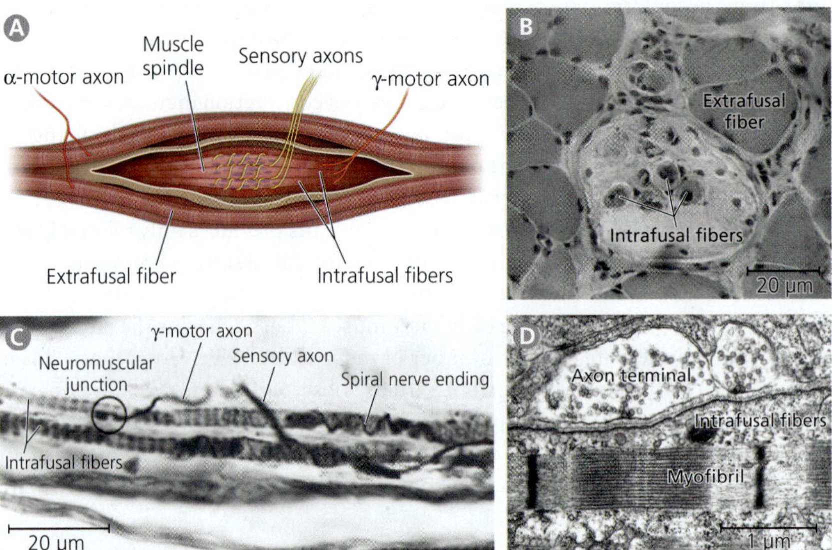

Figure 8.13 Muscle spindles. Shown schematically in (A) is a muscle spindle sandwiched between two extrafusal muscle fibers. The muscle spindle consists of intrafusal muscle fibers that are innervated by spiraling sensory axons and by gamma motor neurons. Panel (B) shows a muscle spindle in cross section. The photograph in (C) depicts several intrafusal muscle fibers that were dissected out of the muscle with their innervation intact. Image (D) is an electron micrograph of an intrafusal fiber that was sectioned longitudinally. One complete sarcomere is visible. [B–D from Ovalle et al., 1999]

more fast muscle fibers. Conversely, rerouting a nerve from a slow muscle to a fast one increased the proportion of slow muscle fibers in the cross-innervated muscle. Thus, there is strong evidence that neural activity can influence fiber type composition, at least to some extent. Neural activity can also influence muscle fiber diameter, which is the main reason why exercise can increase muscle mass.

BRAIN EXERCISE

Alpha motor neurons usually innervate muscle fibers of the same fiber type (expressing the same type of myosin). Why is this useful? What would happen if this weren't the case?

Muscle Spindles

The muscle fibers we have discussed so far in this chapter are called **extrafusal fibers** (*fusus* means "spindle" in Latin). They account for most of the fibers in skeletal muscle. Nestled among the extrafusal fibers are muscle spindles, which are elongated in shape (spindle shaped) and encapsulated by connective tissue (Figure 8.13). As we discussed in Chapter 7, muscle spindles sense muscle stretch and thus provide information about the position of various body parts relative to one another (proprioception). To get a deeper understanding of their function, you need to know some more details about muscle spindle organization.

Intrafusal Fibers and Gamma Motor Neurons

Inside each muscle spindle are a handful of **intrafusal fibers**. They are thinner than the extrafusal fibers (~10 vs. ~60 μm in diameter) and much shorter (4–7 mm vs. 3–40 cm). The intrafusal fibers contain a few myofibrils with contractile sarcomeres (Figure 8.13), but their contractions are much weaker than those of the extrafusal fibers. Spiraling around the center of each intrafusal fiber are the annulospiral endings that we discussed in Chapter 7. They contain mechanosensitive ion channels and therefore are sensitive to intrafusal fiber stretch.

Intrafusal fibers are also innervated by motor axons, which target the two ends of each fiber (Figure 8.13 A). Because these motor axons are myelinated and of medium thickness (2–8 μm diameter), they are classified as Aγ fibers (recall our discussion of fiber type classification in Chapter 7). Accordingly, the motor neurons to which these axons belong are called **gamma (γ) motor neurons**. Like the cell bodies of the **alpha (α) motor neurons** that innervate extrafusal muscle fibers, the cell bodies of γ-motor neurons are located in the ventral (anterior) horn of the spinal cord, although the γ-motor neurons tend to be smaller. Importantly, stimulation of the γ-motor axons causes intrafusal fibers to contract at their two ends. This contraction stretches the fiber's central region, where the annulospiral endings end.

Alpha-gamma Coactivation

Why is it useful to stretch the central region of intrafusal muscle fibers? Consider what happens when a skeletal muscle contracts. Before the contraction, spindle sensory axons that wrap their annulospiral endings around the central region of the intrafusal fibers can accurately signal the extent to which the muscle is stretched (e.g., by a weight applied to one of its ends; Figure 8.14). However, when the muscle contracts and the extrafusal fibers shorten, the intrafusal fibers slacken. As a result, the sensory axons fall silent (Figure 8.14 B) even though the muscle as a whole is still under tension. Moreover, the sensory axons cannot signal further changes in the muscle stretch (e.g., changes in the applied weight) unless the stretch is strong enough to pull the intrafusal fibers taut. In essence, the slackening of the intrafusal fibers makes the spindle sensors useless during each muscle twitch.

An elegant solution to this problem is to activate the γ-motor neurons simultaneously with the α-motor neurons. This **alpha-gamma coactivation** ensures that the intrafusal fibers are pulled taut (by the γ-motor neuron activity) during each muscle contraction (driven mainly by the α-motor neuron activity). As a result, the sensory axons remain capable of signaling muscle stretch even during a muscle twitch (Figure 8.14 C). In other words, alpha-gamma coactivation allows muscle spindles to remain sensitive to unexpected changes in muscle length even as an animal moves. Indeed, some studies have confirmed that α- and γ-motor neurons tend to be coactivated during natural movements. However, the correlation is variable across both studies and behaviors, suggesting that the biological role of γ-motor neurons is more complex than our discussion has implied.

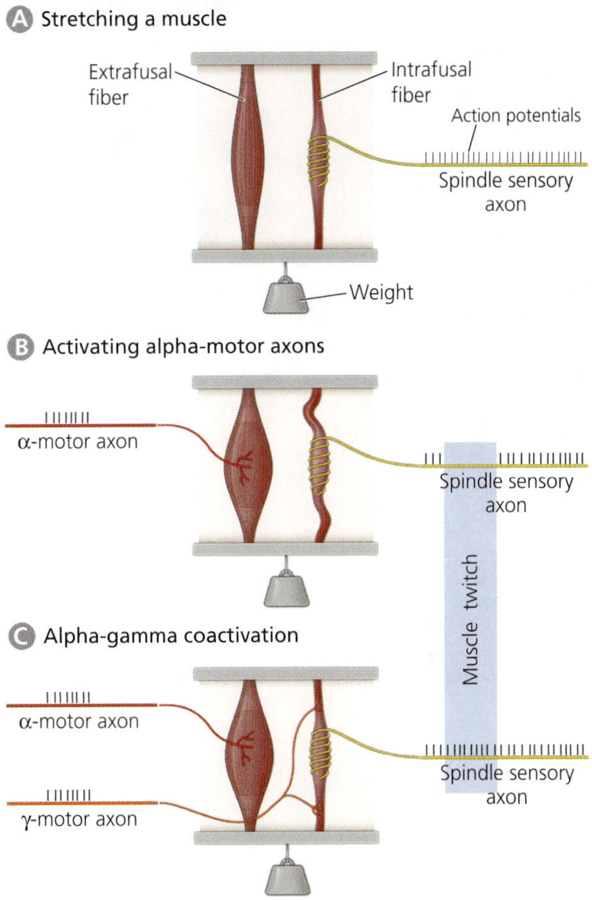

Figure 8.14 Alpha-gamma coactivation. (A) Stretching a muscle activates the sensory axons that spiral around the intrafusal fibers of muscle spindles. (B) Activating the α-motor axons of such a stretched muscle causes extrafusal fibers to contract and the intrafusal fibers to slacken, which makes the latter go silent during the muscle twitch (shaded blue). (C) Coactivation of the α- and γ-motor axons shortens the intrafusal fibers during the muscle twitch, allowing the sensory axons to sense muscle stretch without interruption. [After Hunt and Kuffler, 1951]

BRAIN EXERCISE

Mammalian motor neurons innervate either intrafusal or extrafusal muscle fibers, never both. Why is this useful? What would happen if this weren't the case?

8.2 What Makes the Heart Beat?

The muscle fibers of the heart resemble the extrafusal muscle fibers of skeletal muscle but differ in a few important respects (Table 8.1). For one thing, cardiac muscle fibers are relatively short and thin. For another, they generally contain only one or two cell nuclei per muscle fiber. In addition, cardiac muscle fibers tend to branch at their ends, causing the fibers to form an interlocking web.

An even more distinctive feature of cardiac muscle is that the ends of individual cardiac muscle fibers are joined end-to-end by **intercalated disks** (Figure 8.15), which contain adhesion and linkage proteins that fasten the individual fibers to one another. The intercalated disks also contain gap junctions through which ions and other molecules can freely flow from one cardiac fiber into adjoining fibers. This direct contact causes cardiac fibers to be **electrically coupled**, which means that

	SKELETAL MUSCLE	CARDIAC MUSCLE	SMOOTH MUSCLE
ATTACHMENT	Bone, tendons, and skin	Other cardiac muscle fibers	Other smooth muscle fibers
FIBER DIAMETER	up to 100 μm	~10 μm	2–5 μm
FIBER LENGTH	up to 20 cm	~50 μm	up to 500 μm
FIBER SHAPE	Cylindrical	Branched	Spindle-shaped
NERVE ENDINGS	Motor end plates	Varicosities	Varicosities
NUCLEI PER FIBER	Many	1–2	1
SARCOMERES	Present	Present	Absent
BINDING SITE REGULATION	Tropomyosin	Tropomyosin	Calmodulin
ELECTRICAL COUPLING	Absent	Extensive	Extensive
ENERGY COST	High	High	Low
CONTRACTION SPEED	Fast	Slow	Very slow
ACTION POTENTIAL DURATION	~1 ms	~200 ms	$>$ 200 ms

Table 8.1 Comparison of muscle tissue types.

action potentials can spread between connected cardiac fibers without the need for neurotransmission. If one cardiac fiber fires an action potential, then its neighbors will do so as well.

Action potential duration also differs between skeletal and cardiac muscle fibers. Like most action potentials, cardiac action potentials begin with the rapid opening of voltage-gated sodium channels, which then close again within about a millisecond. However, in cardiac action potentials, the initial membrane depolarization also opens **voltage-sensitive calcium channels**. Calcium ions flow through these open channels into the muscle fiber and generate additional depolarization. Importantly, the voltage-gated calcium channels close more slowly than the voltage-gated sodium channels, thereby extending the period of membrane depolarization. Thus, a typical cardiac action potential lasts for about 100 ms, rather than ~1 ms, as in a typical neuron or skeletal muscle. Because of this difference in duration, a single action potential elicits a more sustained contraction in cardiac muscle than in skeletal muscle.

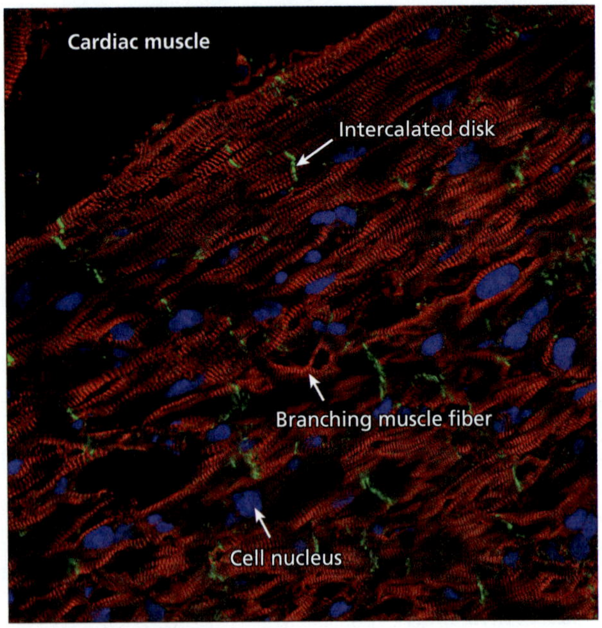

Figure 8.15 Cardiac muscle anatomy. This section through mouse heart muscle was treated with antibodies against myomesin (to which thin filaments attach) to stain parts of each sarcomere (red). It was also treated with anti-beta catenin to reveal intercalated disks (green) and with a stain called DAPI (4′,6-diamidino-2-phenylindole) to label all cell nuclei blue. You can see that cardiac muscle fibers often branch and are joined end-to-end by the intercalated disks. [Courtesy of Elisabeth Ehler]

Generation of the Cardiac Rhythm

The most dramatic difference between skeletal and cardiac muscle is that the latter contracts rhythmically even without neural input. This intrinsic rhythm of the heart is generated by a **cardiac conduction system** (Figure 8.16). The principal pacemaker in this system is the **sinoatrial (SA) node**, which comprises a small clump of modified cardiac muscle fibers in the wall of the right atrium. These cells fire synchronously with one another at a rate of roughly 100 action potentials per minute, even when all neural input has been surgically removed.

Intracellular recordings reveal that the rhythmic firing of the SA node cells involves **hyperpolarization-activated cyclic nucleotide (HCN) channels**. These channels open during the undershoot phase at the end of each action potential when the membrane potential decreases below approximately –50 mV. Once the HCN channels are open, they let positive ions flow into the cell, thereby generating a gradual depolarization that eventually brings the membrane potential to the threshold for firing an action potential (Figure 8.16 B). The current responsible for the gradual depolarization is called the **funny current** (or, more formally, the H-current) because it differs from most other currents in being unusually slow and activated by hyperpolarization (rather than depolarization or ligands). Because the funny current is triggered by the undershoot of each action potential and then triggers another spike, it accounts for the regular, rhythmic production of action potentials in the SA node. It lies at the core of the heart's pacemaking system.

The rhythmically generated action potentials in the SA node spread through gap junctions to the surrounding atrial muscles, causing rhythmic contractions of the atrium. The action potentials also spread along modified muscle fibers to the **atrioventricular (AV) node** at the boundary between the right atrium and ventricle (Figure 8.16 A). These AV cells fire rhythmically when isolated from other cells and thus exhibit their own independent pacemaker activity. However, the intrinsic rhythm of the AV node is slower than that of the SA node, which means that the SA node normally imposes its rhythm on the AV node.

From the AV node, action potentials propagate along **Purkinje fibers** to all regions of the ventricle and cause it to contract. Although these Purkinje fibers function like axons, they are cardiac muscle fibers that have lost their contractile elements and conduct action potentials through gap junctions rather than synaptic transmission. Overall, the system is designed so that the right atrium contracts shortly before the ventricle, which helps move blood from the former into the latter. Within the walls of the ventricle, action potentials are conducted so rapidly that all the muscle fibers contract nearly simultaneously. This synchrony ensures that blood is ejected from the ventricle with maximum force.

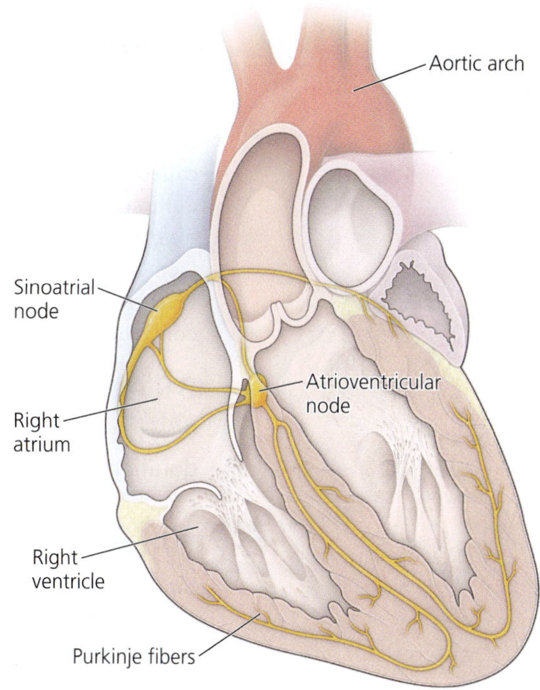

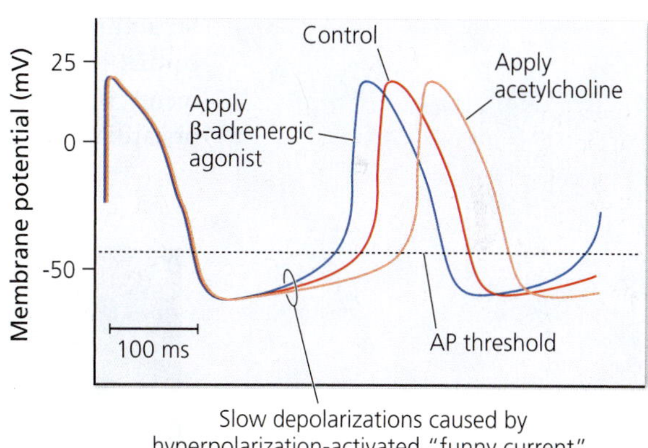

Figure 8.16 The heart's pacemaker system. The heart's primary pacemaker, the sinoatrial (SA) node, lies in the wall of the right atrium (A). From there, action potentials (APs) travel through the atrioventricular node to the ventricles, which contract in response. The graph in (B) is an overlay of three recordings from an SA node, aligned so that the first APs coincide. The interval between successive APs shortens when the SA node is bathed in a β-adrenergic receptor agonist (mimicking norepinephrine) and lengthens after application of acetylcholine. Individual action potentials are triggered by "funny" ion channels that allow positive ions to flow into the cell when the membrane potential drops below about –50 mV. [B after DiFrancesco, 1993]

Modulation of the Cardiac Rhythm

Although the heart can beat independently of neural input, the rate at which it beats is modulated by neural inputs. We will discuss the circuitry underlying this modulation in Chapter 9. For now, let us simply note that the heart is innervated by two sets of nerves that release their transmitter not from neuromuscular junctions but from small swellings along the axon. One set of axons releases **norepinephrine** onto the heart, especially onto the SA node. In response, heart rate speeds up. The other set of axons releases *acetylcholine,* which causes the heart rate to slow down. As you will learn in Chapter 9, it is the balance of acetylcholine versus norepinephrine acting on the SA node that regulates heart rate. Both neurotransmitters modulate heart rate by changing the size of the funny current we discussed earlier. Norepinephrine increases this current, thereby decreasing the time it takes for the membrane potential to rise to the action potential threshold (Figure 8.16 B). Acetylcholine, in contrast, decreases the funny current and thus increases the delay between successive action potentials.

Acetylcholine receptors come in two distinct varieties. The receptors for acetylcholine in the heart are metabotropic, meaning that they modify the postsynaptic cell's internal enzymatic machinery. In contrast, acetylcholine receptors at the neuromuscular junction are ionotropic. The receptors at the neuromuscular junction are called **nicotinic acetylcholine receptors**, whereas the receptors in the heart are called **muscarinic acetylcholine receptors**. As these names suggest, the former are quite sensitive to nicotine, whereas the latter are more sensitive to muscarine, a deadly toxin produced by a few species of mushroom. Curare, a toxin used by some hunters (see Chapter 2, Box 2.4), blocks only nicotinic acetylcholine receptors. Therefore, victims of curare poisoning cannot move their skeletal muscles, but their heart continues to beat.

BRAIN EXERCISE

Why is it important that action potentials in cardiac muscle have such a long duration? What would happen if they were shorter?

8.3 What Is Special about Smooth Muscle?

Smooth muscle lines the walls of the intestines, bladder, blood vessels (especially arterioles), and bronchial passageways (Figure 8.17). Smooth muscles are also found at the base of body hairs, giving you goosebumps when they contract; and in the eye, where they control pupil diameter.

Smooth Muscle Anatomy

Structurally, smooth muscle has even less in common with skeletal muscle than cardiac muscle does (Table 8.1). Smooth muscle does contain actin and myosin filaments, but these do not exhibit the regular sarcomeric arrangement seen in skeletal or cardiac muscle. Instead, the actin and myosin filaments in smooth muscle attach

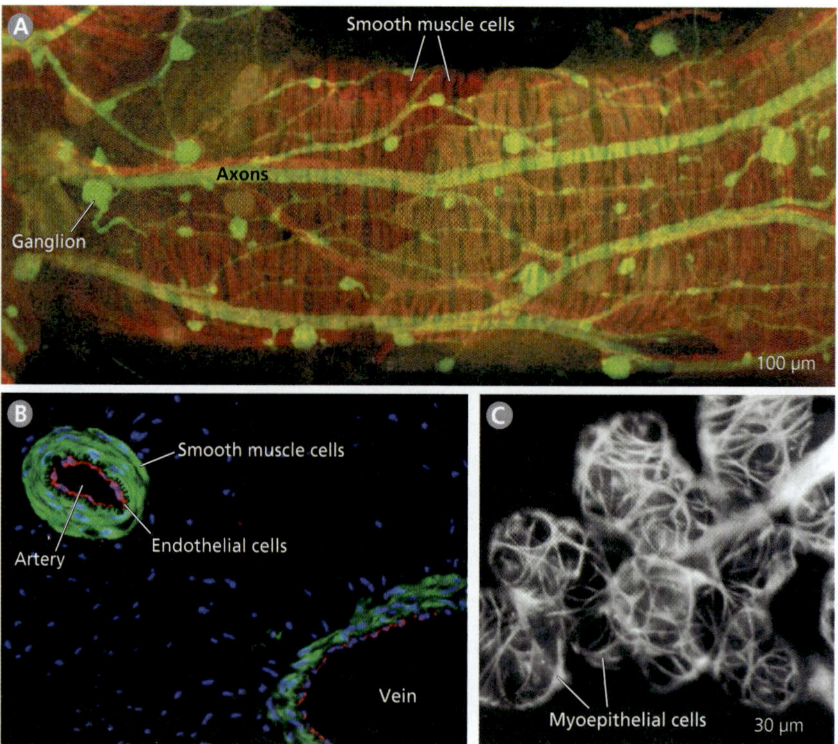

Figure 8.17 Smooth muscle and myoepithelial cells. (A) depicts part of a human lung treated with an antibody against alpha-actin, which reveals smooth muscle fibers (red), and a second antibody to highlight neural tissue (green). Panel (B) shows a transverse section through an artery and a vein in a rat's penis stained with phalloidin (from the poisonous "death cap" mushroom) to reveal smooth muscles (green), with antibodies to label endothelial cells (red), and with DAPI to visualize cell nuclei (blue). The image in (C) shows star-shaped myoepithelial cells surrounding clusters of secretory cells in a rat's salivary gland. The tissue was stained with fluorescent phallacidin, which is similar to phalloidin in that it binds tightly to actin filaments. [A from Sparrow et al., 1999; B from Lin et al., 2011; C from Murakami et al., 1989]

to the internal surface of the cell membrane and crisscross the cell's interior rather haphazardly. As a result of this arrangement, activation of the actin-myosin system causes a smooth muscle cell to assume a more spherical shape. Because relaxed smooth muscle cells tend to be elongated, this rounding up makes the muscle fibers less elongated. In other words, it shortens them. The effect of this shortening depends on how the muscle fibers are arranged. When they are arranged in rings around a cavity, such as a blood vessel, their shortening constricts the cavity.

Smooth Muscle Physiology

Excitation-contraction coupling also differs between smooth muscle and the other two muscle types. Although smooth muscle cells can fire action potentials, these action potentials involve only voltage-gated calcium channels (no sodium channels). Moreover, smooth muscle can contract without firing action potentials, as long as sufficient amounts of calcium can be released from the internal stores. As calcium levels increase inside a smooth muscle cell, calcium ions bind to **calmodulin**, which then activates an enzyme that phosphorylates the myosin heads. This, in turn, makes the myosin heads more likely to bind to the adjacent actin filaments. After binding to actin, the myosin heads in smooth muscle change shape just as they do in skeletal and cardiac muscle, thereby forcing the myosin and actin filaments to slide against one another.

The major point here is that excitation-contraction coupling in smooth muscle depends on **calcium-calmodulin activated phosphorylation** of the myosin heads rather than tropomyosin displacement (which triggers contractions in skeletal muscle). Partly because of this difference, smooth muscle contracts 300–500 times slower than skeletal muscle and takes much longer to relax.

Smooth Muscle Innervation

The innervation of smooth muscle cells involves no neuromuscular junctions. Instead, axons release either norepinephrine or acetylcholine onto smooth muscle cells from small varicosities (swellings) along the axon. The released transmitters then activate their respective receptors on the smooth muscle cells. This much is similar to what you see in cardiac muscle. However, smooth muscle cells respond to neurotransmitter release by increasing or decreasing their degree of contraction rather than accelerating or decelerating their intrinsic rhythm.

Unfortunately, there is no simple rule about which transmitter does what in smooth muscle. The effects depend on which muscles are involved. For example, norepinephrine promotes the contraction of the smooth muscle around arterioles, causing a rise in blood pressure, but the same transmitter dilates the bronchial airways. Smooth muscles can also be influenced by substances other than norepinephrine and acetylcholine, including various hormones. Particularly important is the hormone **epinephrine** (adrenaline), which is secreted from the adrenal gland and can mimic the effects of neurally released norepinephrine by binding to the same (adrenergic) receptors that bind norepinephrine. This explains why asthma inhalers and EpiPens, both of which contain epinephrine, dilate the bronchi and increase blood pressure as well as increasing heart rate.

BRAIN EXERCISE

Why does smooth muscle contract and relax more slowly than skeletal or cardiac muscle? Why might this be useful?

8.4 How Do Muscles Lengthen after Contractions?

Although the various types of muscles differ in numerous respects, they all generate physical force only as they contract. The myosin heads bend back and forth, but they are bound to the actin filaments only in one direction of movement, during the power stroke. Therefore, force is generated only when the muscle contracts. For a muscle to

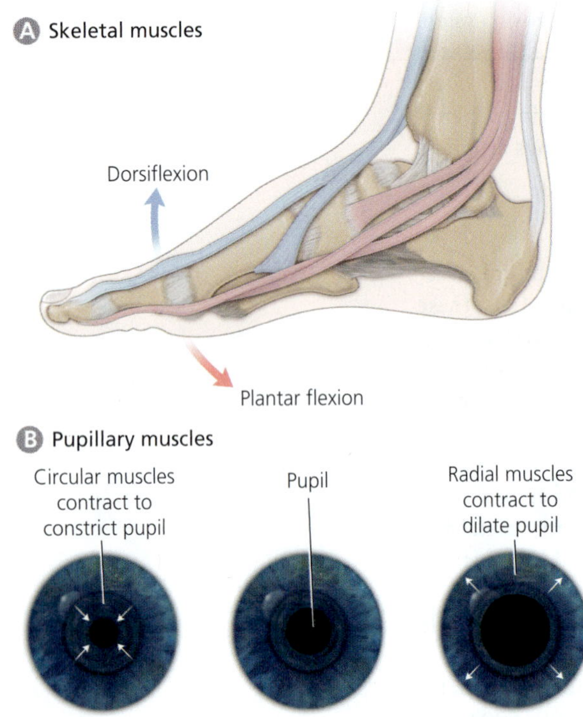

Figure 8.18 Antagonistic muscle pairs. The muscles that dorsiflex the toes (A) are opposed by the muscles that flex the toes downward (plantar flexion). When one set of muscles contracts, the antagonists lengthen. The smooth pupillary dilator and constrictor muscles (B) also operate as an antagonistic pair: when the dilators contract, the constrictors are stretched; conversely, contraction of the pupillary constrictors lengthens the pupillary dilators.

lengthen after a contraction, an external force must be applied. Muscles differ in this respect from some molecular motors (see Box 8.1: Outer Hair Cell Motility) and from most man-made effectors, such as electric and piezoelectric motors, which can push as well as pull.

Muscles Must Be Antagonized

If muscles exert force in only one direction, how can they do useful work? How can they generate back-and-forth movements of body parts? They do so by being arranged into **antagonistic pairs**. For example, contraction of leg flexor muscles forces the leg to bend around the knee, whereas contraction of leg extensor muscles reverses that movement. Similarly, contraction of dorsiflexor muscles that run along the shinbone bends the foot upward around the ankle joint, whereas another set of muscles causes the opposite movement, namely, plantar flexion (Figure 8.18 A). Importantly, contraction of one muscle, or group of muscles, always pulls on its antagonists. Most skeletal muscles are arranged in such antagonistic sets. Some smooth muscles are likewise arranged antagonistically. For example, the ring-shaped muscles that constrict the pupil are antagonized by smooth muscle fibers that attach to the pupillary constrictor muscles, pulling them radially to dilate the pupil (Figure 8.18 B).

However, not all smooth muscles are organized antagonistically. Consider, for example, the **myoepithelial smooth muscle cells** that surround the secretion-filled base of mammary and salivary glands (Figure 8.17 C). When these cells contract, the gland's cavity constricts and the secretions are pushed out. How do the cavities dilate again? There are no muscles to pull them open. Instead, the cavities are reinflated when additional secretions (e.g., saliva) push against the cavity walls. Similarly, the smooth muscle cells around most blood vessels (Figure 8.17 B) are not antagonized by other muscles. When these **perivascular smooth muscle cells** (a.k.a. pericytes) contract, the blood vessels constrict. The work of dilating the blood vessels again is performed not by other muscles but by the blood inside the vessels, which pushes on the vessel walls.

Pumping Blood through the Heart

The mechanisms involved in dilating the heart after a contraction are more complex. Some of the muscles in the wall of the heart's ventricle are arranged in two opposing spirals. Contraction of one spiral twists the ventricle so that its apex and its base rotate in opposite directions (Figure 8.19). This twisting motion squeezes blood out

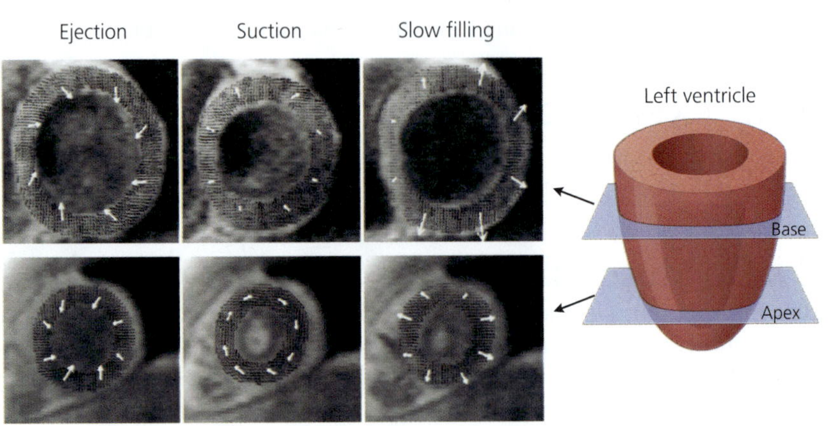

Figure 8.19 Pumping blood. Shown on the left are sections through the left ventricle of a beating human heart at two levels, near the base and the apex. The white arrows indicate the mean direction and velocity of motion in the two indicated planes as determined from MRI scans. During blood ejection, the apex and the base twist in opposite directions, squeezing blood out. When these motions are reversed, blood is sucked into the ventricle. After that, blood flows into the ventricle under weak positive pressure. [From Buckberg et al., 2006]

NEUROBIOLOGY IN DEPTH

Box 8.1 *Outer Hair Cell Motility*

As you may recall from Chapter 6, hearing involves the transformation of acoustic vibrations into basilar membrane vibrations, which drive side-to-side movements of the inner hair cell's stereocilia. What we have not yet discussed are the **outer hair cells**, which outnumber inner hair cells by roughly 3 to 1. Outer hair cells are contacted by relatively few sensory axons and therefore account for only a small fraction of the auditory information that is sent into the brain. Instead, the principal function of outer hair cells is to amplify the sound-induced vibrations of the basilar membrane and thus increase our ability to hear very quiet sounds. To see how this works, let us review some key experiments.

William Brownell and his collaborators in 1985 isolated outer hair cells from guinea pigs and discovered that the cells shorten in response to intracellular depolarization and lengthen when their membrane potential is decreased (hyperpolarized). These findings were later confirmed and extended (Figure b8.1).

Although the changes in outer hair cell length are relatively small (1–5%), they are large enough to amplify basilar membrane vibrations. When an incoming sound makes the basilar membrane vibrate up and down, the outer hair cells are alternately depolarized and hyperpolarized, making them change length in synchrony with the sound stimulus. Because the outer hair cells insert their stereocilia into the cochlea's relatively stiff tectorial membrane, the oscillatory changes in hair cell length amplify the basilar membrane's vibrations. This amplification, in turn, increases inner hair cell activation and thereby improves sound perception.

Changes in outer hair cell length are mediated by *prestin*, a molecule named for its ability to change shape rapidly (presto). Prestin is a transmembrane protein found at high density (~6,000 molecules/μ^2) in the cylindrical walls of mammalian outer hair cells. When prestin is expressed in cells that normally do not contain prestin and do not change their shape when membrane voltage is varied, the prestin-transfected cells exhibit electromotility quite similar to that of outer hair cells.

Prestin is an *incomplete anion transporter*. On membrane depolarization, prestin binds extracellular chloride (Cl^-) and moves it through a central pore toward the cell's interior. However, prestin does not release the Cl^- inside the cell. Instead, the Cl^- remains bound and moves back toward the cell's exterior when the plasma membrane becomes hyperpolarized. As the Cl^- moves back and forth, the prestin protein alters its cross-sectional area in the plane of the membrane, decreasing with membrane depolarization and increasing on hyperpolarization. As you can probably imagine, the synchronized thickening or thinning of millions of prestin molecules in a hair cell's side walls causes the entire hair cell to lengthen or shorten, respectively. An interesting aspect of this mechanism is that both hair cell lengthening and shortening are forceful movements. The extreme speed of these movements comes from the fact that they involve only changes in molecule shape. No transmitters or other signaling mechanisms are involved.

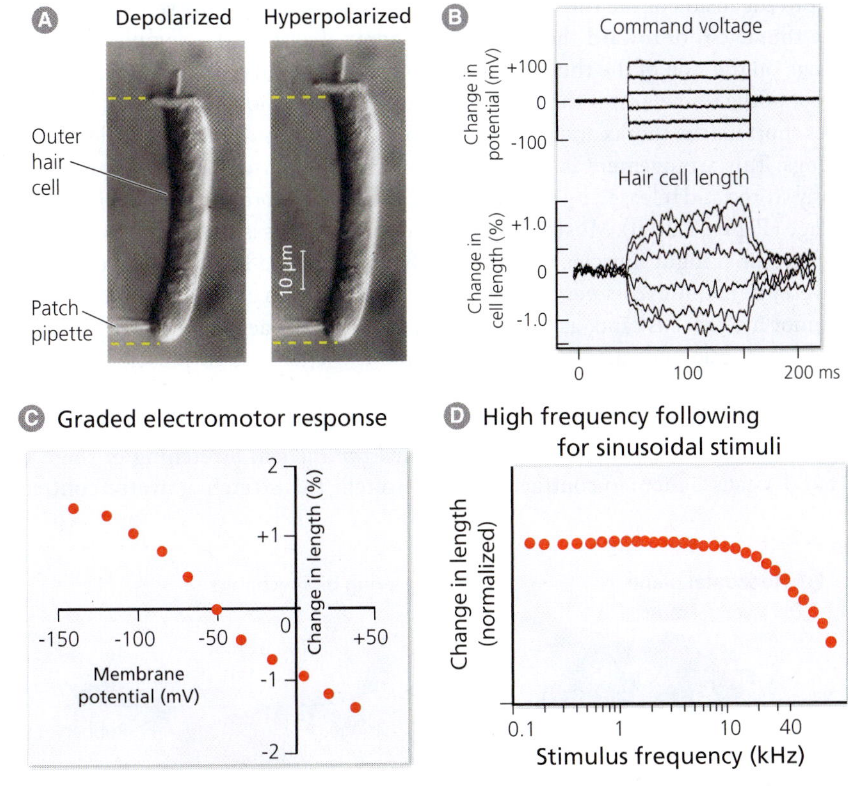

Figure b8.1 Outer hair cell motility. Shown in (A) is an isolated outer hair cell at two different membrane potentials. The graphs in (B) illustrate how outer hair cell length varies as the membrane potential is clamped at 6 different voltages. As shown in (C), cell length varies almost linearly with membrane potential. The changes in cell length remain relatively constant in amplitude with increasing stimulus frequency (D). [A courtesy of Ashmore and Holley, unpublished from 1998; B and C from Ashmore, 1987; D after Frank et al., 1999]

of the heart, much as wringing out a wet towel drains it of water. After the blood is ejected, the opposing spiral contracts, twisting the ventricle the other way. This untwisting creates suction that pulls blood into the ventricle and dilates it. Only after this initial suction does pressure from the inrushing blood expand the ventricle further. Modeling studies suggest that only this combination of active suction and passive filling can account for how much blood our hearts pump through our blood vessels. So next time you see a beating heart, look whether you can see it twist. The general principle at work here is that muscles may be antagonized by one another or by other physical forces (blood pressure), or both. Without this antagonism, contracted muscles cannot be lengthened.

Insect Flight

Another good example of muscle antagonism is found in the machinery of insect flight. In all insects, the flight muscles (whose myofibrils are quite similar to the striated muscle fibers of vertebrates) are organized into antagonistic pairs. In locusts, dragonflies, and a few other species, the flight muscles attach directly to the base of the wing so that one set of muscles elevates the wing when it contracts, whereas the other set lowers the wing. Given this arrangement, the rate at which the flight muscles can contract and relax limits how rapidly the wings can beat. In most species, this limit is just a few cycles per second.

However, several insect lineages (including flies, beetles, and bees) can beat their wings at much higher rates. In these species, the flight muscles attach not to the wing but to the inside of the thoracic exoskeleton (Figure 8.20). One set of muscles pulls the thoracic roof toward the floor of the thorax. Because of a complex mechanical hinge on the side of the thorax, right where the wings attach, this downward movement of the roof elevates the wing. In contrast, contraction of the longitudinal muscles shortens the thorax and pushes the thorax roof upward, which then depresses the wings. This arrangement is energetically efficient because potential energy is cyclically stored and released by accumulations of a rubbery protein (resilin) in the wing's hinge (Figure 8.20 B). Most interesting for our purposes is that the dorsoventral and longitudinal flight muscles can alternately contract and lengthen at frequencies of several hundred cycles per second (up to ~1,000 Hz in some small flies). Neurons cannot fire that fast and, as noted earlier, high rates of action potential firing cause tetanic muscle contractions rather than rapid oscillations in muscle length. How then can flies beat their wings so fast?

The answer lies in the antagonism of the dorsoventral and longitudinal muscles as well as their capacity for *stretch-activated contraction*. Stretching of these flight muscles causes them to contract after the stretch. This stretch-activated contraction

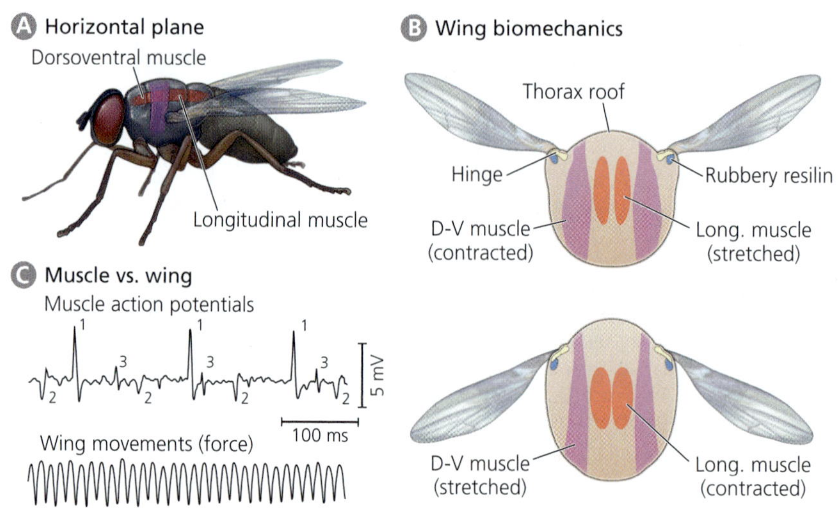

Figure 8.20 Insect flight. On contraction, a fly's dorsoventral and longitudinal flight muscles (A) flatten and shorten the thorax, respectively. Simplified transverse sections through an insect thorax are shown in (B). As the roof of the thorax is pulled down, the wings are raised. When the thorax roof is elevated, the wings are depressed. Accumulations of the elastic protein resilin at the hinge points store and release potential energy, thereby increasing flight efficiency. The graphs in (C) show electromyogram (top) and wing force (bottom) recordings from a tethered beetle during flight. Three different motor units (numbered) are discernible. The crucial point is that in beetles (and flies), the motor units fire well below the wingbeat frequency. [C from Josephson et al., 2000]

then stretches the antagonist, which in turn is triggered by the stretching to contract. This pattern of alternating contractions of antagonists does not require a matching pattern of motor unit activity as long as the motor axons fire often enough to keep the muscle fibers in a state of sustained excitation, characterized by elevated calcium levels. Indeed, as shown in Figure 8.20 C, the motor units in such flight muscles do not fire with every wing beat. They fire just often enough to keep the muscles tonically excited, but this sustained excitation generates oscillatory contractions that have a frequency much higher than the neuronal firing rate.

BRAIN EXERCISE

Why are there no muscles that run from (say) the ribs to the outer wall of the heart's ventricle to increase its volume after a contraction? If you had such muscles, what problems might they create?

8.5 How Do Neurons Control Hormones, and Vice Versa?

As noted at the beginning of this chapter, brains need effectors. The most obvious effectors are the various muscles, especially the skeletal muscles. However, the other major type of effector—the glands—perform a wide variety of important functions ranging from the regulation of development and reproduction to the control of stress and memory. Their actions are generally slow, especially compared to those of skeletal muscles, but the effects of glandular secretions are often long-lasting and powerful. Especially powerful are the **endocrine glands**, which include the pituitary, thyroid, thymus, adrenal, and pineal glands as well as the gonads. All these endocrine glands discharge their secretions—called **hormones**—into the blood. Because the release of several hormones is regulated by neurons, and some hormones affect the brain, neurobiologists should know at least some basics of how hormones work. Even the **exocrine glands**, which eject their secretions through ducts to the skin or into the gastrointestinal tract, are controlled by neurons and therefore related to neurobiology.

Exocrine Glands

The most obvious exocrine glands are sweat, salivary, tear, and mammary glands. Exocrine secretions are also produced by the lining of the stomach, the liver, and most cells in the pancreas; these all aid in digestion. As noted earlier, exocrine secretions accumulate in cavities at the base of the individual exocrine ducts, which are surrounded by a spidery web of myoepithelial smooth muscle cells (see Figure 8.17 C). The regulation of these myoepithelial cells involves both neurons and hormones, but these mechanisms are complex and need not concern us now. Instead, let us focus on endocrine glands, especially on the pituitary gland. It is tightly regulated by the brain and in turn affects the brain in a variety of ways.

The Posterior Pituitary

The pituitary gland protrudes from the inferior surface of the brain just posterior to the optic chiasm (Figure 8.21). It consists of two principal divisions, namely, the **anterior pituitary** (adenohypophysis) and the **posterior pituitary** (neurohypophysis). In some species the pituitary also contains an intermediate lobe, which controls skin pigmentation, but this part of the pituitary degenerates in humans after the fetal period.

Magnocellular Hypothalamic Neurons

The posterior pituitary consists mainly of a dense capillary bed and of axons that terminate there (Figure 8.21). The cell bodies of these axons are located in the **supraoptic nucleus** and the **paraventricular nucleus**. Both of these structures are located in the hypothalamus, above (supra) the optic chiasm and near (para) the third ventricle

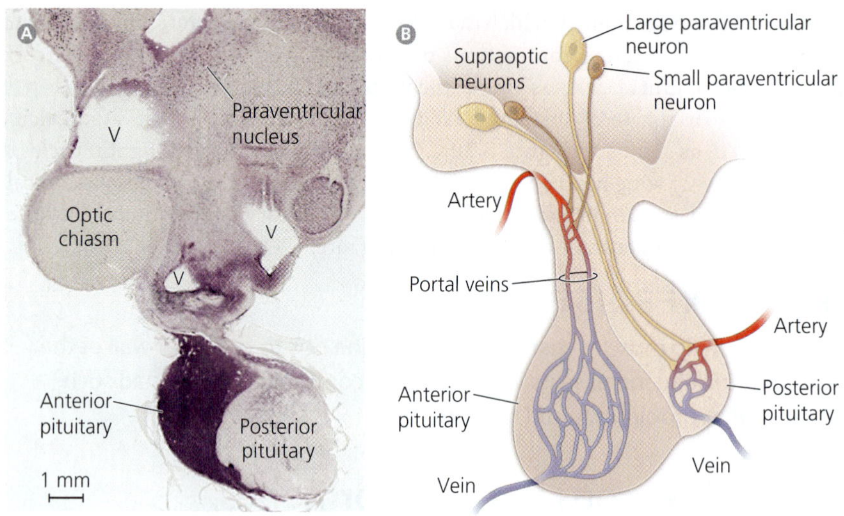

Figure 8.21 The pituitary gland and its innervation. The image in (A) depicts a monkey's pituitary gland and hypothalamus in a sagittal section (anterior is to the left). As shown schematically in (B), magnocellular neurons in the paraventricular and supraoptic nuclei of the hypothalamus send their axons into the posterior pituitary where they release vasopressin and oxytocin into a capillary bed (red-blue lines). Parvocellular neurons release "hormone releasing factors" into capillaries in the upper portion of the anterior pituitary (median eminence). The releasing factors are carried through portal veins to the anterior pituitary where they stimulate pituitary cells to secrete other hormones into the bloodstream. [A from brainmaps.org]

(Figure 8.21 A). The cell bodies of the supraoptic and paraventricular neurons that innervate the posterior pituitary are very large (100–300 μm^3 in volume), which is why they are said to be **magnocellular**.

Why are the cell bodies of these hypothalamic neurons so large? They are voluminous because these neurons synthesize large amounts of hormone. Specifically, they synthesize either **oxytocin** or **vasopressin**. Once synthesized, these peptide hormones are packaged into large vesicles and shipped down to the axon terminals. In response to action potentials invading the axon terminals, some of the vesicles release their contents into the extracellular space just outside of the terminal. From there, the hormones enter the blood through the relatively leaky capillary walls. We will discuss vasopressin in Chapter 9; here, we focus on oxytocin.

Triggering Milk Ejection

One major function of oxytocin is to trigger the ejection of milk from the mammary glands. The mechanisms underlying this **milk ejection reflex** (let-down reflex) have been studied extensively in rats. Specifically, it has been shown that magnocellular supraoptic neurons in female rats fire bursts of action potentials when rat pups suckle on the mother's nipple (Figure 8.22). The bursts of magnocellular activity trigger the release of oxytocin into the blood. Roughly 15 seconds later, the myoepithelial cells in the mother's mammary glands contract, ejecting drops of milk. Additional studies have shown that large amounts of oxytocin are released only when the magnocellular neurons fire in bursts, not when they fire the same number of action potentials in an evenly distributed pattern (Figure 8.22 C). Furthermore, the magnocellular oxytocin neurons fire in bursts only when females have recently given birth (are lactating).

A related observation is that magnocellular oxytocin neurons tend to fire in synchrony with one another only when female rats are lactating. The basis for this synchrony appears to be that the astrocytic processes that normally lie between the magnocellular neurons disappear just after a rat gives birth, causing the neurons to contact one another directly rather than being separated by the astrocytes (Figure 8.23). In correlation with this direct apposition, the magnocellular neurons become linked through gap junctions, as revealed by an increased tendency for injected dyes to spread from one neuron to another. Through these gap junctions, depolarization in one neuron can spread to neighboring neurons without the need for synapses. This current spread, in turn, promotes the tendency of the magnocellular neurons to fire action potentials synchronously (just as gap junctions promote synchronous firing in heart muscle).

Because the magnocellular neurons in lactating rats all fire in synchronized bursts, oxytocin is released from all those neurons simultaneously, causing a brief spike in blood oxytocin concentration. This **pulsatile hormone release** maximizes

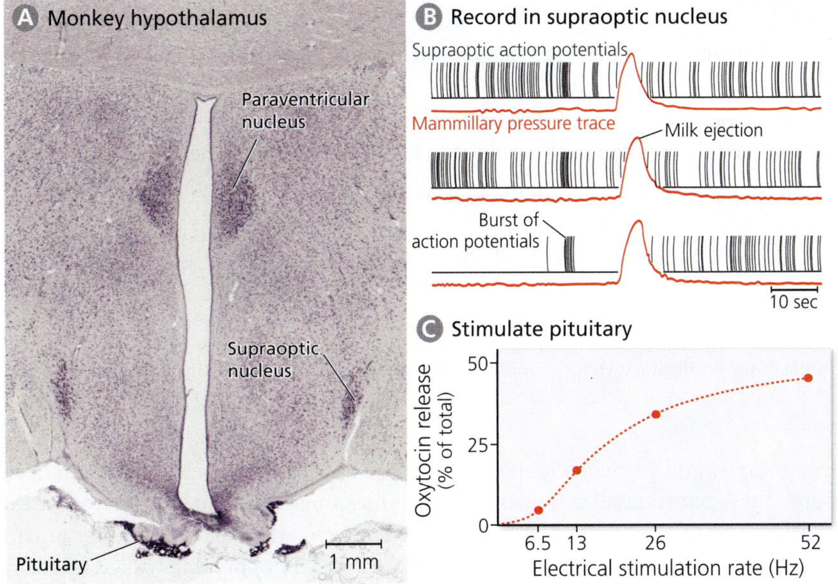

Figure 8.22 Pulsatile oxytocin release. Shown in (A) is a transverse section through the hypothalamus. (B) shows the activity of supraoptic neurons during three milk ejection events, recorded by pressure transducers in the mammary ducts of an anesthetized lactating rat. Bursts of action potentials precede positive pressure deflections (milk ejection) by 10–17 seconds. The graph in (C) shows how much oxytocin is released from an isolated pituitary gland after stimulating it with 156 current pulses at 4 different pulse rates. Because oxytocin release increases sharply with stimulation rate, milk ejection occurs only after supraoptic neurons fire high-frequency bursts of action potentials. [A from brainmaps.org; B from Lincoln & Wakerly, 1974; C after Bicknell, 1988]

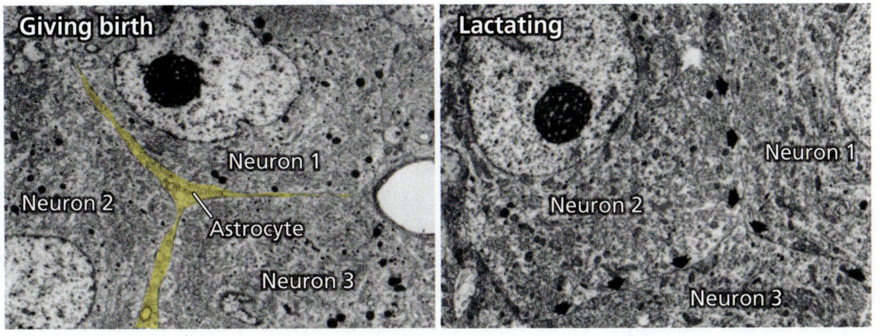

Figure 8.23 Lactation brings supraoptic neurons closer together. These electron micrographs show neurons in the supraoptic nucleus on the day a rat gave birth (left) and 14 days later, when the rat was suckling her pups (right). The three partially visible neurons are separated by astrocytic processes (shaded yellow) on the day of birth. Once the rat is lactating, astrocytes shrink and direct membrane appositions between neurons become common. [From Hatton and Tweedle, 1982]

oxytocin receptor activation in the mammary glands. In contrast, when magnocellular supraoptic neurons fire asynchronously, blood oxytocin levels never rise to the level that is needed for oxytocin receptor activation and milk ejection. Pulsatile hormone release also minimizes receptor desensitization, which occurs when receptors are continually exposed to their ligands (in this case, the oxytocin).

Triggering Contractions of the Uterus

A second important function of oxytocin is to trigger uterine contractions during parturition (giving birth). This **fetus ejection reflex** involves mechanosensory nerve endings in the cervix that increase their firing rate when a fetus pushes against the cervix. This mechanosensory signal is conveyed (through a poorly understood pathway) to the supraoptic magnocellular neurons, which respond with a pulse of oxytocin release. When the oxytocin reaches the smooth muscle in the uterine wall, the uterus contracts and presses the fetus even more firmly against the cervix. The increased pressure from the fetus increases the mechanosensory signal, which then causes even more oxytocin release. The positive feedback loop is broken only when the fetus is delivered. Given this role of oxytocin in parturition, you can understand why labor can be induced with **pitocin**, a synthetic form of oxytocin.

The Anterior Pituitary

The anterior pituitary has been called the "conductor of the endocrine orchestra," but its activity is in turn regulated by neurons in the hypothalamus. In contrast to the posterior pituitary, the anterior pituitary is innervated by relatively small neurons in

	HPA AXIS	HPG AXIS	HPT AXIS
RELEASING HORMONE	CRH (corticotropin releasing hormone)	GnRH (gonadotropin releasing hormone)	TRH (thyrotropin releasing hormone)
PITUITARY HORMONE	ACTH (adrenocorticotropic hormone; a.k.a. corticotropin)	LH and FSH (luteinizing hormone and follicle stimulating hormone)	TSH (thyroid stimulating hormone; a.k.a. thyrotropin)
PRINCIPAL END ORGAN	Cortex of the adrenal gland	Gonads (testes or ovaries)	Thyroid
END ORGAN HORMONES	Cortisol (corticosterone in rodents)	Testosterone or estradiol	Thyroxin and other thyroid hormones

Table 8.2 Major endocrine "axes" passing through the anterior pituitary. HPA = hypothalamic-pituitary-adrenal; HPG = hypothalamic-pituitary-gonadal; HPT = hypothalamic-pituitary-thyroid; a.k.a = also known as.

the supraoptic and paraventricular nuclei as well as several other hypothalamic cell groups. These **parvocellular neurons** (*parvus* meaning "small" in Latin) terminate on a capillary bed in the **median eminence**, which forms the narrow, superior portion of the anterior pituitary. There the axons release a variety of peptides called **releasing hormones** (or releasing factors). These releasing hormones enter the blood and travel through the pituitary capillaries to the inferior portion of the anterior pituitary where they stimulate pituitary cells to synthesize and secrete **pituitary hormones**, which enter the local capillaries and are carried in the blood to distant body parts.

Three Endocrine Axes

The hormones related to the anterior pituitary form three endocrine "axes," namely, the hypothalamic-pituitary-adrenal (HPA) axis, the hypothalamic-pituitary-gonadal (HPG) axis, and the hypothalamic-pituitary-thyroid (HPT) axis (Table 8.2).

The releasing hormone in the HPA axis is called **corticotropin releasing hormone (CRH)**. It stimulates the inferior portion of the anterior pituitary to release *adrenocorticotropic hormone (ACTH)*, which then stimulates the release of *cortisol* (or corticosterone in rodents) from the adrenal gland. The HPG axis, in contrast, begins with **gonadotropin releasing hormone (GnRH)**. It stimulates the release of *luteinizing hormone (LH)*, which ultimately causes the release of *testosterone* or *estradiol* from the testes or ovaries, respectively. GnRH also stimulates anterior pituitary cells to release *follicle stimulating hormone (FSH)*, which promotes the maturation of germ cells (eggs in women, sperm in men). The HPT axis, finally, involves **thyrotropin releasing hormone (TRH)**; this hormone causes the release of *thyrotropin*, which stimulates the thyroid to release *thyroid hormones* (notably thyroxin). The behavioral functions of these three axes are complicated; but, to a first approximation, we can say that the HPA axis regulates stress; the HPG axis regulates reproduction; and the HPT axis regulates metabolism.

Why does the hypothalamus regulate the release of hormones from the anterior pituitary indirectly through the various releasing hormones? The answer is that the releasing hormones amplify the amount of pituitary hormone that enters the blood for a given amount of neural activity. A few action potentials in the hypothalamic neurons trigger the release of a few (thousand) molecules of the releasing hormone, but these activate enzymes in pituitary cells that then facilitate the synthesis and release of many more molecules of the pituitary hormone. Furthermore, a relatively brief pulse of releasing hormone can stimulate pituitary hormone secretion for many minutes. For example, a 60 picogram/ml pulse of GnRH that lasts just a minute or two can trigger an LH pulse of 2 nanogram/ml that lasts for more than an hour. Thus, a weak and brief releasing hormone pulse generates a strong and prolonged release of pituitary hormone. Because of this signal amplification, the hypothalamic neurons that project to the median eminence do not need to synthesize as much hormone as the neurons that secrete oxytocin or vasopressin, which do not function as releasing hormones. This explains why the parvocellular neurons can have much smaller cell bodies than the magnocellular neurons.

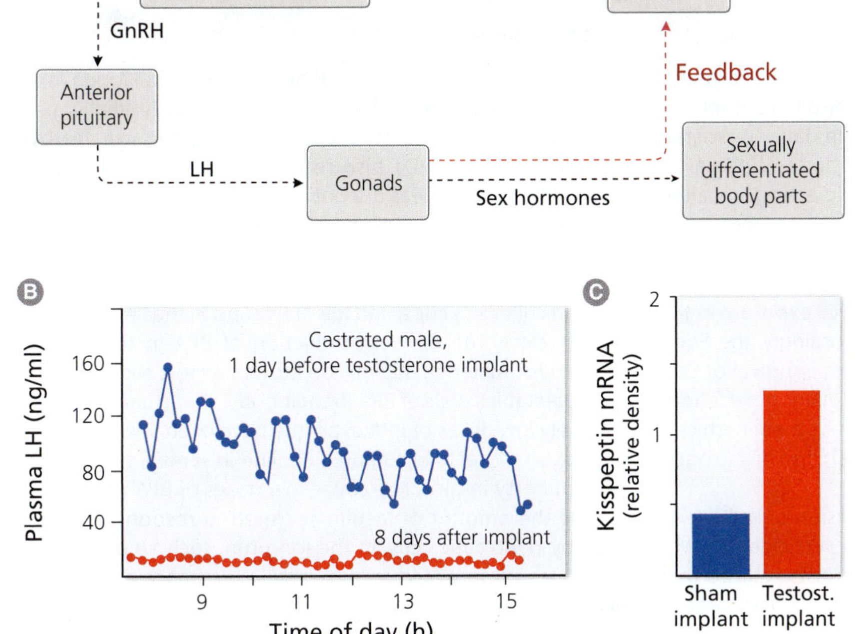

Figure 8.24 Sex hormone regulation. Panel (A) depicts the principal interactions regulating sex hormones. The indicated feedback is always negative in males; in females it is negative except during ovulation, when the feedback turns positive. The graph in (B) shows that artificially supplied testosterone gradually suppresses LH secretion in a castrated male monkey. The histogram in (C) demonstrates that giving testosterone to a castrated male reduces kisspeptin mRNA expression in the basal hypothalamus (which includes the arcuate nucleus). [B after Plant, 1986; C after Shibata et al., 2007]

Neuroendocrine Feedback Loops

Although biologists speak of multiple "axes" that pass through the anterior pituitary, these endocrine pathways are not as linear as the term "axis" suggests. Particularly important is that the pathways contain a **negative feedback loop** that shuts down hormone secretion after a target concentration has been reached. To demonstrate this negative feedback, investigators often remove the end organ on which the anterior pituitary hormones act. For example, after removal of a male's testes, secretion of LH increases dramatically, compared to controls. If the castrated males are then given testosterone implants, which gradually release testosterone into the blood, LH levels decrease over the course of a few days (Figure 8.24). This finding shows that circulating testosterone inhibits LH secretion.

Negative feedback also regulates the concentration of other important hormones. For example, an ovariectomy (removal of the ovaries) causes an increase in LH secretion, which can be reversed by estrogen replacement therapy. Similarly, an adrenalectomy (removal of the adrenal gland) increases CRH and ACTH levels; and thyroid removal boosts TRH and thyrotropin release. These regulatory feedback loops ensure that increases or decreases in anterior pituitary hormone levels tend to last for just a few minutes or hours, rather than days or weeks. This is beneficial because serious problems arise when pituitary hormone levels are elevated chronically (see Box 8.2: Endocrine Disruptors).

The mechanisms underlying the feedback regulation of hormone secretion involves interactions between the circulating hormones and the brain. These interactions are possible because thyroid and steroid hormones (including cortisol, estradiol, and testosterone) are lipophilic, which means that they can easily traverse the blood–brain barrier (see Chapter 5). Once in the brain, the hormones interact directly with matching hormone receptors on the cell or nuclear membranes of neurons. For example, when testosterone enters the brain, it downregulates the production of the neurotransmitter **kisspeptin** (the gene was discovered where Hershey's "chocolate kisses" are made) in neurons of the hypothalamic **arcuate nucleus**. These arcuate neurons use kisspeptin as an excitatory transmitter to activate GnRH-secreting

NEUROLOGICAL DISORDERS

Box 8.2 *Endocrine Disruptors*

Humans and other animals are vulnerable not just to toxins that impair neuron or muscle function but also to compounds that interfere with hormone signaling. The dangers posted by such endocrine disruptors became clear in the 1960s after millions of women were given a synthetic estrogen called *diethylstilbestrol* (DES) in the hope of preventing miscarriages. As it turned out, DES did not function as advertised and instead increased the risk of developing a rare vaginal cancer in the daughters of DES-treated mothers. DES exposure in the womb also decreased female fertility. Accordingly, the Food and Drug Administration (FDA) withdrew its approval of DES for pregnant women in 1971. Since then, many other endocrine disruptors have been identified. Most of them mimic or block the functions of estradiol, the natural form of estrogen in animals.

The most famous synthetic endocrine disruptor is *dichlorodiphenyltrichloroethane* (DDT), a powerful insecticide that was used worldwide until the 1970s, when scientists realized that DDT harms not only insect pests but also vertebrates. For example, they discovered that alligators in a Florida lake heavily contaminated with DDT have significantly smaller penises and lower testosterone levels than alligators in a nearby, less polluted lake. DDT also reduces eggshell thickness in birds. Whereas DDT was discontinued as an agricultural pesticide, other endocrine disruptors continue to be used widely.

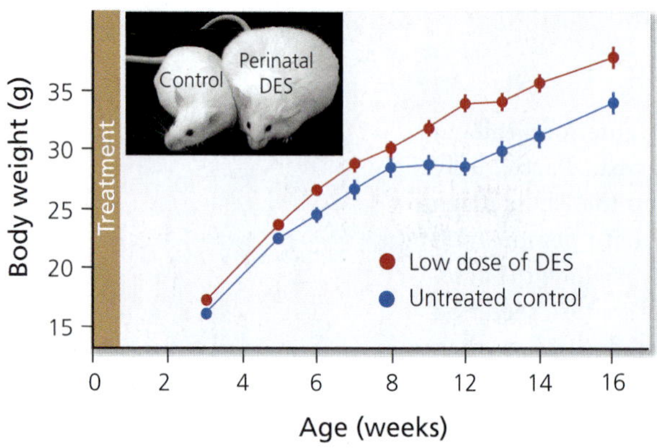

Figure b8.2 Early exposure to the estrogen mimic DES causes adult obesity. Young mice injected with a low dose of diethylstilbestrol (DES) for the first 5 days after birth gradually become obese compared to untreated controls. The difference in body weight becomes statistically significant by 6 weeks of age. The inset shows mice at 6 months. Most of the difference can be attributed to increased body fat in the DES-treated mice. [After Newbold et al., 2005; inset from Newbold et al., 2009]

The best example is *bisphenol A* (BPA). Until the summer of 2012, it was commonly used in the manufacture of baby bottles and children's cups; and it is still found in the lining of aluminum cans. The annual production of BPA in the United States is estimated at 800 million kg, and almost every human has detectable levels of BPA in their body. Unfortunately, even relatively low doses of BPA exposure have been linked to various adverse effects including a decline in semen quality and early puberty in girls. Moreover, low doses of BPA can nearly double the amount of insulin secreted in response to glucose by pancreatic cells. In the long run, such an overactive insulin response would lead to insulin resistance and type 2 diabetes. Extensive recent studies suggest that average levels of BPA exposure in humans remain safe, but continued watchfulness is clearly warranted.

Some plants and some fungi also synthesize endocrine disruptors. For example, soy beans contain the estrogen mimics *genistein* and *coumestrol*. Exposure to these compounds during fetal or juvenile development can interfere with sexual differentiation. For instance, exposing newborn rats to genistein prevents the development of normal estrous cycling in females and alters the size of a sexually dimorphic brain region in males.

Determining the risks associated with exposure to endocrine disruptors is difficult because fetuses and neonates tend to be more sensitive than older animals, and the consequences of early exposure tend not to manifest until later in life. For example, a low dose of DES given to newborn mice has no immediate effect on body weight; but several weeks later, the female DES-treated mice become obese (Figure b8.2). Similarly, perinatal exposure to genistein causes male mice to become obese later in life. The mechanisms underlying these effects remain largely unclear, but increased insulin release and resistance are probably involved. In any case, the available data suggest that the ongoing obesity crisis in many parts of the world may be a consequence, at least in part, of exposure to estrogenic endocrine disruptors.

neurons in other parts of the hypothalamus (Figure 8.24). The overall effect is to reduce GnRH, LH, and testosterone secretion when testosterone levels are high. When testosterone levels are low, the negative feedback loop is inactive, allowing testosterone levels to rise.

Although pituitary hormone regulation is dominated by negative feedback loops, parts of the system can exhibit positive feedback, at least briefly. Specifically, estradiol at some point during the estrous cycle begins to stimulate, rather than inhibit, GnRH release. Increased GnRH levels then boost the secretion of LH, which

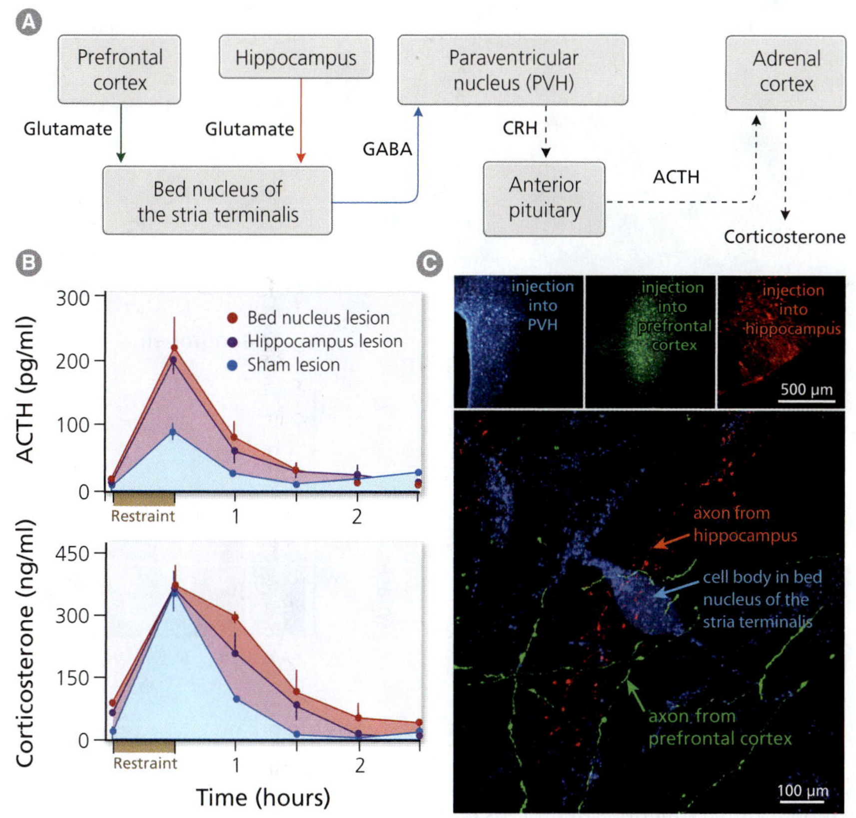

Figure 8.25 Stress hormone regulation. Panel (A) illustrates the principal pathways regulating corticosterone secretion. The graphs in (B) depict blood ACTH (adrenocorticotropic hormone) and corticosterone levels during and after stressing a rat by restraining it in a tightly fitting plastic container. Rats with lesions in the hippocampus (subiculum) or the bed nucleus of the stria terminalis (BNST) exhibit a stronger stress-induced elevation of ACTH and a longer rise in corticosterone. The photographs in (C) illustrate an experiment in which blue, green, and red fluorescent tracers were injected into the paraventricular nucleus (PVH), prefrontal cortex, and hippocampus, respectively. The large image shows labeled cell bodies and axons in the BNST, demonstrating that the hippocampus and prefrontal cortex send converging inputs to the BNST, which in turn projects to the PVH. [From Radley and Sawchenko, 2011]

in turn enhances estradiol secretion from the ovaries. This sets up a positive feedback loop that causes brief but dramatic elevations in GnRH, LH, and estradiol concentrations. These **hormone surges** trigger ovulation. Soon thereafter, the negative feedback is reestablished and estradiol levels drop.

Hormone Set Point Regulation

What mechanisms determine the set points around which pituitary hormone levels are regulated, and how does the system determine what hormone concentrations are "normal"? Answers to these questions remain elusive, but one structure that appears to be involved in setting pituitary hormone levels is the **suprachiasmatic nucleus**, which harbors the brain's "internal clock" (see Chapter 9). This nucleus projects to several of the hypothalamic regions that control pituitary hormone release, and lesions of the suprachiasmatic nucleus abolish the "normal" daily fluctuations in hormone levels, such as the rise of ACTH, cortisol, and testosterone levels in the early morning and the elevation of vasopressin at night.

Another region that controls pituitary hormone levels is the **bed nucleus of stria terminalis** (Figure 8.25). This small nucleus in the ventral telencephalon sends inhibitory projections to the paraventricular nucleus of the hypothalamus and receives excitatory inputs from both the prefrontal cortex and the hippocampus. Although neuroscientists tend to think of the prefrontal cortex and the hippocampus as performing complex cognitive functions (see Chapters 14 and 15), they also help to regulate stress hormone release.

Hippocampal Regulation of Stress Hormones

The mammalian hippocampus is probably the single most intensively studied region of the brain. As you may recall from Chapter 3 (Box 3.2: Hippocampal Structure and Functions), the hippocampus plays a major role in the ability to navigate through space using a remembered "map." It is also involved in the recall of memories about

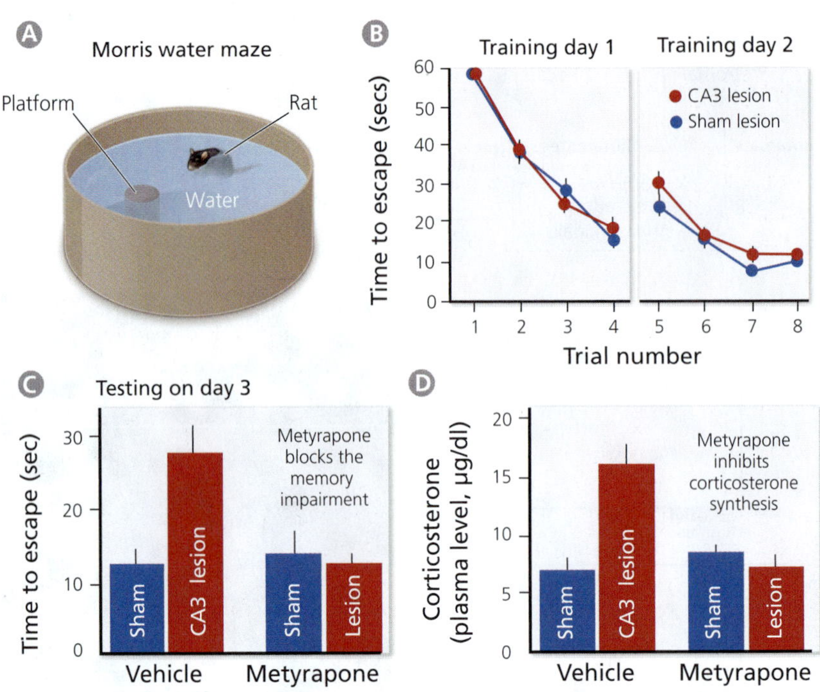

Figure 8.26 The hippocampus, stress, and memory. Rats with bilateral lesions in the hippocampus (specifically in CA3; see Chapter 14) were trained over the course of 2 days to find a hidden platform in a Morris water maze (A). The lesioned rats learned to find the platform as efficiently as sham-lesioned rats (B). However, when the rats were tested on day 3, the CA3-lesioned rats needed significantly longer to find the platform, suggesting that memory recall was impaired (C). Injecting the rats with a corticosterone synthesis inhibitor (metyrapone) abolished the effect. The histogram in (D) confirms that metyrapone blocks the lesion-induced rise in corticosterone. These data imply that hippocampus lesions impair memory recall indirectly by altering stress hormone regulation. [After Roozendaal et al., 2001]

what happened when and where (episodic memories). We will discuss these aspects of hippocampal function later in this book.

Less widely known is that the hippocampus also helps to regulate stress hormone release. Strong evidence for this hypothesis comes from an experiment in which part of the hippocampus was lesioned bilaterally (Figure 8.26). Several days later, the lesioned rats were trained to navigate a **Morris water maze**, which is a circular tub filled with water that hides a small submerged platform. When rats are placed in the tub, they must swim until they find the platform and then climb onto it to get some rest. Over the course of several trials, the rats learn where the platform is and swim to it more directly, reducing their "time to escape" (for more details, see Chapter 11).

The rats in this experiment were given 8 training trials over the course of 2 days. On the first day, the lesioned rats learned to find the platform, and they did so just as well as sham-lesioned rats (which underwent similar surgery but no hippocampus lesions). On the second training day, both groups of rats learned more detailed information about the platform's spatial location, reducing their average time to escape to just over 10 seconds; the hippocampus-lesioned rats took slightly longer to find the hidden platform, but this difference was not statistically significant. However, when the rats were tested again on day 3, the hippocampus-lesioned animals took significantly longer than control rats to find the platform. Indeed, the lesioned rats were no better than they had been on the first trial of the second day, suggesting that they remembered only that there was a platform somewhere in the tank (and thus stopped swimming along the wall, which is what totally naive rats tend to do). In other words, the hippocampus-lesioned rats did not remember the platform's specific location. Overall, these findings indicate that an intact hippocampus is required for some aspect of long-term spatial memory. This much is not controversial, but the scientists performed a few additional experiments.

First, they showed that that blood levels of the stress hormone corticosterone (cortisol in humans) are significantly higher in hippocampus-lesioned rats than in controls. This finding is consistent with the anatomical data indicating that the hippocampus is part of the circuitry that regulates stress hormone levels (Figure 8.25). Next, they showed that the lesion-induced elevation in stress hormone can be reversed by giving the lesioned rats **metyrapone**, a blocker of corticosterone synthesis

(Figure 8.26 D). Finally, they gave some rats metyrapone just before testing their performance on day 3. This treatment eliminated the hippocampus lesion effect on spatial memory. That is, both groups of rats remembered the hidden platform location just fine (Figure 8.26 C).

These data indicate that the hippocampus-lesioned rats had learned where the platform was on days 1 and 2, but could not recall that information on day 3 unless their stress hormone levels were brought into the normal range by metyrapone administration. This finding is consistent with other data showing that high stress levels generally interfere with memory recall rather than memory formation. It is well known, for example, that the stress of taking a test may cause your memory to fail. Once the stress has passed, you remember the information perfectly well. The larger lesson to be learned from this experiment is that hippocampus lesions can alter hormone levels, which then affect some cognitive functions.

Brain Regions Are Multifunctional

An even more general principle emerging from this discussion is that brain regions tend to have multiple functions. The hippocampus is involved in high-level information processing, such as long-term spatial memory, and in the control of stress hormones. Similarly, the pituitary gland is involved in the regulation of many different hormones, which in turn have multiple functions. Because multifunctionality is pervasive at all levels of the nervous system, grasping the brain's functional organization is difficult. As you delve into neurobiology, it is natural to want to know what each brain region "does." However, each brain area has multiple functions, and each function is distributed across multiple brain areas. Therefore, learning about the brain one region at a time is misleading, or at least confusing in the long run. That is why this book is not organized around individual brain regions, but around functional systems. In the next chapter, you will learn about relatively simple vegetative functions such as breathing and temperature regulation. Subsequent chapters will cover progressively more complex functions.

BRAIN EXERCISE

If cortisol in the blood downregulates the release of corticotropin releasing hormone (CRH) in the anterior pituitary, how can stressful situations cause a significant increase in blood cortisol concentration?

SUMMARY

Section 8.1 - Skeletal muscle contains numerous muscle fibers, which consist of multiple myofibrils, each of which contains a series of sarcomeres. Individual sarcomeres contain interleaved actin and myosin filaments.

- Actin and myosin filaments slide past one another as muscles change in length. Contractile force is generated when the heads of myosin molecules bind to actin and, consequently, bend. When a myosin head detaches from actin, a process that requires ATP, the myosin reverts to its straight shape.
- Motor neurons release acetylcholine at neuromuscular junctions. Acetylcholine depolarizes the sarcolemma, triggering an action potential that spreads through the T-tubule system into the muscle fiber's interior where it triggers calcium release from the sarcoplasmic reticulum. The rise in calcium makes actin binding sites accessible to myosin and thus facilitates cross-bridge cycling.
- A motor neuron may innervate just a few muscle fibers or many, but all the fibers it innervates tend to be of the same type. Individual motor units (a motor neuron and its associated muscle fibers) are recruited in an orderly fashion as muscle contractions increase in strength. Small motor units, which by themselves generate little contractile force, fire during most muscle contractions, whereas large and powerful motor units fire only during strong contractions.
- Muscle spindles contain intrafusal muscle fibers, which are innervated by gamma motor neurons. When intrafusal fibers contract in a muscle that has already

shortened, the intrafusal fibers are pulled taut. This restores the muscle spindle's sensitivity to muscle stretch.

Section 8.2 - Some cardiac muscle fibers contain hyperpolarization-activated ion channels that cause them to be rhythmically active, even without neural input. When neurons release norepinephrine onto the sinoatrial node, the heart's primary pacemaker, its rhythm speeds up. In response to acetylcholine, the rhythm slows.

Section 8.3 - Smooth muscle fibers do not contain sarcomeres, and they control actin-myosin binding by calcium-calmodulin-activated phosphorylation of the myosin heads. They typically surround structural cavities, such as the intestines, bronchial airways, blood vessels, and exocrine glands. On contraction of the smooth muscle, the cavities constrict.

Section 8.4 - Because the actin-myosin motor is unidirectional, lengthening a contracted muscle requires an external force, which is usually supplied by antagonistic muscles or, for muscles lining cavities, by internal pressure.

Section 8.5 - Exocrine glands eject their secretions through ducts to the body surface, whereas endocrine glands secrete hormones into the blood.

- Magnocellular neurons in the hypothalamus release oxytocin or vasopressin into capillaries that course through the posterior pituitary. Pulsatile oxytocin release is involved in the ejection of mother's milk and in the fetus ejection reflex.
- Cells in the anterior pituitary secrete hormones that stimulate the secretion of other hormones in the adrenal and thyroid glands as well as the gonads. These hormones all feed back onto hypothalamic neurons that regulate hormone secretion in the anterior pituitary.
- The hippocampus is involved in spatial navigation and memory, but it also participates in regulating stress hormone levels, which in turn affect memory recall.

Box 8.1 - Outer hair cells shorten when they are depolarized, and they lengthen in response to hyperpolarization. This electromotility depends on the shape-changing molecule prestin, and it increases auditory sensitivity.

Box 8.2 - Both synthetic and natural compounds can mimic or block estrogen signaling, which may lead to serious problems such as obesity and abnormal sexual development, especially if the exposure occurs early in development.

KEY TERMS

ADDITIONAL READINGS

8.1 - Skeletal Muscle

Allen DG, Lamb GD, Westerblad H. 2008. Skeletal muscle fatigue: cellular mechanisms. *Physiol Rev* **88**:287–332.

Barbara J-G, Clarac F. 2011. Historical concepts on the relations between nerves and muscles. *Brain Res* **1409**:3–22.

Duchateau J, Enoka RM. 2011. Human motor unit recordings: origins and insight into the integrated motor system. *Brain Res* **1409**:42–61.

Henneman E, Somjen G, Carpenter DO. 1965. Functional significance of cell size in spinal motoneurons. *J Neurophysiol* **28**:560–580.

Lieber RL, Ward SR. 2011. Skeletal muscle design to meet functional demands. *Phil Trans R Soc B* **366**:1466–1476.

Llewellyn ME, Thompson KR, Deisseroth K, Delp SL. 2010. Orderly recruitment of motor units under optical control *in vivo*. *Nat Med* **16**:1161–1165.

Rhee HS, Hoh JFY. 2010. Immunohistochemical analysis of the effects of cross-innervation of murine thyroarytenoid and sternohyoid muscles. *J Histochem Cytochem* **58**:1057–1065.

Schiaffino S, Reggiani C. 2011. Fiber types in mammalian skeletal muscles. *Physiol Rev* **91**:1447–1531.

Slater CR. 2003. Structural determinants of the reliability of synaptic transmission at the vertebrate neuromuscular junction. *J Neurocytol* **32**:505–522.

Spillane J, Beeson DJ, Kullmann DM. 2010. Myasthenia and related disorders of the neuromuscular junction. *J Neurol Neurosurg Psychiatry* **81**:850–857.

Wu H, Xiong WC, Mei L. 2010. To build a synapse: signaling pathways in neuromuscular junction assembly. *Development* **137**:1017–1033.

8.2 - Cardiac Muscle

DiFrancesco D. 2010. The role of the funny current in pacemaker activity. *Circ Res* **106**:434–446.

Fedorov VV, Glukhov AV, Chang R, Kostecki G, Aferol H, Hucker WJ, et al. 2010. Optical mapping of the isolated coronary-perfused human sinus node. *J Am Coll Cardiol* **56**:1386–1394.

8.3 - Smooth Muscle

Burnstock G. 2008. Non-synaptic transmission at autonomic neuroeffector junctions. *Neurochem Int* **52**:14–25.

8.4 - Principles of Muscle Mechanics

Buckberg G, Hoffman JIE, Mahajan A, Saleh S, Coghlan C. 2008. Cardiac mechanics revisited: the relationship of cardiac architecture to ventricular function. *Circulation* **118**:2571–2587.

8.5 - Glands

Christian CA, Moenter SM. 2010. The neurobiology of preovulatory and estradiol-induced gonadotropin-releasing hormone surges. *Endocr Rev* **31**:544–577.

Denver RJ. 2009. Structural and functional evolution of vertebrate neuroendocrine stress systems. *Ann N Y Acad Sci* **1163**:1–16.

Leng G, Moos FC, Armstrong WE. 2010. The adaptive brain: Glenn Hatton and the supraoptic nucleus. *J Neuroendocrinol* **22**:318–329.

Plant TM, Ramaswamy S. 2009. Kisspeptin and the regulation of the hypothalamic-pituitary-gonadal axis in the rhesus monkey (*Macaca mulatta*). *Peptides* **30**:67–75.

Sapolsky RM, Krey LC, McEwen BS. 1984. Glucocorticoid-sensitive hippocampal neurons are involved in terminating the adrenocortical stress response. *Proc Natl Acad Sci U S A* **81**:6174–6177.

Theodosis DT, Poulain DA, Oliet SHR. 2008. Activity-dependent structural and functional plasticity of astrocyte-neuron interactions. *Physiol Rev* **88**:983–1008.

Boxes

Adams NR. 1990. Permanent infertility in ewes exposed to plant oestrogens. *Aust Vet J* **67**:197–201.

Dallos P, Fakler B. 2002. Prestin, a new type of motor protein. *Nat Rev Mol Cell Biol* **3**:104–111.

Guillette LJ, Pickford DB, Crain DA, Rooney AA, Percival HF. 1996. Reduction in penis size and plasma testosterone concentrations in juvenile alligators living in a contaminated environment. *Gen Comp Endocrinol* **101**:32–42.

Rochester JR, Millam JR. 2009. Phytoestrogens and avian reproduction: exploring the evolution and function of phytoestrogens and possible role of plant compounds in the breeding ecology of wild birds. *Comp Biochem Physiol Part A* **154**:279–288.

Sharpe RM. 2010. Is it time to end concerns over the estrogenic effects of bisphenol A? *Toxicol Sci* **114**:1–4.

Tan X, Pecka JL, Tang J, Okoruwa OE, Zhang Q, Beisel KW, He DZZ. 2011. From zebrafish to mammal: functional evolution of prestin, the motor protein of cochlear outer hair cells. *J Neurophys* **105**:36–44.

Veurink M, Koster M, de Jong-van den Berg LTW. 2005. The history of DES, lessons to be learned. *Pharm World Sci* **27**:139–143.

Waye A, Trudeau VL. 2011. Neuroendocrine disruption: more than hormones are upset. *J Toxicol Environ Health B* **14**:270–291.

Regulating Vital Bodily Functions

CHAPTER 9

CONTENTS

FEATURES

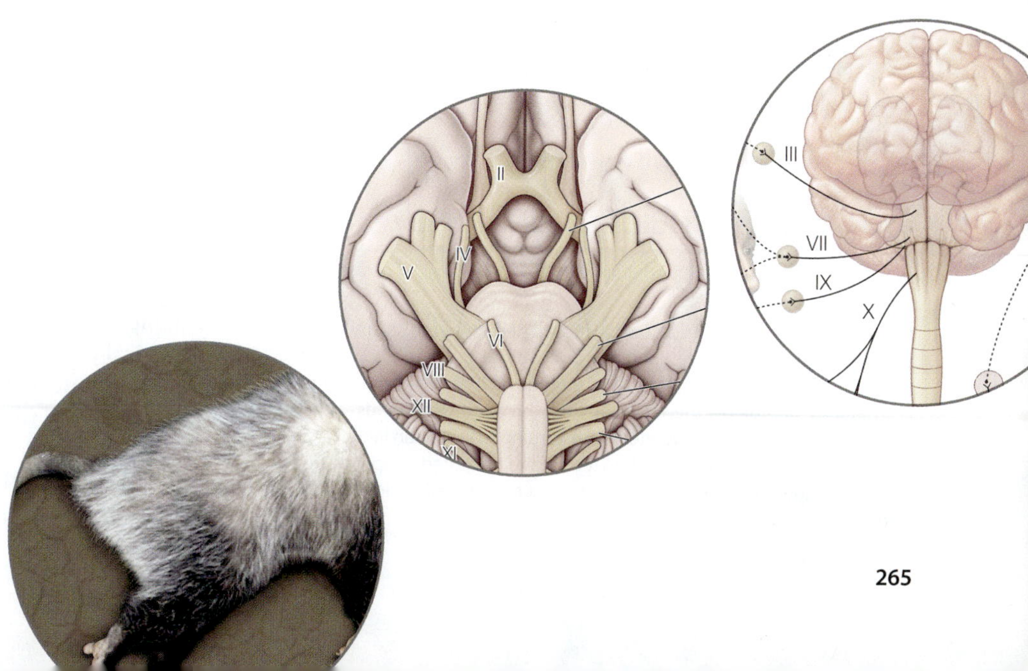

In previous chapters, we discussed the function of individual neurons, sensors, and effectors. Now it is time to consider how these elements are combined into circuits that control behavior. The most obvious sort of behavior involves skeletal muscles and outwardly visible movements, such as walking or turning your head; we will deal with such behaviors in Chapter 10. In the present chapter, we focus on the less obvious behaviors that occur within an organism's body and involve, for the most part, effectors other than skeletal muscle. These less obvious behaviors include the heart's pumping of blood as well as breathing, shivering, and bladder contraction. Such vital bodily functions are often called *vegetative processes* because they require no thought. Indeed, many vegetative processes are well beyond conscious control.

Of course, the lack of voluntary control over our vegetative processes does not imply a lack of neural control. As you will learn in this chapter, the inside of your body is packed with vital processes that are controlled by finely honed neural circuits. Because many of the circuits controlling our vital bodily functions are relatively simple, it is a good idea to study them before we contemplate the more complex circuits that control other behaviors, especially because several key principles of how neurons control behavior are conserved across all kinds of neural circuits and behaviors.

9.1 How Do We Maintain Physiological Stability?

The French physiologist Claude Bernard in the 1850s developed the idea that an animal's internal environment, which he called the internal milieu, is more stable than the external environment and that this internal stability is vital for survival. Eighty years later, in an influential book called *The Wisdom of the Body*, the American physiologist Walter Cannon coined the term **homeostasis** to denote this internal stability. He described how tightly our bodies regulate body temperature, pH, water content, blood sugar, calcium, and many other physiological variables. His major point was that homeostasis is not a happy coincidence but the result of active regulation by many finely tuned homeostatic processes.

Homeostatic control can often be achieved by simple feedback loops, such as processes that shut down pituitary hormone secretion when sex, stress, or thyroid hormone levels get too high (see Chapter 8). These negative feedback systems are functionally analogous to the heating system in your house or apartment where a thermostat compares the temperature of the room to the desired temperature and then turns off the heater if room temperature exceeds the **set point** (the target value or range).

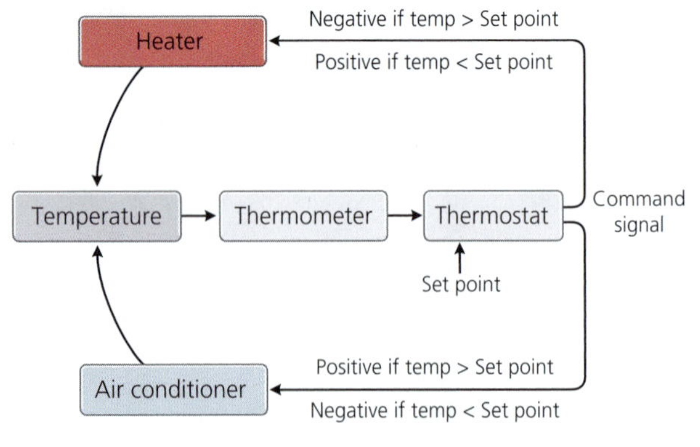

Figure 9.1 Push-pull regulation. Illustrated here is a sophisticated thermostat that regulates temperature by sending opposing commands to antagonistic effectors, namely, a heater and an air conditioner. If the temperature falls below the set point, then the thermostat sends an excitatory command signal to the heater, turning it on, and an inhibitory signal to the air conditioner, shutting it off. If the measured temperature rises above the set point, then the polarity of the command signals is reversed.

However, homeostatic control often involves more than simple negative feedback and stable set points. For example, one of our physiological responses to a drop in body temperature is to increase the set point for thyroid hormone concentration in the blood. The resulting rise in thyroid hormone levels increases the rate at which cells take up carbohydrates and fat, which they then burn for energy. This increase in basal metabolic rate helps to raise body temperature because cellular metabolism generates a significant amount of heat as well as ATP. Thus, regulation of one physiological variable can involve changes in the set point of another variable.

An important principle of homeostatic regulation is that physiological variables may be affected by multiple mechanisms that act antagonistically. One mechanism may push the variable upward while another one actively pulls it down. To visualize this kind of **push-pull regulation**, imagine a high-tech thermostat that controls both a heater and an air-conditioner (Figure 9.1). If the temperature in the room is below the set point, such a thermostat would turn the air-conditioner off and the heater on. Conversely, if the temperature is too high, the thermostat would turn on the air-conditioning and shut down the heater. As you will see, the control of body temperature and many other homeostatic systems involve precisely this sort of push-pull control.

A key feature of push-pull regulation is that the controller sends opposing commands to the antagonistic effectors. This is an efficient design because simultaneous activation of opposing effectors would not alter the variable that is to be controlled or do so inefficiently. The principle of push-pull control of opposing effectors will feature heavily in this chapter, but it is also critical for the control of the muscles that move our limbs (see Chapter 10).

Where in the nervous system are the neurons responsible for all this homeostatic control of vegetative processes? The hypothalamus is one critical player, as it controls the pituitary gland. However, vegetative control involves a wide variety of neurons in many parts of the nervous system. To understand where these neurons are located and how they relate to one another, you need some neuroanatomical background on what is called the autonomic nervous system.

BRAIN EXERCISE

To what extent does the "cruise control" in a modern car employ push-pull regulation?

9.2 What Parts of the Nervous System Control the Vital Bodily Functions?

Claudius Galen, the most famous physician of ancient Rome, realized that an extensive network of nerves connects the body's various internal organs with one another and with the spinal cord. He speculated that this network allows the internal organs to be controlled in a highly coordinated fashion or, as he put it, in sympathy with one another. Consequently, Galen's system of nerves became known as the **sympathetic nervous system**.

Around 1900, the British physiologist John Langley expanded Galen's concept and coined the term **autonomic nervous system** (Figure 9.2), by which he meant "the nervous system of the glands and of the involuntary muscle." Langley chose the word "autonomic" to indicate that neural control of the visceral organs is largely autonomous, or independent of the will. This independence is not as complete as the name implies, but Langley's term is generally appropriate and widely accepted. As we discuss shortly, the autonomic nervous system also contains a parasympathetic division, which tends to function in opposition to the sympathetic nervous system.

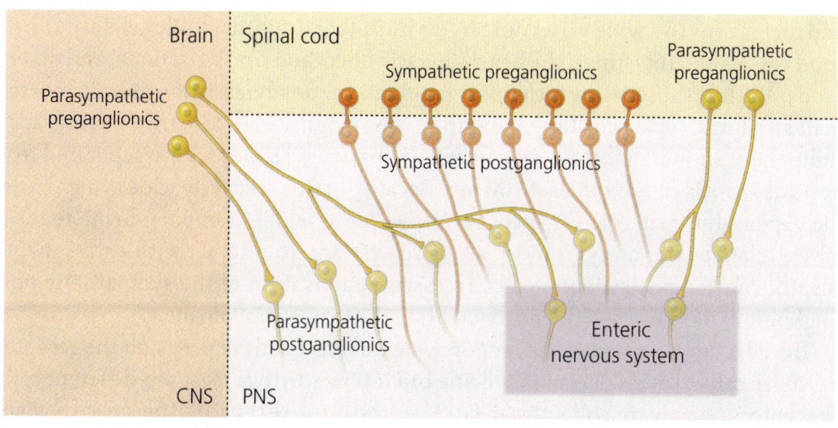

Figure 9.2 The autonomic nervous system in context. This diagram shows the principal components of the autonomic nervous system relative to the central and peripheral nervous systems (CNS and PNS, respectively). The preganglionic neurons of the sympathetic (orange) and parasympathetic (yellow) divisions have their cell bodies within the CNS, mainly within the brainstem and spinal cord, and extend their axons into the PNS. All of the postganglionic neurons lie within the PNS, as does the enteric nervous system.

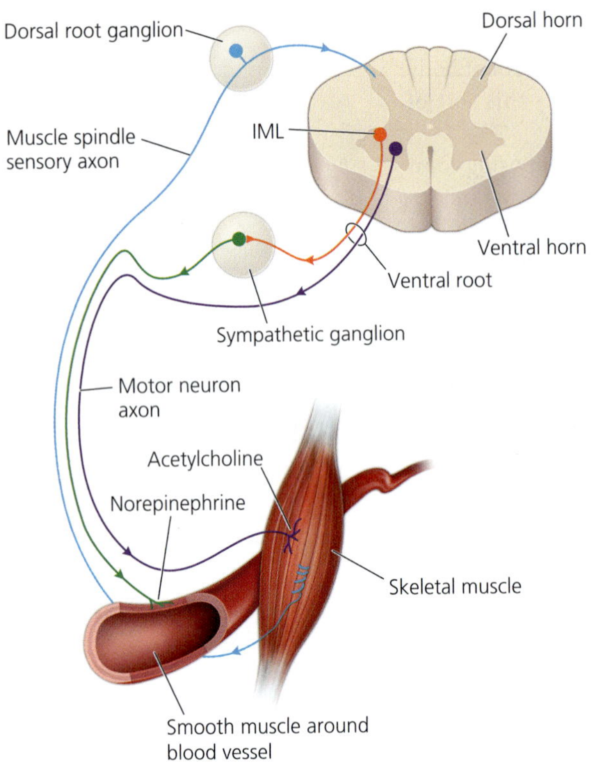

Figure 9.3 Sympathetic pre- and postganglionic neurons. Sympathetic preganglionic neurons in the intermediolateral column (IML) of the spinal cord (orange) innervate sympathetic postganglionic neurons (green) that innervate smooth muscle surrounding blood vessels. In contrast, spinal motor neurons (purple) innervate skeletal muscle directly. Also shown is a peripheral sensory neuron (light blue).

Some parts of the autonomic nervous system lie in the brainstem and spinal cord, but most of the autonomic nervous system lies in the peripheral nervous system (Figure 9.2). The cell bodies of the autonomic neurons that innervate muscles and glands are located in the peripheral nervous system and clustered into *autonomic ganglia*. They are called **postganglionic neurons** because their axons lie downstream of those autonomic ganglia. By extension, the premotor neurons providing input to the postganglionic neurons are referred to as **preganglionic neurons**.

The Sympathetic Division of the Autonomic Nervous System

An important feature of the sympathetic division of the autonomic nervous system is that the cell bodies of its preganglionic neurons are located in a very specific portion of the spinal cord called the **intermediolateral column**. This term reflects the fact that these neurons lie between the spinal cord's dorsal and ventral horns and occupy a lateral position within the spinal gray matter (Figure 9.3). The term "column" is similarly fitting, as the sympathetic preganglionic neurons are found in several consecutive spinal cord segments. Specifically, they extend from the first thoracic segment (T1) to the third lumbar (L3) segment (see Figure 7.4). Thus, they form a slender, elongated column of cells.

The axons of the sympathetic preganglionic neurons exit the spinal cord through the ventral roots of the spinal nerves (Figure 9.3) but do not travel far. Most of them terminate in autonomic ganglia that lie close to the spinal cord, where they contact the dendrites and cell bodies of the sympathetic postganglionic neurons. Many of the **sympathetic ganglia** lie to the left and right of the vertebral column and are interconnected by axons so that they resemble two parallel strings of beads; others form a chain ventral to the spinal cord; and yet others form isolated beads within the neck and abdomen. A common feature is that all the sympathetic ganglia lie close to the spinal cord, far from the organs their neurons innervate. The neurotransmitter released by sympathetic preganglionic neurons inside the sympathetic ganglia is *acetylcholine*. It depolarizes the sympathetic postganglionic neurons and thus functions as an excitatory transmitter (just as it does at the neuromuscular junction).

Norepinephrine as a Transmitter

The postganglionic neurons that have their cell bodies in the sympathetic ganglia send their axons to a wide variety of targets including smooth muscle cells around the blood vessels and intestines, the heart's muscle fibers and sinoatrial node, and diverse glands. Within those tissues, the postganglionic axons release *norepinephrine* from small swellings, or varicosities, along the axons (Figure 9.4). Electron micrographic studies have shown that these *en passant* ("in passing") synapses are separated from the postsynaptic cells by at least 100 nm. Because of this relatively wide synaptic cleft (the gap at glutamatergic synapses is only ~25 nm wide), norepinephrine tends to diffuse relatively far away from its site of release and to interact with extrasynaptic receptors on the postsynaptic cells. In essence, activation of the sympathetic postganglionic axons "bathes" the postsynaptic cells in norepinephrine.

The effect of norepinephrine on postsynaptic target cells depends on the receptors that those cells express. Norepinephrine can interact with at least five different types of receptors, which are collectively called **adrenergic receptors** (because they also interact with adrenaline). These receptors form two subfamilies, referred to as alpha- and

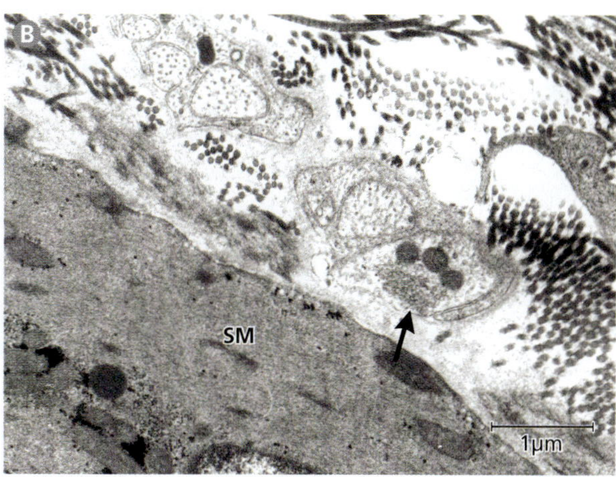

Figure 9.4 Sympathetic postganglionic axons and synapses. Shown in (A) is a scanning electron micrograph of a sympathetic postganglionic axon coursing along a blood vessel ringed with smooth muscle fibers. You can see two varicosities (*en passant* synapses) along the axon. Panel (B) is a transmission electron micrograph of an axon terminal innervating a smooth muscle (SM). [A from Uehara and Suyama, 1978; B from Monos et al., 2001]

beta-adrenergic receptors. All of them are G protein-coupled metabotropic receptors (see Chapter 2), but they activate different intracellular signaling cascades.

Alpha- and beta-adrenergic receptors tend to have opposite effects. For example, activation of alpha-adrenergic receptors tends to constrict the smooth muscle cells surrounding blood vessels, whereas beta-adrenergic activation tends to relax them. However, the effect of adrenergic receptor activation often depends on the identity of the postsynaptic cell. For example, alpha-adrenergic activation in the pancreas inhibits insulin secretion but promotes the secretion of glucagon from other pancreatic cells. This variation in adrenergic receptor effects is confusing until you realize that insulin generally lowers blood sugar, whereas glucagon raises it. Thus, norepinephrine release within the pancreas has opposite effects on antagonistic effectors. It is a good example of the push-pull regulation that we discussed earlier. In general, the varied postsynaptic effects of norepinephrine release become more comprehensible once you realize that sympathetic activation triggers the fight-or-flight response, which involves increased glucose consumption and many other physiological and behavioral effects that help an animal confront a sudden threat (see Section 9.5).

Before we finish this description of the sympathetic nervous system, you should know that one set of sympathetic axons defies the sympathetic stereotype: the sympathetic preganglionic axons that innervate the adrenal gland do not synapse in a sympathetic ganglion but instead project directly to epinephrine-secreting **chromaffin cells** in the adrenal gland. Although this arrangement is unusual, chromaffin cells look like fairly typical sympathetic ganglion cells early in development. Only later do they migrate into the core of the adrenal gland where they mature into epinephrine-secreting cells. Thus, the adrenal chromaffin cells are modified sympathetic postganglionic neurons. Why do mature chromaffin cells release epinephrine rather than norepinephrine? No one knows for sure, but epinephrine is synthesized from norepinephrine by removing a single methyl group. Therefore, changing a norepinephrine-releasing cell into one that secretes epinephrine is a relatively minor developmental modification.

The Parasympathetic Division of the Autonomic Nervous System

The pre- and postganglionic neurons of the parasympathetic division of the autonomic nervous system both use acetylcholine as their main neurotransmitter. No epinephrine

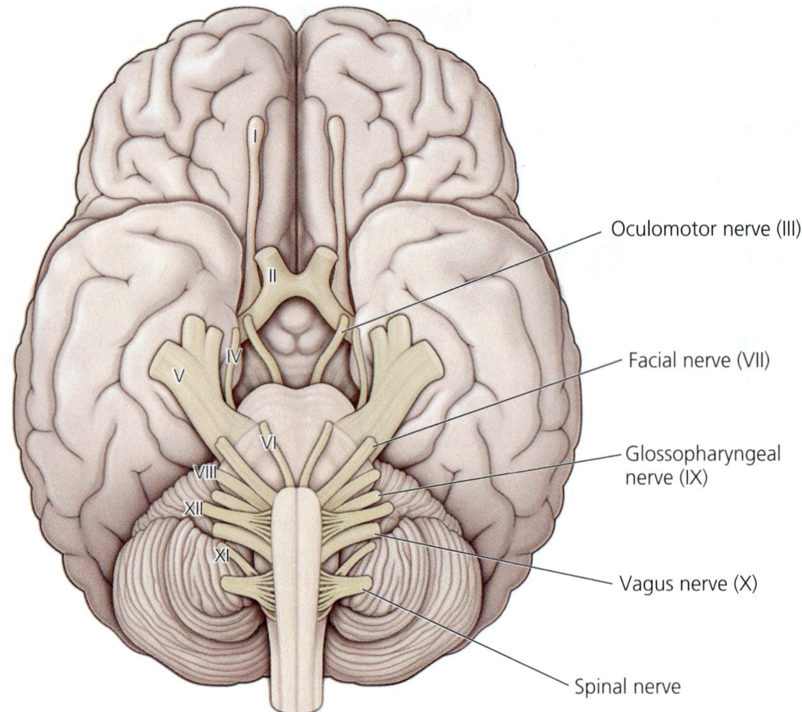

Figure 9.5 Cranial nerves related to the autonomic nervous system. The twelve cranial nerves of humans are numbered in a roughly anterior-to-posterior sequence. Each nerve may be referred to either by its proper name (e.g., vagus nerve) or by its number, expressed in Roman numerals (e.g., nerve X). Autonomic axons are found in cranial nerves III, VII, IX, and X as well as in the spinal nerves.

here! Another difference from the sympathetic division is that the cell bodies of parasympathetic preganglionic neurons are located not just in the spinal cord but also in the midbrain and the medulla (Figure 9.2). The axons of those midbrain and medullary preganglionic neurons exit the brain through **cranial nerves**, rather than spinal nerves. Learning which axons pass through which cranial nerve is very important for budding neurologists who want to diagnose potential nerve damage from a patient's symptoms. However, for us right now it is enough to note that humans have twelve pairs of cranial nerves and that the parasympathetic preganglionic axons travel through cranial nerves III, VII, IX, and X (Figure 9.5).

In contrast to the sympathetic preganglionic neurons we discussed in the previous section, the parasympathetic postganglionic neurons tend to lie close to their targets (Figure 9.6). For example, the parasympathetic axons that innervate the pupillary constrictor muscle have their cell bodies in a small ganglion that sits right behind the eye. Similarly, the parasympathetic neurons that innervate the heart form small clusters on the surface of the heart. In some organs, the parasympathetic postganglionic neurons are actually embedded within the target tissue. In the bladder, for example, small clusters of parasympathetic neurons are intermingled with the bladder's smooth muscle fibers.

Although the parasympathetic and sympathetic divisions of the autonomic nervous system innervate many of the same organs, they tend to have opposite, antagonistic effects on the function of those structures (Figure 9.6). For example, sympathetic stimulation of the heart tends to quicken the heartbeat, whereas parasympathetic activation slows it down. Similarly, sympathetic stimulation dilates the pupil, whereas parasympathetic activation triggers pupillary constriction.

This antagonism is not universal (the liver receives only sympathetic innervation, and the tear glands are innervated only by parasympathetic axons), but sympathetic and parasympathetic activation tend to have opposing effects on the body as a whole. As you will learn in Section 9.5, strong sympathetic activation triggers the fight-or-flight response, which is characterized by arousal, exertion, and fear, whereas parasympathetic activation tends to calm a creature down.

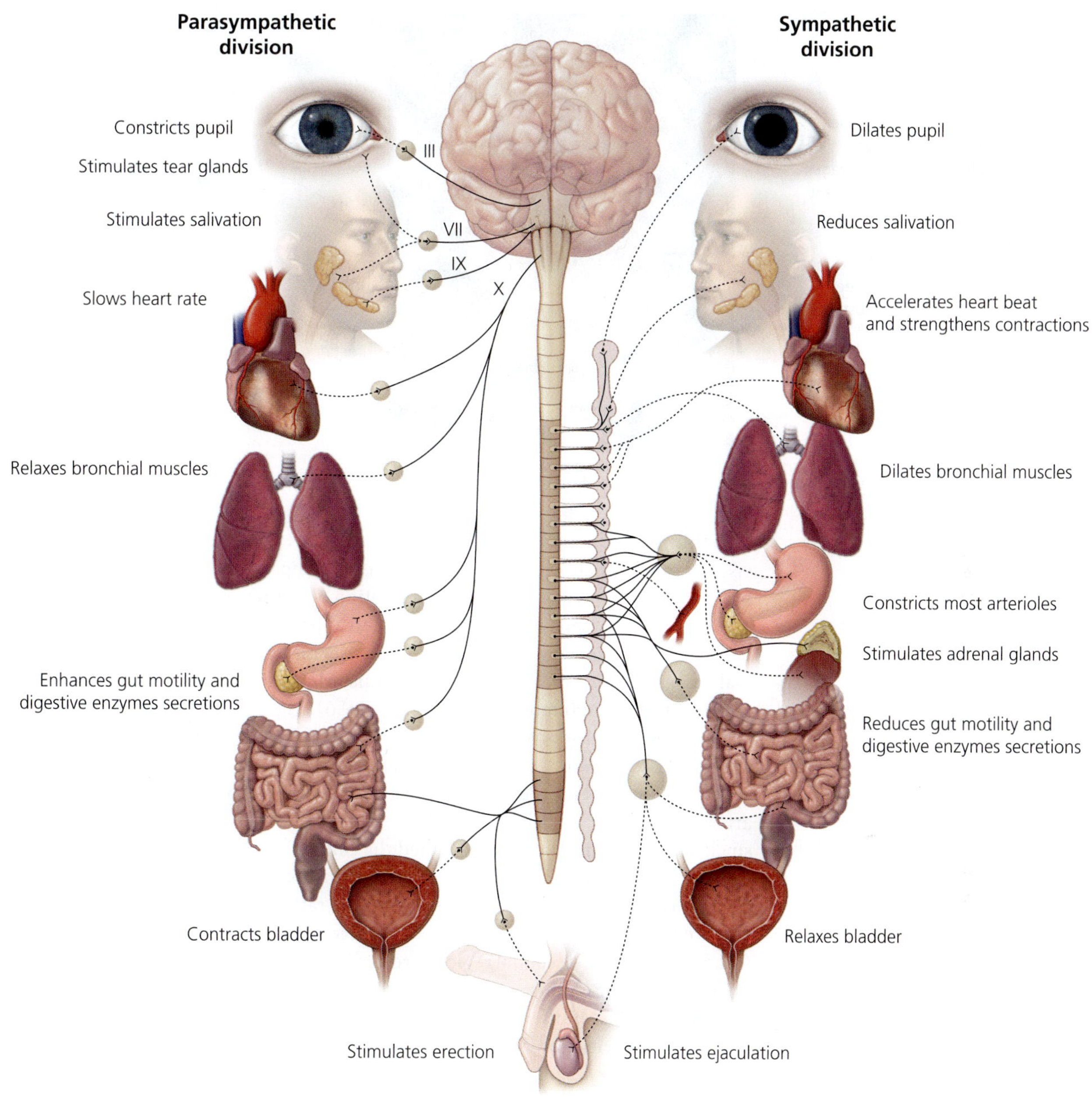

Figure 9.6 Functional antagonism of sympathetic and parasympathetic activation. The eyes, heart, lungs, stomach, intestines, and bladder receive both sympathetic and parasympathetic innervation. In many of these organs, sympathetic activation opposes the effects of parasympathetic innervation. The autonomic ganglia are shown lightly shaded. The parasympathetic ganglia tend to lie near their target organs, whereas the sympathetic ganglia tend to lie near the spinal cord. Cranial nerves are labeled with Roman numerals. For clarity, sympathetic innervation of the arterioles is shown for just one level of the spinal cord.

Sensory Components of the Autonomic Nervous System

According to its traditional definition, the autonomic nervous system contains only *efferent neurons,* which carry impulses toward the target organs. Although neurobiologists had long suspected that the autonomic nervous system also contains *afferent neurons* that transmit information from the internal organs (viscera) to the central nervous system, those afferent neurons proved hard to find. Most of them have thin axons that are difficult to see in anatomical material and hard to impale with recording electrodes. Only after neurobiologists learned how to visualize and record from

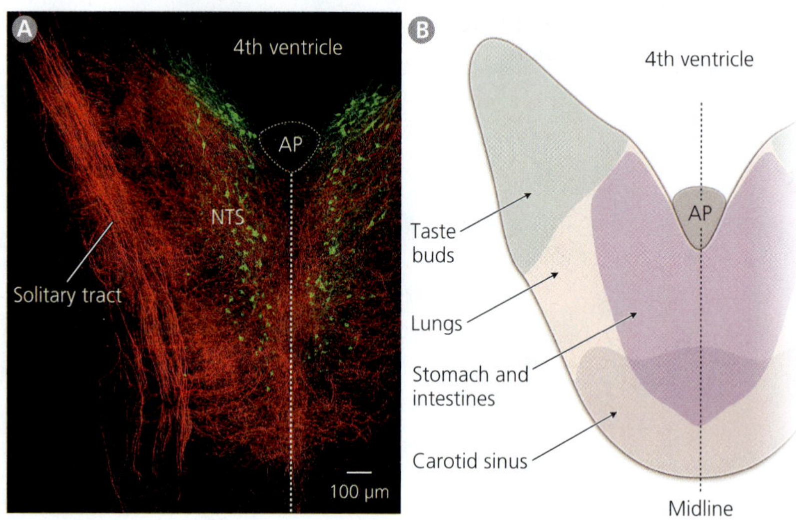

Figure 9.7 Nucleus tractus solitarius. Shown in (A) is a horizontal section through the medulla of a rat whose vagus nerve was injected with a red fluorescent tracer. Labeled axons are evident in the solitary tract and, more medially, the nucleus tractus solitarius (NTS). The section is also stained with a fluorescent green antibody against the hormone aldosterone. The diagram in (B) illustrates schematically which parts of NTS receive sensory information from which types of sensors. Also shown is one of the circumventricular organs, the area postrema (AP). [From Shin et al., 2009]

such thin axons did they realize that **visceral sensory neurons** are plentiful. Because visceral sensory neurons work closely with the sympathetic and parasympathetic divisions, we here include them in our definition of the autonomic nervous system.

Many visceral sensory neurons send their axons through spinal nerves and terminate in the spinal cord's dorsal horn. Other visceral sensory axons travel through cranial nerves and terminate within the medulla. Most important for visceral sensations is the **vagus nerve** (cranial nerve X; Figure 9.6). More than 80% of the axons in the vagus nerve are sensory axons. Some of them innervate taste buds on the tongue and in the throat, but most vagal afferent neurons provide information from internal sensors, including stretch receptors in the gut, pressure sensors in the aorta and carotid artery, and sensors of blood chemistry. These vagal sensory axons terminate in a small but important nucleus of the medulla called the nucleus of the solitary tract or, in the shorter Latin form, *nucleus tractus solitarius* (NTS). This name derives from the fact that NTS is closely associated with a rather isolated, "solitary" axon tract (Figure 9.7 A). The rostral end of NTS receives its major input from taste buds (see Chapter 7). The more caudal regions are dedicated to sensory input from the internal organs, including the heart, major blood vessels, lungs, and gut (Figure 9.7 B).

Although visceral sensory information comes mainly through the vagus nerve, additional information about the body's internal condition is provided by several *circumventricular organs*, including the *area postrema* (Figure 9.7). These circumventricular organs are located at the brain's midline, adjacent to the ventricles (hence their name), and they contain dense capillary beds. Because the capillaries in the circumventricular organs lack a blood–brain barrier (see Chapter 5), neurons and glial cells in the circumventricular organs can sense blood chemistry quite directly. Fortunately, this breach of brain security causes no harm because the circumventricular organs tend to be separated from other brain regions by tight junctions. As you will learn in the following sections, cells in the circumventricular organs respond to changes in the blood's pH and osmolarity as well as to diverse hormones.

The Enteric Nervous System

Each person carries 200–600 million neurons in the lining of their stomach and intestines. This "second brain" is called the **enteric nervous system**, a name derived from the Greek word for intestines ("entrails" is a related English word).

The neurons of the enteric nervous system form two spidery webs sandwiched between layers of smooth muscle and mucous membrane. Some enteric neurons sense chemicals within the stomach and intestines. Others sense when the gut is bloated (stretched). In addition, the enteric nervous system contains motor neurons that release acetylcholine to stimulate smooth muscle contraction or a more complex mix of neurotransmitters to promote smooth muscle relaxation. An important subset of these enteric motor neurons modulates the release of mucus and enzymes into the gut's interior. Aside from sensory and motor neurons, the enteric nervous system contains interneurons that release acetylcholine as well numerous other signaling molecules. Given this diversity of cell types and signaling molecules in the enteric nervous system, it is not surprising that many of our medicines have side effects on digestion. Opioids like morphine, for example, are powerful pain relievers but tend to cause constipation. Although the enteric nervous system can function independently to some extent, its activity is normally modulated by the sympathetic and parasympathetic components of the autonomic nervous system.

BRAIN EXERCISE

Do neuroscientists define the various subdivisions of the autonomic nervous system on the basis of their structure or their function? Can you imagine alternate ways of subdividing the autonomic nervous system?

9.3 How Do Neural Circuits Regulate the Vital Bodily Functions?

Now that you've been introduced to the major components of the autonomic nervous system, we can begin to ask how they control the body's internal activities. Here we focus on the circuits regulating heart rate, blood pressure, respiration, and body temperature.

Adjusting Heart Rate

Without adequate blood supply, brain cells soon stop functioning. Of course, too much blood flow is also bad, as this may burst some blood vessels. Given these considerations, you might expect heart rate to be regulated around some optimum set point. Indeed, resting heart rate usually ranges from 60 to 80 beats/min in adults. However, this range varies considerably with age and physical condition. More important, an individual's heart rate can rise as high as 200 beats/min during strenuous exercise, arousal, joy, or fear. Digestion, too, can increase your heart rate. Even if you simply stand up after lying down, your heart rate increases. This is a useful adjustment because standing requires that blood is pumped against the force of gravity to reach the brain. Without the increase in heart rate, or if you rise too fast, fainting is likely (Box 9.1: Fainting Spells and Vagal Nerve Stimulation). Thus, heart rate is not as tightly regulated as blood pressure or other vital parameters.

Baroreflex Modulation of Heart Rate

One well-studied trigger for changes in heart rate is a sudden change in blood pressure. Particularly effective are blood pressure changes in the *aortic arch*, and the *carotid sinus*, which sits at the origin of the internal carotid artery (right next to the carotid body; see Figure 7.17). Artificially boosting blood pressure at these locations decreases heart rate and causes **vasodilation** (*vas* means "vessel" in Latin); conversely, decreasing blood pressure at either location increases heart rate and stimulates **vasoconstriction**. These adjustments in heart rate and blood vessel diameter synergistically regulate blood pressure, keeping it from rising too high or falling too low. Collectively, they constitute the **baroreflex** (*bar-*, meaning "pressure," as in the word "barometer"). It can be confusing that the baroreflex involves changes in both blood pressure and heart rate. Therefore, we consider the two responses separately, beginning with heart rate control.

The heart rate control circuit (Figure 9.8) begins with viscerosensory neurons that have their cell body in a vagal nerve ganglion. These neurons send the peripheral branch of their axon to the aortic arch and carotid sinus, where they terminate as mechanoreceptive nerve endings that sense blood vessel stretch. Because blood

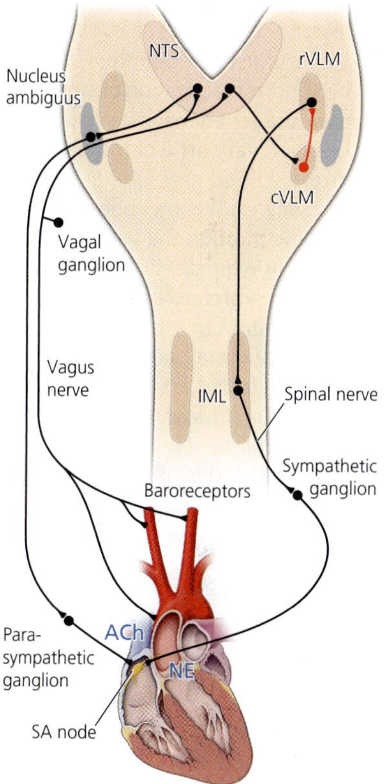

Figure 9.8 Heart rate control. Neurons in the vagus nerve transmit baroreceptor information to nucleus tractus solitarius (NTS). Neurons there project to parasympathetic preganglionic neurons in nucleus ambiguus, which in turn project to postganglionic neurons that release acetylcholine (ACh) onto the heart's sinoatrial (SA) node, slowing its rhythm. NTS neurons also project to inhibitory neurons (red) in the caudal ventrolateral medulla (cVLM). These neurons inhibit tonically active neurons in the rostral ventrolateral medulla (rVLM), which project to sympathetic preganglionic neurons in the spinal cord's intermediolateral column (IML). The sympathetic postganglionic neurons release norepinephrine (NE) onto the SA node, which in response fires more rapidly.

NEUROLOGICAL DISORDERS

Box 9.1 *Fainting Spells and Vagal Nerve Stimulation*

Roughly 35% of people have fainted at least once in their life, and some faint frequently, most commonly in response to extreme fear or pain or other highly emotional situations. This kind of neurally mediated fainting is called *vasovagal syncope* (the latter term is derived from the ancient Greek word for "to cut short" and is related to the musical concept of syncopation, which in essence means "offbeat"). The characteristic features of vasovagal syncope are sudden vasodilation and a sharp decrease in heart rate, which collectively reduce blood pressure so much that the brain no longer receives sufficient blood, causing the subject to pass out.

Scientists used to think that vasovagal syncope is caused by abnormal activation of stretch receptors in the aorta, which then trigger an inappropriate baroreflex. This hypothesis has received very little empirical support. However, the sudden drop in blood pressure during vasovagal syncope is almost certainly caused by abnormal activation of parasympathetic outflow through the vagus nerve and by a simultaneous decrease in sympathetic activity, which is what occurs during the normal baroreflex response to increased blood pressure. What remains unclear are the neural mechanisms that trigger syncope. Abnormal surges of serotonin release and unusual patterns of cortical activity have both been implicated, as have a rise in blood pressure and heart rate just prior to the fainting spell, but little else is known. One intriguing idea is that neurally mediated fainting is related to the normal "play dead" response of many animals to inescapable threats (hence the expression "playing possum"; Figure b9.1). What can you do to prevent an imminent fainting spell? Isometrically contracting your limb and trunk muscles appears to help, presumably because these contractions elevate blood pressure. However, more research into fainting is warranted.

Figure b9.1 An opossum "playing dead." When opossums are severely frightened, they faint and appear dead, often for several hours. During this state they even "smell dead" because of putrid secretions from their anal glands.

Because the vagus nerve carries both parasympathetic axons going to the heart and baroreceptor information coming from there, strong artificial stimulation of the vagus nerve, through implanted stimulating electrodes, causes you to faint (if you are standing up). However, weak repetitive electrical stimulation of the vagus nerve can have beneficial effects.

Best documented is that *vagal nerve stimulation* reduces the frequency of epileptic seizures, which result from abnormal bursts of brain activity, in patients whose epilepsy has been resistant to other treatments. In such patients a helical stimulating electrode is implanted around a cervical segment of the left vagus nerve. Electrical pulses are then applied through this electrode at regular intervals or on demand, whenever the patient senses an impending seizure. No one is quite sure why this form of weak electrical vagal nerve stimulation can prevent seizures, but it probably provides weak activation of the parasympathetic division and a concomitant slight reduction of sympathetic tone. As long as these effects are weak, they do not cause fainting. Instead, they should decrease anxiety and promote behavioral as well as physiological relaxation (the opposite of the fight-or-flight response). If this is true, then one would expect vagal nerve stimulation to influence not just the frequency of epileptic seizures, but a wide variety of behavioral, cognitive, and physiological phenomena.

Indeed, vagal nerve stimulation is increasingly used to battle major depression in patients who have not benefited from standard drug therapies. Several studies have shown such treatments to be effective, although more extensive clinical trials are still ongoing. Preliminary evidence further suggests that vagal nerve stimulation can reduce inflammation (such as inflammation of the heart after a heart attack) and the progression of Alzheimer's disease. It may also alleviate migraines, tinnitus (persistent ear ringing), multiple sclerosis, bulimia, and obesity. One should be skeptical when a single procedure is touted as a treatment for so many different ailments, but the vagus nerve clearly is implicated in a wide variety of normal processes. Further studies will have to demonstrate which of the promised effects of vagal nerve stimulation are real and which are spurious. It will also be interesting to learn what mechanisms lie behind the reported effects. So far, we have only a few intriguing hints, such as elevated epinephrine and serotonin levels in many brain regions after vagal nerve stimulation.

vessel stretch correlates tightly with blood pressure, we can refer to these vagal nerve endings as **baroreceptors**. They increase their rate of firing when blood pressure rises and decrease their firing rate as blood pressure falls.

Once action potentials have been triggered in the baroreceptors, they are conducted along the axons to NTS. Many of the neurons there project to **nucleus ambiguus**

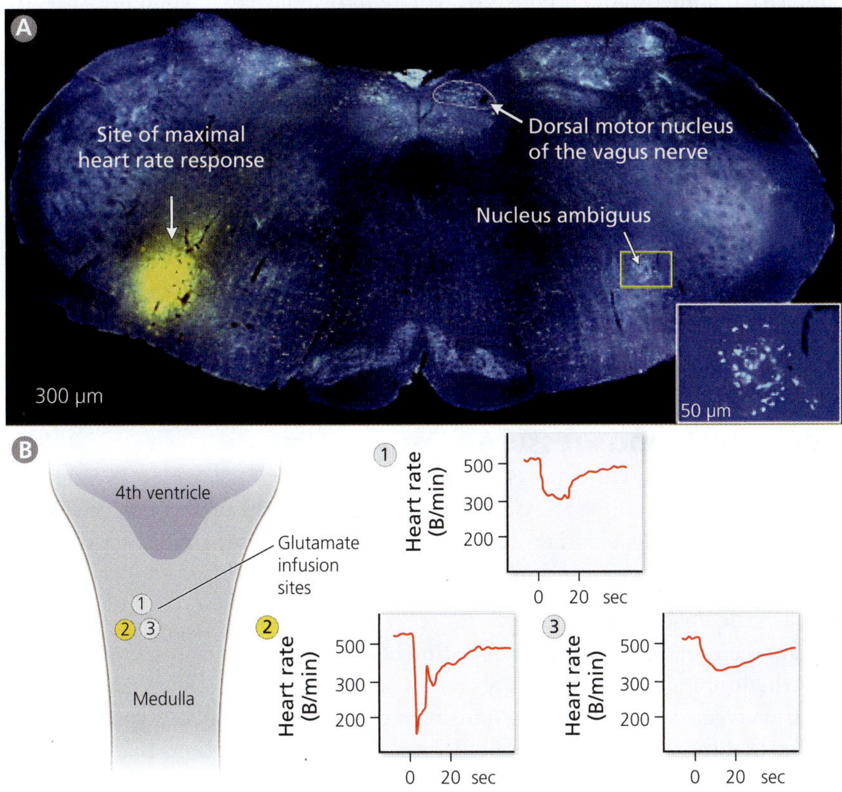

Figure 9.9 Nucleus ambiguus. Experimenters recorded the heart rate of a mouse while they injected glutamate at various locations in the medulla. A fluorescent substance was injected where glutamate evoked the largest decrease in heart rate (yellow area in A). On the right side of this section (yellow rectangle), you can see nucleus ambiguus neurons that were labeled with a retrograde tracer (close-up in inset). Panel (B) shows the location of 3 different injection sites (left) and the associated heart rate responses (1–3). All injections started at t = 0 sec. [From Yan et al., 2009]

(the name is fitting because the nucleus is difficult to see in standard neuroanatomical material). Nucleus ambiguus then sends its axons through the vagus nerve to parasympathetic postganglionic neurons on the surface of the heart. Activation of these neurons causes acetylcholine release, which decreases heart rate (Figure 9.9).

Putting it all together, we find that an increase in blood pressure increases baroreceptor firing; this activates neurons in NTS, which then excites neurons in nucleus ambiguus; nucleus ambiguus, in turn, activates the heart's postganglionic neurons, which slow down the heart. There are four synaptic relays in this circuit (Figure 9.8), but it is nicely linear. From here on out, the circuitry gets more complex.

Sympathetic Regulation of Heart Rate

The principal complication is that the heart rate control circuit has two branches. So far, we have discussed only the parasympathetic branch. What about the sympathetic inputs to the heart? They are part of a slightly longer circuit that also begins with the baroreceptors and goes through NTS. However, some of the neurons in NTS project to a small region in the **caudal ventrolateral medulla**, abbreviated cVLM (Figure 9.8). This region projects to a more rostral portion of the ventrolateral medulla, appropriately named rVLM, which in turn projects to preganglionic sympathetic neurons in the spinal cord that innervate the heart. When those sympathetic postganglionic neurons release norepinephrine, the heart speeds up.

As you can see in Figure 9.8, this sympathetic branch of the heart rate control circuit contains not four but six neuronal links. Importantly, the neurons in cVLM use GABA as their neurotransmitter and are therefore inhibitory. All the other neurons in the heart control circuit are excitatory. Therefore, an increase in the firing rate of the baroreceptors leads to increased inhibition on the neurons in rVLM, which ultimately causes a decrease in heart rate. Conversely, a decrease in baroreceptor firing rate reduces the amount of inhibition on rVLM neurons. This **disinhibition** causes rVLM neurons to become more active, effecting an increase in heart rate. Thus, the sympathetic branch of the heart control circuit is also, like its parasympathetic counterpart, homeostatic: it boosts heart rate in response to a blood pressure drop and decreases heart rate when blood pressure rises.

As you have now learned, any change in baroreceptor activity has opposite effects on the sympathetic and parasympathetic branches of the heart rate control circuit, increasing activity in one and decreasing it in the other. Therefore, the two branches function synergistically (cooperatively) rather than antagonistically. Going back to our high-tech thermostat analogy, a rise in blood pressure activates the parasympathetic neurons and inhibits the sympathetic innervation of the heart, just as a rise in temperature activates the air conditioner and turns off the heater. Conversely, a drop in blood pressure reduces activity in the parasympathetic branch of the circuit and disinhibits (activates) the sympathetic branch, just as a drop in temperature shuts down the air conditioner and fires up the heat. In short, the circuits that control heart rate are a good example of push-pull regulation.

Regulating Blood Pressure

As noted earlier, blood pressure can be also altered by changes in blood vessel diameter. When the body's blood vessels constrict, the space through which the blood is pumped must shrink and, other things being equal, blood pressure rises. Conversely, vasodilation usually causes a drop in blood pressure.

Sympathetic Vasoconstriction

The neural circuit responsible for adjusting blood vessel diameter is identical to the sympathetic branch of the heart rate control circuit until you reach rVLM. There the circuits diverge, as a subset of rVLM neurons projects to sympathetic (preganglionic) neurons that innervate (postganglionic) neurons with projections to blood vessels (Figure 9.10). Activation of these postganglionic neurons promotes vasoconstriction, which causes blood pressure to rise.

Here is how the circuit works: a sudden drop in blood pressure causes the baroreceptors to fall silent, which prompts the neurons in NTS and cVLM likewise to simmer down; decreased activity in cVLM then decreases the inhibition on rVLM neurons, which become more active in response (they are disinhibited). Increased activity in rVLM activates (through IML) the sympathetic postganglionic neurons, which then release norepinephrine onto the smooth muscles around the blood vessels. They, in response, contract. Given this circuit model, it is not surprising that infusing GABA into rVLM lowers blood pressure (Figure 9.11 A and B). The infused GABA mimics GABA release from activated neurons in cVLM.

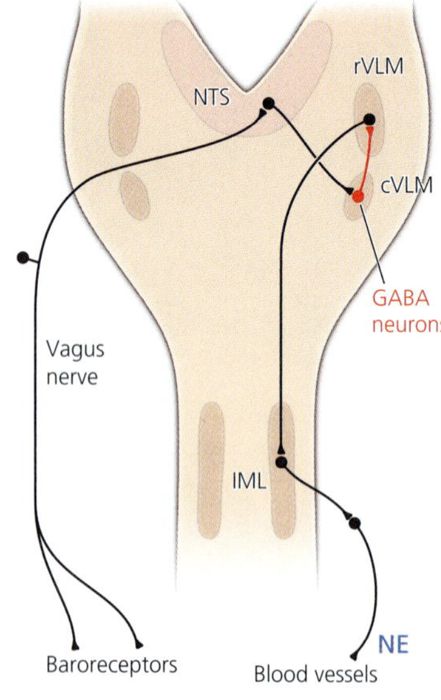

Figure 9.10 Sympathetic control of blood pressure. This circuit is very similar to the circuit that controls heart rate, except that it ends with sympathetic neurons that control blood vessel diameter, rather than heart rate. NTS = nucleus tractus solitarius; rVLM = rostral ventrolateral medulla; cVLM = caudal ventrolateral medulla; IML = intermediolateral column; NE = norepinephrine.

More difficult to understand is why glutamate infusions into rVLM can increase blood pressure (Figure 9.11 C). The circuit diagram shows no excitatory inputs to rVLM that could be mimicked by the glutamate infusion. However, as you may recall, rVLM neurons are tonically active when they are not inhibited by cVLM. This tonic activity is driven by glutamatergic inputs to rVLM from other sources (not shown in the diagram). Therefore, the rVLM neurons must be expressing glutamate receptors, which can be activated experimentally by local infusions of glutamate. In fact, most neurons in the brain can be activated with glutamate, even if their main inputs are GABAergic.

An important aspect of the sympathetic inputs to the blood vessels is that they, too, are tonically active, which means that blood vessels are slightly constricted most of the time. Therefore, if blood pressure suddenly increases, sympathetic background activity can be reduced, leading to a decrease in the activity of the smooth muscles responsible for vasoconstriction. Because of blood pressure inside the vessels, the decrease in vasoconstriction leads to vasodilation. Without the tonic sympathetic activation, vasodilation would be much harder to accomplish. After all, neurons that are already silent cannot be silenced further, and nonconstricted blood vessels are hard to dilate more.

Parasympathetic Vasodilation

Although most blood vessels receive only sympathetic innervation, some vessels in your face, as well as the penis, can be dilated by parasympathetic inputs. The

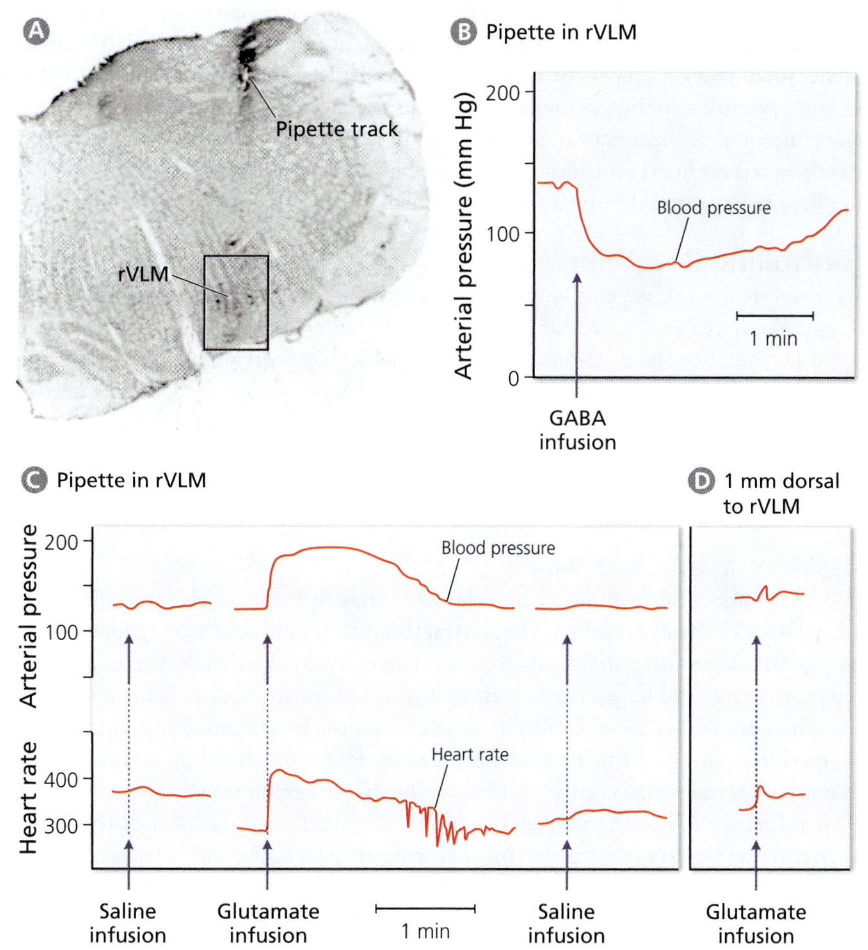

Figure 9.11 Functions of rVLM. Experimenters injected GABA, glutamate, or saline into the rostral ventrolateral medulla (rVLM) of rats. The photograph in (A) shows rVLM in a transverse section through the medulla. The graphs in (B) and (C) show that blood pressure was decreased by injecting GABA, increased by injecting glutamate, and unaltered after control injections of saline. Panel (D) shows that injecting glutamate dorsal to rVLM causes only minimal changes in blood pressure or heart rate. [From Ross et al., 1984]

mechanisms of this dilation remain somewhat unclear, but include the release of **nitric oxide** (from parasympathetic nerve terminals, smooth muscle cells, or both). Nitric oxide is a gaseous transmitter that activates guanylate cyclase, which increases intracellular cGMP levels, which in turn relaxes smooth muscle cells. Consistent with this model, inhibiting cGMP degradation in the penis with a compound called **sildenafil citrate** (Viagra) dilates the vessels through which blood flows into the penis. (Ejaculation, in case you were wondering, is driven mainly by sympathetic activation and by rhythmic contractions of striated muscles that are innervated by a sexually dimorphic nucleus in the spinal cord).

Blood Pressure Set Points

As you now realize, the baroreceptor reflex regulates blood pressure homeostatically, increasing it when blood pressure is low and lowering it when the pressure is high. However, this regulation is not always perfect. A major problem is that many people have **hypertension**, which is defined as chronically high blood pressure (>140/90). Although hypertension can have a variety of causes, including excess salt intake and genetics, it may also involve an elevation in the set point around which blood pressure is regulated. According to this hypothesis, the various aspects of the baroreflex still work in people with high blood pressure, but those people stabilize blood pressure around an abnormally high target value. A good analogy would be a heater's thermostat being turned up too high. Although many different drugs can influence blood pressure and are frequently prescribed to treat hypertension, the neural mechanisms that alter blood pressure set points remain an almost total mystery.

Sometimes the blood pressure set point is altered temporarily. For example, blood pressure must rise when you stand up. This posture-driven increase in blood pressure is caused in part by an increase in heart rate, but it also involves a rapid constriction of blood vessels in large muscles and viscera. Blood pressure also increases rapidly when you get excited, be it from happiness or fear. Such acute changes in blood pressure require a change in the blood pressure set point. Otherwise, any increase in heart rate or vasoconstriction would immediately be neutralized by reflexive commands to reduce heart rate and re-dilate the blood vessels. The mechanisms underlying these temporary set point adjustments also remain largely mysterious.

Controlling Breathing

Breathing is a vital, vegetative process that can go on without conscious control. Nonetheless, we can control our breathing much better than we can control the beating of our heart or the dilation of our blood vessels. Spoken language, in particular, requires exquisite, rapid control over our respiratory muscles. Such rapid control is generally impossible for smooth muscles because their contractions are modulated by relatively slow metabotropic receptors. Accordingly, breathing (respiration) is controlled by striated muscles and skeletal motor neurons rather than smooth muscles and autonomic neurons.

Respiratory Muscles and Neurons

The most important respiratory muscle is the **diaphragm**, which forms the dome-shaped floor of the chest cavity. On contraction, the diaphragm moves down toward the gut, thus expanding the chest. This expansion lowers the chest's internal pressure and pulls air into the lungs. In contrast to inhalation, exhalation is a relatively passive phenomenon. When the diaphragm relaxes, it returns to its domed shape, compressing the lungs and pushing air back out. If more forceful exhalation is needed, then abdominal muscles also contract and push the diaphragm upward.

Another set of respiratory muscles lies in your voice box (larynx). Although the laryngeal muscles are best known for their role in speech, they are also used to close the trachea just after you inhale deeply. This closure prevents the inhaled air from escaping after the diaphragm starts to relax. It lets you hold your breath without keeping your diaphragm forcefully contracted. This is a useful strategy because the laryngeal muscles are much smaller than the diaphragm and, therefore, consume less energy when contracted.

The motor neurons that innervate the diaphragm sit in the spinal cord's ventral horn at levels C3–5, and the abdominal muscles are innervated by ventral horn neurons at thoracic and lumbar levels. In contrast, the laryngeal motor neurons are located in nucleus ambiguus, right next to the neurons that control heart rate. Because the respiratory motor neurons are distributed so widely across the central nervous system, it is difficult to make a general statement about which other neurons project to them.

However, many of the neurons that project to respiratory motor neurons are located in the ventral medulla. Some of these neurons fire during inhalation (inspiration), some fire mainly during exhalation (expiration), and others have more complex rhythms of activity (Figure 9.12). Because these rhythms persist after the vagus nerve (with its visceral sensory axons) is cut, we can conclude that sensory feedback from the lungs or respiratory muscles is not required for generating the rhythm. Instead, the respiratory rhythm must be produced by a **central pattern generator**, defined as a set of neurons that generates patterned activity without requiring sensory input.

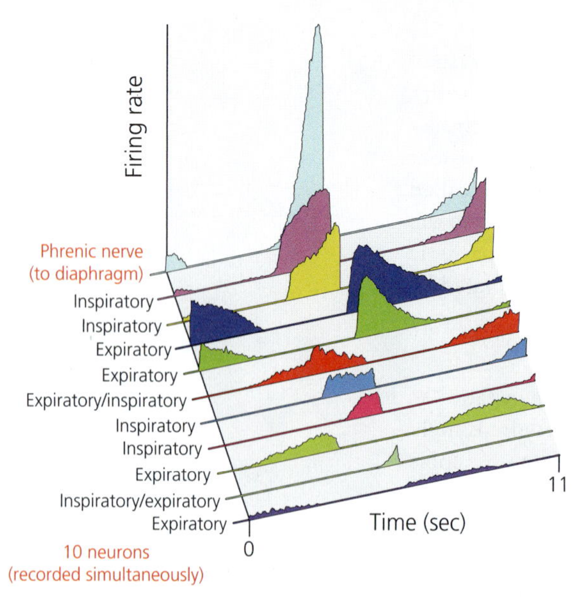

Figure 9.12 Neurons that fire rhythmically with respiration. Shown here are the firing rates of 10 simultaneously recorded neurons in the ventral medulla of a cat. The timing of inspiration (inhalation) is marked by increased activity in the phrenic nerve, which innervates the diaphragm. Some neurons are active during inspiration, others during expiration, and yet others exhibit more complex firing patterns. [After Rybak, 2008]

The Central Pattern Generator for Breathing

Major progress in locating the central pattern generator for breathing came when neuroscientists dissected the medulla

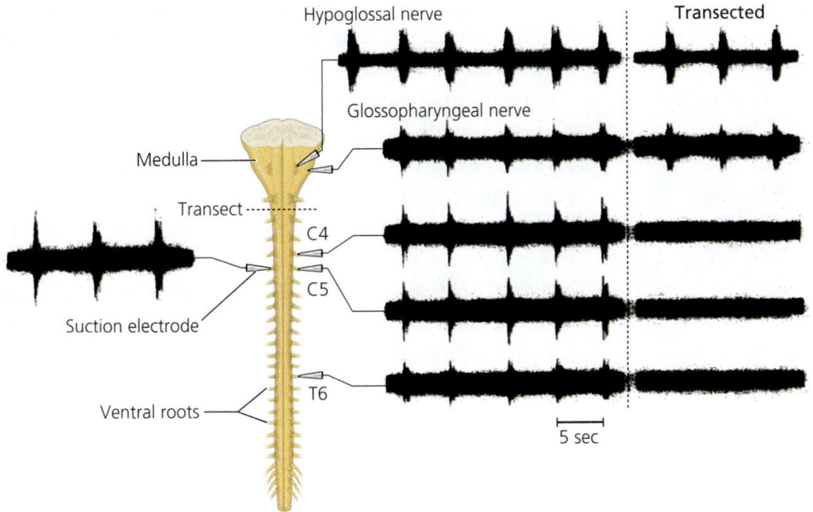

Figure 9.13 Breathing in a dish. Researchers dissected the medulla and spinal cord out of neonatal rats, placed them in a culture dish, and recorded rhythmic neural activity from several cranial and spinal nerves. This rhythmic activity is similar to that which drives normal breathing. When the spinal cord is transected (cut) at upper cervical levels, the respiratory rhythm persists in the cranial nerves but disappears in spinal nerves. Therefore, the rhythm must be generated within the medulla. [After Smith and Feldman, 1987]

and spinal cord out of newborn rats, placed them into a culture dish, and found rhythmic activity in the remaining spinal and cranial nerve roots (Figure 9.13). The researchers went on to show that the rhythmic neural output is linked to rhythmic movements of the chest (breathing). Others built on this discovery by systematically cutting away various parts of the medulla and spinal cord. Eventually, they learned that a small piece of neural tissue in the ventral medulla exhibits rhythmic activity even when it is completely isolated from more rostral and more caudal brain regions. They named this region the **pre-Bötzinger complex,** because it lies just caudal to rhythmically active neurons in the **Bötzinger complex** (named after a winery that hosted a major conference on the neural control of breathing).

Subsequent research revealed that the central pattern generator for breathing includes both the pre-Bötzinger complex and the Bötzinger complex, working together. The cellular and network properties that generate the respiratory rhythm are too complex for us to consider. However, an important principle emerging from this research is that breathing involves several central pattern generators that are normally linked but can act independently of one another. This model explains why lesions in the medulla tend to alter the respiratory rhythm but rarely abolish it. Apparently, the central pattern generator for breathing is designed so that it can continue to function, at least imperfectly, even after many of its constituent neurons have been damaged. An engineer would say that the central pattern generator for breathing is *fault tolerant*. More informally, we can say that the central pattern generator for breathing knows how to take a hit and keep on going.

The Chemoreflex

Although adult humans at rest tend to breathe at a fairly steady rate of ~15 breaths a minute, this rate can vary enormously. Voluntary changes in respiratory rate probably involve projections from the primary motor cortex to the medulla and spinal cord. Most animals also slow down their respiration rate during deep sleep, but the mechanisms of this modulation are poorly understood. Better studied is the involuntary **chemoreflex,** which makes you breathe faster and more deeply when the blood runs low on oxygen or is enriched in carbon dioxide, or both.

Decreases in blood oxygen levels activate *glomus cells* in the carotid body, which we discussed in Chapter 7. The glomus cells release a mix of excitatory neurotransmitters onto the peripheral terminals of axons in the glossopharyngeal nerve (cranial nerve IX), which project to the caudal pole of NTS. Chemoreceptive NTS neurons project to a small brain region in the ventral medulla that is called the **retrotrapezoid nucleus,** which projects to the central pattern generator for breathing and modulates its activity (Figure 9.14). The way this circuit works is relatively straightforward: when glomus cells are activated, NTS and retrotrapezoid neurons increase

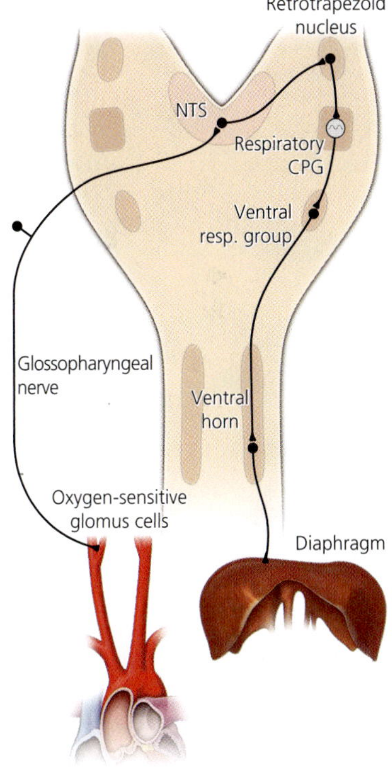

Figure 9.14 The chemoreflex. A drop in blood oxygen level is sensed by glomus cells, which project to nucleus tractus solitarius (NTS). NTS neurons project to neurons in the retrotrapezoid nucleus, which project to the respiratory CPG (central pattern generator) and modulate its activity. As a result, the animal breathes faster and more deeply when blood oxygen levels are low. Only one side of the circuit is shown.

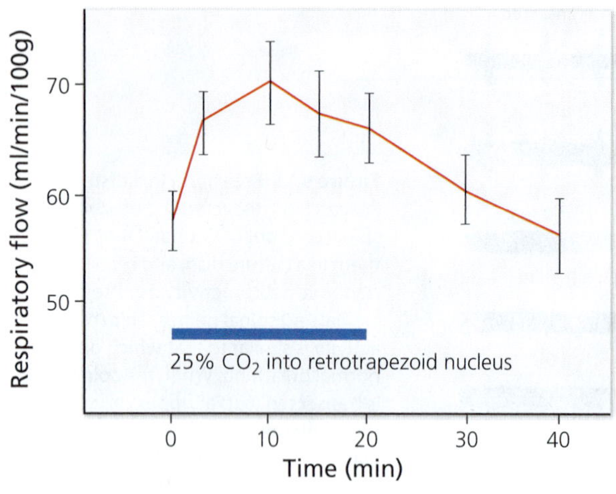

Figure 9.15 CO_2 sensors in the retrotrapezoid nucleus. A 25% solution of CO_2 was pumped through a microdialysis probe whose tip was located in the retrotrapezoid nucleus. In response, the unanesthetized rat breathed more deeply. These data suggest that CO_2 sensors are located in or very near the retrotrapezoid nucleus and that their activation is sufficient to drive the chemoreflex. [After Li et al., 1999]

their firing rates, the central pattern generator speeds up, and the animal breathes more deeply and more rapidly. As a result, blood oxygen levels rise.

The chemoreflex can also be elicited by increases in blood carbon dioxide (CO_2) levels. Breathing harder when CO_2 levels are high is adaptive because CO_2 is a potentially toxic byproduct of cellular respiration that is removed from the blood by gas exchange in the lungs. As discussed in Chapter 7, CO_2 concentration can be measured by tissue or blood acidity sensors. Where are these sensors? At least some of them must be located in the retrotrapezoid nucleus because infusions of CO_2 directly into the retrotrapezoid nucleus make rats breathe more deeply (Figure 9.15).

Because blood oxygen and CO_2 levels tend to be inversely correlated, the O_2- and CO_2-driven versions of the chemoreflex typically work together. However, an important distinction between these two versions of the chemoreflex is that O_2 sensors in the carotid body can trigger the chemoreflex proactively, before the oxygen-poor blood reaches the brain; whereas the CO_2 sensors in the retrotrapezoid nucleus can respond to changes in blood chemistry only reactively, after the brain is already perfused with suboptimal blood. This distinction between proactive and reactive homeostatic reflexes comes up again in the next section, which deals with temperature regulation.

Regulating Body Temperature

Warm-blooded animals regulate their body temperature tightly, which allows them to be relatively independent of environmental temperatures. For example, being warm-blooded allowed early mammals to be much more active during cold nights than a cold-blooded dinosaur could ever be (although some large dinosaurs may well have been somewhat homeothermic).

An important aspect of temperature regulation is that skin temperature is more variable than body core temperature. Brain temperature is even more tightly controlled. Heating the brain by just 1–2°C causes mental confusion, and prolonged brain temperatures above 42°C cause serious brain damage or death. Neurons are more tolerant of cold, but it does silence them. Therefore, brain cooling makes an organism sluggish, which only worsens the **hypothermia** (low body temperature) because muscles generate heat as a byproduct of their contractions. The less you move, the less heat your muscles generate! Death may ensue.

Battling Hypo- and Hyperthermia

To prevent hypothermia, organisms employ a variety of skeletal motor behaviors including shivering; moving to a warmer location; and, in the case of humans, putting on warmer clothes. Hypothermia also leads to a constriction of the superficial blood vessels, which minimizes heat loss across the skin (this is why chilled people look pale). Another autonomic response to cold is the formation of goosebumps. This response is pointless in humans but useful in other mammals, where the goosebumps erect skin hairs, thereby fluffing up the fur. A third autonomic response to cold is the burning of **brown adipose tissue**, which is an unusual type of body fat. Most significant is that brown fat cells synthesize *uncoupling protein 1*, which allows protons to flow back across the mitochondrial membrane during cellular respiration. As a result, brown fat cells generate more heat and fewer molecules of ATP than other types of fat. Adult humans possess relatively little brown adipose tissue, relative to small mammals and human infants, but recent data suggest that brown fat deposits just above the collar bone contribute significantly to thermogenesis even in adult humans.

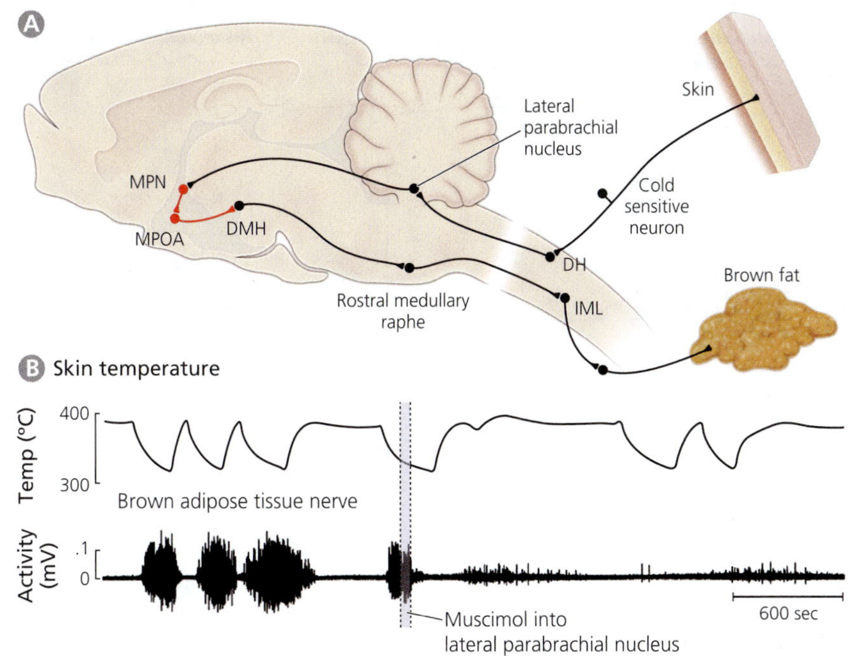

Figure 9.16 Burning brown fat for heat. Many mammals burn brown adipose tissue when they get cold. The neural circuit for this behavior is shown in (A). Excitatory neurons are black, inhibitory neurons red. The graphs in (B) show that activity in the nerve innervating the brown adipose tissue is correlated with decreases in skin temperature. Also, injecting the GABA agonist muscimol into the lateral parabrachial nucleus blocks the neural response to cold. DMH = dorsomedial hypothalamus; DH = dorsal horn; IML = intermediolateral column; MPN = median preoptic nucleus; MPOA = medial preoptic area. [After Nakamura and Morrison, 2008]

The neural circuit responsible for brown fat burning has been studied extensively in rats (Figure 9.16). It begins with cold sensors in the skin, most notably free nerve endings that express TRP-M8 (see Chapter 7). These cold sensors project to the dorsal horn of the spinal cord, from where the temperature information is conveyed to the *lateral parabrachial nucleus*. The parabrachial neurons, in turn, project to part of the **preoptic area**, which comprises several small cell groups just anterior to the optic chiasm. The circuit continues caudally from there until it reaches the sympathetic postganglionic neurons that innervate brown adipose tissue. An analogous pathway from the preoptic area to skeletal motor neurons triggers cold-induced shivering.

An intriguing aspect of these thermoregulatory circuits through the preoptic area is that they include two inhibitory neurons in series (Figure 9.16). The first inhibitory neuron (in the median preoptic nucleus) responds to cold with an increase in firing rate, thereby strengthening its inhibition of the second inhibitory neuron (in the medial preoptic area). As the latter neuron falls silent, its target (in the dorsomedial hypothalamus) is disinhibited. Because the remaining neurons in the circuit are all excitatory, the end result is a cold-induced increase in the rate at which brown fat is burned and skeletal muscles shiver. To reiterate, the inhibition of inhibition causes excitation, just as in mathematics multiplying two negative numbers yields a positive result. In general, if you examine a linear neural circuit with an even number of inhibitory neurons (in series), then you can predict that excitatory input to the circuit will usually cause excitation on the output side. Conversely, if the circuit contains an odd number of inhibitory neurons, then excitatory input will reduce the circuit's output.

Hyperthermia (overheating) is even more dangerous than hypothermia. One of the most obvious responses to heat is the production of sweat, which cools the body when it evaporates. Sweating is driven by sympathetic postganglionic neurons that use acetylcholine as their transmitter. This should seem odd to you because sympathetic postganglionic neurons generally release norepinephrine, but the neurons that innervate sweat glands produce norepinephrine early in life and only later switch to using acetylcholine. The second major autonomic response to hyperthermia is peripheral vasodilation, which promotes convective heat loss by increasing blood flow to the skin. As you learned earlier, vasodilation is caused primarily by a decrease in the activity of sympathetic vasoconstrictor nerves and, for blood vessels in the face,

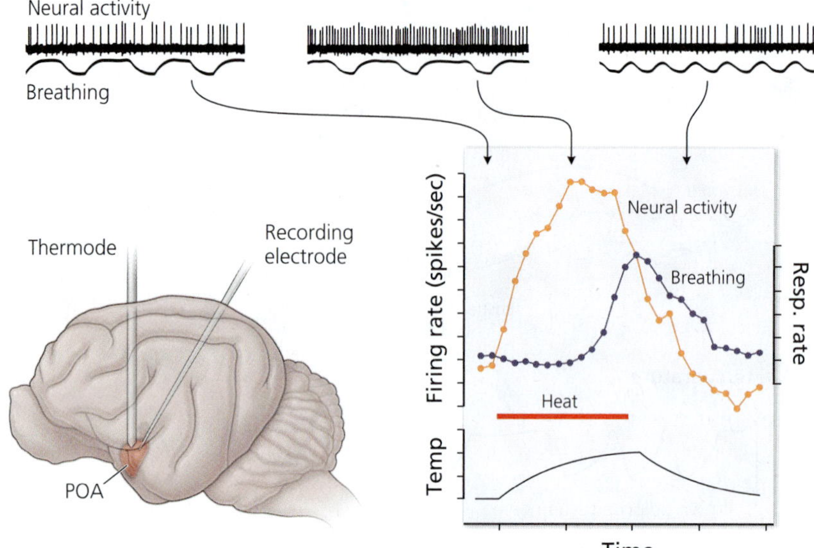

Figure 9.17 Preoptic temperature sensors. Experimenters recorded from neurons in the preoptic area of anesthetized cats while heating the surrounding tissue with an inserted thermode (which heats tissue by circulating warm water through its tip). Roughly 20% of the recorded neurons increased their firing rate as tissue temperature increased. The researchers also observed that heating the preoptic area triggered an increase in respiration rate. Curiously, many of the neurons decreased their firing rate before the heater was turned off; why this happened is not clear. [From Nakayama et al., 1963]

by the activation of parasympathetic nerves. Activation of the latter pathway is what makes your face turn very red when you feel hot (blushing is a related phenomenon, triggered by embarrassment rather than heat).

Temperature Sensors in the Brain

Most responses to heat and cold are triggered by sensors in the skin, but the brain also contains some temperature sensors. The preoptic area is especially sensitive to changes in local temperature, as can be shown by inserting heating or cooling probes into the preoptic area and recording the activity of nearby neurons as local temperature varies. Using this approach, researchers have shown that a significant fraction of preoptic neurons respond to local heating with increases in firing rate (Figure 9.17); a few respond to local cooling. Because neurons in other brain regions do not exhibit such temperature-driven changes in firing rate, the temperature sensors must be located in the preoptic area.

In analogy with the peripherally and centrally triggered versions of the chemoreflex, we can think of peripherally and centrally controlled thermoregulation as providing fault tolerance. If one circuit fails, the other one provides a backup system that allows for at least some temperature regulation. However, the two systems are not completely redundant because when your environment changes in temperature, your skin will follow suit long before brain temperature rises or falls. Accordingly, activation of the peripheral thermoreceptors can adjust body and brain temperature proactively, whereas central thermoreceptors must function reactively.

Fever Modifies the Temperature Set Point

Fever is an adaptive change in the thermoregulatory set point that stimulates the body's immune system and slows down most pathogens. To trigger a fever, immune cells secrete *pyrogenic* (meaning "fire producing") molecules into the blood. These molecules, especially **prostaglandin E2**, are sensed by neurons in the vagus nerve and circumventricular organs, which convey the information to the preoptic area. What happens there remains unclear, but somehow the temperature set point for thermoregulation is increased. This means that shivering, peripheral vasoconstriction, and other warmth promoting behaviors (such as piling on blankets) are triggered at higher than normal body temperatures. As a result, body temperature rises. Once the pathogens have been vanquished, pyrogen levels decrease and the temperature set point returns to normal. When this happens, the brain detects that body temperature is higher than it should be and triggers appropriate countermeasures, which is why profuse sweating usually signals that your fever has broken.

BRAIN EXERCISE

If heart rate, blood pressure, respiration, and body temperature are controlled by homeostatic reflexes, then how can they be modified in times of stress, when you are sick, with medication, or voluntarily?

9.4 How Do Neurons Control Fluid and Energy Balance?

To stay alive, we must stay properly hydrated and maintain adequate levels of energy. Fortunately, we have evolved a variety of mechanisms to achieve both goals.

Balancing the Bodily Fluids

Mammals regulate the *osmolarity* (salt concentration) of their extracellular fluid tightly. If the extracellular fluid becomes too salty, either because too much water is lost or because too much salt is taken in, water diffuses out of the cells and they begin to malfunction. Conversely, if the extracellular fluid becomes too dilute, then cells begin to swell. Such swelling is a serious problem especially within the brain because the brain already fits so tightly in the skull. Any extra pressure causes headaches and may lead to death. For example, a 4-year-old girl died in 2002 after being forced to drink a gallon of water as punishment. This case was extreme, but the point is that drinking too much water can be as much of a problem as drinking too little water. Either way, our bodies must adjust. They do so mainly by regulating urine production and fluid intake.

Vasopressin and Diabetes

Some of our body's responses to fluid loss are purely hormonal, and shall not concern us here. However, dehydration also activates neurons, especially the large neurons in the supraoptic and paraventricular nuclei of the hypothalamus (Figure 9.18). As we discussed in Chapter 8, some of these magnocellular neurons secrete oxytocin into the blood vessels of the posterior pituitary. The remaining magnocellular neurons release **vasopressin**. As its name suggests, vasopressin increases blood pressure through vasoconstriction. In addition, vasopressin combats dehydration by inserting water channels into the distal segments of the kidney's renal tubules, which reduces water loss by allowing more water to flow out of the urine back into the tissue. Even relatively slight changes in the activity of magnocellular vasopressin neurons have significant effects on urine production.

Destruction of the magnocellular vasopressin neurons causes **diabetes insipidus**, which is characterized by a massive overproduction of urine (up to 25 liters/day) and a corresponding need to replenish the lost fluid by drinking. This medical condition

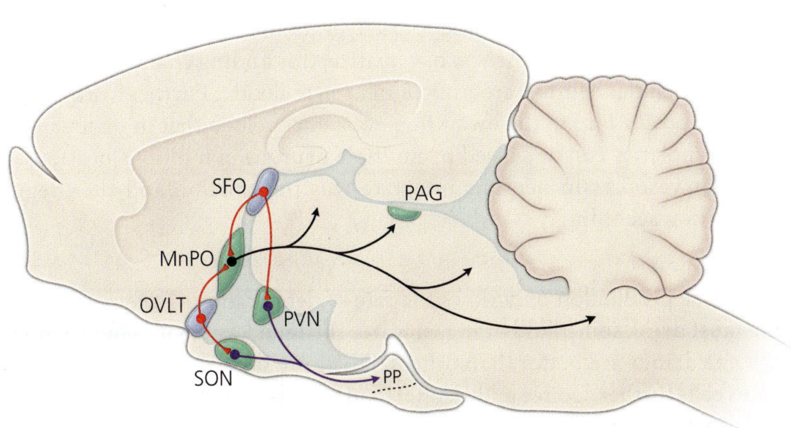

Figure 9.18 Neural circuits involved in regulating fluid balance. In response to dehydration, magnocellular neurons in the supraoptic and paraventricular nuclei (SON and PVN) secrete the antidiuretic hormone vasopressin into the blood vessels of the posterior pituitary (PP). Information about the body's hydration state is conveyed to SON and PVN from two circumventricular organs, namely the subfornical organ (SFO) and the vascular organ of the lamina terminalis (OVLT). Also involved in responding to dehydration is the median preoptic nucleus (MnPO), which projects to multiple brain areas including the periaqueductal gray (PAG). [After Bourque, 2008]

Figure 9.19 Supraoptic neurons are activated by dehydration. Shown here is the response of an isolated (in vitro) magnocellular supraoptic neuron to being bathed in a hypertonic solution. The neuron gradually becomes depolarized and increases its firing rate. We can conclude that this neuron can sense changes in osmolarity directly, without the aid of other brain regions. [From Oliet and Bourque, 1992]

is distinct from the more common **diabetes mellitus**, which results from impaired insulin production or signaling. Both diseases involve an overproduction of urine (diabetes is Greek for "passing through"); but in diabetes mellitus, this overproduction is due to elevated blood glucose, which causes more water to flow out of tissues into the blood. In case you're wondering, the word *mellitus* means "sweetened with honey" and refers to fact that the urine of someone with diabetes mellitus tastes slightly sweet (supposedly). In contrast, the word *insipidus* means "flavorless."

Dehydration Sensors

How does information about the body's hydration state reach the magnocellular vasopressin neurons? Part of the answer is that the magnocellular neurons themselves can sense dehydration. When these cells are placed in a culture dish and bathed in extracellular fluid that contains too few solutes, they swell. This swelling closes a stretch-inactivated cation channel and hyperpolarizes the neurons. Conversely, when the magnocellular neurons are bathed in hypertonic fluid, they shrink, become depolarized, and start firing action potentials (Figure 9.19). These findings indicate that magnocellular neurons can sense a rise in the osmolarity of the extracellular fluid and, through their connections to the posterior pituitary, effect a hormonal response that limits any further rise. They seem to be a single-neuron homeostatic reflex arc!

Additional dehydration sensors are located in two circumventricular organs, namely, the **vascular organ of the lamina terminalis (OVLT)** and the **subfornical organ** (Figure 9.18). Because these two structures lack a blood–brain barrier, they can be activated by blood-borne factors, such as *angiotensin II*, a peptide hormone whose levels in the blood increase when animals experience fluid loss. However, some of the cells in the OVLT and the subfornical organ are capable of sensing osmolarity directly, just like the magnocellular neurons we discussed in the preceding paragraph. Intriguingly, neurons in OVLT provide a major input to the magnocellular neurons and those inputs are required for dehydration-induced vasopressin release by the magnocellular neurons. As you may realize, this finding challenges the notion that the magnocellular neurons are a single-neuron reflex arc. Further work is needed to resolve this puzzle but, in the meantime, we can conclude that in intact animals, changes in osmolarity can be sensed by multiple neurons in multiple locations. The magnocellular vasopressin neurons integrate this information and then regulate urine production accordingly.

Regulating Thirst

Reducing urine production only slows the rate of dehydration. To reverse dehydration, you must drink some fluid or ingest water through food (e.g., a juicy fruit). Of course, some drinks are better than others. Drinking salt water, for example, only dehydrates cells further. Coffee and alcohol tend to worsen dehydration because they promote urination.

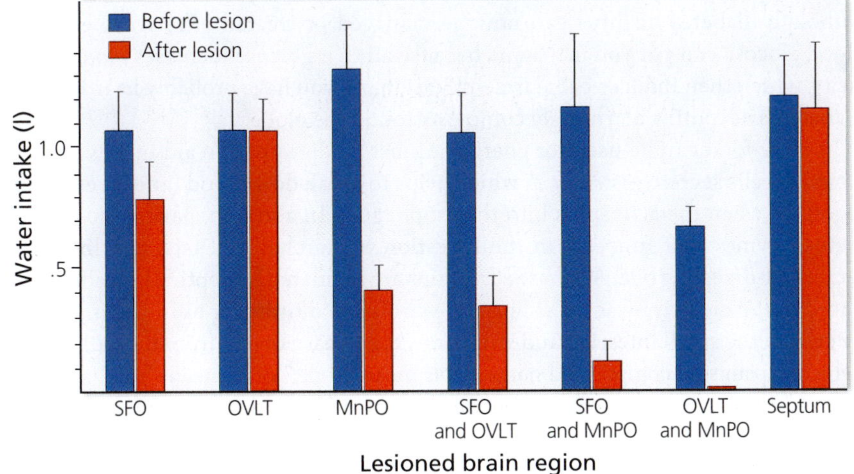

Figure 9.20 Brain regions required for drinking. Shown here are the effects of lesioning various brain regions on drinking by dehydrated sheep. The most dramatic reductions in water intake are observed when the lesions include the median preoptic nucleus (MnPO) and the vascular organ of the lamina terminalis (OVLT) or the subfornical organ (SFO). Significant effects are observed also when MnPO is lesioned by itself or when SFO and OVLT are lesioned together. This pattern of lesion effects is consistent with the known connectivity of these three brain regions (see Figure 9.18). [After McKinley et al., 1999]

How we learn what to drink and where to find it is a complicated question, which we will leave to later chapters. Even the simpler question of what causes thirst is difficult to study experimentally. Comparisons of brain activity in thirsty humans before and after they drink a glass of water have revealed consistent changes in the activity of anterior cingulate and insular cortex. These cortical areas are thought to be required for the conscious experience of thirst, but destroying them experimentally does not prevent drinking. Nor does destroying the magnocellular vasopressin neurons eliminate thirst, as patients with diabetes insipidus drink insatiably. Therefore, thirst-induced drinking must involve some other brain regions as well.

One of these other thirst-related brain regions is the **median preoptic nucleus**, which sits in the midline just rostral to the medial preoptic area (note the subtle difference in names). Lesions of this nucleus reduce water intake in sheep that were injected with hypertonic saline to induce thirst (Figure 9.20). The effects of lesioning the median preoptic nucleus are even more severe when the lesions are combined with lesions of the subfornical organ and OVLT, both of which project to the median preoptic nucleus (Figure 9.18) and presumably provide it with information about the body's hydration state. Projections from the median preoptic nucleus to a variety of lower brain regions, including the *periaqueductal gray* (Figure 9.18), are thought to stimulate drinking.

Regulating Digestion

To support life, we must eat and store some energy for use between mealtimes. Particularly important for normal brain function is the regulation of glucose (see Chapter 5). We normally regulate blood glucose to 0.8–1.1 grams per liter. Given that a normal adult human contains about 5 liters of blood, this means that we normally have about 5 g of glucose floating through our blood vessels. Think about this the next time you drink a 12 oz. can of non-diet soda that contains, on average, 40 g of sugar!

Hormones Involved in Digestion

To minimize fluctuations in blood glucose levels, our bodies employ the hormones glucagon and insulin. **Glucagon** is secreted by the pancreas in response to low glucose levels. It stimulates fat and liver cells to convert glycogen and fat into glucose. **Insulin**, in contrast, is secreted from the pancreas when blood glucose is high. Its effects are largely opposite to those of glucagon. Specifically, insulin stimulates liver and muscle cells to convert glucose into glycogen, and it prompts adipose tissue to make more fat.

The importance of these two hormones for glucose homeostasis becomes apparent when the hormones don't function properly. The loss of insulin-producing pancreatic cells and an insensitivity to insulin cause type I and type II diabetes mellitus, respectively. Both forms of the disease are characterized by elevated blood glucose. Although diabetes mellitus can now be managed or treated effectively, excessive blood glucose can put you in a coma because all that glucose increases blood osmolarity, which then induces cellular dehydration. As you have probably heard, type II diabetes is becoming alarmingly common around the globe.

Before food can be used for energy, it must be digested. To aid in this process, stomach cells secrete *gastric acid*, which helps to break down food (and gives you indigestion when the acid spills into the esophagus). In addition, pancreatic cells secrete enzymes that function in the digestion of starches and fats, and liver cells secrete fat-digesting bile. All these secretions are regulated by peptide hormones produced in the digestive tract. The hormone **cholecystokinin**, for example, is secreted by cells in the small intestine and stimulates the release of bile from the gall bladder (which explains its complicated name: *chole* means "bile," *cysto* means "sac," and *kinin* means "move"). Similarly, the hormone **gastrin** is secreted from stomach cells and triggers the secretion of gastric acid.

This raises an obvious question: what controls the secretion of gastrin, cholecystokinin, and other digestion-promoting hormones? The main trigger is food in the gut because stretch receptors and chemosensors in the intestinal walls activate neurons of the enteric nervous system that, through local enteric circuits, stimulate hormone release. The main advantage of this arrangement is that the digestion-promoting hormones are released when they are needed most.

Moving Food through the Gut

As food is being digested, wave-like **peristaltic contractions** of the intestinal smooth muscle move the food along. Because the intestines can "wriggle" even after they have been completely disconnected from the brain and spinal cord, the enteric neurons in the intestine's walls must harbor some kind of motor pattern generator. However, the rhythm of the peristaltic contractions is modulated by the presence of food inside the intestines. Specifically, stretching a piece of intestine with a bolus of food activates enteric stretch receptors, which then excite enteric motor neurons upstream of the food, resulting in a muscle contraction that moves the food downstream toward the rectum. Once the food has moved, it activates another set of stretch receptors, which then trigger another muscle contraction, and so on. The resulting wave of contraction combines with the intestine's intrinsic tendency for rhythmic contractions to yield repeated waves of contraction that gradually move digested food toward the rectum.

Although gut secretions and movements are controlled by enteric neurons, the enteric nervous system is itself modulated by the brain and spinal cord. Specifically, enteric neurons receive descending inputs from parasympathetic preganglionic neurons in the **dorsal motor nucleus of the vagus nerve** (which lies in the medulla at the same level as nucleus ambiguus; see Figure 9.9) and from sympathetic postganglionic neurons. Activation of the parasympathetic inputs promotes digestion and nutrient absorption by increasing peristalsis, digestive secretions, and intestinal blood flow. In contrast, activation of the sympathetic inputs slows digestion down. In short, the parasympathetic and sympathetic inputs to the enteric nervous system have opposite effects on digestion.

Regulating Appetite

A wide variety of signals govern when we get hungry and when we feel sated. Stimuli that promote eating are called **orexigenic** ("orexis" being the Greek word for appetite). One important orexigenic signal is the sensation of an empty stomach, which is conveyed to the brain by the relative silence of stretch receptors in the stomach wall. Another orexigenic signal is low blood glucose, which is sensed by glucose-sensing neurons in the area postrema (see Figure 9.7).

The best studied orexigenic signal is the peptide hormone **ghrelin** (whose name is derived from the Hindi word for "growth"). Ghrelin is secreted mainly by the stomach

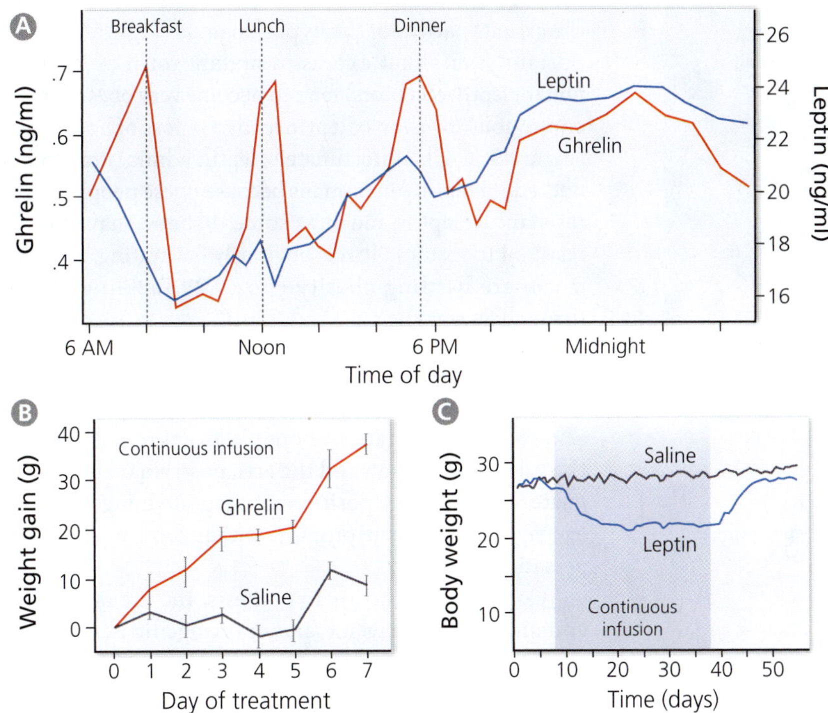

Figure 9.21 Ghrelin stimulates eating, whereas leptin promotes weight loss. The graph in (A) charts human blood plasma levels of ghrelin and leptin, two peptide hormones that are released by the stomach and adipose tissue, respectively. As you can see, ghrelin levels increase before each meal and decrease afterward. In contrast, leptin levels change more slowly and are highest at night. Graph (B) shows that continuous infusions of ghrelin into the cerebral ventricles of rats leads to a gradual increase in body weight compared to saline injected controls. Graph (C) shows that intracerebral infusion of leptin for several weeks causes weight loss in mice. [After Cummings et al., 2001; Tschöp et al., 2000; Halaas et al., 1997]

but also by the intestines when they are empty and insulin levels are low. Circulating ghrelin can be sensed by cells in the hypothalamus and a few other brain regions. Importantly, blood levels of ghrelin tend to rise sharply just before mealtime and to decrease after eating (Figure 9.21). Intravenous injection of ghrelin into healthy human volunteers increases food consumption by ~28% compared to saline-injected controls. Such excessive eating causes obesity in the long run. Unfortunately, if you were hoping to discover an anti-obesity drug, interfering with ghrelin signaling does not prevent normal eating and only causes minor decreases in body weight. This finding is consistent with the idea that eating is controlled by multiple orexigenic signals that can substitute for one another.

Appetite Suppressants

Anorexigenic signals, which suppress appetite, are likewise plentiful. Stomach stretch is one anorexigenic signal, as are high levels of glucose or insulin. Several peptide hormones secreted by the digestive tract also act as appetite suppressants.

One of the most powerful anorexigenic signals is cholecystokinin, the previously mentioned peptide that is secreted by the intestine in response to fat or protein-rich foods. The presence of cholecystokinin in the digestive tract is sensed by axons in the vagus nerve, which convey this signal to NTS. Cholecystokinin also enters the bloodstream and activates neurons in the area postrema, which likewise send their axons to NTS. It remains unclear how the cholecystokinin signal coming out of NTS suppresses the urge to eat, but it evidently does because rats with impaired cholecystokinin signaling become obese. This observation raises the hope that boosting cholecystokinin signaling might reverse obesity, but rats infused with extra cholecystokinin quickly habituate to its anorexigenic effects.

Another highly studied anorexigenic signal is **leptin** (from *leptos*, meaning "thin"). This peptide hormone is secreted by white (not brown!) adipose tissue, and its levels in the blood correlate with levels of white fat. In simple terms, the fatter you are, the more leptin you have in your circulation. In addition, leptin levels rise during the night (Figure 9.21). Circulating leptin enters the brain through the pituitary gland's median eminence, which lacks a blood–brain barrier. From there, the leptin

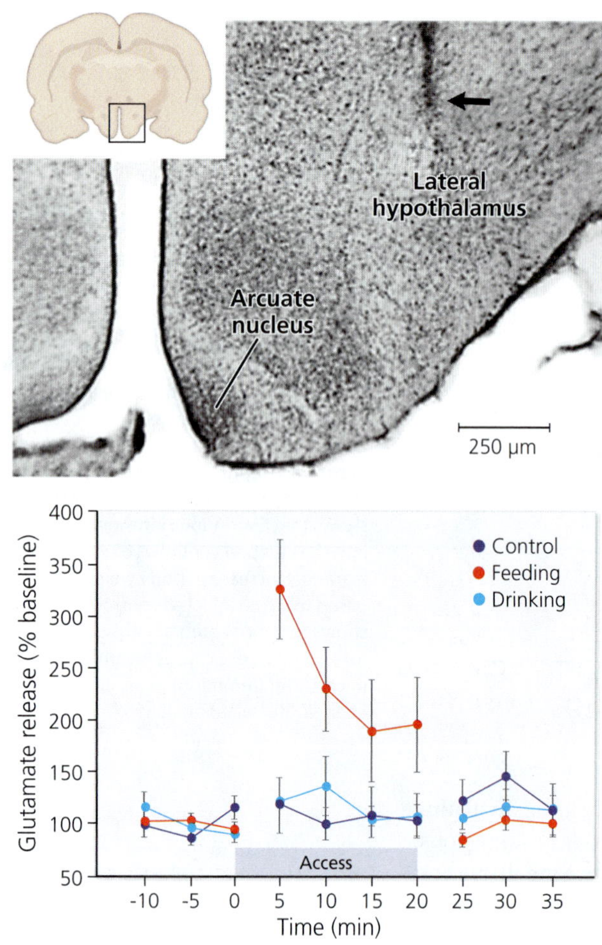

Figure 9.22 Hypothalamic regulation of feeding. Food-deprived rats were given access to food, drink, or a control substance while glutamate levels were measured in the lateral hypothalamus with microdialysis probes. The arrow in the top panel indicates the tip of the microdialysis probe. The graph shows that glutamate levels in the lateral hypothalamus increase only when the rats are given access to food (gray horizontal bar) but not when they are given water or inedible bits of plastic (control). [From Thongkhao-on et al., 2008]

diffuses a short distance to leptin-sensitive neurons in the *arcuate nucleus* of the hypothalamus (Figure 9.22). Importantly, rats that express a mutant form of leptin (or mutant leptin receptors) tend to become very obese, whereas continuous infusion of leptin leads to a loss of body weight (Figure 9.21 C). Unfortunately, leptin administration does not reduce obesity in humans because obese people become resistant to leptin, much as some diabetics have become resistant to insulin. This is obviously frustrating, especially if you are battling obesity (Box 9.2: Tackling Obesity through Surgery).

Integrating Hunger and Satiety Signals

Neurons in the arcuate nucleus respond to leptin, but they also receive information about plasma ghrelin levels from the subfornical organ and the area postrema. Thus, the arcuate nucleus is in a position to integrate hunger and satiety signals and then promote either food intake or the cessation of eating.

Consistent with this hypothesis, the arcuate nucleus contains both orexigenic and anorexigenic neurons. The orexigenic arcuate neurons use **neuropeptide Y (NPY)** as their main neurotransmitter. The anorexigenic ones release **α-melanocortin stimulating hormone (αMSH)**. Because these two types of neurons are chemically distinct from each other and from most other neurons, their activity can be mimicked by infusing NPY or αMSH directly into the cerebral ventricles. Injecting NPY prompts animals to eat ravenously, whereas injecting αMSH inhibits food intake. Importantly, activation of the orexigenic neurons inhibits the anorexigenic neurons. This ensures that when hunger and satiety signals are both present, hunger wins out.

Arcuate neurons send their axons to the **lateral hypothalamus** (Figure 9.22), which is a crucial link in the feeding control circuit. Lesions of this area reduce food intake; injections of glutamate into the lateral hypothalamus promote eating; and glutamate levels in the lateral hypothalamus increase during eating (Figure 9.22). Still, it remains unclear precisely how the lateral hypothalamic area controls eating. Among its many targets are part of the *periaqueductal gray* and the taste-sensitive portion of NTS. It is likely that these regions all work together to promote eating, or to stop it, but the details remain murky.

BRAIN EXERCISE

Given what you learned in this chapter, what strategy would you pursue to find a pharmacological treatment for obesity? How about a treatment for anorexia?

9.5 How Do We Coordinate Our Vegetative Processes?

Thus far we have treated the various vegetative control circuits independently of one another. This approach makes sense when you first learn about vegetative control, but the vegetative control circuits are not really as independent as you might think. For example, we have already seen that dehydration tends to lower blood pressure and that increases in heart rate boost blood pressure, other things being equal. Heart

THERAPIES

Box 9.2 *Tackling Obesity through Surgery*

The most widely used criterion for obesity is based on the *body mass index* (*BMI*), which is calculated by taking your weight (in kg) and dividing it by the square of your height (in meters). People with a BMI > 30 kg/m^2 are considered obese, although very muscular individuals may reach such high values without being fat. People with BMIs higher than 40 kg/m^2 are considered morbidly obese. By these measures, ~32% of the adult men in the United States were obese in 2007–08, and ~5% were morbidly obese. The percentages are even higher for women and for some ethnic or racial groups. This presents a serious problem because obesity correlates with rates of heart disease, diabetes, and cancer. Morbid obesity has been estimated to increase mortality at least threefold and as much as tenfold.

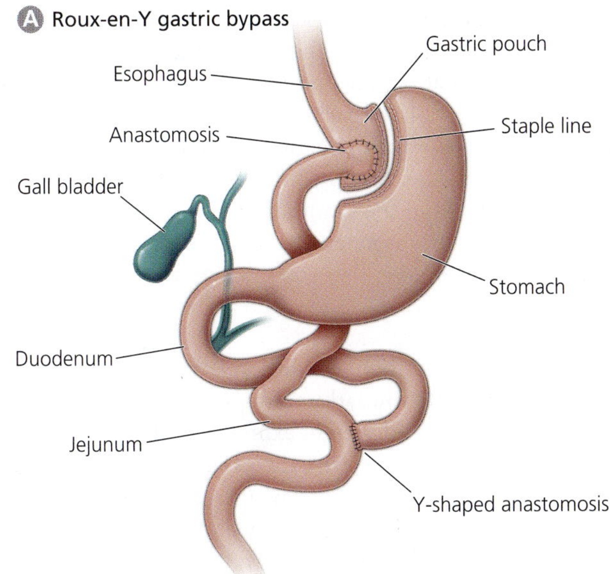

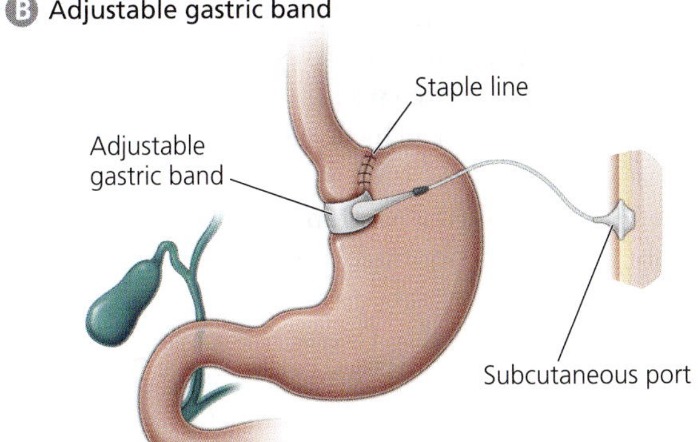

Figure b9.2 Treating obesity through stomach surgery. The two most widely used surgical treatments for morbid obesity are (A) Roux-en-Y gastric bypass (RYGB) surgery and (B) gastric banding. In RYGB, the surgeon creates a small gastric pouch and connects it to the middle of the small intestine's jejunum. The remaining part of the stomach is connected to a lower portion of the jejunum. Gastric banding, which is popular in Europe and Australia, involves threading a band around a portion of the stomach through which food must pass before it can enter the duodenum. Illustrated here is an adjustable gastric band that has a balloon on its inner surface. The balloon can be inflated or deflated by injecting or withdrawing fluid through a silicone "port" implanted under the skin. [After Tessier and Eagon, 2008]

Dieting, exercise, and anti-obesity drugs typically reduce body weight by 5–10%, but much of the lost weight often returns. The average weight loss after at least a year of pharmacological treatment is only about 5 kg. More successful have been *bariatric surgical procedures*, which were developed in the 1960s after doctors realized that patients whose stomachs were removed (because of gastric ulcers) lost a great deal of body weight after the surgery. About 20,000 persons per year received bariatric surgery in the United States in the 1990s; by 2006, this number had risen to 177,000. This enormous increase can be attributed in part to the invention of laparoscopic surgery, which is performed through small holes in the body wall. However, bariatric surgeries became so popular primarily because they seem to be relatively effective. For example, a Swedish study followed 4,047 obese patients and found that two years after a bariatric procedure, the patients' body weights had dropped by as much as 35%, much more than in control subjects. Ten years after the surgery only some of the lost weight had been regained. Diabetes rates and overall mortality were also lower in the treated patients than in untreated controls.

The most popular form of bariatric surgery in the United States is the *Roux-en-Y gastric bypass*, named after the surgeon Cesar Roux. In this technique, a small portion of the stomach is isolated and connected to the middle portion of the small intestine's jejunum (Figure b9.2). Thus, incoming food bypasses ~95% of the stomach, all of the duodenum, and the upper portion of the jejunum. The bypassed portions remain continually empty of food (although their secretions, including gastric acid and bile, drain into the lower part of the jejunum). As a result of these alterations, incoming food reaches the lower jejunum in a much less digested state than it normally does. This change, in turn, reduces the absorption of nutrients through the intestine's wall. Micronutrients like iron, calcium, and vitamin B12 are particularly affected, which is why they must be taken as dietary supplements. However, weight loss after gastric bypass surgery cannot be explained by reduced nutrient absorption alone because the patients after their surgery eat significantly smaller meals, eat less frequently, and shift their food preference away from fatty foods. In general, they have less appetite.

Why does gastric bypass surgery suppress appetite? One possibility is that the small gastric pouch is easily stretched by even small amounts of food, which then sends a satiety signal to the brain via the vagus nerve. The problem with this hypothesis is that food quickly passes out of the gastric

Box 9.2 *Tackling Obesity through Surgery (continued)*

pouch, whereas low appetite persists. Plus, the bypassed portion of the stomach remains empty and thus unstretched. A more likely hypothesis is that the bypass surgery alters intestinal hormone release by letting relatively undigested food reach the small intestine, Supporting this hypothesis is the discovery that gastric bypass surgery in rats increases plasma levels of peptide YY and glucagon-related peptide, which are secreted from the intestine and known to stimulate anorexigenic neurons in the hypothalamus. Furthermore, inhibiting the secretion of these hormones decreases the appetite suppressing effects of gastric bypass surgery. Similar results have now been obtained in humans as well.

rate is also known to increase with each inspiration, especially in young children, and the ability to thermoregulate is affected by various hormones and dehydration.

Along with such low-level coordination, the body's vegetative functions exhibit several forms of more global orchestration. In particular, they are subject to coordinated regulation during the daily sleep-wake cycle; during periods of excitement or acute stress; and during chronic stress. Because the high-level coordination of vegetative processes involves projections from the brain to the spinal cord, people with spinal cord injury frequently suffer from autonomic dysfunction (see Box 9.3: Vegetative Control after Spinal Cord Injury).

Circadian Regulation

When human volunteers live for weeks in bunkers or caves, their daily rhythms of eating, sleeping, and locomotion persist. Similarly, animals that are housed in constant darkness or, more kindly, in constant dim light exhibit activity rhythms with a period close to 24 hours. Such daily rhythms are called **circadian rhythms** (*circa* meaning "approximately" and *diem* being Latin for "day").

Although circadian rhythms are most obvious for sleep and locomotor activity, they also exist for many vegetative functions. For example, core body temperature is highest in the early evening and late morning, and lowest between 3 and 5 AM. Blood pressure is lowest during the early stages of sleep and highest at midday. Hunger, too, has a clear circadian rhythm. In fact, many digestive secretions increase even before you have eaten, shortly before your customary mealtimes (see Figure 9.21 A). This is useful because it ensures that digestion can begin as soon as food arrives in the stomach.

Many hormones are released in a circadian rhythm. For example, the levels of LH and ACTH, which regulate gonadal sex hormones and cortisol, respectively, are highest in the early morning. Growth hormone and vasopressin reach maximal levels at night. Because you already learned about vasopressin, you can infer that the circadian rhythm of this antidiuretic hormone promotes undisturbed sleep by reducing the need to urinate at night. One of the most important hormones with circadian fluctuations is **melatonin**, a small peptide hormone that is released at night from the **pineal gland**, a small endocrine gland sitting on top of the third ventricle. Melatonin is a powerful antioxidant and promotes sleep (see Box 9.4: Jet Lag and Night Shift Work). In many species, melatonin also controls seasonal changes in skin color and behavior.

The Suprachiasmatic Nucleus

All these behavioral, physiological, and hormonal rhythms are regulated by the **suprachiasmatic nucleus**, a tiny brain region immediately dorsal to the optic chiasm. Bilateral lesions of the suprachiasmatic nucleus in rodents abolish or disrupt virtually all circadian and seasonal rhythms. Many of the rhythms that are lost after lesions of the suprachiasmatic nucleus can be restored by replacing the lesioned nuclei with a suprachiasmatic nucleus from another animal of the same species. Because the transplanted suprachiasmatic neurons fail to make their normal neural connections, we can infer that the suprachiasmatic neurons must release some humoral

NEUROLOGICAL DISORDERS

Box 9.3 *Vegetative Control after Spinal Cord Injury*

Spinal cord injury typically destroys neurons in a specific spinal segment and disconnects regions below the site of injury from upper levels of the spinal cord and from the brain. The outwardly most obvious effect of spinal transection is the inability to make voluntary leg movements (paraplegia) or, if the damage is above C5, voluntary arm and leg movements (quadriplegia). However, spinal cord injury also impairs breathing, digestion, urination, defecation, and sexual enjoyment. Right after the injury, most patients with upper spinal cord damage lose vegetative control almost entirely. After this initial period of *spinal shock* some autonomic reflexes return, but the recouped reflexes are often abnormal and uncontrolled (dysreflexia). Moreover, descending control of spinal neurons controlling vegetative functions is often permanently lost.

To understand how spinal cord injury affects urination, you need to know how urination is controlled. When the bladder fills, mechanosensors in the bladder send this information to parasympathetic preganglionic neurons in the sacral spinal cord. These neurons (acting through their associated postganglionic neurons) stimulate bladder contraction and relax the internal sphincter, a smooth muscle looped around the urethra. Relaxation of this sphincter enables urine flow. All this circuitry continues to function in patients with spinal cord damage (after the period of spinal shock has passed), but below the urethra's internal sphincter lies the external sphincter, a skeletal muscle innervated by alpha motor neurons in the spinal cord ventral horn. Because these motor neurons tend to be tonically active, the external sphincter tends to be closed, which means that the reflexive bladder contractions must push the urine out through an external sphincter that is working to keep the urine in. As a result, the purely reflexive bladder contractions in people with spinal cord injury are usually not strong enough to empty the bladder entirely. As a result, the patients often suffer from bladder infections.

Urination also involves a *micturition center* in the brain (*micturition* means "desire to urinate"). Activity in the medial portion of this micturition center promotes urination by exciting the parasympathetic preganglionic neurons that constrict the bladder and relax the internal sphincter. Activity in the lateral portion prevents urination by causing constriction of the two urethral sphincters and inhibition of the parasympathetic postganglionic neurons that innervate the bladder wall. Important inputs to the micturition center come from the midbrain's periaqueductal gray, which in essence "throws the switch" between urinating and holding it in. The periaqueductal switch throwers, in turn, receive input from higher brain regions as well as ascending information about how full the bladder is. None of these circuits are functional in people with spinal cord injury because in such patients, the brain cannot communicate with the spinal urination circuits.

Bowel movements are likewise problematic for people with spinal cord injuries. Although peristalsis is largely controlled by the enteric nervous system, it is slowed down by spinal cord injury, causing constipation. Spinal cord injury also disables voluntary defecation because it too requires pushing against two sphincters. The internal sphincter is made of smooth muscle and relaxes automatically when the rectum is full. The external sphincter, in contrast, is a skeletal muscle that can be contracted or relaxed at will. The circuitry underlying this voluntary control of the external anal sphincter has not been studied thoroughly, but it probably involves a *defecation center* that is similar to the micturition center. In people with spinal cord injury, the defecation center cannot communicate with the relevant spinal circuits, which means that defecation occurs whenever the rectum is full enough to activate the spinal defecation reflex. To manage this problem, the defecation reflex must be triggered at carefully scheduled times by means of mini-enemas or rectal suppositories.

Can people with spinal cord injury have sex? Penile erection and vaginal engorgement are caused by the activation of parasympathetic preganglionic neurons in the sacral spinal cord. Much of the excitatory input to these neurons comes through spinal nerves directly from the genitals. Thus, genital arousal involves a spinal autonomic reflex, which remains intact in people with spinal cord damage above the sacral level. Spinal injury patients can even ejaculate. But can they enjoy sexual stimulation and full-fledged orgasms? Probably not, because in most spinal cord injury patients information from the genitals cannot pass through the spinal cord to the brain, which needs genital sensory information to generate conscious sexual pleasure. It has been suggested that the vagus nerve may provide the brain with the requisite information, but this has not been studied thoroughly. One should also note that the reflexive changes in heart rate and blood pressure that accompany orgasms tend to be abnormally large or even dangerous in people with spinal cord injury.

factor (probably prokineticin) that influences the activity of distant neurons without the benefit of direct axonal connections. However, in intact brains, suprachiasmatic neurons do have axonal connections. Most notably, suprachiasmatic neurons project to the pituitary gland and to the hypothalamic paraventricular nucleus. The latter nucleus projects to many different areas, including sympathetic and parasympathetic neurons that in turn project to the pineal gland.

Individual suprachiasmatic neurons exhibit a circadian rhythm of gene expression and neural activity even when they are isolated from other neurons. The clock-like gene expression patterns involve primarily the genes **period (*per*)** and **cryptochrome (*cry*)**,

NEUROBIOLOGY IN DEPTH

Box 9.4 *Jet Lag and Night Shift Work*

Jet travel has made it possible for humans to move rapidly across multiple time zones. This travel tends to cause *jet lag*, defined as a misalignment between a person's internal circadian rhythm and the external world's day–night cycle. Jet lag causes difficulties sleeping at night, daytime sleepiness, and poor performance on demanding tasks. For example, athletes tend to perform more poorly if they traveled to their competition across multiple time zones. Fortunately, the misalignment usually disappears after a few days in the new location. On average it takes about one day per hour of time difference to overcome jet lag if you traveled eastward. Adjusting your rhythm takes less time if you traveled westward because it is easier to delay your internal clock (by going to bed late) than to advance it (by retiring early).

To manage jet lag well you have to understand that our internal circadian rhythm is influenced by several timing signals, which include external light or darkness, strenuous activity, sleep, and the hormone *melatonin*, which is secreted by the pineal gland at night. The effect of these timing signals depends on where they fall in the ongoing internal rhythm. For example, bright light in the early morning portion of the internal rhythm advances the rhythm, whereas the same light in the internal evening (evening according to the internal clock) delays the rhythm (Figure b9.3). Similarly, taking extra melatonin in the middle of the internal day advances the rhythm, whereas the same melatonin taken in the internal night delays it. Therefore, if you want to use light exposure or melatonin to manage your jet lag, then you have to know the phase of your internal rhythm. This can be difficult if you are in the throes of ongoing jet lag because you do not know how rapidly your rhythm is adjusting.

For brief trips it is usually best not to battle your internal rhythm but to manage the jet lag with caffeine and sleeping pills (don't go overboard). After all, you will soon be back home. However, if you really need to be at your best when you arrive at a distant location—say you were a US athlete traveling to the 2008 Olympics in China—then you should shift your rhythm several days before you leave. For example, for an eastward trip across 5 time zones, you should start 4 days before departure to shift your sleep schedule earlier by an hour a day, always taking melatonin 5 hours before bedtime. This recipe supposedly works well for many people, although it requires considerable discipline.

Night shift work is even more troublesome than jet lag. For one thing, night shift workers must sleep when the sun shines and interact with people whose schedules are not shifted. Second, their rhythm misalignments tend to be larger and more permanent than those associated with jet lag. Third, night shift workers often revert to a daytime schedule on their days off to be with family and friends. Given these challenges, it is not surprising that night shift workers tend to have more ulcers, heart disease, and cancer than people on day shifts.

To minimize these problems, experts recommend that night shift workers sleep in total darkness, are exposed to plenty of artificial light in the early part of their night shift, and avoid bright light on their trip back home to sleep (wear dark sunglasses). Even with these tricks, persistent night shift work remains stressful. If possible, it should be replaced with slowly rotating shifts that maintain a constant schedule for 2 or more weeks at a time. How all of this applies to college life is a question I leave to you.

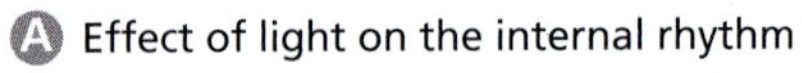

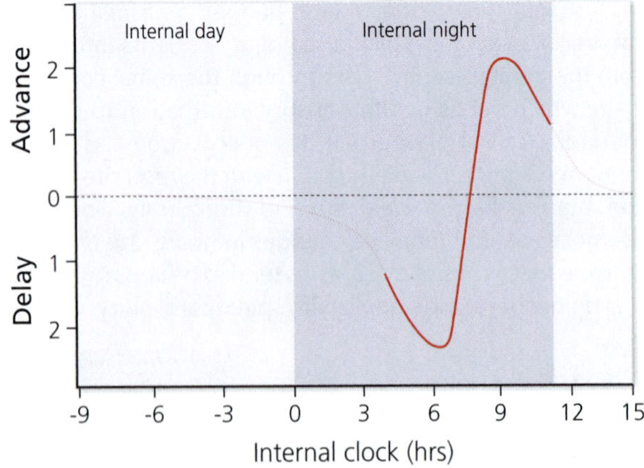

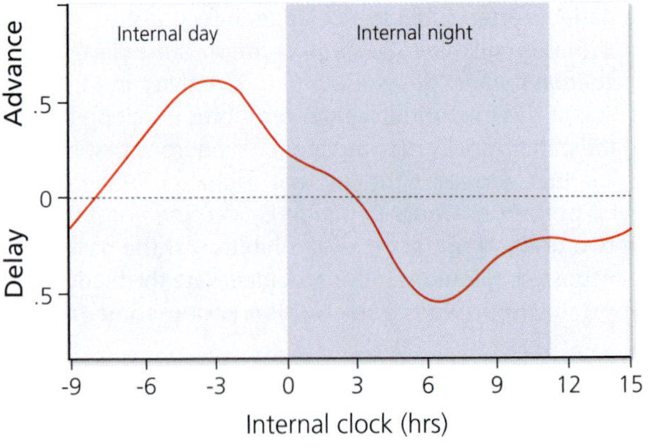

Figure b9.3 Manipulating the circadian rhythm. Light exposure in the first half of the night tends to delay the internal rhythm, whereas light in the last part of the night tends to advance it (A). The solid red line represents actual data; the dashed red line is a reasonable guess; time 0 is defined as the start of endogenous melatonin release (~ 9 PM). Shown in (B) is the analogous graph for people who took melatonin (0.5 mg) at the same time of day for 4 days. Melatonin taken during the internal day advances the internal rhythm, whereas extra melatonin taken at night delays the rhythm. [After Revell and Eastman, 2005]

whose transcription is promoted by the transcription factors **Bmal** and **Clock** (Figure 9.23). Once translated in the cytoplasm, the proteins Per and Cry form heterodimer complexes that can enter the nucleus and prevent Bmal and Clock from driving the expression of *per* and *cry*. Thus, the expression of *per* and *cry* ultimately shuts itself off through negative feedback. Because the Per/Cry complex is degraded slowly, the delay in this negative feedback loop is long enough to support a slow, rhythmic oscillation.

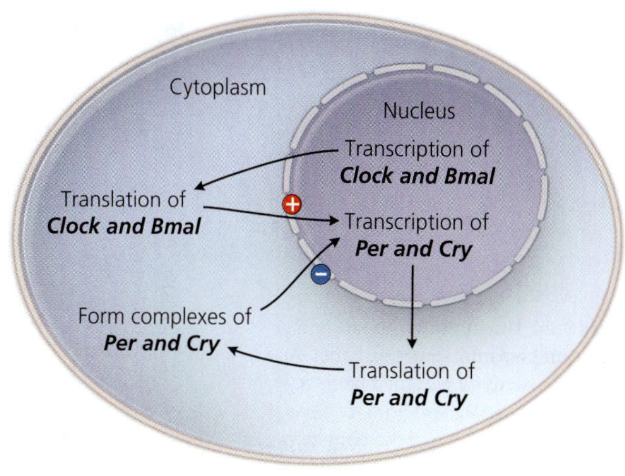

Figure 9.23 The intracellular circadian clock. The transcription factors Clock and Bmal cooperate to promote transcription of the genes period (*per*) and cryptochrome (*cry*), which are translated into Per and Cry protein in the cytoplasm. When Per and Cry bind to one another, they enter the nucleus and prevent Bmal and Clock from promoting the transcription of *per* and *cry*. As Per and Cry proteins are slowly degraded, the inhibition of *per* and *cry* transcription wanes until eventually Per and Cry levels rise again.

Evidence that these regular fluctuations in gene expression are behaviorally significant comes from transgenic mice that are missing (or have defective forms of) one or more clock-related genes. Most dramatically, transgenic mice without a functional *bmal* lack the circadian activity rhythm. When housed in constant darkness, these *bmal* knockout mice sleep and run around at relatively random intervals. Knocking out individual *per* or *cry* genes has less obvious effects, but those genes all come in multiple isoforms (*per1* and *per2*, as well as *cry1* and *cry2*) that can substitute for one another. Indeed, when multiple isoforms of *per* or *cry* are deleted, circadian rhythms are disrupted severely.

For the circadian rhythms of individual neurons to generate circadian rhythms in animal behavior, the activity of the individual neurons must be synchronized. This cellular synchronization appears to be mediated by gap junctions and reciprocal synapses between suprachiasmatic neurons through which action potentials in one neuron can increase the probability of action potentials being triggered in neighboring neurons.

Because the synchronization between the left and right suprachiasmatic nuclei is relatively weak, the two structures can become uncoupled from one another when hamsters are kept in continuous light for several days. As shown in Figure 9.24, this uncoupling of the left and right suprachiasmatic nuclei causes hamsters to express two separate circadian rhythms, meaning that the hamsters become active twice every 24 hours. Using the immediate early gene *c-fos* to monitor neural activity, one can show that in such split-rhythm hamsters, the left suprachiasmatic nucleus is active when the right one is silent and vice versa (Figure 9.24 B).

Entrainment of the Circadian Clock

If you look carefully at Figure 9.24 A, you can see that the hamsters start their active period slightly later each day, even before the rhythm splits. This is a common finding for animals and humans under constant light conditions because the intrinsic rhythm of the suprachiasmatic nucleus has a period slightly longer than 24 hours. If this is the case, what keeps our circadian rhythms from drifting out of phase with the daily cycle of light and dark? As the seasons change and days vary in length, how does the internal rhythm adapt? What happens when you travel across time zones (Box 9.4: Jet Lag and Night Shift Work)?

The answer to these questions is that the daily fluctuations in ambient light levels **entrain** the circadian clock of the suprachiasmatic nucleus so that the two rhythms are matched in phase and frequency. To grasp the general phenomenon of entrainment, consider two metronomes (not digital metronomes but the ones that have a pendulum arm) standing next to each other on a moveable base. One metronome is set to beat at a slightly faster pace than the other, and the two are started out of phase, so that their pendulums swing in opposite directions. After a few cycles, the base will start to move. Due to the transmission of mechanical energy through the movements of the base, the metronomes quickly become synchronized, settling on an intermediate phase and frequency. The entrainment of the suprachiasmatic nucleus by the

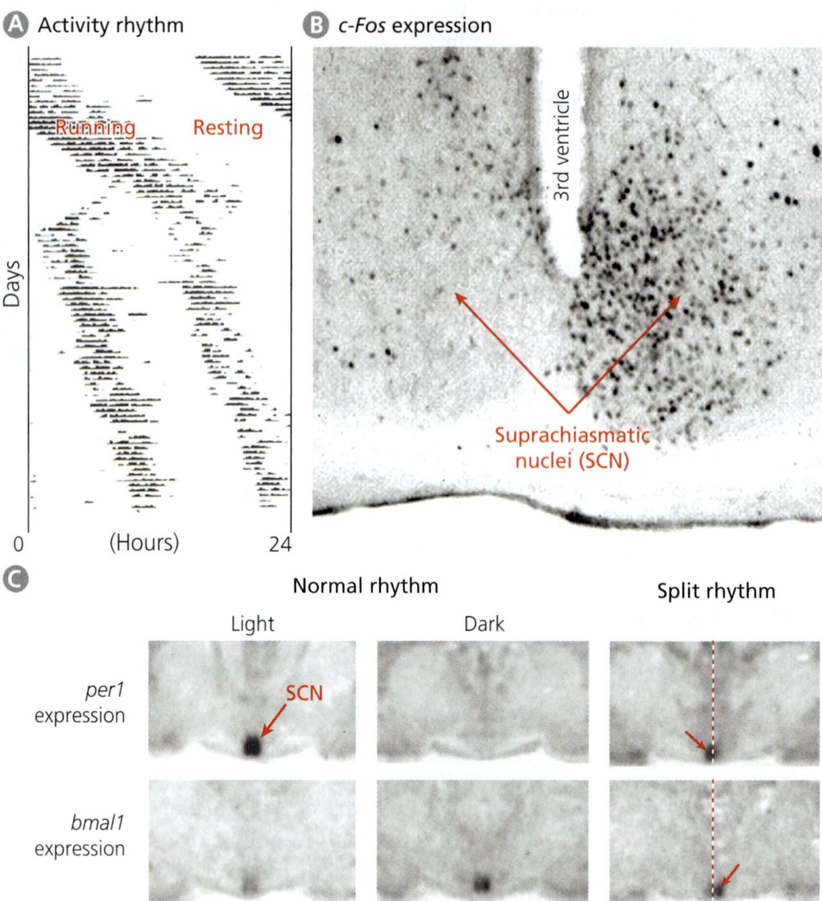

Figure 9.24 Uncoupling the left and right suprachiasmatic nuclei. Panel (A) is a locomotor activity plot for a hamster that was kept in constant light for several weeks. Each revolution of the hamster's running wheel is represented by a black mark. After about 3 weeks the hamster's activity rhythm splits, with the hamster becoming active twice every 24 hours. Panel (B) shows a coronal section through the suprachiasmatic nuclei (SCN) of a split-rhythm hamster, stained to reveal the immediate-early protein c-Fos, an indicator of neuronal activity. One side of the SCN is very active, while the other side is silent. Panel (C) shows analogous sections processed to reveal *per1* and *bmal1* expression (top and bottom row, respectively). Whereas *bmal1* and *per1* expression is bilaterally symmetric in hamsters with normal circadian rhythms, it is asymmetric in the split-rhythm animals. [From de la Iglesia et al., 2000, 2003.]

daily light–dark cycle is similar except that the entrainment in this case is unidirectional rather than mutual. That is, the suprachiasmatic nucleus adjusts its rhythm to match the movements of the sun, but the sun is not so obliging.

Also, the entrainment of the suprachiasmatic nucleus is due to light-sensing retinal neurons rather than mechanical coupling. Specifically, a subset of retinal ganglion cells projects to the suprachiasmatic nucleus. These retinal neurons express the photosensitive pigment *melanopsin*, which makes them sensitive to light even when retinal rods and cones are deleted genetically. When the retinal ganglion cells signal that the external environment is bright, the suprachiasmatic neurons increase their rate of *per* expression. Through this periodic light-induced modulation of *per* expression, the daily light–dark cycle entrains the intracellular circadian clock.

Dealing with Acute Stress

As mentioned at the beginning of this chapter, Walter Cannon had argued that our vegetative control systems are primarily homeostatic in function. However, Cannon also recognized that organisms can change their internal environment to deal with acutely stressful situations, such as threats from predators, fire, earthquakes, floods, or loss of blood. What struck Cannon as interesting is that organisms respond to these rather diverse stressors in remarkably uniform ways. They tend to increase heart rate and respiration, boost blood pressure, increase alertness, slow digestion, and dilate the pupils. Cannon referred to this constellation of responses as the **fight-or-flight response**.

Cannon's notion of a fight-or-flight response simplifies how you think about the high-level neural control of vegetative processes. The physiological changes associated with the fight-or-flight response all seem linked to increased sympathetic activation

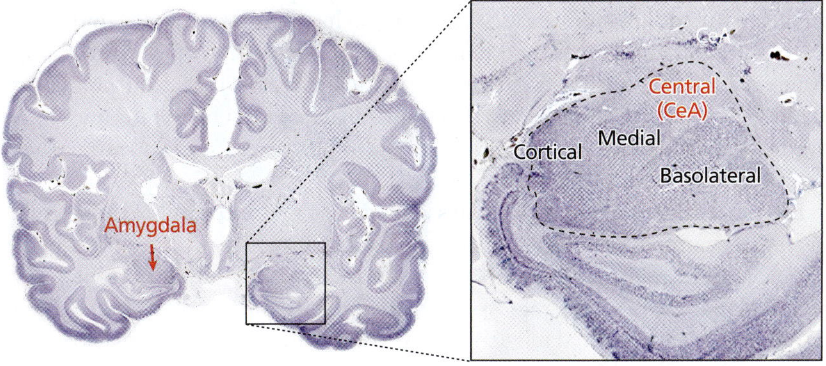

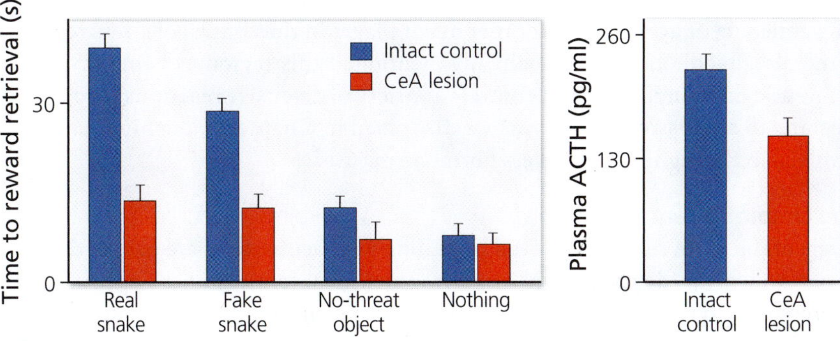

Figure 9.25 Lesions of the central amygdala. Panel (A) presents a Nissl-stained coronal section through a human brain, highlighting the amygdala. Panel (B) illustrates the results of lesioning the central amygdala bilaterally in adult male rhesus monkeys. The investigators measured how long it took each monkey to reach for a food reward on the far side of an open box that contained either a real snake, a plastic snake, a roll of blue masking tape, or nothing. Because the lesioned monkeys took less time than control monkeys to reach for the food when the snakes were present, we can infer that they were less afraid of snakes. The bar graph on the right shows that the monkeys with central amygdala lesions also have lower circulating levels of adrenocorticotropic hormone (ACTH), which stimulates cortisol secretion. [After Kalin et al., 2004]

and decreased parasympathetic activity. Therefore, one can hypothesize that the fight-or-flight response is triggered by a single brain region that, through direct or indirect connections, excites all the sympathetic neurons and inhibits the parasympathetic ones. Conversely, the relaxation that comes after a fight or after reaching safety might occur simply because this hypothetical fight-or-flight "command center" returns to its normal, baseline rate of neural activity. There is no need to postulate a separate brain region that commands **rest-and-digest**, as the opposite of fight-or-flight is often called. A single brain region can do the job, simply by increasing or decreasing its level of activity.

The Central Amygdala

A good candidate for such a fight-or-flight command center is the **central nucleus of the amygdala**. Bilateral lesions of this relatively small and inconspicuous brain area deep in the temporal lobe (Figure 9.25) reduce the behavioral and hormonal responses of monkeys to snakes, which are a natural threat to monkeys in the wild. Conversely, electrical stimulation of the central amygdala in rats and cats increases heart rate, blood pressure, and redirects blood flow from the skin and digestive tract to the muscles. In short, activation of the central amygdala is necessary and sufficient for the major components of the fight-or-flight response.

Given this fight-or-flight function, it is not surprising that the central amygdala projects to many of the vegetative control regions we discussed earlier in this chapter. For example, the central amygdala projects to NTS and rVLM, which regulate heart rate and blood pressure. The central amygdala also projects (at least indirectly) to the hypothalamic paraventricular nucleus, which activates most sympathetic preganglionic neurons and projects to the pituitary gland. Through these pathways, activation of the central amygdala prompts the release of norepinephrine from sympathetic postganglionic axons, of epinephrine from the adrenal medulla, and of cortisol from the adrenal cortex.

Freezing Out of Fear

An important complication to this relatively simple picture is that the responses of animals to threats are more varied than Cannon had suggested. For example, threatened animals often **freeze** rather than run or fight. This freezing behavior is adaptive because it allows the animal to gather information about the threat and perhaps avoid detection. Physiologically, animals that freeze decrease their heart rate, muscle blood flow, and breathing rate. This is very different from what you see in an animal getting ready to fight or run away.

Because the central amygdala plays a major role in the control of threat-induced freezing, we cannot think of the central amygdala as simply triggering an all-or-none, fight-or-flight response. Most likely its effects are modulated by the activity of other brain regions. Those other regions probably include the **septal nuclei**, which are located in the ventromedial telencephalon. This is interesting because lesions of the septal nuclei tend to produce animals that are extremely aggressive. Furthermore, septal neurons are very active when an animal is freezing in response to a threat. These data suggest that neurons in the septal nuclei normally suppress the neurons that trigger fleeing or aggression. Another major player in the control of stress responses is the periaqueductal gray. We already mentioned this region as being involved in eating and drinking, but it also controls a variety of defensive reactions ranging from freezing to aggression. Finally, as we discussed in Chapter 8, the hippocampus is involved in the regulation of stress hormone release.

The Limbic System

Many of the brain regions involved in dealing with acute stress are part of the *limbic circuit*, which was defined by the American neuroanatomist James Papez as a set of interconnected forebrain areas that lie close to the midline and are thought to be involved in regulating emotion. The limbic circuit originally included the hippocampus, the cingulate cortex, the mammillary bodies of the hypothalamus, and part of the dorsal thalamus. Paul MacLean later expanded the circuit to include the amygdala, septum, and ventral striatum.

MacLean's expanded circuit, which he called the **limbic system**, remains an influential idea, but it is problematic. One major problem is that the structures of the limbic system are connected not just with one another but also with a host of other brain regions. Therefore, the limbic system does not delimit a well-defined circuit, which is what the word "system" implies. In addition, the individual components of the limbic system all have multiple functions, not all of which are clearly linked to "emotion," another concept that is tricky to define. Therefore, we generally avoid the term limbic system and instead refer to its components individually.

Effects of Chronic Stress

Responses to acute stress tend to be appropriate for dealing with looming threats. It is adaptive, for example, to send more glucose to your muscles and brain when facing a threat, or to hold still and be alert. However, when stressful situations occur repeatedly, the stress becomes chronic and the responses unhealthy. Common causes of chronic stress are persistent food or water shortages, prolonged temperature extremes, or chronic pain. Similarly stressful are frequent and dramatic environmental changes that you cannot predict or escape. Among the most potent stressors tend to be members of your own species. As Jean-Paul Sartre once remarked, "Hell is other people." Among animals, too, submissive animals that have been housed with a dominant conspecific (a bully) often exhibit severe symptoms of chronic stress such as reduced immune responses, chronically high blood pressure, malnutrition, reduced sex drive, and infertility.

The Stress Hormone Cortisol

The effects of chronic stress are caused primarily by **cortisol**. This hormone is released more slowly than epinephrine in response to a stressor, and its effects on target cells

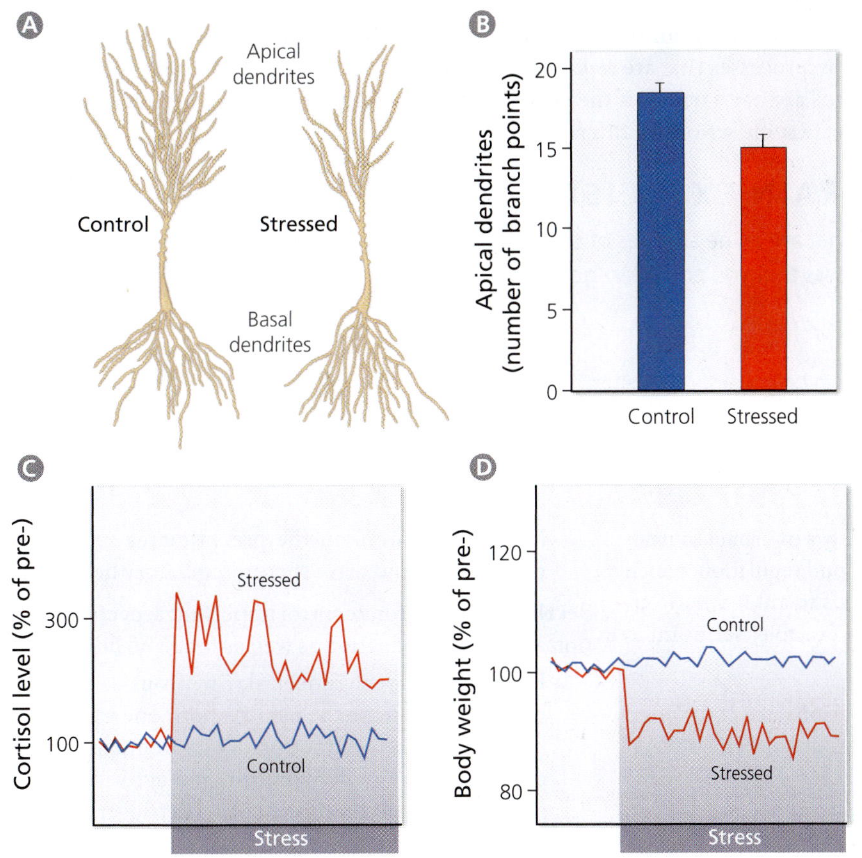

Figure 9.26 Effects of chronic stress. Male tree shrews (close relatives of primates) were stressed by forcing them to live in close quarters with a dominant male (a bully) for 28 days. Panel (A) shows Golgi-stained neurons from the CA3 region of the hippocampus in a stressed tree shrew and a control subject. The bar graph in (B) shows that the apical dendrites contain significantly fewer branches in the stressed animals than in controls. Graph (C) shows that plasma cortisol levels increased after the experimental animals were paired with the bully. As shown in (D), the stressed animals also lose weight. [After Magariños et al., 1996]

are often sluggish. However, the effects of cortisol on target cells tend to be longer lasting than the effects of epinephrine. Furthermore, in contrast to epinephrine, cortisol can cross the blood–brain barrier and affect any neurons expressing cortisol receptors. Particularly interesting is that numerous cortisol-binding neurons are found in the hippocampus, which is part of a negative feedback loop that limits cortisol release (see Chapter 8). This negative feedback ensures that our cortisol response to an immediate threat fades out over the course of an hour or two, as you "calm down" after the threat has passed. However, if the threat is unusually severe, prolonged, or repeated, then the negative feedback circuit breaks down and cortisol levels remain elevated long after the threats have receded.

Effects of Chronic Stress on the Hippocampus

Chronically elevated cortisol causes the dendrites of some hippocampal neurons to shrink. Specifically, giving adult rats daily injections of corticosterone (the rat equivalent of cortisol) reduces the number and length of dendritic branches in the CA3 region of the hippocampus. Shrunken dendrites are also seen in the hippocampus of tree shrews that were stressed for 4 weeks by interactions with a dominant conspecific (Figure 9.26). Given that the hippocampus is involved in learning and memory (see Chapter 14), the observed changes in hippocampal dendrites may explain why you do not remember things so well when you are chronically stressed. However, before you worry too much about the state of your hippocampus, note that stress-induced changes in hippocampal structure are rapidly reversible. Indeed, they may actually protect the hippocampal neurons from excess stimulation. Still, when newborn rats are given daily injections of corticosterone for 30 days, they end up with fewer hippocampal neurons later in life. Thus, chronic stress can cause some irreversible damage if it occurs early in life.

Why is chronic stress so bad when acute stress is beneficial? No one knows for sure. Most likely our responses to acute stress divert some metabolic resources away from processes that are required for long-term health and reproduction. Such diversions are not harmful if they occur occasionally, but repeated diversions eventually create some serious health risks.

BRAIN EXERCISE

What are some sources of chronic stress in your own life? Is there some level of stress that you consider "good for you"?

SUMMARY

Section 9.1 - Homeostasis of the body's internal environment is often achieved through push-pull regulation, which involves sending opposite commands to antagonistic effectors until the desired, and often variable, set point is reached.

Section 9.2 - Most vegetative functions are controlled by the autonomic nervous system, which contains several divisions.

- Sympathetic preganglionic neurons release acetylcholine as their main neurotransmitter and project to sympathetic postganglionic neurons, which release norepinephrine from *en passant* synapses.
- Pre- and postganglionic neurons of the parasympathetic division both use acetylcholine as their neurotransmitter. Some of the parasympathetic preganglionics have their cell bodies in the brain.
- An important subset of visceral sensory neurons projects through the vagus nerve to nucleus tractus solitarius, which processes a wide variety of sensory inputs.
- The enteric nervous system comprises 200–600 million neurons that line our stomach and intestines.

Section 9.3 - Specialized neural circuits regulate heart rate, blood pressure, breathing, and body temperature.

- A drop in blood pressure reduces activity in a parasympathetic circuit that decreases heart rate (the end result being that heart rate increases). A drop in blood pressure also increases, through disinhibition, activity in sympathetic neurons that elevate heart rate.
- A drop in blood pressure activates sympathetic neurons that cause vasoconstriction, which raises blood pressure. Vasodilation in the face and penis involves parasympathetic axons and nitric oxide.
- Rhythmic breathing is controlled by a central pattern generator in the brainstem. It can be modulated by a circuit that includes chemosensitive cells in the carotid sinus and the retrotrapezoid nucleus.
- The burning of brown fat in response to skin cooling is mediated by a long circuit with two inhibitory neurons in series. Neurons in the preoptic area sense brain temperature and drive thermoregulatory behaviors.

Section 9.4 - Neurons control important aspects of digestion and urine production as well as food and fluid intake.

- Hypothalamic magnocellular neurons sense dehydration and release the peptide hormone vasopressin to reduce urine production. Two circumventricular organs also sense dehydration and activate median preoptic neurons that are involved in triggering thirst.
- The autonomic nervous system modulates enteric neurons that, in turn, control the movement of food along the intestines and the secretion of substances that promote digestion.
- Mechanosensory and hormonal signals from the gastrointestinal tract are integrated by neurons in the arcuate nucleus, which projects to appetite-controlling neurons in the lateral hypothalamus.

Section 9.5 - The neural circuits that control vegetative functions are often coordinated with one another.

- Individual neurons in the suprachiasmatic nucleus contain an intracellular circadian clock, are synchronized with one another, and are entrained by light. They control a variety of behavioral and physiological rhythms.
- The central amygdala coordinates many aspects of the fight-or-flight response, but other brain areas, notably the septum, also help coordinate responses to acute stressors.
- Persistent severe stress can damage hippocampal neurons, especially if the chronic stress occurs early in development.

Box 9.1 - Fainting in response to emotional stimuli involves a rapid drop in blood pressure. Electrical stimulation of the vagus nerve can reduce the frequency of epileptic seizures and may have a variety of other beneficial effects.

Box 9.2 - Bariatric surgery helps reduce appetite and body weight for several years, but the patients need vitamin and mineral supplements.

Box 9.3 - After an initial period of spinal shock, patients with spinal cord injury recover many vegetative reflexes. However, these reflexes tend to be poorly controlled and inefficient.

Box 9.4 - When taking melatonin to manipulate your internal circadian clock, timing is important. Night shift work is harder to deal with than jet lag.

KEY TERMS

ADDITIONAL READINGS

9.1 - Principles of Homeostatic Control

Cannon WB. 1932. *The wisdom of the body.* New York: Norton.

9.2 - The Autonomic Nervous System

Blessing WW. 1997. *The lower brainstem and bodily homeostasis.* New York: Oxford University Press.

Blessing WW. 1997. Inadequate frameworks for understanding bodily homeostasis. *Trends Neurosci* **20**:235–239.

Furness JB. 2006. *The enteric nervous system.* Malden, MA: Blackwell.

Jänig W. 2006. *The integrative action of the autonomic nervous system: neurobiology of homeostasis.* New York: Cambridge University Press.

9.3 - Controlling Vital Bodily Functions

Guyenet PG. 2006. The sympathetic control of blood pressure. *Nat Rev Neurosci* 7:335–346.

Lopez-Barneo J, Ortega-Saenz P, Pardal R, Pascual A, Piruat JI. 2008. Carotid body oxygen sensing. *Eur Respir J* **32**:1386–1398.

Nakamura K. 2011. Central circuitries for body temperature regulation and fever. *Am J Physiol Regul Integr Comp Physiol* **301**: R1207–28.

Ouellet V, Labbé SM, Blondin DP, Phoenix S, Guérin B, Haman F, Turcotte EE, Richard D, Carpentier AC. 2012. Brown adipose tissue oxidative metabolism contributes to energy expenditure during acute cold exposure in humans. *J Clin Invest* **122**:545–552.

Smith JC, Abdala APL, Rybak IA, Paton JFR. 2009. Structural and functional architecture of respiratory networks in the mammalian brainstem. *Philos Trans R Soc Lond B Biol Sci* **364**:2577–2587.

9.4 - Controlling Fluid and Energy Balance

Bourque CW. 2008. Central mechanisms of osmosensation and systemic osmoregulation. *Nat Rev Neurosci* **9**:519–531.

Dhillo WS. 2007. Appetite regulation: an overview. *Thyroid* **17**:433–445.

Ferguson AV. 2009. Angiotensinergic regulation of autonomic and neuroendocrine outputs: critical roles for the subfornical organ and paraventricular nucleus. *Neuroendocrinology* **89**:370–376.

Fry M, Hoyda TD, Ferguson AV. 2007. Making sense of it: roles of the sensory circumventricular organs in feeding and regulation of energy homeostasis. *Exp Biol Med* **232**:14–26.

Geerling JC, Loewy AD. 2008. Central regulation of sodium appetite. *Exp Physiol* **93**:177–209.

9.5 - High-level Autonomic Regulation

Albrecht U. 2012. Timing to perfection: the biology of central and peripheral circadian clocks. *Neuron* **74**:246–260.

Buijs RM, La Fleur SE, Wortel J, Van Heyningen C, Zuiddam L, Mettenleiter TC, Kalsbeek A, Nagai K, Niijima A. 2003. The suprachiasmatic nucleus balances sympathetic and parasympathetic output to peripheral organs through separate preautonomic neurons. *J Comp Neurol* **464**:36–48.

Bracha HS, Ralston TC, Matsukawa JM, Williams AE, Bracha AS. 2004. Does "fight or flight" need updating? *Psychosomatics* **45**:448–449.

Dhabhar FS. 2009. A hassle a day may keep the pathogens away: the fight-or-flight stress response and the augmentation of immune function. *Integr Comp Biol* **49**:215–236.

Ko CH, Takahashi JS. 2006. Molecular components of the mammalian circadian clock. *Hum Mol Genet* **15**:R271–R277.

Kollack-Walker S, Watson SJ, Akil H. 1997. Social stress in hamsters: defeat activates specific neurocircuits within the brain. *J Neurosci* **17**:8842–8855.

Kriegsfeld LJ, Silver R. 2006. The regulation of neuroendocrine function: timing is everything. *Horm Behav* **49**:557–574.

McEwen BS. 1999. Stress and hippocampal plasticity. *Annu Rev Neurosci* **22**:105–122.

Mendoza J, Challet E. 2009. Brain clocks: from the suprachiasmatic nuclei to a cerebral network. *Neuroscientist* **15**:477–488.

Saha S. 2005. Role of the central nucleus of the amygdala in the control of blood pressure: descending pathways to medullary cardiovascular nuclei. *Clin Exp Pharmacol Physiol* **32**:450–456.

Vyas A, Mitra R, Shankaranarayana Rao BS, Chattarji S. 2002. Chronic stress induces contrasting patterns of dendritic remodeling in hippocampal and amygdaloid neurons. *J Neurosci* **22**:6810–6818.

Boxes

Arendt J. 2009. Managing jet lag: some of the problems and possible new solutions. *Sleep Med Rev* **13**:249–256.

Beekwilder JP, Beems T. 2010. Overview of the clinical applications of vagus nerve stimulation. *J Clin Neurophysiol* **27**:130–138.

Craggs MD, Balasubramaniam AV, Chung EAL, Emmanuel AV. 2006. Aberrant reflexes and function of the pelvic organs following spinal cord injury in man. *Auton Neurosci* **126–127**:355–370.

Dalcanale L, Oliveira CPMS, Faintuch J, Nogueira MA, Rondó P, Lima VMR, Mendonça S, Pajecki D, Mancini M, Carrilho FJ. 2010. Long-term nutritional outcome after gastric bypass. *Obes Surg* **20**:181–187.

Hainsworth R. 2003. Syncope: what is the trigger? *Heart* **89**:123–124.

Krassioukov A. 2009. Autonomic function following cervical spinal cord injury. *Respir Physiol Neurobiol* **169**:157–164.

Leff DR, Heath D. 2009. Surgery for obesity in adulthood. *BMJ* **339**:b3402–b3402.

Controlling Posture and Locomotion

CHAPTER 10

CONTENTS

FEATURES

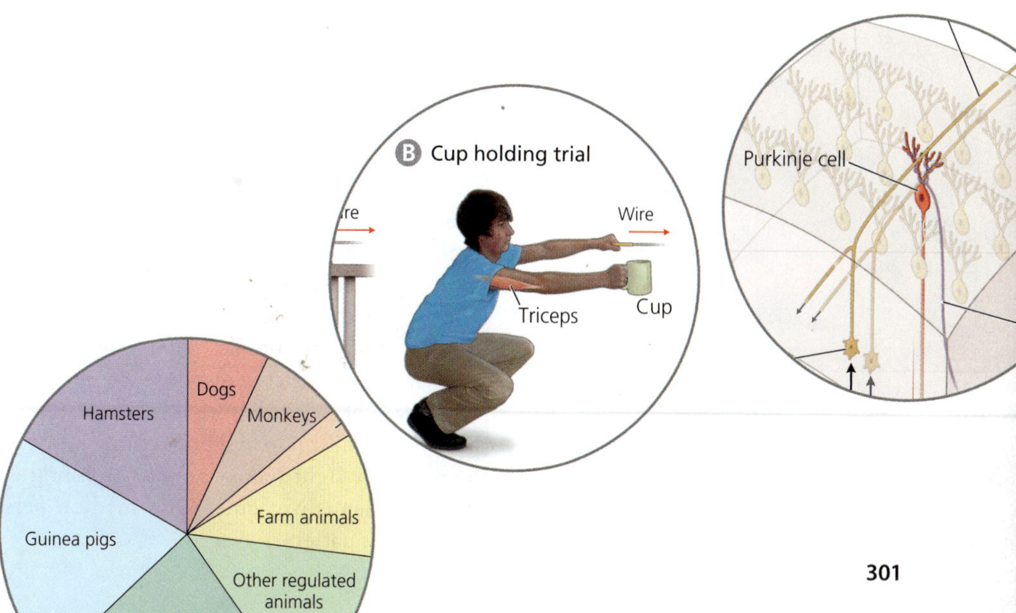

In comparison to the vegetative processes we discussed in Chapter 9, the movements of our limbs and head seem far more voluntary. Indeed, many of them are. However, the movements we make in response to sudden pain or other threats are usually not subject to volitional control; we make them automatically, reflexively. Similarly, most of our movements are accompanied by subtle postural adjustments that are involuntary and, frequently, escape our awareness entirely. Another interesting similarity between postural adjustments and vegetative reflexes is that both are homeostatic in the sense that they stabilize some aspect of our bodies. Whereas the vegetative reflexes stabilize our internal environment, postural reflexes stabilize our bodies in the external environment. A third important similarity between vegetative processes and bodily movements is that both can be driven by central pattern generators, which are subject to sensory modulation but do not require sensory inputs to generate rhythmic activity.

One significant difference between vegetative and skeletal motor control is that the latter involves much more learning. In Chapter 9, we noted how blood pressure and heart rate tend to change with age and how the set point for body temperature rises when you have an infection; but these forms of plasticity are not what most people mean by "learning." In contrast, most motor skills are clearly learned. Furthermore, the neural commands required for a given movement must constantly be recalibrated as the body grows or as muscles fatigue. Two regions heavily involved in motor learning and calibration are the motor cortex and the cerebellum, respectively. We will discuss these regions toward the end of this chapter after we have dealt with the motor functions of the spinal cord and medulla.

10.1 What Is a Reflex?

Although we have already discussed some reflexes in Chapter 9, it is useful now to define the term "reflex" explicitly: A **reflex** is an involuntary, stereotyped response to a specific stimulus. Reflexes usually involve multiple neurons that form a reflex arc (circuit). Many reflex arcs contain a mix of excitatory and inhibitory neurons chained together; and many reflexes have multiple branches, each leading to a different effector. For example, as you already know, the baroreflex has two distinct branches, one that regulates heart rate and one that controls vasoconstriction (see Chapter 9).

Although most reflexes are not learned, in the sense that they emerge without prior experience of the stimulus, it is possible to acquire new reflexes through learning. You can learn, for example, to exhibit a reflexive aversion to food that made you sick (see Chapter 14). In addition, many reflexes can be modified through learning. For example, baby seagulls reflexively peck at the beaks of their parents to obtain food. When the chicks are very young, they readily peck at very simple models of the parent's beak; however, as they grow up, the chicks require ever more precise models to elicit the pecking reflex. Finally, as you will learn toward the end of this chapter, the strength of many reflexes can be adjusted through trial-and-error learning to compensate for changes in the body or the external environment. Thus, we can conclude that reflexes are stereotyped but not always immutable.

Pupillary Reflexes

One reflex that fits our definition well is the constriction of the pupil that occurs when you look into a source of light. This **pupillary light reflex** is involuntary, highly stereotyped, and triggered by a specific stimulus, namely, bright light. When emergency medical technicians shine a flashlight into an unconscious person's eyes, one after the other, the pupillary light reflex is what they want to see. If one or both of the pupils fail to constrict in response to the light, then part of the circuity mediating the pupillary light reflex (Figure 10.1 A) must be impaired. Because this circuitry passes through the midbrain and pretectum (just anterior to the midbrain), impairment of the pupillary light reflex suggests that brain damage is probably widespread and serious, which is why emergency medical personnel routinely test for this reflex in unconscious patients.

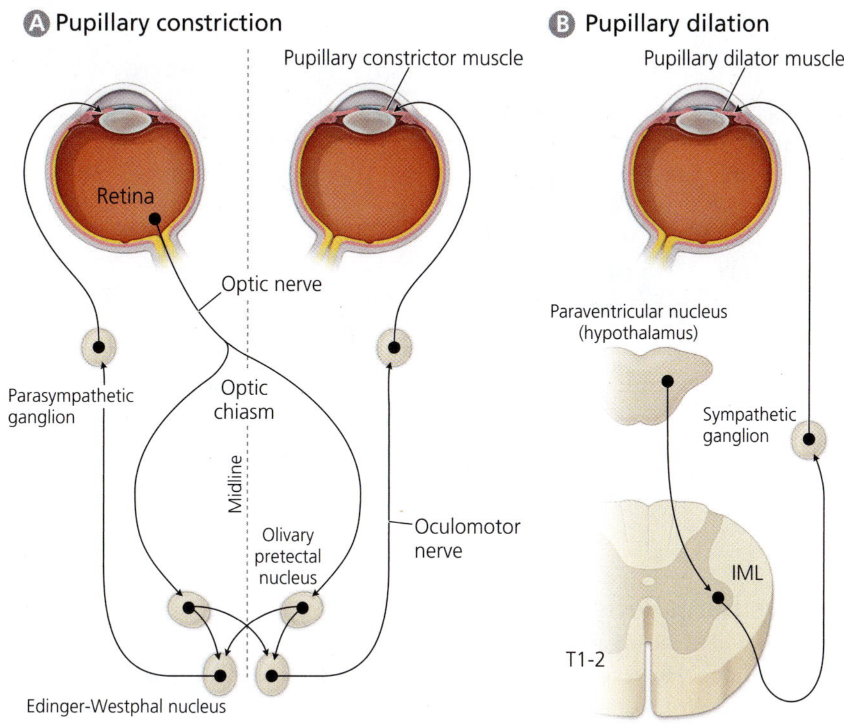

Figure 10.1 Pupillary constriction and dilation. Shown in (A) is the neural circuit for the pupillary light reflex, which constricts the pupil in response to bright light. For clarity, the retinal projections are shown only on one side. Shown in (B) is the circuit for pupillary dilation. Because this circuit is tonically active, the pupillary constrictor circuit always works against a background of commands to dilate the pupil. IML = intermediolateral column.

The counterpart to pupillary constriction is **pupillary dilation**. Because dimming the lights reliably dilates the pupil, this behavior can be considered reflexive. However, the pupils also dilate when you become aroused or perform a challenging task (or take hallucinogenic drugs). Moreover, pupillary dilation is not driven by neurons that detect darkness. Instead, pupillary dilation is driven by part of the sympathetic nervous system (Figure 10.1 B), which increases its activity during a variety of stressful conditions (see Chapter 9).

Because the sympathetic pathway controlling pupillary dilation is tonically active, the pupillary light reflex always occurs against a background of signals that promote pupillary dilation. In a sense, pupillary dilation occurs when the pupillary light reflex diminishes. More accurately, pupil diameter is determined by the relative balance of constrictive and dilatory signals. When bright light enters the eye and sympathetic activation is low, the constrictive signals predominate and pupil size shrinks. When the light dims or sympathetic activation increases, the balance shifts to pupillary dilation.

Once you grasp this mechanism of control, you can appreciate that reflexes occur not independently of other behaviors but in conjunction with them. You can also understand why damage to the pupillary light reflex arc causes not just a lack of pupillary constriction in response to light but abnormally wide pupil dilation.

BRAIN EXERCISE

Is the pupillary light reflex immutable? Can it be influenced by learning?

10.2 How Do Reflexes Protect Us from Harm?

Among the simplest and best studied movements are defensive reflexes, which allow animals to escape threats or otherwise prevent damage. In a way, the pupillary light reflex is defensive because reducing the amount of light striking the retina reduces the risk of light-induced photoreceptor damage (as occurs if you look directly into the sun) and maximizes an organism's ability to see potential threats and other stimuli. However, organisms have many other reflexes that are more obviously defensive.

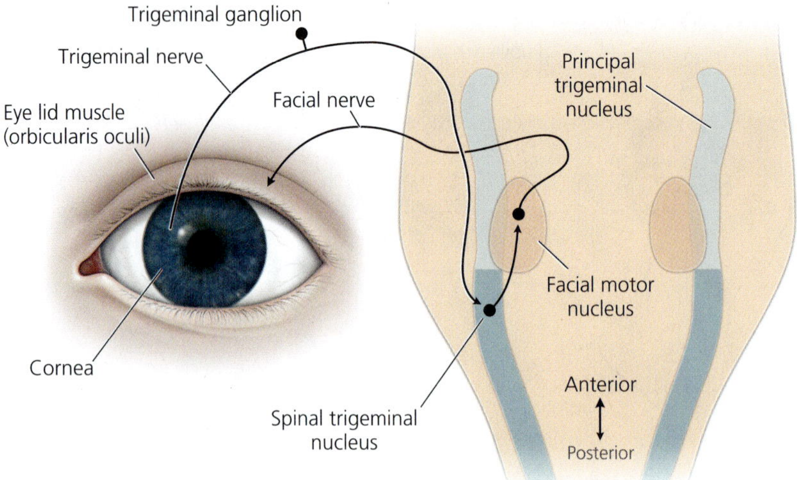

Figure 10.2 The eye blink reflex. The cornea is innervated by pain- and pressure-sensitive neurons that project to the spinal trigeminal nucleus, which is a long column of cells in the medulla (here shown from a dorsal perspective). Neurons in the spinal trigeminal nucleus project to neurons in the facial motor nucleus that innervates the eyelid closer muscles (mainly the orbicularis oculi).

The Eye Blink Reflex

The eye is a delicate organ that is not, in contrast to our inner ear, enclosed within protective bone. However, the eye does have some defenses. Tears, for example, prevent the cornea from drying out. They consist mostly of water but also contain some lipid molecules, which form a thin protective film that slows the rate at which the tears evaporate. Eye blinks help to spread the tears across the surface of the eye and sweep away some irritants. In addition, eye blinks can prevent irritants from entering the eye. Such proactive eye blinks can be triggered by loud sounds, rapidly approaching visual stimuli, or puffs of air hitting the cornea.

Neurobiologists have focused their research primarily on eye blinks triggered by corneal stimulation. They have learned that touch- and pain-sensitive axons project from the cornea through the trigeminal nerve to the *spinal trigeminal nucleus* (Figure 10.2), which generally receives information about noxious stimuli affecting the head (see Chapter 7). The trigeminal neurons send their axons to the **facial motor nucleus**, which in turn projects to the *orbicularis oculi*, a muscle that closes the eyelids when it contracts. Thus, the shortest route by which stimulation of the cornea can elicit an eye blink involves just three neurons, connected by two synapses. Not surprisingly, this disynaptic reflex arc produces a rapid response. Although minimal eye blink latencies vary across species, they can be as short as 10 ms. This is almost 20 times faster than the reaction time for pressing a button in response to a light and fast enough to keep most approaching objects, such as a fly, from entering the eye.

Withdrawal Reflexes

When you touch a hot stove with your fingers, you automatically and rapidly withdraw your hand. Similarly, when you step on a thumbtack, you reflexively withdraw your foot. Indeed, strong stimulation of any part of your skin tends to elicit a rapid withdrawal of the affected body part. Even when you are sleepy, key withdrawal reflexes remain intact, which is why they are commonly used to assess an animal's level of anesthesia. For example, if a rat squirms when you pinch its toes or tail, then the rat is not anesthetized sufficiently for surgery.

The **leg withdrawal reflex**, first studied by Charles Sherrington around 1900 (see Chapter 2), is generated by a short and simple circuit through the spinal cord. Touch-sensitive axons that innervate the foot project to the deep layers of the spinal cord's dorsal horn (Figure 10.3). There the sensory axons synapse on excitatory interneurons that in turn project to ventral horn motor neurons. These motor neurons innervate the hamstring muscles, which flex your leg and pull the foot away from the

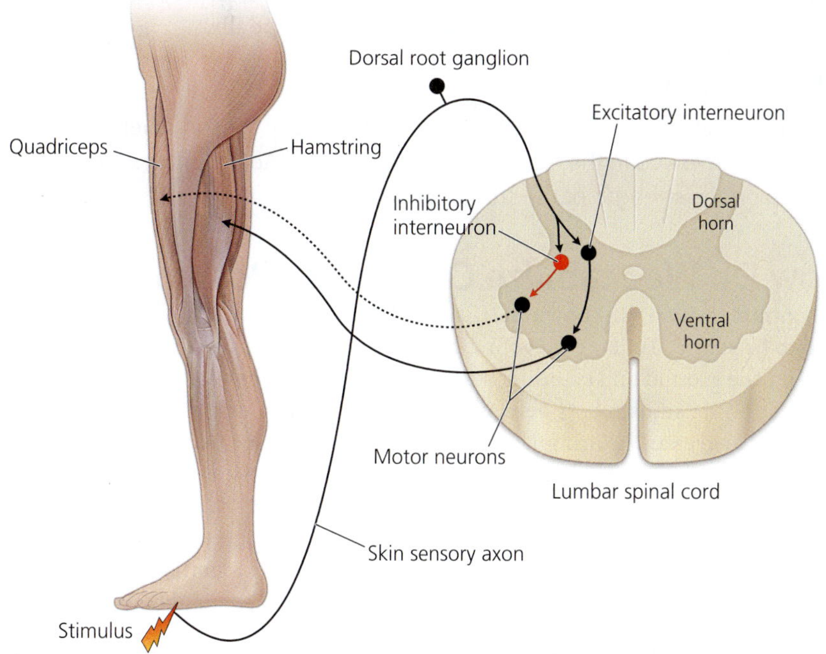

Figure 10.3 The leg withdrawal reflex. When you step on a thumbtack, you reflexively bend your knee and withdraw your foot. This withdrawal reflex is triggered by mechanosensitive neurons that innervate the foot and project into the spinal cord. There they synapse on excitatory interneurons that project to hamstring motor neurons. The sensory neurons also synapse on inhibitory interneurons (red) that project to motor neurons innervating the quadriceps. Thus, strong stimulation of the foot triggers contraction of the leg flexors and inhibits contraction of the leg extensors.

pain's source. Thus, the leg withdrawal reflex is mediated by a minimum of 3 neurons, all of which are excitatory. However, this is not the entire story.

Stimulation of the foot also triggers a slackening of the quadriceps muscles (Figure 10.3). This quadriceps relaxation during the leg withdrawal reflex is adaptive because the hamstring and the quadriceps act antagonistically. The former flexes the knee joint, the latter extends it. Therefore, a fast and efficient withdrawal of the foot requires hamstring contraction to be accompanied by quadriceps relaxation. This **reciprocal innervation** of leg flexors and extensors is accomplished by branches of the incoming sensory axons that project to inhibitory spinal interneurons, which in turn project to quadriceps motor neurons (Figure 10.3).

Thus, a full leg withdrawal reflex involves at least 5 different neurons, including one inhibitory interneuron. The inhibitory interneuron is required because neurons tend not to make excitatory synapses with one of their axon branches and inhibitory ones with another branch. Although it is possible for neurons to use more than one transmitter at a given synapse, vertebrate neurons generally use the same combination of transmitters at all their synapses (recall our discussion of Dale's principle in Chapter 2). Therefore, the easiest way for sensory axons to generate excitation in one motor neuron and inhibition in another is to enlist the help of an inhibitory interneuron.

Another interesting aspect of the withdrawal reflex is that the number of muscles involved in a withdrawal response increases with stimulus strength. For example, a minor prick on the underside of a toe may simply cause the toe to be lifted, whereas a stronger stimulus may cause the whole leg to be withdrawn. This phenomenon is called **irradiation** because the sensory information coming into the spinal cord is thought to involve progressively more spinal neurons as stimulus strength increases. Although the mechanisms underlying irradiation have not been fully described, they almost certainly include axon branches that travel across multiple segments of the spinal cord as well as differences in the activation thresholds of the neurons contacted by those axon collaterals.

In any case, most foot withdrawal responses clearly involve more than the 5-neuron reflex arc that we discussed. This is a general principle: under natural conditions, most reflexes involve neuronal pathways that are much more complex than the minimal circuits that neurobiologists tend to study. Of course, progress in research is usually made by focusing on simple problems, at least initially.

BRAIN EXERCISE

Recalling what you learned in Chapter 7, which type of sensory fibers are most likely to be involved in the leg withdrawal reflex? Aβ, Aδ, or C fibers? Does it depend on the strength of the stimulus?

10.3 How Do We Stabilize Our Body's Position?

In contrast to the well-known defensive reflexes, an organism's stabilizing reflexes tend to be subtle. To increase your awareness of these reflexes, try raising one knee high off the ground and standing still without moving your arms. Can you feel the numerous little adjustments being made by muscles in your standing leg? If not, try closing your eyes and waiting a minute or two, until your leg muscles fatigue. It is not easy to maintain your balance under these conditions, is it? An almost constant stream of postural adjustments is needed to stand upright, walk, or do virtually anything other than lying down. The mechanisms underlying these adjustments are dauntingly complex, but we can simplify the problem by identifying some specific reflexes that stabilize our body parts relative to one another and the outside world.

Muscle Stretch Reflexes

Among the most important and best studied stabilizing reflexes are those that contract a muscle after it has been stretched. The most famous of these stretch reflexes is the **knee jerk response**, which doctors evoke by swinging a little "reflex hammer" at your knee (just below the kneecap) while your leg is bent and off the floor (Figure 10.4). If this is done properly, your foot tends to jerk forward reflexively. Why is the knee jerk response a stretch reflex? Is there a muscle that is stretched when the hammer hits your knee? Actually, there is! The hammer's impact stretches a muscle tendon that attaches your quadriceps to the tibia. By stretching this tendon you also stretch the quadriceps, which then triggers a stretch reflex that leads to quadriceps contraction and, consequently, leg extension.

At first glance, the knee jerk response or, more appropriately, the **quadriceps stretch reflex**, does not appear to be a postural reflex at all. After all, how could a knee jerk in response to a hammer tap possibly help your body maintain a stable position? The opposite would seem to be the case. However, it is decidedly unnatural for muscles to be stretched as abruptly as they are when a tendon is struck with a hammer. A much more natural context for the quadriceps stretch reflex is the act of catching a heavy object, for example. Because catching an object increases the weight that your

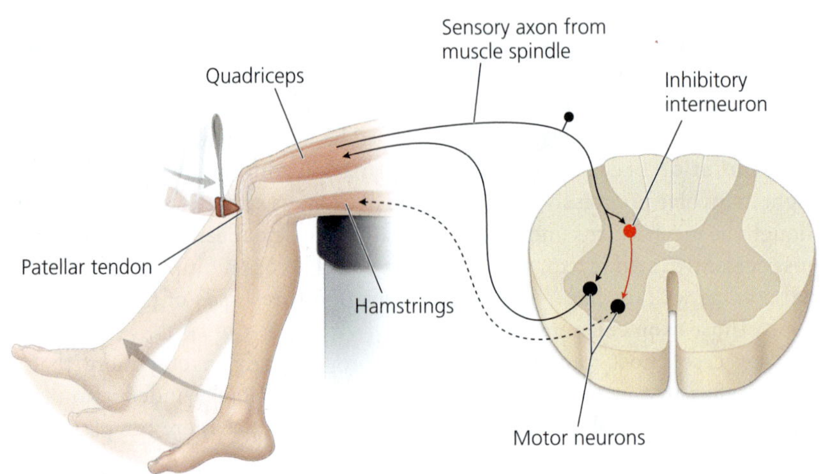

Figure 10.4 The knee jerk response. When a doctor hits you with a small hammer just below the knee cap while your leg is bent and off the ground, your leg straightens reflexively. This happens because the tap on the knee stretches a tendon that in turn stretches the quadriceps muscle. Stretching the quadriceps activates muscle spindle receptors that project directly to spinal cord motor neurons, which project back to the quadriceps. The hammer tap also activates inhibitory interneurons that project to motor neurons innervating the hamstring muscles. In general, stretching any skeletal muscle causes that same muscle to contract and its antagonistic muscle to relax.

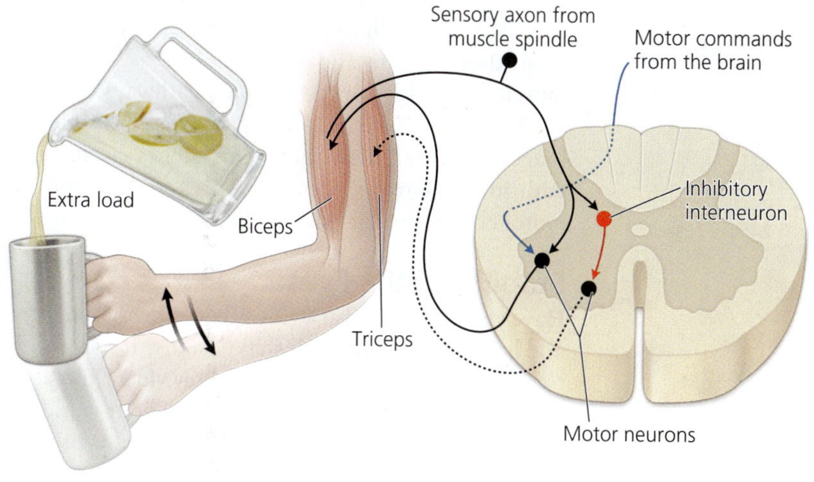

Figure 10.5 The biceps stretch reflex. When someone or something pushes down on your hand while your elbow is flexed, the biceps is stretched. This activates the biceps stretch reflex, which causes the biceps to contract and the triceps to relax. These actions tend to bring the hand back to its original position.

legs must support, your knees will buckle slightly. This stretches the quadriceps, triggers its stretch reflex and, consequently, increases activity in the quadriceps and other leg extensor muscles. The heavier the object, the stronger the muscle stretch and the stronger the leg extension. In other words, the quadriceps stretch reflex increases the force exerted by the quadriceps to compensate for the added load. Thus, the quadriceps stretch reflex helps to maintain the leg in a relatively constant position, regardless of variations in load.

The quadriceps stretch reflex is just one of many stretch reflexes; almost all skeletal muscles contract in response to being stretched. Consider what happens when someone unexpectedly fills a cup you are holding. Your hand and forearm briefly dip toward the floor but then return to their initial position. This reflexive return to the status quo occurs because the downward movement stretches the biceps, which triggers the **biceps stretch reflex** and thus increases biceps contraction (Figure 10.5). If you perform similar experiments with other body parts (it's best to do this with a friend), you'll see that most body parts resist the unexpected application of external forces.

Relaxing the Antagonists

The stretch of a muscle not only triggers the contraction of that same muscle but also relaxes the muscle's antagonists. For the biceps stretch reflex, this means that stretching the biceps causes the triceps to relax. In the case of the quadriceps stretch reflex, the hamstring becomes relaxed.

In essence, each stretch reflex forms a stabilizing, homeostatic feedback loop that is analogous to the homeostatic loops we discussed in Chapter 9. Opposite signals are sent to antagonistic effectors to stabilize the system around a desired set point. Collectively, these stabilizing stretch reflexes make the trunk and limbs act like stiff springs that resist movement away from their equilibrium position. They keep joint angles constant and battle gravity. Without them, your body would resemble a marionette whose puppet strings were cut.

Neural Mechanisms

The neural mechanisms underlying the muscle stretch reflex have been studied extensively. Muscle stretch is sensed by the neurons that innervate muscle spindles, which we discussed in Chapter 7. The axons of these sensory neurons project into the spinal cord's ventral horn, where they synapse directly on large (alpha) motor neurons that project right back to the muscle in which the muscle spindles sit. Thus, the muscle stretch reflex includes a simple monosynaptic reflex arc (Figures 10.4 and 10.5). Relaxing the antagonists requires an inhibitory interneuron that receives direct input from muscle spindles and projects to motor neurons innervating the antagonist muscles. All in all, a muscle stretch reflex involves a minimum of four neurons: one sensory neuron, one inhibitory interneuron, and two motor neurons.

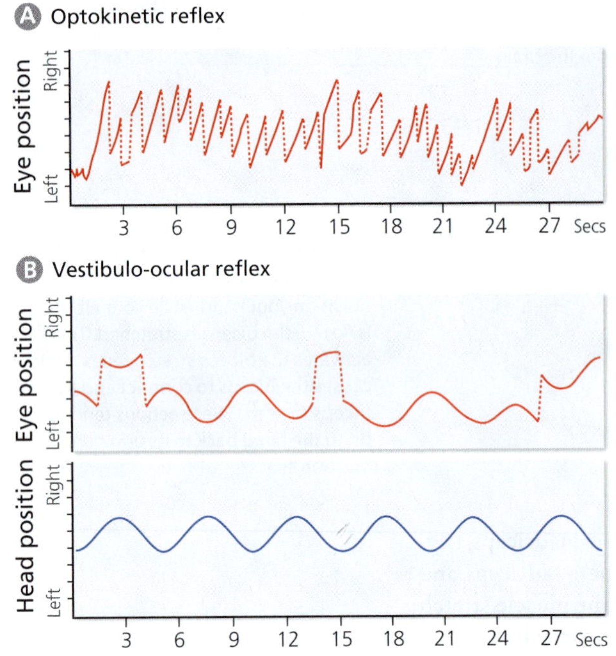

Figure 10.6 Eye movement stabilization. The graph in (A) shows a monkey's eye position while a large visual stimulus is moved at a steady velocity from left to right through the visual field. The eyes follow the stimulus (solid lines) until the eyes have moved far to the right, at which point they rapidly move back toward the left (dashed lines). The graphs in (B) depict eye and head position while a monkey's head is rotated sinusoidally back and forth (in the dark). Except for a few very rapid eye movements (dashed red lines), the eyes compensate for the head movement by rotating in the opposite direction. [After Ilg, 1997]

The stretch reflexes we have discussed so far are extremely rapid, with muscle contractions typically beginning 30–40 ms after the stretch (as little as 10 ms for some muscles). In addition to these fast pathways through the spinal cord, there are secondary stretch reflex circuits with longer response times. These long-latency circuits involve additional neurons, some of which are located in the motor cortex, but they do not concern us here.

Stabilizing the Eyes

Have you ever looked at a video that was shot at high zoom and held by someone with shaky hands? It's very difficult to make out anything in such a wobbly video. The situation is similar when the projected image of your world moves rapidly across your retina. To minimize this problem of **retinal image slip**, animals have evolved several reflexes that help them stabilize the eyes relative to the world.

The Optokinetic Reflex

When you look out the window of a moving train, you see the world passing by. As this happens, you are unconsciously moving your eyes from left to right (or right to left) and back again. A closer look reveals that the eyes tend to focus on a spot in the outside world and track it as it moves. When the eyes have moved as far as they can, they rapidly move in the opposite direction and focus on another spot. Then the cycle repeats. If you plot the changes in eye position during this reflexive behavior, you get a distinctive saw-tooth pattern (Figure 10.6 A). This behavior is called the **optokinetic reflex**. It is most evident for horizontal eye movements, but you also move your eyes up and down repeatedly when the world moves vertically past your head (as in a glass elevator). The principal function of the optokinetic reflex is to minimize retinal image slip. Without it, the outside world would be a blur.

The neural circuit underlying the optokinetic reflex has been described in some detail for horizontal eye movements (Figure 10.7). The shortest circuit begins in the retina with the standard pathway from photoreceptors via bipolar cells to the retinal ganglion cells (see Chapter 6). A subset of the retinal ganglion cells then sends its axons to a small diencephalic nucleus called the **nucleus of the optic tract**. The neurons in this nucleus are exquisitely sensitive to large moving stimuli, including movements of the entire visual image. Importantly, they respond most strongly to visual stimuli that move toward the contralateral side (toward the nose). The optic tract neurons then project to one or more small nuclei in the medulla that, in turn, project to the **vestibular complex**.

The vestibular complex contains many different types of neurons, but some of them have excitatory projections to the contralateral **abducens nucleus**. The abducens nucleus, in turn, is important because it innervates some of the muscles that rotate the eyeball. Specifically abducens neurons innervate the **lateral rectus** muscle, which is attached to the eyeball's lateral edge. Contraction of this muscle rotates the eye away from the nose and toward the temporal bone. So, when the visual world passes from left to right, the visual motion activates the optokinetic reflex arc, which ultimately causes the right eyeball to rotate away from the nose. If the speed of this rotation matches the speed of the visual motion, then the image on the right retina is stabilized.

So far so good, but there is an obvious problem, namely, that the left eye must receive different commands. To stabilize a moving visual stimulus on both retinas, you have to rotate one eye toward the temporal bone while the other eye turns nasally. In our example, you have to rotate the left eye toward the nose and the right eye away from it. To solve this problem, a subset of neurons in the vestibular complex project to the **oculomotor nucleus**, rather than the abducens (Figure 10.7). Neurons in the oculomotor nucleus then innervate the **medial rectus**, which moves the eye toward the nose when it is contracted. Importantly, the projection to the oculomotor nucleus is ipsilateral (uncrossed), whereas the projection to the abducens nucleus is contralateral. Because of this arrangement, a horizontally moving stimulus causes the two eyes to rotate in opposite directions relative to the nose, but in the same direction relative to the outside world.

What about reciprocal innervation in the optokinetic reflex? As we discussed, contraction of one muscle is usually accompanied by relaxation of that muscle's antagonist. In the case of the optokinetic reflex, this means that contraction of the lateral rectus should be accompanied by relaxation of the medial rectus attached to the same eye (and vice versa). This task is accomplished by inhibitory projections from the vestibular complex to the ipsilateral abducens nucleus and by excitatory projections from the abducens to the contralateral oculomotor nucleus. This pathway through the ipsilateral abducens nucleus is not shown in Figure 10.7 for the sake of simplicity, but it is illustrated in Figure 10.8. The pathway ensures that in the optokinetic reflex, as in the stretch reflex, antagonistic muscles receive opposing commands.

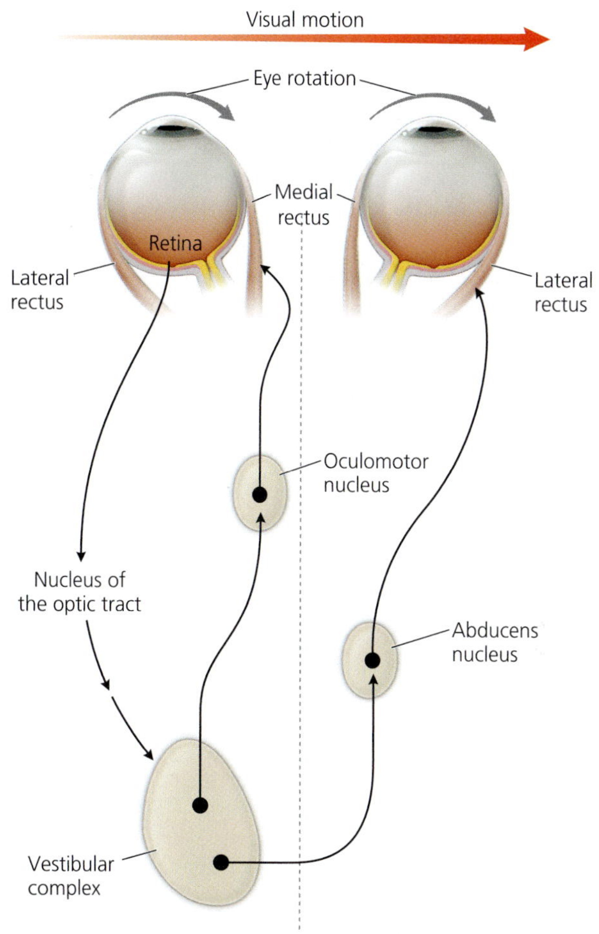

Figure 10.7 The optokinetic reflex arc. Shown here are the neural pathways through which large-scale visual motion toward the right causes a rightward movement of both eyes, thus stabilizing the retinal image. The dashed line represents the midline. Only one side of the reflex circuit is shown. Also omitted are inhibitory projections from the vestibular complex to the ipsilateral abducens nucleus, which relax the lateral rectus when the medial rectus contracts.

The Vestibulo-ocular Reflex

Closely related to the optokinetic reflex is the **vestibulo-ocular reflex (VOR)**. To see this reflex in action, look in the mirror and turn your head from side to side. You'll notice that your eyes don't turn with the head but keep staring straight ahead. Even when you tilt the head up and down, your eyes remain eerily stable. You might think your eyes counterrotate within the head because you are intentionally staring at your eyes to see if they're moving. Fair enough! However, the same counterturns occur in complete darkness (Figure 10.6 B). This finding shows that, in contrast to the optokinetic reflex, the VOR is not driven by visual signals. Instead, it is controlled by signals from the vestibular apparatus.

As you may recall from Chapter 7, the semicircular canals in the vestibular apparatus contain sensory hair cells that fire action potentials in response to head rotations. There are three pairs of semicircular canals, each specialized to sense head rotations in a different plane. All these canals can drive the VOR, but most vestibulo-ocular research focuses on the horizontal canals because they can easily be stimulated by placing experimental subjects in a swivel chair or on a turntable.

The VOR uses the same basic pathways from the vestibular complex to the eye muscles that you just read about. However, the VOR is controlled by inputs that come directly from the horizontal semicircular canals (Figure 10.8). A crucial aspect of this vestibular input is that head turns to the left activate hair cells mainly in the left horizontal canal, whereas head turns to the right activate primarily the right horizontal canal. This asymmetry is important because as we discussed earlier, neural activity in the vestibular complex has opposite effects on each pair of antagonistic eye muscles. The fact that a head turn has opposite effects on the left and right

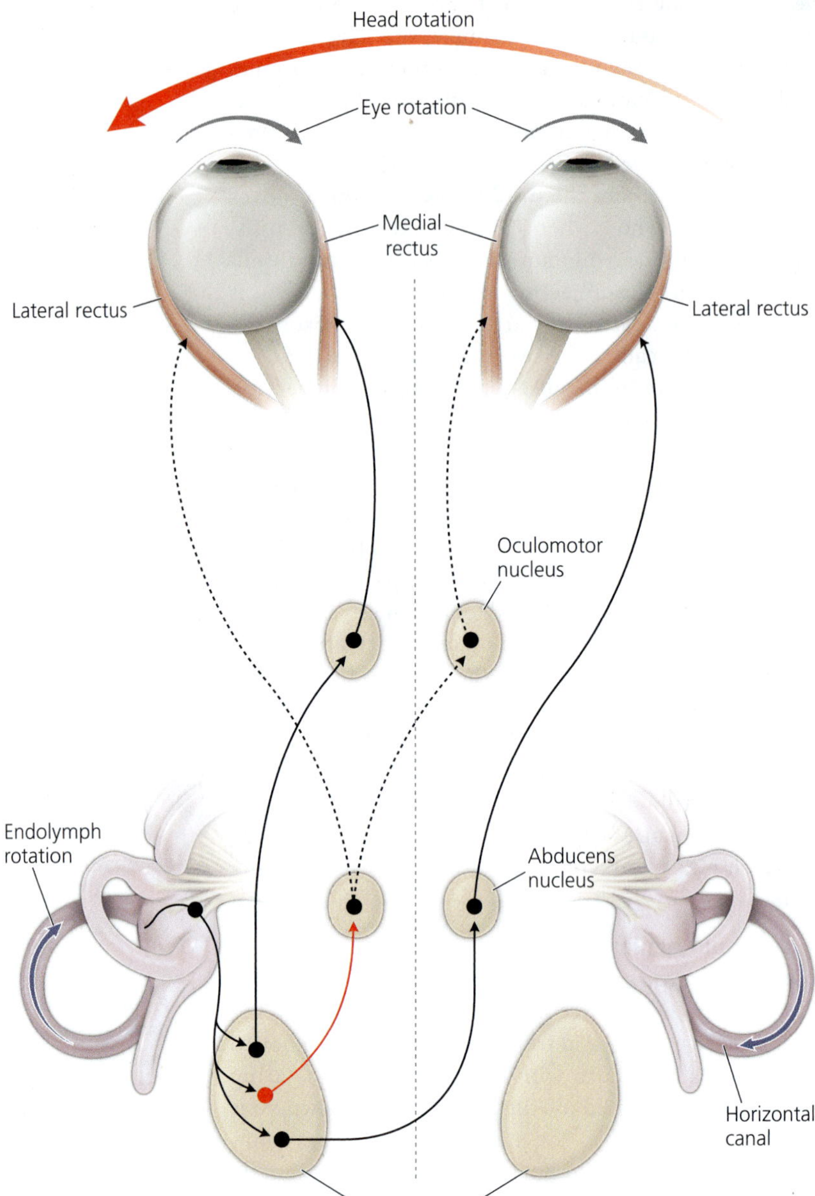

Figure 10.8 The vestibulo-ocular reflex arc. If you turn your head to the left, the endolymph in the horizontal vestibular canals counterrotates due to inertia. This fluid movement stimulates hair cells in the left horizontal canal, which activates vestibular neurons that project to the oculomotor and abducens nuclei, which then trigger contraction of the left medial rectus and the right lateral rectus. The vestibular signal also excites inhibitory neurons (red) that suppress activity in the ipsilateral abducens nucleus, causing relaxation of the left lateral rectus and the right medial rectus (the suppressed pathways are shown as dashed lines). As a result, both eyes rotate toward the right, compensating for the leftward head turn.

semicircular canals and thus on the left and right vestibular complexes ensures that the two eyes turn in the same direction, which is opposite to the head turn. This sounds quite confusing (too many opposites!), but the fog should clear when you examine the circuit diagram in Figure 10.8 and consider what would happen if both horizontal canals were activated simultaneously. The eyes would be pulled in both directions simultaneously, which is clearly not good.

In contrast to its optokinetic cousin, the VOR is not a feedback circuit. Both reflexes minimize retinal slip, but only the optokinetic reflex is driven by a retinal error signal (retinal image slip). To be specific, the optokinetic reflex uses information about how fast the retinal image slips to boost the output of its eye movement commands until the retinal image is stabilized. In contrast, the VOR bases its eye movement commands on the signals from the semicircular canals. This design minimizes the number of synapses involved and allows the VOR to be extremely rapid. However, there is a potential problem. What if the VOR generates eye movement commands that are too strong, or too weak, to eliminate the retinal image slip? Because the

VOR circuit receives no immediate feedback about the retinal image slip, the reflex would seem unable to correct for such overshoots or undershoots. As you will learn toward the end of this chapter, the problem is solved by adjusting the strength of the VOR through trial-and-error learning.

Stabilizing the Head

The **vestibulocollic reflex**, whose name derives from the Latin word for neck ("collum"), tends to stabilize the head as the body turns. To catch a glimpse of this reflex, ask a friend to sit in a swivel chair and then twist the chair back and forth. Alternatively, skip the swivel chair and simply grab the friend's shoulders, twisting them back and forth. You are likely to see that the friend's head stays relatively still even as the rest of the body turns. This happens because your friend's neck muscles are turning the head at the same speed as the shoulders are turning, but in the opposite direction. Because the head also counterturns in complete darkness or when the eyes are closed, the vestibulocollic reflex must resemble the VOR insofar as it is independent of retinal image slip. Indeed, extensive studies have shown that the vestibulocollic reflex is driven, as its name implies, by vestibular signals. However, the vestibular signals for this reflex are routed to the motor neurons responsible for neck movements rather than eye movements.

Another head-stabilizing behavior is the **optocollic reflex**. Have you ever noticed how a pigeon seems to bob its head rhythmically forward and backward while it is walking? This head bobbing, which is even more obvious in walking cranes or other long-necked birds, is a very precise head-stabilizing reflex (Figure 10.9). High-speed photography has shown that walking birds tend to keep their head as stable as possible, relative to the world, for as long as possible. They accomplish this by moving the head backward as the body moves forward. Birds can do this better than we can because they have much longer necks. Eventually, when the head has moved as far back as the neck will allow, the head is rapidly thrust forward, at which point the head's backward movement (relative to the body) begins again. Importantly, birds whose bodies are moved passively (by a human, for example) also keep their heads remarkably stable in space, whereas birds walking in darkness do not head bob. These observations indicate that the optocollic reflex is driven by visual rather than vestibular signals. Just like the optokinetic reflex, the optocollic reflex minimizes retinal image slip.

Of course, a moving animal rarely keeps its head perfectly stable for more than a few seconds. This is okay because the optokinetic and vestibulo-ocular reflexes kick in whenever the head does move relative to the world. All of these reflexes work together to achieve a single goal, namely, to stabilize the retinal image. The evolutionary benefit of this design is apparent. Stabilizing the retinal image allows a moving animal to localize and identify external objects more clearly, especially if those

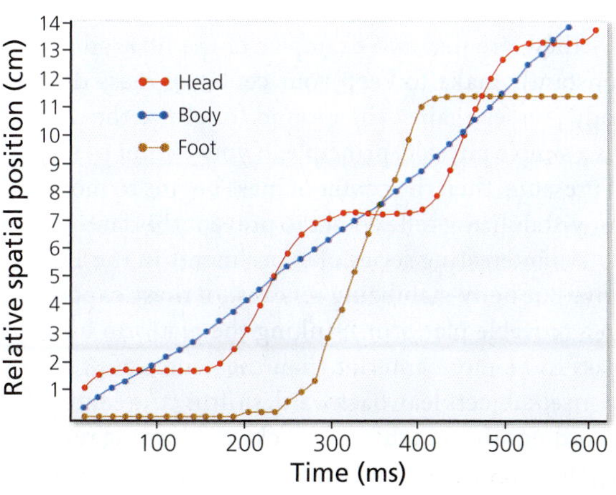

Figure 10.9 The optocollic reflex in birds. When birds walk forward, they hold their head stable relative to the outside world as long as possible and then rapidly thrust it forward, only to keep it stable again (repeating the cycle). This is illustrated in the graph, which plots head, body, and foot position as a pigeon walks forward. Shown at the left are two frames from the kind of video on which the graph is based. The dashed line, arrowhead, and asterisk indicate the positions of the beak, ground foot, and swinging foot, respectively. Next time you see a pigeon, look for this behavior! [Left from Necker, 2007; right after Frost, 1978]

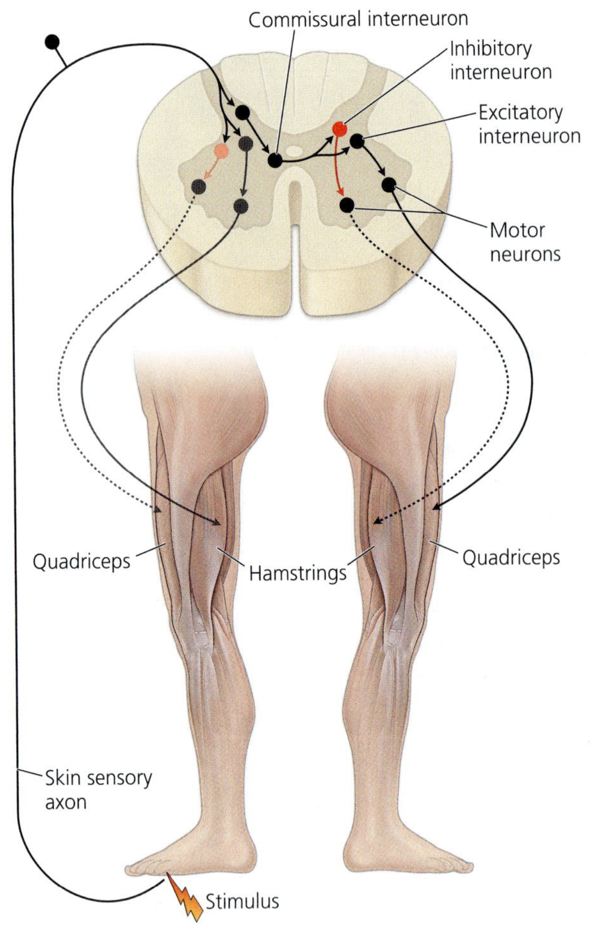

Figure 10.10 The crossed extensor reflex. When you step on a thumbtack and reflexively withdraw your leg (see Figure 10.3), the weight on the other leg is suddenly increased. To counteract the increased load, you reflexively stiffen this other leg by contracting its extensor muscles and inhibiting its flexor muscles. Shown here is the neural circuit underlying this crossed extensor reflex. Inhibitory neurons are red; the neurons of the leg withdrawal reflex are semitransparent.

objects are themselves moving in the world. For example, consider a walking cat in search of a meal. If the image on the cat's retina is stabilized, then a small moving object, such as a mouse, will stand out clearly against an immobile background. On the other hand, if the retinal image is shifting constantly across the retina, then the cat will find it difficult to see the moving mouse. As a rule, making sense of the visual world is significantly easier when that world is stable.

Stabilizing the Body

Reflexes that stabilize the trunk and limbs in the world contribute to stabilizing the retinal image, but their principal function is to keep the body from falling to the ground. Consider, for example, what happens during the leg withdrawal reflex that we discussed previously. If you step on a thumbtack and lift the stimulated foot, you may have gotten the foot out of harm's way, but now you have a different problem. Because the stimulated foot is no longer touching the ground, the other leg must support more of your body's weight. To bear this extra load, the extensor muscles in the standing leg must increase the force of their contractions. This is accomplished in part by the quadriceps stretch reflex, which we discussed earlier.

In addition, you have the **crossed extensor reflex**, which activates the extensor muscles of one leg when the other leg is suddenly withdrawn. The neural circuitry underlying the crossed extensor reflex involves at least 6 kinds of neurons, including one commissural (midline-crossing) interneuron and two inhibitory ones. We don't need to discuss them further, but they are shown in Figure 10.10.

Although the crossed extensor reflex is useful, additional adjustments are needed to keep from falling down when you lift one leg off the ground. Try standing sideways against a wall so that your right foot and right shoulder touch the wall. Now lift your left leg! Were you surprised at how quickly you fell to the left? This happens because the wall prevents you from doing what you normally do when you lift your left leg, namely, shifting your center of body mass toward the right. Now try standing up straight, touching your hands to the front of your chest, and then extending your arms rapidly forward. Can you tell how your body tilts backward as your arms extend? If you are not convinced, try performing the same movements while holding something heavy in your hands. Your legs and trunk will definitely tilt backward.

These are just two examples of the little adjustments in body position that you constantly make to keep your center of mass directly above the point where your body presses against the ground (or above the average of multiple pressure points). It's a simple physical principle: if your center of mass is not directly above the center of pressure, then the center of mass begins to move downward (you fall down). Your body-stabilizing reflexes act to prevent this fate.

An interesting series of experiments in the 1980s clarified which sensory inputs drive our body-stabilizing reflexes. In these experiments people were asked to stand on a movable platform. Yanking the platform backward causes a subject's center of mass to be more anterior than the center of pressure; that is, the body tilts. In response, subjects lean backward, shifting the center of mass back over the feet. Muscle recordings showed that this backward leaning response is accomplished by contracting the calf muscles and knee flexors. In a clever twist, the experimenters then tilted the platform as they pulled it backward so that the subject's foot-shin angle remained constant during the procedure. This manipulation prevented subjects from using sensory information from the ankles to tell them about the body tilt. Although the subjects still leaned backward, they did so more slowly. On other trials, the experimenters prevented the subjects from using vision to tell them about the body tilt. Again, the subjects were slow to lean backward.

Overall, these experiments demonstrated that information from the ankles and the eyes can both be used to compensate for body tilt. Subsequent research revealed that information about body tilt can also be obtained from vestibular sensors, which explains why people with vestibular damage often have enormous trouble maintaining their balance.

Modulation of Postural Reflexes

Did you try some of those little experiments I suggested to demonstrate postural reflexes on yourself or a friend? If you did, then you probably realize that the experiments don't always work. Sometimes the reflexes are clear, but often they are weak or nonexistent. Neurologists are well aware of this when, for example, they are trying to test your knee jerk response. If your knee jerk is weak, the neurologist may ask you to bring your hands up to your chest, hook the fingers, and try to pull the hands apart. If the doctor taps your knee while you perform this so-called *Jendrassik maneuver*, your jerk response will (hopefully) be much larger than before. No one understands exactly why the Jendrassik maneuver facilitates the knee jerk response, or why it inhibits some other reflexes.

Neither do we know the neural mechanisms that modulate various stretch and withdrawal reflexes during walking. However, these modulations are far from random. For example, the stretch reflex for leg extensor muscles is stronger when the leg is in its stance phase (on the ground) than its swing phase. This boosts activity in the leg extensor muscles precisely when they are most needed to counter gravity. Similarly helpful is the fact that your leg withdrawal reflex is boosted during the leg's swing phase, when leg flexion helps you step over obstacles, and minimized during the leg's stance phase, when a withdrawal response might make you fall.

Many reflexes can be inhibited entirely. For example, if you turn your head, your eyes are not compelled to counterturn. You can override the VOR by making voluntary eye movements. Similarly, you can override the muscle stretch reflex to make intentional movements. This is crucial, of course, because without this ability you could never move one body part relative to another. As soon as you tried to move a limb, the stretch reflex would bring that limb back to its starting position. The neural circuitry that overrides the stretch reflex is not well understood but probably involves projections from the motor cortex to the gamma motor neurons that adjust muscle spindle sensitivity (see Chapter 7). Another good example of reflex modulation is the fact that you can intentionally let yourself fall backward without engaging the body-stabilizing reflexes. Once again, projections from the motor cortex are almost certainly involved.

A truly fascinating aspect of the postural reflexes is that what they stabilize may vary across tasks. For example, I once missed a step while walking down the stairs with a baby in my arms. I really hurt myself, but the baby was unharmed because all my reflexes worked to stabilize the baby, rather than my own body. This happened without any

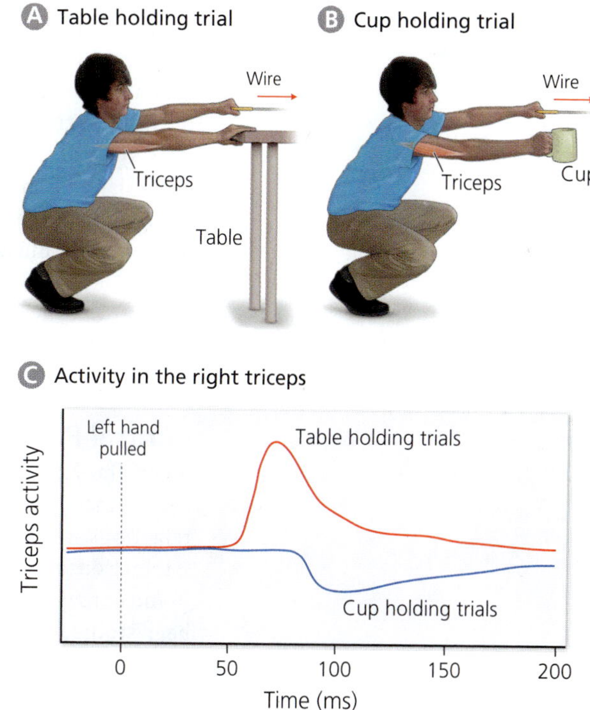

Figure 10.11 Task-dependent modulation of a postural reflex. Subjects were instructed to kneel and extend both arms. A wire was attached to their left hand and unexpectedly pulled forward at some point during each trial. The subjects were asked with their right hands to either (A) grab a table for support or (B) hold a fluid-filled cup (actually a "surrogate cup"). On table-holding trials, the right triceps increased its level of contraction (C), which helps to stabilize the body. The opposite happened on cup-holding trials: the right triceps relaxed, stabilizing the cup. Thus, our stabilizing reflexes vary depending on what we are trying to stabilize. [After Marsden et al., 1981]

explicit preparation or awareness on my part. My nervous system had automatically "reprogrammed" my normal postural reflexes to protect the baby instead of me.

Such reflex reprogramming is relatively common, although rarely appreciated. Consider, for example, what happens when you are walking along with a full cup of coffee in your hand. You try not to spill any liquid, but how much of this effort is conscious? Not much. Somehow your central nervous system "knows" that it should switch from stabilizing your center of body mass to stabilizing the cup. In a classic study of this phenomenon, scientists showed that pulling on a person's left arm elicits a very different pattern of reflexive muscle contractions when that person is holding a cup in their other hand than when they are touching a table (Figure 10.11). How this works is still unclear.

BRAIN EXERCISE

What are some situations in which several of the stabilizing reflexes we have discussed would occur simultaneously, in concert with one another?

10.4 How Do Animals Move through the World?

Locomotor activities such as swimming, crawling, walking, and flying involve highly rhythmic and stereotyped patterns of muscle contractions. As neurobiologists began to examine the neural mechanisms underlying these patterns, they considered two main hypotheses.

The **reflex chain hypothesis** holds that locomotor activity is caused by a chain of sensorimotor reflexes. Sherrington, for example, believed that locomotion occurs because the movement of one muscle stimulates sensors that reflexively activate another muscle, whose contraction then triggers another reflex, and so on until you get back to contracting the first muscle. In contrast, the **central pattern generator (CPG) hypothesis** holds that locomotor rhythms are created within the central nervous system, independently of sensory inputs. Strong support for this second hypothesis came in 1961, when Donald Wilson cut the sensory nerves in the wings of flying locusts and discovered that the locusts continued to send rhythmic commands to their wing muscles. This should not have happened if the rhythmic motor commands depend on sensorimotor reflexes. By now, neurobiologists have identified a number of different CPGs, including the CPG for breathing that we discussed in Chapter 9. Before we deal specifically with the CPGs for locomotion, let us discuss some general principles of rhythmic pattern generation.

Central Pattern Generation

As you may recall from our discussions of the circadian and cardiac rhythms in earlier chapters, rhythmic patterns of neural activity may be generated by molecular interactions within a single cell. In addition, rhythmic patterns may result from the interactions of multiple neurons with one another. Both sorts of mechanisms can be found in the neural circuitry that generates locomotion. This locomotor circuitry has been studied most extensively in **lampreys**, which look and swim like eels but are an ancient group of jawless vertebrates.

An Intracellular Rhythm Related to Locomotion

The rhythmic spiking of some locomotion-related neurons in the lamprey spinal cord has been linked to interactions between NMDA-type glutamate receptors and the neurons' membrane potential (Figure 10.12). When these lamprey neurons are exposed to glutamate (or the glutamate agonist NMDA), they slowly become depolarized. Eventually, the membrane depolarization reaches the level at which most NMDA receptors lose their magnesium block (see Chapter 3). At that point the depolarization accelerates explosively. This strong depolarization opens potassium channels that gradually hyperpolarize the cell membrane until the magnesium block is reinstated, at which point the hyperpolarization accelerates. The cycle then repeats.

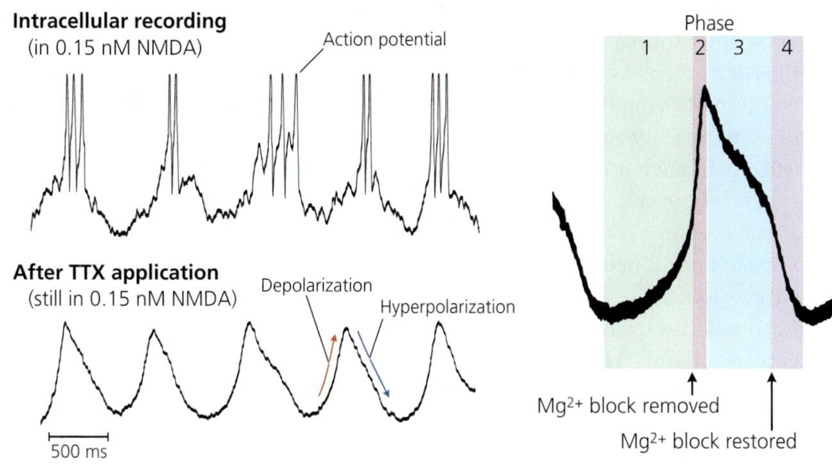

Figure 10.12 NMDA-induced oscillations in isolated neurons. Shown at the top left is an intracellular recording from a neuron in a lamprey's spinal cord bathed in 0.15 mM NMDA. The neuron's membrane potential oscillates and the neuron fires 2–3 action potentials at the height of each depolarization. The lower graph shows that the oscillation persists after applying TTX (Tetrodotoxin), which blocks all spiking activity. Shown on the right is one depolarization-hyperpolarization cycle. In phase 1, the neuron is slowly depolarized until the membrane potential reaches the voltage at which NMDA receptors are no longer blocked by magnesium ions. At that point (phase 2), the influx of positive ions through the open NMDA receptors causes rapid depolarization, until voltage-sensitive K^+ channels open and gradually hyperpolarize the neuron again (phase 3). When the membrane potential drops to a level where the Mg^{++} block is restored, the pace of membrane hyperpolarization accelerates (phase 4). [From Wallén and Grillner, 1987]

To show that this rhythmic activity requires no rhythmic external input but instead originates within each cell, the experimenters bathed the tissue in tetrodotoxin (TTX), which blocks all voltage-gated sodium channels. This procedure eliminates all action potentials and therefore isolates the neurons from one another. Nonetheless, the rhythmic depolarization-hyperpolarization cycle persists (Figure 10.12), indicating that the rhythm is generated intracellularly.

Rhythm-generating Neuronal Networks

Although isolated neurons can exhibit rhythmic patterns of activity, most of the relatively fast rhythms in the nervous system are generated by small networks of interacting neurons. The simplest sort of rhythm generating neural network is illustrated in Figure 10.13. At its core is a **half-center oscillator**, which consists of two neurons that reciprocally inhibit each other. When one of these neurons, say B1 in the figure, is slightly more active than the other, it inhibits the less active neuron B2. After a while, neuron B1 becomes fatigued, most likely because its energy supplies run low,

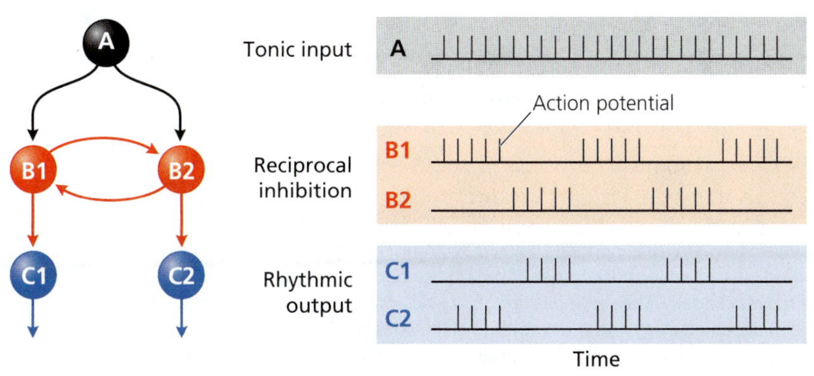

Figure 10.13 Pattern generation through reciprocal inhibition. The diagram on the left shows two inhibitory neurons (red) that receive tonic excitatory input (black) and reciprocally inhibit each other as well as two output neurons (blue). The activity of these hypothetical neurons is shown on the right. Each vertical tick mark represents an action potential. Neuron B1 responds to the tonic excitation before B2 fires and then inhibits B2 until B1 is fatigued. At that point, B2 is disinhibited, fires a set of action potentials and thereby inhibits B1 until B2, in turn, becomes fatigued. The overall result of these interactions is that the network generates rhythmic output (C1 and C2). The two inhibitory neurons in this network form a half-center oscillator. The tonic input is required to get the oscillation going and then maintain it.

and ceases to fire. At that point neuron B2, which receives tonic excitatory input from some other source (cell A in the figure), is disinhibited and begins to fire action potentials of its own. As B2 becomes active, B1 is inhibited. Eventually, B2 becomes fatigued, thus allowing B1, which also receives tonic excitatory input from neuron A, to increase its activity again. Then the cycle repeats. As you can see, the two inhibitory neurons in such a network alternate rhythmically in their activity. The tonic excitation is required to get the rhythm started and then keep it going, but the excitatory input need not itself be rhythmic. Connections from the inhibitory neurons to two sets of output neurons ultimately convey the rhythmic oscillation to the relevant muscles.

As you will discover in the next section, some important CPGs indeed contain half-center oscillators. However, most CPGs are more complicated than the network shown in Figure 10.13. They often contain more than two sets of inhibitory neurons, as well as several types of excitatory interneurons, and individual neurons in the network frequently exhibit the sort of intrinsic rhythmicity we discussed earlier (see Figure 10.12).

Swimming in Fishes

Fishes such as eels or sharks swim mainly by **lateral undulation** (*undula* means "little wave" in Latin). In this form of locomotion, muscles on one side of a body segment contract while the muscles on the other side relax (Figure 10.14 A). Then the side that was relaxed contracts while the other side slackens. Next, the cycle repeats. This pattern of muscle activity bends the body into a C-shape that rhythmically flips around its long axis. The other important feature of lateral undulation is that each C-bend travels down the body like a wave. Thus, lateral undulation is produced by a rhythmic series of traveling waves that steadily push water caudally and thereby generate forward propulsion.

The neural mechanisms of lateral undulation are well documented in lampreys. In a groundbreaking experiment, neuroscientists dissected away a lamprey's head

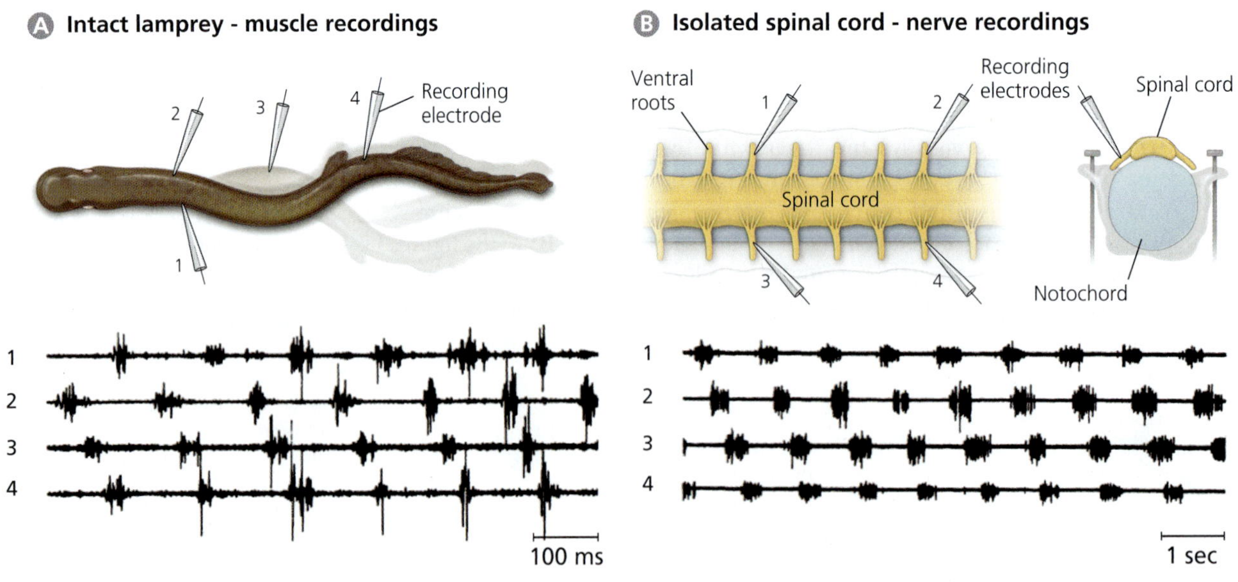

Figure 10.14 Real versus fictive swimming in lampreys. Shown in (A) is a dorsal view of a swimming lamprey equipped with four electrodes that record muscle activity. During swimming, corresponding muscles on opposite sides of the body alternate in their activity. On a given side, the bursts of muscle activity form a "traveling wave" that moves caudally. The same pattern of muscle activity can be recorded from isolated lamprey spinal cords as shown in (B). Here the muscles have been dissected away and the recording electrodes are attached to the spinal cord's ventral roots, which carry the motor axons (rostral is to the left.) Note that "fictive swimming" is much slower than real swimming (compare the scale bars!). [A from Wallén and Williams, 1984; B from Cohen and Wallén, 1980]

and body wall, leaving intact only the spinal cord, the underlying notochord, and the roots of numerous spinal nerves. The investigators then placed this isolated spinal cord in a fluid-filled dish and recorded the electrical activity in the nerve roots. At first, silence prevailed. However, when the researchers added some glutamate to the bath solution, the spinal roots became rhythmically active (Figure 10.14 B).

Importantly, the left and right spinal roots of a single spinal cord segment fired out of phase with one another, meaning that a burst of activity on the left side was accompanied by silence on the right and *vice versa*. Furthermore, when comparing spinal roots on the same side of the body, bursts of activity in rostral spinal roots tended to precede the bursts in more caudal segments. These results are precisely what one would expect if the isolated lamprey spinal cord were sending out commands to swim by lateral undulation. Of course, the lack of muscles prevents any sort of actual movement, which is why the recorded activity pattern is called **fictive swimming**.

The CPG for Swimming in Lampreys

Because fictive swimming involves no actual movement, it cannot be based on a sensorimotor reflex chain. Instead, the rhythmic motor commands coming out of the lamprey's isolated spinal cord must be generated by a CPG. To discover the detailed workings of this CPG, Sten Grillner and others recorded from individual neurons in the lamprey's spinal cord during fictive swimming. One important finding to emerge from this work is that many of the neurons in the swimming CPG are excitatory and tend to fire rhythmically even when all inhibitory connections are blocked pharmacologically. However, after blocking the inhibitory connections, the two sides of the spinal cord no longer fire in alternation. This finding implies that the side-to-side alternation arises because a burst of activity on one side of the spinal cord inhibits inhibitory neurons on the other side until that burst has exhausted itself, which then lets the other side enter its own burst phase. In other words, at the core of the CPG for swimming in lampreys lies a half-center oscillator composed of two sets of neurons that reciprocally inhibit each other (Figure 10.15).

Subsequent studies revealed that sensory input from stretch receptors in the body wall can modulate the swimming CPG. This was shown most clearly in dogfish, a small shark species. The experimenters anesthetized a dogfish, separated its brain from the spinal cord, paralyzed the animal with curare, recorded fictive swimming from the spinal roots, and then moved the paralyzed tail back and forth by means of an external motor. Remarkably, the fictive swimming rhythm changed its speed to match the rhythm of the imposed tail movements. This matching of one rhythm to another is called *entrainment*, which you already read about in Chapter 9. The entrainment of the swimming CPG by the rhythmic tail movements is mediated by sensory inputs from muscle spindles in the tail and from stretch-sensitive nerve endings in the spinal cord. Sensory entrainment of the swimming CPG has also been demonstrated in lampreys and seems to be a general phenomenon that helps fishes compensate for variations in the external forces that push against the tail. Such variations occur, for example, when fishes swim in variable currents or when they burrow into sand.

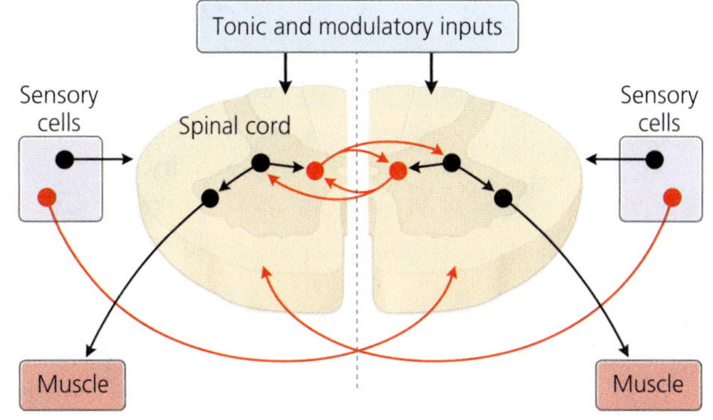

Figure 10.15 The central pattern generator for swimming in lampreys. Excitatory connections are shown in black, inhibitory ones in red. Each circle represents a class of neurons, and all the neurons are repeated in each spinal cord segment. Because the commissural connections are inhibitory, high levels of activity on one side of the spinal cord tend to inhibit the other side (the dashed line represents the midline). If those commissural connections are cut, then each side continues to fire rhythmically, but the two sides no longer alternate in their activity. The illustrated circuit is incomplete and contains a few assumptions that remain to be tested. [After Grillner, 2003]

Walking in Quadrupeds

Walking is more complicated than lateral undulation because it involves more diverse muscles and multiple joints. In particular, walking requires that leg extensors and leg flexors in the same leg be activated alternately. Given this added complexity, one would expect the control of walking

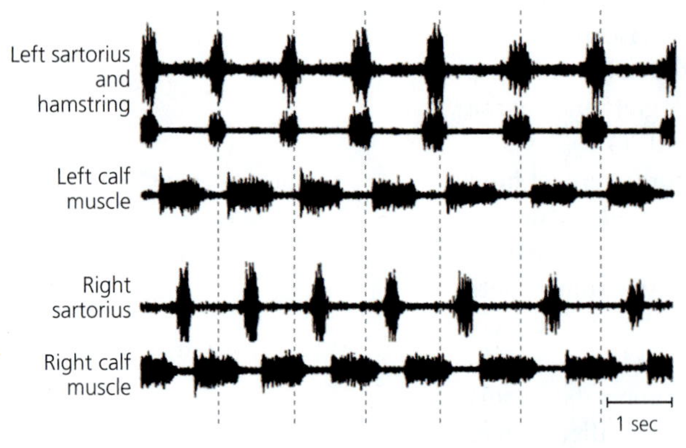

Figure 10.16 The walking CPG. A cat's spinal cord was cut at T13. A few days later the cat was paralyzed pharmacologically and implanted with recording electrodes. Finally, the cat was injected with a norepinephrine agonist and pinched in the perineum. In response, the leg motor nerves became active in the same sort of pattern as is seen during normal walking. Activity alternated between nerves innervating corresponding muscles in the left and right leg and, within a leg, between extensor and flexor muscles. [From Pearson and Rossignol, 1991]

to be more complex than the control of swimming. Indeed, it is. However, many of the concepts derived from the research on swimming in fishes have been invaluable for understanding the neural mechanisms of walking.

A Spinal CPG for Walking in Cats and Rats

Graham Brown, a student of Sherrington's, showed in the early 1900s that *decerebrate cats*, whose forebrains are surgically removed, can be induced to walk on a treadmill as long as their bodies are supported in a sling. Even in *spinal cats*, which had their brain separated from the spinal cord, walking can be elicited by injecting an excitatory neuromodulator into the spinal cord. Most important, Brown could elicit **fictive walking** in spinal cats after cutting the hindlimb sensory nerves as well as most of the motor nerves. The remaining motor nerves exhibited a rhythmic pattern of activity similar to that observed during walking. Thus, Brown showed that the mammalian spinal cord contains a walking CPG. Curiously, Brown's finding received little attention and Brown himself did little further work on this subject. Only in the late 1970s did scientists confirm that fictive walking can be evoked in paralyzed spinal cats (Figure 10.16).

The cat experiments we just discussed were important but ethically challenging, both because they involved cats and because they were quite invasive (see Box 10.1: Using Animals in Research). Therefore, experimenters were keen to study the walking CPG in less intelligent vertebrates and, if possible, in isolated spinal cords (which, presumably, cannot suffer). Their efforts proved successful when in the late 1990s, they began to work with embryonic and newborn rats. Isolated spinal cords from such young rats display the left–right and flexor–extensor alternation that is typical of normal walking. Curiously, the alternating pattern of activity emerges gradually from a synchronous pattern. This developmental progression probably results from a developmental change in the postsynaptic effect of the neurotransmitters GABA and glycine, which are found in spinal cord commissural neurons. These transmitters inhibit postsynaptic neurons in adults but are excitatory in young embryos because GABA and glycine open chloride channels, and intracellular chloride concentrations decrease as development proceeds. As long as the transmitters are excitatory, activity on one side of the spinal cord promotes activity on the other, so that the two sides fire synchronously. After they become inhibitory, the left and right sides alternate in their activity.

Like the CPG for swimming in fishes, the spinal CPG for walking in mammals can be modulated by external inputs. It can, for example, be entrained by leg movements when decerebrate cats are placed on a moving treadmill; such cats adjust their walking speed to that of the treadmill. The walking CPG can also be modulated by descending input from the **midbrain locomotor area**. When this brain region is stimulated electrically, the walking CPG becomes active and, with increasing stimulation intensity, its rhythm accelerates. At very high stimulus intensities, the rhythm changes its pattern so that equivalent muscles on the left and right side of the body are activated synchronously, as they are during galloping.

Do Humans Have a Spinal CPG for Walking?

Humans with a transected spinal cord do not, in general, exhibit rhythmic leg movements when they are placed on a moving treadmill, even when the trauma is long past. Does this mean that the spinal CPG for walking was lost during human evolution? The answer is probably "no." Have you ever noticed how your arms tend to swing back and forth as you walk and that they swing out of phase with the corresponding leg? This swinging of the arms during walking is reminiscent of the fore- and hindlimb

RESEARCH METHODS

Box 10.1 *Using Animals in Research*

If you love animals, some of the experiments discussed in this book are bound to make you feel uncomfortable. This is especially true of some of the experiments on cats and monkeys that are described in this chapter. Although these experiments have contributed significantly to our understanding of the nervous system, it is both fair and appropriate to ask how much the animals suffered in our pursuit of that knowledge. As you can probably imagine, this is a difficult question that people argue heatedly about. In such debates, knowing some history and facts can be useful.

Back in Descartes' day, animals were widely considered mere machines incapable of feeling pain or suffering. Only in the 18th and 19th centuries did animal welfare become a growing concern. The first legislation on animal welfare, Britain's Cruelty to Animals Act, passed in 1875. Since then, new laws have steadily increased the protections afforded to animals. The main law governing animal welfare in the United States is the *Animal Welfare Act* (AWA), which has been amended repeatedly. Two key provisions of the current AWA are that any animals used in research or teaching must receive professional veterinary care and that all procedures involving those animals must be approved by an *Institutional Animal Care and Use Committee* (IACUC). This committee must include not only a veterinarian but also someone not affiliated with the research institution. One of the IACUC's main missions is to assure that the proposed procedures reduce animal discomfort, distress, and pain as much as possible without obstructing the scientific aims. The IACUC must also inspect all research and animal facilities twice a year and report any serious violations to the US Department of Agriculture, which is in charge of enforcing the AWA.

One limitation of the AWA is that it excludes all cold-blooded animals, as well as birds, mice, rats, and any farm animals commonly used in food or clothing production. Does this mean that scientists can do whatever they want with animals not covered by the AWA? No it does not because most research institutions must adhere to a separate set of regulations issued by the National Institutes of Health (NIH), which is the principal funding agency for animal research in the United States. Any institution that wants to receive NIH funding must follow the NIH's guidelines for animal research, which cover all vertebrate animals. Responsibility for enforcing the NIH rules falls to the institution's IACUC, which is required to report any serious violations to the Office of Laboratory Animal Welfare at the NIH. An additional layer of supervision is provided by the Association for Assessment and Accreditation of Laboratory Animal Care (AAALAC), which inspects most research institutions every 3 years. Although AAALAC accreditation is not required by law, it is eagerly sought by most institutions that receive NIH funding because it signifies that an animal research program meets current standards of excellence.

Highly relevant to any debate about animal research is the number of animals involved and the species being examined (Figure b10.1). Back in the 1950s, almost a million monkeys were sacrificed in the successful quest for a polio vaccine. Dogs and cats were also used quite commonly in biological research. Since then, research has shifted away from these species and toward rats and mice as well as zebrafish. Many people believe that performing experiments on mice and fishes is ethically less problematic than working with rabbits, cats, or other "higher" animals; but this is certainly debatable. Indeed, animal rights advocates are generally reluctant to draw any distinctions between different kinds of animals, arguing that all animal research should be halted. If you share this view, you have a right to voice that opinion and demonstrate. However, animal rights advocates sometimes go beyond peaceful protests by vandalizing research laboratories, intimidating researchers, and directly attacking people who support animal research. Since the 2006 passage of the *Animal Enterprise Terrorism Act*, such activities are considered acts of terrorism and have been prosecuted as such.

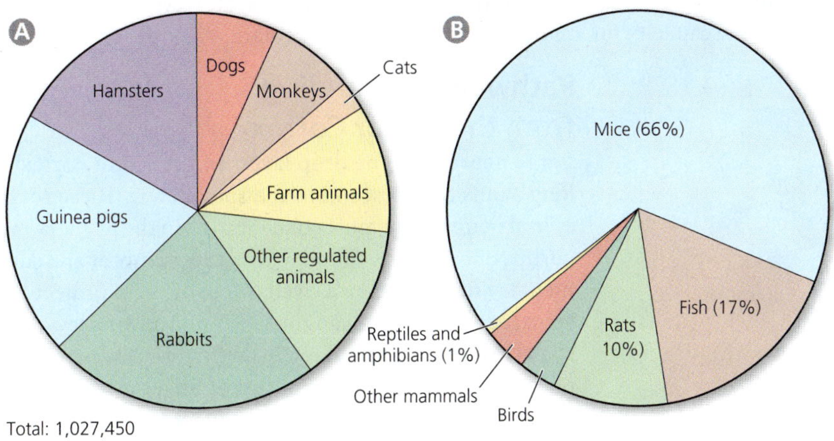

Figure b10.1 Animals in research. (A) shows the distribution of USDA-regulated animals used for research and teaching in the United States during the fiscal year 2007. This chart excludes birds, most rodents, and all cold-blooded animals. Based on the distribution of species used for research in the United Kingdom (B), the excluded animals probably account for roughly 96% of all animals used in teaching and research. Therefore, the total number of vertebrate animals used for research and teaching in the United States during 2007 was probably ~25 million. By comparison, 9.5 billion animals were slaughtered for food in the United States during 2008, 98% of them chickens.

movements observed in walking quadrupeds and suggests that our spinal cords are similar to those of other mammals. In addition, rhythmic leg activity has been observed in a few patients with severed spinal cords, especially if those spinal cords are stimulated with implanted electrodes that tonically activate spinal sensory nerves (so-called epidural electrodes). Collectively, these data suggest that human spinal cords do contain a CPG but that this CPG is more difficult to activate in humans than in other vertebrates.

BRAIN EXERCISE

How might you test whether walking in a cockroach involves a chain of sensorimotor reflexes or a central pattern generator?

10.5 What Does the Motor Cortex Contribute to Motor Control?

As noted in the previous sections, CPGs and motor neurons receive substantial descending inputs that modulate their activity. In primates, many of these descending pathways originate from the motor cortex, which is located in the posterior portion of the frontal lobe, anterior to the central sulcus (Figure 10.17). The motor cortex in primates consists of several subdivisions. Most posteriorly lies a thin strip of tissue called **primary motor cortex (M1)**. Anterior to this lies the **premotor cortex**, which is itself divisible into a several smaller areas. More medially you find the **supplementary motor area**.

Because neocortical lesions impair consciousness, it is tempting to believe that the motor cortex is responsible for all voluntary movements, which occur against a background of involuntary acts controlled by "lower" brain regions. This view has some validity, but not everything that happens in the neocortex is accessible to consciousness. Many unconscious and involuntary reflexes, including the muscle stretch reflexes, involve long-loop circuits through the motor cortex that supplement the subcortical circuits we discussed earlier. Furthermore, experimental removal of the motor cortex in monkeys and most other mammals leaves many voluntary behaviors intact. It is true that humans with motor cortex lesions are massively impaired in their ability to walk, talk, and do things with their hands. However, even in such patients some capacity for voluntary movements may remain.

Pathways Descending from the Motor Cortex

Large neurons in the deep layers of the motor cortex have long axons that descend to the spinal cord. These **corticospinal** neurons are most concentrated in the primary motor cortex, but some are found within the premotor and supplementary motor cortex as well. Most of them project to the spinal cord's intermediate zone (Figure 10.18). A few corticospinal axons terminate directly on motor neurons, especially on those involved in the control of finger movements; but most of the corticospinal axons project to spinal cord interneurons. In addition, some corticospinal axons terminate in the spinal cord's dorsal horn, where they are thought to regulate the flow of movement-related sensory information.

Some parts of the motor cortex project not to the spinal cord but to the medulla. Among the targets of these **corticobulbar** projections (the term "bulbar" refers to the medulla) are regions that control respiration, urination, and other vegetative functions as well as regions that control eye blinks. One small part of the motor cortex projects to motor neurons that innervate the tongue muscles. Another region

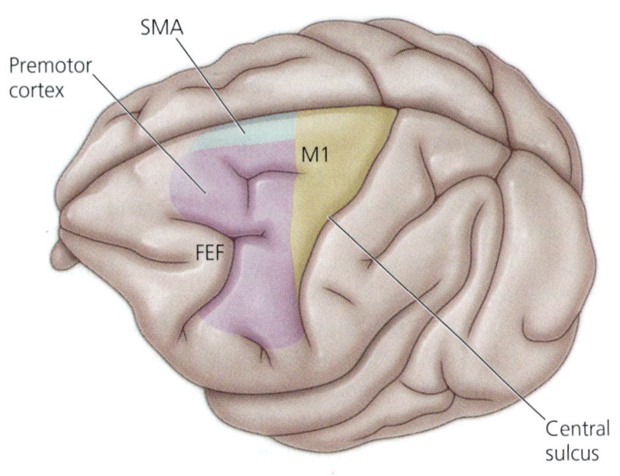

Figure 10.17 Dorsolateral view of macaque motor cortex. The motor cortex includes the primary motor cortex (M1; yellow), premotor cortex (purple), and the supplementary motor area (SMA; green). FEF stands for frontal eye field.

of motor cortex projects to motor neurons (in nucleus ambiguus) that innervate the vocal cords. As you can imagine, both of these pathways are critical for human speech.

Laterality of the Corticospinal and Corticobulbar Projections

Approximately 90% of the corticospinal axons in macaque monkeys cross the midline at the ventral surface of the medulla (in the so-called pyramidal decussation) and then terminate in the contralateral spinal cord (Figure 10.18). Only spinal neurons that control the neck and trunk muscles receive bilateral input from the motor cortex. Because of this arrangement, extensive unilateral damage of the motor cortex severely weakens the arm and leg muscles on the opposite side of the body but has much less severe effects on the control of trunk and neck muscles.

In contrast to the corticospinal projections, the corticobulbar projections are almost entirely bilateral. Therefore, unilateral motor cortex lesions deprive the corticobulbar targets of only half their inputs, leaving them capable of functioning with only minor impairments. The major exception to this rule is that the motor neurons that innervate the muscles of the mouth receive exclusively contralateral input. This explains why patients with unilateral motor cortex damage generally have trouble smiling with the half of the mouth that is opposite the lesion. Of course, if both the corticospinal and the corticobulbar pathways are damaged bilaterally, as can happen with strokes or tumors in the upper medulla, people become incapable of most, if not all, voluntary movements (see Box 10.2: Locked-in Syndrome).

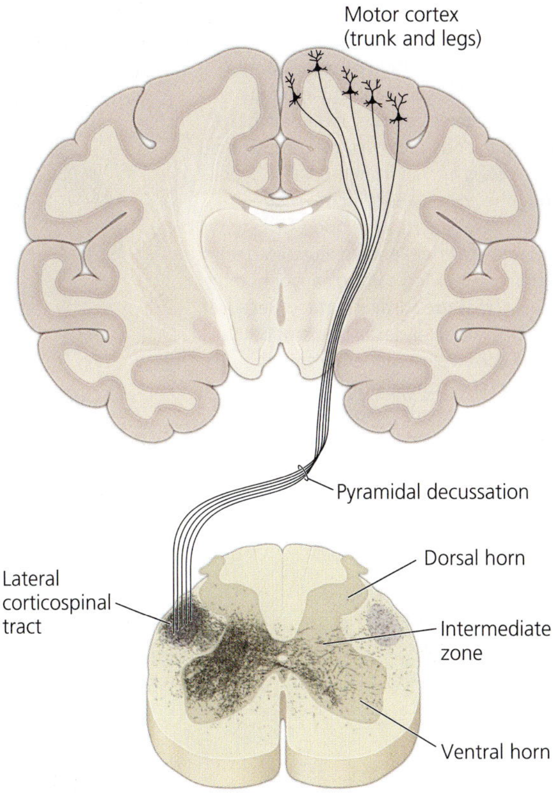

Figure 10.18 The corticospinal tract. Illustrated on these coronal sections through the forebrain and spinal cord of a macaque monkey are projections from the leg area of the primary motor cortex to the spinal cord (lumbar level). Most of the corticospinal axons cross the midline and terminate in the spinal cord's ventral horn and intermediate zone. [After LaCroix et al., 2004]

Cortical Motor Maps

Epilepsy is a brain disorder characterized by repeated episodes of abnormal brain activity called **epileptic seizures** that are commonly (but not necessarily) associated with uncontrollable muscular convulsions. Back in 1863, the neurologist Hughlings Jackson noticed that epileptic convulsions sometimes begin in one part of the body and then "march" across the body in sequence. They might, for example, start in the fingers, move from there up the arm and eventually involve some muscles of the face. Jackson hypothesized that this *Jacksonian march* of seizure activity occurs because the neurons controlling adjacent body parts tend to be located adjacent to one another in the motor cortex and are successively engulfed by a wave of epileptic seizure activity.

Jackson's hypothesis was confirmed and extended in the 1930s by Wilder Penfield, who examined patients with cortical tumors or other conditions that triggered epileptic seizures. To determine where the seizures originated and which cortical areas could be removed without causing catastrophic paralysis, Penfield applied electrical stimuli to the cortex of his patients prior to neurosurgery. Because the patients were awake, they could tell Penfield what they felt in response to the stimulation and report even minor muscle twitches. As Penfield and his colleagues analyzed these responses, they noted that the area of the body affected by the stimulation varied systematically as the stimulating electrodes were moved across the motor cortex. For example, stimulation in the ventral motor cortex evoked small movements of the tongue and mouth whereas more dorsal stimulation elicited finger twitches. Penfield eventually summarized his work by drawing a human figurine across the surface of the motor cortex (Figure 10.19). This **homunculus** (little man) was a clever way to show that adjacent regions in the motor cortex tend to control adjacent body parts. By giving the homunculus an oversized head and hand, Penfield conveyed that twitches

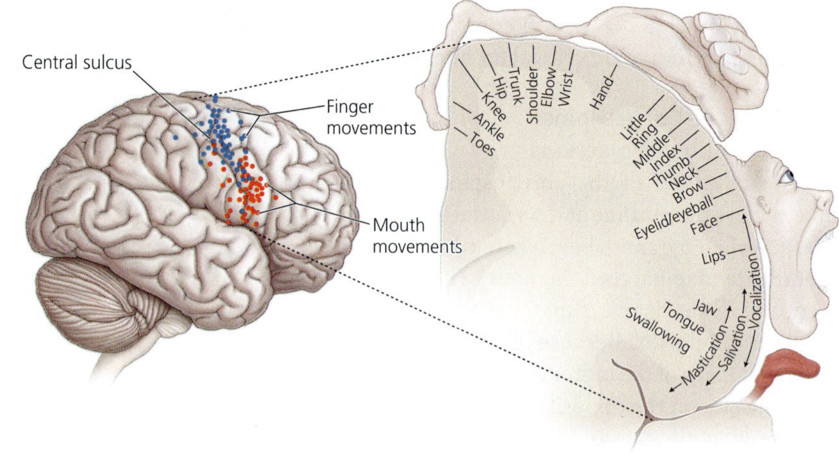

Figure 10.19 Penfield's motor homunculus. Brief bursts of current applied to the motor cortex elicit muscle twitches in specific body parts. As the stimulation electrode is moved from medial to lateral along the cortex, the twitches move systematically across the body, from toe to tongue. Wilder Penfield illustrated this finding by drawing a little man, or homunculus, across the cortical surface. Because disproportionately large portions of the motor cortex elicit movements in the hand and face, those portions of the homunculus are disproportionately large. [Left after Penfield and Boldrey, 1937; right after Penfield and Rasmussen, 1950]

in those body parts can be evoked from disproportionately large regions of the motor cortex.

Modern Views of the Motor Homunculus

Penfield's homunculus implies that the motor cortex contains a fine-grained, continuous motor map in which adjacent neurons control adjacent muscles. When researchers tested this hypothesis by stimulating very small patches of the motor cortex with fine-tipped microelectrodes, they observed that Penfield's notion was only partially correct. Microstimulation of adjacent locations in the motor cortex, especially in the primary motor cortex, do tend to evoke movements of the same or of adjacent body parts; but the motor map is hardly continuous. Instead, it resembles a map that was torn into many pieces and then taped back together by a child (or one of David Hockney's cubist photomontages). Examined from a distance, the map looks smooth; but a closer look reveals many discontinuities. It is a *fractured map* of body movements. A second challenge to Penfield's vision of a smooth and fine-grained motor homunculus was that microstimulation in a small patch of the motor cortex generally activates not just a single muscle but a number of functionally related muscles (such as all muscles that flex the wrist). Thus, what is mapped in the motor cortex are not muscles but movements.

The idea that the motor cortex controls movements rather than muscles was expanded recently when neuroscientists stimulated the motor cortex of monkeys with

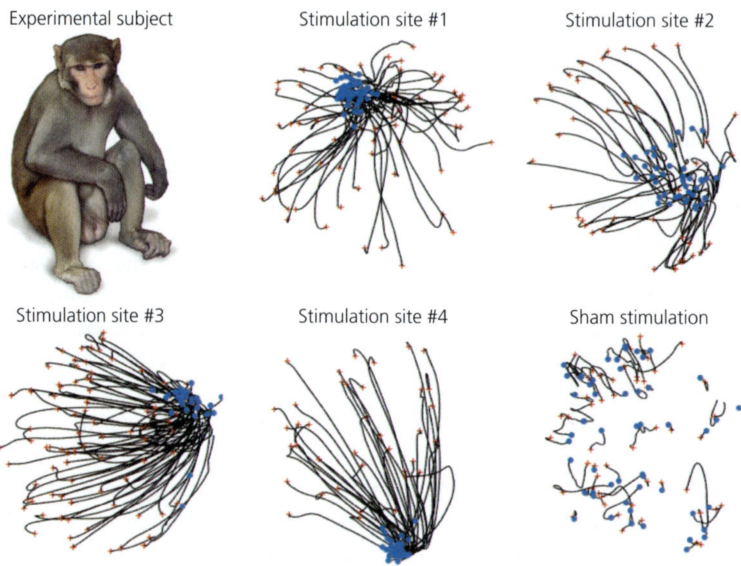

Figure 10.20 Goal-directed hand movements evoked by motor cortex stimulation. Half-second trains of current pulses (25–100 μA each) were applied to the motor cortex of an awake macaque. Shown here are the hand movements evoked by stimulation at 4 sites. The hand's position at the beginning of each trial is indicated by a red plus sign; its final position is marked in blue; the hand's trajectory is drawn as a black curve. Stimulation caused the hand to move toward a specific location in space no matter where the hand had been prior to stimulation, and different sites were associated with hand movements to different locations. On control trials, when no current was passed (sham), hand movements were small and unpredictable. [After Graziano, 2006]

NEUROLOGICAL DISORDERS

Box 10.2 *Locked-in Syndrome*

In the winter of 1995, at the age of 43, Jean-Dominique Bauby suffered a brain stem stroke that put him in a coma for 20 days. Gradually he emerged from this state, only to find himself unable to breathe, speak, or move his limbs. Bauby was fully conscious, his mind as sharp as usual, but he could not tell anyone; he had *locked-in syndrome* (LIS). Sadly, Bauby's doctors thought he had entered a vegetative state. Bauby recalls his feelings at the time:

> I am fading away, slowly, but surely. Like the sailor who watches his home shore gradually disappear, I watch my past recede. My own life still burns within me, but more of it is reduced to the ashes of memory. Since taking up life in my cocoon, I have made two brief trips to the world of Paris medicine to hear the verdict pronounced on me from medical heights. I shed a few tears as we passed the corner café where I used to drop in for a bite. I can weep discreetly, yet the professionals think my eye is watering.

Given that Bauby never recovered from his stroke, how do we know what Bauby felt? Remarkably, an attentive speech therapist noticed that Bauby could move his left eyelid. After asking Bauby to think of a letter, the therapist recited the letters of the alphabet in order of their frequency (in French) and told Bauby to blink when she reached the desired letter. In this manner, letter by letter, Bauby dictated his short but powerful book entitled *The Diving Bell and the Butterfly*. He died two days after the book's original publication. In 2007, the book was made into a film.

Now, computers can help LIS patients communicate. The simplest approach is to use a computer to track a patient's eye movements as they look at specific places on a computer screen. If the patient looks at a specific letter on a virtual keyboard for longer than a second or so, then the software can infer that the patient wants to spell this specific letter. This method allows messages to be spelled out more rapidly than Bauby's eye blink method. Of course, eye tracking only works in LIS patients who remain capable of eye movements.

Patients who cannot move their eyes can sometimes communicate through a brain–computer interface. The least invasive approach is to record a patient's electroencephalogram (EEG; see Chapter 13) with scalp electrodes while the patient is looking at a virtual keyboard whose letters periodically flash brightly. Whenever a letter of interest flashes, the EEG changes its pattern. By detecting this change, a computer can infer which letter the patient was looking at. A related approach is to have patients learn how to move a cursor on the screen by adjusting their own neural signals (see Chapter 3, Box 3.3). In essence, they learn to move the cursor by thinking about it. Simultaneously, a powerful computer learns how to interpret the patient's neural activity patterns. This kind of dynamic collaboration between computers and brains may sound like science fiction, but much of the technology is already in use.

As Bauby pointed out, a major issue for LIS patients is that the condition is frequently misdiagnosed. Of all the people who appear to be in a coma or vegetative state, which ones are actually conscious? This is not an easy question to answer, but functional brain imaging can help. For example, scientists can compare the brain's responses to the patient's own name versus some other name. If the responses are the same, then the patient is unlikely to be conscious. Even better is to examine brain activity when a person is told to imagine playing tennis and then compare it to the activity observed when the same patient is instructed to imagine walking through their house. In humans with intact nervous systems, the thought of playing tennis activates the supplementary motor cortex, whereas the thought of walking through a house activates a different set of brain regions. If a paralyzed patient exhibits the normal instruction-dependent difference in brain activation, then they are likely to be locked in. In contrast, comatose or vegetative patients should exhibit no such difference.

Bauby's experience and more recent brain imaging studies have shown that at least some apparently vegetative patients suffer from LIS. This is disturbing because locked-in patients should not, most people would agree, be taken off life support. Instead, they should be given a chance to share their thoughts with friends and family, or write a book as Bauby did! It's scary stuff to think about. On the other hand, many people who are in a persistent vegetative state really have no mental life and no hope of recovery.

higher current intensities and longer stimulus trains. The stimulated monkeys exhibited not isolated joint movements or muscle twitches but coordinated motor acts. For example, stimulation in one part of the motor cortex caused the monkey's contralateral hand to move to a specific location in space (Figure 10.20). Importantly, the monkey's hand moved to this target region regardless of where the hand had been before the stimulus began, implying that the movement was goal oriented. Stimulation of adjacent cortical regions elicited hand movements to different spatial locations. Stimulation of more distant motor cortical regions elicited different, but still apparently purposeful, motor acts. These findings are controversial because prolonged electrical stimulation activates a fairly large and widely distributed network of cortical neurons. Still, the findings are reliable. At the very least, they demonstrate that the effect of cortical stimulation may vary with the prior location of a limb in space, which implies that sensory information about limb position can dynamically reprogram motor commands.

Encoding Movement Details

To complement the results of stimulating motor cortical neurons, researchers have recorded the activity of individual motor cortex neurons while monkeys are performing diverse hand movements. In a typical experiment, monkeys are trained to grasp the handle of a mechanical arm equipped with force and position sensors and then to move their hand around. To record the activity of neurons during these movements, experimenters use extracellular electrodes inserted into a monkey's brain through special recording chambers that were fastened to the monkey's skull during a previous surgery. After recording a neuron's activity during many different hand movements, the experimenters look for correlations between the neuronal activity and various movement parameters.

Early studies of this type reported that the firing rates of motor cortical neurons correlate with the force or speed of specific movements. Subsequent research revealed that the activity of individual motor cortex neurons also correlates with more global aspects of movement, especially with the direction in which the hand is moved (Figure 10.21). Importantly, the *direction tuning* of most motor cortex neurons is rather broad, making it impossible to predict the direction of a specific hand movement from the activity of a single neuron. However, such predictions are possible if one analyzes the firing rates of many different neurons with diverse direction preferences. This finding implies that movements are encoded not by single neurons but by large populations of neurons working together. In other words, the instructions for performing a movement are encoded in some kind of neuronal **population code**, defined as a code in which information (about a movement or a sensory stimulus) is represented in the pattern of activity across a population of neurons.

As you have now learned, some of the stimulation and recording data indicate that motor cortex neurons encode specific joint movements, whereas other findings imply that the cortical neurons encode more global, purposeful movement

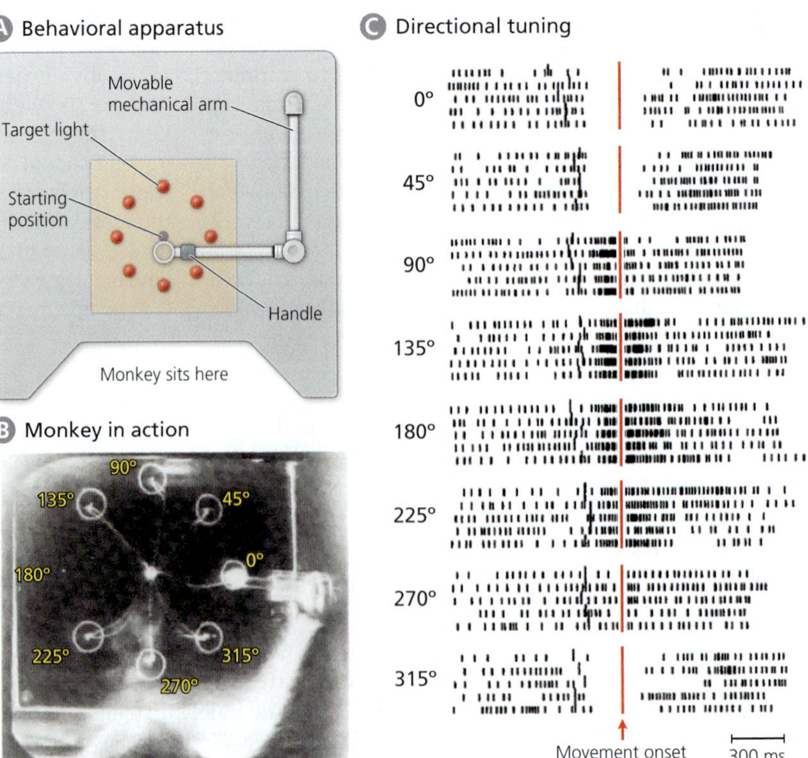

Figure 10.21 Direction tuning in the motor cortex. Monkeys were trained to grasp a mechanical arm and position it at the center of a circle of 8 LEDs (A). When one of the LEDs was illuminated, the monkeys had to move their hand toward the light and then return it to the starting position. The photograph in (B) depicts several such trials superimposed in a long-exposure photograph (the end of the mechanical arm was equipped with a small light). Meanwhile, the experimenters recorded the activity of single neurons in the monkeys' motor cortex (C). Each vertical tick mark represents an action potential, and each trial is represented by a row of ticks. Five trials are shown for each movement direction, and all trials are aligned to movement onset (red vertical lines). The illustrated neuron was most active during (and just before) movements toward 135, 180, and 225 degrees. It was inhibited during movements toward 0, 45, and 315 degrees. [From Georgopoulos et al., 1982]

parameters. These two viewpoints are difficult to reconcile, which is why the field of cortical motor control is rife with controversy and debate. Some functional brain imaging data suggest that one might be able to resolve this problem by proposing that global movement parameters are specified in the premotor cortices, whereas more specific commands are issued from the primary motor cortex. However, neurophysiological studies have not revealed such a definite distinction. For example, neurons with the kind of broad direction tuning shown in Figure 10.21 are found in both premotor and primary motor cortex. Despite these uncertainties, most neuroscientists agree that cortical movement control involves the concerted activity of many different neurons and that individual neurons are activated during many different movements.

Mirror Neurons in the Premotor Cortex

The most complicated part of the primate motor cortex is the ventral premotor region. It contains multiple subdivisions and its neurons exhibit remarkably complex activity patterns. Most of these neurons increase their firing rates shortly before and during arm, hand, lip, or tongue movements.

Best studied are the neurons related to hand movements, especially to reaching and grasping. As Giacomo Rizzolatti and his colleagues discovered in the 1990s, ventral premotor neurons tend to be fairly selective, increasing their firing rate only before and during very specific movements. For example, some neurons fire during whole-hand grasps, whereas others fire during precision grips. This is consistent with the microstimulation data indicating that premotor neurons control purposeful motor acts.

What came as a major surprise, however, was that about 16% of the neurons fire not only during a monkey's own movement but also when the monkey observes human movements. Moreover, different neurons respond preferentially to different sorts of observed movements. For example, some neurons increase their firing when the experimenter grasps an object with his fingers but not when the experimenter grips the object with a pair of pliers (Figure 10.22). Other neurons have different

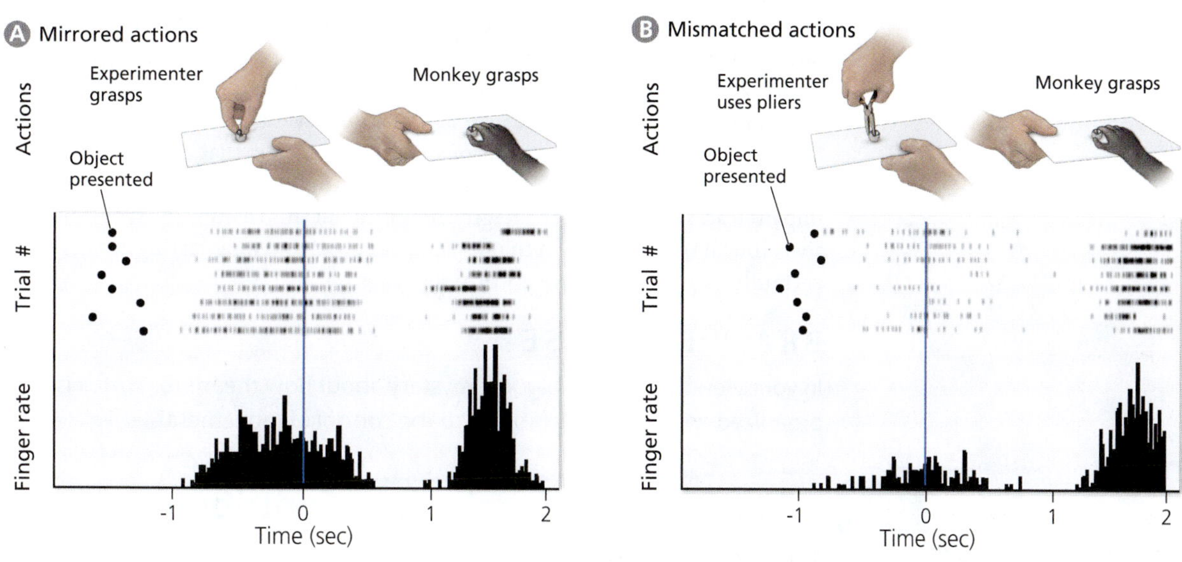

Figure 10.22 A mirror neuron. As shown in (A), mirror neurons increase their firing rate when the experimenter grasps an object in front of the monkey as well as when the monkey grasps a similar object a second or so later. Data from 8 separate trials are shown in the form of a raster plot (each tick is an action potential) and, below that, as a peri-movement time histogram. All trials were aligned to the time at which the experimenter touched the object (blue line). The filled circles indicate when in each trial the stimulus object was first presented. Shown in (B) is the activity of the same neuron when the experimenter grabbed the small object with pliers just before the monkey grasped a similar object with his hand. In this case, the actions do not "mirror" each other; and the neuron fired strongly only during the monkey's own action. [After Gallese et al., 1996]

preferences. Finally, and crucially, 6–11% of all neurons in the area respond selectively to the same sort of observed movement that they are selective for when the monkey makes the movement himself. For example, they fire only when the monkey observes a human grasp an object with the fingertips and only when the monkey makes the same grasping movement himself (Figure 10.22). These neurons are called **mirror neurons**.

The functions of mirror neurons remain intensely debated. The most widely discussed hypothesis is that these neurons facilitate action recognition and understanding. The argument is that a mirror neuron's activity in response to an observed movement can represent or "code for" the movement's purpose because when we perform this same movement ourselves, we know the action's purpose and the same mirror neurons would fire. In essence, the mirror neurons are thought to represent the action's meaning. A natural extension of this core idea is that mirror neurons play a role in language evolution and comprehension. It has also been argued that mirror neurons might endow humans with the ability to feel empathy. Although these hypotheses are intriguing, they have been criticized. Particularly troublesome is that the selectivity of mirror neurons can be modified by experience in adult humans, which suggests that mirror neurons develop as a result of associative sensorimotor learning. If this is true, then the mirror neurons in the ventral premotor cortex may be involved in learning manual skills rather than more lofty forms of cognition. Of course, mirror neurons might well have multiple functions.

Motor Cortex Plasticity

When Wilder Penfield mapped the motor cortex of his patients prior to neurosurgery, he noticed that the map's fine structure varied across individuals. In addition, when patients returned weeks later for a second mapping session, Penfield observed that aspects of the motor map had changed. Subsequent research has confirmed and extended these observations. As we discussed at length in Chapter 3, training on specific motor tasks leads to a gradual expansion of the cortical territory that is dedicated to the task-related movements, especially when the required movements are challenging. Experience can also "rewire" some of the connections within motor cortex.

The idea that brains can rewire themselves drives a great deal of current research on how to maximize recovery from brain damage. Particularly interesting is **constraint-induced movement therapy**, which requires that patients who have lost the function of one arm put their good arm in a sling for several hours at a time and then use the impaired arm to perform various tasks. The idea is that forcing subjects to use their impaired arm will increase plasticity within the brain regions controlling that impaired arm. Few of us can imagine how difficult this therapy must be, but an extensive clinical trial has shown that it produces good and long-lasting results. It almost certainly involves extensive cortical rewiring.

BRAIN EXERCISE

In your view, what is the biggest mystery about how the motor cortex is organized and how it contributes to the control of movement?

10.6 What Does the Cerebellum Contribute to Motor Control?

Another structure that is intimately involved in the control of movements is the cerebellum, which in Latin means "little brain." Back in the mid-1600s, Thomas Willis (see Chapter 5) had suggested that the cerebellum is in charge of our involuntary movements, whereas the motor cortex controls the voluntary ones. This functional dichotomy is simplistic. Nonetheless, the cerebellum's contributions to motor control are primarily the sort of automatic adjustments that tend to go unnoticed until they are lost. The cerebellum also performs some sensory and cognitive functions, which we discuss toward the end of this section.

Cerebellar Anatomy

The largest cerebellar neurons are the **Purkinje cells**, which were named after their discoverer, Jan Purkinje. The cell bodies of the Purkinje cells are arranged into a thin layer that parallels the folded cerebellar surface, and the Purkinje cell dendrites extend from there toward the cerebellar surface (Figure 10.23). As we discussed in Chapter 4, the dendritic tree of a Purkinje cell resembles an open hand whose fingers reach toward the cerebellar surface (see Figure 4.28).

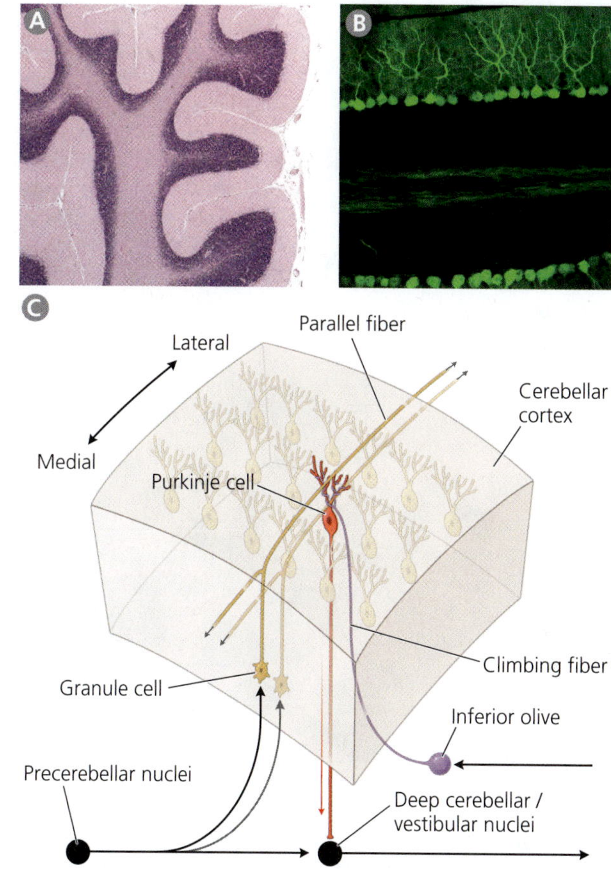

Figure 10.23 Anatomy of the cerebellum. The densely stained granule cell layer is sandwiched between the Purkinje cells and the underlying white matter (A). The photograph in (B) is from a transgenic mouse in which all the Purkinje cells contain green fluorescent protein. The diagram in (C) illustrates the cerebellum's basic circuitry. Several types of cerebellar interneurons have been omitted for the sake of clarity. [A from Comparative Mammalian Brain Collections (brainmuseum.org), funded by NSF; B created by The Gene Expression Nervous System Atlas (GENSAT) Project, funded by NIH]

Parallel and Climbing Fiber Inputs

One of the two major inputs to Purkinje cells comes from the **cerebellar granule cells** that lie just beneath the Purkinje cell layer. These granule cells are small, densely packed, and exceedingly numerous, comprising ~80% of all neurons in adult human brains. The axons of the cerebellar granule cells are called **parallel fibers** because, after ascending to the Purkinje cell dendrites, they split into left and right branches that extend in straight lines parallel to one another and to the cerebellar surface (Figure 10.23). According to some estimates, each Purkinje cell receives parallel fiber input from more than 180,000 granule cells. The granule cells, in turn, get input from a variety of different brain regions that are collectively referred to as precerebellar nuclei.

The second major source of input to Purkinje cells is the **inferior olive**, which lies within the medulla. Each neuron in the inferior olive of adult mammals sends an axon to just a single Purkinje cell, but it forms tens of thousands of synapses onto that cell. To make all those synapses, the axon climbs all over the Purkinje cell's dendritic tree, which is why the inferior olivary axons are called **climbing fibers**. As you might imagine, whenever an inferior olive neuron fires an action potential, the Purkinje cell to which that neuron projects also fires an action potential; in fact, it fires an unusually large and long action potential called a *complex spike*. In contrast, Purkinje cells respond to granule cell activity only when thousands of those cells fire simultaneously. Even then, granule cell activity elicits only *simple spikes* in the Purkinje cells.

Cerebellar Outputs and Divisions

All cerebellar Purkinje cells use GABA as their main neurotransmitter and therefore inhibit their target cells (at least in adulthood). Those target cells lie mainly in the **deep cerebellar nuclei**, which lie "deep" in the cerebellum's interior. However, some Purkinje cells in the most posterior portion of the cerebellum project directly to the vestibular complex, rather than a deep cerebellar nucleus.

This posterior part of the cerebellum (Figure 10.24), called the **vestibulocerebellum**, has been studied in great detail. As you will see shortly, it helps fine-tune the VOR and other reflexes that involve vestibular input. The **spinocerebellum** lies near the cerebellum's midline and is connected primarily to regions that control neck and trunk movements. The cerebellum's third major division is the **cerebrocerebellum**. It forms the cerebellum's lateral portions, which are quite large in primates and connected mainly to the frontal lobe. Within each of these three large cerebellar

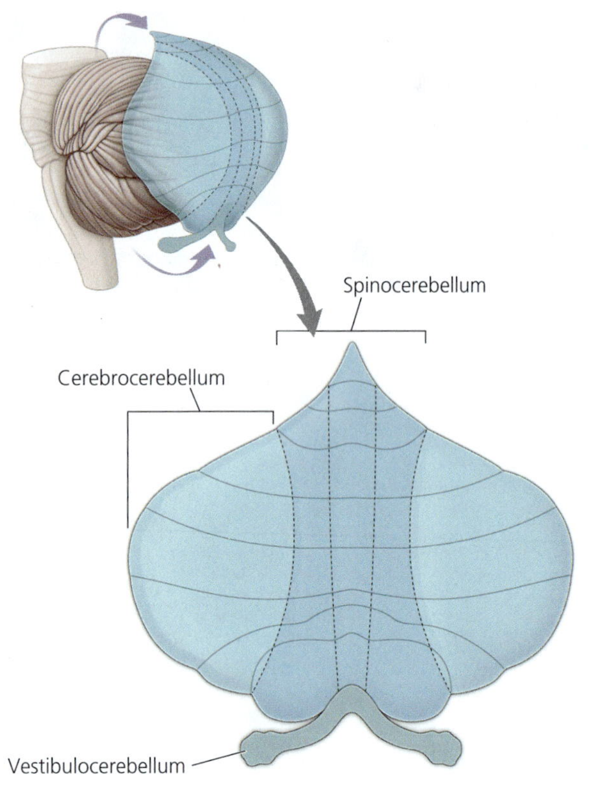

Figure 10.24 Divisions of the cerebellum. The main divisions of the cerebellar cortex are most easily seen when you conceptually unfold the cerebellar cortex and flatten it, as illustrated here.

divisions lie smaller subdivisions that also differ in external connectivity and function, but the details of this organization are controversial and not important for our purposes. The important general concept is that the cerebellum is relatively homogeneous in internal structure but heterogeneous in its external connectivity.

Cerebellar Function: Adaptive Feedforward Control

The question of what the cerebellum "does" has been debated long and hard. Many answers have been proposed (see Box 10.3: The Conditioned Eye Blink Response), but a tentative consensus is gradually emerging: a major function of the cerebellum is **adaptive feedforward control**. The word "adaptive" here simply means capable of being adjusted through learning. To understand the term "feedforward control" you need first to grasp the limitations of feedback control.

Feedback versus Feedforward Control

Feedback is excellent for controlling slow processes, such as heating your house, but it becomes problematic when the process being controlled is faster than the feedback mechanism. For example, if you swerve to avoid an obstacle while driving at highway speeds, then you may oversteer because it takes too long for you to get visual feedback about the car's response to the steering action. Having oversteered, your next response will probably be to countersteer, but this maneuver will likely be exaggerated as well. If so, then you'll careen from one edge of the road to the other. Crucially, the oversteering occurs only when you are driving too fast. At slower speeds you can tell when you have turned the steering wheel enough. In engineering terms, the problem at high speeds is that the process being controlled (the car's motion) is much faster than the feedback mechanism (your ability to detect mismatches between the car's intended and actual trajectories and then to issue appropriate steering commands). In such cases, the controlled variable (the car's direction) begins to oscillate wildly. In other words, control breaks down.

A good solution to the type of problem just described is to employ adaptive feedforward control. This simply means that prior learning adjusts the commands so that the target state is reached with only minimal error. In our example, this amounts to learning from previous mistakes how much you need to turn the wheel, given your current speed, direction, and distance to the obstacle. Having learned to decrease their steering commands as driving speed goes up, experienced drivers are less likely than novices to oversteer. More generally, trial-and-error learning allows a controller to predict the error that would occur if there had been no previous learning and to correct for it before the error arises. This sort of predictive error correction is what engineers call *adaptive feedforward control.*

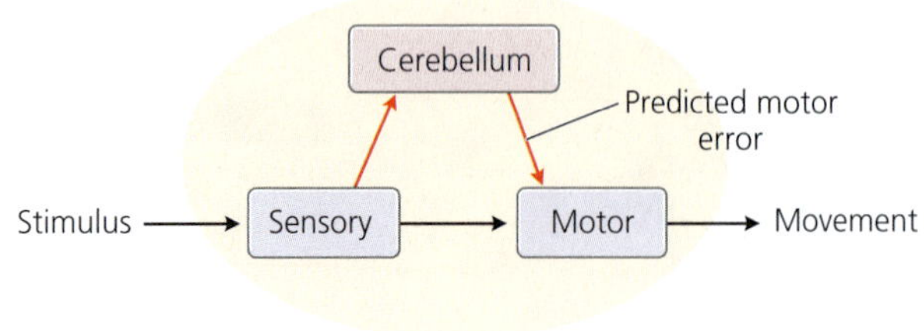

Figure 10.25 The cerebellum fine-tunes reflexes. Parts of the cerebellum form a side-loop (red) to sensorimotor reflex arcs (see also Figure 10.23). Based on past experience, the cerebellum computes the expected motor error and uses this prediction to adjust motor output, preventing the error.

Adaptive Modulation of Sensorimotor Circuits

To understand the cerebellum's role in adaptive feedforward control, consider how the cerebellum forms a side-loop to sensorimotor reflex arcs that pass through precerebellar and deep cerebellar nuclei (Figure 10.25). In essence, the Purkinje cells receive, by way of the granule cells, a copy of the sensory input that drives the sensorimotor reflexes,

NEUROBIOLOGY IN DEPTH

Box 10.3 *The Conditioned Eye Blink Response*

If you repeatedly present an animal with a sound (or other stimulus) ~100 ms before you puff some air into its eye, the animal gradually learns to blink just before the air puff comes. The neural mechanisms underlying this conditioned eye blink response have been studied intensively for over 30 years. The basic circuit for eye blink conditioning to a tone is shown in Figure b10.2. Where in this circuit does the learning take place? In the search for an answer, Richard Thompson and his colleagues arrived at one of the deep cerebellar nuclei, the interpositus nucleus (IPN). Even more interesting than this answer is the series of experiments leading to it.

Especially informative were experiments in which the IPN was inactivated during the conditioning trials. This manipulation prevented the animals from learning the association. After reversing the inactivation, the animals acted as if they had not experienced any previous conditioning trials. Inactivation of the *red nucleus* (Figure b10.2) during the conditioning trials prevented the blinking response; but when the inactivation was reversed, the animals showed an obvious conditioned eye blink response. Therefore, these animals must have learned the association while the red nucleus was inactive. Thompson concluded that the synaptic changes responsible for the conditioned eye blink response must occur either in the IPN or in structures upstream of it.

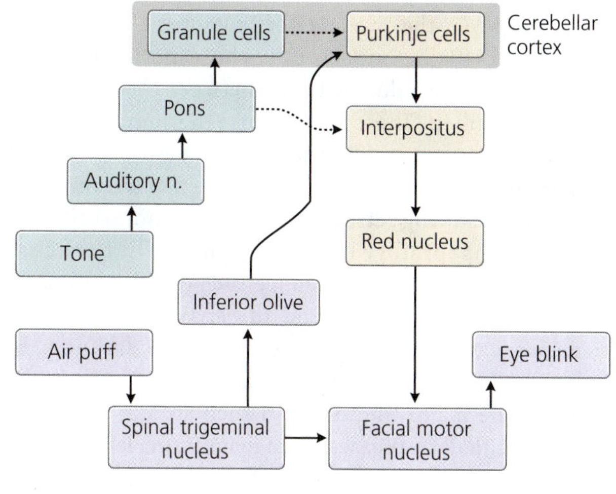

Figure b10.2 The eye blink conditioning circuit. The circuit activated by an air puff is depicted in purple. The pathway responsive to tones is shown in green. The dashed lines indicate connections that are strengthened as a result of the conditioning. Shown in yellow is the pathway by which cerebellar Purkinje cells can activate neurons in the facial motor nucleus.

Additional experiments revealed that the IPN receives information about the tone (the conditioned stimulus) directly from auditory brain regions (see Figure 3.17) and via cerebellar Purkinje cells that project to the IPN. These Purkinje cells also get information about the air puff (the unconditioned stimulus) from a subset of neurons in the inferior olive. Importantly, activation of the inferior olive decreases the strength of any inputs from the granule cells that had just been active. Over many trials, these changes in synapse strength cause the Purkinje cells, which tend to be tonically active, to fall silent in response to the tone. Because the Purkinje cells are inhibitory, their silence leads to an increase in IPN activity, which in turn can trigger an eye blink.

Although these experiments suggest that the memory that gives rise to the conditioned eye blink response is stored in the cerebellar cortex, at the synapses between parallel fibers and Purkinje cells, things are not quite that simple. Several studies have shown that the conditioned eye blink response may persist after lesioning the cerebellar cortex, implying that the conditioned eye blink memory resides in the IPN. Still, without an intact cerebellar cortex, the conditioned eye blink does not emerge. The simplest way to make sense of these data is to suppose that the conditioned eye blink first develops because of plasticity at the parallel fiber synapses but that it then, after a large number of training trials, emancipates itself from the cerebellar cortex because of synaptic changes within the IPN. Such a gradual "transfer" of the memory trace from the cerebellar cortex to the IPN would be consistent with findings on plasticity in the vestibulo-ocular reflex, which also depends on synaptic change in the cerebellar cortex only during the early phases of adaptation.

To reconcile the cerebellum's role in eye blink conditioning with its adaptive feedforward control function, one can think of the air puff signal coming through the inferior olive as an error signal that indicates something went wrong with protecting the eye. In response to the error signal, parallel fiber synapses are adjusted until Purkinje cell activity is modified in such a way that, on future trials, the eye blinks before the air puff comes. The sensorimotor coupling between tone and eye blink may not have been evident before the conditioning, but a weak pathway from the auditory nuclei via the Purkinje cells to the eye blink motor neurons probably existed before conditioning. If this is true, then the training trials may have increased the "gain" of this feedforward pathway to a level where it can trigger a full-blown eye blink in response to the tone.

modify this signal, and then send an inhibitory output "forward" to the motor side of the reflex arc, modulating its output. For simple reflexes, the Purkinje cells can be thought of as controlling the **reflex gain** (how much output you get for a given input). For more complex circuits and behaviors, the Purkinje cells can also alter the temporal pattern of the circuit's activity. In either case, the feedforward circuit through the cerebellum is thought to fine-tune the sensorimotor circuit, optimizing performance.

The feedforward control signal coming from the Purkinje cells is adaptive rather than fixed. It eliminates movement errors by means of trial-and-error learning. The learning in this case is accomplished by adjusting the strength of the synapses between the parallel fibers and the Purkinje cells whenever the system's overall output produced an error. The output error signals are provided by the climbing fiber inputs from the inferior olive. When those climbing fibers fire, the targeted Purkinje cells fire action potentials of their own (complex spikes), which then triggers plasticity at parallel fiber synapses (see Chapter 3).

Importantly, synapse strength is modified only for those parallel fiber synapses that were active shortly before the error signal (the climbing fiber input) arrives at a given Purkinje cell. Thus, change comes only to those synapses that contributed to the error. Over many trials, this (Hebbian) form of synaptic plasticity leads to a set of synapse strengths that eliminates the errors. As long as the system's performance remains error free, the climbing fibers remain silent, no synapses change in strength, and the Purkinje cell output to any given set of parallel fiber inputs remains constant. When errors recur, learning resumes.

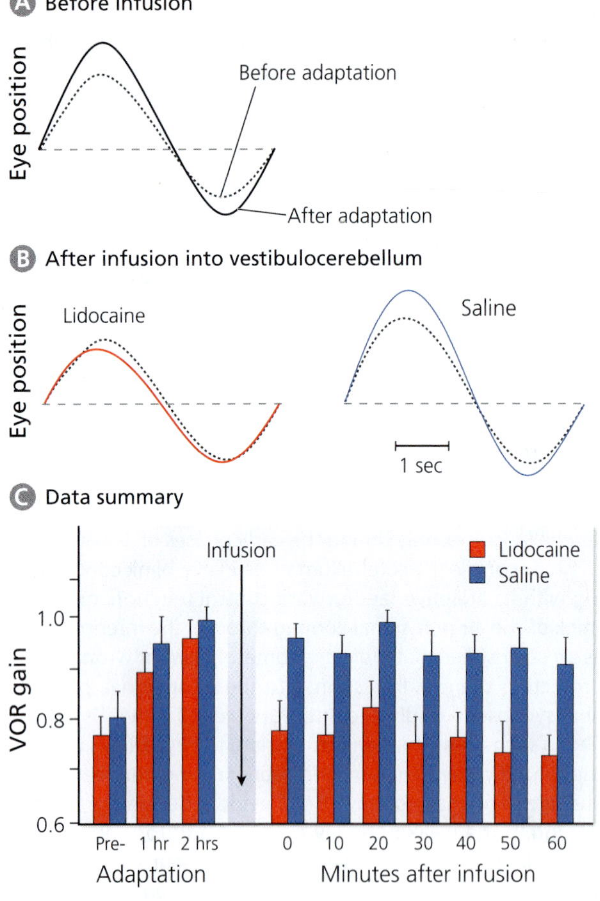

Figure 10.26 Cerebellar modulation of the VOR. Monkeys looked at a checkered stimulus through magnifying lenses while their heads were rotated sinusoidally. Before and after this period of adaptation, the gain of the VOR (vestibulo-ocular reflex) was measured in darkness. As shown in (A), the adaptation increased the speed and amplitude of the eye movements (VOR gain). An infusion of the local anesthetic lidocaine into the vestibulocerebellum of fully adapted animals reversed the adaptation (B). Control injections of saline did not have this effect. The bar graph in (C) provides more detail on the time course of the adaptation and the lidocaine effect. [After Nagao and Kitazawa, 2003]

Fine-tuning the Vestibulo-ocular Reflex (VOR)

The most detailed analyses of cerebellar function have focused on the VOR. As you may recall, the shortest VOR circuit proceeds from the vestibular apparatus through the vestibular complex to the abducens and oculomotor nuclei (see Figure 10.8). The vestibulocerebellum forms a side-loop to this circuit and modulates its activity.

To study VOR modulation experimentally, researchers train monkeys or other animals to look at a patterned screen while the investigators rotate the animal's head back and forth. Importantly, the experimenters place magnifying or reducing lenses between the animal's eyes and the screen. These lenses cause the compensatory eye movements created by the VOR to be too small or too large, respectively. The question is whether the animals will use the resulting retinal error signal to adjust the size of their compensatory eye movements. Will they change the gain of their VOR to reduce and possibly eliminate the retinal image slip? To find the answer, experimenters measure the animal's VOR gain in total darkness, both before the experiment begins and at regular intervals thereafter. The major finding from such studies is that VOR gain gradually increases when magnifying lenses are used (Figure 10.26) and decreases for reducing lenses. When the lenses are removed, the VOR gain gradually returns to normal. Overall, these data indicate that animals recalibrate their VOR whenever its gain is incorrect. This phenomenon is called **VOR adaptation**.

Is the cerebellum required for VOR adaptation? Yes it is, as demonstrated by the observation that cerebellar lesions, specifically of the vestibulocerebellum, abolish VOR adaptation. Is the cerebellum also required to maintain the VOR adaptation after it has been learned? Stated differently, does the cerebellum house the memory of how the VOR gain should be adjusted? To answer this question, scientists trained monkeys to change their VOR gain and then inactivated the cerebellum with lidocaine (a sodium channel blocker). They found that after bilateral inactivation of the vestibulocerebellum, the VOR gain rapidly fell back to its preadaptation level (Figure 10.26). This finding strongly supports the hypothesis that the cerebellum provides a feedforward signal that is required for maintaining the change in VOR gain.

Many questions remain, of course. Particularly interesting is whether adaptive changes in VOR gain can become independent of cerebellar activity. This possibility is raised by the observation that inactivation of the vestibulocerebellum does not affect VOR gain if the animals had been seeing the world through magnifying lenses for several days. Such findings suggest that synaptic plasticity occurs not only in the cerebellar cortex but also, albeit more slowly, at synapses in the direct VOR pathway.

Cerebellar Dysfunction

If the cerebellum provides adaptive feedforward control, then interfering with normal cerebellar function should lead to the sort of oscillations that we discussed earlier in the context of our highway driving example. Indeed, people with cerebellar damage exhibit movement tremors when they try to move their hand toward a target. That is, they move their hand along an oscillating trajectory rather than smoothly. More generally, people with cerebellar damage routinely overshoot or undershoot the goals of their movements (Figure 10.27).

The lack of motor coordination due to cerebellar damage is called **cerebellar ataxia** (*ataxia* means "without order" in Greek). What causes cerebellar damage in humans? There are many possible causes, including hereditary ones, but a particularly interesting one is chronic alcoholism. After all, the symptoms of alcohol intoxication are remarkably similar to those of cerebellar dysfunction. Unfortunately, the effects of alcohol on neurons remain quite poorly understood.

Non-motor Functions of the Cerebellum

Traditionally, the cerebellum has been thought to function mainly in motor control. However, recent studies have shown that the cerebellum is also involved in a variety of non-motor functions. Particularly well supported is the notion that cerebellar circuits help organisms discriminate sensations that are caused by their own movements from sensations that are caused by external factors.

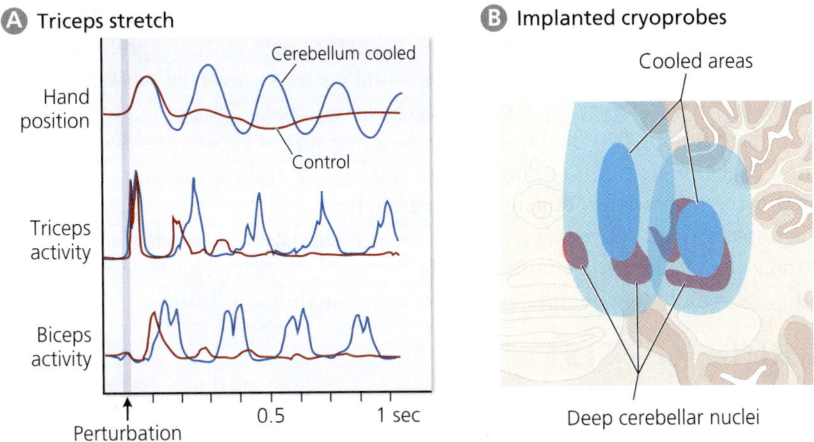

Figure 10.27 Cerebellar inactivation causes movement tremors. Cebus monkeys were trained to hold a handle steady in response to a briefly applied external force (perturbation). The monkeys were also implanted with cryoprobes that could temporarily cool the deep cerebellar nuclei, thus silencing most cerebellar output. Shown in (A) is the response to a perturbation that stretches the triceps and therefore triggers a triceps stretch reflex. When the cerebellum is cooled, the hand oscillates back and forth repeatedly (blue trace), mainly because biceps activation occurs too late to counteract the triceps contraction, and vice versa. Shown in (B) is the location of the two implanted cryoprobes. The tips of the cryoprobes are depicted in dark blue, the cooled region in light blue. [After Vilis and Hore, 1980]

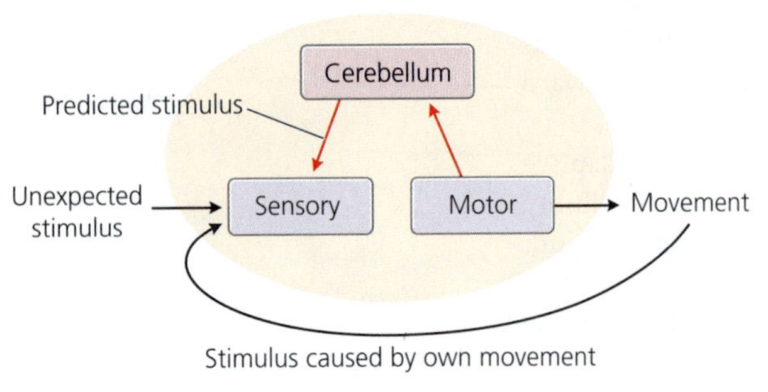

Figure 10.28 Filtering out expected sensory signals. Some parts of the cerebellum receive copies of motor commands, compute what kinds of sensory signals the movements should generate, and then send this information to sensory structures, which use it to filter out the self-generated sensory input. As a result, sensitivity to unexpected signals is enhanced.

Predicting Self-generated Sensory Input

Sharks and a few other fishes use electroreceptors in their skin (see Chapter 6, Box 6.3) to detect weak electric fields that are generated by the muscles of moving aquatic prey. A major problem in detecting these prey-generated signals is that they are contaminated by electrical noise from the predator's own muscular activity, such as the movements of its gills. To solve this signal contamination problem, the fishes send a copy of their motor commands (notably the signals commanding gill movements) to the cerebellum. The cerebellum then uses this **efference copy** to generate a prediction of the sensory signals that should (based on prior experience) result from the self-generated movement (Figure 10.28). The sensory prediction is sent to the electrosensory system where it is subtracted from the actual sensory input to isolate the unpredicted input, which in the case of these fishes is the signal generated by the moving prey.

Humans also seem to use their cerebellum to predict self-generated sensory inputs. Supporting this hypothesis are functional brain imaging data showing that activity in the cerebellar cortex is lower when sensory input matches what subjects expect, given their own movement, than when it does not match. The function of this sensory prediction error signal remains controversial, but it is most likely used to modify sensory predictions on subsequent trials.

At this point you may ask yourself, does my brain really filter out the sensations my own movements will cause? Yes, to a large extent it does. For example, the ability to predict the sensory consequences of your own actions is thought to explain why tickling yourself is difficult. If another person tickles you, the sensations are unpredictable and, therefore, difficult to filter out. In contrast, when you tickle yourself, the sensations are predictable and therefore less intense.

The Cerebellum's Role in Cognition

Although cognitive symptoms after cerebellar lesions are subtle and variable, there is no doubt that the cerebellum is intimately connected with brain regions that are crucial for cognition. Specifically, the lateral part of the cerebrocerebellum has topographic projections to a part of the dorsal thalamus that in turn projects to the prefrontal cortex. As you will learn in Chapter 15, the prefrontal cortex is involved working memory, problem solving, and a variety of other cognitive functions. The cerebellum's contribution to these functions probably involves the same sort of experience-dependent fine-tuning that the cerebellum contributes to sensory and motor functions. Jeremy Shmahmann, an early proponent of the cerebellum's role in cognition, stated the hypothesis well: "In the same way as the cerebellum regulates the rate, force, rhythm, and accuracy of movements, so may it regulate the speed, capacity, consistency, and appropriateness of mental or cognitive processes" (Schmahmann, 1991).

BRAIN EXERCISE

People who grow up without a cerebellum (a rare condition!) may exhibit surprisingly normal behavior. How would you explain such observations, and what kinds of experiments would you perform to test whether such patients have, after all, some subtle deficits?

10.7 How Do the Motor Systems Interact?

It is tempting to think that the spinal cord and medulla are responsible for postural reflexes, that the motor cortex commands all conscious voluntary movements, and that the cerebellum is required for motor learning. However, the division of labor between these regions is hardly precise. Most movements are shaped by interactions between multiple brain areas.

Consider what happens in the brain and spinal cord when you pull on the handle of a heavy door. This is a conscious voluntary movement that activates projections from the motor cortex to hand and arm motor neurons in the spinal cord. However, 60 ms before the arm starts to pull on the door, the motor cortex instructs your calf muscles to contract and tilt your body backward. Without this unconscious backward shift, pulling on the door handle would simply pull you toward the door. The motor cortex also modulates some spinal reflexes as it performs the posture adjustment. In our example, it modifies the calf muscle stretch reflex to stiffen the ankle joints. Meanwhile, the cerebellum may receive error signals, which it can use to adjust the motor commands the next time you want to open this same door.

The importance of the motor cortex and other brain regions in modulating spinal reflexes becomes quite clear after spinal cord injury. Transection of the upper spinal cord removes all descending inputs to spinal neurons below the site of injury, triggering a state of **spinal shock**. During this period, all vegetative functions below the site of injury are lost (see Chapter 9, Box 9.3) and the limbs go limp. The **flaccid paralysis** occurs because the various stretch, withdrawal, and crossed extensor reflexes need descending inputs from the brain to function normally. Those spinal reflexes come back after the spinal shock has passed, but they are exaggerated and uncontrolled because the descending inputs do not regenerate. Exaggeration of the spinal reflexes makes the limbs abnormally rigid (recall that stretch reflexes are homeostatic) and triggers excessive movements in response to limb perturbation or skin stimulation. Neurologists call these symptoms **spasticity** (from the Greek word *spasmos*, which refers to an involuntary muscle contraction). Collectively, these consequences of spinal cord injury illustrate how much control the brain normally exerts over the spinal reflexes.

Motor learning likewise involves multiple interacting systems. When you first learn a challenging skill, your movements tend to be awkward and stiff, and they require focused attention. Functional brain imaging at this early stage of learning reveals increased activity in many brain regions, including the motor and prefrontal cortices. As learning proceeds and the movements become more automatic, the activity in several of these areas goes down. Intriguingly, the cerebellum becomes more important (although not necessarily more active) as the movements become more automatic. For example, people with cerebellar damage find it difficult to learn two complex motor tasks in a row. Most of us can learn a second task once the first one has become automatic, but people with cerebellar damage cannot learn the second task without losing their ability to perform the first task. Each task requires their undivided attention. This difficulty with *movement automatization* is evident from this quote by a patient with damage in the right half of his cerebellum: "The movements of my left arm are done unconsciously, but I have to think out each movement of the right (affected) arm" (Holmes, 1939, p. 22).

We can conclude that motor control involves the coordinated activity of multiple brain regions whose balance of involvement changes as movements become increasingly automatic. As you will learn in Chapter 15, motor control also involves the striatum and several other brain regions that we have not yet discussed. Thus, the neurons that control our movements comprise a widely distributed, but highly interactive, neural system.

BRAIN EXERCISE

When people say that spinal cord injury left a person "paralyzed from the waist down," does this mean that the legs of such a patient cannot move?

SUMMARY

Section 10.1 - A reflex is an involuntary, stereotyped response to a specific stimulus. Pupillary constriction in response to light is a reflex that occurs against a background of sympathetic activation promoting pupillary dilation.

Section 10.2 - Animals are equipped with numerous defensive reflexes. Even tears are protective.

- Blinking in response to corneal stimulation involves a circuit through spinal trigeminal and facial motor nuclei.
- Stepping on a thumbtack triggers leg withdrawal, which causes leg flexors to contract and leg extensors to relax. The circuit goes through the spinal cord and includes one set of inhibitory interneurons.

Section 10.3 - Animals have reflexes that stabilize their body parts relative to one another and the outside world.

- The knee jerk response is one of many muscle stretch reflexes, which cause stretched muscles to contract. Stretch reflexes counteract gravity and stabilize the limbs against varying loads.
- The optokinetic and vestibulo-ocular reflexes use visual and vestibular signals, respectively, to generate eye movement commands that minimize retinal image slip when the head moves relative to the world.
- The optocollic and vestibulocollic reflexes use visual and vestibular signals, respectively, to generate neck muscle commands that stabilize the head when the body moves relative to the world.
- When you lift one leg, the crossed extensor reflex stiffens the other leg, thus helping to stabilize the body. Postural reflexes help to realign the body's center of mass with its center of pressure.
- Many reflexes can be strengthened or weakened, or inhibited entirely, by descending inputs.

Section 10.4 - The rhythmic neural activity that is central to locomotion generally results from the activity of central pattern generators rather than sensorimotor reflex chains.

- Single neurons can exhibit rhythmic activity, as can small networks of neurons. A half-center oscillator consists of two neurons (or sets of neurons) that alternately inhibit each other.
- The central pattern generator for swimming in fish can be entrained by rhythmic tail movements.
- The spinal pattern generator for walking receives modulatory input from the midbrain locomotor area.

Section 10.5 - The motor cortex includes several cortical areas in the posterior portion of the frontal lobe.

- Corticospinal projections are mainly contralateral, whereas corticobulbar projections are more bilateral.
- Stimulating the motor cortex with short stimulus trains elicits muscle twitches and reveals a fractured motor map. Stimulation with longer stimuli reveals more complex, organized movements.
- Most movements are encoded by large populations of cortical neurons, all working together.
- Mirror neurons fire when a specific hand movement is made as well as when the same movement is observed.
- Learning to perform difficult movements is accompanied by motor cortex plasticity.

Section 10.6 - The cerebellum modulates involuntary movements as well as more complex behaviors.

- The cerebellum's internal circuitry is uniform, but its extrinsic connections vary across the cerebellum.
- The cerebellum uses trial-and-error learning to modulate circuits that control movement. Lesions of the cerebellum remove its adaptive feedforward control function, creating tremors and instability.
- The cerebellum helps organisms predict the sensory consequences of their own movement and fine-tune thoughts.

Section 10.7 - Removal of descending inputs to the spinal cord leads to spinal shock, followed by exaggerated reflexes.

Box 10.1 - The use of animals in research is highly regulated, focused mainly on rodents, and minor compared to the use of animals for food.

Box 10.2 - Locked-in syndrome is characterized by almost total paralysis, although the mind continues to work.

Box 10.3 - Learning to blink in anticipation of an air puff to the cornea involves cerebellar plasticity.

KEY TERMS

reflex 302
pupillary light reflex 302
pupillary dilation 303
facial motor nucleus 304
leg withdrawal reflex 304
reciprocal innervation 305
irradiation 305
knee jerk response 306
quadriceps stretch reflex 306
biceps stretch reflex 307
retinal image slip 308
optokinetic reflex 308
nucleus of the optic tract 308
vestibular complex 308
abducens nucleus 308

lateral rectus 308
oculomotor nucleus 309
medial rectus 309
vestibulo-ocular reflex 309
vestibulocollic reflex 311
optocollic reflex 311
crossed extensor reflex 312
reflex chain hypothesis 314
central pattern generator hypothesis 314
lamprey 314
half-center oscillator 315
lateral undulation 316
fictive swimming 317
fictive walking 318
midbrain locomotor area 318
primary motor cortex (M1) 320
premotor cortex 320
supplementary motor area 320
corticospinal 320
corticobulbar 320
epileptic seizure 321
homunculus 321
population code 324
mirror neurons 326
constraint-induced movement therapy 326
Purkinje cell 327
cerebellar granule cell 327
parallel fibers 327
inferior olive 327
climbing fiber 327
deep cerebellar nuclei 327
vestibulocerebellum 327
spinocerebellum 327
cerebrocerebellum 327
adaptive feedforward control 328
reflex gain 329
VOR adaptation 330
cerebellar ataxia 331
efference copy 332
spinal shock 333
flaccid paralysis 333
spasticity 333

ADDITIONAL READINGS

10.1 - Defining a Reflex

Windholz, G. 1995. Pavlov on the conditioned reflex method and its limitations. *Am J Psychol* **108**:575–588.

10.2 - Defensive Reflexes

Pellegrini JJ, Horn AK, Evinger C. 1995. The trigeminally evoked blink reflex. I. neuronal circuits. *Exp Brain Res* **107**:166–180.

Schouenborg J, Kalliomäki J. 1990. Functional organization of the nociceptive withdrawal reflexes. I. Activation of hindlimb muscles in the rat. *Exp Brain Res* **83**:67–78.

10.3 - Stabilizing Reflexes

Massion J. 1992. Movement, posture and equilibrium: interaction and coordination. *Prog Neurobiol* **38**:35–56.

Nashner LM. 1982. Adaptation of human movement to altered environments. *Trends Neurosci* **5**:358–361.

10.4 - Locomotor Control

Dimitrijevic MR, Gerasimenko Y, Pinter MM. 1998. Evidence for a spinal central pattern generator in humans. *Ann N Y Acad Sci* **860**:360–376.

Frigon A, Rossignol S. 2006. Experiments and models of sensorimotor interactions during locomotion. *Biol Cybern* **95**:607–627.

Grillner S, Jessell TM. 2009. Measured motion: searching for simplicity in spinal locomotor networks. *Curr Opin Neurobiol* **19**:572–586.

Grillner S, Wallén P. 1982. On peripheral control mechanisms acting on the central pattern generators for swimming in the dogfish. *J Exp Biol* **98**:1–22.

Kiehn O. 2006. Locomotor circuits in the mammalian spinal cord. *Ann Rev Neurosci* **29**:279–306.

Parker D. 2006. Complexities and uncertainties of neuronal network function. *Philos Trans R Soc Lond B Biol Sci* **361**:81–99.

Spitzer NC. 2010. How GABA generates depolarization. *J Physiol (Lond)* **588**:757–758.

Zehr EP, Duysens J. 2004. Regulation of arm and leg movement during human locomotion. *Neuroscientist* **10**:347–361.

10.5 - The Motor Cortices

Canedo A. 1997. Primary motor cortex influences on the descending and ascending systems. *Prog Neurobiol* **51**:287–335.

Heyes C. 2010. Where do mirror neurons come from? *Neurosci Biobehav Rev* **34**:575–583.

Hickok G. 2009. Eight problems for the mirror neuron theory of action understanding in monkeys and humans. *J Cogn Neurosci* **21**:1229–1243.

Jackson A, Mavoori J, Fetz EE. 2006. Long-term motor cortex plasticity induced by an electronic neural implant. *Nature* **444**:56–60.

Monfils M-H, Plautz EJ, Kleim JA. 2005. In search of the motor engram: motor map plasticity as a mechanism for encoding motor experience. *Neuroscientist* **11**:471–483.

Nakajima K, Maier MA, Kirkwood PA, Lemon RN. 2000. Striking differences in transmission of corticospinal excitation to upper limb motor neurons in two primate species. *J Neurophysiol* **84**:698–709.

Schott GD. 1993. Penfield's homunculus: a note on cerebral cartography. *J Neurol Neurosurg Psychiatry* **56**:329–333.

Scott SH. 2008. Inconvenient truths about neural processing in primary motor cortex. *J Physiol* **586**:1217–1224.

Stewart L. 2008. Do musicians have different brains? *Clin Med* **8**:304–308.

10.6 - The Cerebellum

Bastian AJ. 2006. Learning to predict the future: the cerebellum adapts feedforward movement control. *Curr Opin Neurobiol* **16**:645–649.

Beaton A, Mariën P. 2010. Language, cognition and the cerebellum: grappling with an enigma. *Cortex* **46**:811–820.

Bell CC. 2001. Memory-based expectations in electrosensory systems. *Curr Opin Neurobiol 11*:481–487.

Buckner RL. 2013. The cerebellum and cognitive function: 25 years of insight from anatomy and neuroimaging. *Neuron* **80**:807–815.

Dean P, Porrill J, Ekerot C-F, Jörntell H. 2010. The cerebellar microcircuit as an adaptive filter: experimental and computational evidence. *Nat Rev Neurosci* **11**:30–43.

Ito M. 2006. Cerebellar circuitry as a neuronal machine. *Prog Neurobiol* **78**:272–303.

10.7 - Interactions between Multiple Motor Systems

Jacobs JV, Horak FB. 2007. Cortical control of postural responses. *J Neural Transm* **114**:1339–1348.

Prochazka A. 1989. Sensorimotor gain control: a basic strategy of motor systems? *Prog Neurobiol* **33**:281–307.

Yarrow K, Brown P, Krakauer JW. 2009. Inside the brain of an elite athlete: the neural processes that support high achievement in sports. *Nat Rev Neurosci* **10**:585–596.

Boxes

Animal Care Annual Report of Activities for Fiscal Year 2007. http://www.aphis.usda.gov/publications/animal_welfare/

Statistics of Scientific Procedures on Living Animals: Great Britain 2008. https://www.gov.uk/government/statistics/statistics-of-scientific-procedures-on-living-animals-great-britain-2008

Thompson RF, Steinmetz JE. 2009. The role of the cerebellum in classical conditioning of discrete behavioral responses. *Neuroscience* **162**:732–755.

Wolpaw JR, Birbaumer N, McFarland DJ, Pfurtscheller G, Vaughan TM. 2002. Brain-computer interfaces for communication and control. *Clin Neurophysiol* **113**:767–791.

Localizing Stimuli and Orienting in Space

CHAPTER 11

CONTENTS

FEATURES

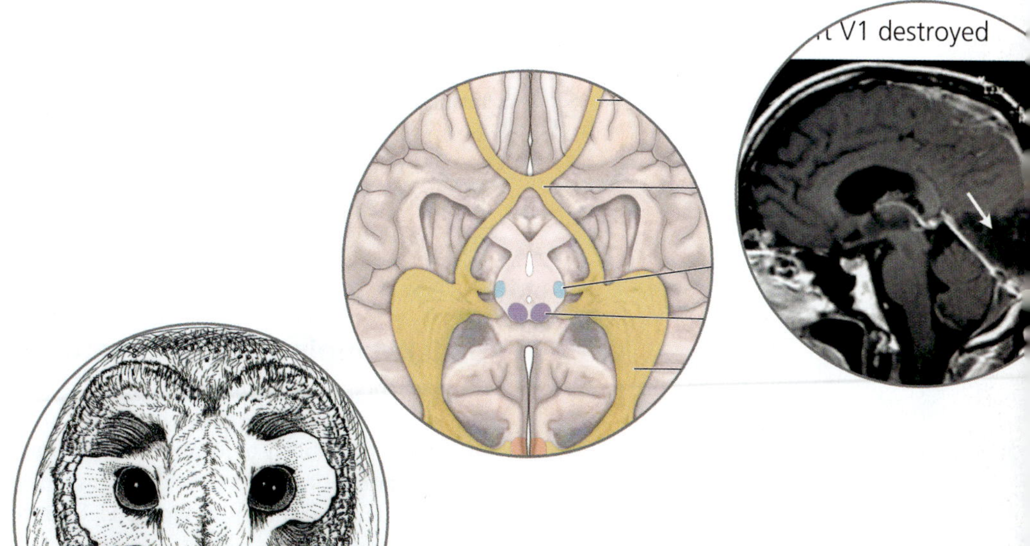

In Chapter 10, you learned about various reflexes and forms of locomotion. The present chapter builds on this knowledge by exploring how movements can be oriented toward or away from stimuli in the external environment. We examine how organisms localize the sources of external stimuli and then make eye, head, or hand movements toward those objects. We will also consider the related problem of how animals navigate through the world to find food or other valuable resources.

The task of localizing stimuli and objects seems simple enough. For example, the optics of the eye ensure that light from an object strikes photoreceptors in a predictable portion of your retina. Given this predictability, you can determine the location of that light source simply by determining which retinal neurons responded to the stimulus. In essence, the visual world is "mapped" onto the retina, although the retinal image is inverted. Somatosensory stimuli likewise seem easy to localize. For example, if sensors in your right big toe become active, then something touchable is most likely down there. In these cases, localization is relatively easy because both the retina and the skin are sensory surfaces that directly encode the spatial origin of stimuli. To localize the stimulus, one needs simply to determine where on the sensory surface the activated sensors are.

On closer inspection, however, the problem of stimulus localization becomes more complicated. In the auditory system, for example, stimulus location is not mapped onto the sensor array. Instead, the position of activated hair cells within the cochlea reflects the pattern of sound frequencies. Therefore, the location of sound sources must be computed in the brain. Other problems arise when organisms move. When you move your eyes, for example, light from a stationary object moves across the retina, but you do not perceive the stimulus object as having moved. Nor do sound sources appear to move whenever you turn your head. Or consider what happens as you walk around town? The world around you appears relatively stable, but the stimuli coming from objects in that world are constantly changing their spatial relationships to you and your sensor arrays. The more you think about this, the more you realize that a great deal of spatial information is not directly accessible to the nervous system. It must be constructed in the brain.

11.1 How Do the Somatosensory and Visual Systems Encode Space?

As you learned in earlier chapters, all sensors "prefer" certain stimuli over others. Photoreceptors, for example, are tuned to different wavelengths of light. What we have not yet discussed in depth is that many sensors and sensory neurons respond to their preferred stimuli much better when those stimuli originate from a specific region of space. For example, retinal neurons respond most strongly when light (or dark) spots are presented at specific locations in space. Similarly, touch-sensitive neurons respond preferentially to touches in specific regions of the body surface.

Neuroscientists quantify this kind of spatial selectivity by determining a cell's **spatial receptive field**, defined as the region of physical space in which stimuli elicit robust neuronal responses. Some spatial receptive fields are small, others are large; some are relatively fixed, while others are quite context dependent. You will learn more about the fine structure of spatial receptive fields in Chapter 12. For now, the important point is that most sensory neurons, especially in the visual and somatosensory systems, have spatial receptive fields, and this spatial selectivity helps organisms localize objects in space.

Spatial Mapping in the Somatosensory System

Touch sensors in the skin (see Chapter 7) are more densely packed in the fingers, toes, and face than in the back. This variation in sensor density across the skin correlates with your ability to localize touch stimuli. When someone touches you on the back, for example, two successive touches spaced less than 10 mm apart feel as if they were applied to the same spot (Figure 11.1). In contrast, spatial separations of 1–2 mm

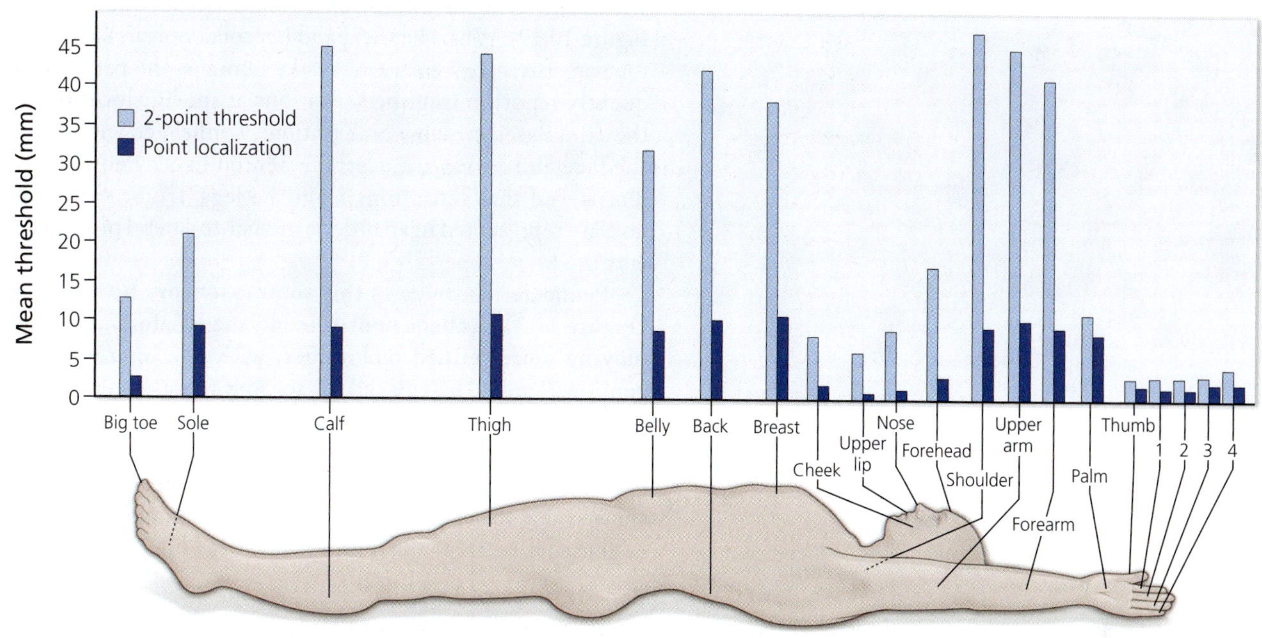

Figure 11.1 Spatial resolution varies across the skin. To measure spatial resolution in the somatosensory system, one can touch the skin twice in succession at the same or different locations and ask how small a spatial separation between the stimuli can be detected (point localization threshold). Alternatively, one can apply two stimuli simultaneously and ask how far they have to be apart to be discriminable (2-point touch threshold). For both measures, the thresholds are much lower in the fingers, face, and toes than in the rest of the body. [After Weinstein, 1968]

are readily discriminable if those touches are applied to the upper lip. In real life, this means that a fly will be much easier to localize if it lands on your face or hands than if it lands on your backside. This is a common theme: our ability to localize sensory stimuli is better in some regions of space than in others. Put differently, sensory systems disproportionately represent external space, overrepresenting some locations at the expense of others. In the human somatosensory system, the most overrepresented regions are the hands and face.

The Ascending Somatosensory Pathway

As we discussed in Chapter 7, touch sensors in the trunk and limbs project into the spinal cord, terminate in the dorsal horn, but also send a long axon branch to the **dorsal column nuclei** of the caudal medulla (see Figure 7.7). This ascending projection is topographically organized so that information from the legs and lower trunk is conveyed to medial parts of the dorsal column nuclei, whereas information from the arms and upper trunk is represented laterally. Moreover, adjacent neurons in the dorsal column nuclei tend to have adjacent spatial receptive fields, forming a somatotopic map. Touch sensors in the face and mouth project to the **principal trigeminal nucleus**, instead of the dorsal column nuclei, but the principal trigeminal nucleus is also somatotopically organized.

Axons coming out of the dorsal column and principal trigeminal nuclei cross the midline and terminate in the contralateral **ventral posterior nucleus** of the dorsal thalamus. This nucleus, in turn, projects to the ipsilateral **primary somatosensory cortex** (S1; Figure 11.2). Thus, somatosensory information from each side of the body is conveyed to the contralateral somatosensory cortex. No one really knows why this pathway is crossed, but the pathways descending from the limb-controlling regions of the motor cortex are likewise crossed. Because of this arrangement, each cerebral hemisphere receives sensory inputs from the side of the body whose movements it controls.

The primary somatosensory cortex contains a map of the entire body surface. This body map was explored systematically by Wilder Penfield, whom you may

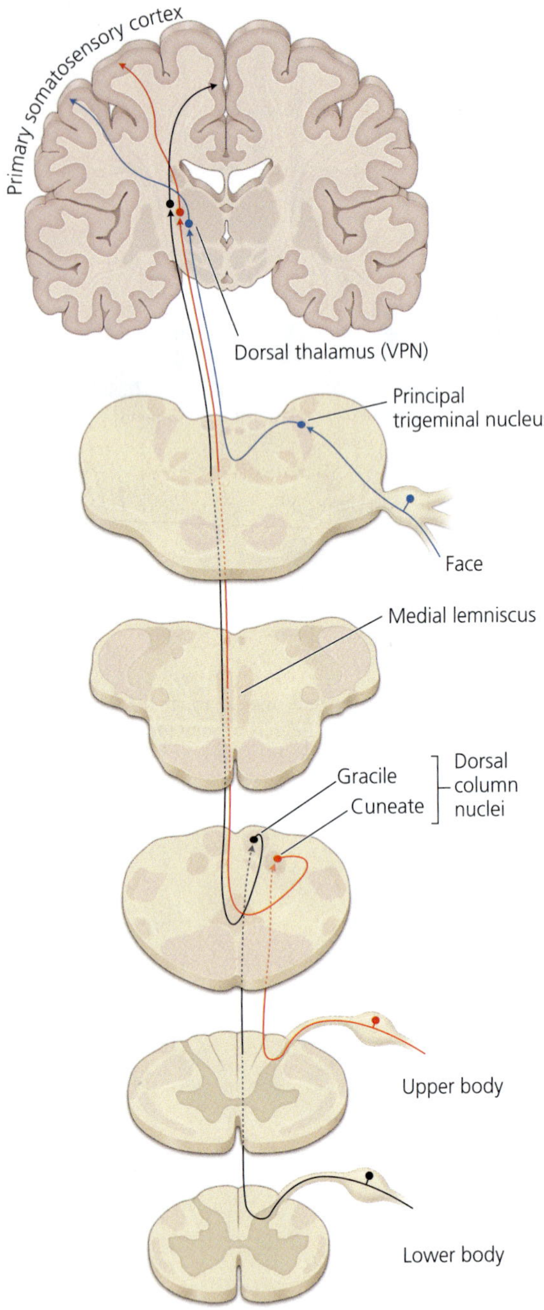

Figure 11.2 The ascending somatosensory pathway. Sensory axons innervating the skin of the trunk and limbs ascend through the spinal cord and terminate in the dorsal column nuclei, whose axons cross to the other side and ascend to the thalamic ventral posterior nucleus (VPN). Axons innervating facial skin project to the principal trigeminal nucleus, which projects to a more medial portion of the contralateral VPN. Thus, information from the trunk, legs, and face all converges in the VPN. From there, the information is sent to the primary somatosensory cortex.

remember from our discussion of the motor homunculus (see Figure 10.19). When Penfield and his collaborators stimulated the somatosensory cortex of awake humans, the patients frequently reported tingling sensations at specific locations on the skin. Based on these observations, Penfield concluded that the face and hands are overrepresented in S1. Penfield also discovered that sensations from the legs, trunk, arms, and face are represented in an orderly medial-to-lateral progression within S1.

Penfield's discovery of this **somatosensory homunculus** (Figure 11.3) has been confirmed by many later studies employing more refined techniques, with one interesting exception. Penfield had placed the sensory representation of the genitals at the most medial edge of the S1 homunculus, just beyond the toes. If you think about it, this placement is anatomically not quite correct. Indeed, subsequent fMRI research showed that the genitals are represented in S1 just where they ought to be, between the hip and upper leg.

Barrel Cortex of Rodents

The most intensively studied portion of S1 is the rodent **barrel cortex**, which is so named because it contains barrel-shaped clusters of cells. These cortical barrels are most obvious when you section the cortex parallel to its surface (Figure 11.4). In such sections, each cortical barrel appears as a hollow ring of neuronal cell bodies, just as a sectioned wine barrel would be a set of hollow rings. Neurophysiological studies in the 1970s established that each cortical barrel processes sensory information from a specific whisker on a rodent's snout. Some of these whiskers are large and are arranged in distinct columns and rows; others are smaller and more randomly distributed. All of them encode information derived from mechanosensory nerve endings that wind around the whisker's base (a hair follicle; see Chapter 7) and fire when the whisker is bent. Rodents and other whisker-bearing animals can use this information to determine a tunnel's width, for example. As you can imagine, this sort of whisker-based sensing is especially useful in total darkness.

Functional imaging data have confirmed that a rat's whiskers are represented topographically in the barrel cortex. Bending a specific whisker tends to activate a specific cortical barrel, and bending adjacent whiskers activates adjacent cortical barrels. Despite this nice topography, it is important to note that the activity elicited in a cortical barrel by bending a single whisker tends quickly to spread to other cortical barrels. Such spreading activity challenges the simple notion that stimuli applied to any body part activate only that body part's corresponding spot in the S1 homunculus. However, the early phase of an activity increase within S1 can still provide valuable information about stimulus location, especially for weak, brief stimuli.

Lateral Inhibition and Topography

As fascinating as it is to discover a little homunculus (or a "rattunculus") in the somatosensory cortex, it is important to ask why such sensory maps exist in the first place. Why is the cortical representation of the skin so nicely somatotopic rather than scrambled like one of Picasso's cubist paintings? The answer probably relates to the

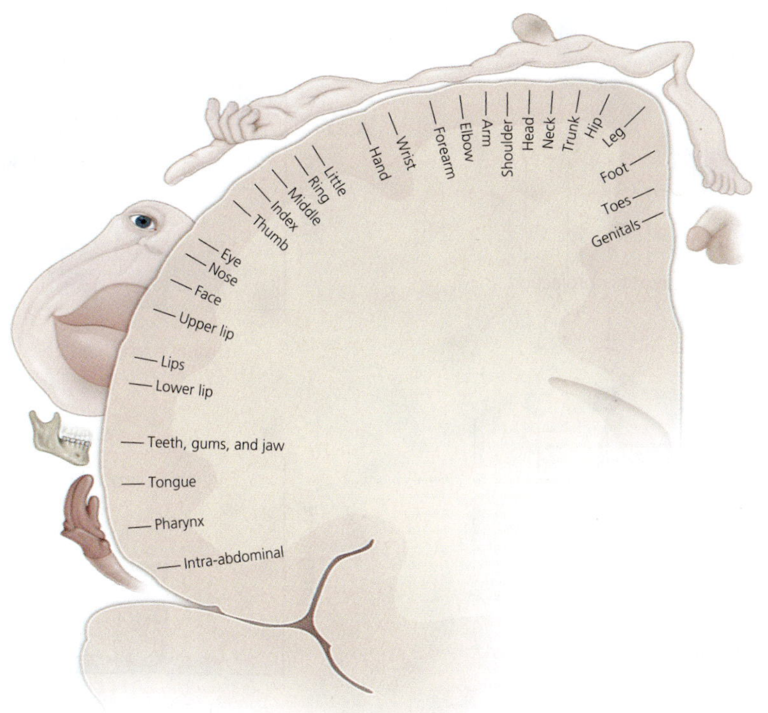

Figure 11.3 The somatosensory homunculus. Penfield and Rasmussen (1950) analyzed skin sensations evoked by electrical stimulation of the somatosensory cortex. They found that the location of tingling sensations varied systematically across the somatosensory cortex and that the face and hands were overrepresented. They summarized their findings by drawing a "somatosensory homunculus" stretched over the somatosensory cortex. [After Penfield and Rasmussen, 1950]

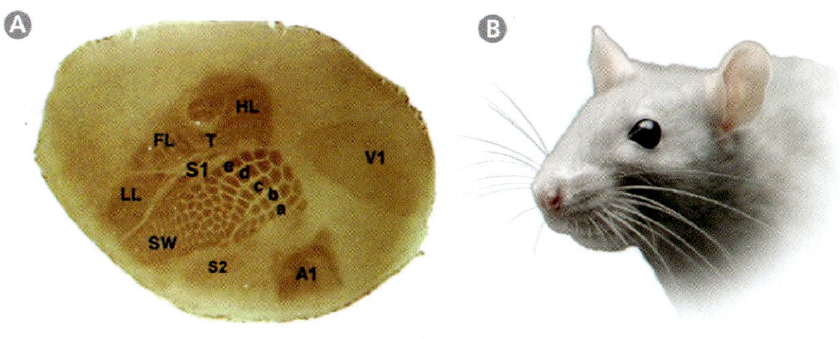

Figure 11.4 Barrel cortex of rodents. Shown in (A) is a flattened and sectioned rat cortex, treated with an antibody against a serotonin transporter to stain sensory cortices brown. You can discern the primary visual cortex (V1), primary auditory cortex (A1), and the secondary somatosensory cortex (S2). The primary somatosensory cortex (S1) comprises regions representing the hindlimb (HL), trunk (T), forelimb (FL), lower lip (LL), small whiskers (SW), and large whiskers (5 rows labeled a–e). An image of a rat's whiskered head is shown in (B). [A from Maier et al., 1999]

phenomenon of **lateral inhibition** (surround inhibition), which is the tendency for excited neurons to inhibit their neighbors.

Lateral inhibition is typically accomplished by the axon collaterals of excitatory neurons that project to nearby inhibitory neurons, which then suppress activity in the excitatory neurons next to them (Figure 11.5). Given this sort of wiring, any excitatory neuron is capable of suppressing other excitatory neurons in its vicinity, but only if it can overcome the reciprocal inhibitory influence of those same neighbors. In essence, the excitatory neurons are always trying to inhibit each other, but only the strongest wins (much as one loudmouth in class can prevent anyone else from speaking up). The end effect is that weak excitation is quickly suppressed if there is a strongly excited neuron nearby. Importantly, lateral inhibition within a somatotopic map, such as S1, allows neurons that are strongly excited by stimulation of one spot

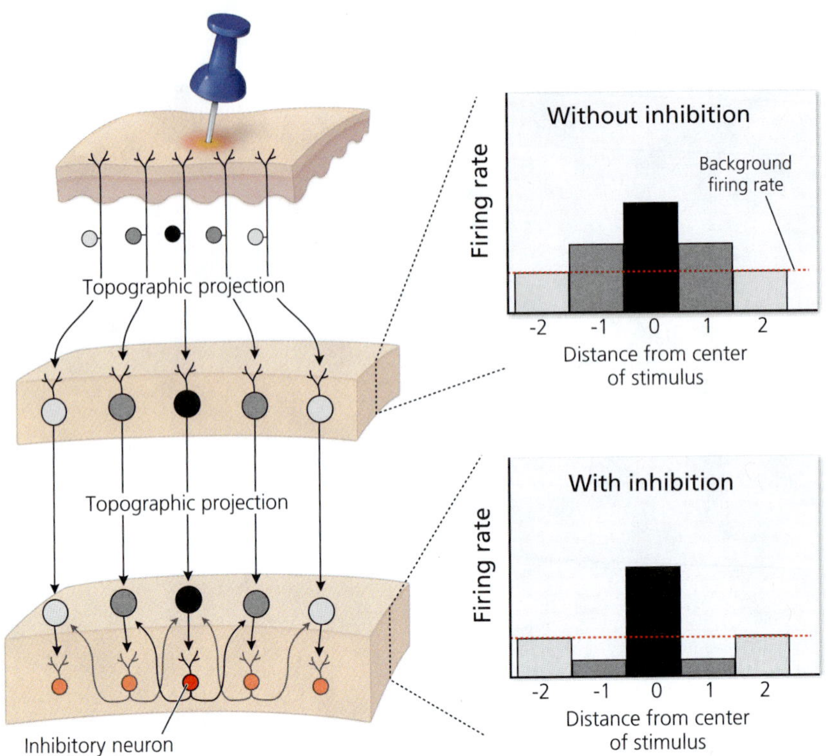

Figure 11.5 Lateral inhibition improves stimulus localization. Shown here is a simplified topographic pathway from the skin to a brain region containing only excitatory neurons and, from there, to a second brain region that contains excitatory neurons as well as inhibitory neurons (red/orange) with laterally directed axons. A thumbtack applied to the skin strongly stimulates nerve endings near the tip of the pin but also triggers weak activity in nearby axons. This pattern of activity is maintained in the first stage of the pathway, as illustrated in the upper bar graph. In the second stage, the strong central excitation activates local inhibitory neurons, which then suppress activity in nearby neurons. This lateral inhibition narrows the stimulus representation (lower bar graph), thereby improving stimulus localization.

on the skin to suppress neighboring neurons that are weakly excited. Thus, the lateral inhibition sharpens the edges of the stimulus representation within the somatotopic map (analogous mechanisms are used for image sharpening in digital photography). This sharpening, in turn, allows for more precise stimulus localization.

If you think about it, lateral inhibition can theoretically exist without topography. However, in a nontopographic representation, long and precisely targeted connections would be needed to interconnect neurons with adjacent spatial receptive fields (recall Figure 6.28). In contrast, in a topographic map, short axon collaterals between neighboring neurons suffice to generate lateral inhibition. This feature minimizes axonal wiring and is developmentally simple to build.

Spatial Mapping in the Visual System

The retina resembles the skin insofar as its sensors, the photoreceptors, directly encode stimulus location and are more tightly packed in some regions than others. Because photoreceptors are packed especially densely in the retina's fovea (see Chapter 6), the region of space at which your fovea is aimed is disproportionately represented in the retina and the rest of the visual system. Defining the **visual field** as the region of space from where visual stimuli can reach your retina, and the center of the visual field as the location where the foveae (plural of fovea) are aimed, we can say that the central few degrees of your visual field are overrepresented in the visual system. This explains why you aim your foveae at locations in space where you want to make out fine details. The peripheral retina has far less spatial resolution.

The Main Ascending Visual Pathways

The retina's output neurons, the retinal ganglion cells, project to multiple targets in the brain. You already read about some of these targets in earlier chapters. Specifically, you learned about the suprachiasmatic nucleus, which regulates circadian rhythms (see Chapter 9); about the olivary pretectal nucleus, which plays a role in pupillary constriction (see Figure 10.1); and about the nucleus of the optic tract, which helps to minimize retinal image slip (see Figure 10.7). For these functions, stimulus

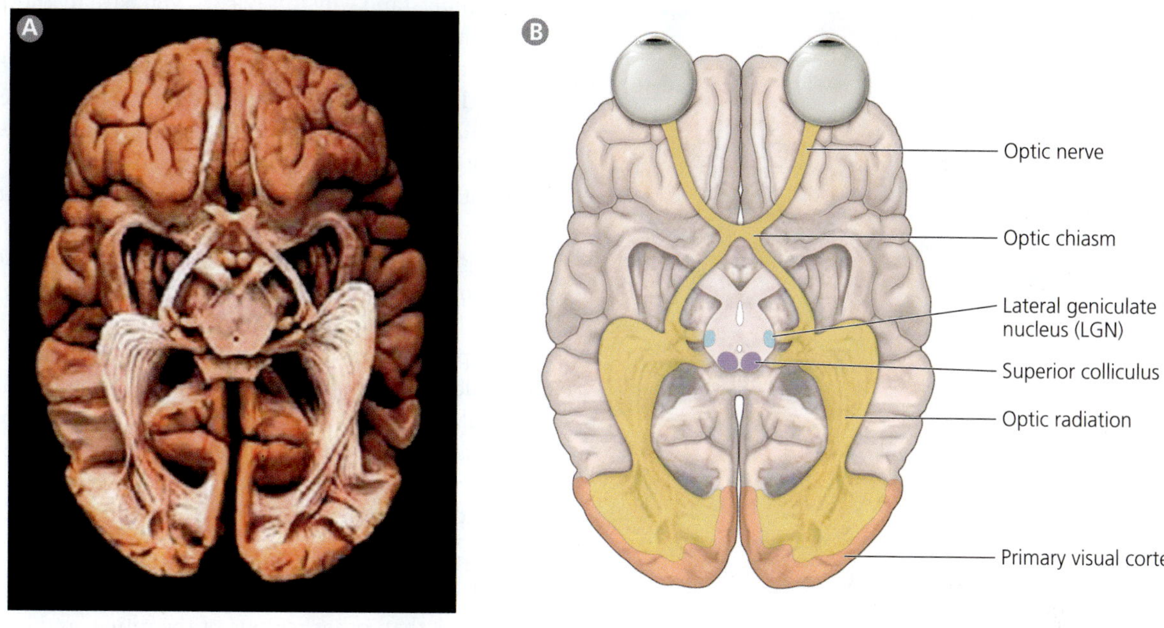

Figure 11.6 The principal ascending visual pathways. The photograph in (A) depicts an inferior view of a human brain from which the brainstem and part of the neocortex have been removed to reveal the optic radiation and other components of the ascending visual pathways. Panel (B) shows the same areas diagrammatically. [From Huxlin, 2008]

localization is not important. In contrast, spatial information is crucial for the proper function of the two largest retinal targets, namely, the **superior colliculus** and the **lateral geniculate nucleus** (**LGN**; Figure 11.6). The former plays a major role in eye movements; the latter is best known for its projections to the **primary visual cortex (V1)**, which is involved in conscious visual perception.

The retina's projections to the superior colliculus and LGN, as well as the projections from the LGN to V1, are topographically organized. In some species, each retina projects to the contralateral superior colliculus and LGN, with the retinal axons crossing in the **optic chiasm** (the Greek word *kiasma* means crossing). However, in primates and other species with forward facing eyes, the ascending visual pathways are more complex.

The clearest indication of this increased complexity is that damage of V1 on the left side of the brain in humans causes blindness not in one eye (the region of space viewed by one eye), but in the right **visual hemifield**, defined as the visible region of space that lies to the right of the **fixation point**, where the foveae are aimed (Figure 11.7). In contrast, lesions of the right V1 impairs vision in the left visual hemifield. Thus, lesions of V1 on one side of the brain cause blindness in portions of both eyes.

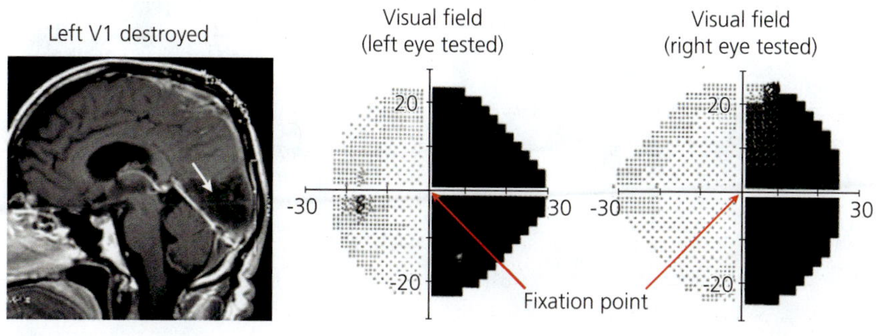

Figure 11.7 Effect of damage to V1. The image on the left depicts a sagittal brain scan from a patient whose left visual cortex (V1) was severely damaged. Also shown are the results of this patient's visual field test; the black regions correspond to portions of the visual field in which the patient was blind. As you can see, damage of the left visual cortex impairs vision in the right visual hemifield, regardless of which eye is tested. [From Huxlin, 2008]

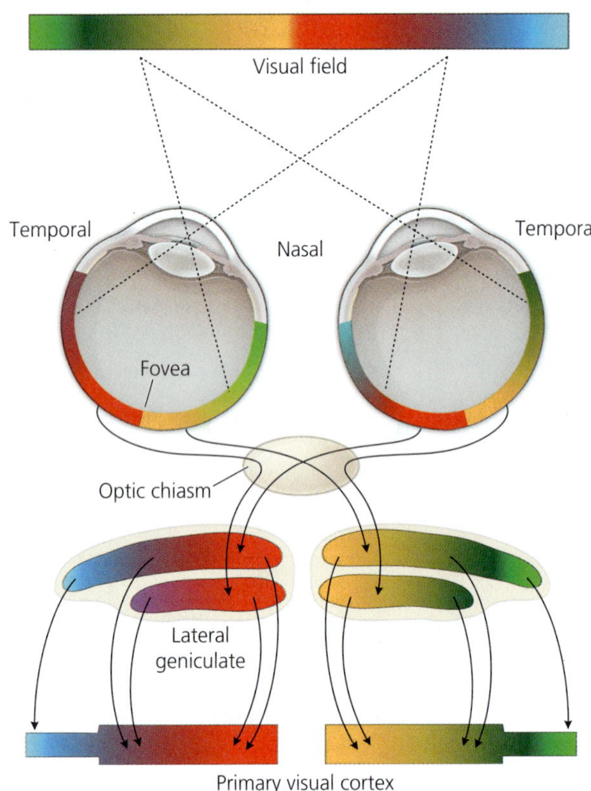

Figure 11.8 Visuotopic mapping in the geniculocortical pathway. The color of a brain region in this schematic diagram indicates which portion of the visual field its neurons respond to best. Information from the right visual hemifield is sent to the left lateral geniculate and visual cortex, while information from the left visual field is sent to the right side of the brain. Information from the two eyes remain anatomically segregated in the lateral geniculate nucleus but converges in the visual cortex. [After Kaas, 2005]

To understand why this occurs, you have to know that in humans and other primates only the axons from the nasal portion of each retina cross to the other side of the brain (Figure 11.8). The axons from the temporal retina enter the optic chiasm but then remain uncrossed. Therefore, visual information from the left nasal retina ends up together with information from the right temporal retina in the right side of the brain. Put differently, information from one visual hemifield ends up represented in the contralateral side of the brain. This sounds complicated, but it is relatively easy to grasp from Figure 11.8. There you can see how the nasal and temporal retinas in each eye project to the contralateral and ipsilateral LGN, respectively, and how the LGN projects ipsilaterally to V1. In essence, the map of the contralateral visual hemifield in V1 is "stitched together" from the retinotopic maps in the ipsilateral temporal retina and the contralateral nasal retina.

A Distorted Map in the Primary Visual Cortex

Because the fovea contains a much higher number of photoreceptors than other regions of the retina, and because the retinogeniculate pathway to the visual cortex is topographically organized, one would expect the central region of the visual field to be overrepresented in V1. A relatively easy way to test this hypothesis is to inject animals with radioactive 2-deoxyglucose (2-DG), which is taken up by active neurons (see Chapter 6), and then to present animals with a patterned visual stimulus that covers the entire visual field (Figure 11.9). Because retinal neurons respond poorly to steady stimuli, the patterned stimulus is dynamic, with local patches flickering on and off.

The results of such an experiment are shown in Figure 11.9 B. They show that that the central (foveal) region of the visual field is vastly overrepresented relative to the periphery. They also show that the visual field is topographically mapped onto the primary visual cortex, although this map is heavily distorted. Specifically, the concentric rings and spokes of the visual stimulus are represented by a roughly rectangular pattern of V1 neurons. Despite this distortion, if you know the rules for those distortions, then you can infer the location of a visual stimulus simply by knowing where in V1 the activated neurons are. In this respect, the map in V1 is very similar to that in S1.

Figure 11.9 A distorted visuotopic map. A patterned stimulus (A) was presented to an anesthetized monkey, whose fovea was aimed at the center of the stimulus. The monkey was injected with 2-deoxyglucose (2-DG), which accumulates in active neurons. The monkey was sacrificed 45 minutes later, and its primary visual cortex (V1) was processed to reveal where the 2-DG had accumulated. As shown in (B), the semicircles and spokes of the visual stimulus are represented by a nearly rectangular grid of active (dark) cells. The central portion of the stimulus, where the monkey's fovea is aimed, is overrepresented. [From Tootell et al., 1982]

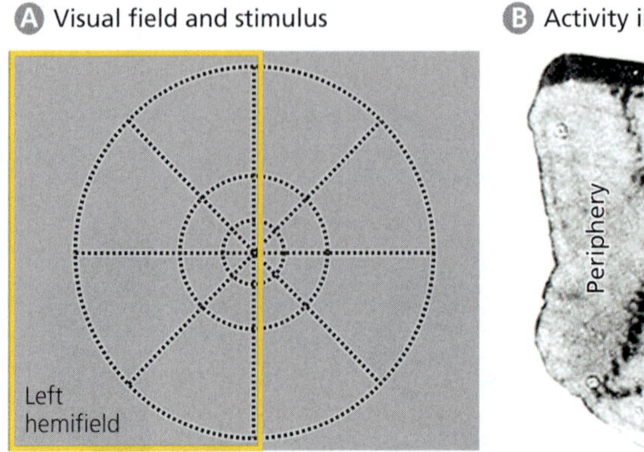

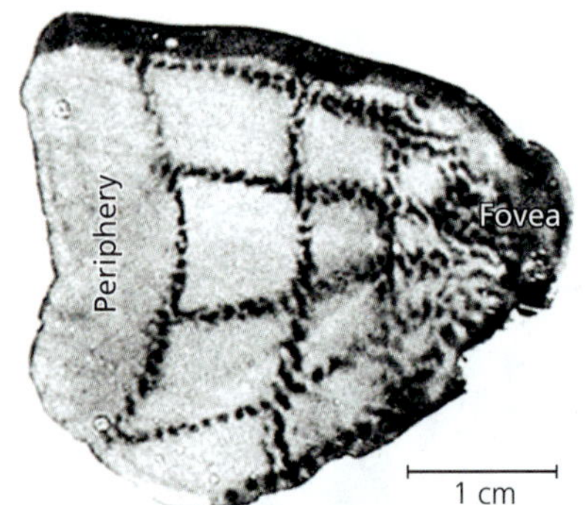

Ocular Dominance Bands

Because both eyes send axons to both sides of the brain, it is interesting to ask whether information from the two eyes is combined in the retinal target structures. In the superior colliculus of primates, most visually responsive neurons receive converging input from both eyes. That is, most of the neurons are **binocular**. In contrast, the neurons of the LGN are strictly **monocular**; they respond to visual inputs only from one eye or the other, but not both. Indeed, LGN neurons receiving input from the left eye are located in different layers of the LGN than the neurons receiving input from the right eye (Figures 11.8 and 11.10). Primates have more of these LGN layers than cats and other mammals do, but the principle is the same: each layer receives inputs from just one eye. As we discussed in Chapter 4, the segregation into eye-specific layers occurs during development and depends on competition for synaptic targets between the axons from the left and right retinas.

Neurons of the LGN project mainly to deep layer 4 of V1. Within that cortical layer, inputs from the two eyes remain strictly segregated, just as they are in the layers of the LGN (Figure 11.10). This segregation forms the basis for the ocular dominance bands (or columns) that one can see in tangential sections through V1 of monkeys whose retinas were injected with a transneuronal axon tracer (see Chapter 4, Figure 4.35). These snaking **ocular dominance bands** are not found in all species, but they are present in humans. This becomes apparent when V1 of humans who have lost one eye is flattened, sectioned tangentially, and stained to visualize cytochrome oxidase, an enzyme that preferentially stains cortical regions with strong sensory inputs. In such preparations, dark bands corresponding to inputs from the intact eye are evident in sections through deep layer 4 (Figure 11.11).

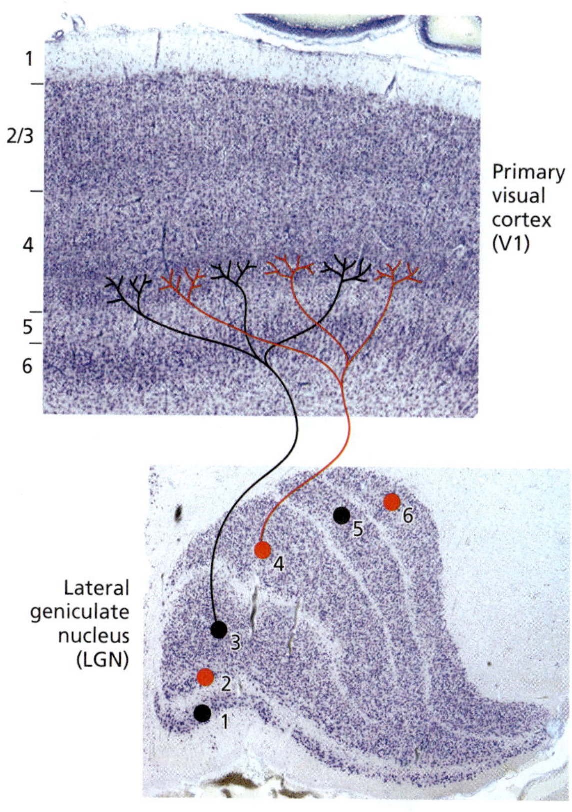

Figure 11.10 Eye-specific segregation in LGN and V1. The connections from the lateral geniculate nucleus (LGN) to the primary visual cortex (V1) are here shown schematically. Neurons from one eye are colored red; those from the other eye are black. Each of the 6 layers in the LGN carries information from only one of the eyes. The LGN axons terminate in an eye-specific manner in deep layer 4. Only the projections from LGN layers 3 and 4 are shown.

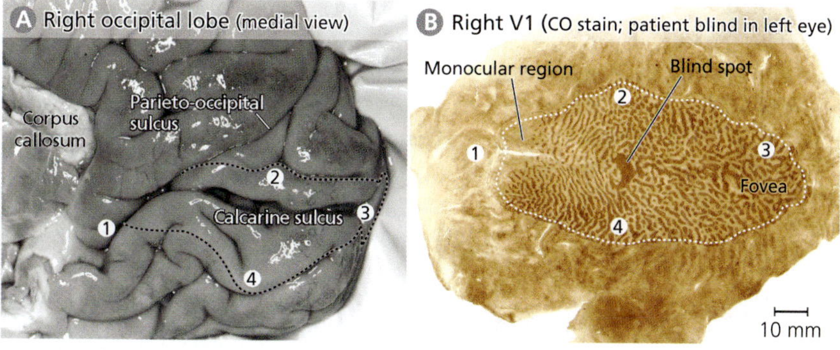

Figure 11.11 Primary visual cortex in humans. The photograph in (A) is a medial view of the right occipital lobe. Much of the primary visual cortex (V1), outlined by the dashed line, is buried within the calcarine sulcus. Panel (B) shows a horizontal section through V1 of a person who was blind in the left eye. The cortex was flattened and sectioned through layer 4. The numbers indicate four points of correspondence with the folded cortex shown in (A). The section was stained for cytochrome oxidase (CO) to visualize, as dark brown patches and bands, the portions of V1 that process information from the remaining, intact eye. [From Adams et al., 2007]

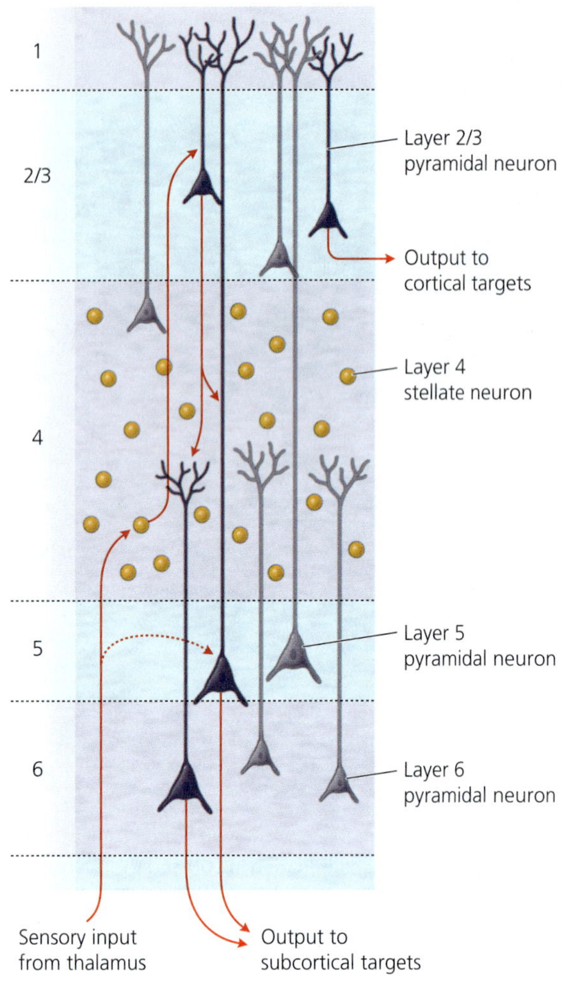

Figure 11.12 Cortical columnar connections. This schematic section through V1 depicts the major neuron types and their axonal connections (red). Pyramidal neurons have pyramid-shaped cell bodies and long radial dendrites. The dashed arrow indicates a "shortcut" through which sensory information can reach layer 5 pyramidal neurons.

Binocular Convergence

In contrast to the monocular neurons in deep layer 4, many of the neurons in the upper and lower layers of V1 do receive converging inputs from both eyes. These binocular neurons function in **binocular depth perception** (stereopsis), which helps us estimate the distance of visible objects.

The key to binocular depth perception is that focusing on a point in space causes any objects in front or behind that point to be projected onto mismatched locations on your two retinas. You can visualize this phenomenon by focusing on an object and then holding your index finger between that object and your nose. The finger will appear in two different locations, one for each eye. We are normally unaware of these *binocular disparities*, but the binocular V1 neurons are highly sensitive to them. By measuring binocular disparities, our visual system can estimate the distance between an object and the retina. This explains why Siamese cats, which tend to be cross-eyed and have few binocular V1 neurons, exhibit poor depth perception.

At this point you may wonder, do the monocular neurons in deep layer 4 project to the binocular neurons in the upper and lower layers of V1? Are those projections direct or indirect? More generally, what sort of circuitry connects the various layers of the neocortex to one another? The answer is complex enough to warrant a few paragraphs.

Columnar Cortical Circuits

The mammalian neocortex is characterized by the presence of numerous **pyramidal neurons**, which have a pyramid-shaped cell body and a long radial dendrite that extends toward the brain surface (see Figure 2.19 D). The largest of these pyramidal neurons have their cell bodies in the deep cortical layers (5 and 6). Smaller pyramidal neurons predominate in the upper cortical layers (2 and 3). Layer 4, in contrast, contains few pyramidal cell bodies and instead is packed with small stellate (star-shaped) neurons. Years of painstaking research has revealed that these cortical neurons are interconnected in specific ways.

As shown in Figure 11.12, sensory input from the dorsal thalamus terminates mainly on the stellate neurons of layer 4. The stellate neurons project mainly to the pyramidal neurons radially above them, in layers 2 and 3. Some of those upper layer pyramidal neurons have long axons that project to other cortical areas, but most of them terminate on the dendrites of the large pyramidal neurons that have their cell bodies in layers 5 and 6. Those deep-layer pyramidal neurons then project to a variety of subcortical areas. Because this basic circuitry has been observed in diverse cortical areas, not just in V1, it is often called the **canonical cortical circuit**.

An important aspect of the canonical cortical circuit is that the interconnected neurons all lie within a relatively narrow radial column (20–50 μm wide). The existence of these **cortical minicolumns** is consistent with the neurophysiological finding that cortical neurons lying radially above or below one another tend to have very similar response profiles. After all, one would expect interconnected neurons to display similar patterns of activity. Collectively, these data support the general notion that the neocortex exhibits a radially columnar architecture in which each radial minicolumn forms a structurally and functionally discrete module. According to this view, the human neocortex contains more than 100 million cortical minicolumns, each performing similar computations on the input it receives. If this is true, then functional differences between cortical minicolumns arise primarily from differences in their inputs.

Although the idea of a canonical cortical circuit is widely accepted, it is subject to many caveats. For example, a recent study showed that layer 5 pyramidal neurons still respond to thalamic input when layers 1–4 are inactivated with lidocaine, implying that some thalamic inputs terminate directly on layer 5 pyramidal neurons (dashed arrow in Figure 11.12). Another complication is that the canonical circuit does not account for inhibitory interneurons, which are plentiful in the neocortex and diverse. An even more important problem is that cortical areas differ from one another, at least in the thickness of their cortical layers. For example, layer 4 is very thin or nonexistent in the motor cortex and large parts of the prefrontal cortex. These areas still receive inputs from the dorsal thalamus, but those inputs terminate primarily in layers 2 and 3. Indeed, thalamic projections to layers 2 and 3 are seen in many cortical areas. This obviously complicates our view of the simplified canonical circuit illustrated in Figure 11.12.

Most important to our discussion of binocular convergence in V1 is that information flow between neighboring minicolumns is probably much more extensive than the canonical circuit implies. As we discussed, information from the two eyes is strictly segregated in deep layer 4 of V1, but binocular neurons are found both radially above and radially below that layer. This observation implies that information must be spreading tangentially as it flows from layer 4 to the other cortical layers. Such tangential spread is not completely inconsistent with the canonical circuit hypothesis, which allows for upper layer pyramidal neurons to have tangential intracortical connections (Figure 11.12), but it does complicate the view of cortical minicolumns. They probably are not as independent of one another as a strictly modular view of the neocortex implies.

BRAIN EXERCISE

Imagine a bug crawling in a straight line and at a constant velocity from your left big toe to your right eyebrow. What spatiotemporal pattern of activity would this bug elicit in your somatosensory cortex? What would the pattern of activity be in the primary visual cortex if you could see the crawling bug (in a tall mirror)?

11.2 How Can Animals Determine Where a Sound Came From?

Localizing stimuli is more difficult in the auditory system than in the visual and somatosensory systems because the spatial location of stimuli is not mapped onto the cochlea as it is onto the retina or skin. As you learned in Chapter 6, the location of hair cells on the cochlea's basilar membrane encodes a sound's frequency spectrum but not its spatial location. Therefore, sound source location must be computed in the brain.

Interaural Comparisons

To determine a sound's location, neurons must compare the sound's timing and intensity between the two ears. As illustrated in Figure 11.13, a sound coming from directly in front of you (or anywhere else in the midline plane) will be equally loud in your two ears. In contrast, sounds coming from the side will be louder in the ear closest to the sound source. Such **interaural level differences** (ILD; *interaural* means "between the ears") are larger for high-frequency sounds than for low-frequency sounds because the latter sounds travel more easily past obstacles, such as your head (which explains why you hear the bass in your neighbor's music so much better than the high-pitched sounds). In addition, sounds coming from the side arrive earlier at the ear closest to the source. Because sound travels at 343 m/sec in air and the extra distance a sound off to the side must travel (mainly along the head's circumference)

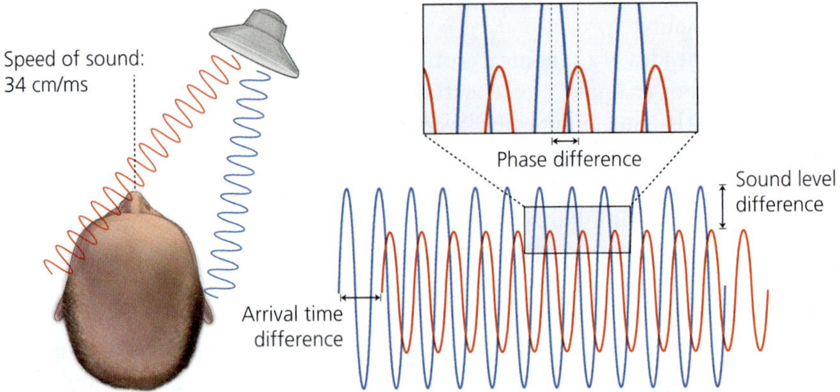

Figure 11.13 Sound localization through interaural comparisons. Sound coming from your right must travel further to reach your left ear than your right ear. This means that the sound waves will reach your right ear a fraction of a millisecond before they reach the left ear. In addition, sounds are louder at the ear closest to the sound source, both because sound intensity falls off with the square of distance and because sounds are attenuated as they travel through or around the head. Such interaural time (phase and/or arrival time) and level differences provide useful cues for localizing sounds in the horizontal plane.

to reach the other side is 25–26 cm, the **interaural time difference (ITD)** for a sound off to your side is about 0.7 msec. Although small, such time differences (including both arrival time and phase differences, as shown in Figure 11.13) yield enough information for you to localize a sound, especially when that sound is low in frequency. Therefore, ITDs are complementary to ILDs: the former are most useful for localizing low-frequency sounds and the latter for high-frequency sounds.

To measure ITDs and ILDs, neurons must receive *binaural* input; that is, they must receive auditory information from both ears. In contrast to binocular convergence, which first occurs in the neocortex, binaural convergence is accomplished in the medulla, specifically the lateral and medial *superior olives*. As you can see in Figure 11.14, the **lateral superior olive** receives excitatory inputs from the ipsilateral cochlear nucleus and indirect inhibitory inputs from the contralateral cochlear nucleus. Whenever the ipsilateral excitation is stronger than the contralateral inhibition, which happens whenever a sound is louder in the ipsilateral ear, neurons in the lateral superior olive increase their firing rate. Thus, lateral superior olive neurons are sensitive to ILDs. The more they fire, the further ipsilateral the sound source must be. More generally, we can say that a sound source's horizontal location is encoded in the firing rate of lateral superior olive neurons, at least for high-frequency sounds that generate detectable ILDs.

Whereas ILDs are processed in the *lateral* superior olive, ITDs are analyzed by the **medial superior olive**. The cellular mechanisms underlying this computation are still controversial (at least in mammals), but they must be incredibly precise because the relevant time differences are much shorter than the duration of a typical action potential (~1 ms). Despite the mysteries that still surround the medial superior olive, few doubt that this structure is critical for the analysis of interaural time differences. Consistent with this hypothesis, albino cats have a shrunken medial superior olive and are quite terrible at localizing sounds.

Given the preceding discussion, it is not surprising that individuals with unilateral hearing loss have trouble localizing sounds, which can cause serious problems, for example, when they must localize the sound of an approaching car. Less obvious is that such individuals also have problems with speech comprehension in noisy environments. Although most of us are unaware of this phenomenon, being able to localize a sound makes it much easier to interpret the meaning of that sound.

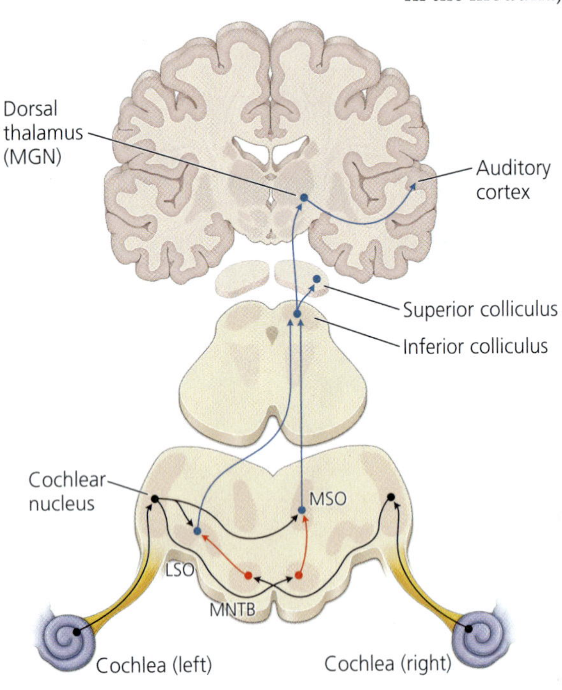

Figure 11.14 The ascending auditory pathway. Auditory nerve axons project to the cochlear nuclei in the medulla, which project ipsilaterally to the lateral superior olivary (LSO) nucleus, bilaterally to the medial superior olive (MSO), and contralaterally to the medial nucleus of the trapezoid body (MNTB). The MNTB has inhibitory projections (red) to the MSO and LSO, which project to the inferior colliculi. The latter structure projects to the superior colliculus and to the medial geniculate nucleus (MGN) of the thalamus, which sends most of its axons to the auditory cortex. For clarity, most connections are shown only on one side of the brain.

As you will learn in Chapter 13, spatial attention in general tends to enhance the perception of stimuli at the attended locations.

Encoding Space in the Auditory Midbrain and Forebrain

Axons from the lateral and medial superior olive ascend to the midbrain's **inferior colliculus** (Figure 11.14). How do inferior colliculus neurons encode sound source location? The answer depends on which species one is examining.

Barn owls excel at localizing sounds, especially those made by mice rustling under leaves (see Box 11.1: Sound Localization in Barn Owls). Therefore, barn owls have long been a favored species for the study of sound localization and its underlying neural mechanisms. These studies have shown that individual neurons in the inferior colliculus of barn owls have relatively small spatial receptive fields, meaning that they respond only to sounds from very limited regions of space. Importantly, different neurons respond to sounds from different locations (Figure 11.15 A). Therefore, sound source location can be inferred by knowing which specific neurons are active against a background of relative silence. This coding strategy is similar to how the visual and somatosensory systems encode stimulus location: each neuron has a clearly circumscribed spatial receptive field. Indeed, the inferior colliculus of barn owls contains a topographic map of auditory space that is analogous to the map of the visual field in V1 and to the map of the body surface in S1.

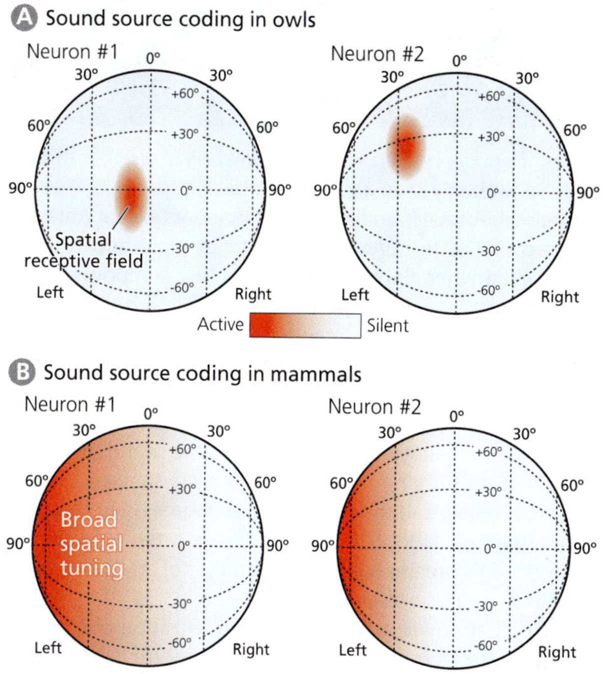

Mammals use a different strategy to encode the location of sounds (Figure 11.15 B). Individual neurons in the mammalian inferior colliculus increase their firing rate quite gradually as a sound source moves from the midline toward the contralateral side. Different neurons increase their firing rate at different rates as the sound moves laterally, but each neuron responds to stimuli from many different locations. Therefore, in the mammalian midbrain, the location of an auditory stimulus is encoded not in the firing rate of single cells but in the firing rates of many different neurons firing at diverse rates. This coding strategy resembles the *population code* you learned about in Chapter 10, when we discussed how the motor cortex represents movements across a population of many different neurons; except now we are talking about a sensory population code. Although population codes are more difficult to comprehend than single cell codes, which allow detailed information to be extracted from the activity of single cells, population codes are common in nervous systems.

What about the auditory cortex? Do the neurons there have circumscribed spatial receptive fields similar to those observed in the midbrain of owls? The short answer is no. Most of the neurons in the **primary auditory cortex**, which lies in the superior portion of the temporal lobe, encode spatial locations just as the mammalian inferior colliculus does. They respond best to sounds from the contralateral side and increase their firing rate the further lateral the sound source moves. In short, the mammalian brain does not contain a nicely topographic map for auditory space.

In one respect, however, the strategy for coding space in the mammalian auditory system does resemble the strategy employed in the visual and somatosensory systems. In all these systems, particular regions of space are overrepresented in the brain. In the auditory system, the favored region is the head's midline plane, where

A Sound source coding in owls
Neuron #1 | Neuron #2
Spatial receptive field
Active | Silent

B Sound source coding in mammals
Neuron #1 | Neuron #2
Broad spatial tuning

Figure 11.15 Sound source coding in owls versus mammals. Shown in (A) are the spatial receptive fields of two auditory neurons in the inferior colliculus of a barn owl. These neurons respond to sounds only if they come from vertically and horizontally restricted locations. By contrast, mammalian neurons exhibit broad spatial tuning (B). As auditory stimuli move contralaterally, different neurons increase their firing at different rates; but a stimulus at any given location activates many different neurons. Therefore, mammalian brains employ a population code, rather than a single neuron code, to represent sound source location.

EVOLUTION IN ACTION

Box 11.1 *Sound Localization in Barn Owls*

Even if you've never seen a barn owl (*Tyto alba*) in the wild, you've probably seen a picture of one (Figure b11.1). These fascinating birds are found across all continents except Antarctica. Barn owls like to hunt small rodents at night and can catch mice even when light levels are very low (a moonless night) or when the mice are hidden under leaf litter or snow, which suggests that barn owls can localize their prey acoustically. This hypothesis was tested by Roger Payne in the late 1950s. He observed that barn owls in a totally darkened room could catch not only live mice but also paper wads that were pulled across the leaf-littered floor, mimicking the sound (but not the smell) of a scurrying mouse. Further tests showed that barn owls can localize sounds to within 1–2 degrees in both azimuth and elevation (about the width of your little finger at arm's length). This precision is better than that of any other species studied to date, including humans.

The barn owl's amazing sound localization ability is due primarily to its frontally directed, asymmetric ears. After removing the acoustically transparent feathers that cover a barn owl's face, you can see an underlying layer of acoustically dense feathers that form a sound collector that looks like a heart-shaped radar-dish (Figure b11.1 B). Careful measurements have shown that the left external ear is directed slightly downward, whereas the right ear aims upward. Because of this asymmetry, high-frequency sounds (>3 kHz) become louder in the right ear and softer in the left ear as a sound source moves upward (conversely as a sound source moves downward).

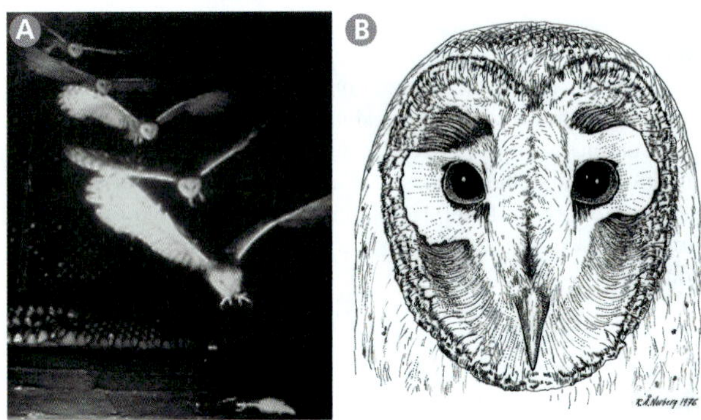

Figure b11.1 Barn owls are nocturnal hunters and have asymmetric ears. Shown in (A) is a barn owl descending from a perch to catch a mouse. The room was pitch black, but five flashes of light captured the owl in action. Shown in (B) is a barn owl's face after the superficial face feathers were removed. The ear flaps, which extend laterally away from the eyes to cover the auditory troughs, are asymmetric, with the right ear aimed upward and the left one downward. This asymmetry creates interaural level differences for high frequency sounds that differ in elevation. [A courtesy of Mark Konishi; B courtesy of Åke Norberg]

Barn owls can use these interaural level differences (ILDs) to determine a sound source's elevation. If you plug a barn owl's right ear, the owl tends to misjudge the location of a rustling mouse and strike short of the target when attacking from a perch. This is what you would expect if the owl is using ILDs to determine sound elevation (ear plugs do not affect interaural time differences). An owl with its right ear plugged also tends to strike to the left of its target, indicating that the owl uses ILDs to localize sounds also in the horizontal plane. Apparently, high frequencies are used to estimate sound elevation, whereas low frequencies are used to calculate sound azimuth (horizontal position). Because a mouse running through leaf litter creates a broadband sound, an attentive barn owl can localize that mouse accurately in both elevation and azimuth, even in total darkness. Owls with symmetric ears cannot perform this feat.

The neural mechanisms underlying sound localization in barn owls have been studied intensively by Mark Konishi, Eric Knudsen, and their collaborators. Much of their neurobiological work was conducted on anesthetized owls that were kept in a dark, soundproof and echo-free room. The experimenters then recorded from neurons in the owl brain while a computer presented sounds from various locations around the room. The first major finding was that neurons in the barn owl's superior colliculus had spatially limited receptive fields and formed an orderly map of auditory space. Subsequent research explored how this auditory space map is generated from interaural level and time differences. Many findings from owls have proven applicable to mammals as well. However, some aspects of the barn owl's auditory system are quite unusual, even compared to other birds. This is not surprising, given how exceptionally good owls are at localizing sounds. Evidently, evolution endowed owls with some highly specialized circuits for sound localization.

ILDs and ITDs are small. For animals with frontal eyes, part of this "acoustic fovea" is aimed at the same region of space as the retinal fovea, at least as long as the eyes are aimed directly ahead.

BRAIN EXERCISE

If two identical sounds are presented simultaneously to your two ears through headphones, where will you perceive the sound to be coming from? (Don't say "from the headphones"!)

11.3 In Which Spatial Coordinate System Should Stimuli Be Localized?

Thus far in this chapter we have discussed only how organisms localize stimuli with immobile sensor arrays. This was a reasonable starting point because most of the research on stimulus localization has been conducted on anesthetized animals whose sensors are, of course, as immobile as the animal. However, organisms normally move about as they explore their environment.

Movable Sensor Arrays

When animals attempt to localize a stimulus, they often move their eyes, head, or other body parts. Horses and cats, for example, turn their ears backward as you approach them from behind. Humans don't move their ears, but they do turn their head toward interesting sounds. This maneuver brings the sound source into the acoustic fovea, where sounds are localized more precisely. What do we do when trying to localize a sound source in the vertical dimension? We often tilt our head sideways, thereby ensuring that sounds above the horizon will be louder and arrive earlier in one ear than the other.

Movable sensor arrays are equally useful in the other sensory systems. If a visual stimulus attracts your attention, you will most likely aim your eyes directly at the stimulus. This moves the retinal image of the stimulus onto the visual fovea, where you have the highest spatial acuity. In the somatosensory system, too, the most sensitive body parts are most frequently used to explore objects of interest. Just think of how you move about in a pitch-black environment. You usually extend your arms. Rats don't use their forelimbs in this manner, but they move their whiskers to scrutinize the space around their nose. Simply put, animals rarely just sit around and wait for stimuli. They explore their surroundings by moving their sensor arrays. Some of these movements are purely exploratory, with the sensor array acting like a rotating radar dish searching for anything interesting. Others serve to investigate a stimulus in more detail. Either way, sensor movements improve spatial localization and stimulus identification.

As wonderful as movable sensors may be, they are not easy to use. If you are looking straight ahead and a stimulus elicits responses in your fovea, then you can infer that the stimulus must be directly in front of you. However, if your eyes are aimed 30° to the left, then a stimulus projected onto your fovea would be to your left, and a stimulus directly in front of you would elicit responses 30° to the left of your fovea (recall that the eye's optics invert the image). Thus, the location of active neurons in the retina provides accurate information about stimulus location only if you know where your eyes are pointing, which in turn requires knowledge about your current eye and head orientations. To make things worse, if you want to know where a stimulus is in the outside world, independently of you, then you must also know where your body is and how it is turned. Nor are these problems limited to the visual system. To localize an object with your hand, for example, you need to know where your hand is relative to your body and where your body is in the world. Amazingly, our nervous system solves such problems routinely, without conscious effort on our part.

Spatial Coordinate Transformations

To understand how our nervous system deals with movable sensors, it is useful to consider that spatial information can be encoded in a variety of different coordinate frames (Figure 11.16). The retina, for example, encodes stimulus location in **eye-centered coordinates**. That is, the activation of neurons or photoreceptors in a specific portion of the retina tells the rest of the brain where a visual system is relative to the eye's current position. To determine where this stimulus is in the outside world, the eye-centered information must be combined with information about eye, head, and body position. Similarly, information about the location of a sound is initially computed in **head-centered coordinates** (at least in organisms with immobile ears).

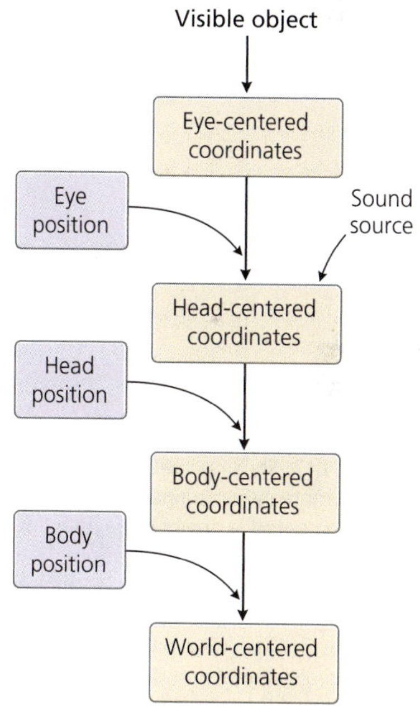

Figure 11.16 Spatial coordinate transformations. In theory, spatial information that is encoded in eye-centered coordinates may be converted to head-centered coordinates by integrating it with information about eye position. Auditory information enters the nervous system in head-centered coordinates (if the ears are immobile). Although this diagram implies a serial progression from eye-centered to head-centered, to body-centered, and ultimately world-centered coordinates, recent research suggests that coordinate transformations are not as neatly serial and progressive. [After Pouget and Sejnowski, 1997]

To know where in the outside world the sound is coming from, we must also consider how our head is turned relative to the rest of our body and how our body is turned relative to external landmarks. As shown in Figure 11.16, we can imagine that spatial sensory information is transformed sequentially from eye- and head-centered coordinate frames into **body-centered coordinates** and, ultimately, into **world-centered coordinates**. Each coordinate transformation involves the integration of additional information about the current orientation of the body and its movable parts. An important question, of course, is whether the brain actually performs the kind of coordinate transformations that the figure depicts. The answer is yes, but the transformations are not as nicely sequential as the diagram implies.

Shifting Spatial Receptive Fields in the Superior Colliculus

Most of the research related to spatial coordinate transformations has been performed on the primate superior colliculus, which receives topographic visual inputs from the retina and auditory inputs from the inferior colliculus. Individual neurons in the deep layers of the superior colliculus respond to both visual and auditory stimuli and, as long as the visual and acoustic foveae are aimed in the same direction, the neurons respond most strongly when the visual and auditory stimuli are coming from the same location in space. That is, the auditory and visual space maps in the superior colliculus are aligned with one another. But what happens when the eyes turn to look aside?

To answer this question, neuroscientists recorded the auditory spatial receptive fields of deep collicular neurons as a head-fixed monkey moved its eyes from one fixation point to another (Figure 11.17). Amazingly, the auditory spatial receptive fields shifted with eye position. If the eyes were aimed to the left of center, then the auditory receptive fields moved to the left as well. This means that the deep superior colliculus neurons do not encode sound source location in purely head-centered coordinates. Instead, they must be using information about eye position to transform the auditory signals from head-centered to eye-centered coordinates. Oddly, the coordinate transformation is incomplete. As eye position changes, the auditory spatial receptive fields shift only about half as much as you would expect if the auditory information were fully transformed into eye-centered coordinates. Why this is so remains unclear.

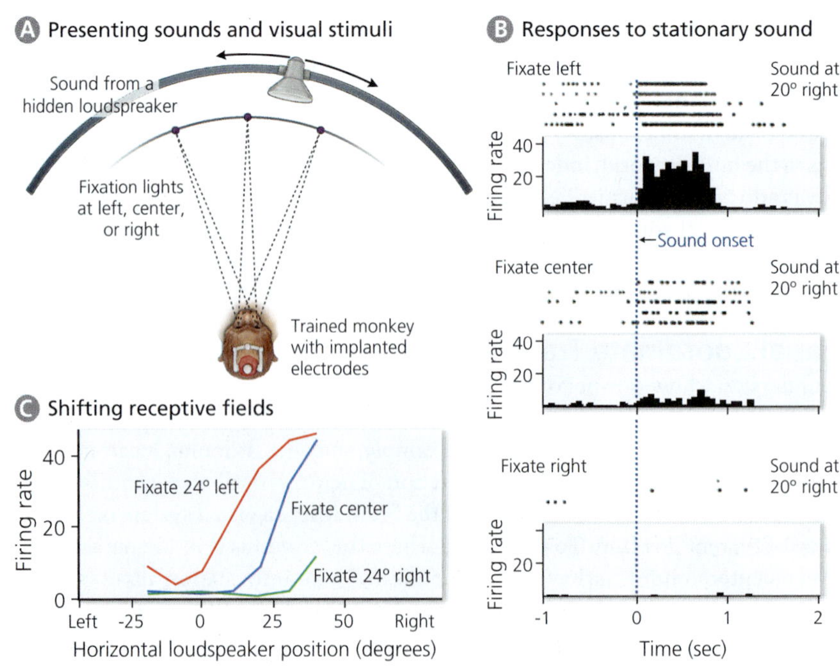

Figure 11.17 Shifting spatial receptive fields. A monkey was trained to fixate on a light presented at one of three different locations (24° left, center, or 24° right). The experimenters then played noises through a moveable, hidden speaker and recorded how individual neurons in the monkey's superior colliculus responded. Crucially, they analyzed how those responses varied with eye position. The diagram in (A) illustrates the experimental setup. The spike rasters and firing rate histograms in (B) illustrate how one neuron's responses to a sound at 20° to the right of center varied with eye position. Panel (C) summarizes the data obtained when sounds were played from a variety of different speaker positions. The auditory receptive field of the illustrated neuron shifted rightward as the eyes moved to the right and leftward as the eyes moved left. This is what you would expect if the superior colliculus neurons transform information about sound location from head-centered coordinates into eye-centered coordinates. [A after Jay and Sparks, 1987a; B and C from Jay and Sparks, 1987b]

IF THE SPATIAL RF DOES THIS:	THEN THE NEURON CODES SPACE IN:
Moves with the eyes	Eye-centered coordinates
Is stable as the eyes move, but shifts with head movements	Head-centered coordinates
Is stable as the eyes or head move, but shifts with body movements	Body-centered coordinates
Is stable as the eyes, head, or body move	World-centered coordinates

Table 11.1 Inferring coordinate transformations from receptive field (RF) shifts.

Posterior Parietal Cortex

Spatial coordinate transformations have also been studied in the **posterior parietal cortex**, which lies just posterior to the somatosensory cortex. As you will learn later in this chapter, the posterior parietal cortex contains several cortical areas involved in the control of hand, head, and eye movements. Important for our present discussion is that posterior parietal cortex receives convergent visual, auditory, and somatosensory input. This raises a question: what spatial coordinate frame is used to represent that multimodal sensory information?

In search of an answer, researchers presented awake monkeys with sounds and visual stimuli at various spatial locations and determined at which locations each neuron responded most robustly. Having mapped the visual and auditory spatial receptive fields, they asked the monkey to move its eyes to a different location (the head was fixed) and examined how those changes in eye position altered the neurons' spatial receptive fields. If an auditory spatial receptive field moves together with the eyes, then the auditory spatial information can be inferred to be encoded in eye-centered coordinates. Alternately, if the auditory receptive field remains stationary, then the auditory information is probably represented in head-, body-, or world-centered coordinates (to discriminate between the latter three alternatives, one must determine how the receptive fields change when the head or body moves in space; see Table 11.1). Similarly, if the visual receptive fields move with the eyes, then the visual information must be encoded in eye-centered coordinates; if it does not, then some other coordinate system must be at work.

How did the experiments come out? No single, simple answer emerged. In the best studied portion of the posterior parietal cortex (the ventral intraparietal area), more than one-third of all neurons encode visual information in eye-centered coordinates, and 28% of them employ head-centered coordinates. Most surprisingly, one-third of the neurons encode visual information in coordinates that are neither eye-centered nor head-centered, but intermediate. The responses of the posterior parietal neurons to auditory stimuli were similarly complex. Forty percent of the neurons encoded the auditory information in head-centered coordinates, but the others employed either eye-centered or intermediate coordinate frames. These data do not fit the model of sequential, stepwise coordinate transformations illustrated in Figure 11.16. However, diverse and intermediate coordinate frames also appear in artificial neural networks trained to process spatial information. Thus, a multiplicity of coordinate frames is not detrimental to spatial processing. It is just difficult to comprehend.

11.4 How Do Animals Orient toward an Interesting Stimulus?

When you see or hear something of interest, you tend to move your eyes toward the stimulus, to turn your head, or both. Even in the dark, you may extend your hands toward shadowy stimuli. The main objective of such orienting movements is to gather additional information about the stimulus, which can then be used to identify the stimulus (see Chapter 12) and to select the most appropriate course of action (Chapter 15).

Targeted Eye Movements

If you are in a dark room and a light suddenly flashes off to the side, your eyes will likely turn toward that location rapidly. Similarly, if you are looking at a static landscape in which something suddenly moves, your eyes will be drawn toward that movement. Such targeted eye movements are called **saccades**. They tend to be extremely swift (>350 degrees/second) and follow a straight trajectory.

Most of the research on saccadic eye movements involves training animals, primarily monkeys, to stare at a central fixation spot and then make a single saccade to a visual target. Such saccades to isolated visual targets are relatively simple. More complicated, multistep saccades occur whenever animals look at an object or scene of interest. Under those conditions, the eyes make saccades from one interesting spot to another (Figure 11.18). We are constantly scanning the world with our foveae, much as one might sweep the beam of a flashlight through a darkened room. Yet our mental image of the outside world is remarkably steady. This intriguing phenomenon probably involves very brief gaps in visual attention during each saccade and anticipatory shifts in visual spatial receptive fields, but it remains quite poorly understood. Therefore, let us here focus on a simpler problem: how does our nervous system convert information about stimulus location into motor commands that turn the eyes toward that location?

Eye Muscles and Saccade Generators

To turn an eyeball you need eye muscles. You already learned about the lateral and medial rectus muscles, which rotate the eyes within the horizontal plane (see Chapter 10). Four additional *extraocular muscles* (as opposed to intraocular muscles, such as the pupillary constrictor) are the superior and inferior rectus and the superior and inferior oblique (Figure 11.19). They rotate the eyes up and down and in the plane of the face. Collectively, the extraocular muscles are innervated by motor neurons in the abducens and oculomotor nuclei, which you read about in Chapter 10, and by the *trochlear nucleus*.

The motor neurons that innervate the extraocular muscles receive major inputs from two **saccade generator regions** in the brainstem. The motor neurons that innervate muscles controlling horizontal eye movements (mainly the abducens nucleus and part of the oculomotor nucleus) receive strong input from a diffuse region called the *pontine and medullary reticular formation* (Figure 11.20). Electrical stimulation of this part of the reticular formation elicits horizontal saccades. In contrast, the motor neurons in charge of vertical eye movements (the parts of the oculomotor nucleus

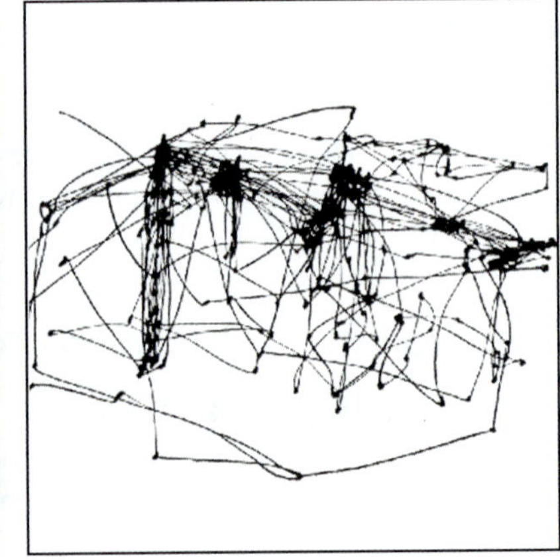

Figure 11.18 Scanning a scene with rapid eye movements. The Russian psychologist Alfred Yarbus in the 1950s and 60s performed classic studies on how we move our eyes when we examine visual objects or scenes. Shown at the left is one of the images Yarbus asked his subjects to examine (a painting by Efimovich Repin titled *They Did Not Expect Him*). Shown on the right is a simultaneous recording of the subject's eye movements. You can see that the eyes moved in straight lines, rested on selected spots, and then moved on. Such jerky eye movements are called saccades. [From Yarbus, 1967]

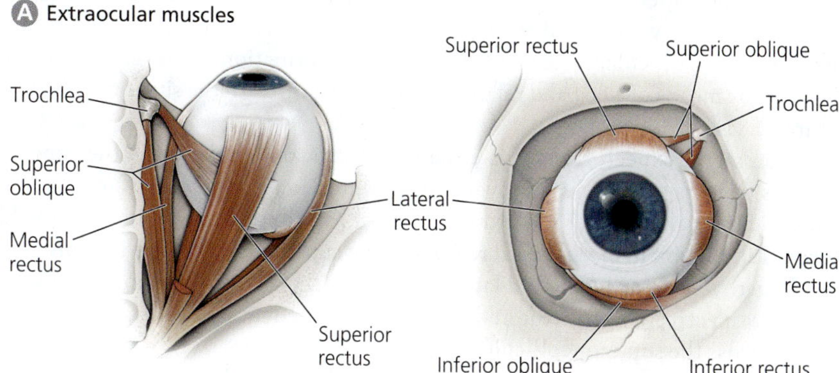

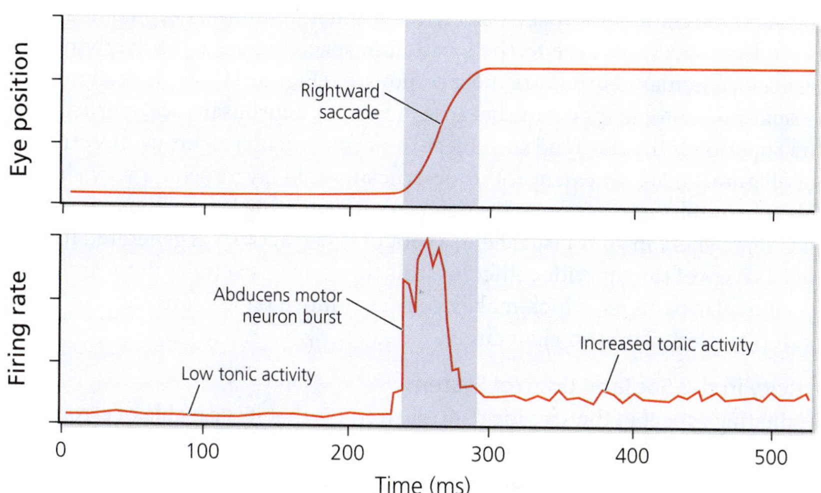

Figure 11.19 Muscles that turn the eyes. Panel (A) depicts superior and frontal views of a human eye and its extraocular muscles, which are arranged in 3 antagonistic pairs. The superior oblique passes through a fibrous loop, called the trochlea, that acts as a pulley to change the muscle's direction of force. The graphs in (B) display eye position during a rightward saccade (top trace) and a simultaneous record of activity in an abducens motor neuron that innervates the lateral rectus (bottom trace). This motor neuron increases its firing rate during the saccade. Its firing rate remains higher after the saccade to prevent the eye from returning to its natural, forward-aiming position. [After Sparks, 2002]

that innervate the superior and inferior rectus muscles) receive strong inputs from the *midbrain reticular formation,* where electrical stimulation triggers vertical saccades.

At this point you may ask yourself, if one saccade generator region controls horizontal saccades and the other one produces vertical saccades, how do you generate saccades that are neither horizontal nor vertical but oblique? The answer is that such slanted saccades occur when reticular neurons commanding vertical saccades fire simultaneously with neurons that generate horizontal saccades, with the precise angle of each saccade being determined by the relative balance of activity in the two pools of reticular premotor neurons. If you like math, you might say that oblique saccades result from vector addition.

The circuits we just discussed are in charge of executing eye movements, but how are those movements targeted to stimuli presented at specific spatial locations? For that, you need descending inputs from the superior colliculus and the cortical eye fields. Let us discuss these two pathways in turn, beginning with the projections from the superior colliculus.

Collicular Control of Eye Movements

As we discussed earlier, the superior colliculus receives topographic inputs from the retina. The retinal axons terminate on

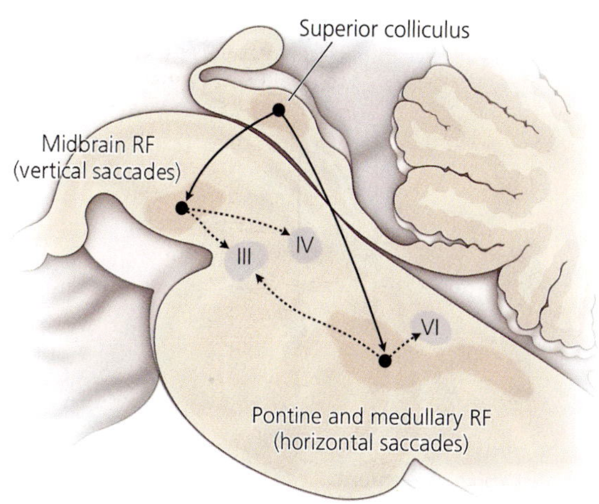

Figure 11.20 Saccade control circuits in the brainstem. The midbrain's superior colliculus projects to neurons in the reticular formation (RF), which in turn project to eye muscle motor neurons in the oculomotor (III), trochlear (IV), and abducens (VI) nuclei. Importantly, neurons in the midbrain RF command vertical eye movements, whereas neurons in the pontine and medullary RF generate commands for horizontal saccades.

neurons in the superficial layers of the superior colliculus, which project to the neurons in deep layers. The deep collicular neurons then project to the saccade generators in the reticular formation. Thus, visual information can reach the saccade generators by means of a fast and relatively direct path through the superior colliculus (Figure 11.21).

As you might expect, given this circuitry, electrical stimulation in the superior colliculus triggers saccadic eye movements. Stimulation in the rostral superior colliculus elicits small saccades toward locations in the contralateral visual hemifield, whereas stimulation in more caudal regions of the superior colliculus elicits progressively larger saccades. Medial and lateral stimulation evoke upward and downward saccades, respectively. In fact, the endpoints of all elicited saccades are systematically mapped across the superior colliculus.

Remarkably, in most of the superior colliculus, this **saccade motor map** is almost perfectly aligned (in register) with the retinotopic map in the more superficial layers of the superior colliculus. That is, electrical stimulation of the deep collicular neurons tends to elicit a saccade to the location in space to which the overlying, more superficial collicular neurons are most responsive (Figure 11.22). This alignment of the visual and motor maps occurs because individual neurons in the superficial layers of the superior colliculus tend to project to deep collicular neurons directly below them (Figure 11.21). In essence, the superficial visual layers send a very short and highly topographic projection to the deep saccade-eliciting layers. Thus, when a light flashes somewhere in your visual field, a spot of visual activity is generated in the superficial layers of the superior colliculus; this then triggers activity in the underlying deep collicular neurons, which makes your eyes move rapidly toward that flashing light. It is an elegant and efficient circuit.

Plasticity in the Saccade Control System

A fascinating aspect of the saccade control circuit is that it is capable of error correction. For example, if saccades consistently overshoot or undershoot their targets, then the size of the saccades is adjusted over the course of several trials. To study this phenomenon, monkeys were trained to make horizontal saccades to visual targets and implanted with "search coils" around the eyes to track all eye movements. Then

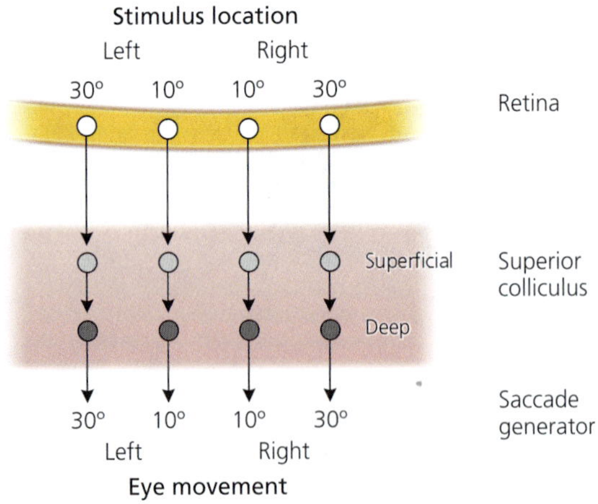

Figure 11.21 A topographic circuit through the superior colliculus. Parallel topographic pathways course from the retina through the superior colliculus to the saccade generators, which trigger saccades to the locations of the stimuli. Because neurons in the superficial layer of the superior colliculus project to neurons directly below them, collicular neurons tend to command saccades toward the locations where they respond to visual stimuli.

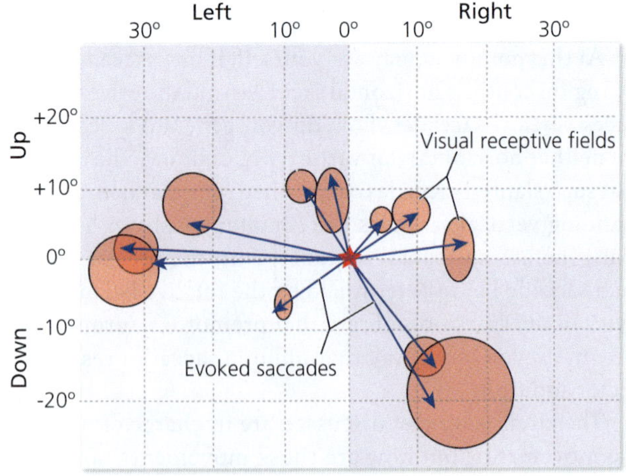

Figure 11.22 Aligned sensory and motor maps. Experimenters measured the saccades (blue arrows) elicited by electrical stimulation at 11 different sites in a monkey's superior colliculus. They then mapped the visual receptive fields (reddish ellipses) of neurons at those sites. The endpoints of each elicited saccade fall within the visual receptive fields, indicating that the visual and saccade maps are almost perfectly in register. [After Schiller and Stryker, 1972]

the lateral and medial rectus muscles were surgically weakened in one eye. The expectation was that the eye with the weakened eye muscles would make abnormally small saccades. Indeed, with an eye patch on the weakened eye, the weakened eye undershot the target consistently, even as the intact eye made a correct saccade (Figure 11.23).

The experimenters then switched the eye patch to the other eye and tested the animal again. The weakened eye still made saccades that were too small. However, when the test was repeated 5 days later, the weakened eye performed accurately, whereas the intact eye now made saccades that were too large. This last observation is important because it means that the plasticity was probably not due to the weakened muscle simply growing stronger. Instead, the motor commands to both eyes (which usually receive matching commands) must have grown in strength. If you remember our discussion of adaptive plasticity in the vestibulo-ocular reflex (see Chapter 10), you might suspect that the cerebellum is involved in this sort of across-trial plasticity. Indeed, in monkeys with large lesions of the posterior cerebellum, the weakened eye never "learns" to reach its target with a single saccade.

The Frontal Eye Field

In addition to the subcortical circuits we have discussed so far, several areas in the neocortex contribute to the control of eye movements. Particularly important and well studied is the **frontal eye field**, which is located just anterior to the premotor cortex (Figure 11.24; see also Figure 10.17). The frontal eye field receives visual input from a variety of cortical areas and projects to the superior colliculus as well as to the saccade generators in the reticular formation. Its neurons tend to increase their firing shortly before saccades and other eye movements, and electrical stimulation in the frontal eye field can trigger eye movements even at low current intensities.

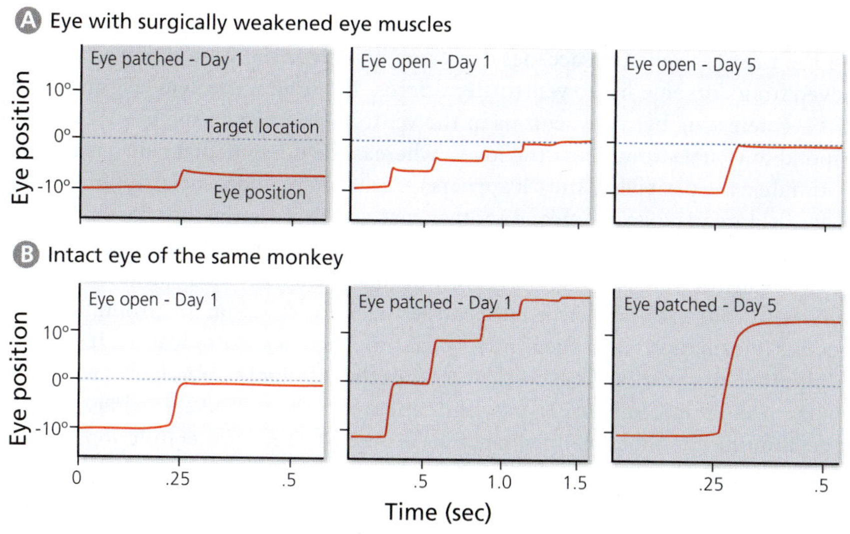

Figure 11.23 Saccade adaptation. Experimenters implanted a monkey with eye-tracking coils and removed a small piece of the medial and lateral rectus muscles in one of the eyes. They then placed a patch over the operated eye and rewarded the monkey for making a horizontal saccade to a visual target. The patched eye consistently undershot the target, whereas the intact eye moved directly to the target (left graphs). After the patch was switched to the intact eye, the weakened eye moved to the target in multiple saccades, whereas the patched eye overshot (middle graphs). These findings were expected because the muscles of the operated eye are now weaker than they used to be, yet continue to get the same motor commands as the intact eye. After 5 days of patching the intact eye, the weakened eye was able to reach the target in a single saccade (right graph in A), whereas the patched eye overshot (right graph in B). Thus, given enough trials, errors in saccade magnitude can be corrected by adjusting the strength of the responsible motor commands. [After Optican and Robinson, 1980]

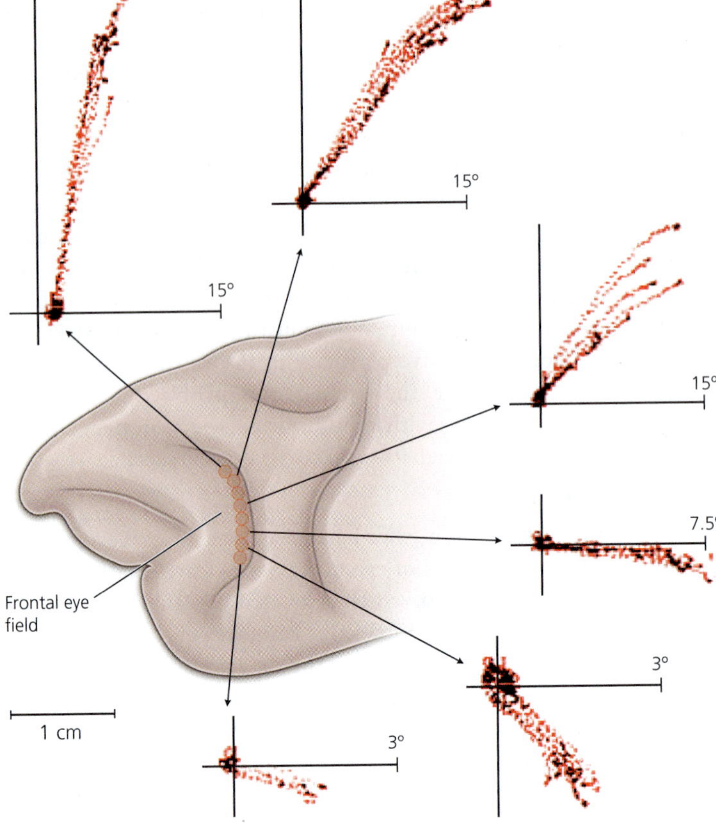

Figure 11.24 Saccades elicited from the frontal eye field (FEF). Shown in the middle of this figure is the left frontal lobe of a macaque monkey (anterior is to the left). The light red circles indicate 8 locations in the FEF where eye saccades were evoked by applying electrical stimuli. The trajectories of the evoked saccades at some of these locations are shown as red dotted lines in the graphs. The elicited saccades became smaller and smaller (note the changes in scale) as the electrode was moved to progressively more ventrolateral locations. All saccades were directed at targets in the right visual field. [From Bruce et al., 1985]

The size of the elicited saccades decreases steadily as the stimulation electrode is moved from dorsomedial to ventrolateral across the frontal eye field (Figure 11.24). This is interesting because neurons in the ventrolateral part of the frontal eye field respond to visual stimuli near the fovea, whereas the dorsomedial neurons respond to stimuli in the periphery. Thus, it appears that the motor and visual maps in the frontal eye field are in register with one another, just as they are in the superior colliculus.

One major function of the frontal eye field is to select a target for the next saccade (or two) when several potential targets are available, such as when you are scanning a scene with your eyes. To perform this target-selection function, the frontal eye field requires information about the features of a stimulus, not just its location. It probably obtains this information from cortical areas in the inferior temporal lobe, which you will learn about in Chapter 12. A second function of the frontal eye field is to retain target information about planned saccades in memory. Yet another function is to prevent saccades under some conditions. This ability to suppress saccades helps prevent the eyes from "getting stuck" on a particular target when scanning a scene and gives us the flexibility to look away from interesting stimuli (perform anti-saccades).

The fourth major function of the frontal eye field is to control **smooth pursuit eye movements**. These are not saccades but rather the sort of automatic eye movements you make when you are tracking a moving stimulus, such as a bird flying across your field of view. Neurons related to these smooth pursuit eye movements tend to be segregated into a special subdivision of the frontal eye field, but they have not been studied as thoroughly as the saccade-related neurons. In any case, you can see that the frontal eye field adds a substantial layer of complexity to the functions of the subcortical eye movement circuits.

This raises an interesting question: does the frontal eye field control eye movements directly, through its connections to the reticular saccade generators, or indirectly, through its projections to the superior colliculus? To address this question, experimenters

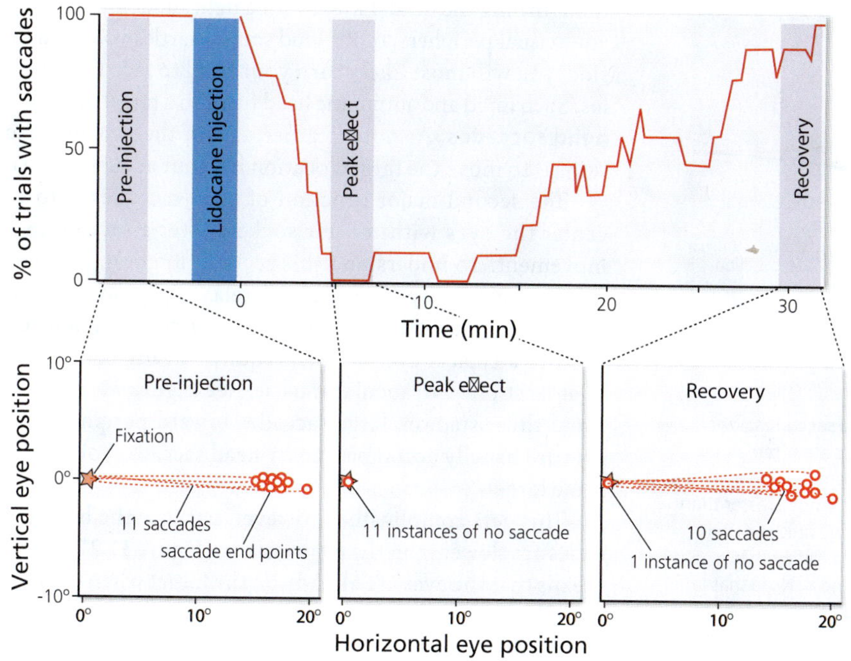

Figure 11.25 The frontal eye field needs the superior colliculus. Macaque monkeys were implanted with a stimulating electrode in the frontal eye field (FEF). They were also implanted with a micropipette through which the local anesthetic lidocaine could be injected into the superior colliculus. As the graphs show, stimulation of the FEF consistently evoked a saccade during the preinjection period. Shortly after the lidocaine injection, the same FEF stimulation no longer elicited saccades. Half an hour later, the evoked saccades returned. Because lidocaine temporarily inactivates neurons by blocking their sodium channels, we can infer that FEF neurons cannot trigger saccadic eye movements without assistance from the superior colliculus. [After Hanes and Wurtz, 2001]

repeatedly stimulated the frontal eye field of monkeys electrically and injected the sodium channel blocker lidocaine into the superior colliculus (Figure 11.25). As the lidocaine took effect and silenced the collicular neurons, frontal eye field stimulation ceased to elicit saccades. Half an hour later, after the lidocaine effects had faded, the evoked saccades returned. These results strongly suggest that the frontal eye field evokes saccades indirectly through its projections to the superior colliculus. Of course, this leaves unresolved the question of what the frontal eye fields might be doing with their direct projections to the saccade generators in the reticular formation.

A Complex Neural Net Controls Targeted Eye Movements

The neocortex also contains *parietal* and *supplementary eye fields.* These additional eye fields are densely interconnected with each other and with the frontal eye field. They probably cooperate in most of their functions. Besides these official "eye fields," several other cortical regions are also capable of influencing eye movements. For example, artificial stimulation of the primary visual cortex can evoke saccades. These saccades persist even if the frontal or parietal eye fields are lesioned, but they disappear after lesions of the superior colliculus. This suggests that the primary visual cortex can trigger saccades by means of its direct projections to the superior colliculus.

In summary, the neocortex contains a variety of areas that can control eye movements. Collectively, these areas form a long and complicated side-loop to the subcortical circuit that passes through the superior colliculus (Figure 11.26). One might think that such a long side-loop would involve unacceptable time delays (causing the descending cortical input to arrive after the subcortical circuits had already completed their work), but this is not the case. Although saccades are the fastest muscle-controlled movements in humans, the movements only start about 100–200 ms after the stimulus appears. During that period, activity in the superior colliculus increases gradually until it reaches the threshold for triggering an eye movement. Some of the cortical side-loops are fast enough to boost or suppress this ramp-up of collicular activity. Thus, they can change the probability of making a saccade to a particular target.

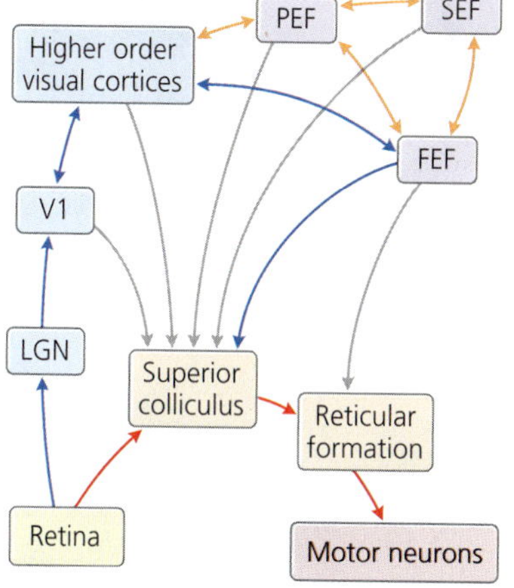

Figure 11.26 Neocortical side-loops in the saccade control system. The most direct pathway from the retina to the extraocular motor neurons is shown in red. The cortical side-loop through the primary visual cortex (V1) and the frontal eye field (FEF) is shown in blue. Additional side-loops through the parietal and supplementary eye fields (PEF and SEF, respectively) are shown in yellow.

Targeted Head Movements

When scanning a visual scene, we typically move only our eyes and keep our heads relatively still. This is efficient because moving the eyes is faster and takes less energy

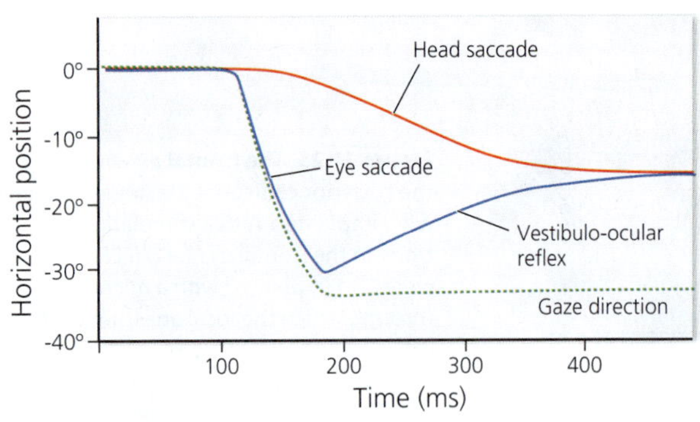

Figure 11.27 Interactions between head and eye saccades. When the head is not restrained, saccadic eye movements are often accompanied by saccadic head movements. The graph depicts such a head-and-eye saccade to a visual target. Because the eyes move faster than the head, the head is still moving when the eyes have finished their saccade. This mismatch triggers the vestibulo-ocular reflex that counterrotates the eyes relative to the head movement. Note that head position in this graph is represented in world-centered coordinates (relative to an external reference point), whereas eye position is represented in head-centered coordinates (relative to the head). [After Freedman, 2008]

than turning the head. However, if a light suddenly flashes in your visual periphery, or if a loud sound startles you from the side, you will most likely turn your head to face the stimulus. Such rapid and automatic head turns to a target are called **head saccades**. As noted earlier, one of their major functions is to move the target location into our acoustic fovea.

The second major function of head saccades is to recenter the eyes within their sockets after a saccadic eye movement. To understand this second function, you need to know that the eyeball's natural, relaxed position is pointing directly forward. Keeping the eye off-center, as it would be after a typical eye saccade, requires steady contraction of at least one extraocular muscle (see Figure 11.19 B). To avoid this situation, large saccades toward peripheral targets are usually accompanied by head saccades toward the same target.

This gets complicated, however, because the head saccades are slower than the eye saccades (Figure 11.27), which means that the eyes are already on the target when the head is still turning. As the head continues to turn, the eyes would overshoot the target unless they counterturn to compensate for the head turn. Indeed, they counterturn with awesome precision, keeping the gaze (the line of sight) locked on target. This counterturning of the eyes results from the operation of the vestibulo-ocular reflex, which generally counterrotates the eyes as the head turns (see Chapter 10). As you might imagine, the vestibulo-ocular reflex must be actively suppressed (through mechanisms that are not yet clear) when you want to turn your head without moving the eyes in their sockets.

The neural circuits underlying head saccades seem to overlap with those controlling eye saccades because electrical stimulation in the superior colliculus and the cortical eye fields can produce head saccades as well as eye saccades. Despite this commingling, the neural mechanisms for head and eye saccades can be uncoupled from one another. For instance, in monkeys trained to make both head and eye saccades, lidocaine inactivation of the superior colliculus impairs only the eye movements, leaving the head movements largely intact. Furthermore, some cortical eye fields can command head saccades even when the superior colliculus is taken off line. Unfortunately, little else is known about the neural mechanisms underlying head saccades. In general, neurobiologists have studied head movements much less than eye movements for the simple reason that it is very difficult to record from neurons in an animal that is moving its head. Such experiments require the use of electrodes that are chronically implanted in the brain (Box 11.2: Recording Neural Activity in Awake Animals).

Targeted Hand Movements

Targeted hand movements, such as reaching and pointing, are not as automatic as saccadic head or eye movements; nor are they as fast. Nonetheless, targeted hand movements can be enormously useful, especially if they lead to grasping and other direct interactions with the stimulus object. Given this functional significance, it is not surprising that the neural mechanisms of reaching, pointing, and grasping have received considerable scrutiny, at least in primates.

One finding to emerge from this research is that the superior colliculus is less involved in hand movements than in the head and eye movements we just discussed. Electrical stimulation in a small part of the superior colliculus does elicit hand movements, but lesions of the superior colliculus do not abolish reaching (although more subtle effects remain a possibility). In contrast, pathways through the cerebral cortex are heavily involved in the control of hand and arm movements. As mentioned in Chapter 10, some neurons in the primary motor and premotor cortices project

RESEARCH METHODS

Box 11.2 *Recording Neural Activity in Awake Animals*

A major disadvantage to recording from neurons in anesthetized animals is that the recorded activity may not reflect what happens when the animal is awake and alert. Moreover, different anesthetics are known to alter brain activity in different ways, raising questions about which anesthetics are least problematic. Yet another limitation is that neural activity related to skeletal movement and cognition cannot be recorded from anesthetized animals. To overcome these shortcomings, experimenters have devised several methods for recording from awake animals. These methods are called *chronic recording* techniques because they involve chronic (persistent) implants on an animal's head.

One chronic recording technique involves stabilizing the animal's head while letting the body move (at least to some extent). In such experiments, most commonly performed with monkeys, a binding post is cemented to the skull while the animal is fully anesthetized. When the animal wakes up, the binding post is clamped to a sturdy metal frame, while the rest of the animal remains relatively free to move. The recording electrode is then advanced into the brain through a small hole in the skull that was drilled while the animal was anesthetized. Importantly, the skull hole is covered with a cylindrical recording chamber that is filled with a transparent oil (similar to Vaseline) to prevent the meninges from drying out. As the animal goes about the task for which it has been trained, the experimenters move the electrode around until its tip is close to a neuron. This causes no pain because there are no pain sensors in the brain. The experimenters then record extracellular action potentials and correlate them with the animal's behavior. At the end of the day's experiment, the electrode is withdrawn and the recording chamber is closed with a tight-fitting cap. At that point, the animal is allowed to move about freely. Although the binding post and recording chamber on top of the animal's head look sinister, they cause minimal discomfort once the skin incisions have healed.

A second chronic recording technique involves permanently implanted electrodes (Figure b11.2). With this approach no binding post is required, and the animal can move its entire body (after recovering from the anesthesia). During a recording session, a cable is used to convey the electrical signals from the animal's head to the recording apparatus. Rats tolerate such tethering quite well, as long as the cable is prevented from twisting or dragging. Monkeys, in contrast, tend to grab the cable and chew on it. To solve this problem, experimenters may transmit the signals wirelessly, but this requires relatively bulky transmitters and batteries that must be attached to the animal. Therefore, chronic recording experiments on monkeys typically avoid implanted electrodes and instead involve head-stabilized monkeys with implanted recording chambers.

A recent trend in chronic recording has been to employ more and more recording electrodes. Particularly popular have been *tetrodes*, which consist of 4 insulated microwires that are twisted together. By comparing the shape of action potentials recorded simultaneously at the tips of each microwire in a tetrode, the experimenters can identify up to a dozen or so distinct neurons (just as you can separate the sounds produced by individual performers when you record a musical performance with multiple, strategically placed microphones). By implanting dozens of tetrodes, the experimenters can record from more than 100 neurons at a time. Although powerful computers are required to handle all this information, the effort is worth it. The ability to "listen in" on hundreds of neurons while an animals is performing an interesting behavior is truly magical. The method allows you to "see" the brain at work.

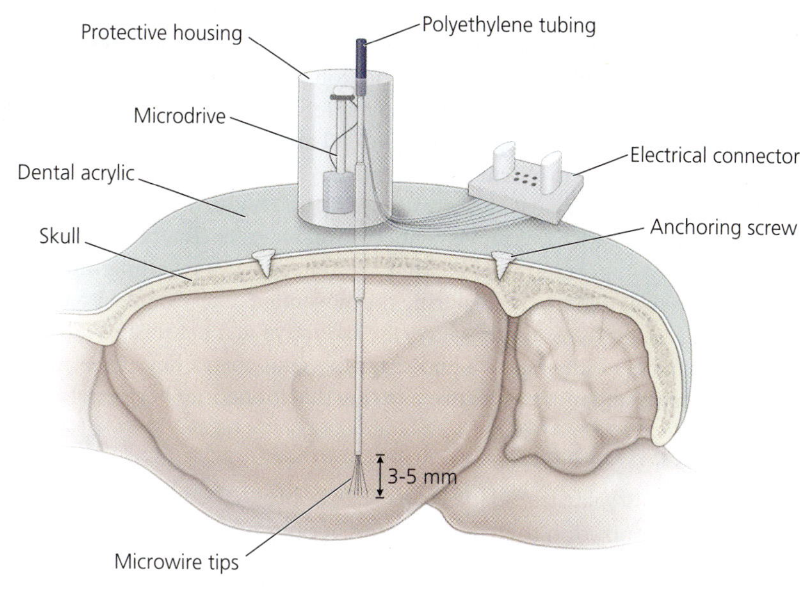

Figure b11.2 Chronic recording assembly for a rat. This diagram depicts a typical implant used to record neural activity from an awake rat. Several microwires are threaded through a guide tube (cannula) so that they protrude by a few millimeters into the targeted brain region. The other ends of the microwires are soldered to an electrical connector that is affixed to skull with dental acrylic. The guide tube can be raised or lowered, together with the microwires, by means of a microdrive, which is also cemented to the skull. For each recording session, a lightweight cable from a multichannel amplifier is plugged into the connector on the skull. [After Szymusiak and Nitz, 2003]

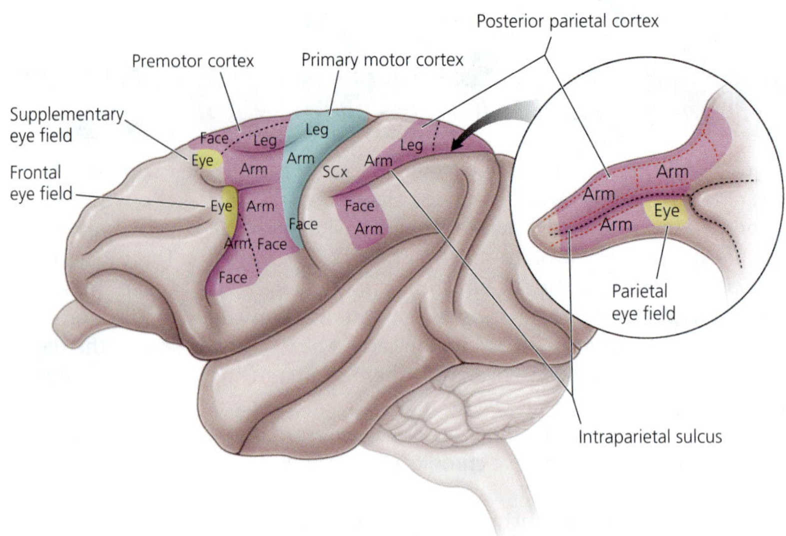

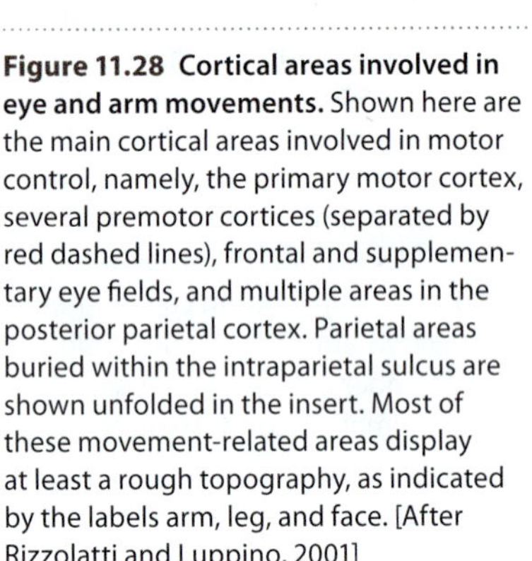

Figure 11.28 Cortical areas involved in eye and arm movements. Shown here are the main cortical areas involved in motor control, namely, the primary motor cortex, several premotor cortices (separated by red dashed lines), frontal and supplementary eye fields, and multiple areas in the posterior parietal cortex. Parietal areas buried within the intraparietal sulcus are shown unfolded in the insert. Most of these movement-related areas display at least a rough topography, as indicated by the labels arm, leg, and face. [After Rizzolatti and Luppino, 2001]

to subcortical motor and premotor neurons controlling the contralateral arm, and electrical stimulation of these areas can elicit targeted hand movements (see Figure 10.20). Moreover, large lesions of the premotor and primary motor cortices cause weakness (paresis) or outright paralysis of the contralateral arm. These findings show that the motor and premotor cortices can control hand movements.

Which parts of the neocortex are involved in directing the hand toward a specific target? The answer lies in the posterior parietal cortex, in a set of cortical areas just anterior to the parietal eye field (Figure 11.28). As noted earlier, posterior parietal neurons encode stimulus location in a variety of complex coordinate systems. One likely reason for this complexity is that these neurons transform sensory information into signals that the motor system can use (no matter how hard they are for neuroscientists to comprehend). Indeed, posterior parietal areas have strong projections to the premotor areas that control arm movements.

Lesions of the posterior parietal cortex cause **optic ataxia**, which is defined as an inability to make accurate hand movements to visual targets. The posterior parietal cortex is especially important for adjusting the trajectory of reaching movements when the target moves unexpectedly during the movement. Consistent with this idea, many neurons in the posterior parietal cortex are active just before and during targeted hand movements. In short, the posterior parietal cortex lies at the interface between the sensory and motor systems, blurring the boundaries between the two.

BRAIN EXERCISE

How does the circuitry illustrated in Figure 11.26 relate to the Meynert-James model of brain organization that we discussed in Chapter 1?

11.5 How Do Animals Navigate through Space?

You have now learned that the brain encodes stimulus locations in a variety of different coordinate systems. Making sense of this diversity becomes easier once you consider that neurons encode spatial information for specific functions, all related to the guidance of movement. So far we have discussed how spatial information is used to target eye, head, and arm movements. Now let us ask, what kind of spatial information do you need to guide your whole body as it moves through the environment and toward a target?

What information do you need when driving to a new restaurant or taking a novel shortcut? You need information about the location of various objects and places relative to one another and relative to where you are right now. For that, you must encode spatial information in world-centered coordinates, independently of your own body's position. Such world-centered representations are sometimes called *allocentric* to distinguish them from *egocentric* (eye-, head-, or body-centered) representations of space. They can also be called **cognitive maps**, a term that the American psychologist Edward Tolman coined in 1948 to indicate that animals may learn not only how to get to a specific place from a specific starting point but also where that place is relative to other places and objects.

If animals possess a world-centered cognitive map, then they should be able to use it to navigate their world more flexibly and more efficiently. In support of this idea, the psychologist Karl Lashley once observed that a rat trained to navigate a complex maze climbed over the maze wall right at the beginning of the maze and scrambled directly over to the place with the food. This was certainly more efficient than making all those silly turns inside the maze. In general, learning where a place is relative to other places (having a cognitive map) is more efficient than simply learning a sequence of left and right turns, especially when the learned route is needlessly long or suddenly blocked. A cognitive map can help an animal identify shortcuts and chart detours. We certainly use our own cognitive map in this manner. But do non-human animals really navigate by means of allocentric cognitive maps, or did Lashley's impatient, maze-escaping rat just get lucky?

Testing for Allocentric Navigation in Animals

In today's age of Google Maps and GPS-enabled devices, it is tempting to think that all animals carry inside their brains a cognitive map that guides them through the world, but this need not be so. Even humans can find their way from one place to another if they are given a stepwise series of instructions such as turn left at the light, then right at the stop sign, and go until you see the school. Similarly, animals might find their way to a food source by learning when to turn in which direction and how far to go relative to various landmarks in the environment. Egocentric representations of space are sufficient for this kind of navigation. Indeed, egocentric navigation is used to great effect by many invertebrates. Honeybees are the only invertebrates for whom these is some evidence of allocentric navigation, but even this remains controversial. How, then, can you prove that an animal is navigating allocentrically rather than egocentrically?

Confronted with this question, Edward Tolman designed an elegant experiment. He put rats in one arm of a cross-shaped maze (Figure 11.29), blocked the opposite arm, and hid food at the end of one of the open arms. Over multiple trials in which the food was always placed in the same arm, the rats learned to turn into the baited arm and retrieve the food. Then Tolman changed the task. He moved the block to the opposite arm and started the rat from the previously blocked arm. The food remained in the same arm. Tolman's question was whether the animals in these test trials would turn away from the food or toward it. If they turn away, one can infer that they are using egocentric **habit learning** to navigate the maze; they simply make the same sort of turn (left or right, relative to their body axis) that they had learned to make during training.

In contrast, if the rats turn toward the food, then they must have learned where in the room the food was located. They might have learned, for example, that the food is consistently near the wall with the big red poster and away from the window (the walls of a rat maze are usually low enough so that the rat can look out!). This kind of **place learning** differs from egocentric habit learning in that it involves learning about the spatial relationships of environmental features relative to one another rather than relative to the animal's body. In other words, it involves an allocentric representation of space, a cognitive map.

When researchers examined how rats performed in Tolman's cross maze, they made an interesting observation. After one week of training, most rats exhibited the allocentric place learning strategy. After an additional week of training, the same rats were using the egocentric habit strategy; that is, when tested on day 16, they turned

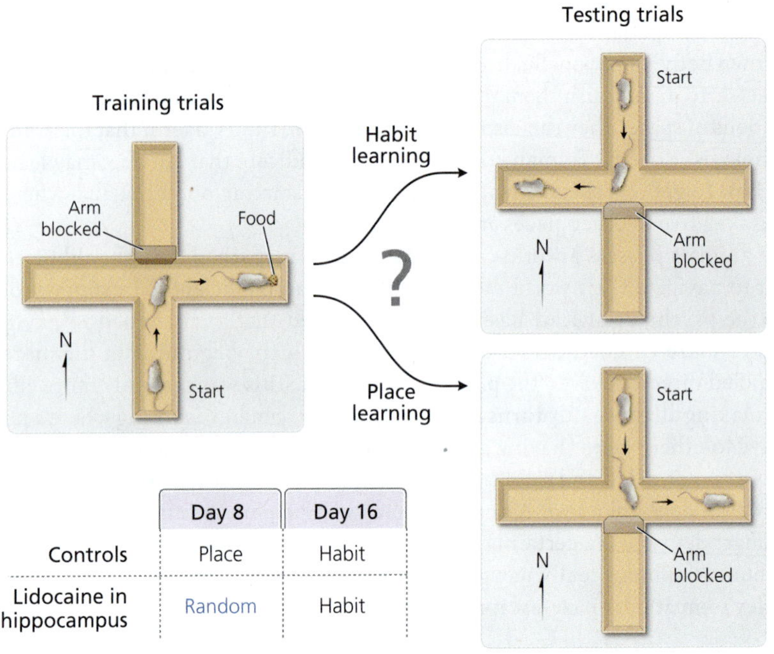

	Day 8	Day 16
Controls	Place	Habit
Lidocaine in hippocampus	Random	Habit

Figure 11.29 Tolman's cross maze. Shown at the top left is a cross-shaped maze in a room with various distinct features (e.g., a door and a poster on one of the walls). During training trials, the Northern arm is blocked so that the maze becomes T-shaped. Rats are then placed in the Southern arm and trained to obtain food at the East end. During test trials, the Southern arm is blocked and the rats start their trials from the previously blocked Northern arm. The important question is how the rats behave on these test trials. If they consistently turn West, then they formed a habitual response that requires only egocentric representations of space. On the other hand, if the rats turn East, then they must have learned which place inside the room contained the food. They must have formed a "cognitive map" of the room. The data table shows that normal rats generally exhibit the allocentric place learning strategy if they were trained for 7 days (4 trials/day). After 7 additional days of training, the same rats switch to the egocentric habitual response. Lidocaine injections into the hippocampus selectively impair place learning. [After Packard and McGaugh, 1996]

away from the place where the food was typically located (Figure 11.29). Thus, rats are capable of using allocentric as well as egocentric strategies to navigate; but they tend to use the egocentric strategy only after they have learned the task really, really well. This resembles what we do. For example, if you routinely walk the same route to school or work but plan to make a little detour just today (to stop by a friend's house), you are likely to forget about the detour and go home directly, as if you were on autopilot. How many of your routine tasks have become almost entirely automatic, ruled by what psychologists would call habit learning? Very likely, most of them have. In any case, experiments using Tolman's cross maze clearly confirmed that rats are capable of allocentric place learning as long as they have not been trained to the point where habits prevail (overtrained).

Another widely used test for allocentric navigation is the **Morris water maze**, named after its inventor Richard Morris. In this task, a rat is forced to swim in a large water-filled circular tank until it finds a submerged (hidden) platform on which to rest (Figure 11.30). The platform is kept in a constant location, but the rat's starting position is varied from trial to trial. This makes it impossible for the rat to find the platform by means of an egocentric strategy (e.g., turn 30° left and swim forward 1 m). Instead, the rat must learn where the platform is relative to landmarks outside of the maze. It must construct for itself a cognitive map of the room, which it can then use to plot a course for the platform from any arbitrary starting position. Rats can learn the water maze quite well, as demonstrated by a steady reduction across trials in the

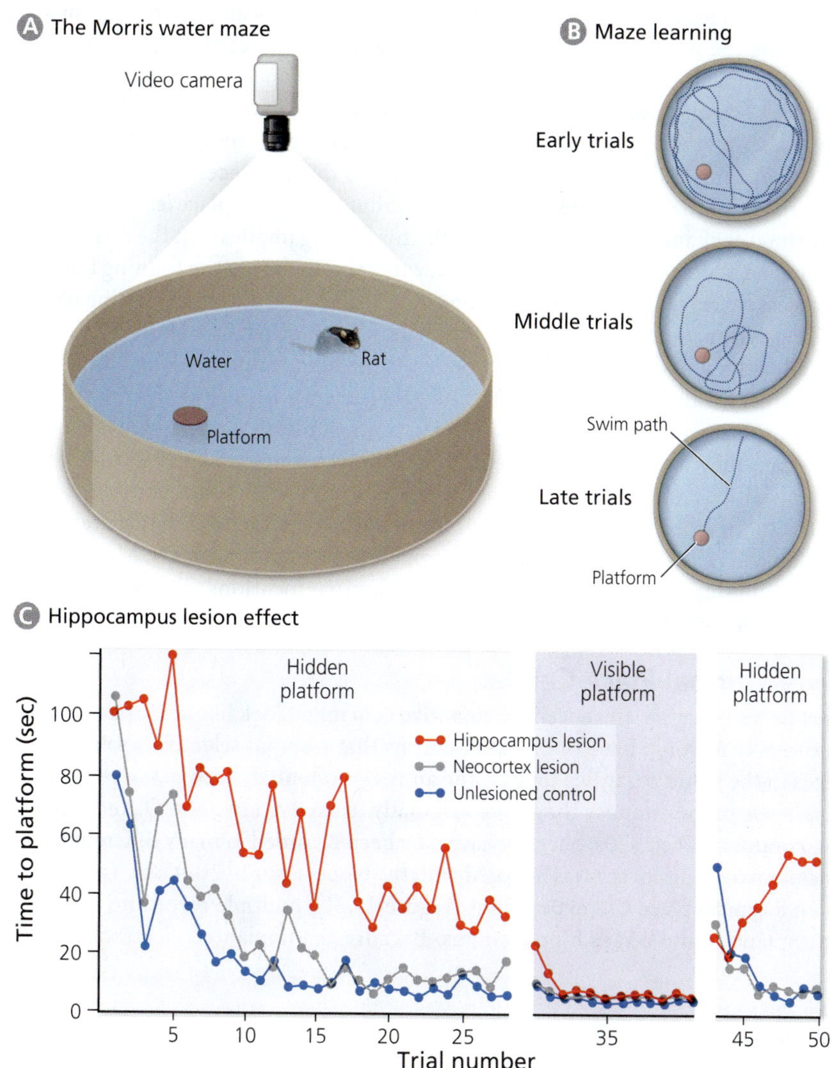

Figure 11.30 Spatial learning in the Morris water maze. A Morris water maze (A) consists of a circular tub filled with opaque water. Individual rats are placed into the water at random starting positions and must swim until they get to a submerged platform, where they can rest. As training progresses, swim paths shorten (B). The graph in (C) shows that rats with large bilateral lesions of the hippocampus (red line) solve the water maze more slowly than control rats or rats with small neocortical lesions. The hippocampus-lesioned rats are unimpaired if the platform is visible (blue shaded region of the graph), proving that hippocampus lesions do not simply reduce a rat's swimming ability or motivation to escape from the water. [B after Morris, 2008; C after Morris et al., 1982]

time it takes the rats to swim to the platform (Figure 11.30 C). If the maze is rotated within the room, trained rats swim to the place in the room where the platform used to be, confirming their use of allocentric cues.

Hippocampal Lesions Impair Allocentric Navigation

To determine where in the brain an allocentric cognitive map might be, neurobiologists have lesioned diverse brain regions and asked which of these lesions disrupt allocentric navigation in tasks such as the Morris water maze. Overwhelmingly, those studies showed that the hippocampus is a critical player. As we discussed in earlier chapters, the hippocampus has multiple functions, some of which may differ across species (see Box 3.2: Hippocampal Structure and Functions). In rodents, at least, its major and best studied function is spatial navigation, specifically navigation by means of a cognitive map.

Strong evidence for this hypothesis was obtained by Richard Morris and his collaborators when they showed that large bilateral lesions of the hippocampus slow the rate at which rats learn the water maze. Moreover, the lesioned rats never learn the maze as well as control rats (Figure 11.30 C). Importantly, the lesioned rats are not impaired if the platform is visible rather than submerged, proving that the lesioned rats can swim just fine, are motivated to get out of the water, and can use an egocentric method of navigation (moving toward a landmark). Morris also showed that rats with lesions in the neocortex overlying the hippocampus are not impaired at learning the water maze. These basic findings have been confirmed by numerous later experiments.

If hippocampus lesions impair allocentric navigation in the water maze, then they should also interfere with allocentric navigation in Tolman's cross maze. Indeed, when the hippocampus is temporarily inactivated in rats that had received 1 week of training in the cross maze, and should therefore exhibit allocentric place learning, the rats performed at chance levels (Figure 11.29). This finding implies that the rats had constructed an allocentric cognitive map inside their brain during the training but could not access this map when the hippocampus was inactivated during the testing phase.

The vast majority of studies on the neural bases of navigation have been performed with rats, which makes you wonder: is the hippocampus needed for allocentric navigation also in humans? The principal problem with answering this question is that experimenters cannot lesion human brains unless it is medically necessary. Therefore, they usually study patients whose brains were damaged accidentally or through disease (or, in a few cases, through surgery). Such brain damage is rarely limited to the brain region of interest, but a handful of patients with relatively selective damage to the hippocampus have been identified over the years. Most of them have trouble learning the spatial layout of unfamiliar locations. This finding is consistent with the data from rats.

Hippocampal Place Cells

What do you imagine an allocentric cognitive map might look like at the cellular level? Presumably it would have to contain neurons that respond selectively to particular places in the world regardless of how the animal is oriented. Such **place cells** indeed exist in the hippocampus. They were originally discovered by John O'Keefe and his collaborators in the 1970s, but they have now been recorded in many different laboratories, and their discovery was honored with the Nobel Prize in 2014 (see Table 11.2).

In a typical place cell experiment (Figure 11.31), multiple recording electrodes are implanted into a rat's hippocampus. Because the implanted electrodes cannot

Year	Laureates
2014	John **O'Keefe**, May-Britt **Moser**, and Edvard I. **Moser** "for their discoveries of cells that constitute a positioning system in the brain"
2004	Richard **Axel** and Linda B. **Buck** "for their discoveries of odorant receptors and the organization of the olfactory system"
2000	Arvid **Carlsson**, Paul **Greengard**, and Eric R. **Kandel** "for their discoveries concerning signal transduction in the nervous system"
1998	Robert F. **Furchgott**, Louis J. **Ignarro**, and Ferid **Murad** "for their discoveries concerning nitric oxide as a signaling molecule in the cardiovascular system"
1991	Erwin **Neher** and Bert **Sakmann** "for their discoveries concerning the function of single ion channels in cells"
1986	Stanley **Cohen** and Rita **Levi-Montalcini** "for their discoveries of growth factors"
1981	Roger W. **Sperry** "for his discoveries concerning the functional specialization of the cerebral hemispheres" David H. **Hubel** and Torsten N. **Wiesel** "for their discoveries concerning information processing in the visual system"

Table 11.2 Nobel Prizes in physiology or medicine (in the area of Neurobiology and Behavior). All quotations from http://www.nobelprize.org/nobel_prizes/medicine/laureates/

1977	Roger **Guillemin** and Andrew V. **Schally** "for their discoveries concerning the peptide hormone production of the brain" Rosalyn **Yalow** "for the development of radioimmunoassays of peptide hormones"
1973	Karl **von Frisch**, Konrad **Lorenz**, and Nikolaas **Tinbergen** "for their discoveries concerning organization and elicitation of individual and social behavior patterns"
1970	Sir Bernard **Katz**, Ulf **von Euler**, and Julius **Axelrod** "for their discoveries concerning the humoral transmittors in the nerve terminals and the mechanism for their storage, release, and inactivation"
1967	Ragnar **Granit**, Haldan Keffer **Hartline**, and George **Wald** "for their discoveries concerning the primary physiological and chemical visual processes in the eye"
1963	Sir John Carew **Eccles**, Alan Lloyd **Hodgkin**, and Andrew Fielding **Huxley** "for their discoveries concerning the ionic mechanisms involved in excitation and inhibition in the peripheral and central portions of the nerve cell membrane"
1961	Georg **von Békésy** "for his discoveries of the physical mechanism of stimulation within the cochlea"
1957	Daniel **Bovet** "for his discoveries relating to synthetic compounds that inhibit the action of certain body substances, and especially their action on the vascular system and the skeletal muscles"
1949	Walter Rudolf **Hess** "for his discovery of the functional organization of the interbrain as a coordinator of the activities of the internal organs" Antonio Caetano de Abreu Freire Egas **Moniz** "for his discovery of the therapeutic value of leucotomy in certain psychoses"
1944	Joseph **Erlanger** and Herbert Spencer **Gasser** "for their discoveries relating to the highly differentiated functions of single nerve fibers"
1938	Corneille Jean François **Heymans** "for the discovery of the role played by the sinus and aortic mechanisms in the regulation of respiration"
1936	Sir Henry Hallett **Dale** and Otto **Loewi** "for their discoveries relating to chemical transmission of nerve impulses"
1935	Hans **Spemann** "for his discovery of the organizer effect in embryonic development"
1932	Sir Charles Scott **Sherrington** and Edgar Douglas **Adrian** "for their discoveries regarding the functions of neurons"
1914	Robert **Bárány** "for his work on the physiology and pathology of the vestibular apparatus"
1906	Camillo **Golgi** and Santiago Ramón y **Cajal** "in recognition of their work on the structure of the nervous system"
1904	Ivan Petrovich **Pavlov** "in recognition of his work on the physiology of digestion, through which knowledge on vital aspects of the subject has been transformed and enlarged"

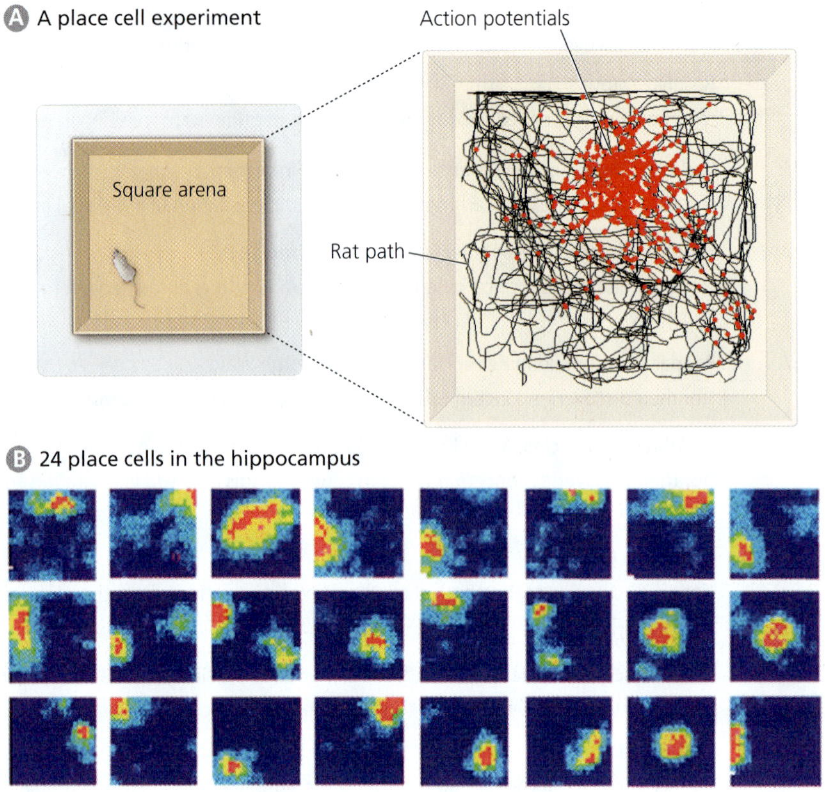

Figure 11.31 Hippocampal place cells. A rat with implanted recording electrodes is placed into a square box (A). As the rat wanders around the arena, the experimenter records from the implanted electrodes and notes where along the rat's path a neuron fires (red dots on the black lines). Shown in (B) are the spatial receptive fields of 24 neurons in the dorsal hippocampus of a rat exploring such a square arena. The warmer the color (yellow, red), the higher the neuron's firing rate at that location in the box (normalized to the number of times the rat was in each place). Note that the vast majority of hippocampal neurons have only a single "place field" within this arena. [A after Moser et al., 2008; B from Jung et al., 1994]

move and the brain is not exposed, the rat can move around without harming itself (Box 11.2: Recording Neural Activity in Awake Animals). The experimenters then use a long and flexible cable to connect the electrodes to a set of amplifiers, which allow them to record action potentials from multiple neurons as the rat is wandering about. The arena in which the rat can move is typically rectangular and placed inside a room with multiple distinctive features on the walls. In such a setup, many hippocampal neurons increase their firing rate when the rat is in a specific portion of the arena (Figure 11.31). Importantly, neurons that increase their firing rate at a particular location do so regardless of the direction in which the rat is facing. By definition, such neurons encode spatial information in allocentric, rather than egocentric, coordinates.

Detailed analyses have shown that hippocampal place cells tend to have just one or two spatial receptive fields (as long as the test arena is small). Furthermore, the place fields tend to be stable for weeks or even months if the environment remains unchanged. They also persist when the lights are turned off or if one or two features of the room have been removed. If the arena walls have no distinctive features, then rotating this arena inside the room tends to leave most place fields where they were, relative to the room. If the whole environment is scaled up or down, the place fields tend to change in size, getting bigger in a bigger arena and room. All of these findings support the notion that hippocampal place cells are part of the brain's cognitive map. It is important to note, however, that adjacent place cells almost never encode adjacent spatial locations. Therefore, they do not really form a map; the topological relationships between places in the world are not reflected in the topological relationships between the cells that code for those places.

Recent studies have shown that the firing rate of many place cells can be influenced by factors unrelated to the rat's current location, such as where the rat has been, where it is going, and what task it is trying to accomplish. At first blush, these findings seem incompatible with the idea that place cells encode an animal's spatial location. However, a better way to think about these data is to hypothesize that

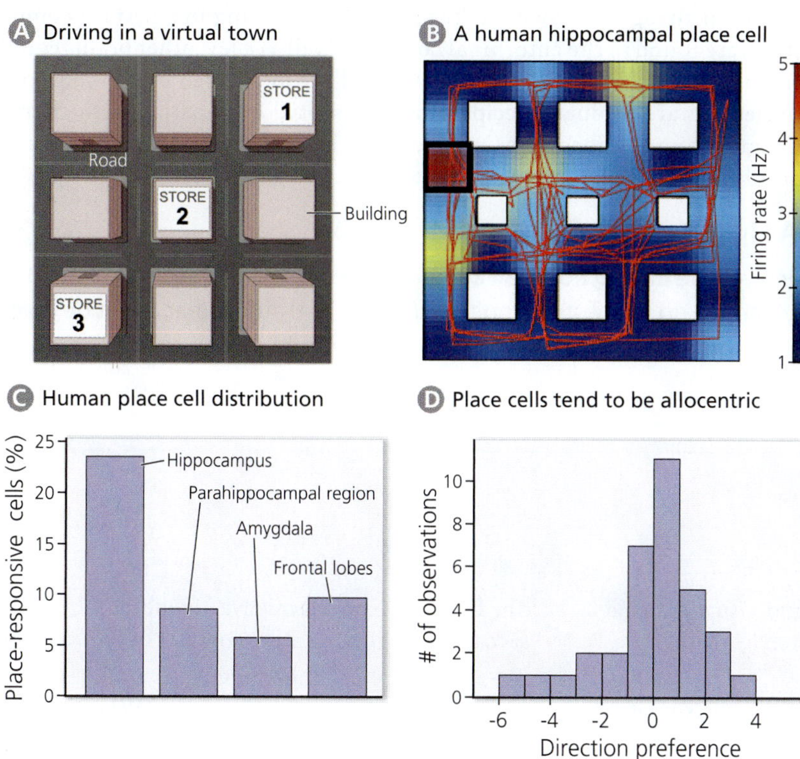

Figure 11.32 Place cells in humans. Epilepsy patients were implanted with recording electrodes in the hippocampus and several other brain regions. The patients then played a computer game in which they were driving a taxi through a virtual town (A). They were instructed to pick up passengers and drop them off at specific stores. Shown in (B) is the color-coded firing rate of a neuron in the right hippocampus of one subject. The red square (outlined in black) indicates a location in which this cell increased its firing rate consistently. The histogram in (C) shows that such place-responsive cells were most common in the hippocampus. As shown in (D), the place-sensitive cells in the hippocampus tend to fire in their preferred location regardless of the direction of travel, indicating that they probably encode spatial location in allocentric, rather than egocentric, coordinates. [From Ekstrom et al., 2003]

hippocampal neurons encode multiple types of information, superimposed on one another. For example, the animal's current location may be represented by the set of place cells that are activated above their background firing rate; whereas the location where the rat is planning to go may be encoded in the specific firing rate of individual place cells. According to this view, hippocampal place cells function in both spatial orientation and in some forms of memory that can be used to plan future actions.

Place Cells in Primates

Place cells are rarely studied in monkeys because monkeys don't appreciate being tethered; they are liable to grab the recording cable and chew on it! However, place cell experiments have been conducted in more cooperative primates, namely humans. Specifically, place cells have been recorded from humans who were being prepared for brain surgery (Figure 11.32). The patients were not allowed to move around, but they used virtual reality equipment (computer games) to navigate through a simulated town, picking up passengers and delivering them to various storefronts. Using this approach, almost a quarter of the recorded cells in the human hippocampus were place sensitive, meaning that they increased their firing rate at a specific place in the virtual town. Most of the neurons were relatively insensitive to the direction in which the patient moved through the place field, implying that they encode space allocentrically. Although place cells can also be recorded outside of the human hippocampus, such cells are relatively rare (Figure 11.32 C).

The cellular computations that generate place cells remain unclear, but an important hint comes from the recent discovery of **grid cells** in the entorhinal cortex. These neurons have multiple place fields that are arranged in a triangular grid-like pattern. Theoretical models suggest that hippocampal place cells may obtain their single, unique place fields by combining inputs from multiple grid cells with different grid spacings, but this hypothesis remains speculative. In any case, it simply pushes our question back a step: what kind of cellular computations generate grid cells? Again, the question remains open, but a reasonable guess is that grid cells integrate information about external landmarks with information about head direction and running speed. It is interesting to note, therefore, that **head-direction cells**, which increase their firing rate whenever the rat's head is facing in a certain compass direction, are found in the entorhinal cortex (as well as a few other brain regions) and that the activity of these head-direction cells is modulated by running speed. Thus, scientists are gradually deciphering how spatial information in the brain is transformed from an egocentric (eye-, head-, or body-centered) coordinate frame into an allocentric (world-centered) one.

BRAIN EXERCISE

Imagine you are driving home on autopilot when, suddenly, you remember that you wanted to go to the store and therefore pull a U-turn. What might be going on inside your brain during that "override" of the habitual behavior?

SUMMARY

Section 11.1 - In the somatosensory and visual systems the sensor arrays (skin and retina, respectively) are represented topographically inside the brain.

- Topographic circuits from the skin to the somatosensory cortex create a map-like representation of the body surface that is called a homunculus. Fingers and lips are overrepresented, and lateral inhibition sharpens the map.
- The ascending visual pathways are topographically organized, forming a highly distorted map of the visual field in the primary visual cortex (V1). Inputs from the two eyes converge in the upper layers of V1. The binocular neurons provide information about stimulus distance.

Section 11.2 - The location of sound sources is not mapped onto the auditory sensor array and must therefore be computed inside the brain.

- Localizing a sound source is usually accomplished by comparing the intensity and timing of the sound between the two ears. These comparisons are carried out by dedicated circuits in the medulla.
- The brains of owls contain a topographic map of auditory space in which each neuron has a small spatial receptive field. Mammalian brains do not have such a map and instead use a population code to specify the locations of auditory stimuli.

Section 11.3 - With mobile sensors, stimulus localization requires knowing sensor position. Spatial information can be encoded in eye-, head-, body-, or world-centered coordinates as well as intermediate coordinate frames.

Section 11.4 - Orienting toward stimuli involves some reflex arcs, especially through the superior colliculus, as well as long side-loops through cortical areas.

- Saccadic eye movements are generated by brainstem saccade generators. Saccades to visual targets are driven by a topographic circuit through the superior colliculus. The frontal eye field is involved in inhibiting reflexive saccades and guiding more complex eye movements.
- The neural circuits underlying head saccades overlap with those controlling eye saccades. When head and eye saccades occur together, the vestibulo-ocular reflex prevents the eyes from overshooting their target.
- Targeted hand movements, such as reaching or pointing, involves projections from the posterior parietal cortex to the motor cortex. Lesions of the posterior parietal cortex cause optic ataxia.

Section 11.5 - Navigating through space, animals may either memorize a habitually traveled route or form an allocentric (world-centered) cognitive map of their external environment.

- Tolman's cross maze and the Morris water maze have both been used to demonstrate that rats are capable of allocentric navigation. Overtrained animals tend to navigate along habitual routes, ignoring their cognitive map.
- Bilateral lesion or inactivation of the hippocampus disrupts a rat's capacity for allocentric navigation. Humans with hippocampal damage have trouble learning to navigate through novel environments.
- Hippocampal place cells encode specific spatial locations in allocentric coordinates. Place cells have been studied most extensively in rats, but are found also in humans.

Box 11.1 - Barn owls have asymmetric ears that allow them to localize broadband sounds with exceptional accuracy, both in azimuth and vertically.

Box 11.2 - To record neural activity in awake behaving animals, researchers implant electrodes into the brain and then record from those electrodes after the animal wakes up. Alternatively, they may implant a recording chamber onto the skull through which electrodes can be advanced into the brain after the animal has recovered from surgery.

KEY TERMS

spatial receptive field 338
dorsal column nuclei 339
principal trigeminal nucleus 339
ventral posterior nucleus 339
primary somatosensory cortex (S1) 339
somatosensory homunculus 340
barrel cortex 340
lateral inhibition 341
visual field 342
superior colliculus 343
lateral geniculate nucleus (LGN) 343
primary visual cortex (V1) 343
optic chiasm 343
visual hemifield 343
fixation point 343
binocular 345
monocular 345
ocular dominance bands 345
binocular depth perception 346
pyramidal neuron 346
canonical cortical circuit 346
cortical minicolumn 346
interaural level difference (ILD) 347
interaural time difference (ITD) 348
lateral superior olive 348
medial superior olive 348
inferior colliculus 349
primary auditory cortex 349
eye-centered coordinates 351
head-centered coordinates 351
body-centered coordinates 352
world-centered coordinates 352
posterior parietal cortex 353
saccade 354
saccade generator regions 354
saccade motor map 356
frontal eye field 357
smooth pursuit eye movement 358
head saccade 360
optic ataxia 362
cognitive map 363
habit learning 363
place learning 363
Morris water maze 364
place cell 366
grid cell 369
head-direction cell 369

ADDITIONAL READINGS

11.1 - Encoding Space in the Visual and Somatosensory Systems

Adams DL, Horton JC. 2009. Ocular dominance columns: enigmas and challenges. *Neuroscientist* **15**:62–77.

Constantinople CM, Bruno RM. 2013. Deep cortical layers are activated directly by thalamus. *Science* **340**:1591–1594.

Ferezou I, Haiss F, Gentet LJ, Aronoff R, Weber B, Petersen CCH. 2007. Spatiotemporal dynamics of cortical sensorimotor integration in behaving mice. *Neuron* **56**:907–923.

Parker AJ. 2007. Binocular depth perception and the cerebral cortex. *Nat Rev Neurosci* **8**:379–391.

Wandell BA, Dumoulin SO, Brewer AA. 2007. Visual field maps in human cortex. *Neuron* **56**:366–383.

11.2 - Encoding Space in the Auditory System

Grothe B. 2003. New roles for synaptic inhibition in sound localization. *Nat Rev Neurosci* **4**:540–550.

Werner-Reiss U, Groh JM. 2008. A rate code for sound azimuth in monkey auditory cortex: implications for human neuroimaging studies. *J Neurosci* **28**:3747–3758.

11.3 - Movable Sensors and Spatial Coordinate Frames

Avillac M, Denève S, Olivier E, Pouget A, Duhamel J-R. 2005. Reference frames for representing visual and tactile locations in parietal cortex. *Nat Neurosci* **8**:941–949.

Diamond ME, Von Heimendahl M, Knutsen PM, Kleinfeld D, Ahissar E. 2008. "Where" and "what" in the whisker sensorimotor system. *Nat Rev Neurosci* **9**:601–612.

Schlack A, Sterbing-D'Angelo SJ, Hartung K, Hoffmann K-P, Bremmer F. 2005. Multisensory space representations in the macaque ventral intraparietal area. *J Neurosci* **25**:4616–4625.

11.4 - Orienting Movements

Elsley JK, Nagy B, Cushing SL, Corneil BD. 2007. Widespread presaccadic recruitment of neck muscles by stimulation of the primate frontal eye fields. *J Neurophysiol* **98**:1333–1354.

Luppino G, Rizzolatti G. 2000. The organization of the frontal motor cortex. *News Physiol Sci* **15**:219–224.

Pisella L, Gréa H, Tilikete C, Vighetto A, Desmurget M, Rode G, Boisson D, Rossetti Y. 2000. An "automatic pilot" for the hand in human posterior parietal cortex: toward reinterpreting optic ataxia. *Nat Neurosci* **3**:729–736.

Ramat S, Leigh RJ, Zee DS, Optican LM. 2007. What clinical disorders tell us about the neural control of saccadic eye movements. *Brain* **130**:10–35.

Schall JD. 2002. The neural selection and control of saccades by the frontal eye field. *Philos Trans R Soc Lond B Biol Sci* **357**:1073–1082.

Scudder CA, Kaneko CS, Fuchs AF. 2002. The brainstem burst generator for saccadic eye movements: a modern synthesis. *Exp Brain Res* **142**:439–462.

Thier P, Ilg UJ. 2005. The neural basis of smooth-pursuit eye movements. *Curr Opin Neurobiol* **15**:645–652.

11.5 - Navigating Through Space

Goodrich-Hunsaker NJ, Livingstone SA, Skelton RW, Hopkins RO. 2010. Spatial deficits in a virtual water maze in amnesic participants with hippocampal damage. *Hippocampus* **20**:481–491.

Griffin AL, Hallock HL. 2013. Hippocampal signatures of episodic memory: evidence from single-unit recording studies. *Front Behav Neurosci* **7**:54.

Maguire EA, Nannery R, Spiers HJ. 2006. Navigation around London by a taxi driver with bilateral hippocampal lesions. *Brain* **129**:2894–2907.

Moser EI, Kropff E, Moser M-B. 2008. Place cells, grid cells, and the brain's spatial representation system. *Annu Rev Neurosci* **31**:69–89.

O'Keefe J, Nadel L. 1978. *The hippocampus as a cognitive map.* Oxford, England: Clarendon Press.

Tolman EC. 1948. Cognitive maps in rats and men. *Psychol Rev* **55**:189–208.

Wehner R, Boyer M, Loertscher F, Sommer S, Menzi U. 2006. Ant navigation: one-way routes rather than maps. *Curr Biol* **16**:75–79.

Boxes

Knudsen EI. 1981. The hearing of the barn owl. *Sci Am* **245**:112–125.

Konishi M. 2003. Coding of auditory space. *Annu Rev Neurosci* **26**:31–55.

Szymusiak R, Nitz D. 2003. Chronic recording of extracellular neuronal activity in behaving animals. *Curr Protoc Neurosci.* Ch. 6, Unit 6.16.

Identifying Stimuli and Stimulus Objects

CHAPTER 12

CONTENTS

FEATURES

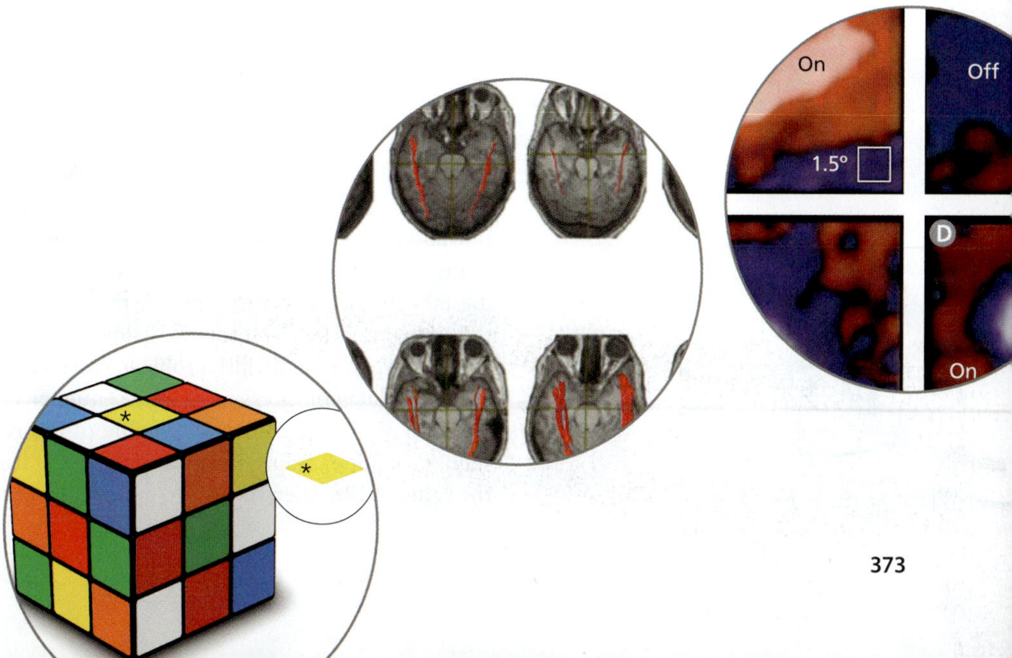

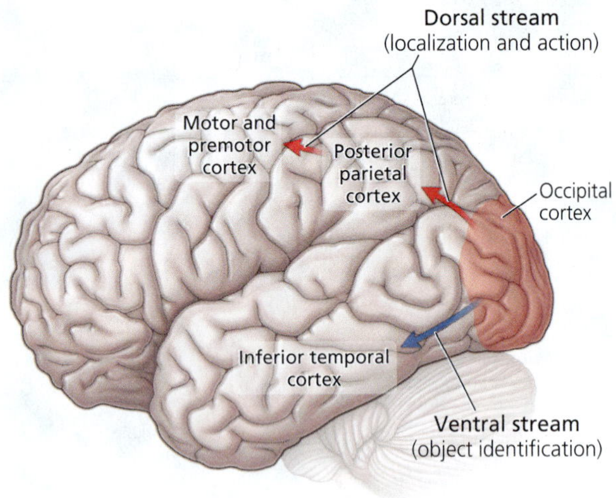

Figure 12.1 Dorsal and ventral visual streams. After processing in the occipital cortex, visual information is conveyed either into the parietal cortex or the inferior temporal lobe. These two "streams" of information flow are called the dorsal and ventral streams. After integration with auditory and somatosensory signals, information in the dorsal stream is passed on to motor and premotor cortices to guide eye, hand, and head movements. Information in the ventral stream is used primarily for object identification.

In the primate visual system, stimulus localization and stimulus identification are carried out in two largely separate cortical circuits called the dorsal and ventral streams (Figure 12.1). The **dorsal stream** originates in the *primary visual cortex (V1)* of the occipital lobe and then passes through the posterior parietal cortex to the premotor and motor cortices. As we discussed in Chapter 11, this pathway uses spatial information to guide eye, head, and hand movements. In contrast, the **ventral stream** proceeds from V1 to the inferior portion of the temporal lobe and is concerned primarily with stimulus identification. Although the dorsal and ventral streams both originate from V1, they also receive some auditory and somatosensory information. Neurobiologists sometimes debate whether the dorsal and ventral streams are truly divergent, as some of their areas are interconnected, but most experiments support the notion that they are functionally distinct.

Why do our brains segregate the tasks of stimulus localization and identification from one another? Why don't we have the same neurons perform both tasks? In theory, the two sorts of information could be processed together. One could easily imagine neurons that respond to specific stimulus objects (objects from which stimuli originate) only when those objects are in a specific location. Such neurons would encode stimulus location and identity at the same time.

There is a problem, however, with such a coding scheme. What happens when the stimulus object moves, or when the organism moves relative to the object? Then you would need a different neuron to represent the object at its new location. After all, objects don't change their identity as they change location. Indeed, you would need a vast number of neurons to code for each perceptible object at every possible location. It is much more efficient to have one group of neurons respond to specific objects regardless of where they are, and another group of neurons respond whenever any kind of stimulus appears at a specific location.

Of course, this division of labor creates a different problem, namely, how to integrate information about stimulus location with information about stimulus identity. The solution to this second problem probably involves connections between the ventral and dorsal streams as well as a convergence of the two streams in the prefrontal cortices, but these integrative mechanisms are poorly understood. We will return to the problem of perceptual integration at the end of the chapter. For now, we focus just on the problem of stimulus identification; it is challenging enough. Before we begin, let us briefly explore some general principles of sensory coding.

12.1 What Coding Strategies Do Sensory Systems Employ?

If your sensors were to feed your brain a complete, uncompressed stream of information about the outside world, your brain would rapidly be overwhelmed. The problem is analogous to what happens when you take a photograph with a high-resolution digital camera and try to send the raw file over a wireless connection. Most likely, the connection will be intolerably slow.

To solve this bandwidth problem, you must compress the file. In the simplest form of digital data compression, a long sequence of identical pixel values is encoded as just two numbers, namely, the value of the repeating pixel and the length of the sequence. For example, if an image file contains 500 successive pixels that all have the value of 128, then those 500 data points can be compressed into just two values, namely, 128 and 500. This *run-length encoding scheme* is a form of lossless data compression because the original data can be reconstructed without any information having been lost. A more aggressive compression algorithm might lump nearby pixel values together, for example, treating all values between 126 and 130 as 128. Such an algorithm would lose some of the original information but would retain information about major variations in pixel value. To compress videos, rather than static images, you can identify any pixels that do not change their value across multiple video frames and then encode the original time series of pixel values with just one pixel value and the number of video frames for which it is invariant.

Sparse and Efficient Sensory Coding

Sensory systems also compress information by filtering out redundant (repeating) data. As we discussed in earlier chapters, most sensors quickly adapt to constant stimuli but respond robustly when a stimulus is suddenly changed. As you will discover in this chapter, most sensory neurons also respond better to spatial variations in stimuli than to spatially uniform stimuli. Many visual neurons, for example, respond more strongly to a black outline on a white background than to a blank sheet of paper. By focusing on spatial and temporal variations, rather than constancies, sensory neurons pass on to the next stage of sensory processing only the most informative information. This neural data compression, or **sparse coding**, drastically reduces the number of action potentials that must be sent from the sense organs to the higher levels of sensory processing. As you learned in Chapter 2, such a reduction in firing rate saves precious metabolic energy.

To identify objects and other features in the external environment, compressed sensory data must be analyzed. This analysis requires neurons that respond selectively to features of the stimulus such as its color, texture, or direction of movement. Of course, stimulus objects may have an almost infinite number of possible features, and coding for them all would require an enormous number of neurons. To solve this problem, nervous systems tend not to encode all possible features but instead to focus on the features that occur most frequently in natural stimuli.

This principle of **efficient sensory coding** is supported by computer modeling studies, which show that artificial neural networks trained to extract the statistically most common features of natural images (or sounds) end up containing simulated neurons with the same kind of stimulus preferences as real sensory neurons. For visual stimuli, this means that both the simulated and the real neurons respond most strongly to edges and lines, which are common features in natural images. This matching of neural stimulus preferences to natural stimulus features saves both energy and space, as it would be wasteful to devote many sensory neurons to coding for features that rarely or never occur in the real world.

Grandmother Cells versus Combinatorial Coding

To encode complex features and entire objects, rather than just isolated features, the brain uses neurons that are activated only by specific combinations of features, such as a dark line of a particular orientation that is moving in a specific direction. Such combination-sensitive neurons receive converging input from multiple feature-sensing neurons but increase their firing rate only when all (or most) of those lower level neurons are active simultaneously. By chaining several such combination-sensitive neurons together, you can, at least in theory, end up with neurons that respond only to a very specific object, such as your grandmother.

Such a **grandmother cell coding scheme** (Figure 12.2) is appealing but problematic. Does your brain also contain grandfather neurons, and neurons that encode your neighbor down the street, or people you have never met? Surely, you do not have enough neurons to encode all possible objects, as long as each object is encoded by a specific cell. Furthermore, the grandmother cell coding scheme is vulnerable to neuron death, as each deceased neuron would make you incapable of perceiving the

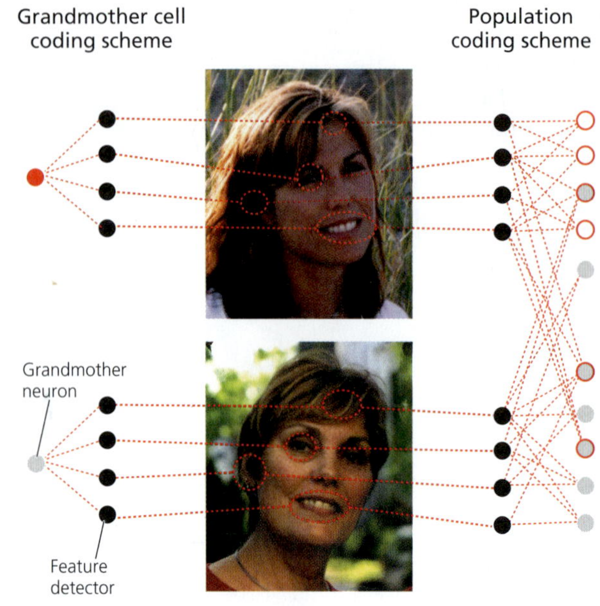

Figure 12.2 Coding schemes for complex stimuli. Depicted on the left is the grandmother cell coding scheme, in which complex stimuli, such as specific faces, are represented by the activity of single neurons that receive convergent input from a unique set of lower level feature detectors. The illustrated red neuron responds only to the face of the woman at the top of the figure, whereas the gray neuron responds specifically to the face of the woman at the bottom of the figure. Shown on the right is population coding, which represents the two faces by the activity of partially overlapping sets of neurons. The six neurons outlined in red are activated by the upper woman's face, whereas the seven neurons shaded gray respond to the lower woman's face.

object that the neuron encodes. You might counter this argument by claiming that the brain contains multiple grandmother neurons to mitigate the loss of just a single cell, but such redundancy would exacerbate the problem of not having enough neurons to encode all possible objects.

How does the brain solve the problems associated with the grandmother cell coding scheme? It does so by encoding specific objects (such as your grandmother) in the activity of many neurons that receive inputs from combination-sensitive neurons but do not themselves converge onto a single grandmother cell. For example, seeing your grandmother activates a specific combination of neurons that, collectively, represent your grandmother. Importantly, some of those neurons may also be active when you see someone else (Figure 12.2). The code resides not in the activity of the individual neurons but in their combination. It is a *population code*, conceptually analogous to the population codes for movement and sound localization that we discussed in Chapters 10 and 11, respectively.

BRAIN EXERCISE

Which sensory coding scheme (grandmother cell or population coding) involves a larger number of active neurons? Which scheme is more efficient? Why?

12.2 How Does the Visual System Identify Objects?

The visual system has been studied more intensively than any other sensory system because humans tend to be highly visual creatures and because the visual system is relatively easy to study. The retina is simpler to access with recording electrodes than the cochlea or olfactory epithelium, and well-controlled stimuli can be delivered to specific retinal locations by projecting them onto a screen. Furthermore, visual stimuli can be presented to anesthetized animals as long as their eyes are propped open and the corneas kept moist. Indeed, research on anesthetized monkeys and cats has shaped much of our current understanding of visual processing. Fortunately, many of the discoveries made in anesthetized animals have been confirmed and extended by more recent work in awake animals.

One significant obstacle to understanding the processing of visual information in mammals is that cat, monkey, and human brains contain a large number of visual cortical areas that are linked to one another through a large number of connections. The complexity of this network can be overwhelming, but a simplified schema suffices for our purposes (Figure 12.3). In this simplified model, visual information is sent from the retina to the thalamus and from there to the V1, which is also known as "striate cortex" because it exhibits an obvious striation (stripe) with some stains. From V1, information travels to a series of visual cortical areas in the occipital lobe that are collectively referred to as the **extrastriate cortex**. These areas include V2, V3, V4, and the middle temporal area. As visual information passes through these extrastriate cortices, it segregates into the dorsal and ventral streams (Figure 12.1).

One potential problem with the idea that visual information flows from the thalamus to V1, and from there toward the higher visual areas in the parietal and temporal lobes, is that most of the connections between cortical areas are reciprocal (Figure 12.3). This reciprocity implies that information can flow both up and down the cortical hierarchy. However, neurons at the lower levels of the cortical hierarchy tend to have shorter response latencies than those at the higher levels and, as you will see, they tend to be activated by simpler stimulus features. Given these data, it is safe to assume that the visual cortices are hierarchically organized, at least to a first approximation, and that visual information tends to flow primarily from lower to higher visual areas. Therefore, it is reasonable to begin our discussion of visual processing at the very bottom of the hierarchy, in the retina and thalamus. Still, information flowing back down the cortical hierarchy is likely to be crucial for generating perceptual expectations, such as those that generate perceptual illusions and "fill in" our blind spot; for voluntary attention; and for the recall of memories (see Chapters 13 and 14).

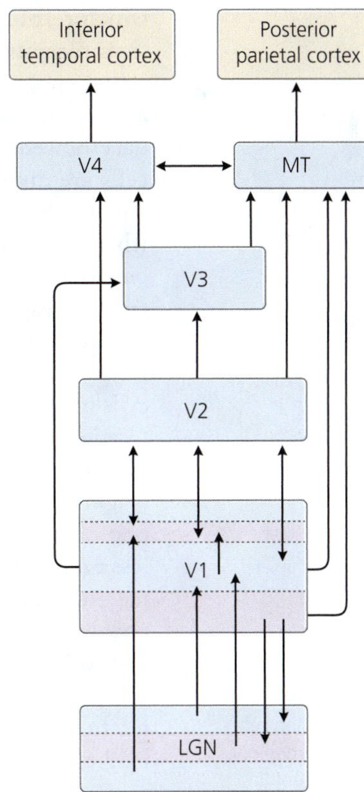

Figure 12.3 A simplified model of the visual cortical hierarchy. At the bottom of the visual cortical hierarchy is the primary visual cortex (V1), which receives visual inputs from the lateral geniculate nucleus (LGN). At the top of the hierarchy are the inferior temporal and posterior parietal areas. Most of the connections are reciprocal, but connections from lower areas to higher areas tend to be more influential during stimulus perception than those in the opposite direction. Different cortical layers in V1 are demarcated by dashed lines, as are the major cell types in the LGN. [After Rokszin et al., 2010]

Retinal Receptive Fields

The concept of a **spatial receptive field** was developed in the 1930s and 40s by Keffer Hartline, whose work was honored with a Nobel Prize. Hartline projected small spots of light onto the retinas of frogs (and horseshoe crabs) and recorded the responses of retinal ganglion cells. He discovered that many retinal ganglion cells respond to such stimuli only when the stimulus is presented at a specific location in space, which he called the cell's spatial receptive field. Hartline also observed that some retinal ganglion cells respond when a light is suddenly turned on at the center of the cell's spatial receptive field, whereas other cells respond only when that light is turned off or, equivalently, a dark spot suddenly appears. These retinal ganglion cells are called **on-center cells** and **off-center cells**, respectively (Figure 12.4). You can think of them as bright and dark spot detectors. As you may recall from Chapter 6, these two types of retinal ganglion cells receive input from ON and OFF bipolar cells, which respond in opposite ways to glutamate release from cones because they express metabotropic and ionotropic glutamate receptors, respectively.

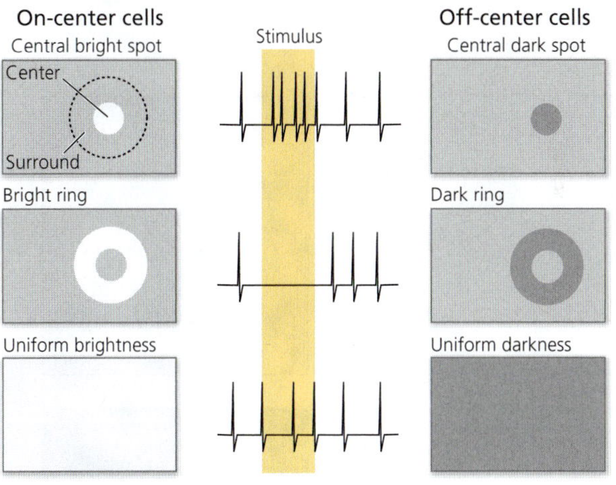

Figure 12.4 Center-surround receptive fields. On-center retinal ganglion cells (left) respond with an increase in firing rate to bright spots presented at the center of their spatial receptive field. If these cells are, instead, stimulated with a ring of light surrounding the center of their receptive field, they lower their firing rate. Stimulating these cells with uniform light yields no response (no change in firing rate). The retina also contains off-center retinal ganglion cells, which respond with an increased firing rate to dark spots in the receptive field's center (right). They, too, exhibit surround inhibition and do not respond to uniform stimulation. [After Schiller, 1995]

Center-surround Receptive Fields

Hartline went on to discover that the response of retinal ganglion cells to stimuli presented in the center of their receptive field can often be inhibited by a second stimulus that is presented simultaneously at a second, nearby location. More detailed studies then revealed that responses to a spot of light can be inhibited by light presented anywhere within a ring around the spot that generates excitation. Indeed, illumination of the surrounding ring alone yields a robust inhibitory response, and diffuse illumination of both the center spot and the surrounding ring produces little to no change in firing rate (Figure 12.4). Additional research has extended these findings to retinal ganglion cells with off responses,

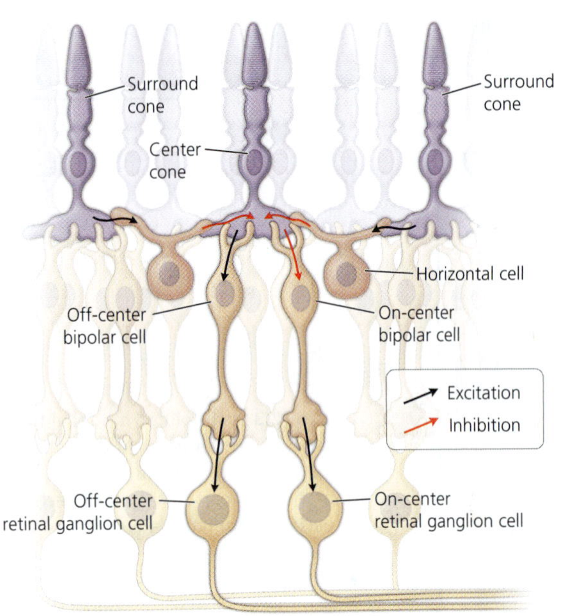

Figure 12.5 Retinal surround inhibition. Activation of the "center cone" in this diagram leads to activation of a retinal ganglion cell (on- or off-center depending on the stimulus). However, when one of the "surround cones" is also activated, horizontal cells inhibit the center cone, reducing its effect on the bipolar cells and thus the retinal ganglion cell response. This explains why uniform light (or uniform darkness) are less effective stimuli than spots of light (or dark).

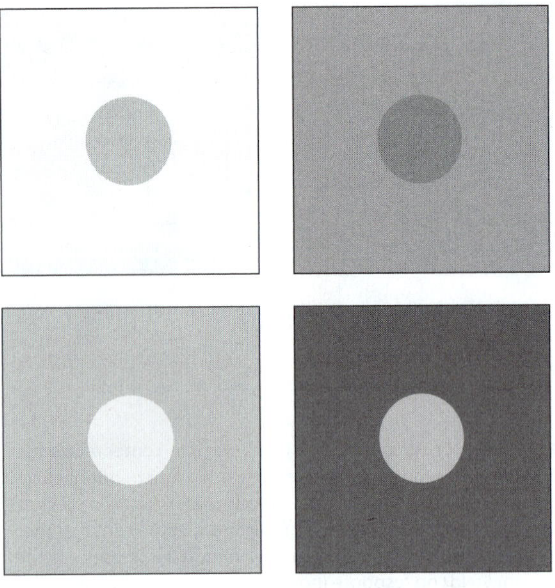

Figure 12.6 Local contrast versus absolute luminance. We perceive the circles in the top two panels as similarly dark even though the circle on the right has a lower luminance. Similarly, the circles in the bottom row appear equally bright even though they differ in luminance. In fact, the top left and bottom right circles have equal luminance. We do not perceive this equality because our visual system is more sensitive to local luminance contrast than to absolute luminance.

which respond briskly to small dark spots in the center of their receptive field, are inhibited by a dark ring around the center spot, and remain unfazed when both the central spot and the surround are covered in darkness. Collectively, these data showed that the spatial receptive fields of retinal ganglion cells are divisible into an **excitatory center**, in which focal stimuli elicit robust responses, and an **inhibitory surround** (this definition of spatial receptive fields is broader than the one we used in earlier chapters where we focused only on excitatory responses).

The surround's inhibitory influence is due primarily to retinal **horizontal cells**, which connect nearby photoreceptor cells to one another. Importantly, the input from the photoreceptors onto horizontal cells is excitatory, whereas the output of horizontal cells back onto photoreceptors is inhibitory (Figure 12.5). As a result of this arrangement, strongly excited photoreceptors in one retinal location tend to inhibit their less excited neighbors (much as one highly vocal person in a meeting tends to inhibit the others from speaking up). This phenomenon of **surround inhibition** is functionally equivalent to *lateral inhibition* in the somatosensory system, which we discussed in the previous chapter. In essence, surround inhibition sharpens the somewhat blurry retinal image detected by the photoreceptors.

Detecting Spatial and Temporal Change

Surround inhibition makes the retina's output an unreliable sensor of **luminance**, the technical term for absolute brightness. For example, an off-center retinal ganglion cell may respond to a gray spot on a white background as strongly as it responds to a black spot on a gray background (Figure 12.6). What such a neuron "cares about" is that the stimulus is darker in the center of the cell's spatial receptive field than in the surround. Conversely, on-center cells care only that the stimulus is brighter in the center than in the surround. Absolute levels of luminance are much less important. This is a useful property because objects don't change their properties as ambient light levels vary: a rose is a rose, no matter how bright the light. In contrast, spatial *variations* in luminance often delineate objects and other visual features that are of interest to an animal.

Retinal ganglion cells are also highly sensitive to temporal variations in luminance. Specifically, on- and off-center cells respond more strongly to the sudden appearance of bright or dark spots, respectively, than to persistent spots in the same location. This phenomenon is related to background adaptation in the photoreceptors, which reduces visual responses to steady illumination (see Chapter 6). Overall, the insensitivity of the visual system to stable stimuli, combined with its exquisite sensitivity to spatial variation, explains why relatively immobile animals, such as frogs, are almost blind to what is constant in their world yet remain sensitive to moving objects such as flies. It also explains why experimenters who record visual responses from anesthetized animals use flashing or moving stimuli. Most neurons simply do not respond to steady stimuli.

Recalling our earlier discussion of sparse and efficient coding in sensory systems, you can now appreciate that surround inhibition and background adaptation are mechanisms for eliminating, or at least reducing, the spatially and temporally redundant (unchanging) information in visual signals. They compress the information that retinal neurons send to the brain.

BRAIN EXERCISE

How would you define a sensory neuron's spatial receptive field? How well do the data on retinal responses to visual stimuli fit your definition?

Thalamic Receptive Fields

The axons of retinal ganglion cells terminate in a variety of different brain regions. One of their principal targets is the *lateral geniculate nucleus* (LGN), which provides the major visual input to V1. As you may recall from Chapter 11, the primate LGN has 6 distinct layers. The four thickest layers contain relatively small (parvocellular) neurons and receive input from small retinal ganglion cells (midget cells) that have small receptive fields and convey color as well as luminance information. The two thinner LGN layers contain large (magnocellular) neurons and receive input mainly from large retinal ganglion cells (parasol cells) that have large receptive fields and are color insensitive. Various other, less common, types of retinal ganglion cells project sparsely to both sets of LGN layers and the cell-free zones between them.

To map the spatial receptive fields of LGN neurons, experimenters usually record the activity of individual geniculate neurons while presenting a random sequence of light and dark spots in the animal's visual field. The spots can be presented one at a time or simultaneously in a randomly flickering pattern. The important thing is that the experimenters can correlate the neuron's spiking activity with where a spot was presented and whether it was light or dark. The experimenters then construct a map that shows how a recorded neuron responds to bright and dark spots at each location in space. To make these spatial receptive field maps easy to understand, the maps are usually color coded, with one color to represent the responses to bright spots (red in Figure 12.7) and a different color (blue) to denote responses to dark spots.

Given this method, what do LGN spatial receptive fields look like? Some resemble retinal on-center cells in that their spatial receptive field contains a central region where bright spots increase the neuron's firing rate and a ring-shaped surround where bright spots generate inhibition and, conversely, dark spots elicit excitation (Figure 12.7 B). Other geniculate neurons resemble retinal off-center cells (Figure 12.7 D). Given these similarities between retinal and LGN receptive fields, one can conclude that individual geniculate neurons receive converging input from just a few retinal ganglion cells that are of the same type (either on-center or off-center) and have nearly identical spatial receptive fields. The observation that geniculate receptive fields tend to be much larger in newborn animals than in adults (Figure 12.7) suggests that the connections between retinal neurons and the LGN become more precise with development. As we discussed in Chapter 4, this kind of developmental refinement in connectivity probably involves the activity-dependent pruning of axon branches and dendrites.

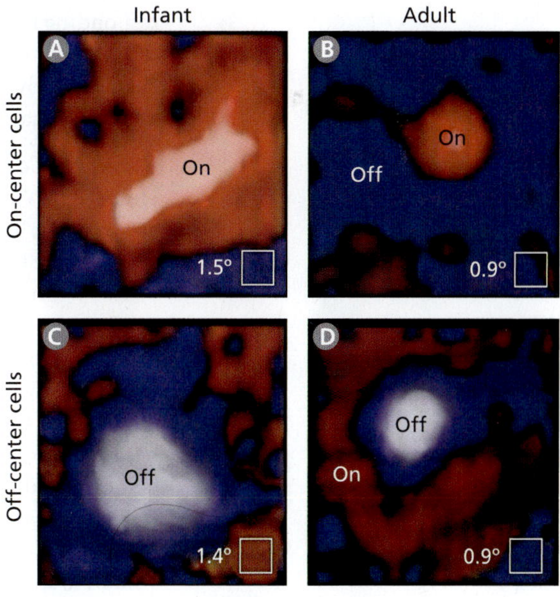

Figure 12.7 Receptive fields in the lateral geniculate nucleus (LGN). Shown here are receptive field maps for LGN neurons in infant and adult ferrets (left and right, respectively). Red areas correspond to locations where bright stimuli produce an excitatory neuronal response; blue regions show where dark stimuli elicit responses. The paler the red or blue, the greater the response (the pale pink area elicits maximal excitation in A, the pale blue area maximal inhibition in B). The white squares and numbers indicate the scale of each visual field map (in degrees of visual angle). [From Tavazoie & Reid, 2000]

Because the spatial receptive fields of adult LGN neurons are so similar to the receptive fields of retinal ganglion cells, it is tempting to consider the LGN a passive relay in the path from retina to visual cortex, performing little or no information processing. However, LGN neurons also receive descending projections from V1. These descending inputs target the distal dendrites of geniculate neurons and are less powerful than the retinal inputs, which consist of large synapses on geniculate cell bodies and proximal dendrites. Nonetheless, the cortical inputs are massive, accounting for roughly 70% of all synapses in the LGN. It has been proposed that these descending projections alter the antagonism between a geniculate neuron's receptive field center and its antagonistic surround, but this idea remains debatable.

BRAIN EXERCISE

What kind of experiment would you design to determine the function(s) of the descending projection from the primary visual cortex to the LGN?

Edge and Line Detectors in V1

Major advances in our understanding of stimulus coding in V1 were made by David Hubel and Torsten Wiesel, who shared the 1981 Nobel Prize in Physiology and Medicine. In the late 1950s Hubel and Wiesel were postdoctoral researchers in the laboratory of Stephen Kuffler, who had been instrumental in showing that retinal ganglion cells have circular center-surround receptive fields. At Kuffler's urging, Hubel and Wiesel sought to determine whether neurons in V1 also have center-surround receptive fields. At first, their efforts were disappointing because most cortical neurons respond only weakly to the spots or rings of light (or dark) that so effectively stimulate retinal and LGN neurons.

Then Hubel and Wiesel made an exciting discovery. They had been projecting stimuli onto a screen with an old-fashioned slide projector. Some of the stimuli were black spots painted onto a transparent glass slide. As Hubel and Wiesel slid one of these glass slides into the slide projector, the neuron they had been recording from fired a burst of action potentials. Closer inspection revealed that this neuron was not responding to the black spot at all but to the moving gray shadow cast by the edge of the glass slide as it was pushed into the projector. Hubel and Wiesel pounced on this chance discovery and soon learned that most V1 neurons respond far better to lines and edges than to circular spots.

V1 Simple Cells

Hubel and Wiesel discovered that many V1 neurons respond best to lines presented at a specific location and in a particular orientation. If the stimuli are rotated or moved away from the preferred location, the neurons no longer respond. Hubel and Wiesel referred to such neurons as **simple cells**.

Some of the simple cells respond to bright lines on a dark background; their spatial receptive fields consist of an elongated on-responsive central region that is flanked by two antagonistic off-responsive sidebands (Figure 12.8 A). Other simple cells exhibit the opposite pattern: they respond best to dark lines, and their receptive fields consist of an elongate off-responsive center flanked by on-responsive sidebands. This dichotomy between on-center and off-center simple cells is analogous to the distinction between on- and off-center cells in the retina and LGN. Simple cells also resemble retinal and geniculate neurons insofar as they prefer flashing or moving stimuli to static ones. However, no self-respecting retinal or geniculate neuron would respond more strongly to a line than to a spot of the same width presented in the center of the neuron's receptive field.

How do V1 cells become sensitive to lines rather than spots? Hubel and Wiesel proposed an elegant answer. They hypothesized that each simple cell receives converging input from multiple geniculate neurons whose receptive field centers are all of the same type (e.g., on-center) and arranged in a line (Figure 12.8 B). If the geniculate receptive fields are all of the on-center type, then the V1 simple cell receiving the converging geniculate inputs should, according to the model, respond to a bright line that falls across those on-centers. Conversely, if the geniculate receptive fields are all of the off-center variety, then the simple cell should respond best to a dark line that

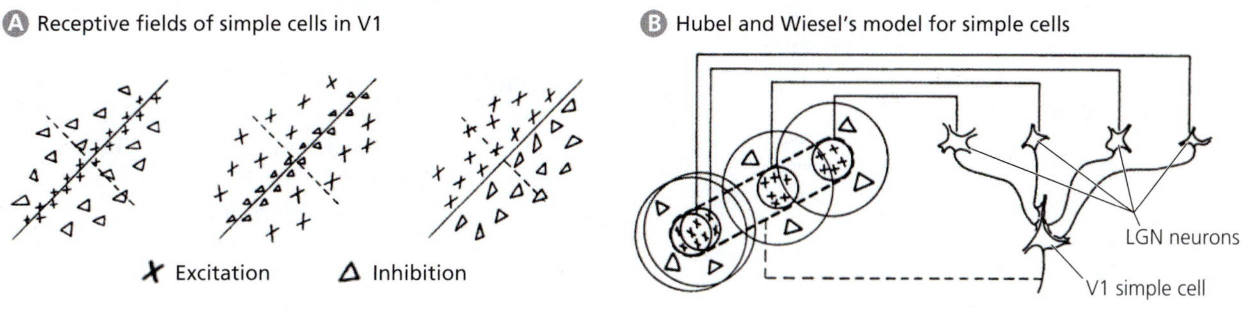

Figure 12.8 V1 simple cells. Some neurons, called simple cells, respond best to narrow slits of light, dark lines, or edges at specific orientations. Three such simple cell receptive fields are shown in (A). Regions where light elicits excitatory responses are marked with X's; regions where light evokes inhibition are labeled with triangles. As shown in (B), Hubel and Wiesel suggested that simple receptive fields might result from the convergence of multiple LGN neurons onto a single V1 neuron. The converging LGN neurons must be of the same type (either on-center or off-center) and have their receptive field centers arranged in a straight line. According to this model, the antagonistic surrounds of the geniculate receptive fields generate the antagonistic sidebands of V1 simple cells. [From Hubel and Wiesel, 1962]

covers those off-centers. Importantly, Hubel and Wiesel's model explains why simple cells respond weakly (at best) when the line stimuli are moved to a different location, rotated, or replaced by wider lines. These non-optimal stimuli encroach on the antagonistic surrounds of the converging geniculate receptive fields.

Although this model was elegant, Hubel and Wiesel soon realized that it was too simple. To make their model more realistic, Hubel and Wiesel proposed a more elaborate model in which the receptive fields of simple cells are generated primarily by the spatial arrangement of the geniculate receptive field centers, whereas the ring-shaped surrounds of the geniculate receptive fields play no substantial role in shaping the simple cell receptive fields (Figure 12.9).

In contrast to the simple model, the elaborate model allows for simple cells with asymmetric receptive fields, such as the one illustrated on the right in Figure 12.8 A. Such cells, which respond better to edges than to lines, are common in V1. According to the elaborate model, their on-responsive region is formed by the convergence of on-center geniculate neurons, whereas their off-responsive region results from the convergence of off-center geniculate neurons (right side of Figure 12.9). Again, the antagonistic surrounds in this model play at best a minor role.

Strong support for the elaborate version of Hubel and Wiesel's model comes from a study in which experimenters recorded simultaneously from pairs of geniculate neurons and V1 simple cells. For each cell, the scientists mapped the on- and off-responsive subregions of the receptive field. They also used correlations in the neuronal firing patterns to identify pairs of geniculate and V1 cells that were directly (monosynaptically) connected with one another. The study's major finding was that for interconnected pairs of cells, the geniculate neuron's receptive field center is reliably located in the predicted subregion of the simple cell's receptive field. Thus, the center of an on-center geniculate neuron invariably lies in the on-responsive region of the simple cell's receptive field, and the center of off-center geniculate neurons consistently overlaps with the simple cell's off-responsive region. This overlap of geniculate and V1 receptive fields was not observed

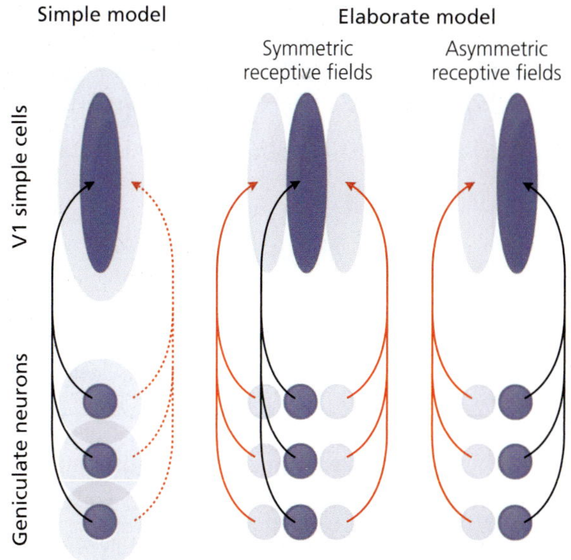

Figure 12.9 Simple and elaborate versions of Hubel and Wiesel's model. Shown on the left is the simple Hubel and Wiesel model (Figure 12.8) in which antagonistic sidebands of a V1 simple cell receptive field result from the convergence of the antagonistic surrounds (dashed black circles). In a more elaborate model (middle and right) the sidebands of the simple cell receptive fields result from the convergence of receptive field centers. This elaborate model can explain simple cells with asymmetric receptive fields (right). For clarity, the inhibitory surrounds are not shown for the lateral geniculate neurons in the elaborate model.

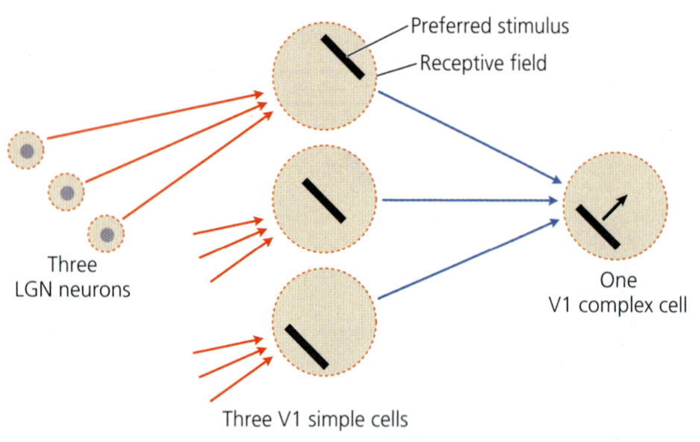

Figure 12.10 Types of convergence in V1. Simple cells are thought to receive converging input from multiple LGN neurons whose receptive fields are aligned with one another. In this type of convergence, the postsynaptic neuron fires only when most of its input neurons are active simultaneously. V1 complex cells are also thought to arise through converging input from multiple neurons with similar receptive fields. However, in this second type of convergence, activity in any one of the input lines generates a postsynaptic response. Therefore, complex cells respond to line-shaped stimuli at multiple locations within, or moving through, their receptive field (dashed orange line).

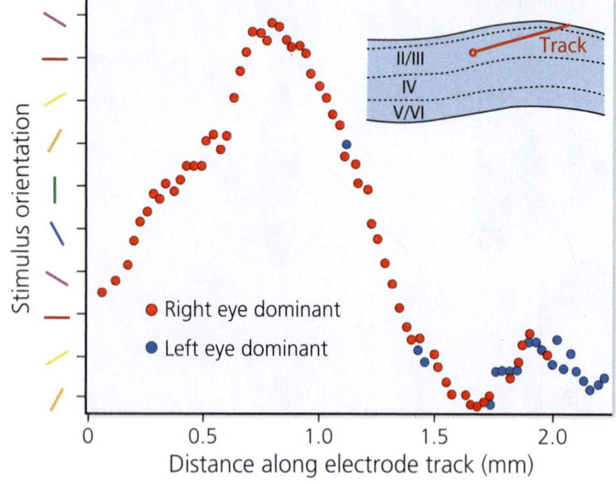

Figure 12.11 Orientation tuning varies systematically across V1. The orientation selectivity of neurons is recorded along a single electrode track roughly parallel to the cortical surface (inset at top right). The orientation of the most effective stimuli is shown along the y-axis. As you can see, the best orientation varies gradually along the track. Also shown is that the recorded neurons tend to be right-eye dominant until about 1.7 mm into the track, at which point they become primarily left-eye dominant (despite the dominance, most neurons were binocular). [After Hubel and Wiesel, 1977]

in pairs of neurons that were not directly connected. Hubel and Wiesel had predicted as much.

V1 Complex Cells

In addition to simple cells, Hubel and Wiesel described V1 neurons with more complex receptive fields. These **complex cells** are similar to simple cells in that they also respond preferentially to bright or dark lines and edges in a particular orientation. However, their responses do not depend on where in the receptive field the stimulus is located, and their receptive fields are not divisible into separate on- and off-responsive subregions.

To explain these properties, Hubel and Wiesel proposed that complex cells receive convergent input from multiple V1 simple cells (Figure 12.10). According to this model, V1 complex cells come after V1 simple cells in the visual processing hierarchy. You can think of the complex cells as "surveying" many different simple cells with overlapping receptive fields and similar **orientation sensitivity**, and firing whenever any one of these simple cells fires a burst of action potentials. This complex cell model is similar to the simple cell model described earlier except that the complex cells respond to activity in any one of their converging input lines, whereas the simple cells fire only when a particular combination of geniculate neurons is excited.

Orientation Preference Maps

Another of Hubel and Wiesel's important discoveries was that the orientation sensitivity of simple and complex cells varies systematically across V1. If a recording electrode is advanced radially through V1 (perpendicular to the cortical surface), then all the encountered neurons tend to have very similar stimulus orientation preferences (as you might expect, given our discussion of the canonical cortical circuit in Chapter 11). However, if an electrode is advanced tangentially through V1, then the preferred stimulus orientation of successive neurons rotates clockwise or counterclockwise, until, after several hundred microns, the pattern reverses direction (Figure 12.11).

An excellent method for studying these systematic changes in orientation preference is **intrinsic signal optical imaging**, which uses differences in the light reflectance of oxygenated versus deoxygenated blood to map neural activity patterns (Box 12.1: Intrinsic Signal Optical Imaging). To map the orientation preference of cortical neurons with this technique, experimenters present the animal with line-shaped stimuli in various orientations and observe how the activated cortical patches shift across the cortex as the stimuli change. When the results are color coded so that each preferred orientation is represented in a different color, it becomes apparent that orientation preference is mapped across V1 in a "pinwheel" pattern (Figure 12.12). Each pinwheel is less than 1 mm in diameter, and all the neurons in a single pinwheel respond to stimuli in the same general region of space. This is important because it means that for each location in visual space, V1 has neurons capable of responding to lines and edges of all possible orientations.

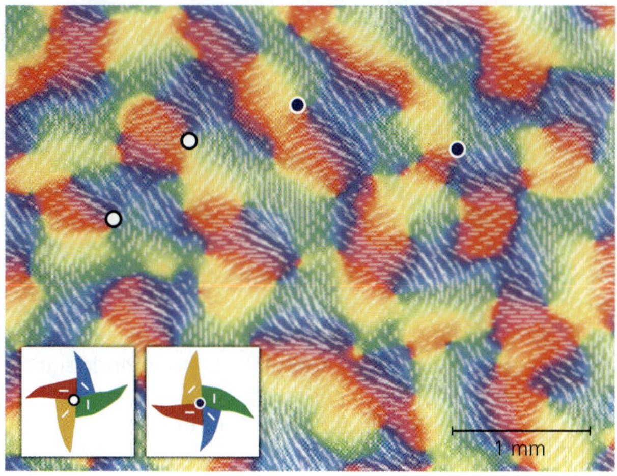

Figure 12.12 Orientation selectivity "pinwheels." Intrinsic signal optical imaging was used to map the responses of macaque V1 to visual stimuli of various orientations. The color coding reveals a "pinwheel" pattern of orientation selectivity. The white and dark blue circles indicate the centers of four relatively obvious pinwheels. If you examine the orientation preferences of neurons along a circle surrounding each pinwheel center, you can see that they rotate through a full 180°. In some of the pinwheels the orientation preference rotates clockwise; in others it rotates counterclockwise. One schematic pinwheel of each type is shown in the two insets. [From Blasdel, 1992]

Figure 12.13 V1 neurons function like edge detectors in image processing. Shown here is a picture of two elephants before and after it was passed through a digital edge detection filter. Clearly, much of the information needed to identify the elephants is contained in the edge-filtered image. [From Blumberg and Kreiman, 2010]

Detecting Object Contours

Because of the response properties we just discussed, the act of looking at an object tends to activate all V1 simple and complex cells whose receptive fields and orientation preferences match some segment of the object's contour (its edges). Furthermore, because V1 contains a map of visual space (see Chapter 11), the V1 neurons that are activated when you look at an object form a distorted, yet predictable, spatial representation of the object's contour. In essence, the object's outline is encoded in the spatial pattern of activated V1 neurons.

In contrast, the interior of an object's visual contour elicits much less of a response from the V1 neurons we have discussed. To use an analogy, V1 cells are like the edge-finding filters in image processing software, which convert natural images to line drawings (Figure 12.13). In the process, much information is thrown away. However, most of the information required to identify visual objects is contained in such edge-filtered line drawings. In comparison to an object's contours, the interior of an object's image is, as David Hubel once remarked, "boring." It is fair to say, therefore, that V1 neurons encode visual information efficiently from the standpoint of what is needed for object identification. They are a good illustration of the efficient coding principle at work.

BRAIN EXERCISE

How much information can you get from visual contours, and what are you missing if that is all you have in an image? As you answer this question, consider the advantages and limitations of sketches and cartoons.

Identifying Visual Motion

Movement plays an important role in visual object identification because any lines and edges in a visual scene that move at the same time and in the same direction

RESEARCH METHODS

Box 12.1 *Intrinsic Signal Optical Imaging*

An important method for visualizing patterns of neural activity is intrinsic signal optical imaging (ISOI). This method estimates neural activity by measuring changes in tissue oxygenation. The core principle is that deoxyhemoglobin absorbs light with a wavelength of ~600 nm (orange) much better than oxyhemoglobin does. Therefore, deoxygenated blood looks darker under orange light than oxygenated blood. Blood oxygenation levels, in turn, reflect neural activity because active neurons consume oxygen, which is replenished by diffusion out of capillary beds. The diffusion of oxygen out of the blood vessels occurs within a few hundred milliseconds of the increase in neural activity. A few seconds later, extra blood rushes into the area to resupply the active neurons with glucose and oxygen. This new incoming blood is highly oxygenated. Therefore, an increase in neural activity causes an initial dip in blood oxygenation followed by a slow rise in blood oxygenation levels. ISOI with orange light typically measures the initial dip. In contrast, the standard fMRI technique focuses on the delayed influx of oxygenated blood (see Chapter 5, Figure 5.25).

In practice, ISOI requires shining a constant amount of light at the specified wavelength onto the brain and then recording how much of that light is reflected back into the microscope (Figure b12.1). To provide access to the brain, experimenters usually drill a hole in the animal's skull and cut away the dura mater (the toughest of the three meninges). Alternatively, they can scrape the skull with a drill until it becomes transparent. This skull thinning technique leaves the brain and meninges intact and allows repeated imaging across multiple days, weeks, or even months.

Whether or not the skull is left intact, the intrinsic signal produced by activity-dependent changes in blood oxygenation is very small, on the order of 0.01–0.1% of the total reflected light. Therefore, an extremely sensitive low-noise camera is needed to capture the optical images. Furthermore, multiple image frames must usually be averaged to remove at least some of the noise. Then, the averaged image obtained in one condition (during stimulus presentation) is compared to the image obtained in another condition (no stimulus). The pixel-by-pixel difference between the two images is interpreted as the difference in neural activity between the two conditions. This subtractive image analysis method is also at the core of most fMRI experiments.

Compared to the fMRI technique, ISOI has several advantages. Because the initial dip in oxygenation is more localized than the activity-dependent blood flow response, ISOI has better spatial resolution than conventional fMRI (~0.05 vs. >1 mm for BOLD fMRI). This makes the ISOI technique ideal for studying mouse and rat brains, which are commonly used in neuroscience research but too small for most fMRI research. ISOI imaging is also cheaper than fMRI, which requires an expensive magnet and a team of physicists to keep it operational. However, a major drawback of ISOI imaging is that the orange light penetrates only a few hundred microns into the brain. This explains why most ISOI studies focus on the upper layers of the mammalian neocortex. Its sheet-like organization makes it ideal for ISOI. In contrast, fMRI can target structures deep in the brain.

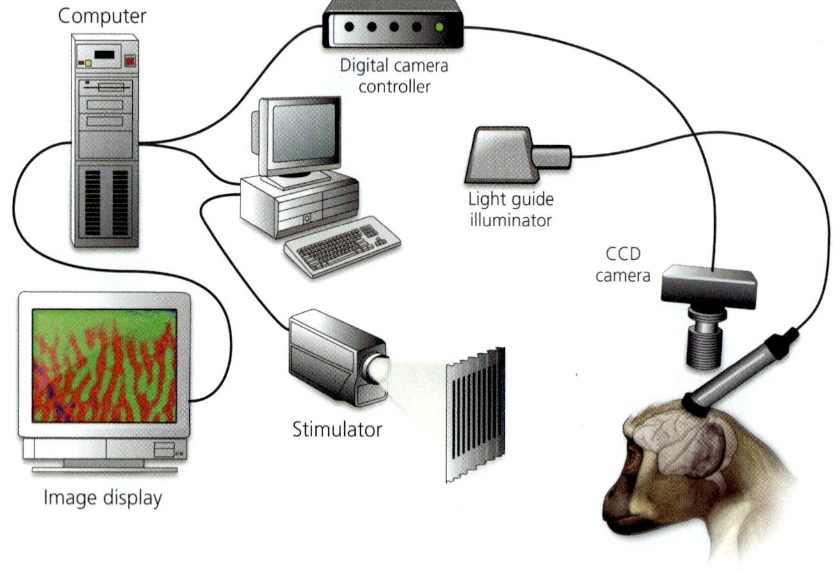

Figure b12.1 Recording intrinsic signals in response to visual stimulation. Orange light is directed at an animal's exposed cortex, which is then imaged with a digital camera. Simultaneously a computer presents the animal with visual stimuli, such as moving stripes. Images obtained under different stimulus conditions are digitally subtracted from one another and displayed on a monitor as pseudocolored images. [This drawing was prepared by A. Grinvald and D. Ts'o]

(coherently) are very likely to belong to a single object. To appreciate the importance of this fact, consider that even well camouflaged animals "pop out" of the background as soon as they move against that background. It is the motion that gives them away. Therefore, you will not be surprised that many animals have motion- and direction-sensitive neurons.

As mentioned earlier, most retinal and geniculate neurons prefer dynamic stimuli to static ones. However, very few of them are sensitive to the direction of stimulus movement (Figure 12.14). Therefore, we can infer that **direction sensitivity** is generated mainly in the visual cortex (and, separately, in the nucleus of the optic tract; see Chapter 10). Most models that attempt to explain direction sensitivity are based on the idea that input from one region of space arrives at a visual cortical neuron more slowly than input from another region. Because of this conduction delay, the cortical neuron receives simultaneous input from both regions of space only if a stimulus appears earlier in the location with the longer transmission delay. As we discussed in Chapter 2, such simultaneous input causes a greater neural response than non-coincident input. Alternative models propose that visual cortical neurons inhibit other cortical neurons with receptive fields to one side (but not the other) of their own receptive field. Such asymmetric inhibitory projections would suppress responses to any stimuli that move in the same direction as the inhibition. Both kinds of models are supported by some evidence.

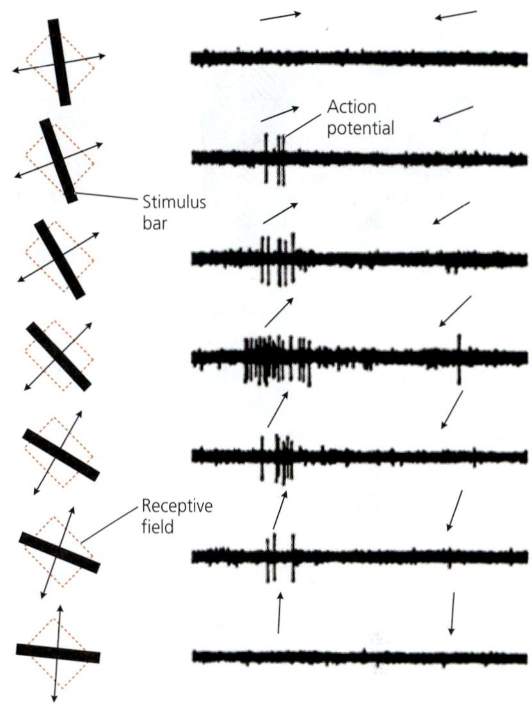

Figure 12.14 A direction sensitive neuron. Hubel and Wiesel discovered that many V1 neurons respond preferentially to light or dark bars that move in a specific direction. Shown here are the responses of one such neuron to moving dark bars. The stimuli are shown on the left, along with the neuron's spatial receptive field (dashed red rectangle). The arrows indicate movement direction. [From Hubel and Wiesel, 1968]

The Aperture Problem

Whatever the mechanism, roughly half of all V1 cells are directionally sensitive. They respond more strongly when oriented lines or edges are moving one way rather than the other (Figure 12.14). However, all V1 neurons are imprecise in their encoding of movement direction. To understand this imprecision, imagine that you are looking at a moving line through a small hole (an aperture). As long as the line extends across the aperture, you will always think it is moving in one of the two directions that are perpendicular to the line's long axis, regardless of the line's actual direction of movement (Figure 12.15 A). V1 neurons have this **aperture problem** because their receptive fields cover only a small portion of most lines in an image. For half of all possible movement directions, a V1 neuron will signal that the line in its receptive field is moving left, relative to the line's long axis; for all other possible directions, the neuron will signal movement to the right. This is inaccurate, of course.

How does the brain solve the aperture problem? By combining information from multiple V1 neurons whose receptive fields are aimed at different segments of a moving contour (Figure 12.15 B). This computation is performed outside of V1, in the **middle temporal area (MT)**, which is located between the occipital and temporal lobes and receives inputs from V1, V2, and V3 (see Figure 12.3). MT neurons take longer than V1 neurons to respond to visual stimuli (~110 ms vs. 60 ms), and their receptive fields are an order of magnitude larger. Most notably, they are highly sensitive to the direction of stimulus motion. Some MT neurons have the same aperture problem that is characteristic of V1 neurons, but others have solved it by integrating inputs from several different cortical neurons that each "look" at different small portions of the large moving object.

Awake Brain Stimulation

Some of the strongest evidence for MT's role in visual motion processing comes from elegant experiments that use chronically implanted electrodes (Box 11.2) to stimulate small groups of neurons in awake, behaving monkeys. Such **awake brain stimulation**

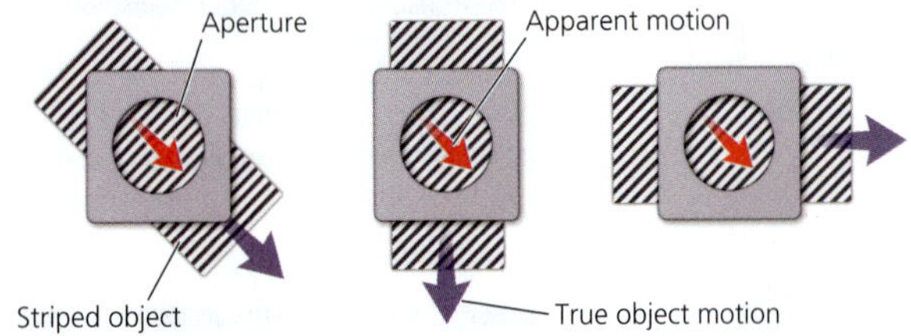

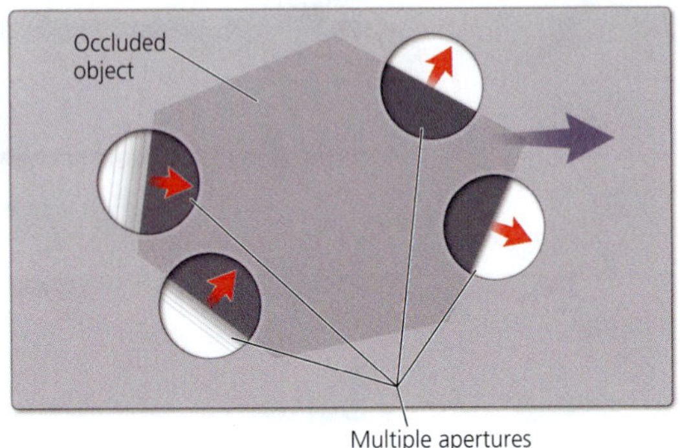

Figure 12.15 The aperture problem in motion perception. If you look at a moving straight edge (or line) through a small hole (an aperture), then the edge always seems to move at right angles to its long axis, regardless of the underlying object's true direction of motion (A). The aperture problem can be resolved if you look at multiple edges of the moving object through multiple apertures. Each visible edge appears to move in a different direction, but taking the average of all these apparent motions (weighted by their velocity) reveals the object's true direction of motion.

experiments are powerful because they can demonstrate causal links between neuronal activity and perception.

To appreciate the value of awake brain stimulation experiments, consider an experiment conducted to test area MT's role in motion processing. In this study, a monkey was trained to look at a screen that displays a random array of dots, each of which is moving around the screen. The monkey's task is to determine whether the average motion of all the dots presented during a "sample period" matches the direction of another set of dots that is presented during a later "test period" (Figure 12.16). Whenever the monkey gives the right response (pushing one button if the motions matched, another if they didn't), it gets a juice reward. During the test period, the individual dots all move in the same direction, but during the sample period they may move independently of one another, in a variety of directions. As you can imagine, this task is easy when most of the individual dots presented during the sample period are moving in the same direction, but it becomes increasingly difficult as the experimenters dial up a more balanced distribution of movement directions. When the number of dots moving in any one direction is precisely counterbalanced by the number of dots moving in the opposite direction, the task becomes impossible, and the monkey should be correct on just 50% of the trials (chance performance).

Once the monkey is well trained, the experimenters implant electrodes and map the receptive fields and directional preferences of neurons in a small patch of a monkey's MT area. They then stimulate these neurons electrically while the monkey is shown the moving dot stimuli in the neuron's receptive field and asked to perform the direction discrimination task. The question is whether the stimulation of those neurons will influence the monkey's perceptual judgment and thus its performance on the task. To maximize

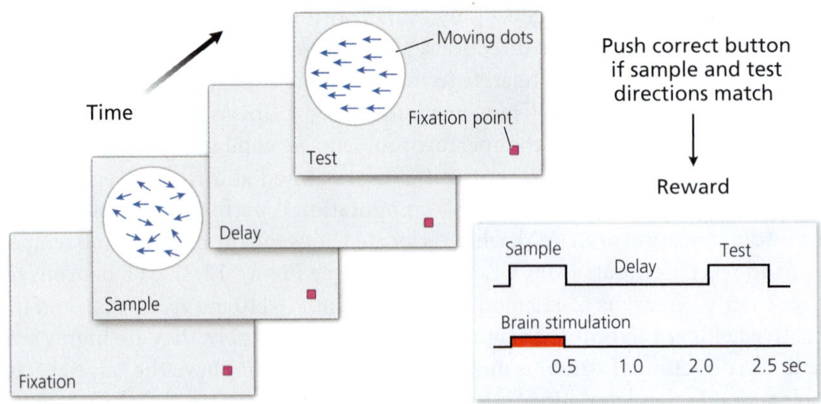

Figure 12.16 Design of an awake brain stimulation experiment. During a sample period, a monkey is shown a set of independently moving dots (blue arrows) that, on average, move left or right. During a later test period, the monkey is shown similar dots that move coherently in one of the two directions. The monkey is trained to push one button if the sample and test directions matched and another button if they did not. On some trials, the experimenters apply electrical stimulation to specific neurons during the sample period (B). The question is whether the brain stimulation affects the monkey's judgment of movement direction. For results, see Figure 12.17.

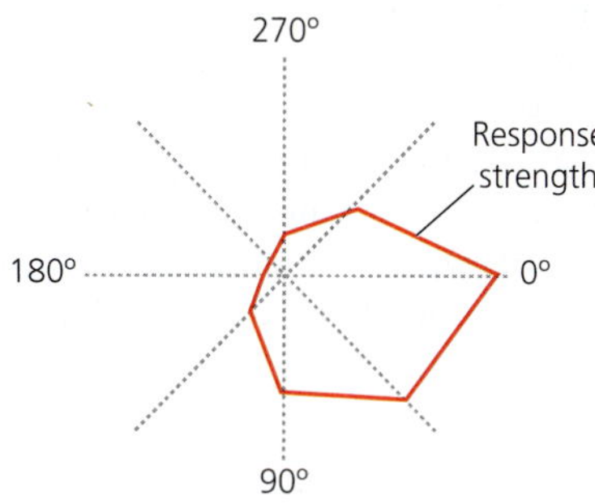

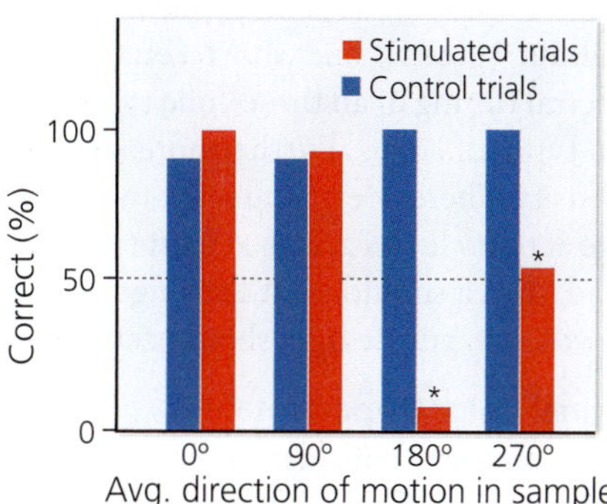

Figure 12.17 Microstimulation in MT. Monkeys were trained on the task illustrated in Figure 12.16. During some trials, the experimenters applied electrical stimulation to a small set of MT neurons. Shown at the left is the directional tuning curve of neurons at one stimulation site. Response strength is plotted against movement direction on a polar plot. The illustrated neuron responds much better to rightward stimulus motion (0°) than to leftward motion (180°). The results for stimulation at that site are shown on the right. When stimulation was applied during trials in which the stimuli moved to the left, the monkey almost never reported leftward motion (chance is 50%). Thus, boosting the activity of MT neurons tuned to rightward stimulus motion can bias the monkey's perception against leftward movement. [After Bisley et al., 2001]

the chance of seeing a stimulation effect, the experimenters make the task challenging by presenting stimuli in which the movement directions are nearly counterbalanced but not quite.

The major finding to emerge from this study (and several other, similar studies) is that the monkeys fail to identify the movement direction correctly when the preferred direction of the stimulated neurons differs significantly from that of the dots presented during the sample period (Figure 12.17). More specifically, the electrical stimulation biases the monkey's perception of movement direction in favor of the direction to which the stimulated neurons are tuned (their preferred direction) and against the opposite direction. The monkey's performance is not impaired if the preferred direction at the stimulated site matches the direction of the moving stimulus, or if the dots are presented outside of the stimulated neurons' receptive field.

Overall, these data allow us to conclude that MT neurons do not merely encode stimulus motion but are causally involved in generating motion perception. They are not epiphenomena but part of the brain's perceptual machinery!

BRAIN EXERCISE

Why is it important that the task in the MT microstimulation experiment is relatively difficult? Do you think the experiment would work if it were very easy for the monkey to tell which way the dots were moving during the sample period?

Identifying Color

Just as coherent motion can be used to separate an object from the rest of the image, so can color. Specifically, color can help identify an object's edges whenever those edges differ in color from the adjacent background. This color-based edge detection is especially useful when an object's edges are not definable by luminance contrasts (when the object and the background are equally bright). This explains, for example, why some species of octopus match the color of their skin to that of the background; it makes them virtually invisible to a potential predator. Color can also provide useful information about an object's identity and properties. It helps you determine, for example, how ripe a tomato is.

As we discussed in Chapter 6, color processing begins with retinal cone photoreceptors. The L-, M-, and S-cones of primates are most sensitive to long (yellow/red), medium (green), and short (blue/violet) wavelengths, respectively. Because the spectral tuning of all three cone types is relatively broad, most colors activate at least two types of cones. Furthermore, any cone's level of activation varies with stimulus intensity. Therefore, if you want to determine the color of a stimulus, you must compare activity levels across at least two different cone types. For example, if a stimulus is red, then it should elicit a stronger response from L-cones than from M- or S-cones. Conversely, purple light should activate only the S-cones.

Retinal Color Opponent Cells

Based on these considerations, it is not surprising that many retinal ganglion cells are specialized for comparing activity levels across multiple cone types. **Red-green opponent cells** compare the activation levels of L-and M-cones. They are either excited by L-cone activation and inhibited by M-cone activation, or they exhibit the inverse pattern. Another class of color-sensitive retinal ganglion cells compares activity of the S-cones to the combined activity of the L- and M-cones; these are the **blue-yellow opponent cells**. Thus, the retina contains at least some of the circuits required to identify stimulus color independently of light intensity.

Although the red-green and blue-yellow opponent cells we just discussed are sensitive to stimulus color, they provide little information about spatial variations in color, such as the edge of a red object on a green background. The retinal color opponent cells do have antagonistic center-surround receptive fields, but the surrounds usually respond to the same colors as the receptive field centers. Consequently, these opponent cells are sensitive to luminance contrasts (such as a dark red spot on a light red background) but insensitive to color contrasts (e.g., a bright red spot on an equally bright green background). Even in the lateral geniculate nucleus, few neurons are sensitive to such **color contrast**.

Color Contrast Neurons in V1

In contrast to the neurons of the retina and LGN, many neurons in V1 are sensitive to color contrast, meaning that they respond to colored edges in their receptive field even if the stimuli on either side of the edge are equally bright. Best studied are the **double-opponent cells**, which comprise 5–10% of all neurons in a macaque monkey's V1. These cells exhibit one type of cone opponency in their receptive field center and another in the surround. For example, the double-opponent cell illustrated in Figure 12.18 prefers L-cone activation over M-cone activation in its receptive field center (making it a red-center cell) but exhibits the opposite preference in its surround. As a result, this neuron responds strongly to red spots on a green background, even if the red and green are matched for luminance. Other double-opponent cells are sensitive to other types of color contrast.

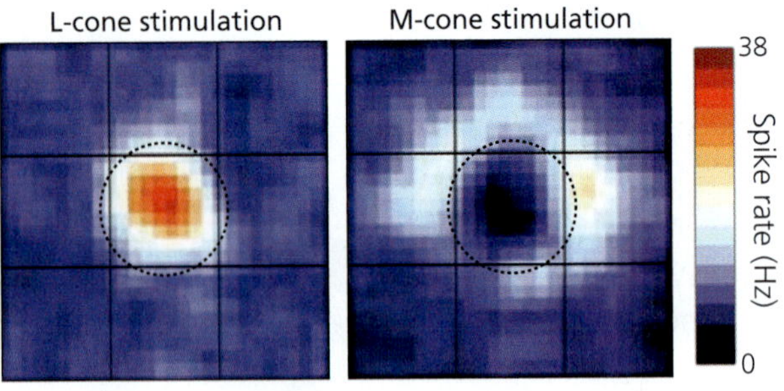

Figure 12.18 A color sensitive V1 neuron. The image on the left shows how a neuron in V1 of an awake monkey changes its firing rate in response to spots of light that preferentially activate L-cones. Shown on the right are the responses of the same neuron to stimuli that isolate M-cone activity. For stimuli in the center of the neuron's receptive field, the neuron is excited by L-cone stimulation but inhibited by M-cone stimulation. The neuron is also excited by M-cone stimulation in the surround region of its receptive field. Such a neuron is called a double-opponent cell because it has opposite responses to L- and M-cones and opposite color preferences in its center and surround. [From Conway and Livingstone, 2006]

The surrounds of double opponent cells are often asymmetric (sickle shaped), which makes these neurons sensitive to the orientation of colored lines and edges that cross through their receptive fields. However, double-opponent cells tend to have relatively large receptive fields and are not as orientation selective as V1 simple cells. This low degree of orientation selectivity probably explains why the double-opponent cells tend to be located at the centers of the orientation pinwheels that we discussed earlier. Their orientation selectivity is poor compared to that of the surrounding cells. Because the color-sensitive pinwheel centers appear as small dark spots when the cortex is stained for cytochrome oxidase (a metabolic enzyme), they are named *V1 blobs*.

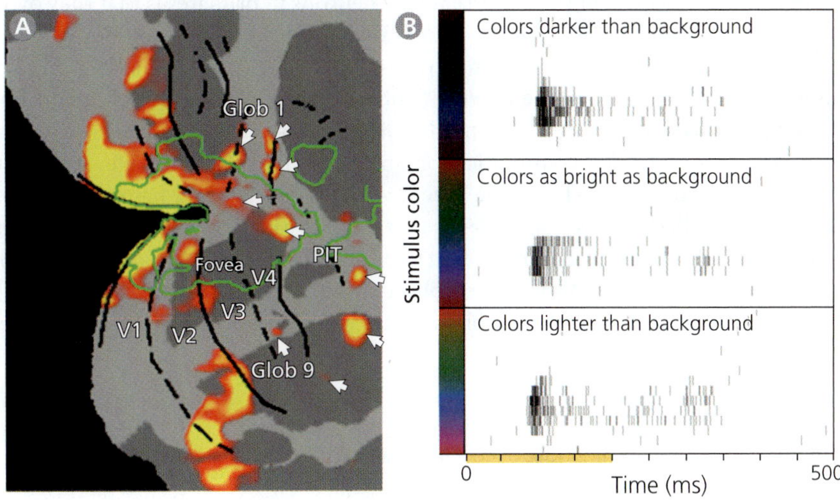

Figure 12.19 Color-responsive areas in the monkey cortex. Monkeys were shown a moving pattern of stripes that were either black-and-white or red and blue (with the two colors being equally bright). The cortical flat-map (A) reveals strong color sensitivity in the foveal regions of V1 and V2 as well as in nine "globs" (white arrows) scattered across the extrastriate cortex, including V4 and the posterior inferior temporal cortex (PIT). The data in (B) were obtained by recording extracellularly from a single neuron in a color-sensitive glob. Shown along the vertical axis are the color and brightness of the stimuli. Darker tick marks represent higher firing rates. The illustrated neuron responds to blue stimuli regardless of how bright they are relative to the background. [From Conway et al., 2007]

Color Sensitivity in the Extrastriate Visual Cortex

Functional brain imaging has shown that color processing continues past V1, in V2, V4, and several regions near the boundary between the temporal and occipital lobes. Specifically, brain imaging has revealed several clusters of neurons that respond more strongly to local color contrasts (colored stripes) than to luminance contrasts (black-and-white stripes). These color-sensitive areas do not respect the traditional boundaries of cortical areas and are, instead, scattered across multiple cortical areas (Figure 12.19).

By analogy to V1 blobs, color sensitive patches in the anterior occipital and posterior temporal lobes are called **globs**. Individual neurons within a glob tend to be tuned to specific colors independently of how bright or dark those colors are. This is important because most other color-sensitive neurons in the visual system change their color preference as light levels vary, which is not consistent with our subjective impression of color (Figure 12.20). After all, our perception of an apple's color does not change as the lights are dimmed (unless it gets so dark that the cones stop functioning and we no longer see colors at all). However, only a few neurons in our brains, especially within the globs, exhibit such **hue constancy**.

Figure 12.20 Hue constancy. Our perception of an object's color (or hue) barely changes when incident light levels are decreased (left vs. right). The asterisks mark two yellow patches that appear similar in color when they are part of the larger Rubik's cubes but different when they are isolated from the cubes.

An intriguing aspect of the color-sensitive neurons in primates is that they overrepresent the red end of the color spectrum. This bias toward red, together with the primate-specific evolution of red-sensitive L-cones, is probably related to the fact that many primates like to dine on ripe fruits and young leaves, which tend to be highly nutritious and red. In addition, the ability to discriminate various shades of red is useful for nonverbal communication among primates. Have you ever noticed how sick or fearful people look pale green, whereas healthy and happy ones have rosy cheeks? Well, there is some evidence that monkeys use the color of hairless skin on the face and rump of other monkeys to gauge the condition (or mood!) of their potential mates and competitors. These observations suggest that an increased ability to discriminate red from other colors would have been adaptive during primate evolution.

BRAIN EXERCISE

When you look at a color photograph and then look at the same image in black-and-white, how does neural activity in your visual cortex change? Do all the color sensitive neurons fall silent? Why or why not?

Identifying Complex Visual Objects

The perception of complex visual objects involves cortical areas in the ventral stream, which reaches into the temporal lobe. Early studies revealed that neurons in the temporal cortex respond poorly to simple stimuli, such as circles, lines, or rectangles. However, some neurons in the temporal lobe respond well to monkey and human faces (Figure 12.21).

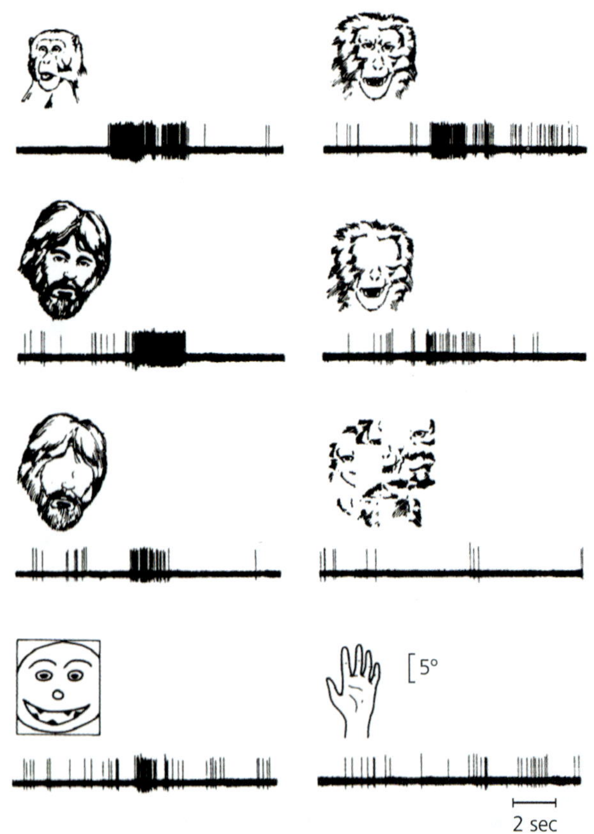

Figure 12.21 Face cells in the monkey's temporal lobe. Bruce, Desimone, and Gross reported in 1981 that a few neurons in the temporal cortex of anesthetized monkeys responded strongly to monkey and human faces but not to other visual stimuli, including scrambled faces and hands. The responses weakened considerably when the faces had no eyes or were drawn as cartoons. [From Bruce et al., 1981]

Face-selective Cells

Neurons in the temporal cortex that respond well to faces respond poorly to scrambled faces in which the eyes or mouth are displaced (as in a cubist painting). They also respond poorly to faces without eyes or to various other complex visual stimuli. Therefore, they are called **face-selective neurons.** In contrast to V1 neurons, which respond only to small and simple image features, the face-selective neurons detect a complex constellation of features that, collectively, define a face. They clearly occupy a higher level of the visual processing hierarchy than the line and edge detectors in V1.

To determine whether face-selective neurons respond to just one face or to faces more generally, researchers recorded from single neurons while showing monkeys a variety of different faces. These studies showed that face-selective neurons generally respond better to some faces than to others, although all respond to more than just one face (Figure 12.22). Therefore, the face-selective neurons are not "grandmother cells." Instead, each face is encoded by a combination of simultaneously activated neurons; and different faces are encoded by different, although potentially overlapping, combinations of neurons (recall Figure 12.2). As noted earlier, an advantage of such a population code is that a relatively small number of face-selective neurons can encode a vast variety of faces. Another benefit is that the loss of a few face-selective neurons should cause only minor impairments in the ability to recognize any specific face.

Early studies on visual responses in the temporal lobe had been unclear about the precise location of the face-selective cells. In monkeys, such neurons are most commonly

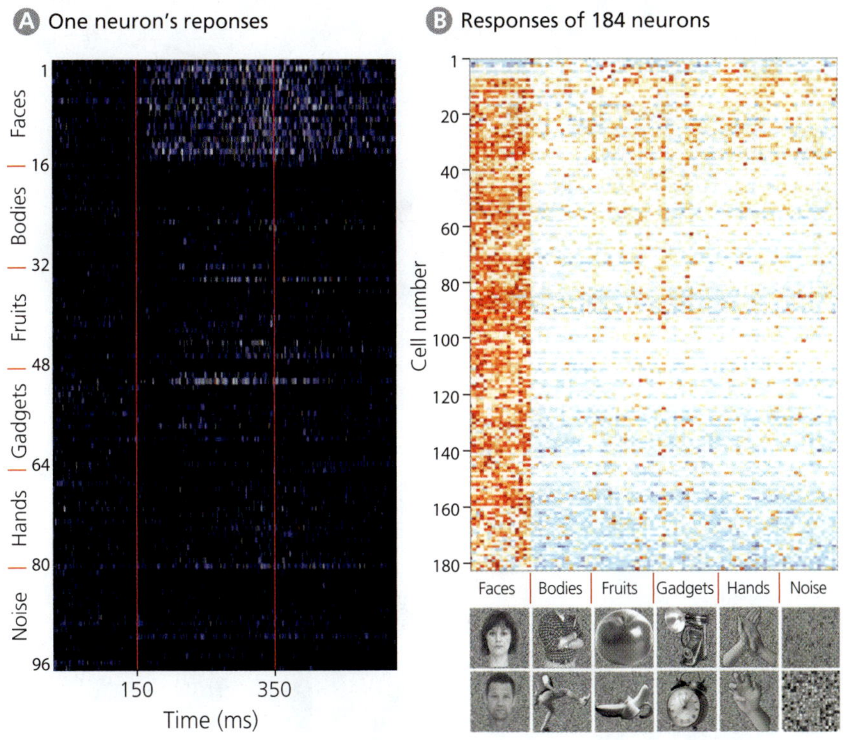

Figure 12.22 Face-selective neurons. Experimenters used fMRI in monkeys to identify multiple patches of the temporal lobe cortex that are activated more strongly by faces than by other visual stimuli. They then recorded the responses of individual neurons in those patches to 16 different (familiar or unfamiliar) faces, 16 headless bodies, 16 fruits, 16 gadgets, 16 hands, and 16 random noise patterns. The responses of one such neuron are shown in (A), where each white tick mark represents an action potential. The responses of 184 neurons from a single monkey are summarized in panel (B). Pixel color represents response rates (orange/red = strong response). Two examples of each stimulus type are shown at the bottom of panel (B). [From Tsao et al., 2006]

found in the **inferior temporal cortex**, which occupies the lateral and inferior surfaces of the temporal lobe. The analogous (and probably homologous) region in humans includes the anterolateral portion of the occipital lobe and the posterior portion of the temporal lobe. For an extensive study of the face-selective neurons in monkeys, scientists used functional brain imaging to identify multiple face-selective patches, the location of which varied slightly from monkey to monkey. The investigators then targeted individual patches with microelectrodes and confirmed that nearly every neuron in a face-selective patch responds more strongly to faces than to other complex stimuli, such as fruits, hands, or headless bodies (Figure 12.22). Thus, the inferior temporal cortex in monkeys contains a relatively small number of face-selective neurons that are clustered into discrete face-selective patches.

Linking Face-selective Cells to Face Perception

To demonstrate that face-selective neurons in the inferior temporal cortex are causally involved in face perception, scientists stimulated face-selective neurons while monkeys were discriminating faces from other sorts of stimuli. This brain stimulation experiment is conceptually similar to the experiment we discussed earlier, in which experimenters stimulated MT neurons to manipulate movement perception.

Specifically, the experimenters identified a group of face-selective neurons in a monkey's inferior temporal cortex and stimulated these neurons while the monkey discriminated faces from flowers and other non-face stimuli (Figure 12.23). Whenever the monkey perceived a face, it was trained to move its eyes to a specific location for a juice reward. Importantly, the faces and the non-face stimuli were embedded in visual noise to varying degrees. When the noise level was low, the monkeys could perceive the faces and the non-face stimuli quite easily and give the appropriate eye movement response. However, when the stimuli were swamped in noise, the monkeys found the task difficult (50% correct equals chance performance in Figure 12.23). The crucial finding was that when noise levels were high, electrical stimulation of the face-selective neurons biased the monkeys toward perceiving faces rather than non-face stimuli. Thus, manipulating the number of active face-selective neurons in the inferior temporal cortex can change a monkey's perception.

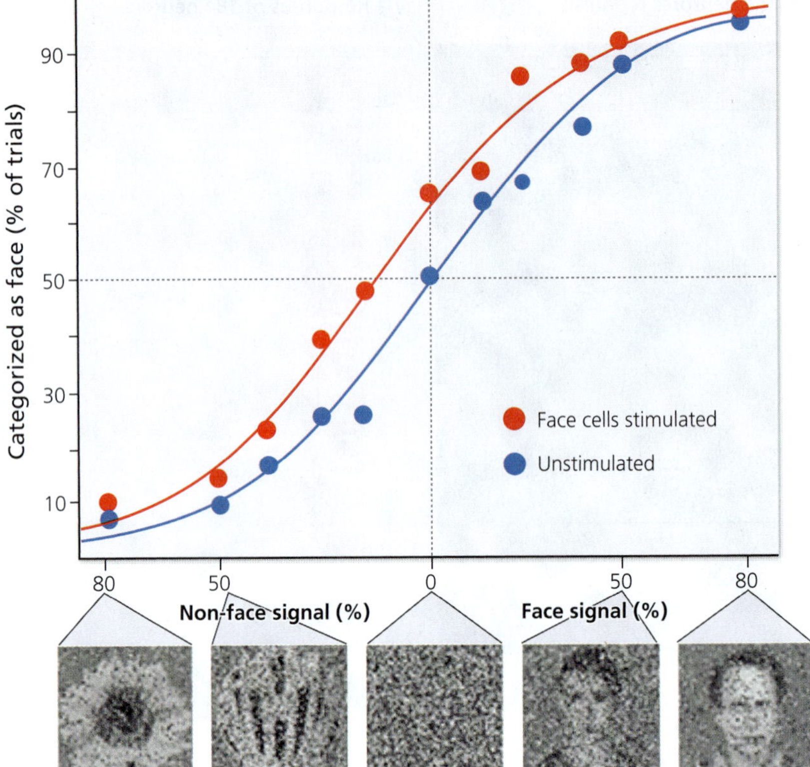

Figure 12.23 Stimulation of face-selective cells can bias perception. A small patch of face-selective neurons in a monkey's inferior temporal cortex was stimulated electrically immediately after the monkey saw pictures of a face or a non-face stimulus embedded in varying amounts of visual noise. The monkey was trained to move its eyes to a specific target after detecting a face. The major finding was that stimulation of the face-selective cells increased the likelihood that the monkeys would categorize the noisy, relatively ambiguous stimuli as faces. This effect was observed at 19 of 31 face-selective sites and at only 1 site where neurons did not respond to faces. The effect was largest when the noise levels were high. [From Afraz et al., 2006]

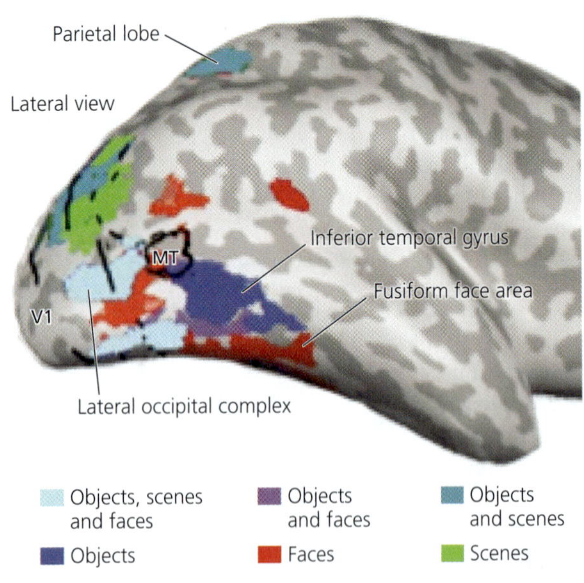

Figure 12.24 Patchy representation of visual objects. Investigators compared fMRI responses to pictures of novel objects, human faces, four-legged mammals, houses, and natural scenes to the responses to visual noise. They found that the lateral occipital complex is more strongly activated by objects, scenes, and faces than by noise. Parts of the inferior temporal gyrus are activated more strongly by novel objects than by other sorts of stimuli. The fusiform face area is activated most strongly by faces. Parts of the parietal lobe are activated by visual objects and natural scenes. [From Grill-Spector, 2003]

Patchy Encoding of Complex Visual Stimuli in Humans

Human brain imaging research has confirmed the existence of face-selective patches in the temporal cortex. The largest of these patches lies in the inferior portion of the temporal lobe and is called the **fusiform face area** (Figure 12.24). Other face-selective patches can be found in more posterior and superior regions of the temporal lobe.

Interspersed among the face-selective patches are patches of neurons that preferentially respond to other types of complex visual objects such as tools, animals, fruits, or natural scenes (such as houses and landscapes). These non-face object selective neurons have not been studied as extensively as the face-selective neurons, but some of them are similar to face-selective neurons insofar as they are relatively insensitive to changes in stimulus size or rotation. Some respond to their preferred stimulus object even when that object is partially obscured.

Neural Correlates of Categorical Perception

The discovery that neurons in the inferior temporal cortex are relatively tolerant of variations in the size, rotation, and completeness of their preferred stimulus raises an important question: how do these neurons draw the line between stimuli that they get excited about and those that they ignore? Do their responses gradually fade out as preferred stimuli are modified, or do the responses vanish more abruptly?

To answer this question, monkeys were shown computer-generated images of various cats and dogs as well as

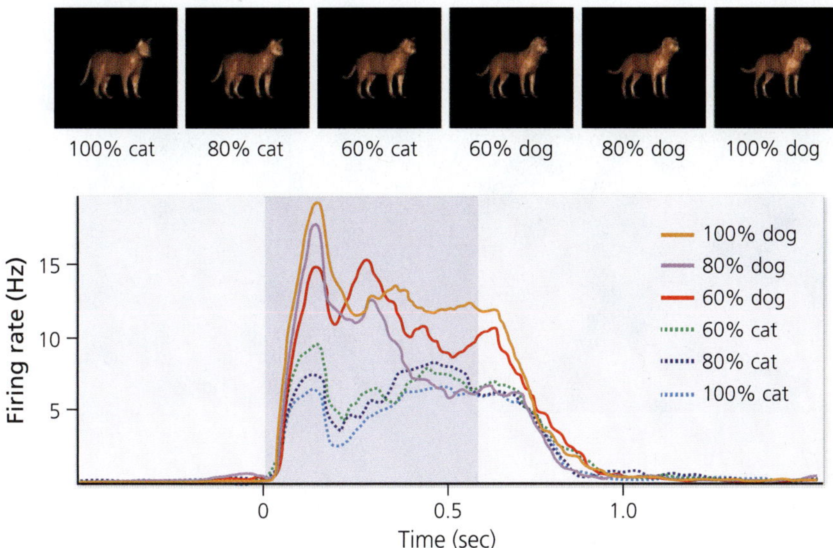

Figure 12.25 Category-sensitive neurons. Monkeys were shown two successive pictures of cats, dogs, or digitally morphed intermediate creatures and trained to indicate whether or not the stimuli were of the same category (e.g., two cats). The illustrated graph shows the responses of one neuron in the inferior temporal cortex to the 6 types of stimuli shown at the top. If a stimulus was >50% dog, the neuron responded as if it had been a 100% dog; if the stimulus was >50% cat, the neuron acted as if had been a perfectly good cat. This dichotomous response pattern reflects what we do when we categorize animals as being either cats or dogs but never intermediate creatures. [Top from Freedman et al., 2001; bottom after Freedman et al., 2003]

intermediate morphs (Figure 12.25). The monkeys were shown two such pictures in succession and trained to indicate whether the depicted animals belonged to the same category (e.g., two dogs) or to two different categories. Simultaneously, the experimenters recorded neuronal activity. They found many neurons that respond more strongly to pictures of cats than dogs, or vice versa. They even found a few neurons whose responses do not change gradually as one type of animal is morphed into another. Instead, these neurons respond robustly as long as the stimulus is more than 50% cat or, for other neurons, more than 50% dog. Such neurons behave as if they recognize a category boundary between dogs and cats and are probably involved in categorical perception. However, such **category-sensitive neurons** are relatively rare in the inferior temporal cortex. They are found more frequently in the prefrontal cortex, which receives input from the inferior temporal lobe and is involved in a variety of complex cognitive processes (see Chapter 15).

In summary, neurons in the inferior temporal cortex of monkeys and humans respond preferentially to images of complex visual objects. Different neurons respond to different kinds of objects; and neurons responding to the same major type of object, such as faces or animals, tend to be clustered into discrete patches. However, object-selective neurons in the inferior temporal cortex rarely respond to only one specific object or to a distinct object category. Instead, most of them respond to a multitude of similar objects, such as various faces or pictures of animals that share some similar features. To find more selective neurons, neuroscientists must look downstream of the inferior temporal cortex, in the hippocampus and its related cortices (see Chapter 14) and in the prefrontal cortex (see Chapter 15).

BRAIN EXERCISE

Do you think people who prefer dogs to cats might have more dog-selective neurons than cat-selective neurons in their inferior temporal cortex? What about people who've never seen a dog (or a cat)?

12.3 How Do Neurons Encode Non-Visual Objects?

As mentioned earlier, the vast majority of research on object identification has been conducted with visual stimuli. Significantly less is known about how we identify non-visual objects, such as the smell of a rose, the meow of a cat, or the feel of a coffee mug. Therefore, our discussion of non-visual object identification is relatively short. Let us begin with olfaction.

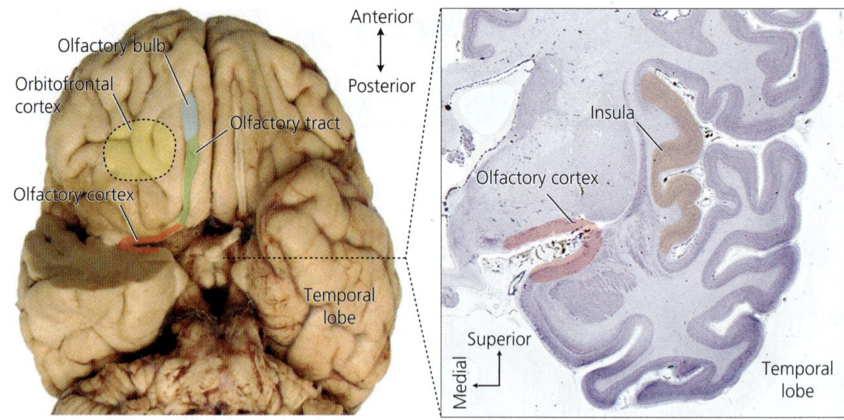

Figure 12.26 Cortical areas involved in olfaction and taste. Shown on the left is an inferior view of a human brain from which the anterior part of the right temporal lobe was cut away, revealing the olfactory cortex (shaded red). Also indicated are the olfactory bulb, the olfactory tract, and part of the orbitofrontal cortex that receives olfactory and taste information (dashed outline). Depicted on the right is a coronal section through the olfactory cortex. This section also shows the insular cortex, which contains taste-sensitive neurons. [After Gottfried and Zald, 2005]

Object Identification in the Olfactory System

Back in Chapter 6, we reviewed how olfactory sensory neurons in the olfactory epithelium detect odorant molecules and send their axons to specific glomeruli in the olfactory bulb. You also learned that mitral cells, the principal projection cells in the olfactory bulb, extend their dendrites into the glomeruli and project to the olfactory cortex. Now it is time to examine this projection in more detail.

Projections to the Olfactory Cortex

The **olfactory cortex** in humans is located near the junction of the temporal and frontal lobes (Figure 12.26). It is composed of multiple subdivisions, the largest of which is called the *piriform cortex* (which means pear-shaped cortex, although the structure hardly resembles a pear). The piriform cortex is sometimes called primary olfactory cortex, but all divisions of the olfactory cortex receive direct projections from the olfactory bulb, which makes them all equally primary. As you can tell by comparing Figures 12.26 and 12.27, the olfactory cortex is much smaller in humans than in mice, at least relative to the neocortex.

An important feature of the connections from the olfactory bulb to the olfactory cortex is that they are non-topographic and highly divergent. Each mitral cell gets input from a single glomerulus and sends a long axon to the olfactory cortex. There the axon branches and meanders extensively (Figure 12.27), synapsing onto a large number of pyramidal neurons along the way. Because each mitral cell contacts so many pyramidal neurons throughout so much of the olfactory cortex, we can infer that each pyramidal neuron in the olfactory cortex receives converging inputs from many different mitral cells associated with many different glomeruli. Indeed, physiological studies have shown that olfactory pyramidal neurons require the simultaneous activation of multiple glomeruli before they themselves are excited enough to increase their own firing rate.

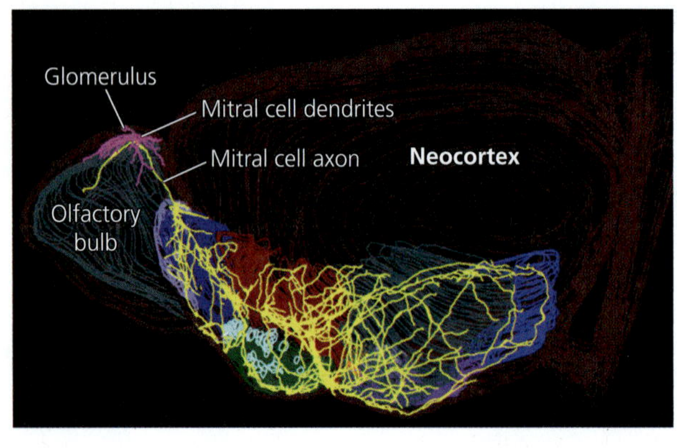

Figure 12.27 Projections of one mitral cell to the olfactory cortex. Labeling of a single mitral cell reveals that this neuron receives input from just one glomerulus (dendrites are shown in light purple) and sends a meandering and highly branched axon (yellow) to most of the olfactory cortex (colored areas). This lateral view of the mouse's forebrain was reconstructed from many sections. [From Igarashi et al., 2012]

Sensory Coding Schemes in the Olfactory System

To understand the implications of this circuitry for olfactory coding, we must briefly review how odorants are represented in the olfactory bulb. As you learned in Chapter 6, each olfactory sensory neuron expresses a specific olfactory receptor molecule, which recognizes a specific epitope (submolecular feature) in odorant molecules. Because most molecules possess multiple epitopes, and most natural odors are mixtures of multiple odorants (clam broth, for example, emits a mix of 49 different odorants), most odors activate a wide variety of olfactory sensory neuron types. These olfactory sensory neurons, which are spread across the olfactory epithelium, converge onto a specific set of glomeruli in the olfactory bulb (Figure 12.28; also Figure 6.17). Each natural odor activates a different set of glomeruli, but each glomerulus may be activated by a variety of different odors. Therefore, odor identity is encoded in the pattern of activated glomeruli. It is a beautiful example of population coding.

How does the encoding of odorants change as you go from the olfactory bulb to the olfactory cortex? The main difference is that the olfactory cortex contains neurons that are activated only by specific mixtures of epitopes and odorants. They only fire when several of their input lines are active simultaneously. In a sense, they are "grandmother odor" cells. However, each odorant or odorant mixture activates a relatively large number of neurons scattered across the olfactory cortex.

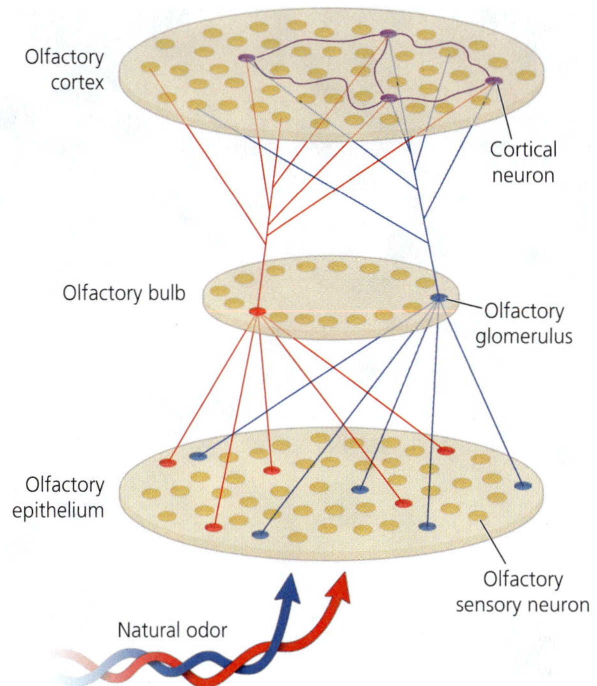

Figure 12.28 Encoding natural odors. Most natural odors are mixtures of two or more odorants (red and blue) that activate multiple types of olfactory sensory neurons in the olfactory epithelium. The outputs of these neurons converge onto specific glomeruli in the olfactory bulb, which then project divergently to the olfactory cortex. Within the olfactory cortex, natural odors activate widely dispersed neurons (highlighted in purple) that receive coincident input from the activated glomeruli. With repeated odor exposure, those cortical neurons become "wired together" (purple lines).

Cell Assemblies in the Olfactory Cortex

An unusual aspect of the olfactory cortex is that each of its principal neurons is connected, at least weakly, to a very large number of other olfactory cortex neurons. Moreover, these connections exhibit Hebbian LTP (see Chapter 3), which means that simultaneous activation of a subset of olfactory cortex neurons will tend to strengthen the connections between those neurons. Therefore, repeated exposure to a specific odorant or odorant mixture causes the olfactory cortex neurons that are activated by this stimulus to become "wired together" into what Donald Hebb called a **cell assembly**. We will discuss the formation and significance of Hebbian cell assemblies more thoroughly in Chapter 14.

For now, the important point is that repeated exposure to a specific olfactory stimulus forges the olfactory cortex neurons activated by that stimulus into a coherent entity—an assembly of interconnected cells. When one part of such a cell assembly is later activated by a partial stimulus (such as a familiar odor from which a few components are missing), the excitatory connections between the activated assembly members can activate the whole assembly, "filling-in" the missing components. Thus, with experience, the olfactory cortex comes to contain a multitude of different (although often overlapping) cell assemblies that each encode a specific odor.

BRAIN EXERCISE

The olfactory system is unusual among the sensory systems for exhibiting almost no topography in its pathways. What are the functional consequences of this difference? Why do you think the difference exists?

Identifying Sounds

Although we humans are highly visual creatures, we can identify many objects from their sounds. Just think of a car alarm, the sound of a chainsaw, a dog's bark, or the voice of your mother. Relatively little is known about the neural mechanisms underlying auditory object identification, but we do know that the task is not simple.

You may remember that auditory information ascends to the neocortex through a series of medullary, midbrain, and thalamic nuclei (see Figure 11.14). Within the neocortex, the major target of auditory information is the **auditory cortex**, which lies along the superior edge of the temporal lobe and is divisible into core, belt, and parabelt regions (Figure 12.29). The core region includes the **primary auditory cortex (A1)**

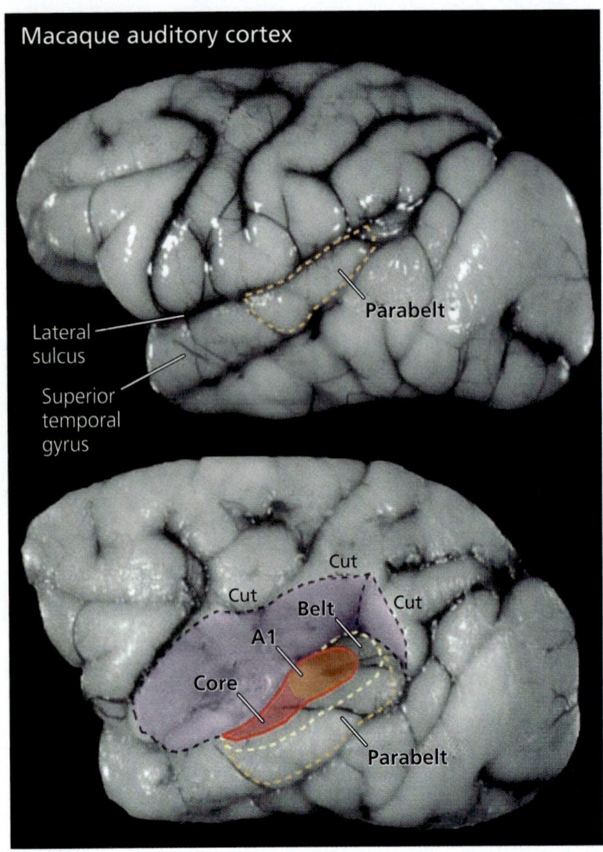

Figure 12.29 Auditory cortex. Primate auditory cortex is divisible into a core, which includes the primary auditory cortex (A1), a belt region surrounding the core, and a parabelt region adjacent to the belt. Only the parabelt region is visible in a lateral view of the intact macaque brain (top). The rest of the auditory cortex lies on the inferior bank of the lateral sulcus. It becomes visible when brain regions superior to the lateral sulcus are cut away (bottom). [After Kaas and Hackett, 2000]

as well as two other cortical areas. The belt and parabelt regions can also be subdivided. Although A1 is recognized across all mammals, many of the other auditory areas are difficult to compare between rodents, primates, bats, and cats—the taxonomic groups in which they have been studied best. Because of these species differences, it is difficult to make general statements about the structural and functional organization of the mammalian auditory cortex.

Nonetheless, it is clear that the core auditory areas are tonotopically organized, just like the cochlea. In contrast, the belt and parabelt regions of auditory cortex are not tonotopically organized, and many of their neurons respond more strongly to complex sounds than to pure tones. For example, some neurons in these areas respond selectively when a low-frequency tone is followed by a higher frequency (a rise in pitch). Others respond best to species-typical communications sounds; in humans, such sounds include speech.

Spectrotemporal Receptive Fields

To understand how neurons encode those complex sounds, researchers often employ the concept of a **spectrotemporal receptive field** (Figure 12.30). You already learned that many auditory neurons tend to respond most strongly to a specific range of sound frequencies; by analogy with spatial receptive fields, you can think of those preferred sound frequencies as the neuron's *spectral receptive field* (the word "spectrum" here refers to the spectrum of sound frequencies). Once you extend this concept to sound stimuli that vary in frequency (pitch) over time, you end up with a neuron's spectrotemporal receptive field (STRF). In theory, at least, this STRF tells you how a neuron will respond to complex sounds whose pitch varies with time (such as music or speech).

To determine a neuron's STRF, researchers record how the cell responds to a wide variety of sounds (often randomly

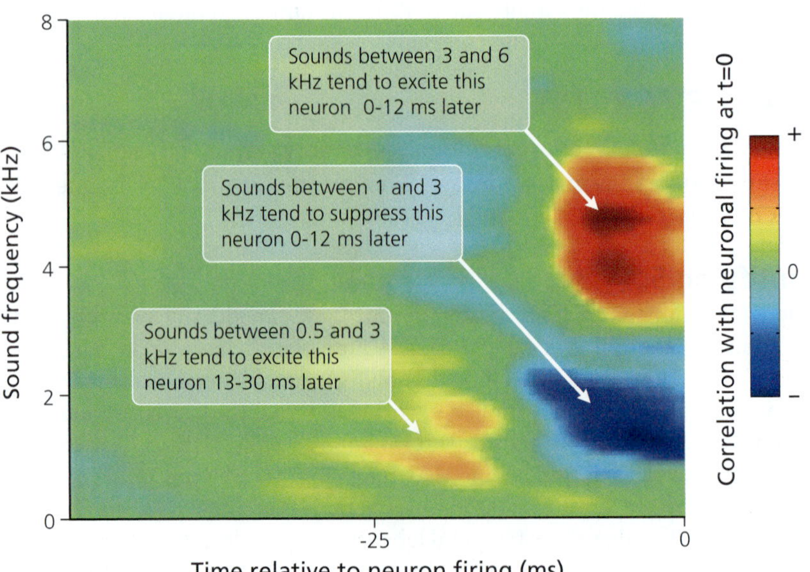

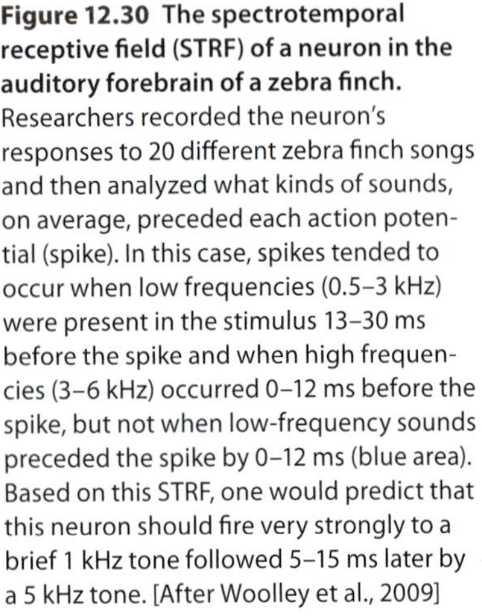
Figure 12.30 The spectrotemporal receptive field (STRF) of a neuron in the auditory forebrain of a zebra finch. Researchers recorded the neuron's responses to 20 different zebra finch songs and then analyzed what kinds of sounds, on average, preceded each action potential (spike). In this case, spikes tended to occur when low frequencies (0.5–3 kHz) were present in the stimulus 13–30 ms before the spike and when high frequencies (3–6 kHz) occurred 0–12 ms before the spike, but not when low-frequency sounds preceded the spike by 0–12 ms (blue area). Based on this STRF, one would predict that this neuron should fire very strongly to a brief 1 kHz tone followed 5–15 ms later by a 5 kHz tone. [After Woolley et al., 2009]

varying or complex natural sounds). They then use computer algorithms to perform spike-triggered averaging (plus a statistical normalization procedure), which identifies the stimulus frequencies that were most prominent, on average, at various intervals before each of the many action potentials (spikes) that the neuron fired during the testing procedure. The output of such an analysis is generally a pseudo-colored plot as shown in Figure 12.30. Assuming a neuron's response to long, time-varying stimuli is just the (linear) sum of the responses to the individual bits and pieces of those long stimuli, such STRF plots should allow you to predict a neuron's responses to other complex sounds. For example, the neuron illustrated in Figure 12.30 should respond very strongly to a 1 kHz tone that is followed 10 ms later by a 5 kHz tone. In contrast, its response to a sustained 1 kHz tone should be much more subdued because the neuron is both suppressed and excited (with different delays) by this sound frequency.

Although STRF analyses are widely used, the results of these analyses are mixed. Neurons at lower levels of the auditory pathway, notably the midbrain's inferior colliculus, tend to respond to complex sounds as the STRF analyses predict. However, many neurons in the auditory cortex defy those expectations, mainly because the response of those neurons to a snippet of sound tends to depend on its acoustic context, much as your interpretation of a spoken word often depends on the preceding words (e.g., you interpret the word "bus" quite differently when it occurs in a sentence about dirty tables at a restaurant than when you hear it in a conversation about schoolchildren). More concretely, the neuron in Figure 12.30 might not, despite our predictions, increase its firing rate when a low-frequency tone is followed 10 ms later by a high-frequency tone, even though it would respond with excitation to both elements individually. We still know very little about the context dependencies (nonlinearities) that shape auditory processing at higher levels of the cortical hierarchy, but the observed mismatch between actual responses and those predicted from STRF analyses implies that those dependencies exist.

Why do we know so little about stimulus encoding in the auditory cortex, when our knowledge of stimulus coding in the visual cortex seems so incredibly detailed? Part of the answer is that the visual cortex has been studied much more intensively, by far more researchers, than the auditory cortex. However, it is also true that studies on the visual cortex rarely use the kind of complex, time-varying stimuli that STRF analyses work with. For example, very few studies have asked how neurons in the visual cortex fire during the viewing of movies. A major conclusion from those studies is that the responses of V1 neurons to movies are not fully predictable from how those same neurons respond to simple stimuli, such as the lines and edges we discussed earlier. Thus, our visual cortex is more mysterious than you might think.

Identifying Things by Touch or Taste

Aside from using vision, smell, or sound to identify the objects around us, we can also use touch. As we reviewed in Chapter 7, our skin harbors a wide variety of sensors that are activated by touch or other mechanical disturbances. Information from these sensors is conveyed through the dorsal thalamus to the primary somatosensory cortex (S1; see Figure 11.2). Because this area is somatotopically organized, contact between an object and our skin activates S1 in a spatial pattern that reflects the object's shape. The cortical activation is especially pronounced at the time of initial contact, before the receptors habituate, or when the object is moving. From S1, touch information passes anteriorly to the primary motor cortex and posteriorly to the secondary somatosensory cortex as well as the posterior parietal cortex. All of these areas are activated when we explore objects by touch.

To identify more precisely which cortical areas function in touch-based object identification, experimenters have used fMRI to compare the pattern of cortical activity observed while subjects are identifying three-dimensional objects by touch to the activity observed when the subjects are merely touching textured surfaces (such as sandpaper). This analysis revealed small patches of cortex related to touch-based object identification in the posterior parietal and the lateral occipital cortex. The latter area is interesting because its neurons also respond better to visual objects than to visual noise. Thus, the lateral occipital cortex is one of the few brain areas where multimodal information about objects converges.

Perhaps the most fascinating question about touch-based object identification is how we analyze the complex flow of information that is generated when we actively explore an object with our hand. To determine an object's shape by touch, our nervous systems must combine complex spatial and temporal patterns of skin sensor activity with information about where our fingers are in space and how they are moving. Unfortunately, very little is known about these surely awesome computations.

Taste is another sense that we use to determine object identities; babies, in particular, love to identify objects by taste! When an object contacts the taste cells in our mouth, the gustatory information is sent to a special part of nucleus tractus solitarius. From there, gustatory information is relayed through the dorsal thalamus to the **insular cortex** (see Figure 12.26). The insular cortex, in turn, projects to the **orbitofrontal cortex** (Figure 12.26), where tastes are integrated with olfactory information. Little is known about how different tastes are represented in the insular and orbitofrontal cortices. However, the coding scheme is probably quite similar to that used in the olfactory cortex because the available data reveal that different flavors tend to generate overlapping patterns of cortical activation.

BRAIN EXERCISE

For each of your major senses, imagine how it might be used to identify an object. Which senses are best for identifying which kinds of objects? If you have a dog, would it agree with your answer?

12.4 Are We Born with All the Neural Circuitry We Use to Identify Stimulus Objects?

As noted in Chapter 4, neural circuits that formed during embryogenesis are frequently refined during later stages of development, especially early postnatal life. You also read in Chapter 3 that neural circuits can be modified in adulthood as organisms learn new information and skills. Therefore, you will not be surprised to learn that we are far from being born with all the neural circuitry we need to identify complex objects. Much of this machinery develops as we grow up and, importantly, is shaped by our experiences during that time. In general, experience makes neural circuits more efficient at encoding the experienced stimuli and, conversely, less capable of encoding the stimuli that were encountered rarely.

Sensory Deprivation Experiments

A common approach to studying neural plasticity is to deprive young animals of normal sensory experiences. Back in Chapter 4, you learned that raising kittens with one eye shut leads to an expansion of the cortical ocular dominance bands for the open eye (see Figure 4.35). Now we may ask, what happens when both eyelids are shut, depriving the kitten of all pattern vision? Under those conditions, most V1 neurons become unresponsive to visual stimuli. Importantly, the deficit persists after the eyes are opened, implying that a lack of early pattern vision causes permanent visual deficits. This seems to be true also in humans because children born with cloudy lenses in their eyes (cataracts) are unlikely ever to develop effective pattern vision unless those lenses are replaced early in development.

These age-dependent **deprivation effects** are consistent with the general notion that experience-dependent neural plasticity occurs most easily during early development, during a **sensitive period** (also known as the critical period). The onset and duration of the sensitive period may vary from feature to feature. Furthermore, the end of a sensitive period may be delayed under some conditions. For example, when cats are raised in total darkness, their visual cortex remains plastic for an unusually long time. Even after the end of a supposed sensitive period, some neural plasticity is possible. For example, one girl whose congenital cataracts were removed at age 12, which is unusually late, developed good pattern vision. According to her mother, it took the girl about 6 months after the surgery to recognize her brothers by sight and another 6 months to visually identify objects around the house.

Instructive Effects of Early Experience

Besides permitting normal development, experience can guide development in specific directions. Evidence for such **instructive effects** comes from kittens that were raised so that they could see only lines of a particular orientation (by fitting them with prism goggles or raising them in striped enclosures). Such rearing conditions increase the fraction of V1 neurons that are tuned to the experienced orientation (Figure 12.31).

Similar effects have been observed in the auditory system. In one study, young rat pups were exposed to tones of a specific frequency for 10–16 hours a day. When the experimenters later examined the frequency tuning of auditory cortical neurons in these tone-reared rats, they found twice the normal number of neurons tuned to the experienced frequency. In addition, the neurons that were tuned to the experienced frequency had become more sharply tuned, meaning that they responded less

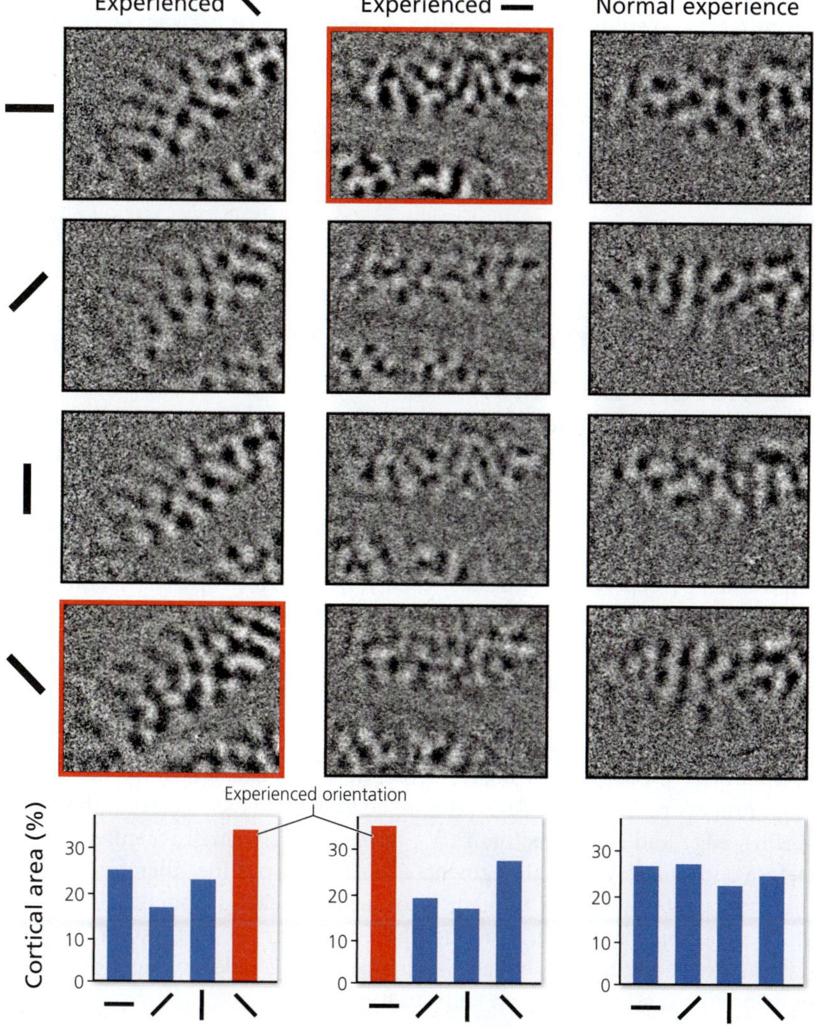

Figure 12.31 Early experience can alter orientation tuning. Kittens were raised in a tall cylinder with walls covered in black and white stripes of a single orientation (e.g., horizontal). The experimenters then used intrinsic signal optical imaging to record the orientation preference of neurons in V1. Shown here are results from 3 kittens that grew up with diagonal stripes (left column), horizontal stripes (middle column), or normally (right column). The first 4 rows show optical imaging data collected while moving stripes were presented at 4 different orientations (indicated at the left). The black patches represent neurons that were activated by the stimulus. In the kittens with the abnormal visual experience, a disproportionate amount of cortical territory was activated by stimuli of the experienced orientation (red image frames). The bar graphs in the bottom row present this finding quantitatively. [From Sengpiel et al., 1999]

strongly to other frequencies (their spectral receptive field had narrowed). Later studies showed that the sensitive period for inducing such changes in frequency tuning is only 11 days long, starting on day 9 of a rat's life.

Overall, these findings indicate that auditory and visual cortical neurons can be shaped by early sensory experiences in such a way that they end up responding disproportionately well to the experienced stimuli. Thus, experience during development increases the efficiency with which the brain encodes its natural environment (see Section 12.1).

Experience Effects in Higher Cortical Areas

The studies we have discussed thus far show experience dependent plasticity in primary visual and auditory cortices, but analogous plasticity exists also in higher order sensory cortices. For example, functional brain imaging studies have shown that expert bird watchers have an unusually large portion of the lateral occipital cortex that is activated preferentially by images of birds. In contrast, car experts have an abundance of car-responsive neurons in the same cortical region. Another study showed that monkeys trained to identify entirely novel objects end up with neurons in the inferior temporal cortex that respond selectively to familiar views of those objects.

These findings suggest that expertise in identifying a particular type of object leads to an increase in the number of neurons coding for those objects. If this is true, then the preponderance of face-responsive neurons in the inferior temporal cortex might be the result of primates having extensive experience with faces as they grow up. This hypothesis is hotly debated, however. To resolve the controversy, one would have to study the visual cortex of monkeys that grew up without seeing monkey or human faces. Sadly, such monkeys would be abnormal in many ways, confounding the results.

BRAIN EXERCISE

Can you think of a way in which the sensory experiences you had as a child influenced your current perceptual abilities? What might be the neural basis of that influence?

12.5 Why Do We Perceive Objects as Coherent Entities?

As you have now learned, neurons in diverse cortical areas are specialized to detect specific features of objects as well as specific combinations of features. In some cortical areas, the neurons are specialized to detect motion; in others, they focus on color, contours, sounds, tactile information, or smell. This raises an important question: how can neural activity distributed over so many different areas be integrated to create unitary percepts of real-world objects? When you pick up a baby, hold it in your arms, gaze on its face, smell it, and listen to it cry, how many babies are you perceiving? Probably just one! Somehow, you integrate all the information gathered about this baby by your various senses into a single percept. How does this work? How does your nervous system "bind together" the activity of the many neurons that encode the baby's disparate features?

This **binding problem** of object perception also exists at lower levels of the sensory processing hierarchy. As we discussed earlier (in the context of the aperture problem), edge and line detectors in V1 have relatively small receptive fields and therefore respond only to small segments of an object's outline. Therefore, looking at an object activates a diverse assortment of neurons, each responding to a specific portion of the object's outline. How does the rest of your nervous system "know" that the activity of those neurons collectively represents a single, coherent object? How does it "bind" the activity of these neurons together?

Before you answer that this problem is easily solved because only the neurons activated by the object will be active, consider that real objects must be perceived against cluttered backgrounds, to which the neurons are likely to respond as well. For example, talking with someone at a party can be difficult because their voice is probably embedded in a complex auditory scene replete with other sounds. We solve this **cocktail party problem** by grouping together any sounds that have similar frequencies, timbre, rhythm, or location in space. This "perceptual binding" of acoustic stimuli allows us to pick out a distinctive voice at a noisy party, the bass line in a jazz piece, or the sound of a distant highway. However, the neural mechanisms underlying this *acoustic stream segregation* remain largely unknown. Nor do we know much about the neural mechanisms underlying perceptual binding in other sensory modalities. In neural terms, the question is, how does the widespread neural activity triggered by a single, coherent object get bound together into a distinct representation of that object?

One solution to the binding problem is to have all the neurons that code for parts of an object send converging projections to a higher level neuron, or small set of neurons, whose activity then codes for the entire object. This binding-through-convergence model may be implemented in some parts of the sensory processing hierarchy, but few parts of the brain exhibit the kind of massive multimodal convergence that the model predicts. A second problem for the binding-through-convergence model is that it requires an implausibly large number of neurons. What if someone hands you something that you have never seen, touched, or smelled? Is it plausible that you would be equipped with a set of neurons that is wired specifically to identify this strange object? Probably not. Faced with these challenges to the binding-through-convergence model, neuroscientists have sought another solution to the binding problem.

Binding through Temporal Correlation

The most influential hypothesis proposed to solve to the binding problem is the **temporal correlation hypothesis**. Its core idea is that all the neurons activated by an object's various features will tend to be activated simultaneously whenever the object moves. This is certainly true for the neurons that encode bits and pieces of an object's contour, but it is also true for many of an object's internal features (such as its color or texture). Even the sounds made by an object tend to correlate with the object's movement; just think of a mouse running through leaf litter, making a rustling sound. This simultaneity (temporal correlation) could bind the activated neurons together through Hebbian synaptic plasticity (neurons that fire together wire together) and thus form the kind of Hebbian cell assembly that we discussed previously. In other words, the temporal correlation hypothesis would allow an object to be represented by a Hebbian cell assembly.

Unfortunately, there are several problems with this hypothesis. First, Hebbian cell assemblies form as a result of experience and should therefore have problems coding for entirely novel objects. Second, neocortical neurons do not form the sort of densely interconnected mesh that is ideal for creating cell assemblies. As you may recall from Chapter 3, Hebbian plasticity can only strengthen connections that already exist (albeit weakly). Therefore, Hebbian cell assemblies form most readily in networks where most neurons are connected to many of the other neurons in the area. The olfactory cortex contains such an auto-associative network and so does the hippocampus (see Chapter 14). However, neocortical areas are less densely interconnected and much more reluctant to form Hebbian cell assemblies (we will come back to this in Chapter 14).

Another significant problem for the temporal correlation hypothesis is that Hebbian plasticity requires fine-grained synchrony on a millisecond timescale. It strengthens synapses only if they were active a few milliseconds before the postsynaptic neuron spikes. There is no guarantee that the distributed neural activity caused by a moving object would be so tightly synchronized. As an object moves, the responsive neurons might all increase their firing rates at roughly the same time, but there is no a priori reason to expect tight synchrony among the individual action potentials within those bursts of increased activity.

Experimental Evidence for Binding through Temporal Correlation

Yet, some influential experiments by Wolf Singer and his colleagues suggest that coherent stimulus motion does create precisely the sort of fine-grained synchrony needed to create Hebbian cell assemblies. These investigators recorded simultaneously from two neurons with different receptive fields in different parts of a cat's visual cortex (Figure 12.32). They then stimulated these neurons by moving dark bars across their receptive fields. The bars always moved at the same speed, but on some trials the bars moved in opposite directions across the two receptive fields (Figure 12.32 A). In other trials the bars moved in the same direction. In a third type of trial, the two stimulus bars were replaced with a single long bar that swept across both receptive fields. In all these trials, both neurons responded to the stimulation with robust increases in firing rate. This was not surprising. The remarkable discovery was that the individual action potentials of the two neurons were highly synchronized when the neurons were stimulated with the long bar, but asynchronous when the two small bars moved in opposite directions. Moving the two bars in the same direction generated an intermediate degree of synchrony (Figure 12.32 C).

The takeaway message from this experiment is that coherent motion of the sort generated by a moving object causes widely separated neurons to synchronize their action potentials on a timescale of 1–3 milliseconds, precisely what is needed to support the formation of Hebbian cell assemblies. This kind of fine-grained synchrony would also support binding-by-convergence because synchronous action potentials would maximize the depolarization of downstream neurons because of spatial summation (see Chapter 2). Thus, the fine-grained action potential synchrony

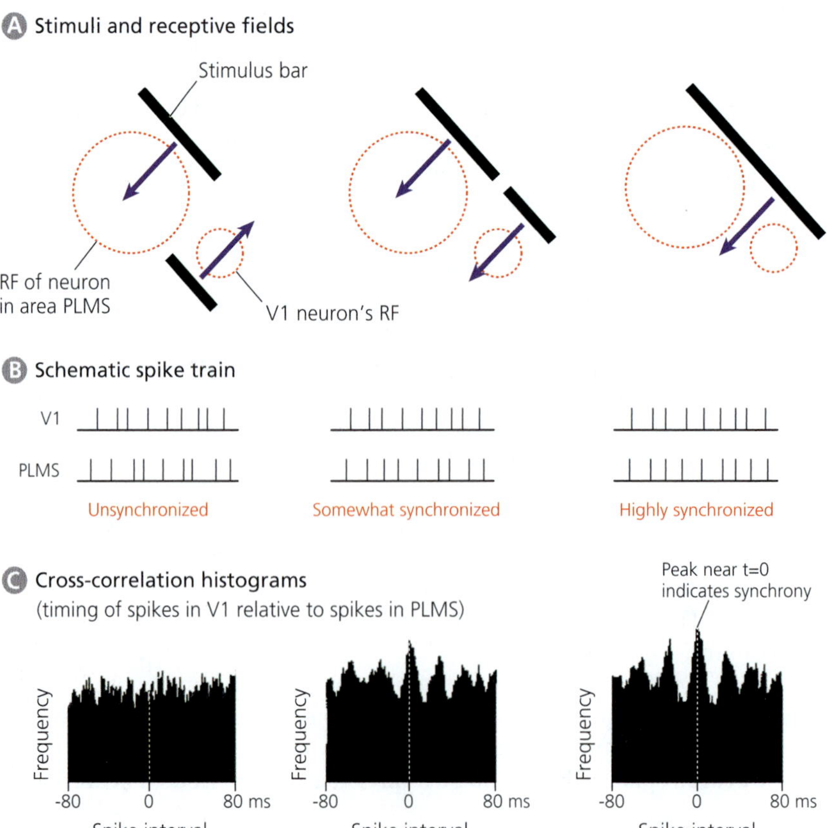

Figure 12.32 Synchronization of neuronal firing by coherent motion. Two neurons with different receptive fields (red dashed circles) were recorded in different areas of a cat's visual cortex (V1 and PMLS [posteromedial lateral suprasylvian area]). The neurons were stimulated by moving dark bars across the two receptive fields (A) in opposite directions (left), in the same direction (middle), or coherently, as part of one long bar (right). Both neurons responded to all stimuli with robust bursts of action potentials, but the temporal structure of their responses varied. As shown schematically in (B), the individual action potentials (spikes) in the response bursts were more synchronous between the two neurons when stimulus motion was coherent. This observation was quantified by correlating the timing of each spike in one neuron with the timing of the spikes in the other neuron (C). When such cross-correlation histograms exhibit strong peaks at 0 ms, then the spike timing is highly synchronized. Side-peaks are evidence of synchronized rhythmic activity. [After Singer and Gray, 1995]

NEUROLOGICAL DISORDERS

Box 12.2 *Deficits in Object Identification: Agnosias*

Deficits in object identification, called agnosias, come in several varieties. *Apperceptive agnosia* is characterized by the inability to perceive a collection of features as an object. People with this kind of agnosia have trouble seeing objects as coherent entities, distinct from other objects. To them, the objects seem no more than a bunch of individual features. The second major type of agnosia is called *associative agnosia*. People with this condition can perceive objects, but they cannot recognize specific objects, determine how they might be used, or otherwise interpret their meaning.

Agnosias can also be categorized according to the kinds of objects or attributes that can no longer be identified. For example, people may have tactile, auditory, or visual agnosia, and each of these modality-specific agnosias can be subdivided further. Auditory agnosia, for example, includes *amusia* (an inability to identify pieces of music), *pure word deafness* (an inability to interpret the sounds of words), and *nonverbal auditory agnosia* (an inability to identify nonverbal sounds, such as the sound of a cat, a bell, or a whistle). Importantly, people with these conditions are not deaf. They hear the sounds just fine; they just cannot determine what they mean.

The most common form of visual agnosia is face blindness, or *prosopagnosia* (from the Greek word *prosopon*, meaning "face" or "mask"). People with this condition have trouble recognizing familiar persons by sight. As you can imagine, it can be quite embarrassing not to recognize your best friend's face, your mother's face, or even the faces of famous individuals. Fortunately, people with prosopagnosia can identify familiar persons as soon as they begin to speak. Often they can also identify them by their posture, by the way they walk, or by some distinctive visual feature such as a large hooked nose or a unique hairstyle. This shows that people with prosopagnosia are not blind, unintelligent, or amnesic (unable to remember things). They simply cannot perceive a face as a face or, in the case of associative prosopagnosia, identify a specific face as one that means something to them. If you are curious about prosopagnosia, read *The Man Who Mistook His Wife for a Hat and Other Clinical Tales* by Oliver Sacks, who suffers from prosopagnosia himself (as does the famous primatologist Jane Goodall).

The cause of prosopagnosia and other visual agnosias is usually a stroke or a traumatic brain injury that damages the neocortex near the junction between the temporal and occipital lobes, most often bilaterally. This observation is consistent with the finding that neurons in this general area tend to be activated selectively by faces and other visual objects. Because brain damage is usually widespread, visual agnosias are often less specific than their names suggest. For example, most people with prosopagnosia have trouble identifying visual objects other than faces. In addition, many of them have cortical color blindness (achromatopsia) as well as motion blindness (akinetopsia). These deficits are likely due to damage that extends into the color-sensitive globs of the occipito-temporal cortex and the motion-sensitive neurons of area MT.

Although prosopagnosia usually develops in adulthood, some people are born with it. The congenital form of prosopagnosia is usually specific to the recognition of familiar faces, but it may also be associated with an inability to recognize other sorts of familiar objects. Intriguingly, congenital prosopagnosia does not involve the sort of dramatic damage that characterizes acquired prosopagnosia. Indeed, people with congenital prosopagnosia show remarkably normal activation of the face-selective cortical regions. However, diffusion tensor imaging, which reveals the orientation and structural integrity of large axon tracts, has shown that the microstructure of a major pathway connecting the posterior temporal lobe and occipital cortex to the anterior temporal lobe is perturbed in people with congenital prosopagnosia (Figure b12.2). This finding is consistent with the hypothesis that recognition of specific faces requires robust connections between posterior cortical structures that identify faces as faces and more anterior temporal lobe structures that compare the perceived faces to memory. Despite this satisfying conclusion, many aspects of prosopagnosia and, for that matter, other agnosias, remain mysterious.

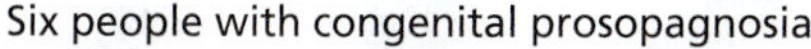

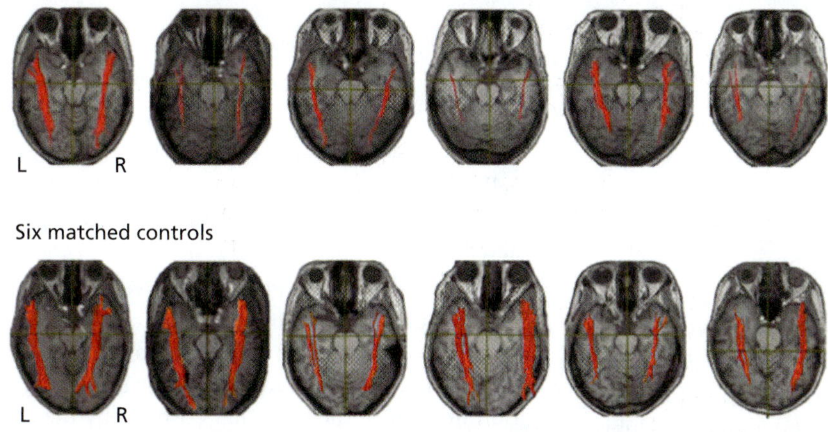

Figure b12.2 Neural correlates of congenital prosopagnosia. Subjects had their brains examined with diffusion tensor imaging, a variant of structural MRI that reveals large axon tracts. Shown in red are the likely trajectories of the inferior longitudinal fasciculus (ILF), which connects the occipital and posterior temporal lobes to the anterior temporal lobe. In the top row are horizontal scans through the brains of 6 people with congenital prosopagnosia (CP). The bottom row depicts analogous scans for six age- and sex-matched control subjects. On average, the ILF is smaller in the individuals with CP, especially on the right side of the brain (R). [From Thomas et al., 2009]

discovered by Singer and his colleagues can, at least in theory, solve the binding problem both by promoting the formation of cell assemblies and by driving the activation of "grandmother cells" that receive converging inputs from lower level feature detectors.

Although Singer's work has been influential, it remains controversial. The most important concern is that the mechanisms underlying the proposed spike synchrony remain unclear. What causes action potentials in disparate neurons to become synchronized when those neurons are activated by features of the same object, and to remain asynchronous when the features belong to different objects? No answer is obvious. Spike synchrony may be a signal that perceptual binding occurred, but it is not a mechanism for that binding. Thus, the binding problem remains largely unsolved. Nonetheless, Singer's work has stimulated much valuable discussion about the mechanisms of perceptual binding, which is one of the most important unsolved aspects of perception.

Disorders of Perceptual Binding

The importance of perceptual binding is apparent when you consider people who are impaired in that capacity (Box 12.2: Deficits in Object Identification: Agnosias). Most dramatically, people with **apperceptive agnosia** can see the individual features of visual objects just fine, but they cannot see the objects as coherent entities. Their world is full of features, rather than objects.

In contrast, people with **synesthesia** (from the Greek words for "together" and "sensation") seem to have overactive perceptual binding machinery. In the most common form of synesthesia, specific sounds or words are accompanied by the perception of specific colors. In other forms, various tastes or musical sounds trigger the perception of specific colors or shapes. The neural mechanisms underlying synesthesia remain enigmatic, but imaging data suggest that synesthesia results from the abnormal strengthening of weak cross-modal connections or from the formation of intracortical pathways that never exist in people without synesthesia. Either way, synesthesia probably results from neural connections that bind together neurons that, in most of us, remain unbound.

SUMMARY

Section 12.1 - Sensory systems compress sensory information by ignoring spatially and temporally invariant data. They selectively encode stimulus features that occur commonly in natural scenes.

Section 12.2 - The visual system is hierarchically organized and, in the primate neocortex, diverges into dorsal and ventral streams. The ventral stream is specialized for object identification.

- The spatial receptive fields of retinal ganglion cells are divisible into an excitatory center and an inhibitory surround. Retinal ganglion cells respond preferentially to spatial and temporal variations in luminance.
- Lateral geniculate neurons have relatively simple receptive fields similar to those of retinal ganglion cells. They receive substantial cortical input, but the function of these descending projections remain unclear.
- Simple cells in V1 respond best to lines and edges of a particular orientation in a specific location. Complex cells are less finicky about stimulus location. Preferred stimulus orientation varies in a pinwheel pattern across V1.
- Many V1 neurons are motion sensitive but, due to the aperture problem, do not encode movement direction reliably. In contrast, many MT neurons do. Electrical stimulation of MT neurons can bias motion perception in favor of the direction encoded by the stimulated neurons.
- Double-opponent cells in V1 respond to color contrasts independently of luminance contrasts. Color sensitive neurons are also found in the "globs" of the extrastriate cortex. Some of these neurons exhibit hue constancy, meaning that their color preference is independence of luminance.

- The inferior temporal (IT) cortex contains neurons that respond selectively to faces or other kinds of visual objects. Neurons with similar response profiles tend to be clustered into patches in IT cortex. Electrical stimulation of face-selective neurons can bias perception in favor of faces.

Section 12.3 - The more one learns about the encoding of stimulus objects, the more blurry becomes the boundary between grandmother cell and combinatorial population coding. Still, the distinction is conceptually useful.

- Pyramidal neurons in the olfactory cortex receive converging inputs from multiple glomeruli in the olfactory bulb. Natural odors activate a far-flung combination of neurons in both the olfactory bulb and olfactory cortex.
- The auditory cortex in primates contains both tonotopic and non-tonotopic areas. The complexity of stimulus encoding in the auditory cortex is apparent from the fact that many neurons there don't respond to complex stimuli as one would expect from the neuron's spatiotemporal receptive field.
- Gustatory information is analyzed in the insular cortex and integrated with olfactory information in the orbitofrontal cortex.

Section 12.4 - Experience-dependent neural plasticity is needed for normal development and can modify circuits so that they preferentially encode the experienced stimuli. This form of plasticity is greatest in early development, during sensitive periods, but can persist to adulthood.

Section 12.5 - To perceive objects as coherent entities, their individual features must be bound together perceptually. The perceptual binding problem may be solved by high-level neurons that receive converging input from multiple feature detectors or by temporally correlated activity that leads to the formation of Hebbian cell assemblies.

Box 12.1 - Because orange light is absorbed better by deoxyhemoglobin than oxyhemoglobin, intrinsic signal optical imaging can be used to measure neuronal oxygen consumption.

Box 12.2 - Deficits in object identification, called agnosias, come in several varieties. Among the most intriguing is prosopagnosia, the inability to perceive or recognize faces. Agnosias may result from localized cortical lesions or from abnormally weak intracortical connections.

KEY TERMS

ADDITIONAL READINGS

12.1 - Principles of Sensory Coding

Milner AD, Goodale MA. 2008. Two visual systems reviewed. *Neuropsychologia* **46**:774–785.

Olshausen B, Field D. 2004. Sparse coding of sensory inputs. *Curr Opin Neurobiol* **14**:481–487.

12.2 - Object Identification in the Visual System

Alonso JM, Usrey WM, Reid RC. 2001. Rules of connectivity between geniculate cells and simple cells in cat primary visual cortex. *J Neurosci* **21**: 4002–4015.

Bair W. 2005. Visual receptive field organization. *Curr Opin Neurobiol* **15**: 459–464.

Carandini M, Demb JB, Mante V, Tolhurst DJ, Dan Y, Olshausen BA, Gallant JL, Rust NC. 2005. Do we know what the early visual system does? *J Neurosci* **25**:10577–10597.

Changizi MA, Zhang Q, Shimojo S. 2006. Bare skin, blood and the evolution of primate colour vision. *Biol Lett* **2**: 217–221.

Conway BR. 2001. Spatial structure of cone inputs to color cells in alert macaque primary visual cortex (V-1). *J Neurosci* **21**:2768–2783.

Clifford CWG, Ibbotson MR. 2002. Fundamental mechanisms of visual motion detection: models, cells and functions. *Prog Neurobiol* **68**: 409–437.

Desimone R. 1991. Face-selective cells in the temporal cortex of monkeys. *J Cogn Neurosci* **3**:1–8.

Ferster D, Miller KD. 2000. Neural mechanisms of orientation selectivity in the visual cortex. *Annu Rev Neurosci* **23**: 441–471.

Freedman DJ, Miller EK. 2008. Neural mechanisms of visual categorization: insights from neurophysiology. *Neurosci Biobehav Rev* **32**:311–329.

Gross CG. 2002. Genealogy of the "grandmother cell." *Neuroscientist* **8**:512–518.

Nassi JJ, Callaway EM. 2009. Parallel processing strategies of the primate visual system. *Nat Rev Neurosci* **10**:360–372.

12.3 - Identifying Non-visual Objects

Bandyopadhyay S, Shamma SA, Kanold PO. 2010. Dichotomy of functional organization in the mouse auditory cortex. *Nat Neurosci* **13**:361–368.

Davison IG, Ehlers MD. 2011. Neural circuit mechanisms for pattern detection and feature combination in olfactory cortex. *Neuron* **70**:82–94.

Gottfried JA. 2010. Central mechanisms of odour object perception. *Nat Rev Neurosci* **11**:628–641.

Hsiao S. 2008. Central mechanisms of tactile shape perception. *Curr Opin Neurobiol* **18**:418–424.

Leinwand SG, Chalasani SH. 2011. Olfactory networks: from sensation to perception. *Curr Opin Genet Dev* **21**:806–811.

Shamma SA, Micheyl C. 2010. Behind the scenes of auditory perception. *Curr Opin Neurobiol* **20**:361–366.

Wilson DA, Sullivan RM. 2011. Cortical processing of odor objects. *Neuron* **72**:506–519.

12.4 - Experience-dependent Cortical Plasticity

Hoffman KL, Logothetis NK. 2009. Cortical mechanisms of sensory learning and object recognition. *Philos Trans R Soc Lond B Biol Sci* **364**:321–329.

Keuroghlian AS, Knudsen EI. 2007. Adaptive auditory plasticity in developing and adult animals. *Prog Neurobiol* **82**:109–121.

12.5 - Solving the Binding Problem

Tarr MJ. 1999. News on views: pandemonium revisited. *Nat Neurosci* **2**:932–935.

Hebb DO. 1949. *The organization of behavior: a neuropsychological theory.* New York: Wiley & Sons.

Shadlen MN, Movshon JA. 1999. Synchrony unbound: a critical evaluation of the temporal binding hypothesis. *Neuron* **24**:67–77.

Singer W, Gray CM. 1995. Visual feature integration and the temporal correlation hypothesis. *Annu Rev Neurosci* **18**:555–586.

Varela F, Lachaux JP, Rodriguez E, Martinerie J. 2001. The brainweb: phase synchronization and large-scale integration. *Nat Rev Neurosci* **2**:229–239.

Boxes

Frostig RD, Masino SA, Kwon MC, Chen CH. 1995. Using light to probe the brain: intrinsic optical imaging. *Int J Imaging Syst Technol* **6**:216–224.

Sacks O. 1985. *The man who mistook his wife for a hat and other clinical tales.* New York: Summit Books.

Behrmann M, Avidan G, Marotta JJ, Kimchi R. 2005. Detailed exploration of face-related processing in congenital prosopagnosia: 1. behavioral findings. *J Cogn Neurosci* **17**:1130–1149.

CHAPTER 13

Regulating Brain States

CONTENTS

FEATURES

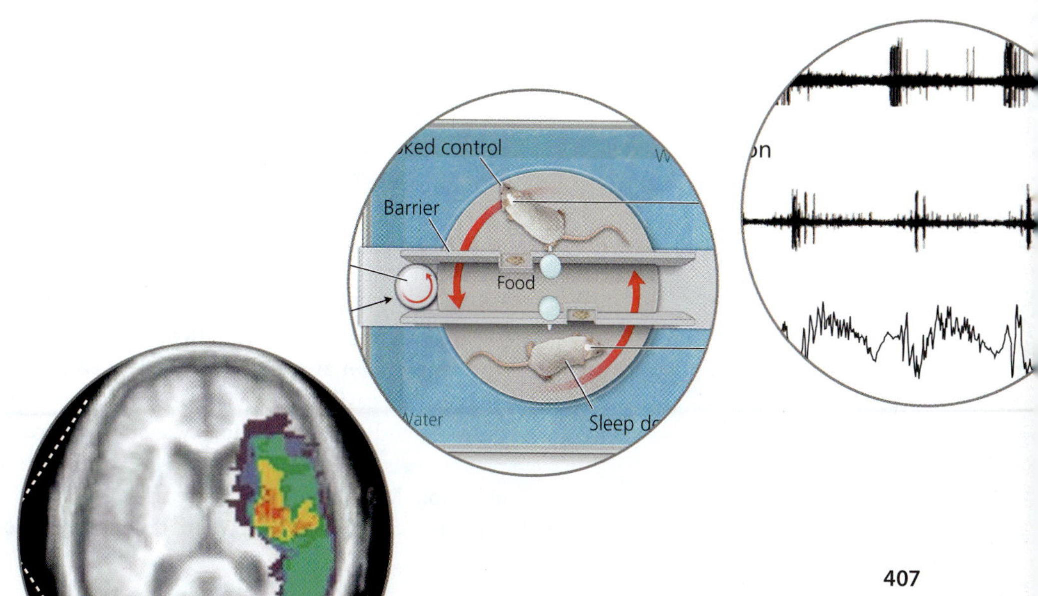

Can you recall a dripping faucet in your kitchen or bathroom? Perhaps you didn't hear the drips until you settled down and tried to sleep; then, the sound became quite audible. Perhaps the dripping bothered you to the point where you got up and tried, in vain, to turn the faucet off. Presumably, your weariness eventually won out and sleep took hold, the sounds of the external world receding with your consciousness. Then, let's speculate, some loud and unexpected sound roused you. Startled, you listened alertly, your heart racing briefly. Again, you heard the faucet's steady drip, but now you were also aware of other sounds, some rather faint. Because none of those sounds seemed threatening, you gradually relaxed and fell asleep again. If elements of this story sound familiar, then you already know that what you hear is influenced by attention, arousal, and sleep. More generally, the message is, what you perceive depends on your behavioral state.

This chapter deals with the mechanisms that change our behavioral states and thereby modulate our perceptions. We begin with the neurobiology of attention and then discuss, in turn, arousal and sleep. These are all distinct behavioral states, but they are tricky to define. For example, when you say that someone snaps to attention, you probably mean that their general state of arousal is heightened. To ease potential confusions, each major section in this chapter starts with a description of the state being discussed.

13.1 How Does the Brain Generate and Direct Attention?

The psychologist William James (see Chapter 1) famously declared

> Everyone knows what attention is. It is the taking possession by the mind, in clear and vivid form, of one out of what seem several simultaneously possible objects or trains of thought. Focalization, concentration of consciousness are of its essence. It implies withdrawal from some things in order to deal effectively with others, and is a condition which has a real opposite in the confused, dazed, scatter-brained state which in French is called *distraction*, and *Zerstreutheit* in German.

This characterization of attention as the focusing of the mind on specific objects or trains of thought is reasonable and widely accepted. However, James himself went on to describe several different forms of attention, and later generations of psychologists have offered even more detailed analyses. Therefore, let us consider the major forms of attention before we grapple with their neural underpinnings.

Psychological Aspects of Attention

If we accept that attention implies mental focus, then it is selective by definition. The focusing of attention on one set of stimuli, thoughts, or actions comes at the expense of attention to other things. You may think that you can pay attention to many things at once (such as texting and driving), but this is very difficult. Air traffic controllers can, after extensive training, learn to keep track of many planes; star athletes may know where all their teammates are; and orchestra conductors can keep track of multiple instruments. However, these masters of attention are probably scanning their environment rapidly, attending to one object at a time, and constructing from this information a mental model of where those objects are and how they are changing. This skill requires extensive training, but it does not negate our definition of attention as being limited in scope.

A good analogy for attention is the beam of a flashlight, which can illuminate only one region at a time. The beam can be wide or narrow, but it cannot be split (at least not easily). Magicians are well aware of this limitation, for they try hard to draw the audience's attention away from where the secret moves are made. Similarly, pickpockets know that they can lift your wallet easily if they first misdirect your attention. Most of the time these tricksters make you think that something important is happening in one place, when it is occurring someplace else. Once your *attentional spotlight* is drawn to the extraneous location, the rest of space is plunged into attentional darkness.

Of course, the flashlight or spotlight analogy is limited because attention need not be visual. We can focus our attention on one voice out of many at a party, on the smell of faint perfume, or on the texture of an intriguing surface. Nor does attention have to be spatial. For example, we can pay attention to an object's color rather than its shape, and we can pay attention to a person's body language, rather than their words. Despite these considerations, neurobiologists have focused most of their research on *visual spatial attention*. This is reasonable because humans tend to be highly visual and because most forms of attention have a strong spatial component.

Overt versus Covert Spatial Attention

When your attention is drawn to a particular location in space, what do you do? Most of the time you turn your head, or at least your eyes, toward the location of interest. These orienting movements are sensible because they aim your retinal and acoustic foveae at the attended location, thereby maximizing your ability to gather information from that site. Sometimes, of course, we choose to inhibit these orienting movements and shift our attention to areas in our peripheral vision. Because such shifts of attention are invisible to outside observers (only "the mind's eye" moves), they are said to be covert (secret). Although **covert spatial attention** is relatively rare in everyday life, it is important for researchers because it allows shifts in attention to be studied without the confounding effects of orienting movements.

Involuntary Attention

Some stimuli draw our attention automatically, without involvement of our will and sometimes even against it. This kind of **involuntary attention** is also known as bottom-up, stimulus-driven, or exogenous attention. Its key characteristic is that the target of our attention is determined by some external stimuli and not by our goals or intentions.

What kind of stimuli can drive involuntary attention? As William James put it, in cases of involuntary attention

> The stimulus is a sense-impression, either very intense, voluminous, or sudden—in which case it makes no difference what its nature may be, whether sight, sound, smell, blow, or inner pain—or else it is an instinctive stimulus, a perception which, by reason of its nature rather than its mere force, appeals to some of our normal congenital impulses and has a directly exciting quality. . . . These stimuli differ from one animal to another, and what most of them are in man: strange things, moving things, wild animals, bright things, pretty things, metallic things, words, blows, blood, etc., etc., etc.

As you can see, many of the stimuli that attract our involuntary attention indicate potential threats or benefits. However, our attention is also drawn to stimuli that have no intrinsic significance but simply stand out from the background. A single yellow flower on a green lawn, for example, will command our involuntary attention. So will a fly on a white wall or a red circle in a sea of blue (Figure 13.1 A). Why are such stimuli so **salient** (attracting attention)? Probably because they, too, represent at least potential threats or benefits. They are worth investigating further, just in case. By aiming our attention, if not our eyes and heads as well, toward salient stimuli, we can assess more thoroughly their true significance. If we then determine that the stimuli are not interesting after all, our attention moves on.

Voluntary Attention

We are also capable of **voluntary attention**. For example, you can intentionally focus your attention on a specific person at a party, on the sound of specific instrument in a piece of music, or on the words in front of you. Voluntary attention is also used to search for a specific item among many distractors. For example, you might look for a young man wearing a red-and-white-striped sweater in a painting full of similar people (looking for Waldo), or you might look for the red circle in Figure 13.1 B. The red circle in this figure does not immediately "pop out" from the background because

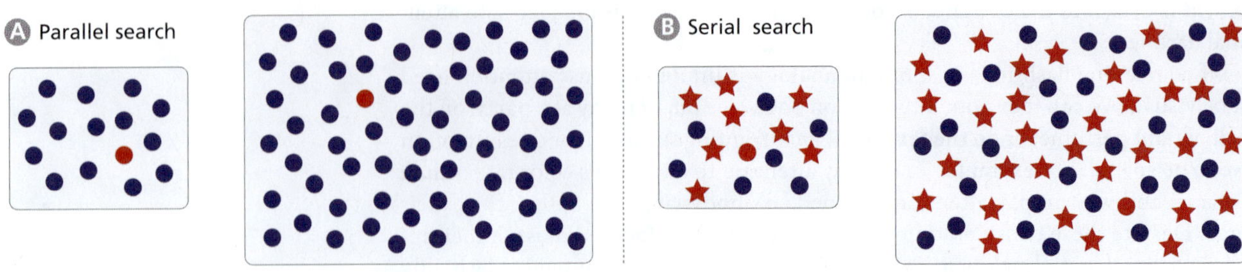

Figure 13.1 Parallel versus serial search. When you look at the images in (A), the red circle rapidly "pops out" of the background, even when the number of blue circles is increased (left vs. right). In such tasks your visual system analyzes all stimuli in parallel. In contrast, it takes more time to find a red circle among a mix of blue circles and red stars (B). In such a task, you must examine each stimulus in turn (serially), and the time to find the target increases with the number of distractors.

its difference from the other objects is evident only when you consider both color and shape. This sort of conjunctive analysis requires the voluntary movement of attention from one object to another until the "search image" is found. Neurobiologists refer to the intentional roving of your attentional spotlight as *serial search* and distinguish it from *parallel search*, in which salient stimuli automatically pop out of the background (Figure 13.1 A).

A historically important study of voluntary spatial attention was performed by Hermann von Helmholtz at the end of the 19th century. As part of a series of experiments on depth perception, Helmholtz painted letters on a piece of cardboard, which he then placed inside a light-tight box (Figure 13.2). A small hole in the center of the card let light shine through from behind the card. Therefore, when Helmholtz looked inside the box, he could see only a spot of light at the center of the card. Helmholtz then set off a flash of light inside the box. The flash was too brief to read the letters on the card. However, when Helmholtz decided before the flash to attend to a specific portion of the card, then the flash allowed him to read the letters there. This task involved covert attention because Helmholtz's eyes were always aimed on the central fixation point (the spot of light on which his eyes were fixed); only his attention moved.

The most important conclusion Helmholtz drew from this experiment is that even covert attention improves perception. As mentioned earlier, overt attention improves perception because the retinal and acoustic foveae are aimed at the attended location.

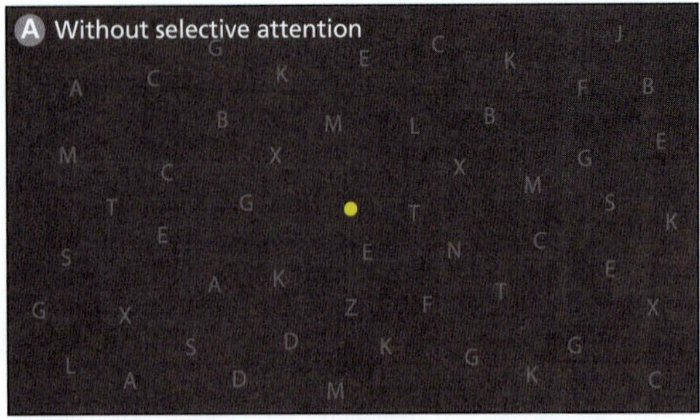

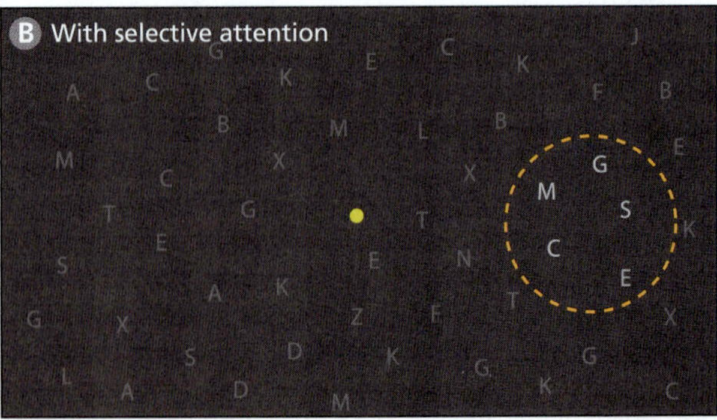

Figure 13.2 Covert attention improves perception. Helmholtz painted letters on a piece of cardboard inside a light-tight box. Peeking inside the box, Helmholtz could see only a spot of light coming through a pinhole (yellow). Keeping his eyes fixed on the light, he triggered a flash inside the box to illuminate the card. None of the letters were readable (A). However, when Helmholtz decided before the next flash to focus his attention on a specific region (dashed orange circle), he could read the letters in the targeted region (B). Because Helmholtz kept his eyes fixed on the pinhole, the shift in his attention was covert.

In covert attention, the eyes and ears don't move, but the perception of stimuli at the attended location is nonetheless enhanced. This is important because it shows that covert spatial attention serves a valuable function. Furthermore, experimenters can use this spatially selective enhancement of perception to determine whether their test subjects are covertly attending to specific locations. Without these effects on perception, covert attention would not just be outwardly invisible, it would be completely undetectable and impossible to study experimentally.

BRAIN EXERCISE

Can you think of some examples in your life where you intentionally used covert spatial attention? Do you ever use it involuntarily?

Neural Correlates of Involuntary Attention

The time it takes for a salient stimulus to capture our attention and pop out of the background does not increase significantly as the number of background distractors goes up (Figure 13.1 A). This finding suggests that the salience of stimuli that catch our involuntary attention is being computed in parallel rather than serially. Specifically, it suggests that separate populations of neurons are scrutinizing each region of space for the presence of salient stimuli. This conclusion, in turn, implies that stimulus salience is computed by neurons with relatively small spatial receptive fields.

Another clue to how neurons compute stimulus salience is that pop-out stimuli can vary from the background in many different aspects, including color, luminance, shape, and movement. Furthermore, attention-grabbing objects usually differ from the background in more than one respect. The black fly on the white wall, for example, becomes even more salient once it begins to move. Therefore, overall stimulus salience is probably computed by neurons that receive converging information about multiple stimulus features (color, motion, etc.). In our example, information about the fly's movement and its relative brightness are likely to converge onto the same set of neurons.

The Saliency Map

Based on such considerations, neuroscientists have built computer models of how visual salience is computed. In these models, an image is analyzed separately for local color, intensity, orientation contrasts, or other stimulus features (Figure 13.3). This information is then combined to generate a single **saliency map** that indicates for each part of the image how different it is from neighboring regions in the analyzed

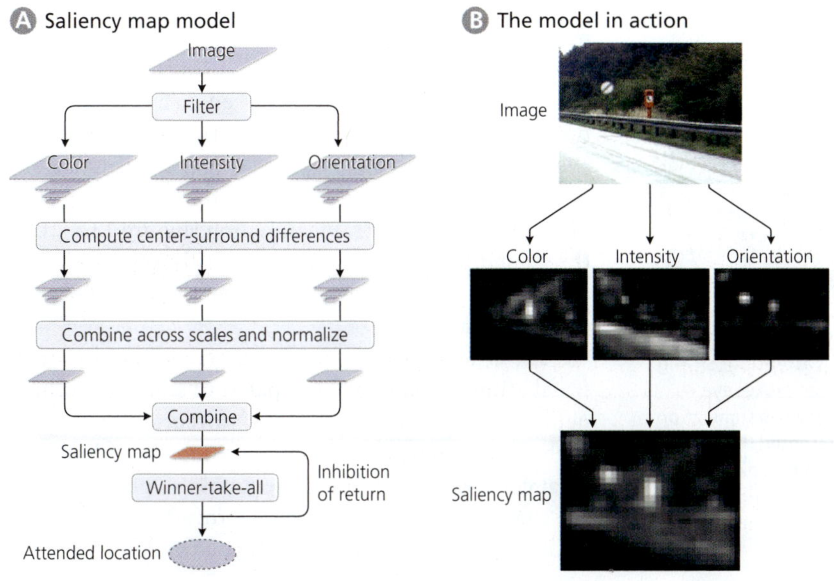

Figure 13.3 A computational model of spatial attention. According to an influential model, all parts of a visual scene are analyzed in parallel to identify local differences in color, intensity, and orientation, as well as other stimulus features (A). This information is then combined across the various feature dimensions to generate a single "saliency map" that specifies all the attention-deserving locations. A competitive algorithm determines the most salient location, to which attention is then drawn. In addition, "inhibition of return" decrements the salience of winning locations, ensuring that attention does not get stuck. Shown in (B) is an example of the computer model applied to an actual image. The emergency call box is the most salient item, followed closely by the traffic sign. [After Itti et al., 1998]

features. Different parts of the saliency map then compete with one another, using a winner-take-all algorithm, to determine which region is most salient. The winning region is where attention is drawn. An additional mechanism, called **inhibition of return**, makes a winning location temporarily less salient after its victory, thus ensuring that attention does not "get stuck" on just a single location (we will return to this shortly). Compared to what humans consider most interesting in an image, and where they aim their eyes, the computer models perform well, suggesting that they are realistic.

Where in the brain might a saliency map be located? This question has no simple answer. Much research has focused on the parietal and frontal eye fields (which you may recall from Chapter 11), but these regions are also involved in voluntary attention. We will discuss them in the next section. Another region that appears to be involved in computing the sort of involuntary, bottom-up saliency that generates pop-out effects is the **superior colliculus**. As you learned in Chapter 11, the superior colliculus receives topographic retinal input, as well as topographic projections from multiple cortical areas. These converging inputs are consistent with the saliency map model. In addition, the superior colliculus is heavily involved in targeted eye and head movements. This, too, is consistent with the saliency map hypothesis, given that involuntary attention is normally associated with orienting reflexes. Finally, the superior colliculus receives somatosensory and auditory inputs, which could mediate involuntary shifts of attention to salient touches or sounds.

A Strong current: elicit saccades

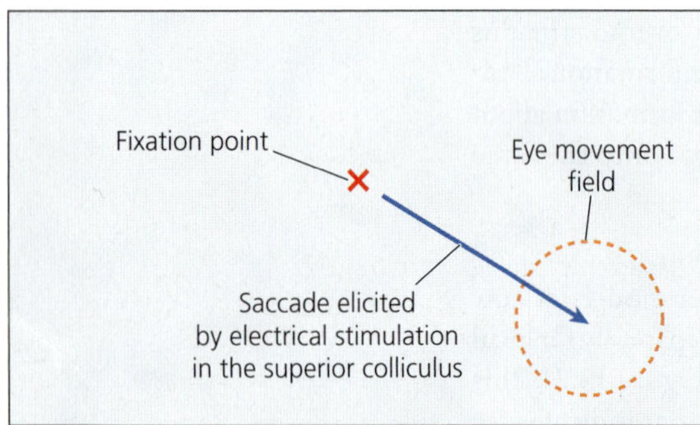

B Weak current: covert attention test

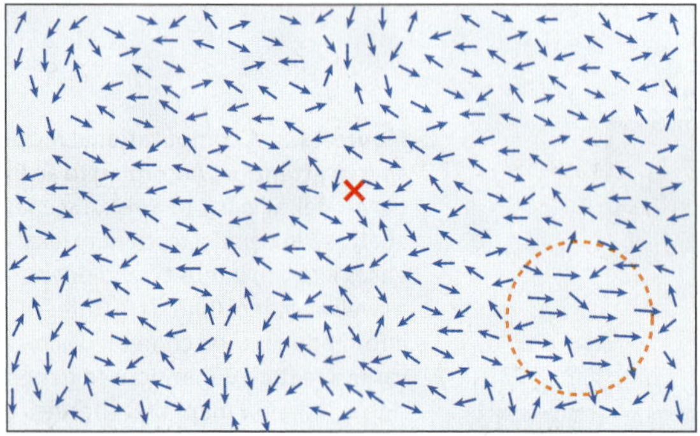

Figure 13.4 Can electrical stimulation shift covert attention? Strong electrical stimulation in a monkey's superior colliculus evokes eye movements to a specific location (A). To test whether weak stimulation can shift covert spatial attention, experimenters presented the monkey with dots that were moving in random directions, except in the eye movement field, where the majority of dots moved in a single direction. The monkey was rewarded if it identified this average movement correctly (without moving its eyes).

Subthreshold Stimulation of the Superior Colliculus

Strong evidence that the superior colliculus is involved in covert, as well as overt, spatial attention comes from an experiment in which researchers stimulated an awake, behaving monkey's superior colliculus electrically (Figure 13.4). This experiment is conceptually similar to the awake brain stimulation experiments we discussed in Chapter 12, except that the experimenters in this case were trying to alter perception indirectly by activating neurons that effect a change in covert spatial attention.

The first step in this experiment was to train a monkey on the same sort of direction discrimination task that we discussed in Chapter 12. Once the monkey was good at reporting the average direction of the moving dots, the task was made more difficult by having the dots move totally randomly throughout most of the screen. Only in one relatively small portion of the visual field did most of the dots move coherently (in the same direction). The monkey managed to perform this task correctly, using just its peripheral vision (the eyes had to stay focused on a central fixation point). However, the experimenters then made the task even more challenging by decreasing the coherence of the dots in the selected region so that, for example, only 70% of them moved in the same direction (Figure 13.4). Under these conditions, the task was essentially impossible unless the monkey was covertly paying attention to that particular region of space.

The crucial experimental question was whether electrical stimulation of a small patch of superior colliculus neurons could shift the monkey's attention to the region of the screen in which the dots were moving at least somewhat coherently. If so, then the monkey's performance on the challenging task should be improved. Of course, given what you learned in Chapter 11, you would expect strong stimulation of the superior colliculus to elicit saccades toward a specific region in space (the eye movement field). Such an effect would represent an *overt* shift of attention, bringing the eye movement field into the retina's fovea. By itself, this result would be uninteresting because it doesn't tell you whether the superior colliculus can shift attention independently of eye movements. The experimenters solved this problem by reducing current intensity until the electrical stimulation no longer elicited an eye movement. They then asked whether the weak stimulation can shift the monkey's *covert* attention toward the eye movement field (to which the monkey would saccade if current intensity were high).

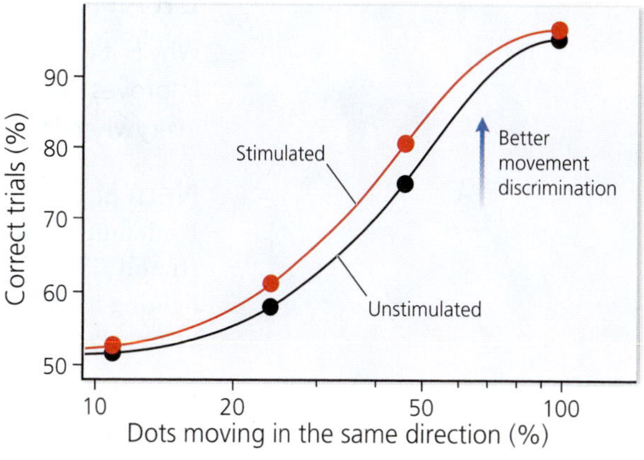

Figure 13.5 Collicular stimulation can shift covert attention. A small patch of the superior colliculus was stimulated with weak currents (that elicited no overt eye movements) while the monkey tried to perform the movement direction discrimination task illustrated in Figure 13.4 B. As long as the task was not too easy (>90% of the dots in the eye movement field are moving coherently) or too hard (<20% coherent motion), the electrical stimulation boosted the monkey's odds of reporting average movement direction correctly. [After Müller et al., 2005]

The short answer is "yes." When the task was moderately difficult, weak electrical stimulation in the superior colliculus improved the monkey's performance as long as the coherently moving dots were presented in the eye movement field (Figure 13.5). On trials in which the task is too easy (100% of the dots in the eye movement field are moving coherently), the monkey was already performing so well that further improvements were impossible. On the other hand, when only 12% of the dots were moving coherently, the task was just too difficult for a shift in attention to make a difference. Because the perceptual improvement on the moderately difficult trials was evident only when the moving dots were presented in the eye movement field, rather than elsewhere, we can conclude that the boost in performance resulted from a specific shift in covert attention rather than some more general mechanism (such as increased arousal). Overall, this experiment revealed that stimulation of the superior colliculus can guide covert, as well as overt, spatial attention.

The idea that both overt and covert spatial attention are controlled by the superior colliculus is consistent with the observation that both eye movements and covert shifts of visual attention exhibit inhibition of return. As we scan a scene with our eyes, we look first at the most salient feature; but then our eyes move on. Similarly, our attention may be attracted to the most interesting location in a scene, but it does not get stuck there forever. Once our attention has shifted, it tends not to return to its earlier target for at least a few hundred milliseconds. The mechanisms underlying this inhibition of return remain obscure, but they probably include a temporary suppression of neural activity at the "winning" location in the saliency map. Similarly, it remains unclear how activated neurons in the superior colliculus compete with one another for dominance. Recent studies with barn owls suggest an important role for inhibitory neurons in the *lateral tegmental nucleus*, which are reciprocally interconnected with the superior colliculus. These tegmental neurons probably ensure that only one winner at a time is selected, but how this winner is "chosen" and then suppressed remains unclear.

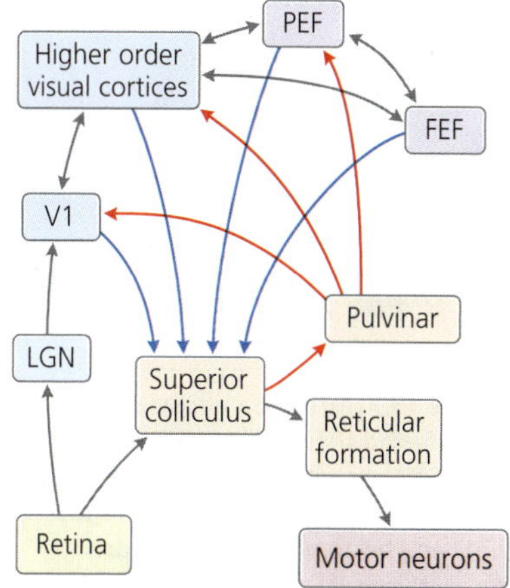

Figure 13.6 Pathways related to visual spatial attention. This figure depicts the major pathways controlling eye movements and visual attention, including pathways that carry information from the superior colliculus through the pulvinar nucleus to V1 and to higher order visual cortices and the parietal eye field (PEF). The frontal eye field (FEF) helps direct voluntary spatial attention. LGN = lateral geniculate nucleus.

The Pulvinar's Role in Visual Spatial Attention

How does the superior colliculus create the perceptual enhancements that result from shifts in covert attention? This question remains unanswered, but the mechanism probably involves projections from the superior colliculus to the **pulvinar nucleus** of the thalamus. As shown in Figure 13.6, this nucleus receives input from the superior colliculus and then projects to most visual cortices, as well as the parietal eye field. Because these projections are excitatory and topographically organized, increased activity in one part of the superior colliculus may, by way of the pulvinar, enhance neural activity in multiple visual cortical areas, and do so in a spatially specific manner. This idea remains to be tested, but pulvinar lesions are known to cause deficits in visual spatial attention.

BRAIN EXERCISE

Why is it important that weak stimulation of neurons in the superior colliculus improves movement perception only in a specific region of space (rather than everywhere)?

Neural Correlates of Voluntary Attention

By definition, voluntary attention is controlled by the subject, rather than external stimuli. This form of attention is sometimes referred to as *top-down attention*, highlighting its dependence on "executive" control, and contrasted with bottom-up, involuntary attention.

Attention Selectively Enhances Neural Responses

In the laboratory, voluntary attention is commonly studied in **cued spatial attention tasks**. In these tasks, subjects are trained to look at a central fixation point, where they are presented with an instructive cue (such as an arrow) that instructs them to direct their attention to a specific location on a computer screen (often just to the left or right of the fixation point). Importantly, the subject's eyes are not supposed to move. Then a visual stimulus is presented either at the location where the subject was instructed to attend or, in control trials, at some other location on the screen. When monkeys are used in such experiments, they are trained to make a specific hand or eye movement as soon as they identify the stimulus. If they respond correctly, a juice reward ensues. Although the monkeys in such experiments are highly trained, the shifts in attention are voluntary (top-down) because the monkeys are choosing to cooperate.

The key finding to emerge from such experiments is that monkeys generally react more rapidly when the stimuli appear at the attended location. This is consistent with our own experience, as paying attention to a specific location allows us to respond more rapidly to what is happening there, compared to non-attended locations. Overall, such experiments have shown that monkeys are as capable of covert voluntary spatial attention as we are. Importantly, monkeys performing these cued spatial attention tasks can be used to study the neural bases of voluntary attention in considerable detail.

In a typical experiment, monkeys are trained to direct their attention either to the left or to the right of the fixation point, as directed by a centrally presented cue (Figure 13.7). The experimenters then record from individual neurons in area V4 of the visual cortex (see Figure 12.3), determining both the location of their receptive field and their stimulus preferences. The key question is whether the strength of those responses vary with the location of the monkey's covert attention. Are the responses

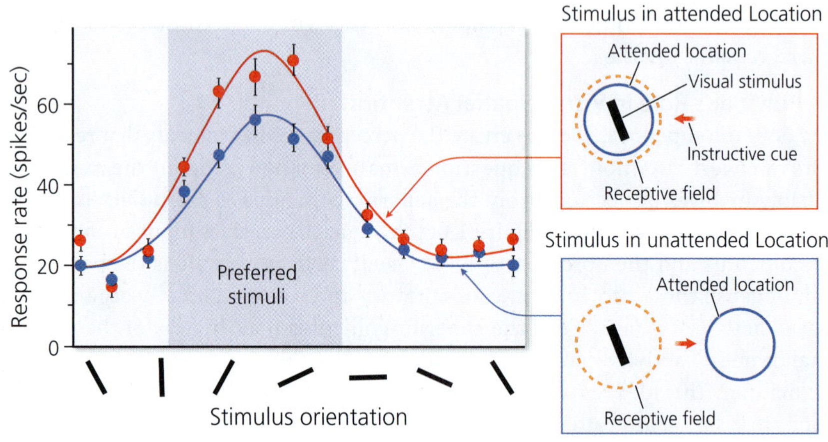

Figure 13.7 Attention boosts responses to preferred stimuli. Monkeys were trained to direct their attention either to the left of the fixation point or to the right (top right and bottom right of the figure, respectively). If they correctly identified the stimuli at the attended location, they received a juice reward. Simultaneously, the experimenters recorded the orientation preferences of V4 neurons to stimuli in their receptive fields. The major finding was that attending to a stimulus location increases the responses of neurons with receptive fields in the attended area (the graph shows the results for one such neuron), but that this attention-driven boost occurs only for responses to preferred stimuli. [After McAdams and Maunsell, 1999]

stronger when the monkey's attention is directed toward the neuron's receptive field, where the stimuli are being presented? The general answer is "yes," but with an interesting caveat.

Covert voluntary attention only strengthens responses to the neuron's preferred stimuli (such as lines of a specific orientation); the responses to non-preferred stimuli tend to be unaffected (Figure 13.7). To see why this is important, consider what would happen if attention simply increased neural responses to all stimuli. Such an indiscriminate increase in response strengths would add more noise to the neural representation of the stimulus and decrease its efficiency. By boosting responses selectively, covert attention makes it easier for animals to discriminate between similar stimuli, a feature that would often be useful.

The Frontal Eye Fields in Attention

You learned in Chapter 11 that the frontal eye fields (FEF) play a major role in the control of voluntary eye and head movements. They also function in controlling attention, specifically the voluntary shifts of visual spatial attention that we discussed in the previous section. Particularly interesting is that weak electrical stimulation of the FEF significantly enhances visual responses of V4 neurons if, and only if, the receptive fields of the V4 neurons coincide with the eye movement fields of the stimulated FEF neurons. This selectivity implies that low-intensity FEF stimulation triggers a shift in covert spatial attention rather than a general arousal of the animal.

Additional evidence for a prominent role of the FEF in voluntary spatial attention comes from an experiment in which monkeys were trained to find a target stimulus (a line of a particular color and orientation) in an array of other stimuli using a serial search strategy (Figure 13.8). Because the monkeys had to keep their eyes trained on a central fixation point, they were forced to search for the target stimulus covertly rather than overtly. When they found the target, the monkeys had to make an eye movement straight to the target to obtain a reward. Behavioral reaction time data indicated that the monkeys typically began their covert serial search on the right side of the stimulus array and then searched in a clockwise direction. As the monkeys performed the task, the experimenters recorded the activity of neurons in the FEF.

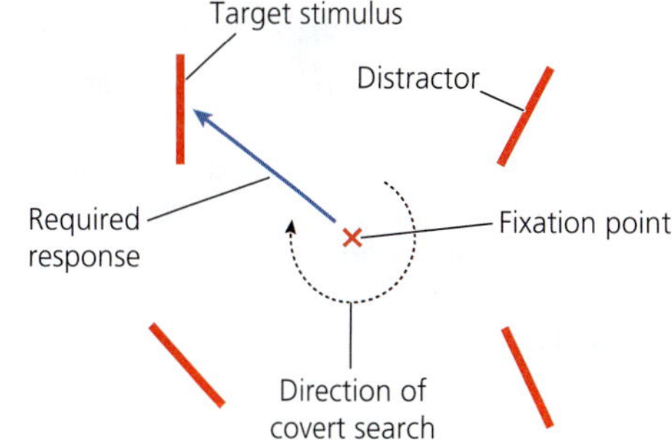

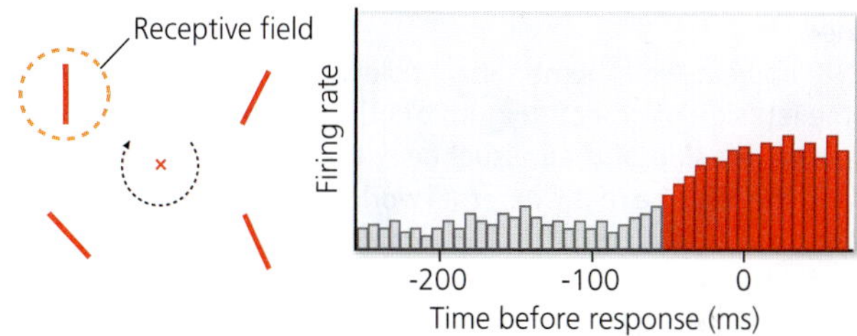

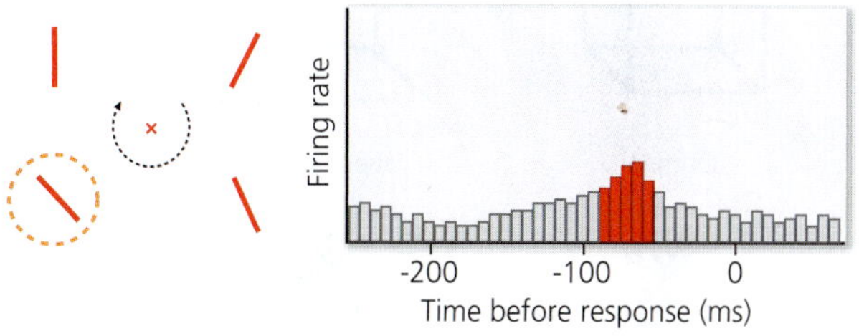

Figure 13.8 Frontal eye field (FEF) activity correlates with covert voluntary attention. Monkeys were trained to search for a line of a specific orientation among an array of distractors and then to move their eyes directly from the fixation point to the target stimulus (A). The behavioral data revealed that the monkeys searched for the target covertly in a clockwise direction. FEF neurons increased their firing rates whenever the attentional spotlight reached their receptive field. For neurons whose receptive field coincided with the target stimulus (B), firing rates increased ~50 ms before the eye movement response. For neurons whose receptive field was counterclockwise to the target stimulus (C), firing rates increased roughly 80 ms before the eye movement response, when the attentional spotlight would have been aimed at the neuron's receptive field. [After Buschman and Miller, 2009]

The major finding of this study was that neurons whose receptive field contained the target stimulus showed an increase in firing rate roughly 50 ms before the monkey responded overtly (Figure 13.8 B). This increase in activity might be the result of the attentional spotlight reaching the target stimulus, or it could be some sort of premotor activity commanding the trained eye movement response. Strong evidence in favor of the former hypothesis is that FEF neurons with receptive fields located immediately counterclockwise to the target increased their firing rate approximately 80 ms before the monkey's response and then reduced it again by the time the eye movement response began (Figure 13.8 C). The simplest way to explain these data is that the FEF neurons increase their firing rate whenever the attentional spotlight enters their receptive field.

NEUROLOGICAL DISORDERS

Box 13.1 *Attention Deficits*

Brain lesions that impair attention tend also to impair other functions. Lesions of the posterior parietal cortex, for example, interfere with motor attention as well as with the execution of visually guided hand movements. Similarly, lesions of the frontal eye field or superior colliculus lead to deficits in visual attention, but they also interfere with eye and head movements. Because spatial attention and the control of orienting movements are implemented by overlapping neural circuits, most brain lesions impair both functions equally. An important exception to this rule are brain lesions that cause *hemispatial neglect.*

Subjects with hemispatial neglect syndrome often neglect the left side of objects (Figure b13.1). They may also ignore the entire left side of their visual field, the left side of their body, and the left side of the external world, independently of where their eyes are aimed. In other words, the deficits may occur in a variety of spatial coordinate frames (see Chapter 11), but are usually associated with the left side of the affected spatial frame.

Important for our discussion here is that the deficits in hemispatial neglect tend to be neither sensory nor motor but attentional (as the term "neglect" implies). For example, when hemispatial neglect patients are tested in a peripheral cueing task, they can see stimuli in the left half of their visual field, but they are slow to react when targets appear there, especially when their attention had been focused on the right. These data indicate that hemispatial neglect results, at least in part, from a decreased ability to shift attention from the right side to the left.

What kind of brain damage causes hemispatial neglect? The lesions in subjects with this syndrome are usually in the right cerebral hemisphere and involve large portions of the parietal and temporal lobes. Giving a more precise answer is difficult because the lesions vary considerably across patients. However, lesion overlap analyses suggest that most hemispatial neglect patients have damage in the inferior parietal cortex, the superior temporal cortex, and the temporo-parietal junction (Fig. b13.1).

This finding is somewhat surprising because the equivalent brain regions in the left cerebral hemisphere function primarily in language-related processes. Furthermore, you may wonder how lesions in the inferior parietal and temporal lobes can impact the attention-related circuits that pass through the more dorsally located frontal and parietal eye fields. This question remains open, but functional brain imaging has shown that right inferior parietal and superior temporal lesions reduce activation of more dorsal frontal and parietal areas in attention-demanding tasks. Forty weeks after such lesions, activity in the frontal and parietal regions returns to relatively normal levels, and the hemispatial neglect usually resolves. This finding suggests that hemispatial neglect results, at least in part, from disturbances in the normal activity of the dorsal parietal cortex, which results from disruptions of activity in more ventral (i.e., more inferior) cortical areas.

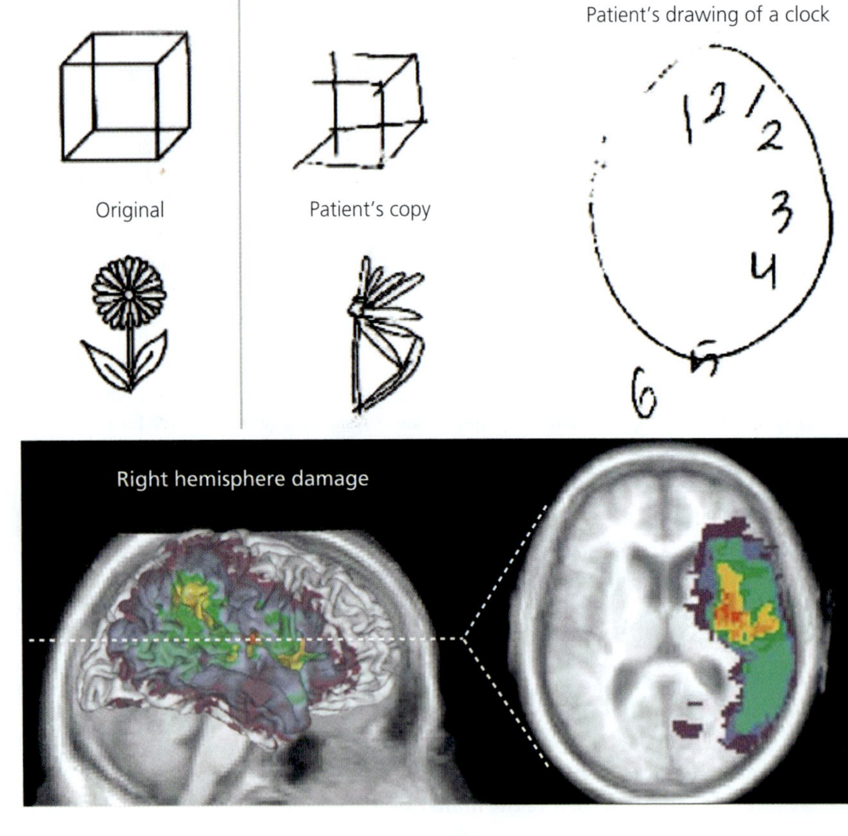

Figure b13.1 Hemispatial neglect. When patients with hemispatial neglect are asked to copy simple drawings, such as a flower or a clock, they tend to copy only the right side of the drawing (top panel). Although neglect syndrome may sometimes include perceptual as well as motor problems, the principal problem appears to be attentional. When regions of damage for multiple neglect patients are mapped onto a reference brain (bottom panel), overlap tends to be maximal in the inferior parietal and superior temporal cortices and at the junction between the two (the more overlap, the warmer the color in this figure). Importantly, the damage is usually limited to the right side of the brain. [Top from Parton et al., 2004; bottom from Corbetta et al., 2005]

Box 13.1 *Attention Deficits (continued)*

Can *attention deficit/hyperactivity disorder (ADHD)*, which you have surely heard about, also be linked to brain damage? This question remains hotly contested, but good evidence suggests that the brains of boys with ADHD are 3–4% smaller than the brains of age-matched controls. These differences in overall brain size are largely due to difference in the size of the prefrontal cortex, striatum, and cerebellum. Of these three structures, only the prefrontal cortex has clear connections to the neural circuitry of attention. Indeed, patients with ADHD perform normally in standard tests of attention. They may shift their attention more frequently than others do, but this symptom is part of a more general inability to suppress inappropriate actions, which leads to impulsivity and hyperactivity. Therefore ADHD is not, despite its name, primarily a deficit of attention.

Collectively, these data indicate that FEF neurons are causally involved in guiding covert voluntary spatial attention. Consistent with this hypothesis, inactivation of the FEF with the GABA agonist muscimol impairs voluntary shifts of covert spatial attention.

Network Interactions in Attention

It is interesting to note that weak activity in the superior colliculus and FEF generate covert shifts of attention, whereas higher levels of activity in the same brain regions cause overt orienting movements. Also intriguing is that voluntary attention is linked most closely to the FEF, which is involved in voluntary eye movements; whereas involuntary attention is linked to the superior colliculus, which is more closely linked to reflexive, involuntary orienting movements. It is important to recognize, however, that the superior colliculus and FEF are part of a complex neural web (see Figure 13.6) and that voluntary and involuntary attention can interact.

It has been suggested, for example, that the parietal eye fields integrate information from the saliency map in the superior colliculus with top-down information from the FEF. Such a melding of bottom-up saliency with top-down attention would explain why you are more likely to find a target stimulus in a serial search paradigm when that stimulus suddenly moves, flashes, or dims. In such instances, bottom-up salience grabs our voluntary attention. The notion that the parietal cortex is involved in the control of spatial attention is further supported by the observation that lesions of the right parietal cortex often cause **hemispatial neglect**, a syndrome in which patients have trouble shifting attention to the left half of their world (see Box 13.1: Attention Deficits).

Although this section has focused exclusively on visual spatial attention, humans are certainly capable of attending to non-visual and non-spatial features. Indeed, subjects with hemispatial neglect ignore not only visual stimuli in the left half of their world but also stimuli in other sensory modalities. Auditory, somatosensory, and other non-visual forms of attention have not been studied as thoroughly as visual spatial attention, but both the parietal and the lateral prefrontal cortex appear to be involved.

BRAIN EXERCISE

What would happen to your perceptual abilities if attention were to increase the responses of cortical neurons to all stimuli and not just those that the neurons already prefer?

13.2 What Mechanisms Generate Behavioral Arousal?

So far we have discussed the mechanisms that control an organism's attention, selectively enhancing the perception of stimuli with the attended attributes (notably stimuli at the attended location). Sometimes, of course, a stimulus is so startling, thrilling, or threatening—so highly salient—that the animal gets really excited, ready for anything, prepared to act. Some people refer to this state of heightened

behavioral arousal as a state of "heightened attention," but attention is selective, whereas arousal is not. When arousal levels rise, all sorts of stimuli become more likely to elicit a response. For that reason, the term arousal is often applied to the transition from sleeping to the waking state.

The term "arousal" is also applied to sexual excitement. Although sexual arousal can involve a heightened general awareness, as well as focused attention, it refers primarily to penile erection or vulval engorgement. These physiological responses are generated by parasympathetic activation, which dilates blood vessels and, thus, allow more blood to flow into the genitals. As you may recall from Chapter 9, parasympathetic activation is generally associated with relaxation rather than heightened behavioral arousal. Indeed, men often get erections in their sleep precisely because sleep involves decreased sympathetic tone and increased parasympathetic activity. Orgasms, in contrast, are associated with an increase in heart rate and other signs of sympathetic activation. Thus, sexual arousal involves both parasympathetic and sympathetic activation, one after the other. In these respects, it is quite different from the behavioral arousal that will occupy us in this chapter.

The Electroencephalogram (EEG)

Neurobiologists usually identify behavioral arousal by looking for its correlate in the **electroencephalogram (EEG)**. Obtaining an EEG involves placing multiple recording electrodes on a person's scalp and then recording voltage differences between those electrodes and a reference electrode on the neck. Because the observed voltage changes are tiny (think microvolts), the first EEG recordings were regarded skeptically. It probably didn't help that Hans Berger, who recorded the first human EEGs in the 1920s, was doing so because he wanted to study the neural basis of telepathy. In any case, later researchers showed that EEG waves are indeed generated by cortical neurons and that changes in the EEG reflect changes in an organism's behavioral state.

As individuals go from being alert to being drowsy, the EEG increases slightly in amplitude and starts to exhibit fairly clear, rhythmic oscillations (Figure 13.9). If the individual then falls asleep, EEG amplitude increases further and the fast oscillations are gradually replaced by slower ones. We will discuss these sleep-related changes later in this chapter. For now, the important point is that the EEG has a lower amplitude when individuals are aroused (wide awake) than when they're drowsy or relaxed.

Neurobiologists usually refer to the decrease in EEG amplitude that accompanies behavioral arousal as **EEG desynchronization**. This term is appropriate because large deflections in the EEG are caused by the synchronous firing of many cortical neurons, whereas small undulations occur whenever those neurons fire asynchronously. The link between neuronal synchrony and EEG amplitude arises because the dendrites of neocortical pyramidal neurons tend to be arranged in parallel to one another and perpendicular to the neocortical surface (see Figure 11.12). As a result of this radial alignment, synchronous activation causes the dendritic potentials generated by the individual pyramidal neurons to sum, rather than cancel, when they are recorded by large electrodes placed either on the scalp or on the neocortical surface. The larger the sum of these individual potentials, the larger the deflections in the EEG.

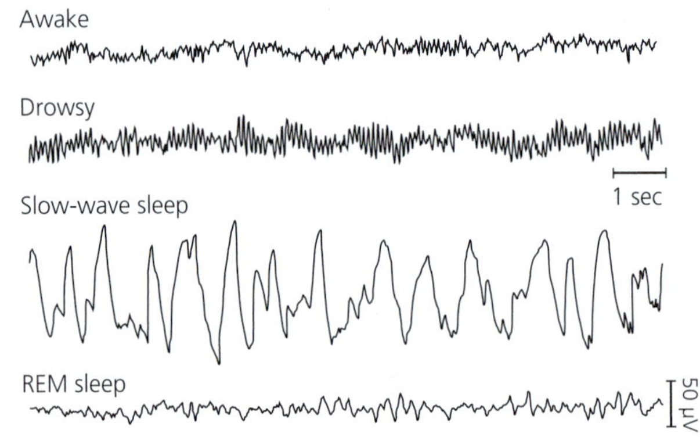

Figure 13.9 The electroencephalogram (EEG) reflects behavioral states. As people become drowsy and then fall asleep, their EEG becomes dominated by large-amplitude slow oscillations (waves). These slow waves disappear during rapid eye movement (REM) sleep. [After Moore, 1999]

A good analogy for EEG synchronization and desynchronization is the marching of soldiers across a bridge. If they are marching in lockstep, then the forces exerted by their footsteps on the bridge all sum, causing the bridge to oscillate. If the bridge is pliable, then the oscillations can become quite large and dangerous. In contrast, if the soldiers step across the bridge asynchronously, then the stepping forces do not sum and the bridge is safe. Soldiers aside, the

EEG becomes desynchronized when mammals are aroused. This observation has allowed researchers to discover the neuronal mechanisms underlying arousal.

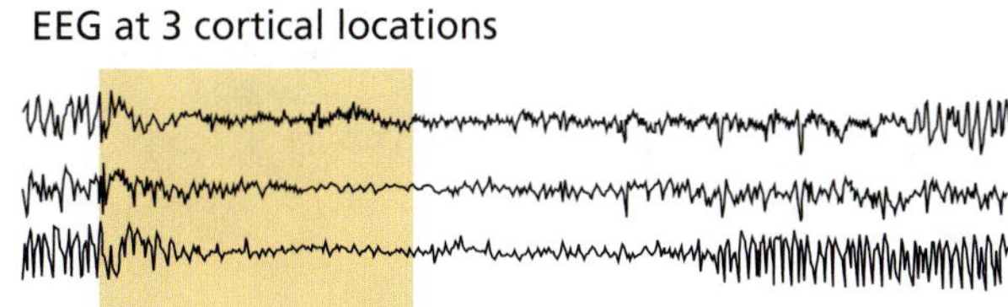

Figure 13.10 Brainstem stimulation can desynchronize the EEG. Shown here are EEG traces recorded by Moruzzi and Magoun from three cortical areas in an anesthetized cat. Shortly after the onset of electrical stimulation of the brainstem (yellow shading), all three EEG traces decreased in amplitude, indicating widespread EEG desynchronization. A few seconds after the end of the brainstem stimulation, the large-amplitude oscillations returned. [After Moruzzi and Magoun, 1949]

Ascending Arousal Systems

Guiseppe Moruzzi and Horace Magoun in 1949 published an influential study showing that electrical stimulation in the reticular formation of the upper pons and midbrain can desynchronize the EEG of anesthetized cats. Because the desynchronization was observed simultaneously at multiple recording electrodes (Figure 13.10), Moruzzi and Magoun inferred that the stimulation causes general arousal rather than a selective activation of specific cortical areas.

Moruzzi and Magoun never stimulated the reticular formation of awake and intact animals, but they studied some cats whose brains had been surgically separated from the spinal cord. Such cats are paralyzed but generally awake. Moruzzi and Magoun were able to synchronize the EEG of these cats by stimulating the thalamus (which somehow made the cats sleepy) and then to desynchronize the EEG again by presenting the cats with strong sensory stimuli (blowing on their face). Importantly, stimulating the reticular formation had the same desynchronizing effect as the arousing sensory stimuli. Moruzzi and Magoun inferred that stimulation of the reticular formation causes EEG desynchronization not only in sleeping cats, but also in awake (but drowsy) cats.

Moruzzi and Magoun used crude stimulation methods and probably stimulated multiple cell groups. Subsequent research has clarified that there are two principal **ascending arousal systems**. One involves cholinergic neurons, which are discussed in the next section because they have been studied primarily in the context of sleep regulation. The other major arousal system centers on *locus coeruleus*, a small group of neurons near the fourth ventricle.

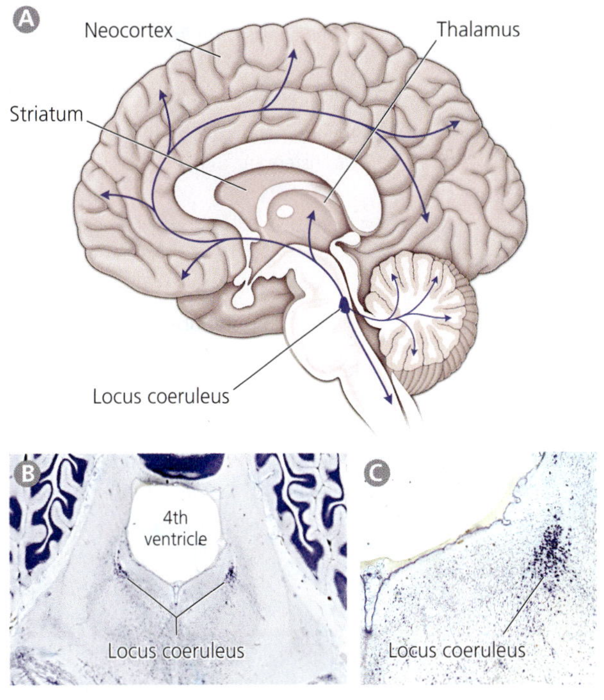

Figure 13.11 Locus coeruleus. The diagram in (A) shows the approximate location of locus coeruleus and its projections to most of the central nervous system, except the striatum. The photographs in (B) and (C) depict locus coeruleus neurons in Nissl-stained horizontal sections through a human brain at low and high magnifications, respectively. [B and C courtesy of the Yakovlev-Haleem Collection]

The Locus Coeruleus System

The neurons of **locus coeruleus** tend to be pigmented (*locus coeruleus* means "blue spot" in Latin) and use norepinephrine as their main neurotransmitter. As you may recall, norepinephrine is also released by postganglionic neurons of the sympathetic nervous system. In the brain, however, most of the noradrenergic (norepinephrine-using) neurons have their cell bodies in locus coeruleus.

Widespread Noradrenergic Projections

Although the human locus coeruleus contains only about 20,000–23,000 neurons per side, the axons of these neurons project widely throughout the brain (Figure 13.11). Because it is technically difficult to visualize norepinephrine directly, neuroscientists usually examine the projections of locus coeruleus by staining brains with antibodies against **tyrosine hydroxylase**, an enzyme needed for the synthesis of norepinephrine. Using this technique, stained axons are visible throughout the brain and spinal cord (although their density is low in primary sensory cortices). However, as you can infer from the biosynthetic pathway for norepinephrine (Figure 13.12), staining for tyrosine hydroxylase labels both noradrenergic and dopaminergic axons.

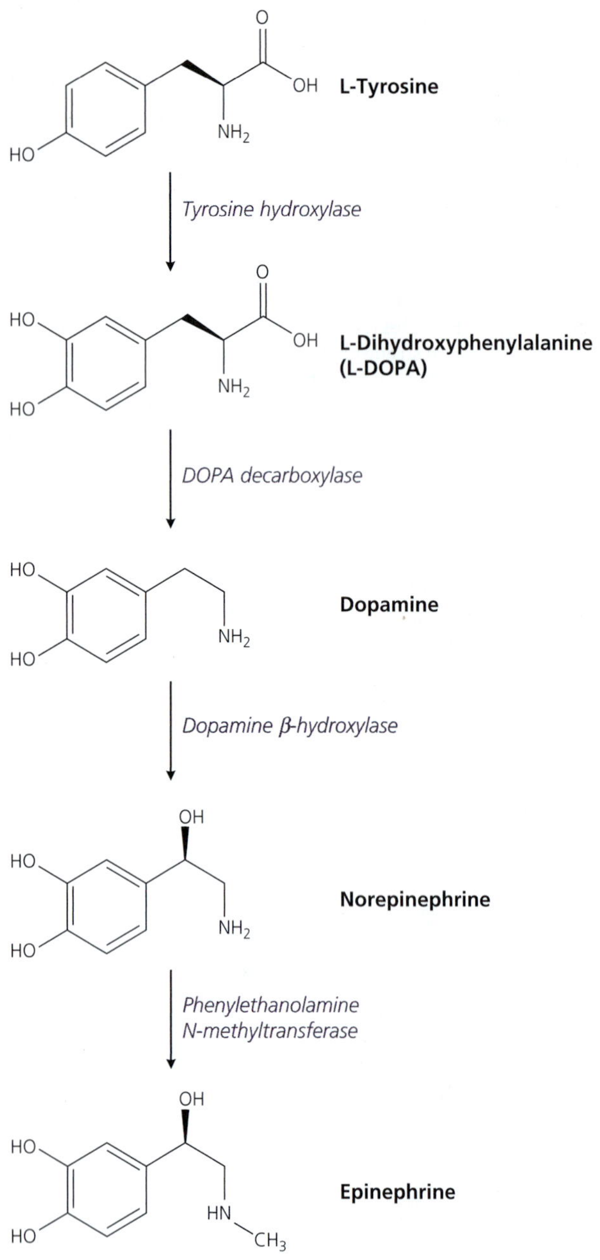

L-Tyrosine

Tyrosine hydroxylase

L-Dihydroxyphenylalanine (L-DOPA)

DOPA decarboxylase

Dopamine

Dopamine β-hydroxylase

Norepinephrine

Phenylethanolamine N-methyltransferase

Epinephrine

Figure 13.12 Catecholamine biosynthesis.

To discriminate between these two types of axons, one can use antibodies against **dopamine beta-hydroxylase**, the enzyme that converts dopamine to norepinephrine. This more selective staining approach yields results that are very similar to those obtained with antibodies against tyrosine hydroxylase, suggesting that most of the stained axons synthesize norepinephrine. Furthermore, staining with antibodies against the enzyme that converts norepinephrine to epinephrine reveals only a few labeled axons and cell bodies in the medulla. Therefore, we can conclude that most tyrosine hydroxylase-positive axons use norepinephrine as their main neurotransmitter rather than dopamine or epinephrine. Only in the striatum does staining with antibodies against dopamine beta-hydroxylase fail to yield labeled axons, indicating that the tyrosine hydroxylase-positive inputs to the striatum are dopaminergic rather than noradrenergic (we will discuss the dopamine system at length in Chapter 15).

The axons of locus coeruleus neurons are thin, unmyelinated, and highly branched, with individual neurons sending axon collaterals to diverse brain regions; some even project to both sides of the brain. Because noradrenergic synapses lack a glial barrier, released norepinephrine can diffuse out of the synapse to more distant receptors, influencing a large number of postsynaptic neurons. A related observation is that locus coeruleus neurons are electrically coupled, which means that depolarization in one neuron spreads through gap junctions to its neighbors. This electrical coupling decreases in adult animals, but adult locus coeruleus neurons still tend to fire synchronously. Overall, these data show that any inputs that can activate locus coeruleus will end up influencing a very large number of widely dispersed neurons, which is consistent with the idea that locus coeruleus functions as a brain arousal system.

Control of Locus Coeruleus Activity

What kinds of situations cause locus coeruleus neurons to become excited and release norepinephrine throughout so much of the brain? One clear answer is naturally arousing stimuli. For example, when awake rats are startled by strong sensory stimuli, such as loud unexpected sounds, locus coeruleus neurons fire brief bursts of action potentials (Figure 13.13). Similar increases in locus coeruleus firing are seen when a rat receives a mild foot shock or tail pinch.

Locus coeruleus can also be activated by initially boring stimuli that animals have learned to associate with rewards or punishments. In one experiment a monkey was trained to perform a specific movement in response to a vertical stimulus and to ignore a horizontal bar (Figure 13.14). Correct responses produced a reward; errors elicited a punishment. The monkey soon learned this visual discrimination task; but then the reward contingencies were switched so that now, only responses to the horizontal bar were rewarded. As you would expect, the monkey made numerous errors right after the task reversal until it learned the new stimulus-response association. Meanwhile, the experimenters recorded the activity of neurons in the monkey's locus coeruleus.

The study's principal finding was that the locus coeruleus neurons increased their firing rate soon after the monkey's error rate increased (Figure 13.14). This observation implies a correlation with behavioral arousal because the monkey probably became excited when it was punished instead of receiving the expected reward. Furthermore, a detailed trial-by-trial analysis revealed that before the task reversal, locus coeruleus firing rates increased in response to the vertical bar but not the horizontal bar. After the reversal (and after the error rate had once again declined), the same neurons responded only to the horizontal bar. Thus, the neurons responded to whichever stimulus was linked to the reward. This is consistent with the idea that locus coeruleus activation occurs after the presentation of salient, arousing stimuli. After all, reward-predicting stimuli are likely to be salient and arousing. Collectively, these data show that locus coeruleus neurons are activated not only by innately arousing stimuli (such as a tail pinch) but also by stimuli that have acquired their arousing quality through learning.

Another stimulus that activates locus coeruleus is **epinephrine** (adrenaline). This hormone does not cross the blood–brain barrier, but its levels in the blood are sensed by neurons in the vagus nerve. These neurons project to the nucleus of the solitary tract, which in turn projects to nucleus coeruleus. Indeed, locus coeruleus neurons increase their firing rate as blood epinephrine levels rise, which they do when animals are excited or afraid. Locus coeruleus activation stimulates the central amygdala, which activates a variety of other neurons that ultimately cause the release of cortisol and epinephrine from the adrenal gland (see Chapter 9). In short, locus coeruleus is part of a positive feedback loop that leads to ever more epinephrine release and ever more locus coeruleus activation.

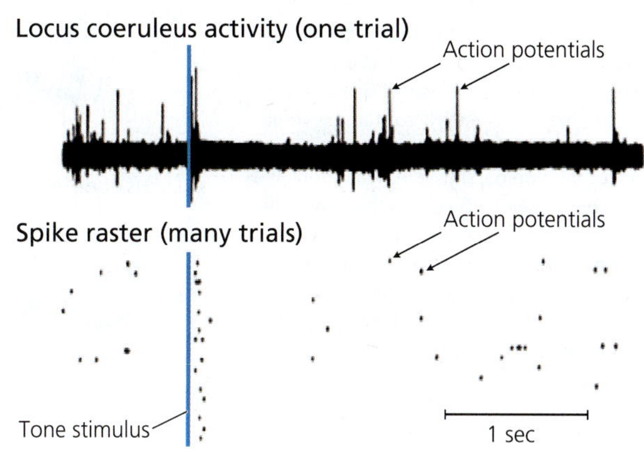

Figure 13.13 Locus coeruleus (LC) is activated by arousing stimuli. The upper trace depicts an extracellular recording from a group of LC neurons before, during, and after the presentation of a short, loud sound (vertical blue bar). The lower diagram shows the timing of individual action potentials (tick marks) for a single LC neuron across multiple trials. [From Aston-Jones and Bloom, 1981]

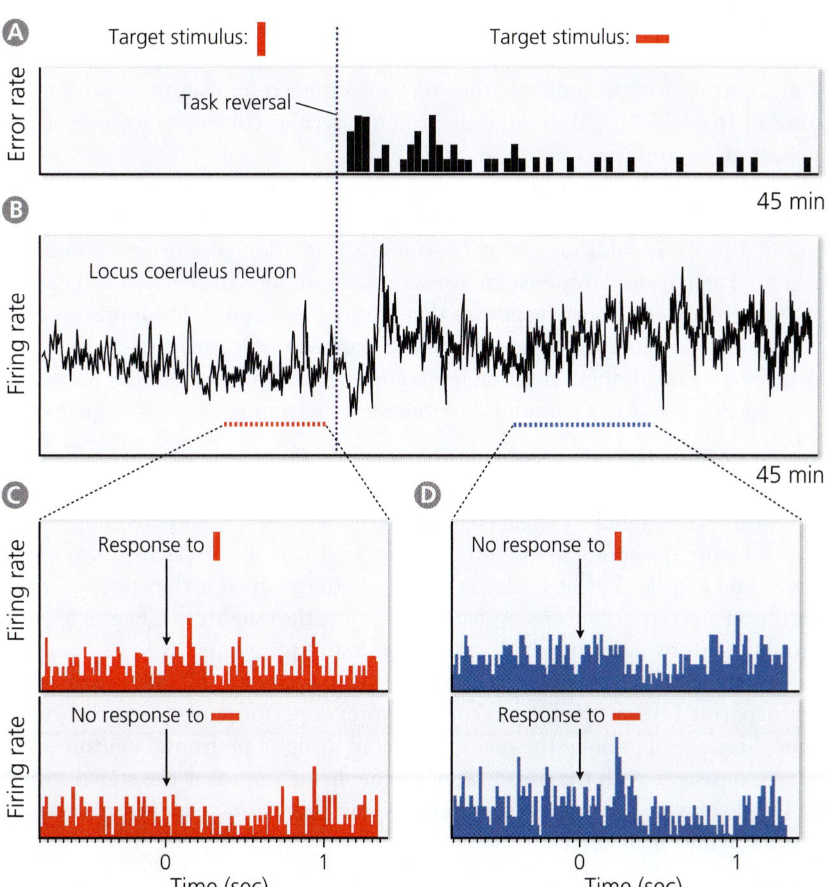

Figure 13.14 Locus coeruleus neurons change their tune after task reversals. Monkeys were trained to release a lever in response to seeing a vertical bar (target stimulus). Responses to non-target stimuli were punished by a "time-out." After monkeys learned this task, the target and non-target stimuli were switched, causing the monkey's error rate to jump (A). Simultaneous recordings from locus coeruleus neurons revealed increased firing rates shortly after task reversal (B), as one would expect if the errors (and associated punishments) increased behavioral arousal. A more fine-grained analysis revealed that, prior to task reversal, only the vertical bar elicited a response from the illustrated locus coeruleus neuron (C). After reversal, the formerly effective stimulus became ineffective, whereas the new target stimulus elicited a neuronal response (D). [After Aston-Jones et al., 1997]

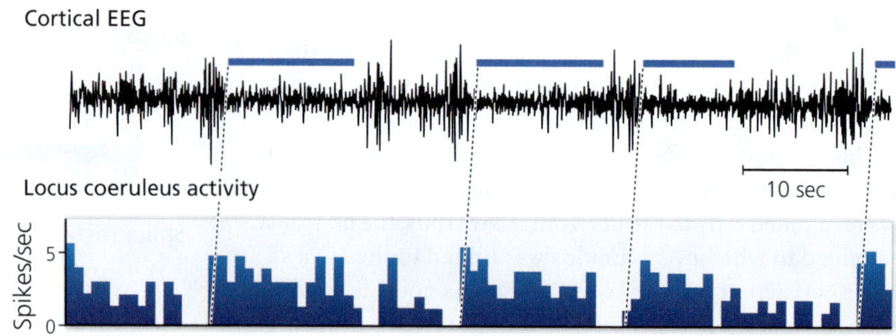

Figure 13.15 Locus coeruleus activity correlates with EEG desynchronization. The upper trace is a cortical EEG recording from an awake monkey. It contains several brief periods of EEG desynchronization (horizontal blue bars) that indicate behavioral arousal. Below the EEG is the firing rate histogram for a simultaneously recorded neuron in locus coeruleus. You can see that the neuron's firing rate increases just before and during the desynchronized periods. [After Foote et al., 1980]

If this is true, then how can you ever calm down after a stressful, arousing incident? One answer to this question is that locus coeruleus neurons project onto themselves and express α-2 adrenergic receptors that inhibit postsynaptic cells (more on that shortly). Therefore, locus coeruleus activity is kept in check by a very short and simple feedback loop. This negative feedback helps you calm down after an arousing event. Unfortunately, it also makes it difficult to stay alert for sustained periods of time.

Effects of Norepinephrine Release in Sensory Cortices

Experimental activation of locus coeruleus leads to EEG desynchronization and can wake a sleeping animal. However, locus coeruleus also controls behavioral arousal within the waking state, as illustrated by the observation that locus coeruleus activity in waking animals correlates with the degree of EEG desynchronization (Figure 13.15). Whenever locus coeruleus neurons increase their firing rate substantially, the cortical EEG becomes desynchronized; when they fall silent, EEG amplitude goes up.

In addition to altering the EEG, locus coeruleus activity affects how thalamic and cortical neurons respond to sensory stimuli. For example, infusing norepinephrine directly onto a set of thalamic neurons increases their responses to sensory stimuli (Figure 13.16 A). Electrical stimulation of locus coeruleus likewise increases sensory responses of thalamic neurons.

A similar effect is seen in sensory neocortex. Shortly after electrical stimulation of locus coeruleus, cortical responses to sensory stimuli are markedly enhanced (Figure 13.16 B). In addition, locus coeruleus stimulation suppresses spontaneous background activity and responses to weak thalamic inputs. In effect, locus coeruleus activation enhances sensory responses that are already strong and suppresses everything else. It increases what engineers call the **signal-to-noise ratio** of the neocortex. Another way to think about it is that locus coeruleus activation reduces the number of neurons that fire after a stimulus, but those few neurons respond with gusto.

Norepinephrine Effects on Motor and Prefrontal Cortices

Locus coeruleus projects more heavily to motor and prefrontal cortices than to primary sensory cortex, but its effects outside of the sensory cortices are poorly understood. Infusing norepinephrine into the motor cortex improves performance on complex motor tasks, but little else is known. Better studied is the effect of norepinephrine on prefrontal neurons. As we discuss more thoroughly in Chapter 15, many prefrontal neurons fire robustly during tasks that require animals to retain a piece of information for several seconds before the task can be fulfilled. Infusions of norepinephrine into the prefrontal cortex improves performance on such "working memory" tasks and prolong the memory-related firing of prefrontal neurons. These data are consistent with the idea that norepinephrine improves the signal-to-noise ratio in all sorts of neuronal computations, ranging from sensory processing to problem solving and memory.

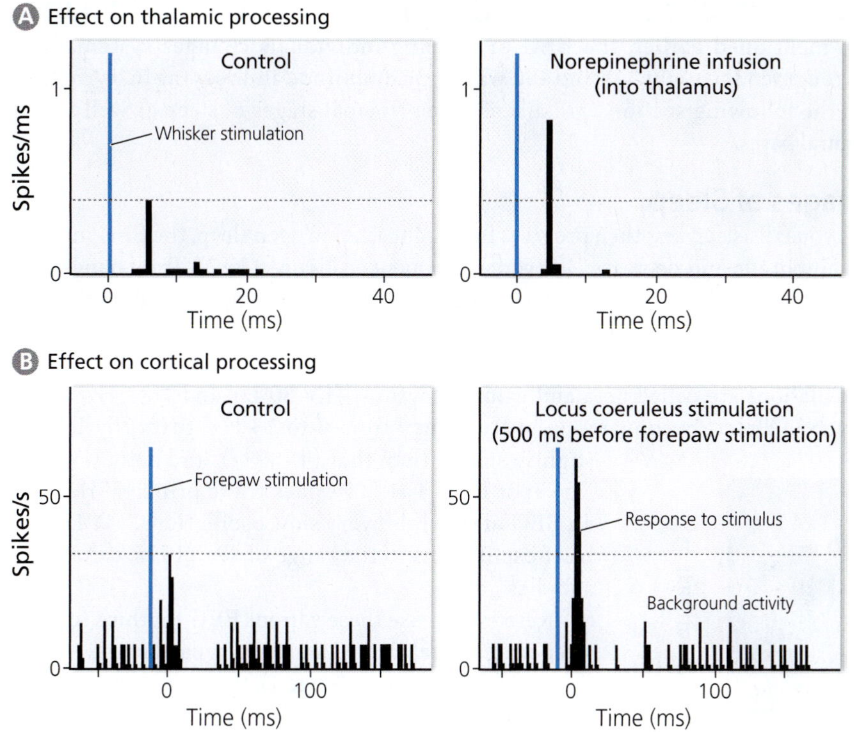

Figure 13.16 Locus coeruleus activation enhances sensory processing. Shown in (A) are data from a neuron in the somatosensory thalamus of an anesthetized rat. The neuron responds with a brief delay to mechanical stimulation of a contralateral whisker. The neuron responds even more robustly after norepinephrine was infused into the thalamus, mimicking locus coeruleus activation. Panel (B) shows a similar enhancement, except that the recording electrode was positioned in the somatosensory cortex; the sensory stimulus consisted of a touch to the forepaw; and the locus coeruleus was stimulated electrically. This manipulation not only enhances the sensory response but also suppresses background activity. [A from Hirata et al., 2006; B from Waterhouse et al., 1998]

However, the response of prefrontal neurons depends on the amount of norepinephrine that is released. These neurons become more active as norepinephrine levels rise and then shut down as norepinephrine levels increase further. To understand this unusual (inverse U-shaped) dose-response curve, you need to know that there are two main types of **adrenergic receptors** (whose effects on muscles we discussed in Chapter 9). As a general rule, neurons expressing α-1 or β adrenergic receptors tend to become more active in response to norepinephrine; whereas neurons that express α-2 adrenergic receptors (including locus coeruleus neurons themselves) tend to be inhibited by norepinephrine. This rule is fairly general, but prefrontal neurons do not obey. They are excited by α-2 receptor activation and inhibited by α-1 activation. Because α-2 receptors bind norepinephrine at lower concentrations than α-1 receptors do, prefrontal neurons are excited at low levels of norepinephrine and inhibited as more norepinephrine is released.

Why does the prefrontal cortex shut down at levels of locus coeruleus activation that boost processing in other cortical areas? As Amy Arnsten and her colleagues have pointed out, high levels of locus coeruleus activity are often evoked in highly stressful situations, such as confronting an imminent threat. At such times, it makes sense to improve processing in sensory and motor structures. In contrast, boosting prefrontal cortical functions may get in the way of rapid reactions because prefrontal functions are relatively slow to execute and tend to inhibit rapid, automatic responses. Therefore, taking the prefrontal cortex off-line may be adaptive during extremely stressful situations when immediate actions are more useful than long-term plans. This would explain why we sometimes look back on highly stressful episodes in our life and ask, what was I thinking? Insofar as thinking involves the prefrontal cortex, we were probably not thinking much at all. According to this view, panicking under stress is adaptive, at least in the short term.

BRAIN EXERCISE

Which anatomical and physiological attributes of locus coeruleus neurons make them well suited for controlling behavioral arousal? Why?

13.3 Why Do We Sleep, and What Helps Us Wake Up?

As mentioned earlier, the EEG of sleeping individuals changes systematically throughout the night, waxing and waning in amplitude and varying in its rhythms. In the following sections, we discuss the principal stages of sleep as well as their neural bases.

Stages of Sleep

As you fall asleep and then progress from light sleep to deep sleep, the EEG increases in amplitude and decreases in average frequency (Figure 13.17). This change is deceptively simple, however, because the EEG contains multiple rhythms that vary in amplitude, as well as frequency, and come and go throughout the night. Most of these rhythms are given Greek letter names (Figure 13.17). The fast, low-amplitude oscillations are called beta and gamma rhythms (13–30 Hz and >30 Hz, respectively). Alpha rhythms range in frequency from 8 to 13 Hz, theta rhythms are slightly slower than that (4–8 Hz), and delta waves are slower still (1–4 Hz). Besides these principal rhythms, an EEG may exhibit very slow oscillations (<1 Hz) and brief oscillations in the range of 12–14 Hz, called sleep spindles.

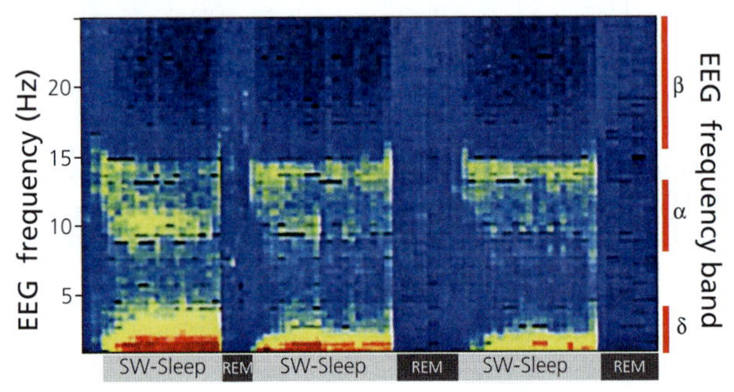

Figure 13.17 Changes in the EEG during one night of sleep. Time is represented along the x-axis, EEG frequency along the y-axis. The color indicates how much power the EEG contains at each frequency and point in time (dark blue represents minimum power; red is maximum power). You can see that low frequencies dominate during slow-wave (SW) sleep and largely vanish during REM. Indicated by red lines along the right edge of the image are three important EEG frequency bands. [From Achermann and Borbély, 1998]

Researchers use these various EEG rhythms to divide sleep into several stages. During a typical night, sleepers descend from *light sleep* (stage 1), through a couple of brief intermediate stages, to *deep sleep* (stage 4), which is characterized by prominent delta waves. Sleepers then return in the reverse order to light sleep, from which they either wake up or enter a special state (stage 5) in which the EEG is as desynchronized as it is during waking (Figure 13.18), even though the individual is clearly still asleep. During such periods of "paradoxical sleep" (the paradox being that the individual is sleeping while the EEG suggests they are awake), a person's eyes move jerkily, which is why this stage is usually called **rapid eye movement (REM) sleep**.

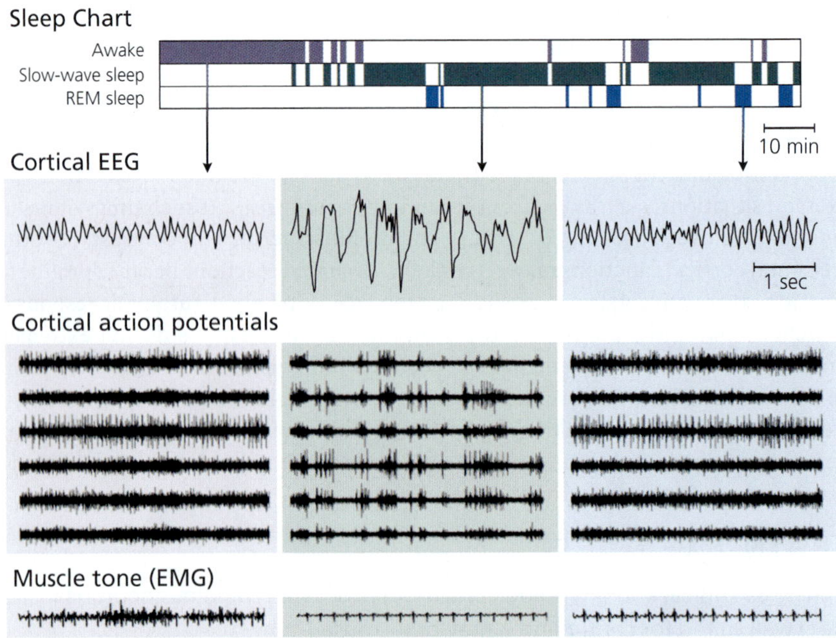

Figure 13.18 Neural correlates of sleep. A sleeping rat spends most of its time in slow wave sleep, interrupted by brief bouts of REM sleep and occasional transitions to wakefulness (top). The cortical EEG during slow-wave sleep exhibits large, low-frequency oscillations, which arise when cortical neurons in an area burst synchronously, as illustrated here by recordings of extracellular action potentials at 6 sites in the rat's somatosensory cortex. The bottom traces show that muscle tone, as measured with EMG electrodes, is very low during both slow-wave and REM sleep. The regular spikes in the EMG are due to the heart, which (thankfully) continues to beat during REM sleep. [From Vyazovskiy et al., 2009]

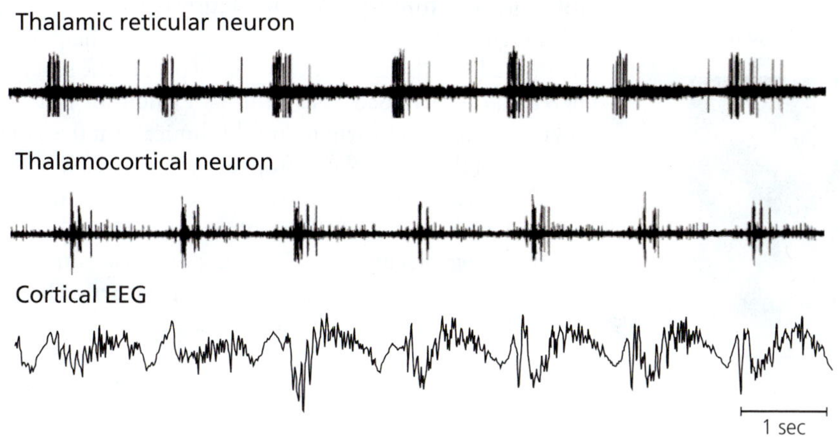

Figure 13.19 Rhythmic bursting in the thalamus. Shown here are simultaneous recordings from a thalamic reticular neuron, a thalamocortical neuron in the dorsal thalamus, and the cortical EEG in a sleeping (deeply anesthetized) cat. You can see that the thalamic reticular neuron fires in rhythmic bursts, followed 100–300 ms later by bursts of activity in the thalamocortical neuron. This pattern of activity is consistent with the hypothesis that bursts of thalamic reticular activity inhibit thalamocortical neurons; when this inhibition ends, the thalamocortical neurons rebound with a burst of action potentials of their own. [From Contreras and Steriade, 1997]

We will discuss REM sleep more thoroughly shortly. For now, suffice it to say that people who are awoken from REM sleep often report having had vivid dreams.

After REM sleep, people either wake up or descend again into the other stages of sleep, which are collectively called non-REM sleep. Another name for non-REM sleep is *slow-wave sleep* (Figure 13.18), although some scientists restrict this term to the deeper stages of non-REM sleep, when the slow (low-frequency) waves become very large. A typical night of sleep for an adult human contains 3–4 episodes of REM sleep, most of them in the second half of the night.

The Origins of EEG Rhythms

Where do the large EEG oscillations of slow-wave sleep come from? We can say that the oscillations arise whenever many cortical neurons are active rhythmically and synchronously so that their synaptic potentials tend to sum, but what mechanisms generate this synchrony and what accounts for the rhythms?

Thalamocortical Network Rhythms

A major role in generating the slow rhythms of the cortical EEG is played by neurons in the dorsal thalamus. Physiological recordings from deeply anesthetized or sleeping animals have shown that neurons in the dorsal thalamus with projections to the neocortex (thalamocortical neurons) tend to fire rhythmic bursts of action potentials in synchrony with the slow oscillations of the cortical EEG (Figure 13.19). This dorsal thalamic rhythm probably drives the neocortical rhythm because the dorsal thalamus provides strong excitatory input to most neocortical areas; but what mechanisms generate the dorsal thalamic rhythm?

To answer this question you have to know that the dorsal thalamus receives a major GABAergic input from the **thalamic reticular nucleus** (Figure 13.20). Furthermore, thalamic reticular neurons tend to fire bursts of action potentials just before each burst of activity in the thalamocortical neurons (Figure 13.19). The simplest explanation for this alternating pattern of activity is that each burst of activity in the reticular nucleus inhibits the thalamocortical neurons. When this inhibition stops, the disinhibited thalamocortical neurons exhibit a strong **postinhibitory rebound**, which means that they fire a burst of action potentials whenever they are released from strong inhibition. Because thalamocortical neurons project to the reticular nucleus, as well as to the neocortex, the thalamocortical rebound spikes trigger a burst of activity in the thalamic reticular neurons. Then the cycle repeats, with the reticular neurons once again inhibiting the thalamocortical neurons. In short, thalamocortical and thalamic reticular neurons form an oscillatory feedback loop that can explain, at least to some extent, the rhythmic firing in both sets of neurons during slow-wave sleep.

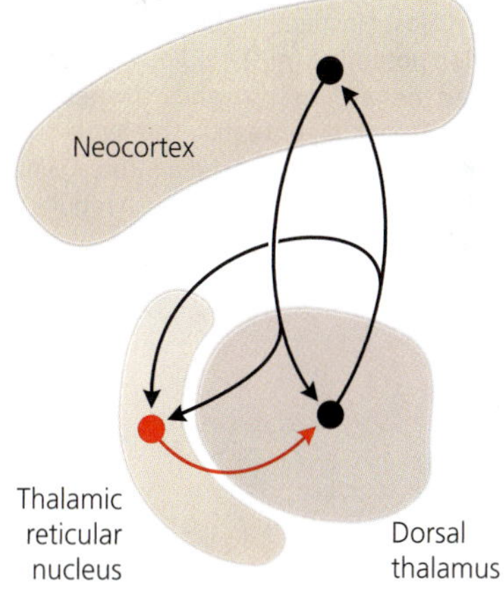

Figure 13.20 Loops between corticothalamic and thalamic reticular neurons. Neurons in the thalamic reticular nucleus have inhibitory projections (red) to thalamocortical neurons in the dorsal thalamus, which send excitatory projections (black) to the neocortex and back to the thalamic reticular nucleus. The neocortical neurons are excitatory and project back to the dorsal thalamus as well as to the thalamic reticular nucleus.

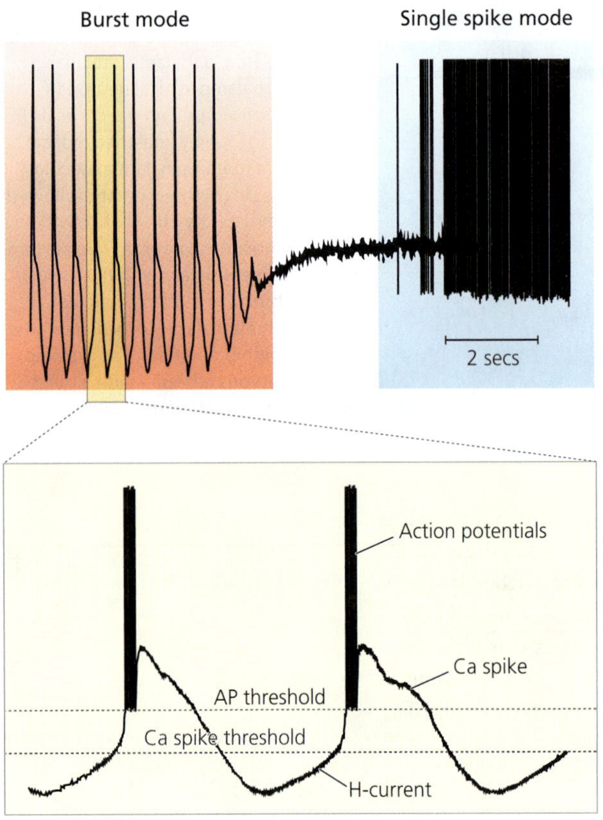

Figure 13.21 An intracellular rhythm in thalamocortical neurons. In this intracellular recording from a thalamocortical neuron, the red shaded area outlines a period during which the neuron bursts rhythmically. Closer inspection (bottom) reveals that each burst of regular action potentials (APs) is riding atop a much slower calcium (Ca) spike. When the calcium spike ends, the neuron becomes hyperpolarized below its resting potential. This hyperpolarization activates an h-current that slowly depolarizes the neuron again until it fires another calcium spike, and so on and so forth in a regular rhythm. [After McCormick and Bal, 1997]

Intrinsic Rhythms in Thalamic Neurons

Although feedback loops within the thalamus help to explain the EEG rhythms, they are only part of the story. In particular, thalamocortical neurons exhibit an intrinsic rhythm that causes them to fire rhythmically at 0.5–4 Hz even when they are isolated from other neurons.

This intracellular rhythm arises because the thalamocortical neurons tend to fire not only the typical action potentials we discussed in Chapter 2, but also action potentials in which the depolarizing current is carried by calcium ions (rather than sodium). These **calcium spikes** are much slower than traditional action potentials and occur at significantly lower thresholds (Figure 13.21). Near the peak of each calcium spike, the thalamocortical neurons fire a burst of regular action potentials. They then fall silent as the membrane potential reverts to the hyperpolarized state it displayed before the calcium spike. The falling membrane potential, in turn, opens hyperpolarization-activated channels. Once these channels are open, they generate an *h-current* (the same "funny current" we discussed in Chapter 8) that slowly depolarizes the neuron until it fires another calcium spike, thus completing the cycle. Overall, these calcium spikes strongly influence the period (frequency) of the thalamocortical rhythm.

Rhythm Synchronization

So far we have discussed only how thalamic and cortical neurons come to fire rhythmically. We have not yet discussed how rhythmically active neurons become synchronized with one another, which they must be if they are to cause large oscillations in the EEG (recall the soldiers on the bridge analogy).

One possibility is that cortical neurons with similar rhythms synchronize each other by means of excitatory intracortical connections. Such connections would increase the likelihood that a burst of action potentials in one neuron would be accompanied by bursts of activity in neighboring neurons. Supporting this hypothesis is the observation that silencing the connections between adjacent cortical regions with lidocaine desynchronizes their slow activity rhythms. Neuronal synchronization may also be accomplished by gap junctions between reticular thalamic neurons. Given the circuits shown in Figure 13.20, synchrony among the thalamic reticular neurons should synchronize the corticothalamic and neocortical neurons.

Consequences of Rhythmic Activity for Sensory Processing

An important consequence of the pronounced thalamic rhythms during sleep is that ascending inputs arriving at the dorsal thalamus have trouble getting through to the neocortex. Although sensory detection and processing can occur during sleep, especially for unexpected stimuli, thalamocortical neurons respond poorly to ascending inputs during slow-wave sleep. They are too busy bursting or being hyperpolarized. Therefore, the neocortex in deep sleep is largely cut off from the world (especially in species with few predators, such as humans). That is why people in deep sleep are difficult to wake.

BRAIN EXERCISE

In what ways is a "drum circle," defined as a group of people playing drums in a circle, analogous to the rhythmic firing of thalamic neurons during sleep?

Brain Systems That Wake Us Up

Although locus coeruleus neurons can wake a sleeping animal, they do so in collaboration with neurons in the basal forebrain and upper pons that use acetylcholine as their transmitter. These *cholinergic neurons* are relatively silent in deep sleep and active during the waking state. Their activation tends to desynchronize the EEG and rouse a sleeping animal.

Cholinergic Peribrachial Neurons

Cholinergic **peribrachial neurons** (whose name derives from their position around a major fiber tract, or *brachium*, that connects the cerebellum to the rest of the brain) project to both the basal forebrain and the thalamus (Figure 13.22). Within the thalamus, they innervate both the dorsal thalamus and the thalamic reticular nucleus (see Figure 13.20). Acetylcholine release in the dorsal thalamus slowly depolarizes the postsynaptic neurons (by closing some potassium channels) until they are so depolarized that they stop firing the calcium spikes we discussed earlier. In contrast, acetylcholine release in the thalamic reticular nucleus causes only a brief depolarization, followed by long-lasting hyperpolarization that reduces overall activity of the reticular neurons, which in turn decreases or eliminates their rhythmic inhibition of the dorsal thalamus.

The overall result of these effects is that dorsal thalamic neurons stop bursting rhythmically and become more capable of responding to ascending inputs with individual action potentials. This explains why stimulation of the cholinergic peribrachial neurons desynchronizes the EEG and makes the neocortex more responsive to external stimuli.

Did Moruzzi and Magoun stimulate the cholinergic peribrachial axons in their famous discovery of the reticular activating system? Probably yes, but it remains unclear precisely which neurons Moruzzi and Magoun reached with their rather large stimulating electrodes.

(A) Cholinergic brain systems

Neocortex

Thalamus

Basal forebrain

Peribrachial nuclei

(B) Basal forebrain activity

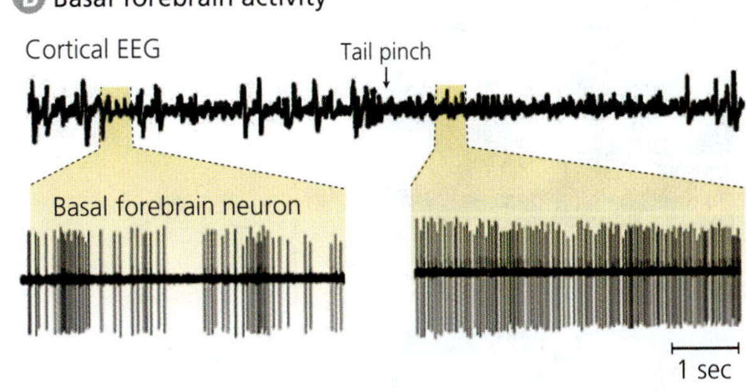

Figure 13.22 Cholinergic neurons help desynchronize the EEG. As shown in (A), cholinergic neurons (purple) in the basal forebrain have widespread projections to the neocortex and receive inputs from glutamatergic (green) neurons in the peribrachial nuclei. The upper trace in (B) shows the cortical EEG of a rat awakened by a tail pinch. The lower traces show the simultaneously recorded activity of a cholinergic basal forebrain neuron as the rat wakes up. [B after Manns et al., 2000]

Basal Forebrain Cholinergic Neurons

The second major group of cholinergic neurons involved in arousal from sleep (or drowsiness) is the **cholinergic basal forebrain**. These neurons project to the neocortex (Figure 13.22) as well as to the hippocampus and amygdala. All these projections are non-topographic, meaning that neighboring neurons often project to very different target areas. Therefore, even highly localized stimulation of the basal forebrain tends to cause widespread acetylcholine release.

The main effect of acetylcholine on neocortical neurons is similar to its effect on dorsal thalamic neurons, namely a slow depolarization, which reduces the tendency of the neurons to fire rhythmic bursts of action potentials. Neocortical acetylcholine release also hyperpolarizes astrocytes. This finding is interesting because astrocyte depolarization has been linked to the generation of slow oscillations in the neocortex during sleep, but the idea requires more testing. In any case, the overall effect of acetylcholine release in the cortex is clearly to reduce the slow rhythms in the EEG.

What activates the basal forebrain cholinergic neurons? A major excitatory input to these neurons comes from peribrachial neurons that use glutamate, rather than acetylcholine, as their main neurotransmitter (Figure 13.22 A). This is interesting because it means that electrical stimulation of the peribrachial region can desynchronize the EEG through two different pathways: one pathway reduces the tendency of thalamic neurons to oscillate; the other pathway activates the basal forebrain

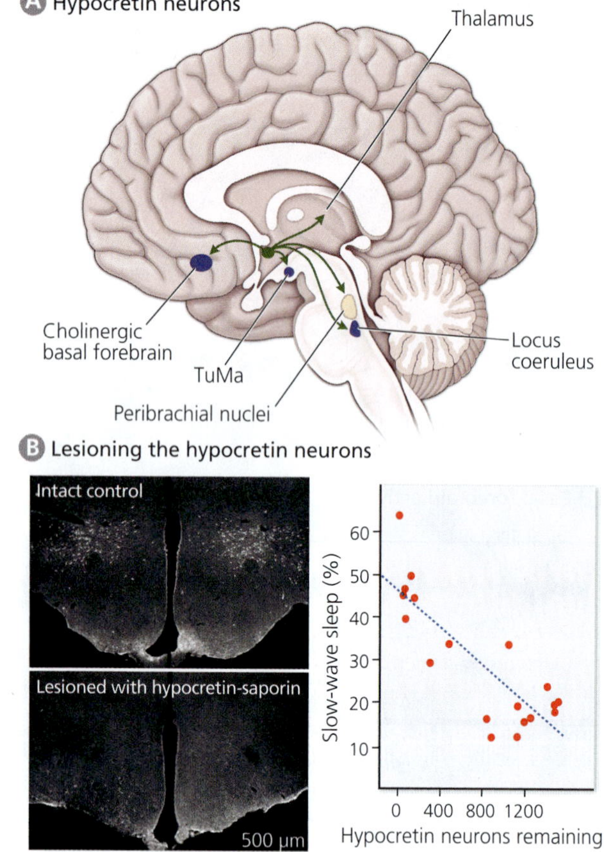

Figure 13.23 Hypocretin neurons. Neurons using hypocretin as their transmitter (green) project to many brain regions that are involved in waking up an animal (A). Panel (B) shows what happens when rats are given hypocretin-saporin, a toxin that selectively kills hypocretin neurons. The photomicrographs show the lateral hypothalamus of lesioned and control rats stained with antibodies against hypocretin. As illustrated in the graph, the more hypocretin neurons have died, the more time a rat spends in slow-wave sleep. [B from Gerashchenko et al., 2001]

cholinergic neurons, which then reduce the tendency of neocortical neurons to fire rhythmically. A second important excitatory input to the basal forebrain cholinergic neurons comes from the locus coeruleus, which suggests that the ability of the locus coeruleus to wake a sleeping rat depends, at least in part, on its ability to activate cholinergic neurons in the basal forebrain. As you are starting to see, the neurons involved in "waking up the brain" all tend to be connected to one another.

Histamine and Hypocretin

Aside from cholinergic and noradrenergic neurons, the brain contains a third set of neurons that become active as animals awake. These neurons use **histamine** as their main neurotransmitter and are located in the **tuberomammillary nucleus** of the posterior hypothalamus. Although tuberomammillary neurons are few in number, they project to many different brain regions, including the neocortex, amygdala, and striatum. Their role in promoting the waking state is supported by the observation that injecting histamine into the ventricle causes strong behavioral arousal. Conversely, drugs that interfere with histamine signaling make you drowsy (unless those antihistamines can't cross the blood–brain barrier; see Chapter 5). Lesions of the tuberomammillary region cause excessive sleep in cats (but oddly not in rats). Taken together, these data indicate that tuberomammillary neurons help animals wake up and stay awake.

A fourth group of neurons that fire at relatively high rates in waking animals, and fall silent when the animals sleep, uses the neuropeptide **hypocretin** (orexin) as its main neurotransmitter. These neurons are scattered across the lateral hypothalamus (Figure 13.23), where they intermingle with other types of cells. To test their role in sleep, researchers lesioned the hypocretin cells by infusing the toxin hypocretin-saporin into the lateral hypothalamus. This compound binds selectively to, and ultimately kills, any neurons that contain the hypocretin receptor, including the hypocretin neurons themselves. When more than 60% of the hypocretin neurons are destroyed, rats become excessively sleepy, showing increased REM as well as slow-wave sleep (Figure 13.23 B). This finding is consistent with the observation that dogs with mutations in the hypocretin receptor sleep more than normal. Furthermore, humans with **narcolepsy**, who are abnormally sleepy during the day and easily fall into REM sleep, have lost on average ~90% of their hypocretin neurons. Given these data, we can conclude that hypocretin cells are needed to maintain the waking state and suppress sleep, especially REM sleep.

How do the hypocretin neurons desynchronize the EEG? Mainly by activating the other wakefulness-promoting brain regions we discussed earlier, namely the locus coeruleus, the cholinergic basal forebrain neurons, and the tuberomammillary region (Figure 13.23 A). These target areas contain hypocretin receptors and increase their activity in response to hypocretin infusions. Thus, the hypocretin cells desynchronize the EEG and rouse sleeping animals indirectly by prompting other wakefulness-promoting brain regions to spring into action. Which leaves one more question: what activates the hypocretin neurons? No one is sure, but some hypocretin neurons are known to receive inputs from the suprachiasmatic nucleus, which controls the sleep–wake cycle and other circadian rhythms (see Chapter 9).

Brain Systems That Induce Sleep

It has long been suggested that slow-wave sleep may be the brain's default state, which it naturally adopts when the wakefulness-promoting systems fall silent. This idea is consistent with the observation that the locus coeruleus, cholinergic basal forebrain, tuberomammillary, and hypocretin neurons are all active when animals are awake and silent during slow-wave sleep (we will deal with REM sleep in the next section). Moreover, blocking cholinergic activation in the brain and boosting GABAergic inhibition with barbiturates tends to induce deep anesthesia (see Box 13.2: Anesthesia and the Death Penalty).

Still, one might wonder whether any neurons exhibit the opposite pattern, being more active during sleep than in the waking state. To search for such neurons, neurobiologists compared the expression of the gene *c-fos*, which increases shortly after neurons have been highly active, between rats that had been sleeping and rats that had been awake just before being sacrificed. Only the suprachiasmatic nucleus and a small cell group called the **ventrolateral preoptic area** (vlPOA) exhibited more *c-fos* positive cells in the sleeping animals. Subsequent research revealed that the vlPOA is activated even when rats are sleep deprived during the day and then allowed to sleep shortly before sacrifice during the night (when rats are normally awake). This shows that ventrolateral preoptic neurons become active during sleep, whenever it occurs. To test whether the vlPOA is required for sleep, researchers lesioned it in rats. Sure enough, the lesioned animals slept significantly less than the controls. In humans, too, preoptic lesions tend to cause insomnia.

A Sleep–Wake Switch Inside the Brain

Ventrolateral preoptic neurons induce sleep primarily by suppressing activity in the wakefulness-promoting brain regions we discussed earlier. This hypothesis is based on the observation that vlPOA neurons are GABAergic and project to the locus coeruleus as well as to the tuberomammillary and hypocretin neurons (Figure 13.24). Other studies have shown that the inhibitory projection from the vlPOA to locus coeruleus is reciprocated by inhibitory noradrenergic projections from the locus coeruleus to vlPOA. This is interesting because it means that locus coeruleus and the vlPOA, two brain regions that promote wakefulness and sleep, respectively, inhibit each other (Figure 13.24). This mutual inhibition creates clean and stable transitions between the waking state and sleep. It ensures that animals spend most of their time either awake or asleep, not in some intermediate state. Of course, throwing the *sleep–wake toggle switch* may take some time, especially in the morning.

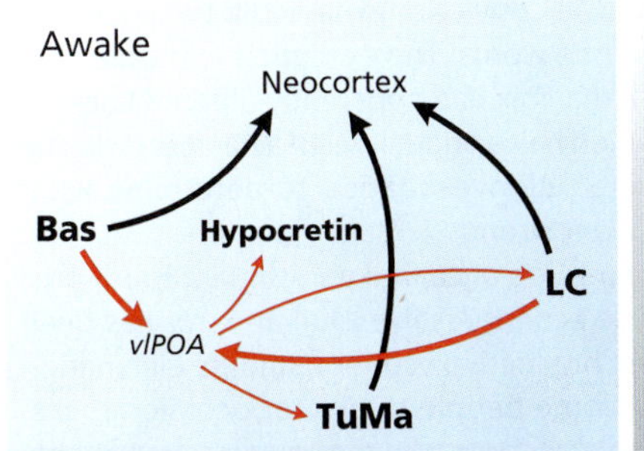

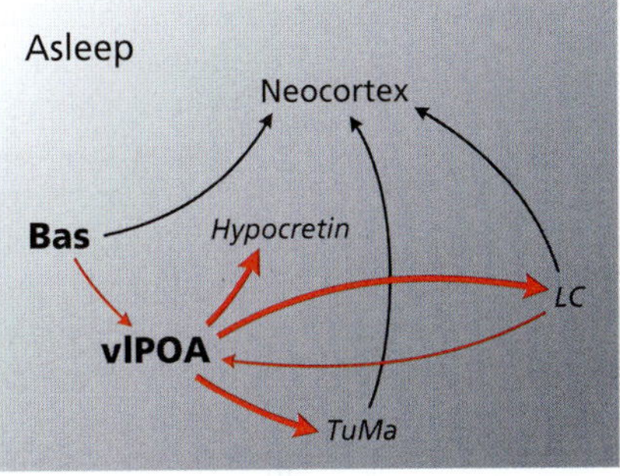

Figure 13.24 Reciprocal inhibition creates a sleep–wake toggle switch. The red arrows represent inhibitory connections, and the thickness of each arrow indicates its level of activity. Similarly, a large bold font indicates that a brain region is highly active, whereas a small italic font indicates reduced activity. Comparing the awake and sleeping states, you can see that the ventrolateral preoptic area (vlPOA) is active during sleep and inhibits the areas that promote wakefulness, namely, the cholinergic basal forebrain neurons (Bas), and the locus coeruleus (LC), tuberomammillary (TuMa), and hypocretin neurons. Conversely, vlPOA is inhibited during the waking state.

NEUROBIOLOGY IN DEPTH

Box 13.2 *Anesthesia and the Death Penalty*

Public hanging was the principal form of capital punishment in the United States until the late 1880s, when employees of Thomas Edison developed the electric chair. The first convict was electrocuted in New York State in 1890. When the first shock (700V for 17 seconds) failed to kill him, he was given a second, stronger, and longer shock that caused smoke to pour from his head. Although the use of the electric chair was later refined, botched executions continued to occur and generate concerns that they are "cruel and unusual punishment," which is forbidden under the Eighth Amendment of the US Constitution.

To address this concern, legislators in the 1970s developed *lethal injection* as a "more humane" form of capital punishment. Indeed, the US Supreme Court in 2007 ruled that execution by lethal injection does not violate the Constitution. The method has, however, run into some serious problems. To understand these problems, you need to know that lethal injection typically involves a sequence of three drugs, each given for a specific reason.

The first drug injected into a convict's vein has traditionally been *thiopental*, a fast-acting barbiturate that induces general anesthesia by enhancing GABA receptor activity and inhibiting neuronal acetylcholine receptors (as well as a variety of other receptors). Thiopental is also used to prepare patients for major surgeries, such as C-sections to deliver babies. In such surgeries, the induction of anesthesia is immediately followed by the insertion of a breathing tube into the trachea, through which the doctors then administer an anesthetic gas (such as sevoflurane) to maintain a surgical level of anesthesia. This is important because the thiopental is not just quick acting; it also wears off rapidly, at least at doses that are safe.

The second drug given during a typical lethal injection procedure is not a maintenance anesthetic but a muscle relaxant, most commonly *pancuronium*. These drugs block acetylcholine receptors at the neuromuscular junction and therefore paralyze the subject. Paralytic agents are also administered during major surgeries to facilitate the tracheal intubation, prevent vomiting, and eliminate the chance of patient movement during a surgery. In executions, their sole function is to prevent the subject from moving during death, which observers would find unsettling.

The third component of the lethal injection "cocktail" is *potassium chloride*. Low doses of this drug are sometimes given to replenish a patient's electrolytes, but rapid elevation of extracellular potassium prevents the repolarization of cardiac muscle fibers, which causes the heart to get stuck in a sustained contraction. Death rapidly ensues.

Once you know about these drugs and what they do, it's easy to foresee some potential problems. Most vexing is the question of what happens if the thiopental does not, for whatever reason, anesthetize the subject deeply enough or long enough. After the paralytic agent has taken effect, the subject would not be able to tell you that he has woken up. Moreover, he (most executed prisoners are male) would slowly suffocate because the diaphragm and other respiratory muscles would be paralyzed. Finally, the potassium chloride would prevent repolarization not only in cardiac muscle but also in pain-sensing axons that innervate the blood vessels. If the subject were awake during this stage of the procedure, he would feel as if his veins were "on fire."

To prevent these problems, one would need to monitor the subject's level of anesthesia throughout the procedure. Because the subject is paralyzed, such monitoring would require EEG recording. Making and interpreting such recordings has become easier since the development of *bispectral EEG analyzers*, but it is rarely done and still requires well-trained personnel. Indeed, a major problem with capital punishment is that the executions are usually carried out by prison personnel, rather than medical professionals, who tend to interpret their oath to "do no harm" as meaning that they should not intentionally kill another person.

A relatively recent problem for the lethal injection procedure is that the only US manufacturer of thiopental has stopped making the drug, and the remaining European manufacturer is not allowed to sell the drug to customers who use it for executions. US authorities have tried to use alternative drugs, such as *pentobarbital* (a close relative of thiopental) and the fast-acting benzodiazepine *midazolam*, but now the manufacturers of those drugs are coming under pressure as well. Indeed, anesthesiologists in the United States are becoming increasingly concerned that the persistent controversies surrounding lethal injections are making it difficult to obtain the anesthetics they need for surgeries.

Another strategy state governments are pursuing to execute their prisoners is to obtain anesthetics from secret suppliers, including lightly regulated "compounding pharmacies" in the United States. Of course, this strategy is likely to increase the incidence of botched executions. On March 29, 2014, Oklahoma used such a secret cocktail to execute Clayton Locket, who had been convicted of the brutal rape and murder of a young woman. Shortly after the executioner declared Locket unconscious, the convict unexpectedly twitched and gasped and spoke some words. The execution was called off, but enough of the lethal mix had apparently entered Locket's system because he died half an hour later of a heart attack. The governor promised a full investigation to determine what part of the process went wrong.

Ultimately, the question of capital punishment is one that science and medicine cannot resolve. Human error has been responsible for most botched executions, and its elimination seems implausible. Some people think that occasional mishaps are an acceptable cost because the convicts presumably "had it coming." Others consider such incidents abhorrently "cruel and unusual," especially because several death row inmates have in the past been falsely convicted. Even if the legal system were foolproof and science could develop a fail-safe execution protocol, some would object to the taking of human life on general, principled grounds. You, too, will have to find and possibly defend your stance on this issue. Knowing what issues are involved may help you in this task.

What controls the sleep–wake switch? We have already discussed one structure that regulates the timing of sleep and wakefulness, namely the suprachiasmatic nucleus (see Chapter 9). Its circadian signals ensure that you're awake when it is productive to be awake and sleep when it is not. In addition, the observation that you get sleepier and sleepier the longer you have been awake suggests that some sort of sleep-inducing factors may accumulate while you're awake and dissipate when you're asleep. One such sleep-inducing molecule is **adenosine**, which is a byproduct of energy metabolism (because ATP is converted to ADP, which is then degraded to adenosine). Adenosine inhibits basal forebrain cholinergic neurons and increases sleep duration. The fact that **caffeine** is a potent adenosine receptor blocker largely explains why caffeine is so effective at keeping you awake.

BRAIN EXERCISE

Based on what you've learned, which one involves a greater number of active neurons: going to sleep or waking up? Why do you think this is the case?

13.4 What's Happening During REM Sleep?

Researchers have known since the late 1930s that slow-wave sleep is interrupted by periods during which the EEG becomes desynchronized (Figure 13.25). Because these periods of EEG desynchronization are accompanied by relatively rapid, jerky eye movements, they were named rapid eye movement (REM) sleep.

The EEG desynchronization during REM sleep is caused primarily by the basal forebrain cholinergic neurons, which are quiet during slow-wave sleep but tend to increase their firing rate during REM sleep. In contrast, the locus coeruleus and tuberomammillary neurons remain silent during REM. Serotonergic neurons may also be involved in REM-associated EEG desynchronization, but this hypothesis is not yet strongly supported.

When people are suddenly awoken from REM sleep, they generally report having had vivid dreams (less vivid dreams occur also in slow-wave sleep). Because vivid dreams and rapid eye movements occur at the same time, it is reasonable to suppose that dreamers move their eyes to scan the dream environment. However, the eye movements in REM sleep are slower than normal saccades and tend to follow looping trajectories. Unfortunately, no good alternative hypotheses have been proposed to explain why REM-associated eye movements exist.

Muscle Atonia

If dreams occur primarily during REM sleep, and if the brain's activity during REM sleep resembles the waking state, why don't we act out our dreams as we are having them? The answer is that during REM sleep most of our body is paralyzed. The cardiac and smooth muscles continue to work, as do the diaphragm and eye muscles, but the body's main striated muscles lose their muscle tone during REM sleep. This **muscle atonia** occurs because the skeletal motor neurons are tonically hyperpolarized during REM sleep (Figure 13.25 A).

Crucial for this inhibition of the motor neurons is the **subcoeruleus region**, which lies ventral to the locus coeruleus. Its glutamatergic neurons project to inhibitory neurons with direct projections to skeletal motor neurons in the brainstem and spinal cord. Moreover, lesions of the

(A) Key features of REM sleep

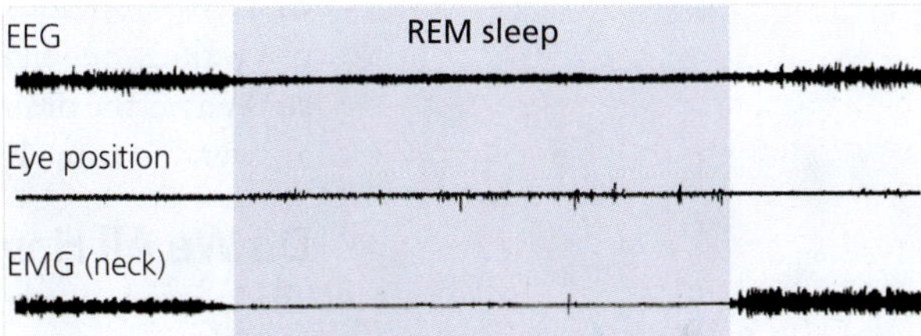

(B) Switching to REM and back

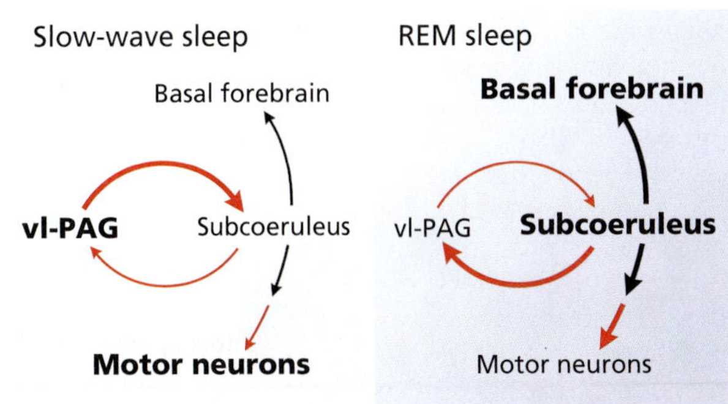

Figure 13.25 Rapid eye movement (REM) sleep. REM sleep is accompanied by EEG desynchronization, jerky eye movements, and loss of skeletal muscle tone (A). The muscle atonia is due to motor neuron hyperpolarization, as shown here for a jaw motor neuron. Panel (B) shows the neural network responsible for switching between slow-wave and REM sleep. The ventrolateral periaqueductal gray (vl-PAG) and the subcoeruleus region are thought to inhibit one another, forming a toggle switch. [A from Chase and Morales, 2000; B after Saper et al., 2010]

subcoeruleus region cause **REM without atonia**, which means that animals enter REM sleep without paralysis. Loosely speaking, such animals are acting out their dreams. REM without atonia has also been reported in humans, although it is rare. In case you're wondering, the condition is distinct from sleepwalking, which occurs during slow-wave sleep.

A REM Sleep Switch?

What triggers REM sleep? According to one hypothesis, REM begins when glutamatergic neurons in the peribrachial region become active and excite the basal forebrain cholinergic neurons. Another hypothesis emphasizes the **ventrolateral periaqueductal gray (vl-PAG)**, which lies ventrolateral to the cerebral aqueduct in the midbrain. Its neurons fall silent during REM sleep, thereby disinhibiting their target neurons in the peribrachial and subcoeruleus regions (Figure 13.25 B). An interesting aspect of this second hypothesis is that the subcoeruleus region contains GABAergic neurons that project back to the vl-PAG. Thus, the subcoeruleus neurons and the vl-PAG neurons may mutually inhibit one another, forming a toggle switch that determines whether the animal will be in REM or slow-wave sleep.

BRAIN EXERCISE

Have you ever woken up and felt unable to move? What might cause such "sleep paralysis"?

13.5 Why Does the Brain Have Discrete States?

A central theme in this chapter has been that a brain's physiological state varies over time. Most obviously, the brain of a sleeping organism is very different from the brain of a waking one in both intrinsic rhythms and responsiveness to stimuli. Why is this so? Why is the brain not constantly in the same state? These questions are difficult to answer, but some limited answers are possible.

Do We All Have to Sleep?

The longer you are awake, the sleepier you get; and if you reach the age of 70, you will have spent roughly 20 years asleep. These observations suggest that sleep performs a valuable function. To test this hypothesis, scientists have examined the effect of chronic sleep deprivation in rats (Figure 13.26). They found that severely sleep-deprived rats lose weight compared to control rats (even though the deprived rats eat more) and deteriorate in physical appearance and health. Most of them die within 3 weeks. Depriving rats selectively of REM sleep, rather than all sleep, is also lethal, but the rats' health declines less rapidly (Figure 13.26 B).

In humans, too, prolonged sleep deprivation impairs performance on difficult tasks and causes irritability, hallucinations, and disorientation. One young man was able to go for 11 days without sleep, but he was exceptional. Most people become severely delusional after about a week without sleep and then must end their strange experiment. Fortunately, the symptoms of sleep deprivation disappear rather quickly after recovery sleep.

How about species other than rats and humans? Must they also sleep? Although most species exhibit recurring periods of inactivity, slow-wave and REM sleep are well documented only in mammals and birds. Within those two taxonomic groups, all studied species sleep. However, some birds can go for many days without sleep during their annual migrations; and penguins do not sleep while they are defending their brood.

Aquatic mammals have a special problem with sleep because they must come up for air regularly. To solve this dilemma, many aquatic mammals tend to sleep near the water's surface. Intriguingly, dolphins and whales (cetaceans) sleep with one hemisphere at a time. That is, the two cerebral hemispheres in these species can enter

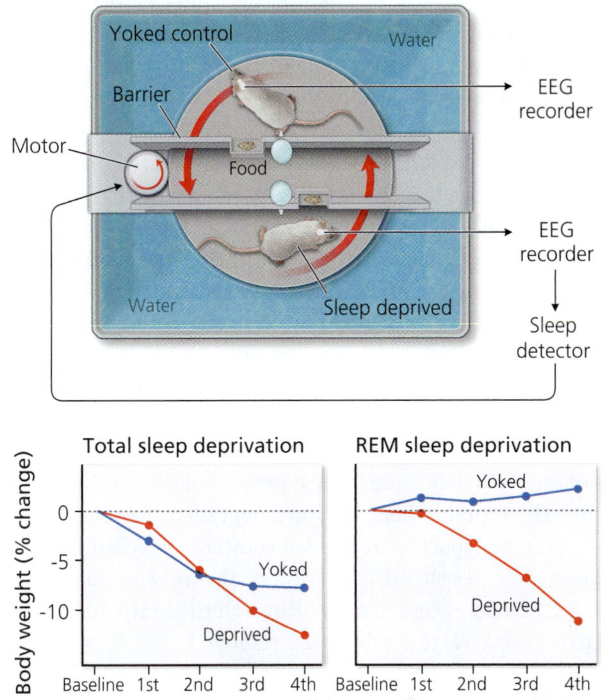

Figure 13.26 Sleep deprivation in rats. Two rats are placed on opposite sides of a disk over a shallow pool of water. One of the rats is targeted for sleep deprivation. Whenever it falls asleep or enters REM, as determined from the EEG, the disk rotates and forces the rat to walk or fall in the water. The other rat must also walk when the disk rotates, but will often be awake when this happens. The graphs show changes in body weight when rats were deprived of all sleep (left) or just of REM sleep (right). The sleep-deprived rats needed to be euthanized after 17 days, whereas the REM-deprived rats lived for 36 days. [Top after Bergman et al., 1989; bottom after Everson et al., 1989, and Kushida et al., 1989]

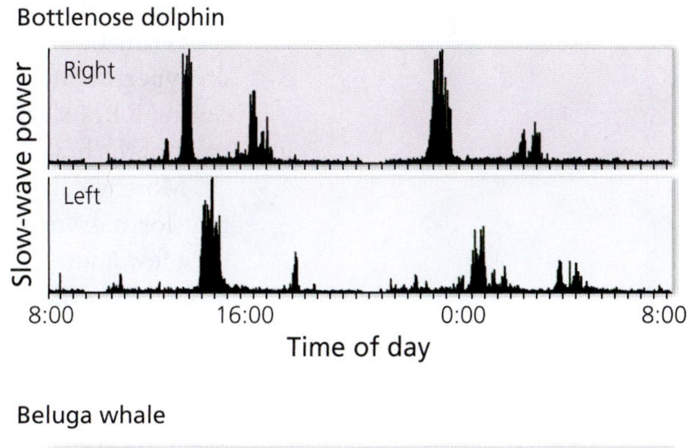

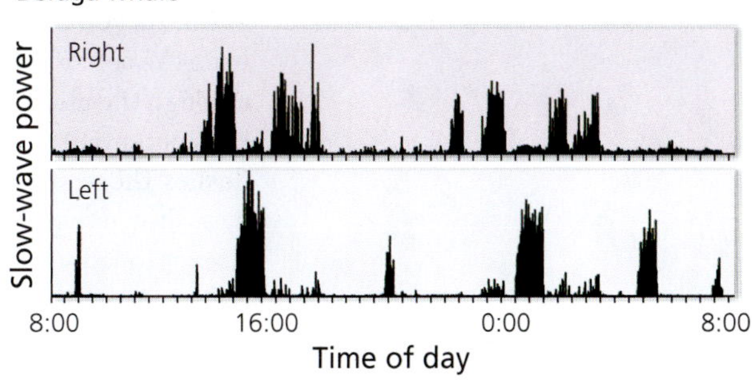

Figure 13.27 Cetaceans can sleep with one hemisphere at a time. These graphs show how the EEG's power in the slow-wave range (1.2–4 Hz) varies over a 24-hour period in the left and right cerebral hemispheres of a dolphin (top) and a whale (bottom). You can see that in both species, the left and right hemispheres exhibit slow-wave sleep alternately. [From Lyamin et al., 2008]

slow-wave sleep alternately (Figure 13.27). Whether dolphins and whales possess REM sleep is controversial but unlikely, given that prolonged muscle atonia would cause these animals to drown. Overall, we can conclude that slow-wave sleep is so important for warm-blooded species (birds and mammals) that evolution has found ways to accommodate it even in species that cannot afford to be inactive for long.

Energy Conservation, Memory Consolidation, and Toxin Removal

One crucial function of slow-wave sleep is the conservation of metabolic energy. Although the brain is hardly silent during slow-wave sleep, it is considerably less active during slow-wave sleep than during waking or REM sleep. Its oxygen consumption decreases by about 10% during deep slow-wave sleep (compared to waking), and its rate of glucose utilization drops 30–40%. Given that the brain accounts for roughly 20% of the body's entire energy budget at rest, these savings are substantial. Additional savings accrue from a ~1°C drop in core body temperature during slow-wave sleep. This may not seem like much, but it significantly reduces the rate of heat loss in small animals when it is cold outside. Reduced heat loss, in turn, allows small animals to lower the amount of metabolic energy that they must burn just to keep warm.

If a principal function of sleep is to conserve energy, why not silence the brain completely during sleep? In particular, why does the sleeping brain remain so active during REM? One likely answer is that a totally silent brain would be incapable of triggering thermoregulatory behaviors, which would increase the risk of lethal hypo- or hyperthermia. Some scientists have even speculated that the eye movements during REM sleep may warm the brain (because the eye muscles are very close to parts of the brain).

More widely accepted is the idea that REM and slow-wave sleep are both important for **memory consolidation**. Indeed, many studies have shown that sleeping for a few hours after you learn something improves retention, compared to control subjects who stayed awake. Furthermore, parts of the sleeping brain seem to "replay" patterns of activity that occurred during the earlier waking period. Such replay activity, which we discuss in Chapter 14, is likely to increase the strength of memories.

Yet another benefit of sleep is enhanced clearance of potential toxins from the brain. As you may recall from Chapter 5, cerebrospinal fluid (CSF) enters the brain from the cerebral ventricles and eventually drains into blood vessels. On this journey through the brain, along the **interstitial space** between brain cells, the CSF accumulates potentially harmful metabolites (such as amyloid β protein; see Box 5.2). It then flushes those substances into the blood, where they are degraded. Recent studies have shown that the brain's interstitial space increases substantially in volume during sleep, probably because some astrocytes shrink at that time. The increase in interstitial space, in turn, allows the CSF to flow much faster through the brain, increasing the rate at which potential toxins and waste products are removed.

Costs and Benefits of Arousal and Attention

Just as a sleeping brain can be in diverse states, so can the waking brain. As you have learned, how a brain processes stimuli is modulated by arousal and attention. Why is that? If processing improves with arousal, why is this state not permanent? The likely answer is that an optimal response in one behavioral context need not be optimal in a different context. For example, as we discussed, panicking may be adaptive at times. In addition, putting the brain into a low-arousal state whenever an animal feels safe (or otherwise relaxed) is likely adaptive because the relaxed state consumes less metabolic energy. Arousal is metabolically expensive not only because neurons respond more robustly to stimuli in aroused brains but also because it increases heart rate, respiration, and the mobilization of energy reserves (see Chapter 9). Given these considerations, arousal should be limited to times when it is most useful (such as when facing a threat or a chance to procreate). At other times, rest is more efficient.

What about attention? Does it also reflect the need for energy conservation? Yes it does because you can think of attention as the selective arousal of some neurons over others. By increasing activity and responsiveness in just a few neurons, you can keep energy costs down for the rest of the brain. If this is true, then why have general arousal at all as opposed to having only attention? Isn't it a waste of energy to make ALL neurons more responsive to stimuli? No it isn't, because organisms cannot always know where important stimuli are likely to appear or what those stimuli might be. In those situations, it is best to be ready for anything, anywhere, no matter the cost.

In general, attention should be employed when an organism has prior expectations about what features, objects, and locations are likely to be important. Those expectations might be the result of prior learning (for voluntary attention), or they might have been built into the nervous system by the evolutionary process (for involuntary attention). Either way, organisms can focus on what is likely to deserve attention and neglect everything else. According to this view, organisms continuously weigh the costs of increased brain activity against its benefits. Of course, they rarely do so consciously. Instead, evolution has created brains that automatically adjust their states to match the current needs and conditions.

SUMMARY

Section 13.1 - Attention generally improves performance on perceptual (or motor) tasks. It differs from behavioral arousal in being selective to particular stimuli, stimulus features, or regions of space.

- Visual spatial attention can be voluntary, such as when you're searching for a specific object in a cluttered space, or "grabbed" against your will by salient stimuli. Although spatial attention is often accompanied by orienting movements, it can also be covert.
- Salient stimuli may activate neurons in the superior colliculus, which then direct attention toward the stimulus. With strong stimulation, the shift in attention is accompanied by orienting movements toward the stimulus location; with weak stimulation, the shift is covert.
- Voluntary spatial attention, which involves the frontal eye fields, makes neurons with receptive fields in the attended area respond more strongly to their preferred stimuli (but not to other stimuli).

Section 13.2 - Behavioral arousal improves perception generally, rather than selectively.

- Wakefulness and behavioral arousal are associated with EEG desynchronization (smaller, more irregular EEG traces).
- Electrical stimulation in the reticular formation can trigger arousal. This effect probably results from the activation of noradrenergic axons from locus coeruleus and cholinergic axons from peribrachial neurons.
- Locus coeruleus contains a relatively small number of neurons with widespread axonal projections that use norepinephrine as their transmitter. These neurons are activated by various arousing stimuli.
- Most neurons become more responsive to stimuli and reduce their background firing rate in response to noradrenergic stimulation. The prefrontal cortex is unusual in that its neurons are inhibited by high levels of norepinephrine.

Section 13.3 - Sleep is good; it rests your body and brain, although the brain is not shut off entirely. After all, you can be roused from sleep by your own internal rhythm and by highly salient stimuli.

- Sleep is divisible into slow-wave and rapid eye movement (REM) sleep. During REM sleep, the EEG is as desynchronized as it is during the waking state, but muscle tone is minimal.
- The rhythmic activation of cortical neurons during slow-wave sleep results from intrinsic neuronal rhythms as well as looping interactions between the neocortex, the dorsal thalamus, and the thalamic reticular nucleus.
- Waking from sleep involves the activation of basal forebrain cholinergic neurons, which receive input from glutamatergic peribrachial neurons, locus coeruleus, and hypocretin neurons in the hypothalamus. Loss of the hypocretin neurons causes narcolepsy.
- A small group of neurons in the ventrolateral preoptic area is more active during sleep than during waking. These neurons inhibit wakefulness promoting neurons and are, in turn, inhibited by them.

Section 13.4 - The muscle atonia of REM sleep is caused by motor neuron hyperpolarization, which involves activation of neurons in the subcoeruleus region.

Section 13.5 - Slow-wave sleep saves energy, promotes memory consolidation, and facilitates the removal of potential toxins from the brain. Prolonged sleep deprivation is lethal.

Box 13.1 - Damage to the right inferior parietal cortex often results in hemispatial neglect. Patients with this syndrome generally have trouble shifting attention the left side of their external world.

Box 13.2 - Lethal injections typically involve a fast-acting anesthetic, followed by a paralytic agent, followed by potassium chloride to stop the heart. The level of anesthesia is rarely monitored after the second drug is administered.

KEY TERMS

covert spatial attention 409
involuntary attention 409
salient 409
voluntary attention 409
saliency map 411
inhibition of return 412
superior colliculus 412
pulvinar nucleus 413
cued spatial attention task 414
hemispatial neglect 417
behavioral arousal 418
electroencephalogram (EEG) 418
EEG desynchronization 418
ascending arousal system 419
locus coeruleus 419
tyrosine hydroxylase 419
dopamine beta-hydroxylase 420
epinephrine 421
signal-to-noise ratio 422
adrenergic receptor 423
rapid eye movement (REM) sleep 424
thalamic reticular nucleus 425
postinhibitory rebound 425
calcium spike 426
peribrachial neurons 427
cholinergic basal forebrain 427
histamine 428

tuberomammillary nucleus 428
hypocretin 428
narcolepsy 428
ventrolateral preoptic area 429
adenosine 431
caffeine 431
muscle atonia 431
subcoeruleus region 431
REM without atonia 432
ventrolateral periaqueductal gray (vl-PAG) 432
memory consolidation 434
interstitial space 434

ADDITIONAL READINGS

13.1 - Attention

Awh E, Armstrong KM, Moore T. 2006. Visual and oculomotor selection: links, causes and implications for spatial attention. *Trends Cogn Sci* **10**:124–130.

Boehnke SE, Munoz DP. 2008. On the importance of the transient visual response in the superior colliculus. *Curr Opin Neurobiol* **18**:544–551.

Jans B, Peters JC, De Weerd P. 2010. Visual spatial attention to multiple locations at once: the jury is still out. *Psychol Rev* **117**:637–684.

Kuhn G, Findlay JM. 2010. Misdirection, attention and awareness: inattentional blindness reveals temporal relationship between eye movements and visual awareness. *Q J Exp Psychol* **63**:136–146.

Maunsell JHR, Treue S. 2006. Feature-based attention in visual cortex. *Trends Neurosci* **29**:317–322.

Mysore SP, Knudsen EI. 2013. A shared inhibitory circuit for both exogenous and endogenous control of stimulus selection. *Nat Neurosci* **16**:473–478.

Rushworth MFS, Johansen-Berg H, Göbel SM, Devlin JT. 2003. The left parietal and premotor cortices: motor attention and selection. *Neuroimage* **20 Suppl 1**:S89–100.

Shipp S. 2004. The brain circuitry of attention. *Trends Cogn Sci* **8**:223–230.

13.2 - Arousal

Benarroch EE. 2009. The locus ceruleus norepinephrine system: functional organization and potential clinical significance. *Neurology* **73**:1699–1704.

Ramos BP, Arnsten AFT. 2007. Adrenergic pharmacology and cognition: focus on the prefrontal cortex. *Pharmacol Ther* **113**:523–536.

Sara SJ. 2009. The locus coeruleus and noradrenergic modulation of cognition. *Nat Rev Neurosci* **10**:211–223.

13.3 and 13.4 - Sleep

Amzica F, Massimini M. 2002. Glial and neuronal interactions during slow wave and paroxysmal activities in the neocortex. *Cereb Cortex* **12**:1101–1113.

Constantinople CM, Bruno RM. 2011. Effects and mechanisms of wakefulness on local cortical networks. *Neuron* **69**:1061–1068.

Fuentealba P, Steriade M. 2005. The reticular nucleus revisited: intrinsic and network properties of a thalamic pacemaker. *Prog Neurobiol* **75**:125–141.

Fuller P, Sherman D, Pedersen NP, Saper CB, Lu J. 2011. Reassessment of the structural basis of the ascending arousal system. *J Comp Neurol* **519**:933–956.

Haas H, Panula P. 2003. The role of histamine and the tuberomamillary nucleus in the nervous system. *Nat Rev Neurosci* **4**:121–130.

Hassani OK, Lee MG, Henny P, Jones BE. 2009. Discharge profiles of identified GABAergic in comparison to cholinergic and putative glutamatergic basal forebrain neurons across the sleep-wake cycle. *J Neurosci* **29**:11828–11840.

Hungs M, Mignot E. 2001. Hypocretin/orexin, sleep and narcolepsy. *Bioessays* **23**:397–408.

Lu J, Sherman D, Devor M, Saper CB. 2006. A putative flip–flop switch for control of REM sleep. *Nature* **441**:589–594.

Ruby P, Caclin A, Boulet S, Delpuech C, Morlet D. 2008. Odd sound processing in the sleeping brain. *J Cogn Neurosci* **20**:296–311.

Siegel J. 2004. Brain mechanisms that control sleep and waking. *Naturwissenschaften* **91**:355–365.

Steriade M. 2006. Grouping of brain rhythms in corticothalamic systems. *Neuroscience* **137**:1087–1106.

13.5 - Functional Considerations

Diekelmann S, Born J. 2010. The memory function of sleep. *Nat Rev Neurosci* **11**:114–126.

Mignot E. 2008. Why we sleep: the temporal organization of recovery. *PLoS Biol* **6**:e106.

Orzeł-Gryglewska J. 2010. Consequences of sleep deprivation. *Int J Occup Med Environ Health* **23**:95–114.

Xie L, Kang H, Xu Q, et al. 2013. Sleep drives metabolite clearance from the adult brain. *Science* **342**:373–377.

Boxes

Corbetta M, Kincade MJ, Lewis C, Snyder AZ, Sapir A. 2005. Neural basis and recovery of spatial attention deficits in spatial neglect. *Nat Neurosci* **8**:1603–1610.

Kerkhoff G. 2001. Spatial hemineglect in humans. *Prog Neurobiol* **63**:1–27.

Olk B, Hildebrandt H, Kingstone A. 2010. Involuntary but not voluntary orienting contributes to a disengage deficit in visual neglect. *Cortex* **46**:1149–1164.

CHAPTER

14

Remembering Relationships

CONTENTS

FEATURES

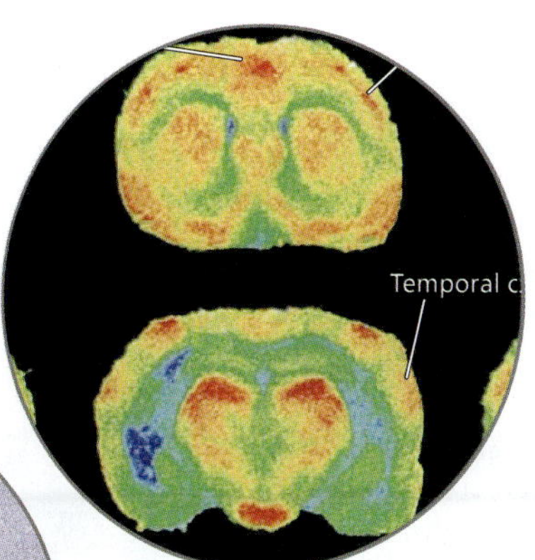

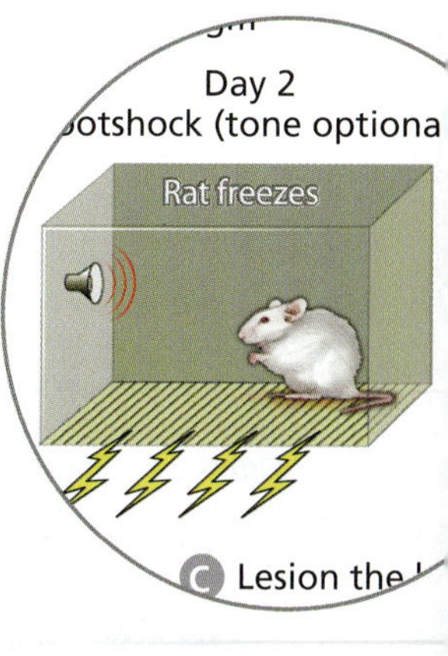

The ability to learn from experience is a crucial function of our brains. Without it, we would live exclusively in and for the moment, unable to benefit from earlier interactions with the environment or plan for the future. Such a life on the fly might be fun for a while, but it would probably be short. Given the importance of learning and memory, it is not surprising that scientists have studied them intensively. However, learning and memory cannot be studied directly. The only way to know that an animal has learned from an experience is to show that afterward, the animal behaves as if it had learned. That is, one must compare the behavior of animals that had the potential learning experience to the behavior of naive control animals.

Importantly, the design of the experiments constrains our inferences. For example, the observation that a dog salivates to the sound of a bell after repeated pairings of sound and food (Figure 14.1) tells you that the dog learned to associate the sound with salivation, but it does not tell you what else the dog might have learned. The dogs in Pavlov's famous experiment were strapped into the apparatus and, thus, limited in how they could respond. Therefore we cannot say that Pavlov's dogs learned ONLY to salivate in response to the sound. If they had been released from the harness, they might well have run around excitedly on hearing the telltale bell, indicating that they learned more than just a single, limited response. This may seem obvious, but it becomes important when one tests for the loss or retention of memories by looking for specific conditioned responses. The lack of a response need not imply a loss of memory! Perhaps the animal simply cannot show you what it knows, or isn't motivated to do so.

It might seem simpler to study learning and memory in humans, rather than dogs or other animals, because you can ask humans what they have learned. However, humans sometimes learn things that the experimenter did not think to ask about. Plus, humans may be unaware of what they've learned, and some of them may lie. Thus, discovering what individuals have learned always requires carefully designed experiments, regardless of which species is studied.

14.1 How Many Forms of Learning and Memory Are There?

Learning is not a unitary process. It comes in a variety of forms, as do the memories that it creates. Consider what you learn when playing a video game in which your character must navigate a complex virtual environment and shoot as many enemies as possible. As you play the game, you learn which buttons to push to run, turn, jump, aim, and shoot. With practice, you get faster at these skills and more precise. Neuroscientists refer to this process as **procedural learning**. In addition, you remember

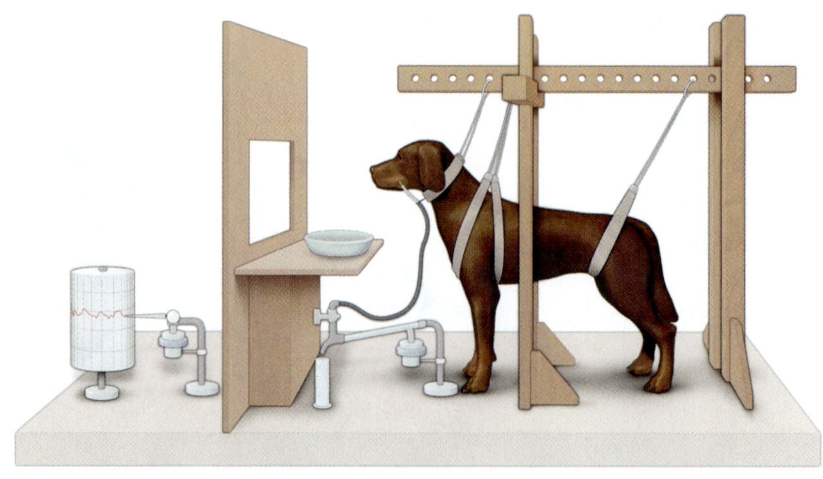

Figure 14.1 Pavlovian conditioning. Pavlov's famous dogs were restrained and connected to equipment that measures salivation. In a typical experiment, a specific sound was repeatedly presented just before the food. After several trials, the dog learned to salivate in response to the sound, even when no food was presented. [After Yerkes and Morgulis, 1909]

the most interesting things that happened during specific games. In particular, you'll probably remember where and when those interesting events occurred. As neuroscientists put it, you formed a series of **episodic memories**.

What else do you learn while playing the video game? You may remember that the game is rated M, for mature audiences, even though you cannot recall when and where you saw this rating displayed. Neuroscientists refer to this kind of learning as the creation of *semantic memories* (they may use the term "declarative memory" to refer to semantic and episodic memory collectively). Furthermore, if you play the game repeatedly, you'll gradually learn the spatial layout of its virtual environment, which means that you are performing the kind of *place learning* that we discussed in Chapter 11. Playing the game may also improve your ability to perceive small objects hidden in clutter, which implies some *perceptual learning*. In addition, you learn which stimuli predict imminent threats, rewards, or punishments; this is **Pavlovian (or classical) conditioning**. As if your brain was not already busy enough, you also learn the rules of the game, the strategies, the most efficient ways to reach your goals; we will cover this kind of *instrumental learning* in Chapter 15. Finally, you may habituate to all the violence or perhaps be sensitized by it. *Habituation* and *sensitization* are nonassociative (you are not learning about relationships), but they are forms of learning nonetheless.

Some scientists have tried to organize these diverse forms of learning and memory into a single, overarching scheme, but these efforts have remained controversial. Therefore, let us just recognize that learning and memory come in a variety of forms. As you will see in this chapter, the diverse forms of learning and memory are implemented by distinct but overlapping brain systems and can function independently of one another. This explains why some individuals excel at some forms of learning and memory (Box 14.1: Memory Specialists) while being average in other realms.

BRAIN EXERCISE

Which form(s) of learning and memory are you good at (compared to your peers) and which one(s) do you struggle with? How has this shaped your life?

14.2 What's Wrong with H.M.?

The most influential study on the neurobiology of learning and memory involved a patient known to most neuroscientists only by his initials, H.M. By the time he was 27, H.M. suffered more than 50 full-blown epileptic seizures per year and many smaller ones each day (Box 14.2: Epilepsy). Medication could not control them.

H.M.'s Amnesia

At that point, neurosurgeon William Scoville offered to remove the anterior medial portion of H.M.'s temporal lobes, including both the hippocampus and the amygdala bilaterally. The rationale for this treatment was that epileptic seizures often begin in the medial temporal lobe and then spread to the remaining brain. If the seizure's site of origin is gone, the seizures should abate. Scoville's proposal was daring, but similar procedures had already been used with some success in other epilepsy patients. Therefore, in the fall of 1953, H.M. had the anterior portion of his medial temporal lobes surgically removed (Figure 14.2). His seizure frequency indeed decreased, but it soon became apparent that the surgery also destroyed a crucial part of H.M.'s memory.

As Brenda Milner and Scoville reported in 1957, H.M. lost the ability to form new long-term memories. For example, H.M. only briefly remembered the names and faces of people he met after the surgery, forgetting them as soon as he was distracted. This **anterograde amnesia** extended across all sensory modalities and affected memories for faces, places, events, facts, and words. In H.M.'s words, this state was "like waking from a dream . . . every day is alone in itself." This sort of amnesia severely impairs life because it doesn't just impede the forming of new memories; it also prevents the formation and pursuit of plans for the future.

NEUROBIOLOGY IN DEPTH

Box 14.1 *Memory Specialists*

Some people have exceptional memories for certain types of information. One such memory specialist was described by Aleksandr Luria in *The Mind of a Mnemonist* (1968). This mnemonist, called Shereshevsky or simply S., could remember long lists of numbers or words, as well as nonsense words and complex mathematical formulas, after seeing or hearing them just once. Fifteen years after hearing several stanzas from Dante's epic poem *Divina Commedia,* S. was able to recite those stanzas "not only with perfect accuracy of recall, but with the exact stress and pronunciation" (Luria, 1968, p. 45). Part of what made this feat so impressive was that S. did not speak Italian, which means that the poem sounded like nonsense to him.

Shereshevsky's mnemonic ability was linked to synesthesia (see Chapter 12), which meant that words, numbers, and even nonsense syllables evoked vivid images that S. could then manipulate. In particular, he could arrange the images evoked by items in a list along an imaginary route. To recall the list, S. traversed the route again and noted what he saw along the way. By taking his mental walk in the opposite direction, he could recite the list in reverse order. One of his rare mistakes was omitting "egg" from a remembered list because he had placed his mental image of the white egg against a white wall. On his later mental walk, S. overlooked the egg. Thus, S. used the mental walks as a *mnemonic device*, a memory-aiding technique. Still, his ability was truly outstanding. He remembered hundreds of these lists, usually for many years.

It is tempting to be envious of S's incredible memory, but he was not a happy man. Before he became a professional mnemonist, S. changed jobs dozens of time, considering them all to be inferior to what he had imagined. Most importantly, the synesthesia made it difficult for S. to read a book or focus on conversations. Images of individual words continually arose in his mind, distracting him. Although S. had a wife and son, he perceived them "as through a haze."

A different kind of memory specialist was described by James McGaugh and his colleagues in 2006. They reported that A.J. (Jill Price) has *highly superior autobiographical memory.* In A.J.'s words

> My memory has ruled my life . . . There is no effort to it . . . It's like a running movie that never stops. It's like a split screen. I'll be talking to someone and seeing something else. . . . Like we're sitting here talking and I'm talking to you and in my head I'm thinking about something that happened to me in December 1982, December 17, 1982, it was a Friday, I started to work at Gs (a store). (Parker et al., 2006, p. 35)

In several memory tests, A.J. performed astoundingly. For example, she could recall the date of every Easter Sunday between 1980 and 2003, which is difficult because the timing of Easter varies from year to year. A.J. also recalled what she did on those days. Much of what AJ remembered about her life was verified from her extensive collection of diaries. However, in contrast to Luria's mnemonist, A.J. is not particularly good at memorizing lists. She freely admits to having struggled in school. Nor does A.J. record all her experiences in their totality (as a video camera might); she remembers only what is interesting to her. Still, A.J. does remember details from virtually every day of her life, going back at least to when she was 14.

The basis of A.J.'s superior autobiographical memory remains a mystery. She is compulsive about remembering her past experiences but does not memorize her diaries. A.J. is superb at matching dates and days of the week, but she differs from other calendar specialists (including some autistic savants) in that her ability is limited to the days of her life. She does not have the sort of synesthesia where words elicit images, but she probably has time-space synesthesia, which lets her think of dates as being arranged in space.

A substantial number of other people with highly superior autobiographical memory have now been identified (Figure b14.1). Working with this fascinating population, rather than just one or two subjects, will help scientists identify the mechanisms underlying their extraordinary abilities. Some of their brains are being scanned, but the results are still coming in and difficult to interpret.

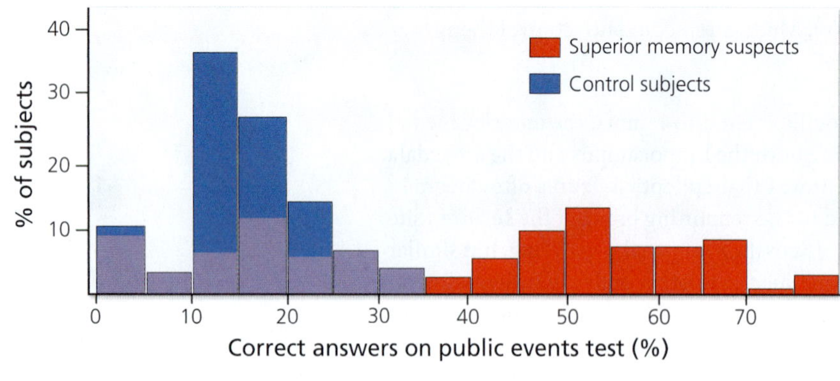

Figure b14.1 Screening humans for superior autobiographical memory. To identify people with superior autobiographical memory, researchers gave potential subjects (who claimed to have superior autobiographical memory) a public events test with 30 questions such as "What year, date, and day of the week was John Lennon killed?" Control subjects can typically answer 10–25% of these questions (blue bars). In contrast, people who think they have superior autobiographical memory exhibit a bimodal distribution of test scores: roughly half score like control subjects; the others score significantly better than controls. [After LePort et al., 2012]

NEUROLOGICAL DISORDERS

Box 14.2 *Epilepsy*

Approximately 0.5–1% of the people in North America have some form of epilepsy, which is characterized by repeated, spontaneous *epileptic seizures*. Seizures, in turn, are defined as episodes of highly synchronized, excessive neuronal activity. Roughly half of all epilepsy patients have seizures that begin in a specific brain region, especially the medial temporal lobe, and then spread to other brain regions. In other patients, the seizures appear simultaneously in many different brain regions. The behavioral symptoms of epileptic seizures range from whole-body convulsions to tremors to blank stares. When a seizure spreads, the behavioral symptoms vary systematically as seizure activity invades new brain regions. In many epileptic seizures, consciousness is eventually lost and memory of the event impaired. Young children often suffer fever-induced seizures, but this does not mean that they have epilepsy. Still, some evidence suggests that febrile (fever-induced) seizures increase the risk of developing epilepsy later in life. Many forms of childhood epilepsy resolve with age.

Unfortunately, this is not the case for *medial temporal epilepsy*, which is what H.M. had; this form of epilepsy usually worsens over time. Medial temporal epilepsy is thought to be caused by some sort of traumatic trigger, such as a severe concussion, a stroke, or an intense febrile seizure. Weeks after this initial trigger, the hippocampus and adjacent cortices (usually on just one side of the brain) become increasingly prone to spontaneous seizures. The seizures themselves are thought to arise from an imbalance of excitation and inhibition. According to one hypothesis, the imbalance is created by the death of tonically active neurons that normally excite inhibitory neurons in the dentate gyrus. Without this tonic excitation, the inhibitory neurons fall silent, causing the rest of the hippocampus to become overly excitable. Other evidence suggests that the initial trauma causes dentate granule cells in the hippocampus to sprout abnormal connections that create an excitatory intrahippocampal feedback loop. One way or the other, hippocampal neurons receive too much excitatory input, which then leads to cell death (excitotoxicity). With every seizure, more neurons die. According to this hypothesis, medial temporal epilepsy is a progressive degenerative disease. This idea remains controversial but is supported by some evidence (Figure b14.2).

The early stages of medial temporal epilepsy can usually be managed pharmacologically. Eventually, however, medial temporal lobe epilepsy tends to become intractable and must be treated by removing the hippocampus and adjacent cortices on the side of the brain where the seizure begins. To identify a seizure's site of origin (its focus), neurosurgeons sometimes implant recording electrodes in the temporal lobes of epileptic patients, hoping to catch a seizure's onset "in the act." Fortunately, surgical removal of the hippocampus and adjacent cortices virtually eliminates seizures in 80% of the patients. Because the surgery is unilateral, the effects on memory tend to be much less severe than they were in H.M.

Many forms of epilepsy have a genetic component. This is evident from the preponderance of epilepsy in some (rare) families. Furthermore, if one monozygotic twin has epilepsy, the chance of the other twin having it, too, is 40–50%. In contrast, the concordance rate for dizygotic twins is 10–15%. Based in part on the study of such epilepsy-prone families and twins, researchers have identified a variety of genetic mutations that are associated with epilepsy. Many of these mutations affect voltage-gated sodium and potassium channels; others change GABA receptors. It is easy to see how these mutations might alter the balance between neuronal excitation and inhibition. However, other epilepsy-associated genes are less obviously linked to neuronal excitability.

A general principle emerging from this genetic work is that a specific mutation can be associated with several different forms of epilepsy and that mutations in different genes are linked to seemingly identical epilepsies. This makes it difficult to establish causal linkages between specific mutations and specific epilepsy symptoms. Most likely, many epilepsy-associated mutations do not cause epilepsy directly but increase a person's susceptibility to the disease.

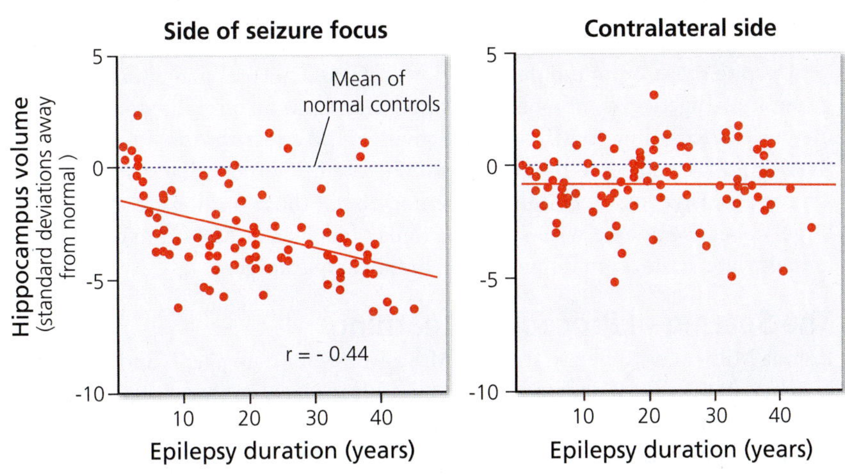

Figure b14.2 Hippocampus shrinkage in epilepsy. MRI scans were used to measure hippocampus volume in patients after their medial temporal epilepsy became intractable. The graphs reveal that hippocampus volume decreases with epilepsy duration (years since diagnosis) on the side of the seizure focus but not on the contralateral side. [After Bernasconi et al., 2005]

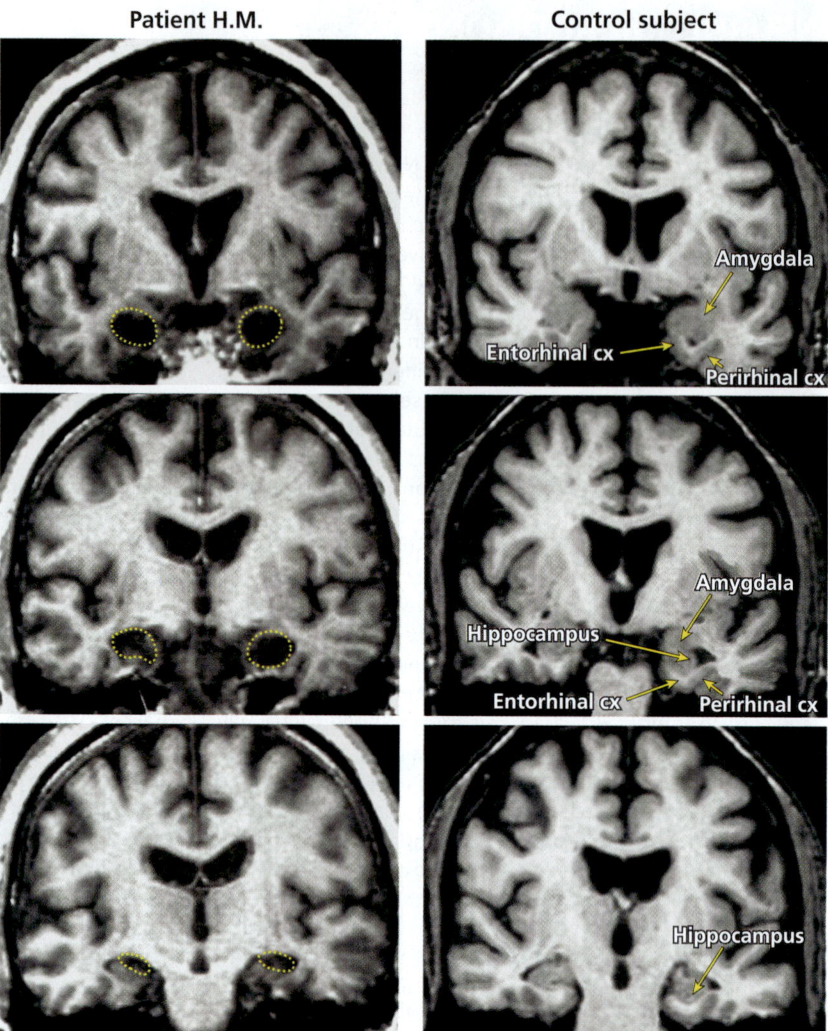

Figure 14.2 Patient H.M.'s brain. Shown along the left are three coronal MRI scans of H.M.'s brain taken 40 years after his surgery. Shown at the right are corresponding scans from an age-matched control subject. The damaged area in H.M.'s brain is outlined by the dashed yellow line. It includes the amygdala and most of the hippocampus, as well as the entorhinal cortex and portions of the perirhinal cortex. [From Corkin et al., 1997]

Despite these profound deficits, H.M. retained normal perceptual abilities and general intelligence as measured by an IQ test. H.M also retained many of his old memories. According to Milner and Scoville, H.M's **retrograde amnesia** was relatively mild, extending to roughly 3 years before his surgery. Later studies revealed a more extensive loss of childhood memories, but those findings remain contested, largely because people with severe epilepsy (such as H.M.) often develop severe amnesia late in life, even if they had no hippocampal surgery.

The Sparing of Procedural Learning

Brenda Milner continued to study H.M.'s amnesia for many years and in 1962, made another significant discovery: H.M. could learn new motor skills. For example, H.M.'s performance on a mirror tracing task (in which he had to trace a drawing that was visible only in a mirror) improved over the course of several days, even though H.M. reported no memory of having done the task before. Similarly, H.M.'s ability to track a moving spot with his hand improved over the course of several training sessions (Figure 14.3), even though he could not remember across training sessions that he had ever done the task before. H.M. could also learn new cognitive skills. For example, he got better with practice at recognizing objects from incomplete, fragmented drawings. Thus, H.M.'s brain lesion destroyed his ability to form new long-term episodic memories but largely spared his ability to learn new perceptual,

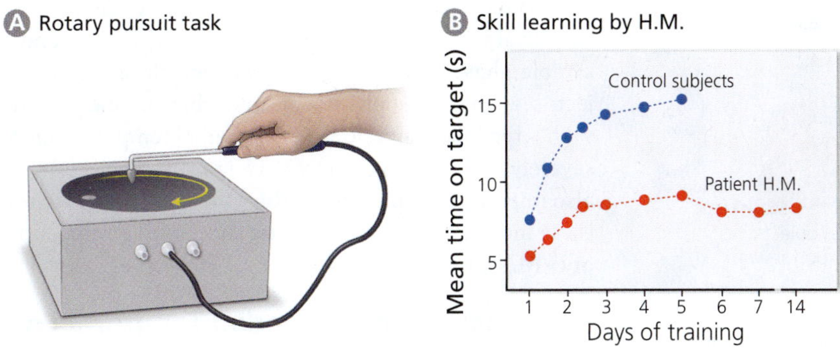

Figure 14.3 H.M. could learn a motor skill. In the rotary pursuit task, a subject must keep a probe on a target as it moves in a circle (A). Control subjects rapidly improve their performance on this task (B). H.M. also improved, although his performance plateaued after about 4 training sessions. Curiously, he could never remember having performed the task before. [After Corkin, 1968]

cognitive, and motor skills. As the neuroscientists put it, he remained capable of procedural learning.

This finding came as a significant surprise. Although the existence of multiple memory types had long been hypothesized, it had not been clear that the neural mechanisms underlying those forms of memory are segregated in the brain such that taking out one set of brain regions can interfere selectively with a subset of learning types. H.M.'s case also proved that learning and memory can be uncoupled from basic perception and intelligence. Again, this had not been so clear before. Finally, the study of H.M. revealed that some important types of memories can be affected by lesions that are largely limited to the hippocampus and amygdala. Earlier studies had focused on the neocortex as being the principal "seat" of memory.

In short, H.M.'s amnesia profoundly altered how neurobiologists think about learning and memory. H.M died in 2008. At that point, his full name was revealed to be Henry Gustave Molaison. He was not a neurobiologist, but he inspired many of them.

BRAIN EXERCISE

If you met an amnesiac resembling H.M., what would you want to know about his or her brain and memories?

14.3 Can H.M.'s Amnesia Be Reproduced in Non-humans?

After learning of H.M.'s amnesia, many neurobiologists attempted to create a similar amnesia in non-humans where more refined studies are possible. In particular, they sought to determine whether H.M.'s amnesia was caused by destruction of the hippocampus or the amygdala, both of which were damaged in H.M. This seemed like an easy question to answer, but monkeys and rats with large bilateral lesions of the hippocampus and amygdala showed only minor deficits in standard learning tasks. For example, the lesioned animals could learn which objects were associated with rewards and which were not. This was puzzling, until researchers realized that the standard learning tasks did not require the sort of memories that H.M. was unable to form. Specifically, the lesioned animals might learn object-reward associations over the course of many trials without remembering explicitly that they had encountered these objects before.

Neurobiologists therefore devised new learning tasks. The most famous of these is the **delayed non-match to sample task (DNMTS)**. Despite the intimidating name, the task is straightforward (Figure 14.4). During a sample phase, monkeys are shown a single object. After a variable delay, the monkeys are shown the same object again, as well as a novel object. Over the course of many trials, the monkeys are trained always to lift up the novel object (the one that does not match the sample) and retrieve food hidden underneath. Because each trial involves a different set of objects,

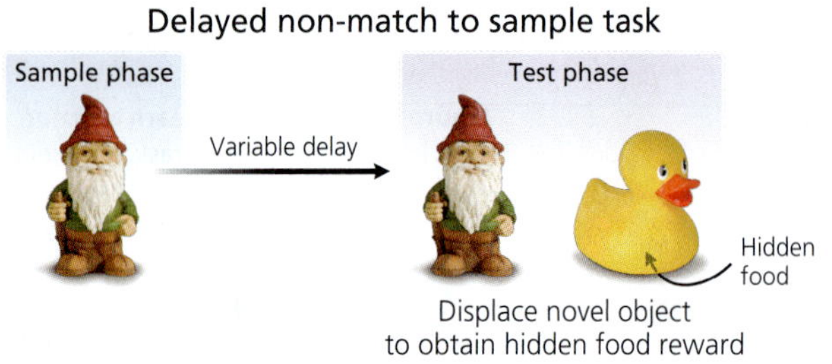

Delayed non-match to sample task

Sample phase

Variable delay

Test phase

Hidden food

Displace novel object to obtain hidden food reward

Figure 14.4 A recognition memory test. The delayed non-match to sample task requires animals to displace an object that they did not see during an earlier "sample phase" of the experiment. Each trial uses a new pair of objects. After extensive training, monkeys will learn that food is always hidden under the unfamiliar object.

the monkeys can solve the task only if they remember on each trial which object they saw during the corresponding sample phase. They must recognize which of the two objects seen in the test phase is old and which is new. Monkeys with large bilateral lesions of the medial temporal lobe are severely impaired in the DNMTS task. Because H.M. was also impaired in this kind of **object recognition memory**, these monkeys provided the first successful animal model of H.M.'s amnesia.

Subdivisions of the Medial Temporal Lobe

Having established that monkeys with large lesions of the medial temporal lobe have a memory impairment similar to that of H.M., researchers set out to determine which divisions of the medial temporal lobe are causing the deficit. To that end, they selectively lesioned the hippocampus and the amygdala, sparing the cortical areas adjacent to the hippocampus. These areas include the **entorhinal cortex** and the **perirhinal cortex** (Figure 14.5) as well as the posterior extension of the perirhinal cortex, which is called the *parahippocampal cortex* in primates and the *postrhinal cortex* in rats. Substantial portions of these areas had been damaged in H.M. Would bilateral lesions that spared these cortices, but took out both the hippocampus and the amygdala, impair a monkey's ability to perform the DNMTS?

No, they did not! Monkeys with large bilateral lesions of the amygdala and the hippocampus (including the dentate gyrus, the CA fields, and the subiculum) performed the DNMTS task as well as intact monkeys did (Figure 14.6). In contrast, monkeys with lesions limited to the perirhinal and entorhinal cortices were severely impaired, at least at long delays between sample presentation and memory testing. These findings were startling because most neuroscientists had assumed that H.M.'s amnesia was caused primarily by damage to the hippocampus. However, studies in rats soon confirmed the results.

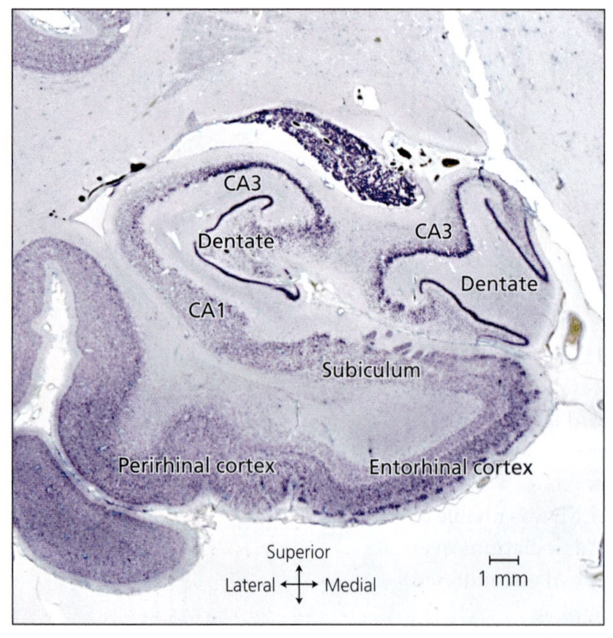

Figure 14.5 The hippocampus and its neighbors. In this Nissl-stained coronal section through the left hippocampal region of an adult human, you can see the hippocampus (dentate, CA1, CA3, and subiculum) as well as the adjacent cortices (entorhinal and perirhinal). The dentate and CA3 are bent in such a way that this section cuts through each of them twice. [Courtesy of the Yakovlev-Haleem Collection]

Object Recognition Tests in Rats

For reasons that are not entirely clear, rats have a harder time in the DNMTS task than monkeys do. However, researchers in the late 1980s discovered a much simpler test for object recognition memory in rats. First, an experimenter places two identical objects into a rat's enclosure and lets the rat explore them both for several minutes. The objects are then removed. After a variable delay, one of the old objects is placed back into the cage, together with a novel object. When this is done (controlling carefully for object location), the rats preferentially explore the novel object, ignoring the familiar one. Importantly, the rats need not be trained to exhibit this preference; they reveal it spontaneously.

Experiments using this **spontaneous novel object recognition task** have shown that rats with lesions limited to the hippocampus still exhibit a reliable preference for novel, rather than familiar, objects. In contrast, rats with lesions that extend across both the perirhinal and postrhinal cortex, but leave the hippocampus untouched, show little of this preference (Figure 14.7). These data indicate that the peri- and postrhinal cortices, but not the hippocampus, are needed for object recognition. This hypothesis remains controversial because some studies do show object recognition deficits in monkeys with selective

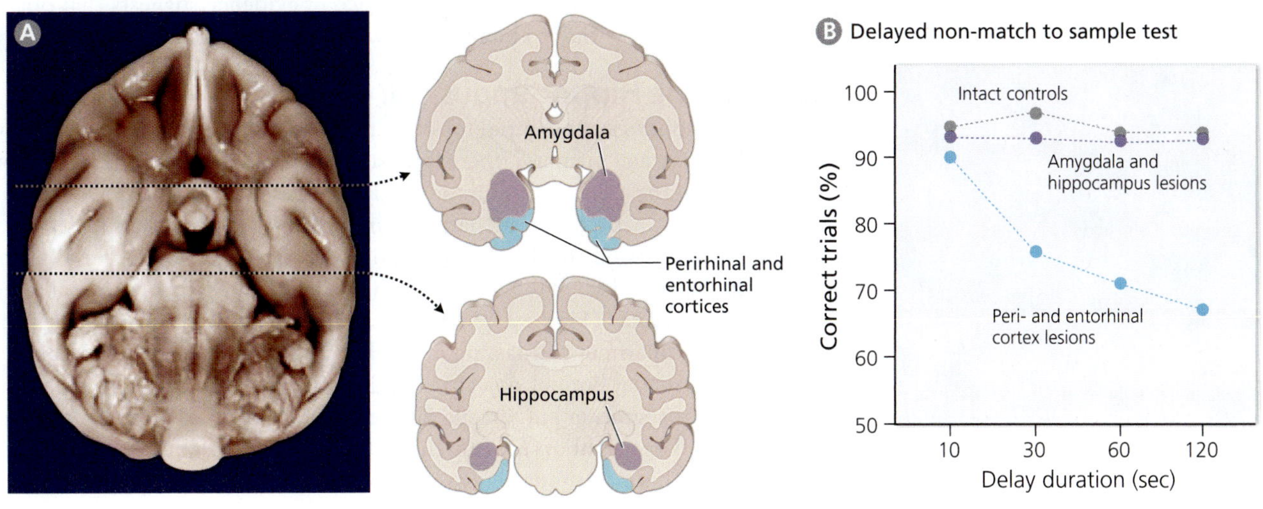

Figure 14.6 The importance of the peri- and entorhinal cortices. Monkeys with large bilateral lesions of both the amygdala and the hippocampus (A) were unimpaired in the delayed non-match to sample task (B). In contrast, combined lesions of the perirhinal and entorhinal cortices created a profound impairment, as long as the interval between sample presentation and testing was longer than 10 seconds. [After Murray and Mishkin, 1998, and Meunier et al., 1993; brain image from brainmuseum.org]

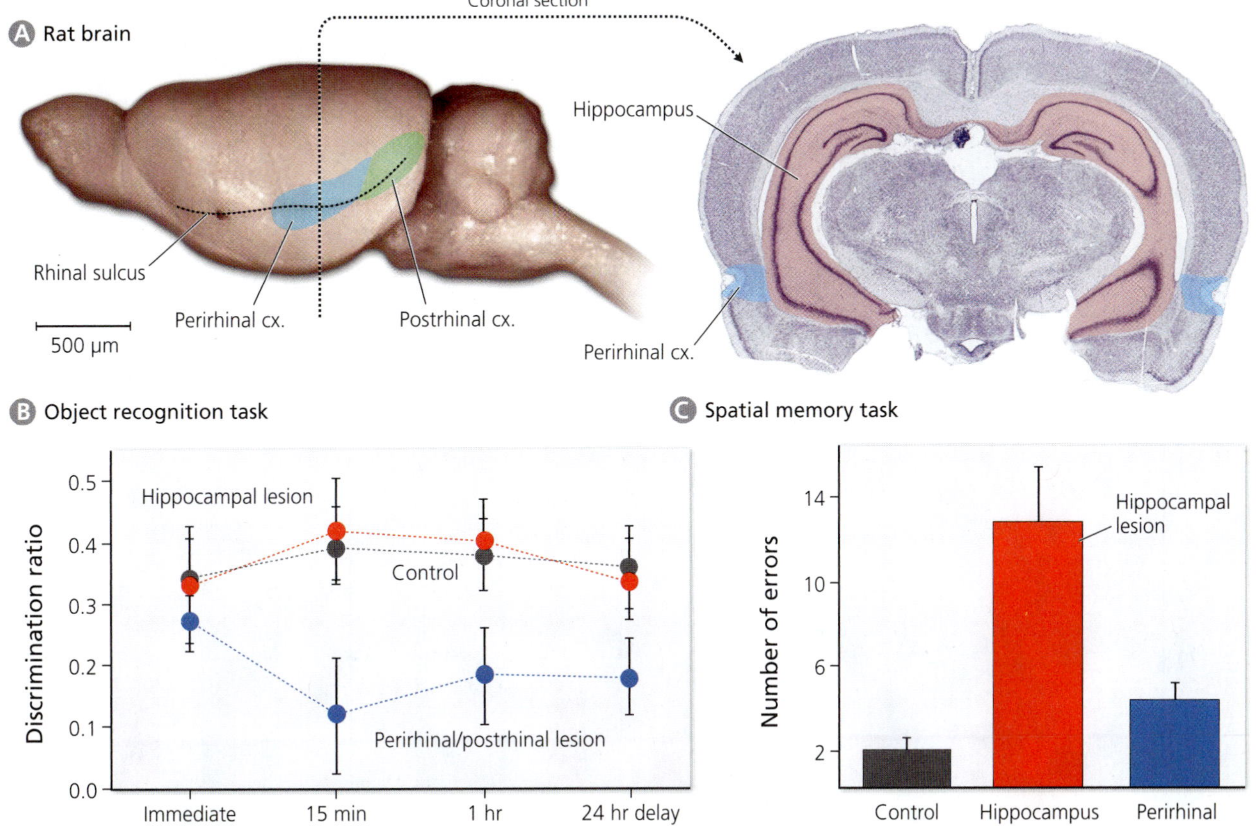

Figure 14.7 Double dissociation of hippocampus and perirhinal cortex functions. Panel (A) illustrates the locations of the hippocampus, perirhinal cortex, and postrhinal cortex in a rat brain. The graph in (B) shows that lesions of perirhinal plus postrhinal cortex impair performance on the spontaneous object recognition task, whereas hippocampus lesions do not. Panel (C) shows that hippocampus lesions, but not perirhinal cortex lesions, impair performance on a spatial memory task (in which rats must remember which arms of an 8-arm radial maze they had already visited). [After Winters et al., 2004; brain image from brainmuseum.org; brain section from brainmaps.org]

hippocampal damage. However, a preponderance of evidence suggests that object recognition is possible without a hippocampus.

What Does the Hippocampus Do?

If the hippocampus is not essential for object recognition, then what is its major function in learning and memory? There is no simple answer to this question, but you already know that the hippocampus is required for learning to construct a "cognitive map" of space (see Chapter 11). Furthermore, in the study we just discussed, the rats with large bilateral lesions of the hippocampus were severely impaired in a spatial memory task that asked them to remember which arms of an eight-arm radial maze they have already visited (Figure 14.7 C). In contrast, the rats with perirhinal lesions showed no such impairment. This kind of **double dissociation** (with hippocampus lesions affecting spatial but not object recognition memory, and perirhinal lesions affecting object recognition but not spatial memory) is powerful evidence that the hippocampus and its neighbors are functionally distinct. In particular, the evidence suggests that the hippocampus plays some sort of special role in spatial memory (at least in rats).

However, the hippocampus is clearly concerned with more than just spatial relationships. For example, humans with damage limited to the hippocampus (usually because of temporary hypoxia, such as that induced by near drowning) cannot recall a simple story after a ten-minute delay, and they have trouble copying complex drawings from memory. In rats as well, the hippocampus is involved in more than spatial

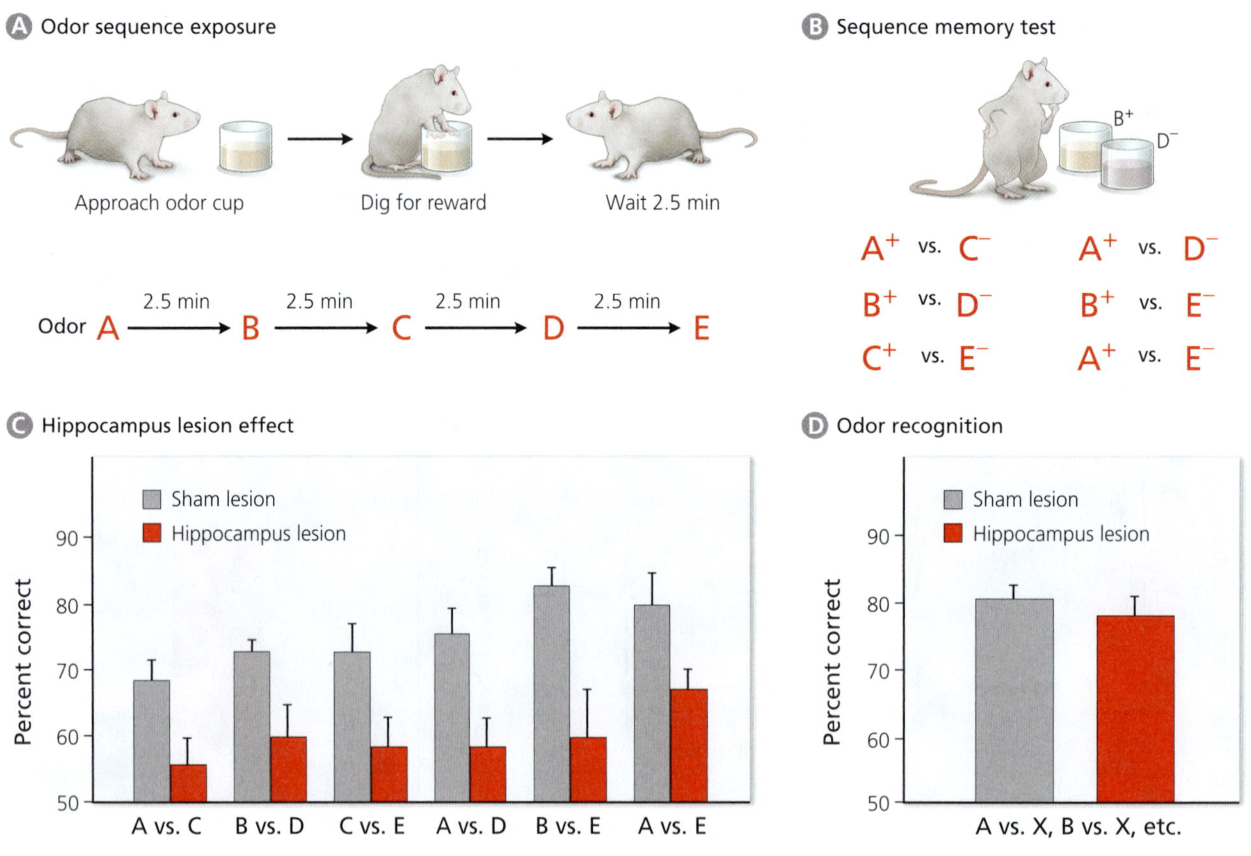

Fig. 14.8 Hippocampus lesions impair sequence memory. Rats were trained to dig for rewards in cups filled with scented sand. They were then exposed to a sequence of different odors (A). To test for sequence memory (B), the rats were presented with two cups and trained to dig for a reward in the cup that had been presented earlier in the sequence (+ symbols mark the rewarded odor in each pair). Bilateral hippocampus lesions impair performance on this odor sequence memory test (C). In contrast, hippocampus lesions do not impair performance on a simple odor recognition test in which rats are trained to select a novel odor (D). [After Fortin et al., 2002]

memory. For instance, selective hippocampus lesions impair a rat's ability to remember the sequence in which various odorants were presented (Figure 14.8).

Grappling with these diverse findings, researchers have proposed that the perirhinal and postrhinal cortices may be sufficient for forming memories of individual objects, but that the hippocampus, in both humans and non-humans, is needed to connect those objects to one another, through space and across time. In other words, the hippocampus is involved in learning about relationships, be they spatial, temporal, or even logical. This **relational memory hypothesis** would explain why the hippocampus is involved in both spatial and episodic memories. After all, both forms of memory entail complex relationships. Although this hypothesis is widely accepted, not everyone agrees on what, specifically, the hippocampus "does." Part of the problem is that it appears to have a variety of different functions, many of which are intertwined (see also Chapter 3, Box 3.2: Hippocampal Structure and Functions).

BRAIN EXERCISE

Milner and Scoville had argued that H.M.'s amnesia was due to hippocampal, rather than amygdalar, damage. Were they correct?

14.4 How Are Hippocampus-dependent Memories Created, and How Are They Recalled?

It is satisfying to know that the hippocampus is required for relational memory, but how are these memories formed, and how are they later recalled? These questions require answers at several levels of analysis.

At the cellular and molecular levels, memory formation generally involves changes in synaptic strength and, probably, the growth of some new connections. We discussed these mechanisms back in Chapter 3. Particularly important is the phenomenon of *long term potentiation (LTP)*, defined as the persistent growth or strengthening of synapses that were active just before or during postsynaptic depolarization (a.k.a. Hebbian LTP). As you may recall, LTP is thought to involve the activation of NMDA receptors and various intracellular mechanisms that enhance synaptic transmission. After LTP has been induced, a variety of additional molecular mechanisms, some of which require protein synthesis, stabilize the changes in synaptic strength.

Although much has been learned about the molecular mechanisms underlying changes in synaptic strength, this information is not sufficient to explain how memories are formed and how they are recalled. We also need to understand which synapses are strengthened in which brain regions and how those changes alter the interactions between neurons in those areas. That is, we have to think about neural circuits and systems. The following sections will introduce you to the "standard model" of memory formation and recall. It is centered on the hippocampus and its interactions with the neocortex.

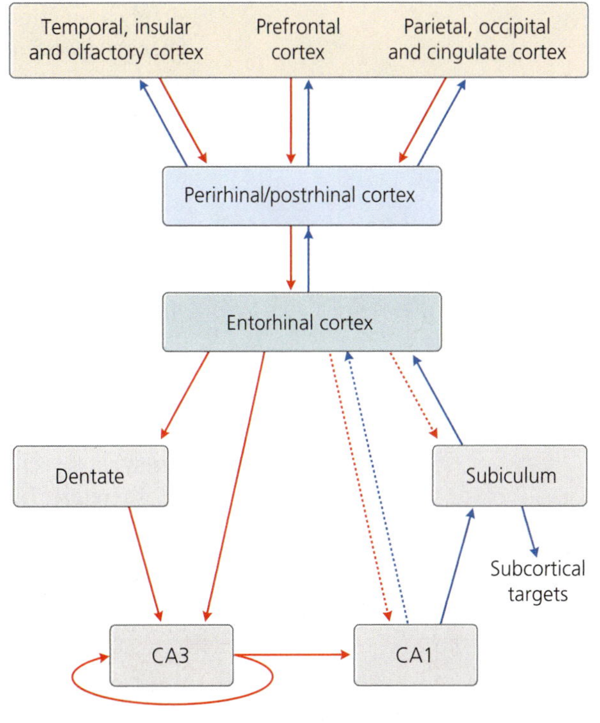

Figure 14.9 Pathways through the hippocampus. The hippocampus consists of the dentate gyrus, CA1, CA3, and the subiculum (see Figure 14.5). Information generally flows from higher order neocortical areas through the perirhinal/postrhinal and entorhinal cortices into the hippocampus and back out. Minor connections are depicted as dashed lines. Even so, the diagram is highly simplified.

Hippocampal Circuits and Synaptic Plasticity

The principal circuits connecting the hippocampus to the neocortex are shown in Figure 14.9. As you can see, various neocortical areas are reciprocally connected with the perirhinal and postrhinal cortices, which are reciprocally

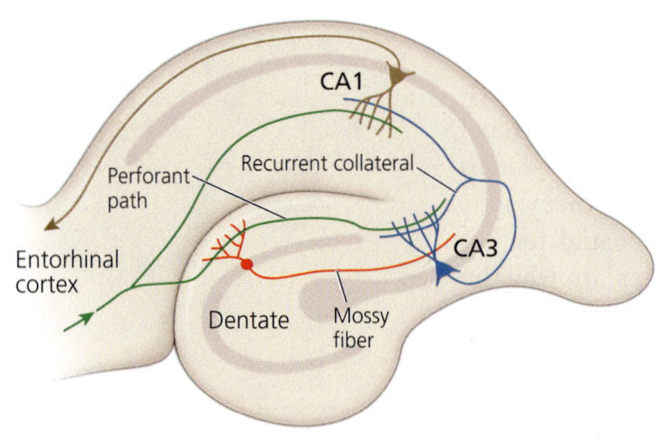

Figure 14.10 Intrahippocampal circuitry. The major pathways through the hippocampus are illustrated here in a schematic coronal section. The projection from the entorhinal cortex to neurons in the dentate gyrus and CA3 is called the perforant path because it supposedly "perforates" the boundary between the dentate and the entorhinal cortex. The mossy fiber pathway connects the dentate gyrus to CA3. The neurons of CA3 project to CA1 and, via recurrent collaterals, to other CA3 neurons. [After Treves and Rolls, 1992]

connected with the entorhinal cortex. Neurons in the entorhinal cortex then project into the hippocampus. In particular, they project to the **dentate gyrus** (Figure 14.10), whose output cells project to one of the CA fields (CA stands for *cornu ammonis*, meaning Ammon's horn). Specifically, they project to **CA3**, which projects to **CA1**. The CA1 neurons in turn project to the **subiculum**, which lies sandwiched between CA3 and the entorhinal cortex. The subiculum, finally, projects out of the hippocampus to the entorhinal cortex and to several subcortical targets, including the amygdala and hypothalamus.

An important aspect of this circuitry is that the connections within the hippocampus tend to be unilinear rather than reciprocal. Thus, information tends to flow linearly through the hippocampus, from dentate to CA3 to CA1 to the subiculum (Figures 14.9 and 14.10). Furthermore, the entorhinal neurons that project into the hippocampus (to the dentate) are distinct from those that receive projections back from the subiculum. Therefore, information tends to flow from various neocortical areas through the entorhinal cortex into the hippocampus, traces a looping circuit through the hippocampus, and then flows, at least potentially, back out to the neocortex. This scheme is somewhat simplified, as there are several "shortcuts" across the intrahippocampal loop and connections from CA3 back to itself (more on these shortly). However, the basic scheme of unilinear information flow through the hippocampus is compelling.

Many of the synapses within the hippocampal circuitry are malleable. In the mossy fiber pathway, which connects the dentate gyrus to CA3, synaptic plasticity is simply a function of presynaptic activity. That is, repeated activation makes mossy fiber synapses more powerful for several hours, but this boost in synaptic strength is not dependent on postsynaptic depolarization and, therefore, not LTP as we defined it in Chapter 3. However, LTP (following Hebb's rule) is observed at synapses connecting CA3 to CA1 as well as at the synapses connecting CA3 neurons to one another through recurrent axon collaterals (Figure 14.9).

Because the synapses that CA3 neurons make with one another play an important role in memory formation and recall, let us consider their plasticity in more detail. As shown in Figure 14.11, synapses onto CA3 neurons can be strengthened by applying a strong, high-frequency stimulus (a long tetanic stimulus) to the axons of other CA3 neurons. Such strong stimulation is sufficient to depolarize the postsynaptic cell to the point where it fires some action potentials. With weaker stimuli, the postsynaptic cell is depolarized much less, and LTP is not observed. However, if weak stimulation is coupled with intracellular current injections that depolarize the postsynaptic cell, then LTP is evident (Figure 14.11 B).

Therefore, LTP at the synapses that CA3 neurons make on other CA3 neurons obeys Hebb's postulate: "When an axon of cell A is near enough to excite a cell B and repeatedly or persistently takes part in firing it, some growth process or metabolic change takes place in one or both cells such that A's efficiency, as one of the cells firing B, is increased" (Hebb, 1949, p. 62; see Chapter 3). Put more simply, the pre- and postsynaptic cells become more tightly "wired together" because they are strongly depolarized (and hence likely to fire) at the same time (or nearly so). When such Hebbian plasticity is embedded in the right kind of neural circuitry, neurons become capable of storing information about their own patterns of activity. How this might work shall occupy us in the next section.

Pattern Learning within the Hippocampus

As mentioned earlier, CA3 neurons have recurrent collaterals that project to other CA3 neurons. Because this recurrent projection is dense, with each neuron connecting

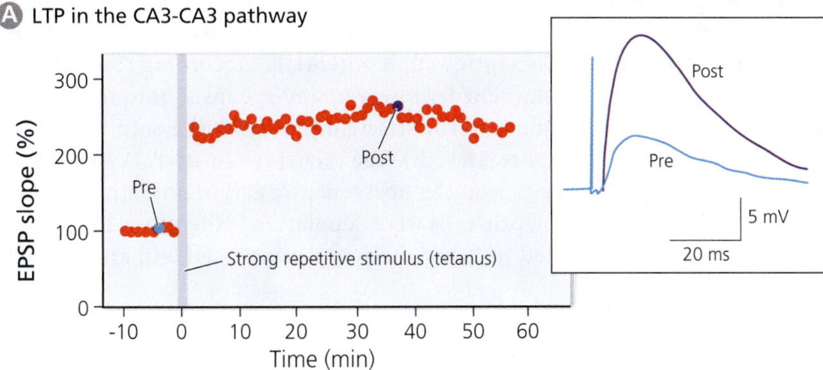

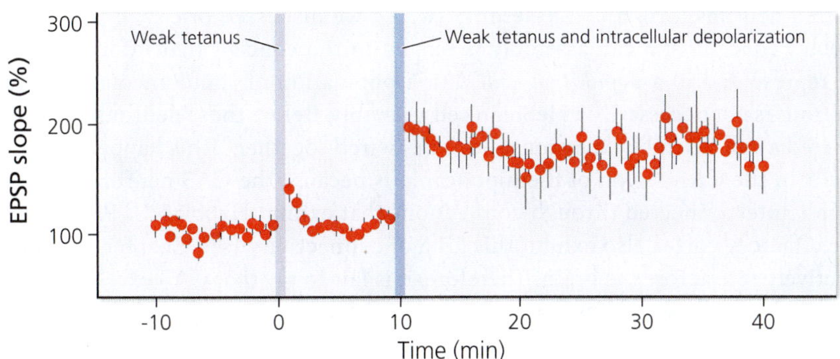

Figure 14.11 LTP in CA3 requires postsynaptic depolarization. An intracellular electrode was used to record from individual CA3 neurons while electrical stimuli were applied to axons coming from other CA3 neurons. Strong, high-frequency (tetanic) stimulation potentiates the activated synapses for at least 1 hour (A). Specifically, it increases the amplitude and rising slope of the EPSP (excitatory postsynaptic potential) recorded in response to individual test stimuli. The graph in (B) shows that weak tetanic stimulation does not elicit LTP (long-term potentiation), unless it is coupled with intracellular depolarization of the postsynaptic cell. [After Zalutsky and Nicoll, 1990]

to a substantial fraction of all other CA3 neurons, CA3 is said to form an **autoassociative network** (*auto* meaning "self" in Latin). In addition, CA3 neurons receive strong inputs from the entorhinal cortex, both directly and indirectly by way of the dentate gyrus.

To see how LTP makes it possible for such a network to store information about its own activity, consider what happens when a set of active entorhinal neurons triggers action potentials in a set of CA3 neurons (Figure 14.12). The action potentials

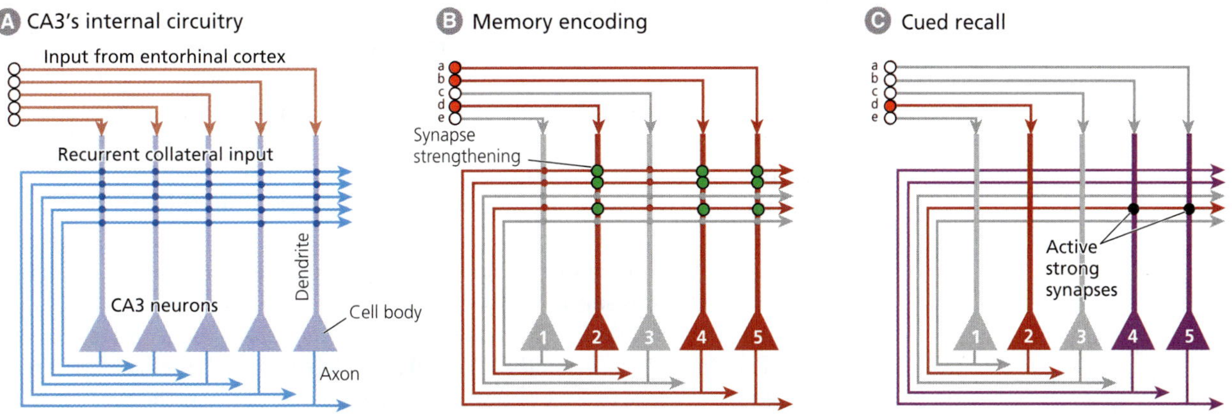

Figure 14.12 Hebbian memory storage and recall. Panel (A) depicts five CA3 neurons with their recurrent collaterals, which are here shown as synapsing weakly onto all other CA3 neurons. The neurons also receive inputs from the entorhinal cortex. Panel (B) illustrates a hypothetical situation in which three entorhinal neurons fire action potentials and trigger action potentials in three CA3 neurons (numbered 2, 4, and 5). The green circles mark synapses that are strengthened according to Hebb's rule. Panel (C) shows what happens when this network is later activated by a subset of the original entorhinal input. Initially, the partial input activates only neuron 2, but this neuron then activates neurons 4 and 5 (through previously strengthened synapses; black circles). Thus, partial input can activate all of the CA3 neurons that had been activated by the original, complete input.

rapidly propagate along the recurrent collaterals of the CA3 neurons and initiate transmitter release at all their synapses. Most of these synapses are too weak, even in aggregate, to trigger postsynaptic action potentials. According to Hebb's postulate, they will not be strengthened. However, any synapses onto neurons that fire action potentials in response to the strong inputs from the entorhinal cortex will increase in strength (Figure 14.12 B). The recurrent collateral synapses onto these neurons may not have caused the postsynaptic action potentials, but they were active while the postsynaptic cells were depolarized. Therefore, they do obey Hebb's rule; they participated in firing the postsynaptic cell and are, therefore, strengthened.

Hebbian Cell Assemblies in CA3

The result of all this synapse strengthening is that the activated CA3 neurons become more strongly connected to one another. Eventually all the activated CA3 neurons become capable of activating each other. To use Donald Hebb's term, these interconnected neurons form a **cell assembly** (which we discussed briefly in Chapter 12). Hebb's concept of a cell assembly is related to the concept that information may be represented in a *population code,* but a population of simultaneously activated neurons is not necessarily a Hebbian cell assembly. Before those neurons can be considered a cell assembly, they must become "wired together." This happens relatively easily in the CA3 region of the hippocampus because the CA3 neurons tend to be highly interconnected through connections that exhibit Hebbian LTP. Neurons in the olfactory cortex also exhibit this kind of connectivity (see Chapter 12), but it is relatively rare across the brain. Therefore, it is fair to say that CA3 excels at forming the kinds of cell assemblies that Hebb had envisioned.

Each cell assembly, once formed, is a potential *engram,* or memory trace. Consider what happens when only a subset of the neurons in a fully formed cell assembly are activated by some external inputs (Figure 14.12 C). As a result of prior LTP, the activated CA3 neurons have become strongly connected to the other neurons in the same assembly. Through these previously strengthened connections, neural activity spreads from the initial subset of active neurons to the entire assembly. This process is called **pattern completion.** It is analogous to cued recall in memory in which a partial cue can trigger the recall of an entire, complex memory. Indeed, given the importance of the hippocampus in memory for relationships among multiple stimuli (or stimulus objects), pattern completion in CA3 is probably part of the mechanism that enables the recall of memories that depend on an intact hippocampus.

Learning and Remembering Temporal Sequences

So far we have discussed cell assemblies as storing information about an instantaneous pattern of neuronal activity. However, cell assemblies can also encode temporal patterns of neuronal activity. The key notion here is that LTP preferentially strengthens not only synapses that were active simultaneously with a postsynaptic action potential but also synapses that were active a few milliseconds earlier. In contrast, synapses that become active after the postsynaptic spike remain unchanged or are weakened. This temporal asymmetry in LTP (spike timing-dependent plasticity; see Chapter 3) makes **sequence learning** possible.

Consider, for example, two weakly interconnected neurons with Hebbian plasticity. If neuron A is consistently active 10 ms before neuron B fires an action potential, then the synapse from A to B will be strengthened, whereas the synapse from B to A will remain weak (or become weaker). Afterward, activation of A can trigger a spike in B, assuming the LTP was strong enough. In contrast, activation of cell B cannot trigger a spike in A. Therefore, our 2-neuron network has stored information about the temporal sequence of the original firing pattern (namely, that A fired before B). Simply put, autoassociative networks with spike timing-dependent LTP are capable of storing sequences.

Keeping Assemblies Apart

As you might suspect, the ideas we just discussed are simplified. A more detailed model would require, for example, inhibitory neurons that can dampen overall network activity.

The most serious problem with our simple model is that different cell assemblies, representing different experiences, must remain segregated from one another. Although individual neurons may participate in multiple cell assemblies, the overlap between different assemblies must be small. Otherwise the assemblies may fuse and become incapable of representing separate memories. The best way of solving this problem is to ensure that even similar experiences activate different sets of CA3 neurons from the outset, during the initial encoding of the memory.

The mechanisms underlying this **pattern separation** in the hippocampus remain hotly debated. However, the dentate gyrus is thought to play a major role. Specifically, it is thought that the projections from the neocortex to the dentate are nontopographic and designed to ensure that even very similar patterns of neocortical activity cause relatively random, nonoverlapping sets of neurons in the dentate to fire. These distinct sets of dentate neurons then influence the CA3 neurons in such a way that they, too, respond with very different patterns of activity to similar patterns of neocortical input. You can think of the dentate gyrus as helping to "pull apart" the cell assemblies in CA3. Evidence for this general hypothesis comes, for example, from the observation that blocking NMDA receptor function in the dentate gyrus makes it more difficult for rats to discriminate between two slightly different environments.

Memory Recall

If memories are stored in CA3 of the hippocampus, then what happens during memory recall? Is the reactivation of a cell assembly in CA3 sufficient to generate a full-blown memory experience? Although this hypothesis is plausible, the evidence suggests that neurons outside of the hippocampus are also involved in memory recall.

Neocortical Involvement in Memory Recall

One way to examine which neurons are involved in memory recall is to record from individual neurons while subjects are actively experiencing a memory. This approach is technically challenging, but experimenters have implanted electrodes in the brains of a few people with intractable epilepsy. The main purpose of these electrodes was to localize where in the brain the seizures start so that their site of origin can be surgically removed. Because seizures often start in the medial temporal lobe, many of the implanted electrodes were located there. With the electrodes already in place, the experimenters decided to pursue some exciting basic research, namely to search for neural correlates of memory recall (of course, the patients consented).

One particularly striking finding was that a few neurons in the human entorhinal cortex respond selectively to specific video clips, both when those clips are displayed on a screen and when those clips are later remembered (Figure 14.13). Such recall-activated neurons have also been observed in the hippocampus, but their presence in the entorhinal cortex implies that memory recall involves at least some neocortical neurons.

Further support for neocortical involvement in memory recall comes from functional brain imaging. In one influential study, subjects were trained to associate images or sounds with printed words on a screen. One day later, the subjects were again presented with the images or sounds and their associated words. Then they were asked to remember for each word whether it had been paired with a sound or an image. At the end of the experiment, the researchers compared fMRI scans taken from each subject during the perception and recall phases of the experiment. They discovered that recalling images activates a subset of the neocortical areas that were

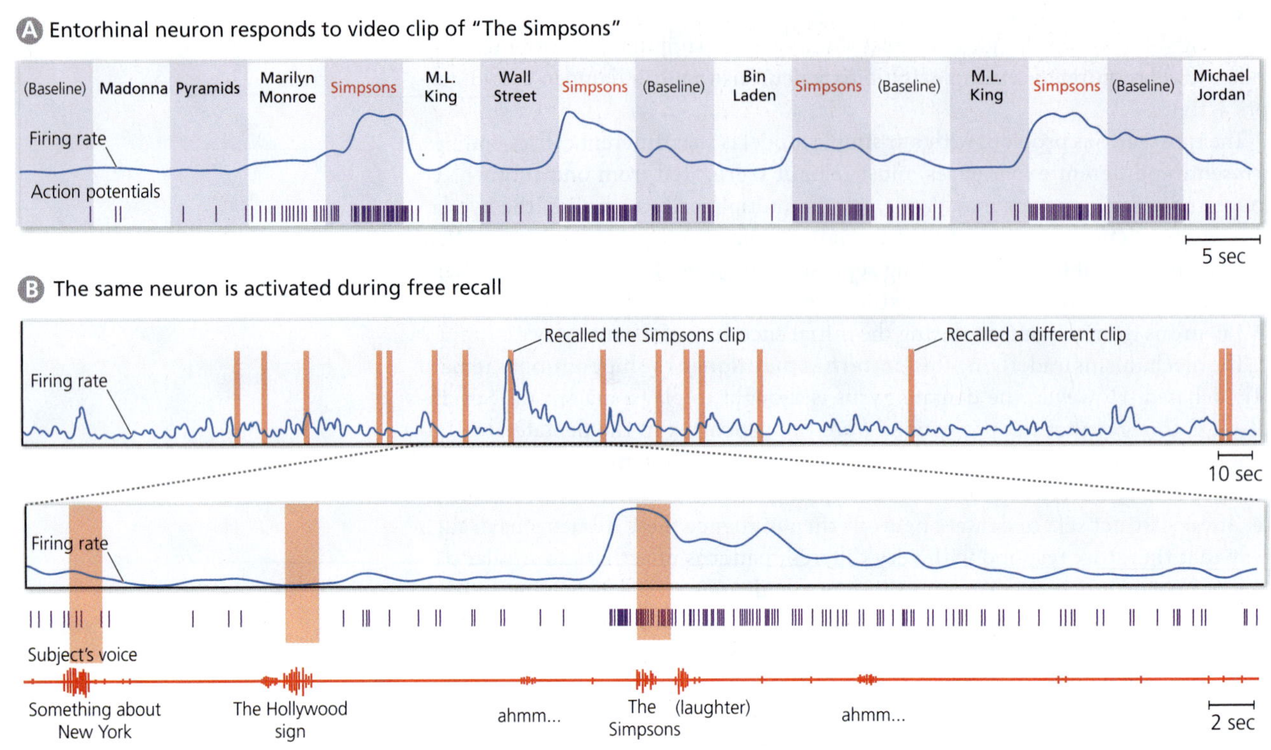

Figure 14.13 Recalling "The Simpsons." Recording electrodes were implanted in the hippocampus and entorhinal cortex of patients awaiting epilepsy surgery. One entorhinal neuron (A) increased its firing rate whenever the patient was watching brief (5-sec) clips of a "Simpsons" episode. The graphs in (B) show the same neuron's activity when the patient was later asked to recall which clips he viewed. Remarkably, the neuron increased its firing rate just before the patient reported remembering "The Simpsons." Similar recall-related activity has also been recorded in the hippocampus. [After Gelbard-Sagiv et al., 2008]

active during perception of those images, whereas recalling sounds activates a subset of the brain regions active during perception of the sounds (Figure 14.14).

Subsequent research confirmed that the recall of specific objects activates at least some of the same areas that are involved in the perception of the objects. Some studies even report that recalling one's own movements tends to increase activity in the motor cortex. Thus, memory recall appears to reinstate at least part of the neocortical activity patterns that existed when the memory was formed. This is the **reinstatement hypothesis** of memory recall.

Reactivation of the Neocortex by the Hippocampus

How can neocortical activity be reinstated during memory recall? For recent memories, at least, the reinstatement of neocortical activity is thought to be driven by hippocampal activity. As you saw in Figure 14.9, the pathways from the neocortex into the hippocampus are paralleled by pathways going in the opposite direction, from hippocampus to entorhinal cortex, perirhinal/postrhinal cortex, and, ultimately, a variety of other cortical areas. Thus, we can think of information flowing from the neocortex to the hippocampus during memory encoding and in the opposite direction during subsequent memory recall (Figure 14.15).

This model of memory formation and recall is difficult to test, but it is consistent with a variety of evidence. For example, it has been shown that neurons in the inferior temporal cortex of monkeys are active shortly *before* the hippocampal neurons during stimulus perception but *after* the hippocampal neurons during recall of the remembered stimulus. These data are consistent with visual information flowing in opposite directions during memory formation and recall.

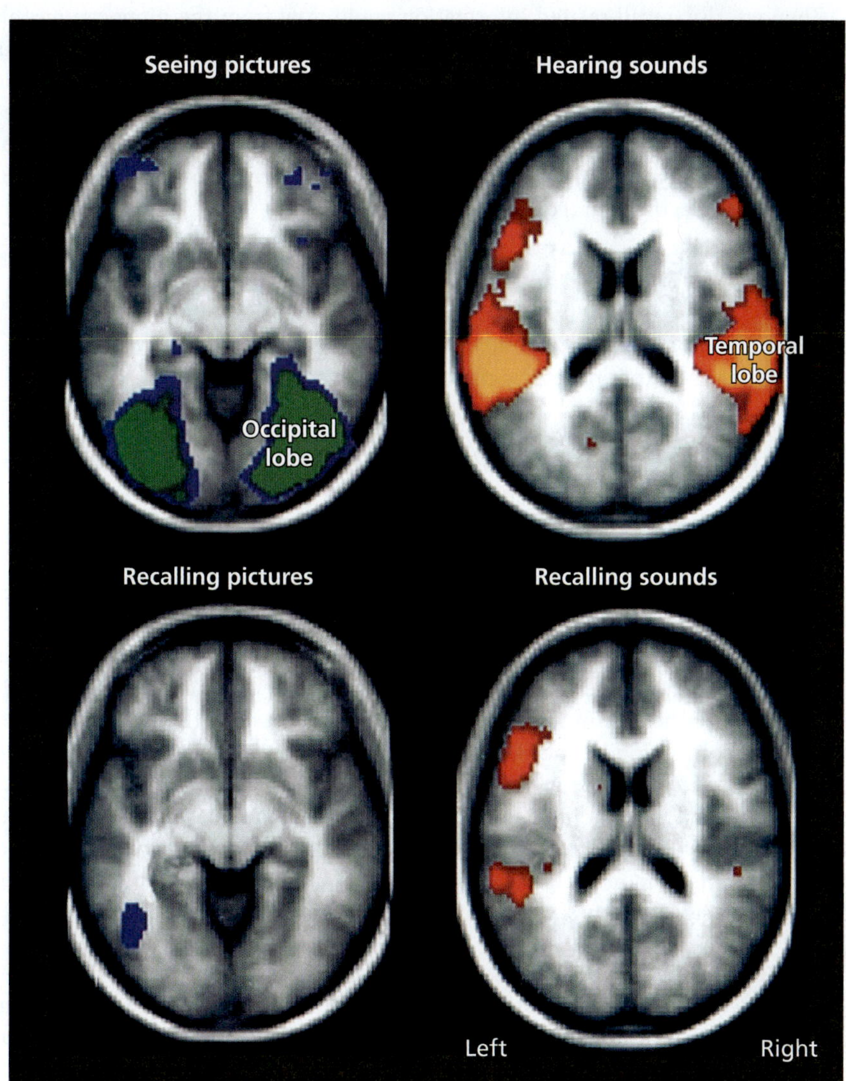

Figure 14.14 Recall activates a subset of the regions activated during perception. People were asked to memorize 20 pictures and 20 sounds. Functional MRI was performed as they viewed the pictures, listened to the sounds, or tried to recall those stimuli as vividly as possible. Recalling pictures activated a subset of the same brain regions that were activated during perception of the pictures, and recalling sounds activated a subset of the regions activated during listening. [From Wheeler et al., 2000]

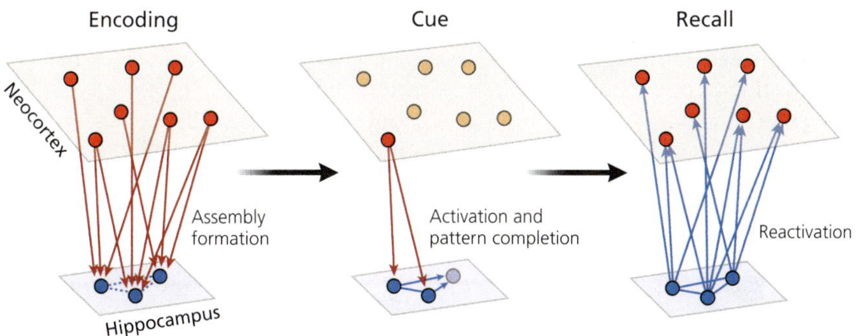

Figure 14.15 Memory formation and recall. According to a widely accepted model, memories are formed when a set of simultaneously active neocortical neurons drives the formation of a corresponding cell assembly in the hippocampus. During cued recall, activation of just a few of the previously active neocortical neurons activates the hippocampal cell assembly, which then "reinstates," at least to some extent, the pattern of neocortical activity that had been active during the original experience.

Theoretically, recall-related information could flow from the hippocampus all the way back down to the dorsal thalamus; the requisite anatomical connections exist. However, the functional brain imaging data indicate that memory recall tends to activate mainly the higher order sensory cortices rather than the primary cortices or subcortical areas. As noted earlier, recall involves only a partial reinstatement of the original activity. This may explain why we can distinguish vivid memories from actual experience; the neural correlates are not identical.

BRAIN EXERCISE

What experiments might you do to test the hypothesis that memory recall is based on hippocampus-driven reinstatement of neocortical activity patterns? Why would those experiments be difficult?

14.5 What Happens to Memories as They Grow Old?

Patient H.M. exhibited some retrograde amnesia, losing the memories that had formed in the 3 (or so) years before the surgery. Older memories were spared, at least when H.M. was tested shortly after his surgery.

To study this **retrograde amnesia gradient** in non-humans, researchers presented monkeys with 100 different object pairs, teaching them that only one object in each pair was associated with a food reward. They then lesioned the hippocampus and some adjacent cortices (but not the perirhinal cortex) of the monkeys 2, 4, 8, 12, or 16 weeks after the training period. The major finding was that the lesions impaired memories for object-food associations that had been learned 2 and 4 weeks before the surgery but spared the older memories (Figure 14.16). In a reversal of the usual pattern, the lesioned monkeys remembered the older associations better than the recently learned ones. Thus, medial temporal lobe damage causes temporally limited retrograde amnesia in monkeys as well as humans, although the temporal extent of the amnesia is significantly shorter in monkeys (for reasons that are unclear).

Systems Consolidation

To explain why hippocampus lesions interfere selectively with recent memories, one can hypothesize that memories are initially stored in the hippocampus but then somehow transferred (relocated) to the neocortex. How might this relocation work?

The most plausible answer is that the hippocampus-driven reinstatement of neocortical activity will, if it occurs repeatedly, cause the formation of neocortical cell assemblies, which can then be reactivated even without a functioning hippocampus (Figure 14.17). According to this model, the relocation of memory traces from the hippocampus to the neocortex is like the transmission of knowledge from teacher to student; the hippocampus "teaches" the neocortex until the latter, too, remembers what happened (although, perhaps, less precisely). Once the neocortex has been

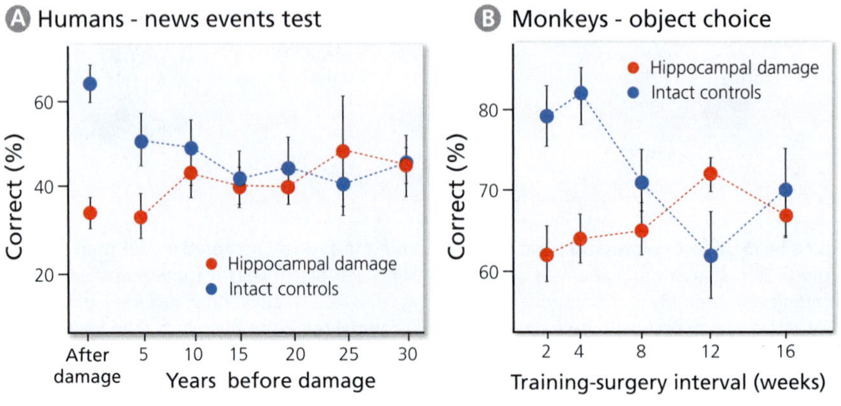

Figure 14.16 Time-limited retrograde amnesia. Humans with bilateral damage to the hippocampus have impaired memory for major news events that happened after their injury (anterograde amnesia) or 1–5 years before the brain damage (retrograde amnesia); older memories are not impaired (A). Panel (B) shows similar data for monkeys that had learned 100 different 2-choice object discrimination problems (only one member of each pair was rewarded) at various intervals prior to brain surgery. Hippocampal lesions weakened only the 2- and 4-week old memories. [After Bayley et al., 2006, and Zola-Morgan and Squire, 1990]

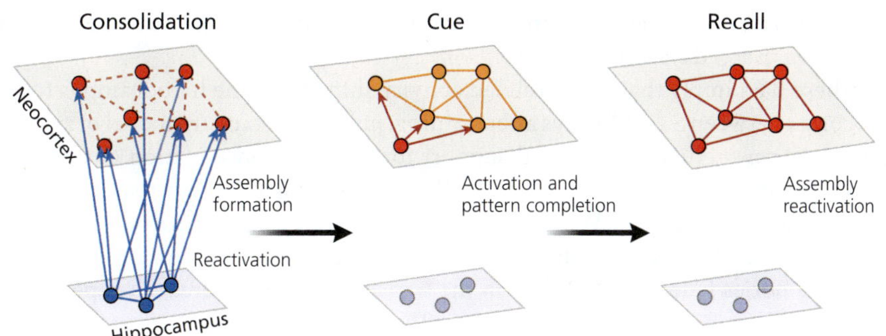

Figure 14.17 Systems consolidation. According to the standard model of memory consolidation, repeated reactivation of the neocortical neurons is thought to strengthen the connections between them, leading to the formation of a neocortical cell assembly that can then be activated without involving the hippocampus.

trained, the hippocampus is no longer required to reinstate the neocortical activity patterns that existed when the memory was formed.

An important aspect of this hypothesis is that the memory traces become stronger and less susceptible to disruption (by electroconvulsive shock, for example) once they have become established in the neocortex. This gradual strengthening of memories is called *memory consolidation*. The idea that memory consolidation involves the relocation of memories to the neocortex is called **systems consolidation**.

Experimental Support for Systems Consolidation

Experimental support for systems consolidation comes from an experiment using radioactive 2-deoxyglucose (2-DG), which you learned about in Chapter 6. Because 2-DG accumulates inside of active neurons, experimenters can infer rates of neuronal activity from accumulated 2-DG levels. Using this method to compare levels of neuronal activity in multiple brain regions at various stages of memory consolidation (Figure 14.18), neurobiologists have found that neuronal activity in the hippocampus

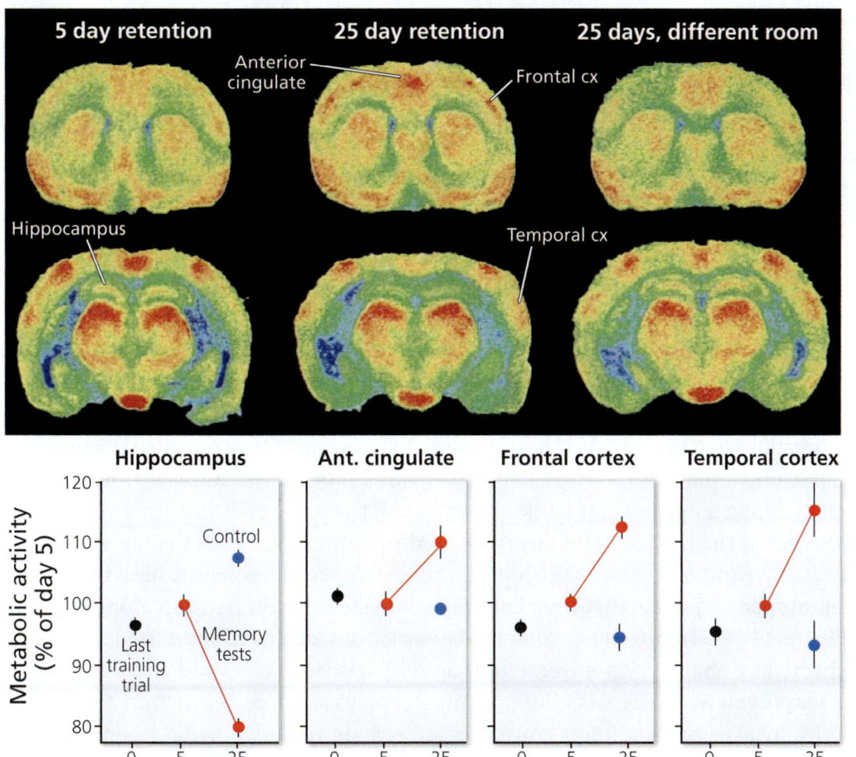

Figure 14.18 Evidence for systems consolidation. Mice were trained to remember which arms of a radial maze contain food. Their memory was tested 5 or 25 days after training; control mice were tested in a different room. Just before the memory test (or before the last training trial), the mice were injected with radioactive 2-DG, which accumulates in active neurons. Shown at the top are representative autoradiographs (two sections per brain) in which warm colors indicate high metabolic activity. The hippocampus was less active during the memory test on day 25 than on day 5. The opposite pattern is evident in the anterior cingulate, frontal, and temporal cortices. The four graphs summarize the data quantitatively. [From Bontempi et al., 1999]

is significantly *lower* during a memory test at 25 days after training (when systems consolidation is largely complete in mice) than 5 days after training. Conversely, activity levels in several neocortical areas are significantly *higher* during recall at 25 days than at 5 days after training.

These data imply that recall-related activity shifts from the hippocampus to the neocortex between 5 and 25 days after training, just as the standard model of systems consolidation predicts. Additional support for systems consolidation comes from transgenic mice in which neocortical LTP is severely impaired, whereas hippocampal LTP appears normal. When these transgenic mice are tested in a variety of memory tasks, they learn normally and can remember what they learned. However, those memories fade after 1–3 days in the transgenic mice, whereas they persist for at least 10 days in wild-type control mice. These data imply that neocortical LTP is required for memories lasting longer than a day or two in mice.

The Formation of Neocortical Assemblies

An important assumption of the systems consolidation model is that neocortical cell assemblies form more slowly than hippocampal cell assemblies. Indeed, LTP can be induced more easily (with fewer rounds of stimulation) in the hippocampus than the neocortex. Why is neocortical LTP so sluggish when change at hippocampal synapses can be rapid?

No one is sure, but it has been suggested that the neocortex is specialized for the gradual learning of statistical regularities in its input patterns. Computer modeling has shown that this kind of statistical pattern learning works best if it occurs slowly and gradually, while the network is presented with a wide variety of input patterns. The hippocampus, in contrast, seems specialized for the rapid storage of unique input patterns. According to the modeling studies, this kind of rapid learning is incompatible with gradual pattern learning.

Viewed from this perspective, systems consolidation is an efficient compromise. The hippocampus serves as an initial, temporary store for rapidly acquired information. Some of this information is then transmitted to the neocortex, where it is gradually integrated with information that was learned earlier. Importantly, the information that the neocortex stores is not identical to what was stored in the hippocampus. What the neocortex stores tends to be less detailed and better integrated with previously acquired information.

Unconscious "Replay" of Memories

When and how is the hippocampus "teaching" the neocortex? The most obvious answer is that the hippocampus can teach the neocortex every time you recall a memory by reinstating a neocortical activity pattern. However, if neocortical plasticity is sluggish, then many recall episodes would be required to "burn" a memory trace into the neocortical network. This is problematic because we can remember some things for years even if we don't recall them frequently.

The most likely solution to this puzzle is that hippocampal cell assemblies can reinstate neocortical activity patterns not only during ordinary memory recall, when we are awake, but also during sleep. This hypothesis is based on a remarkable set of observations first made by Matthew Wilson and colleagues. They recorded from multiple hippocampal place cells (see Chapter 11) while rats were running, rather monotonously, along a figure-eight track. As expected, the place cells fired in a stereotyped sequence as the rats ran. The surprising finding came later, when the rats were sleeping after a running episode: the hippocampal place cells frequently fired in the same stereotyped sequence that they had followed when the rat had run along the track (Figure 14.19). Importantly, such **replay episodes** occurred more frequently than expected by chance. Moreover, replay activity has been observed also in the visual cortex, as you would expect if hippocampal replay can drive neocortical activity.

Many aspects of systems consolidation remain to be worked out. For example, the standard model assumes that neocortical neurons are interconnected at a level that is high enough to allow for the formation of cell assemblies. As we discussed, dense interconnectivity is found in CA3 and, to some extent, in the olfactory cortex. However, the connections between neocortical neurons are much less dense, raising doubts about their ability to form Hebbian cell assemblies. After all, traditional LTP can only strengthen weak connections; it cannot strengthen nonexistent connections. This problem would be lessened if intracortical axons regularly sprout (and lose) new axon branches, as suggested by the discovery of dendritic spine turnover in adult brains (see Chapter 3). Still, doubts remain. One should also note that not all data are consistent with the standard consolidation model. For example, some rich and detailed memories require an intact hippocampus, even if those memories are fairly old.

BRAIN EXERCISE

What is the relationship between systems consolidation and synaptic consolidation? Which one is more important for long-term memory formation?

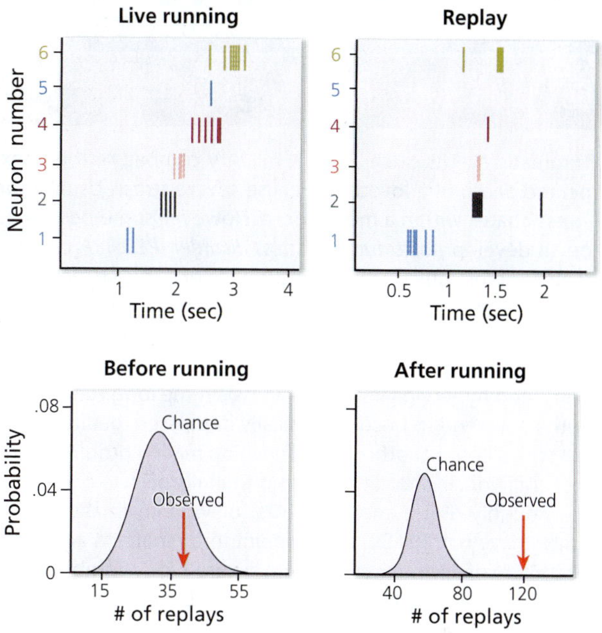

Figure 14.19 Replay in the hippocampus during sleep. The activity of 6 hippocampal place cells was recorded while a rat was running along a figure-8 shaped maze (live running). Also shown is an instance of replay activity during slow-wave sleep just after the running. During this replay, the 6 neurons fire in roughly the same sequence as they did during live running (although the sequence is compressed). The bottom graphs show that replay activity after running occurred more frequently than expected by chance. [After Ji and Wilson, 2007]

14.6 What Makes Some Memories Stronger Than Others?

Most of what you remember for a day or two is relatively trivial and can safely be forgotten, but some information is worth retaining for years. How can your brain anticipate which memories are likely to be useful down the road? There are two strategies. The first involves enhancing the initial input to the memory forming system.

Boosting the Initial Experience

As we discussed in Chapter 13, attention may be voluntary or involuntary. Either way, the stimuli that attract our attention tend to be those that provide useful information about the current environment. Importantly, attention enhances the neural signals carrying the attended information. The neural enhancement makes attended stimuli easier to detect and identify than unattended stimuli. It also increases the likelihood that attended information will be remembered long term. This is adaptive in evolutionary terms because things that were important to an organism once are likely to be important again. The underlying neural mechanisms are also straightforward: the stronger a sensory input pattern, the more likely it is to cause the formation of robust hippocampal cell assemblies. Strong hippocampal cell assemblies, in turn, are likely to drive neocortical neurons strongly enough to form assemblies of their own.

Behavioral arousal also enhances sensory encoding and makes the information easier to remember. As you may recall from Chapter 13, behavioral arousal involves the activation of locus coeruleus, which then boosts any sensory signals that are being processed at the time while suppressing background activity. Therefore, behavioral arousal should boost memories for whatever happens during the period of arousal. You probably know this from personal experience. If you have ever been severely frightened, or geared up for a fight, you probably remember the experience well. In fact, such emotional memories can become so strong that they interfere with normal life, as they do in posttraumatic stress disorder (Box 14.3).

Post-training Memory Enhancement

Behavioral arousal can also enhance memories for things that happened earlier, shortly before the arousal got underway. This kind of **post-training memory enhancement** is normally beneficial because the better an organism can remember what led up to a

NEUROLOGICAL DISORDERS

Box 14.3 *Posttraumatic Stress*

Traumatic events, such as rape, military combat, or the unexpected death of a loved one, cause severe stress. Usually the stress abates within a month or so. However, some people go on to develop *posttraumatic stress disorder (PTSD)*. A cardinal feature of this affliction is the recurrence of intrusive memories of the trauma, including flashbacks and nightmares. In addition, PTSD involves a tendency to avoid thinking about the traumatic event and hyperarousal, including insomnia, irritability, and hypervigilance (jumpiness). In the long run, people with PTSD tend to become clinically depressed, get into trouble with alcohol or other drugs, develop marital problems, and lose their job. In short, their lives often fall apart.

Roughly 5% of men and 10% of women in the United States have had PTSD at some point in their life. In any given year, 1.3% of all adults suffer from severe PTSD. Combat veterans are even more likely to develop PTSD, with lifetime rates typically in the range of 15–20% (depending on sampling methodology and PTSD criteria). The fact that PTSD is twice as common in women than in men suggests that women may be more susceptible to PTSD. However, women are raped more than ten times as often than men, and 55% of rape victims develop PTSD. Indeed, rape is the leading cause of PTSD in women. Importantly, when men are raped, they are just as likely as female rape victims to develop PTSD. Furthermore, both sexes are equally likely to develop PTSD after the sudden loss of a loved one, which the two sexes experience with roughly equal probability. These data suggest that men and women do not differ in their susceptibility to PTSD, but the matter is not settled yet.

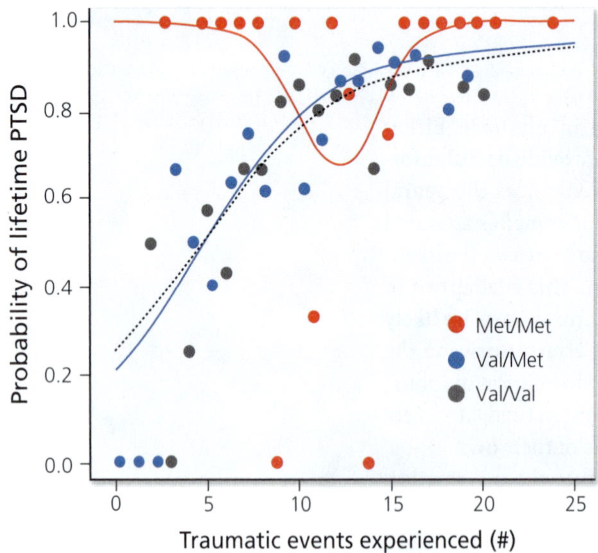

Figure b14.3 A mutation that increases PTSD susceptibility. This study examined refugees of the 1994 genocide in Rwanda. Many of them had experienced multiple traumatic events, including rape and witnessing murder. The graph shows the distribution in this population of a genetic polymorphism that causes valine (Val) at codon 158 catechol-O-methyltransferase to be replaced with methionine (Met). For subjects with the Val/Val and Val/Met genotypes, the probability of having severe PTSD (at some point in their life) increases with the number of traumatic events the subject has experienced. In contrast, subjects with the Met/Met genotype have a very high probability of developing PTSD even if their trauma load was low. [After Kolassa et al., 2010]

The extent to which genetic factors influence PTSD susceptibility is difficult to determine. For example, a shared mutation might make monozygotic twins highly susceptible to PTSD; but if only one of the twins experienced significant trauma, then only one of them will develop the disorder. Thus, differences in trauma exposure can obscure genetic factors underlying PTSD. After correcting for such complications, scientists estimate that about 30% of the variation in PTSD symptoms is due to genetic factors.

One study found that the lifetime probability of developing PTSD is influenced by a single nucleotide substitution in the gene for *catechol-O-methyltransferase*, an enzyme involved in the inactivation of catecholamines (notably dopamine and norepinephrine). Survivors of the Rwandan genocide who are homozygous for this mutation were found to have an increased likelihood of developing PTSD, even if their trauma exposure was relatively low (Figure b14.3). This is an important discovery, especially because norepinephrine has been implicated in arousal-based memory enhancement. More recent work has revealed additional PTSD susceptibility genes.

PTSD can be treated effectively with *exposure therapy*, which requires patients to confront their trauma-related feelings by various techniques, including mental imagery, writing, and visiting the scene of the trauma. The treatment tends to work because the patients learn to associate the trauma with more positive feelings. The procedure is analogous to extinction training in rats. If a rat that had learned to fear a shock-predicting tone is repeatedly presented with the tone alone (without the shock), then it will gradually stop freezing to the tone. At first glance, this extinction of the old conditioned response (freezing to the tone) seems like simple forgetting. However, a single trial in which the tone is once again paired with the shock suffices to reinstate the freezing response, implying that the rat did not really forget. It still remembered that the tone used to predict shock, but it also learned that the tone eventually stopped being predictive of bad things. Similarly, exposure therapy does not erase a patient's traumatic memories, but it teaches them that the trauma is no longer as threatening as it once was. This new knowledge can reduce and eventually eliminate the PTSD.

To make exposure therapy even more effective, it can be coupled with therapies that help patients deal with the stigma, sense of isolation, and anger that often accompany PTSD. Exposure therapy can also be improved by giving patients *D-cycloserine*, an NMDA agonist that enhances extinction learning in rats. Finally, serotonin reuptake inhibitors and other antidepressants can alleviate some symptoms of PTSD. Despite all this good news, roughly a third of all PTSD patients never fully recover.

negatively arousing experience, the better it can avoid similar experiences in the future. Similarly, strengthening the memories for what happened shortly before a positive experience makes it easier for organisms to find that happiness again.

At this point you may be doubtful: does getting excited or scared really strengthen my memory of what happened a few minutes earlier? Yes, it does, in both humans and rats. As many studies have shown, rats injected with stimulants (such as amphetamine) or stress hormones (such as adrenaline) remember better than control rats what happened shortly before the injections. In humans, too, adrenaline injections improve the memory of previously viewed images. One recent, rather dramatic study showed that skydiving improves the retention of information learned shortly before the jump (at least in men and for at least some kinds of information).

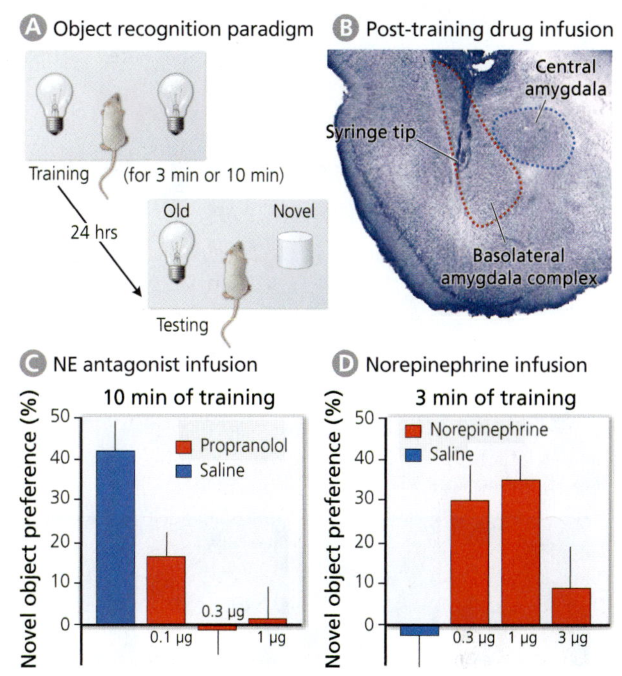

Figure 14.20 Manipulating memory consolidation. In the spontaneous object recognition paradigm (A), rats are exposed to 2 identical objects for various lengths of time. One day later they are presented with the old object as well as a novel object. The better they remember the old object, the less they explore it. To study the mechanisms of memory consolidation, researchers infused drugs bilaterally into the basolateral complex of rats (B) immediately after training. Post-training infusions of propranolol (a beta-adrenergic receptor blocker) interfere with memory consolidation (C), whereas post-training infusions of norepinephrine boost memory, at least at low and intermediate doses (D). In this second experiment, only 3 minutes of training were given, which is too short for control rats to remember the object. [From Roozendaal et al., 2008]

Memory Modulation by the Basolateral Amygdala

The neural mechanisms underlying arousal-driven, post-training memory enhancement have been studied extensively. A key player is the **basolateral amygdala complex** (Figure 14.20). This relatively large part of the amygdala lies inferior to the central nucleus of the amygdala, which we discussed in Chapter 9. Its neurons increase their firing rate after behaviorally arousing events, be they negative or positive. This arousal response is probably driven by noradrenergic inputs from the locus coeruleus because norepinephrine levels in the basolateral amygdala correlate with arousal. Consistent with this hypothesis, blocking norepinephrine signaling in the basolateral complex right after exposing a rat to a novel, exciting object impairs the rat's recognition of that object on a subsequent test trial (Figure 14.20 C). Conversely, infusing norepinephrine directly into the basolateral complex can strengthen an otherwise weak memory (Figure 14.20 D).

An important element in the design of these studies is that the drug infusions occur after training but well before the memory tests so that the drug has worn off before the test. Therefore, the manipulations can be interpreted as affecting memory consolidation rather than perception, memory formation, or recall. In other words, the experiments allow us to conclude that noradrenergic activation of the basolateral complex is necessary and sufficient for post-training memory enhancement.

Studies on brain slices have shown that activation of the lateral nucleus of the amygdala, which is part of the basolateral complex, facilitates the flow of information from the neocortex into the hippocampus. More specifically, it appears that area 35 of the perirhinal cortex acts as a "gate" for information flow into the hippocampus (Figure 14.21). Neuronal recordings from intact animals support this hypothesis.

There is currently no direct evidence that activation of the basolateral complex facilitates the flow of information in the reverse direction, from the hippocampus back to the neocortex, during memory consolidation or recall. However, post-training activation of the basolateral complex increases hippocampal levels of the protein Arc, which is linked to synaptic plasticity; whereas post-training inactivation of the basolateral complex decreases hippocampal Arc levels. These findings imply that activity in the basolateral complex can influence hippocampal activity not only during a learning episode but also afterward, during the memory consolidation phase. The details of this influence remain to be worked out.

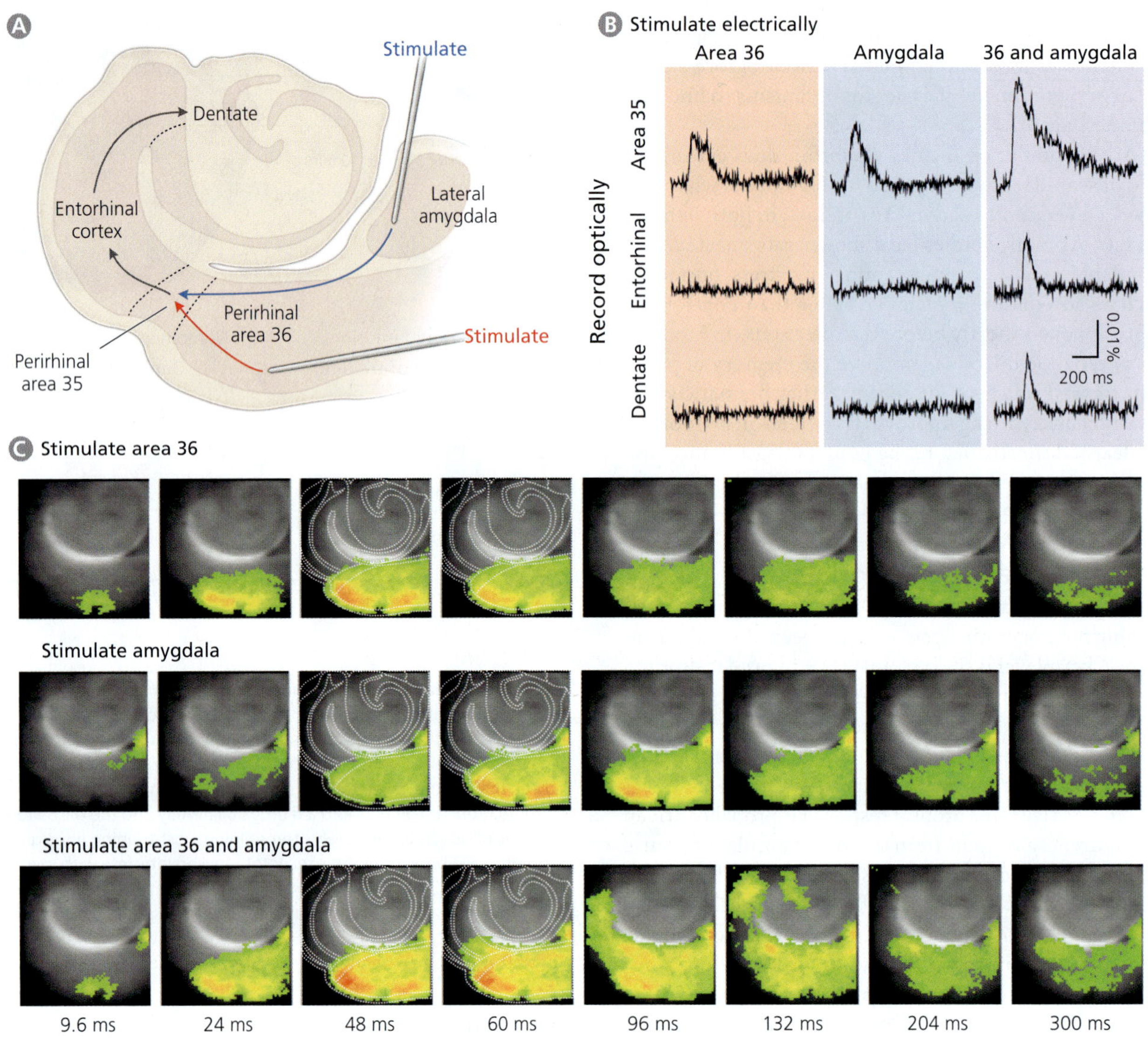

Figure 14.21 Regulating the flow of information into the hippocampus. The diagram in (A) depicts a hippocampal slice with stimulating electrodes in the lateral amygdala (part of the basolateral complex) and area 36 (part of the perirhinal cortex). Optical imaging with a voltage-sensitive dye was used to visualize the spread of neuronal activity after stimulation at those two locations. The graphs in (B) show that electrical stimulation of area 36 and the lateral amygdala, but not of either area by itself, allows activity to spread into the entorhinal cortex and dentate gyrus. The same results are shown as pseudocolored time-lapse images in panel (C). [From Kajiwara et al., 2003]

Function of the Human Basolateral Amygdala

Does the basolateral complex play the same memory-modulating role in humans as in rats? This question is difficult to answer because in humans one cannot simply lesion the basolateral complex or inject it with diverse drugs. However, researchers have identified two humans with large amygdala lesions that include the basolateral complex but exclude the hippocampus. In contrast to control subjects, these patients do not remember emotionally arousing images, presented as part of a narrated slide-show, better than emotionally neutral images.

This finding is consistent with the hypothesis that part of the amygdala modulates memory for arousing experiences in humans, as it does in rats. Further support comes from functional brain imaging studies, which show that the degree of amygdala activation at the time of memory formation (encoding) correlates positively with memory strength when subjects are tested several weeks later. Intriguingly, this correlation holds only for the right amygdala in men and only for the left amygdala in women. Such sex differences and brain asymmetries seem to be more common in humans than scientists had expected (see Chapter 16).

BRAIN EXERCISE

Given the standard model of systems consolidation (and your own vivid imagination), what kinds of mechanisms would suffice to strengthen memory consolidation? How might the basolateral amygdala affect those mechanisms?

14.7 How Do Animals Learn What's Dangerous?

Animals of most species are quick to learn what's dangerous. Because this type of learning is easily elicited under laboratory conditions, neurobiologists have studied it extensively. Most of the research has focused on auditory fear conditioning, contextual fear conditioning, and inhibitory avoidance training. We consider each of these subjects in turn.

Auditory Fear Conditioning

The most intensively studied form of learning about threats is **Pavlovian fear conditioning** with tones as the conditioned stimuli. In this paradigm, an animal is placed into an experimental chamber and habituated to this environment (Figure 14.22). The animal, typically a rat, is then presented with an auditory stimulus, usually a tone. As the tone ends, the rat is given a foot shock. Sometime later, the rat is placed into a novel experimental chamber and presented with the previously heard tone minus the foot shock. The crucial question is whether the rat remembers that the tone signals a threat.

One way to answer this question is to measure the animal's **freezing response**, which is defined as immobility (except for breathing) while having all feet on the ground. Rats that were exposed to the tone-shock pairing during training freeze when the tone is presented by itself more frequently than rats that did not experience the tone-shock pairing. Because the experimental chamber used for testing differs significantly from the training chamber, whereas the tone remains constant, we can infer that the animals were conditioned to fear the tone rather than the chamber.

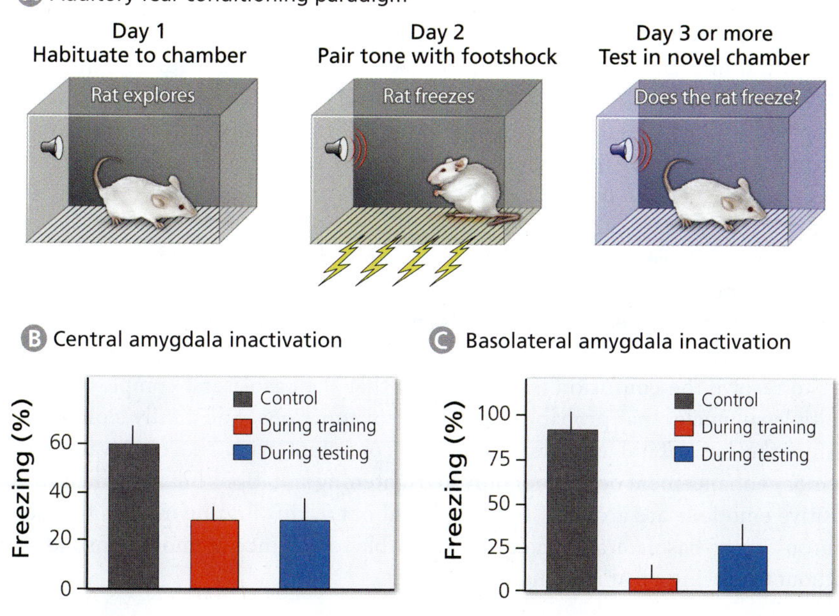

Figure 14.22 Auditory fear conditioning. In auditory fear conditioning, testing is performed in a novel chamber to control for the possibility that the rat learned to fear the chamber, rather than the tone. Inactivating the central nucleus of the amygdala with muscimol (a GABA agonist) during training or testing reduces the freezing response (B). Injecting muscimol into the basolateral complex likewise reduces the freezing response (C). [After Muller et al., 1997, and Ciocchi et at., 2010]

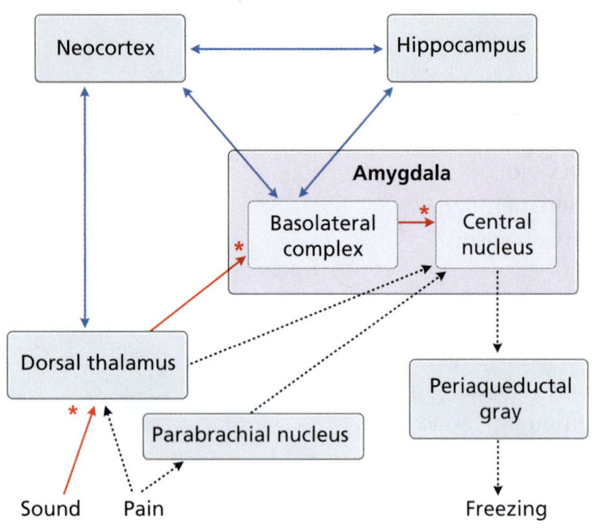

Figure 14.23 Circuits underlying auditory fear conditioning. Information about sounds is conveyed to the amygdala directly from the thalamus (red arrows) as well as indirectly through the neocortex (blue arrows). Information about the painful stimulus can trigger a freezing response (dashed arrows) and converges with the auditory information at several locations. The red asterisks mark sites at which synaptic plasticity has been inferred.

Pathways through the Central Amygdala

The neural circuits underlying auditory fear conditioning have been studied intensively. A key player is the *central nucleus of the amygdala,* which projects to the autonomic control regions involved in the fight-or-flight response (see Chapter 9). The central amygdala also projects to the midbrain's *periaqueductal gray,* which coordinates the freezing response. Consistent with these pathways, stimulating the central amygdala elicits freezing, whereas inactivating it reduces freezing, both during the initial training and during memory testing (Figure 14.22 B).

Information about foot shocks or other fear-evoking stimuli can reach the central nucleus of the amygdala through the parabrachial nuclei (which are distinct from the peribrachial nuclei you learned about in Chapter 13), the dorsal thalamus, and the basolateral complex (Figure 14.23). Nobody knows why mammals have three different pathways for sending pain information to the central nucleus of the amygdala, but this multiplicity of **unconditioned fear pathways** makes the system fault tolerant. If one of the three pathways fails, the others will ensure that an appropriate response occurs. Also worth noting is that the pain pathways to the central nucleus are all relatively fast, compared to the pathways that go through the neocortex. When responding to pain, speed can be a life saver!

Conveying Auditory Information to the Central Amygdala

The unconditioned fear pathways prompt animals to freeze in response to painful stimuli, such as a foot shock, but how can animals learn to fear a harmless sound after it has been paired with a foot shock? For this to happen, information about the sound must somehow be conveyed to the unconditioned fear pathways through the central amygdala. Indeed, research has shown that pairing a sound with pain strengthens previously weak connections from the auditory pathway to the unconditioned fear pathway. Because of this strengthening, the sound alone can, after training, evoke the freezing response.

Where specifically does auditory information enter the unconditioned fear pathway, and where does the synaptic strengthening take place? One answer is the basolateral amygdala complex (especially its lateral nucleus). Single neurons there respond to sounds as well as to foot shock, and their auditory responses are strengthened by training. Auditory information also converges with pain information in the central nucleus of the amygdala and in the dorsal thalamus, both of which exhibit synaptic plasticity (Figure 14.23). Therefore, auditory fear conditioning involves synaptic strengthening in multiple brain areas. This, too, promotes fault tolerance. If synaptic plasticity in one location is insufficient to accomplish the learning, then plasticity at other sites can do the job.

As you may have noticed, the basolateral amygdala complex is involved in both auditory fear conditioning and memory modulation. This may be confusing, but the two functions are related: if a stimulus is frightening enough to evoke freezing, then it probably pays to remember what happened before and after this stimulus. Another way to resolve the confusion is to point out that the basolateral complex contains multiple subnuclei that probably have different functions. Such a division of labor within the basolateral complex would be consistent with the observation that memory enhancement occurs not only in frightening situations but also when strong positive emotions are aroused. If you think about it, this finding implies that some neurons in the basolateral complex must be able to enhance memory consolidation without triggering a fear response.

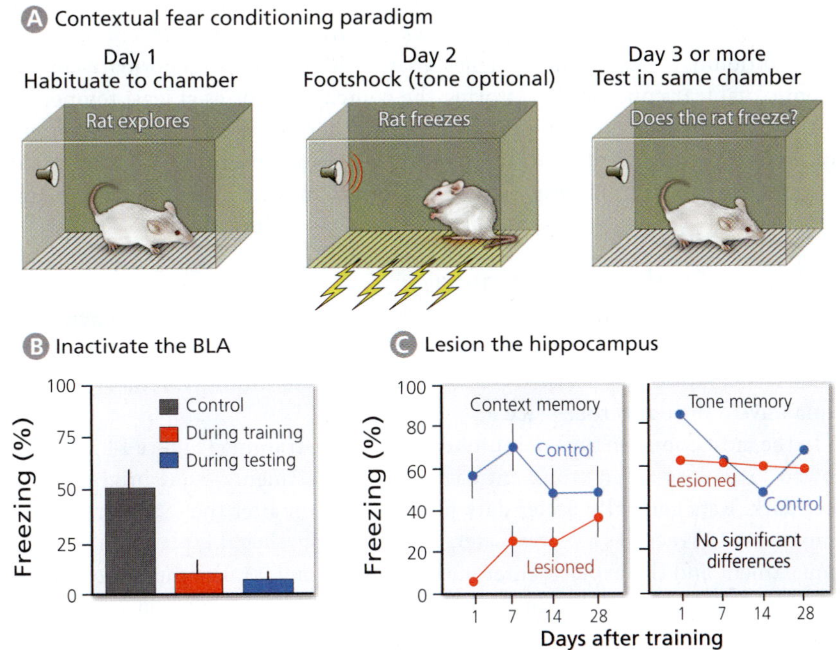

Figure 14.24 Contextual fear conditioning. In contextual fear conditioning, a rat is asked whether it remembers the chamber (the context) in which it experienced a footshock (A). Inactivating the basolateral amygdala (BLA) complex with muscimol, either during training or during testing, impairs contextual conditioning, as assessed by the freezing response (B). The expression of contextual fear is also impaired by hippocampal lesions, as long as the lesions are made within 14 days after training (C). Auditory fear conditioning is not affected by those same lesions. [After Muller et al., 1997, and Kim and Fanselow, 1992]

Contextual Fear Conditioning

A second experimental paradigm frequently used to examine what animals have learned about a threat is **contextual fear conditioning**. The training procedure in this paradigm is similar to that in auditory fear conditioning. For testing, however, the trained animals are placed into the same experimental chamber in which they experienced the shock (Figure 14.24). If the foot shock had been paired with a tone during training, then the tone is omitted during testing because the question now is not whether the animals learned to fear the tone but whether they learned to fear the chamber (the context in which the tone had been presented). To quantify an animal's degree of fear in the contextual fear conditioning paradigm, experimenters usually measure freezing, as they do in most rodent studies on fear. They then compare the degree of freezing between animals that had received a shock (the trained animals) and those that had not (untrained controls).

In this paradigm, rats can learn to fear the experimental chamber after just a single training trial, as long as they had a few minutes to explore the chamber before the shock occurs. The function of contextual fear conditioning is fairly obvious: it puts animals on guard in places where bad things have happened in the past.

Connecting Context to Fear

The neural circuitry underlying contextual fear conditioning includes, again, the basolateral complex. Bilateral inactivation of the basolateral complex reduces freezing in the contextual fear conditioning paradigm, regardless of whether the inactivation occurs during training or testing. In this respect, auditory and contextual fear conditioning are similar.

However, contextual fear conditioning is also impaired by hippocampus lesions, which do not affect auditory fear conditioning (Figure 14.24 C). More specifically, hippocampal lesions impair contextual fear memory as long as they are made within 2 weeks after training; after that, the context memories become independent of the hippocampus, just as our model of systems consolidation predicts. The overall conclusion is that the neural circuits involved in contextual fear conditioning overlap with those for auditory fear conditioning but include some extra elements, notably the hippocampus.

Because hippocampal lesions *after training* impair contextual fear memory, it is surprising that hippocampal lesions made *before training* do not prevent contextual fear conditioning. These findings suggest that the hippocampus is normally involved in contextual fear conditioning, storing the context memories at least temporarily, but that animals can learn about contexts by other means if the hippocampus is unavailable. This kind of neural flexibility, where one brain region functionally substitutes for another, represents yet another way in which brains tend to be fault tolerant.

Inhibitory Avoidance Training

The third paradigm commonly used to study fear learning is **inhibitory avoidance training**. In this paradigm, animals are taught to fear a place; but in contrast to contextual fear conditioning, learned fear is assessed by determining whether the trained animals avoid the dangerous place.

In the most common type of inhibitory avoidance training (Figure 14.25), a rat is allowed to explore an environment with two compartments—one brightly lit, the other dark. Rats generally prefer dark places, but soon after the rat enters the dark compartment, it receives a shock. Later, during testing, the rat is placed in the bright compartment and the experimenter measures how long it takes the rat to enter the dark side. Rats usually learn in a single trial to avoid the place where they were shocked. This form of learning is highly adaptive because it's usually a good idea to avoid dangerous places when possible. By comparison, animals in the contextual fear conditioning paradigm cannot exhibit avoidance. They are stuck in a bad place and have to act accordingly.

The neural circuits underlying inhibitory avoidance overlap extensively, although not entirely, with those involved in contextual fear conditioning. Inactivating the hippocampus and most of the amygdala one day after inhibitory avoidance decreases the time it takes for the experimental rats to enter the shock area (Figure 14.26). In contrast, when the same manipulation is made 30 days after training, inhibitory avoidance is intact. Therefore, the memory of the bad place becomes independent of the hippocampus within a month of training. Most likely the neocortex has by then formed its own cell assemblies to represent the bad experience. Consistent with this hypothesis, inactivation of the entorhinal cortex 30 days after training impairs inhibitory avoidance, and inactivation of the parietal cortex is deleterious for even longer periods of time (Figure 14.26).

The role of the basolateral amygdala complex in inhibitory avoidance is complex. In the standard paradigm, pre-training lesions of the basolateral complex decrease the time it takes for rats to enter the shock compartment at the time of memory testing. However, when rats with such lesions are given a choice between the place where they were shocked and another, novel place, they prefer the novel location. The simplest interpretation of these data is that lesions of the basolateral complex impair a rat's freezing response but spare at least some memories of the "bad place," which are

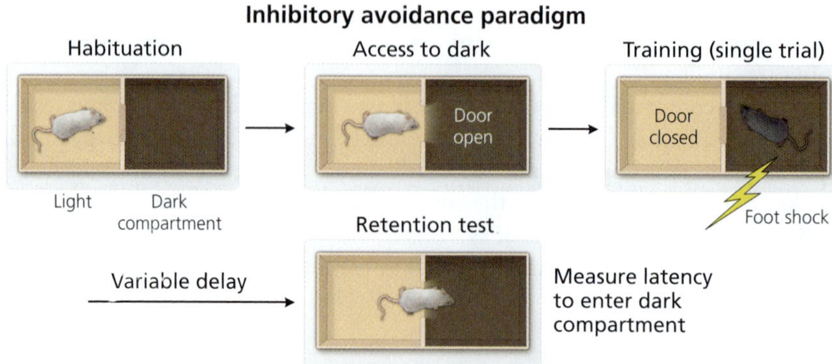

Figure 14.25 Inhibitory avoidance. An animal is given a foot shock in the dark half of a conditioning chamber. Experimenters later put the animal into the light half of the chamber and measure how long it takes the animal to enter the dark half. The longer it takes, the stronger the fear memory.

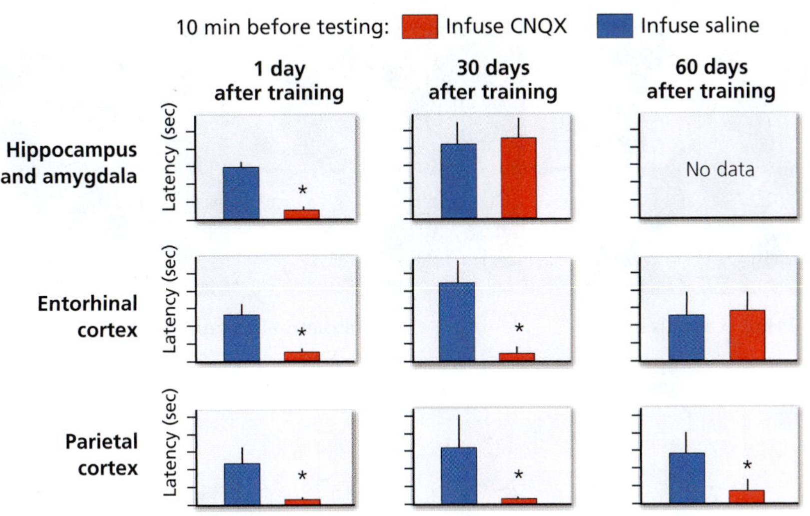

Figure 14.26 Temporal changes in the neural substrates of inhibitory avoidance. The drug CNQX (6-cyano-7-nitroquinoxa-line-2,3-dione), which blocks AMPA-type glutamate receptors (see Chapter 2), was infused into various structures just prior to testing at variable intervals after training. The graph reveals that a functional hippocampus and amygdala are required for rats to show inhibitory avoidance on post-training day 1 but not on day 30; that the entorhinal cortex remains required for at least 30 days; and that the parietal cortex is needed to express the memory for at least 60 days. These data are consistent with the standard model of memory consolidation. [After Izquierdo et al., 1997]

sufficient to guide avoidance behavior. The location of these residual fear memories is debatable, but the hippocampus is a good candidate.

BRAIN EXERCISE

Studies of auditory fear conditioning usually measure an animal's freezing response. What else might you measure to examine whether an animal has learned to fear the tone? Might the hippocampus be needed for showing those other signs of fear?

14.8 How Do We Learn What to Eat or Not to Eat?

Knowing what is safe to eat—and what is not—is very important. To handle this problem, organisms have evolved specialized taste and olfactory receptors that warn them about food that is potentially harmful. This is the principal function of sensors for bitter tastes and rotten smells (see Chapters 6 and 7). In addition, organisms tend to stay away from novel, unfamiliar foods; and they avoid the foods that previously have made them sick. This avoidance of novel or illness-causing foods obviously requires some sort of learning and memory.

Learning from Others

Many animals are capable of learning from other individuals what foods are good to eat, and which should be avoided at all costs. Rats, in particular, develop a preference for the foods that they can smell on the breath of other rats who ate the food and "lived to tell the tale" (Figure 14.27). If the food didn't kill those other individuals, then it is likely safe to eat.

Large hippocampal lesions impair this **social transmission of food preferences** in rats, at least if the lesions are made shortly after the learning experience (Figure 14.27 B). The basolateral complex must also be intact for rats to learn about a novel food from other rats. However, once the preference has been established, the basolateral amygdala is not required to exhibit the food preference (Figure 14.27 C). These data imply that the basolateral complex is involved in the initial learning of socially transmitted food preferences but that those memories are stored elsewhere (although no one is quite sure of their specific location).

Learning from Nausea

Another strategy for learning about food is **conditioned taste aversion**, which is defined as learning not to eat (again) food that made you sick. This phenomenon was

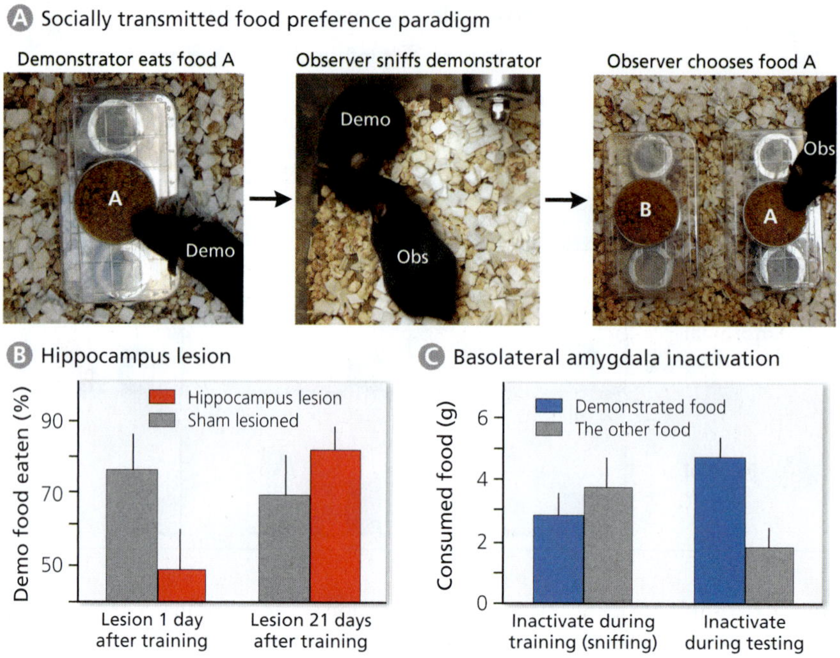

Figure 14.27 Socially Transmitted Food Preferences (STFP). After sniffing a specific food on another rodent's breath, the observer generally prefers to eat the food eaten by the demonstrator (demo). Large hippocampus lesions impair STFP when performed 1 day after training but not at 21 days after training (B), as predicted by the standard model of memory consolidation. Inactivating the basolateral amygdala during the sniffing phase, but not during testing, eliminates STFP (C). [A from Munger et al., 2010; B and C after Ross and Eichenbaum, 2006 and Wang et al., 2006]

discovered in the 1950s by John Garcia and his colleagues, who noticed that rats avoided foods they had eaten before falling ill with radiation poisoning. Nowadays conditioned taste aversion is usually studied by giving rats fluids, rather than solid foods, and sickening them with lithium chloride rather than radiation (which is interesting because, at lower doses, lithium alleviates the manic symptoms of bipolar disorder).

In a typical experiment, thirsty rats are allowed to drink water sweetened with saccharin. Thirty minutes later, they are injected with lithium to make them nauseous. After the rats have recovered, they are given access to both regular water and the sweetened solution. Intact rats usually need only a single trial to associate the sweet taste with nausea and, subsequently, show a strong preference for regular water. The most remarkable aspect of conditioned taste aversion is that it works even when the nausea comes several hours after the drink. This differs from Pavlovian conditioning, in which the unconditioned stimulus must be presented at the end of the conditioned stimulus (or very soon thereafter). However, the long delay between conditioned and unconditioned stimuli in conditioned taste aversion (food and sickness, respectively) makes good biological sense because food poisoning takes time to develop.

Neural Substrates of Conditioned Taste Aversion

Pretraining lesions of the basolateral complex impair conditioned taste aversion. Indeed, it appears that the association between a taste and nausea is formed within the basolateral amygdala.

Evidence for this hypothesis comes from a study that examined the expression of *Arc* mRNA in the basolateral complex (Figure 14.28). Because the Arc protein is implicated in synaptic plasticity, Arc expression can reveal which neurons were activated by, and likely "changed their tune" in response to, particular stimuli. Moreover, the fact that mRNA is initially transcribed in the nucleus but then transported to the cytoplasm allows experimenters to infer when, in relation to a stimulus, the *Arc* gene was turned on (Figure 14.28 B). Using this approach, neuroscientists have discovered that some neurons in the basolateral complex are activated by *both* the saccharin

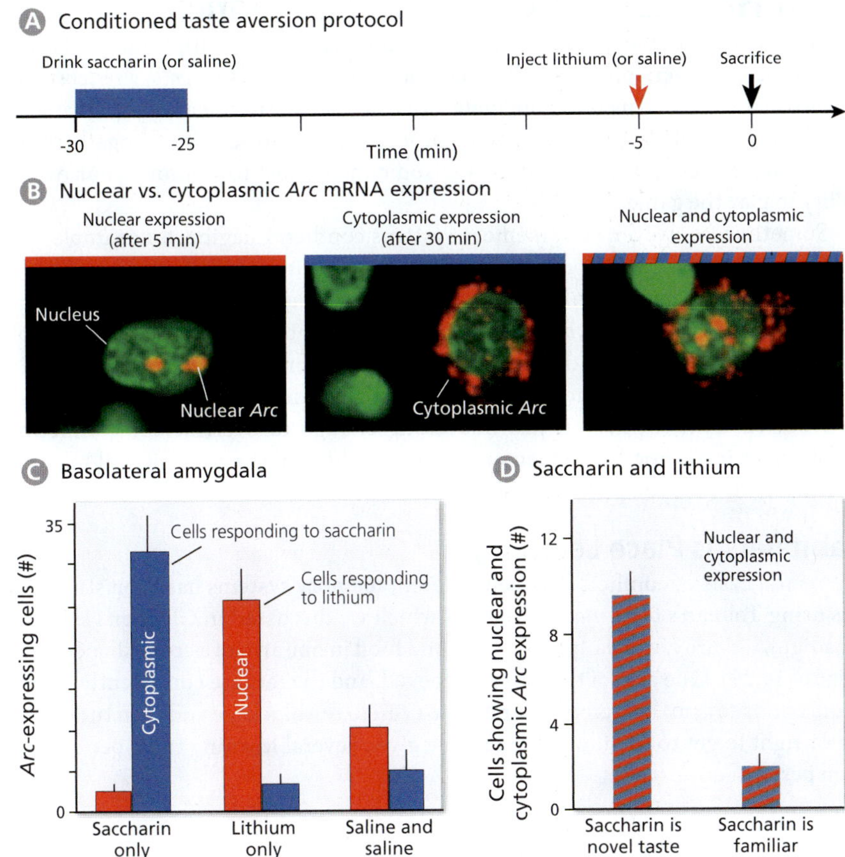

Figure 14.28 Conditioned taste aversion. Twenty minutes after drinking saccharin, rats were injected with lithium chloride, which makes them sick. Five minutes after that, the rats were sacrificed to analyze *Arc* gene transcription (A). As shown in (B), *Arc* mRNA is first seen (as red fluorescence) in cell nuclei but moves to the cytoplasm within ~30 minutes after neuronal activation. Cells exhibiting both nuclear and cytoplasmic *Arc* mRNA must have been activated ~30 minutes and ~5 minutes prior to sacrifice. Some cells in the basolateral amygdala express *Arc* mRNA in response to saccharin; others respond to lithium (C). A few cells in the saccharin-plus-lithium group exhibit both nuclear and cytoplasmic *Arc* expression, especially in rats that never tasted saccharin before (D). This is important because conditioned taste aversion occurs preferentially with novel foods. [From Barot et al., 2008]

and the subsequent illness (Figure 14.28 D). These neurons are probably part of the neural machinery underlying conditioned taste aversion, especially because such jointly activated neurons are rarely seen in rats that were already used to drinking saccharin and therefore unlikely to develop an aversion to it after just one nauseating experience.

Does the hippocampus also play a role in conditioned taste aversion? The answer seems to be: not usually. However, if the delay between the taste and the nausea is increased to 3 hours, then hippocampus lesions do prevent conditioned taste aversion. This is intriguing because other studies have shown that the hippocampus becomes necessary also in auditory fear conditioning when tone and shock are separated by a relatively long interval (30 seconds). Similarly, the hippocampus becomes necessary for eye blink conditioning (see Box 10.3) when the tone and the air puff are separated by more than a few milliseconds.

No one is sure why the hippocampus becomes involved in these forms of associative learning when the conditioned and unconditioned stimuli are separated by long intervals. The mystery is deepened by the fact that the intervals between unconditioned and conditioned stimuli that trigger hippocampal involvement are so much longer in conditioned taste aversion than in auditory fear or eye blink conditioning. Perhaps, some have argued, the hippocampus gets involved whenever a learning task is difficult.

BRAIN EXERCISE

Why it is difficult to exterminate rats with toxic bait? How might you circumvent those difficulties?

14.9 What Happens When Memories Conflict?

As you have now learned, humans and other animals exhibit multiple forms of learning that involve overlapping but distinct neural circuits. These *memory systems* tend to work in parallel and usually cooperate. For example, the motor and cognitive skills you learn while playing a video game (procedural memories) work together with your memories of specific situations (episodic memories) to improve your overall ability to play the game.

Sometimes, however, your memory systems conflict. Imagine, for example, that someone snuck into your home and surreptitiously switched the hot and cold water faucets of your bathroom sink. When you now go to brush your teeth, you'll get only hot water when you open what used to be the cold water faucet. You will probably remember this episode, but the next day, when you return to brush your teeth again, you will almost certainly reach for the wrong faucet again. As a neurobiologist would say, your episodic memory is just not strong enough to override the conflicting, deeply engrained procedural memory. If you are like most people, it will take you several days before you pick the desired faucet reliably.

Habit versus Place Learning

The neural basis of conflicts between different memory systems has been studied in rats using Tolman's *cross-maze paradigm*, which we discussed in Chapter 11. In this paradigm, rats are given daily training to find food in one arm of a cross-shaped maze (Figure 14.29). One arm of the maze is blocked, and the rats are consistently started in the opposite arm. Therefore, the rats must run to the blockage and then turn either left or right to get to the food. The rats are given several learning trials per day and soon perform quite well.

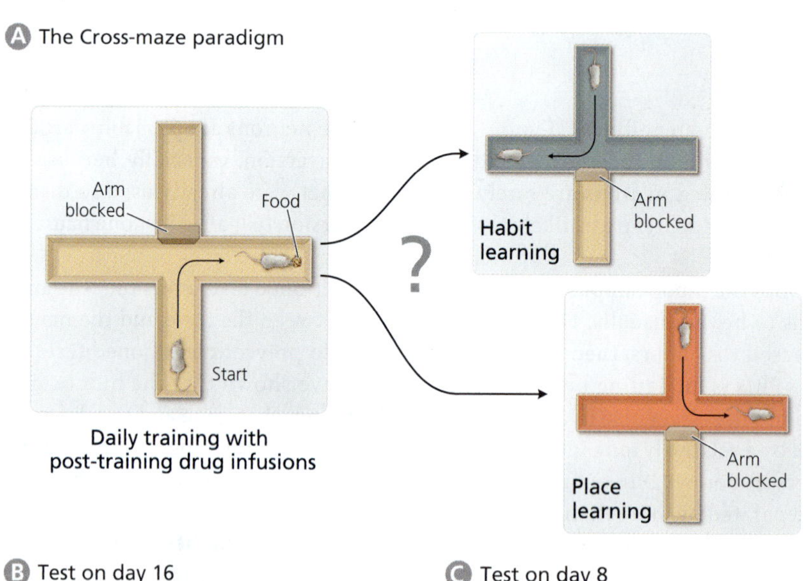

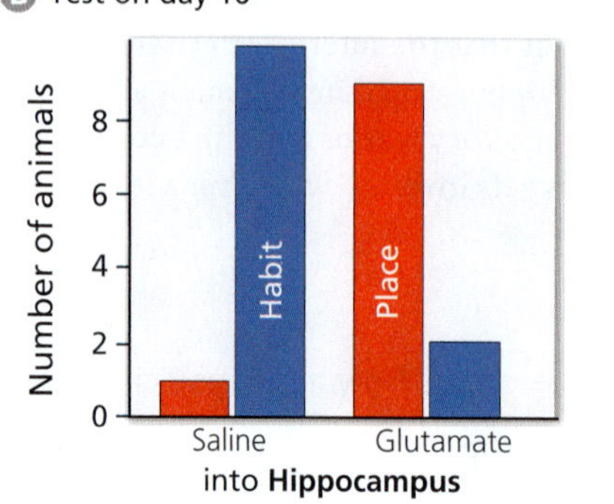

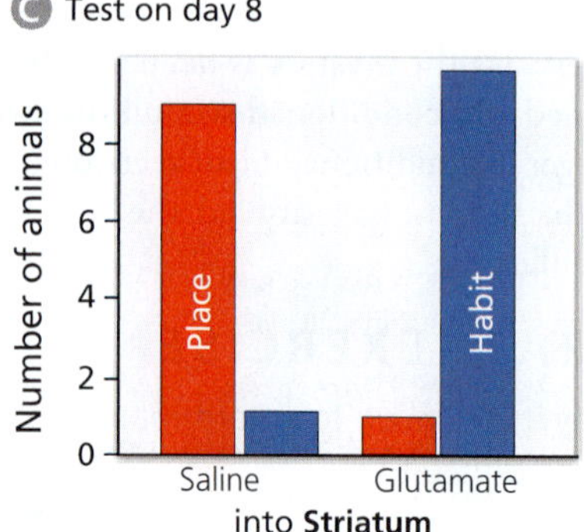

Figure 14.29 Competition between two memory systems. The cross-maze paradigm (A) is used to test whether animals have learned to find a specific place in the room or are navigating by habit (see Figure 11.29). Infusing glutamate into the hippocampus after each training trial causes the experimental rats to exhibit place learning on day 16, when normal (saline-infused) rats have switched to being guided by the force of habit (B). In contrast, animals receiving glutamate infusions into the striatum exhibit habit learning already on day 8, when control rats normally exhibit place learning (C). [After Packard, 1999]

On days 8 and 16 of the experiment, the rats are given special test trials, during which the blockage is moved to the opposite arm and the rat is started from the previously blocked arm. The crucial question is whether the rats now make the habitual turn (e.g., turn right if they always turned right during training) or turn toward the place where the food had been during training. As you learned in Chapter 11, normal rats generally exhibit "place learning" on day 8 and "habit learning" on day 16. Apparently, the motor habit takes at least 8 days to become strong enough to override the rat's allocentric place memory.

You also learned in Chapter 11 that allocentric place memory requires an intact hippocampus. Given this information, what do you think would happen if you briefly make a rat's hippocampus more excitable after each training trial in the cross-maze paradigm? According to our discussion in this chapter, you would expect such a manipulation to facilitate the formation and reactivation of hippocampal cell assemblies, which should then lead to stronger hippocampal memory traces. Behaviorally, you would expect this manipulation to boost a rat's place memory.

Indeed, when low doses of glutamate were infused into the hippocampus after each cross-maze training trial, the rats exhibited place learning when tested on day 16, which is when control rats usually exhibit habit learning (Figure 14.29 B). The simplest interpretation of this finding is that the post-training glutamate infusions strengthen the hippocampal memory traces to the point where the rats (during the test trials on day 16) cannot ignore them and, instead, ignore the information being provided by the habit learning system. According to this view, place learning and habit learning compete with one another for control over the animal's behavior.

At this point, you might ask, which brain regions control habit learning? We will explore this question in Chapter 15. For now, suffice it to say that the *striatum*, which is the largest subcortical component of the mammalian telencephalon, is critical for habit formation. Given this information, what do you predict would happen if you infuse glutamate into the striatum, rather than the hippocampus, after each cross-maze training trial? Such rats exhibit habit learning already on day 8 when control rats are still using place learning to find the food (Figure 14.29 C). Apparently, the glutamate infusions strengthen striatal memory traces, which allows the striatum to outcompete the hippocampus for control over the animals' behavior.

BRAIN EXERCISE

Can you think of instances in your own life where different memories (or learned behavior patterns) competed with one another in guiding your actions? What do you think was going on in your brain at the time?

SUMMARY

Section 14.1 - Learning and memory come in a variety of forms. In this chapter, we focused mainly on the neural bases of episodic memory and Pavlovian conditioning.

Section 14.2 - Patient H.M. had his amygdala, hippocampus, and parts of the adjacent cortices surgically removed. He developed anterograde amnesia for events in his life but retained some ability to learn new skills.

Section 14.3 - Research on monkeys and rats revealed that the perirhinal/postrhinal cortex, but not the hippocampus, is critical for object recognition memory. In contrast, the hippocampus is needed for remembering relationships (spatial or otherwise).

Section 14.4 - The hippocampus is important for several kinds of learning and memory.

- The hippocampus is reciprocally connected to the entorhinal cortex, which is reciprocally connected to the perirhinal/postrhinal cortex, which has reciprocal connections with many other neocortical areas.
- Area CA3 of the hippocampus is an autoassociative network that is capable of forming Hebbian cell assemblies and thus potential memory traces.

- It is hypothesized that reactivation of a hippocampal cell assembly can reinstate, at least in part, the pattern of neocortical activity that occurred during the original experience, leading to memory recall.

Section 14.5 - According to the standard model of systems consolidation, repeated reactivation of hippocampal memory traces leads to the creation of neocortical cell assemblies that can be activated without the hippocampus.

Section 14.6 - Behavioral arousal activates the basolateral complex of the amygdala, which then enhances memories while they are formed, while they are consolidated, or both.

Section 14.7 - Animals have a variety of rapid and robust mechanisms for learning what is dangerous.

- In auditory fear conditioning, rats learn to freeze in response to a tone previously paired with an aversive stimulus. The underlying neural circuits run through the basolateral and central nuclei of the amygdala.
- In contextual fear conditioning, animals learn to fear environments in which bad things happened to them. Recalling recent, but not old, contextual fear memories requires an intact hippocampus.
- It usually takes just one trial for a rat to avoid a place where it was scared. The ability to express inhibitory avoidance is hippocampus dependent for a few weeks but then becomes hippocampus independent.

Section 14.8 - Animals can learn after just one trial not to eat what made them sick. They can also learn from other individuals what's safe to eat. The hippocampus and basolateral amygdala are involved in these forms of learning.

Section 14.9 - Although the brain's memory systems usually collaborate, they sometimes compete for control of the organism's behavior. Localized drug infusions can influence who wins these competitive interactions.

Box 14.1 - Memory specialists (mnemonists) vary in what kinds of things they can remember exceptionally well.

Box 14.2 - Epilepsy is characterized by seizures, which are brief periods of highly synchronized and excessive neuronal activity that often begin in one place and then spread. Repeated seizures can shrink the hippocampus.

Box 14.3 - Posttraumatic stress disorder (PTSD) is a good example of a syndrome that has a genetic component but, in most people, requires special environmental influences (severe trauma) to develop.

KEY TERMS

ADDITIONAL READINGS

14.1 - Multiple Forms of Learning and Memory

Henke K. 2010. A model for memory systems based on processing modes rather than consciousness. *Nat Rev Neurosci* **11**:523–532.

Rescorla RA. 1988. Pavlovian conditioning. it's not what you think it is. *Am Psychol* **43**:151–160.

14.2 - Patient H.M. and the Medial Temporal Lobe

Burgess N, Maguire EA, O'Keefe J. 2002. The human hippocampus and spatial and episodic memory. *Neuron* **35**:625–641.

Milner B, Corkin S, Teuber H-L. 1968. Further analysis of the hippocampal amnesic syndrome: 14-year follow-up study of H.M. *Neuropsychologia* **6**:215–234.

Scoville WB, Milner B. 1957. Loss of recent memory after bilateral hippocampal lesions. *J Neurol Neurosurg Psychiatry* **20**:11–21.

14.3 - Animal Models of Anterograde Amnesia

Dere E, Kart-Teke E, Huston JP, De Souza Silva MA. 2006. The case for episodic memory in animals. *Neurosci Biobehav Rev* **30**:1206–1224.

Mumby DG, Gaskin S, Glenn MJ, Schramek TE, Lehmann H. 2002. Hippocampal damage and exploratory preferences in rats: memory for objects, places, and contexts. *Learn Mem* **9**:49–57.

Winters BD, Saksida LM, Bussey TJ. 2010. Implications of animal object memory research for human amnesia. *Neuropsychologia* **48**:2251–2261.

Wixted JT, Squire LR. 2010. The role of the human hippocampus in familiarity-based and recollection-based recognition memory. *Behav Brain Res* **215**:197–208.

14.4 - Memory Formation and Recall

Danker JF, Anderson JR. 2010. The ghosts of brain states past: remembering reactivates the brain regions engaged during encoding. *Psychol Bull* **136**:87–102.

Harris KD, Csicsvari J, Hirase H, Dragoi G, Buzsáki G. 2003. Organization of cell assemblies in the hippocampus. *Nature* **424**:552–556.

McHugh TJ, Jones MW, Quinn JJ, Balthasar N, Coppari R, Elmquist JK, et al. 2007. Dentate gyrus NMDA receptors mediate rapid pattern separation in the hippocampal network. *Science* **317**:94–99.

Myers CE, Scharfman HE. 2011. Pattern separation in the dentate gyrus: a role for the CA3 backprojection. *Hippocampus* **21**:1190–1215.

Naya Y, Yoshida M, Miyashita Y. 2001. Backward spreading of memory-retrieval signal in the primate temporal cortex. *Science* **291**:661–664.

O'Reilly RC, Rudy JW. 2000. Computational principles of learning in the neocortex and hippocampus. *Hippocampus* **10**:389–397.

14.5 - Retrograde Amnesia and Systems Consolidation

Frankland PW, Bontempi B. 2005. The organization of recent and remote memories. *Nat Rev Neurosci* **6**:119–130.

McGaugh, J. L. (2000). Memory—a century of consolidation. *Science* **287**:248–251.

Stickgold R, Walker MP. 2005. Memory consolidation and reconsolidation: what is the role of sleep? *Trends Neurosci* **28**:408–415.

Wiltgen BJ, Zhou M, Cai Y, Balaji J, Karlsson MG, Parivash SN, Li W, Silva AJ. 2010. The hippocampus plays a selective role in the retrieval of detailed contextual memories. *Curr Biol* **20**:1336–1344.

Winocur G, Moscovitch M, Bontempi B. 2010. Memory formation and long-term retention in humans and animals: convergence towards a transformation account of hippocampal-neocortical interactions. *Neuropsychologia* **48**:2339–2356.

14.6 - Memory Modulation

Adolphs R, Cahill L, Schul R, Babinsky R. 1997. Impaired declarative memory for emotional material following bilateral amygdala damage in humans. *Learn Mem* **4**:291–300.

de Curtis M, Paré D. 2004. The rhinal cortices: a wall of inhibition between the neocortex and the hippocampus. *Prog Neurobiol* **74**:101–110.

McGaugh JL. 2004. The amygdala modulates the consolidation of memories of emotionally arousing experiences. *Annu Rev Neurosci* **27**:1–28.

Paz R, Pelletier JG, Bauer EP, Paré D. 2006. Emotional enhancement of memory via amygdala-driven facilitation of rhinal interactions. *Nat Neurosci* **9**:1321–1329.

Wolf OT. 2009. Stress and memory in humans: twelve years of progress? *Brain Res* **1293**:142–154.

14.7 - Learning about Threats

Choi J-S, Kim JJ. 2010. Amygdala regulates risk of predation in rats foraging in a dynamic fear environment. *Proc Natl Acad Sci U S A* **107**:21773–21777.

Raybuck JD, Lattal KM. 2011. Double dissociation of amygdala and hippocampal contributions to trace and delay fear conditioning. *PLoS ONE* **6**:e15982.

Wilensky AE, Schafe GE, LeDoux JE. 2000. The amygdala modulates memory consolidation of fear-motivated inhibitory avoidance learning but not classical fear conditioning. *J Neurosci* **20**:7059–7066.

14.8 - Learning about Food

Koh MT, Wheeler DS, Gallagher M. 2009. Hippocampal lesions interfere with long-trace taste aversion conditioning. *Physiol Behav* **98**:103–107.

Morris R, Frey S, Kasambira T, Petrides M. 1999. Ibotenic acid lesions of the basolateral, but not the central, amygdala interfere with conditioned taste aversion: evidence from a combined behavioral and anatomical tract-tracing investigation. *Behav Neurosci* **113**:291–302.

Winocur G, McDonald RM, Moscovitch M. 2001. Anterograde and retrograde amnesia in rats with large hippocampal lesions. *Hippocampus* **11**:18–26.

14.9 - Interacting Memory Systems

Biedenkapp JC, Rudy JW. 2009. Hippocampal and extrahippocampal systems compete for control of contextual fear: role of ventral subiculum and amygdala. *Learn Mem* **16**:38–45.

Oliveira AMM, Hawk JD, Abel T, Havekes R. 2010. Post-training reversible inactivation of the hippocampus enhances novel object recognition memory. *Learn Mem* **17**:155–160.

Poldrack RA, Packard MG. 2003. Competition among multiple memory systems: converging evidence from animal and human brain studies. *Neuropsychologia* **41**:245–251.

Wiltgen BJ, Sanders MJ, Anagnostaras SG, Sage JR, Fanselow MS. 2006. Context fear learning in the absence of the hippocampus. *J Neurosci* **26**:5484–5491.

Boxes

Ben-Ari Y, Dudek FE. 2010. Primary and secondary mechanisms of epileptogenesis in the temporal lobe: there is a before and an after. *Epilepsy Curr* **10**:118–125.

Helbig I, Scheffer IE, Mulley JC, Berkovic SF. 2008. Navigating the channels and beyond: unravelling the genetics of the epilepsies. *Lancet Neurol* 7:231–245.

Kindt M, Soeter M, Vervliet B. 2009. Beyond extinction: erasing human fear responses and preventing the return of fear. *Nat Neurosci* **12**:256–258.

Myers KM, Carlezon WA, Davis M. 2011. Glutamate receptors in extinction and extinction-based therapies for psychiatric illness. *Neuropsychopharmacology* **36**:274–293.

Parker ES, Cahill L, McGaugh JL. 2006. A case of unusual autobiographical remembering. *Neurocase* **12**:35–49.

CHAPTER 15

Selecting Actions, Pursuing Goals

CONTENTS

FEATURES

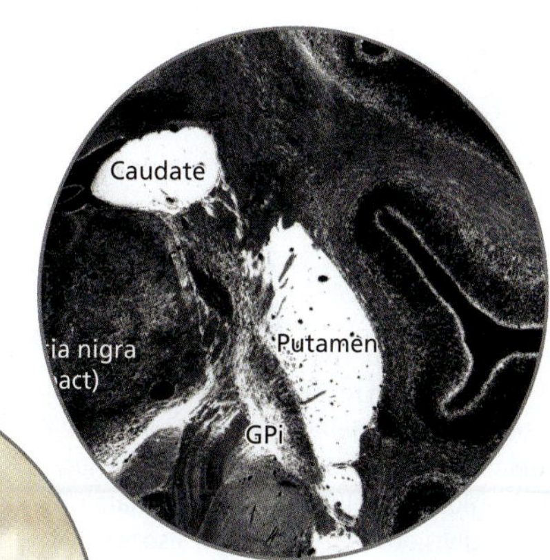

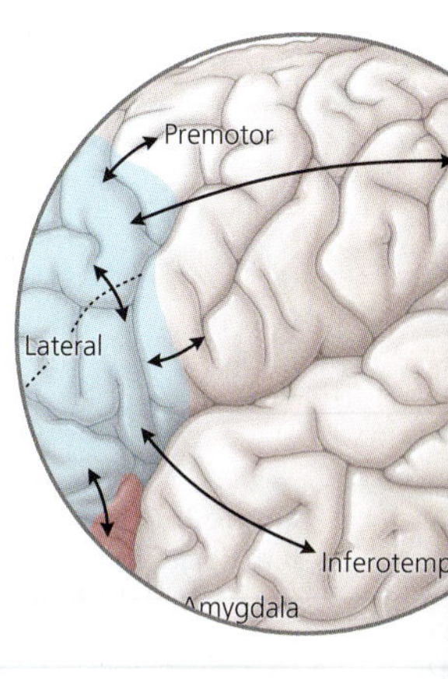

In Chapter 14 we discussed multiple forms of learning and memory, but we only scratched the surface of habit learning, which generates stereotyped sequences of behavior. In essence, habits are formed when you repeatedly pursue a goal in the same way. With enough repetition, you eventually perform the series of actions automatically, perhaps entirely subconsciously. As you will learn in this chapter, the selection of goals, actions, and movements involves a far-flung system of interconnected brain regions called the **frontostriatal system**.

15.1 What Is the Frontostriatal System?

The frontostriatal system includes parts of the neocortex as well as several subcortical structures, which are collectively referred to as the **basal ganglia**. The word "basal" in this name indicates that the basal ganglia lie deep underneath (basal to) the neocortex; and the term "ganglia" refers to their nuclear, rather than laminar, organization (*ganglion* means "gathered into a ball" in ancient Greek).

The largest component of the basal ganglia is the striatum, which is divisible into a large **dorsal striatum** and a smaller **ventral striatum** (Figure 15.1). The ventral striatum is generally involved in generating cravings for food, water, sex, and drugs. It receives strong inputs from the *orbital prefrontal cortex*, which occupies the inferior surface of the frontal lobe, just behind the orbits (eye sockets). In contrast, the dorsal striatum receives major inputs from posterior regions of the frontal lobe, including the motor and premotor cortices, and is involved in the learning and control of stereotyped movements.

The outputs of the striatum target primarily the pallidum. Specifically, the dorsal striatum projects to the **dorsal pallidum**, whereas the ventral striatum projects to the **ventral pallidum**. Both divisions of the pallidum project to the thalamus, with the dorsal and ventral pallidum targeting distinct thalamic nuclei. These thalamic areas then project to the frontal lobe. Roughly speaking, the parts of the thalamus that get input from the ventral pallidum project to the anterior prefrontal cortex, whereas the thalamic nuclei receiving input from the dorsal pallidum project to more posterior portions of the frontal lobe. Thus, these connections collectively form two large, parallel loops. Some of these loops can be subdivided further, and the loops are not as parallel as they at first appear; but the core idea is clear: the frontal cortices and the striatum are connected to each other through a series of looping connections. Collectively, they form the core of the frontostriatal system.

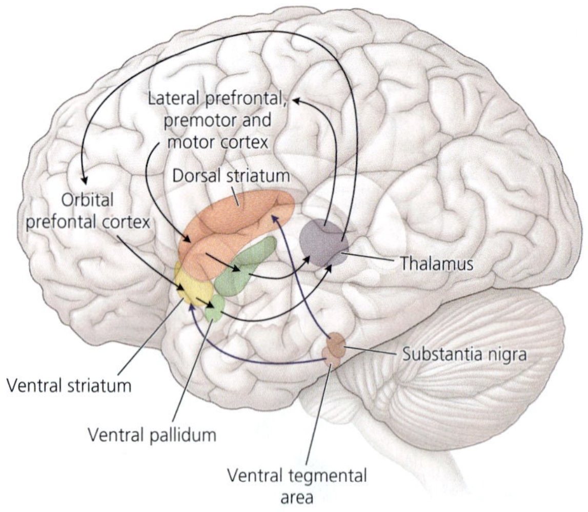

Figure 15.1 The frontostriatal system. The striatum is divisible into dorsal and ventral divisions, which receive inputs from different parts of the frontal lobe. The two striatal divisions project, through the pallidum (dorsal and ventral divisions) and thalamus, back to their cortical input regions, forming large frontostriatal loops. The dorsal and ventral divisions of the striatum also receive dopaminergic inputs from the substantia nigra and ventral tegmental area, respectively.

Complexities of Basal Ganglia Nomenclature

Before proceeding, let us briefly discuss some terminological issues involving the basal ganglia. Most of these issues need not worry us, but if you read other texts, then you will come across some terms that we largely refrain from using in this book. Knowing how the various terms relate to one another can help avoid unnecessary confusion.

Two potentially confusing terms are *caudate* and *putamen*. They refer to two parts of the primate dorsal striatum that are separated by a fiber tract (Figure 15.2). These divisions are not recognized in rodents because in these species, the comparable fiber tract is broken up into many small axon bundles, which give the striatum its "striated" appearance but make a simple division into caudate and putamen impossible.

The largest component of the ventral striatum is called the *nucleus accumbens*, but we don't need to use this term. Similarly, we do not need the term *globus pallidus*, which refers to two very different, although adjoining, structures. The *internal segment*

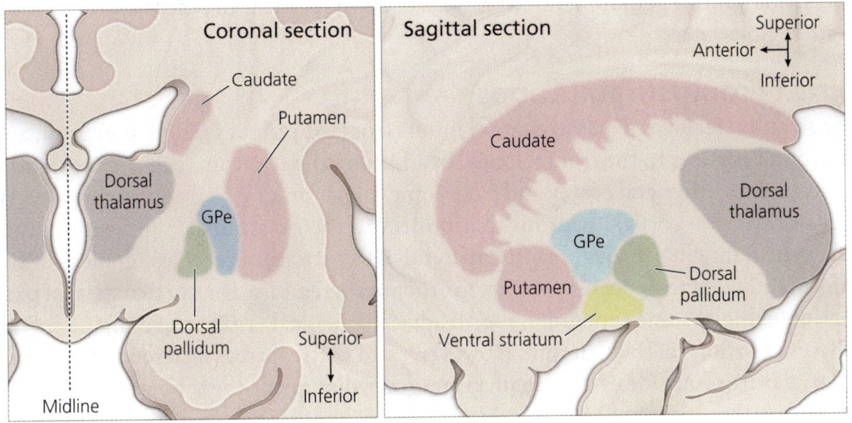

Figure 15.2 Schematic sections through the human striatum and pallidum. The panel on the left shows a coronal section through the dorsal striatum (caudate plus putamen) and the dorsal thalamus. The external globus pallidus (GPe) is sandwiched between the putamen and the dorsal pallidum. The right panel depicts a sagittal section through the same structures, plus the ventral striatum.

of the globus pallidus is part of the dorsal pallidum, as we define it here. The *external segment of the globus pallidus* contains very different neurons and is not part of the dorsal pallidum at all.

The term *substantia nigra* likewise refers to two adjacent but very different structures. The *reticulated division of the substantia nigra* is functionally and anatomically quite similar to the internal segment of the globus pallidus, which is why we refer to these two areas by just one name: the dorsal pallidum. The **substantia nigra, compact division (SNc)**, is very different from the reticulated division. Instead of being part of the dorsal pallidum, it resembles an adjoining structure called the **ventral tegmental area (VTA)**. Both structures are located in the ventral midbrain and contain dopaminergic neurons. The SNc projects mainly to the dorsal striatum, whereas the VTA sends its axons to the ventral striatum (Figure 15.1). We refer to them collectively as **SNc/VTA**.

An Overarching Function for the Frontostriatal System

A potentially confusing aspect of the frontostriatal system is that it has been linked to a broad variety of functions, ranging from sex and drugs to Parkinson's disease, which is generally considered to be mainly a movement disorder. To manage this complexity, we here consider the frontostriatal system as performing just one very general function, namely, the selection of context-appropriate behaviors from competing alternatives. This overarching function includes the selection of goals, of actions to attain those goals, and of specific movements to perform the actions.

This view of the frontostriatal system as being critical for behavior selection is relatively new, first emerging in the mid-1990s, but it has gained acceptance steadily. If you think about it, selecting appropriate behaviors is extremely important. After all, organisms have many different behaviors they could perform at any given time. How are they to decide what to do when? The frontostriatal system helps them make those often vital decisions.

BRAIN EXERCISE

Can you image a set of neural mechanisms that would help animals make decisions about how to behave in various situations? Would the mechanisms you imagine be hardwired or capable of learning from experience?

15.2 What Are the Direct and Indirect Pathways through the Striatum?

To understand how the frontostriatal system mediates the selection of goals, actions, and movements, we must take a closer look at its circuits. Let us begin with the shortest, most direct pathway from the frontal cortex through the basal ganglia and back. To keep it simple, we do not distinguish between dorsal and ventral striatum, or dorsal and ventral pallidum, for now; we will consider the differences between these brain regions later.

Direct Frontostriatal Loops

The striatum receives dense excitatory inputs from the primary motor, premotor, and prefrontal cortices in the frontal lobes. In addition, the striatum receives substantial inputs from the parietal cortex and from some higher order visual and auditory cortices. Because the cortical neurons that project to the striatum also project to other subcortical targets (through axon collaterals), the striatum receives "copies" of the information that the neocortex sends to other brain regions. The **corticostriatal projection** exhibits some topography, with adjacent cortical areas projecting mainly to adjacent portions of the striatum. However, most cortical areas also send a few axons to striatal areas outside of their main target zone, disrupting the topography. Overall, we can say that each part of the striatum receives input mainly from a single region of the neocortex but integrates this input with information from other cortical areas.

The principal neurons of the striatum are the **medium spiny neurons** (Figure 15.3). As the name implies, these neurons are covered with numerous dendritic spines, which is where most of the corticostriatal synapses are located. It has been estimated that each medium spiny neuron receives converging input from at least 1,000 different cortical neurons. Many of these cortical inputs must be active simultaneously to trigger action potentials in a medium spiny neuron, which explains why these neurons tend to fire only 2–10 action potentials per second. The axons of medium spiny neurons branch locally within the striatum (Figure 15.3 C) and send long axon branches to the pallidum. The pallidal neurons then project to the dorsal thalamus, which projects to the neocortex (Figure 15.4).

Disinhibition of the Pallidum

An important aspect of this frontostriatal loop is that it contains two excitatory and two inhibitory synapses. The thalamic and cortical neurons use glutamate and excite

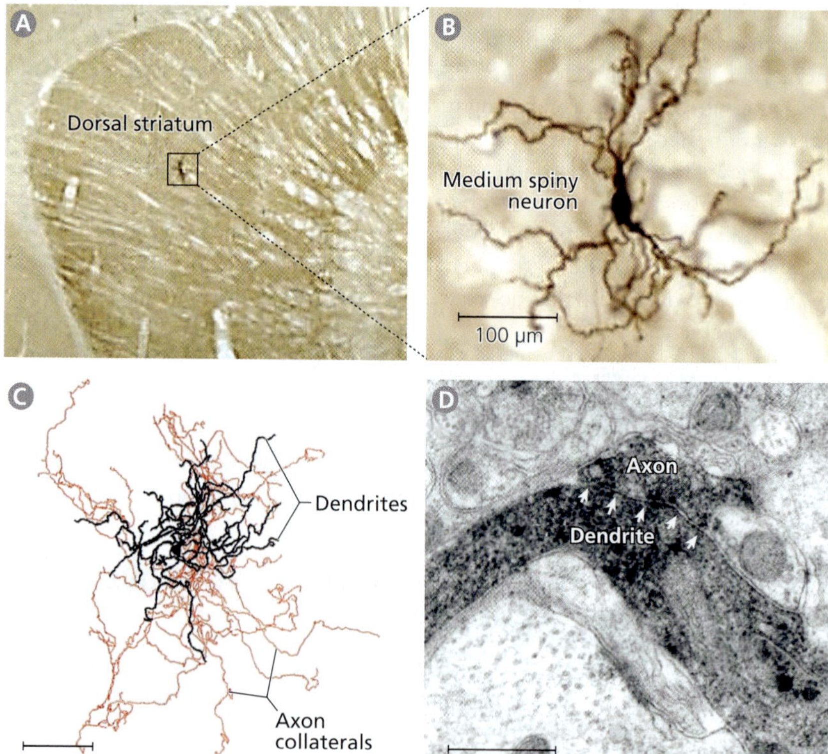

Figure 15.3 Striatal medium spiny neurons. A medium spiny neuron in a rat's dorsal striatum was filled intracellularly with biocytin to color it brown (A). You can see the cell body and dendrites (B). Shown in (C) is another medium spiny neuron. Its axon (red) has many collateral branches that contact other medium spiny neurons. The electron micrograph in (D) shows a synapse between two intracellularly labeled medium spiny neurons. [A and B from Gerfen, 2003; C and D from Tepper et al., 2004]

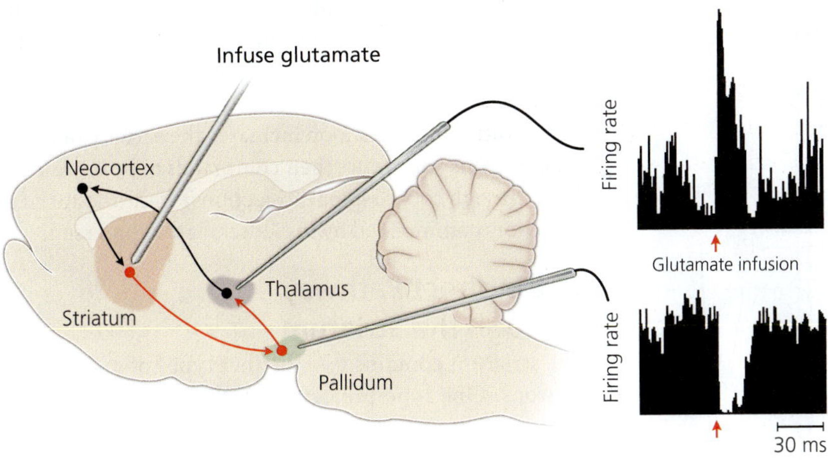

Figure 15.4 Disinhibition in the frontostriatal loop. A small amount of glutamate was infused into the striatum of a rat. Simultaneously, recording electrodes were placed in the rat's pallidum and thalamus. As the glutamate was injected, the pallidal neurons temporarily fell silent and the thalamic neurons briefly increased their firing rate. These data are consistent with the principal neurons in both the striatum and the pallidum being inhibitory (red). Excitation of the striatal neurons increases inhibition of the pallidal neurons, which in turn removes inhibition from (disinhibits) the thalamic neurons. [After Deniau and Chevalier, 1985]

their postsynaptic cells. In contrast, the medium spiny neurons of the striatum use GABA as their transmitter and are inhibitory. The same is true of the pallidal neurons. Given these facts, can you predict how thalamic and neocortical neurons would react to a burst of striatal activity?

The answer is shown in Figure 15.4: a burst of action potentials in the striatum increases inhibition of the target neurons in the pallidum, reducing their firing rate from about 60 spikes/second to almost nil. This decrease in pallidal activity *disinhibits* (removes inhibition from) the thalamic neurons, causing them to fire a strong burst of action potentials, which then provides additional excitation to the neocortex. Thus, the loop through the basal ganglia can function as a positive feedback loop. Whenever a group of neocortical neurons is active enough to excite the striatum medium spiny neurons, the loop causes the neocortical neurons to become even more active.

Competition between Frontostriatal Loops

To understand the function of this positive feedback loop you have to consider that the medium spiny neurons in the striatum inhibit each other (Figure 15.3 C, D). Although the inhibitory connection between any two medium spiny neurons is relatively weak, their collective effect is to create a **winner-take-all competition** in which the most active medium spiny neurons inhibit their less active neighbors (Figure 15.5).

This kind of competition is analogous to what happens in a classroom when one student is eager to answer the teacher's questions but, in doing so, inhibits the others. Moreover, just as highly vocal students sometimes take the silence of the other students as a signal that they themselves should continue talking, the positive feedback loop through the striatum provides neurons that are more active than their neighbors with the excitation that they need to continue suppressing the other neurons. Without the positive feedback, the winning neurons would soon fall silent, giving up their hard-earned victory.

What happens when you combine this idea of a winner-take-all competition in the striatum with the observation that the striatum gets excitatory input from behavior-controlling neurons in the frontal lobe? You get a system that is well-suited for behavior selection. The idea is that cortical neurons commanding different behaviors activate different sets of striatal neurons, to varying degrees. The most active set of

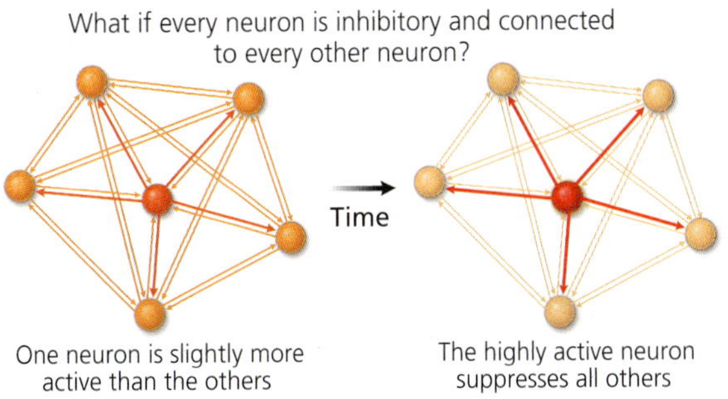

Figure 15.5 Intrastriatal winner-take-all competition. Shown here is a hypothetical network of six medium spiny neurons that all inhibit each other (arrows). Shown at the left is a situation in which the central neuron is slightly more active than the other neurons (as indicated by its darker color and thicker arrows). Because it is more active, the central neuron tends to inhibit its neighbors more than they inhibit it back. Therefore, the network soon transitions to the state shown on the right in which only the central neuron is active. It has won the winner-take-all competition.

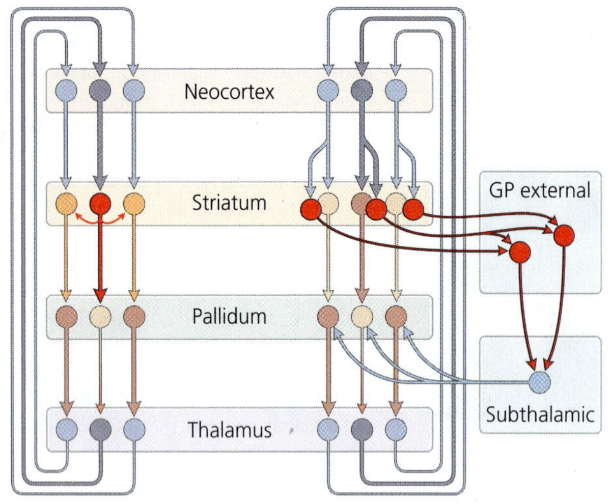

Figure 15.6 Direct and indirect pathways through the basal ganglia. Illustrated at the left is the so-called direct pathway through the basal ganglia. Excitatory neurons are colored blue; inhibitory ones are orange or red. Highly active neurons are colored dark blue or red, whereas relatively silent neurons are shown in light blue or orange. The inhibitory axon collaterals of medium spiny neurons are shown as laterally directed arrows within the striatum. The diagram on the right depicts the indirect pathway, which originates from a distinct set of inhibitory striatal neurons and ends with a relatively diffuse excitatory projection from the subthalamic nucleus to the pallidum.

striatal neurons then inhibits the others and wins the intrastriatal competition. The winning neurons transmit the news of their victory through the pallidum and thalamus to the neocortical neurons that activated them in the first place. This positive feedback loop increases the activity of the winning neurons and allows their commands to be implemented (e.g., through corticospinal projections) while the alternative behaviors commanded by the "losers" are suppressed.

The Indirect Pathway through the Striatum

The striatum contains two distinct types of medium spiny neurons. One type projects directly to the pallidum. The other type projects to the **external segment of the globus pallidus (GPe)**, which contains inhibitory neurons and projects to the **subthalamic nucleus**. The latter nucleus is relatively small and contains glutamatergic neurons that are tonically active at about 20 spikes per second. These subthalamic neurons project to the pallidum. Together, GPe and the subthalamic nucleus form a side-loop to the direct pathway from the striatum to the pallidum. This side-loop is sometimes called the **indirect pathway** through the basal ganglia (Figure 15.6).

Interactions between the Direct and Indirect Pathways

A crucial difference between the direct and indirect pathways is that the latter employs two inhibitory neurons to pass information from the striatum to the pallidum, whereas the former uses one. Thus, a burst of activity in the striatum briefly inhibits pallidal neurons through the direct pathway but excites (disinhibits) them through the indirect pathway. In other words, the direct and indirect pathways oppose one another.

Do the two pathways cancel each other out? They do to some extent, but not exactly, because the indirect pathway is topographically much less precise than the direct pathway. In particular, the projection from the subthalamic nucleus to the pallidum is highly divergent (Figure 15.6). Therefore, activation of a small cluster of medium spiny neurons in the striatum will, through the indirect pathway, excite a relatively large group of pallidal neurons. In contrast, the same pattern of striatal activation will, through the direct pathway, inhibit only a small cluster of pallidal neurons. The net result is that the direct pathway provides the pallidum with tightly focused inhibition, whereas the indirect pathway provides it with diffuse excitation. The functional significance of this arrangement remains debatable, but the diffuse excitation coming from the subthalamic nucleus should make it harder for the direct pathway to activate (disinhibit) thalamic neurons. Therefore, you can think of the indirect pathway as regulating how easily signals pass through the direct pathway to set up the positive feedback loops we discussed earlier.

Huntington's Disease

What happens to behavior when the indirect pathway through the subthalamic nucleus and GPe becomes dysfunctional? The answer can be gleaned from patients with **Huntington's disease**. This disease, named after the man who first described its clinical features in 1872, afflicts about 1 person out of every 10,000 and is caused by a mutation in the gene for the **huntingtin protein**. It emerges when a 3-nucleotide repeated sequence (a CAG [cytosine/adenine/guanine] repeat) in the huntingtin gene expands to more than 35–39 copies. At that point, the huntingtin protein becomes neurotoxic. Although the abnormally expanded huntingtin gene is expressed in all neurons, it preferentially kills the striatal medium spiny neurons of the indirect pathway.

According to the model we have been discussing, losing the medium spiny neurons in the indirect pathway should increase the firing rate of GPe neurons (due to the removal of striatal inhibition), decrease the activity of subthalamic neurons, and decrease the average firing rate of neurons in the pallidum. Collectively, these changes ought to make it difficult for Huntington's patients to suppress unwanted, functionally inappropriate behaviors. Indeed, Huntington's disease is characterized by excessive involuntary movements. These motor symptoms are commonly referred to as **chorea** (after the Greek word for "dance"), which is why the disease is sometimes called Huntington's chorea.

Sadly, late stage Huntington's patients also develop serious cognitive deficits. These probably result from massive striatal degeneration and from neuron loss in the neocortex, thalamus, and hippocampus. Treatments for Huntington's disease are currently severely limited; most patients die within 15 years of their diagnosis.

BRAIN EXERCISE

How would the function of the frontostriatal system be affected if the inhibitory connections between medium spiny neurons were eliminated?

15.3 What Is the Influence of Dopamine on the Frontostriatal Loops?

Aside from cortical input, the striatum receives substantial input from neurons that use dopamine as their main neurotransmitter (Figure 15.7). These dopaminergic neurons are located in the compact division of the substantia nigra (SNc) and the ventral tegmental area (VTA). As noted earlier, these two structures project to the dorsal and ventral striatum, respectively, but for now we can think of them as a single structure called SNc/VTA.

Dopaminergic Modulation of the Striatum

The neurons in the SNc/VTA typically fire at a slow and steady rate of about 2 spikes per second. Individual SNc/VTA neurons branch profusely within the striatum, and much of the dopamine that they release diffuses out of the synapses before it is taken up by dopamine transporter molecules. Thus, the dopaminergic projection from SNc/VTA to the striatum is nonspecific and diffuse.

The postsynaptic effect of dopamine released within the striatum depends on the kind of dopamine receptors that the target neurons express. **D1-type dopamine receptors** tend to promote depolarization in their postsynaptic targets, whereas **D2-type dopamine receptors** promote hyperpolarization (both receptor types are G protein-coupled receptors, but they work through different G proteins). As you might expect, given these opposing effects, striatal neurons generally express only one of these two receptor types, not both. Specifically, medium spiny neurons that project directly to the pallidum express D1 receptors, whereas medium spiny neurons with projections to GPe express D2 receptors (Figure 15.8).

Therefore, dopamine tends to excite the direct pathway and inhibit the indirect pathway (Figure 15.9). Because the direct and indirect pathways have roughly opposing effects

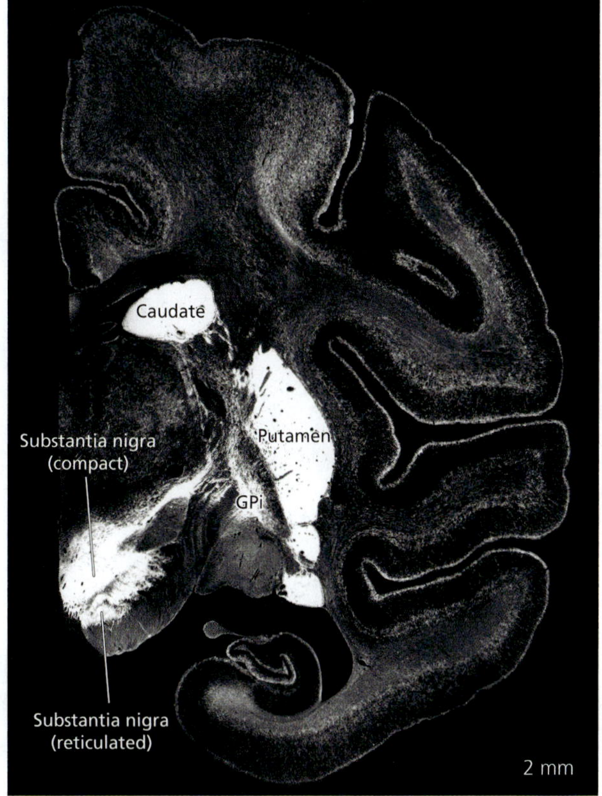

Figure 15.7 Dopamine in a monkey brain. This coronal section was stained with an antibody against the dopamine transporter, which is expressed almost exclusively in dopaminergic cell bodies, dendrites, and axons. The section was photographed using darkfield illumination, which highlights labeled structures in white against a black background. The caudate and putamen collectively comprise the dorsal striatum. Dopaminergic neurons in the compact division of the substantia nigra extend numerous dendrites into the reticulated substantia nigra. [From Lewis et al., 2001]

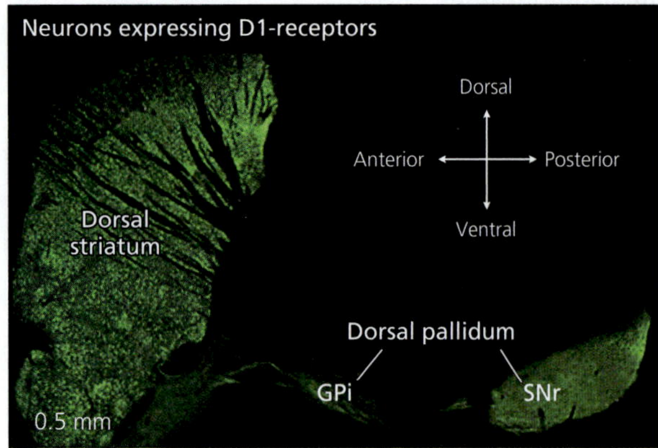

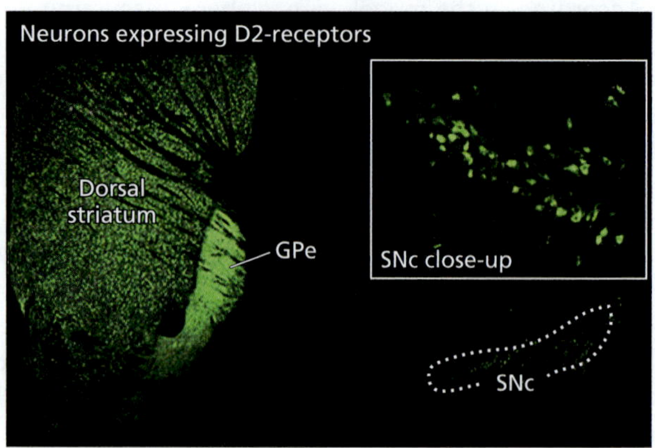

Figure 15.8 Striatal neurons expressing D1 and D2 dopamine receptors. The left image is from a mouse that expresses green fluorescent protein (GFP) in all cells that normally express the D1-type dopamine receptor; both the axons and the cell bodies are fluorescent green. The labeled axons project to the dorsal pallidum (GPi plus SNr). The right image is from a mouse that expresses GFP in neurons with D2-type receptors. The cell bodies of these neurons are found in the striatum and the compact division of the substantia nigra (SNc). The striatal D2 neurons project to the external globus pallidus (GPe). [From Valjent et al., 2009]

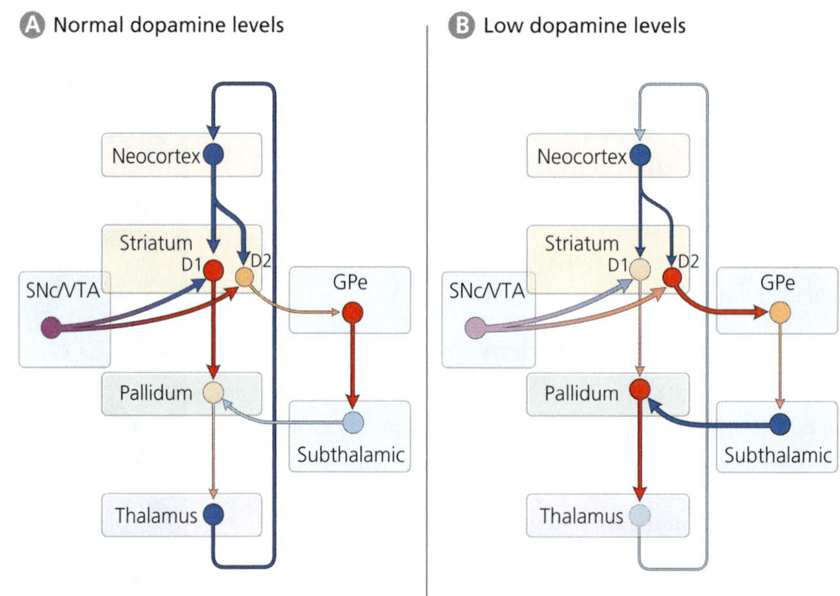

Figure 15.9 Effects of dopamine. Dopamine released from SNc/VTA neurons tends to excite (blue arrows) striatal neurons that express D1 receptors and project directly to the pallidum. It also inhibits (red/orange arrows) the D2-expressing neurons that project to the GPe (the indirect pathway). With chronically low levels of dopamine (as in Parkinson's disease), the pallidal neurons become overly active, putting a brake on thalamic and neocortical activity.

on the pallidum, striatal dopamine release tends to decrease the firing rate of pallidal neurons and thus disinhibit the thalamus. Conversely, decreases in striatal dopamine levels increase activity in the pallidum and suppress the firing of thalamic neurons.

Parkinson's Disease

A concrete illustration of what happens when dopamine levels are low is provided by **Parkinson's disease**. Named after James Parkinson, who in 1917 provided the first clinical description of what was then called "the shaking palsy," the disease afflicts about 1% of the world's population over 60 years of age.

Parkinson's disease is characterized by the progressive degeneration of dopaminergic neurons in the SNc (Figure 15.10) and, in late stages, the VTA. One common symptom of Parkinson's disease is a shaking tremor of the hand. In addition, Parkinson's patients tend to have slow reaction times, reduced facial expressions, and a sluggish, shuffling gait. They also tend to have abnormally rigid limbs and a hunched posture. At later stages, Parkinson's disease also affects a person's moods, memories, and other cognitive functions.

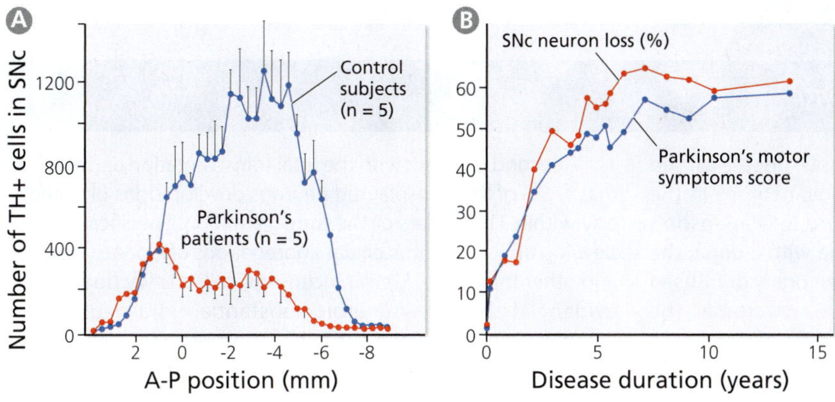

Figure 15.10 Causes and consequences of Parkinson's disease. The number of neurons that express tyrosine hydroxylase (TH), an enzyme needed for dopamine synthesis, in the compact substantia nigra (SNc) is significantly lower in Parkinson's patients than in healthy controls (A). SNc neuron loss and motor symptoms become more severe the longer a patient has suffered from Parkinson's disease (B). [After Damier et al., 1999, and Greffard et al., 2006]

No single theory can account for all these symptoms. However, most Parkinson's symptoms can be viewed as impairments in the ability to select actions (and the movements that those actions entail). According to the model we have been discussing, the loss of dopaminergic inputs makes striatal neurons in the direct pathway less active and those in the indirect pathway more active. Both effects increase the activity of pallidal neurons, which then inhibit the thalamus, thereby decreasing excitatory input to the neocortex. Thus, the death of the dopaminergic neurons in Parkinson's disease disrupts the positive feedback loops that promote the execution of the cortically generated action commands. Without those positive feedback loops, no actions can be selected.

This model can explain why Parkinson's patients are slow to move and ultimately become incapable of voluntary movements. The cognitive and mood effects are more difficult to explain but probably involve problems with the selection of goals, which we'll discuss shortly. The origins of limb tremors are even more obscure. However, the observation that tremors can be reduced by manipulation of the subthalamic nucleus (see Box 15.1: Treating Parkinson's Disease) suggests that tremors may arise from oscillatory interactions between the subthalamic nucleus and the GPe, which are reciprocally interconnected.

Animal Models of Dopamine Depletion

To learn more about the cause of Parkinson's disease and test potential therapies, it helps to have animal models of the disease. One such model is based on the discovery that a few human drug addicts had accidentally given themselves Parkinson's disease.

MPTP Toxicity in Primates

In the fall of 1976, a graduate student injected himself with a home-cooked batch of MPPP, a synthetic opioid similar to heroin. Shortly thereafter, he became unable to speak or move his limbs. A few years later, similar symptoms appeared in at least six heroin addicts in San Francisco. Clever detective work revealed that all of these "frozen addicts" had injected MPPP contaminated with a chemical called **MPTP** (1-methyl-4-phenyl-1,2,3,6-tetrahydropyridine). Injecting this compound into rats does little harm, but in primates it rapidly induces full-blown Parkinson's disease.

Further research revealed that MPTP is converted in the brain to MPP+ and that the latter molecule is extremely toxic to primate dopamine neurons. Current research is focused on understanding why dopaminergic neurons in rats are relatively resistant to MPP+ toxicity and whether Parkinson's disease might be linked to environmental toxin exposure.

Dopamine-deficient Mice

A rodent model of Parkinson's disease was developed by creating transgenic mice in which all the normally dopaminergic neurons fail to make dopamine because they lack tyrosine hydroxylase, the enzyme needed to convert L-tyrosine into **L-dopa**,

THERAPIES

Box 15.1 *Treating Parkinson's Disease*

Parkinson's disease is a devastatingly progressive disease caused mainly by the degeneration of dopaminergic neurons in the substantia nigra. Although there is no cure for Parkinson's disease, some of its symptoms are treatable with L-dopa, the precursor of dopamine. L-dopa can be taken orally or infused through a catheter into the small intestine. It crosses the blood–brain barrier and increases the amount of dopamine that is produced by dopaminergic neurons that have not yet degenerated. L-dopa can alleviate many Parkinson's disease symptoms, and, at moderate doses, it has relatively few side effects, especially if given together with carbidopa, which blocks the conversion of L-dopa to dopamine in the peripheral nervous system. Unfortunately, most Parkinson's patients need to increase their dose of L-dopa as the disease progresses. This becomes a problem because those high doses can trigger spontaneous involuntary movements similar to those observed in Huntington's disease. High doses of the drug can also promote impulsive behaviors and risk taking, including compulsive gambling.

Starting in the late 1980s, some Parkinson's patients were treated by transplanting fetal dopamine neurons into their striatum. This *fetal transplantation* approach raised serious ethical concerns because the transplanted cells were isolated from the ventral midbrain of aborted fetuses. However, cell transplantation therapy has met with some success. For example, it has been shown that the transplanted cells survive for more than 10 years, develop long processes, and release dopamine into the striatum. Most important, the procedure reduces the symptoms of Parkinson's disease in at least some patients when they are off medication (Figure b15.1). Such patients can reduce their L-dopa dose and lead more normal lives. Unfortunately, transplant patients older than 60 years of age experienced no statistically significant benefits. This is an obvious concern, given that 60 is the average age at which people are diagnosed with Parkinson's disease.

A second problem with the fetal transplantation approach is that 1–5% of the transplanted neurons develop signs of pathology within 11–16 years of the transplantation. Specifically, they develop abnormal intracellular aggregations of alpha-synuclein and other molecules. Similar inclusions, called Lewy bodies, are evident in the dopaminergic substantia nigra neurons of Parkinson's patients and probably contribute to their death. These findings suggest that some of the transplanted neurons have been "infected" by whatever kills the substantia nigra neurons in Parkinson's disease. The rate of putative "infection" is relatively low, but the phenomenon is troubling nonetheless. After all, what is the point of grafting in young, healthy dopamine cells if those transplanted cells also catch the disease?

A third significant problem with the fetal transplantation approach is that the grafted cells are placed in the striatum, rather than the ventral midbrain, where the substantia nigra is normally located. Therefore, the grafted cells do not receive the normal complement of substantia nigra inputs, which means that they cannot generate the phasic dopamine bursts that facilitate striatal plasticity. Intriguingly, recent research indicates that fetal dopaminergic neurons transplanted into the ventral midbrain of adult mice (with lesions of the substantia nigra) can grow axons that reach the striatum. Of course, the distance between the ventral midbrain and the striatum is significantly greater in humans than in mice, but the rodent work is nonetheless quite promising. Recent studies have also explored the possibility of transplanting stem cells that were derived from adult tissue, rather than embryos. Again, technical hurdles remain, but the prospects are good.

Currently, the most widely used surgical treatment for Parkinson's disease is *deep brain stimulation*. For this therapy, stimulating electrodes are implanted bilaterally into the subthalamic nuclei. The patients are also equipped with a battery pack and a programmable stimulator that passes brief current pulses through the implanted electrodes. How can stimulation of the subthalamic nucleus counteract the symptoms of Parkinson's disease, given that this nucleus is thought to be hyperactive (rather than hypoactive) in this disease? The likely answer is that electrical stimulation at frequencies higher than 100 Hz can reduce (rather than increase) neuronal activity. This would explain why deep brain stimulation typically ameliorates Parkinson's disease symptoms only after the stimulation parameters have been fine-tuned for individual patients. Although the surgical risks associated with deep brain stimulation are low, the implants can get infected or dislodged. Furthermore, this treatment is not a cure, as the disease continues to progress.

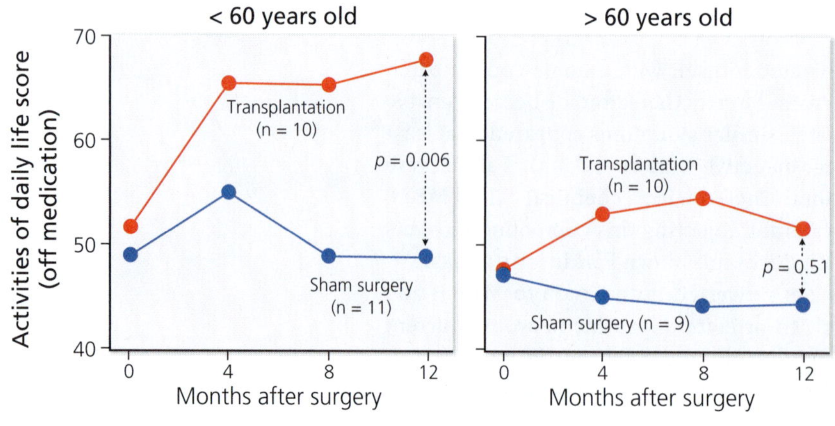

Figure b15.1 Dopamine neuron grafts. Dopaminergic neurons from human embryos were transplanted into the striatum of Parkinson's patients. Other Parkinsons's patients received only sham surgeries (no transplants). The two graphs show that patients younger than 60 years of age improve after such transplantation but that older patients don't do as well. [After Freed et al., 2001]

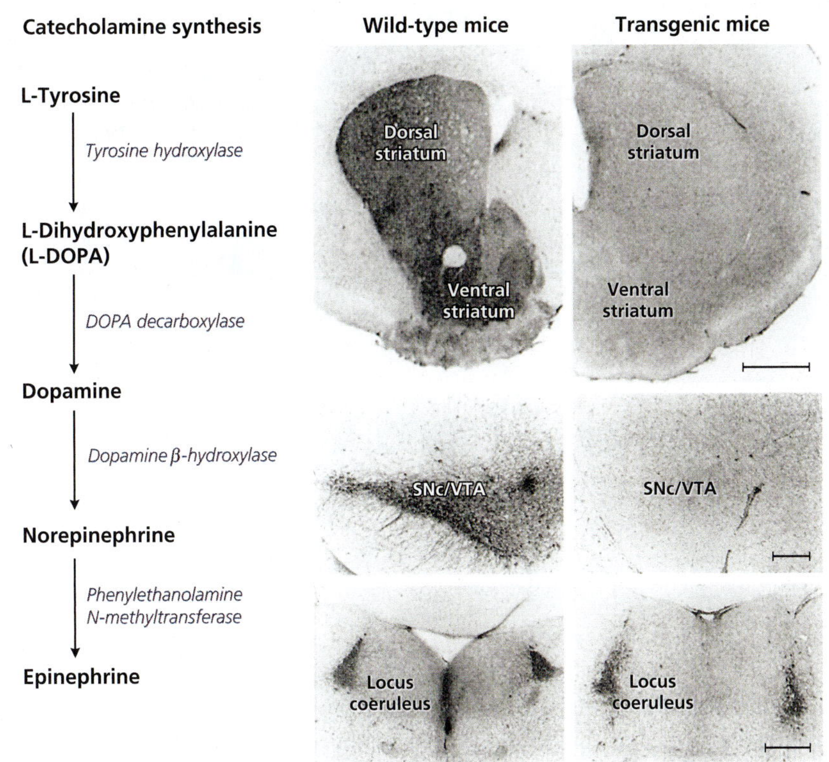

Figure 15.11 Dopamine-deficient mice. Scientists created transgenic mice with a defective tyrosine hydroxylase (TH) gene, which they then crossed with another strain engineered to express functional TH in any neurons that normally express dopamine beta hydroxylase (DBH). The resultant double transgenic mice can make norepinephrine in any neurons that normally make it (locus coeruleus). However, SNc/VTA (compact substantia nigra/ventral tegmental area) neurons do not express DBH and therefore lack TH and the capacity for making dopamine in the transgenic strain (unless those mice are fed the dopamine precursor L-dopa). The right side of the figure confirms that in the transgenic mice only locus coeruleus neurons stain with antibodies against TH. Importantly, the dopamine-deficient neurons do not die; they simply live without making dopamine. [From Zhou and Palmiter, 1995]

dopamine's immediate precursor (Figure 15.11). Such mice normally die shortly after birth, but they can be kept alive by feeding them L-dopa. While on the L-dopa diet, these mice behave like normal mice. However, when the L-dopa is withdrawn, the transgenic mice exhibit symptoms of advanced Parkinson's disease. Most obviously, the mice become much less active than normal and reduce their food and water intake more than 10-fold. They seem as "frozen" as the addicts who had injected themselves with MPTP.

Although these transgenic dopamine-deficient mice are not paralyzed, they are not motivated to do anything. This observation is consistent with the idea that striatal dopamine is required for goal and action selection. Without a goal, why should the mice do anything? Even if they had a goal, why should they engage in any actions? Why move?

Now think about the opposite problem. What would happen if striatal dopamine levels become abnormally high? One way to find an answer to this question is to give dopamine-deficient mice L-dopa in combination with **carbidopa**, a compound that prevents the L-dopa from being converted to dopamine outside the brain. This combined L/C-dopa treatment creates roughly normal levels of dopamine in the otherwise dopamine-deficient mice. However, the D1 receptors in these animals are hypersensitive because, before treatment, dopamine levels had been chronically low (most receptors become more sensitive, or exhibit higher levels of expression, when they are chronically deprived of their ligand). Therefore, normal amounts of dopamine levels in the C/L-dopa treated mice generate excessive levels of dopamine receptor activation, causing the mice to act very strangely (Figure 15.12). Most dramatically, they engage in a highly stereotyped behavior called "taffy-pulling" in which the mice sit back and repeatedly draw their front legs up to and away from their mouth.

Given the model we have been working with, this **behavioral stereotypy** probably results from excessive positive feedback within the basal ganglia loop. The uncontrolled positive feedback traps the mice in a single repetitive behavior, unable to select alternatives.

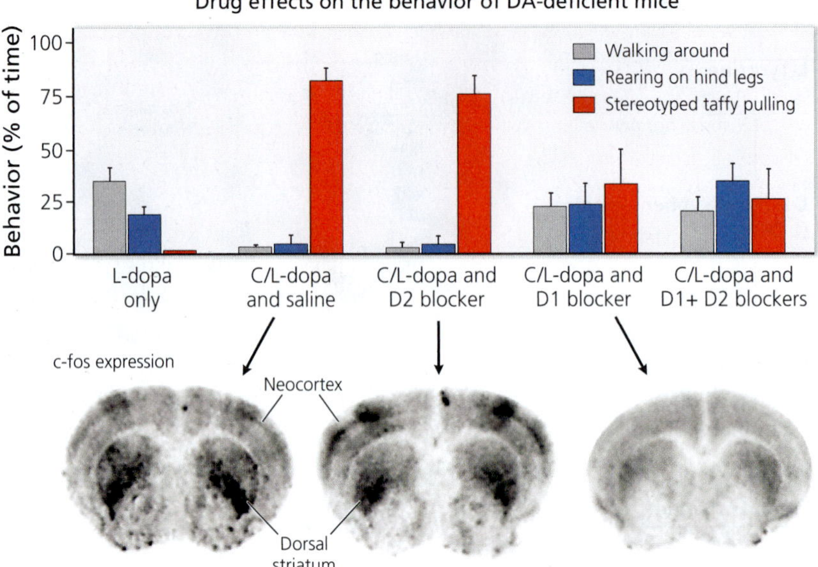

Figure 15.12 Excessive dopamine receptor activation causes stereotypy. When genetically dopamine-deficient mice are given L-dopa, they behave like typical mice. However, when the L-dopa is given in conjunction with carbidopa (C/L-dopa), further boosting dopamine levels, the mice exhibit a very odd, repetitive, and highly stereotyped behavior called "taffy-pulling." This behavior is unaffected by D2 receptor antagonists but reduced by D1 antagonists. The brain sections at the bottom of the figure illustrate expression of the immediate early gene *c-fos* in mice from three of the groups. [From Chartoff et al., 2001]

Dopamine and Drugs of Abuse

Elevated levels of striatal dopamine are also found in people high on **cocaine** or **amphetamines**. These drugs block dopamine reuptake and cause dopamine to be released from presynaptic terminals, respectively. The resulting elevation of dopamine within striatal synapses creates a feeling of euphoria and the desire to take the drugs again (we will return to these topics in the next section). In addition, the increase in striatal dopamine causes hyperactivity and, often, repetitive and stereotyped behaviors. Again, this is consistent with the general idea that the dopamine promotes goal and action selection.

While on the topic of drugs of abuse, you should know that repeated use of amphetamines depletes striatal dopamine and triggers the degeneration of dopaminergic terminals. Indeed, it may kill off entire dopaminergic neurons, not just their terminals. These destructive effects explain, at least in part, why amphetamine addicts require ever higher doses of the drugs to get the expected effects and why, without the drugs, the addicts tend to be unmotivated and sluggish.

BRAIN EXERCISE

How can transgenic dopamine-deficient mice be used as an animal model for both Parkinson's disease and dopamine receptor hyperactivity?

15.4 How Do We Learn What to Do When?

So far, we have discussed dopamine signaling as playing a major role in selecting movements, actions, and behavioral goals. This view is well established in the research literature, but dopamine is also widely discussed as signaling rewards or, in the popular media, "pleasure." How can one reconcile these seemingly divergent views? The link is actually quite straightforward.

As you know, deciding which behavior to select in a particular context gets easier with experience. Key in this shaping of behavior is the process of **instrumental conditioning**, which involves animals trying out various behaviors to learn which ones result in positive reinforcement (and minimal negative consequences). Given enough repetitions, the animals learn to repeat the most effective behaviors whenever they are in similar situations. As we now discuss, dopamine release within the striatum is key to this form of learning.

Dopamine Bursts Can Follow or Precede Rewards

Earlier in this chapter we discussed the effects of relatively slow and gradual changes in dopamine levels. Now let us examine the effects of rapid increases in striatal dopamine, which typically result from rapid-fire bursts of action potentials in SNc/VTA neurons. Such **phasic dopamine bursts** increase the excitability of the medium spiny neurons that express D1 receptors, making it easier for cortical neurons to trigger action potentials in those striatal neurons.

Measuring Dopamine Release

What triggers phasic dopamine bursts? It has been known since the late 1980s that striatal dopamine levels increase temporarily in response to rewarding stimuli, such as palatable food or a receptive mate. This discovery was originally made by implanting *microdialysis probes* into the striatum of rats and then measuring dopamine concentrations in the extracellular fluid sampled by these probes. When the rats obtained rewards, dopamine levels increased, at least slightly. However, the microdialysis technique has a temporal resolution of several minutes, which means that it cannot detect individual dopamine bursts.

A more suitable technique for measuring phasic dopamine release is **fast-scan cyclic voltammetry**, which involves a carbon fiber microelectrode whose voltage is varied rapidly and cyclically while the experimenter measures the current flowing through the electrode (Figure 15.13). At a specific voltage dopamine is oxidized, creating a small current that is proportional to the local dopamine concentration. Importantly, the temporal resolution of fast-scan cyclic voltammetry is approximately 100 ms, which allows it to detect brief rises in dopamine. Several studies using this technique have now confirmed that dopamine is often phasically released in the striatum when organisms are presented with rewards.

Reward-predicting Bursts of Dopamine Release

However, phasic dopamine bursts do not always follow rewards; sometimes they precede them. This is shown in Figure 15.14. In this experiment, rats repeatedly received a food reward a few seconds after the onset of an audiovisual stimulus (CS+). Through implanted voltammetry electrodes, striatal dopamine levels were monitored as the rats learned that the audiovisual stimulus reliably predicts the food reward. The crucial finding was that, after training, the phasic dopamine signals no longer occurred in response to the reward but in response to the reward-predicting stimulus. Therefore, phasic bursts of dopamine do not signal rewards per se. In well-trained

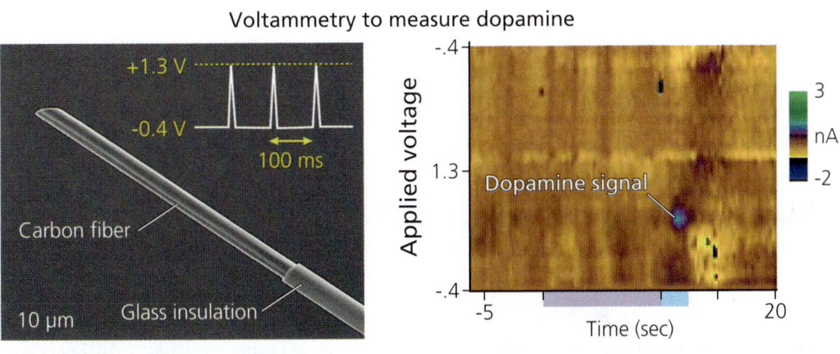

Figure 15.13 Fast-scan cyclic voltammetry. As the electrical potential of a carbon fiber electrode is varied from −.4 to 1.3V and back, dopamine is oxidized at a characteristic voltage (around +0.6V as the voltage becomes more positive). This redox reaction induces a small current, measured in nanoamps (nA), that can be subtracted from the larger current induced by the voltage changes and is proportional to dopamine concentration. The purple and light blue bars along the bottom indicate the presentation of a conditioned stimulus and a reward, respectively (see Figure 15.14). [From Robinson et al., 2003, and Day et al., 2007]

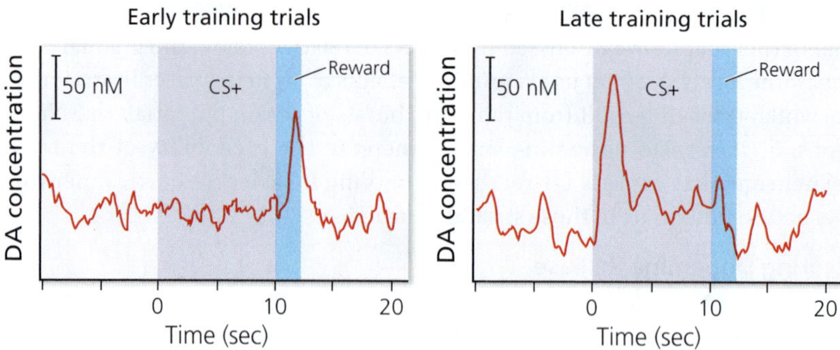

Figure 15.14 Dopamine bursts may anticipate rewards. Shown here are representative voltammetry traces from electrodes in the ventral striatum of rats as they learn to associate an audiovisual stimulus (CS+) with a food reward. In early training trials, dopamine levels increase sharply just after the reward is obtained. Later on, the dopamine response occurs just after stimulus onset, several seconds before the reward. [After Day et al., 2007]

animals, dopamine bursts occur when the animal detects some cue that, based on prior experience, predicts a reward. The predicted rewards themselves elicit little response.

Dopamine in Drug Addiction

The tendency of dopamine bursts to predict rewards in highly trained animals was originally discovered by Wolfram Schultz and his collaborators in the late 1990s, but it has now been confirmed in diverse studies and species. Many of these studies involve drugs of abuse, such as cocaine. One of them is illustrated in Figure 15.15. The rats in this experiment were trained to self-administer cocaine. That is, they learned to give themselves cocaine through an implanted catheter by pressing a lever. Importantly, the cocaine was consistently accompanied by specific sensory cues (a tone and dimmed lights). Using implanted voltammetry electrodes, the experimenters discovered that striatal dopamine levels in trained rats increased shortly before the rats received cocaine, while they were approaching the lever. These data are consistent with dopamine predicting the cocaine delivery. This predictive quality is also evident from the observation that striatal dopamine levels increase sharply to the cocaine-associated sensory cues, even when no cocaine is given. In untrained rats, no such increase is seen.

If phasic bursts of dopamine can predict a reward and do not occur when predicted rewards occur, then they cannot be the simple "pleasure signals" that one occasionally

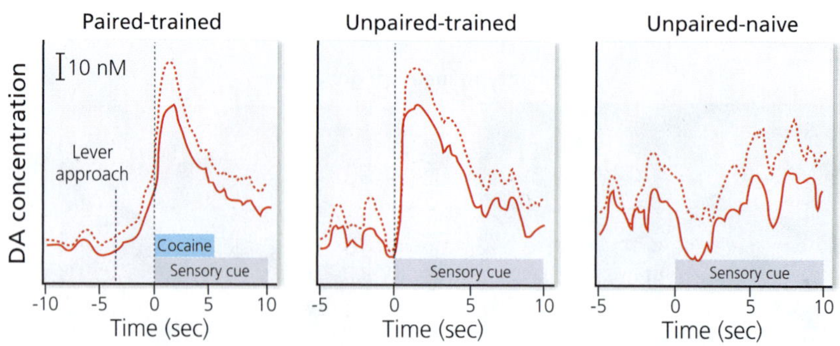

Figure 15.15 Dopamine and drug-seeking behavior. Rats were taught to self-administer cocaine (through an implanted catheter) by pressing a lever. Every time a rat received cocaine, it was simultaneously presented with a sensory cue (a tone and a change in room lighting). The left graph shows that, after training, striatal dopamine levels increase even before rats press the lever for cocaine. In the trained animals, the cocaine-associated sensory cues also prompt a burst of dopamine release, even if no cocaine is given (middle). This response does not occur in naive, untrained rats (right). The dashed red lines indicate 1 standard error above the mean for the 6 studied rats. The dashed blue and black lines indicate the timing of lever approach and bar press, respectively. [After Phillips et al., 2003]

reads about. Instead, they represent anticipatory pleasure. Moreover, this anticipation seems to be accompanied by the urge to do whatever it takes to obtain the pleasure-providing stimulus (the reward); we call that a **craving**.

To test the hypothesis that bursts of dopamine trigger cravings, one can electrically stimulate the dopaminergic neurons in the VTA of rats that had learned to self-administer cocaine by pressing a lever. When this was done, the rats tended to go to the lever and press it shortly after each stimulation event; in contrast, unstimulated rats did not exhibit this behavior (Figure 15.16). Thus, the artificially elicited dopamine signals caused the rats to perform the behavior that had previously been associated with dopamine bursts, namely, pressing the lever.

Putting it all together, we can conclude that reward-predicting dopamine signals trigger actions that previously led to the predicted reward. This is interesting because recovering drug addicts tend to relapse when they are exposed to sensory cues that had been associated with drug consumption (e.g., drug paraphernalia). Most likely, seeing the drug-associated stimulus triggers a burst of dopamine deep in the addict's brain, which then prompts him or her to perform behaviors that, in the past, led to receiving the drug. Understanding the neural basis of such *cue-induced drug seeking* may help neurobiologists find treatments for drug addiction and relapse.

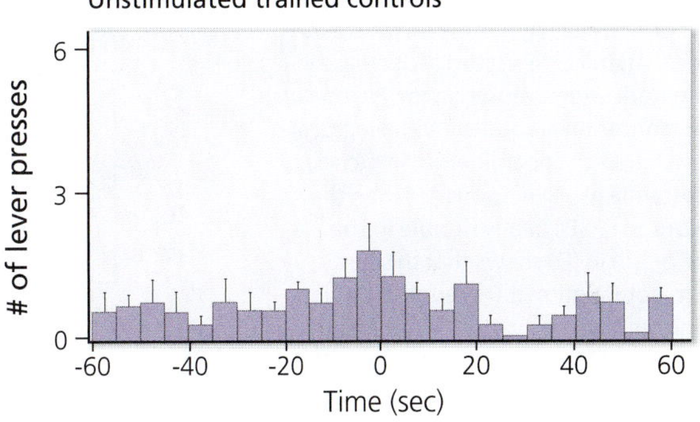

Figure 15.16 Dopamine can trigger a craving. Rats were taught to self-administer cocaine by pressing a lever. Electrical stimulation of the ventral tegmental area (VTA) causes such trained rats to press the lever for cocaine. [After Phillips et al., 2003]

Phasic Dopamine Bursts as Teaching Signals

So far you have learned that phasic dopamine signals are shaped by learning and can trigger behaviors that animals have learned to associate with rewards, but we have not yet discussed the role of dopamine signals in learning to obtain rewards in the first place. According to most models, phasic bursts of dopamine serve as *teaching signals* that tell the frontostriatal system which actions immediately preceded the dopamine burst and should therefore be selected again when a similar context recurs in the future. This is useful because those actions must have either led to a reward or increased the likelihood that a reward is coming soon (led to a predicted reward). The key idea here is that phasic dopamine signals cause a form of synaptic plasticity within the striatum that increases the likelihood that actions immediately preceding the dopamine signal will recur in the future when similar circumstances arise again.

Using Dopamine to Learn a Sequence of Actions

Consider what happens when a naive animal receives an unexpected reward. On receiving the reward, the animal's dopamine neurons fire a burst of action potentials that, according to the teaching signal model, increases the likelihood that the action preceding the reward will be selected again whenever a similar situation recurs. If the reward was food at the end of a maze and the animal turned left at the last choice point, then the burst of dopamine would "teach" the animal to turn left again at that choice point on subsequent trials.

Once the animal has learned to make that last turn consistently, the dopamine neurons stop firing in response to the food and instead begin to fire earlier, in anticipation of the food. For example, they might fire when the animal reaches the last choice point before the food. This burst of dopamine should, according to the model, reinforce whatever behavior the animal performed shortly before it reached that location. For example, it might "teach" the animal which way to turn at the second-to-last choice point.

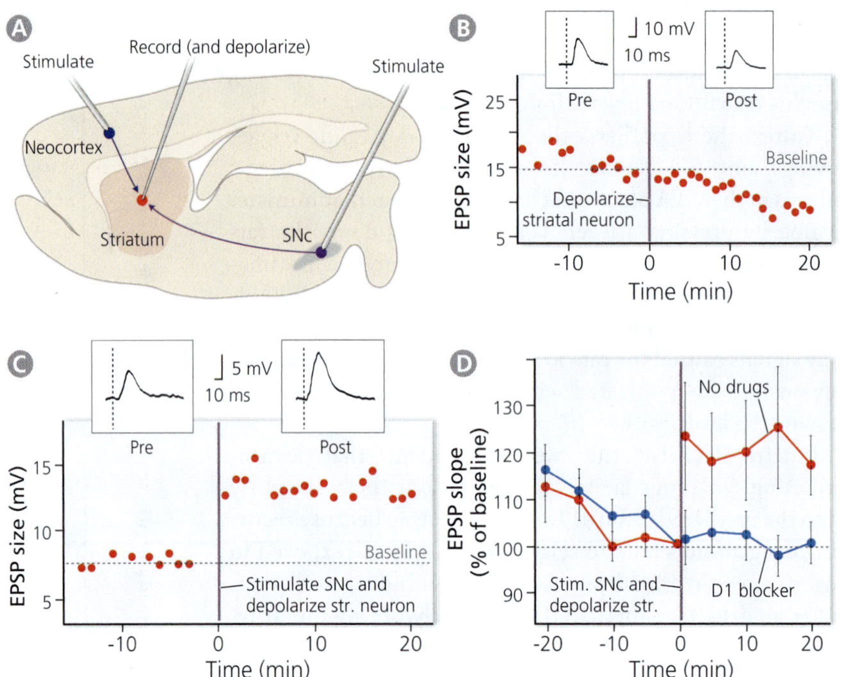

Figure 15.17 Dopamine facilitates corticostriatal LTP. EPSPs (excitatory postsynaptic potentials) were recorded intracellularly from striatal neurons in response to neocortical stimulation before and after brief sessions during which striatal neurons were prompted by intracellular depolarization to fire action potentials (A). The neocortex was unstimulated during these sessions, but its neurons provided spontaneous input to the striatal neurons. Therefore, corticostriatal synapses were active while the striatal cells were firing. The graph in (B) shows that, in this paradigm, striatal EPSPs steadily decrease in amplitude. However, when the intracellular depolarization is coupled with electrical stimulation of dopaminergic SNc (compact substantia nigra) neurons (C), then the recorded EPSPs nearly double in size. The graph in (D) shows that this long-term potentiation (LTP) requires D1 dopamine receptor activation. [After Reynolds et al., 2001]

If you think this through, you'll realize that, over time, the predictive dopamine signals can teach the animal an ever longer sequence of behaviors that ultimately lead to the reward. The fundamental insight is that, just as a reward can reinforce a preceding behavior, so can the prediction of a future reward. Computer algorithms that use reward-predicting signals to reinforce behavior are, indeed, highly effective. Therefore, the dopamine-as-teaching-signal model is highly plausible.

Plasticity at Corticostriatal Synapses

An important prediction of the model is that phasic bursts of dopamine should facilitate synaptic plasticity within the striatum. Much evidence to that effect is now available.

Particularly influential was a study that measured how medium spiny neurons in anesthetized rats respond to cortical stimulation before and after dopamine bursts. As shown in Figure 15.17, striatal responses to repeated cortical stimulation get weaker across trials, even when the striatal neurons are depolarized by intracellular current injection. However, when the intracellular depolarization is accompanied by SNc stimulation and, thus, striatal dopamine release, then the cortical inputs to the striatum are strengthened for at least 20 minutes (Figure 15.17 C). Because this long-term potentiation (LTP) is not observed when the animals are injected with a D1-type dopamine receptor antagonist (Figure 15.17 D), we can conclude that **corticostriatal LTP** is dopamine dependent.

If dopamine strengthens a specific set of corticostriatal inputs, then the goals, actions, or movements commanded by those inputs have an increased chance of reoccurring in the future. After all, in our model of striatal function, the winner-take-all competition rewards the strongest, most active inputs. If this is correct, then bursts of striatal dopamine should make it more likely that the goals, actions, or movements that immediately preceded or accompanied a dopamine burst will reoccur, given the same behavioral context. In doing so, bursts of dopamine facilitate instrumental conditioning; they "teach" animals how to obtain rewards.

BRAIN EXERCISE

If someone asks you "what does dopamine do," what would your answer be?

15.5 How Do the Dorsal and the Ventral Striatum Relate to One Another?

In this chapter's opening section, you learned that the frontostriatal system contains at least two parallel loops: one courses through the ventral striatum, the other through the dorsal striatum (see Figure 15.1). Given their similarities in connectivity and physiology, the two loops probably perform a similar function, namely, the selection of context-appropriate behaviors. However, the two loops are specialized for different aspects of behavior selection.

The ventral striatal loop seems to be involved mainly in the selection of high-level behavioral goals—such as the procurement of food, drink, sex (Box 15.2: Sex Drive and Sexual Conditioning) and, in some cases, drugs. It is involved in learning under which conditions to select which goal. For example, activation of this loop by a stimulus associated with delicious food can, over time, generate a learned craving for that food.

In contrast, the major function of the dorsal striatal loop is to learn which actions and movements are most likely to accomplish the chosen goal. Obviously, craving some goal is not enough. You need to know what to do to satisfy the craving. This, in essence, is the task of the dorsal striatal system. Given this hypothesis, you will not be surprised to learn that the striatal lesions that impair learned motor habits in Tolman's cross maze (see Figure 14.29) all involve the dorsal, rather than the ventral, striatum.

Drug-conditioned Place Preference

Further evidence for a division of labor between the dorsal and ventral striatal loops comes from studies on **drug-conditioned place preference**. In these experiments, an animal, usually a rat, is allowed to explore two experimental chambers connected by a smaller middle chamber (Figure 15.18). The two main chambers are colored differently, have floors of different textures, and are scented with different odors. Over the next few days the animals are confined alternately to one chamber or the other (for about half an hour at a time). Importantly, the animals are injected with a pleasure-inducing drug (usually cocaine, morphine, or amphetamine) just before they are put into one of the chambers, and injected with saline before being placed in the other chamber. Thus, they tend to be "high" in one of the chambers and "sober" in the other.

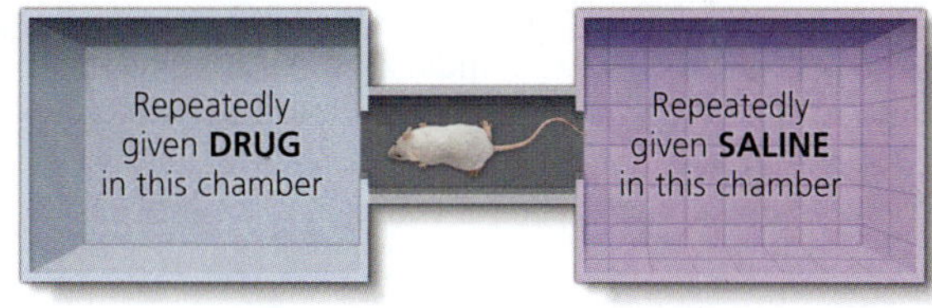

Figure 15.18 Drug-conditioned place preference. A rat is repeatedly given an addictive drug (such as cocaine, morphine, or amphetamine) in one of two easily distinguishable chambers. Later, with no drugs in its system and free to explore both chambers, the conditioned animal prefers spending time in the drug-associated chamber.

After several days of these conditioning trials, the rats are given a test trial, during which they are drug free and allowed to roam freely between the two chambers. Which chamber do you think the rats prefer? Of course, they preferentially visit the chamber where they had been "on drugs." That is, they show a strong drug-conditioned place preference.

To test whether the ventral striatum is involved in acquiring drug-conditioned place preferences, experimenters have infused diverse drugs directly into the ventral striatum. For example, infusing amphetamine directly into the ventral striatum causes massive dopamine release in that brain area, and it causes the rats to prefer the compartment in which they received those amphetamine infusions (Figure 15.19). By contrast, infusing amphetamine into the dorsal striatum does not cause such a preference. Furthermore, interfering with ventral striatal plasticity (by disrupting *fos* signaling) prevents the development of drug-conditioned place preferences, whereas interfering with dorsal striatal plasticity does not have this effect. Collectively, these data indicate that drug-induced dopamine release in the ventral striatum, but not the dorsal striatum, is critical for the development of drug cravings.

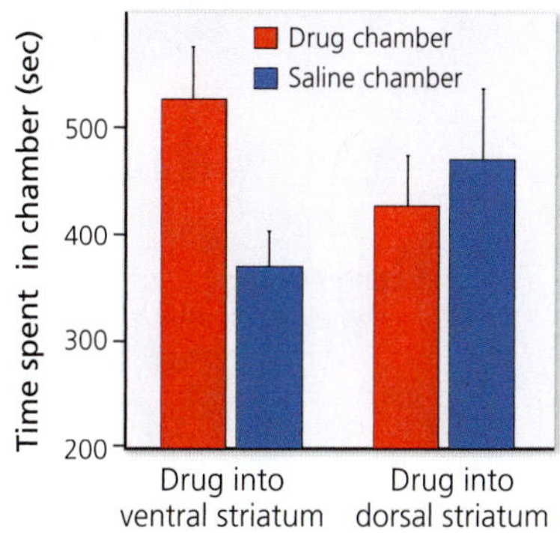

Figure 15.19 Ventral striatal dopamine in conditioned place preference. Rats receiving infusions of amphetamine directly into the ventral striatum (causing massive dopamine release) develop a preference for the "drug den." This place preference does not emerge when the amphetamine is infused into the dorsal striatum. [After Carr and White, 1983]

Connections between the Dorsal and Ventral Striatal Loops

If the ventral striatal loop is involved in learning to select high-level behavioral goals (such as seeking a drug), whereas the dorsal striatal loop is needed for learning how

NEUROBIOLOGY IN DEPTH

Box 15.2 *Sex Drive and Sexual Conditioning*

During early development, circulating sex hormones are involved in sexual differentiation, which leads to sex differences in the size of several brain regions (see Chapter 16, Section 16.4). Later in life, sex hormones affect the sex-related circuits in more subtle ways. In species that breed seasonally, sex hormones generate seasonal changes in sexual arousability and receptivity (sex drive). Humans do not breed seasonally, but castration and ovariectomy (without hormone replacement) dramatically reduce the sex drive in humans. In addition, women experience small fluctuations in their sex drive in association with the menstrual cycle. In case you're wondering, testosterone levels do vary across men, but they have little correlation with sex drive as long as they exceed some rather low threshold concentration. Consistent with this finding, testosterone supplements do little to alter sexual arousability in adult men.

Because sex is fundamental to the survival of most vertebrate species, one might expect it to require no learning. However, sexually naive rodents take significantly longer to copulate with a receptive partner, and they are distracted more easily than sexually experienced animals. Precisely what the animals are learning from their sexual experiences is often difficult to tell, but research has shown that both males and females learn where to go for sex (sex conditioned place preferences) and who is most likely to be a suitable partner. For example, in one study, male marmoset monkeys were repeatedly allowed to copulate with ovulating females that were released from a small "stimulus box" shortly after the males had smelled a lemon scent (Figure b15.2). Later, when the conditioned males were exposed to lemon scent, they became sexually aroused and explored an empty stimulus box with great enthusiasm. Without the whiff of lemon scent, the males showed no such interest. Moreover, the males had not expressed much interest in the box before conditioning. Based on these data, we can infer that the males had been sexually conditioned to associate lemon scent with copulation.

The phenomenon of sexual conditioning raises interesting questions about the causal basis of homosexuality. Is it "in the genes" or is it learned? The question remains open and is enormously controversial, but the data on sexual conditioning suggest that, even if some learning is involved, the resulting preferences would probably be deeply felt and quite resistant to change. Indeed, sexual conditioning is probably at least as powerful as the development of some drug addictions and food aversions. At a minimum, we can say that sexual conditioning probably explains much of the remarkable diversity among humans in what it is that turns us on.

The neural mechanisms underlying sexual conditioning have been studied extensively, primarily in rats. Much of this research has pointed to dopamine release in the ventral striatum as playing a major role. Genital stimulation is rewarding to both male and female rats and can cause dopamine release in the ventral striatum. This burst of dopamine probably strengthens synapses that were active just before copulation, causing dopamine to be released earlier and earlier during sexual encounters. Indeed, the mere sight or smell of a receptive female rat can increase dopamine levels in the ventral striatum of a sexually experienced male rat.

Sexual conditioning also involves endogenous opioids, which are released during male orgasm, cross the blood–brain barrier, and bind to opioid receptors in the ventral striatum. The details of how opioid receptor binding influences ventral striatal neurons remain unclear, but ventral striatal infusions of naloxone, an opioid antagonist, block sexual conditioning. This is interesting because opioid receptors in the ventral striatum have also been linked to cravings for sweet and fatty food, as well as cravings for morphine and heroin (exogenous opioids). Another interesting observation is that endogenous opioids are released during vigorous exercise. This finding suggests that ventral striatal opioid receptors may also be involved in exercise addiction, which is defined as an unhealthy craving for excessive exercise.

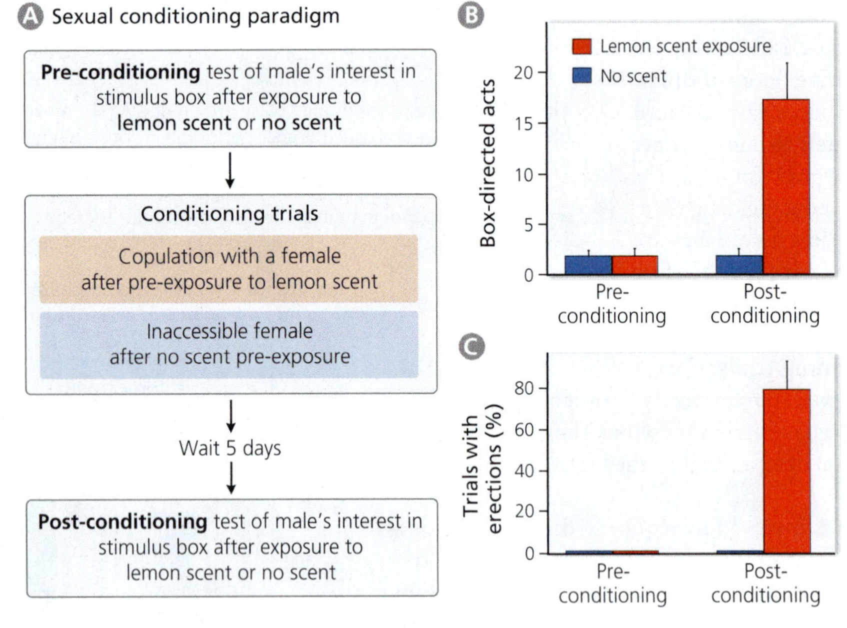

Figure b15.2 Sexual conditioning. Across multiple conditioning trials (A), male marmoset monkeys were pre-exposed to lemon scent and then allowed to copulate with an ovulating female that emerged from a small "stimulus box." On other trials, the same males were given no scent pre-exposure and presented with an ovulating female that remained locked in the box. Both before and 5 days after the conditioning trials, the males were tested by exposing them to lemon scent, or to no special scent, and then measuring their interest in the stimulus box. As shown in (B), the conditioned males directed more behavior (sniffing, touching) at the stimulus box if they had just smelled lemon scent. This lemon scent-induced interest in the stimulus box did not exist before conditioning. The conditioned males frequently got erections after smelling the lemon scent (C). [After Snowdon et al., 2011]

to achieve that goal, then one might expect the dorsal striatal loop to receive inputs from the ventral striatal loop. After all, if you are going to pursue a goal, it helps to know what that goal is. Indeed, the ventral striatum and ventral pallidum have projections to the dorsal pallidum and to the SNc, which projects mainly to the dorsal striatum (Figure 15.20). Through these pathways information may flow from the ventral striatal loop into the dorsal striatal loop.

To test this hypothesis, experimenters monitored dopamine release in the ventral and dorsal striatum of rats every day for several weeks as the rats gradually developed a cocaine habit. In the ventral striatum, dopamine was released reliably as the rats self-administered cocaine. This ventral striatal dopamine was strongest at the beginning of the experiment, when the cocaine was still unexpected, and then decreased in amplitude over successive weeks, as the cocaine became routine. By contrast, dopamine was not released in the dorsal striatum until one week after the cocaine self-administration had begun. Moreover, dorsal striatal dopamine release emerged only if the ventral striatum was intact. Collectively, these findings are consistent with the idea that the ventral striatum is involved in selecting and driving goal-oriented behavior, whereas the dorsal striatum is more involved in learned habitual actions, and that the functions of the dorsal striatum require input from the ventral striatal loops.

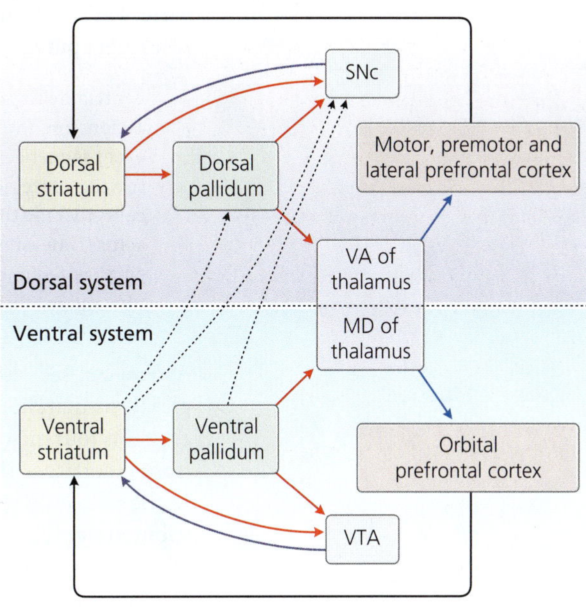

Figure 15.20 Dorsal versus ventral frontostriatal loops. Although the circuits through the ventral striatum are similar to those through the dorsal striatum, structures in the ventral system project to the dorsal system (dashed lines) but not vice versa.

Unfortunately, not all the data fit so nicely into our simple model of how the ventral and dorsal striatal loops are related to one another. For example, the ventral striatum has strong inhibitory projections to the lateral hypothalamus, which is involved in feeding behavior (see Chapter 9). Disrupting this pathway causes rats to increase their food consumption dramatically. This finding is consistent with the ventral striatum being involved in goal selection, but it is not consistent with the dorsal striatum being necessary for learning how to implement the goal. Apparently, the ventral striatum can control some behaviors directly through the hypothalamus, without much assistance from the dorsal striatal loop.

Despite such complications, most data indicate that the dorsal striatum is functionally "downstream" of the ventral striatum. As you will see in the next section, this functional division of labor is also reflected in the organization of the frontal lobe, which is the one component of the frontostriatal system that we have not yet discussed. In humans, the largest portion of the frontal lobe is the prefrontal cortex.

BRAIN EXERCISE

Amphetamine causes dopamine release by acting directly on dopaminergic terminals, whereas food rewards cause dopamine release indirectly, by increasing the firing rate of dopaminergic VTA neurons. Does this difference matter? If so, how?

15.6 What to Do with the Prefrontal Cortex?

The portion of the frontal lobe anterior to the motor and premotor areas is called the **prefrontal cortex**. For many years this brain region was considered to be "silent cortex" because electrical stimulation there had no discernible effects. Nor did lesions of the prefrontal cortex alter behavior dramatically. Since those early days, however, data on prefrontal cortex functions have accumulated rapidly. Indeed, the variety of functions now attributed to the prefrontal cortex is overwhelming. As one researcher put it, the prefrontal cortices

> contribute to prospective coding; sensory working memory; attention to objects, places, actions and intentions; response inhibition; the categorization of objects, sounds and event sequences; the selection, generation and sequencing of multiple actions; problem solving strategies; and the selection of behavior-guiding rules or task sets. These concepts include the preparatory set for future actions, in addition to processing events across time gaps, decision making and goal selection. Also included are the ideas emphasized in contemporary neuroeconomics such as reward expectation, the valuation and revaluation of actions and stimuli, and the effect of reinforcement history on decisions, choices and actions. (Steven P. Wise, 2008, p. 605)

As you can see, these functions are complex and interrelated. This functional complexity is matched by anatomical complexity. For starters, the prefrontal cortex contains four major divisions, which are called lateral, orbital, medial, and polar prefrontal cortices. Their locations and major connections are shown in Figure 15.21. In the following sections, we will focus on just a few key experiments that have contributed significantly to our understanding of prefrontal cortex functions. At the end of the chapter, we will return to the question of how the various components of the frontal lobe are related to the dorsal and ventral striatum.

Prefrontal Lobotomies

Carlyle Jacobsen in the 1930s lesioned the prefrontal cortex of non-human primates and observed that his animals became deficient on tasks in which they had to remember information for more than a few seconds. We will come back to this later. Almost as an aside, Jacobsen also reported that one of his chimpanzees, called Becky, became much calmer after the prefrontal lesions. Before the surgery, Becky had often thrown temper tantrums when she made a mistake and failed to receive a reward. At some point she refused to participate in the experiments entirely. After the surgery Becky no longer got upset, even though her ability to solve the task had been reduced to chance. As Jacobsen put it, Becky seemed to have joined a "happiness cult."

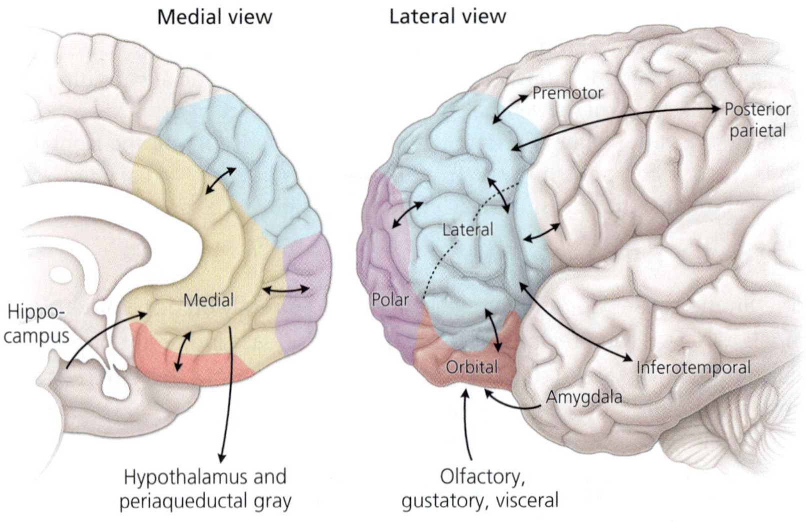

Figure 15.21 Major divisions and connections of the prefrontal cortex. Shown here are the medial, polar, lateral, and orbital divisions of the human prefrontal cortex. The arrows indicate some of their major connections. The lateral prefrontal cortex is divisible into dorsolateral and ventrolateral divisions, here separated by a dashed line. The diagram on the left is a medial view of the frontal lobe, viewed from the midsagittal plane.

On hearing Jacobsen's report of these results at a scientific meeting in 1935, a Portuguese psychiatrist named Egas Moniz asked if such a surgery might help mentally disturbed and hotheaded patients. Jacobsen was skeptical, but Moniz soon teamed up with a neurosurgeon and began performing **prefrontal lobotomies** (large lesions in the prefrontal cortex) in humans suffering from diverse mental disorders, ranging from schizophrenia to depression. How much these patients improved was never clear because Moniz quickly became a fan of his own method and failed to perform rigorous postoperative analyses. Nonetheless, Moniz was an energetic promoter of his work and was rewarded with the Nobel Prize in Physiology or Medicine in 1949.

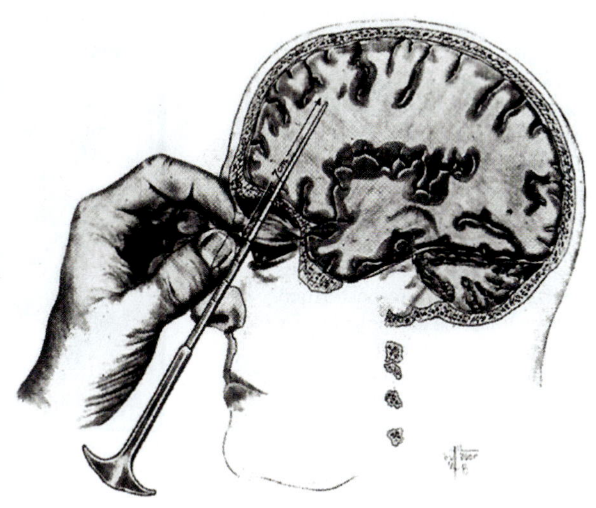

Figure 15.22 Transorbital prefrontal lobotomy. This drawing illustrates the surgical approach used by Walter Freeman and James Watts to perform transorbital lobotomies (a.k.a. leukotomies). In essence, they inserted an ice pick above the patient's eye and through the orbital bone into the prefrontal cortex. They then rotated this ice pick mediolaterally to sever many of the axons that connect the prefrontal cortex with more posterior brain regions. This procedure was relatively rapid and easy to perform, but it left many patients severely impaired. [From Freeman and Watts, 1950]

One of the people who heard about Moniz's work was the American physician Walter Freeman. He tried Moniz's method on an agitated woman with clinical depression and found her to be much calmer after the surgery. Freeman then collaborated with a neurosurgeon to improve the procedure. Their new approach was to insert a metal ice pick–like rod above the eyeball and to hammer it through the skull until it penetrates the brain (Figure 15.22). They then swung the "ice pick" side to side (mediolaterally) to sever the axons that connect the prefrontal cortex with more posterior brain regions. The advantage of this **transorbital lobotomy** technique was that it could be performed in a few minutes by people without surgical training on patients who were minimally anesthetized (usually by electroconvulsive shock). Enamored of his method, Freeman performed about 3,500 lobotomies at mental hospitals throughout the United States. His approach became popular and was performed widely, mostly on schizophrenics.

Powerful Drugs Replace Lobotomies

In retrospect, prefrontal lobotomies are a dark chapter in the history of neurosurgery. Although some patients were indeed calmer after their lobotomies, many who did not receive lobotomies likewise improved. Furthermore, lobotomies often caused severe side effects, including unresponsiveness, lack of motivation, decreased attention span, inappropriate or blunted emotions, and the loss of behavioral inhibitions. Given such tragic consequences, why did medical professionals in the 1940s and 50s collectively perform more than 40,000 lobotomies? The best explanation is that mental hospitals in the 1940s were overflowing with patients, many of whom were difficult to manage. Lobotomies promised to calm these patients so that they could leave the mental institutions and, perhaps, reenter society. This promise was not fulfilled.

Fortunately, the popularity of prefrontal lobotomies declined precipitously after 1952, when clinical trials showed that the drug chlorpromazine (thorazine) can be used to treat a broad range of psychotic disorders. In contrast to lobotomies, chlorpromazine is reversible and trivially easy to administer. Within a few years, it was given to millions of patients, many of whom were able to live outside of mental hospitals. Nowadays, a wide range of antipsychotic drugs is available to manage patients who, in the past, might have received a prefrontal lobotomy.

The Story of Phineas Gage

Although the problems associated with prefrontal lobotomies took many years to become clear, hints of their potential harm had existed since the mid-1800s when Phineas Gage (Figure 15.23) gave himself a prefrontal lobotomy (of sorts) by accident. As foreman of a rail line construction crew, part of Gage's job had been to blast large rocks that were obstructing the planned railroad tracks. This involved drilling

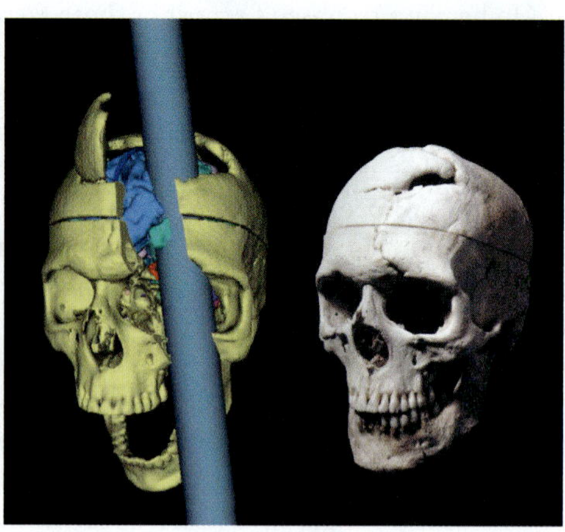

Figure 15.23 Phineas Gage suffered an accidental prefrontal lobotomy. The image on the left is an early photograph of Phineas Gage after his accident, holding the tamping iron that accidentally shot through his head. Shown on the right is Gage's skull with a prominent hole at the top and a fissure across the forehead. Also shown is a digital reconstruction of this skull and the tamping iron as it passed through the brain. The damage was almost certainly confined to the left prefrontal cortex. [From Wilgus and Wilgus, 2009, and Ratiu and Talos, 2004]

holes into rock and filling them with explosives. For maximum effectiveness, the explosives had to be tamped down using the blunt end of a tamping iron, which is a metal rod about 1 meter long, 3 cm in diameter, and tapered at one end (Figure 15.23). On September 13, 1848, Gage became distracted during such a tamping procedure and struck the rock with his iron, creating a spark that ignited the explosives. The tamping iron was propelled upward, passed through Gage's skull, and landed about 20 feet behind him, covered with bits of brain.

Although Gage initially lay on the ground, quivering with convulsions, he regained consciousness within a few minutes. His colleagues sat Gage down in the back of an oxcart and drove him to a nearby inn, where a skilled physician stemmed the bleeding and dressed the wound. Amazingly, Gage survived both the initial trauma and the ensuing infection. He emerged from the hospital a few months later and lived for another 11 years.

Gage's brain was not examined at autopsy, but a detailed model of his skull and the tamping iron (which Gage carried around with him after the accident) revealed that the tamping iron probably destroyed most of Gage's left anterior prefrontal cortex. Despite this massive injury, Gage's behavior was surprisingly normal. He could speak and move as he did before the accident. His memory and intellectual capacity appeared normal. However, Gage's friends and family noticed some profound changes in Gage's personality after the accident. The formerly polite and responsible man had become fitful, irreverent, grossly profane, impatient, obstinate, and capricious. He made plans for the future, but abandoned them almost immediately. In the words of John Harlow, the physician who saved his life, Gage had become "a child intellectually" but with "the animal passions of a strong man."

Gage never regained his foreman's job and, by 1850, earned his living as a circus exhibit. However, Gage did eventually recover some of his lost capacities and managed to hold down a stagecoach driver's job in South America.

Modern Cases Resembling Phineas Gage

Nowadays, prefrontal lobotomies are performed only rarely (typically to remove a tumor), but those few cases are studied intensively. In particular, neurologists have studied numerous patients with damage that is relatively limited to the orbital prefrontal cortex, which is likely to have been the most extensively damaged portion of the prefrontal cortex in Phineas Gage.

Patients with orbital prefrontal damage perform normally on many tasks but tend to make bad decisions both in the real world and in laboratory settings. For example, in a simulated gambling task, patients with large orbital prefrontal lesions are unable to pass up short-term rewards, even if this means that they incur some long-term costs. They also tend not to get upset at having made bad decisions and, more generally, have trouble generating emotions. These findings suggest that orbital prefrontal cortex is needed for using "gut feelings" in making decisions. Because of this deficit, patients with orbital prefrontal lesions tend to deliberate excessively before making a decision and then have trouble sticking to the decisions they made. They often end up losing money, respect, and friends. In many respects, they resemble Phineas Gage, at least during the first few years after his injury.

Response Inhibition

The reports of Phineas Gage becoming grossly profane, impatient, and capricious after his self-inflicted prefrontal lobotomy suggest that the prefrontal cortex is needed for impulse control. Subsequent studies have confirmed that the prefrontal cortex indeed plays a major role in the inhibition of **prepotent responses**, which we may define as responses that an individual performs habitually and automatically. Let us review some relevant experiments.

In one influential study, experimenters presented marmoset monkeys with pairs of stimuli consisting of black lines superimposed on blue polygons that varied in shape (Figure 15.24). The monkeys were trained to select one of the polygons for a reward and to ignore the patterns of black lines. Then the monkeys underwent surgery to lesion either the lateral or the orbital prefrontal cortex. After the surgery, the monkeys could remember the previously learned association. They were also able to learn a new association, between a different blue polygon and the reward (Figure 15.24 B). Therefore, the lesions did not impair visual discrimination or the learning of novel stimulus-reward associations.

However, the experimenters then changed the reward contingencies so that the animals had to ignore the blue polygons and instead select one of the black line patterns. The animals with lateral prefrontal lesions took much longer than the other animals to learn this novel kind of stimulus-reward association (Figure 15.24 C). The most likely interpretation is that the monkeys with lateral prefrontal lesions had trouble shifting their attention away from the blue polygons and toward the black lines. This was interesting, but the investigators had one more trick up their sleeve.

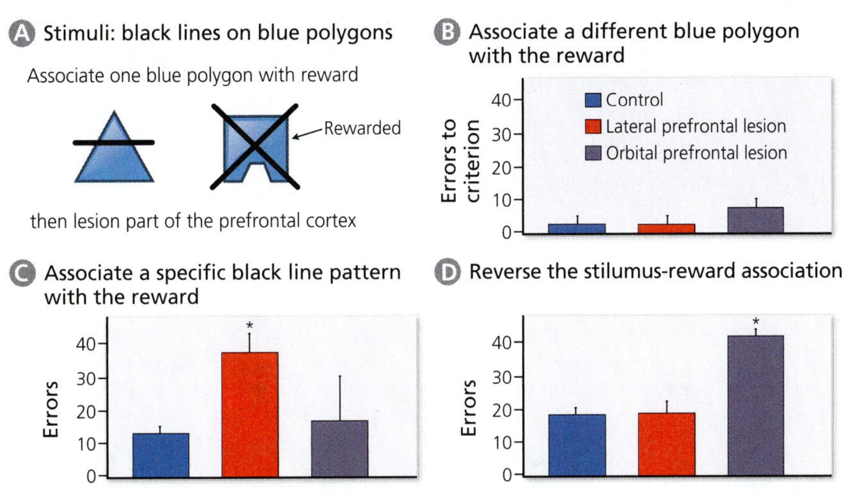

Figure 15.24 Prefrontal inhibition of behavior. Marmoset monkeys were shown pairs of stimuli comprised of black lines superimposed on blue polygons and trained to associate one of the polygons with a reward (A). Then the experimenters lesioned either the lateral or the orbital divisions of the prefrontal cortex; control animals were sham lesioned. Postoperatively, all animals remembered the learned stimulus-reward association. Next, the monkeys were trained to associate a different, new polygon with the reward. Again, all groups learned this new association readily (B). After that, the reward contingencies were changed again, but now one of the black line patterns (rather than a polygon) was positively reinforced. This time the monkeys with the lateral prefrontal lesions took longer than the other groups to learn the task (C). Finally, the stimulus-reward association was reversed so that the monkeys had to select the stimulus that had not been rewarded on the previous trials. The monkeys with the orbital prefrontal lesions were very slow to learn this final task (D). [After Dias et al., 1997]

Once the animals had learned which black line pattern was associated with the reward, and which was not, the experimenters reversed the association: the monkeys now had to ignore the previously reinforced pattern and, instead, select the other one. This task was relatively easy for the animals with lateral prefrontal or control lesions, but animals with orbital prefrontal cortex lesions found it extremely difficult. They had trouble letting go of the behavior that had previously earned them rewards.

Overall, these data indicate that the orbital and lateral divisions of the prefrontal cortex perform subtly different functions (we will come back to this). More important for our discussion, we can conclude that the lateral and orbital divisions of the prefrontal cortex are both needed to suppress responses that have become well learned and habitual. They are needed for the inhibition of prepotent responses.

The Wisconsin Card Sort Task

Complementary data have been obtained in humans, using primarily the **Wisconsin card sort task** (Figure 15.25). This task employs a deck of 128 cards with geometric symbols that differ in shape, color, or number. Four cards differing in all three respects are presented face-up as reference cards. The subject then draws successive cards from the remaining deck and assigns them to one of the four reference cards (much as you do in the card game Solitaire). Importantly, the subject does not know whether the cards should be matched (sorted) by color, shape, or number of symbols. However, the experimenter provides feedback after each card, allowing the subject to deduce the sorting rule by trial and error. An interesting twist is that, after several correct trials, the experimenter surreptitiously changes the sorting rule. This causes unexpected errors and forces the subject to deduce the new rule.

Patients with extensive damage to the prefrontal cortex have trouble with the Wisconsin card sort task, mainly because they are slow to give up on sorting rules that used to work. They make a large number of **perseverative errors**, defined as responses that used to be correct but no longer are (the term is derived from "perseverance").

Using functional brain imaging, researchers have shown that large parts of the prefrontal cortex and several other cortical areas are more active during the card sort task than a less demanding control task, such as matching identical cards. More detailed results emerged from studies using **event-related fMRI**, which involves taking brain scans every 2–3 seconds while a subject is performing a task and then looking for correlated changes in the blood oxygenation signal. Such an analysis reveals that oxygenation levels in the lateral prefrontal cortex tend to increase 5–9 seconds after each rule change in the Wisconsin card sort task (Figure 15.25 B). Because local

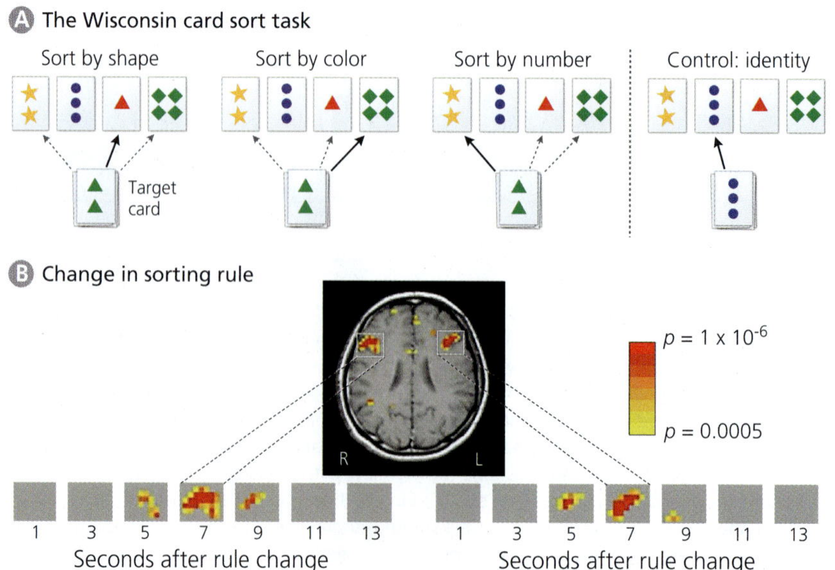

Figure 15.25 The Wisconsin card sort task activates the lateral prefrontal cortex. Subjects are instructed to sort a stack of cards into four piles, matching each new target card to one of four reference cards according to the shape of the symbols on the cards, the color of the symbols, or the number of symbols (A). The subjects are not told which sorting rule to use, but they are told when they made a mistake. After several correct responses, the sorting rule is changed, forcing the subject to abandon the old rule and learn a novel one. The fMRI scans in (B) show that the flow of oxygenated blood to the lateral prefrontal cortex increases 5–9 seconds after changes in the sorting rule. [From Konishi et al., 1998]

changes in blood oxygenation lag changes in neural activity by roughly this amount of time, we can conclude that neurons in the lateral prefrontal cortex increase their activity whenever the sorting rule is changed. This finding is consistent with the notion that the prefrontal cortex is involved in response inhibition because changes in the sorting rule require subjects to inhibit their old (prepotent) pattern of responses and develop a new behavioral strategy.

The Stroop Task

A second task that has been widely used to study prefrontal cortex functions in humans is the **Stroop task**, named after its inventor Ridley Stroop. In this task, subjects are asked to name the color of the ink in which a color word (such as "red") is printed (Figure 15.26). This exercise is challenging because the subjects usually cannot help reading the word, which then interferes with the task of naming the printed color when the two are incongruent (e.g., when the word "blue" is printed in red).

Brain activity during the Stroop task is usually studied by administering the task in individual trials, telling the subject before each trial whether they should name the printed color or the word. In "congruent trials," the word and its color match; on "incongruent trials," they don't. The investigators then scan the brain every 2.5 seconds during each trial to look for changes in blood oxygenation as an indicator of brain activity. Such studies have shown that activity in the lateral prefrontal cortex increases shortly after subjects are instructed to name the color of the printed word rather than the word itself (Figure 15.27). Moreover, the medial prefrontal cortex (specifically the anterior cingulate cortex) is preferentially engaged when subjects

Name the colors of the rectangles

Name the colors of the printed words

Red	Blue	Green	Brown	Purple
Blue	Brown	Red	Purple	Green
Brown	Red	Purple	Green	Blue
Green	Purple	Brown	Blue	Red
Purple	Green	Blue	Red	Brown

Figure 15.26 The Stroop task requires inhibition of a prepotent response. It is easier to name the color of a filled rectangle (top) than to name the printed color of a word that names a different color (bottom). Naming the color of mismatched (incongruent) color words is difficult mainly because we process word meanings so rapidly that they interfere with our intention of naming the printed colors.

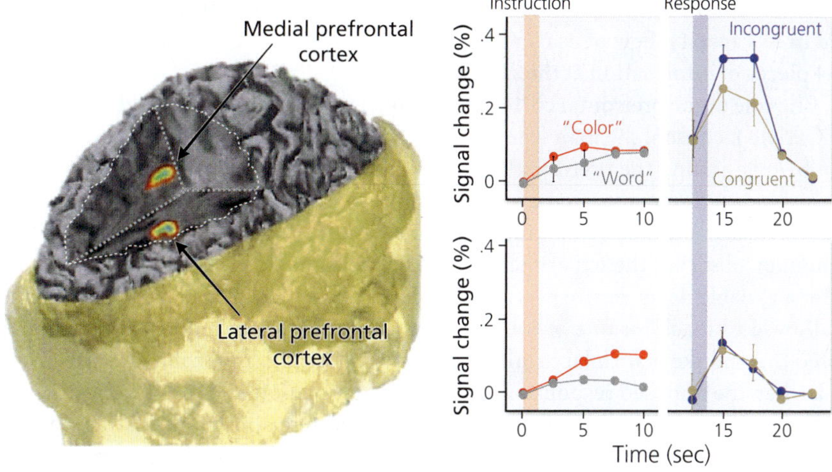

Figure 15.27 Event-related fMRI during Stroop task performance. Subjects performed the Stroop task while their brain was scanned 10 times per trial. Each trial begins with an instruction, telling the subject whether to read the color word or name the color that the word is printed in. Eleven seconds later, a word is flashed for 1.5 seconds, and the subject is asked to respond as instructed. The top graphs show oxygenation level changes in the right medial prefrontal cortex (anterior cingulate). You can see that this region becomes active during the response period, especially on incongruent trials (when word and font color are mismatched). In contrast, the lateral prefrontal cortex exhibits no differential activity during the response period (bottom graphs). However, the lateral prefrontal cortex is more active during the instruction period whenever subjects are asked to name the font color rather than read the word. [From MacDonald et al., 2000]

actively suppress their impulse to name the word (during the response period on incongruent trials). These data suggest that the lateral prefrontal cortex is involved in preparing for a challenging behavior, whereas the medial prefrontal cortex is involved in monitoring ongoing behavior and suppressing unwanted responses.

Thus, a considerable amount of data indicates that the various divisions of the prefrontal cortex are involved in, and required for, the inhibition of behaviors that are not helpful in the current behavioral context. The orbital prefrontal cortex is probably concerned mainly with inhibiting urges to obtain rewards. In contrast, the lateral prefrontal cortex is more involved with shifting attention to task-related stimuli or stimulus features. The medial prefrontal cortex probably performs a more general function related to performance monitoring. The polar prefrontal cortex has been studied much less than the other prefrontal regions. It is very small or nonexistent in rodents and disproportionately large in humans compared to other primates (see Chapter 16). Its functions are controversial, but the polar prefrontal cortex appears to be involved in handling multiple goals, keeping one on ice while temporarily pursuing another.

Despite all these divisions of labor, it is probably simplistic to think of the various prefrontal divisions as working independently of one another. They are densely interconnected and almost certainly collaborate in most situations. Therefore, it is probably best just to conclude, for now, that the prefrontal cortex is involved whenever you are actively thinking about what you're doing, as opposed to performing the actions automatically. This deliberate attention to your own behavior takes mental effort and self-control.

Working Memory

As noted earlier, the prefrontal cortex is involved not only in response inhibition but also in a wide array of other functions. One of these other functions is **working memory**, defined as the temporary storage of information that was just experienced or retrieved from long-term memory but is no longer accessible in the external environment. A good example of working memory is the ability to remember a phone number you have just heard. Another instructive example is reading a recipe and then trying to remember both the list of ingredients and what you are supposed to do with them. In both instances, you quickly discover that working memory is limited in its capacity. Few of us can remember and manipulate (work with) more than 3–4 pieces of information at the same time.

The role of the prefrontal cortex in working memory was discovered in the 1930s by Carlyle Jacobsen, the man who inspired Moniz's prefrontal lobotomies. Jacobsen had been lesioning the prefrontal cortex of non-human primates and observed that the animals became impaired on tasks that require holding information in memory for several seconds before making a response. In one such **delayed response task**, a monkey observes the experimenter hiding food under one of two inverted cups. After a variable delay, during which the food cups are hidden from view, the monkey is allowed to reach for one of the two cups and retrieve the hidden food. Jacobsen noted that monkeys with large bilateral prefrontal lesions are unable to perform this task when the imposed response delays are longer than a few seconds.

Subsequent research confirmed this basic finding using more abstract tasks in which monkeys are given sensory cues indicating which of two alternative behaviors they should perform after a variable delay. For example, a monkey might be trained to push a button on its left several seconds after seeing a red light on a screen, and to push a button on the right several seconds after seeing a blue light. Again, monkeys with prefrontal lesions tend to be impaired on such tasks, especially if they are distracted by other stimuli during the delay. These findings indicate that the prefrontal cortex is required for working memory.

A Neurophysiological Correlate of Working Memory

Additional support for a causal link between the prefrontal cortex and working memory comes from neurophysiological recordings of prefrontal activity during delayed response tasks. In most of these studies, monkeys are rewarded for making an

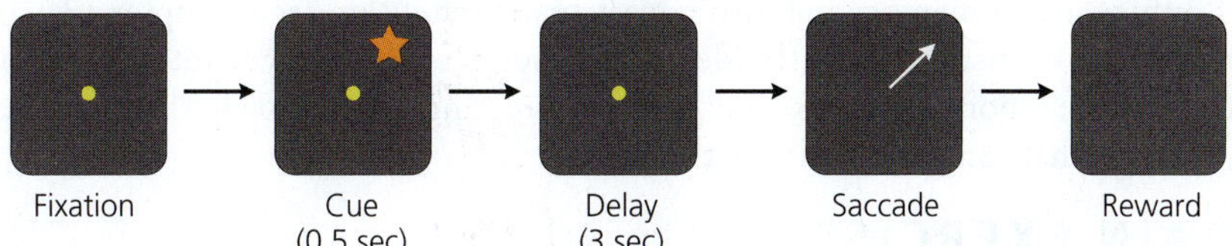

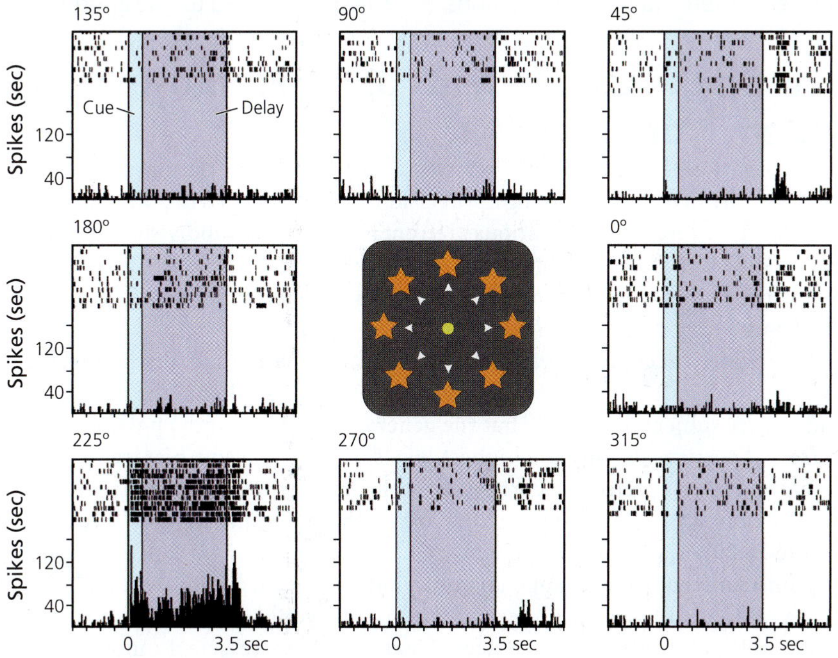

Figure 15.28 Delay activity in lateral prefrontal cortex. Monkeys were trained to make a saccadic eye movement toward a location where a target cue (the orange star) had been 3 seconds earlier (A). The spike rasters and firing rate histograms in (B) illustrate the activity of a representative neuron in the dorsolateral prefrontal cortex that increased its firing rate only on trials in which the cue was presented at the bottom left of the screen (at 225°). The increase in firing rate began with the presentation of the cue, continued during the delay, and ended shortly after the saccade. [After Chafee and Goldman-Rakic, 1998]

eye movement to a location where a stimulus appeared briefly (Figure 15.28). Because the cue disappears several seconds before the monkey is allowed to make the eye movement, the animal must hold the target location in working memory. As monkeys perform this task, the experimenters record from neurons in the prefrontal cortex.

Some of the recorded neurons increase their firing rate when the cue appears; others respond to the reward. Most intriguing is that many prefrontal neurons maintain an elevated firing rate during the delay period, when the monkey is remembering the target location. Because this **delay activity** occurs only when the cue is presented at a specific location (Figure 15.28), it cannot be due to increased arousal. Instead, it seems to be a correlate of working memory. Although such "working memory neurons" are most common in the lateral prefrontal cortex, they can also be found in other parts of the prefrontal cortex (and even in the striatum).

Linking Working Memory to Response Inhibition

Because the prefrontal cortex is involved in both working memory and response inhibition, it is important to ask how these two functions relate to one another. Although they are different by definition, working memory is often needed for inhibitory control. For example, in a typical delayed response task, the animals must not just remember an instruction; they must also suppress the impulse to respond before the end of the delay period.

In general, as individuals perform a complex task, they must remember their goal. As you know from your own life, pursuing a goal requires not only inhibitory control but also a good memory of what it is that you are trying to accomplish. Therefore, response inhibition and working memory often go hand in hand.

Nor are these two functions all you need for handling complex tasks. For example, considerable evidence suggests that delay activity in the prefrontal cortex reflects not only working memory but also covert spatial attention (see Chapter 13). This observation is consistent with the idea that pursuing a goal involves not only working memory and response inhibition but also the directing of attention to those locations and stimuli that can help you reach the goal.

BRAIN EXERCISE

The quotation from Steven Wise at the beginning of this section lists a large number of prefrontal cortex functions. How are they related to one another?

15.7 How Do the Components of the Frontostriatal System Work Together?

As you have now learned, both the prefrontal cortex and the striatum, as well as their associates, are involved in making the decisions that are needed to pursue a goal. You have also discovered that the various prefrontal and striatal subdivisions are specialized for different aspects of this overarching "executive" function. It remains a serious challenge, however, to synthesize all this information into a coherent scheme. Still, let us try.

One widely accepted idea about the frontal lobe is that it is hierarchically organized and displays an anteroposterior functional gradient. The details of this gradient are subject to dispute, but the general notion is that the polar and orbital prefrontal cortices occupy the highest levels of the executive hierarchy; that the medial and lateral prefrontal cortices are concerned with mid-level decisions; and that the premotor and motor cortices are in charge of making low-level choices (Figure 15.29).

Consider what would happen if you decide to put this book away and instead, search for food. This is a high-level decision that would most likely activate your orbital and polar prefrontal cortices. Next, you have to select among multiple strategies for getting food. This task requires an evaluation of your current context, recall of relevant experiences, and then selecting among the various potential strategies. It also requires suppressing alternative courses of action (suppressing the urge to get sidetracked). All of these activities are likely to engage the lateral and medial prefrontal cortices. Finally, you have to select a series of movements that culminate in eating the food. You might have to walk down the hall, open a cupboard, unwrap some candy, and stick it in your mouth. Selection of these movements involves the premotor and motor cortices. Thus, your pursuit of food begins with the selection of a high-level goal, such as the mental image of a candy

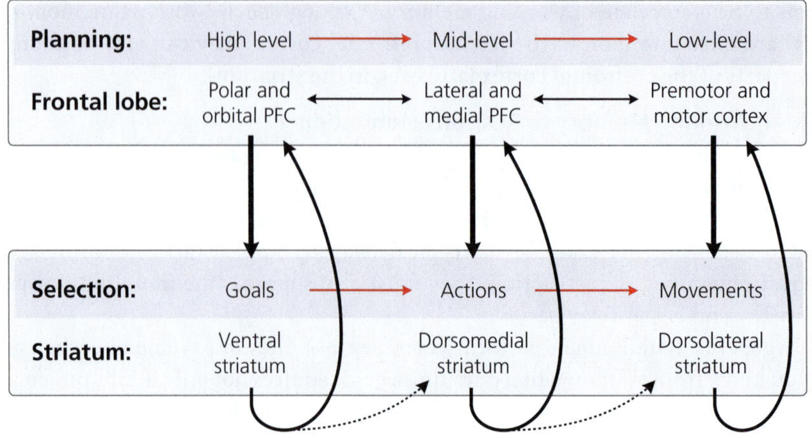

Figure 15.29 Hierarchies of behavioral control. This simplified diagram emphasizes the hierarchical organization of behavioral control through both the frontal lobe and the striatum. The dashed lines indicate pathways through which information can pass from high-level frontostriatal loops to lower level loops. PFC = prefrontal cortex.

bar or a more diffuse vision of food, which is followed by a series of lower level decisions that help you reach this goal.

It is tempting to think of this hierarchy of decisions as forming a temporal sequence, with one occurring after the other; but the goal of getting food must remain active throughout the process, as must the chosen strategy. Therefore, it is better to think of the high-level executive functions occurring over a longer time scale than the low-level decisions, which are anchored in the present. You can also think of the high-level decisions as being more abstract than lower level decisions.

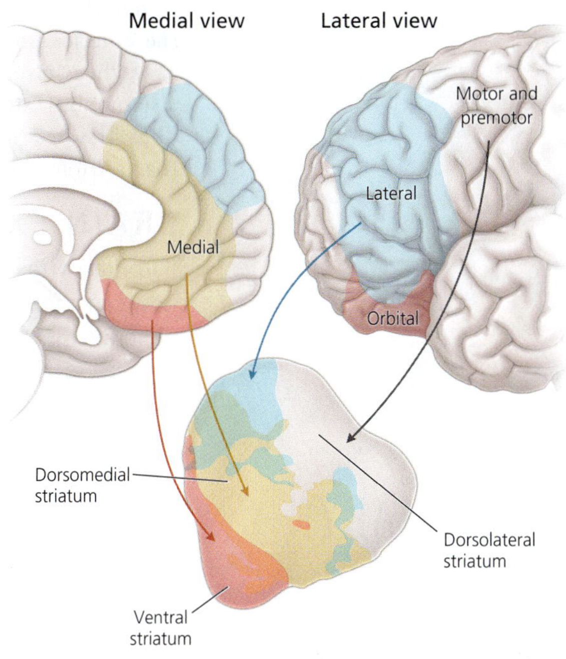

Figure 15.30 Topography in the frontostriatal projections. This diagram shows that the orbital prefrontal cortex projects predominantly to the ventral striatum, whereas the medial and lateral prefrontal cortices project to the dorsomedial part of the dorsal striatum. The motor and premotor cortices project primarily to the dorsolateral portion of the dorsal striatum. [After Haber and Knutson, 2010]

Frontostriatal Interactions

To understand how the gradient of executive functions within the frontal lobe relates to the remainder of the frontostriatal system, you have to know that the frontal lobe projects topographically to the striatum. As diagrammed in Figure 15.30, the orbital prefrontal cortex projects selectively to the ventral striatum, the lateral and medial prefrontal cortices project to the ventromedial portion of the dorsal striatum (called dorsomedial striatum), and the premotor and motor cortices project most heavily to the dorsolateral striatum.

Integrating this anatomical information with what you've learned about functional divisions within the frontal lobe, we can propose that the striatum contains three (rather than just two) major subdivisions that are each linked to a different component of the frontal lobe (Figure 15.29). The ventral striatum is involved in goal selection, which is a high-level process that requires input from the polar and orbital prefrontal cortices. The dorsomedial striatum is involved in mental processes of intermediate complexity, such as deciding what to hold in working memory and what strategies to use in pursuit of a goal; it collaborates mainly with the lateral and medial prefrontal cortices. The dorsolateral striatum, finally, is charged with selecting specific actions and movements, which it does in cooperation with the motor and premotor cortices.

Because the frontal lobe is linked so tightly to the striatum, you may wonder whether these two brain regions are functionally distinct. To some extent, the answer is yes because the frontal lobe has many connections that the striatum lacks, and vice versa. For example, the prefrontal cortex (but not the striatum) has direct reciprocal connections with the posterior parietal and inferior temporal cortices. These connections are thought to mediate top-down, voluntary attention by biasing sensory processing in favor of "search images" and other stimuli related to the organism's currently selected goal.

Another important difference between the striatum and the frontal lobe is that the output neurons of the striatum (the medium spiny neurons) are inhibitory, whereas those of the frontal cortex are excitatory. Striatal neurons also change their firing patterns more rapidly than prefrontal neurons as an individual learns a new task. Because of these anatomical and physiological differences, the striatum and the frontal cortex surely differ in their contributions to brain function and behavior. Nonetheless, the striatum and the frontal lobe collaborate so tightly that their functional differences should not be overstated.

Historically, neuroscientists have viewed the frontal lobe as being responsible for high-level, complex cognition and the striatum as controlling simpler behaviors, especially the learning of motor habits. This view is no longer tenable. Instead of thinking of the striatum as functionally "lower" than the frontal cortices, it is better to view the ventral and intermediate portions of the striatum as working *together* with the anterior prefrontal cortex to select high-level goals and devise strategies for achieving those goals. Similarly, the learning of motor habits involves not just the striatum but interactions between the dorsolateral striatum and the motor cortices. Thus, the striatum is not obviously "lower" than the frontal lobe in anything other than anatomical position (the striatum having a more inferior location).

BRAIN EXERCISE

If frontal cortex and the striatum are connected through looping pathways, how can information get out of these loops to the motor neurons that ultimately generate movements?

SUMMARY

Section 15.1 - The frontostriatal system consists of roughly parallel loops that course from the frontal lobe to the striatum and then return through the pallidum and thalamus. The system also includes dopaminergic neurons in the substantia nigra and ventral tegmental area.

Section 15.2 - The overarching function of the frontostriatal system is to select context-appropriate behaviors.

- The core of the frontostriatal system is a set of topographically arranged positive feedback loops that each contain two inhibitory neurons. Inhibitory connections within the striatum ensure that these loops inhibit each other, setting up a winner-take-all competition.
- An indirect side-loop to the direct pathway through the striatum regulates behavior selection. This indirect pathway is damaged in Huntington's disease, causing uncontrolled actions.

Section 15.3 - Parkinson's disease is linked to deficits in striatal dopamine release.

- Dopamine increases activity in the striatal neurons that project directly to the pallidum, but inhibits striatal neurons in the indirect (side-loop) pathway. By depleting dopamine, Parkinson's disease increases pallidal activity, which reduces the likelihood of actions being selected.
- Dopamine-deficient mice lack motivation to do anything unless they're given L-dopa, the precursor to dopamine. Excessive activation of D1 dopamine receptors leads to behavioral stereotypy.
- Cocaine or amphetamines directly affect brain dopamine levels.

Section 15.4 - The striatum is involved in learning what to do when to reach a goal. More formally, it is involved in instrumental conditioning.

- Phasic increases in striatal dopamine levels signal unexpected rewards or, after training, predict that a reward is coming soon.
- Bursts of dopamine increase the likelihood of LTP in corticostriatal synapses. This in turn increases the likelihood that rewarded behaviors will be selected again, given similar contexts.

Section 15.5 - The ventral striatum is involved in the pursuit of behavioral goals, including drug seeking. The dorsal striatum is more involved in learning motor habits. Information from the ventral striatal loop can pass to the dorsal striatal loop.

Section 15.6 - The prefrontal cortex is involved in a variety of "executive" functions related to goal pursuit and decision making.

- Prefrontal lobotomies were conducted with great enthusiasm in the 1940s and 50s until chlorpromazine was approved as an antipsychotic drug. Phineas Gage and other lobotomy patients often act impulsively, unable to pursue long-term goals.
- The ability to inhibit prepotent responses is impaired by lesions of the prefrontal cortex, which is activated during the Wisconsin card sort and Stroop tasks. The various parts of the prefrontal cortex are specialized for somewhat different functions, but most of them are related to self-control.
- Monkeys with prefrontal cortex lesions tend to be impaired on delayed response tasks that require them to remember information for more than a few seconds. Many prefrontal neurons maintain elevated firing rates during the delay period, suggesting that they are a neural substrate of working memory.

Section 15.7 - The functions of the frontal lobe are hierarchically organized and arranged in an anteroposterior gradient, which is paralleled by an analogous gradient across the striatum.

Box 15.1 - Currently, there are no cures for Parkinson's disease, but its symptoms can be alleviated with L-dopa, cell transplantation, or stimulation of the subthalamic nucleus. All of these therapies have limitations.

Box 15.2 - Males and females can be conditioned to prefer places where they had pleasant sexual experiences and to prefer partners with whom they copulated successfully. Dopamine and endogenous opioids are both involved.

KEY TERMS

frontostriatal system 474
basal ganglia 474
dorsal striatum 474
ventral striatum 474
dorsal pallidum 474
ventral pallidum 474
substantia nigra, compact division (SNc) 475
ventral tegmental area (VTA) 475
SNc/VTA 475
corticostriatal projection 476
medium spiny neuron 476
winner-take-all competition 477
external segment of the globus pallidus (GPe) 478
subthalamic nucleus 478
indirect pathway 478
Huntington's disease 478
huntingtin protein 478
chorea 479
D1-type dopamine receptor 479
D2-type dopamine receptor 479
Parkinson's disease 480
MPTP 481
L-dopa 481
carbidopa 483
behavioral stereotypy 483
cocaine 484
amphetamine 484
instrumental conditioning 484
phasic dopamine bursts 485
fast-scan cyclic voltammetry 485
craving 487
corticostriatal LTP 488
drug-conditioned place preference 489
prefrontal cortex 491
prefrontal lobotomy 493
transorbital lobotomy 493
prepotent response 495
Wisconsin card sort task 496
perseverative errors 496
event-related fMRI 496
Stroop task 497
working memory 498
delayed response task 498
delay activity 499

ADDITIONAL READINGS

15.1 - Overview of the Frontostriatal System

Kropotov JD, Etlinger SC. 1999. Selection of actions in the basal ganglia-thalamocortical circuits: review and model. *Int J Psychophysiol* **31**:197–217.

Mink JW. 1996. The basal ganglia: focused selection and inhibition of competing motor programs. *Prog Neurobiol* **50**:381–425.

Redgrave P, Prescott TJ, Gurney K. 1999. The basal ganglia: a vertebrate solution to the selection problem? *Neuroscience* **89**:1009–1023.

Verstynen TD, Badre D, Jarbo K, Schneider W. 2012. Microstructural organizational patterns in the human corticostriatal system. *J Neurophysiol* **107**:2984–2995.

15.2 - Pathways through the Striatum

Chuhma N, Tanaka KF, Hen R, Rayport S. 2011. Functional connectome of the striatal medium spiny neuron. *J Neurosci* **31**:1183–1192.

Gurney K, Prescott TJ, Redgrave P. 2001. A computational model of action selection in the basal ganglia. I. A new functional anatomy. *Biol Cybern* **84**:401–410.

McHaffie JG, Stanford TR, Stein BE, Coizet V, Redgrave P. 2005. Subcortical loops through the basal ganglia. *Trends Neurosci* **28**:401–407.

Tripp G, Wickens JR. 2009. Neurobiology of ADHD. *Neuropharmacology* **57**:579–589.

15.3 - Dopaminergic Modulation

Langston JW, Ballard P, Tetrud JW, et al. 1983. Chronic Parkinsonism in humans due to a product of meperidine-analog synthesis. *Science* **219**:979–980.

Matsuda W, Furuta T, Nakamura KC, Hioki H, Fujiyama F, Arai R, Kaneko T. 2009. Single nigrostriatal dopaminergic neurons form widely spread and highly dense axonal arborizations in the neostriatum. *J Neurosci* **29**: 444–453.

Palmiter RD. 2008. Dopamine signaling in the dorsal striatum is essential for motivated behaviors. *Ann N Y Acad Sci* **1129**:35–46.

15.4 - Learning What to Do When

Calabresi P, Picconi B, Tozzi A, Di Filippo M. 2007. Dopamine-mediated regulation of corticostriatal synaptic plasticity. *Trends Neurosci* **30**:211–219.

Cheer JF, Aragona BJ, Heien MLAV, Seipel AT, Carelli RM, Wightman RM. 2007. Coordinated accumbal dopamine release and neural activity drive goal-directed behavior. *Neuron* **54**:237–244.

Galvan A, Hare TA, Davidson M, Spicer J, Glover G, Casey BJ. 2005. The role of ventral frontostriatal circuitry in reward-based learning in humans. *J Neurosci* **25**:8650–8656.

Schultz W. 2007. Behavioral dopamine signals. *Trends Neurosci* **30**:203–210.

Tolliver BK, Sganga MW, Sharp FR. 2000. Suppression of c-fos induction in the nucleus accumbens prevents acquisition but not expression of morphine-conditioned place preference. *Eur J Neurosci* **12**:3399–3406.

15.5 - Dorsal versus Ventral Striatum

Barbano MF, Cador M. 2007. Opioids for hedonic experience and dopamine to get ready for it. *Psychopharmacology* **191**:497–506.

Belin D, Everitt BJ. 2008. Cocaine seeking habits depend upon dopamine-dependent serial connectivity linking the ventral with the dorsal striatum. *Neuron* **57**:432–441.

Berridge KC. 2009. "Liking" and "wanting" food rewards: brain substrates and roles in eating disorders. *Physiol Behav* **97**:537–550.

Bromberg-Martin ES, Matsumoto M, Hikosaka O. 2010. Dopamine in motivational control: rewarding, aversive, and alerting. *Neuron* **68**:815–834.

Humphries MD, Prescott TJ. 2010. The ventral basal ganglia, a selection mechanism at the crossroads of space, strategy, and reward. *Prog Neurobiol* **90**:385–417.

Kelley AE, Baldo BA, Pratt WE, Will MJ. 2005. Corticostriatal-hypothalamic circuitry and food motivation: integration of energy, action and reward. *Physiol Behav* **86**:773–795.

Willuhn I, Burgeno LM, Everitt BJ, Phillips PEM. 2012. Hierarchical recruitment of phasic dopamine signaling in the striatum during the progression of cocaine use. *Proc Natl Acad Sci U S A* **109**:20703–20708.

15.6 - Prefrontal Cortex

Acharya HJ. 2004. The rise and fall of the frontal lobotomy. *Proceedings of the 13th Annual History of Medicine Days* (WA Whitelaw, ed.) Calgary: Health Sciences Centre; pp. 32–41.

Cato MA, Delis DC, Abildskov TJ, Bigler E. 2004. Assessing the elusive cognitive deficits associated with ventromedial prefrontal damage: a case of a modern-day Phineas Gage. *J Int Neuropsychol Soc* **10**:453–465.

Chudasama Y. 2011. Animal models of prefrontal-executive function. *Behav Neurosci* **125**:327–343.

Coutlee CG, Huettel SA. 2012. The functional neuroanatomy of decision making: prefrontal control of thought and action. *Brain Res* 1428:3–12.

Curtis CE, D'Esposito M. 2003. Persistent activity in the prefrontal cortex during working memory. *Trends Cogn Sci* 7:415–423.

Hazy TE, Frank MJ, O'Reilly RC. 2006. Banishing the homunculus: making working memory work. *Neuroscience* **139**:105–118.

Lebedev MA, Messinger A, Kralik JD, Wise SP. 2004. Representation of attended versus remembered locations in prefrontal cortex. *PLoS Biol* **2**: e365.

MacLeod CM. 1991. Half a century of research on the Stroop effect: an integrative review. *Psychol Bull* **109**:163–203.

Macmillan M. 2008. Phineas Gage—unravelling the myth. *Psychologist* **21**:828–831.

Passingham RE, Wise SP. 2012. *The neurobiology of the prefrontal cortex: anatomy, evolution, and the origin of insight.* New York: Oxford University Press.

Rossi AF, Bichot NP, Desimone R, Ungerleider LG. 2007. Top down attentional deficits in macaques with lesions of lateral prefrontal cortex. *J Neurosci* **27**:11306–11314.

Swayze VW. 1995. Frontal leukotomy and related psychosurgical procedures in the era before antipsychotics (1935–1954): a historical overview. *Am J Psychiatry* **152**:505–515.

15.7 Synthesis of Frontostriatal Functions

Ashby FG, Turner BO, Horvitz JC. 2010. Cortical and basal ganglia contributions to habit learning and automaticity. *Trends Cogn Sci* **14**:208–215.

Diekhof EK, Gruber O. 2010. When desire collides with reason: functional interactions between anteroventral prefrontal cortex and nucleus accumbens underlie the human ability to resist impulsive desires. *J Neurosci* **30**:1488–1493.

Haber SN, Knutson B. 2010. The reward circuit: linking primate anatomy and human imaging. *Neuropsychopharmacology* **35**:4–26.

Boxes

Benabid AL. 2003. Deep brain stimulation for Parkinson's disease. *Curr Opin Neurobiol* **13**:696–706.

Gaillard A, Jaber M. 2011. Rewiring the brain with cell transplantation in Parkinson's disease. *Trends Neurosci* **34**:124–133.

Paredes RG, Agmo A. 2004. Has dopamine a physiological role in the control of sexual behavior? A critical review of the evidence. *Prog Neurobiol* **73**:179–226.

Pfaus JG, Kippin TE, Coria-Avila GA, Gelez H, Afonso VM, Ismail N, Parada M. 2012. Who, what, where, when (and maybe even why)? How the experience of sexual reward connects sexual desire, preference, and performance. *Arch Sex Behav* **41**:31–62.

CHAPTER 16

Being Different from Others

CONTENTS

FEATURES

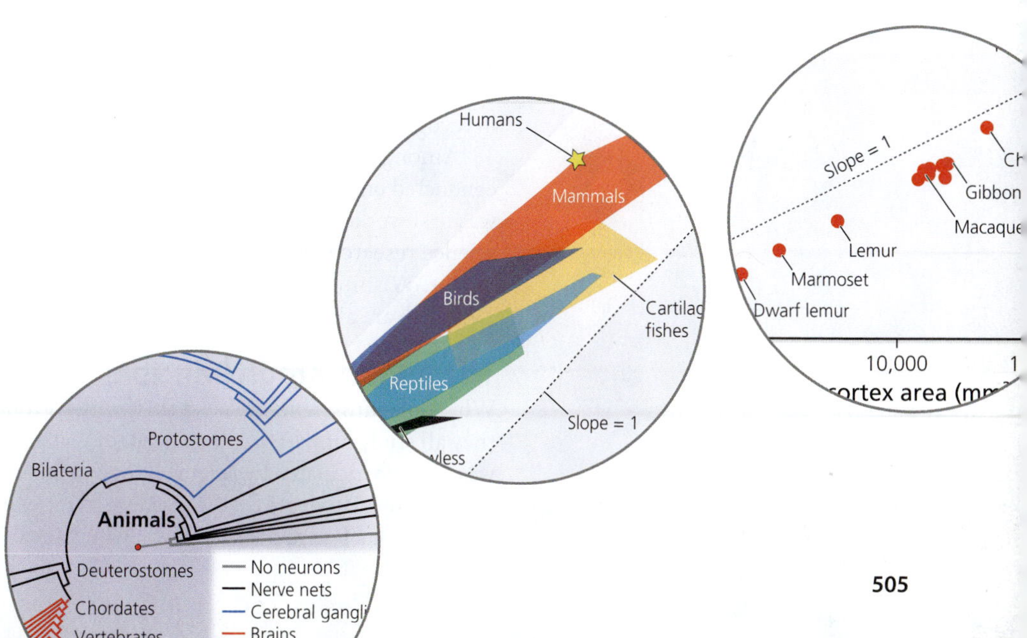

Throughout this book, we have been concerned mainly with learning about human brains. However, our discussions have included data from diverse species including monkeys, rodents, cats, and some invertebrates. Therefore, you might ask, does it matter from which species the data were obtained? Can what we learn from mice or squid be generalized to more complex creatures, including us? If this is the case, why do neuroscientists study so many different non-human species rather than just one? Alternatively, if we cannot extrapolate from non-humans to us, then why study non-human species at all? As you will see, these questions are complex but important. To answer them, we need to consider the kinds of similarities and differences in brain organization that evolution has brought forth. This task will occupy us in the first half of this chapter.

In the second half of the chapter, we shall turn our attention to the variability in brain organization that exists within the human species. Historically, neuroscientists have focused their research on relatively young adults, especially undergraduates. Most of these studies were conducted in Europe or North America, and the majority focused on males rather than females. Therefore, our knowledge of human brains comes from a non-random sample of humans. Is this a problem? The answer depends on how well the data from this biased sample can be generalized to the rest of the population. Are male and female brains significantly different? How much do human brains vary with age? As you will see, our understanding of human brain variability, both between the sexes and within a sex, is still severely limited.

A major theme in this chapter is that the analysis of model species and non-random samples of humans (model humans) has been enormously successful in neuroscience. It has led to the discovery of many fundamental principles that are broadly conserved, both within and across species. Nonetheless, not all findings can easily be generalized. Therefore, researchers must select their subjects with great care, and readers of the published literature must likewise be on guard against the assumption that all brains are alike. It is important to note, however, that the variation is not merely a nuisance. It is what makes us unique as individuals and as a species. It can also provide novel insights into how neural systems work.

16.1 Which Species Should Neuroscientists Study, and Why?

Structural and functional brain imaging techniques have enabled neuroscientists to study living human brains in unprecedented detail. Nonetheless, this work is based on a vast foundation of research on non-humans (Figure 16.1). For example, most of what we know about membrane potentials was discovered by recording from the giant axons of squid (see Chapter 2), and most of the early work on synaptic plasticity was carried out on the marine snail *Aplysia californica* (see Chapter 3). Another great example of groundbreaking invertebrate research is the discovery of lateral inhibition in horseshoe crab retinas. All of these studies were honored with Nobel Prizes in recognition of their broad impact.

Among the vertebrates, historically important studies using brain lesions were conducted on pigeons, rabbits, dogs, and monkeys. The physiology of the visual cortex was first worked out in cats (Chapter 12), and rats have long been a staple of neuroscience research, especially for studies on learning, memory, and reproductive behavior. Within the last decade, rats are increasingly being replaced with mice. Zebrafish are also on the rise. Meanwhile, the use of cats and dogs has plummeted.

The August Krogh Principle

Why did neuroscientists select this motley group of species for research? Why didn't they all study human brains instead, or just a single non-human species? The answer was clearly articulated by August Krogh, winner of the 1920 Nobel Prize in Physiology or Medicine. He wrote that "for a large number of problems there will be some animal of choice, or a few such animals, on which it can be most conveniently studied."

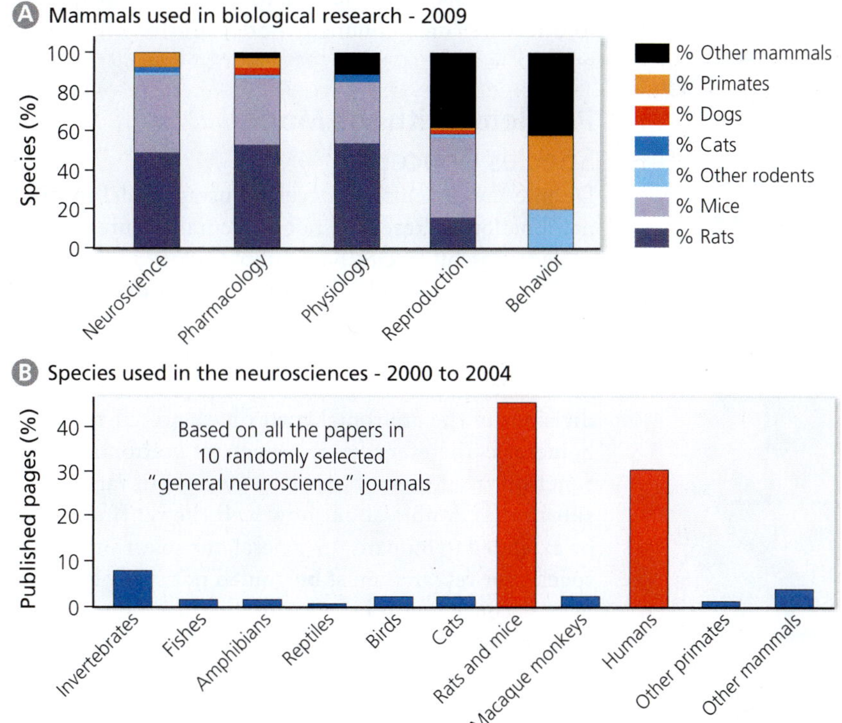

Figure 16.1 Non-random species selection in biological research. The data presented in (A) were generated by analyzing the first 20 research articles published in 2009 and dealing with organismal mammalian biology in 4 journals in each of 5 biological subdisciplines. For the neuroscience category, the journals were *Journal of Neuroscience, Neuroscience, The Journal of Comparative Neurology,* and *Nature Neuroscience.* The graph shows that neuroscientists and most other biologists study primarily rats and mice. Studies on humans were not included in this graph. Panel (B) illustrates the distribution of species studied in 10 randomly selected neuroscience journals from 2000 to 2004. The vast majority of studies focused on rats, mice, or humans. [After Beery and Zucker, 2011, and Manger et al., 2008]

Convenience, of course, can take a variety of forms. For example, squid giant axons are ideal for studying membrane potentials because their enormous size allows researchers to insert recording electrodes into the axons and to manipulate intracellular ion concentrations (Figure 16.2). Performing such experiments in thin mammalian axons is much more difficult. Cats were studied mainly because their visual cortex is relatively large; it is also easier to access surgically in cats than in primates. What makes rats such good subjects for neuroscience research? Mainly the fact that rats have been available from large commercial breeders since the early 1900s and are relatively cheap to house in animal facilities, compared to primates. In addition, the size of a rat brain is convenient for both histology (sectioning and staining) and stereotaxic surgery, in which specific brain regions are targeted for lesioning or physiological recording. Mice are even cheaper to house and, more important, relatively easy to manipulate genetically. By now, a plethora of transgenic mouse strains can be obtained commercially, and the techniques for creating "designer mice" are practically routine.

A critical assumption underlying **Krogh's principle**, as his view came to be called, is that discoveries made in one especially convenient species can be generalized to other, less convenient species. This assumption has held up well within the neurosciences. For example, the mechanisms underlying membrane potentials in squid are conserved across all animals with only minor modifications; and lateral inhibition is found not only in the eyes of horseshoe crabs but also in mammalian retinas and other brain regions. Similarly, an enormous number of anatomical and physiological findings obtained in macaques, the most commonly studied non-human primates, have been confirmed in human subjects, at least in outline.

Thus, Krogh's principle embodies a successful research strategy. If the basic principles of nervous system structure and function are broadly conserved, then it makes good sense to conduct experiments on whatever species are most convenient for the research. Most often, the most convenient species will not be our own, as it will always be more difficult, both practically and ethically, to perform experimental manipulations (such as brain lesions, single cell recording, or pharmacological

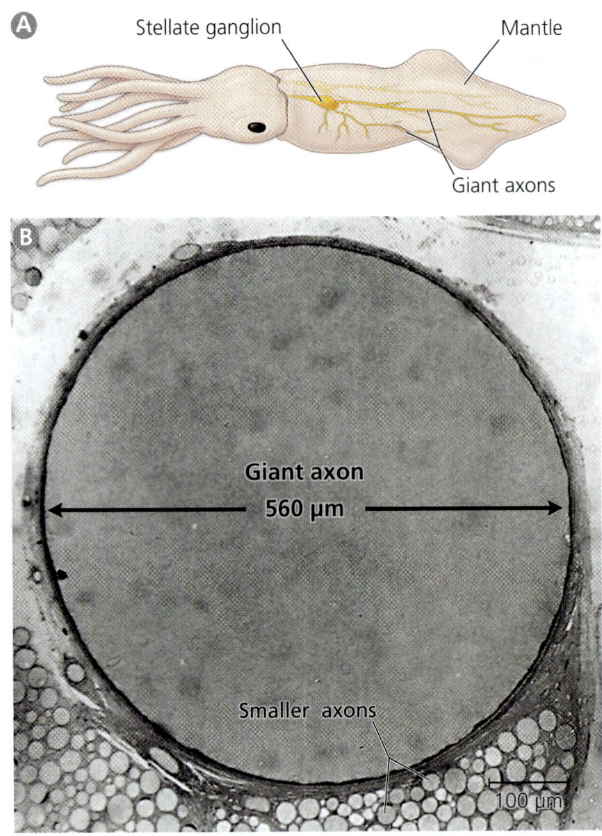

Figure 16.2 The giant axon of a squid. Several long axons emerge from the stellate ganglion of a loliginid squid. The giant axons innervate circular muscles toward the tail end of the animal's mantle (A). The photomicrograph in (B) depicts one of these giant axons in cross section, next to many smaller axons. When the giant axons fire, the squid's muscular mantle contracts nearly synchronously throughout its length, ejecting water from the mantle's front end and propelling the squid backward. [From Lee et al., 1994]

manipulations) in humans than in non-humans. Therefore, research on non-human **model species** will certainly persist.

Problems with the Model Species Concept

Despite the undeniable success of using model species in neurobiological research, not all neural features are conserved across all species. For example, color vision is trichromatic in humans and other primates, but dichromatic in rodents (see Chapter 6). Therefore, rats and mice are not good model species for studies aimed at understanding color vision in humans. Similarly, the lateral and polar subdivisions of the prefrontal cortex have no self-evident rat homologs. Therefore, some aspects of prefrontal cortical function in humans cannot be studied in rats without substantial doubts about how well the rat findings can be extended to humans. In general, the selection of model species for research must be guided not solely by experimental convenience but also by the likelihood that any findings will apply to other species, including us.

The likelihood that a discovery made in one species can be extrapolated to another generally decreases with the phylogenetic distance between the species. Therefore, scientists who are interested primarily in humans often want to study monkeys (if they cannot study humans directly). Of course, this desire must be balanced against ethical concerns, cost, and other aspects of experimental convenience (ah, Krogh's principle again!).

A good general strategy is to study a phenomenon initially in the most convenient species possible, regardless of how closely it is related to humans. Then, after some knowledge has been gained, the studies can be replicated in other species to see how general the findings are. This approach is often employed by researchers who seek new cures for human diseases. They often begin their research on mice, or even on fruit flies. Once a promising new therapy has been identified, they may test it on monkeys and, ultimately, on humans in a series of clinical trials. Sometimes the strategy succeeds; sometimes it fails. Predicting which findings will generalize from model species to humans is currently a serious challenge. Still, from a basic science point of view, most of what we've learned from animal research has helped illuminate the human condition.

Studying Non-human Species for Their Own Sake

Although neuroscientists often study non-human species to learn about humans, many study non-humans for their own sake. For example, a substantial number of neuroscientists study the neural mechanisms of bird song—not because of its close similarity to human speech (although such similarities exist) but because bird song is a fascinating biological phenomenon. Why do only male birds sing in many songbird species, and why is this song seasonal? How do young birds learn their song, and how can they produce such complex sounds so precisely? Answering these question does not reveal how humans speak, nor cure a specific disease. However, understanding how birds sing goes a long way toward understanding how brains produce complex behaviors.

Moreover, the study of bird song has led to several important but totally unexpected discoveries. For example, Fernando Nottebohm's effort to explain why birds sing more in spring than during other seasons led him to discover that the neural circuitry for song expands and contracts throughout the year. This finding, in turn, led to the discovery of adult neurogenesis, which has now been described also in other species, including humans (see Chapter 5). Similarly, the chance discovery of a gynandromorph (half male and half female) songbird has led to major advances in our general understanding of sexual differentiation. You will learn more about this strange bird and its significance later. For now, suffice it to state that the study of non-human brains can reveal fundamental principles of brain structure, function, and development, whether those principles were sought explicitly or discovered by chance.

BRAIN EXERCISE

What role do you think the cost of buying and housing research animals plays (or should play) in selecting a study species? What role do you think cost plays in studying human subjects?

16.2 Who Evolved the Largest and Most Complex Brains?

Neurons probably evolved early in animal phylogeny. Even sponges have neuron-like cells, as well as many of the genes that in more complex animals are used to construct synapses. Echinoderms (such as starfish) and cnidaria (including jellyfish) have proper neurons, but these neurons are distributed throughout the body, forming diffuse **nerve nets**. Most other groups of animals possess a highly centralized nervous system (i.e., a CNS). In some taxa, this CNS includes enlarged, often fused ganglia in the animal's head. Many scientists refer to these **cerebral ganglia** as brains, although it is probably better to reserve the latter term for the larger and more complex collections of neurons that are found in chordates (including vertebrates), arthropods (e.g., insects and spiders), mollusks (e.g., octopuses), and some annelid worms. Given this definition and taxonomic distribution, we can conclude that brains probably evolved at least three times independently (Figure 16.3).

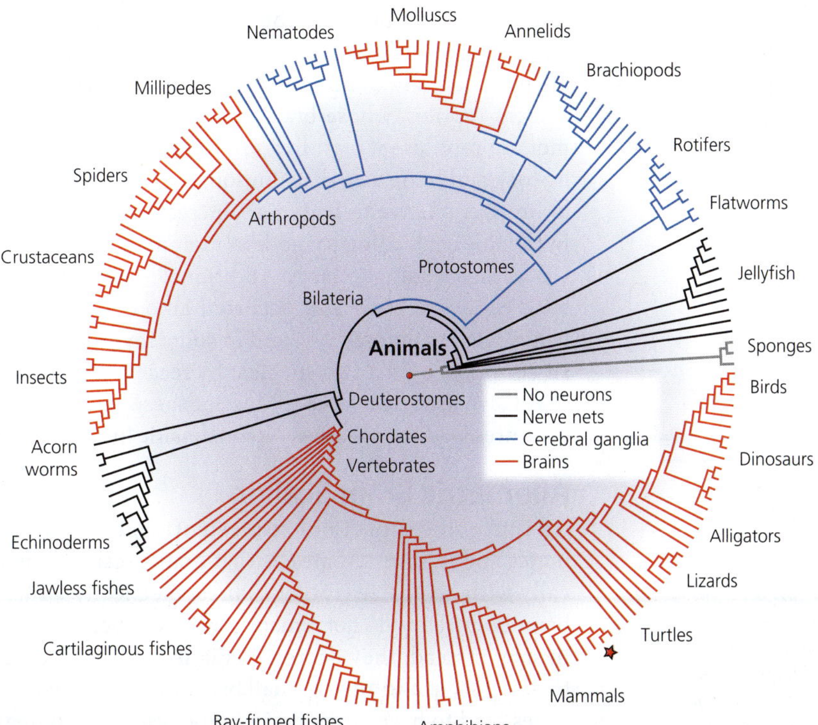

Figure 16.3 The evolutionary history of nervous systems. The root of this phylogenetic tree (or tumbleweed) is at its center. Modern and extinct taxa are represented as the tips of the various branches. The colors indicate, at least roughly, what kinds of nervous systems are found in which taxa. The last common ancestor of bilaterian animals probably had a diffuse nervous system (black) or relatively simple cerebral ganglia (blue). Complex brains (red) evolved at least three times independently: namely, in chordates, arthropods, mollusks, and annelids. Humans are represented by the red star. [After Jenner and Wills, 2007, and Northcutt, 2012]

Although brains evolved repeatedly, the molecular mechanisms underlying neural development are remarkably similar across all **bilateria** (animals with bilateral symmetry, at least at some point in their life). This fundamental similarity suggests either that the last common ancestor of bilateria had a simple brain (cerebral ganglia) or that ancient mechanisms that predated centralized nervous systems were co-opted to build cerebral ganglia and brains in multiple lineages. As both hypotheses are difficult to test, they remain hotly debated.

Evolutionary Increases in Brain Size and Complexity

Less controversial is the idea that brain size and complexity increased repeatedly within the major lineages that have definite brains. Among invertebrates, the largest brains are found in **cephalopods** (squids and octopuses). The brain of a large octopus, for example, contains approximately 380 million neurons. However, not all mollusks have large brains. For example, *Aplysia californica* has less than 20,000 neurons in its brain. Clams and other bivalves have even simpler nervous systems. Therefore, brain size and complexity must have increased within mollusks, specifically in the lineage leading to modern cephalopods. Paralleling this evolutionary increase in brain size and complexity is a substantial increase in behavioral complexity. For instance, octopuses can learn to solve problems by observing the behavior of other individuals.

Arthropods (including spiders, millipedes, crustaceans, and insects) also have large and complex brains, as well as intricate behavior. The brainiest arthropods are the social insects. The brain of a honeybee, for example, contains nearly 1 million neurons. By comparison, a fruit fly's brain consists of only about 100,000 neurons. These data imply that brain size increased at least once within insects, in the lineage leading to bees. Recent studies have shown that the cognitive abilities of bees are remarkably complex.

Brain size and complexity also increased in the lineage leading to **vertebrates**. Even the jawless fishes (lampreys and hagfishes) have significantly larger brains than their closest invertebrate relatives. Within the vertebrates, brain size increased again, and it did so repeatedly. To visualize this variation in brain size, neuroscientists often plot brain size versus body size and then draw polygons around the data points from each major vertebrate taxon (Figure 16.4).

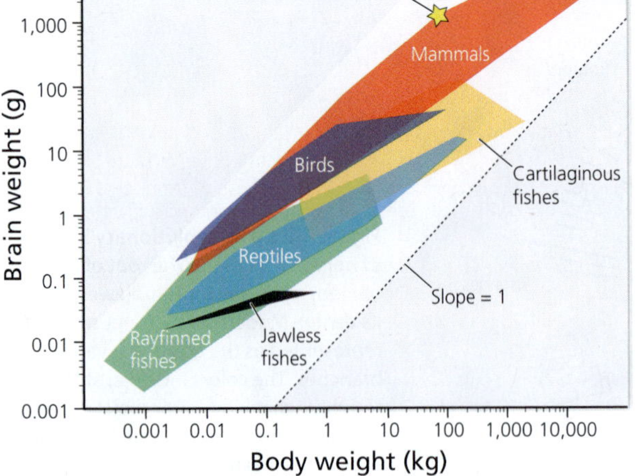

Figure 16.4 Brain-body scaling. This graph shows how brain size varies with body size in the 6 major groups of vertebrates. Minimum convex polygons are drawn around the individual data points. You can see, for example, that birds and mammals have relatively large brains for their body size compared to reptiles. Brain:body ratios are highest in small birds and lowest in large cartilaginous fishes (such as whale sharks). [After Striedter, 2005]

This kind of graph quickly reveals that brain size tends to increase with body size but is, at any given body size, larger in mammals and birds than in reptiles. This is interesting because both birds and mammals descended from ancient reptiles, whose brains likely resembled those of modern reptiles (at least in size). Given this evolutionary history and brain size distribution, we can conclude that brain size, relative to body size, increased independently in the lineage leading to modern mammals and, again, in the lineage leading to birds. Additional increases in brain size relative to body size occurred among cartilaginous fishes (e.g., in manta rays) and in some ray-finned fishes. In contrast, only a few species decreased their relative brain size as they evolved. Most of these small-brained creatures lead very passive, often parasitic lives.

Allometric Brain Scaling

An intriguing aspect of brain evolution, both in vertebrates and in invertebrates, is that brains scale allometrically with body size (rather than isometrically). That is, brain-body ratios do not remain constant as brains increase in size; instead, they decrease (Figure 16.4). Because of this **negative allometry**, small birds and a few ray-finned fishes tend to have larger brain-body ratios than humans, whose ratio hovers near 2.5%.

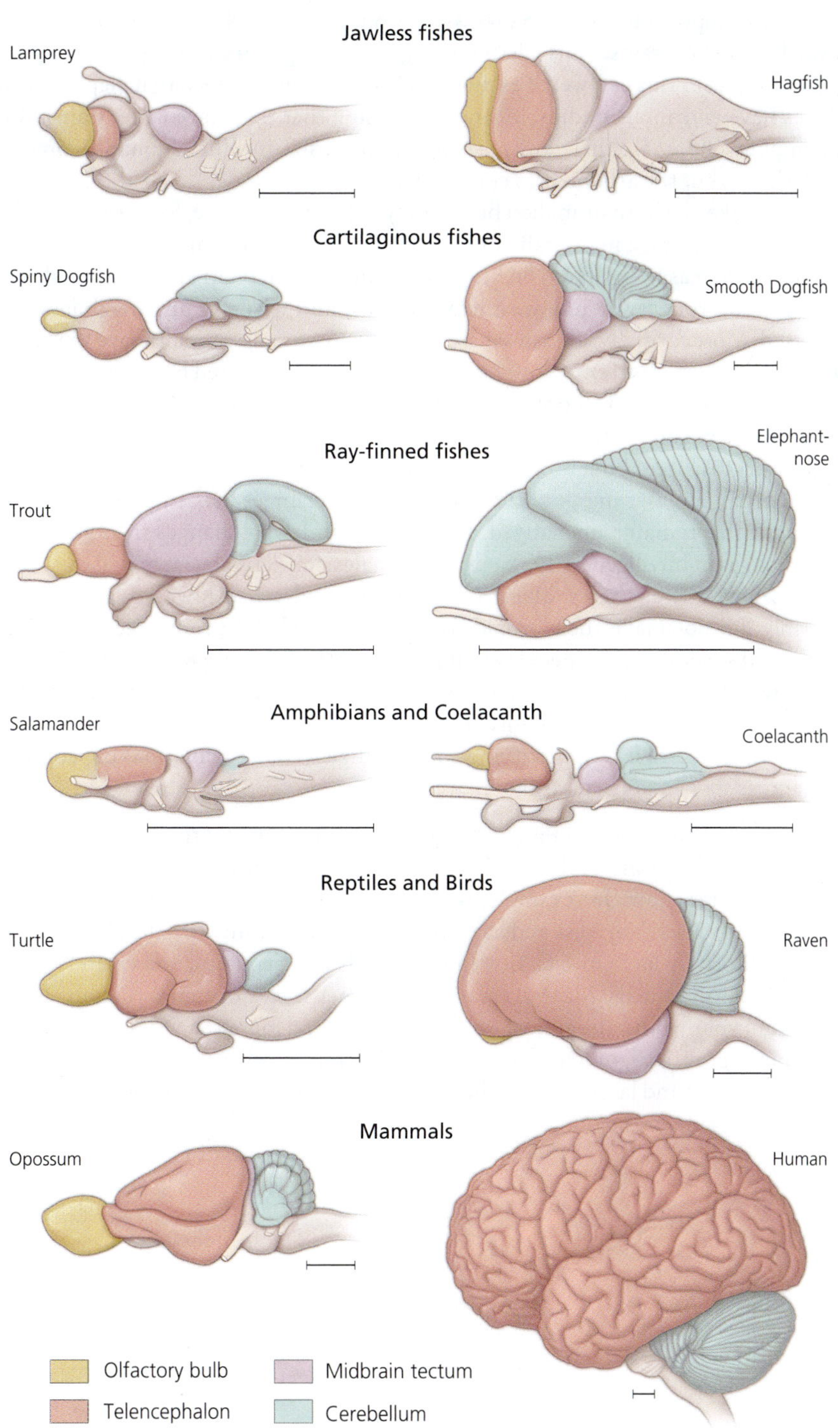

Figure 16.5 Similarities and differences among vertebrate brains. Virtually all vertebrate brains contain the same major brain divisions (although jawless fishes lack an obvious cerebellum). However, these divisions vary in size, both relative to one another and in absolute terms. All scale bars equal 1 mm.

Particularly impressive is the brain of the little elephant-nose fish (Figure 16.5). These fishes tend be less than 25 cm long, but their brain-body ratio is ~3%, greater than our own. Even more amazing is that in elephant-nose fish, the brain accounts for roughly 60% of the entire body's oxygen consumption; for humans, the equivalent fraction is ~20%. How these little fish can afford to spend so much energy on their brain, and how they use those brains, remains largely unknown. We can speculate that the ability of elephant-nose fish to generate and sense electric fields (see Chapter 6, Box 6.3) may have favored the evolution of large brains, but other fish with similar behavioral skills (glass knifefish and their relatives) do not have such enormous brains.

A third important lesson to take away from Figure 16.4 is that humans do not have the largest brains. This distinction goes to elephants (see Figure 1.19) and whales, whose brains can weigh as much as 7 kg. Of course, given that elephants and whales have enormous bodies, it is not surprising that they have large brains. Still, you might wonder: are there any measures of brain size in which humans "come out on top"? Yes, but the argument is complicated.

If you take all the mammalian brain-body data and calculate, for every body size, how large the average mammalian brain should be, then human brains are roughly 3 times as large as they should be, given our body size. In contrast, brain weights for elephants are roughly in line with expectations; and whale brains weigh less than expected. More formally, we can say that, among mammals, humans have the largest **encephalization quotient** (EQ = (actual brain size / expected brain size) × 100).

Whether EQ always correlates with intelligence (IQ) remains a subject of considerable debate. Although many animals with high EQs seem relatively smart, researchers disagree about how best to quantify intelligence. Moreover, evolutionary increases or decreases in body size can cause large changes in the EQ, even as absolute brain size remains unchanged. Do such body-driven changes in EQ correlate with changes in intelligence? Probably not. Whales, for example, increased their body size dramatically as they evolved an aquatic lifestyle. This lowered their EQ but probably did not make them stupid. Indeed, some whales produce long, complex songs, and several whale species exhibit cooperative hunting behavior. Some dolphins, which belong to the same order as whales (Cetacea), put sponges on their snouts and then use them as tools. Thus, dolphins and whales appear remarkably intelligent. Nonetheless, quantifying this intelligence is difficult, making comparisons across species persistently controversial.

Variations in Brain Region Proportions and Other Characteristics

Brains vary not only in size, but also in the proportional sizes of their various subdivisions. The cerebellum, for example, is gigantic in elephant-nose fish but essentially absent in the jawless lampreys and hagfishes (Figure 16.5). Similarly, the olfactory bulb is proportionately large in rats, tiny in most birds and primates, and entirely absent in all aquatic mammals. Most dramatically, the neocortex occupies roughly 20% of total brain volume in bats, almost 80% in humans, and more than 85% in whales. As shown in Figure 16.6, much of this variation correlates with absolute brain size. As you get to larger and larger brains, the neocortex becomes disproportionately large.

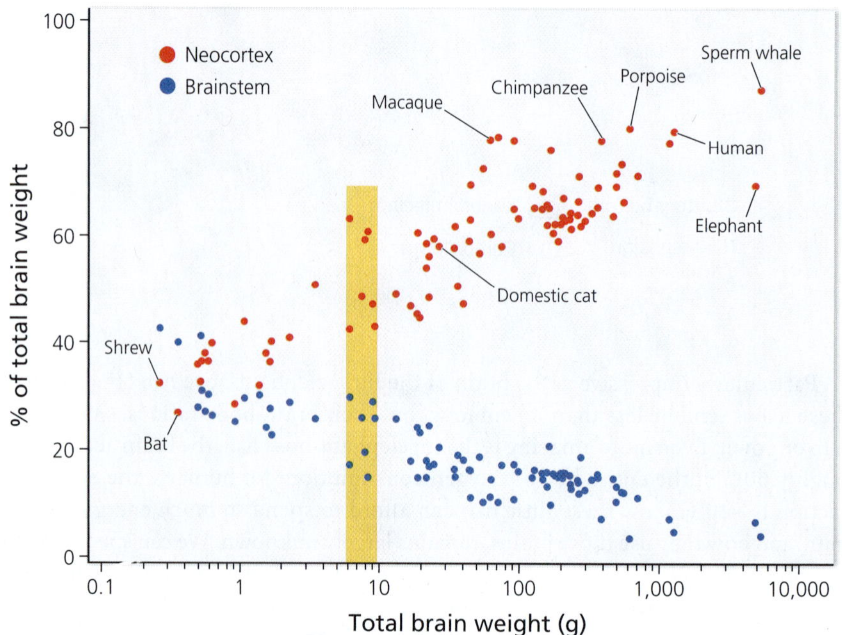

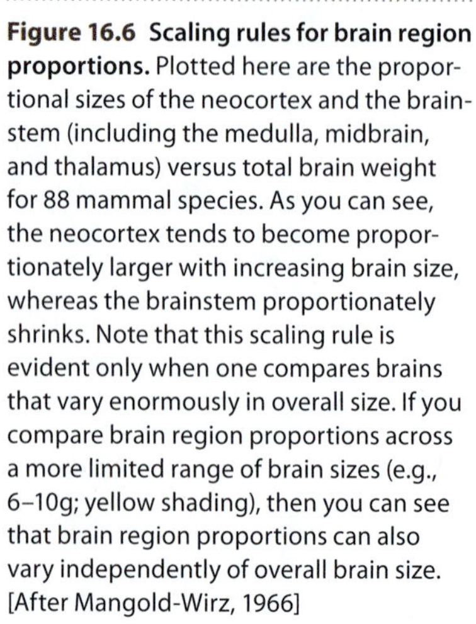
Figure 16.6 Scaling rules for brain region proportions. Plotted here are the proportional sizes of the neocortex and the brainstem (including the medulla, midbrain, and thalamus) versus total brain weight for 88 mammal species. As you can see, the neocortex tends to become proportionately larger with increasing brain size, whereas the brainstem proportionately shrinks. Note that this scaling rule is evident only when one compares brains that vary enormously in overall size. If you compare brain region proportions across a more limited range of brain sizes (e.g., 6–10g; yellow shading), then you can see that brain region proportions can also vary independently of overall brain size. [After Mangold-Wirz, 1966]

In addition, the brains of different species vary in the types of neurons they contain (including variations in neurotransmitter use and receptor expression), in how those neurons are arranged (layers or nuclei), and in their connections. We will not survey all this variation. Instead, let us focus on the major features that distinguish human brains from those of other species.

BRAIN EXERCISE

Why is it incorrect to say that humans have the largest brain-body ratios? What would be a more correct statement?

16.3 What Makes Human Brains Unique?

Scientists have long wondered what, if anything, makes human brains different from other brains? If one accepts the theory of evolution, then one would expect human brains to share many features with other brains, especially if the analysis is limited to our close relatives, the other primates. On the other hand, our cognitive and language skills far exceed those of other animals. Presumably, these behavioral differences are, somehow, reflected in our brains. To explore this complex web of similarities and differences, let us begin with a general discussion of primate brain evolution.

Primate Brain Evolution

Primates originated roughly 65 million years ago as relatively small creatures that were nocturnal and lived primarily in trees. During subsequent evolution, some primates became more diurnal and less arboreal. Many, especially the apes, also increased in body size. As one would expect, the size of their brains likewise increased.

However, at any given body size, primates have roughly twice as much brain tissue as rodents, rabbits, and hares, all of which are closely related to primates (see Figure 16.7). Furthermore, the haplorhine primates (monkeys and apes) tend to have slightly larger brains than prosimian primates. Human brains stand out because they are roughly three times as large as one would expect given the average human body size. Thus, among primates, we are clearly the most encephalized species.

The Hominid Fossil Record

To reconstruct when in evolution our ancestors became so highly encephalized, we must look to fossil specimens. Although brains do not fossilize, one can estimate brain size from fossil skulls by measuring the volume of the **endocranial cavity** in which the brain is housed. As shown in Figure 16.8, endocranial volume in australopithecines (including the specimen widely known as Lucy), ranges between 400 and 500 cm^3. In contrast, early specimens of the genus *Homo*, including *Homo habilis* and *Homo erectus*, have endocranial volumes as high as 1,000 cm^3. Thus, absolute brain size increased roughly 1.5–2 million years ago, soon after the genus *Homo* first appeared. Brain size then increased again roughly 150,000 years ago, as *Homo sapiens* emerged.

Therefore, the fossil record suggests that absolute brain size in our closest ancestors increased in two relatively brief growth spurts, separated by a long period of relative stability. The brain growth spurt in early hominids correlates, at least roughly, with the taming of fire, which allowed

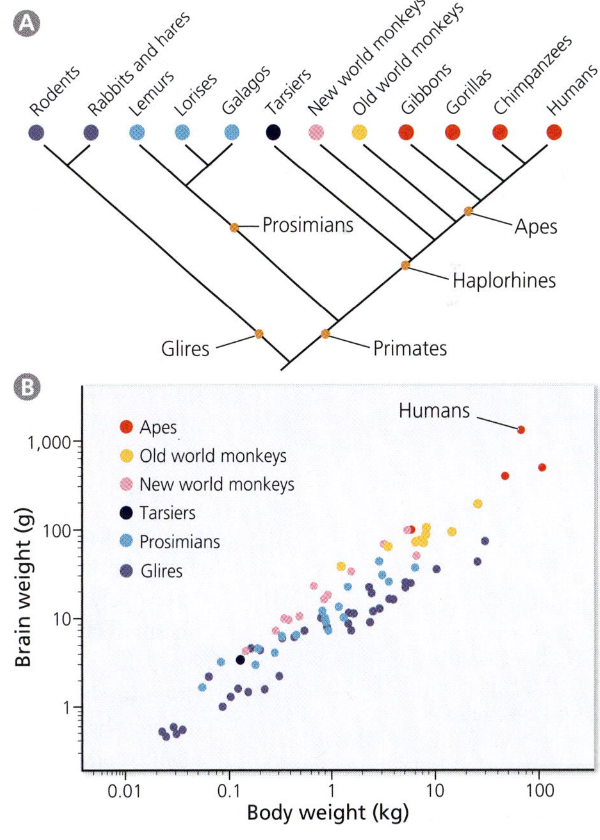

Figure 16.7 Brain-body scaling in primates and their close relatives. Panel (A) presents the generally accepted phylogeny of primates and glires (rodents, rabbits, and hares). Plotted in (B) are the brain and body sizes of representatives from all major primate taxa, as well as diverse glires. You can see that brain size, relative to expectations based on body size, increases progressively in the lineage leading to humans. [After Mangold-Wirz, 1966, Stephan et al., 1981, and Pirlot and Kamiya, 1982]

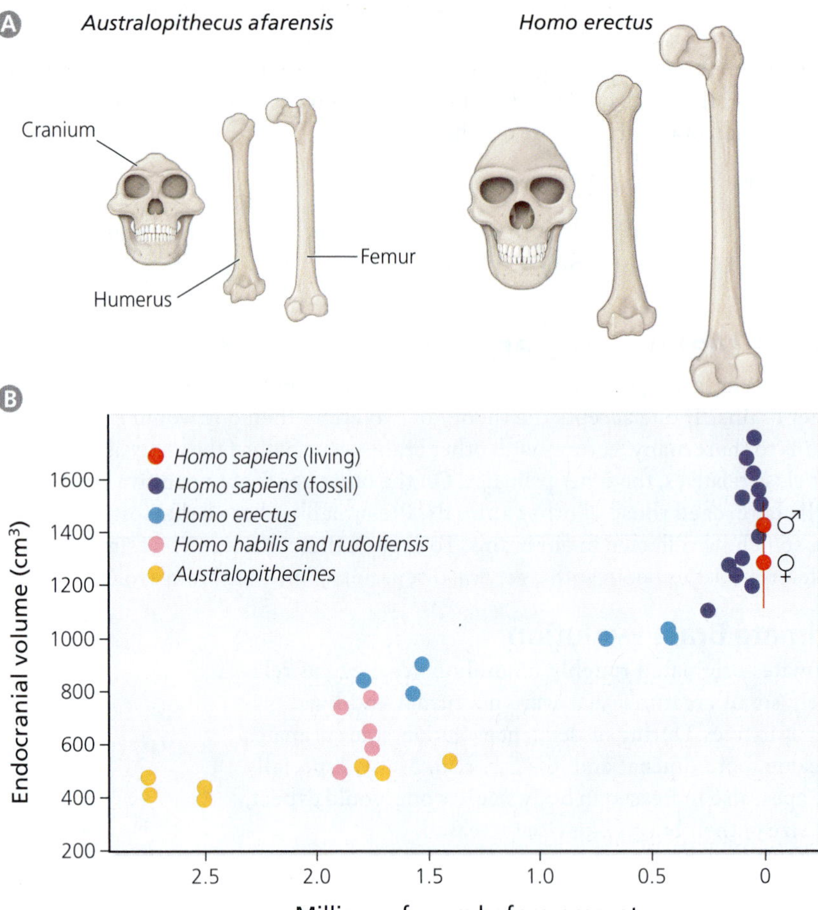

Figure 16.8 Absolute brain size more than tripled as modern humans evolved. *Homo erectus* was much taller and had a larger cranium than a typical australopithecine (A). The graph (B) reveals that endocranial volume was quite small in the australopithecines, increased dramatically with the evolution of the genus *Homo*, and then increased again as *Homo sapiens* appeared. The data for modern humans represent averages (calculated separately for women and men), whereas the other data points represent individual fossil specimens. [B after Hofman, 1983, and Kappelman, 1996]

Homo erectus to cook its food and, thus, obtain more nutrients. In contrast, the increase in the brain size of *Homo sapiens* was most likely linked to intraspecific competition for resources and mates (including social maneuvering, which requires "Machiavellian intelligence"—a term based on a famous old book by Niccolò Machiavelli that describes a recipe for leadership success based on the ruthless manipulation of others). However, both of these hypotheses are controversial and difficult to test.

Genetic Correlates of Changes in Hominid Brain Size

The genetic changes underlying increases in hominid brain size are being studied intensively. This research is facilitated by the fact that DNA is relatively stable, allowing small chunks to be extracted from fossilized bones up to 400,000 years old.

Using such data, scientists have assembled a draft of the nuclear genome for **Neanderthals**, a line of hominids that lived in Europe and parts of Asia from 400,000 to roughly 30,000 years ago. For about half that time, Neanderthals coexisted with *Homo sapiens*. This is fascinating, but comparisons between Neanderthal and human genes will not tell us much about how brains increased in size because Neanderthal brains were just as large as ours are today. Moreover, there currently is little hope of being able to extract useful DNA from fossil australopithecines or *Homo erectus*.

Fortunately, additional information can be gleaned from comparing our own genome to that of living non-human primates. Employing this approach, scientists have identified several genes involved in brain development that have changed more rapidly in the human lineage than in non-human primates or rodents. For example, the gene *abnormal spindle-like microcephaly associated* (*ASPM*) causes microcephaly (abnormally small brains) when it is mutated. Moreover, this gene was modified in the human lineage after it split from that of chimpanzees. Unfortunately, variations in ASPM do not appear to correlate with variations in brain size among modern humans; and even if changes in ASPM did play a role in boosting hominid brain size, it surely was just one of many genes involved in doing so.

Predictable Changes in Neuron Numbers

To determine how the number of neurons in human brains compares to that in other species, researchers have estimated neuron numbers in a variety of primates and rodents (Figure 16.9). In both taxonomic groups, neuron number increases predictably with increasing brain weight. However, the rate of this increase is significantly faster in primates than in rodents. Therefore, large primate brains contain more neurons than equally large rodent brains. Intriguingly, human brains contain almost exactly the number of neurons you would expect, given the primate scaling rule. This is true not only for the entire human brain but also for its two largest divisions: the neocortex and cerebellum (Figure 16.9 B). It is also worth noting that the neocortex contains more neurons in humans than in whales or elephants. Therefore, in terms of absolute neocortical neuron number (but not absolute brain weight), we come out "on top" in the dubious competition for neuroanatomical superiority.

Predictable Changes in Brain Region Proportions

Human brains follow the primate scaling rules not only for neuron numbers but also for brain region proportions. As mentioned earlier, proportional neocortex size increases predictably as overall brain size goes up, and the human neocortex follows this scaling rule nicely (see Figure 16.6). Of the various neocortical areas, the temporal and occipital lobes are proportionately smaller in humans than in non-human primates, but the prefrontal cortex is disproportionately large in human brains. According to Korbinian Brodmann, who compared cortical anatomy across a wide variety of species in the early 1900s, the prefrontal cortex comprises nearly 30% of the entire neocortex in humans, 11% in chimpanzees, and less than 8% in small lemurs. Although this shift in proportions is large, it is in line with allometric expectations. Across primates, prefrontal cortex volume scales predictably with

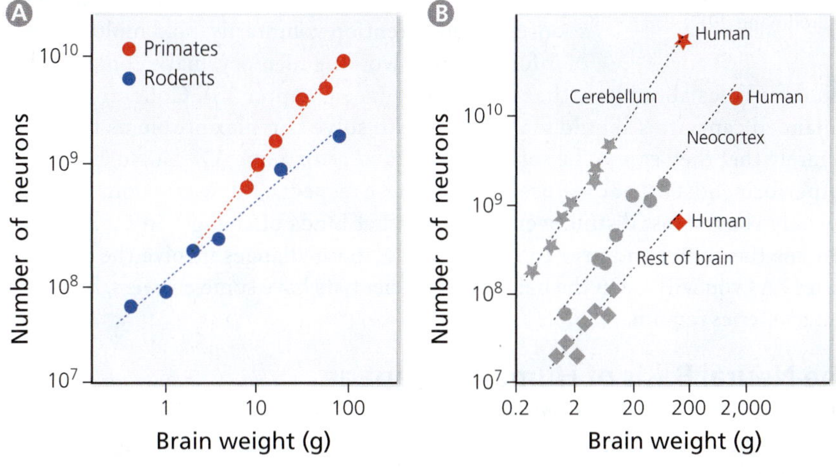

Figure 16.9 Neuron number scaling rules. To estimate the number of neurons in whole brains or brain regions, researchers homogenized neural tissue and then counted the number of cell nuclei in samples of the resulting suspension. Neurons were distinguished from other cells by staining with an immunohistochemical marker. The graph in (A) illustrates that total neuron number increases with brain weight more rapidly in primates than in rodents. Panel (B) plots data only for primates and shows that the major brain regions in human brains contain as many neurons as one would expect, given our brain size and the primate scaling rules. Note that 80% of all the neurons in the human brain are found in the cerebellum. [After Herculano-Houzel et al., 2006, 2007, and Azevedo et al., 2009]

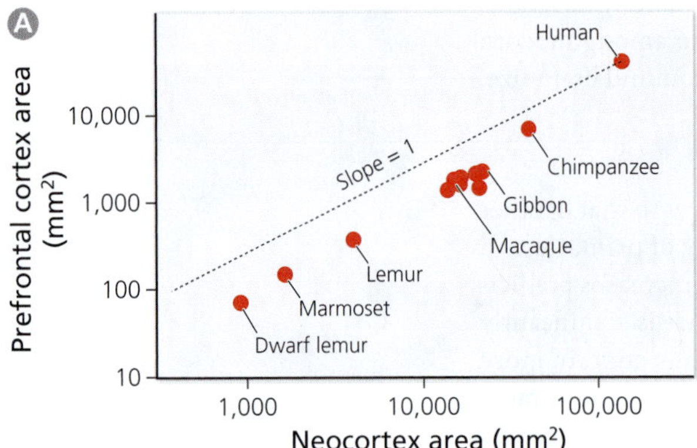

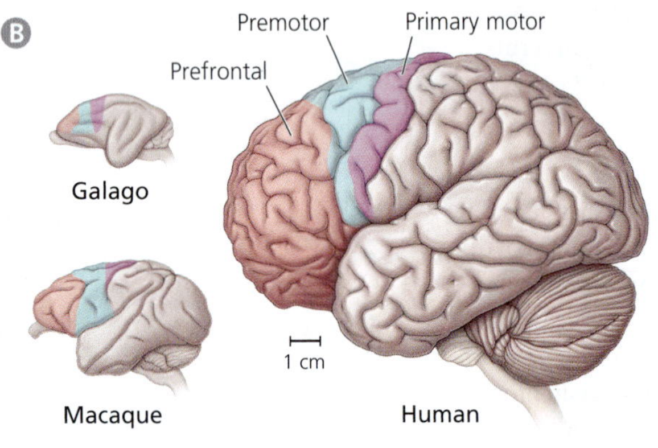

Figure 16.10 The human prefrontal cortex is predictably enlarged. The size of the prefrontal cortex in primates increases predictably with total neocortex size (A). The best-fit line through the illustrated data points has a slope greater than 1, which means that the prefrontal cortex becomes disproportionately large as neocortex size increases. This enlargement is evident in (B), which presents lateral views of three primate brains, all drawn to the same scale. [A after Brodmann, 1912]

neocortex size and does so with a slope greater than one (Figure 16.10). Therefore, it is not surprising that humans, with their enormous neocortices, have a disproportionately large prefrontal cortex.

The observation that the human neocortex and, within that, the human prefrontal cortex, are just about as large as they should be, given our brain size, implies that natural selection did not increase their size selectively. Nonetheless, the shifts in proportions may well have been functionally significant. For example, the expansion of the human neocortex, relative to the remaining brain and spinal cord, may have been linked to an expansion of direct neocortical projections to subcortical regions, which would explain why humans tend to have more precise control over their hand movements than their close relatives. It might also explain why cortical damage produces more severe and long-lasting paralysis in humans than in non-humans. After all, if subcortical regions are "accustomed" to receiving strong inputs from the neocortex, then they are likely to be impaired when those projections are destroyed.

Whether the disproportionate enlargement of the human prefrontal cortex altered human behavior and cognition remains a contentious issue. Some neuroscientists have argued that because the prefrontal cortex is of the size one would expect (given our brain size and the primate scaling rule), the functions of our prefrontal cortex must be quite similar to those of other primate species. Other neuroscientists suspect that making the prefrontal cortex larger, relative to other brain regions, is bound to increase its importance within the brain and for behavior.

Following the latter line of argument, the proportional enlargement of the human prefrontal cortex should correlate with an increased ability to inhibit impulsive responses, orient attention voluntarily, hold multiple pieces of information in working memory, make complex plans, and infer rules about how the world works (see Chapter 15). Collectively, these enhanced capacities should make it easier to solve complex problems that occur so rarely that they cannot be solved by trial and error alone. Of course, humans do outperform most other creatures in all of these respects. But what about language? It surely is our most distinctive attribute. What kinds of changes in the brain give humans the ability and urge to speak? Might those changes involve the prefrontal cortex? As you will see in the next section, scientists have some answers, but important mysteries remain.

The Neural Basis of Human Language

The most influential early work on the neural basis of human language was performed by Paul Broca in the 1860s. Working in Paris, Broca examined the brains of several patients with language deficits. Broca's first patient, Monsieur Leborgne, was given the nickname Tan because a brain lesion had made him incapable of saying anything other than "tan." Importantly, Tan was able to comprehend speech and could move his lips and tongue; he just could not produce articulate speech. Today, we call this kind of language deficit **Broca's aphasia**. When Broca examined Tan's brain at autopsy, he found a lesion centered on the posterior portion of the left inferior prefrontal cortex (Figure 16.11). This part of the brain contains **Broca's area**, which consists mainly of Brodmann areas 44 and 45 (named by the same Korbinian Brodmann we discussed in the preceding section).

Broca's 1st aphasic patient: M. Leborgne (a.k.a. Tan)

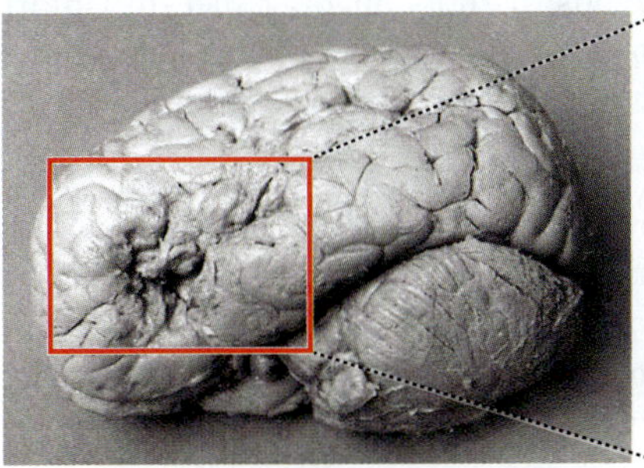
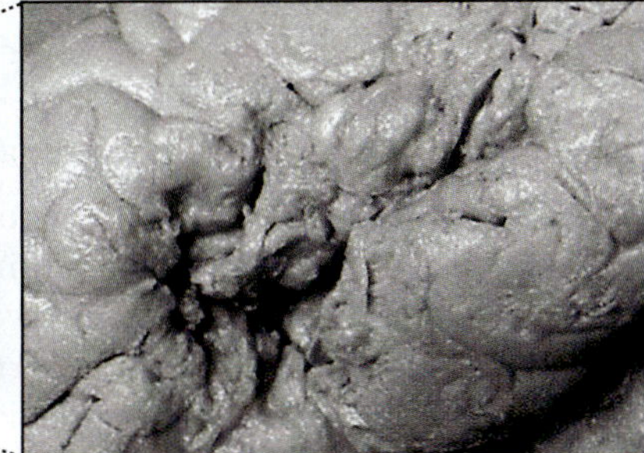

Broca's 2nd aphasic patient: M. Lelong

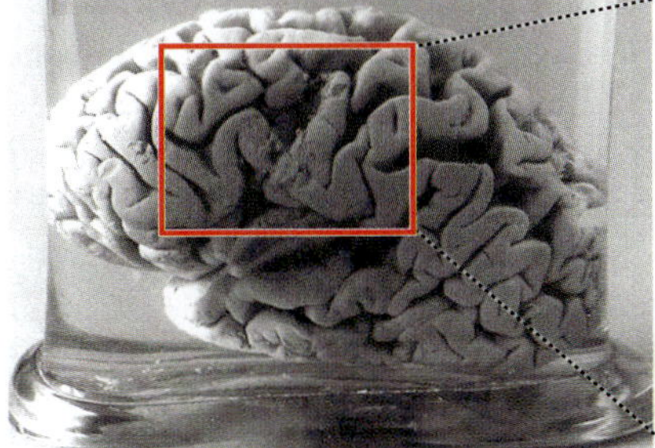
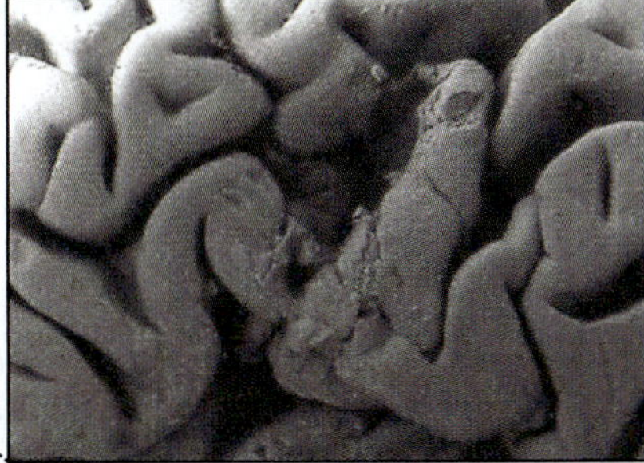

Figure 16.11 The brains of Paul Broca's first aphasic patients. Shown at the top is the brain of Broca's first aphasic patient called Leborgne and nicknamed Tan. His lesion is in the posterior portion of the left inferior prefrontal cortex. Shown at the bottom are analogous views of the brain of Broca's second patient, Lelong. Again, the posterior portion of the left inferior prefrontal cortex is seriously damaged. Both patients could comprehend language but were incapable of fluent speech. [From Dronkers et al., 2007]

A few months after publishing his paper on Tan, Broca studied a second aphasic patient called Monsieur Lelong. This patient could only utter five different words. A postmortem inspection revealed that, like Tan, Lelong had incurred damage to the posterior portion of the left inferior prefrontal cortex (Figure 16.11). After Broca examined several additional patients with similar symptoms and brain damage, he became convinced of an important pattern: in all of these aphasic patients the damage was on the left side of the brain. This insight explained why the earlier literature on prefrontal cortex lesions and language had been so full of confusion: those earlier studies had all combined data from left-sided and right-sided lesions. As Broca recognized, only left-sided lesions in the posterior inferior prefrontal cortex produce Broca's aphasia.

Diverse Aphasias

A decade after Broca's initial work, Carl Wernicke discovered that damage in more posterior brain regions can also cause language deficits. Specifically, he noted that damage to the posterior portions of the superior temporal lobe, in what we now call **Wernicke's area**, impairs language comprehension without impairing speech. Wernicke's patients were able to speak fluently, but what they said made little or no sense. Importantly, Wernicke noted that this form of aphasia, called **Wernicke's aphasia**, occurs preferentially after damage to the left hemisphere, just as Broca had noted for Broca's aphasia.

Although Broca's and Wernicke's areas are the best known "language areas," they are hardly the only brain regions involved in language control and processing. It has long been known, for example, that lesions in the inferior parietal lobe cause **conduction aphasia**, which is characterized by deficits in word repetition and object naming. It is also clear that lesions in prefrontal cortex outside of Broca's area can lead to language deficits (so-called transcortical aphasias) that are distinct from Broca's aphasia.

Further support for language being supported by a widespread network of cortical areas comes from electrical stimulation of the cortex during speech. As you may recall from Chapters 10 and 11, Wilder Penfield and his colleagues stimulated the neocortex of awake humans to localize the primary somatosensory and motor cortices.

Using the same approach, researchers in the late 1980s discovered that electrical stimulation in many different cortical regions disrupts ongoing speech. Unfortunately, it was difficult from these data to construct a detailed model of how language is processed in the brain. For that, researchers turned to functional brain imaging.

Brain Imaging of Language Functions

Hundreds of studies have used fMRI or other functional brain imaging techniques to examine changes in brain activity during a wide variety of language tasks. These imaging studies confirm that language involves an array of cortical areas that extends well beyond the areas that Broca and Wernicke had focused on. For example, listening attentively to complex sentences activates parts of the anterior temporal lobe, as well as Broca's area and more posterior regions near and in Wernicke's area.

A second major finding to emerge from the fMRI research is that listening to spoken sentences activates many cortical areas bilaterally rather than only on the left (Figure 16.12A). This finding was surprising given the strong left hemisphere dominance for language functions suggested by the brain lesion data. However, subtle language deficits can be observed after right hemisphere lesions. Specifically, the language-related areas in the right hemisphere appear to be specialized for the *prosodic* (musical) aspects of language including rhythm, stress, and intonation. Another fascinating finding is that perceiving and understanding sign language activates many of the same brain areas as listening to speech (Figure 16.12B). This shows that the brain treats sign language as what it is, a proper language.

It is difficult to synthesize the wealth of data on language-related deficits and brain activity into a coherent model of how language is implemented in the brain. However, Gregory Hickok and David Poeppel have developed a model of speech processing that is widely discussed (Figure 16.13). A central feature of their model is the division of the language-related circuitry into dorsal and ventral "streams" (in analogy to the dorsal and ventral streams in the visual system; see Chapter 12). The dorsal stream passes through the inferior parietal lobe and is involved in the production of speech. In contrast, the ventral stream passes toward the anterior temporal lobe and is involved in complex speech comprehension. Both streams probably converge in

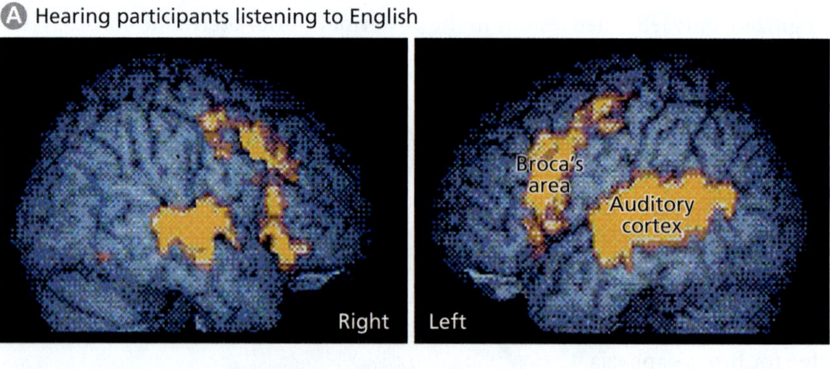

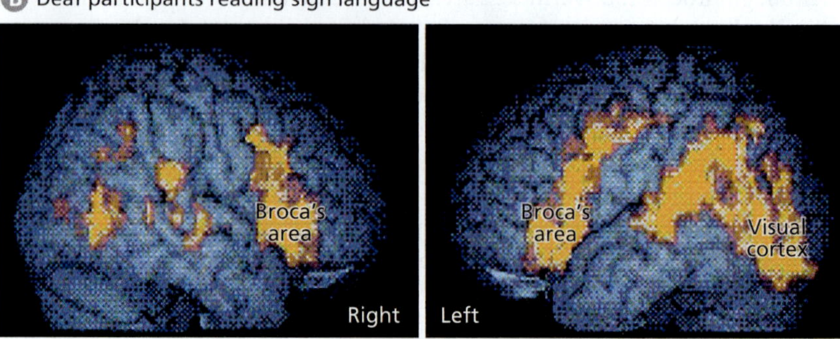

Figure 16.12 Spoken and sign language activate similar circuits bilaterally. Shown in (A) are the brain regions that become activated when hearing participants watch and hear a video of a person speaking a sentence (vs. attending to a video of the same person doing nothing). Shown in (B) are analogous images for deaf participants who grew up using sign language. Not surprisingly, the auditory cortex is more active in (A) than in (B), whereas the visual cortex exhibits the opposite pattern. Except for these differences, sign language and spoken language activate very similar cortical areas. In both conditions, activation of the right hemisphere is substantial. [From MacSweeney et al., 2002]

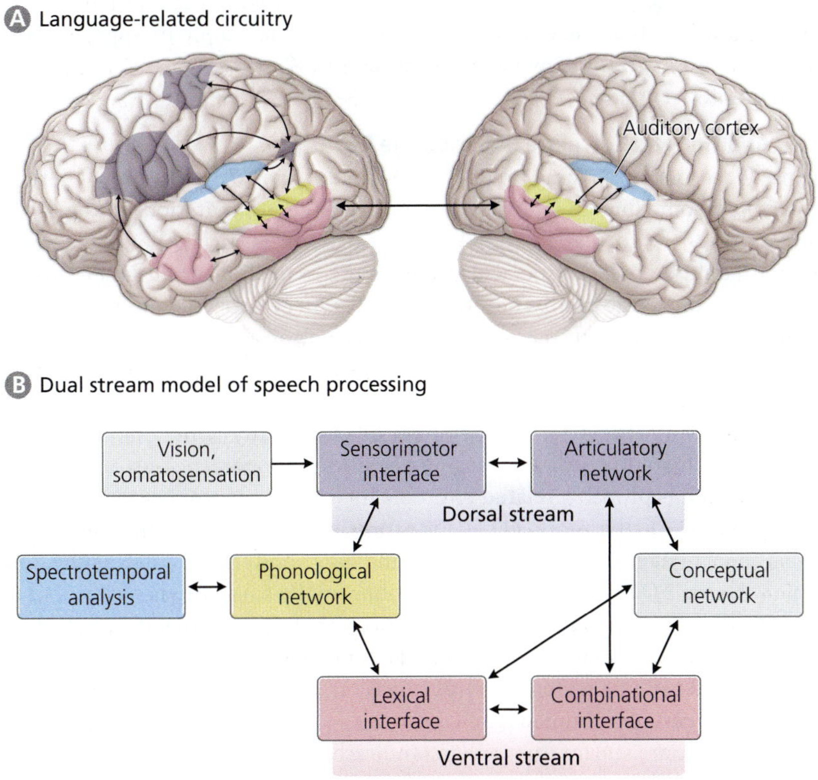

Figure 16.13 **A neurobiological model of language functions.** Shown in (A) are the principal language-related circuits of the human neocortex. The diagram in (B) illustrates an influential model of speech processing that proposes that auditory information is initially analyzed in the auditory cortex (light blue) and then processed further along two divergent pathways. The so-called ventral stream is concerned primarily with language comprehension, whereas the dorsal stream is needed for speech production. According to this model, only the late stages of speech processing are left-hemisphere dominant; the others are bilateral. [After Hickok et al., 2011]

Broca's area, which would be consistent with the relatively recent observation that lesions of Broca's area cause deficits not only in speech production but also in the comprehension of complex speech. Still, many aspects of this dual stream model remain to be fleshed out and are controversial.

The Evolution of Language-related Circuitry

Although chimpanzees and gorillas are capable of using hand gestures or plastic tokens to communicate with humans, these forms of communication do not (by most accounts) amount to proper language as we recognize it in humans. How, then, did language evolve?

For comparative neurobiologists, the first step in answering this question is to ask whether the language-related circuits in humans have homologs in non-humans. The answer is clearly yes. Using histological criteria, one can identify homologs of areas 44 and 45 (which together comprise Broca's area) in all great apes and even in macaques. Similarly, a major component of Wernicke's area, called *Area Tpt*, has been identified in chimpanzees, macaques, and some prosimian primates. Remarkably, Area Tpt is larger on the left than on the right in chimpanzees, just as Wernicke's area is larger on the left in us. Connectional studies have further shown that the dorsal and ventral "streams" connecting the posterior temporal lobe to the ventral prefrontal cortex exist not only in humans but also in macaques. The pathways are smaller in non-humans than in us, but they are nonetheless quite similar.

These similarities leave us with a dilemma: if the language-related circuitry in human brains exists in non-humans, why do only humans have language? Nobody knows for sure. The increased size of the language-related areas in humans is probably part of the answer. In addition, humans are likely to have some language-related connections that non-humans just lack. For example, it has been proposed that humans have evolved more direct connections from the neocortex to the motor neurons that innervate the lips, the tongue, the larynx (which harbors the vocal folds), and the muscles that control breathing. Such connections would have given humans more fine-grained, voluntary control over their vocalizations. It seems unlikely, however, that the evolution of language is due simply to improved motor control. It almost certainly involved new intracortical connections as well, but these remain unknown.

FOXP2 as a Genetic Correlate of Language Evolution?

The mechanisms underlying language evolution can also be studied genetically. For example, some human language deficits are linked to mutations in the gene for a transcription factor called FOXP2. This protein is highly conserved across mammals, but two of its amino acids were altered in the lineage leading to humans. Moreover, in one large family, half the members carry a specific point mutation in the *FOXP2* gene. All family members with this mutation are language impaired. This discovery raised hopes that *FOXP2* might be "the human language gene," but reality is more complex.

For one thing, the affected members of this family generally have trouble making coordinated mouth and face movements. Therefore, at least some of their language deficit is probably related to problems with fine motor control rather than language per se. Furthermore, *FOXP2* is expressed in many different brain regions, not just the language areas, as well as in some other body parts (such as the lungs). People with the *FOXP2* mutation do have abnormally small Broca's areas, but they also differ from controls in other brain regions that are less specifically linked to language such as the striatum and the cerebellum. Overall, these findings imply that *FOXP2* is involved in a wide variety of functions, only some of which are likely to have a direct bearing on human language.

Although we are still far from understanding how human language evolved, there is real hope that such an understanding can be reached. This sentiment stands in stark contrast to that of the Linguistic Society of Paris in 1866, which had grown so frustrated with speculations on the origins of language that it formally banned the topic from public discussion. Modern neuroscientists no longer feel that such discussions are futile.

BRAIN EXERCISE

In what sense are human brains typical primate brains, and how are they unique?

16.4 Do Brains Differ between the Sexes?

Males and females of most species differ in anatomy, physiology, and behavior; they also differ in the structure and function of their brains. For many years, neural sex differences were thought to be limited to a few regions that are involved in reproductive behaviors. Other neural circuits were thought to be quite similar between the two sexes, especially for our own species.

However, men and women differ behaviorally in more than just a few reproductive behaviors. Multiple studies have shown, for example, that men tend to perform better on visuospatial tasks, whereas women have the edge in diverse verbal tasks. It has also been established that men are more likely to be diagnosed with Parkinson's disease, autism, drug addiction, and schizophrenia; whereas women suffer disproportionately from depression and anxiety. The neural mechanisms underlying these behavioral sex differences are poorly understood. However, it is now clear that the brains of men and women differ, on average, in a variety of ways. Before we discuss these differences, let us briefly review the mechanisms that cause males and females to develop along some diverging trajectories.

Mechanisms of Sexual Differentiation

In humans, sex is determined by the two sex chromosomes called X and Y. Females have two X chromosomes, whereas males have both an X and a Y. The Y chromosome carries the *SRY* gene, which causes gonadal precursor tissue to develop into testes rather than ovaries.

Testosterone and Estrogens

Near the middle of embryonic development and shortly after birth, a male's testes produce a surge of **testosterone**. This testosterone binds to **androgen receptors** (testosterone being one of several hormones in the androgen family) in the external genitalia, which in response enlarge to turn the embryonic phallus into a penis. Testosterone also helps to *masculinize* the brain by binding to neuronal androgen receptors. However, in many species much of the testosterone is converted within the brain to **estradiol** (the body's main form of estrogen during the reproductive years) by the enzyme **aromatase**. Estradiol binds to estrogen receptors and causes the cells expressing those receptors to develop in the male-typical direction.

This may surprise you, as we tend to think of estrogens as female sex hormones. However, during fetal and early postnatal development, females express very little estrogen, and whatever estrogen they do express is sequestered by *alpha-fetoprotein*, which does not cross the blood–brain barrier. Thus, during early development, androgens masculinize the brain indirectly through estradiol. This is certainly true for rodents; in humans, testosterone seems to affect brain development more directly, but this remains controversial. In any case, without testosterone, the human fetus develops along the female (default) trajectory. Estrogens do not become important in females until puberty, when they induce the development of breasts and (together with progesterone) regulate the timing of female reproductive behavior (notably the menstrual cycle).

Strong support for this model of sexual differentiation comes from the observation that neonatal castration causes genetic males to become phenotypically female in many (but not all) respects. Conversely, treating genetic females with high levels of testosterone (or estradiol) during fetal development causes them to develop a number of male-typical features.

Sex Differences that Develop Independently of Sex Hormones

Recent research has shown that some sexually dimorphic characteristics can emerge independently of circulating sex hormones. This independence was underscored by the existence of a rare, mutant zebra finch that looked like a male on the right side of its body and like a female on the left (Figure 16.14). Such **bilateral gynandromorphism** should be impossible if all sex differences were caused by circulating sex hormones, as blood flow is not strictly divided between the right and left sides of the body.

Indeed, further study showed that most of the cells on the right side of the gynandromorph finch contained the male mix of sex chromosomes, whereas most of the cells on the left side contained the female sex chromosomes (presumably due to abnormal sex chromosome segregation during one of the first embryonic cell divisions). Therefore, the left–right differences in the plumage of this finch were most likely caused by genes that are located on only one of the sex chromosomes. Ultimately, we can conclude that the normal sex difference in zebra finch plumage results from a direct genomic effect rather than circulating sex hormones.

The situation in the brain is more complex. Male zebra finches sing complex songs and possess a network of brain regions that help them learn and produce those songs (the song system). In contrast, female zebra finches don't sing and have only a very small song system. In the gynandromorph zebra finch, the song system was larger on the right than on the left (Figure 16.14 B), consistent with a direct genomic effect. However, the song system on the left side of the gynandromorph finch was significantly larger than it is in normal females. Most likely, the left side of the song

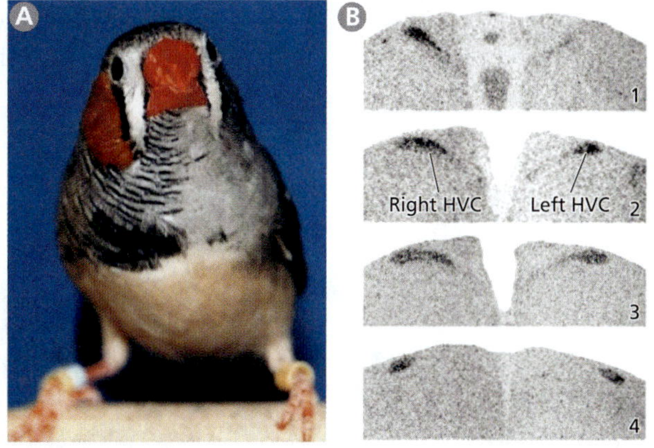

Figure 16.14 A bilateral gynandromorph finch. Depicted in (A) is an extremely rare, mutant zebra finch that exhibits male-typical plumage (orange cheek patch and black neck stripes) on the right side of its body, and female plumage on the left. Internally, this bird had one ovary (on the left) and one testis. Panel (B) shows four transverse sections through this finch's brain, highlighting the song system nucleus "high vocal center" (HVC; stained for androgen receptor RNA). You can see that the HVC is larger on the right side than the left. This difference is reminiscent of the normal sex difference in HVC size, although normal females have an even smaller HVC. [From Agate et al., 2003]

system was partially masculinized by testosterone coming from the bird's single testis (and converted to estradiol in the brain).

We can conclude that the normal sex differences in the song system of zebra finches result from both direct genomic effects and the action of circulating sex hormones. Direct genomic effects also contribute to some sex differences in mice; but in most species, the genomic contributions to sexual differentiation are difficult to disentangle from hormone effects.

Sex Differences in the Spinal Cord, Hypothalamus, and Midbrain

The spinal motor neurons that innervate muscles at the base of the penis (as well as nearby muscles) are more numerous and occupy a larger volume in men than in women. These neurons form **Onuf's nucleus** (Figure 16.15). A homologous nucleus is found in most mammals, although its name and degree of sexual dimorphism varies across species. Because injecting newborn females with testosterone increases the number of neurons in Onuf's nucleus, we can conclude that the sex difference in the volume of this spinal nucleus is driven, at least indirectly, by circulating sex hormones.

Sex Differences in the Hypothalamus

In the brain, the best studied sex differences involve the anterior hypothalamus. Specifically, the **sexually dimorphic nucleus (SDN)** of the preoptic area is 3–7 times larger in male rats than in female rats. This neuronal sex difference probably contributes to sex differences in sexual behavior, as SDN neurons are known to increase their activity during a variety of male sexual behaviors (notably ejaculation). The human homolog of SDN appears to be the **third interstitial nucleus of the anterior hypothalamus (INAH-3)**. This nucleus is likewise larger in men than in women, although the size difference is less pronounced than for the rodent SDN.

Intriguingly, several studies have shown that INAH-3 has a smaller volume in homosexual men than in heterosexual men, although it contains a roughly equal number of neurons (Figure 16.16). Interpreting these data is problematic. Homosexual men might be predisposed to homosexuality because they have a relatively small INAH-3; alternatively, they might have a smaller INAH-3 because they are homosexual. Neither hypothesis has been tested. What does seem reasonably clear is that this difference between heterosexual and homosexual men is not attributable to differential rates of HIV infection, as this variable was controlled for in the research.

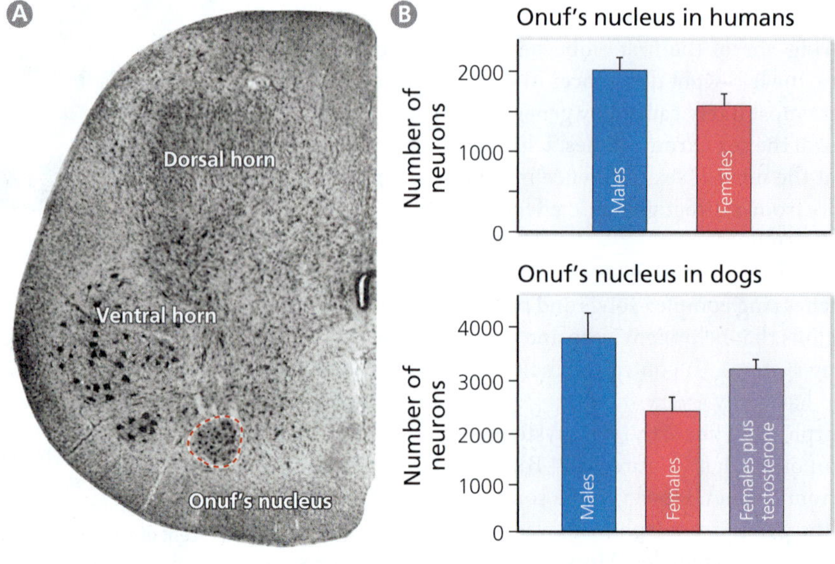

Figure 16.15 Onuf's nucleus is larger in males. Onuf's nucleus lies in the ventral horn of the spinal cord and appears circular in cross section (A). As shown in (B), Onuf's nucleus contains more neurons in men than in women. The bottom graph shows that female dogs exposed to high levels of testosterone in utero and shortly after birth end up with an increased number of Onuf's motor neurons, implying that the sexual dimorphism results, at least in part, from sex differences in testosterone exposure during early development. [From Forger and Breedlove, 1986]

Ⓐ The human anterior hypothalamus

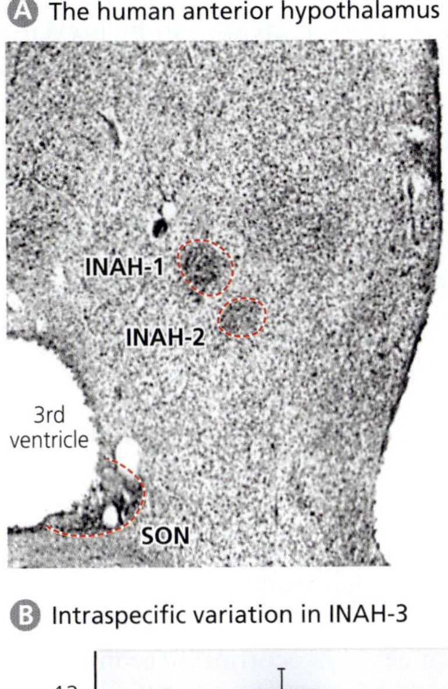

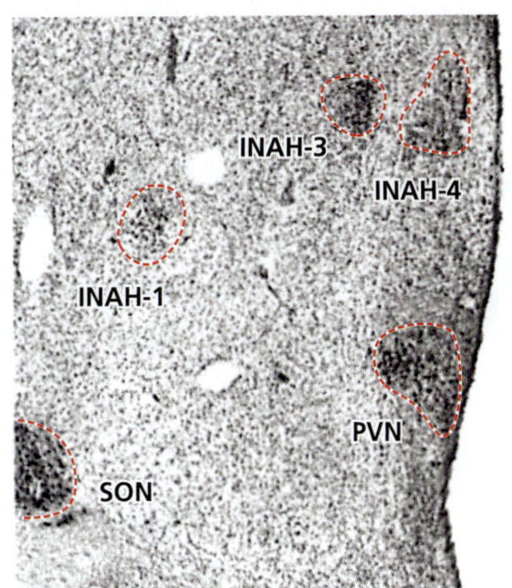

Ⓑ Intraspecific variation in INAH-3

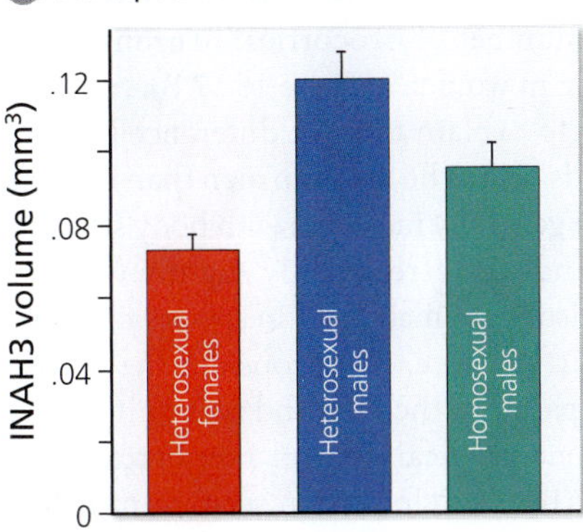

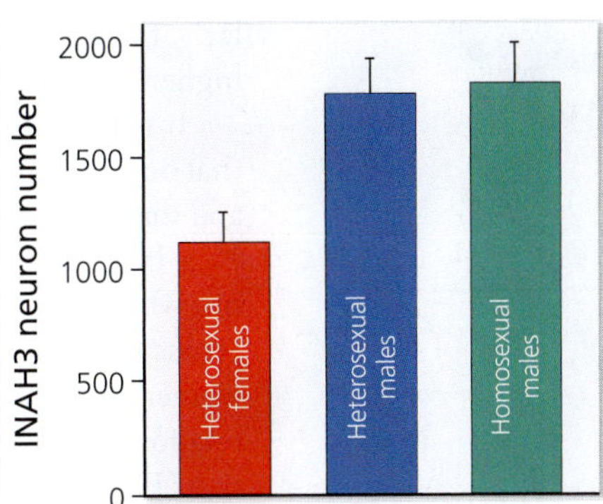

Figure 16.16 Sex differences in the human anterior hypothalamus. The photographs in (A) depict two sections through the right anterior hypothalamus of a 59-year-old man (the section on the right is more posterior). You can see the supraoptic and paraventricular nuclei (SON and PVN, respectively), as well as the 4 interstitial nuclei of the anterior hypothalamus (INAH). INAH-3 is significantly larger in heterosexual men than in heterosexual women, both in volume and neuron number (B). Intriguingly, this nucleus has the same number of neurons in homosexual males as in heterosexual males, but its volume is smaller in the former group. [A from Byne et al., 2000; B after Byne et al., 2001]

In contrast to SDN and INAH-3, which are larger in males than in females, the **anteroventral periventricular nucleus (AVPV)** of the preoptic area is significantly larger in females. This makes sense, given that AVPV helps to trigger ovulation by driving a surge in luteinizing hormone release from the pituitary gland (see Chapter 8).

The sex differences in SDN and AVPV size are dependent on the action of sex hormones during the perinatal period (around the time of birth). Castration of newborn male rats causes SDN to shrink, whereas giving testosterone or estradiol to newborn females causes the SDN to attain male-like proportions. Similarly, perinatal testosterone or estradiol treatments decrease the number of AVPV neurons in female rats. These effects of circulating sex hormones are accomplished mainly by the modulation of developmental cell death. Specifically, perinatal testosterone increases the rate of apoptosis in AVPV and decreases it in SDN. Furthermore, blocking cell death eliminates the sex differences in SDN and AVPV.

This is interesting because one might have expected sex differences in neuron number to be due to differences in rates of proliferation rather than death. However, for small cell groups at least, creating sex differences by differential cell death seems to be the rule. This hypothesis is supported by the observation that Onuf's nucleus exhibits higher rates of cell death in females than in males.

Sex Differences in the Substantia Nigra

The sex differences we have discussed so far all relate to reproductive or sexual behaviors, but some sex differences do not. For example, the dopaminergic neurons in the substantia nigra (SN) are more numerous in females than in males, even though they do not function specifically in sexual behavior. Intriguingly, this sex difference is sensitive to circulating sex hormones in adulthood. Castrating an adult male rat

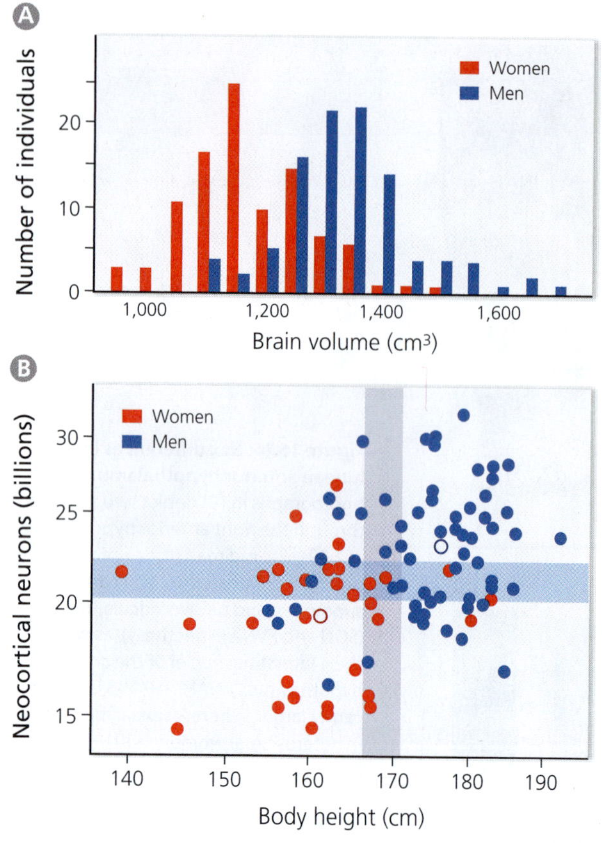

Figure 16.17 Sex differences in human brain and body size. Absolute brain sizes, as determined from structural MRI scans, are significantly smaller, on average, in women than in men ($p < .0001$). The graph in (B) shows that men tend to be taller than women and have more neocortical neurons (open symbols indicate population means). However, within a narrow range of body heights (light purple rectangle), men tend to have more neocortical neurons than women. Conversely, within a narrow range of neuron numbers (light blue rectangle), men tend to be taller. Therefore, the sex differences in human brain size and height appear to be independent of one another. [A after Leonard et al., 2008; B after Pakkenberg and Gundersen, 1997]

increases the number of dopaminergic SN neurons, relative to control rats, and this increase can be prevented by testosterone replacement therapy. Conversely, removing the ovaries from an adult female rat reduces the number of dopaminergic SN neurons, and this reduction is preventable by estrogen replacement. Overall, these data suggest that testosterone is slightly toxic to dopaminergic SN neurons in adulthood, whereas estrogen is protective. Assuming these findings can be generalized from rats to humans, they would help to explain why men are more likely than women to develop Parkinson's disease.

Sex Differences in the Telencephalon

The brains of men and women overlap in size, but, on average, men's brains are larger (Figure 16.17). Most of this difference is due to a sex difference in the volume of the cerebral hemispheres, which are 14% larger, on average, in men than in women (excluding the ventricles). Similarly, the average number of neocortical neurons is ~16% higher in men than in women (Figure 16.17 B).

It is tempting to explain this sex difference by noting that body size tends also to be larger in men than in women and that brain size generally increases with body size. However, brain size tends to increase only slightly with body size when comparisons are made within species rather than across species. Furthermore, if we consider the brains of women or men separately, then we find no significant correlation between neocortical neuron number and body size (Figure 16.17 B). Therefore, men seem to have larger brains *and* larger bodies than women, but it is not clear that they have larger brains *because* they have larger bodies. Instead, it appears that the sex difference in human brain size evolved *independently* of the sex difference in body size. This conclusion is extremely controversial because it forces us to explain the sex difference in human brain size independently of the sex difference in human body size.

Why do men have larger bodies than women (on average, even at the same brain size)? Probably because their ancestors were more involved in predator defense, hunting, or aggression toward other men. Women, in turn, might have preferred to mate with larger men, at least on average and at some time during human evolution. But why did men evolve larger brains than women (on average, even at the same body size)? The observation that men consistently perform better than women on visuospatial ability suggests that men might have experienced natural selection for visuospatial processing. Increasing neocortex size might have improved this skill somehow. In contrast, the skills in which women tend to outperform men (such as verbal ability) might have been boosted in women by mechanisms that do not involve neocortical enlargement (such as earlier brain maturation). But this is all speculation. At this point, we know only that men and women differ in average brain size. We have no clear idea why this is so.

Predictable Sex Differences in Telencephalic Organization

Because it is so difficult to link differences in brain or neocortex size to specific behaviors or cognitive capacities, most neuroscientists focus on more specific sex differences in telencephalic organization. Specifically, they point out that neocortical gray matter, the corpus callosum (the fiber tract connecting the two cerebral hemispheres), and several subcortical areas are proportionately larger in women than in men.

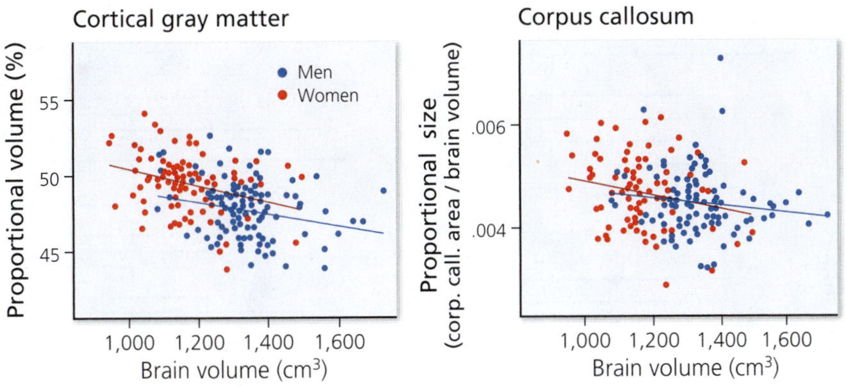

Figure 16.18 Sex differences in gray matter and corpus callosum size. Compared to men, women tend to have proportionately more cortical gray matter (left) and a proportionately larger corpus callosum (right). However, gray matter fraction and proportional corpus callosum size decrease predictably with increasing brain size. [After Leonard et al., 2008]

The problem with these findings is that differences in proportional size are often linked to variation in absolute brain size. Thus, the sex difference in proportional gray matter volume is rather slight, and the sex difference in corpus callosum size disappears entirely, once you control for the sex difference in absolute brain size (Figure 16.18). Similarly, the greater size of various subcortical regions in women compared to men is predictable, given that men have a larger brain, and larger brains generally have a proportionately larger neocortex (see Figure 16.6).

Of course, differences in proportional brain region size may be functionally significant even if they are **allometrically predictable** (just as the proportionate enlargement of the prefrontal cortex in humans compared to non-humans may be functionally significant despite being predictable from the change in brain size). For example, the proportionately smaller corpus callosum in men may cause the two cerebral hemispheres to function more independently of one another than they do in women.

Sex Differences in Functional Lateralization

Indeed, some functional brain imaging data suggest that left hemisphere dominance for language processing is less pronounced in women than in men (Figure 16.19). Although these studies showed statistically significant sex differences, many other studies failed to confirm this finding (Figure 16.20). Part of the problem is that the studies varied in which specific aspects of language processing they examined. Maybe sex differences in laterality exist for some language-related functions but not for others. In any case, the available data suggest that the two sexes vary only slightly, if at all, in their degree of language lateralization. By comparison, the sex difference in handedness is much more robust (right side of Figure 16.20).

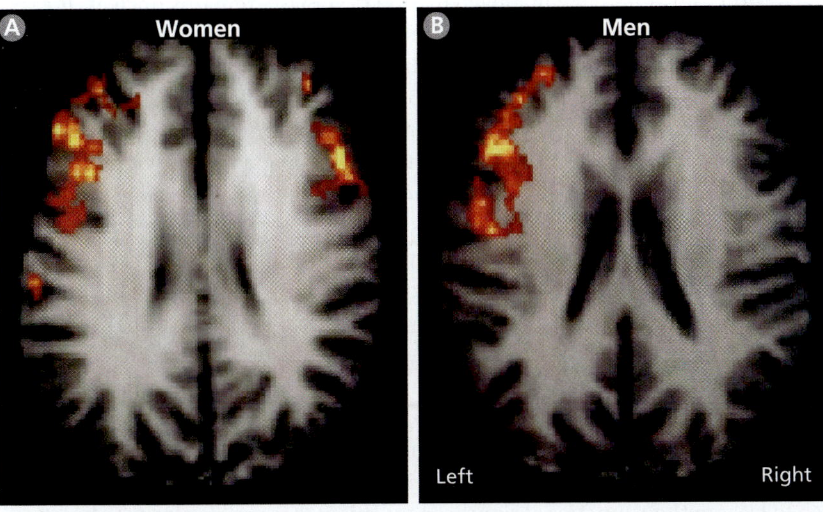

Figure 16.19 Some language functions appear more lateralized in men than in women. Shown here are fMRI data obtained when 19 men and 19 women evaluated whether two sets of nonsense words rhymed. In both sexes, activity increased in the prefrontal cortex (centered on areas 44 and 45), but the activity increase was more bilateral in women than in men. [From Shaywitz et al., 1995]

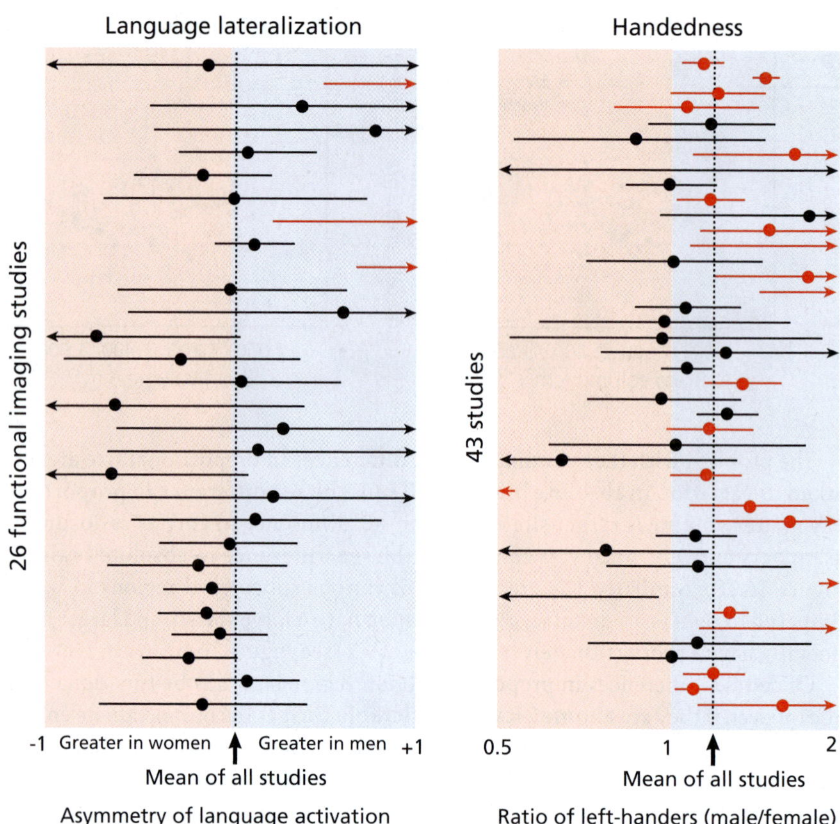

Figure 16.20 Meta-analysis of sex differences in language lateralization and handedness. Summarized on the left are the results of 26 studies that used functional brain imaging to look for sex differences in language lateralization. Each circle represents the degree of asymmetry observed in one study. Statistically significant effects are shown in red. The horizontal lines indicate the 95% confidence interval for each study, with longer lines indicating less confidence in the result (arrowheads indicate out-of-range confidence intervals). Only 3 studies showed one or more language functions to be significantly more lateralized in men than in women. Across all studies, the mean sex difference is virtually nil. Shown on the right are analogous results from 43 studies looking for sex differences in handedness. They show that men are 25% more likely than women to be left-handed or ambidextrous. [After Sommer et al., 2008]

A more convincing sex difference in functional lateralization involves the amygdala. As we discussed in Chapter 14, people tend to remember emotionally arousing images better than neutral images. Subsequent studies have shown that the degree of amygdala activation at the time of memory formation correlates positively with increased memory for the arousing images when subjects are tested several weeks later. However, this correlation exhibits a statistically significant sex difference. For women, the correlation is more robust for the left amygdala; for men, it is more significant on the right side of the brain (Figure 16.21). This is a surprising discovery because the left and right amygdalae are of roughly equal size in men and women and seem structurally identical. However, this sex difference in functional lateralization has been replicated (Figure 16.21).

Why this sex difference in amygdala function exists remains unclear. It may be linked to a general left–right difference between the two cerebral hemispheres, with the left preferentially processing detailed information and the right focused on more global aspects. Given this context, one might predict that emotional arousal should boost the memory for details in women but increase memory for the overall gist of an event in men. This is indeed the case. However, the matter is not yet fully resolved, and the general topic of left–right differences between the two cerebral hemispheres is notoriously controversial.

As you can see, controversy and uncertainty surround the topic of sex differences in human brains, especially when one moves beyond the hypothalamic and spinal cell groups that are linked to reproductive behaviors. Part of the challenge is that discussions of sex differences easily degenerate into debates about which sex is "better" along some imaginary dimension (such as general intelligence). However, given the existence of neurobiological sex differences in nonhumans, such as zebra finches, it is a priori likely that male and female human brains exhibit some divergent specializations. The challenge for neuroscientists is to identify those differences and then to fathom their functional significance. This task is very far from completion.

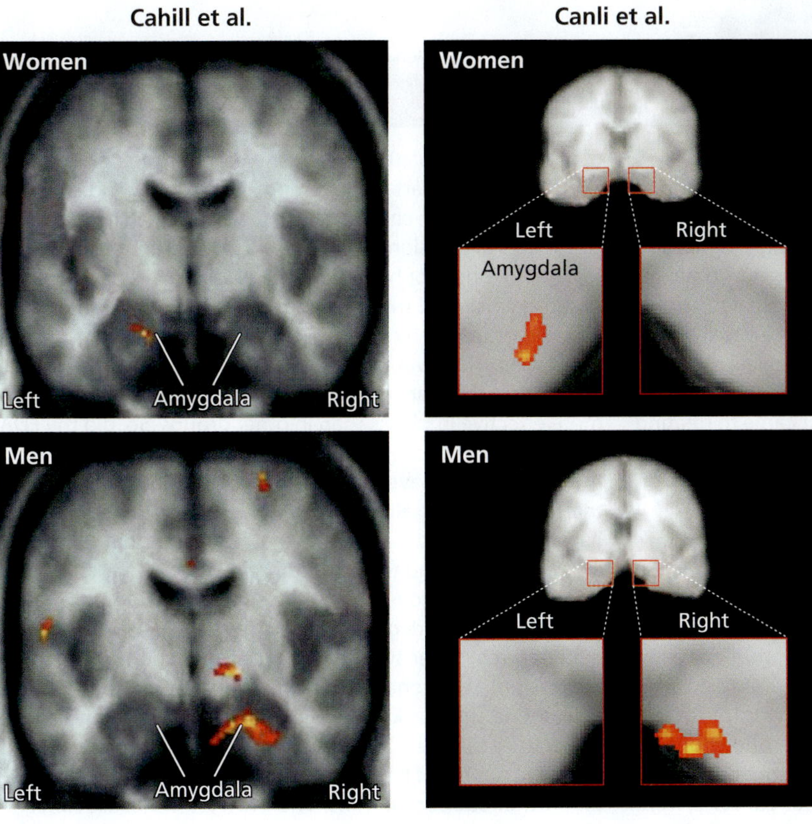

Figure 16.21 Sex differences in the laterality of amygdala activation. Amygdala activation, as measured with fMRI, correlates with enhanced memory for emotionally arousing images. Intriguingly, the correlation between amygdala activation and memory for highly emotional pictures is more robust in the left hemisphere for women and more robust in the right hemisphere for men. The red and orange areas indicate the regions with the statistically most robust correlations. This basic finding has been confirmed repeatedly (left vs. right). [From Cahill et al., 2004, and Canli et al., 2002]

BRAIN EXERCISE

Why do you think the topic of sex differences in human brains is so controversial? What kind of evidence for or against sex differences in human brains would you like to see?

16.5 Within a Sex, How Much Do Human Brains Vary?

Brains vary not only between species and between the two sexes, but also across individuals of the same sex and species. Much of this interindividual variability remains poorly explored. For example, evidence suggests that even relatively fundamental cognitive traits may vary between ethnic or racial groups (Box 16.1: The Challenges of Human Subjects Selection), but few studies have explored the neural correlates of that variation. Moreover, this research area is even more controversial than the study of sex differences, in part because most ethnic and racial groups are so heterogeneous that their very definition becomes problematic.

In addition, questions about the role of genes versus experience (including nutrition) complicate the interpretation of intraspecific variation in brains. For example, one study has reported that the hippocampus is significantly smaller in female Australian aborigines than in female Caucasians, but this size difference emerges late in development and may well be linked to differences in health and nutrition.

Once we move away from racial and ethnic differences, the research becomes less problematic. For example, it is widely accepted that humans vary in visual ability, as measured in diverse visual processing tasks. Humans also vary more than twofold in the size of their primary visual cortex and lateral geniculate nucleus, even within a single sex. These differences in visual system size probably correlate with the behavioral differences in visual ability, although this has not yet been established. Other studies have shown that the shape of the hippocampus differs significantly between

RESEARCH METHODS

Box 16.1 *The Challenges of Human Subjects Selection*

The Food and Drug Administration (FDA) in the 1970s banned all women of child-bearing age in the United States from participating in clinical trials because it worried that these women might be pregnant, or become pregnant, during the trial and that the tested drug might harm the fetus. Even in basic research, female subjects were woefully underrepresented, mainly because the estrous cycle is thought to increase variability in data obtained from females. In the 1980s, however, people began to realize that the exclusion of women from biomedical research had disadvantaged women when it came to treating them. Because the drugs had not been tested on women, little was known about how well the drugs would work in women or what kinds of side effects they might trigger.

By protecting women from potential harm during a drug's testing phase, society inadvertently exposed them to more risk after the drug had been officially approved. According to a study published in 2008, women were 1.5 times as likely as men to develop adverse reactions to prescription drugs. In addition, researchers became increasingly aware that many diseases differ between men and women in symptoms, incidence, severity, and age of onset. Therefore, one could no longer assume that studying diseases and treatments in men would aid women as much as men. Analogous concerns were raised about the paucity of racial and ethnic minorities in clinical trials. Was medical research neglecting them?

In response to these concerns, the US Congress in 1993 required the National Institutes of Health (NIH), the principal sponsor of biomedical research in the United States, to ensure that women and members of minorities and their subpopulations are included in all human subjects research. Furthermore, all phase 3 clinical trials, which aim to establish a drug's effectiveness, must be designed so that data from the two sexes and racial/ethnic groups can be analyzed separately. Part of the justification for these new regulations was that "because the population of the United States is heterogeneous, the health needs and responses to treatment of individuals in the country must be assumed *a priori* to be just as heterogeneous" (Hohmann and Parron, 1996). Although it took a number of years, by 2014, women accounted for roughly half of the participants in NIH-funded clinical research. Having achieved some progress on the clinical front, the NIH recently expanded its policy of inclusion to preclinical (basic science) research. Since October 2014, it requires that all grant applicants "report their plans for the balance of male and female cells and animals in preclinical studies in all future applications, unless sex-specific inclusion is unwarranted, based on rigorously defined exceptions" (Clayton and Collins, 2014, p. 283).

Still, a larger question looms. If the research uncovers substantial sex differences, or other sorts of intraspecific variation, will we have to find different treatments for different subsets of the overall population? Probably yes, but developing a therapy for just one part of humanity is practically and ethically challenging. Those challenges are well illustrated by the FDA's controversial decision in 2005 to approve a drug called BiDil for the treatment of heart disease in blacks. This decision was based on a clinical trial showing that BiDil reduced mortality from heart failure by 43% in self-described blacks or African Americans. However, the trial did not include other racial groups, which means that it could not show that the drug was more effective in blacks than in other groups. So, did the FDA unfairly deprive non-blacks of a potentially life-saving treatment? Should it have required another, more costly, clinical trial that included all recognized racial groups? Would that have been fair to the company that was developing the drug and would have had to pay for the trial?

Another controversial aspect of this episode is that self-described blacks are genetically extremely diverse. As one geneticist put it, "After 400 years of social disruption, geographic dispersion, and genetic intermingling, there are no alleles that define the black people of North America as a unique population or race." Yet, self-described black people in the United States die of heart disease at a significantly higher rate than other subpopulations. Therefore, it is at least conceivable that the ideal treatment for heart disease may likewise vary across the population.

Currently, both scientists and the public are ambivalent about how to handle diversity in biomedical research. On one hand, most people are excited about the prospect of "personalized medicine" in which treatments are selected for each individual based on their genotype or other biomarkers. In theory, this is the ideal way to deal with human heterogeneity in disease susceptibility and treatment responses. On the other hand, the more subgroups that are included in a clinical trial or other kind of biomedical experiment, the larger the total number of subjects must be. Otherwise the risks of obtaining false-positive or negative results increase dramatically. But increasing the number of subjects in a study increases its cost. The NIH regulations state explicitly that increased cost cannot be an excuse for excluding women or minorities from a subject pool, but cost considerations are hard to ignore in this era of limited budgets. Furthermore, the biomedical industry, which is responsible for many clinical trials, need not be bound by NIH regulations. Indeed, it is difficult to see how companies can make money developing "personalized" treatments for multiple subpopulations unless they focus on subgroups that are relatively large.

Despite these caveats and procedural complexities, the rise in awareness of human heterogeneity among biomedical researchers is an important development that should make medicine more equitable. Clearly the status quo is not ideal.

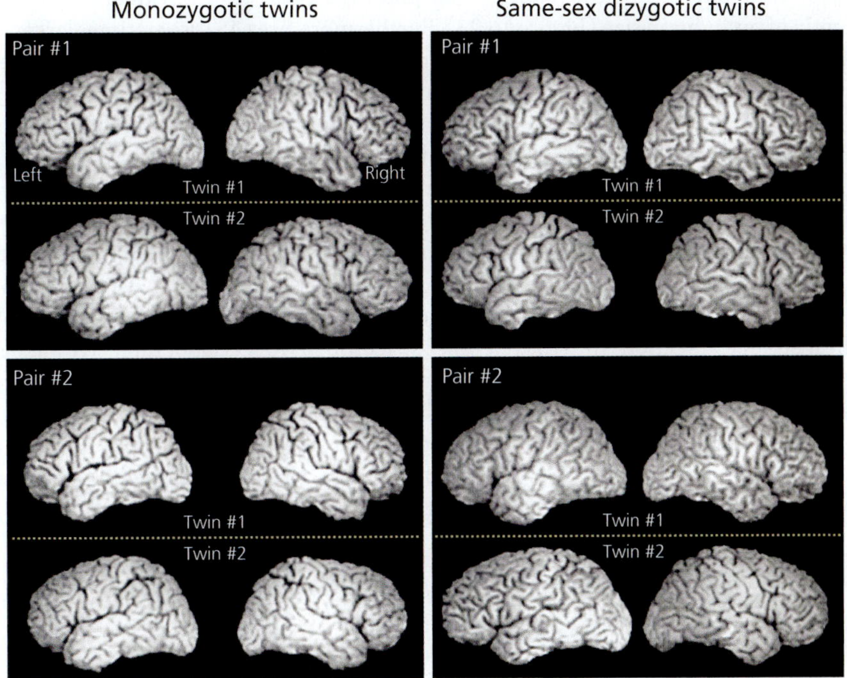

Figure 16.22 Heritability of brain size and shape. The left side of this figure shows left and right cerebral hemispheres from two pairs of monozygotic twins. Shown on the right are analogous images for 2 pairs of same-sex dizygotic twins. You can see that brain size, general brain shape, and the major folds of the neocortex are more similar between monozygotic twins than dizygotic twins. In contrast, the smaller folds vary considerably even between corresponding hemispheres of monozygotic twins. [From Bartley et al., 1997]

experienced London taxi drivers (London is a very difficult city to navigate!) and control subjects, including London bus drivers, who drive more constrained routes.

Well documented are variations in the shape and neocortical folding pattern of human brains. Even some of the major sulci and gyri (troughs and ridges, respectively) can be difficult to compare across multiple brains. Some of this variation is heritable, as the brains of monozygotic twins (derived from a single fertilized egg) are significantly more similar than the brains of same-sex dizygotic twins (Figure 16.22). Whether the remaining variation is due to chance or some kind of experience remains unclear.

Implications of Brain Variability for Functional Brain Imaging

The variability in human brains poses a challenge for neuroscientists who conduct functional brain imaging research. How can they average data across multiple subjects if the brains of those subjects vary in size, shape, and folding pattern?

To solve this problem, researchers digitally stretch and squeeze each subject's brain scan into a standard size and shape. The reference brain used to guide this morphing procedure is most often that of a 60-year-old French woman, whose left cerebral hemisphere was the basis for an influential brain atlas published by Talairach and Tournoux in 1988. After transforming each subject's brain into a standard size and shape, it becomes relatively easy to average fMRI activity patterns across subjects.

However, because individual brains often differ in shape, the morphing procedure shrinks or expands different brain regions to varying degrees. This can cause artifacts in the location of averaged fMRI results when they are presented in *Talairach space* (or any other standard reference coordinate system). For example, spots of increased brain activity may come to lie between cortical folds or in the ventricles. Such artifacts usually involve only the edges of the observed activity patterns and therefore do not cause serious interpretive problems. Nonetheless, they indicate that human brains vary much more than most brain imaging studies suggest.

Probabilistic Brain Atlases

A related problem for functional brain imaging research is that individual cortical areas may vary in size and shape. For example, cortical areas 44 and 45 (the two major components of Broca's area) vary enormously across individuals, independently of sex or brain size (Figure 16.23). Furthermore, area 45 is sometimes larger in the left cerebral hemisphere, and sometimes larger on the right. This variability remains when the data are transformed into a standard reference space (Figure 16.24). Similarly, mapping the location of the primary visual cortex in 10 different individuals onto a standard brain reveals imperfect overlap (Figure 16.24 B).

Because of this variability, one can rarely be sure that any spot of fMRI activity, presented in standardized coordinates, corresponds to a specific cortical area. Instead, we can only say that a given location in the reference space *probably* represents a specific cortical area. To quantify these probabilities, researchers have to examine multiple human brains histologically (to identify the cortical areas) and then morph the histological sections into the reference space. Such work is labor intensive but has now been performed for many cortical areas. The ultimate goal of this research is to generate a complete probabilistic atlas of the human brain or, more accurately, human brains. In summary, when neuroscientists report that any specific spot of imaged brain activity is located in a particular cortical area, they are usually giving you an educated guess. Most of the time, such estimates are good enough.

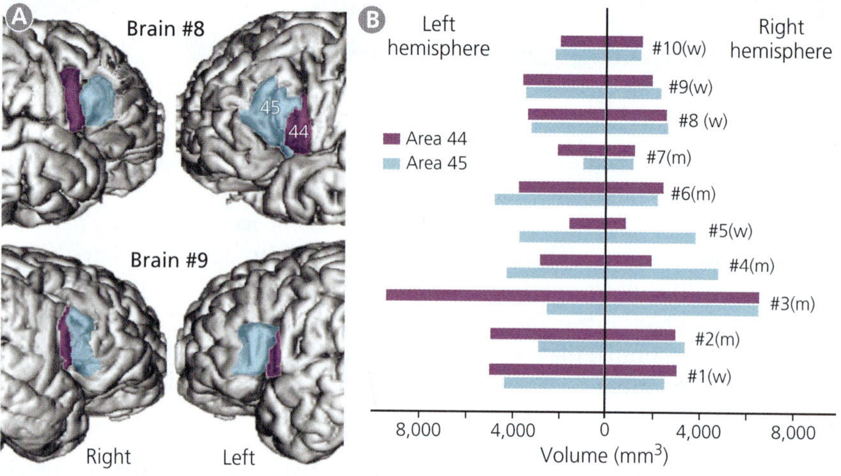

Figure 16.23 Variation in the size of Broca's area. The sizes of areas 44 and 45 were measured for 5 men (m) and 5 women (w). Shown in (A) are left and right hemispheres from 2 representative brains. The graph in (B) shows that significant variability exists both across brains and between hemispheres. On average, area 44 was significantly larger in the left hemisphere. Area 45 was sometimes larger on the left and sometimes larger on the right. Variation in the size of these two areas correlates only weakly with overall brain weight and is independent of sex. [After Amunts et al., 1999]

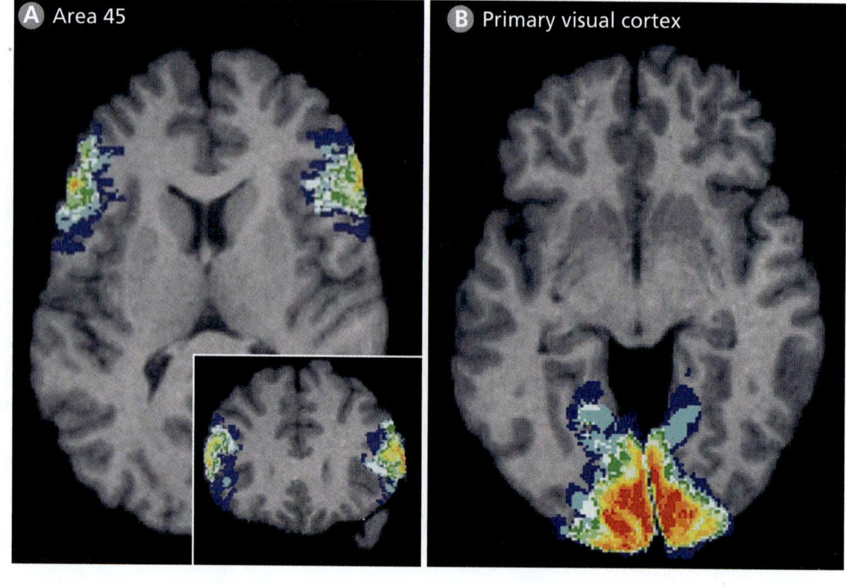

Figure 16.24 Probabilistic maps of two cortical areas. Panel (A) summarizes the location of area 45 in ten different human brains that were morphed into a standard reference brain. The overlap between the areas in different subjects is color coded, with red indicating overlap in all 10 specimens and dark blue meaning that Area 45 extends into that location in just one brain (no overlap). The insert shows a coronal section. Panel (B) shows an analogous probabilistic map for the primary visual cortex (in horizontal section). The cortical areas in this study were delineated in stained sections (postmortem) and reconstructed in 3 dimensions before being mapped into the standard reference space. [From Uylings et al., 2005]

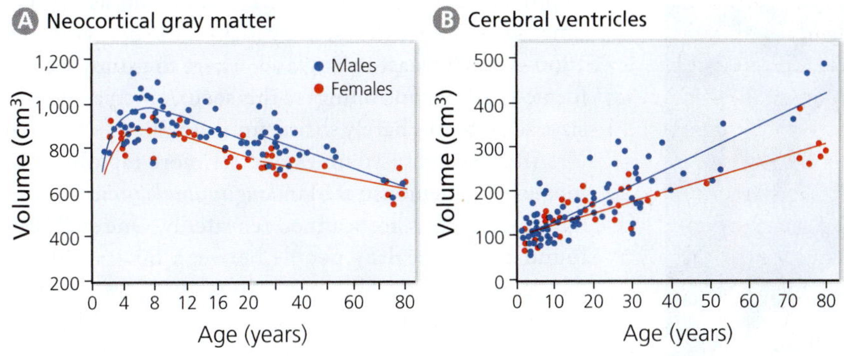

Figure 16.25 Gray matter grows, then shrinks with age. The graphs in (A) and (B) show neocortical gray matter and ventricle volumes, respectively, for 116 subjects. On average, gray matter growth peaks earlier in girls than boys. By age 70, men and women have similar gray matter volumes, although men have larger ventricles. [After Courchesne et al., 2000]

Age-related Variability in Brains

Although most neurons in the human brain are born before the birth of the person, human brains continue to grow for several years after birth (Figure 16.25). Neocortical surface area, for example, triples between birth and young adulthood. Much of this expansion is due to increased myelination of the long axons that course through the neocortical white matter. However, gray matter also increases in volume, mainly due to the addition of more synapses and the expansion of dendritic trees.

Delayed Maturation of High-level Cortical Areas

Neocortical expansion and maturation are delayed in the prefrontal, parietal, and lateral temporal lobes, relative to other cortical areas. These areas also show the greatest expansion when humans are compared to non-humans; and, as you have learned, they are involved in a variety of complex cognitive processes. The delayed maturation of high-level cortical areas might explain why children differ from adults behaviorally. In particular, it has been argued that the late maturation of the prefrontal cortex explains why teenagers tend to take more risks than older adults. This suggestion is consistent with the evidence that the prefrontal cortex is involved in decision making and impulse inhibition, but risk taking in juveniles may be more adaptive than older people tend to imagine. It helps young people discover new solutions to old problems and often attract a mate.

Although these ideas are fun to contemplate, the implications can be far ranging and serious. For example, lawyers have argued that delayed brain maturation makes young people less responsible than adults when they commit a crime. Not surprisingly, this is a controversial argument.

Sex Differences in Brain Growth and Maturation

Similarly controversial is the observation that boys lag girls in brain growth and maturation. On average, neocortical gray matter volume peaks in girls at 8.5 years of age; in boys this peak is reached 2 years later. Within the frontal lobe, peak gray matter volumes are reached at age 11 in girls and at age 12 in boys.

How should we interpret these developmental sex differences? An interesting possibility is that earlier cortical maturation is causally linked to earlier language development. Indeed, girls typically lead boys in language development, an advantage they retain to adulthood (perhaps because of the head start). It is important to note, however, that both the behavioral and the neural differences between the two sexes are evident only when data are averaged across a large number of subjects. For any given person, knowing their sex does not allow you to make specific predictions about either their brain or their language abilities.

Brains Shrink in Old Age

Once brains have reached their peak volume, they remain relatively stable in size for many years. Using cross-sectional study designs in which contemporaneous individuals of various ages are examined, scientists find that brain volumes decrease by 0.5–1% per year (Figure 16.25). However, much of this decrease is likely due to

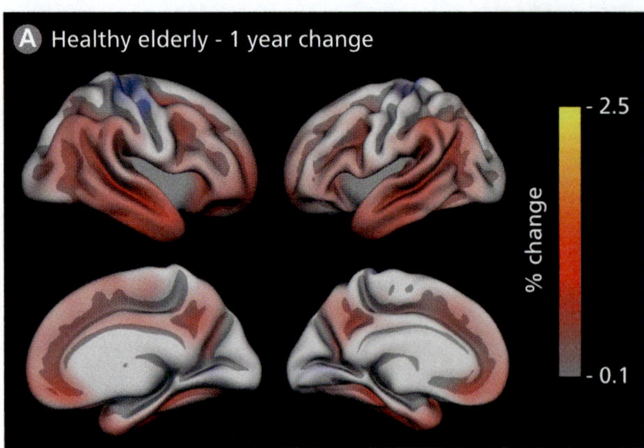

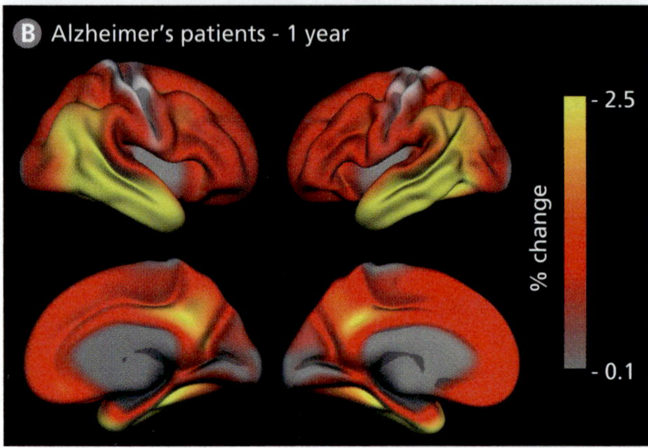

Figure 16.26 Brain shrinkage in health and disease. Many cortical regions become thinner with age, even in healthy 60–90 year olds (A). The rate of cortical thinning is significantly faster in subjects with probable Alzheimer's disease than in age-matched controls. The accelerated tissue shrinkage is most apparent in the temporal lobe. [After Fjell et al., 2009; images by Anders M Fjell, University of Oslo, Norway]

improvements in human health and nutrition, as brain volumes generally increased by 0.3–0.7% per year from 1860–1940 (at least in England, where the study was performed). After controlling for this secular increase in brain size, adult brains barely shrink until age 50.

After 50, brain volumes decline more rapidly. This is most clearly demonstrated in *longitudinal studies* in which individual brains are scanned repeatedly. One such study found that in healthy people between 60 and 90 years of age, many cortical regions shrink by 0.5% a year, on average (Figure 16.26 A). This may not seem like much, but it adds up! The shrinkage is most dramatic in the frontal, parietal, and temporal lobes—the same areas that mature late. Outside of the neocortex, age-related tissue shrinkage is most significant in the amygdala and the hippocampus. What takes the place of all that lost tissue? The answer is mainly the ventricles, which expand with age, especially in men (Figure 16.25 B).

As aging brains shrink in volume, they lose substantial amounts of white matter, as well as some neurons and synapses. Between age 20 and 90, neocortical white matter volume decreases by a whopping 28%, even in healthy humans. Whether this white matter loss is linked to neuron loss remains controversial. However, studies using rigorous counting procedures have shown that the number of neocortical neurons in healthy humans decreases by about 10% between 20 and 90 years of age. Compared to the age-related loss of white matter, the age-related change in neuron number is, therefore, relatively mild. The loss of synapses with age is even more difficult to determine, in part because it varies across brain regions. For example, the *perforant path*, which connects the entorhinal cortex to CA3 and the dentate gyrus, loses a significant number of synapses as rats get old; but other hippocampal pathways are less affected by age.

Alzheimer's Disease

A major challenge in the study of human brain aging is that old people often suffer from **Alzheimer's disease**. This devastating form of dementia currently affects roughly 13% of humans over the age of 60 and 43% of people over 85. It usually begins with a mild inability to form new episodic memories, but then progresses to the point where even old memories are lost, mood swings become severe, and thoughts become confused. The terminal stage of Alzheimer's disease is characterized by a loss of motor and autonomic coordination, which generally leads to infections and death.

Neurologically, Alzheimer's disease is characterized by the presence of beta-amyloid deposits (amyloid plaques) and neurofibrillary tangles (see Chapter 5, Box 5.2: Immune Responses in Alzheimer's Disease). According to the **amyloid cascade hypothesis**, these plaques and tangles result from the faulty processing of the amyloid precursor protein. By means of processes that still remain controversial, the products of this faulty processing destroy synapses and, eventually, kill some neurons. As a result, the brains of Alzheimer's patients shrink relatively rapidly (Figure 16.26 B).

An interesting aspect of Alzheimer's disease is that it devastates some brain regions more than others. Particularly vulnerable are the hippocampus and entorhinal cortex as well as the cholinergic neurons of the basal forebrain and the noradrenergic neurons in locus coeruleus. As we discussed in earlier chapters, these regions play crucial roles in attention and memory.

Is Alzheimer's disease merely an accelerated form of normal aging or does it entail a distinct pattern of neuron and synapse loss? This question is difficult to answer because the brains of both Alzheimer's patients and healthy age-matched controls are highly variable. Thus, one group of neuroscientists found that CA1 neurons tend to die in Alzheimer's disease and not in healthy aging, but another group reported opposite results. Separating Alzheimer's disease from healthy aging is also complicated by the fact that, presently, Alzheimer's can be diagnosed with certainty only after a postmortem examination of the brain for tangles and plaques. This means that even a seemingly healthy elderly individual might, unbeknownst to anyone, have early stage Alzheimer's disease. Furthermore, some people with senile dementia have few plaques in their brains at autopsy; others who died with lots of plaques had only mild cognitive deficits.

Despite these complications, most researchers agree that Alzheimer's disease, as classically defined, impairs cognition and memory by destroying many synapses and some neurons in brain regions that are required for those processes. The quest to understand this destruction is ongoing and intense.

BRAIN EXERCISE

How do you think your own brain might differ from the brains of your closest friends, your parents, or your grandparents?

16.6 What Can We Learn by Comparing Diverse Brains?

The fact that brains vary both within and across species can complicate research, but it can also be useful. Consider, for example, the study of Alzheimer's disease. Although age-related beta-amyloid deposits and dementia are seen in several species, including dogs, only humans develop full-blown Alzheimer's disease. This species difference in the incidence of Alzheimer's disease may be related to the fact that contemporary humans live much longer, on average, than they did when *Homo sapiens* originally evolved; but other factors are probably involved as well.

Working with Animal Models

Regardless of the cause, the lack of Alzheimer's disease in non-humans creates a problem for neuroscientists. How can they study the mechanisms underlying this disease, or explore possible therapies, if animals do not get Alzheimer's disease? To get around this problem, researchers have made transgenic mice that develop something akin to human Alzheimer's disease. Early efforts to create such mice were stymied by the fact that mouse brains can tolerate very high levels of beta-amyloid production. Therefore, researchers had to make mice containing multiple human genes with mutations that are all linked to Alzheimer's disease in humans. Some of these double- or triple-transgenic mice develop amyloid plaques and neurofibrillary tangles, exhibit hippocampal cell death, and become demented in old age. Therefore, they are excellent models for exploring the causal links between amyloid processing, neuropathology, and cognitive decline. In a way, they are the exception that proves Krogh's principle. Nature did not provide a convenient species in which to study Alzheimer's disease, so scientists created one.

Looking beyond transgenic animals, wild-type rodents turn out to be a good model for normal human brain aging. For example, the perforant path shrinks with age not only in rodents but also in humans (Figure 16.27). Specifically, high resolution **diffusion tensor imaging**, which allows investigators to infer the orientation of myelinated axons at specific locations in the brain, has revealed that the perforant path is smaller in 70-year-old human subjects than in 21-year-old college students.

NEUROLOGICAL DISORDERS

Box 16.2 *Complex Mental Disorders: Autism and Schizophrenia*

Autism is characterized by a combination of symptoms including deficits in social interaction, repetitive behaviors or unusually restricted interests, and often some language impairments. Autism typically manifests at 2–3 years of age and afflicts boys more than twice as often as girls. The rate of autism in the population is difficult to estimate, as it depends in part on subjective criteria. However, a reasonable estimate is that 1–2% of all people have autism, broadly defined. Because some people with autism can function well in their society, whereas others are severely impaired, autism is often referred to as *autism spectrum disorder.* Some high-functioning autistics, such as Temple Grandin, have made important contributions to society.

Autism is highly heritable, as indicated by concordance rates of almost 90% for monozygotic twins and 30% for dizygotic twins. Even non-twin siblings of autistic children are ~25 times as likely to develop autism as the general population. Indeed, dozens of genes (hundreds, by some criteria) have been linked to autism. The problem is that most of these genes, when mutated, increase the risk of developing autism by less than 1%. Therefore, autism is caused by the aggregation of mutations in many different autism susceptibility genes. This probably explains why autism is so variable in its severity and constellation of symptoms; it all depends on the combination of genes a person is carrying.

Further complicating the picture is the recent discovery of rare chromosomal rearrangements that dramatically elevate the risk of autism but often arise *de novo* rather than being inherited. These rearrangements generally increase the copy number of several genes and are linked not only to autism but also to several other psychiatric disorders, including schizophrenia. Thus, it remains unclear how they contribute to the most common forms of autism. Finally, the panoply of causes underlying autism includes several environmental factors, including complications during pregnancy and the age of the baby's father. There is, in other words, no single smoking gun for autism. Instead, there is a battery of them, each shooting small pellets.

The brains of autistic toddlers are significantly larger than those of typically developing children (by 5–12%, depending on the study; Figure b16.1). This overgrowth is most pronounced in the prefrontal cortex and is accompanied by an even larger increase in neuron numbers. By adulthood, however, the brains of autistic subjects are similar in size to those of controls. Furthermore, most autistics did not have unusually large brains at birth (as estimated from pediatric head circumference data). Therefore, we can conclude that autism involves accelerated brain expansion during the first 2–4 years of life, followed by a premature cessation of brain growth and, more tentatively, some neuronal shrinkage.

Autism is also linked to reduced white matter development. In particular, the long axon tracts that link the prefrontal cortices to each other (through the corpus callosum) and to other parts of the brain are smaller in autistics than in controls. One can only speculate about the behavioral impact of this neuronal "disconnection," but one would expect it to decrease the level of functional integration across the brain. Because these deficits arise during early childhood, they are likely to affect subsequent development in diverse ways. Individual variation in these downstream effects (e.g., in how an individual responds to the initial neuron overproduction) may explain at least some of the enormous variation that is seen across adults with autism.

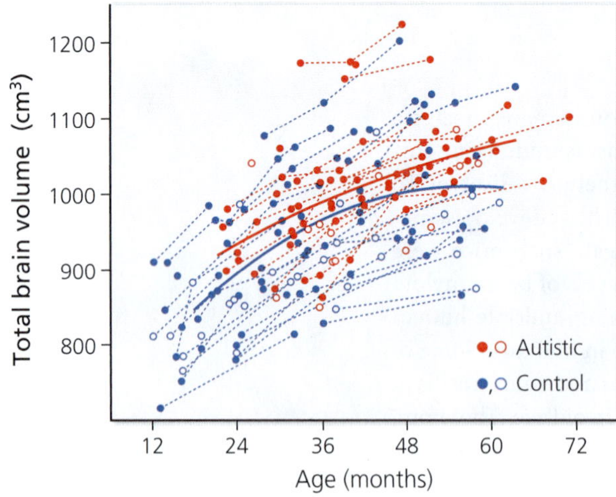

Figure b16.1 Altered brain development in autism. Total brain volume, as determined from MRI scans, is plotted for 41 autistic children and 44 typically developing controls (A). Most subjects underwent multiple scans (dashed lines connect repeated measures). Data points from boys and girls are represented by filled and open symbols, respectively. The solid lines are best-fit brain growth functions for the autistic (red) and control (blue) groups. [After Schumann et al., 2010]

Schizophrenia is very different from autism, but it is similarly complex. It afflicts slightly less than 1% of the population, beginning in adolescence or early adulthood, and is characterized by hallucinations, delusions, social withdrawal, and a flat affect (reduced emotional expressiveness). Schizophrenia has been called the "worst disease affecting mankind"; and before the 1950s, schizophrenics crowded mental hospitals throughout the world. Nowadays, the psychotic symptoms of schizophrenia (delusions and hallucinations) can be managed with antipsychotic drugs. However, schizophrenia remains a life-long disease. On average, the lives of schizophrenics are 10–15 years shorter than those of the general population, partly due to an increased rate of suicide, but also because of a predilection for smoking cigarettes, which has been linked to lung cancer. Indeed, an astonishing 85% of schizophrenics smoke, and they consume significantly more cigarettes than the average smoker. Because nicotine has been shown to lessen some of the cognitive symptoms of schizophrenia, it appears that many schizophrenics are effectively self-medicating with nicotine.

The heritability of schizophrenia is lower than that of autism but still significant. If one monozygotic twin has schizophrenia, the chance of the other twin having it as well

Box 16.2 *Complex Mental Disorders: Autism and Schizophrenia (continued)*

is ~50%; and if one of your blood relatives is schizophrenic, then your odds of having or developing this terrible disease increases tenfold, relative to the general population.

As in the case of autism, large number of genetic mutations have been linked to schizophrenia. Many of them affect neural development, but most increase the risk of schizophrenia only slightly. The risk of developing schizophrenia can also be increased by a variety of environmental factors, such as complications during fetal development and living in an urban environment. One study of 45,570 Swedish military recruits showed that individuals who had smoked marijuana (cannabis) at least once before they entered military service were 2.6 times as likely to develop schizophrenia (over the next 15 years) as those who never smoked the drug. Those who smoked marijuana more than 50 times increased their risk sixfold. Even more interesting is that people with specific allelic combinations of the gene for catechol-O-methyltransferase (COMT) are especially sensitive to the schizophrenia-inducing effects of marijuana. The mechanisms underlying this multiplicative gene-environment interaction remain unknown.

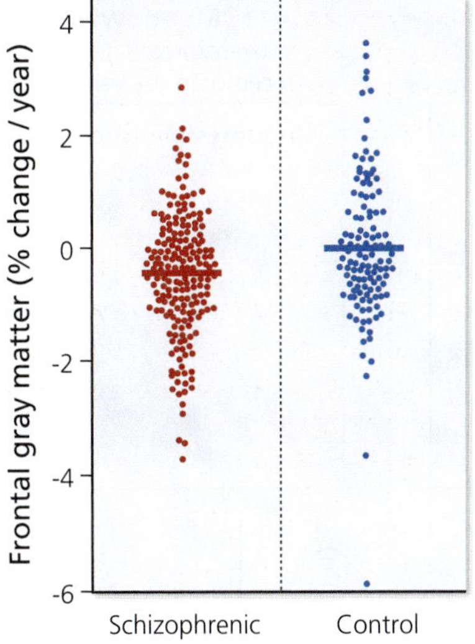

Figure b16.2 The brain in schizophrenia. A pool of 202 schizophrenics (18–45 years old; 148 male, 54 female) and 125 control subjects had their brains scanned twice, once just after being diagnosed with schizophrenia, and then again 2 years later. You can see that, on average, frontal gray matter shrank in the schizophrenics but not in the controls. [After Andreasen et al., 2011]

In the brain of schizophrenics, the cerebral ventricles enlarge excessively, whereas neural tissue shrinks (Figure b16.2). This shrinkage in tissue volume is due mainly to the loss of synapses and the shrinking of individual neurons rather than cell death. Most heavily affected are the prefrontal cortex, the hippocampus, and parts of the thalamus. Some axon tracts are also disrupted in schizophrenia, but these effects are relatively minor and inconsistent across studies.

Schizophrenia is known to change some of the brain's neurotransmitter systems. In particular, some schizophrenia susceptibility genes affect glutamatergic or GABAergic transmission. In addition, the high rate of smoking among schizophrenics suggests that cholinergic transmission (through nicotinic acetylcholine receptors) may be altered in schizophrenia. Finally, the fact that virtually all the drugs used to treat schizophrenia are dopamine antagonists suggests some sort of deficit in dopaminergic transmission.

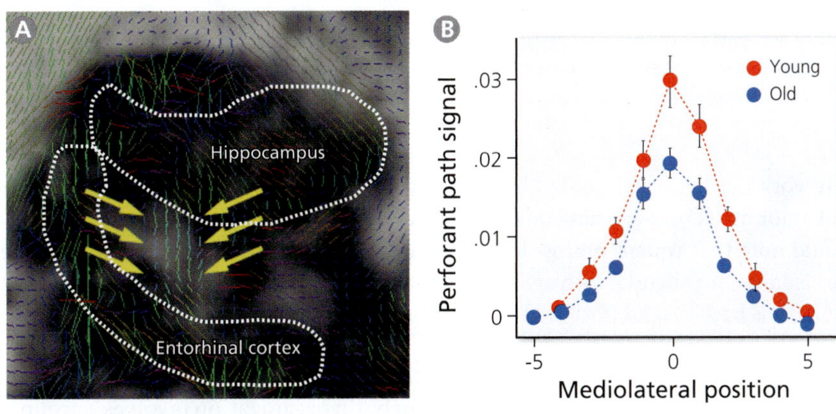

Figure 16.27 The perforant path in humans shrinks with age. The image in (A) shows a diffusion tensor imaging (DTI) scan of a human medial temporal lobe. The short colored lines indicate the orientation of the local tensor signal, which correlates with the dominant orientation of the axons at that location. The area between the yellow arrows contains the perforant path, which connects the entorhinal cortex to the hippocampus. The graph in (B) summarizes data obtained through such scans from 17 young adults and 19 older individuals (21 vs. 70 years of age, on average). The data indicate that the perforant path is more robust in the young than in the old. [From Yassa et al., 2010]

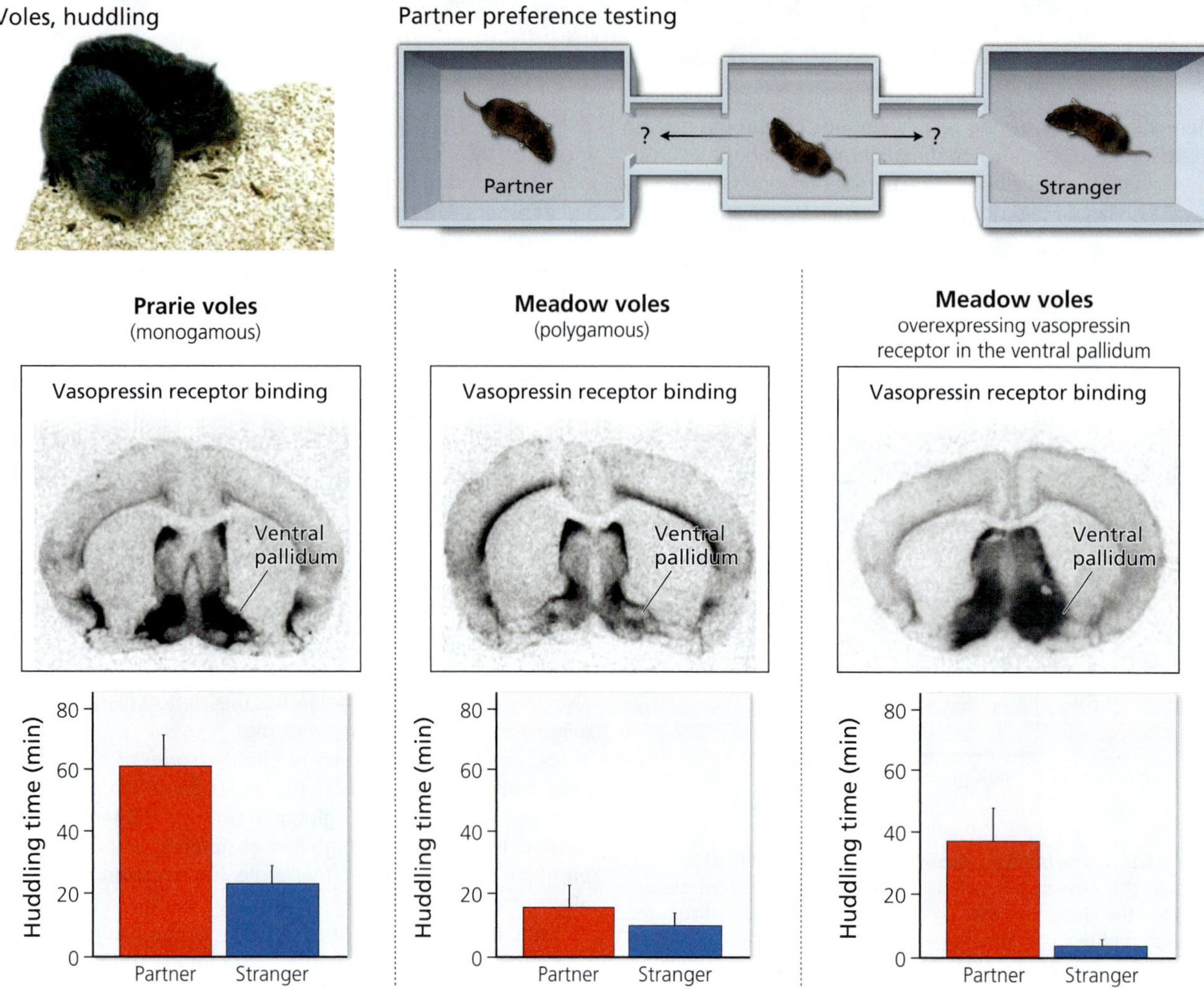

Figure 16.28 Pair bonding in voles. Prairie voles tend to be monogamous, whereas meadow voles are polygamous. To quantify this behavioral difference, male voles of either species are placed into an apparatus that allows them to interact either with a previous sexual partner or an unfamiliar female (of the same species). In this paradigm (top), prairie vole males consistently huddle more with the partner than the stranger, whereas the meadow voles exhibit no partner preference. Prairie voles also express higher levels of the vasopressin receptor gene in the ventral pallidum (middle). To test whether this neural difference accounts for the behavioral species difference, researchers used a virus to overexpress the vasopressin receptor in the ventral pallidum of meadow voles. Sure enough, those genetically modified meadow voles preferentially huddled with their partner, just as prairie voles do (bottom). [From Lim et al., 2004]

This work was explicitly guided by the research on rodents, and it revealed a significant amount of cross-species conservation. Despite this propitious conservation, one should note that rodent brains do not age exactly like human brains. For example, the visual cortex shrinks with age in rats, not in thickness but in areal extent. No such reductions have been reported in humans or other species.

Taking Advantage of "Natural Experiments"

One intensively studied species difference in brain organization involves a group of small rodents called *voles*. Some vole species are monogamous (males and females form stable pairs after mating), whereas others are polygamous. Comparative neuroanatomical studies have shown that the monogamous voles generally express higher levels of the receptor for *vasopressin* in the ventral pallidum (part of the basal ganglia; see Chapter 15). This seems odd at first, because we already know that vasopressin is a circulating hormone involved in fluid retention (see Chapter 9).

However, vasopressin is also released in the ventral pallidum by axons from parts of the amygdala.

Because high levels of vasopressin receptor expression in the ventral pallidum correlate (at least roughly) with monogamy across several species of voles, mice, and primates, neuroscientists suspected that the two traits were causally linked. To test this hypothesis, researchers used a genetically engineered virus to overexpress vasopressin receptors in the ventral pallidum of meadow voles, which are normally polygamous. As predicted, the manipulated meadow voles exhibited more pair bonding than the controls, as determined in a partner preference test (Figure 16.28). Thus, in the research on voles, species differences in brain organization and behavior suggested a causal hypothesis that was then tested by experimental manipulation. In essence, the experimenters used the species differences as "natural experiments" to inspire a laboratory experiment.

Scientists commonly employ an analogous approach in studying human disease. They correlate genetic variation among humans with the incidence of a disease or other abnormality, and then use these correlations to bolster claims about causal hypotheses. This approach helped link beta-amyloid to Alzheimer's disease, FOXP2 to language skills, and catechol-O-methyltransferase to PTSD. Unfortunately, the comparative approach has been less successful for autism and schizophrenia, which apparently involve a multitude of different genes. Even in these cases, however, comparisons between affected individuals and unaffected controls are likely to provide important clues about what causes the diseases (Box 16.2: Complex Mental Disorders: Autism and Schizophrenia).

BRAIN EXERCISE

Many scientists study transgenic mice expressing human Alzheimer's disease-related genes with the aim of developing treatments for human Alzheimer's disease. What assumptions underlie this research strategy?

SUMMARY

Section 16.1 - Many problems are more easily studied in some species than in others. Indeed, much of what we know about human brains was first discovered in non-humans, although species differences also exist.

Section 16.2 - Complex brains evolved at least three times among animals. Within vertebrates, brain size increased repeatedly, even when changes in body size are factored out. Brains also vary in the proportional sizes of their subdivisions.

Section 16.3 - As modern humans evolved, their brains and behavioral capacities diverged from those of other primates. Still, in many ways, humans brains are fairly typical primate brains.

- As brain size increased in primate evolution, neuron numbers increased predictably, as did the proportion of the neocortex that is occupied by the prefrontal cortex. These changes probably made humans more flexible in their behavior. Neanderthal brains were similar in size to modern human brains.
- Lesions of Broca's area and Wernicke's area in the left hemisphere cause profound language deficits. However, language also involves other cortical areas, including some in the right hemisphere.
- Apes and monkeys possess homologs of Broca's and Wernicke's areas, and the pathways connecting them. It is unclear what changes in the brain give us our unique language capacity.

Section 16.4 - Nervous systems differ between males and females, even in humans. However, many of these sex differences are small, controversial, or unclear, especially when they involve forebrain structures.

- Sexual differentiation results from developmental differences in sex hormone levels and from direct genomic effects.
- Some cell groups in the brain and spinal cord are larger in males than in females, whereas others are larger in females. Many, but not all, of these sex differences are linked to differences in reproductive behaviors.

- On average, men have a larger neocortex than women do. This difference appears to be independent of the sex difference in body size, but its functional significance remains unclear. A reported sex difference in the degree of language lateralization has not been replicated consistently.

Section 16.5 - Even within a sex, human brains vary considerably. Least controversial are variations due to developmental maturation, old age, and disease.

- When functional brain imaging data are averaged across subjects, individual variation in brain size, shape, and internal organization creates uncertainties about the precise anatomical location of the imaging signals.
- Human brains grow after birth, their gray matter volume peaking slightly later in boys than girls. In old age, brains shrink, especially in people with Alzheimer's disease. Rates of brain growth and shrinkage vary across brain regions, both in health and in disease.

Section 16.6 - Neuroscientists frequently use transgenic mice as "models" for the study of human diseases. They also use naturally occurring species differences to generate hypotheses about brain-behavior relationships.

Box 16.1 - In recognition of intraspecific differences, human subjects research must nowadays include both women and men as well as diverse ethnic groups. Animal and tissue culture research is also becoming more inclusive.

Box 16.2 - Autism and schizophrenia are complex diseases involving many susceptibility genes and diverse neural correlates, including abnormalities in neural connections.

KEY TERMS

Krogh's principle 507
model species 508
nerve net 509
cerebral ganglia 509
bilateria 510
cephalopods 510
arthropods 510
vertebrates 510
negative allometry 510
encephalization quotient (EQ) 512
endocranial cavity 513
Neanderthals 514
Broca's aphasia 516
Broca's area 516
Wernicke's area 517
Wernicke's aphasia 517
conduction aphasia 517
testosterone 521
androgen receptor 521
estradiol 521
aromatase 521
bilateral gynandromorphism 521
Onuf's nucleus 522
sexually dimorphic nucleus (SDN) 522
third interstitial nucleus of the anterior hypothalamus (INAH-3) 522
anteroventral periventricular nucleus (AVPV) 523
allometrically predictable 525
Alzheimer's disease 532
amyloid cascade hypothesis 533
diffusion tensor imaging 535

ADDITIONAL READINGS

16.1 - Model Species in Neuroscience

Logan CA. 2001. "[A]re Norway rats . . . things": diversity versus generality in the use of albino rats in experiments on development and sexuality. *J History Biol* **34**:287–314.

Martin B, Ji S, Maudsley S, Mattson MP. 2010. "Control" laboratory rodents are metabolically morbid: why it matters. *Proc Natl Acad Sci U S A* **107**:6127–6133.

16.2 - The Evolution of Large Brains

Butler AB, Hodos W. 2005. *Comparative vertebrate neuroanatomy: evolution and adaptation.* Second Ed. New York: Wiley & Sons.

Striedter GF. 2005. *Principles of brain evolution.* Sunderland, MA: Sinauer Associates.

Strausfeld NJ. 2012. *Arthropod brains: evolution, functional elegance, and historical significance.* Cambridge, MA: Harvard University Press.

Yopak KE, Lisney TJ, Darlington RB, Collin SP, Montgomery JC, Finlay BL. 2010. A conserved pattern of brain scaling from sharks to primates. *Proc Natl Acad Sci U S A* **107**:12946–12951.

16.3 - Distinctively Human Brains

Konopka G, Geschwind DH. 2010. Human brain evolution: harnessing the genomics (r)evolution to link genes, cognition, and behavior. *Neuron* **68**:231–244.

Petrides M, Pandya DN. 2009. Distinct parietal and temporal pathways to the homologues of Broca's area in the monkey. *PLoS Biol* **7**:e1000170.

Scharff C, Petri J. 2011. Evo-devo, deep homology and FoxP2: implications for the evolution of speech and language. *Phil Trans Roy Soc B* **366**:2124–2140.

Spocter MA, Hopkins WD, Garrison AR, Bauernfeind AL, Stimpson CD, Hof PR, Sherwood CC. 2010. Wernicke's area homologue in chimpanzees (*Pan troglodytes*) and its relation

Glossary

2-deoxyglucose a form of glucose that is taken up by cells but cannot be processed through glycolysis and, therefore, accumulates inside the cells; it can be used to quantify how (metabolically) active neurons have been

abducens nucleus a small group of motor neurons that innervate the lateral rectus muscle (which rotates the eyeball laterally when it is contracted) and receives input from the vestibular complex

acetylcholine an important neurotransmitter used at neuromuscular junctions, in much of the autonomic nervous system, and by a few neurons in the brain

acetylcholinesterase the enzyme that degrades acetylcholine

acid-sensing ion channel (ASIC) an ion channel that allows positive ions to flow into a cell when the extracellular pH is low (acidic); it plays a major role in sensing tissue acidity, typically caused by CO_2 buildup

actin a filamentous protein that interacts with myosin to generate contractile forces in muscle

actin filaments very thin chains of actin subunits that grow and shrink in length as subunits are added or removed from the filament's two ends

action potential an all-or-none electrochemical signal generated by neurons in response to above-threshold stimulation

action potential threshold the membrane potential at which voltage-gated sodium channels tend to open and, consequently, an action potential is triggered

activity-related cytoskeletal protein (Arc) a protein whose expression is upregulated rapidly after synapses have been highly active; it is involved in stabilizing long-term potentiation (LTP)

adaptive feedforward control using prior learning (from mistakes) to adjust motor commands in such a way that performance errors are minimized

adaptive immune system the relatively slow response of the body to foreign invaders involving the production of antibodies and other processes

adenosine a byproduct of ATP metabolism that can induce sleep by inhibiting cholinergic basal forebrain neurons

adenosine 5′-triphosphate (ATP) a key source of metabolic energy but also used as a neurotransmitter by taste cells

adenylate cyclase an enzyme that promotes the conversion of ATP to cyclic AMP

adrenergic receptors a family of G protein-coupled receptors that bind norepinephrine as well as epinephrine (adrenaline); they are commonly divided into alpha- and beta-adrenergic receptors

agrin a protein released from motor axons when they contact a muscle; it causes postsynaptic acetylcholine receptors to cluster at the location of the developing synapse

allometrically predictable a difference in proportional size of a structure that is consistent with how that structure generally scales relative to other structures (such as absolute brain or body size)

alpha (α) motor neuron a large motor neuron that innervates the extrafusal fibers in skeletal muscle

alpha-gamma coactivation the idea that gamma motor neurons are activated simultaneously with alpha motor neurons to ensure that the intrafusal muscle fibers remain taught and sensitive to muscle stretch during a muscle contraction

Alzheimer's disease a common neurodegenerative disease that is associated with progressive dementia and the loss of synapses (and eventually neurons) in diverse brain regions, especially in and around the hippocampus

amacrine cell a diverse class of retinal neurons, some of which receive input from rod bipolar cells and synapse onto OFF and ON cone bipolar cells with inhibitory and electrical synapses, respectively

AMPA receptor an ionotropic type of glutamate receptor that, in contrast to the NMDA receptor, can be opened by glutamate when the postsynaptic cell is near its resting potential

amphetamine a highly addictive drug of abuse that causes dopamine to be released in large amounts from presynaptic terminals; it may disrupt the blood–brain barrier

amygdala a collection of cell groups in the anterior pole of the human temporal lobe; its functions include the control of strong emotions and the emotion-related modulation of memory.

amyloid cascade hypothesis the idea that Alzheimer's disease is caused primarily by the faulty processing of amyloid precursor protein

anastomoses the point at which two branches of a blood vessel reconnect such that blood may flow from one branch into the other

androgen receptor a type of receptor that binds testosterone as well as similar compounds (which are collectively called androgens)

annulospiral ending a sensory nerve ending that wraps around the central region of intrafusal muscle fibers (inside muscle spindles) and is activated when those fibers are stretched; the faster the stretch, the stronger the response

anorexigenic acting as an appetite suppressant

antagonistic pairs pairs of mechanisms or processes that exert opposite influences on some other process

anterior toward the front of the body when standing upright; for the brain: toward the front of the skull

anterior neural ridge an embryonic signaling center at the anterior end of the neural tube; it secretes fibroblast growth factor 8 (Fgf8)

anterior pituitary the anterior part of the pituitary gland including the median eminence, a capillary bed, and a large inferior portion that secretes a variety of pituitary hormones into the blood

anterograde amnesia the inability to form new long-term memories

anterograde signals signals that travel from the dendrites or the cell body toward the tips of the axon

anteroventral periventricular nucleus (AVPV) a cell group in the preoptic area that is larger in females than in males; it is involved in triggering ovulation in females

antihistamines drugs that interfere with the action of histamine, which is used as a neurotransmitter at some synapses; if they cross the blood–brain barrier, antihistamines can make subjects drowsy

aperture problem any long, moving line or edge that is viewed through a small hole (aperture) will appear to move perpendicular to its long axis (just two possible directions), regardless of its actual direction of motion

Aplysia californica a large, marine gastropod mollusk (snail), commonly known as a sea hare, that has been used extensively by scientists studying the neural and molecular mechanisms underlying learning and memory

apoptosis a form of neuron death that proceeds in an orderly fashion, recycling some neuronal components and minimizing harm to neighboring cells

apperceptive agnosia the inability to perceive visual objects as coherent entities, perceiving them instead as a disconnected set of individual features

aquaporins a family of transmembrane proteins that allow water molecules to flow through their central pore (water channels)

arachnoid barrier a very thin layer of cells between the subarachnoid space and the veins or sinuses, usually formed by tight junctions between its cells

arachnoid granulations mushroom-like protrusions of the arachnoid membrane into veins or sinuses though which cerebrospinal fluid (CSF) is recycled into the blood

arachnoid membrane the middle one of the three meninges; it encloses the subarachnoid space

arcuate nucleus a group of neurons in the hypothalamus that is involved in a variety of functions, including the secretion of gonadotropin releasing hormone (GnRH) and the integration of hunger and satiety signals

aromatase an enzyme expressed by some brain cells that converts testosterone into estradiol

arthropods a large and highly successful group of animals that includes insects, crustaceans, spiders, and millipedes

ascending arousal system a set of neural pathways that, when stimulated electrically, can elicit EEG desynchronization; it includes the axons of locus coeruleus and peribrachial neurons

association fibers axonal connections that are both divergent and convergent so that each neuron projects to (and receives input from) many other neurons

astrocytes star-shaped glial cells that perform a variety of functions including secretion of inflammatory molecules, control of local blood flow, recycling of neurotransmitters, and forming glial scars

atrioventricular (AV) node a small clump of modified muscle fibers at the boundary between the right atrium and the right ventricle; its cells fire rhythmically and can generate rhythmic cardiac contractions

auditory cortex a set of cortical areas on the superior aspect of the temporal lobe that are involved in auditory processing

autoassociative network a neural network in which most neurons project to many other neurons in the same network; if such a network exhibits synaptic plasticity that follows Hebb's rule, it can form associative memories

autonomic ganglia collections of cell bodies in the peripheral nervous system; they project to a variety of targets, including smooth muscle, the heart, and various glands

autonomic nervous system a complex, widespread network of neurons that regulate the body's vital processes, including digestion, heart rate, and blood pressure; it acts fairly independently (autonomously) of the CNS and has sympathetic as well as parasympathetic divisions

autoradiograph a 2-dimensional image showing the spatial distribution of radioactivity in, for example, a brain tissue section

awake brain stimulation the technique of applying a series of short electrical pulses to a set of neurons in an awake animal, usually to examine how this stimulation affects behavior

axon a long and thin neuronal process that remains relatively constant in thickness along its length; its main function is to conduct action potentials over long distances

axon collateral a major branch of an axon, distinct from the smaller branches that axons typically form near their targets

axon guidance molecules a broad class of molecules that influence the direction of axon outgrowth through a variety of mechanisms

axon hillock a small mound (hillock) on neuronal cell bodies where the axon typically emerges; it typically contains a high density of voltage-gated ion channels, making it a spike initiation zone

axoplasmic transport the active intracellular transport of molecules along an axon, either away from the cell body (anterograde) or toward the cell body (retrograde)

Aβ fibers myelinated sensory axons that have a relatively large diameter (7–12 μm)

background adaptation the idea that sensors gradually stop responding to steady, persistent stimuli

background rate the rate at which a neuron fires action potentials when it is not activated or suppressed

back-propagation the propagation of action potentials from the axon or cell body into the dendrites

baroreceptor vagal nerve endings that are activated when the walls of blood vessels are stretched, which occurs when blood pressure rises

baroreflex a reflexive response that is triggered by a change in blood pressure and tends to restore blood pressure to its set point; it typically involves changes in heart rate and blood vessel diameter (e.g., vasoconstriction)

barrel cortex part of a rodent's primary somatosensory cortex that represents the whiskers, with each whisker corresponding to one of the cortical "barrels" (neuronal cell bodies arranged in the shape of a barrel)

basal ganglia a collection of subcortical structures that are part of the **frontostriatal system**; they include the dorsal and ventral striatum, the dorsal and ventral pallidum, and several groups of dopaminergic neurons

basilar membrane a thin membrane inside the cochlea, on top of which the inner and outer hair cells sit

basolateral amygdala complex part of the amygdala that is activated by strongly arousing stimuli and is involved in both memory modulation and some forms of fear learning

bed nucleus of the stria terminalis a small cell group in the ventral telencephalon that helps control hormone secretions through inhibitory projections to the paraventricular nucleus of the hypothalamus

behavioral arousal a condition in which all sorts of stimuli become more likely to elicit a response, often because the organism has experienced a threatening stimulus

behavioral stereotypy the prolonged repetition of a stereotyped (often quite meaningless) behavior

beta-catenin a protein with diverse functions, one of which is to control the timing of neurogenesis (the timing of when progenitors stop dividing and begin to differentiate)

biceps stretch reflex a reflexive contraction of the biceps muscle when it is stretched

bilateral gynandromorphism rare individual in whom cells on one side of the body are genetically male while cells in the other half are genetically female

Bilateria the ancient taxonomic group that includes all animals with bilateral symmetry (although some of its members have lost that symmetry; e.g., starfish)

binding problem the problem of how to perceive the various features of a stimulus object as belonging to (coming from) a single, coherent object (i.e., how to "bind" those features together perceptually)

binocular involving both eyes at once

binocular depth perception the ability to estimate the distance to a visual object by comparing the inputs to both eyes

birth date the time at which a cell stopped dividing and began to differentiate (develop toward a mature cell fate)

blastocyst the stage of embryonic development when the embryo consists of a hollow (fluid-filled) ball of cells

blind spot the location in the visual field where you cannot detect a visual stimulus (as long as you are looking through just one eye) because the corresponding location on the retina (the optic nerve head) lacks photoreceptors

blood–brain barrier a barrier to the flow of large hydrophilic (not lipophilic) molecules from the blood into the brain; formed by tight junctions between cells that line the blood vessels

blood-CSF barrier a barrier to the flow of large hydrophilic (not lipophilic) molecules from the blood into the cerebrospinal fluid (CSF); formed by tight junctions between ependymal cells in the choroid plexus

blood oxygen level dependent (BOLD) signals changes in the degree of blood oxygenation that vary with brain activity and can be detected with functional magnetic resonance imaging (fMRI)

blue-yellow opponent cell a type of neuron that compares the activity of S-cones to the combined activity of L- and M-cones in its spatial receptive field; some fire most when the S-cones predominate, whereas others exhibit the opposite preference

Bmal one of the "clock genes" involved in generating intracellular circadian rhythms; it promotes the transcription of period (*per*) and cryptochrome (*cry*), but it suppresses those genes when combined with Clock protein

body-centered coordinates a spatial coordinate system that is locked to the position of the body and thus moves as the body moves (regardless of how the eyes or the head are moving)

bone morphogenetic protein (BMP) a diffusible protein secreted by cells on the ventral side of the early embryo; it stimulates the ectoderm to develop into skin, but it can be inhibited by chordin

Bötzinger complex a small group of neurons in the ventral medulla that is part of the central pattern generator for breathing

Braille alexia the inability to read Braille, the touch-based writing system for the blind and visually impaired

brain-in-a-vat the idea of an isolated brain, floating in a fluid-filled container (a vat)

brain nuclei clusters of neuronal cell bodies with similar features, typically segregated from cell bodies that exhibit different sets of features

brain slice a slice of living brain tissue that can be maintained *in vitro* for many hours and is, therefore, convenient to study experimentally

brainstem defined differently by different people; in this book the term refers to medulla, pons, and tegmentum

bridging veins veins that "bridge" the gap between the dura mater and the brain; they may rupture when the brain moves excessively within the skull

Broca's aphasia an inability to produce articulate speech, despite being able to move the lips and tongue, that is caused by lesions in the left inferior prefrontal cortex; deficits in language comprehension are typically minor

Broca's area two cortical areas (called Brodmann areas 44 and 45) in the general area where left-sided lesions cause Broca's aphasia; homologs of these two areas exist in non-human primates

bromodeoxyuridine (BrdU) a thymidine analog that can be detected immunohistochemically; it is used to determine cellular birth dates in a way that is analogous to the use of tritiated thymidine

brown adipose tissue a type of fatty deposit that generates more heat (and less metabolic energy) than regular fat when it is "burned" by the body

Brown-Séquard syndrome a clinical condition in which subjects with damage to one side of the spinal cord lose the ability to sense pain on the contralateral side of the body (below the level of the injury) and the ability to sense touch and vibration on the ipsilateral side (also below the injury)

CA1 the abbreviation of "*cornu ammonis* area 1," which is part of the mammalian hippocampus; it receives strong inputs from CA3 and has robust projections to the subiculum

CA3 the abbreviation of "*cornu ammonis* area 3," which is part of the mammalian hippocampus; it has robust projections to CA1

caffeine the ingredient in coffee and tea that promotes the waking state, at least in part by inhibiting adenosine receptors in the cholinergic basal forebrain

caged glutamate molecules of the neurotransmitter glutamate that are enclosed in a molecular "cage" that prevents them from binding to their receptors (until they are released from that cage by, e.g., a laser pulse)

calcium-activated calcium channels ion channels that let calcium flow through their central pore when calcium levels are relatively high, thus further increasing calcium concentration

calcium-calmodulin activated phosphorylation the mechanism of excitation-contraction coupling in smooth muscle; phosphorylation of the myosin heads promotes cross-bridge cycling

calcium/calmodulin kinase II (CaMKII) a protein kinase that is activated by calcium/calmodulin complexes and (among other things) helps to insert AMPA receptors into postsynaptic membranes

calcium spike an action potential (spike) that is based on the influx of calcium, rather than sodium, ions; such action potentials are significantly longer in duration than typical sodium-based action potentials

calmodulin a calcium-binding protein that, when bound to calcium, activates calcium/calmodulin kinase II

cAMP response element binding protein (CREB) a transcription factor that binds to cAMP response elements (CREs) in DNA and regulates the expression of downstream genes; it regulates some forms of synaptic plasticity

canonical cortical circuit the idea that all cortical areas contain a fundamentally conserved (canonical) intracortical circuit; although there is some truth to this idea, intracortical circuits do vary somewhat across areas

capsaicin the molecule that causes chili peppers to taste "hot" (spicy)

carbidopa a drug that prevents L-dopa from being converted to dopamine outside the brain, which means that more of the circulating L-dopa is available for the neurons that need to synthesize dopamine

cardiac conduction system a set of modified (not contractile) muscle fibers that conduct action potentials to cardiac muscle fibers throughout the heart's ventricles

carotid body a highly vascularized collection of glomus cells and other cells located where the carotid artery divides; it is involved in sensing blood oxygenation

category-sensitive neurons neurons that exhibit category boundaries in their responses to stimuli that fall on a continuum; e.g., neurons that respond to morphed cat-dog stimuli only when the stimulus is >50% dog

caudal toward the tail in four-legged animals; not well defined in humans

caudal ventrolateral medulla a group of GABAergic neurons in the caudal medulla that is involved in regulating heart rate and blood pressure

caudalize to change the developmental trajectory of cells so that they will adopt a more caudal (posterior) cell fate

cell assembly the idea that a weakly interconnected group of neurons that is repeatedly coactivated will over time, as a result of synaptic plasticity that obeys Hebb's rule, become strongly interconnected so that the activation of just some of those neurons can activate all the others in the group

cell fate the kind of adult cell that a young cell will become (its adult phenotype)

central canal the extension of the cerebral ventricles into the spinal cord

central nervous system (CNS) in vertebrates, the brain and spinal cord; the CNS also includes the eye, which develops as an outpocketing of the diencephalon

central nucleus of the amygdala a group of neurons in the amygdala that projects to various targets in the hypothalamus and medulla (especially those related to the sympathetic nervous system) and can trigger a fight-or-flight response

central pattern generator a set of neurons that generate a rhythmic pattern of activity, and control some sort of rhythmic behavior, without requiring rhythmic external input

central sulcus one of the most prominent sulci in primate brains; it lies between the primary somatosensory cortex and the primary motor cortex, separating the frontal lobe from the parietal lobe.

centralization the idea that the majority of neurons (in many but not all animals) tend to be concentrated in a few specific locations, notably the brain, rather than being scattered throughout the body

cephalopods a taxonomic subgroup of mollusks that includes octopus and squid

cerebellar ataxia a lack of motor coordination brought about by cerebellar damage

cerebellar granule cell a small and extremely numerous type of neuron in the cerebellar cortex; these cells have unusual axons called parallel fibers, which terminate on the dendrites of Purkinje cells

cerebellum a highly folded part of the hindbrain (in many vertebrates) that is involved in fine-tuning movements (and probably some cognitive processes) based on trial-and-error learning

cerebral aqueduct the thin tubular portion of the cerebral ventricles that connects the third ventricle to the fourth ventricle

cerebral cortex the layered (laminated) part of the telencephalon, including neocortex, olfactory cortex, and hippocampus

cerebral fatigue the idea that neurons become less active when they run out of metabolic energy

cerebral ganglia large clusters of neurons in invertebrates that some researchers consider to be a brain (a cerebrum)

cerebrocerebellum the lateral portions of the cerebellum, which are very large in humans; they are indirectly connected with, and functionally related to, the neocortex

cerebrospinal fluid (CSF) the fluid that fills the cerebral ventricles and subarachnoid space

chemoaffinity hypothesis Roger Sperry's idea that each axon expresses a unique combination of molecular markers and finds it target by growing until it finds cells that express the complementary set of molecules

chemoreflex a reflexive increase in breathing rate that is triggered when the blood runs low on oxygen or is enriched in carbon dioxide, or both

chemotaxis in the context of axon guidance, the idea that developing axons grow toward or away from a source of a diffusible substance (following its concentration gradient)

chemotopy the idea that similar odorants are represented in adjacent locations (notably within the olfactory bulb)

cholecystokinin a peptide hormone secreted by cells in the lining of the small intestine that stimulates the gall bladder to release bile, which aids in the digestion of fats

cholinergic basal forebrain neurons in the basal forebrain that use acetylcholine as their main neurotransmitter and project diffusely to the neocortex; they play a role in regulating behavioral arousal and maintaining the waking state

chordin one of the diffusible proteins that is secreted by the dorsal blastopore lip and capable of inducing the ectoderm to adopt a neural cell fate

chorea a disorder characterized by uncontrolled movements, as in Huntington's disease (also known as Huntington's chorea)

choroid a capillary bed that covers the retina's pigment epithelium, supplying it with metabolic energy

choroid plexus a thin, highly folded and vascularized tissue that extends into the cerebral ventricles and produces much of the cerebrospinal fluid (CSF)

chromaffin cells modified sympathetic postganglionic neurons that, during development, migrate into the core of the adrenal gland where they secrete epinephrine (adrenaline)

chronic traumatic encephalopathy (CTE) inflammation, cell death, and dementia caused by repeated blows to the head

circadian rhythm a behavioral or physiological rhythm with a period of about (circa) one day

circle of Willis a set of small vessels that link the major arteries feeding the brain and form a "circle" at the base of the brain; if an upstream artery is clogged, blood can be rerouted through those small vessels to minimize potential brain damage

circumventricular organs locations, generally near the midline and close to the ventricles, where the blood–brain barrier is leaky

climbing fiber the axons of neurons (in the inferior olive) that each make a multitude of synapses onto a very small number of cerebellar Purkinje cells

Clock one of the "clock genes" involved in generating intracellular circadian rhythms; it promotes the transcription of period (*per*) and cryptochrome (*cry*), but it suppresses those genes when combined with Bmal protein

cocaine a highly addictive drug of abuse that blocks dopamine reuptake, which means that dopamine signaling is enhanced (at least in the short term)

cochlea the spiraling, tubular part of the inner ear that is responsible for sensing sounds

cochlear nucleus a set of neurons in the upper medulla that receives input from the cochlea and projects to the inferior colliculus

cocktail party problem the challenge of focusing on a single stream of sounds embedded in a background of other sounds (e.g., the voice of one person at a busy party); it is an auditory version of the binding problem

cognitive map the idea that organisms carry "in their head" a map-like representation of the external world (in world-centered coordinates) that they can use to calculate novel routes (e.g., shortcuts, detours)

collateral branches small blood vessels that connect arteries on the surface of the brain to one another, allowing blood to be rerouted around blockages

colliculus the dorsal/superior part of the midbrain, usually divided into superior and inferior colliculi. The superior colliculus is called the optic tectum in non-mammalian vertebrates

color contrast the difference between two stimuli that is due to a difference in color rather than luminance (e.g., the difference between a bright red spot and an equally bright green background)

color-blindness an inability to discriminate some colors that the majority of people can discriminate; it comes in a variety of forms

combinatorial *Hox* code model the idea that the rostrocaudal identity of cells in the hindbrain and spinal cord is determined by the combination of *Hox* genes that those cells express

communicating arteries small arteries that link larger arteries, allowing blood to flow between them, as occurs in the circle of Willis

complex cell a type of neuron in the primary visual cortex that responds preferentially to lines or edges of a particular orientation anywhere in its spatial receptive field

computed tomography (CT) scan a three-dimensional X-ray, computed from multiple X-ray "slices"

concentration gradient in the context of resting and action potentials, a difference in the concentration of ions (or other substances) across the two sides of a cell membrane

conditioned taste aversion the tendency to avoid eating foods that may have made you nauseous in the past (i.e., you got quite sick a few hours after you ate that food last time); it is most effective for previously unfamiliar food

conduction aphasia a deficit in the ability to repeat words or name objects; it is thought to result from lesions that disconnect Wernicke's area from Broca's area

congenital insensitivity to pain a medical condition that renders subjects insensitive to pain (they can feel the noxious stimuli but don't perceive them as painful)

constraint-induced movement therapy the idea of improving function of a neurologically impaired limb by constraining use of the intact limb, forcing the subject to use the impaired limb

contextual fear conditioning a form of Pavlovian fear conditioning in which the threat-associated (conditioned) stimulus is the context in which the threat is presented (e.g., the room in which a rat is shocked)

contralateral on the opposite side of the midline

convergence the idea that neurons may receive input from two or more other neurons

coronal one of several planes that cut through a human body transversely, at right angles to the body's long axis; for the brain: a plane that is parallel to the face and perpendicular to the midsaggital and horizontal planes.

cortical area a region of the cerebral cortex that is functionally and structurally distinct from other cortical areas (e.g., in its pattern of lamination, connections, and/or physiological responses)

cortical minicolumn the idea that neocortex is divisible into thin radial (perpendicular to the cortical surface) columns, within which the neurons have similar physiological responses

corticobulbar a term to describe projections from the neocortex (especially primary motor cortex) to the medulla; they help control movements of muscles in the head (e.g., jaws, tongue, eye lids) as well as several vegetative functions (e.g., respiration)

corticospinal a term to describe projections from the neocortex (especially the primary motor cortex but also some other areas) to the spinal cord; they help control movements of the limbs and trunk

corticospinal axons axons that project from the cortex (mainly the motor cortex and some parts of the somatosensory cortex) to the spinal cord

corticostriatal LTP long-term potentiation (LTP) at corticostriatal synapses

corticostriatal projection a set of axons that project from the neocortex to the dorsal and/or ventral striatum

corticotropin releasing hormone (CRH) a releasing hormone secreted into the capillaries of the median eminence; it stimulates pituitary cells to secrete adrenocorticotrophic hormone (also known as corticotropin)

cortisol the principal human "stress hormone"; it is corresponds to corticosterone in non-humans

covert spatial attention a form of visual spatial attention that can be shifted from one location to another without movement of the eyes (or head)

cranial nerve a large bundle of axons that connect the brain (rather than the spinal cord) to rest of the body; in mammals, the principal cranial nerves are numbered 1–12 in Roman numerals (I–XII)

cranial vault the space inside the skull that is occupied by the brain

craving a strong desire for something combined with the motivation to do something about it

cross-bridge cycling the idea that contractile force in muscle fibers is generated when myosin binds to actin, bends, releases from the actin, straightens out, and binds again, repeating the cycle

crossed extensor reflex reflexive extension of the leg when the opposite leg is withdrawn from a noxious stimulus; it allows the standing leg to carry the extra weight, thus stabilizing the body

cryptochrome (*cry*) one of the "clock genes" involved in generating intracellular circadian rhythms; its transcription is promoted by Bmal and Clock, but heterodimer complexes of those two proteins ultimately suppress *cry* transcription

cued spatial attention task a behavioral task in which the subject is given a signal (a cue) indicating where they should direct their spatial attention

cupula a large gelatinous mass that sits on top of the hair cells in each semicircular canal; when the fluid inside a canal moves, the cupula bends, which then activates the associated hair cells (depending on movement direction)

cyclic adenosine monophosphate (cAMP) a second messenger derived from ATP that plays a role in diverse forms of cellular plasticity

D1-type dopamine receptor a class of G protein-coupled receptors that, when activated by dopamine, depolarize (excite) the postsynaptic cells; in the striatum, they are found on neurons that project directly to the pallidum

D2-type dopamine receptor a class of G protein-coupled receptors that, when activated by dopamine, hyperpolarize (inhibit) the postsynaptic cells; they are found on striatal neurons that project into the "indirect pathway"

Dale's principle the idea that neurons always release the same set of neurotransmitters at all of their synapses

dark adaptation a set of processes that makes photoreceptors more sensitive to light after they have been in darkness (or low-light conditions) for a while

deep cerebellar nuclei a set of cell groups deep in the cerebellum (beneath the cerebellar cortex) that receive input from cerebellar Purkinje cells and project to various targets outside of the cerebellum

defasciculation the process through which growing axons exit from a bundle of other, similar axons (an axon fascicle)

delay activity an increase in neuronal firing rate (above the background rate) that occurs while the subject is holding some sort of information in working memory

delayed non-match to sample task (DNMTS) a task in which subjects are presented with one item (the sample) and then, after a delay, presented with a pair of objects that includes the old item as well as a new one; to get a reward, the subject must select the novel (non-match) item; this task tests object recognition memory

delayed response task a task in which the subject gets a clear signal about what to do but must then wait (for a variable delay) before giving the desired response

delta-1 a protein that helps initiate neurogenesis

dendrite parts of a neuron that taper as they extend away from the cell body and often branch repeatedly; they tend to be covered with synapses

dendritic filopodia slender, often transient processes on dendrites that may develop into dendritic spines

dendritic protein synthesis the idea that some proteins are synthesized in the dendrites rather than the cell body

dendritic spine small, often mushroom-shaped protrusions on dendrites; dendritic spines tend to receive one or more synaptic inputs

dendritic tiling the idea that the dendrites of neurons (of the same type) repel one another during development and therefore tend not to overlap

dentate gyrus a part of the hippocampus that, in contrast to most other mammalian brain regions, exhibits a high degree of adult neurogenesis

depolarize making a cell's membrane potential more positive or less negative (e.g., changing it from –70 to –40 mV)

deprivation effects the effects on the nervous system (and behavior) of depriving an organism of some aspect of normal (species-typical) experiences (e.g., growing up with a cataract that deprives you of pattern vision)

design principles physical or engineering principles that can be used to argue that a biological feature is "well designed," even though natural selection is not guided by any kind of "designer"

developmental neuron death the death of neurons as part of normal development rather than injury or disease

diabetes insipidus a relatively rare form of diabetes associated with excessive urination and insatiable fluid intake; it is caused by damage to vasopressin-secreting magnocellular neurons

diabetes mellitus a medical condition that results from impaired insulin production or reduced sensitivity to insulin that is being produced; it interferes with the body's ability to regulate blood glucose levels

diaphragm the dome-shaped skeletal muscle forming the floor of the chest cavity; when it contracts, air is drawn into the lungs

diencephalon the part of the forebrain that is closest to the midbrain; it includes the thalamus and hypothalamus

diffusion tensor imaging a form of structural MRI that allows investigators to determine the predominant orientation and structural integrity of fiber tracts in human brains

digital signal a signal that is either on or off (a 1 or a 0); in contrast, analog signals vary continuously in amplitude

direction sensitivity the property of responding preferentially to stimuli that are moving in a particular direction (i.e., being sensitive to the direction of movement)

disinhibition the release of neurons from inhibition, which often causes the disinhibited neurons to generate a burst of action potentials

divergence the idea that axons may split into two or more branches that project to different targets

dopamine an important neurotransmitter used mainly by neurons that project to the striatum

dopamine beta-hydroxylase the enzyme that converts dopamine to norepinephrine

dorsal toward the back of the body or the top of the head in four-legged animals when they hold their head high; ambiguous in humans

dorsal blastopore lip a critical portion of the late blastocyst stage where tissue invaginates as gastrulation begins; it produces molecules that induce adjacent parts of the ectoderm to adopt a neural cell fate

dorsal column a longitudinal column of axons running along the dorsomedial aspect of the spinal cord; it contains sensory axons that convey mainly somatosensory information from the body surface to the caudal medulla

dorsal column nuclei cell groups in the caudal medulla that receive somatosensory information from the body through the dorsal column pathway

dorsal horn the dorsal (posterior in humans) portion of the spinal cord's gray matter

dorsal motor nucleus of the vagus nerve a small cell group in the dorsal medulla containing parasympathetic preganglionic neurons that project to the enteric nervous system

dorsal pallidum a relatively small brain region (also known as the internal segment of the globus pallidus and reticular division of the substantia nigra) that receives inputs from the dorsal striatum and projects to part of the thalamus

dorsal root a bundle of axons that exits a segment of the spinal cord dorsally (posteriorly in humans) and consists mainly of sensory axons; it joins with the corresponding ventral root to form a spinal nerve

dorsal root ganglion a small swelling in each dorsal root that contains the cell bodies of sensory neurons that project through spinal nerves to the dorsal horn of the spinal cord

dorsal stream a set of interconnected cortical areas that extends from the primary visual cortex over the posterior parietal cortex into the motor and premotor cortex; it is involved in stimulus localization and targeted movements

dorsal striatum a large subcortical structure that receives inputs mainly from posterior regions of the frontal lobe and is involved in the learning and control of stereotyped movements (e.g., habit learning); in primates, it is divided by a fiber tract into two separate structures called the caudate and the putamen

double dissociation the set of experimental findings showing that one sort of manipulation affects one function but not another, whereas a second type of manipulation has the opposite effect

double-opponent cell a type of neuron that exhibits one type of cone opponency (e.g., red-green opponent) in the center of their receptive field and another type of opponency in the surround; such neurons can detect edges or lines defined only by color contrast

drug-conditioned place preference the behavioral preference of an animal to be in a place where it has previously received a pleasure-inducing drug (versus some other place)

dual reflex arc model the hypothesis, attributed to William James and Theodore Meynert, that long (generally transcortical) reflex arcs modulate the activity of short (usually subcortical) reflex arcs

dura mater the thickest and toughest outermost layer of the meninges

ear canal the tubular passage from the outer ear to the tympanic membrane

ectoderm the most external of the three germ layers; it develops mainly into neural tissue and skin

EEG desynchronization a decrease in the amplitude and increase in the frequency of the electroencephalogram (EEG) that is typically associated with behavioral arousal

effector structures through which the nervous system can effect change, notably muscles and glands

efference copy a copy of motor (efferent) commands that is sent to other (non-motor) brain regions for various purposes including the canceling out of sensory inputs that result from one's own movements

efficient sensory coding the idea that neurons preferentially encode the stimulus features that are most common in their environment

electrical potential gradient in the context of resting and action potentials, a difference in electrical potential between the inside of a cell and its surrounding, extracellular space

electrically coupled the idea that two cells are coupled by gap junctions so that electrical activity in one cell can spread to the other without employing synapses

electroencephalogram (EEG) recordings of cortical activity made though relatively large electrodes (most often on the scalp), each of which averages the electrical activity of thousands of neurons (rather than single neurons)

encapsulated nerve ending the peripheral tip of a sensory axon that is embedded in (encapsulated by) one or more specialized skin cells

encephalization quotient (EQ) a fraction (quotient) that quantifies the size of an individual's brain, relative to what one would expect for individuals of that body size; it is usually employed in cross-species comparisons

end-feet the stubby endings of long processes extended by some astrocytes onto blood vessels in the brain; they help regulate vessel diameter and thus blood flow

endocranial cavity the cavity within the skull in which the brain is housed; it is frequently used to estimate brain size for fossil hominids (and other large-brained species) because large brains tend not to fossilize

endocrine gland a gland that secretes its contents (hormones) into the blood

endogenous opioids chemical compounds that are related to opium and are produced by the organism's own body (e.g., endorphin)

endolymph the fluid inside the scala media and, therefore, the fluid in which the stereocilia are bathed; it contains an unusually high concentration of potassium ions

end plate potential the postsynaptic potential generated by transmitter release at a neuromuscular junction (whose postsynaptic side is known as the end plate)

enteric nervous system roughly 200–600 million neurons (in humans) in the lining of the gastrointestinal tract that regulate digestion

entorhinal cortex a cortical area that serves as a bottleneck for information flowing into our out of the hippocampus, with which it is reciprocally interconnected

entrain to impose one rhythm on another rhythm, synchronizing the two

ependymal cells the cells that line the cerebral ventricles

ephrin family a large family of membrane-bound proteins that are involved in axon guidance

ephrin gradient hypothesis the idea that retinotectal axons grow along a gradient of ephrin expression until ephrin levels get so high that the axons (which are repulsed by ephrin) cannot grow further

ephrin receptor receptors expressed on the surface of a cell that bind to ephrins expressed on the surface of other cells; involved in axon guidance

epigenetic change a change in DNA structure that does not involve a change in DNA sequence

epileptic seizure an episode of abnormal, highly rhythmic brain activity that often starts in one location (often the medial temporal lobe) and then spreads to other brain regions, temporarily disrupting their normal functions

epinephrine a hormone (chemically related to norepinephrine) secreted from the adrenal gland and involved in the stress response; also known as adrenaline

episodic memory the memory for what happened when and where

epithelial sodium channel an ion channel that is expressed in a subset of taste cells and is involved in sensing salt

epitope a molecular feature; in the context of olfaction, a feature of an odorant to which an olfactory receptor may bind selectively

equilibrium potential the electrical potential at which the flow of a particular ion into the cell is counterbalanced by the flow of ions out of the cell

estradiol the body's main form of estrogen during the reproductive years; it plays a crucial role in the development of secondary female body features (e.g., breasts) and the menstrual cycle

event-related fMRI a form of functional magnetic resonance imaging (fMRI) in which experimenters look for correlations between specific events and changes in the fMRI signal at various times after that event

excitatory center the central region of a neuron's receptive field where optimal stimuli tend to generate an increase in the neuron's firing rate (i.e., excitation)

excitatory postsynaptic current (EPSC) the same phenomenon as an excitatory postsynaptic potential (EPSP) except that EPSCs measure current rather than changes in voltage

excitatory postsynaptic potential (EPSP) a brief membrane depolarization that is generated by activity at an excitatory synapse and may (or may not) trigger an action potential

excitotoxicity excessive stimulation of neurons (e.g., by glutamate) that leads to excessive neuronal firing and, in some cases, cell death

exocrine gland a gland that secretes its contents to the external or internal body surfaces (the skin or the gastrointestinal tract)

exogenous opioids chemical compounds that are related to opium and are NOT produced by the organism's own body (e.g., morphine or heroin)

external segment of the globus pallidus (GPe) a group of inhibitory neurons that receive inputs from the striatum and project to the subthalamic nucleus; it is part of the indirect pathway through the basal ganglia

extrafusal fiber the main muscle fibers in skeletal muscle; they are thicker and longer than intrafusal muscle fibers

extrastriate cortex a set of visual areas surrounding the primary visual cortex (a.k.a. striate cortex) in the occipital lobe

extrasynaptic receptor a receptor that is located outside of a typical synapse, generally activated by neurotransmitters that are released outside of synapses (see **volume transmission**)

exuberant projections axons or axon branches that exist early in development but then disappear as part of normal development

eye-centered coordinates a spatial coordinate system that is locked to the position of the eye and thus moves as the eyes move (even if the head remains stationary)

face-selective neurons neurons that respond more strongly to faces (or a subset of faces) than to other visual stimuli

facial motor nucleus a group of motor neurons in the caudal medulla that innervate the eye lid and other facial muscles; it is a critical link in the eye blink reflex

facial nerve cranial nerve seven (VII), which carries taste information for the anterior portion of the tongue and the oral cavity; it also contains some motor axons

fast-scan cyclic voltammetry a technique used to measure concentrations of dopamine (and other chemicals that are easily oxidized); it involves rapidly changing the voltage at a carbon fiber electrode and measuring the oxidation-induced currents

fault tolerant the idea that a system may lose one of its elements (experience a "fault") but still be capable of functioning, at least to a considerable extent

fetus ejection reflex a reflex that causes uterine contractions in pregnant mammals when pressure is exerted on the cervix (by the head of the fetus that is about to be born)

fiber tracts large bundles of axons, usually myelinated

fibroblast growth factor 8 (Fgf8) a diffusible protein secreted from the midbrain–hindbrain boundary and the anterior neural ridge, involved in rostrocaudal patterning of the neocortex and other brain regions

fictive swimming the rhythmic output of an isolated piece of neural tissue (disconnected from most sensory and motor structures) that resembles the rhythmic output observed during real, actual swimming

fictive walking the rhythmic output of an isolated piece of neural tissue (disconnected from most sensory and motor structures) that resembles the rhythmic output observed during real, actual walking

fight-or-flight response a coordinated set of behaviors and physiological mechanisms that prepare an organism for dealing with a threat

filopodia slender, often transient extensions of an axon's growth cone or developing dendrites

fixation point the point in space at which the fovea is aimed

flaccid paralysis an inability to move that is accompanied by a loss of muscle tone (i.e., the muscles are limp, rather than stiff)

floor plate a column of cells along the ventral midline of the spinal cord and caudal brain regions involved in dorsoventral patterning

flower-spray ending a sensory nerve ending that terminates on the ends of intrafusal muscle fibers (inside muscle spindles) and is activated when those fibers are stretched; the stronger the stretch, the stronger the response

forebrain the most rostral/anterior of the brain's major divisions; it includes the *telencephalon* and *diencephalon*

fourth ventricle the part of the cerebral ventricles that is located in the medulla; its roof has holes in it through which cerebral spinal fluid can enter the subarachnoid space

fovea a small region of the retina that (in humans) contains a very high density of cone photoreceptors and very few rod photoreceptors

free nerve ending the peripheral tip of a sensory axon that is not associated with one or more acessory sensory cells

freeze in the context of responses to external threats, freezing is characterized by almost total immobility; it may help a threatened animal to avoid detection by a predator

freezing response an almost total lack of movement in response to a threatening stimulus (breathing is maintained); it is a frequently used measure of fear, especially in work with rodents

frontal eye field a cortical area that lies at the anterior edge of the premotor cortex (and the posterior edge of the prefrontal cortex) and is involved in the control of eye movements including *smooth pursuit eye movements*

frontal lobe the most rostral/anterior lobe of the neocortex; it includes the motor cortex and the prefrontal cortex

frontostriatal system a set of interconnected brain regions that includes the frontal lobes and the striatum; it is involved in selection of movements, actions, and goals

functional magnetic resonance imaging (fMRI) any kind of magnetic resonance imaging that is aimed at visualizing brain activity

funny current a transmembrane current in cells of the sinoatrial node (and some other rhythmically active cells); it is also called h-current because it is activated by hyperpolarization

fusiform face area a cortical region in the inferior temporal cortex that contains a large number of face-selective neurons

gain-of-function experiment an experiment in which a causal process is enhanced so that the experimenters can observe how this enhancement (gain of function) affects the system

gamma (γ) motor neuron one of the smaller motor neurons that innervate the intrafusal fibers in muscle spindles

gamma-aminobutyric acid (GABA) a widely used neurotransmitter that tends to be inhibitory in its effect (at least in adult animals)

gastrin a peptide hormone secreted by cells in the stomach lining that stimulates the secretion of gastric acid, which aids in digestion

gate control theory of pain the idea that nociceptive signals can be blocked (gated out) at lower levels of the central nervous system so that they do not reach the cerebral cortex

germ layers three distinct groups (layers) of cells in early embryos that develop into very different adult structures

ghrelin a peptide hormone secreted by cells in the stomach and intestines when those structures are relatively empty and insulin levels are low; in response, organisms tend to get hungry

giant vacuoles large organelles inside the cells of the arachnoid barrier that carry cerebrospinal fluid (CSF) from the subarachnoid space into the venous blood

gill withdrawal reflex a reflexive behavior in which *Aplysia* withdraws its gills (and siphon) in response to noxious stimuli applied to its skin

glial cell one of several non-neuronal cell types that are prominent components of the nervous system; they include astrocytes, microglia, oligodendrocytes, and Schwann cells.

glial scar a barrier of reactive astrocytes that forms around damaged brain tissue, limiting the spread of the damage but also interfering with axon regeneration

globs in the context of color vision, globs are patches of neurons spread across several visual cortical areas that contain many neurons capable of sensing color contrast independently of luminance contrast

glomeruli one of many spherical structures in the olfactory bulb where the axons of olfactory sensory neurons synapse onto the dendrites of mitral cells

glomus cells specialized oxygen-sensing cells in the carotid body

glossopharyngeal nerve cranial nerve nine (IX); it carries taste information from the back of the tongue and information about blood oxygen levels

glucagon a hormone secreted by the pancreas when blood glucose levels are low; it stimulates fat and liver cells to convert glycogen and fat into glucose

glucose transporter a membrane-spanning protein that can transport glucose into cells

glutamate in the form of L-glutamate, it is the brain's principal excitatory neurotransmitter

glycogen a complex carbohydrate that many cells (but not neurons) can synthesize from glucose; it can be converted back to glucose when other sources of glucose run low

glycolysis the chemical conversion of glucose into pyruvate, resulting in a net gain of 2 ATPs per glucose molecule

Golgi stain an anatomical technique, invented by Camillo Golgi, that stains a relatively small fraction of all neurons but stains those neurons completely, often revealing both dendrites and axons

Golgi tendon organ sensory nerve ending embedded in the tendons through which muscles attach to bone; they provide information about the force with which a muscle pulls on a tendon

gonadotropin releasing hormone (GnRH) a releasing hormone secreted into the capillaries of the median eminence; it stimulates pituitary cells to secrete luteinizing hormone or follicle stimulating hormones (two gonadotropins)

grandmother cell coding scheme the hypothetical idea that specific stimulus objects (e.g., your grandmother) are represented in the brain by single neurons, one for each object

granule cell a neuron with a very small cell body that is tightly packed together with other, similar cells; granule cells in the olfactory bulb are inhibitory and interconnect mitral cells; most granule cells reside in the cerebellum

gray matter contains neuronal cell bodies and dendrites; it also contains unmyelinated axons and most synapses

grid cell a type of neuron that increases its firing rate (above the background rate) when the organism is in any one of several spatial locations, which are arranged in a hexagonal "grid" pattern

growth cone the tip of a growing axon; it extends and retracts actin-filled filopodia

Guillain-Barre syndrome a neurological disorder characterized by the loss of myelin in parts of the peripheral nervous system

gustation the sense of taste

gustatory system a system of neurons and sensory cells that provide an organism with information about the taste of diverse stimuli

gustducin a G protein that can be activated by taste receptors and then activates intracellular signaling cascades that can alter a cell's membrane potential

gyrus/gyri an outward fold (or ridge) of the cerebral cortex; "gyri" is the plural of "gyrus"

habit learning having learned some action "by habit," meaning that the learned task is accomplished automatically; in the context of spatial learning, it means repeating a well-learned series of movements

hair cell sensory cells that sit on top of the basilar membrane of the cochlea and extend stereocilia into the scala media or, in the case of the outer hair cells, the tectorial membrane; they are also found in the vestibule and semicircular canals

half-center oscillator two neurons (or two groups of neurons) that generate a rhythmic output because they reciprocally (and alternately) inhibit each other

head saccade a rapid, jerky turning of the head, usually toward a specific visual or auditory target

head-centered coordinates a spatial coordinate system that is locked to the position of the head and thus moves as the head moves (regardless of how the eyes or the rest of the body are moving)

head-direction cell a type of neuron that increases its firing rate (above the background rate) when the organism's head is facing in a specific set of compass directions

Hebb's rule the idea that synapses are strengthened only when their (presynaptic) activity coincides with postsynaptic activation (depolarization); as a result, neurons that "fire together" tend to become "wired together"

hemispatial neglect a clinical condition in which the subject has trouble paying attention to stimuli on the left side of the external world (usually in body-centered or world-centered coordinate frames)

heterosynaptic synaptic plasticity is said to be heterosynaptic when activity at one synapse causes a change in the strength of another synapse, the plasticity is said to be heterosynaptic

hindbrain the part of the brain between the spinal cord and the midbrain; it includes medulla, cerebellum, and pons

hippocampus part of the cerebral cortex; it is involved in spatial memory and in the memory of what happened when and where

histamine a neurotransmitter used by neurons in the tuberomammillary nucleus that, when activated, maintain the waking state; antihistamines that cross the blood–brain barrier interfere with these neurons, making you sleepy

homeostasis in the context of neurobiology, the tendency of the body to keep its internal environment (e.g., temperature, pH, and water content) relatively constant

homolog one biological feature is the homolog of another if both features can be traced back, along a continuous history, to a corresponding feature in a common ancestor

homosynaptic synaptic plasticity is said to be homosynaptic when activity at one synapse causes a change in the strength of the same synapse, the plasticity is said to be homosynaptic

homunculus a drawing of a "little man" in which the size of various body parts is modified to reflect the degree to which that part of the body is over- or underrepresented in somatosensory or motor regions of the brain

horizontal planes that cut through an animal's brain or body parallel to the ground, when the head and body are held in their normal, standing position.

horizontal cell a type of retinal neuron that is largely responsible for surround inhibition in the retina; using reciprocal dendrodendritic synapses, horizontal cells are excited by the photoreceptors and, in turn, inhibit them

hormone a signaling molecule secreted from endocrine glands and carried to its targets by the blood

***Hox* genes** a family of genes involved in the rostrocaudal patterning of the hindbrain, the spinal cord, and other body parts

hue constancy the ability to detect the color of a stimulus independently of its brightness (luminance)

huntingtin protein the protein that, when mutated sufficiently (by an excessive accumulation of CAG repeat sequences in the huntingtin gene), causes Huntington's disease; its normal function is still unclear

Huntington's disease a lethal neurodegenerative disorder caused by changes in the huntingtin gene that kill striatal neurons projecting into the indirect pathway; it is characterized by uncontrolled movements

hydrocephalus enlargement of the cerebral ventricles during development, causing potentially lethal brain enlargement

hyperalgesia a heightened reaction to normally painful stimuli and the perception of pain in response to stimuli that do not normally activate nociceptive axons

hyperpolarization a decrease in membrane potential to below (more negative than) the resting potential

hyperpolarization-activated cyclic nucleotide channel an ion channel that is responsible for the "funny current" in heart pacemaker cells; it opens when a cell has become hyperpolarized and then gradually depolarizes that cell

hypertension the condition of having chronically high blood pressure (>140/90)

hyperthermia the condition of having a dangerously high body temperature

hyperventilation excessive breathing (e.g., after hard exercise) that may decrease blood CO_2 levels to the point where it triggers a reflex that reduces blood flow to the brain, causing the subject to pass out

hypocretin a neuropeptide (also known as orexin) used by a subset of neurons in the lateral hypothalamus; damage to these neurons causes narcolepsy

hypothalamus the ventral/inferior part of the diencephalon; it is involved in regulating many vital bodily functions

hypothermia the condition of having an abnormally low body temperature

immune privileged the now outdated idea that foreign substances inside the brain do not generate an immune response

incus one of the middle ear bones; it connects the malleus to the stapes

indirect pathway a pathway through the basal ganglia that starts with neurons in the striatum that project to the external segment of the globus pallidus, which projects to the subthalamic nucleus, which projects to the pallidum

inferior toward the bottom of the feet in humans, when standing upright; for the brain: toward the neck

inferior colliculus a major auditory region in the midbrain that receives inputs from several brainstem auditory areas and projects to the medial geniculate nucleus as well as the superior colliculus

inferior olive a group of neurons that project via climbing fiber axons to cerebellar Purkinje cells, which respond by firing unusually large and long action potentials called complex spikes

inferior temporal cortex a set of cortical areas on the inferior aspect of the primate temporal lobe containing neurons that respond to complex visual stimuli (e.g., face-selective neurons)

inhibition of return the tendency of visual spatial attention (both covert and overt) to not return to locations on which it had just been focused; it prevents attention from getting stuck at the most salient location

inhibitory avoidance training a fear learning paradigm in which a subject (typically a rat) is given an electric shock when it enters a distinctive compartment of the experimental apparatus; learning and memory are tested by measuring how long it takes the animal to enter the threat-associated compartment on a subsequent trial

inhibitory postsynaptic potential (IPSP) a brief membrane depolarization that is generated by activity at an inhibitory synapse and tends to pull the membrane potential toward its resting value

inhibitory surround a region around the excitatory center of a neuron's receptive field where stimuli that trigger an excitatory response in the "excitatory center" instead generate an inhibitory response

innate immune response the relatively rapid and nonselective response of the body to foreign invaders; in the nervous system, it includes inflammation and the phagocytic function of microglia

inner ear the part of the ear that lies deepest within the skull; it includes both the cochlea and the vestibular apparatus (semicircular canals, saccule, and utricle)

inner hair cell the hair cells closest to the center of the spiraling cochlea; they sense vibrations of the basilar membrane

***in situ* hybridization** a technique that uses labeled pieces of single-stranded DNA to visualize the expression of specific RNA molecules

instructive effects developmental factors that guide development in specific directions, promoting the emergence of specific features

instrumental conditioning a form of learning in which behaviors that led to positive reinforcement become more likely to be repeated (in similar circumstances), whereas behaviors that led to negative reinforcement become less likely to recur

insular cortex a portion of the neocortex that is covered by parts of the frontal (and other) lobes; it has a variety of functions, one of which is the processing of gustatory (taste) information

insulin a hormone secreted by the pancreas when blood glucose levels are high; it stimulates muscle and liver cells to convert glucose into glycogen and fat

integrate-and-fire device the idea that neurons sum (integrate) their synaptic inputs and then fire action potentials if the summed potential exceeds the action potential threshold

interaural level difference (ILD) the difference in sound intensity measured at the two ears

interaural time difference (ITD) the difference in the time at which sound waves reach the two ears; it includes differences in sound arrival time (sound onset) as well as phase differences (the time between individual sound wave peaks)

intercalated disk membranous disks that connect the ends of cardiac muscle fibers to one another; they contain gap junctions that allow ions to flow between the connected fibers

intermediolateral column a column of cells in the lateral portion of the spinal cord gray matter, in between the dorsal and ventral horns; it contains sympathetic preganglionic neurons

interneuron a neuron that connects mainly to nearby neurons

interstitial space the space between cells, which in the brain is filled by cerebrospinal fluid (CSF)

intrafusal fiber the muscle fibers inside of muscle spindles; they are thinner and shorter than extrafusal muscle fibers

intrinsic signal optical imaging using the fact that deoxygenated blood looks darker under orange light than oxygenated blood to visualize patterns of oxygen consumption and, by inference, neuronal activity

involuntary attention a form of attention in which external stimuli "grab" a person's attention against their will or, in any case, without them having conscious control

inward current a net influx of positive ions into a cell, causing the membrane to depolarize

ion channel a transmembrane protein with a central pore through which ions can flow

ionotropic receptor an ion channel that opens when it is activated by a specific neurotransmitter (or other ligand)

ipsilateral on the same side of the midline

irradiation the idea that stimuli of increasing strength cause increasingly extensive reflex responses, involving an increasing number of neurons and body parts

ischemic stroke an arterial blockage that may kill brain cells dependent on the blocked vessel for oxygen and nutrients

isometric contraction a muscular contraction that generates a pulling force but does not involve a change in muscle length

joint sensors sensory nerve endings in the joints that are activated mainly when a joint is overextended

kinocilium a long cilium that sits next to the longest stereocilium of a vestibular hair cell

kisspeptin a neurotransmitter used by some neurons in the arcuate nucleus of the thalamus; it helps excite the neurons responsible for secreting gonadotropin releasing hormone (GnRH)

knee jerk response the reflexive extension of the knee triggered by rapid stretching of the patellar tendon, which passes under the knee cap and attaches to the quadriceps muscle; also known as the quadriceps stretch reflex

Krogh's principle the idea that "for a large number of problems there will be some animal of choice, or a few such animals, on which it can be most conveniently studied"

labeled line the idea that activity in a neural pathway is interpreted by the nervous system according to a fixed "label"; if axons normally carry a specific type of visual information, that information becomes their label

laminae layers, typically of the cerebral or cerebellar cortex; in general, a neuron is said to be located in whichever lamina contains its cell body

lamprey an ancient group of jawless vertebrates with an elongate (eel-like) body; they swim by lateral undulation

lateral toward the left or right sides of the body; for the brain: toward the left or right sides of the skull

lateral geniculate nucleus (LGN) a thalamic nucleus that receives projections from the retina and is reciprocally connected with the primary visual cortex

lateral hypothalamus cells in the lateral portion of the hypothalamus that receive inputs from the arcuate nucleus and regulate feeding behavior (as well as other processes)

lateral inhibition the process by which nearby neurons inhibit each other

lateral rectus a small muscle that, when contracting, rotates the eyeball laterally (i.e., temporally); it is antagonized by the medial rectus muscle

lateral superior olive a cell group in the medulla that measures interaural level differences; it receives excitatory and inhibitory inputs from the ipsilateral and contralateral cochlear nuclei, respectively

lateral undulation a mode of locomotion in which a wave of lateral body flexion propagates along the body (e.g., the way that snakes typically move around)

lateral ventricles the parts of the cerebral ventricles that extend into the left and right sides of the telencephalon

L-cones cone photoreceptors that are maximally sensitive to long (L) wavelengths of light

L-dopa the immediate precursor of dopamine that can cross the blood–brain barrier; it is commonly prescribed as a treatment for Parkinson's disease (although side effects often arise)

leak channel a type of ion channel that is permeable to potassium and opens near a neuron's resting potential, allowing potassium ions to "leak" out of a neuron

leg withdrawal reflex a reflexive bending of the leg to move it away from a noxious stimulus, mediated by a relatively simple reflex arc through the spinal cord

length constant a mathematical constant that describes how far a change in membrane potential will travel along an axon by purely passive conduction (without involving voltage-gated ion channels)

leptin a peptide hormone released by white fat cells; high levels of leptin suppress appetite, unless those levels are chronically high

ligand-gated ion channel a type of ion channel that opens when a specific type of molecule (the ligand) binds to it

limbic system a set of structures (including the amygdala, hippocampus septal nuclei, and ventral striatum) that are involved in regulating emotions; the concept of a limbic system is controversial because it is not well defined

locus coeruleus a small group of neurons near the fourth ventricle that use norepinephrine as their transmitter and project throughout most of the brain; they are involved in regulating behavioral arousal

long-term depression (LTD) a form of synaptic weakening (depression) that can last for many hours; it is best studied in the cerebellum but also occurs at synapses in other brain regions

long-term potentiation a form of synaptic strengthening (potentiation) that is observed after tetanic stimulation and lasts for many minutes, hours, or even days

loperamide an anti-diarrhea drug related to morphine but unable to cross the blood–brain barrier

loss-of-function experiment an experiment in which a causal process is disrupted so that the experimenters can observe how this disruption (loss of function) affects the system

luminance absolute brightness of a visual stimulus

lymphocytes another name for white blood cells

macula a small, highly pigmented region of the retina that contains the fovea

macular degeneration degeneration of retinal cells in the region of the macula; it often develops in old age and can cause major vision problems

magnesium block near a neuron's resting potential, the ion-conducting pore of the NMDA receptor tends to be blocked by magnesium ions

magnocellular having a very large cell body

malleus one of the middle ear bones; it attaches to the tympanic membrane

mantle zone a layer of cells that lies superficial to the ventricular zone and consists mainly of cells that have stopped dividing

M-cones cone photoreceptors that are maximally sensitive to medium (M) wavelengths of light

mechanosensory the ability to sense a mechanical disturbance and turn it into an electrical signal that neurons can process

medial toward the organism's midline

medial rectus a small muscle that, when contracting, rotates the eyeball medially (i.e., nasally); it is antagonized by the lateral rectus muscle

medial superior olive a cell group in the medulla that measures interaural time differences

median eminence the superior part of the anterior pituitary; it contains a capillary bed into which parvocellular hypothalamic neurons secrete releasing hormones, which are then ferried via blood vessels to the inferior portion of the anterior pituitary

median preoptic nucleus a midline nucleus in the preoptic area that is involved in regulating drinking behavior

medium spiny neuron the most common type of neuron in the striatum, these medium-sized neurons are covered with dendritic spines and use GABA as their neurotransmitter

medulla the part of the brain that is closest to the spinal cord

Meissner's corpuscle an encapsulated nerve ending in which the sensory axon is embedded in a stack of glia-like "lamellar cells"; it is specialized for sensing low-frequency vibrations (e.g., an object moving across the skin)

α-melanocortin stimulating hormone (αMSH) a neurotransmitter used by anorexigenic neurons in the arcuate nucleus

melanopsin a visual pigment expressed by a subset of light-sensitive retinal ganglion cells

melatonin a small peptide hormone released from the pineal gland in a circadian rhythm; it is a powerful antioxidant and promotes sleep

membrane transporters membrane-spanning proteins that transport substances into or out of cells (or both)

memory consolidation the idea that newly formed memories grow stronger over time (at least for a few hours) because memory traces in the brain take time to stabilize

meninges the membranes surrounding the brain and spinal cord, including the dura mater, pia mater, and arachnoid membrane

Merkel cell mechanosensory skin cells most common in highly touch-sensitive areas such as the fingertips; they adapt slowly to sustained stimuli and help us sense the shape and texture of stimulus objects

metabotropic receptor a molecule that, when activated by the appropriate neurotransmitter (or other ligand), triggers an intracellular signaling cascade that ultimately opens (or closes) a nearby ion channel

metyrapone a drug that blocks the synthesis of corticosterone (cortisol in humans)

microglia small star-shaped glial cells that multiply in response to brain injury, migrate to the site of injury, secrete inflammatory molecules, and engulf cellular debris

midbrain the region between the hindbrain and the forebrain

midbrain–hindbrain boundary a signaling center that secretes fibroblast growth factor 8 (Fgf8) and helps to pattern nearby brain regions

midbrain locomotor area a group of neurons in the midbrain that projects to the central pattern generator for locomotion in the spinal cord and modulates its activity (e.g., turning it on, speeding it up, changing its pattern)

middle ear the portion of the ear that lies between the ear drum and the inner ear (cochlea); it contains the middle ear bones (hammer, anvil, and stapes)

middle ear muscles small muscles that affect movements of the tympanic membrane and middle ear bones, thereby affecting acoustic sensitivity

middle temporal area (MT) a visual cortical area in the posterior part of the temporal lobe that is specialized for the detection of visual motion

midsagittal the plane that bisects the body or the brain into left and right halves

milk ejection reflex a reflex that causes contraction of the mammary glands (and, thus, milk ejection) in lactating female mammals when their nipples are stimulated by suckling offspring

mirror neurons neurons that fire selectively during a specific action AND when the subject is observing another individual performing that same action; they seem to be most common in the ventral premotor cortex

mitral cells the neurons in the olfactory bulb that receive input from olfactory sensory neurons and project to the olfactory cortex

model species a species that serves as a model for another species that is more difficult to study but is assumed to be fundamentally similar (with regard to the aspects that are being investigated)

monoamine a class of compounds with one amino group, including serotonin, dopamine, and norepinephrine

monocular involving just one eye

monosodium glutamate (MSG) a potent flavor enhancer that activates the T1R1/T1R3 taste receptor combination, which generates umami taste

Morris water maze a behavioral test of spatial memory and navigation, in which a rodent is trained to swim to a submerged platform from diverse starting positions; it tests for the presence of a *cognitive map*

motor map the idea that movements are represented in a map-like fashion across a brain region such that adjacent neurons are involved in similar movements

motor unit one motor neuron, together with all the muscle fibers it innervates

MPTP a powerful toxin that kills dopaminergic neurons in primates (but not rodents) and can therefore be used to create primate models of Parkinson's disease; it occasionally contaminates home-cooked batches of the synthetic opioid MPPP

Müller cell a type of glial cell in the retina that acts as a light guide, channeling light from the front of the retina to the photoreceptors at the back of the retina

multiple sclerosis a neurological disorder characterized by the loss of myelin in parts of the central nervous system

muscarinic acetylcholine receptor a type of acetylcholine receptor that is sensitive to muscarine, a compound found in some poisonous mushrooms; the acetylcholine receptors on the heart's pacemaker cells are of this type

muscle atonia the loss of muscle tone during REM sleep that causes most muscles to go limp (as in flaccid paralysis); it is caused by the hyperpolarization of most skeletal motor neurons

muscle spindle specialized structures that look like small seed pods, sandwiched between the principal muscle fibers of skeletal muscle; their major function is to sense muscle stretch

myelin sheath a multilayered wrapper of myelin-rich glial cell membranes around an axon

myasthenia gravis an autoimmune disease that destroys acetylcholine receptors, causing muscular weakness or paralysis

myoepithelial smooth muscle cells smooth muscle cells surrounding the cavities of glands; when they contract, the contents of the glands are secreted

myofibril a single muscle fiber

myoglobin a molecule that delivers oxygen to the mitochondria in muscle cells; it is structurally and functionally similar to hemoglobin

myosin a filamentous protein that interacts with actin to generate contractile forces in muscle

myosin head the part of the myosin molecule that attaches to binding sites on actin filaments

myosin isoforms molecular variants of myosin that differ in their contractile properties (e.g., whether they generate slow or rapid contractions)

Na^+/K^+-ATPase an enzyme that moves 2 potassium ions (K^+) into a cell for every 3 sodium ions (Na^+) that it moves out of the cell, using one molecule of ATP to do so

narcolepsy the tendency to fall asleep more often than normal, often falling into REM sleep without prior warning

natural selection Darwin's theory that heritable traits that boost an individual's chance of survival and reproduction will, over the course of many generations, become more common in the population

Neanderthals a group of ancient *Homo sapiens* that lived in Europe and parts of Asia 400,000 to 30,000 years ago and had brains that were similar in size to those of modern humans

necrosis a form of cell death in which the cell bursts, spilling its contents

negative allometry a form of scaling in which the structure of interest becomes proportionately smaller and smaller as body size increases

negative feedback loop a control system in which the system's output is used to modify the process giving rise to that output in such a way that, overall, less output is produced

neocortex the largest part of the cerebral cortex in humans; it exhibits numerous folds and interacts extensively with the underlying striatum and the thalamus

Nernst equation an equation that allows you to calculate the equilibrium potential for an ion if you know the concentration of that ion on either side of a cell membrane (that is permeable to this ion)

nerve growth factor the first trophic factor to be discovered (by Levi-Montalcini and Cohen in 1954)

nerve net a distributed network of neurons that exhibits no centralization and thus no brain or cerebral ganglion

netrin a diffusible protein, secreted by floor plate cells, that attracts some growing axons in the spinal cord

neural crest a groups of cells that originates at the lateral edges of the neural folds and migrates away from there to form a variety of tissues, including large portions of the peripheral nervous system

neural groove an early stage of neural development when the originally flat neural plate has begun to fold lengthwise

neural plate an early stage of neural development when the prospective nervous system is still a flat sheet

neural tube the stage of neural development when the left and right edges of the neural folds meet along the top of the embryo and fuse, forming a tubular structure

neurexin a protein expressed by presynaptic terminals of developing axons; it interacts with postsynaptic receptors (neuroligin), which then causes postsynaptic proteins to accumulate at the postsynaptic site

neuroethologist a scientist who studies the neural mechanisms underlying natural behavior

neurogenesis the process of generating neurons (and some glial cells) from progenitor cells

neuromodulation the changing of a neuron's response to one transmitter, released from a specific set of synapses, by other substances released from other sites

neuromuscular junction the large synapse between a motor axon and a skeletal muscle fiber

neuron doctrine the idea, most prominently promoted by Santiago Ramón y Cajal, that each neuron is a distinct cell (not continuous with other neurons)

neuropeptide a class of neurotransmitters that consist of several amino acids chained together; they tend to be co-released with other transmitters

neuropeptide Y (NPY) a neurotransmitter released by orexigenic neurons in the arcuate nucleus

neuropsychologist a scientist who explores the neural basis of psychological processes

neurotransmitter one of many different chemicals that neurons release in order to transmit information from one cell to another

nicotinic acetylcholine receptor a type of acetylcholine receptor that is sensitive to nicotine; found, for example, at neuromuscular junctions

Nissl substance a type of rough endoplasmic reticulum found in neurons and stained selectively by so-called Nissl stains

nitric oxide a gaseous neurotransmitter with a variety of poorly understood functions, including the promotion of vasodilation

NMDA-type glutamate receptor an ionotropic type of glutamate receptor that, in contrast to the AMPA receptor, can be opened by glutamate only when the postsynaptic cell is depolarized well above its resting potential

nociceptor a cell that is specialized for the detection of noxious (potentially harmful) stimuli

Node of Ranvier a small gap in the myelin sheath, usually recurring at regular intervals along the axon

Nogo a molecule secreted by oligodendrocytes that tends to inhibit axon growth

norepinephrine an important neurotransmitter used mainly by some neurons in the sympathetic division of the autonomic nervous system and by neurons in locus coeruleus

nucleus ambiguus a set of neurons in the medulla that projects to parasympathetic postganglionic neurons that, in turn, innervate the heart; it also contains neurons that innervate the muscles of the larynx

nucleus of the optic tract a small cell group in the diencephalon whose neurons receive retinal input, respond to large-scale image motion (notably retinal image slip), and are involved in generating the optokinetic reflex

nucleus tractus solitarius (NTS) a cell group in the dorsal medulla that receives taste information mainly through cranial nerves VII and IX as well as visceral sensory information mainly through the vagus nerve (cranial nerve X)

numerical matching the idea that the number of neurons in one brain region can be "matched" by trophic factor-dependent cell death to the number of neurons in other, interconnected brain regions

object recognition memory the ability to recognize an object as being familiar (e.g., something that one has seen before)

occipital lobe the most caudal/posterior lobe of the neocortex; it includes primary visual cortex and several higher visual cortices

ocular dominance bands snaking bands in layer 4 of the primary visual cortex, defined by the fact that they receive visual information from one eye or the other but not both (also known as ocular dominance columns)

oculomotor nucleus a group of motor neurons that innervate several muscles that can rotate the eye, including the medial rectus

odorant binding protein proteins that bind odorant molecules and shuttle them through the mucus covering the olfactory epithelium to the olfactory receptors

OFF bipolar cell a type of retinal neuron that receives input from cone photoreceptors and is depolarized when a light stimulus turns OFF; it expresses ionotropic glutamate receptors (which elicit depolarization in response to glutamate)

off-center cells neurons that increase their firing rate in response to a dark spot presented at the center of their receptive field (against a bright background)

olfactory bulb the most rostral part of the telencephalon in most vertebrates, located on the inferior surface of the frontal lobe in humans; it receives inputs from the olfactory epithelium and projects to olfactory cortex

olfactory cortex a set of cortical areas that receives highly divergent input from neurons in the olfactory bulb; it is involved in odor discrimination

olfactory epithelium the layer of cells in the roof of the nasal cavity that houses the olfactory sensory neurons

olfactory receptor molecule a large family of G protein-coupled receptors that are expressed in the cilia of olfactory sensory neurons and bind to odorants

olfactory sensory neuron neurons in the olfactory epithelium that express olfactory receptor molecules and project to the olfactory bulb

oligodendrocyte a type of glial cell that wraps itself around axons and produces the myelin sheath around myelinated axons in the central nervous system

ON bipolar cell a type of retinal neuron that receives input from cone photoreceptors and is depolarized when a light stimulus comes ON; it expresses metabotropic glutamate receptors (which elicit hyperpolarization in response to glutamate)

on-center cells neurons that increase their firing rate in response to a bright spot presented at the center of their spatial receptive field (against a dark background)

Onuf's nucleus a group of motor neurons in the spinal cord that is larger (and contains more cells) in males than in females; it innervates muscles near the genitals

opioid receptors receptors that are activate by chemicals related to opium

opsin a type of G protein-coupled receptor that is found in photoreceptors and helps make them light-sensitive

optic ataxia a clinical condition marked by an inability to make accurate hand movements to visual targets, often caused by damage to the posterior parietal cortex

optic chiasm the location (in the hypothalamus) where the left and right optic nerves cross (at least partially) and enter the brain

optic nerve head the location where the axons of retinal ganglion cells cross through the retina to exit the eye and enter the optic nerve; it is responsible for creating the blind spot

optic tectum the non-mammalian homolog of the superior colliculus, often used to study the mechanisms of axon guidance

optocollic reflex reflexive movements of the head (by neck muscles) that stabilize the visual image on the retina; it is triggered by neural signals that measure retinal image slip

optokinetic reflex when the outside world moves past the open eyes, the eyes rotate to track the moving image as far as possible, move rapidly in the opposite direction, and then repeat the cycle (creating a saw-tooth pattern of repetitive eye movements)

orbitofrontal cortex a cortical area on the inferior aspect of the frontal lobe that integrates information about taste and smell

orexigenic acting as an appetite stimulant

orientation sensitivity the property of responding preferentially to lines or edges of a particular orientation (i.e., being sensitive to the orientation of the stimulus)

otoconial membrane a gelatinous membrane containing numerous carbonate crystals; when this membrane moves relative to the underlying hair cells, some of the hair cells become depolarized (depending on movement direction)

outer hair cell the three rows of hair cells furthest away from the center of the spiraling cochlea; their main function is to amplify vibrations of the basilar membrane and thus increase acoustic sensitivity

outer segment the part of a photoreceptor cell that faces toward the back of the eye and contains most of the photosensitive pigments (rhodopsin or one of the cone opsins)

oval window a thin membrane covering the entrance to the cochlea

oxidative phosphorylation the conversion of pyruvate into carbon dioxide and water, generating 30 ATP in the process

oxytocin a hormone released in the posterior pituitary by magnocellular neurons; among other functions, it can trigger both the milk ejection reflex and the fetus ejection reflex

P2X an ion channel that is activated by adenosine 5′-triphosphate (ATP) and expressed at postsynaptic sites in gustatory axons

Pacinian corpuscle an encapsulated nerve ending in which the sensory axon is embedded in an onion-like cluster of cells; they usually lie deep in the skin, are highly sensitive, and respond best to high-frequency vibrations

pallidum part of the ventral/inferior telencephalon that receives its main input from the striatum; it contains dorsal and ventral subdivisions

parallel fiber the axons of cerebellar granule cells that run parallel to the cerebellar surface and synapse onto many different Purkinje cells

parallel processing the idea that information need not flow along a linear, unbranched circuit but may instead diverge along multiple pathways as it passes through the brain

paraventricular nucleus a hypothalamic nucleus that contains some of the large (magnocellular) neurons that synthesize vasopressin or oxytocin, in the posterior pituitary as well as some smaller (parvocelluar) neurons

parenchyma in the context of neurobiology, the parenchyma is the tissue of the brain, excluding ependymal cells and blood vessels

parietal lobe the region of the neocortex that lies caudal/posterior to the frontal lobe and anterior/rostral to the occipital lobe; it includes somatosensory cortex and areas involved in the control of targeted eye, head, and hand movements

Parkinson's disease a neurodegenerative disease caused mainly by the loss of dopaminergic neurons from the substantia nigra (compact division); it is characterized by an inability to select (initiate) movements and actions

parvocellular neurons neurons with small cell bodies compared to those of the larger (magnocellular) neurons

pattern completion the idea that presenting just one part of a complex stimulus to a neural network will cause that network to "complete the pattern," responding as it would to the entire (complete) complex stimulus

pattern separation the idea that very similar patterns of sensory input are made less similar (by some neural process) before they are stored in memory, thus making the patterns easier to distinguish in memory

Pavlovian (or classical) conditioning a form of learning in which the organism, after one or more exposures to a neutral stimulus followed by a reward or a threat, learns to associate that stimulus with the reward or threat, and acts accordingly

Pavlovian fear conditioning a form of Pavlovian (classical) conditioning in which the organism learns to predict a threat; it is typically quantified by measuring a rodent's tendency to "freeze" in response to the threat-associated stimulus

penetrating arterioles the arterioles (small arteries) that enter the brain from its surface

perforant path a set of axons that project from the entorhinal cortex to various parts of the hippocampus

peribrachial neurons a cell group that contains cholinergic neurons with projections to the thalamus, as well as non-cholinergic neurons that project to cholinergic basal forebrain neurons; both regulate behavioral arousal

perilymph the fluid surrounding the cell bodies of cochlear hair cells; it is has a much lower concentration of potassium ions than the endolymph

period (*per*) one of the "clock genes" involved in generating intracellular circadian rhythms; its transcription is promoted by Bmal and Clock, but heterodimer complexes of those two proteins ultimately suppress *per* transcription

peripheral nervous system axons and neurons that lie outside of the central nervous system, including most sense organs and nerves, large parts (but not all) of the autonomic nervous system, and the enteric nervous system

perirhinal cortex a cortical area that has reciprocal connections with the entorhinal cortex and is needed for object recognition memory

peristaltic contractions rhythmic waves of muscular contraction that travel along the intestines and push ingested food toward the rectum

perivascular smooth muscle cells smooth muscles that surround blood vessels; when they contract the blood vessels constrict

perseverative errors doing the same thing over and over (perseverating), despite clear signals that the behavior is ineffective (an error); such errors are often associated with prefrontal damage

phantom limb pain pain that is felt in a region of the body that is no longer there (e.g., due to an amputation) or is no longer innervated by sensory axons (e.g., after spinal cord damage)

phasic dopamine bursts a brief increase in dopamine release within the striatum, generally associated with rewards; to be distinguished from the much slower changes in dopamine levels associated with Huntington's and Parkinson's disease

phosphodiesterase an enzyme that can be activated by activated transducin and degrades cyclic GMP

phototransduction cascade the series of event that starts with photons activating the molecule retinal and ends with a change in the membrane potential of a photoreceptor cell

photoreceptor biological sensors of light, including both rod and cone photoreceptors

phrenologist followers of Joseph Gall who believes that the brain is composed of multiple distinct "cerebral organs," the size of which is reflected in the shape of the skull

pia mater the thinnest of the three meninges, it is in direct contact with neural tissue

pigment epithelium a highly pigmented layer of cells covering the outer segments of retinal photoreceptors; it is involved in retinoid recycling

pineal gland a small endocrine gland situated on top of the third ventricle; one of its major functions is to secrete melatonin

pinna the outer ear (ear flap)

pioneer axons axons that grow to their target very early in development and are then used as guides by later axons, which simply grow along the earlier axons to find the same general target area

piriform cortex the largest part of the olfactory cortex; it receives input from the olfactory bulb.

pitocin a synthetic form of oxytocin that is used to induce birth by stimulating uterine contractions

pituitary hormones hormones that are secreted into the blood by cells in the anterior pituitary; they include adrenocorticotropic hormone (ACTH), luteinizing hormone (LH), and thyroid stimulating hormone (TSH)

PKD-2L1 a polycystic kidney disease-like (PKD) TRP channel involved in sensing the carbon dioxide (CO_2) in carbonated drinks

place cell a type of neuron that increases its firing rate (above the background rate) when the organism is in a specific spatial location, regardless of how the animal is oriented or in which direction it is moving

place learning the process of learning where in space some object is located and how to navigate to that location regardless of one's starting location

pons clusters of neurons lying ventral/inferior to the cerebellum and providing much of its input

population code the idea that specific sensory or motor information is represented in the activity of many different neurons (each of which does not encode the information with as much precision as the population does)

positive feedback loop a control system in which the system's output is used to modify the process giving rise to that output in such a way that, overall, more output is produced

posterior toward the back side of the body when standing upright; for the brain: toward the back of the skull

posterior parietal cortex a collection of cortical areas posterior to primary somatosensory cortex that receives visual, somatosensory, and auditory input and is involved in the control of hand, head, and eye movements

posterior pituitary the posterior part of the pituitary gland; it contains a capillary bed into which magnocellular neurons of the hypothalamus secrete the hormones oxytocin and vasopressin

posterior prevalence model the idea that caudally expressed *Hox* genes are more powerful than rostrally expressed *Hox* genes in determining the rostrocaudal identity of cells in the hindbrain and spinal cord

postganglionic neurons neurons that have their cell bodies in the autonomic ganglia

postinhibitory rebound the tendency of some neurons to fire a burst of action potentials after they have been "freed" from inhibition (disinhibited)

post-tetanic potentiation a form of synaptic plasticity that is observed after tetanic stimulation but lasts only for a few minutes

post-training memory enhancement the observation that the strength of some memories can be enhanced by experimental manipulations after the initial learning event (post-training)

potassium buffering the idea that extracellular potassium concentrations must be kept low for neurons to function properly; it is thought to be a major function of astrocytes

pre-Bötzinger complex a small group of neurons in the ventral medulla that is part of the central pattern generator for breathing; it lies caudal to the Bötzinger complex

prefrontal cortex the most anterior (rostral) portion of the neocortex, which is proportionately enlarged in humans, relative to non-humans; it plays a major role in many cognitive functions, including planning and decision making

prefrontal lobotomy a surgical procedure, frequently used in the 1940s and 1950s, that involves large lesions of the prefrontal cortex or disconnecting it from the rest of the brain; it was supposed to make patients more manageable

preganglionic neurons neurons in the brain or spinal cord that project to the autonomic ganglia

premotor cortex a collection of cortical areas immediately rostral (anterior) to the primary motor cortex; its neurons help control movements, at least in part through projections to the primary motor cortex

preoptic area a small set of hypothalamic cell groups that lies rostral (anterior) to the optic chiasm and is involved in regulating body temperature, fluid balance, sexual behavior, and a variety of other important functions

prepotent response a behavioral response that an individual performs habitually and automatically

primary auditory cortex (A1) a core region of auditory cortex (in the superior portion of the temporal lobe) that receives direct projections from the medial geniculate nucleus (of the thalamus) and is tonotopically organized; it is essential for the identification of complex sounds (notably speech)

primary motor cortex a thin strip of cortex just anterior to the central sulcus that has relatively direct projections to motor neurons in the medulla and spinal cord

primary somatosensory cortex (S1) a thin strip of cortex just posterior to the central sulcus that receives input from the ventral posterior nucleus (of the thalamus); it is organized as a somatosensory homunculus

primary visual cortex (V1) the part of mammalian neocortex that has direct reciprocal connections with the lateral geniculate nucleus (LGN) and is needed for conscious visual perception; also known as striate cortex

principal trigeminal nucleus a cell group in the medulla that receives topographically organized information about skin touch and vibration from to the face and mouth (though the trigeminal nerve)

procedural learning various forms of skill learning (e.g., motor skills, perceptual skills) that can occur without conscious awareness; in patient H.M., many forms of procedural learning and memory were spared

progenitor a cell that is still dividing

projection neuron a neuron with a long axon that connects mainly to distant neurons

proprioception the sense of body position and movement of body parts relative to one another

prostaglandin E2 a molecule secreted into the blood (by cells of the immune system) that can be sensed by neurons in the vagus nerve and circumventricular organs; elevated levels of prostaglandin E2 trigger a fever

protein kinase A (PKA) a family of enzymes whose level of activity is regulated by cAMP; their functions vary across cell types

protein kinase C a calcium-activated enzyme that is involved (among other things) in the removal of AMPA receptors from postsynaptic membranes

protein kinase M-zeta (PKM-zeta) an enzyme that has been implicated in the stabilization of long-term potentiation (LTP) but does not appear to be necessary for the formation of LTP

pseudogene a gene that has become nonfunctional during the course of evolution

pseudounipolar a type of neuron with a single process emerging from its cell body, but this process then splits into two major branches, as exemplified by neurons of the dorsal root ganglia

pulsatile hormone release the idea that hormones are released in pulsatile bursts, rather than gradually; this type of release makes it easier to activate the hormone receptors and prevents their habituation

pulvinar nucleus a thalamic nucleus that receives visual inputs from the superior colliculus and projects to a variety of extrastriate cortices; it is involved in the control of visual spatial attention

pupillary dilation a reflex that excites the pupillary dilator muscle, thereby dilating the pupil, when an organism is behaviorally aroused (e.g., frightened)

pupillary light reflex a reflex that excites the pupillary constrictor muscle, and thus constricts the pupil, in response to bright light

Purkinje cell very large neurons in the cerebellar cortex that feature a distinctive, largely two-dimensional dendritic tree

Purkinje fiber modified muscle cells that have lost their contractile properties and help conduct action potentials to cardiac muscle fibers; part of the cardiac conduction system

push-pull regulation the regulation of some variable by sending opposing commands to effectors that move the variable in opposite directions (e.g., turning up a heater while turning down an air conditioner)

pyramidal neuron the largest neurons of the cerebral cortex, they have a pyramid-shaped cell body and long, radially oriented dendrites

quadriceps stretch reflex a reflexive contraction of the quadriceps muscle when it is stretched

quantal varying in discrete steps rather than continuously

radial cell a type of neural progenitor cell that has a radial processes that extends to the brain surface

radioactive carbon (^{14}C) a radioactive isotope of carbon that can be used for birth-dating brain cells because its levels in the atmosphere increased dramatically as a result of nuclear testing in the 1960s and then declined steadily

rapid eye movement (REM) sleep a phase of sleep in which subjects tend to exhibit jerky eye movements and a desynchronized EEG; subjects that suddenly wake from REM sleep often report vivid dreams

reactive astrocyte a kind of astrocyte that can multiply in response to injury and form a glial scar around the damaged tissue

reciprocal dendrodendritic synapse synapses between dendrites that are arranged so that a synapse that transmits signals in one direction (from one dendrite to the other) lies adjacent to a synapse going in the opposite direction

reciprocal innervation the idea that commands to move a limb generally involve sending opposing commands to opposing muscles (flexors vs. extensors)

recurrent collateral an axonal branch that projects back onto the neuron from which the axon originates

red-green opponent cell a type of neuron that compares the activation levels of L- and M-cones in its spatial receptive field; some fire most when the L-cones predominate, whereas others exhibit the opposite preference

reelin a protein that regulates the migration of young cortical neurons along the radial processes of radial cells

referred pain pain that is perceived as coming from one part of the body (usually part of the skin) when in actuality it is caused by noxious stimuli elsewhere (usually some internal organ)

reflex an involuntary, stereotyped response to a specific stimulus; most reflexes are not learned, but it is possible to acquire new reflexes through learning

reflex chain hypothesis the idea that locomotor activity is caused by a chain of sensorimotor reflexes where a movement triggers a reflexive movement, which triggers another reflexive movement, and so on

reflex gain the magnitude (e.g., strength) of the reflexive response to a specific stimulus

refractory period a period after each action potential when the neuron cannot generate another action potential, mainly because the voltage-gated sodium channels are still inactivated

reinstatement hypothesis the idea that memory recall involves the reactivation of the neural activation patterns that existed at the time of memory encoding (at least to some extent)

relational memory hypothesis the idea that the general function of the hippocampus is to form memories of relationships, which may be spatial, temporal, or even logical

releasing hormones hormones that are secreted into the median eminence and trigger the secretion (release) of other hormones from cells in the inferior portion of the anterior pituitary gland

REM without atonia a clinical condition in which REM sleep is not accompanied by muscle atonia, causing the sleeping subject to "act out their dreams"

remapping the idea that map-like representations in the brain may be altered (remapped) as a result of learning or injury

replay episodes the observation that some patterns of neuronal activation (especially in the hippocampus) recorded during sleep recapitulate (replay) patterns of neuronal activation that had occurred during waking

repolarization the process that brings the membrane potential back toward its resting value after a strong depolarization

response profile a summary description of how a neuron responds to a broad range of stimuli

rest-and-digest a state that is generally opposite to that of the fight-or-flight response, promoting relaxation and the restoration of the body

resting potential the voltage difference (electrical potential) across the cell membrane that a cell exhibits when it is not receiving external inputs (resting)

retina the multilayered neural structure at the back of the eye that contains photoreceptors, retinal interneurons, and neurons that project through the optic nerve to the brain

retinal a small molecule that dissociates from opsin molecules when it is "activated" by light (i.e., converted from the *cis* isomer to the *all-trans* isomer, causing a change in the molecule's shape)

retinal activity waves waves of depolarization that spread across the developing retina and are involved in refining the topography of retinal projections to the brain

retinal ganglion cell a type of retinal neuron that receives input from bipolar cells and sends its axon through the optic nerve into the brain

retinal image slip the gradual drifting across the retina of the visual image that is being projected onto the retina from the outside world; when retinal image slip is zero, the projected image is stable

retinoic acid a diffusible molecule that, at high concentrations, can caudalize cells of the embryonic brain

retinoid recycling the conversion of all-*trans* retinal back to *cis*-retinal, which can reassociate with opsins and make them light sensitive

retinotectal system the set of axons that connect the retina to the optic tectum (in non-mammalian vertebrates)

retinotopic the idea that adjacent neurons in a visually responsive brain region respond to stimulation of adjacent locations in the retina

retrograde amnesia the loss of old memories

retrograde amnesia gradient the observation that some brain lesions (notably of the hippocampus) impair recently formed long-term memories more severely than very old memories

retrograde signals signals that travel from the tips of an axon toward the cell body

retrotrapezoid nucleus a group of neurons that sense and process information about CO_2 levels in the blood and modulate breathing rate through projections to the central pattern generator for breathing

reversal potential the electrical potential at which ionic currents flip their polarity (inward currents become outward currents, or vice versa)

reverse engineering the process of trying to understand the inner workings of a system by starting with observations about its behavior

rhodopsin the photosensitive pigment in rod photoreceptors

ribbon synapse a type of chemical synapse in which numerous synaptic vesicles are arranged around a central "ribbon"; they are found in hair cells, allowing them to release large amounts of glutamate over long periods of time

rigor mortis muscle stiffness caused by ATP depletion after death and the release of calcium from ruptured internal calcium stores

robo a receptor for the axon guidance molecule slit, which is secreted by floor plate cells

rod bipolar cell a type of retinal neuron that receives input from rod photoreceptors, expresses metabotropic glutamate receptors (which elicit hyperpolarization in response to glutamate), and synapses onto amacrine cells

rod photoreceptor a type of photoreceptor cell that is more light sensitive than cone photoreceptors but is not used in color vision

rostral toward the snout in four-legged animals; not well defined in humans

round window a thin membrane covering the basal end of the scala tympani; it always vibrates together with the oval window but in the opposite direction

ryanodine receptor a type of voltage-sensitive calcium channel that links the T-tubule system to the sarcoplasmic reticulum, triggering internal calcium release when the muscle fiber fires an action potential

saccade a small and rapid (jerky) movement, most often of the eyes

saccade generator regions two areas of the reticular formation in the pons and medulla that cooperate to generate saccadic eye movements; they project directly to motor neurons that innervate the eye muscles

saccade motor map a systematic representation in the brain of saccade endpoints, such that activation of adjacent neurons triggers saccades to adjacent targets in the external world

sacculus a patch of hair cells in the vestibule that is covered with an otoconial membrane and is involved in sensing upward or downward acceleration (or deceleration) of the head

saliency map the idea of a map-like neural representation of external space in which each neuron's firing rate correlates with the salience of the stimuli at the encoded location

salient grabbing (or deserving) attention

saltatory conduction the idea that, in myelinated axons, action potentials can "jump" from one node of Ranvier to another node further down the axon, thereby increasing the speed of action potential propagation

sarcolemma the cell membrane of muscle cells

sarcomere one of many repeating segments in a striated or cardiac muscle fiber

sarcoplasmic reticulum the endoplasmic reticulum of muscle cells

scala media a fluid-filled compartment in the cochlea, sandwiched between the scala vestibuli and the scala tympani

scala tympani the half of the cochlear tube that extends from the top (apex) of the cochlea to the round window

scala vestibuli the half of the cochlear tube that extends from the oval window to the top (apex) of the cochlea

Schwann cell a type of glial cell that produces myelin sheaths surrounding axons in the peripheral nervous system

S-cones cone photoreceptors that are maximally sensitive to short (S) wavelengths of light

semaphorin a protein that tends to repel growing axons by causing the filopodia of their growth cones to collapse

semicircular canals a bilateral set of three semicircular tubes, oriented roughly at right angles to one another, that are involved in sensing head tilt and rotation

semi-intact preparation a living animal from which most of the body has been dissected away, such that only some parts of the nervous system remain intact and attached to parts of the body that they innervate

sensitive period a period of time, usually during early development, when the nervous systems is most easily modified by experience

sensitization a form of nonassociative learning in which a strong (usually noxious) stimulus enhances the response to a variety of subsequent stimuli

sensory map the idea that sensory representations in the brain resemble geographic maps insofar as adjacent locations in the world correspond to adjacent locations in the neural representation

sensory modality the classification of sensations and perceptions into distinct categories; for example, vision is a sensory modality, as are hearing and touch

sensory transduction the process by which an external stimulus (e.g., a sound) is converted into a change in the electrical activity of sensory neurons or other sensory cells

septal nuclei paired structures in the ventromedial telencephalon that are involved in regulating freezing and (on the other end of the behavioral spectrum) aggression

sequence learning the learning of a temporal sequence, such that the organism can recall the order in which events have previously occurred

serotonin an important neurotransmitter involved mainly in regulating mood

set point the target value or range to which some variable is regulated (e.g., the temperature to which a thermostat is set)

sexually dimorphic nucleus (SDN) a cell group in the preoptic area of rodents that is significantly larger in males than in females; its neurons are involved in a variety of male sexual behaviors

signal-to-noise ratio the ratio between the amplitude of a signal and the noise in which that signal is embedded; the higher the ratio, the “cleaner” the signal

sildenafil citrate marketed as Viagra, this compound inhibits cGMP degradation in the vasculature of the penis, thereby promoting erections

silent synapses glutamatergic synapses that contain no AMPA receptors and are, therefore, “silent” even when glutamate is released (unless the postsynaptic cell becomes depolarized enough to open NMDA receptors)

simple cell a type of neuron in the primary visual cortex that responds preferentially to lines or edges of a particular orientation in a specific location

sinoatrial (SA) node a small clump of modified muscle fibers in the wall of the right atrium; its cells fire rhythmically and are the principal pacemakers of the cardiac rhythm

size principle of motor unit recruitment the idea that, during a contraction that gradually increases in strength, small motor units (containing a small number of muscle fibers) become active before the large motor units

skeletal muscle the type of muscle that attaches to skeletal structures (bones) and a few other structures (e.g., the eyeball); its fibers are unbranched and striated

sliding filament model the idea that actin and myosin filaments slide across one another when muscles change in length

slit a diffusible protein, secreted by floor plate cells; it repulses axons in the spinal cord that express Robo receptors

smooth muscle cells a type of muscle cell, distinct from cardiac and striated muscle, that surrounds blood vessels, the intestines, and other body cavities; it tends to contract slowly

smooth pursuit eye movement an eye movement that smoothly tracks a relatively small, moving visual stimulus

SNc/VTA the collective term for the compact division of the substantia nigra (SNc) and the ventral tegmental area (VTA), two sets of neurons that lie next to one another; both use dopamine as their main neurotransmitter

social transmission of food preferences the tendency of some animals (notably rats) to learn what's safe to eat by smelling what their social companions have eaten (i.e., smelling their breath); if it didn't kill them, it's probably safe

sodium channel inactivation voltage-gated sodium channels automatically close (inactivate) shortly after they have opened, even when the depolarization is prolonged

somatosensory homunculus a drawing of a "little man" in which the size of various body parts is modified to reflect the degree to which that part of the body is over- or under-represented in the somatosensory cortex

somatosensory system a multifaceted system of neurons and sensory cells that provides an organism with information about the physical state of its body including temperature, limb position, and pressure on the skin

somatotopic organized in such a way that adjacent parts of the body (usually the skin) are represented at adjacent locations in the brain

sonic hedgehog (shh) a secreted protein that can ventralize cells in the embryonic spinal cord and brain

sparse coding the idea that neurons filter out redundant information (analogous to data compression)

spasticity abnormal limb rigidity, coupled with exaggerated spinal reflexes (excessive reflex gain)

spatial receptive field the set of locations where stimuli must be presented in order to increase (or suppress) a neuron's firing rate significantly

spatial summation the idea that synaptic potentials generated in different parts of a neuron will be additive in their overall effect on the neuronal membrane potential

spectrotemporal receptive field a summary of how an auditory neuron responds to sounds of various frequencies, combined with a description of the (temporal) delay of those responses; it can sometimes be used to predict responses to complex sounds

spike rate code the idea that neurons transmit information by increasing or decreasing the rate at which they fire action potentials (spikes)

spike timing the time at which an action potential (a spike) occurs

spike timing-dependent plasticity the idea that a synapse may be strengthened if it was active shortly *before* a postsynaptic action potential (spike) but weakened if it was active shortly *after* a postsynaptic spike

spinal nerve nerves that carry signals to or from the spinal cord (or both)

spinal nucleus of the trigeminal nerve a cell group in the medulla that receives nociceptive information from the region of the face (through axons in the trigeminal nerve)

spinal shock a temporary reduction in the strength (or elimination) of spinal cord reflexes that is caused by damage to descending, modulatory inputs from the brain

spine turnover the idea that dendritic spines may disappear and form anew (i.e., are not stable) even in adult brains

spinocerebellar projections axons that project from the spinal cord to the cerebellum; they play a major role in proprioception and motor coordination

spinocerebellum the portion of the cerebellum that lies closest to the midline, receives sensory inputs from the spinal cord, and is involved in stabilizing the body

spinothalamic tract a fiber tract that ascends from the substantia gelatinosa of the spinal cord to the contralateral thalamus; it carries mainly nociceptive signals

spiral ganglion a spiraling string of cell bodies inside the cochlea; its neurons innervate cochlear hair cells and project to the cochlear nuclei in the brain

spontaneous novel object recognition task a task in which subjects are presented with one item and then, after a delay, presented with a pair of objects that includes the old item as well as a new one; most rodents will spontaneously explore the novel item, thereby demonstrating that they recognize which item is familiar

squid giant axons extremely large diameter axons in common squid that help the squid escape from threats and helped neuroscientists discover the mechanisms underlying neuronal resting and action potentials

staggerer a mutant strain of mice in which all cerebellar Purkinje cells die early in development

stapes one of the middle ear bones; it attaches to the oval window

stereocilia the "hairs" of the hair cells; they are actually microvilli, rather than cilia (nor, for that matter, hairs)

striated muscle skeletal or cardiac muscle, which exhibit prominent transverse striations (stripes) as a result of the regular arrangement of sarcomeres

striatum a large ventrolateral/inferior (non-cortical) component of the telencephalon; it is the largest component of the basal ganglia

stria vascularis specialized epithelial cells in the outer wall of the cochlea that pump potassium ions into the scala media

stripe assay cell membranes are deposited in a series of stripes on the bottom of a cell culture dish; experimenters then note whether specific axons grow preferentially along those stripes or avoid them

Stroop task a task that involves subjects either reading a "color word" or naming the color of the ink in which a word is printed; it tests for several abilities related to prefrontal function, including response inhibition, strategy selection, and voluntary attention

subarachnoid space the space between the pia mater and the arachnoid membrane; it is filled with cerebrospinal fluid (CSF)

subcoeruleus region a cluster of cells ventral to the locus coeruleus that is a critical link in the circuit controlling muscle atonia; lesions of this area cause REM without atonia

subdural hematoma an accumulation of blood that leaked from ruptured vessels into the *subarachnoid space*

subfornical organ one of the circumventricular organs; its cells are capable of sensing blood osmolarity (dehydration) and are involved in regulating fluid intake

subiculum part of the hippocampus; it receives strong projections from CA1 and projects to several subcortical targets as well as the entorhinal cortex

substantia gelatinosa the most dorsal layer of the spinal cord's dorsal horn, where unmyelinated nociceptive axons tend to terminate

substantia nigra, compact division (SNc) a group of neurons in the midbrain that use dopamine as their main neurotransmitter and project mainly to the dorsal striatum; loss of these neurons leads to Parkinson's disease

subthalamic nucleus a cell group ventral to (below) the dorsal thalamus that is part of the indirect pathway through the basal ganglia; lesions or electrical inactivation of this cell group may be used to treat Parkinson's disease

subventricular zone a layer of dividing cells that lies between the ventricular zone and the mantle zone in some parts of the embryonic nervous system

sulcus/sulci an inward fold (or groove) of the cerebral cortex; "sulci" is the plural of "sulcus"

superior toward the top of the head in humans

superior colliculus a major midbrain region that receives visual as well as auditory information and projects to the pulvinar and brainstem saccade generators; it is called the optic tectum in non-mammalian vertebrates

suprachiasmatic nucleus a small group of rhythmically firing neurons in the hypothalamus (close to the optic chiasm) that help to generate the body's daily (circadian) and seasonal rhythms

supraoptic nucleus a hypothalamic nucleus that (just like the paraventricular nucleus) contains some of the large (magnocellular) neurons that synthesize vasopressin or oxytocin, as well as some smaller (parvocelluar) neurons

surround inhibition the idea that neurons in the center of a spatial receptive field suppress neurons in the area surrounding that center, and vice versa; it is a special case of lateral inhibition

sympathetic ganglia autonomic ganglia that receive input from sympathetic preganglionic neurons (in the intermediolateral column)

sympathetic nervous system one of the major divisions of the autonomic nervous system; its activity generates a variety of physiological effects that, collectively, prepare an animal to deal with an acute stressor (e.g., a threat)

synapse the site where a presynaptic neuron releases neurotransmitter molecules onto a postsynaptic cell or (for electrical synapses) where a presynaptic neuron comes into direct cytoplasmic contact with a postsynaptic cell

synapse formation the process through which synapses are created

synapse maturation the process by which newly formed synapses grow and accumulate a greater variety of pre- and postsynaptic proteins

synaptic cleft the narrow gap between the pre- and post-synaptic elements of a synapse

synaptic scaling the idea that neurons decrease (scale down) the average strength of all synaptic inputs if the neurons have been highly active and increase the strength of those inputs if the neurons have been relatively quiet

synaptic tagging hypothesis the idea that neurons use special molecules to "tag" synapses that are later recognized by proteins (made in the cell nucleus and transported into the dendrites) that modify those synapses

synaptic vesicles membranous vesicles filled with neurotransmitter molecules, located inside the presynaptic component of a chemical synapse

synesthesia a condition in which some stimuli evoke perceptions that most people do not associate with those stimuli (e.g., colors being associated with words, or sounds being associated with shapes)

systems consolidation the idea that memory traces become stronger and less susceptible to disruption once they have become established in the neocortex (as well as, or instead of, the hippocampus)

T1 receptor family a family of 3 taste receptors (T1R1, T1R2, and T1R3) that form heterodimers (two different molecules assembled into one compound receptor); their sensitivity depends on which receptors are combined

T2 receptors a family of taste receptors that, collectively, respond to a variety of bitter-tasting stimuli

tangential migration the migration of young neurons (and glial cells) parallel to the external brain surface

tapetum lucidum a reflective layer in the retina of many nocturnal species; it accounts for the "eye shine" that arises when you shine a light straight into those animals' eyes

taste bud clusters of 50–150 taste cells on the tongue, the roof of the mouth, and the back of the throat

taste cell a cell that expresses taste receptors; these cells tend to die and be replaced in adulthood

taste receptor a molecule that can bind to one or more tastants and trigger an intracellular response that ultimately causes a cell expressing this receptor to change its membrane potential

tectorial membrane a relatively thick membrane that lies on top of the hair cells of the basilar membrane

tegmentum the ventral/inferior part of the midbrain

telencephalon the most rostral part of the forebrain and its largest division in most vertebrates

temporal correlation hypothesis the idea that the binding problem is solved by having neurons that respond to different features of the same stimulus object fire action potentials in tight synchrony (being temporally correlated)

temporal lobe the most ventral/inferior portion of the primate neocortex, bordered caudally/posteriorly by the occipital lobe; it includes auditory cortex and, deep within the temporal lobe, the hippocampus and amygdala

temporal summation the idea that synaptic potentials that overlap in time will be additive in their overall effect on the neuronal membrane potential

terminal arborization a set of small branches that axons typically form near their target area

testosterone a hormone that is produced by a male's testes and important in sexual differentiation; much of it is converted in the brain to estradiol, which helps to masculinize the brain during development

tetanic stimulation a form of electrical stimulation in which many short electrical pulses are delivered (usually to a set of axons) at a high frequency (high rate of repetition)

tetanus a potentially fatal medical problem characterized by uncontrolled and persistent, spasm-like contractions of some skeletal muscles

tetanus toxin a powerful bacterial toxin that blocks GABA release from inhibitory synapses and causes tetanus

thalamic reticular nucleus a thalamic cell group that inhibits neurons in the dorsal thalamus and receives input from both the dorsal thalamus and neocortex; it is involved in generating rhythmic patterns of thalamocortical activity

thalamus the dorsal/superior part of the diencephalon; it includes the medial geniculate nucleus and the lateral geniculate nucleus.

third interstitial nucleus of the anterior hypothalamus (INAH-3) a hypothalamic cell group in humans that is smaller (by volume) in women and homosexual men than in heterosexual men

thyrotropin releasing hormone (TRH) a releasing hormone secreted into the capillaries of the median eminence; it stimulates pituitary cells to secrete thyroid stimulating hormone, also known as thyrotropin

tight junction a place where the cell membranes of two cells come into contact and are held together by special molecules; tight junctions restrict the flow of large hydrophilic molecules through extracellular space

tip link a thin filament that connects the tip of one stereocilium to the adjacent, taller stereocilium; it is linked to an ion channel that opens when the stereocilia are bent toward the tallest stereocilium

tonotopic map the idea that a sound's frequency (~ pitch) is represented in a map-like fashion across a brain region such that adjacent neurons are tuned to similar sound frequencies

tonotopy the idea that neurons tuned to similar sound frequencies are located adjacent to one another such that, collectively, they form a neural "map" of sound frequency

topographic projection an axonal projection in which neighboring neurons project to neighboring neurons in the target region

transcription factor a protein that binds to short, specific sequences of DNA and regulates the expression of downstream genes

transcutaneous electrical nerve stimulation (TENS) a form of pain therapy that involves transcutaneous (across the skin) electrical stimulation of touch-sensitive axons, which reduces the ability of nociceptive signals to reach the brain

transducin a G protein that can be activated by light-activated opsins

transient receptor potential (TRP) channels a family of about 30 different channels, named after a channel that, when mutated, causes fruit fly photoreceptors to exhibit abnormally transient responses to light

transorbital lobotomy a form of prefrontal lobotomy in which an ice pick–like instrument was inserted above the eye, through the orbital bone, and into the brain and used to disconnect the prefrontal cortex from the rest of the brain

trigeminal nerve the fifth cranial nerve (nV); it contains somatosensory axons coming from the face as well as some motor axons

tritiated thymidine a radioactive form of thymidine that can be used to determine cellular birth dates because it is incorporated into the DNA of dividing cells and then retained, unless it is diluted by further cell divisions

trophic factors a class of molecules that promote neuron survival and growth; it includes nerve growth factor (NGF) and brain derived neurotrophic factor (BDNF)

tropomyosin a filamentous molecule that obstructs the myosin binding sites on actin filaments when calcium levels are low; as calcium levels rise, the tropomyosin moves out of the way and allows cross-bridge cycling

TRP-A1 a member of the family of transient receptor potential (TRP) channels that begins to open at very low, noxious temperatures

TRP-M5 a calcium-activated ion channel that is expressed in taste cells and, in its open state, allows positive ions to flow into the cell; without this channel, sensitivity to sweet, umami, and bitter tastants is lost

TRP-M8 a member of the family of transient receptor potential (TRP) channels that opens at temperatures below normal mammalian body temperatures and is used in thermoregulation

TRP-V1 a member of the family of transient receptor potential (TRP) channels that can be activated by noxious heat as well as capsaicin (which is why this channel is sometimes called the capsaicin receptor)

T-tubule system a set of membranous tubes that extend from folds beneath each neuromuscular junction deep into a muscle fiber and thus allow action potentials to spread deep into the fiber

tuberomammillary nucleus a small group of neurons in the posterior hypothalamus that project to multiple targets, using histamine as their main neurotransmitter; their activity promotes the waking state

tuned a neuron is said to be "tuned" to a particular set of stimuli if it responds significantly better to those stimuli than to others

tympanic membrane the thin membrane separating the ear canal from the middle ear, also known as the ear drum; it vibrates in response to airborne sounds

tyrosine hydroxylase an enzyme that is needed for the synthesis of dopamine, norepinephrine, and epinephrine because it converts L-tyrosine into L-dopa (the immediate precursor of dopamine)

ubiquitin hydrolase an enzyme that (among other things) degrades the regulatory subunit of protein kinase A, causing the latter enzyme to become persistently active

unconditioned fear pathways neural circuits that cause animals to exhibit innate (unlearned) fear responses in response to various threatening stimuli

undershoot a period at the end of an action potential when the membrane potential dips below the cell's resting potential

utricle a patch of hair cells in the vestibule that is covered with an otoconial membrane and is involved in sensing acceleration (or deceleration) of the head forward, backward, or sideways

vagus nerve cranial nerve ten (X); it innervates taste buds at the back of the throat, conveys sensory information from most internal organs, and contains parasympathetic efferent axons

vascular organ of the lamina terminalis (OVLT) one of the circumventricular organs; its cells are capable of sensing blood osmolarity (dehydration) and are involved in regulating fluid intake

vasoconstriction the constriction of blood vessels (especially the smaller ones); it usually causes a rise in blood pressure

vasodilation the expansion (dilation) of blood vessels, especially the smaller ones; it usually causes a drop in blood pressure

vasopressin a hormone secreted into the capillaries of the posterior pituitary by the axons of magnocellular hypothalamic neurons; among other functions, it increases blood pressure through vasoconstriction, which promotes fluid retention

ventral toward the abdomen (underside) of the body or the underside of the head in four-legged animals when they hold their head high; ambiguous in humans

ventralize to change the developmental trajectory of cells so that they will adopt a more ventral cell fate

ventral horn the ventral (anterior in humans) portion of the spinal cord's gray matter

ventral pallidum a cell group that receives inputs from the ventral striatum and projects to a part of the dorsal thalamus that projects to cortical areas that, in turn, project heavily to the ventral striatum (forming a loop)

ventral posterior nucleus a thalamic nucleus that receives somatosensory information from the dorsal column nuclei and the principal trigeminal nucleus; it projects to the primary somatosensory cortex

ventral root a bundle of axons that exits a segment of the spinal cord ventrally (anteriorly in humans) and consists mainly of motor axons; it joins with the corresponding dorsal root to form a spinal nerve

ventral stream a set of interconnected cortical areas that extends from the primary visual cortex into the inferior portion of the temporal lobe; it is involved in stimulus identification

ventral striatum a relatively small brain region (a.k.a. nucleus accumbens) that receives strong inputs from parts of the prefrontal cortex and is involved in generating cravings for food, water, sex, and drugs

ventral tegmental area (VTA) a group of neurons in the midbrain that use dopamine as their neurotransmitter and project mainly to the ventral striatum; they play a major role in learning through positive reinforcement

ventricular zone a layer of cells in embryonic nervous systems that lies adjacent to the cerebral ventricles and consists almost exclusively of densely packed progenitors

ventrolateral periaqueductal gray (vlPAG) a cell group near the cerebral aqueduct whose neurons fall silent during REM sleep

ventrolateral preoptic area a small cell group in the preoptic area whose neurons exhibit increased activity during sleep and are thought to promote the sleeping state

vertebrates the group of animals that has a vertebral column, including jawed fishes, amphibians, reptiles, and mammals

vestibular complex a collection of several cell groups that receive sensory input from the vestibular apparatus (the semicircular canals, utricle, and sacculus) and convey this information to a variety of other brain regions

vestibular system a system of sensory structures and neural pathways concerned with sensing movements of the head, specifically its angular and linear acceleration

vestibule a fluid-filled bony cavity in the inner ear that contains the utricle and the sacculus

vestibulocerebellum the relatively small, most caudal (posterior) portion of the cerebellar cortex; it is involved in fine-tuning the vestibulo-ocular reflex and other reflexes that involve vestibular input

vestibulocochlear nerve cranial nerve number eight (VIII), which innervates the inner ear and vestibular apparatus (semicircular canals, sacculus, and utricle)

vestibulocollic reflex a reflexive counterrotation of the head when the body turns (using neck muscles); it stabilizes the visual image on the retina and is triggered by head-rotation signals from the semicircular canals

vestibulo-ocular reflex a reflexive counterrotation of the eyes when the head rotates; it serves to stabilize the visual image on the retina; it is triggered by head-rotation signals from the semicircular canals

visceral pain sensations of pain that originate from internal organs (viscera)

visceral sensory neurons neurons that convey sensory signals from the internal organs (viscera)to the brain and spinal cord; most of them have their axons in the vagus nerve

visual field the region of space that is visible through the eyes

visual hemifield the half of the visual field that lies either to the left or to the right of the fixation point

voltage-clamp recording a technique that allows experimenters to measure ionic currents flowing across a membrane while the electrical potential difference (voltage) across that membrane is held constant (clamped)

voltage-gated calcium channel a calcium channel that greatly increases its probability of being open when the cell is depolarized above a threshold value

voltage-gated potassium channel a type of potassium channel that greatly increases its probability of being open when the cell is depolarized beyond a threshold value

voltage-gated sodium channel a sodium channel that greatly increases its probability of being open when the cell is depolarized above a threshold value

volume transmission the idea that transmitters may be released outside of synapses, diffuse through extracellular space, and act on extrasynaptic receptors

voluntary attention attention that can be directed at will; for example, when you are looking for a specific item in a cluttered scene or listening for a specific sound embedded in other sounds

VOR adaptation the process of having adjusted the gain of the vestibulo-ocular reflex (after having learned from earlier mistakes) in such a way that retinal image slip is minimized when the head turns

Wernicke's aphasia a deficit in language comprehension (without reducing the fluidity of speech) that typically results from damage to Wernicke's area

Wernicke's area an area in the left posterior portion of the superior temporal lobe that is involved in language comprehension

white matter tissue that contains almost exclusively myelinated axons (see **myelin sheath**)

winner-take-all competition the idea that a set of neurons inhibit each other in such a way that the initially most active neuron will inhibit all the others, effectively "winning the competition," at least temporarily

Wisconsin card sort task a task in which subjects must learn to sort cards according to an unknown rule, using feedback about errors; once the rule is learned, the experimenter changes the rule, forcing a shift in sorting strategy

working memory the temporary storage of information that was just experienced or retrieved from long-term memory but is no longer accessible in the external environment

world-centered coordinates a spatial coordinate system that is anchored to the external world and thus remains immobile as the body moves through the world; it is also known as an allocentric coordinate system

zeta inhibitory peptide (ZIP) a peptide that can reverse well-established long-term potentiation (LTP) through mechanisms that remain controversial

Z-line a dense, zig-zagging zone that separates adjacent sarcomeres

References

Publications Referenced in Text

Preface

Ramos RL, Fokas GJ, Bhambri A, Smith PT, Hallas BH, Brumberg JC. 2011. Undergraduate neuroscience education in the US: an analysis using data from the National Center for Education Statistics. *J Undergrad Neurosci Ed* **9**:A66.

Chapter 1

Crick FH. 1979. Thinking about the brain. *Sci Am* **241**:219–232.

Hippocrates. From *The genuine works of Hippocrates*, translated by F. Adams (1886), Vol. 2, New York: William Wood & Co.

Chapter 2

Fatt P, Katz B. 1952. Spontaneous subthreshold activity at motor nerve endings. *J Physiol* **117**:109–128.

Miledi R. 1973. Transmitter release induced by injection of calcium ions into nerve terminals. *Proc Roy Soc Lond B* **183**:421–425.

Ramón y Cajal S. 1917. *Recuerdos de mi vida, Vol. 2, Historia de mi labor científica*. Madrid: Moya.

Chapter 3

Cajal Ry. 1894. The Croonian lecture: la fine structure des centres nerveux. *Proc Roy Soc Lond* **55**:444–468.

Hebb DO. 1949. *The organization of behavior: a neuropsychological theory*. New York: Wiley.

Chapter 4

Ramón y Cajal S. 1917. *Recuerdos de mi vida, Vol. 2, Historia de mi labor científica*. Madrid: Moya.

Ramón y Cajal S. 1894. *Die retina der wirbelthiere*. Wiesbaden, Bergmann-Verlag, Thorpe and Glick (trans.), Springfield, IL: Thomas.

Tessier-Lavigne M, Placzek M, Lumsden AG, Dodd J, Jessel TM. 1988. Chemotropic guidance of developing axons in the mammalian central nervous system. *Nature* **336**:775–778.

Chapter 6

Kalmijn AJ. 1971. The electric sense of sharks and rays. *J Exp Biol* **55**:371–383.

Chapter 8

Brownell WE, Bader CR, Bertrand D, de Ribaupierre Y. 1985. Evoked mechanical responses of isolated cochlear outer hair cells. *Science* **227**:194–196.

Chapter 10

Bauby J-D. 1997. *The diving bell and the butterfly*. New York: Knopf.

di Pellegrino G, Fadiga L, Fogassi L, Gallese V, Rizzolatti G. 1992. Understanding motor events: a neurophysiological study. *Exp Brain Res* **91**:176–180.

Schmahmann JD. 1991. An emerging concept. The cerebellar contribution to higher function. *Archives Neurol* **48**:1178–1187.

Wilson DM. 1961. The central nervous control of flight in a locust. *Journal Exp Biol* **38**:471–490.

Chapter 12

Hartline HK. 1969. Visual receptors and retinal interaction. *Science* **164**:270–278.

Chapter 13

James W. 1890. *The principles of psychology*. Volumes I and II. New York: Holt.

Chapter 14

Garcia J, Hankins WG, Rusiniak KW. 1974. Behavioral regulation of the milieu interne in man and rat. *Science* **185**:824–831.

Luria AR. 1968. *The mind of a mnemonist*. New York: Basic Books, Inc.

Chapter 15

Wise SP. 2008. Forward frontal fields: phylogeny and fundamental function. *Trends Neurosci* **31**:599–608.

Jacobsen CF. 1939. The effects of extirpations on higher brain processes. *Physiol Reviews* **19**:303–322.

Tolliver BK, Sganga MW, Sharp FR. 2000. Suppression of c-fos induction in the nucleus accumbens prevents acquisition but not expression of morphine-conditioned place preference. *Eur J Neurosci* **12**:3399–3406.

Chapter 16

Talairach J, Tournoux P. 1988. *Co-planar stereotaxic atlas of the human brain.* Stuttgart: Georg Thieme Verlag.

Hohmann AA, Parron DL. 1996. How the new NIH guidelines on inclusion of women and minorities apply: efficacy trials, effectiveness trials, and validity. *J Consult Clin Psychol* **64**:851–855.

Publications Referenced in Figures

Chapter 1

Aponte Y, Atasoy D, Sternson SM. 2011. AGRP neurons are sufficient to orchestrate feeding behavior rapidly and without training. *Nat Neurosci* **14**:351–355.

Brooks R. 1999. *Cambrian intelligence.* Boston: MIT Press.

Deacon T. 1990. Rethinking mammalian brain evolution. *Am Zool* **30**:629–705.

Fuster JM. 2003. *Cortex and mind: unifying cognition.* New York: Oxford University Press.

Goodson JL, Rinaldi J, Kelly AM. 2009. Vasotocin neurons in the bed nucleus of the stria terminalis preferentially process social information and exhibit properties that dichotomize courting and non-courting phenotypes. *Horm Behav* **55**:197–202.

James W. 1890. *The principles of psychology.* New York: H. Holt & Co. [see http://psychclassics.yorku.ca/James/Principles/]

Morgan JL, Dhingra A, Vardi N, Wong ROL. 2006. Axons and dendrites originate from neuroepithelial-like processes of retinal bipolar cells. *Nat Neurosci* **9**:85–92.

Sereno M, Tootell R. 2005. From monkeys to humans: What do we now know about brain homologies? *Curr Opin Neurobiol* **15**:135–144.

Sivak J. 1976. Optics of the eye of the "four-eyed fish" (*Anableps anableps*). *Vision Res* 16:531–534.

Swanson LW. 2005. Anatomy of the soul as reflected in the cerebral hemispheres: Neural circuits underlying voluntary control of basic motivated behaviors. *J Comp Neurol* **493**:122–131.

Wells SR. 1870. *How to read character: an illustrated hand-book of phrenology and physiognomy for students and examiners; with a descriptive chart.* New York: Samuel R. Wells.

Chapter 2

Attwell D, Laughlin SB. 2001. An energy budget for signaling in the grey matter of the brain. *J Cereb Blood Flow Metab* **21**:1133–1145.

Ramón y Cajal S. 1894. The Croonian lecture: la fine structure des centres nerveux. *Proc Roy Soc Lond* **55**:444–468.

Catterall WA. 2000. From ionic currents to molecular mechanisms: the structure and function of voltage-gated sodium channels. *Neuron* **26**:13–25.

Hamill OP, Marty A, Neher E, Sakmann B, Sigworth FJ. 1981. Improved patch-clamp techniques for high-resolution current recording from cells and cell-free membrane patches. *Pflügers Archiv* **391**:85–100.

Hatzidimitriou G, McCann UD, Ricaurte GA. 1999. Altered serotonin innervation patterns in the forebrain of monkeys treated with (+/-)3,4-methylenedioxy-methamphetamine seven years previously: factors influencing abnormal recovery. *J Neurosci* **19**:5096–5107.

Hodgkin A, Horowicz P. 1959. The influence of potassium and chloride ions on the membrane potential of single muscle fibres. *J Physiol* **148**:127–160.

Hursh JB. 1939. Conduction velocity and diameter of nerve fibers. *Am J Physiology* **127**:131–139.

Laughlin SB, Sejnowski TJ. 2003. Communication in neuronal networks. *Science* **301**:1870–1874.

McCusker EC, Bagnéris C, Naylor CE, Cole AR, D'Avanzo N, Nichols CG, and Wallace BA. 2012. Structure of a bacterial voltage-gated sodium channel pore reveals mechanisms of opening and closing. *Nat Commun* **3**:1102.

Roberts TF, Tschida KA, Klein ME, Mooney R. 2010. Rapid spine stabilization and synaptic enhancement at the onset of behavioural learning. *Nature* **463**:948–952.

Segev I. 1998. Sound grounds for computing dendrites. *Nature* **393**:207–208.

Wollmuth LP, Sobolevsky AI. 2004. Structure and gating of the glutamate receptor ion channel. *Trends Neurosci* **27**:321–328.

Chapter 3

Bi GQ, Poo MM. 1998. Synaptic modifications in cultured hippocampal neurons: dependence on spike timing, synaptic strength, and postsynaptic cell type. *J Neurosci* **18**:10464–10472.

Bliss TV, Lømo T. 1973. Long-lasting potentiation of synaptic transmission in the dentate area of the anaesthetized rabbit following stimulation of the perforant path. *J Physiol* **232**:331–356.

Boele H-J, Koekkoek SKE, De Zeeuw CI, Ruigrok TJH. 2013. Axonal sprouting and formation of terminals in the adult cerebellum during associative motor learning. *J Neurosci* **33**:17897–17907.

Cracco JB, Serrano P, Moskowitz SI, Bergold PJ, Sacktor TC. 2005. Protein synthesis-dependent LTP in isolated dendrites of CA1 pyramidal cells. *Hippocampus* **15**:551–556.

Dash PK, Hochner B, Kandel ER. 1990. Injection of the cAMP-responsive element into the nucleus of *Aplysia* sensory neurons blocks long-term facilitation. *Nature* **345**:718–721.

Finch EA, Tanaka K, Augustine GJ. 2012. Calcium as a trigger for cerebellar long-term synaptic depression. *Cerebellum* **11**:706–717.

Kandel ER. 2001. The molecular biology of memory storage: a dialogue between genes and synapses. *Science* **294**:1030–1038.

Kleim JA, Barbay S, Cooper NR, Hogg TM, Reidel CN, Remple MS, Nudo RJ. 2002. Motor learning-dependent synaptogenesis is localized to functionally reorganized motor cortex. *Neurobiol Learn Mem* **77**:63–77.

Matsuzaki M, Honkura N, Ellis-Davies GCR, Kasai H. 2004. Structural basis of long-term potentiation in single dendritic spines. *Nature* **429**:761–766.

Roberts TF, Tschida KA, Klein ME, Mooney R. 2010. Rapid spine stabilization and synaptic enhancement at the onset of behavioural learning. *Nature* **463**:948–952.

Rosenzweig MR, Krech D, Bennett EL, Zolman JF. 1962. Variation in environmental complexity and brain measures. *J Comp Physiol Psychol* **55**:1092–1095.

Serrano P, Yao Y, Sacktor TC. 2005. Persistent phosphorylation by protein kinase Mzeta maintains late-phase long-term potentiation. *J Neurosci* **25**:1979–1984.

Steward O, Levy WB. 1982. Preferential localization of polyribosomes under the base of dendritic spines in granule cells of the dentate gyrus. *J Neurosci* **2**:284–291.

Turner AM, Greenough WT. 1985. Differential rearing effects on rat visual cortex synapses: I. Synaptic and neuronal density and synapses per neuron. *Brain Res* **329**:195–203.

Turrigiano GG, Leslie KR, Desai NS, Rutherford LC, Nelson SB. 1998. Activity-dependent scaling of quantal amplitude in neocortical neurons. *Nature* **391**:892–896.

Weinberger NM. 2007. Associative representational plasticity in the auditory cortex: a synthesis of two disciplines. *Learn Mem* **14**:1–16.

Chapter 4

Anger, EM, et al., 2004. Ultrahigh resolution optical coherence tomography of the monkey fovea. Identification of retinal sublayers by correlation with semithin histology sections. *Exp Eye Res* **78**:1117–1125.

Assimacopoulos S, Kao T, Issa NP, Grove EA. 2012. Fibroblast growth factor 8 organizes the neocortical area map and regulates sensory map topography. *J Neurosci* **32**:7191–7201.

Bray D. 1982. Filopodial contraction and growth cone guidance. In: *Cell behavior.* Bellair R, Curtis A, Dunn G, eds. Cambridge: Cambridge University Press, pp. 299–317.

Briscoe J, Pierani A, Jessell TM, Ericson J. 2000. A homeodomain protein code specifies progenitor cell identity and neuronal fate in the ventral neural tube. *Cell* **101**:435–445.

Brown A, Yates PA, Burrola P, Ortuño D, Vaidya A, Jessell TM, Pfaff SL, O'Leary DD, Lemke G. 2000. Topographic mapping from the retina to the midbrain is controlled by relative but not absolute levels of EphA receptor signaling. *Cell* **102**:77–88.

Burns L. 1911. *Studies in the osteopathic sciences.* Volume 2. Cincinnati: Monfort & Co. [free e-book].

De Robertis EM, Kuroda H. 2004. Dorsal-ventral patterning and neural induction in *Xenopus* embryos. *Annu Rev Cell Dev Biol* **20**:285–308.

Dehay C, Kennedy H. 2007. Cell-cycle control and cortical development. *Nat Rev Neurosci* **8**:438–450.

Emoto K, He Y, Ye B, Grueber WB, Adler PN, Jan LY, Jan Y-N. 2004. Control of dendritic branching and tiling by the Tricornered-kinase/Furry signaling pathway in *Drosophila* sensory neurons. *Cell* **119**:245–256.

Garner CC, Waites CL, Ziv NE. 2006. Synapse development: still looking for the forest, still lost in the trees. *Cell Tissue Res* **326**:249–262.

Guidato S, Prin F, Guthrie S. 2003. Somatic motoneurone specification in the hindbrain: the influence of somite-derived signals, retinoic acid and Hoxa3. *Development* **130**:2981–2996.

Hamburger V. 1934. The effects of wing bud extirpation on the development of the central nervous system in chick embryos. *J Exp Zool* **68**:449–494.

Hamburger V, Hamilton HL. 1992. A series of normal stages in the development of the chick embryo. 1951. *Dev Dyn* **195**:231–272.

Heidemann SR, Lamoureux P, Buxbaum RE. 1990. Growth cone behavior and production of traction force. *J Cell Biol* **111**:1949–1957.

Hollyday M, Hamburger V. 1976. Reduction of the naturally occurring motor neuron loss by enlargement of the periphery. *J Comp Neurol* **170**:311–320.

Holmes GP, Negus K, Burridge L, Raman S, Algar E, Yamada T, Little MH. 1998. Distinct but overlapping expression patterns of two vertebrate slit homologs implies functional roles in CNS development and organogenesis. *Mech Dev* **79**:57–72.

Huberman AD, Feller MB, Chapman B. 2008. Mechanisms underlying development of visual maps and receptive fields. *Annu Rev Neurosci* **31**:479–509.

Kennedy TE, Serafini T, de la Torre JR, Tessier-Lavigne M. 1994. Netrins are diffusible chemotropic factors for commissural axons in the embryonic spinal cord. *Cell* **78**:425–435.

LeVay S, Wiesel T, Hubel D. 1980. The development of ocular dominance columns in normal and visually deprived monkeys. *J Comp Neurol* **191**:1–51.

Lemons D, McGinnis W. 2006. Genomic evolution of *Hox* gene clusters. *Science* **313**:1918–1922.

Li Q, Martin JH. 2002. Postnatal development of connectional specificity of corticospinal terminals in the cat. *J Comp Neurol* **447**:57–71.

Maden M. 2002. Retinoid signalling in the development of the central nervous system. *Nat Rev Neurosci* **3**:843–853.

Masdeu JC, Pascual B, Bressi F, Casale M, Prieto E, Arbizu J, Fernández-Seara MA. (2009). Ventricular wall granulations and draining of cerebrospinal fluid in chronic giant hydrocephalus. *Arch Neurol* **66**:262–267.

Ming GL, Song HJ, Berninger B, Holt CE, Tessier-Lavigne M, Poo MM. 1997. cAMP-dependent growth cone guidance by netrin-1. *Neuron* **19**:1225–1235.

Monschau B, Kremoser C, Ohta K, Tanaka H, Kaneko T, Yamada T, Handwerker C, Hornberger MR, Löschinger J, Pasquale EB, Siever DA, Verderame MF, Müller BK, Bonhoeffer F, Drescher U. 1997. Shared and distinct functions of RAGS and ELF-1 in guiding retinal axons. *EMBO J* **16**:1258–1267.

Noctor S, Flint A, Weissman T, Dammerman R. 2001. Neurons derived from radial glial cells establish radial units in neocortex. *Nature* **409**:714–720.

Portera-Cailliau C, Pan DT, Yuste R. 2003. Activity-regulated dynamic behavior of early dendritic protrusions: evidence for different types of dendritic filopodia. *J Neurosci* **23**:7129–7142.

Rakic P. 1974. Neurons in rhesus monkey visual cortex: systematic relation between time of origin and eventual disposition. *Science* **183**:425–427.

Rapaport DH, Rakic P, LaVail MM. 1996. Spatiotemporal gradients of cell genesis in the primate retina. *Perspect Dev Neurobiol* **3**:147–159.

Ramón y Cajal S. 1890. À quelle époque apparaissent les expansions des cellules nerveuses de la moelle épinière du poulet. *Anatomischer Anzeiger.* Volumes 21 and 22.

Walter J, Kern-Veits B, Huf J, Stolze B, Bonhoeffer F. 1987. Recognition of position-specific properties of tectal cell membranes by retinal axons in vitro. *Development* **101**:685–696.

Wikramanayake AH, Hong M, Lee PN, Pang K, Byrum CA, Bince JM, Xu R, Martindale MQ. 2003. An ancient role for nuclear beta-catenin in the evolution of axial polarity and germ layer segregation. *Nature* **426**:446–450.

Chapter 5

Altman J, Das GD. 1966. Autoradiographic and histological studies of postnatal neurogenesis: I. A longitudinal investigation of the kinetics, migration and transformation of cells incorporating tritiated thymidine in neonate rats, with special reference to postnatal neurogenesis in some brain regions. *J Comp Neurol* **126**:337–389.

Bush TG, Puvanachandra N, Horner CH, Polito A, Ostenfeld T, Svendsen CN, Mucke L, Johnson MH, Sofroniew MV. 1999. Leukocyte infiltration, neuronal degeneration, and neurite outgrowth after ablation of scar-forming, reactive astrocytes in adult transgenic mice. *Neuron* **23**:297–308.

Choudhari KA, Sharma D, Leyon JJ. 2008. Thomas Willis of the "Circle of Willis." *Neurosurgery* **63**:1185–1190.

Ek CJ, Dziegielewska KM, Stolp H, Saunders NR. 2006. Functional effectiveness of the blood-brain barrier to small water-soluble molecules in developing and adult opossum (*Monodelphis domestica*). *J Comp Neurol* **496**:13–26.

Ek CJ, Habgood MD, Dziegielewska KM, Saunders NR. 2003. Structural characteristics and barrier properties of the choroid plexuses in developing brain of the opossum (*Monodelphis domestica*). *J Comp Neurol* **460**:451–464.

Eriksson PS, Perfilieva E, Björk-Eriksson T, Alborn AM, Nordborg C, Peterson DA, Gage FH. 1998. Neurogenesis in the adult human hippocampus. *Nat Med* **4**:1313–1317.

Haines DE. 1991. On the question of subdural space. *Anat Rec* **230**:3–21.

Hamilton R, Keenan JP, Catala M, Pascual-Leone A. 2000. Alexia for Braille following bilateral occipital stroke in an early blind woman. *Neuroreport* **11**:237–240.

Han H, Tao W, Zhang M. 2007. The dural entrance of cerebral bridging veins into the superior sagittal sinus: an anatomical comparison between cadavers and digital subtraction angiography. *Neuroradiology* **49**:169–175.

Jain N, Qi H-X, Collins CE, Kaas JH. 2008. Large-scale reorganization in the somatosensory cortex and thalamus after sensory loss in macaque monkeys. *J Neurosci* **28**:11042–11060.

Lee JE, Liang KJ, Fariss RN, Wong WT. 2008. Ex vivo dynamic imaging of retinal microglia using time-lapse confocal microscopy. *Invest Ophthalmol Visual Sci* **49**:4169–4176.

Masdeu JC, Pascual B, Bressi F, Casale M, Prieto E, Arbizu J, Fernández-Seara MA. 2009. Ventricular wall granulations and draining of cerebrospinal fluid in chronic giant hydrocephalus. *Arch Neurol* **66**:262–267.

Nishimura N, Schaffer CB, Friedman B, Lyden PD, Kleinfeld D. 2007. Penetrating arterioles are a bottleneck in the perfusion of neocortex. *Proc Natl Acad Sci U S A* **104**:365–370.

Oberheim NA, Takano T, Han X, He W, Lin JHC, Wang F, Xu Q, Wyatt JD, Pilcher W, Ojemann JG, Ransom BR, Goldman SA, Nedergaard M. 2009. Uniquely hominid features of adult human astrocytes. *J Neurosci* **29**:3276–3287.

Oddo S, Billings L, Kesslak JP, Cribbs DH, LaFerla FM. 2004. A-beta immunotherapy leads to clearance of early, but not late, hyperphosphorylated tau aggregates via the proteasome. *Neuron* **43**:321–332.

Pardridge WM. 2003. Gene targeting in vivo with pegylated immunoliposomes. *Methods Enzymol* **373**:507–528.

Potts DG, Reilly KF, Deonarine V. 1972. Morphology of the arachnoid villi and granulations. *Radiology* **105**:333–341.

Raichle ME, Mintun MA. 2006. Brain work and brain imaging. *Annu Rev Neurosci* **29**:449–476.

Sadato N, Okada T, Honda M, Yonekura Y. 2002. Critical period for cross-modal plasticity in blind humans: a functional MRI study. *NeuroImage* **16**:389–400.

Schaffer CB, Friedman B, Nishimura N, Schroeder LF, Tsai PS, Ebner FF, Lyden PD, Kleinfeld D. 2006. Two-photon imaging of cortical surface microvessels reveals a robust redistribution in blood flow after vascular occlusion. *PLoS Biol* **4**:e22.

Schenk D, Barbour R, Dunn W, Gordon G, Grajeda H, Guido T, Hu K, Huang J, Johnson-Wood K, Khan K, Kholodenko D, Lee M, Liao Z, Lieberburg I, Motter R, Mutter L, Soriano F, Shopp G, Vasquez N, Vandevert C, Walker S, Wogulis M, Yednock T, Games D, Seubert P. 1999. Immunization with amyloid-beta attenuates Alzheimer-disease-like pathology in the PDAPP mouse. *Nature* **400**:173–177.

Spalding K, Bhardwaj R, Buchholz B, Druid H, Frisen J. 2005. Retrospective birth dating of cells in humans. *Cell* **122**:133–143.

Stark CE, Squire LR. 2001. When zero is not zero: the problem of ambiguous baseline conditions in fMRI. *Proc Natl Acad Sci U S A* **98**:12760–12766.

Takano T, Tian G-F, Peng W, Lou N, Libionka W, Han X, Nedergaard M. 2006. Astrocyte-mediated control of cerebral blood flow. *Nat Neurosci* **9**:260–267.

Viswanathan A, Freeman RD. 2007. Neurometabolic coupling in cerebral cortex reflects synaptic more than spiking activity. *Nat Neurosci* **10**:1308–1312.

Wrigley PJ, Press SR, Gustin SM, Macefield VG, Gandevia SC, Cousins MJ, Middleton JW, Henderson LA, Siddall PJ. 2009. Neuropathic pain and primary somatosensory cortex reorganization following spinal cord injury. *Pain* **141**:52–59.

Chapter 6

Abaffy T, Malhotra A, Luetje CW. 2007. The molecular basis for ligand specificity in a mouse olfactory receptor: a network of functionally important residues. *J Biol Chem* **282**:1216–1224.

Deeb SS. 2006. Genetics of variation in human color vision and the retinal cone mosaic. *Curr Opin Genet Dev* **16**:301–307.

Franze K, Grosche J, Skatchkov SN, Schinkinger S, Foja C, Schild D, Uckermann O, Travis K, Reichenbach A, Guck J. 2007. Müller cells are living optical fibers in the vertebrate retina. *Proc Natl Acad Sci U S A* **104**:8287–8292.

Gilad Y, Man O, Pääbo S, Lancet D. 2003. Human specific loss of olfactory receptor genes. *Proc Natl Acad Sci U S A* **100**:3324–3327.

Glowatzki E, Fuchs PA. 2002. Transmitter release at the hair cell ribbon synapse. *Nat Neurosci* **5**:147–154.

Hackney CM, Furness DN. 1995. Mechanotransduction in vertebrate hair cells: structure and function of the stereociliary bundle. *Am J Physiol* **268**:C1–13.

Johnson BA, Farahbod H, Xu Z, Saber S, Leon M. 2004. Local and global chemotopic organization: general features of the glomerular representations of aliphatic odorants differing in carbon number. *J Comp Neurol* **480**:234–249.

Kachar B, Parakkal M, Kurc M, Zhao Y, Gillespie PG. 2000. High-resolution structure of hair-cell tip links. *Proc Natl Acad Sci U S A* **97**:13336–13341.

Kleene SJ, Gesteland RC. 1981. Dissociation of frog olfactory epithelium with N-ethylmaleimide. *Brain Res* **229**:536–540.

Knaapila A, Zhu G, Medland SE, Wysocki CJ, Montgomery GW, Martin NG, Wright MJ, Reed DR. 2012. A genome-wide study on the perception of the odorants androstenone and galaxolide. *Chem Senses* **37**:541–552.

Leal M, Losos JB. 2010. Evolutionary biology: communication and speciation. *Nature* **467**:159–160.

Loizou PC. 1998. Mimicking the human ear: an overview of signal-processing strategies for converting sound into electrical signals in cochlear implants. *IEEE Signal Proc Mag* **15**:101–130.

Matthews HR. 1990. Messengers of transduction and adaptation in vertebrate photoreceptors. In: *Light and life in the sea*. Herring PJ, Campbell AK, Whitfield M, Maddock L, eds. Cambridge, England: Cambridge University Press.

Meyer AC, Frank T, Khimich D, Hoch G, Riedel D, Chapochnikov NM, Yarin YM, Harke B, Hell SW, Egner A, Moser T. 2009. Tuning of synapse number, structure and function in the cochlea. *Nat Neurosci* **12**:444–453.

Mombaerts P, Wang F, Dulac C, Chao SK, Nemes A, Mendelsohn M, Edmondson J, Axel R. 1996. Visualizing an olfactory sensory map. *Cell* **87**:675–686.

Nayagam BA, Muniak MA, Ryugo DK. 2011. The spiral ganglion: connecting the peripheral and central auditory systems. *Hear Res* **278**:2–20.

Osterberg G. 1935. Topography of the layer of rods and cones in the human retina. *Acta Ophthalmol* (Suppl)**6**:1–103.

Polyak SL. 1957. *The vertebrate visual system*. Chicago: University of Chicago Press.

Russell IJ, Legan PK, Lukashkina VA, Lukashkin AN, Goodyear RJ, Richardson GP. 2007. Sharpened cochlear tuning in a mouse with a genetically modified tectorial membrane. *Nat Neurosci* **10**:215–223.

Spinelli KJ, Gillespie PG. 2009. Bottoms up: transduction channels at tip link bases. *Nat Neurosci* **12**:529–530.

Sugiyama K, Gu ZB, Kawase C, Yamamoto T, Kitazawa Y. 1999. Optic nerve and peripapillary choroidal microvasculature of the rat eye. *Invest Ophthalmol Vis Sci* **40**:3084–3090.

Witmer MT, Kiss S. 2013. Wide-field imaging of the retina. *Surv Ophthalmol* **58**:143–154.

Yokoi M, Mori K, Nakanishi S. 1995. Refinement of odor molecule tuning by dendrodendritic synaptic inhibition in the olfactory bulb. *Proc Natl Acad Sci U S A* **92**:3371–3375.

Chapter 7

Banks RW, Hulliger M, Saed HH, Stacey MJ. 2009. A comparative analysis of the encapsulated end-organs of mammalian skeletal muscles and of their sensory nerve endings. *J Anat* **214**:859–887.

Bautista DM, Siemens J, Glazer JM, Tsuruda PR, Basbaum AI, Stucky CL, Jordt S-E, Julius D. 2007. The menthol receptor TRPM8 is the principal detector of environmental cold. *Nature* **448**:204–208.

Boyd IA, Davey MR. 1968. *Composition of peripheral nerves.* Edinburgh & London: E. & S. Livingstone.

Caterina MJ, Schumacher MA, Tominaga M, Rosen TA, Levine JD, Julius D. 1997. The capsaicin receptor: a heat-activated ion channel in the pain pathway. *Nature* **389**:816–824.

Chandrashekar J, Hoon MA, Ryba NJP, Zuker CS. 2006. The receptors and cells for mammalian taste. *Nature* **444**:288–294.

Devor M. 1999. Unexplained peculiarities of the dorsal root ganglion. *Pain* **82**:S27–S35.

Dib-Hajj SD, Yang Y, Black JA, Waxman SG. 2013. The Na(v)1.7 sodium channel: from molecule to man. *Nat Rev Neurosci* **14**:49–62.

Guinard D, Usson Y, Guillermet C, Saxod R. 2000. PS-100 and NF 70-200 double immunolabeling for human digital skin Meissner corpuscle 3D imaging. *J Histochem Cytochem* **48**:295–302.

Haeberle H, Lumpkin E. 2008. Merkel cells in somatosensation. *Chemosens Percept* **1**:110–118.

Huang AL, Chen X, Hoon MA, Chandrashekar J, Guo W, Tränkner D., et al. 2006. The cells and logic for mammalian sour taste detection. *Nature* **442**:934–938.

Ifediba MA, Rajguru SM, Hullar TE, Rabbitt RD. 2007. The role of 3-canal biomechanics in angular motion transduction by the human vestibular labyrinth. *Ann Biomed Eng* **35**:1247–1263.

Lee MWL, McPhee RW, Stringer MD. 2008. An evidence-based approach to human dermatomes. *Clin Anat* **21**:363–373.

Li A, Xue J, Peterson EH. 2008. Architecture of the mouse utricle: macular organization and hair bundle heights. *J Neurophysiol* **99**:718–733.

Liu Z, Wang W, Zhang T-Z, Li G-H, He K, Huang J-F, et al. 2014. Repeated functional convergent effects of NaV1.7 on acid insensitivity in hibernating mammals. *Proc R Soc B* **281**:20132950. doi:10.1098/rspb.2013.2950

Matsumoto A, Mori S. 1975. Number and diameter distribution of myelinated afferent fibers innervating the paws of the cat and monkey. *Exp Neurol* **48**:261–274.

Pawson L, Prestia LT, Mahoney GK, Güçlü B, Cox PJ, Pack AK. 2009. GABAergic/glutamatergic-glial/neuronal interaction contributes to rapid adaptation in Pacinian corpuscles. *J Neurosci* **29**:2695–2705.

Peers C, Wyatt CN, Evans AM. 2010. Mechanisms for acute oxygen sensing in the carotid body. *Respir Physiol Neurobiol* **174**:292–298.

Todd AJ. 2010. Neuronal circuitry for pain processing in the dorsal horn. *Nat Rev Neurosci* **11**:823–836.

Tominaga M, Caterina MJ. 2004. Thermosensation and pain. *J Neurobiol* **61**:3–12.

Woodbury CJ, Koerber HR. 2007. Central and peripheral anatomy of slowly adapting type I low-threshold mechanoreceptors innervating trunk skin of neonatal mice. *J Comp Neurol* **505**:547–561.

Yang S, Xiao Y, Kang D, Liu J, Li Y. 2013. Discovery of a selective Na_v1.7 inhibitor from centipede venom with analgesic efficacy exceeding morphine in rodent pain models. *Proc Natl Acad Sci U S A* **110**:17534–17539.

Yokota T, Eguchi K, Hiraba K. 2014. Topographical representations of taste response characteristics in the rostral nucleus of the solitary tract in the rat. *J Neurophysiol* **111**:182–196.

Yoshida R, Ohkuri T, Jyotaki M, Yasuo T, Horio N, Yasumatsu K, et al. 2010. Endocannabinoids selectively enhance sweet taste. *Proc Natl Acad Sci U S A*, **107**:935–939.

Chapter 8

Ashmore J. 2008. Cochlear outer hair cell motility. *Physiol Rev* **88**:173–210.

Ashmore JF. 1987. A fast motile response in guinea-pig outer hair cells: the cellular basis of the cochlear amplifier. *J Physiol* **388**:323–347.

Bicknell RJ. 1988. Optimizing release from peptide hormone secretory nerve terminals. *J Exp Biol* **139**:51–65.

Buckberg GD, Mahajan A, Jung B, Markl M, Hennig J, Ballester-Rodes M. 2006. MRI myocardial motion and fiber tracking: a confirmation of knowledge from different imaging modalities. *Eur J Cardiothorac Surg* **29**:S165–177.

Burke RE, Dum RP, Fleshman JW, Glenn LL, Lev-Tov A, O'Donovan MJ, Pinter MJ. 1982. An HRP study of the relation between cell size and motor unit type in cat ankle extensor motoneurons. *J Comp Neurol* **209**:17–28.

Burke RE, Levine DN, Tsairis P, Zajac FE. 1973. Physiological types and histochemical profiles in motor units of the cat gastrocnemius. *J Physiol* **234**:723–748.

Dauber W, Meister A. 1986. Ultrastructure of junctional folds of motor end plates in *extensor digitorum longus* muscles of mice. *J Ultrastruct Mol Struct Res* **97**:158–164.

De Luca CJ, Contessa P. 2012. Hierarchical control of motor units in voluntary contractions. *J Neurophysiol* **107**:178–195.

Desmedt JE, Godaux E. 1977. Fast motor units are not preferentially activated in rapid voluntary contractions in man. *Nature* **267**:717–719.

DiFrancesco D. 1993. Pacemaker mechanisms in cardiac tissue. *Annu Rev Physiol* **55**:455–472.

Frank G, Hemmert W, Gummer AW. 1999. Limiting dynamics of high-frequency electromechanical transduction of outer hair cells. *Proc Natl Acad Sci U S A* **96**:4420–4425.

Hatton GI, Tweedle CD. 1982. Magnocellular neuropeptidergic neurons in hypothalamus: increases in membrane apposition and number of specialized synapses from pregnancy to lactation. *Brain Res Bull* **8**:197–204.

Hunt CC, Kuffler SW. 1951. Stretch receptor discharges during muscle contraction. *J Physiol* **113**:298–315.

Josephson RK, Malamud JG, Stokes DR. 2000. Asynchronous muscle: a primer. *J Exp Biol* **203**:2713–2722.

Keller-Peck CR, Walsh MK, Gan WB, Feng G, Sanes JR, Lichtman JW. 2001. Asynchronous synapse elimination in neonatal motor units: studies using GFP transgenic mice. *Neuron* **31**:381–394.

Lin G, Qiu X, Fandel TM, Albersen M, Wang Z, Lue TF, Lin C-S. 2011. Improved penile histology by phalloidin stain: circular and longitudinal cavernous smooth muscles, dual-endothelium arteries, and erectile dysfunction-associated changes. *Urology* **78**:970.e1–e8.

Lincoln DW, Wakerley JB. 1974. Electrophysiological evidence for the activation of supraoptic neurones during the release of oxytocin. *J Physiol* **242**:533–554.

McGarvey C, Cates PA, Brooks A, Swanson IA, Milligan SR, Coen CW, O'Byrne KT. 2001. Phytoestrogens and gonadotropin-releasing hormone pulse generator activity and pituitary luteinizing hormone release in the rat. *Endocrinology* **142**:1202–1208.

Murakami M, Nagato T, Tanioka H, Uehara Y. 1989. Morphological changes in the myoepithelial cells of the rat sublingual salivary gland during differentiation as shown by the nitrobenzoxadiazole-phallacidin fluorescent method. *Arch Oral Biol* **34**:143–145.

Newbold RR, Padilla-Banks E, Snyder RJ, Jefferson WN. 2005. Developmental exposure to estrogenic compounds and obesity. *Birth Defects Res A Clin Mol Teratol* **73**:478–480.

Newbold RR, Padilla-Banks E, Jefferson WN. 2009. Environmental estrogens and obesity. *Mol Cell Endocrinol* **304**:84–89.

Ovalle WK, Dow PR, Nahirney PC. 1999. Structure, distribution and innervation of muscle spindles in avian fast and slow skeletal muscle. *J Anat* **194**:381–394.

Plant TM. 1986. Gonadal regulation of hypothalamic gonadotropin-releasing hormone release in primates. *Endocr Rev* **7**:75–88.

Radley JJ, Sawchenko PE. 2011. A common substrate for prefrontal and hippocampal inhibition of the neuroendocrine stress response. *J Neurosci* **31**:9683–9695.

Roozendaal B, Phillips RG, Power AE, Brooke SM, Sapolsky RM, McGaugh JL. 2001. Memory retrieval impairment induced by hippocampal CA3 lesions is blocked by adrenocortical suppression. *Nat Neurosci* **4**:1169–1171.

Shibata M, Friedman RL, Ramaswamy S, Plant TM. 2007. Evidence that down regulation of hypothalamic KiSS-1 expression is involved in the negative feedback action of testosterone to regulate luteinising hormone secretion in the adult male rhesus monkey (*Macaca mulatta*). *J Neuroendocrinol* **19**:432–438.

Sparrow MP, Weichselbaum M, McCray PB. 1999. Development of the innervation and airway smooth muscle in human fetal lung. *Am J Respir Cell Mol Biol* **20**:550–560.

Vale RD, Milligan RA. 2000. The way things move: looking under the hood of molecular motor proteins. *Science* **288**:88–95.

Young B, Heath JW. 2000. *Wheater's functional histology*, Fourth edition. Edinburgh, UK: Churchill Livingstone.

Zuo Y, Bishop D. 2008. Glial imaging during synapse remodeling at the neuromuscular junction. *Neuron Glia Biology* **4**:319–326.

Chapter 9

Bourque CW. 2008. Central mechanisms of osmosensation and systemic osmoregulation. *Nat Rev Neurosci* **9**:519–531.

Cummings DE, Purnell JQ, Frayo RS, Schmidova K, Wisse BE, Weigle DS. 2001. A preprandial rise in plasma ghrelin levels suggests a role in meal initiation in humans. *Diabetes* **50**:1714–1719.

de la Iglesia HO, Meyer J, Carpino A, Schwartz WJ. 2000. Antiphase oscillation of the left and right suprachiasmatic nuclei. *Science* **290**:799–801.

de la Iglesia HO, Meyer J, Schwartz WJ. 2003. Lateralization of circadian pacemaker output: activation of left- and right-sided luteinizing hormone-releasing hormone neurons involves a neural rather than a humoral pathway. *J Neurosci* **23**:7412–7414.

Halaas JL, Boozer C, Blair-West J, Fidahusein N, Denton DA, Friedman JM. 1997. Physiological response to long-term peripheral and central leptin infusion in lean and obese mice. *Proc Natl Acad Sci U S A* **94**:8878–8883.

Kalin NH, Shelton SE, Davidson RJ. 2004. The role of the central nucleus of the amygdala in mediating fear and anxiety in the primate. *J Neurosci* **24**:5506–5515.

Li A, Randall M, Nattie EE. 1999. $CO_{(2)}$ microdialysis in retrotrapezoid nucleus of the rat increases breathing in wakefulness but not in sleep. *J Appl Physiol* **87**:910–919.

Magariños AM, McEwen BS, Flügge G, Fuchs E. 1996. Chronic psychosocial stress causes apical dendritic atrophy of hippocampal CA3 pyramidal neurons in subordinate tree shrews. *J Neurosci* **16**:3534–3540.

McKinley M, Mathai M, Pennington G, Rundgren M, Vivas L. 1999. Effect of individual or combined ablation of the nuclear groups of the lamina terminalis on water drinking in sheep. *Am J Physiol Regul Integr Comp Physiol* **276**:R673–R683.

Monos E, Lóránt M, Fehér (2001) Influence of long-term experimental orthostatic body position on innervation density in extremity vessels. *Am J Physiol Heart Circ Physiol* **281**:H1606–H1612.

Nakamura K, Morrison SF. 2008. A thermosensory pathway that controls body temperature. *Nat Neurosci* **11**:62–71.

Nakayama T, Hammel H, Hardy J, Eisenman J. 1963. Thermal stimulation of electrical activity of single units of the preoptic region. *Am J Physiol* **204**:1122–1126.

Oliet SH, Bourque CW. 1992. Properties of supraoptic magnocellular neurones isolated from the adult rat. *J Physiol* **455**:291–306.

Revell VL, Eastman CI. 2005. How to trick mother nature into letting you fly around or stay up all night. *J Biol Rhythms* **20**:353–365.

Ross CA, Ruggiero DA, Park DH, Joh TH, Sved AF, Fernandez-Pardal J, Saavedra JM, Reis DJ. 1984. Tonic vasomotor control by the rostral ventrolateral medulla: effect of electrical or chemical stimulation of the area containing C1 adrenaline neurons on arterial pressure, heart rate, and plasma catecholamines and vasopressin. *J Neurosci* **4**:474–494.

Rybak IA, O'Connor R, Ross A, Shevtsova NA, Nuding SC, Segers LS, et al. 2008. Reconfiguration of the pontomedullary respiratory network: a computational modeling study with coordinated *in vivo* experiments. *J Neurophys* **100**:1770–1799.

Shin J-W, Geerling JC, Loewy AD. 2009. Vagal innervation of the aldosterone-sensitive HSD2 neurons in the NTS. *Brain Res* **1249**:135–147.

Smith JC, Feldman JL. 1987. In vitro brainstem-spinal cord preparations for study of motor systems for mammalian respiration and locomotion. *J Neurosci Methods* **21**:321–333.

Tessier DJ, Eagon JC. 2008. Surgical management of morbid obesity. *Curr Probl Surg* **45**:68–137.

Thongkhao-on K, Wirtshafter D, Shippy SA. 2008. Feeding specific glutamate surge in the rat lateral hypothalamus revealed by low-flow push-pull perfusion. *Pharmacol Biochem Behav* **89**:591–597.

Tschöp M, Smiley DL, Heiman ML. 2000. Ghrelin induces adiposity in rodents. *Nature* **407**:908–913.

Uehara Y, Suyama K. 1978. Visualization of the adventitial aspect of the vascular smooth muscle cells under the scanning electron microscope. *J Electron Microsc* **27**:157–159.

Yan B, Li L, Harden SW, Epstein PN, Wurster RD, Cheng ZJ. 2009. Diabetes induces neural degeneration in nucleus ambiguus (NA) and attenuates heart rate control in OVE26 mice. *Exp Neurol* **220**:34–43.

Chapter 10

Cohen AH, Wallén P. 1980. The neuronal correlate of locomotion in fish. "Fictive swimming" induced in an *in vitro* preparation of the lamprey spinal cord. *Exp Brain Res* **41**:11–18.

Frost BJ. 1978. The optokinetic basis of head-bobbing. *J Exp Biol* **74**:187–195.

Gallese V, Fadiga L, Fogassi L, Rizzolatti G. 1996. Action recognition in the premotor cortex. *Brain* **119**:593–609.

Georgopoulos AP, Kalaska JF, Caminiti R, Massey JT. 1982. On the relations between the direction of two-dimensional arm movements and cell discharge in primate motor cortex. *J Neurosci* **2**:1527–1537.

Graziano M. 2006. The organization of behavioral repertoire in motor cortex. *Annu Rev Neurosci* **29**:105–134.

Grillner S. 2003. The motor infrastructure: from ion channels to neuronal networks. *Nat Rev Neurosci* **4**:573–586.

Ilg UJ. 1997. Slow eye movements. *Prog Neurobiol* **53**:293–329.

Lacroix S, Havton LA, McKay H, Yang H, Brant A, Roberts J, Tuszynski MH. 2004. Bilateral corticospinal projections arise from each motor cortex in the macaque monkey: a quantitative study. *J Comp Neurol* **473**:147–161.

Marsden CD, Merton PA, Morton HB. 1981. Human postural responses. *Brain* **104**:513–534.

Nagao S, Kitazawa H. 2003. Effects of reversible shutdown of the monkey flocculus on the retention of adaptation of the horizontal vestibulo-ocular reflex. *Neuroscience* **118**:563–570.

Necker R. 2007. Head-bobbing of walking birds. *J Comp Physiol A* **193**:1177–1183.

Pearson KG, Rossignol S. 1991. Fictive motor patterns in chronic spinal cats. *J Neurophysiol* **66**:1874–1887.

Penfield W, Boldrey E. 1937. Somatic motor and sensory representation in the cerebral cortex of man as studied by electrical stimulation. *Brain* **60**:389–443.

Penfield W, Rasmussen T. 1950. *The cerebral cortex of man*. New York: Macmillan.

Vilis T, Hore J. 1980. Central neural mechanisms contributing to cerebellar tremor produced by limb perturbations. *J Neurophysiol* **43**:279–291.

Wallén P, Grillner S. 1987. N-methyl-D-aspartate receptor-induced, inherent oscillatory activity in neurons active during fictive locomotion in the lamprey. *J Neurosci* **7**:2745–2755.

Wallén P, Williams TL. 1984. Fictive locomotion in the lamprey spinal cord in vitro compared with swimming in the intact and spinal animal. *J Physiol* **347**:225–239.

Chapter 11

Adams DL, Sincich LC, Horton JC. 2007. Complete pattern of ocular dominance columns in human primary visual cortex. *J Neurosci* **27**:10391–10403.

Bruce CJ, Goldberg ME, Bushnell MC, Stanton GB. 1985. Primate frontal eye fields. II. Physiological and anatomical correlates of electrically evoked eye movements. *J Neurophysiol* **54**:714–734.

Ekstrom AD, Kahana MJ, Caplan JB, Fields TA, Isham EA, Newman EL, Fried I. 2003. Cellular networks underlying human spatial navigation. *Nature* **425**:184–188.

Freedman EG. 2008. Coordination of the eyes and head during visual orienting. *Exp Brain Res* **190**:369–387.

Hanes DP, Wurtz RH. 2001. Interaction of the frontal eye field and superior colliculus for saccade generation. *J Neurophysiol* **85**:804–815.

Huxlin KR. 2008. Perceptual plasticity in damaged adult visual systems. *Vision Res* **48**:2154–2166.

Jay MF, Sparks DL. 1987a. Sensorimotor integration in the primate superior colliculus. I. Motor convergence. *J Neurophysiol* **57**:22–34.

Jay MF, Sparks DL. 1987b. Sensorimotor integration in the primate superior colliculus. II. Coordinates of auditory signals. *J Neurophysiol* **57**:35–55.

Jung MW, Wiener SI, McNaughton BL. 1994. Comparison of spatial firing characteristics of units in dorsal and ventral hippocampus of the rat. *J Neurosci* **14**:7347–7356.

Kaas JH. 2005. Serendipity and the Siamese cat: the discovery that genes for coat and eye pigment affect the brain. *ILAR J* **46**:357–363.

Maier DL, Mani S, Donovan SL, Soppet D, Tessarollo L, McCasland JS, Meiri KF. 1999. Disrupted cortical map and absence of cortical barrels in growth-associated protein (GAP)-43 knockout mice. *Proc Natl Acad Sci U S A* **96**:9397–9402.

Morris RGM. 2008. Morris water maze. *Scholarpedia* **3**:6315.

Morris RG, Garrud P, Rawlins JN, O'Keefe J. 1982. Place navigation impaired in rats with hippocampal lesions. *Nature* **297**:681–683.

Moser EI, Kropff E, Moser M-B. 2008. Place cells, grid cells, and the brain's spatial representation system. *Annu Rev Neurosci* **31**:69–89.

Optican LM, Robinson DA. 1980. Cerebellar-dependent adaptive control of primate saccadic system. *J Neurophysiol* **44**:1058–1076.

Packard MG, McGaugh JL. 1996. Inactivation of hippocampus or caudate nucleus with lidocaine differentially affects expression of place and response learning. *Neurobiol Learn Mem* **65**:65–72.

Penfield W, Rasmussen T. 1950. *The cerebral cortex of man*. New York: Macmillan.

Pouget A, Sejnowski TJ. 1997. Spatial transformations in the parietal cortex using basis functions. *J Cogn Neurosci* **9**:222–237.

Rizzolatti G, Luppino G. 2001. The cortical motor system. *Neuron* **31**:889–901.

Schiller PH, Stryker M. 1972. Single-unit recording and stimulation in superior colliculus of the alert rhesus monkey. *J Neurophysiol* **35**:915–924.

Sparks DL. 2002. The brainstem control of saccadic eye movements. *Nat Rev Neurosci* **3**:952–964.

Szymusiak R, Nitz D. 2003. Chronic recording of extracellular neuronal activity in behaving animals. *Curr Protoc Neurosci*, Chapter 6, Unit 6.16.

Tootell RB, Silverman MS, Switkes E, De Valois RL. 1982. Deoxyglucose analysis of retinotopic organization in primate striate cortex. *Science* **218**:902–904.

Weinstein S. 1968. Intensive and extensive aspects of tactile sensitivity as a function of body part, sex, and laterality. In: *The skin senses*. Kenshalo DR, ed. Springfield, IL: Thomas, pp. 195–222.

Yarbus AL. 1967. *Eye movements and vision*. New York: Plenum Press.

Chapter 12

Afraz S-R, Kiani R, Esteky H. 2006. Microstimulation of inferotemporal cortex influences face categorization. *Nature* **442**:692–695.

Bisley JW, Zaksas D, Pasternak T. 2001. Microstimulation of cortical area MT affects performance on a visual working memory task. *J Neurophysiol* **85**:187–196.

Blasdel GG. 1992. Orientation selectivity, preference, and continuity in monkey striate cortex. *J Neurosci* **12**:3139–3161.

Blumberg J, Kreiman G. 2010. How cortical neurons help us see: visual recognition in the human brain. *J Clin Invest* **120**:3054–3063.

Bruce C, Desimone R, Gross CG. 1981. Visual properties of neurons in a polysensory area in superior temporal sulcus of the macaque. *J Neurophysiol* **46**:369–384.

Conway BR, Livingstone MS. 2006. Spatial and temporal properties of cone signals in alert macaque primary visual cortex. *J Neurosci* **26**:10826–10846.

Conway BR, Moeller S, Tsao DY. 2007. Specialized color modules in macaque extrastriate cortex. *Neuron* **56**:560–573.

Freedman DJ, Riesenhuber M, Poggio T, Miller EK. 2001. Categorical representation of visual stimuli in the primate prefrontal cortex. *Science* **291**:312–316.

Freedman DJ, Riesenhuber M, Poggio T, Miller EK. 2003. A comparison of primate prefrontal and inferior temporal cortices during visual categorization. *J Neurosci* **23**:5235–5246.

Gottfried JA, Zald DH. 2005. On the scent of human olfactory orbitofrontal cortex: meta-analysis and comparison to non-human primates. *Brain Res Rev* **50**:287–304.

Grill-Spector K. 2003. The neural basis of object perception. *Curr Opin Neurobiol* **13**:159–166.

Hubel DH, Wiesel TN. 1962. Receptive fields, binocular interaction and functional architecture in the cat's visual cortex. *J Physiol* **160**:106–154.

Hubel DH, Wiesel TN. 1968. Receptive fields and functional architecture of monkey striate cortex. *J Physiol* **195**:215–243.

Hubel DH, Wiesel TN. 1977. Ferrier lecture. Functional architecture of macaque monkey visual cortex. *Proc R Soc B* **198**:1–59.

Igarashi KM, Ieki N, An M, Yamaguchi Y, Nagayama S, Kobayakawa K, et al. 2012. Parallel mitral and tufted cell pathways route distinct odor information to different targets in the olfactory cortex. *J Neurosci* **32**:7970–7985.

Kaas JH, Hackett TA. 2000. Subdivisions of auditory cortex and processing streams in primates. *Proc Natl Acad Sci U S A* **97**:11793–11799.

Rokszin A, Márkus Z, Braunitzer G, Berényi A, Benedek G, Nagy A. 2010. Visual pathways serving motion detection in the mammalian brain. *Sensors* **10**:3218–3242.

Schiller P. 1995. The ON and OFF channels of the mammalian visual system. *Prog Retin Eye Res* **15**:173–195.

Sengpiel F, Stawinski P, Bonhoeffer T. 1999. Influence of experience on orientation maps in cat visual cortex. *Nat Neurosci* **2**:727–732.

Singer W, Gray CM. 1995. Visual feature integration and the temporal correlation hypothesis. *Annu Rev Neurosci* **18**:555–586.

Tavazoie SF, Reid RC. 2000. Diverse receptive fields in the lateral geniculate nucleus during thalamocortical development. *Nat Neurosci* **3**:608–616.

Thomas C, Avidan G, Humphreys K, Jung K-J, Gao F, Behrmann M. 2009. Reduced structural connectivity in ventral visual cortex in congenital prosopagnosia. *Nat Neurosci* **12**:29–31.

Tsao DY, Freiwald WA, Tootell RBH, Livingstone MS. 2006. A cortical region consisting entirely of face-selective cells. *Science* **311**:670–674.

Woolley SMN, Gill PR, Fremouw T, et al. 2009. Functional groups in the avian auditory system. *J Neurosci* **29**:2780–2793.

Chapter 13

Achermann P, Borbély AA. 1998. Temporal evolution of coherence and power in the human sleep electro-encephalogram. *J Sleep Res* **7(Suppl 1)**:36–41.

Aston-Jones G, Bloom FE. 1981. Norepinephrine-containing locus coeruleus neurons in behaving rats exhibit pronounced responses to non-noxious environmental stimuli. *J Neurosci* **1**:887–900.

Aston-Jones G, Rajkowski J, Kubiak P. 1997. Conditioned responses of monkey locus coeruleus neurons anticipate acquisition of discriminative behavior in a vigilance task. *Neuroscience* **80**:697–715.

Bergmann BM, Kushida CA, Everson CA, Gilliland MA, Obermeyer W, Rechtschaffen A. 1989. Sleep deprivation in the rat: II. Methodology. *Sleep* **12**:5–12.

Buschman TJ, Miller EK. 2009. Serial, covert shifts of attention during visual search are reflected by the frontal eye fields and correlated with population oscillations. *Neuron* **63**:386–396.

Chase MH, Morales FR. 2000. Control of motoneurons during sleep. In: *Principles and practice of sleep medicine.* Third Edition. Kryger MH, Roth T, Dement WC, eds. Philadelphia: WB Saunders, pp. 155–168.

Contreras D, Steriade M. 1997. Synchronization of low-frequency rhythms in corticothalamic networks. *Neuroscience* **76**:11–24.

Corbetta M, Kincade MJ, Lewis C, Snyder AZ, Sapir A. 2005. Neural basis and recovery of spatial attention deficits in spatial neglect. *Nat Neurosci* **8**:1603–1610.

Everson CA, Bergmann BM, Rechtschaffen A. 1989. Sleep deprivation in the rat: III. Total sleep deprivation. *Sleep* **12**:13–21.

Foote SL, Aston-Jones G, Bloom FE. 1980. Impulse activity of locus coeruleus neurons in awake rats and monkeys is a function of sensory stimulation and arousal. *Proc Natl Acad Sci U S A* **77**:3033–3037.

Gerashchenko D, Kohls MD, Greco M, Waleh NS, Salin-Pascual R, Kilduff TS, Lappi DA, Shiromani PJ. 2001. Hypocretin-2-saporin lesions of the lateral hypothalamus produce narcoleptic-like sleep behavior in the rat. *J Neurosci* **21**:7273–7283.

Hirata A, Aguilar J, Castro-Alamancos MA. 2006. Noradrenergic activation amplifies bottom-up and top-down signal-to-noise ratios in sensory thalamus. *J Neurosci* **26**:4426–4436.

Itti L, Koch C, Niebur E. 1998. A model of saliency-based visual attention for rapid scene analysis. *IEEE Trans Pattern Anal Mach Intell* **20**:1254–1259.

Kushida CA, Bergmann BM, Rechtschaffen A. 1989. Sleep deprivation in the rat: IV. Paradoxical sleep deprivation. *Sleep* **12**:22–30.

Lyamin OI, Manger PR, Ridgway SH, Mukhametov LM, Siegel JM. 2008. Cetacean sleep: an unusual form of mammalian sleep. *Neurosci Biobehav Rev* **32**:1451–1484.

Manns ID, Alonso A, Jones BE. 2000. Discharge properties of juxtacellularly labeled and immuno-histochemically identified cholinergic basal forebrain neurons recorded in association with the electro-encephalogram in anesthetized rats. *J Neurosci* **20**:1505–1518.

McAdams CJ, Maunsell JH. 1999. Effects of attention on orientation-tuning functions of single neurons in macaque cortical area V4. *J Neurosci* **19**:431–441.

McCormick DA, Bal T. 1997. Sleep and arousal: thalamocortical mechanisms. *Annu Rev Neurosci* **20**:185–215.

Moore RY. 1999. Circadian timing. In: *Fundamental neuroscience*. Zigmond MJ, Bloom FE, Landis SC, Roberts JL, Squire LR, eds. San Diego: Academic Press, pp. 1189–1227.

Moruzzi G, Magoun HW. 1949. Brain stem reticular formation and activation of the EEG. *Electroencephalogr Clin Neurophysiol* **1**:455–473.

Müller JR, Philiastides MG, Newsome WT. 2005. Microstimulation of the superior colliculus focuses attention without moving the eyes. *Proc Natl Acad U S A* **102**:524–529.

Parton A, Malhotra P, Husain M. 2004. Hemispatial neglect. *J Neurol Neurosurg Psychiatry* **75**:13–21.

Saper CB, Fuller PM, Pedersen NP, Lu J, Scammell TE. 2010. Sleep state switching. *Neuron* **68**:1023–1042.

Vyazovskiy VV, Olcese U, Lazimy YM, Faraguna U, Esser SK, Williams JC, Cirelli C, Tononi G. 2009. Cortical firing and sleep homeostasis. *Neuron* **63**:865–878.

Waterhouse BD, Moises HC, Woodward DJ. 1998. Phasic activation of the locus coeruleus enhances responses of primary sensory cortical neurons to peripheral receptive field stimulation. *Brain Res* **790**:33–44.

Chapter 14

Barot SK, Kyono Y, Clark EW, Bernstein IL. 2008. Visualizing stimulus convergence in amygdala neurons during associative learning. *Proc Natl Acad Sci U S A* **105**:20959–20963.

Bayley PJ, Hopkins RO, Squire LR. 2006. The fate of old memories after medial temporal lobe damage. *J Neurosci* **26**:13311–13317.

Bernasconi N, Natsume J, Bernasconi A. 2005. Progression in temporal lobe epilepsy: differential atrophy in mesial temporal structures. *Neurology* **65**:223–228.

Bontempi B, Laurent-Demir C, Destrade C, Jaffard R. 1999. Time-dependent reorganization of brain circuitry underlying long-term memory storage. *Nature* **400**:671–675.

Ciocchi S, Herry C, Grenier F, Wolff SBE, Letzkus JJ, Vlachos I, Ehrlich I, Sprengel R, Deisseroth K, Stadler MB, Müller C, Lüthi A. 2010. Encoding of conditioned fear in central amygdala inhibitory circuits. *Nature* **468**:277–282.

Corkin S. 1968. Acquisition of motor skill after bilateral medial temporal-lobe excision. *Neuropsychologia* **6**:255–265.

Corkin S, Amaral DG, González RG, Johnson KA, Hyman BT. 1997. H. M.'s medial temporal lobe lesion: findings from magnetic resonance imaging. *J Neurosci* **17**:3964–3979.

Fortin NJ, Agster KL, Eichenbaum HB. 2002. Critical role of the hippocampus in memory for sequences of events. *Nat Neurosci* **5**:458–462.

Gelbard-Sagiv H, Mukamel R, Harel M, Malach R, Fried I. 2008. Internally generated reactivation of single neurons in human hippocampus during free recall. *Science* **322**:96–101.

Izquierdo I, Quillfeldt JA, Zanatta MS, Quevedo J, Schaeffer E, Schmitz PK, Medina JH. 1997. Sequential role of hippocampus and amygdala, entorhinal cortex and parietal cortex in formation and retrieval of memory for inhibitory avoidance in rats. *Eur J Neurosci* **9**:786–793.

Ji D, Wilson MA. 2007. Coordinated memory replay in the visual cortex and hippocampus during sleep. *Nat Neurosci* **10**:100–107.

Kajiwara R, Takashima I, Mimura Y, Witter MP, Iijima T. 2003. Amygdala input promotes spread of excitatory neural activity from perirhinal cortex to the entorhinal-hippocampal circuit. *J Neurophysiol* **89**:2176–2184.

Kim JJ, Fanselow MS. 1992. Modality-specific retrograde amnesia of fear. *Science* **256**:675–677.

Kolassa I-T, Kolassa S, Ertl V, Papassotiropoulos A, De Quervain DJ-F. 2010. The risk of posttraumatic stress disorder after trauma depends on traumatic load and the catechol-o-methyltransferase Val(158)Met polymorphism. *Biol Psychiatry* **67**:304–308.

LePort AKR, Mattfeld AT, Dickinson-Anson H, Fallon JH, Stark CEL, Kruggel F, Cahill L, McGaugh JL. 2012. Behavioral and neuroanatomical investigation of highly superior autobiographical memory (HSAM). *Neurobiol Learn Mem* **98**:78–92.

Meunier M, Bachevalier J, Mishkin M, Murray E. 1993. Effects on visual recognition of combined and separate ablations of the entorhinal and perirhinal cortex in rhesus-monkeys. *J Neurosci* **13**:5418–5432.

Muller J, Corodimas KP, Fridel Z, LeDoux JE. 1997. Functional inactivation of the lateral and basal nuclei of the amygdala by muscimol infusion prevents fear conditioning to an explicit conditioned stimulus and to contextual stimuli. *Behav Neurosci* **111**:683–691.

Munger SD, Leinders-Zufall T, Mcdougall LM, Cockerham RE, Schmid A, Wandernoth P, Wennemuth G, Biel M, Zufall F, Kelliher KR. 2010. An olfactory subsystem that detects carbon disulfide and mediates food-related social learning. *Curr Biol* **20**:1438–1444.

Murray E, Mishkin M. 1998. Object recognition and location memory in monkeys with excitotoxic lesions of the amygdala and hippocampus. *J Neurosci* **18**:6568–6582.

Packard MG. 1999. Glutamate infused posttraining into the hippocampus or caudate-putamen differentially strengthens place and response learning. *Proc Natl Acad Sci U S A* **96**:12881–12886.

Roozendaal B, Castello NA, Vedana G, Barsegyan A, McGaugh JL. 2008. Noradrenergic activation of the basolateral amygdala modulates consolidation of object recognition memory. *Neurobiol Learn Mem* **90**:576–579.

Ross RS, Eichenbaum H. 2006. Dynamics of hippocampal and cortical activation during consolidation of a nonspatial memory. *J Neurosci* **26**:4852–4859.

Sutherland RJ, Weisend MP, Mumby D, Astur RS, Hanlon FM, Koerner A, Thomas MJ, Wu Y, Moses SN, Cole C, Hamilton DA, Hoesing JM. 2001. Retrograde amnesia after hippocampal damage: recent vs. remote memories in two tasks. *Hippocampus* **11**:27–42.

Treves A, Rolls ET. 1992. Computational constraints suggest the need for two distinct input systems to the hippocampal CA3 network. *Hippocampus* **2**:189–199.

Wang Y, Fontanini A, Katz DB. 2006. Temporary basolateral amygdala lesions disrupt acquisition of socially transmitted food preferences in rats. *Learn Mem* **13**:794–800.

Wheeler ME, Petersen SE, Buckner RL. 2000. Memory's echo: vivid remembering reactivates sensory-specific cortex. *Proc Natl Acad Sci U S A* **97**:11125–11129.

Winters BD, Forwood SE, Cowell RA, Saksida LM, Bussey TJ. 2004. Double dissociation between the effects of peri-postrhinal cortex and hippocampal lesions on tests of object recognition and spatial memory: heterogeneity of function within the temporal lobe. *J Neurosci* **24**:5901–5908.

Yerkes RM, Morgulis S. 1909. The method of Pawlow in animal psychology. *Psychol Bull* **6**:257–273.

Zalutsky RA, Nicoll RA. 1990. Comparison of two forms of long-term potentiation in single hippocampal neurons. *Science* **248**:1619–1624.

Zola-Morgan SM, Squire LR. 1990. The primate hippocampal formation: evidence for a time-limited role in memory storage. *Science* **250**:288–290.

Chapter 15

Carr GD, White NM. 1983. Conditioned place preference from intra-accumbens but not intra-caudate amphetamine injections. *Life Sci* **33**:2551–2557.

Chafee MV, Goldman-Rakic PS. 1998. Matching patterns of activity in primate prefrontal area 8a and parietal area 7ip neurons during a spatial working memory task. *J Neurophysiol* **79**:2919–2940.

Chartoff EH, Marck BT, Matsumoto AM, Dorsa DM, Palmiter RD. 2001. Induction of stereotypy in dopamine-deficient mice requires striatal D1 receptor activation. *Proc Natl Acad Sci U S A* **98**:10451–10456.

Damier P, Hirsch EC, Agid Y, Graybiel AM. 1999. The substantia nigra of the human brain. II. Patterns of loss of dopamine-containing neurons in Parkinson's disease. *Brain* **122**:1437–1448.

Day JJ, Roitman MF, Wightman RM, Carelli RM. 2007. Associative learning mediates dynamic shifts in dopamine signaling in the nucleus accumbens. *Nat Neurosci* **10**:1020–1028.

Deniau JM, Chevalier G. 1985. Disinhibition as a basic process in the expression of striatal functions. II. The striato-nigral influence on thalamocortical cells of the ventromedial thalamic nucleus. *Brain Res* **334**:227–233.

Dias R, Robbins TW, Roberts AC. 1997. Dissociable forms of inhibitory control within prefrontal cortex with an analog of the Wisconsin Card Sort Test: restriction to novel situations and independence from "on-line" processing. *J Neurosci* **17**:9285–9297.

Freed CR, Greene PE, Breeze RE, Tsai WY, DuMouchel W, Kao R, Dillon S, Winfield H, Culver S, Trojanowski JQ, Eidelberg D, Fahn S. 2001. Transplantation of embryonic dopamine neurons for severe Parkinson's disease. *New Eng J Med* **344**:710–719.

Gerfen CR. 2003. D1 dopamine receptor supersensitivity in the dopamine-depleted striatum animal model of Parkinson's disease. *Neuroscientist* **9**:455–462.

Greffard S, Verny M, Bonnet A-M, Beinis J-Y, Gallinari C, Meaume S, Piette F, Hauw J-J, Duyckaerts C. 2006. Motor score of the Unified Parkinson Disease Rating Scale as a good predictor of Lewy body-associated neuronal loss in the substantia nigra. *Arch Neurol* **63**:584–588.

Haber SN, Knutson B. 2010. The reward circuit: linking primate anatomy and human imaging. *Neuropsychopharmacology* **35**:4–26.

Konishi S, Nakajima K, Uchida I, Kameyama M, Nakahara K, Sekihara K, Miyashita Y. 1998. Transient activation of inferior prefrontal cortex during cognitive set shifting. *Nat Neurosci* **1**:80–84.

Lewis DA, Melchitzky DS, Sesack SR, Whitehead RE, Auh S, Sampson A. 2001. Dopamine transporter immunoreactivity in monkey cerebral cortex: regional, laminar, and ultrastructural localization. *J Comp Neurol* **432**:119–136.

MacDonald AW, Cohen JD, Stenger VA, Carter CS. 2000. Dissociating the role of the dorsolateral prefrontal and anterior cingulate cortex in cognitive control. *Science* **288**:1835–1838.

Phillips PEM, Stuber GD, Heien MLAV, Wightman RM, Carelli RM. 2003. Subsecond dopamine release promotes cocaine seeking. *Nature* **422**:614–618.

Ratiu P, Talos I-F. 2004. Images in clinical medicine: the tale of Phineas Gage, digitally remastered. *N Engl J Med* **351**:e21.

Reynolds JN, Hyland BI, Wickens JR. 2001. A cellular mechanism of reward-related learning. *Nature* **413**:67–70.

Robinson DL, Venton BJ, Heien MLAV, Wightman RM. 2003. Detecting subsecond dopamine release with fast-scan cyclic voltammetry in vivo. *Clin Chem* **49**:1763–1773.

Snowdon CT, Tannenbaum PL, Schultz-Darken NJ, Ziegler TE, Ferris CF. 2011. Conditioned sexual arousal in a nonhuman primate. *Horm Behav* **59**:696–701.

Tepper JM, Koós T, Wilson CJ. 2004. GABAergic microcircuits in the neostriatum. *Trends Neurosci* **27**:662–669.

Uchino A, Kato A, Yuzuriha T, Takashima Y, Kudo S. 2001. Cranial MR imaging of sequelae of prefrontal lobotomy. *Am J Neuroradiol* **22**:301–304.

Valjent E, Bertran-Gonzalez J, Hervé D, Fisone G, Girault J-A. 2009. Looking BAC at striatal signaling: cell-specific analysis in new transgenic mice. *Trends Neurosci* **32**:538–547.

Wilgus J, Wilgus B. 2009. Face to face with Phineas Gage. *J Hist Neurosci* **18**:340–345.

Zhou QY, Palmiter RD. 1995. Dopamine-deficient mice are severely hypoactive, adipsic, and aphagic. *Cell* **83**:1197–1209.

Chapter 16

Agate RJ, Grisham W, Wade J, Mann S, Wingfield J, Schanen C, Palotie A, Arnold AP. 2003. Neural, not gonadal, origin of brain sex differences in a gynandromorphic finch. *Proc Natl Acad Sci U S A* **100**:4873–4878.

Amunts K, Schleicher A, Bürgel U, Mohlberg H, Uylings HB, Zilles K. 1999. Broca's region revisited: cytoarchitecture and intersubject variability. *J Comp Neurol* **412**:319–341.

Andreasen NC, Nopoulos P, Magnotta V, Pierson R, Ziebell S, Ho B-C. 2011. Progressive brain change in schizophrenia: a prospective longitudinal study of first-episode schizophrenia. *Biol Psychiatry* **70**:672–679.

Azevedo FAC, Carvalho LRB, Grinberg LT, Farfel JM, Ferretti REL, Leite REP, Jacob Filho W, Lent R, Herculano-Houzel S. 2009. Equal numbers of neuronal and nonneuronal cells make the human brain an isometrically scaled-up primate brain. *J Comp Neurol* **513**:532–541.

Bartley AJ, Jones DW, Weinberger DR. 1997. Genetic variability of human brain size and cortical gyral patterns. *Brain* **120**:257–269.

Beery AK, Zucker I. 2011. Sex bias in neuroscience and biomedical research. *Neurosci Biobehav Rev* **35**:565–572.

Brodmann K. 1912. Neue Ergebnisse über die vergleichende histologische Lokalisation der Grosshirnrinde mit besonderer Berücksichtigung des Stirnhirns. *Anat Anz* **41**:157–216.

Byne W, Lasco MS, Kemether E, Shinwari A, Edgar MA, Morgello S, Jones LB, Tobet S. 2000. The interstitial nuclei of the human anterior hypothalamus: an investigation of sexual variation in volume and cell size, number and density. *Brain Res* **856**:254–258.

Byne W, Tobet S, Mattiace LA, Lasco MS, Kemether E, Edgar MA, Morgello S, Buchsbaum MS, Jones LB. 2001. The interstitial nuclei of the human anterior hypothalamus: an investigation of variation with sex, sexual orientation, and HIV status. *Horm Behav* **40**:86–92.

Cahill L, Uncapher M, Kilpatrick L, Alkire MT, Turner J. 2004. Sex-related hemispheric lateralization of amygdala function in emotionally influenced memory: an FMRI investigation. *Learn Mem* **11**:261–266.

Canli T, Desmond JE, Zhao Z, Gabrieli JDE. 2002. Sex differences in the neural basis of emotional memories. *Proc Natl Acad Sci U S A* **99**:10789–10794.

Courchesne E, Chisum HJ, Townsend J, Cowles A, Covington J, Egaas B, Harwood M, Hinds S, Press GA. 2000. Normal brain development and aging: quantitative analysis at in vivo MR imaging in healthy volunteers. *Radiology* **216**:672–682.

Dronkers NF, Plaisant O, Iba-Zizen MT, Cabanis EA. 2007. Paul Broca's historic cases: high resolution MR imaging of the brains of Leborgne and Lelong. *Brain* **130**:1432–1441.

Fjell AM, Walhovd KB, Fennema-Notestine C, McEvoy LK, Hagler DJ, Holland D, Brewer JB, Dale AM. 2009. One-year brain atrophy evident in healthy aging. *J Neurosci* **29**:15223–15231.

Forger NG, Breedlove SM. 1986. Sexual dimorphism in human and canine spinal cord: role of early androgen. *Proc Natl Acad Sci U S A* **83**:7527–7531.

Herculano-Houzel S, Collins CE, Wong P, Kaas JH. 2007. Cellular scaling rules for primate brains. *Proc Natl Acad Sci U S A* **104**:3562–3567.

Herculano-Houzel S, Mota B, Lent R. 2006. Cellular scaling rules for rodent brains. *Proc Natl Acad Sci U S A* **103**:12138–12143.

Hickok G, Houde J, Rong F. 2011. Sensorimotor integration in speech processing: computational basis and neural organization. *Neuron* **69**:407–422.

Hofman MH. 1983. Encephalization in hominids: evidence for the model of punctualism. *Brain Behav Evol* **22**:102–117.

Jenner RA, Wills MA. 2007. The choice of model organisms in evo-devo. *Nat Rev Genet* **8**:311–319.

Kappelman J. 1996. The evolution of body mass and relative brain size in fossil hominids. *J Hum Evol* **30**:243–276.

Lee PG, Turk PE, Yang WT, Hanlon RT. 1994. Biological characteristics and biomedical applications of the squid *Sepioteuthis lessoniana* cultured through multiple generations. *Biol Bull* **186**:328–341.

Leonard CM, Towler S, Welcome S, Halderman LK, Otto R, Eckert MA, Chiarello C. 2008. Size matters: cerebral volume influences sex differences in neuroanatomy. *Cereb Cortex* **18**:2920–2931.

Lim MM, Wang Z, Olazábal DE, Ren X, Terwilliger EF, Young LJ. 2004. Enhanced partner preference in a promiscuous species by manipulating the expression of a single gene. *Nature* **429**:754–757.

MacSweeney M, Woll B, Campbell R, McGuire PK, David AS, Williams SCR, Suckling J, Calvert GA, Brammer MJ. 2002. Neural systems underlying British Sign Language and audio-visual English processing in native users. *Brain* **125**:1583–1593.

Manger P, Cort J, Ebrahim N, Goodman A, Henning J, et al. 2008. Is 21st century neuroscience too focussed on the rat/mouse model of brain function and dysfunction? *Front Neuroanat* **2**:1–7.

Mangold-Wirz K. 1966. Cerebralisation und Ontogenesemodus bei Eutherien. *Acta Anat* **63**:449–508.

Northcutt RG. 2012. Evolution of centralized nervous systems: two schools of evolutionary thought. *Proc Natl Acad Sci U S A* **109**(s1):10626–10633.

Pakkenberg B, Gundersen HJ. 1997. Neocortical neuron number in humans: effect of sex and age. *J Comp Neurol* **384**:312–320.

Pirlot P, Kamiya T. 1982. Relative size of brain and brain components in three gliding placentals. *Can J Zool* **60**:565–572.

Schumann CM, Bloss CS, Barnes CC, Wideman GM, Carper RA, Akshoomoff N, Pierce K, Hagler D, Schork N, Lord C, Courchesne E. 2010. Longitudinal magnetic resonance imaging study of cortical development through early childhood in autism. *J Neurosci* **30**:4419–4427.

Shaywitz BA, Shaywitz SE, Pugh KR, Constable RT, Skularski P, Fulbright RK, Bronen RA, Fletcher JM, Shankweiler DP, Katz L, Gore JC. 1995. Sex differences in the functional organization of the brain for language. *Nature* **372**:607–609.

Sommer IE, Aleman A, Somers M, Boks MP, Kahn RS. 2008. Sex differences in handedness, asymmetry of the planum temporale and functional language lateralization. *Brain Res* **1206**:76–88.

Stephan H, Frahm H, Baron G. 1981. New and revised data on volumes of brain structures in insectivores and primates. *Folia Primatol* **35**:1–29.

Uylings HBM, Rajkowska G, Sanz-Arigita E, Amunts K, Zilles K. 2005. Consequences of large interindividual variability for human brain atlases: converging macroscopical imaging and microscopical neuroanatomy. *Anat Embryol* **210**:423–431.

Yassa MA, Muftuler LT, Stark CEL. 2010. Ultrahigh-resolution microstructural diffusion tensor imaging reveals perforant path degradation in aged humans in vivo. *Proc Natl Acad Sci U S A* **107**:12687–12691.

Credits

Chapter 1

1.1 A Tom McHugh / Science Source; **1.1 B** Adapted from Sivak, 1976; **b1.1** from Morgan et al., 2006; **1.3** Brain images reproduced with permission from http://www.brains.rad.msu.edu and http://brainmuseum.org, supported by the US National Science Foundation; **1.5** MRI scan courtesy of Dr. Craig Stark, PhD; **1.6** Material from the Yakovlev-Haleem Collection; **1.7** Adapted from Sereno and Tootel, 2005; Image courtesy of Martin Sereno, PhD; **1.10** Adapted from Deacon, 1990; **1.13** Adapted from James, 1890; **b1.2** from Goodson et al., 2009; **b1.3** Adapted from Aponte et al., 2011; brain image from brainmaps.org; **1.19** Reproduced with permission from http://www.brains.rad.msu.edu, and http://brainmuseum.org, supported by the US National Science Foundation.

Chapter 2

2.1 A Courtesy of Cajal Legacy; Instituto Cajal; Madrid (Spain); **B** Courtesy of Annie Vogel-Ciernia; **C** from Cajal, 1894; **2.6** Adapted from Hodgkin and Horowicz, 1959; **b2.2 B & C** McCusker, E.C. et al., 2012. Structure of a bacterial voltage-gated sodium channel pore reveals mechanisms of opening and closing. Nature communications, 3, p.1102; **2.16** SynapseWeb, Kristen M. Harris, PI, http://synapses.clm.utexas.edu/; **2.19** Idan Segev; **2.20** Reprinted by permission from Macmillan Publishers Ltd: NATURE Vol. 463, Issue 7283, Marguerita E. Klein; **2.23** Reprinted from Trends in Neurosceinces 27.6 , Lonnie P. Wollmuth and Alexander Sobolevsky 'Structure and the gating of the glutamate receptor ion channel', pp. 321–328. Copyright 2006, with permission from Elsevier; **b2.3 B** McCusker, E.C. et al., 2012. Structure of a bacterial voltage-gated sodium channel pore reveals mechanisms of opening and closing. Nature communications, 3, p.1102; **C** Hatzidimitriou G, McCann UD, and Ricaurte GA. 1999. Altered serotonin innervation patterns in the forebrain of monkeys treated with (+/−)3,4-methylenedioxymethamphetamine seven years previously: factors influencing abnormal recovery. J Neurosci 19: 5096–5107.

Chapter 3

b3.1 Images courtesy of John H. Byrne; **3.5** Adapted from brainmaps.org; **3.9 A** Reprinted by permission from Macmillan Publishers Ltd: NATURE Vol. 429, Issue 6993, Masanori Matsuraki, Naoki Honkura, Graham C.R. Ellis-Davies, Haruo Kasai. 'Structural basis of long-term potentiation in single dendritic spines'. Copyright © 2004; **3.10** Steward O., Levy W.B., "Preferential localization of polyribosomes under the base of dendritic spines in granule cells of the dentate gyrus", Journal of Neurosciencei 2, pp. 284–291. © Society for Neuroscience. Republished by permission of the Society for Neuroscience. Permission conveyed through Copyright Clearance Center Inc.; **3.15** Adapted by permission from Macmillan Publishers Ltd: NATURE Vol. 391, Issue 6670, Gina G. Turrigiano, Kenneth R. Leslie, Niraj S. Desai, Lana C. Rutherford1 & Sacha B. Nelson 'Activity-dependent scaling of quantal amplitude in neocortical neurons' Fig.1a; **3.16** Reprinted by permission from Macmillan Publishers Ltd: NATURE Vol. 463, Issue 7283, Todd F. Roberts, Katherine A. Tschida, Marguerita E. Klein & Richard Mooney 'Rapid spine stabilization and synaptic enhancement at the onset of behavioural learning'. Copyright © 2010; **3.17 B** Boele, H.-J. et al., 2013. Axonal sprouting and formation of terminals in the adult cerebellum during associative motor learning. The Journal of neuroscience : the official journal of the Society for Neuroscience, 33(45), pp. 17897–17907; **3.18** Learning and Memory 14 (January 3, 2007) Norman M. Weinberger, 'Associative representational plasticity in the auditory cortex: A synthesis of two disciplines'. © 2007 Cold Spring Harbor Laboratory Press; **3.19** Reprinted from Neurobiology of Learning and Memory 77.1 , Jeffrey A. Kleim, Scott Barbay, Natalie R. Cooper, Theresa M. Hogg, Chelsea N. Reidel, Michael S. Remple, Randolph J. Nudo. 'Motor Learning-Dependent Synaptogenesis Is Localized to Functionally Reorganized Motor Cortex', pp. 63–77. Copyright 2002, with permission from Elsevier; **b3.3** University of Pittsburgh Medical Center.

Chapter 4

4.1 Wikramanayake, A.H. et al., 2003. An ancient role for nuclear beta-catenin in the evolution of axial polarity and germ layer segregation. Nature, 426(6965), pp. 446–450; **4.3** © 2004, Annual Reviews. De Robertis, E.M. & Kuroda, H., "Dorsal-ventral patterning and neural induction in Xenopus embryos" Annual Review of Cell and Developmental Biology 20 (November 2004) pp. 285–308; **4.5** from Marvin Sodicoff; **4.7** Guidato, S., Prin, F. and Guthrie, S., 'Somatic motoneurone specification in the hindbrain: the influence of somite-derived signals, retinoic acid and Hoxa3'. Development 130:13, pp. 2981–2996. © Company of Biologists. Reprinted with permission; **4.13** Assimacopoulos, S. et al., 2012. Fibroblast Growth Factor 8 Organizes the Neocortical Area Map and Regulates Sensory Map Topography. The Journal of neuroscience : the official journal of the Society for Neuroscience, 32(21), pp. 7191–7201; **4.15 C** from Noctor et al., 2001; **4.16** from Dehay and Kennedy, 2007; **4.17 A** Courtesy of Pasko Rakic, MD, PhD; **B** From Rapaport et al., 1996; retina image from Anger, E.M. et al., 2004. Ultrahigh resolution optical coherence tomography of the monkey fovea. Identification of retinal sublayers by correlation with semithin histology sections. Experimental eye research, 78(6), pp. 1117–1125; **4.18A** from Cajal, 1890; **C** Courtesy of Paul Letourneau, PhD; **4.19** ©1990 Heidemann et al. The Journal of Cell Biology. 111: 1949–1957. doi:10.1083/jcb.111.5.1949; **4.20** Adapted from Ming et al. 1997; **4.21 A** f Kennedy TE, Serafini T, de la Torre JR, Tessier-Lavigne M. 1994. Netrins are diffusible chemotropic factors for commissural axons in the embryonic spinal cord. Cell 78:425–435; **B** Holmes GP, Negus K, Burridge L, Raman S, Algar E, Yamada T, Little MH. 1998. Distinct but overlapping expression patterns of two vertebrate slit homologs implies functional roles in CNS development and organogenesis. Mechs Development 79:57–72; **4.24 A&B** Walter J., Kern-Veits B., Huf J., Stolze B., Bonhoeffer F., "Recognition of position-specific properties of tectal cell membranes by retinal axons in vitro", Development 101, pp. 685–696. © Company of Biologists. Reprinted with permission; **C** Monschau B, Kremoser C, Ohta K, Tanaka H, Kaneko T, Yamada T, Handwerker C, Hornberger MR, Löschinger J, Pasquale EB, Siever DA, Verderame MF, Müller BK, Bonhoeffer F, Drescher U. 1997. Shared and distinct functions of RAGS and ELF-1 in guiding retinal axons. EMBO 16:1258–1267; **4.26** Portera-Cailliau C, Pan DT, Yuste R. 2003. Activity-regulated dynamic behavior of early dendritic protrusions: evidence for different types of dendritic filopodia. J Neurosci 23:7129–7142; **4.27** Emoto K, He Y, Ye B, Grueber WB, Adler PN, Jan LY, Jan Y-N. 2004. Control of dendritic branching and tiling by the Tricornered-kinase/Furry signaling pathway in Drosophila sensory neurons. Cell 119:245–256; **4.30 A, left** Hamburger V, Hamilton HL. 1992. A series of normal stages in the development of the chick embryo. 1951. Dev Dynamics 195:231–272; **right** from Hamburger, 1934; **B** Hollyday, M. & Hamburger, V., 1976. Reduction of the naturally occurring motor neuron loss by enlargement of the periphery. The Journal of Comparative Neurology, 170(3), pp. 311–320; **4.32** Li Q, Martin JH. 2002. Postnatal development of connectional specificity of corticospinal terminals in the cat. J Comp Neurol 447:57–71; **4.35** Le Vay S, Wiesel T, Hubel D. 1980. The development of ocular dominance columns in normal and visually deprived monkeys. J Comp Neurol 191:1–51; **4.3** Copyright 2013, Susan J. Astley, PhD, University of Washington.

Chapter 5

5.2 Altman J, Das GD. 1966. Autoradiographic and histological studies of postnatal neurogenesis. I. A longitudinal investigation of the kinetics, migration and transformation of cells incorporating tritiated thymidine in neonate rats, with special reference to postnatal neurogenesis in some brain regions. J Comp Neurol 126:337–389; **5.3** Eriksson PS, Perfilieva E, Björk-Eriksson T, Alborn AM, Nordborg C, Peterson DA, Gage FH. 1998. Neurogenesis in the adult human hippocampus. Nature Med 4:1313–1317; **5.4** Spalding K, Bhardwaj R, Buchholz B, Druid H, Frisen J. 2005. Retrospective Birth Dating of Cells in Humans. Cell 122:133–143; **5.5** Haines DE. 1991. On the question of subdural space. Anat Rec 230:3–21; **5.6 B** Brain images reproduced with permission from http://www.brains.rad.msu.edu and http://brainmuseum.org, supported by the US National Science Foundation; **5.7** from a supplement to Masdeu JC, Pascual B, Bressi F, Casale M, Prieto E, Arbizu J, Fernández-Seara MA. Ventricular wall granulations and draining of cerebrospinal fluid in chronic giant hydrocephalus. Arch Neurol. 2009;66:262–267, with permission; **5.8** Han, H., Tao, W. & Zhang, M., 2007. The dural entrance of cerebral bridging veins into the superior sagittal sinus: an anatomical comparison between cadavers and digital subtraction angiography. Neuroradiology, 49(2), pp. 169–175; **5.9** Feuillet L, Dufour H & Pelletier J, 2007. Brain of a white-collar worker. The Lancet, 370:262–262; **5.10** Ek CJ, Dziegielewska KM, Stolp H, Saunders NR. 2006. Functional effectiveness of the blood-brain barrier to small water-soluble molecules in developing and adult opossum (Monodelphis domestica). J Comp Neurol 496:13–26; **b5.1 B** Pardridge WM. 2003. Gene targeting in vivo with pegylated immunoliposomes. Methods Enzymol 373:507–528; **5.11** Ek CJ, Habgood MD, Dziegielewska KM, Saunders NR. 2003. Structural characteristics and barrier properties of the choroid plexuses in developing brain of the opossum (Monodelphis domestica). J Comp Neurol 460:451–464; **5.13** Oberheim NA, Takano T, Han X, He W, Lin JHC, Wang F, Xu Q, Wyatt JD, Pilcher W, Ojemann JG, Ransom BR, Goldman SA, Nedergaard M. 2009. Uniquely hominid features of adult human astrocytes. J Neurosci 29:3276–3287; **5.14** from Lee et al., 2008; **b5.2 A and B** Schenk D, et al. 1999. Immunization with amyloid-beta attenuates Alzheimer-disease-like pathology in the PDAPP mouse. Nature 400:173–177; **C and D** Oddo S, Billings L, Kesslak JP,

Cribbs DH, LaFerla FM. 2004. Abeta immunotherapy leads to clearance of early, but not late, hyperphosphorylated tau aggregates via the proteasome. Neuron 43:321–332; **5.15** Bush TG, Puvanachandra N, Horner CH, Polito A, Ostenfeld T, Svendsen CN, Mucke L, Johnson MH, Sofroniew MV. 1999. Leukocyte infiltration, neuronal degeneration, and neurite outgrowth after ablation of scar-forming, reactive astrocytes in adult transgenic mice. Neuron 23:297–308; **5.16** Hamilton R, Keenan JP, Catala M, Pascual-Leone A. 2000. Alexia for Braille following bilateral occipital stroke in an early blind woman. Neuroreport 11:237–240; **5.17** Sadato N, Okada T, Honda M, Yonekura Y. 2002. Critical period for cross-modal plasticity in blind humans: a functional MRI study. NeuroImage 16:389–400; **5.18** Jain N, Qi H-X, Collins CE, Kaas JH. 2008. Large-scale reorganization in the somatosensory cortex and thalamus after sensory loss in macaque monkeys. J Neurosci 28:11042–11060; **5.19 A** Wrigley P.J., Press S.R., Gustin S.M., Macefield V.G., Gandevia S.C., Cousins M.J., Middleton J.W., Henderson L.A., Siddall P.J., 'Neuropathic pain and primary somatosensory cortex reorganization following spinal cord injury', Pain 141 (1–2), pp. 52–59. Copyright © 2009 Lippincott Williams & Wilkins, Inc.; **5.22 B** courtesy of Harald H. Quick, University of Erlangen, Erlangen, Germany; **5.23 B** from Schaffer et al. 2006, PLoS Biology, 4:e22; **b5.3 A** Courtesy of Kasuga Huang; **B** from Stark and Squire, 2001. Copyright National Academy of Sciences, U.S.A.; **5.26** Takano T, Tian G-F, Peng W, Lou N, Libionka W, Han X, Nedergaard M. 2006. Astrocyte-mediated control of cerebral blood flow. Nature Neurosci 9:260–267.

Chapter 6

6.1A Courtesy of National Eye Institute; **6.2A** Witmer, M.T. & Kiss, S., 2013. Wide-field imaging of the retina. Survey of Ophthalmology, 58(2), pp. 143–154; **6.4B** Rhodopsin model courtesy of Roland Deschain; **6.7** Courtesy of Nicolás Cuenca, www.retinalmicroscopy.com; **6.10** Deeb SS. 2006. Genetics of variation in human color vision and the retinal cone mosaic. Curr Opin Gen & Dev 16:301–307; **6.12 B & C** from Franze et al., 2007; Copyright (2007) National Academy of Sciences, U.S.A.; **6.13 C** Kleene SJ, Gesteland RC. 1981. Dissociation of frog olfactory epithelium with N-ethylmaleimide. Brain Res 229:536–540; **6.14** from Abaffy et al., 2007; **6.15** from Gilad et al., 2003; Copyright National Academy of Sciences, U.S.A.; **6.17 A & B** Mombaerts P, Wang F, Dulac C, Chao SK, Nemes A, Mendelsohn M, Edmondson J, Axel R. 1996. Visualizing an olfactory sensory map. Cell 87:675–686; **6.18** from Johnson et al., 2004; **6.22 A** Picture by Mireille Lavigne-Rabillard, from 'Promenade around the cochlea', by R. Pujol, S. Blatrix, T.Pujol and V. Reclar-Enjalbert, (www.iurc.montp.inserm.fr/cric/audition/index.htm) CRIC, University Montpelli; **6.19 & 6.20** Yokoi, M., Mori, K. & Nakanishi, S., 1995. Refinement of odor molecule tuning by dendrodendritic synaptic inhibition in the olfactory bulb. Proceedings of the National Academy of Sciences of the United States of America, 92(8), pp. 3371–3375; **6.23 A** © The American Physiological Society (APS). Hackney C.M., Furness D.N., "Mechanotransduction in vertebrate hair cells: structure and function of the stereociliary bundle", (1995) American Journal of Physiology 268, pp. C1–13; **B & C** from the Proceedings of the National Academy of Science U.S.A. 97.24 (November 21, 2000), pp. 13336–13341. Kachar B., Parakkal M., Kurc M., Zhao Y., Gillespie P.G. 'High-resolution structure of hair-cell tip links.' Copyright (2000) National Academy of Sciences, U.S.A.; **6.25 A** Meyer AC, Frank T, Khimich D, Hoch G, Riedel D, Chapochnikov NM, Yarin YM, Harke B, Hell SW, Egner A, Moser T. 2009. Tuning of synapse number, structure and function in the cochlea. Nature Neurosci 12:444–453; **B & C** Glowatzki E, Fuchs PA. 2002. Transmitter release at the hair cell ribbon synapse. Nature Neurosci 5:147–154; **D** courtesy of Elisabeth Glowatzki; **b6.2** Adapted from Cochlear Implants: Auditory Protheses and Electric Hearing, Fan-Gang Zeng, Arthur N. Popper, and Richard R. Fay, eds. 2004, pp. 101–148, 'Anatomical Considerations and Long-Term Effects of Electrical Stimulation', Patricia A. Leake. © Springer Science + Business Media New York 2004. With kind permission from Springer Science and Business Media; **b6.3** Leal, M. & Losos, J.B., 2010. Evolutionary biology: Communication and speciation. Nature, 467(7312), pp. 159–160.

Chapter 7

7.2 A & B Woodbury CJ, Koerber HR. 2007. Central and peripheral anatomy of slowly adapting type I low-threshold mechanoreceptors innervating trunk skin of neonatal mice. J Comp Neurol 505:547–561; **7.3 A** Guinard D, Usson Y, Guillermet C, Saxod R. 2000. PS-100 and NF 70–200 double immunolabeling for human digital skin meissner corpuscle 3D imaging. J Histochem Cytochem 48:295–302; **B** Pawson L, Prestia LT, Mahoney GK, Güçlü B, Cox PJ, Pack AK. 2009. GABAergic/glutamatergic-glial/neuronal interaction contributes to rapid adaptation in pacinian corpuscles. J Neurosci 29:2695–2705; **7.5 A** after Devor, M., 1999. Unexplained peculiarities of the dorsal root ganglion. Pain, 82, pp. S27–S35; **7.6 A** Matsumoto, A. & Mori, S., 1975. Number and diameter distribution of myelinated afferent fibers innervating the paws of the cat and monkey. Experimental neurology, 48(2), pp. 261–274; **7.7** Todd, A.J., 2010. Neuronal circuitry for pain processing in the dorsal horn. Nature Reviews Neuroscience, 11(12), pp. 823–836; **7.9 A** Dib-Hajj, S.D. et al., 2013. The Na(V)1.7 sodium channel: from molecule to man. Nature Reviews Neuroscience, 14(1), pp. 49–62; **7.10** after Tominaga M, Caterina MJ. 2004. Thermosensation and pain. J Neurobiol 61:3–12; **7.11** A Bautista DM, Siemens J, Glazer JM, Tsuruda PR, Basbaum AI, Stucky CL, Jordt S-E, Julius D. 2007. The menthol receptor TRPM8 is the principal detector of environmental cold. Nature 448:204–208; **7.12** Caterina MJ, Schumacher MA, Tominaga M, Rosen TA, Levine JD, Julius D. 1997. The capsaicin receptor: a heat-activated ion channel in the pain pathway. Nature 389:816–824; **7.14 & 7.15** Chandrashekar J, Hoon MA, Ryba NJP, Zuker CS. 2006. The receptors and cells for mammalian taste. Nature 444:288–294; **7.16** © The American Physiological Society (APS). Yokota, T., Eguchi, K., and Hiraba, K., "Topographical representations of taste response characteristics in the rostral nucleus of the solitary tract in the rat", Journal of Neurophysiology 111:1 (January 2014), pp. 182–196; **7.18 A & B** Banks, R.W. et al., 2009. A comparative analysis of the encapsulated end-organs of mammalian skeletal muscles and of their sensory nerve endings. Journal of anatomy, 214(6), pp. 859–887; **7.21** Li A, Xue J, Peterson EH. 2008. Architecture of the mouse utricle: macular organization and hair bundle heights. J Neurophys 99:718–733.

Chapter 8

8.1 Photographs courtesy of Thomas Caceci; **8.3** from Young and Heath, 2000; **8.4** Vale R.D., Milligan R.A., 'The way things move: looking under the hood of molecular motor proteins' Science 288 (April 2000) pp. 88–95. © The American Association for the Advancement of Science. Reprinted with permission from AAAS. Figure prepared by Graham Johnson; **8.5** Images taken by Yi Zuo in Wes Thompson's lab at UT Austin; **8.6 A & B** Dauber W, Meister A. 1986. Ultrastructure of junctional folds of motor end plates in extensor digitorum longus muscles of mice. J Ultrastructure & Mol Structure Res 97:158–164; **8.7** Photograph from Keller-Peck CR, Walsh MK, Gan WB, Feng G, Sanes JR, Lichtman JW. 2001. Asynchronous synapse elimination in neonatal motor units: studies using GFP transgenic mice. Neuron 31:381–394; **8.10** Desmedt JE, Godaux E. 1977. Fast motor units are not preferentially activated in rapid voluntary contractions in man. Nature 267:717–719; **8.11** © The American Physiological Society (APS). De Luca C.J., Contessa P., "Hierarchical control of motor units in voluntary contractions" Journal of Neurophysiology 107 (2012), pp. 178–195; **8.13 B, C, & D** Ovalle WK, Dow PR, Nahirney PC. 1999. Structure, distribution and innervation of muscle spindles in avian fast and slow skeletal muscle. J Anatomy 194:381–394; **8.15** Courtesy of Dr. Elisabeth Ehler; **8.17 A** parrow M.P., Weichselbaum M., McCray P.B. "Development of the innervation and airway smooth muscle in human fetal lung" in the American Journal of Respiratory and Molecular Biology 20.4 (April, 1999), pp. 550–560. © American Thoracic Society. Reprinted with permission; **B** Lin G, Qiu X, Fandel TM, Albersen M, Wang Z, Lue TF, Lin C-S. 2011. Improved penile histology by phalloidin stain: circular and longitudinal cavernous smooth muscles, dual-endothelium arteries, and erectile dysfunction-associated changes. Urology 78:970.e1–e8; **C** Murakami M, Nagato T, Tanioka H, Uehara Y. 1989. Morphological changes in the myoepithelial cells of the rat sublingual salivary gland during differentiation as shown by the nitrobenzoxadiazole-phallacidin fluorescent method. Arch Oral Biol 34:143–145; **8.19** Buckberg G.D., Mahajan A., Jung B., Markl M., Hennig J., Ballester-Rodes M., 'MRI myocardial motion and fiber tracking: a confirmation of knowledge from different imaging modalities,' European Journal of Cardio-thoracic Surgery 29 (April Supplement 2006), pp. S165–177. ©European Association for Cardio-Thoracic Surgergy. By permission of Oxford University Press; **b8.1 A** Courtesy of Ashmore and Holley, unpublished from 1998; **B & C** from Ashmore, 1987; **8.21 A** from brainmaps.org; **8.22 A** from braimaps.org; **B** from Lincoln & Wakerly, 1974; **8.23** Hatton GI, Tweedle CD. 1982. Magnocellular neuropeptidergic neurons in hypothalamus: increases in membrane apposition and number of specialized synapses from pregnancy to lactation. Brain Res Bull 8:197–204; **b8.2** Inset from Newbold RR, Padilla-Banks E, Jefferson WN. 2009. Environmental estrogens and obesity. Molec Cell Endocrinol 304:84–89; **8.25 C** Radley JJ, Sawchenko PE. 2011. A common substrate for prefrontal and hippocampal inhibition of the neuroendocrine stress response. J Neurosci 31:9683–9695.

Chapter 9

9.4 A Uehara, Y., Suyama, K., 'Visualization of the adventitial aspect of the vascular smooth muscle cells under the scanning electron microscope'. Journal of Electron Microscopy 27.2 (January 1, 1978), pp. 157–159. © The Japanese Society for Microscopy. By permission of Oxford University Press; **B** Monos E, Lóránt M, Fehér (2001) Influence of long-term experimental orthostatic body position on innervation density in extremity vessels. Am J Physiol Heart Circ Physiol 281:H1606–H1612.; **9.7** Shin J-W, Geerling JC, Loewy AD. 2009. Vagal innervation of the aldosterone-sensitive HSD2 neurons in the NTS. Brain Res 1249:135–147; **b9.1** Miguel A Landestoy, 2007; http://www.anoleannals.org/ 2011/10/11/have-you-seen-anoles-play-dead/; **9.9** Yan B, Li L, Harden SW, Epstein PN, Wurster RD, Cheng ZJ. 2009. Diabetes induces neural degeneration in nucleus ambiguus (NA) and attenuates heart rate control in OVE26 mice. Exp Neurol 220:34–43; **9.11 A** Ross CA, Ruggiero DA, Park DH, Joh TH, Sved AF, Fernandez-Pardal J, Saavedra JM, Reis DJ. 1984. Tonic vasomotor control by the rostral ventrolateral medulla: effect of electrical or chemical stimulation of the area containing C1 adrenaline neurons on arterial pressure, heart rate, and plasma catecholamines and vasopressin. J Neurosci 4:474–494; **9.12** © The American Physiological Society (APS). Rybak I.A, O'Connor R, Ross A., Shevtsova N.A., Nuding S.C., Segers L.S., Shannon R., Dick T.E., Dunin-Barkowski W.L., Orem J.M., Solomon I.C., Morris K.F., Lindsey B.G., "Reconfiguration of the pontomedullary respiratory network: A computational modeling study with coordinated in vivo experiments" in Journal of Neurophysiology 100:4 (October 2008); **9.13** Smith JC, Feldman JL. 1987. In vitro brainstem-spinal cord preparations for study of motor systems for mammalian respiration and locomotion. J Neurosci Methods 21:321–333; **9.16** Nakamura, K. & Morrison, S.F., 2008. A thermosensory pathway that controls body temperature. Nature Neuroscience, 11(1), pp. 62–71; **9.19** Oliet S.H.,

Bourque C.W., 'Properties of supraoptic magnocellular neurones isolated from the adult rat'. © 1992 The Physiological Society. Reprinted by permission of John Wiley and Sons; **9.22** Thongkhao-on K, Wirtshafter D, Shippy SA. 2008. Feeding specific glutamate surge in the rat lateral hypothalamus revealed by low-flow push-pull perfusion. Pharm Biochem Behav 89:591–597; **9.24 A&B** de la Iglesia HO, Meyer J, Schwartz WJ. 2003. Lateralization of circadian pacemaker output: Activation of left- and right-sided luteinizing hormone-releasing hormone neurons involves a neural rather than a humoral pathway. J Neurosci 23:7412–7414; **C** de la Iglesia, Horacio O., Jennifer Meyer, Alan Carpino Jr., William J. Schwartz, 'Antiphase oscillation of the left and right suprachiasmatic nuclei', Science 290 (5492), October 2000. © American Association for the Advancement of Science. Reprinted with permission from AAAS. **9.26** Magariños AM, McEwen BS, Flügge G, Fuchs E. 1996. Chronic psychosocial stress causes apical dendritic atrophy of hippocampal CA3 pyramidal neurons in subordinate tree shrews. J Neurosci 16:3534–3540.

Chapter 10

10.9 Springer Journal of Comparative Physiology A: Neuroethology, Sensory, Neural, and Behvioral Phsyiology 193.12, December 2007, pp. 1177–1183, 'Head-Bobbing of Walking Birds' Reinhold Necker, © Springer-Verlag 2007. With kind permission from Springer Science and Business Media; **10.12** Wallén P, Grillner S. 1987. N-methyl-D-aspartate receptor-induced, inherent oscillatory activity in neurons active during fictive locomotion in the lamprey. J Neurosci 7:2745–2755; **10.14 A** Wallén P., Williams T.L., 'Fictive locomotion in the lamprey spinal cord in vitro compared with swimming in the intact and spinal animal', in the Journal of Physiology 347, John Wiley and Sons. © 1984 The Physiological Society; **B** Cohen AH, Wallén P. 1980. The neuronal correlate of locomotion in fish. "Fictive swimming" induced in an in vitro preparation of the lamprey spinal cord. Exp Brain Res 41:11–18; **10.16** 'Fictive motor patterns in chronic spinal cats', K. G. Pearson, S. Rossignol in Journal of Neurophysiology Published 1 December 1991 Vol. 66 no. 6, Copyright © 1991 the American Physiological Society; **10.17** brainmuseum.org **10.18** Lacroix S, Havton LA, McKay H, Yang H, Brant A, Roberts J, Tuszynski MH. 2004. Bilateral corticospinal projections arise from each motor cortex in the macaque monkey: a quantitative study. J Comp Neurol 473:147–161; **10.19** "Penfield and Rasmussen, 1950. Reprinted with permission of Macmillan Publishing Company from "The cerebral cortex of man" by Wilder Penfield and Theodore Rasmussen. Copyright 1950 Macmillan Publishing Company; copyright renewed 1978 Theodore Rasmussen."; **10.20** Copyright © 2006, Annual Reviews. Michael Graziano, "The organization of behavioral repertoire in motor cortex" Annual Review of Neuroscience 29, pp. 105–134; **10.21 B** Georgopoulos AP, Kalaska JF, Caminiti R, Massey JT. 1982. On the relations between the direction of two-dimensional arm movements and cell discharge in primate motor cortex. J Neurosci 2:1527–1537; **10.22** Gallese, V. et al., 1996. Action recognition in the premotor cortex. Brain : a journal of neurology, 119.2 (Pt 2), pp. 593–609; **10.23 A** brainmuseum.org; **B** created by the Gene Expression Nervous System Atlas (GENSAT) Project, funded by NIH.

Chapter 11

11.1 after S. Weinstein, 1968, Intensive and extensive aspects of tactile sensitivity as a function of body part, sex, and laterality. In D. R. Kenshalo (Ed.), The skin senses (pp. 195–222). Springfield, IL: Thomas; **11.3** after Penfield and Rasmussen, 1950. Reprinted with permission of Macmillan Publishing Company from "The cerebral cortex of man" by Wilder Penfield and Theodore Rasmussen. Copyright 1950 Macmillan Publishing Company; copyright renewed 1978 Theodore Rasmussen; **11.4 A** from Maier et al., 1999; **11.6 & 11.7** Huxlin KR. 2008. Perceptual plasticity in damaged adult visual systems. Vision Res 48:2154–2166; **11.9** Tootell RB, Silverman MS, Switkes E, De Valois RL. 1982. Deoxyglucose analysis of retinotopic organization in primate striate cortex. Science 218:902–904; **11.11** Adams DL, Sincich LC, Horton JC. 2007. Complete pattern of ocular dominance columns in human primary visual cortex. J Neurosci 27:10391–10403; **b11.1 A** courtesy of Mark Konishi; **B** courtesy of Ake Norberg; **11.17** © The American Physiological Society (APS). Jay M.F., Sparks D.L., "Sensorimotor integration in the primate superior colliculus. I. Motor convergence", Journal of Neurophysiology 57, pp. 22–34; **11.18** Alfred L. Yarbus, Eye Movements and Vision. © 1967 Springer-Verlag US. With permissions of Springer Science + Business Media; Ilya Efimovich Repin, Unexpected Return. 1884–1888, oil on canvas, 160.5x167.5 cm. Courtesy The State Tretyakov Gallery, Moscow; **11.22** © The American Physiological Society (APS). Schiller, P.H. and M. Stryker, 'Single-unit recording and stimulation in superior colliculus of the alert rhesus monkey', Journal of Neurophysiology 35:6 (November 1972), pp. 915–924; **11.24** Bruce C.J., Goldberg M.E., Bushnell M.C., Stanton G.B., 'Primate frontal eye fields. II. Physiological and anatomical correlates of electrically evoked eye movements' in the Journal of Neurphysiology 54:3 (1985), pp. 714–734. © The American Physiological Society; **11.31 A** Modified with permisson from the Annual Reivew of Neuroscience, Volume 31 © 2008 by Annual Reivews, http://www.annualreviews.org; **B** Jung MW, Wiener SI, McNaughton BL. 1994. Comparison of spatial firing characteristics of units in dorsal and ventral hippocampus of the rat. J Neurosci 14:7347–7356; **11.32** Ekstrom AD, Kahana MJ, Caplan JB, Fields TA, Isham EA, Newman EL, Fried I. 2003. Cellular networks underlying human spatial navigation. Nature 425:184–188.

Chapter 12

12.7 Tavazoie SF, Reid RC. 2000. Diverse receptive fields in the lateral geniculate nucleus during thalamocortical development. Nature Neurosci 3:608–616; **12.8 & 12.11** Hubel DH, Wiesel TN. 1962. Receptive fields, binocular interaction and functional architecture in the cat's visual cortex. J Physiol 160:106–154; **12.12** Blasdel GG. 1992. Orientation selectivity, preference, and continuity in monkey striate cortex. J Neurosci 12:3139–3161; **12.13** Blumberg J, Kreiman G. 2010. How cortical neurons help us see: visual recognition in the human brain. J Clin Invest 120:3054–3063; **b12.1** Springer-Verlag: Grinvald, A. et al., ' In-vivo optical imaging of cortical architecture and dynamics', Figures 1, in Modern Techniques in Neuroscience Research, U. Windhorst and H. Johansson (Editors). Copyright © 1999, Springer-Verlag Berlin Heidelberg. With kind permission from Springer Science and Business Media; **12.14** Hubel, D.H., Wiesel T.N., 'Receptive fields and functional architecture of monkey striate cortex' in the Journal of Physiology 195.1 (March 1, 1968), John Wiley and Sons. © 1968 The Physiological Society; **12.18** Conway BR, Livingstone MS. 2006. Spatial and temporal properties of cone signals in alert macaque primary visual cortex. J Neurosci 26:10826–10846; **12.19** Conway BR, Moeller S, Tsao DY. 2007. Specialized color modules in macaque extrastriate cortex. Neuron 56:560–573; **12.21** Bruce C, Desimone R, Gross CG. 1981. Visual properties of neurons in a polysensory area in superior temporal sulcus of the macaque. J Neurophys 46:369–384; **12.22** Tsao DY, Freiwald WA, Tootell RBH, Livingstone MS. 2006. A cortical region consisting entirely of face-selective cells. Science 311:670–674; **12.23** Afraz S-R, Kiani R, Esteky H. 2006. Microstimulation of inferotemporal cortex influences face categorization. Nature 442:692–695; **12.24** Grill-Spector K. 2003. The neural basis of object perception. Curr Opin Neurobiol 13:159–166; **12.25** Freedman DJ, Riesenhuber M, Poggio T, Miller EK. 2003. A comparison of primate prefrontal and inferior temporal cortices during visual categorization. J Neurosci 23:5235–5246; **12.26** Gottfried JA, Zald DH. 2005. On the scent of human olfactory orbitofrontal cortex: meta-analysis and comparison to non-human primates. Brain Res Rev 50:287–304; **12.27** Igarashi, K.M. et al., 2012. Parallel Mitral and Tufted Cell Pathways Route Distinct Odor Information to Different Targets in the Olfactory Cortex. The Journal of neuroscience : the official journal of the Society for Neuroscience, 32(23), pp. 7970–7985; **12.29** adapted from Hackett, T.A., Stepniewska, I. & Kaas, J.H., 1999. Prefrontal connections of the parabelt auditory cortex in macaque monkeys. Brain Research, 817(1–2), pp. 45–58; **12.30** Woolley, S.M.N. et al., 2009. Functional groups in the avian auditory system. The Journal of neuroscience : the official journal of the Society for Neuroscience, 29(9), pp. 2780–2793; **12.31** Sengpiel F, Stawinski P, Bonhoeffer T. 1999. Influence of experience on orientation maps in cat visual cortex. Nature Neurosci 2:727–732; **12.32** Singer W, Gray CM. 1995. Visual feature integration and the temporal correlation hypothesis. Ann Rev Neurosci 18:555–586; **b12.2** Thomas C, Avidan G, Humphreys K, Jung K-J, Gao F, Behrmann M. 2009. Reduced structural connectivity in ventral visual cortex in congenital prosopagnosia. Nature Neurosci 12:29–31.

Chapter 13

13.3 Itti L, Koch C, Niebur E. 1998. A model of saliency-based visual attention for rapid scene analysis. IEEE Transact Pattern Analysis Machine Intell 20:1254–1259; **b13.1, top** Parton A, Malhotra P, Husain M. 2004. Hemispatial neglect. J Neurol Neurosurg Psych 75:13–21; **bottom** Corbetta M, Kincade MJ, Lewis C, Snyder AZ, Sapir A. 2005. Neural basis and recovery of spatial attention deficits in spatial neglect. Nat Neurosci 8:1603–1610; **13.9** Reprinted from Fundamental Neuroscience, Terry L. Powley. Copyright 2013, with permission from Elsevier; **13.11 B and C** Courtesy of the Yakovlev-Haleem Collection; **13.13** Aston-Jones G, Bloom FE. 1981. Norepinephrine-containing locus coeruleus neurons in behaving rats exhibit pronounced responses to non-noxious environmental stimuli. J Neurosci 1:887–900; **13.14** Aston-Jones G, Rajkowski J, Kubiak P. 1997. Conditioned responses of monkey locus coeruleus neurons anticipate acquisition of discriminative behavior in a vigilance task. Neuroscience 80:697–715; **13.15** Foote SL, Aston-Jones G, Bloom FE. 1980. Impulse activity of locus coeruleus neurons in awake rats and monkeys is a function of sensory stimulation and arousal. Proc Natl Acad Sci USA 77:3033–3037; **13.16 A** from Hirata et al.; **B** from Waterhouse et al., 1998; **13.17** Achermann, P. & Borbély, A.A., 1998. Temporal evolution of coherence and power in the human sleep electroencephalogram. Journal of sleep research, 7 Suppl 1, pp. 36–41; **13.18** Vyazovskiy VV, Olcese U, Lazimy YM, Faraguna U, Esser SK, Williams JC, Cirelli C, Tononi G. 2009. Cortical firing and sleep homeostasis. Neuron 63:865–878; **13.19** Contreras D, Steriade M. 1997. Synchronization of low-frequency rhythms in corticothalamic networks. Neuroscience 76:11–24; **13.21** McCormick DA, Bal T. 1997. Sleep and arousal: thalamocortical mechanisms. Ann Rev Neurosci 20:185–215; **13.22** Manns ID, Alonso A, Jones BE. 2000. Discharge properties of juxtacellularly labeled and immunohistochemically identified cholinergic basal forebrain neurons recorded in association with the electroencephalogram in anesthetized rats. J Neurosci 20:1505–1518; **13.23 B** Gerashchenko D, Kohls MD, Greco M, Waleh NS, Salin-Pascual R, Kilduff TS, Lappi DA, Shiromani PJ. 2001. Hypocretin-2-saporin lesions of the lateral hypothalamus produce narcoleptic-like sleep behavior in the rat. J Neurosci 21:7273–7283; **13.25 A** Chase and Morales: Control of motoneurons during sleep. In: Principles and practice of Sleep Medicine, 3rd ed., ed by Kryger, MH, Roth, T., and Dement, W.C. Philadelphia: WB Saunders., pp 155–168. cited in Lydic, 2005; **13.27** Lyamin OI, Manger PR, Ridgway SH, Mukhametov LM, Siegel JM. 2008. Cetacean sleep: an unusual form of mammalian sleep. Neurosci Biobehav Rev 32:1451–1484.

Chapter 14

14.2 Corkin S., Amaral D.G., González R.G., Johnson K.A., Hyman B.T., "H. M.'s medial temporal lobe lesion: findings from magnetic resonance imaging", Journal of

Neuroscience 17 (1997), pp. 3964–3979. © Society for Neuroscience; **14.5** Courtesy of the Yakovlev-Haleem Collection; **14.6 A** after Murray, E. & Mishkin, M., 1998. Object recognition and location memory in monkeys with excitotoxic lesions of the amygdala and hippocampus. The Journal of neuroscience : the official journal of the Society for Neuroscience, 18(16), pp. 6568–6582; Brain image from brainmuseum.org; **14.7 A** Brain image from brainmuseum.org; **14.8** Fortin, N.J., Agster, K.L. & Eichenbaum, H.B., 2002. Critical role of the hippocampus in memory for sequences of events. Nature Neuroscience, 5(5), pp. 458–462; **14.14** Wheeler ME, Petersen SE, Buckner RL. 2000. Memory's echo: vivid remembering reactivates sensory-specific cortex. Proc Natl Acad Sci USA 97:11125–11129; **14.18** Bontempi B, Laurent-Demir C, Destrade C, Jaffard R. 1999. Time-dependent reorganization of brain circuitry underlying long-term memory storage. Nature 400:671–675; **14.20** Roozendaal B, Castello NA, Vedana G, Barsegyan A, McGaugh JL. 2008. Noradrenergic activation of the basolateral amygdala modulates consolidation of object recognition memory. Neurobiol Learn Mem 90:576–579; **14.21** Kajiwara R, Takashima I, Mimura Y, Witter MP, Iijima T. 2003. Amygdala input promotes spread of excitatory neural activity from perirhinal cortex to the entorhinal-hippocampal circuit. J Neurophys 89:2176–2184; **14.27 A** Munger SD, Leinders-Zufall T, Mcdougall LM, Cockerham RE, Schmid A, Wandernoth P, Wennemuth G, Biel M, Zufall F, Kelliher KR. 2010. An olfactory subsystem that detects carbon disulfide and mediates food-related social learning. Curr Biol 20:1438–1444; **14.28 B** Barot SK, Kyono Y, Clark EW, Bernstein IL. 2008. Visualizing stimulus convergence in amygdala neurons during associative learning. Proc Natl Acad Sci USA 105:20959–20963.

Chapter 15

15.3 A & B Gerfen CR. 2003. D1 dopamine receptor supersensitivity in the dopamine-depleted striatum animal model of Parkinson's disease. Neuroscientist 9:455–462; **C & D** Tepper JM, Koós T, Wilson CJ. 2004. GABAergic microcircuits in the neostriatum. Trends Neurosci 27:662–669; **15.4** after Deniau JM, Chevalier G. 1985. Disinhibition as a basic process in the expression of striatal functions. II. The striato-nigral influence on thalamocortical cells of the ventromedial thalamic nucleus. Brain Res 334:227–233; **15.7** Lewis DA, Melchitzky DS, Sesack SR, Whitehead RE, Auh S, Sampson A. 2001. Dopamine transporter immunoreactivity in monkey cerebral cortex: regional, laminar, and ultrastructural localization. J Comp Neurol 432:119–136; **15.8** Valjent E, Bertran-Gonzalez J, Hervé D, Fisone G, Girault J-A. 2009. Looking BAC at striatal signaling: cell-specific analysis in new transgenic mice. Trends Neurosci 32:538–547; **15.11** Zhou QY, Palmiter RD. 1995. Dopamine-deficient mice are severely hypoactive, adipsic, and aphagic. Cell 83:1197–1209; **15.12** Chartoff EH, Marck BT, Matsumoto AM, Dorsa DM, Palmiter RD. 2001. Induction of stereotypy in dopamine-deficient mice requires striatal D1 receptor activation. Proc Natl Acad Sci 98:10451–10456; **15.13 left** Robinson DL, Venton BJ, Heien MLAV, Wightman RM. 2003. Detecting subsecond dopamine release with fast-scan cyclic voltammetry in vivo. Clin Chem 49:1763–1773; **right** Day JJ, Roitman MF, Wightman RM, Carelli RM. 2007. Associative learning mediates dynamic shifts in dopamine signaling in the nucleus accumbens. Nat Neurosci 10:1020–1028; **15.22** Uchino A, Kato A, Yuzuriha T, Takashima Y, Kudo S. 2001. Cranial MR imaging of sequelae of prefrontal lobotomy. Am J Neuroradiol 22:301–304; **15.23 left** Wilgus J, Wilgus B. 2009. Face to face with Phineas Gage. J History Neurosci 18:340–345; **right** Ratiu P, Talos I-F. 2004. Images in clinical medicine. The tale of Phineas Gage, digitally remastered. New Eng J Med 351:e21; **15.25** Konishi S, Nakajima K, Uchida I, Kameyama M, Nakahara K, Sekihara K, Miyashita Y. 1998. Transient activation of inferior prefrontal cortex during cognitive set shifting. Nat Neurosci 1:80–84; **15.27** MacDonald AW, Cohen JD, Stenger VA, Carter CS. 2000. Dissociating the role of the dorsolateral prefrontal and anterior cingulate cortex in cognitive control. Science 288:1835–1838; **15.28** Chafee MV, Goldman-Rakic PS. 1998. Matching patterns of activity in primate prefrontal area 8a and parietal area 7ip neurons during a spatial working memory task. J Neurophys 79:2919–2940.

Chapter 16

16.2B Lee PG. 1994. Biological characteristics and biomedical applications of the squid Sepioteuthis lessoniana cultured through multiple generations. Biol Bull 186:328–341; **16.11** Dronkers NF, Plaisant O, Iba-Zizen MT, Cabanis EA. 2007. Paul Broca's historic cases: high resolution MR imaging of the brains of Leborgne and Lelong. Brain 130:1432–1441; **16.12** MacSweeney M, Woll B, Campbell R, McGuire PK, David AS, Williams SCR, Suckling J, Calvert GA, Brammer MJ. 2002. Neural systems underlying British Sign Language and audio-visual English processing in native users. Brain 125:1583–1593; **16.13** after Hickok G, Houde J, Rong F. 2011. Sensorimotor integration in speech processing: computational basis and neural organization. Neuron 69:407–422; **16.14** Agate RJ, Grisham W, Wade J, Mann S, Wingfield J, Schanen C, Palotie A, Arnold AP. 2003. Neural, not gonadal, origin of brain sex differences in a gynandromorphic finch. Proc Natl Acad Sci USA 100:4873–4878; **16.15 A** Forger NG, Breedlove SM. 1986. Sexual dimorphism in human and canine spinal cord: role of early androgen. Proc Natl Acad Sci USA 83:7527–7531; **16.16 A** Byne W, Lasco MS, Kemether E, Shinwari A, Edgar MA, Morgello S, Jones LB, Tobet S. 2000. The interstitial nuclei of the human anterior hypothalamus: an investigation of sexual variation in volume and cell size, number and density. Brain Res 856:254–258; **16.19** Shaywitz B, Shaywltz S, Pugh K, Constable R. 1995. Sex differences in the functional organization of the brain for language. Nature 372:607–609; **16.21 left** Cahill L, Uncapher M, Kilpatrick L, Alkire MT, Turner J. 2004. Sex-related hemispheric lateralization of amygdala function in emotionally influenced memory: an FMRI investigation. Learn Memory 11:261–266; **16.21 right** Canli T, Desmond JE, Zhao Z, Gabrieli JDE. 2002. Sex differences in the neural basis of emotional memories. Proc Natl Acad Sci USA 99:10789–10794; **16.22** Bartley AJ, Jones DW, Weinberger DR. 1997. Genetic variability of human brain size and cortical gyral patterns. Brain 120:257–269; **16.24** Amunts K, Schleicher A, Bürgel U, Mohlberg H, Uylings HB, Zilles K. 1999. Broca's region revisited: cytoarchitecture and intersubject variability. J Comp Neurol 412:319–341; **16.25** Uylings HBM, Rajkowska G, Sanz-Arigita E, Amunts K, Zilles K. 2005. Consequences of large interindividual variability for human brain atlases: converging macroscopical imaging and microscopical neuroanatomy. Anat Embryol 210:423–431; **16.26** Figure by Anders M Fjell, University of Oslo, Norway; **16.27** Yassa MA, Muftuler LT, Stark CEL. 2010. Ultrahigh-resolution microstructural diffusion tensor imaging reveals perforant path degradation in aged humans in vivo. Proc Natl Acad Sci USA 107:12687–12691; **16.28** Lim MM, Wang Z, Olazábal DE, Ren X, Terwilliger EF, Young LJ. 2004. Enhanced partner preference in a promiscuous species by manipulating the expression of a single gene. Nature 429:754–757.

Index

D

M

Q

R